THE LAKES HANDBOOK
Volume 1

The Lakes Handbook

VOLUME 1

LIMNOLOGY AND LIMNETIC ECOLOGY

EDITED BY

P.E. O'Sullivan

AND

C.S. Reynolds

Blackwell
Publishing

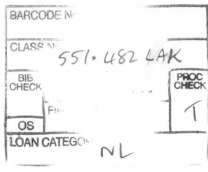
© 2004 by Blackwell Science Ltd
a Blackwell Publishing company

BLACKWELL PUBLISHING
350 Main Street, Malden, MA 02148-5020, USA
9600 Garsington Road, Oxford OX4 2DQ, UK
550 Swanston Street, Carlton, Victoria 3053, Australia

The rights of P.E. O'Sullivan and C.S. Reynolds to be identified as the Authors of the Editorial Material in this Work have been asserted in accordance with the UK Copyright, Designs, and Patents Act 1988.

First published 2004 by Blackwell Science Ltd

2   2005

Library of Congress Cataloging-in-Publication Data

The lakes handbook / edited by P.E. O'Sullivan and C. S. Reynolds.
p. cm.
Includes bibliographical references and index.
ISBN 0-632-04797-6 (hardback, v.1: alk. paper)
1. Limnology. 2. Lake ecology. I. O'Sullivan, P. E. (Patrick E.) II. Reynolds, Colin S.
QH96.L29 2003
551.48'2—dc21
2003000139

ISBN-13: 978-0-632-04797-0 (hardback, v.1: alk. paper)

A catalogue record for this title is available from the British Library.

Set in 9 on 11.5 pt Trump Mediaeval
by SNP Best-set Typesetter Ltd., Hong Kong
Printed and bound in the United Kingdom
by TJ International, Padstow, Cornwall

The publisher's policy is to use permanent paper from mills that operate a sustainable forestry policy, and which has been manufactured from pulp processed using acid-free and elementary chlorine-free practices. Furthermore, the publisher ensures that the text paper and cover board used have met acceptable environmental accreditation standards.

For further information on
Blackwell Publishing, visit our website:
www.blackwellpublishing.com

Contents

Blackwell Publishing is grateful to the various copyright holders who have given their permission to use copyright material in this volume. While the contributors to this volume have made every effort to clear permission as appropriate, the publisher would appreciate being notified of any omissions.

Contributors

Jürg Bloesch *Swiss Federal Institute for Environmental Science and Technology (EAWAG), CH-8600 Dübendorf, Switzerland*

Z. Maciej Gliwicz *Department of Hydrobiology, University of Warsaw, ul. Banacha 2, PL 02-097 Warszawa, Poland*

Dieter M. Imboden *Zürichstrasse 128, CH-8700 Küsnacht, Switzerland*

Pétur M. Jónasson *Freshwater Biological Laboratory, University of Copenhagen, DK-3400 Hillerød, Denmark*

Juha Karjalainen *Department of Biological and Environmental Science, University of Jyväskylä, FIN-40351 Jyväskylä, Finland*

Jan Květ *Faculty of Biological Sciences, University of South Bohemia, CZ-37005 Česke Budejoviče, Czech Republic Institute of Botany, Academy of Sciences of Czech Republic, CZ-379 82 Třeboň, Czech Republic*

Heinz Löffler *Institute of Limnology, University of Vienna, Althanstrasse 14, A-1090 Wien, Austria*

Patrick O'Sullivan *School of Earth, Ocean and Environmental Sciences, University of Plymouth, Plymouth, PL4 4AB, UK*

Judit Padisák *Institute of Biology, University of Veszprém, H-8200 Veszprém, Hungary*

Claudia Pahl-Wostl *Institute of Environmental Systems Research, University of Osnabrück, Albrechtstrasse 28, D-49069 Osnabrück, Germany*

Jan Pokorný *Institute of Botany, Academy of Sciences of Czech Republic and ENKI o.p.s. Dukelská 145, CZ-379 82 Třeboň, Czech Republic*

Martti Rask *Finnish Game and Fisheries Research Institute, Evo Fisheries Research, FIN-16900 Lammi, Finland*

Colin S. Reynolds *Centre for Ecology and Hydrology Windermere, The Ferry House, Ambleside, Cumbria LA22 0LP, UK*

Jouko Sarvala *Department of Biology, University of Turku, FIN 20500 Turku, Finland*

Christian E.W. Steinberg *Leibniz-Institut für Gewässerökologie und Binnenfischerei, Müggelseedamm 310, D-12587 Berlin, Germany*

Werner Stumm *late of EAWAG/ETH, CH-8600 Dübendorf, Switzerland*

Thomas Weisse *Institute of Limnology, Austrian Academy of Sciences, A-5310 Mondsee, Austria*

Ian J. Winfield *Centre for Ecology and Hydrology Windermere, The Ferry House, Ambleside, Cumbria LA22 0LP, UK*

Thomas C. Winter *Denver Federal Center, Box 25046 Denver, Colorado 80225, USA*

1 Lakes, Limnology and Limnetic Ecology: Towards a New Synthesis

COLIN S. REYNOLDS

1.1 INTRODUCTION

From the beginnings of modern science, lakes have fulfilled a focus of attention. Doubtless, this has something to do with the lure that water bodies hold for most of us, as well as for long having been a source of food as well as water. Authors, from Aristotle to Izaak Walton, committed much common knowledge of the freshwater fauna to the formal written record, so it is still a little surprising to realise that the formal study of lakes—limnology (from the Greek word, λιμνοσ, a lake)—is scarcely more than a century in age (Forel 1895). When, yet more recently, the branch of biology concerned with how natural systems actually function (ecology) began to emerge, ponds and lakes became key units of study. The distinctiveness of aquatic biota together with the tangible boundaries of water bodies lent themselves to the quantitative study of the dynamics of biomass (Berg 1938) and energy flow (Lindeman 1942). One of the leading contemporaneous exponents of limnology, August Thienemann, was quick to realise the distinctive properties of individual lakes and the nature of the crucial interaction of lakes with their surrounding catchments, in the broader context of what we now refer to as landscape ecology (Thienemann 1925). Moreover, as the science has developed, it has been recognised that lakes cannot be studied without some appreciation of their developmental history (Macan 1970). Today, we have no difficulty in accepting that the biology of water bodies is influenced by geography, physiography and climate; in the morphometry of basins; in the hydrology and the hydrography of the impounded water; the

hydrochemistry of the fluid exchanges; and the adaptations, dynamics and predilections of the aquatic biota. We may also accept that, just as no two water bodies are identical, we cannot assume that they function in identical ways. Yet the search for underlying patterns and for the underpinning processes continues apace, moving us towards a better understanding of the ecology of lakes and their biota.

The ostensible purpose of *The Lakes Handbook* is, plainly, to provide a sort of turn-of-the-century progress report which brings together the most recent perspectives on the interactions among the properties of water, the distinguishing features of individual basins and the dynamic interactions with their biota. Such reviews are not a new idea, especially not in limnology where the great *Treatise* commenced by Hutchinson (1957, 1967 and later volumes) remains a foundation block in the limnologist's firmament. In these two volumes, we have attempted to bring together limnologists and hydrobiologists, each a recognised and respected authority within a specialist sub-division of limnetic science, and invited from each an up-to-date overview of pattern–process perceptions within defined areas of the knowledge spectrum. The contributed chapters address aspects of the physics, chemistry and biological features of selected (usually phylogenetic) subdivisions of the biota, while several topics (such as structural dynamics and system regulation) are addressed under general headings. The limnetic biologies of selected lakes and systems feature in the second volume.

The immediate inspiration for this book—for

the publishers, Blackwell Science, as well as for us as its editors—has been the excellent *Rivers Handbook* (Calow & Petts 1992). We have sought to select a similar balance of in-depth reviews by leading practitioners in the field, to cover all aspects of limnology. Within the two volumes, we have attempted also a balance between theoretical and applied topics: our objective has been to provide a point of reference for students and professionals alike. We have been conscious, of course, of the challenges these ambitions entail. Neither the encyclopaedic thoroughness of Hutchinson's *Treatise* nor the robust utility of Wetzel's *Limnology* textbooks (1975, 1983, 2001), nor the accessibility of Kalff's (2002) text, nor even the explicative empiricism of Uhlmann's (1975) *Hydrobiologie*, is yet capable of emulation. We do not pretend to rival the more specialist compendia, such as Lerman *et al.* (1995) on the physics or chemistry of lake waters. Nevertheless, by drawing on the talents of our respective contributors, we believe that we have been able to present a contemporary and accurate reflection of current understanding about how lakes and lake ecosystems function.

1.2   THE LURE OF LIMNOLOGY

We have also been keen to mirror two further modern perceptions underpinning current attitudes to limnology. One of these is the importance of the freshwater resource. Although over 70% of our 'blue planet' is covered by sea, lakes and rivers occupy only a tiny percentage of the (c.150 million km²) terrestrial surface. Nobody can say for certain just how many water bodies currently populate the surface of the Earth. The series beginning with the world's largest lakes (Tables 1.1–1.3) progresses through smaller and smaller water bodies, eventually to collapse fractally into a myriad of sumps, melt-water, flood-plain, delta- and other wetland pools, puddles and 'phytotelmata' of rainwater retained in the foliage of terrestrial plants (especially of Bromeliads). Even restricting ourselves to natural (i.e., no artificial ponds), permanent (or seasonally enduring), stillwater-filled bodies,

wholly surrounded by land and exceeding an arbitrary cut off point (say 0.1 km²), then the total number of lakes in the world may be estimated to be in excess of 1.25 million, having an aggregate surface area of 2.6 million km² (Meybeck 1995). For smaller lakes, Meybeck (1995) used intensive regional censuses to extrapolate that there are likely to be a further 7.2 million water bodies with areas in the range 0.01–0.1 km², contributing a further 0.2 million km² of water surface. Of particular interest, however, is that the numbers of lakes in successive logarithmic bands (0.1–1 km², 1–10 km², etc., up to >100,000 km²) diminish by a factor of about 10 at each step; nevertheless, the aggregate of lake areas within each is broadly similar between bands (0.35 ± 0.15 million km²: see Table 1.1). Only the second category of great tectonic and glacial scour lakes—group (b) in Table 1.1—lies outside this generalisation, but the disparity is not vast. It is a reasonable deduction that the present planetary distribution of standing stillwaters (i.e., all lakes by definition, discounting extreme differences in salinity) is very evenly dispersed across the spectrum of lake areas.

The impression of evenness disappears when the volume of lakes is considered. Again, the abundance of water on the planet (nearly 1390 million km³) notwithstanding, barely 225,000 km³ (i.e., <0.016% of the total) is estimated to be contained in lakes and rivers. The amount discharged annually to the sea, c.29,000 km³, suggests an average residence of terrestrial surface water of eight years or so. True, groundwaters boost the planetary aggregate of non-marine liquid water (8 million km³) but the point remains that, if the resource is to be maintained, its sustainable exploitation is necessarily restricted to the interception of the seaward flux. Even then, far from all of the terrestrial storage is potentially potable, owing in part to the excessive salinity or alkalinity of a large proportion of it (see Williams; Chapter 8, Volume 2) and in part to anthropogenic despoliation (see Chapter 2, Volume 2). Variability in the flux rate about the average determines that the supply is far from evenly distributed (Meybeck 1995), either in space or in time, so that problems vary locally, according to drainage, climate, season, usage demands and the

Table 1.1 Global distribution of the world's aggregate area of lake water among area classes (in km², based on data in Meybeck, 1995), with the individual areas of those in the first two classes, as presented by Herdendorf (1982). See also Beeton (1984).

Category	Name of lake	Area (km²)	Number in category	Total area of category (km²)
(a) Inland waters >100,000 km²	Kaspiyskoye More*,†	374,000	1	374,000
(b) Inland waters 10,000–100,000 km²	Lake Superior	82,100	18	624,000
	Aralskoye More†,‡	64,500		
	Lake Victoria	62,940		
	Lake Huron	59,500		
	Lake Michigan	57,750		
	Lac Tanganyika	32,000		
	Ozero Baykal	31,500		
	Great Bear Lake	31,326		
	Great Slave Lake	28,568		
	Lake Erie	25,657		
	Lake Winnipeg	24,387		
	Lake Malawi*	22,490		
	Ozero Balkhash§	19,500		
	Lake Ontario	19,000		
	Ladozhskoye Ozero	18,130		
	Lac Tchad§	18,130		
	Tonle Sap§	16,350		
	Lac Bangweolo§	10,000		
(c) Inland waters 1000–10,000 km²			124	327,000
(d) Inland waters 100–1000 km²			1380	359,000
(e) Inland waters 10–100 km²			12,300	319,000
(f) Inland waters 1–10 km²			127,000	323,000
(g) Inland waters 0.1–1 km²			1,110,000	288,000
(h) Inland waters 0.01–0.1 km²			7,200,000	190,000
Total (a–h)			8.45×10^6	2,804,000

*   The Caspian Sea (see Chapter 2). Here, lakes are identified by the latinised local names recommended by Herdendorf (1982), except that Lake Nyasa is now more generally recognised as Lake Malawi.

†   As terrestrially enclosed water bodies these saline waters are included as 'lakes'.

‡   The area cited, from Herdendorf (1982), refers to the full recent extent of the lake. It has diminished greatly in the last decade owing to exploitative abstraction (see Williams & Aladin 1991).

§   Shallow lakes in typically arid and semi-arid areas are subject to large fluctuations in extent. The area quoted is the arithmetic mean of the range cited by Herdendorf (1982).

extent of defilement. Thus, the world-wide distribution of lakes and of the rivers that feed and drain them is of crucial social and economic importance. Reservoirs provide a means of resource conservation and of balancing ongoing demands against temporal vagaries of flows, but there needs still to be a hydraulic flux. Thus, access to adequate supplies of clean, fresh water is already proving to be a severe social and developmental constraint in several arid nations (e.g., the middle and northeast USA versus Mexico). The risks of a drying local climate or, simply the impact of an abnormally dry year, precipitate immediate and widespread public concern.

However, between-basin differences in the allocation of the standing water is impressively

Table 1.2   The world's ten largest lakes and inland waters, by volume retained (in km^3, based on Herdendorf (1982) and Beeton (1984)).

Kaspiyskoye More	78,200
Ozero Baykal	22,995
Lac Tanganyika	18,140
Lake Superior	12,100
Lake Malawi	6140
Lake Michigan	4920
Lake Huron	3540
Lake Victoria	2700
Great Bear Lake	2240
Great Slave Lake	2088
Ozero Issyk-kul	1740

Table 1.3   The world's ten deepest lakes (maximum depths in metres; based on Herdendorf (1982) and Beeton (1984), but modified after Reynolds *et al.* (2000)).

Ozero Baykal	1741
Lac Tanganyika	1471
Kaspiyskoye More	1025
Lake Malawi	706
Ozero Issyk — kul	702
Great Slave Lake	614
Lago Gral.Carrera/Buenos Aires	590
Danau Matano	590
Crater Lake	589
Danau Toba	529

skewed: the volume stored in Ozero Baykal, Russia (c.23,000 km^3), alone represents over 10% of the total water quantity in the world's lakes and rivers. Discounting the salt waters of the Kaspiyskoye More, the other nine lakes in Table 1.2 together account for 25% of the surface store of fresh water. So it is that, although most lake habitats are really rather small (each <0.1 km^2), most of the aggregate volume resides in large, deep lakes. All this accentuates the need to understand the role of the standing reserves (lakes, including reservoirs and groundwaters), how they function and how they sustain life, within and beyond their confines. There is an urgent need to grasp the productive dependence of fisheries as a source of food, not as a factor of the carbon fixation capacity of the primary producers but on the extent to which physical and chemical processes govern the metabolic transformation into useful food and in the context of the materials supplied by the entire hydraulic catchment.

This brings us to the second of the two perceptions. It is that lakes hold practical attractions for the ecological study of systems. This is not to resurrect the quaint and rather discredited early notion about lakes providing closed, microcosmic model ecosystems. Nevertheless, the substantial integrity of function in the face of seemingly overwhelming environmental constraints set by the high density, high viscosity and high specific heat of the medium is striking. Moreover, despite the alleged transparent colourlessness and universal solvent properties of water, the poverty of underwater light and the extreme dilution of dissolved resources are recurrent, dominating architectural features of limnoecology (Lampert & Sommer 1993). Ecosystems are generally acknowledged to be complex; those of the majority of lakes provide no exceptions. They do offer, as many terrestrial ecosystems do not, the advantage of relevant environmental fluctuations and responses at the population and community levels on timescales convenient to the observer and experimenter. For instance, a great many international exhortations and initiatives have been implemented to protect species diversity and to develop protocols for the sustainable exploitation of ecosystems. These are well motivated but have been offered, as a rule, without an ecosystem-based view of the management objective and, very often, without a clear idea of whence the diversity derives or precisely how it is maintained. Ecology is still beset with what are little more than unproven hypotheses. Lakes (especially smaller ones), on the other hand, are amenable to the study of such high-order ecosystem processes as population recruitment, community assembly, competitive exclusion, niche differentiation and changing diversity indices. The timescales of ecological interest, which in aquatic environments occupy the range seconds to decades, are equivalent to minutes to hundreds of millennia on land. Though similar scales apply as in lakes, they are logistically too difficult or too expensive to research in the oceans.

Common to water bodies of all sizes and shapes is their island quality—to aquatic organisms they function as patches of suitable habitat in a sea of an inhospitable, terrestrial environment. The distribution and relative isolation of the aquatic islands are especially challenging in the context of species dispersion and endemism, while patch size and the variability to which patches may be subject are key factors in viability and survival. Of course, not all lakes are mutually isolated, but fluvial connectivity merely adds to the fascination of freshwater systems and the importance to ecological understanding of their study (Tokeshi 1994).

Limnologists have perhaps advanced furthest in the development of realistic models of species-specific successions and their interaction with the relatively straightforward properties of the limnetic environment (Steel 1995; Reynolds & Irish 1997). This might be reason enough for advocating the study of lakes but we believe that, because freshwater systems are sufficiently representative of all ecosystems, the analogous processes are adequately simulated in the functioning of limnetic biota. Through the pages of the two volumes of this handbook, we seek to project something of the extraordinary diversity of lake basins and their ecologies. However, we strive to reveal the common constraints that link the properties of water, the movements generated in lakes, the sources of input water and its chemical composition, the structure of the pelagic biota, the littoral influence and the significance of benthic processes. We identify the fundamental knowledge required to uphold the management of quality and biotic outputs from lakes and reservoirs, drawing on the assessment of case studies and applications from around the world. Of course, the relevant knowledge emanates from our contributors, through whose writings we are able to convey the excitement felt by all limnologists about the habitats we are fortunate enough to study.

1.3   THE ORGANISATION AND STRUCTURE OF THIS BOOK

This might have been the point at which to conclude this introductory chapter. The organisation of the book is relatively self-evident, the first part being dedicated to the physical and chemical features of lakes and their main biota, the second much more at the practical applications of limnology. The beginning of a book is also the customary place in which to acknowledge the generous efforts of expert authors, the patience and professionalism of the publisher and the interest of readers. We take this opportunity to do just that but we are bound to do something more in this introduction. The book's inordinately long period in gestation—some five years from conception to its entry into the bookshops—is a matter of deep regret and embarrassment to the editors. We apologise profoundly to the contributors whose submissions we received comfortably within the original schedule. Hindsight confirms that, whether out of our naivety, incompetence or as a consequence of other pressures upon us, we did not provide the assistance and exhortation required by a minority of contributors who, for whatever personal difficulties they experienced, were unable to keep to the schedule. Alternative contributors had to be found in more than one instance.

Not one but three tragic setbacks further confounded the slow progress. We learned with great sadness of the death, on 14 April 1999, of Professor Werner Stumm. As the undisputed father of 'aquatic chemistry', his profound contributions and the fundamental research he led in the field of mineral surface reactions and aqueous phase equilibria and kinetics made him a natural choice as a contributory author. This was a task to which he readily committed himself. The chapter here is close to his original draft; however, we are extremely grateful to Dr Laura Sigg for her careful and skilful attention to the final manuscript.

Professor W. Thomas Edmondson died on 10 January 2000. Tommy had been passionate about freshwater life from childhood and his development as a scientist had been encouraged by such great limnologists as G.E. Hutchinson, Chancey Juday and Edward Birge. His pioneering work on rotifers continues to be held in very high regard but he became best known for his long-term study of the eutrophication of Lake Washington, and for his successful public campaign to reverse it. The case is a telling example of how sound limnological understanding, good communication and collective appreciation of a natural asset might be harnessed to bring about one of the most successful exercises in lake restoration. We feel deeply honoured to be able to include Tommy's own perspective on the unfolding story.

The death of Milan Straškraba, on 26 July 2000, also represents a severe loss to the scientific study of lakes. Milan had stepped in at a late date to write the chapter on reservoirs. We had discussed with him the topics to be covered in his contribution and he had submitted a first draft for our consideration. Illness slowed his further progress but he remained determined to complete the work. Sadly, circumstances did not allow him to do so: we are in possession only of the first draft. It was necessary to edit it but we have done so as lightly as we reasonably could.

Further tragedy struck close to the final preparation of Volume 1 for the publishers when we learned with enormous sadness that Professor W.D. (Bill) Williams had finally lost his long battle with myeloid leukaemia and had passed away in Brisbane on Australia Day, 26 January 2002. His eminent contribution to the study of lakes, especially of those saline and temporary waters that he considered to be as important, on areal grounds, as the more familiar subjects of the northern temperate zone, is familiar to all. He was enthusiastic in his support for this project and we are proud to be able to include his Chapter with the text as he agreed it.

The delays are not easily excused and we do not seek this. When it became evident just how long it would be before the book was finished, all the authors of submissions received on time were given the opportunity to update their contributions. In most instances, the landmark quality of the original review endures unscathed but topical revision has brought each of the contributions to the level of a contemporary overview of its subject.

Nevertheless, it is fortuitous that the gestation of the Handbook has coincided with the significant paradigm shift that is transforming limnetic ecology and the consensus view of just how lake ecosystems work. It is becoming increasingly recognised that the level of net aquatic primary production in small lakes does not always generate sufficient organic carbon to sustain the recruitment of heterotrophs at higher trophic levels. Moreover, the amount that is transferred through the planktic food chain seems to be altogether too small to support the measured quantities of fish biomass produced. Systems that are exclusively dependent upon pelagic primary and secondary producers (those of the open water of the oceans and of large, deep lakes) are, characteristically, able to maintain only the very low biomass densities diagnostic of resource-constrained oligotrophic systems. In contrast, the maintenance of productive food webs culminating in high areal densities of fish or macrophytes relies, to varying extents, on the supplement of inorganic nutrients and the residues of organic biomass carbon from the terrestrial catchments. It now seems that, with varying shortfalls, the functioning of many lake systems is not self-sufficient upon autochthonous primary production but is supported by the heterotrophic assimilation of primary products originating from the catchment (Wetzel 1995).

This deduction appears to be applicable to a wide range of smaller lakes where, in fact, gross primary production cannot be shown to exceed community respiration and the net production of the lake appears to be negative. The balance is met by terrestrial primary products, imported from the adjacent lands directly or in the hydraulic inflow. Enhancing *in situ* primary production through the direct fertilisation of the water with nutrients does not prevent the imported materials so that, even if internal production is raised absolutely and relatively, the system continues to function with a distinct heterotrophic component (Cole *et al.* 2000).

Some of the individual chapters represent these ideas and interpretations of system dynamics. Others have emphasised the contemporary problems in limnetic ecology that are propagated by the recognition and acceptance that wholesome and supposedly pristine sites may be, on balance, heterotrophic. The whole concept of the ecosystem health of lakes, as well as the way that they can be managed in order to achieve and maintain a sustainable condition, must now undergo careful revision.

In more ways than one, we both feel more than a little wiser. We have experienced, in more than adequate amount, the tribulations of editing a book of this scale and these ambitions. On the other hand, we have gained a great deal of new knowledge from the accumulated wisdom of our contributors. We take this opportunity of congratulating all of them for their respective submissions, for the deep knowledge and experienced judgement that they reveal, and of thanking them for their co-operation and collaboration with us in bringing the project to fruition. To the majority of them, we express our grateful thanks and appreciation for prompt responses and extreme patience; but our gratitude is no less for persistence and endeavour where these have been the distinguishing attributes.

We take pleasure in acknowledging the support afforded to us by Dr Helen Wilson, of the University of Plymouth, copy editor Harry Langford and the understanding advice and guidance initially provided by Susan Sternberg and, later, by Ian Francis and Delia Sandford at Blackwell. The finished book is a team effort. We are proud to acknowledge an excellent team.

REFERENCES

Beeton, A.M. (1984) The world's great lakes. *Journal of Great Lakes Research*, **10**, 106–13.

Berg, K. (1938) Studies on the bottom animals of Esrom Lake. *Kongelige Danske Videnskabernes Selskabs Skrifter*, **8**, 1–255.

Calow, P. & Petts, G.E. (1992) *The Rivers Handbook*. Blackwell Scientific Publications, Oxford (2 vols.), 536 + 522 pp.

Cole, J.J., Pace, M.L., Carpenter, S.R. & Kitchell, J.F. (2000) Persistence of net heterotrophy in lakes during nutrient addition and food-web manipulations. *Limnology and Oceanography*, **45**, 1718–30.

Forel, F.A. (1895) La limnologie, branche de la Geographie. *Comptes Rendues du sixième Congrès international de Geographie*, 1–4.

Herdendorf, C.E. (1982) Large lakes of the World. *Journal of Great Lakes Research*, **8**, 106–13.

Hutchinson, G.E. (1957) *A Treatise on Limnology*, Vol. 1, *Geography, Physics, Chemistry*. Wiley, New York, 1016 pp.

Hutchinson, G.E. (1967) *A Treatise in Limnology*, Vol. II, *Introduction to Lake Biology and the Limnoplankton*. Wiley, New York, 1115 pp.

Kalff, J. (2002) *Limnology — Inland Water Systems*. Prentice Hall, Upper Saddle River, New Jersey, 592 pp.

Lampert, W. & Sommer, U. (1993) *Limnoökologie*. Georg Thieme Verlag, Stuttgart, 440 pp.

Lerman, A., Imboden, D.M. & Gat, J.R. (1995) *Physics and Chemistry of Lakes* (2nd edition). Springer-Verlag, Berlin, 352 pp.

Lindeman, R.L. (1942) The trophic dynamic aspect of ecology. *Ecology*, **23**, 399–418.

Macan, T.T. (1970) *Biological Studies of the English Lakes*. Longman, London, 260 pp.

Meybeck, M. (1995) Global distribution of lakes. In: Lerman, A., Imboden, D.M. & Gat, J.R. (eds), *Physics and Chemistry of Lakes* (2nd edition). Springer Verlag, Berlin, 1–35.

Reynolds, C.S. & Irish, A.E. (1997) Modelling phytoplankton dynamics in lakes and reservoirs; the problem of *in-situ* growth rates. *Hydrobiologia*, **349**, 5–17.

Reynolds, C.S., Reynolds, S.N., Munawar, I.F. & Munawar, M. (2000) The regulation of phytoplankton population dynamics in the world's great lakes. *Aquatic Ecosystem Health and Management*, **3**, 1–21.

Steel, J.A. (1995) Modelling adaptive phytoplankton in a variable environment. *Ecological Modelling*, **78**, 117–27.

Thienemann, A. (1925) *Die Binnengewässer, Band I*. E. Schweizerbart'sche, Stuttgart, 255 pp.

Tokeshi, M. (1994) Community ecology and patchy freshwater habitats. In: Giller, P.S., Hildrew, A.G. & Raffaelli, D.G. (eds), *Aquatic Ecology: Scale, Pattern and Process*. Blackwell Science, Oxford, 63–91.

Uhlmann, D. (1975) *Hydrobiologie*. G. Fisher, Stuttgart, 345 pp.

Wetzel, R.G. (1975) *Limnology*. Saunders, Philadelphia, 743 pp.

Wetzel, R.G. (1983) *Limnology* (2nd edition). Saunders, Philadelphia, 859 pp.

Wetzel, R.G. (2001) *Limnology and Lake Ecosystems* (3rd edition). Academic Press, San Diego, 1006 pp.

Wetzel, R.G. (1995) Death, detritus and energy flow in aquatic ecosystems. *Freshwater Biology*, **33**, 83–9.

Williams, W.D. & Aladin, N.V. (1991) The Aral Sea: recent limnological changes and their conservation significance. *Aquatic Conservation*, **1**, 3–23.

2 The Origin of Lake Basins

HEINZ LÖFFLER

2.1 LAKE DEFINITION

Given the great variety of bodies of standing waters, it is not surprising that all lake definitions must be arbitrary. It is sometimes even difficult to distinguish between flowing (**lotic**) and standing (**lentic**, **lenitic**) waters—for instance lengthy, shallow swellings in a river channel with only short (weekly) retention times. Many examples of this kind are known from northern Sweden (e.g. the Ångermanälven System).

To begin with the definition of Forel (1901), the founder of limnology, that a lake is 'a body of standing water occupying a basin and lacking continuity with the sea' (pp. 2–3), one can think immediately of many lakes subject to marine influence. The most remarkable example is Lake Ichkeul in northern Tunisia near Bizerta (Fig. 2.1), where periodic changes in water supply occur. During winter and spring fresh water from six inflowing rivers predominates, but during summer and autumn, seawater enters from the Bizerta Lagoon, when its outflow (the Tinja River) becomes an inflow of Lake Ichkeul. Most recently, this remarkable system has been altered by the damming (and irrigation) activities in three of its most important inflows, which has led to a dramatic rise in salinity of the lake and its marshes (together about 115 km^2) and the destruction of large *Phragmites* stands.

Dictionary definitions (Webster 1970; in Timms 1992) define lakes as 'large or considerable bodies of standing (still) water either salt or fresh surrounded by land'. This still leaves the question of a lake's dimensions. More recently, Bayly & Williams (1973, p. 50) considered that 'a typical pond is shallow enough for rooted vegetation to be established over most of the bottom, whereas a typical lake is deep enough for most of the bottom to be free of rooted vegetation'. They add a further distinction: 'Most lakes are permanent and many ponds are temporary' (p. 50). This again is unsatisfactory since shallow lakes completely covered by emergent vegetation are conveniently considered as marshes (Kvet *et al.* 1990), and many alkaline (e.g. Lake Nakuru, Kenya) and saline lakes (e.g. Lake Niriz, Iran) are devoid of vegetation for ecological (physiological) reasons. Moreover, the term 'pond' is normally used in connection with certain types of artificial bodies (e.g. fish ponds, farm ponds, etc.) rather than shallow lakes. Timms (1992), in his *Lake Geomorphology* is clearly satisfied with the definition given by Riley *et al.* (1984; in Timms 1992)—that 'lakes are defined as areas where vegetation does not protrude above the water surface' (p. 2), and 'swamps are defined as areas where vegetation, usually emergent rooted macrophytes, dominate the surface' (p. 2).

All these attempts clearly demonstrate that difficulties with definitions arise only from the case of shallow lakes, which are considered by many authors to be wetlands, with frequent transitions towards marshes, tree swamps, minerogenic bogs or shallow swellings in a river channel. Shallow lakes are, however, merely bodies of water which are easily mixed down to the bottom by casual wind, sometimes by evaporation or irradiation, and with respect to their critical depth, are dependent on their location. Periods of stratification may be induced by ice cover (inverse stratification), by freshwater tributaries of saline lakes, or by floating, dense vegetation cover (e.g. *Azolla*, *Eichhornia*, *Nuphar*, *Salvinia*, *Stratiotes*, etc.) which may minimise wind fetch but much less so evaporation.

The size of the lake then remains the only question which needs to be considered. Very local des-

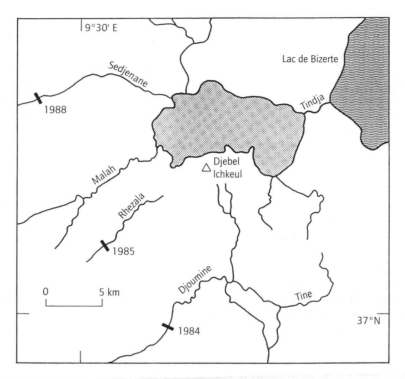

Fig. 2.1 Lake Ichkeul, Tunisia, and its connection with Lac de Bizerte (maritime lagoon) by the Tindja River. During the rainy season (winter) the Tindja River used to flow into the lagoon carrying almost fresh water from the lake, whereas, with low water level during the dry season, the river used to supply Lake Ichkeul with lagoon water. Owing to construction of reservoirs (bold bars with year of construction) for irrigation and water supply for municipalities, the most important rivers no longer reach Lake Ichkeul. Consequently the lake has become highly saline throughout the year and during the last two decades its formerly luxuriant emergent vegetation belt has disappeared.

ignations—often arbitrary or connected with historical events—may frequently be used. For such local labelling, the plain east of the Neusiedlersee (Lake Fertö) in Austria, with about 40 shallow bodies of water, offers an example. The largest lake in this group is about 2 km² in area and is called 'Lange Lacke' ('Long Lake**let**'), whereas much smaller bodies in the same area are given the designation 'See' (i.e. 'Lake'). This distinction is founded on the past occurrence of large inundations by the Neusiedlersee, which could amount up to 500 km² and include such shallow basins as distinct parts of the coastal area of the larger lake. All of these bodies are also called lake.

2.2 LAKES IN THE PAST

Lakes are transitory landscape features. Sometimes they are born of catastrophes (volcanic eruptions, floods, landslides and avalanches, meteoric impacts and major human interventions), sometimes they evolve quietly and over a long period of time. Most often, they pass imperceptibly away as they turn into bogs, marshes or tree swamps, or become filled with permanent sediment (see section 2.12 below). Equally, they may also empty via catastrophic eruptions, or again, as a result of adverse human activities. The eruption of Palcacocha (Cordillera Blanca, Peru) which destroyed a large part of the city of Huaraz in 1941, and the progressive desiccation of the Aral Sea due to human mismanagement (see sections 2.4 and 2.5.1, also Fig. 2.6 and section 2.7, and Volume 2, Chapter 7), are examples of such events.

Ever since precipitation began to collect on the terrestrial surface of our planet, lakes must have been in existence. It is likely that lake basin formation was enhanced by the Precambrian and Palaeozoic glaciations. Likewise, the subsequent formation of mountains during the Silurian (Old and Young Caledonian orogenies) should have contributed to such processes. Although *Opabinia*, often discussed as a Cambrian precursor of the Anostracan Crustacea, has been removed from this order (Hutchinson 1967), it may be the Branchiopoda (Negrea *et al.* 1999) which provide the earliest evidence for the possible existence of lakes. The appearance of the Dipnoi and Ostracoda during the Devonian (as marine groups known from the Cambrian), gives safe testimony of their existence.

Evidence for a variety of lakes, and probably the first wetlands, increases rapidly during the Carboniferous when the first aquatic insects (the Palaeodictyoptera, precursors of the Odonata, and possible ancestors of the Ephemeroptera), Notostraca and Amphibia appear. A great variety of 'tree' species contributed to the well known limnic coal-beds. The formation of the Old and Middle Variscan mountains and, in addition, the long-lasting Permo-Carboniferous glaciation (as evidenced from the Southern Hemisphere), must have created a great many lakes. The Odonata, Ephemeroptera, Plecoptera, Cladocera and Malacostraca all appear during the Permian, and inland lake fish become abundant. Fossil lakes from the early Permian have been identified in central Europe and elsewhere. Many fossil lakes and wetlands are also known from the Mesozoic, when, during the Jurassic, the first Trichoptera and aquatic Rhynchota appeared. The Late Jurassic saw the first teleosts, which became abundant during the Cretaceous. This last period is also characterised by the appearance of the diatoms (Bacillariophyta).

The oldest lakes existing today came into existence during the Tertiary. These are mainly restricted to Eurasia, Africa and perhaps Australia (e.g. Lake Eyre). The Tertiary is also the era of extremely large transient inland waters which were largely the remnants of the Tethys and Parathethys Seas, and which like the Mediterranean (the 'Lago Mare', about 6–5.5 million yr BP), or the Black Sea, were cut off over long periods from the sea. The latter, after a changeable history of almost two million years, regained its marine connections to the Mediterranean via the Bosporus only about 7500 years ago (Fig. 2.2). Other sites were located in the Amazonian Basin, the largest sedimentary area in the world. Many fossil lakes have been reported from the Tertiary, among them several classic sites with excellent preservation of vertebrates. Such examples include an Eocene lake near Halle (Germany) and a selection of Oligocene, Miocene and Pliocene sites.

During the Pleistocene (about 1.7 million until 10,000 yr BP) most of the world's present lake basins (at least 40% of them in Canada) came into existence within the areas of continental glaciation. In addition, many older lake basins became reshaped, although very little is known so far about their extent during the interglacials. This is evidenced by most of the present Alpine piedmont lakes. Many of these basins (e.g. Lake Constance, or the Bodensee) came into existence during the Early Pleistocene, if not the late Tertiary and must have undergone major changes during the glacials. Only recently, results from a profile near Mondsee (Austria) demonstrate a much higher lake level than at present. During the pre-Pleistocene, this lake must therefore have covered an area of about $30\,km^2$, in contrast to its present $14\,km^2$.

At the end of each glaciation, large proglacial lakes developed at the ice fronts. After the most recent glaciation, the Laurentian Ice Lake (with an area of more than $300,000\,km^2$), the Baltic Ice Lake (Fig. 2.3) and the West Siberian Ice Lake, expanded along the line of withdrawal of the Northern Hemisphere continental glaciers. In contrast to the first two, which left behind the Great Lakes and the Baltic Sea, the West Siberian Ice Lake, as

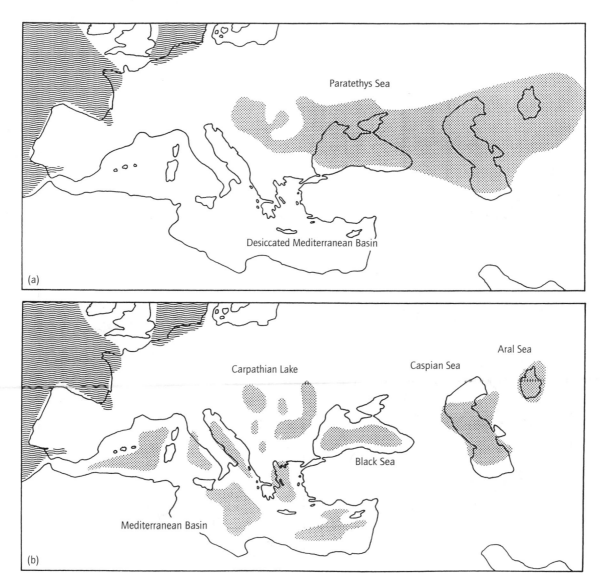

Fig. 2.2 The extent of the Sarmatian Sea (a) some 15 million years BP and (b) the remnants of the Tethys and the Paratethys 6.0–5.5 million years BP. (Modified from Hsü (1972) and Rögl & Steininger (1983).)

well as Lake Agassiz in North America, disappeared completely. In addition, at this time, many volcanic lakes (see section 2.5.2) came into existence. The crater lakes Lago di Monterosi (formed about 26,000 yr BP) and Lago di Monticchio (about 75,000 yr BP) in Italy can be mentioned as such examples. Finally, as another of the many large-scale events of the Pleistocene, the Red Sea, part of the Great Rift Valley system, should be given attention. Owing to repeated eustatic sea-level changes,

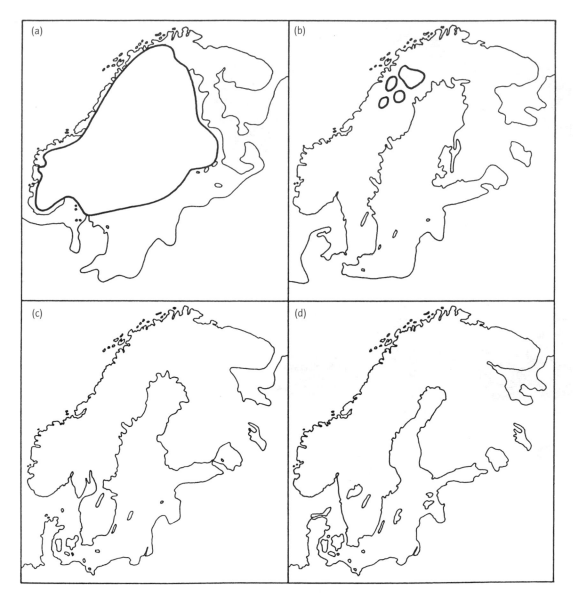

Fig. 2.3 The post-Weichselian development of the Baltic Sea. (a) As isolated ice-lake until c.10,000 years BP, just before the *Yoldia* stage. (b) *Ancylus* Lake until c.8000 years BP. (c) *Littorina* Sea until 3000 years BP. This is followed by (d), the present less saline stage of the *Lymnaea–Mya* Sea. The two driving forces of this development are (i) the melting of the Fennoscandian ice sheet and (ii) isostatic recovery of the Fennoscandian Shield.

it became separated from the Indian Ocean and during such phases was a hypersaline saltwater basin which lost most of its organisms (Thenius 1977).

2.3 HYDROLOGICAL ASPECTS AND GLOBAL BUDGET

With respect to the global water budget (Fig. 2.4)

three extreme conditions may be distinguished. One scenario would be when the oceans preponderate, leaving terrestrial surfaces mainly as desert. Conditions of this kind, though not worldwide, occurred during the Devonian, and perhaps also during the Late Permian. A second, when maximum humidity for terrestrial surfaces is available, could be presumed for the early Tertiary. Finally a third scenario with maximum amounts of water bound as glaciers, ice and snow may be distinguished. The well known examples for these states comprise the Pleistocene, and earlier glaciation periods.

The present condition of our planet demonstrates a complex situation with the impact of the Pleistocene leading to the formation of millions of lakes, mainly as a result of the melting and retreat of the continental and mountain glaciers (Table 2.1). Accumulation of ice began in Antarctica during the late Oligocene, long before the Pleistocene,

when that continent became separated from South America and moved south towards its present polar position. An independent circum-Antarctic current system then developed, and Antarctica cooled rapidly. Only much later, when Arctic ice formation was initiated, some 2.5–3.0 million years ago, did the recognised world-wide Pleistocene glaciation come about. This eventually ended some 10,000 years ago.

Obviously, plate tectonics and, hence, the shift of ocean currents, were greatly involved in these events. It should be kept in mind that, compared with the Permo-Carboniferous glaciation some 280 million years ago, the Pleistocene has lasted only a fraction of time. During the last glaciation (and also earlier ones) many of the present dry and desert areas (such as the Sahara) experienced wet 'pluvial' periods. These contributed greatly to existing 'fossil' groundwater deposits.

Since then, many climatic changes have oc-

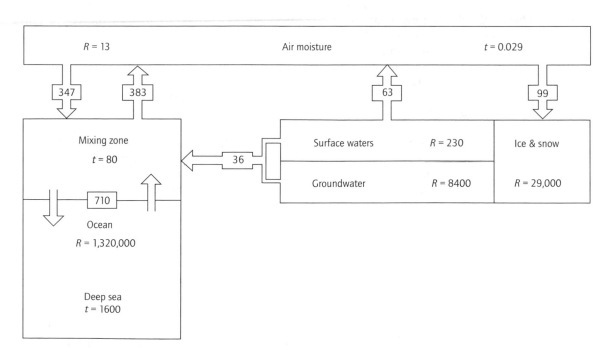

Fig. 2.4 Global water budget: t, retention time (yr); R, reservoirs [km$^3 \times$ 1000]. Arrows denote annual water flow [km$^3 \times$ 1000] (Modified from Stumm & Matter-Müller (1984) and other sources.)

Table 2.1 The existing mass of global water resources ($km^3 \times 1000$).

Ocean	1,319,800	
Ice and snow	29,000	
Air moisture	12.9	(instantaneous mean)
Rivers	1.1	(instantaneous mean)
Lakes (fresh)	125	
Lakes (salt)	104	
Soil	6	?
Groundwater, more than 800 m	4200	?
Groundwater, less than 800 m	4200	?

curred regionally and globally and most recently global warming and increased desertification have contributed to the shrinking of many lakes, especially in Africa. An example of this recent development is Lake Turkana which most probably reached high levels between 9500 and 7500 years BP. Between this pluvial peak and 3000 years BP, the lake underwent two regressions and two major expansions before the final period of falling level (Richardson & Richardson 1972). Since 1977 this decrease has amounted to approximately 1 m annually (Källquist *et al.* 1988). Even more dramatic is the shrinking of the shallow Lake Chad from its former area of 25,000 km^2 (during the 1960s) to less than 1000 km^2 at present (Fig. 2.5). This body drains the largest watershed of all African lakes (about 2.5 million km^2) which extends through six countries.

Apart from the present configuration of the continents, the pattern of ocean currents and their irregularities and other stochastic events, global precipitation (Table 2.2) is greatly influenced by large atmospheric circulation systems such as the trade winds and the westerlies. In addition, orographic features and secondary patterns are of great regional importance. Among the latter, the monsoon system, also recognised elsewhere, is most paradigmatically presented by the thermal regime of the interior of the Asiatic continent. The phenomenon of a wet monsoon blowing across the Indian Ocean during the summer provides for summer rain in areas which would otherwise be desert.

As mentioned above, short-interval fluctuations between 'normal climate' and different climatic conditions (within one or over several decades) occur, which most often follow an irregular pattern. Among the most forceful events of this kind, with almost a global influence, the 'El Niño' of the Pacific region should be mentioned. During the past 40 years, nine El Niño events have increasingly affected the Pacific coasts of North and South America, with development of warm ocean currents off Peru and Ecuador and as far afield as the Galapagos Islands where normally cold surface waters are found. Losses in fishery, imprints on the regional marine life and impacts on the climatic conditions around the globe, such as the disturbance of the Asian monsoon, may be consequences of strong events of this kind such as that which occurred in 1982–1983 (Wallace & Vogel 1994). During this exceptionally warm interval, coastal deserts in northern Peru experienced more than 2000 mm of rain transforming them into grasslands dotted with lakes. Further to the west, abnormal wind patterns deflected typhoons off their natural tracks towards islands such as Hawaii and Tahiti, which are unaccustomed to such catastrophes. They also caused monsoon rains to fall over the central Pacific instead of in its western parts, which led to droughts in Indonesia and Australia. Winter storms in southern California caused widespread flooding across the southern USA while northern areas experienced unusually mild weather and lack of snow. Overall, the loss in economic activity in 1982–1983 as a result of the climatic change amounted to over $1 billion.

In principle, the phenomenon of El Niño originates in oscillations of high and low barometric pressure between the eastern and the western sides of the Pacific. Upwelling cold-ocean water along the South American (Chile to Ecuador) coast caused by wind, induces high pressure in the east whereas weak 'easterlies' cause the upwelling to slow down; the result is low air pressure. The resulting change in ocean temperature causes the major rain zone off the western Pacific to shift east-

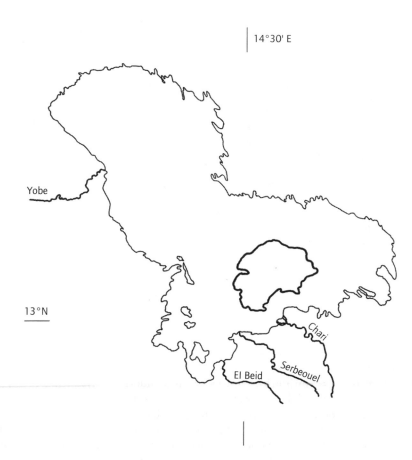

14°30' E

Yobe

13°N

Chari

Serbeouel

El Beid

Fig. 2.5 Lake Chad, which during the 1960s covered 25,000 km², has recently shrunk to a remnant of less than 1000 km².

ward and the related adjustments in the atmosphere cause the rain to fall over the central and eastern Pacific and a rise of barometric pressure over Indonesia and Australia, which results in a further weakening and eastward retreat of the easterlies. At present, our improving understanding of the wind–sea processes allows for better prediction of El Niño years.

2.4 THE PRESENT CONDITION OF LAKES

The total volume of water located in natural and artificial lakes amounts to 229,000 km³ (Margalef 1983), although some other estimates are as high as 280,000 km³. A large fraction of this volume, including that of the Caspian Sea (78,700 km³), is saline. The catastrophic decline of several large lakes (e.g. the Aral Sea from 69,000 km² during the 1960s to less than 30,000 km² at present, Lake Chad from 25,000 km² during the 1970s to about 1000 km²), is easily compensated for by the rapidly increasing number of reservoirs and artificial ponds (e.g. Lake Volta about 8000 km² and each of the two dams, Lake Kariba and the Aswan Lake, with about 5000 km²). Total global lake volume amounts to only 0.017% of the total global water volume (Table 2.1).

Estimates of total global lake surface vary

Table 2.2 Water balance of oceans and continents ($10^3 \, km^3 \, yr^{-1}$). Obtained from sources such as the International Lake Environment Committee and United Nations Environment Programme. The question mark indicates that the figure is still a matter of discussion and object of conflicting opinions. Global values of soil moisture are still missing. Regionally they have been given considerable attention only in the former Soviet Union. Since, with increasing warming, global precipitation values have also risen, the total global annual mean may at present be close to $1000 \, mm \, m^{-2}$.

Precipitation on ocean surfaces	347	
Evaporation from ocean surfaces	383	
Precipitation on land surfaces	99	
Evaporation from land surfaces	63	
Annual runoff (rivers)	28	(Amazonas 6)
Annual runoff (surface, subsurface groundwater)	8	?

greatly and most figures are too low. A value of about 3 million km^2 seems most appropriate. The figure of 1.56 million km^2 given by Herdendorf (1990) for natural lakes only is certainly too low, as it is based upon lakes >500 km^2 in area, which total 1.4 million km^2. This implies that the total surface area of the remainder is only 160,000 km^2. However, the estimated surface area of Canadian lakes, excluding the large ones, amounts to 400,000 km^2. Hammer (1988) mentions that 8% of the total area of Canada is covered by lakes, and that wetlands increase this value to 20%. The total area of Chinese lakes, again excluding the large ones, comes to 724,000 km^2. This means that Herdendorf's sum for the remainder is by far surpassed by only these two countries alone.

At present, the number of lakes on the globe must be of the order of at least 2 million. Figures are, however, available for only a few countries. There are more than 2300 lakes in China with a surface area exceeding 1 km^2, a total which does not include Tibet. The number of natural lakes in Austria, indicated on maps of a scale of 1 : 50,000, surpasses 6000; figures for countries such as Finland and Sweden exceed 100,000 by far, and for Canada a figure close to one million or more

should be expected. Similarly, Russia including Siberia should contain more than 500,000 lakes. Of lakes greater than 500 km^2 in area, 40% occur in North America, mainly in the area covered by the last glaciation.

In addition to natural lakes, artificial lakes and reservoirs abound in many parts of the world. Artificial lakes came into existence more than 6000 years ago (e.g. Kosheish Dam, 4900 yr BP) during the reign of the first Pharaoh Menes, and the first retention basin likewise used for irrigation was constructed during the Middle Empire of Egypt (Amenemhet III, 1830–1801 BC). Its name was originally 'Piom-en-mêré' (flood lake) and, for some unknown reason, later called 'Moeris' Lake by the Greeks.

Any figure for existing artificial ponds, dams and reservoirs is purely a guess, but should be in the order of one million or more. Sri Lanka, with a history of more than 2000 years of damming and pond tradition, contains more than 9000 bodies of water, all of them artificial. In many semi-arid and arid regions (e.g. North America, South Africa, Argentina, etc.) farm ponds and reservoirs are the main type of standing waters, whereas reservoirs (dammed rivers or valleys) used for hydroelectric power have been in existence for about 100 years. In addition to dammed lakes, in 1970 there were 300 reservoirs with areas between 100 and more than 8000 km^2 and more than 100 dams over 150 m in height (Goldsmith & Hildyard 1985). Many artificial lakes came about by excavation of materials such as peat, coal, ore, gravel, loam, etc. Strip mining of certain types of brown coal has produced (sulphuric) acid lakes with pH values below 3. Finally, in many countries, artificial lakes for recreation purposes are rapidly increasing in number.

2.5   LAKE CLASSIFICATION BY ORIGIN

Classification of lake basins by origin has a long tradition and began with the consideration of geological processes which Davis (1882) classified as constructive (e.g. crater lakes), destructive (e.g. the scouring of basins by glaciers) and obstructive (e.g.

natural damming events by lava, landslides, glaciers, etc.). Other schemes were developed by Penck (1894), Forel (1901) and Halbfass (1923). The most elaborated overview was given by Hutchinson (1957) who recognised 76 lake types grouped under 11 kinds of formation. Bayly & Williams (1973) have offered a slightly modified system which is adapted to Australia and New Zealand and recently Timms (1992) delivered a synthesis in which special attention is given to the lakes formed by wind action. Again, it presents a thorough overview of Australian and New Zealand lakes. Hutchinson's scheme, which is followed here with few modifications, allows one to allocate additional lake types with regard to their origin under the 11 headings. One of them ('glacial lakes') should perhaps be replaced by 'lakes formed by glacial, permafrost and ice activity', which could then comprise additional types.

2.5.1 Tectonic basins

Movements of Earth's crust by plate tectonics, regional volcanic activities or local subsidence may lead to the formation of many different morphological features of varying age. It is safe to say that at present the oldest lakes on our planet belong to this category. One of the large regional types comprises basins which became isolated from the sea by epeirogenetic earth movements and other vertical uplift by plate tectonics. Here, the foremost example is presented by the Ponto-Caspian region (Fig. 2.2) which, during the early Tertiary, was still a submerged part of the Tethys Sea. The collision of Africa with Europe brought about the closing of this sea some 20 million years ago, when it was already divided into two great arms. The southern arm was ancestral to the modern Mediterranean, the northern arm—the Paratethys—comprised the Ponto-Caspian region. The Alpine mountain formation eventually isolated the Paratethys which then became the brackish Sarmatian Sea. In its early stage it still retained its marine character.

Although lacking stenohaline organisms such as the echinoderms, it became increasingly fresh, a process which proceeded from west towards east. During the late Miocene (about 6 million yr ago),

mountain building severed the connection of the Mediterranean with the Atlantic and the entire sea evaporated and became a desert basin (Hsü 1972). This condition allowed the migration of large African mammals to Mediterranean islands. During the dry period which lasted for about half a million years, the Paratethys drained into the Mediterranean. As a result, both the Sarmatian and the Mediterranean basin became a network of lakes. The Pannonian descendants of the former Sarmatian Sea—many of them not well defined with regard to their extent—comprise, among others, the Euxine (Black) Sea, the Caspian and the Aral Sea. With the refilling of the Mediterranean by the Atlantic, the Black Sea and, possibly, a few other Pannonian basins became marine environments. However, the connection of the Euxine Sea with the Mediterranean was soon severed again and the Black Sea entered a long interval as an isolated freshwater lake. As recently as 9000 years BP, following post-glacial eustatic sea-level rise, it once again received water from the Mediterranean via the Bosporus.

The problem of the varying extent and the causes of the changes exhibited by the water level of the Caspian Sea has not yet been solved. Likewise, connections with the Black Sea ('Baku Lake' stage) during the earlier glaciations in the Pleistocene are still a matter of discussion. Since 1830, the level of the Caspian has been monitored, and the data obtained show that between 1933 and 1941 a fall of 170 cm occurred whereas between 1978 and 1994 a rapid rise of 225 cm took place. The latter was of great concern to surrounding countries. In the coastal zone flooding has ruined or damaged buildings, engineering structures, farmland and threatens the sturgeon-group fishery. As already mentioned, knowledge of the causes of the Sea's rise and fall is, despite many years of study, still limited (IHP, 1996).

The history of the Aral Sea (Fig. 2.6; see also Volume 2, Chapter 8) is even less well known. From the striking faunistic similarities between the Caspian and the former Aral Sea, especially with regard to the unique diversity of endemic Cladocera, it can be presumed that, during its highest glacial stands, the Aral Sea drained into the

Caspian. Among the endemic Cladocera which abound in the Caspian, and less so in the Aral Sea, the Podonidae most likely gave rise to the world-wide marine genera *Podon* and *Evadne*, probably during the Tertiary. Owing to human intervention with large-scale irrigation activities in the upper watershed of the two tributaries, Amu-Darja and Syr-Darja, and the loading of the Aral Sea with pesticides from agriculture, the lake has shrunk to less than half its area in the 1960s (69,000 km^2) and has lost its unique fauna.

With respect to its changeable history, but on a smaller temporal and spatial scale, the Baltic (Fig. 2.3) exhibits similarities to the Euxine Sea, in that it has alternated between fresh and saline stages during the latest Pleistocene and throughout the Holocene. Little is known about its earlier Pleistocene history. As with the West Siberian Ice Lake, the Baltic began as a proglacial freshwater lake dammed by the Fennoscandian ice sheet in about 12,000 years BP. As melting progressed, it was invaded in about 10,000 years BP in the Billingen area

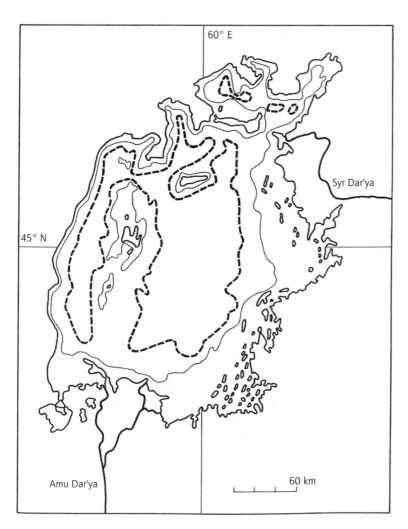

Fig. 2.6 The Aral Sea before 1960 and in 1995 (dashed bold line). (Modified from different sources.) Since 1997 (?), the lake has separated into two bodies.

of southern Sweden by marine waters, to become the Yoldia Sea (named after a mussel which at present is designated *Portlandia arctica*). The southern part of the now deglaciated Scandinavia then rose isostatically more rapidly than the eustatic rise in sea level, and in about 8500 years BP the Baltic became once again a lake, during the *Ancylus* stage (coined after a freshwater snail). Later, isostatic recovery slowed, and the eustatic rise continued with increasing velocity, and in about 8000 years BP the Baltic again became connected to the sea in the area of the Danish Sounds, and entered the Littorina Sea stage (named after the snail *Littorina littorea*). Further isostatic rise finally weakened the connection with the ocean and in about 3000 years BP resulted in the *Peregra (ovata)—Mya (arenaria)* Sea—named after a freshwater snail and a brackish water mussel respectively—with its northern portion almost fresh. Recent weak eustatic rise is reported slightly to influence the southern Baltic.

The Lake Eyre Basin (area $1.3 \times 10^6 \, km^2$) and its highly astatic shallow lake, with an area of 9690 km^2 and a maximum depth of 5.7 m when full, has been endorheic since drainage to the sea was blocked by upwarping across its southern boundary (Timms, 1992). Therefore this basin is a good example of the reversal of hydrographic pattern by tilting or folding. Moreover it represents a cryptodepression with the lowest point located some 15 m b.s.l. (below sea level). It fills only after periods of heavy rainfall (e.g. 1950 and 1974) when the lake changes its configuration, owing to massive shoreline erosion. Lake Eyre is a descendant of Pleistocene Lake Dieri which, according to Timms (1992), may have originated during the latest Tertiary. The most prominent example of large-scale basins formed by warping is represented by Lake Victoria with an area of 68,000 km^2 and a maximum depth which recently was reported to be 81 m. It lies in the basin of an uplifted area between the eastern and the western Rift Valley and is of early Pleistocene origin. In about 10,000 years BP it was (almost?) dry.

Lakes may also occupy sections of syncline, and thus be reasonably attributed to folding. Fählensee in Switzerland is cited as a classic example of this kind (Hutchinson 1957). It has been proposed that at least the northern part of Lake Turkana (see below, this section) may also belong to this category. Old peneplain surfaces which may form intermontane basins during the process of mountain faulting are relatively rare. Such basins may be deepened by subsequent local block faulting. The most remarkable example of this kind is the Altiplano which extends through the central Andes. Apart from smaller and shallow lakes in Peru, Lake Titicaca (with an area of 8300 km^2 and a maximum depth of 304 m) is the foremost example, and, like Lake Victoria, came into existence during the earlier Pleistocene. It is part of an endorheic watershed and drains into the highly saline Lago Poopo. Timms (1992) mentions Lake Buchanan in the central Queensland highlands as an Australian example.

Following earthquakes, local subsidence may produce small to medium lake basins. Hutchinson lists examples from Italy, Tennessee, Missouri and Arkansas. The most important type of tectonic basins, however, very often both in terms of extent and depth, are associated with faulting, either with lakes formed against a tilted fault block, or in down-faulted troughs and in grabens between two fault blocks. Hutchinson (1957) considered Albert Lake in Oregon to be the finest example of a lake lying along a fault block, and Timms (1992) refers to the highly astatic Lake George in New South Wales. Much more common are lakes in grabens. Lake Baikal, in eastern Siberia, the oldest (Oligocene) and deepest (1637 m) lake in the world, with an area of about 31,500 km^2 (ILEC 1992) and a volume of 23,000 km^3, is also unique with respect to its endemic fauna (Fig. 2.7). With its 'coral-like' sponges, its abundant amphipod species and endemic fish families and genera, it belongs to the group of Tertiary lakes with the highest diversity. In spite of its very great depth, Baikal contains bottom oxygen concentrations of more than 75% of the air-saturated value.

With respect to morphological features, such as its long configuration and its division into three basins, Lake Tanganyika (area 31,900 km^2; depth

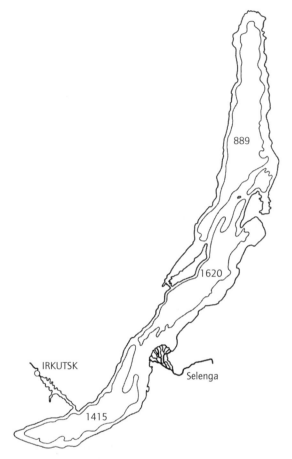

Fig. 2.7 Lake Baikal, the deepest lake in the world (1620 m) which is divided into three basins by ridges. The central basin presents the deepest part, followed by the southern (1415 m) and the northern (889 m). (Modified from different sources.)

ostracods, etc.) is the evidence for living conditions in the lake which came into existence sometime during the Miocene. Lake Malawi, another remarkable lake in the Rift Valley, is, most likely, much younger (early Pleistocene?) and is outstanding because of its fish diversity of more than 500 species (mainly cichlids), which is far greater than that of any lake on our planet. In contrast to these drainage lakes, Lake Turkana in northern Kenya and extending into Ethiopia is a closed lake which has experienced a fast declining water level — about 1 m per year — and increasing salinity since 1977. As a consequence annual fish catches of about 17,000 tons in 1976 have fallen to approximately 1000 tons in 1994 (Källquist *et al.* 1988).

The most renowned example of a graben in North America is Lake Tahoe which has been thoroughly studied by Goldman and his students. It is located in the Sierra Nevada, west of the Great Basin (once Lake Bonneville), and deepened by a lava dam (Hutchinson 1957). A puzzling morphometric feature is its extremely flat bottom, at a depth of about 500 m. Finally, the large lakes of central Asia, Balkhash and Issyk-Kul (with a maximum depth of more than 700 m), lie in down-faulted troughs (Mandych 1995).

On a smaller scale, the astatic Neusiedlersee (Lake Fertö) with an area of about 320 km², and a maximum depth of less than 2 m, represents an example of a lake in an area of tectonic subsidence (Küpper 1957) which took place, however, in two steps. Its basin is embedded in 400–800 m thick Pannonian sediments on top of igneous rocks of the Leitha Mountain range and once possessed a large extension (now the Hanság Plain) to the southeast, which most likely was the precursor of the present lake. The modern lake lost much of its flood-plain early last century, via drainage activities and construction of a channel which functions as an artificial outlet. It came into existence about 12,000–13,000 years BP, only some 1000 years after the Hanság Plain (Fig. 2.9) as is indicated by the maximum depth of the two basins. Moreover, the ostracod assemblage of the Hanság sediments corresponds to a cold water fauna whereas the late Pleistocene assemblage of the present lake indicates a slightly warmer stage. Since its formation,

1435 m) is the other body most similar to Lake Baikal. At present, it is the largest lake within the western branch of the Great Rift Valley which extends through Ethiopia, the Red Sea and into Asia (Dead Sea, Fig. 2.8). However, it differs greatly in its ecology from Lake Baikal. Being meromictic, with an oxygenated upper layer of only 80–100 m, its remarkable fauna is restricted to a volume of about 3000 km³. At the same time the outstanding diversity of fish and molluscs ('thalassoid' snails,

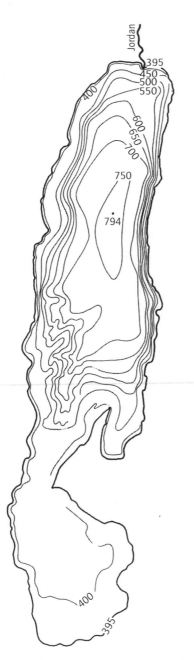

Fig. 2.8 The Dead Sea, a descendant of the much larger Lisan Lake which extended into the Kinneret Basin during the Pleistocene. The Dead Sea represents the deepest cryptodepression in the world. (Modified from Neev & Emery, 1967.)

the Neusiedlersee has been dry many times and on other occasions has expanded into the Hanság area whence it eventually overflowed into tributaries of the Danube.

2.5.2 Volcanic lakes

The great variety of volcanic processes which contribute to lake basin formation has been carefully described by Hutchinson (1957). In order to provide a hierarchical classification, Timms (1992) divides volcanic activities into vents, volcanic–tectonic depressions and lava fields. This classification fails somewhat when faced with an almost unlimited variety of more complex events (e.g. the combination of maars with cinder damming, etc.). Among true crater lakes, lake basins occupying unmodified cinder cones are obviously rare and hard to identify (Hutchinson 1957). In contrast, explosion craters and maars, which are shaped by a single volcanic explosion, often in combination with other volcanic events, are widely distributed and recognised by their frequent small size, their circular shape and very often considerable depth. Examples comprise the classic Eifel Maars (Thienemann 1913), the Lago di Monterosi in Italy (Cowgill & Hutchinson 1970), and lakes in the Auvergne (Lac Pavin, France), Indonesia (Ruttner 1931) and Uganda. Timms (1992) lists many examples from Australia.

Lake Nyos in Cameroon (Fig. 2.10), which became notorious in 1986 for its outburst of CO_2, which caused more than 1600 casualties, is a maar with an area of 2.75 km^2 and a maximum depth of 220 m. Since it is dammed by a cinder barrier, its level is raised and its shape is irregular. Outbursts of CO_2 have been reported from other lakes in the area and the native people are familiar with this kind of catastrophe. They have never settled in valleys below any of the lakes from which there is risk of gas outbursts. Therefore, those who became victims at Lake Nyos were mainly recent immigrants to the area.

The caldera, carefully described by Hutchinson (1957), is probably the most variable and complex volcanic basin, and is connected with catastrophic outbursts such as Krakatau in 1883. Here, and in

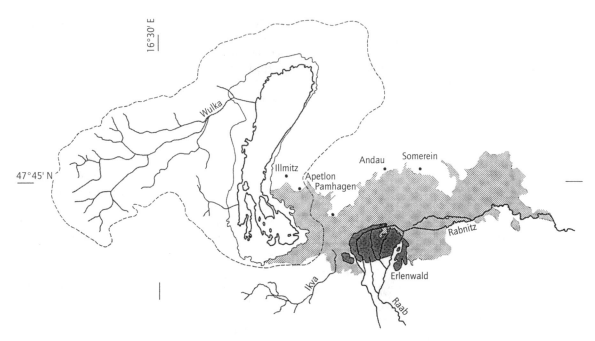

Fig. 2.9 Modern Neusiedlersee with its present watershed (shaded) and its *Phragmites* belt (area between the thick and thin lines). The dashed boundary marks the extent of the Hanság during periods of maximum water level (before its regulation early this century).

other types of eruption, the central area is liable to collapse, so producing the caldera. Among the various types of calderas, Crater Lake in Oregon, the second deepest lake in North America (608 m) and examples from Japan (e.g. Tazawako, Honshu) may be mentioned. A modified type of caldera, a depression formed by several eruptions from craters occupying the central part and later from new craters to the west is exemplified by the Conca di Bolsena (Italy), in which lies Lago di Bolsena, with an area of 114 km² and a maximum depth of 151 m. As a consequence of later eruptions, the material thrown out of the central region has collapsed in stages to form a basin with gently sloping sides interrupted by step faults, which produced terraces (Hutchinson 1957). The term **conca** (plural conche) is used for calderas of this type and is exemplified by yet another Italian lake, Lago di Bracciano.

The volcano-tectonic basin of Lake Toba (Indonesia), with an area of 1150 km² and a maximum depth of 529 m, represents the largest caldera on our planet. During the Tertiary the Toba region was a mountainous landscape with numerous volcanoes. Their eruptions ejected an immense quantity of pumice and ash which cover an area of approximately 20,000 km² in the vicinity of the lake. As a consequence, a large collapse along existing fault lines formed this volcano-tectonic basin. Further volcanic activities resulted in an island. Together with other Indonesian lakes, Toba became a foremost object of early limnological research in 1929 (Ruttner 1931). Since then, this large lake has been adversely influenced by a number of activities.

Lake types connected with eruptive materials such as ash, cinder, lava and rocks exhibit a large variety of origin. The most renowned example of a

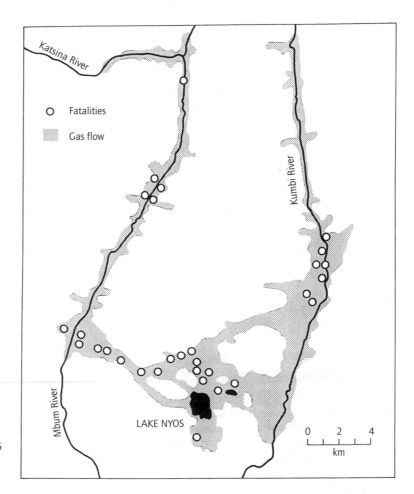

Fig. 2.10 Emission of CO$_2$ from Lake Nyos (Cameroon) in August 1986 which caused the death of 1746 people. (Modified from Löffler, 1988.)

lake dammed by volcanoes is Lake Kivu located in a former valley of a tributary of the Nile. During the Pleistocene, it was dammed by the still active Virunga volcanoes, and its outlet, the River Ruzizi, became an important tributary of Lake Tanganyika, which, after a period of endorheic condition of unknown duration, then regained an outflow towards the Zaire Basin.

Of lakes on or between lava fields many examples can be found world-wide. Timms (1992) cites many cases from eastern Australia. He believes that the largest permanent lake in Victoria (Corangamite, with an area of 250 km^2 and with a maximum depth of 5 m) lies in hollows between lava flows whereas Bayly & Williams (1973) suggest that it may be the result of a volcano-tectonic collapse. Lac d'Aydat in the Auvergne (France) represents a classic case of a lake dammed by a lava flow.

2.5.3 *Lakes formed by landslides*

Landslides, often released by earthquakes or volcanic events and of world-wide distribution in mountain areas, may be considered next. Like those formed by mudflows or avalanches, many of the basins produced in this way are short-lived and can give rise to catastrophic outbursts. Quite a few, however, are stable, some of which came into existence during the Pleistocene. One of the most interesting examples, cited by Hutchinson (1957), is Lake Chaillexon on the French–Swiss border, which was probably dammed by both pre- and post-Würm landslides. Niriz Lake, east of Shiraz in southern Iran, at present an end-lake (see section 2.7), may have lost its outlet owing to a Late Pleistocene landslide (Professor Hans Bobek pers. comm.). From Australia only two examples have so far been reported but there are many more cases described from New Zealand, some of them with areas greater than 10 km^2 (Timms 1992). There are also many landslide basins known from central Asia, such as Murgab in the Pamir Mountains, which came into existence only in 1911 (Hutchinson 1957). Numerous basins of this kind have also been reported from the western USA.

2.5.4   Lakes formed by glacial, nival activities and by permafrost and ice

These represent the most diverse group of basins (at least 23 types). Moreover, the Pleistocene glaciation has contributed to at least 70% of all the lakes now in existence. It must be assumed that earlier periods of glaciation, like the long-lasting Permo-Carboniferous one, have been efficient in basin forming as well.

A large group of lakes still in contact with glaciers comprises the small bodies lying on their surfaces and those which remain dammed by glaciers. The latter, however, may develop channels through crevasses or sub-glacial tunnels, and eventually may empty, sometimes catastrophically. One of the most remarkable lakes dammed by a glacier is Tsola Tso, the largest lake in the Mt Everest area of Nepal. During the monsoon, it acquires an outlet over the glacier, whereas between monsoon periods sub-glacial outflow predominates, and the lake may completely dry up. The difference in lake levels between the two states exceeds 16 m (Löffler 1969; Fig. 2.11).

Classic cases of lateral lakes dammed by glaciers are found in the Märjelensee valley of Valais (Switzerland) and also the Shyok tributary of the upper Indus watershed in the Karakorum. Large and very large lakes have been or may still be formed against continental or regional ice sheets and are designated as 'proglacial' lakes. Amongst these, the most prominent examples are the North American (Laurentian and Agassiz), the Baltic and the West Siberian ice lakes (Fig. 2.3). Lake Warren, the precursor of the Finger Lakes of New York State, is considered a modification of this type. Small examples of proglacial lakes occur when valley glaciers retreat behind solid terminal moraines or non-eroded rock barriers. Many such lakes have come into existence especially in the Andes, the Himalayas and in other mountain ranges. Several have collapsed by catastrophic emptying as mentioned above. About the same time as the eruption of the Cordillera Blanca lake in Peru, other lakes in the surrounding Cordillera Huayhuash emptied by catastrophic outbreaks (Löffler 1988). Similar catastrophes have been reported earlier from the Alps in Switzerland. Sometimes, ephemeral lakes may be dammed by avalanches and become major hazards for the valley below.

A remarkable example of a proglacial lake held by a resistant rock formation is Lewis Tarn on Mt Kenya (4574 m above sea level (a.s.l.)) which came into existence in about 1930. In 1960 it was still in contact with the Lewis Glacier (Löffler 1968) but only a few years later (1976) the ice had completely retreated from it. A few proglacial lakes may become completely ice-covered over some period. This was exemplified by the Curling Pond, the highest lake in Africa on Mt Kenya (4782 m a.s.l.) during the 1960s (Fig. 2.12). Other examples have been reported from Peru and from New Zealand (Timms 1992) and, more recently, Lake Vostok, Antarctica. Although the above-mentioned basins have arisen via glacial erosion, it seems convenient to restrict typical glacial rock lakes, such as ice-scour lakes, to a separate group which would include a vast number of small lakes on the Canadian shield, in Fennoscandia and in mountain regions such as the Alps and the Highlands of Scotland. It is, however, not always possible to distinguish between glacial rock lakes and lakes produced by valley glaciers.

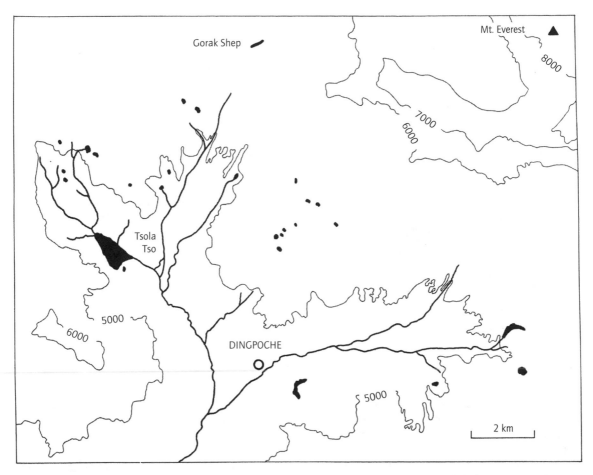

Fig. 2.11 Tsola Tso in the Mt Everest area is dammed by a glacier. During the dry season the lake loses its water by seepage beneath the glacier. Monsoon rains during summer fill the basin again until it attains a surface outflow. The annual fluctuation in lake level exceeds 16 m. (Modified from Löffler, 1969.)

Another abundant group consists of the generally small and often shallow cirque lakes which occur in varying number in practically every mountain range which has once been glaciated or which still possesses active glaciers. In some mountains, where, during the Pleistocene, only small glaciers existed, cirques may be the only or the main type of glacial basins. The Bohemian Mountain Range, and the elevated area near Mt Kosciusko in Australia, are examples (Timms 1992). Very often, cirques are restricted to the head of the valley, but in long valleys at the appropriate altitude a series of cirques may occur which are conveniently referred to as 'paternoster' lakes. Among the processes which can be involved in cirque formation, glacial excavation, and more often frost-riving due to freezing and thawing at the rock face of a névé-filled concavity on the mountain surface, play the major role.

In the northern and southern calcareous zone of the Alps it is often difficult to distinguish between cirques *sensu stricto* and basins formed

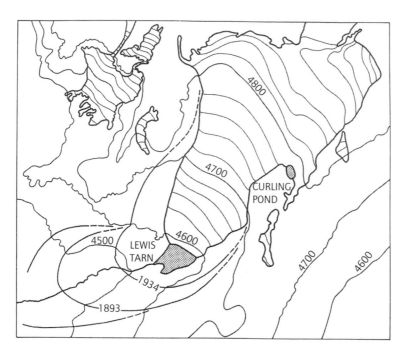

Fig. 2.12 The Lewis Glacier area (Mt Kenya) in 1960. At this time Curling Pond was still permanently frozen over and Lewis Tarn, behind a rock barrier, was still in close contact with the glacier. At present the glacier lies far behind Lewis Tarn and Curling Pond is no longer an amictic lake. The glacier stages (terminal moraines) in 1893 and 1934 are indicated. (Modified from Löffler, 1969.)

in combination with solution processes. Among many examples of this kind, the Lunzer Obersee in Austria is of special interest. Situated at the head of the Seebach Valley it possesses the characteristics of a cirque but also the features of karstic processes. It is thought originally to have possessed a sub-lacustrine outlet which became obstructed during the formation of a peat bog. As a consequence, the lake rose until it reached its present surface outlet, which after a short distance runs for several kilometres in a subterranean river. This matter, however, needs further investigation.

If ice accumulates in the large depressions of wide and open valleys with narrow outlets, glacial excavation may lead to the formation of deep and large rock basins above the exit of such a valley. According to a Norwegian local term (**glint** = boundary), basins of this kind are labelled **glint lakes**. Often these basins (such as Lake Torneträsk in Swedish Lapland, and Scottish examples listed by Hutchinson (1957)) are not formed purely on rock.

In Austria the Enns Glacier excavated a large and deep lake above a prominent gorge ('Gesäuse') which, however, owing to rapid retreat during the Late Pleistocene, filled with gravel.

The large glaciers of mountain ranges such as the Alps, which occupied long valleys during the Pleistocene, descended to the foot-hill elevation or to the plains beyond, and excavated large trough-like basins, conveniently classified as piedmont lakes. Among this category, the English Lake District is a remarkable example for which Hutchinson (1957) offers a detailed description. A variation of this kind are the fjord lakes which are often associated with extremely indented coastlines and which descend to sea level. Lakes in Norway provide classic examples of such Pleistocene processes, as do the fjord lakes of British Columbia. Ongoing glacial activities from southern Chile and the southern island of New Zealand should also be mentioned. The large piedmont lakes along the southern fringe of the Alps (the Lago Maggiore, Lugano, Como, Iseo and Garda) were once fjord lakes, but lost their connection to the Adriatic when, during the Pleistocene, alluvial deposition by the River Po shifted the original coastline eastwards. These lakes were given their final configuration during the last glaciation. In conclusion it should be stressed that many piedmont and fjord basins were pre-shaped tectonically and therefore may have originated during the late Tertiary. During the Pleistocene, they then experienced extensive transformation. Obviously, Lake Constance belongs to this group of old basins.

A large number of lakes are formed in glacial deposits in valleys, such as terminal and lateral moraines, whereas basins on ground moraines are more common in areas of continental glaciation, and are conveniently called drift basins. The most common type is the basin held behind a terminal moraine which comprises lake types already mentioned in connection with pre-glacial, cirque and piedmont lakes. As described earlier they may become sites of catastrophic emptying, if located in steep valleys.

Another very common type of basin formed by glaciers is represented by kettles (or kettle holes), which are produced by melting of stranded ('stagnant') ice masses during periods when glaciers retreat. Melting may occur among terminal moraines or on ground moraines. Ice-blocks may also be washed away and then become stranded in outwash deposits. In many cases it is rather difficult to define the origin of kettles accurately. In the Austrian Alps, basins formed by solution are often confused with kettles.

Sometimes, as in Minnesota, USA (Hutchinson 1957), kettles are arranged in rows. They are most often of moderate size ($0.01–5\,km^2$ in area) but rather deep (up to 50 m and more) and therefore exhibit a strong tendency to meromictic condition. Within the former glaciated region of North America east of the Rocky Mountains, they represent a major portion of the lakes. Among the well known examples, Linsley Pond in Connecticut, USA, and Klopeiner See in Carinthia, Austria, may be mentioned as suitable topics of further limnological research.

Glacial tunnel lakes belong to a regional group of basins which came into existence in areas of the large pre-glacial ice lakes. Examples of such meltwater channels which were excavated under the ice often occupy elongated basins. They are abundant in the once glaciated plain of the Baltic, where they extend from Denmark and northern Germany (Großer Plöner See) into Poland.

Within the permafrost region, which comprises the tundra, and part of the taiga biome, innumerable lakes are formed by thawing during the summer. Complex situations are connected within areas at the border of permafrost. Along with the withdrawal of permafrost during recent decades, the long-term growth of large lakes, such as those from eastern Alaska described by Hutchinson (1957), has come about. It has also been suggested that, in critical transition areas, any damage to vegetation may lead to melting of ice wedges, and the formation of lakes.

Within permafrost and periglacial zones, during periods of frost, both now and during the late Pleistocene, water accumulates in shallow impermeable basins to form **pingos** (ice lenses) during periods of frost. As elevated structures, they may prevent covering of the landscape with material such as gravel. Some of the shallow bodies of water east of the Neusiedlersee (see section 2.1), which are embedded in a large area of Pleistocene gravel from the Danube, are supposed to be remnants of such pingos. Similarly, the **palsa** type of peat bog, developed on elevated ice cones in the permafrost region, may produce basins after thawing. Apart from pingos and palsas, a third type of basin, again formed by ice, is found in the Neusiedlersee (see sections 2.1, 2.5.1, 2.5.7, 2.7, 2.11 and 2.12), which, on average, is covered by ice annually for about one month. During the spring thaw, ice floes driven by winds pile up to more than 10 m height and are driven against the slightly sloping eastern shore. Owing to numerous events of this kind, a levée has been formed along the eastern shore of the lake which has given origin to a series of long, narrow shallow groundwater basins. In contrast to other basins east of the Neusiedlersee, which partly may be derived from pingos, these shallow bodies to the west are probably only a few thousand years old. This is suggested by archaeological findings at the base of the levées.

2.5.5 Lake basins formed by karstic events and solution processes

These kinds of lake are widely distributed in calcareous regions, and are conveniently designated as karstic lakes. This term is derived from the Dinaric Karst with an area of about 70,000 km^2 (Bonacci 1987). The power of water solubility in contact with rocks depends on the water temperature and its chemical composition, with the dominant component being CO_2. In addition, rock salt and gypsum may contribute to solution lakes. In some parts of the Dinaric Karst region (e.g. Istria) there is an extraordinarily dense concentration of dolines (funnel-shaped depressions), which are, however, not always filled with water. If fused, such a group of dolines is labelled an **uvala**. Similar tectono-karstic depressions in Dalmatia are called **poljes**.

Many bodies of water in dolines, uvalas and poljes are astatic, but there are also perennial examples such as Lake Vrana on the island of Cres (Cherso) off the coast of Istria. Situated in a large, fresh cryptodepression with an area of about 5.7 km^2, its surface lies some 14 m above sea level but its maximum depth some 70 m below. In contrast to earlier statements, which suggested that water sources to this lake include an underground component from the mainland, according to recent calculations, its water budget is now considered to be balanced (Bonacci 1993). Very recent decline of its level is explained by coincidence of increased demand for water supply and periods of unusual dryness.

Scutari, the largest lake of the Balkans, with an area of 372 km^2, is of tectonic origin and over most of its area only about 8 m deep. However, it contains sub-lacustrine dolines with a maximum depth of almost 50 m. Obviously, some of these function as a limnokrene. Among the temporary lakes of the region, the Zirknitzer See, a lake which occupies a polje, deserves special mention. This lake, described as early as 1688 by Valvasor, is fed by a number of streams and springs and drained by a great number of sinks. Obviously, water disappears from the lake whenever the level of the local water table falls below that of the lake floor.

Elsewhere, the Lago di Trasimeno in Italy represents an example of a very large (126 km^2) but shallow ($z_{max} = 6.3$ m) solution lake. Numerous perennial and astatic solution lakes are located in the calcareous zones of the Alps. Muttensee (Glarus, Switzerland), at 2448 m altitude represents an **uvala** remodelled by glacial action. Like the Lunzer Obersee (Fig. 2.13) its outlet sinks below ground in a tunnel almost immediately after leaving the lake. One basin which fills only after heavy rainfall is the Halleswies See in Upper Austria. Afterwards, it empties quickly underground into the Attersee. In America, the most important solution basins occur in Florida; although less numerous, they are also known in Central America. From Tasmania, Lake Chrisholm is mentioned by Timms (1992) as a good example.

Other lakes are formed through solution of rock salt and gypsum. Many examples have been described from France and Germany. Small basins in Lower Austria (Gaming) were for a long time thought to be kettles but there is now good evidence that they present solution basins related to an area underlain by Permian sediment known for its high gypsum content (the Haselgebirge). Finally, the numerous basins located in caves (such as Postojna, Slovenia) should be mentioned. Many of these provide environments for endemic and rare species. *Proteus anguineus* (Proteidae: Amphibia), which has been known for more than 300 years from Postojna and other sites in Istria and Dalmatia, is just one example.

2.5.6 Fluvial lakes, lakes in flood-plains and deltaic areas

Basins shaped by rivers may be either (i) **rhithral** (i.e. representing the upper course of the river, where much gravel and other coarse material is present), (ii) **potamal** (characteristic of those middle stretches, where the river carries sand and mud only), or (iii) **deltaic** (i.e. found at the mouth of the river, where there is the additional influence of the sea and the activity of the wind). If the river is fed by a glacier, the section close to the spring is called the **cryal**, and may be occupied by a glacial lake (see section 2.5.4). Otherwise it is often distinguished

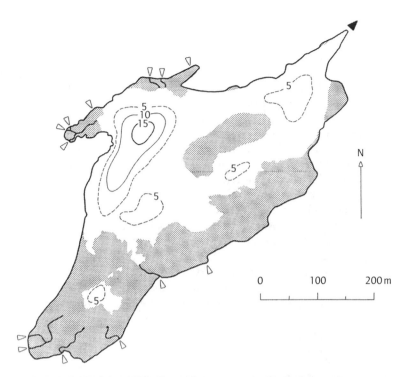

Fig. 2.13 Lunzer Obersee (Lower Austria), a cirque with karstic features. Shaded areas denote floating mats of bog vegetation.

as **crenal**, which can be replaced by a limnocrene, such as mentioned for the Lunzer Obersee (section 2.9).

With its changing but generally distinct inclination, and the coarse material present (which may be missing in rock beds if schistose material prevails) the **rhithron** facilitates the formation of plunge pools, evorsion basins, and gravel levées as multiple barriers for basins along such river sections. Rock or mountain bars may reduce the inclination of the rhithron and extended flood-plains. Here, meandering courses with the formation of oxbow lakes may be observed, which otherwise are typical of the potamon. Generally, however, rhithron rivers, or the rhithron sections of rivers, exhibit a tendency to produce furcation systems within their flood-plains and, therefore, are more likely to create abandoned river channels (**bras mort**). All the types mentioned are widely distributed. Famous plunge pools corroded by waterfalls are well known as tourist sites in warm regions

such as one remarkable example in Jamaica. Evorsion basins (*Auskolkungs-Becken*) can exceed a depth of 10 m and are due to the existence of obstacles in flooded rhithron plains. Typical examples are located within the flood-plain of the Danube east of Vienna. In connection with a mountain ramp, oxbow lakes abounded within the rhithron of the mid-section of the Austrian Enns River before it was regulated.

As a result of regulation of rivers, lakes behind levées have now become rare in industrial countries, but can often be identified on old maps. They are still, however, a prominent feature along the Yangtse Kiang where both large **embarkment** lakes and lateral (blocked valley) lakes are rather common. In addition to the basin types mentioned, unusually large inputs of gravel from a lateral tributary may eventually dam the main river. The Goggau See in Carinthia (Austria) is an example of a lake of this kind, which came into existence during the late Pleistocene and until most recently—when the tributary supplying gravel became regulated—was a slowly growing basin. If such an eroding tributary builds an alluvial fan into a lake, it may not only modify the shoreline but divide the lake. The Wolfgangsee in Upper Austria and Salzburg is one of the notable examples of an almost divided lake (see also p. 54).

The potamon sections of a river, or exclusively potamon rivers, lack most of the rhithron structures such as the evorsion basins and abandoned river channels, whereas plunge pools may occur if a rock barrier crosses the valley. Apart from types such as basins behind levées and lateral lakes, the oxbow basin is the most typical feature of potamon rivers. Classical examples abound along the Mississippi, the lower section of the Amazon, the Mekong and the Tisza in Hungary. In Australia they are called 'billabongs' and are abundant along the Murray and Darling River. Within the inner part of a meander, as part of the river or former river loops, ephemeral lakes and narrow basins due to levées (concave bench lakes) may develop. Examples of these have been described from many sites, especially from the Pantanal in Brazil (Tundisi 1994). In addition, insignificant depressions of the flood-plain may become inundated for short periods only.

Owing to the recent (post-glacial, Holocene) eustatic rise in global sea level, many deltaic areas (e.g. the Danube delta), only came into existence after the Pleistocene. Of more recent origin are some which have been created by erosion processes within watersheds owing to deforestation: the delta of Kizil Irmak (Turkey, Black Sea) belongs to this group. Development of deltas almost always implies formation of levées and hence of basins distributed accordingly. In addition, continual subsidence of alluvial deposits of the delta contributes to the creation of lake basins. Since salt layers are often represented in deltaic areas, and since marine influence there is strong, deltaic lakes are frequently brackish, euhaline or even hyperhaline. An example is the Camargue region of the Rhone Delta, with its wide range of fresh, brackish and highly concentrated salt lagoons (Fig. 2.14). Similarly, the Danube Delta contains a great variety of lagoons of different salinity (Fig. 2.15).

The Volga Delta is of special interest since in this region the Caspian Sea, with only 12–13‰ salinity, is strongly influenced by the large river, so that fresh water prevails. Moreover, since the late 1970s, the level of the Caspian Sea has risen by 1 m per year, which contributes to lower salinity. The reason for this water-level rise is still a matter of discussion (see above). Coastal lakes (see below) may sometimes be connected to deltas if ocean currents provide for a bar which eventually may close a bay in the vicinity of a deltaic river mouth. Examples of this kind are paradigmatically presented by the Vistula 'delta' with the longest bars in Europe, and to a lesser extent by the Danube delta.

2.5.7 Coastal lakes

Maritime coastal lakes, formed by deposition from sea currents of material such as sand, but not connected with deltaic areas, have formed along the western coast of France, large sections of the southern and eastern coasts of Australia, and also the coast of the Gulf of Mexico. Bars between the continental coast and islands, though rare, have also contributed to the formation of lagoons, and similarly, small lakes may become separated by bars from large lake basins. Many examples of this kind are found along the shores of the Laurentian (Great) Lakes. The bar and dam formation is more common in shallow lakes which tend much more to split into parts (see the example of the Neusiedlersee, section 2.5.4). There, wind fetch, often in association with ice floes, can contribute greatly to the shifting of bottom material.

2.5.8 Lakes formed by deflation

Hutchinson (1957) distinguished four types of basins created by wind action, but Timms (1992) proposes seven and classifies them as coastal basins, and basins of arid zones. The latter category, however, includes lakes in important coastal areas in North and South America, southwest Africa and Australia. Deflation processes are dependent on wind speeds $\geq 10\,\mathrm{m\,s^{-1}}$ and average length of duration of wind. Wind-eroded basins in

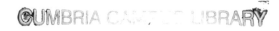

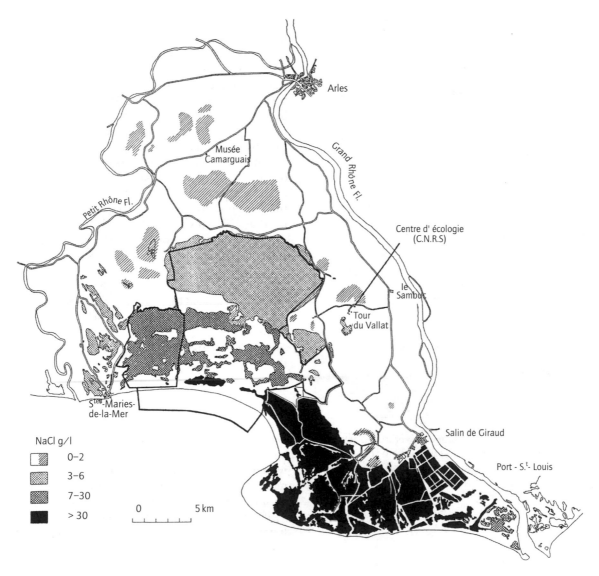

Fig. 2.14 The Rhône Delta and its sodium chloride concentration ranges. Areas with >30 g NaCl L^{-1} comprise the hypersaline brine pans for salt production. Hatched areas along the Petite Rhone and the Grand Rhône represent the still remaining riverine forests.

coastal dune areas are normally just exposures of superficial water tables of varying depth, and are mainly oval shaped or with downwind concave shores. As a special type, Timms (1992) mentions 'parabolic' dune lakes, formed between hairpin-formed dunes. Dunes may also form barrages across depressions and valleys, and basins of this kind can vary greatly with respect to position, shape and limnological character. Basins between parallel dunes are often relatively little affected by

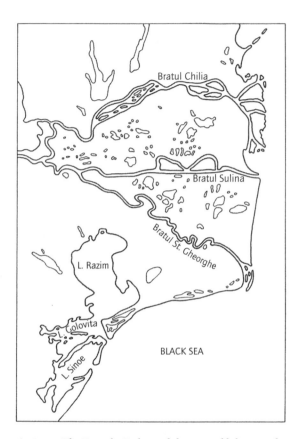

Fig. 2.15 The Danube Delta and the coastal lakes south of St George's Branch. After a critical period of destruction the delta has recently become protected and declared a Biosphere Reserve by UNESCO.

a highly impermeable horizon. Some of these lakes, described from the east coast of Australia, are formed behind barrages of Pleistocene dunes. Dunes may also be connected to shallow lake basins, and the most outstanding example of this kind until the early 1970s was represented by Lake Chad with an area of 20,000–25,000 km^2 but a maximum depth of only about 5 m. During that time its eastern portion was characterised by an extended dune archipelago with dunes oriented NW–SE. Obviously, existence of this dynamic dune system contributed not only to the formation of short-lived separated bodies of water but also (by infiltration of salts) to the low salinity of this large endorheic lake. Owing to on-going desertification and the increasing use of the lake's inflows for irrigation, the area of Lake Chad has shrunk to less than 1000 km^2 during the past 30 years (Fig. 2.5) and the lake has, therefore, lost its connection with the dunes. The catastrophic shrinking appears even more dramatic if it is considered that during the period 5000–6000 years BP the lake area was more than 300,000 km^2 (Leveque 1976).

In arid regions (including their coastal zones), large deflation basins denoted as **playas** and small examples labelled as **pans** are common and not necessarily combined with dunes. Many examples have been described from South Africa and western North America (Hutchinson 1957) and more recently from Australia (Timms 1992). The strongest evidence of the process of deflation is deflated material piled up as a curved mound or lunette. This dune is asymmetrical in cross-section and according to Timms (1992) with the steep side facing windward if composed of clay, or the reverse if sandy. Finally, it should be stressed that during dry periods deflation tends to deepen and extend existing basins, and possibly also move them downwind.

2.5.9 *Lakes formed by plant accumulation and by animals*

Lakes may also be formed by plant accumulation and by the activities of animals. Outbursts of raised peat bogs and dams of *Sphagnum* may be in-

deflation and therefore may persist over long periods. They occur not only in coastal zones but also in arid regions and have been described from North America, Australia and the Tarim Basin (Hedin 1942).

A special type of perched dune lake (Timms 1992) is related to old and highly siliceous dune systems stabilised by vegetation. Their existence depends upon accumulation of organic material on the lake bottom in addition to deflation or damming by advancing dunes. By podsolisation of the underlying sand, the organic material forms volved in the formation of modest basins. Raised peat bogs frequently possess an outer fringe of water, locally denoted as the 'lagg'. In addition, owing to unequal growth rates of peat, elevated plateau mires can contribute to the formation of shallow basins. Much less often, macrophytes may block the drainage outlets of shallow lakes, as mentioned for Lake Okeechobee in Florida (Hutchinson 1957). Quite different from these examples, is the formation of coral atolls which may lead to the development of complete ring structures whose apertures are choked with sand or other material. Hutchinson (1957) cites examples from the Pacific Ocean, such as Washington Island and Chipperton Island (off the western coast of Mexico). A few in the Marshall Islands, amongst other types, were described by the renowned poet Chamisso when he participated in the Russian 'Rurik' scientific expedition (1815–1818), organised by the Count Romanow.

In contrast to the scheme proposed by Hutchinson (1957) animal activities are also grouped among this section. The European beaver (*Castor fiber*) produces dams not more than 15 m long, whereas its American relative (*C. canadensis*) is responsible for basins up to 50 m length. As a result of activities of many generations, beaver lakes attaining 650 m length have been reported.

2.5.10 Artificial lakes and reservoirs

These will be given closer attention elsewhere (see Volume 2, Chapter 12). On this occasion a table of use, and period of origin of such bodies is presented (Table 2.3).

2.5.11 Meteoric impacts

In contrast to large-scale but regional volcanic events, meteoric impacts may be of global dimensions. There is good evidence that the end of the Mesozoic, some 65 million years ago, an event which brought about the extinction of the dinosaurs, the ammonites and certain groups of foraminifera, etc., was caused by a large meteoritic impact in the region of the Yucatan. Since then, many small events of this kind have taken place

Table 2.3 The origins of artificial lakes and reservoirs.

From antiquity:	Drinking water
	Irrigation
	Retention
	Mills
	Fish ponds
	Strip mining
	Defence and means of war
Since the Middle Ages, and modern times:	Fire ponds
	Ice ponds
	Farm ponds
	Hydroelectric power reservoirs
	Recreation
	Purification basins

and numerous sites have been detected. Many of the craters produced in this way are dry, such as the large Arizona Crater. So far about 120 meteoritic lakes, very often with a small area and considerable depth, are known; in several cases their origin is still doubtful. One remarkable example of a meteoric lake is Chubb Lake in Canada (Quebec) with a diameter of 3.3 km and a depth which amounts to 251 m.

2.6 MORPHOMETRY AND MORPHOLOGICAL TYPES OF LAKES

The form of a lake basin, and of the lake which partly or completely occupies it, depends on the forces which produced the basin, and on the events which occur in the lake basin and in its drainage area, after it has been formed. In order to work with lake morphology, it is necessary to obtain an accurate outline of the surface shape, and information on lake bathymetry. With these parameters available, essential morphometric data can be obtained as follows.

The **length** (l) of a lake is defined as the shortest distance, through the water or on the water surface, between the most distant points on the lake-shore (Hutchinson 1957). The **breadth** or the maximum width (b_m) is the maximum distance between shores at right angles to the maximum length.

The **shoreline development** (D_s) expresses the degree of irregularity of the shoreline. Essentially, it is the ratio of the length of the shoreline (s) to the length of the circumference of a circle with the same area as the lake, and expressed by the formula

$$D_S = s/(2\sqrt{A\pi})$$

where A is the lake area (see below). According to this formula, perfectly circular lakes possess a D_s of 1.0. The value for branched lakes may exceed 10, and for irregular lakes, 20.

Lake **area** is determined by planimetry from the outline of an accurate map of appropriate scale with respect to the size of the lake. In computing volumes, and for certain other purposes such as the water bottom relation $(\Delta V/\Delta A)$, it is necessary to measure the area of each contour on the bathymetric map. The area circumscribed by contour z is designated Az.

Estimation of **maximum depth** (z_m), bearing in mind the often very slight (e.g. seasonal) variation in water level, should be made with reference to some independent datum. Considerable variations may occur in the case of large floods, and even more so in astatic lakes. In such cases it seems convenient to report the range of z_m within a certain period. The mean depth (z) is obtained by dividing the volume of the lake by its area.

The **relative depth** $(z_r$ in per cent) is defined by the ratio of maximum depth (in metres) to the mean diameter of the lake; as this it is expressed as

$$z_r = z_m \sqrt{\pi}/(20\sqrt{\pi})$$

The **depth of cryptodepression** (z_c) is defined by the maximum depth of that part or whole of the lake lying below sea level.

The **volume** of a lake (V) is defined by

$$z = z_m$$

$$V = \int_{z=0}^{z=z_m} A_z \times d_z$$

$$Z = 0$$

Standard procedures for evaluation of the integral concern measurement of the maximum possible area of each contour. If plotted against z, the area of the curve so obtained may be measured planimetrically. Alternatively, the volume of a series of defined layers is summed.

The volume between the planes $z = a$ and $z = b$ of successive contours corresponds to

$$Vb-a = 1/3(Aa + Ab + \sqrt{AaAb})(b-a)$$

The volume between the surface and a horizontal plane of depth z is designated V_z.

The **bottom contact area**, which also provides information on the basin's inclination, can be defined by $\Delta V/\Delta A$. Berger (1971) recommends this relation as more illustrative and advantageous for the graphic presentation than the quotient dA/dV used by Hutchinson (1957). In his paper, Berger (1971) calculated the equations $\Delta V/\Delta A$ and dV/dA for a sphere and a cone section.

2.6.1 The morphological types of lake basins

As stressed in section 2.1, shallow lakes, whether of primary or of secondary origin, are potentially mixed throughout by wind—the exceptions being those protected by ice or by floating, emergent or even submerged vegetation. Secondary shallow lakes, which represent the final stage of the **ontogeny** of formerly deep lakes (see section 2.12), may still reflect the former contours of the once deep water basin, or may be only a remnant of the former large lake. Generally, primary shallow lakes vary greatly in their shape, and comprise dendritic and irregular basins (see below) connected to floodplains, large plain wetlands and deltaic landscapes.

The shape of many lakes often indicates their origin and history (see e.g. section 2.5.2). The following types can be considered more or less as such examples.

Circular lakes, such as explosion craters (**maars**) and crater lakes, often result from volcanic events. Similarly, meteorite craters and, less frequently, doline lakes and certain pans produced by deflation, may possess values of D_s corresponding almost to 1. **Subcircular lake basins** with D_s approaching 1.5 belong also to the types just mentioned, but have normally been remodelled by shoreline processes and by deflation. In addition, many kettles, cirque lakes and pans may belong to this category.

Elliptical lakes are exemplified by basins connected with deflation, such as some lakes which lie in coastal parabolic dunes in Queensland (Timms 1992). Here, D_s may exceed 2.0. **Subrectangular elongate basins** are represented by most lakes in grabens and fjords, and by lakes of overdeepened valleys which appear as elongated widening of rivers (see section 2.1). Their shoreline development may reach D_s values of more than 5.0. **Lunate lakes** are exemplified by oxbows, maars and volcanic basins with asymmetrically placed secondary cones, and as such exhibit a crescentic moon shape. **Triangular lakes** arise from flooding of non-dissected valleys behind bars such as sand dunes, spits, levées or artificial dams.

Dendritic basins are represented by drowned, but not overdeepened, valleys with their lower ends blocked by damming or tilting. Lake Weikaremoana in New Zealand, retained behind a landslide, is a striking example of a lake resulting from a damming event (Hutchinson 1957; Timms 1992). Lake Kioga, north of Lake Victoria in Africa, occupies a large basin formed by tectonic tilting. Lake Mälaren, a complex basin between Stockholm, Arboga and Uppsala (Sweden), has come into existence by uplift of the Baltic coast within historical time. Many large artificial reservoirs, such as Volta Lake in Ghana, and many reservoirs in Brazil, Spain, etc., are typical examples of the dendritic lake type.

Irregular lakes are the most complicated morphological type, and are restricted to areas where fusion of basins has occurred. The most extraordinary examples are, amongst others, known from northern Finland, Wisconsin and Northwest Canada, where D_s may, in some cases, exceed 20 (Hutchinson 1957).

2.6.2   The origins of lake islands

Islands in lakes may arise by various processes. In connection with tectonic events, subsidiary fault scarps or other structural features may lead to island formation. They are not common and labelled as **structural islands**. Olkhan Island and the small Ushkani Islands in Lake Baikal are examples of this kind. Much more common are **islands of volcanic origin**, which may themselves contain craters which sometimes enclose lakes. Mokoia Island in Lake Roturua (New Zealand) and Wizard Island in Crater Lake (Oregon, USA) are examples of this type.

The most common type concerns **islands in glacially formed lakes**. Occasionally (e.g. the Saimaa region of southeastern Finland), such islands are so abundant that the insulosity exceeds 30% of the lake area. They may represent areas of rock resistant to glacial erosion, or deposits common in glacial drift areas.

Coastal lakes, if large and of dendritic shape, may allow for **islands forming by erosion of promontories**. Apart from the Ushkani Islands, all the other small islands of Lake Baikal are of this kind. Other examples are known from Lake Superior and the dendritic Lake Weikaremoana in New Zealand (Timms 1992). **Depositional islands** arise when spits or delta-arms become cut-off, or when waves and currents form shoals offshore. Such islands may frequently change their size and shape. Long Point in Lake Erie is mentioned by Hutchinson (1957) and examples from Australia (Lake Wyara, Lake Macquarie) have been reported by Timms (1992).

Ice islands, and islands of ice and frozen sediment, have so far been reported from the southern Altiplano in Bolivia and northern Chile (Hurlbert & Chang 1984). Among ten relevant lakes observed, Laguna Colorada (4278 m) is shallow and saline and contains the largest island with a total area of 0.8 km² and a length of 5 km. According to the authors the ice blocks may have formed during the 'Little Ice Age' which lasted from the mid-15th until the mid-19th century. First recorded observations of the ice exist from 1935. Possibly, an isopleth diagram of the area, as proposed by Troll (1959), could be useful for a better understanding of the formation of ice islands.

Floating islands consist mainly of organic matter, and may occur when gases produced by decomposition buoy up parts of the lake floor. Frequently, such an event concerns transitional mats of algae and, much less often, stands of emergent vegetation such as *Phragmites*, *Typha* spp., *Cyperus papyrus*, etc., with their rhizomes. These may become long lasting and often form floating and drifting islands. Mechanical events (e.g. floods) may also contribute to lifting them from the lake bottom. Likewise, parts of fringing peat bogs may break free and turn into (floating) islands. In some cases there is reason to assume that in basins originally covered by peat bogs an increase in the water-level may cause large vegetation mats to float. In karstic regions this could happen if an underground outflow becomes blocked by organic material and the basin finally obtains a surface outflow.

It has been suggested that Lunzer Obersee in Austria (1117 m a.s.l., $A = 0.14 \, km^2$, $Z_m = 15.5 \, m$) offers an example of an island of this kind. For at least 100 years its surface has been covered by an extended mat of *Sphagnum* transitional peat bog (Fig. 2.13). Although this mat is submerged in calcareous water by considerable loads, and every year between October and June by snow, its acid character is quickly renewed by ion exchange processes performed by the *Sphagnum*. Similar mats of the peat-bog type are known from other Alpine lakes, from Bohemia, and from Minnesota (USA), and have been reported from Loch Lomond during the Middle Ages and from the Roman Empire (Italy; Hutchinson 1957).

Finally, mention should be made of the small artificial islands of Lake Titicaca (Northcote *et al.* 1989). They consist of **kille** or **quili**, a mixture of mud, rhizomes (of *Scirpus tatora*) and decaying organic matter. When kille is dry it is very light and floats on the surface of the lake. Such islands, which are made by groups of the local Uros people, serve as sites of aquaculture. In order to keep the bottom of the platform created dry, **totora amarilla** (dry *Scirpus tatora*) is used as floor material.

2.7   THE HYDROLOGICAL TYPES OF LAKES

Given the global distribution of arid, semi-arid, semi-humid and humid regions (Fig. 2.16), three

hydrological types of **drainage basins**, **catchments** or **watersheds** can be distinguished (Martonne & Aufrère 1928; Hutchinson 1957). **Exorheic watersheds**, which empty into the sea, represent the major part of the drainage of all of the continents except Australia. **Endorheic watersheds**, which discharge inland, into closed lake basins or basins of 'internal drainage', cover approximately 42 million km^2 and are mainly (but not exclusively) restricted to arid and semi-arid regions. However, one of the largest of these, the Caspian Basin, into which drain the River Volga and the Ural River, is, in terms of its climatic and hydrological relations, more typical of exorheic areas. From a climatic point of view, it may be regarded as a 'geographical accident' (Hutchinson 1957) that these rivers run into the closed and relict cryptodepression of the Caspian Sea.

After a series of wet years, endorheic watersheds may become exorheic. Until construction of an artificial outlet early last century, the Neusiedlersee overflowed almost every 100 years via the Hanság (Fig. 2.9) towards the Danube. In one map, produced during such a high water period (which could last for several years), the lake is denominated as 'Flumen Ferteu' (River Ferteu = Fertö, the Hungarian name of the Neusiedlersee).

During periods of high water level, Lake Hamun, Afghanistan (Fig. 2.17) periodically attained an area of more than 4000 km^2 when it overflowed into another closed basin, the Gaud-e Sirreh Plain. More recently, the large tributary of Lake Hamun, the Hilmand River, which drains a large portion of Afghanistan, has been increasingly used for irrigation. Lake Hamun, and with it its old, well adapted human cultures, may now disappear forever.

The history of Lake Tanganyika is more complex. Owing to Pleistocene volcanic events in the Virunga Mountains, which led to obstruction of an upper tributary of the Nile, and formation of Lake Kivu and its outflow Ruzizi (see section 2.5.2), Lake Tanganyika obtained an additional important influx. This transformed its endorheic basin to exorheic. From the time of its initial exploration by Europeans in 1854, until 1878, lake level rose, after which it finally discharged over a dam of silt

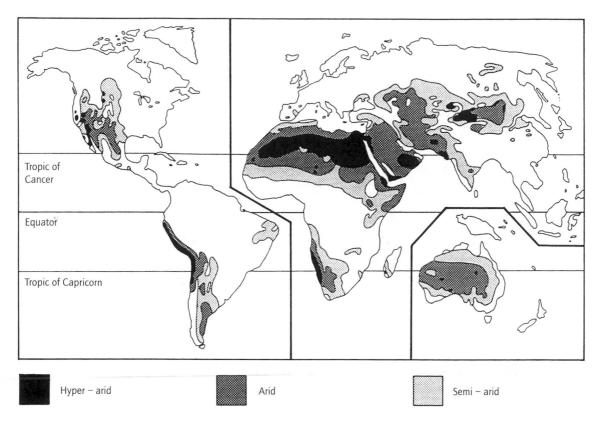

Fig. 2.16 The distribution of global hyper-arid, arid and semi-arid zones. The first two correspond to more or less arheic, the latter to endorheic regions. (Modified from Margalef, 1994.)

which had blocked its outflow, the Lukuga River. As soon as this new discharge was established, the obstructing deposit was very rapidly cut away, resulting in a dramatic fall of the lake level which probably exceeded some 15 m.

Finally, a third type of hydrological region, within which no rivers arise, comprises the **arheic regions** of the globe, which amount to about 28 million km². They may still be crossed, however, by very large streams. The lower part of the Nile and the Oranje, as well as the Niger near Timbuktu, all in Africa, provide extreme and well known examples of this category of basin.

Generally, the endorheic parts of Earth can be seen to lie between the two arheic zones (which are found mainly in the latitude of the trade winds)

and the three exorheic regions (which represent the north and south temperate zones, and the zone of equatorial rains). Most **endorheic lakes** tend to be highly astatic and to possess elevated salt content, depending on the petrographic properties of their watersheds, or on the long-term accumulation of salts within their water column (if not mitigated by deflation—see section 2.12). According to their major anion concentrations, lakes which contain chlorides, bicarbonates and(or) carbonates and sulphates as the most abundant anion may be distinguished as **thalassohaline** if they possess sodium and chloride as the commonest ions (e.g. Lake Niriz, Iran; Fig. 2.18) and **athalassohaline** (see section 2.8) if magnesium occurs in abundance and in combination with chloride and/or sulphate.

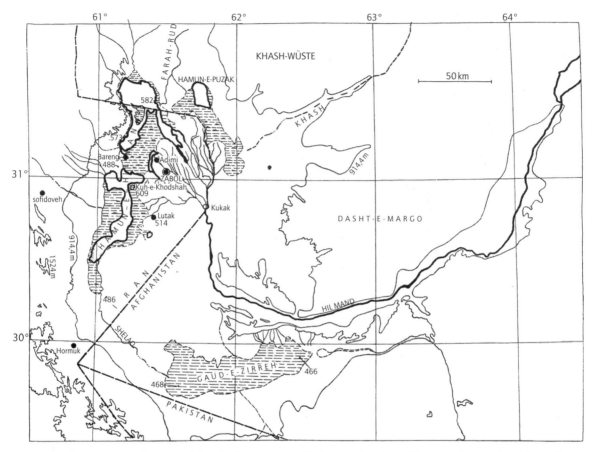

Fig. 2.17   Lake Hamun on the Iran–Afghan border. At critical high water levels the lake gained an outflow (the Shelaq River) towards the Gaud-e Zirreh Basin. Owing to increasing use of water from the only inflow (the Helmand River) for irrigation, the lake faces permanent desiccation. (Modified from Löffler, 1960.)

They may also be placed in the second category if they are sodium (or rarely potassium) alkaline lakes (e.g. Lake Nakuru, Lake Magadi, Lake Natron in East Africa), or sodium sulphate lakes (e.g. Tso Kar, Indian Tibet; Hutchinson, 1957).

The relative proportions of the three major anions and cations of saline lakes (counting Na^+ and K^+ together) are conveniently presented in triangular diagrams (Fig. 2.19) of the type also used by petrologists. Here, each apex of the diagram corresponds to a value of 100% of any particular anion or cation. Exceptions to these standard compositions can be represented by borate, which excep-

tionally may be the most abundant anion in a few lakes.

The sources of incoming water, to both open and closed lakes, are (i) precipitation falling on the lake surface—less often so in endorheic lakes, (ii) surface water inflows, (iii) groundwater seeping through the floor of the lake and (iv) groundwater entering by discrete springs (limnocrenic inflow and sub-lacustrine channels, most often connected with karst environments). Relative contributions from each source may vary greatly, however, according to prevailing hydrological and climatic regime. Thus, whereas Lake Victoria re-

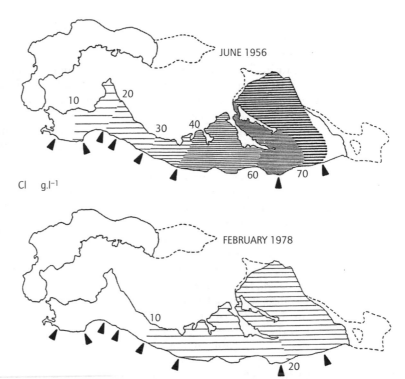

Fig. 2.18 With respect to its range of salinity, Lake Niriz, east of Shiraz (Iran), is probably the most spectacular shallow lake in the world. Along its 100 km axis, salinity varies from fresh at the western end near the mouth of the Kur River, to saturation concentrations in the east during summer. In winter and early spring, fresh to low-salinity water replaces the more saline summer water. Arrows indicate sites of investigation in 1956 and 1978. Values given as chlorinity (g L^{-1}). (Modified from Löffler, 1981.)

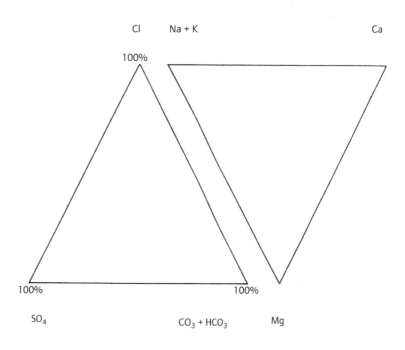

Fig. 2.19 Diagrams used for the presentation of the anionic and cationic composition in inland waters.

ceives more than 50% of its water from precipitation, in the case of the Dead Sea this proportion
would be practically zero compared with that contributed by its main inflow, the River Jordan. In
most large lakes of the temperate zone, direct precipitation represents only a few per cent of the
total water budget. Some lakes, such as Vrana (see
section 2.5.4), receive their water perhaps from
sub-lacustrine channels, and a few are fed solely by
glaciers and snow.

Normally, lake levels in extratropical exorheic
regions vary according to seasonal events such as
rainfall, snow and ice melting, whereas in the tropics precipitation occurs mainly during the rainy
season(s); there, snow and ice are rarely relevant on
a regional basis (Asia, South America) but often
play a local role, such as on several high tropical
mountains in Africa and New Guinea (Irian).

The extent of fluctuations of lakes in exorheic
areas depends on lake area (A), the area of the
drainage basin (D), the mean precipitation over the
lake (Pr) and over the rest of the drainage basin (Pr'),
as well as the mean evaporation from the lake (Ev)
and from the rest of drainage basin (Ev'). In addition, mean discharge by the outflow per unit area
of the lake (Ef) is an important factor, since Ef and A
are monotonic functions of z (maximum depth),
which both increase as z increases, with $\dot{z}$ being
the mean rate of increase of z (Hutchinson 1957).
Therefore

$$A\dot{z} = A(Pr - Ev) + (D - A)(Pr' - Ev') - AEf$$

The condition for $\dot{z}$ to be positive is that

$$(Pr - Ev) + [(D - A)/A](Pr' - Ev') - Ef > 0$$

Variation in discharge with varying hydraulic head
at the outlet may cause lakes of otherwise similar
conditions to rise and to fall at different rates
and/or intervals. In addition, the features of a
drainage basin such as the kind of vegetation, the
presence of wetlands, and influence of human activities may contribute to the water balance of a
lake.

During their earlier history, many exorheic
lakes, such as Lake Geneva, underwent major
changes. Terraces dating from the Oldest Dryas,
and from the Bölling and the Alleröd interstadials,
are now located 10–30 m above the lake level.
Similarly, studies of the sediments of the Neuenburgersee and the Bielersee (both in Switzerland)
reveal major fluctuations (Ammann 1982) which
occurred during the late Pleistocene (Bielersee)
and the Holocene (Bieler- and Neuenburgersee).
The reasons for these events include major climatic changes, hydrogeology (e.g. karstic conditions),
glacial phenomena (see Tsola Tso, Fig. 2.11) and
landslides in the area of the outlet or major erosion
by the outflow. Obviously, forest clearance during
the later Holocene, and even more recently the use
of lakes as recipients of urban and industrial efflu-
ents, or of their waters for hydropower production,
have greatly upset the hydrological balance of
many lakes.

With respect to retention time, the climate and
morphometry of exorheic lakes play a predominant role. Therefore large lake basins in semi-arid
regions tend to exhibit the longest times. There
are, however, exceptional relatively small lakes,
such as the Arendsee of northern Germany $(A = 5.1 \text{ km}^2, z = 49.5 \text{ m})$, which possess long mean residence time, in this case 114 years. Table 2.4 shows
the mean residence time of selected lakes as well
as some large reservoirs.

In closed lakes, in which $Ef = 0$, the required in-
equality becomes

$$(Pr - Ev) + [(D - A)/A](Pr' - Ev') > 0$$

In such a lake $(Pr - Ev)$ is negative, whereas if any
water reaches the lake, then $(Pr' - Ev')$ must be positive (Hutchinson 1957). The majority of closed
lakes, such as Bonneville, Lahontan (USA) and the
Eyre basin (Australia), being usually strongly
astatic, exhibit short (weekly, monthly) and also
longer term fluctuations which are almost exclusively due to climatic events. This is mainly
because hydrographic variability increases with
aridity, the result being that small climatic
changes become amplified by aquatic systems. Another feature of closed drainage basins is accumulation of salt and hence the high frequency of saline
lakes within endorheic (closed) regions. There are,

Table 2.4 Mean residence time of selected exorheic lakes and reservoirs.

Exorheic lakes	Residence time (yr)	Volume (km³)
Titicaca (Peru/Bolivia)	1343	866
Tahoe (USA)	700	156
Baikal (Russia)	380	22,995
Superior (Canada/USA)	191	12,230
Great Bear (Canada)	124	2381
Arendsee (Germany)	114	
Michigan (USA)	99.1	5760
Vättern (Sweden)	55.9	74
Victoria (Uganda, Kenya, Tanzania)	23	2700
Geneva (France, Switzerland)	11.5	89
Lugano (Italy, Switzerland)	12	4.7
Attersee (Austria)	6.8	3.9
Biwa (Japan)	5.5	28
Constance (i.e. Bodensee Austria, Germany, Switzerland)	4.6	48.4
Maggiore (Italy, Switzerland)	4	37.5
Balaton (Hungary)	2	1.8
Windermere (England)	0.75	0.35
Traunsee (Austria)	0.7	2.3
Tai-Hu (China)	0.65	4
Lunzer Untersee (Austria)	0.4	0.013
Reservoirs		
Volta (Ghana)	4.3	
Kariba (Zambia / Zimbabwe)	3	130
Tucurui (Brazil)	0.14	

however, lakes which are exceptions to this rule, which stay fresh during most of their existence owing to special features of their drainage patterns, to precipitation, or to deflation of salt. Lake Chad (Fig. 2.5) and to a lesser extent Lake Hamun (Iran, Afghanistan, Fig. 2.17) are examples of this kind of lake.

Among endorheic drainage lakes whose water-level fluctuations are little understood, the Caspian Sea and its basin (see also sections 2.5.1, 2.7, 2.11 and 2.12) represents one of the most complex. The Caspian was first separated from the Black Sea during the mid-Pliocene. Since that time, development of the Caspian and the Black Sea basins has proceeded independently, although with occasional transient and unilateral links of the Caspian with the Black Sea (Kosarev & Yablonska 1994). The size and salinity of the Caspian have varied both during the Pliocene (mainly as a result of tectonic processes) and the Quaternary, when cyclical climatic changes have given rise to repeated transgressions and regressions, and variations in the level of the surface of the Caspian in relation to world mean sea level (w.m.s.l.).

Amongst transgressions, two major events which led to connection with the Black Sea are known from the Upper Pliocene. During the Pleistocene, three principal transgressions, called the Baku, Khazar and Khvalyn Periods, took place during the interval after 500,000 years BP. Amongst these, the Early Khvalyn (70–40,000 yr BP), which reached more than 45 m **above** w.m.s.l., was the most significant Quaternary transgression, and the last time that a **unilateral(?)** connection existed between the Caspian and the Black Sea. In contrast, the Late Khvalyn transgression, which reached its maximum approximately 16,000 years BP, attained only 0–2 m below w.m.s.l., and was followed by a regression when the Caspian fell to –100 m below w.m.s.l.

In contrast to the Pleistocene, with its fluctuations of up to 190 m in the level of the Caspian, the Holocene (10,000–0 yr BP) has been marked by only moderate changes. After an initial rise, around 9000 years BP, to about –25 m below w.m.s.l., a maximum of –19 m was reached about 8000 years ago. Further fluctuations, which are fairly well investigated, and dated, remained essentially within a range of only 5 m. Between AD 1830 and 1929, they did not even exceed 1 m about a general level of –26 m. After this, the level began to fall steadily, to its lowest value for the last 500 years of –29 m in 1977. This decline was caused partly by a drought in the Volga basin, but was also greatly due to utilisation and regulation of that river (Hutchinson 1957). Owing to reasons which are not fully understood, but which are thought to be climatic, the level began to rise again in 1978, and by 1994 was approaching –26.5 m, causing inundations and considerable economic damage (Golubev 1996). According to recent records, this most recent rise of the Caspian came to a halt in 1998.

Before catastrophic human intervention in the water budget of the Aral Sea (see Volume 2, Chapter 8), its level earlier in the nineteenth century had fluctuated within 2–3 m (Suslov 1947). Until 1960, about half of the flow of **Amu Darya** (the River Oxus) and of **Syr Darya** (the Iaxartes) was used for irrigation. The lake had for a long period experienced stable morphometry, with an area of about 68,000 km^2 (including its islands), a maximum depth of approximately 69 m and a volume of 1000 km^3. Under these conditions, salinity amounted to 10‰.

With the decision at the end of the 1950s to increase the irrigation area within the Aral Sea basin from 5 to 7.4 million ha, practically all of the inflow water became used up, mainly for the sake of monoculture of low quality cotton. During the 1980s, rivers did not reach the Aral Sea every year and only during the period 1992–1994, when flow was above average, was the downward trend of the

Aral Sea slowed. At present only half or even less of the former area, and about 25% of the former volume, remains, with the resulting salinity exceeding even oceanic concentrations. In addition, general pollution, with large amounts of dissolved pesticides, has removed most of the Aral Sea's former organisms and left the three million human population around and near the lake-shore without any of their former fishing resources, which, before the disaster, amounted to c.40,000 t per annum.

At present, salinisation of catchment soils is progressing, and land loss has reached 10,000 km^2 (Golubev 1996). In order to improve conditions, a long-term programme of land and water-resource management, such as the removal of low productivity lands from irrigation, drastic reduction of water applied per unit of cropland, and optimal use of fertilisers and pesticides, is needed. In addition, social and economic problems, such as improvement of farming methods, need also to be addressed. As expressed by Golubev (1996), only a long-term sustainable development strategy for the whole watershed with its 34 million inhabitants may lead to improvement in the hydrology and water quality of the Aral Sea.

'Aral Syndrome' can also be identified for other endorheic, closed lakes. Apart from Hamun Lake (Iran, Afghanistan) already mentioned, Mono Lake (California), the Dead Sea (Israel, Jordan), Lake Corangamite (Australia) and possibly also Balkhasch (Kazakhstan) can be classified amongst endorheic lakes adversely affected by unbalanced water use within their drainage basins. Many such smaller lakes have disappeared for ever.

2.8 CHEMICAL CLASSIFICATION OF SALINE LAKES

Although the majority of saline lakes occur within endorheic regions, there are also many located in exorheic drainage basins. Their salinities derive from underground deposits (e.g. the Permian Zechstein in Europe, the Paratethyan Aralo-Caspian sediments, etc.), often associated with mining.

As early as 1935, Bond referred to inland salt lakes rich in (i) anions other than chloride and (ii) cations other than sodium (mainly magnesium; potassium being the exception) as **athalassohaline**. In contrast, lakes 'simulating' marine salt composition (or very similar in their salinity to seawater) he designated **thalassohaline**. Generally, alkaline, sulphate lakes and lakes with those cations listed above are placed among athalassohaline salt lakes. Very exceptionally borate may become an additional abundant anion, as in Borax Lake, California, and as reported from some Tibetan lakes. The relative proportions of the major anions and major cations (calcium, magnesium and the sum of sodium and potassium) are conveniently presented in triangular diagrams (as shown in Fig. 2.19) in which the data of given lakes are plotted. In addition the diameter of each circle (lake) can be used as a measure of salinity (expressed as a logarithm or as V^-).

More recently, Bayly (1967) proposed the term **thalassic**, which should include only those coastal marine embayments whose features are closer to marine environments than the rest of inland salt lakes, which are then described as **athalassic** (inland salt) lakes. This proposal is supported (although not unequivocally) by other authors (e.g. Margalef 1994; Williams 1996). It should, however, be kept in mind that in large deltaic areas with their mainly Holocene dynamics, such a differentiation may be as difficult as in many coastal zones. Moreover, inland salt lakes of the thalassohaline type, with concentrations less than, or not far beyond, the salinity of the oceans, are often the homes of erstwhile marine organisms such as some foraminifera, which have been found in Eurasian localities, of which Lake Niriz, Iran (Fig. 2.18) may be mentioned.

In addition, there remains the question of old (Tertiary) to recent lakes of marine origin (see sections 2.5.1 and 2.7) which are cited as relict lakes. Examples of the Tertiary Aralo-Caspian Sea region, with its variety of marine-related organisms such as *Cardium*, the Cumacea and the Mysidacea, are obviously more thalassic than recent coastal lakes which have become isolated from the sea by sand and shingle bars produced by sea currents. Instances of the latter are numerous along the western coast of France, the Mediterranean and the Baltic coast (see section 2.5.7). Their salinity may range from fresh to hypersaline, and with respect to chemical composition they can sometimes even be strongly alkaline.

Therefore it seems advisable to reserve the term thalassic exclusively for bodies of water which still bear close hydrographical, hydrological and biological relationships to the sea. Everything else then comprises the athalassic (inland) lakes, with the wide range of anion and cation composition mentioned above. Only to these should the terms thalassohaline and athalassohaline as used by Bond (1935) be applied, and this classification might be useful, amongst others, for regional comparisons.

As a concluding remark, it should be mentioned that highly alkaline lakes, with pH values far above that necessary for the precipitation of carbonate, are distinguished by a small number of specialist organisms which not only tolerate this extreme environment but are adapted to it. Among the algae, the diatom *Surirella peisonis* can be mentioned, and among the animals the rotifer *Hexarthra jenkinae*, the crustaceans *Arctodiaptomus spinosus* and *Branchinecta orientalis*, and the remarkable cichlid fish species *Tilapia grahami* (endemic to Lake Magadi in Kenya) are noteworthy. Molluscs are completely absent in more concentrated alkaline lakes. The physiological explanation for the remarkable adaptation of only a small group of organisms to such conditions is still unknown.

Among freshwater lakes, those with watersheds located over sedimentary rocks may be generally distinguished from lakes lying within igneous regions by their major cation (and less so by their major anion) equivalent proportions. Calcareous and dolomitic watersheds frequently produce equivalent proportions of $Ca^{2+} > Mg^{2+} > Na^+ > K^+$, with Mg^{2+} close to Ca^{2+} in dolomitic lakes. Among the major anions, the sequence $[HCO_3]^- > SO_4^{2-} > Cl^-$ prevails.

In contrast, lakes in igneous rock environments frequently exhibit low conductivity and poor buffer capacity. With regard to major cations each may occupy the most abundant position, albeit that this is most often the case with Na^+. Sometimes (e.g. in some lakes of Mt Kenya), magnesium may fall to trace element concentrations, and in rocks rich in mica, K^+ becomes more abundant than Na^+.

Owing to their frequent lack of buffer capacity, lakes in areas of igneous rocks are strongly influenced by acid rain, and consequently by metal pollution, with many examples recorded from the Northern Hemisphere. Acidification may also be caused by organic acids exported by peat bogs. Close to the coast, and in the vicinity of salt plains, such lakes may become more easily influenced even at distance. Finally, a great variety of alkaline, acid and extremely dilute conditions is documented from lakes in volcanic regions.

2.9   MEROMICTIC LAKES

Findenegg (1935) introduced the term **meromictic** to denote those lakes where, throughout the main circulation period(s), owing to the presence of a vertical (chemical) density gradient, conveniently called a **chemocline**, some water remains partly or wholly unmixed. Most other lakes are at some point in the year **holomictic**. All forms of transitions between holomictic and meromictic lakes are known. The perennial—sometimes periodically—stagnant layer of a meromictic lake, the so-called **monimolimnion**, is separated from the remaining upper part, the **mixolimnion**, in which free circulation occurs.

The most rapid process which renders a lake meromictic, is supply of salt water to a freshwater body, or fresh water to a saline lake. These cases are called respectively **ectogenic** and **crenogenic meromixis**. Ectogenic meromixis is most often observed in lakes close to the sea, which may become exposed to an extraordinary flood. An example of this kind is provided by the Hemmelsdorfersee (Germany), which occupies a cryptodepression near the Baltic Sea. Eventually, such lakes may become fresh and holomictic again if the intervals between episodes of marine flooding are long enough, or if the chemocline gradually descends to the lake bottom. Many examples of such marine-influenced lakes are known along the USA Atlantic coast, and from Japan and Greenland (Hutchinson 1957).

One of the most prominent examples of meromictic conditions brought about by ectogenic events is the Black Sea which, after a long, 'lake' phase, began to receive water from the Mediterranean some 9000 years ago. At present the chemocline, between the mixo- and the monimolimnion, with a salinity of about 18 to 22‰, makes this the largest meromictic body of water on Earth. A much more complex situation with respect to origin of meromixis is provided by the Rift Valley lakes Tanganyika and Malawi, the presence of whose immensely rich endemic faunas suggests that, since they developed their large and anoxic monimolimnia, these lakes have never fully circulated.

Ectogenic meromixis may also be the consequence of industrial activities. For more than 100 years, the alkali works of the Solway Company has released hundreds of tonnes of soluble salts into the Traunsee (Austria). Before appropriate improvement of the outlet of the lake, and despite its short retention time (see Table 2.3), the Traunsee had become meromictic. In addition, ectogenic meromixis is promoted by road salting for de-icing. Most recently, Kjensmo (1997) reported such a case (Svinsjøn) from the Oslofjord area of Norway.

An entirely different type of ectogenic meromixis is seen in saline lakes flooded by fresh water. Owing to the effects of irrigation, the local groundwater table in the vicinity of Big Soda Lake (Hutchinson 1937, 1957) caused a rise in the level of this crater lake between 1905 and 1925. Since the new water supply entered the lake basin at its surface, the lake became meromictic.

In both cases, the **meromictic stability** (defined as the amount of work needed to mix a meromictic closed lake to uniform concentration) may attain remarkable values. For example, depending on transparency, the monimolimnion can often be heated to more than 50°C. Inspired by the case of Solar Lake, close to the Gulf of Eilat, Sinai (Serruya & Pollingher 1983), scientists in Israel have tried to use artificial 'solaric' lakes as a regional source of energy. Owing to many shortcomings, however, they have so far not been able to achieve satisfactory results.

An example of crenogenic meromixis (Lake Ritom, Switzerland) is given by Hutchinson (1957). Prior to establishment of a hydroelectric plant in 1918, the lake mixed only down to 13 m. Other examples from France and Germany (Ulmener Maar: Thienemann 1913; Scharf 1991) have been described in some detail.

Sometimes solution processes in the profundal zone of a lake may be involved in meromixis, which Hutchinson (1957) places in the ectogenic category. They are included here in the crenogenic group because it is often quite difficult to distinguish between saline sediment leaching and possible sources of upwelling saline water (e.g. Brunskill *et al.* 1969).

In both types of meromixis, it is obvious that the relation between lake area and depth, as well as the retention time, are of crucial importance. Therefore, lakes with an area of only a few hectares but a depth which exceeds 20 m, and with an extended retention time, should develop meromictic conditions if the lake basin provides for a minimum of soluble material. Karstic lakes and kettles provide the most likely cases of this morphological meromictic condition. On a small scale, 'crenogenic' conditions were described by Ruttner (1955) when he investigated the funnels of the Lunzer Mittersee, a small limnocrenic system (area = 2.5 ha, z_{max} = c.4 m) with inactive springs which evolved to meromictic sub-systems in a low temperature but almost thermostatic system. According to Hutchinson (1937, 1957), the accumulation of salts liberated from the sediments, primarily of biochemical origin, is considered as a special case of biogenic meromixis which he thinks applies to numerous lakes in Carinthia (Austria).

Biogenetic (biogenic) meromixis in temperate zones is connected with relatively great depth, weak inflows, shelter from wind, and more continental climate. During prolonged winter periods under ice, the deep water of such lakes may acquire sufficient electrolytes, predominantly from biochemical processes in the mud, to increase significantly in density. If spring comes late but is warm, the ice may melt and the lake begin to warm at the surface before the chemically stabilised deep water has been fully mixed with the main body of the lake. The lake thus enters the period of summer stagnation with excess electrolyte already present in the deep water. During summer stagnation, there will be considerable further accession of dissolved material by the deeper part of the hypolimnion (Hutchinson 1957). Thus, winter and summer accumulation may finally lead to permanent meromictic conditions.

Recent investigations in several Carinthian lakes (e.g. Längsee, Löffler 1997; Kleinsee, Löffler 1977; Goggausee, Löffler *et al.* 1975; Table 2.5), and in other lakes such as Lobsigensee, Switzerland (Löffler 1986), show that meromixis had already begun during the pre-Bölling stage, and sometimes (Lobsigensee) even under oligotrophic conditions. In these cases it was the climatic switch from very cool short summers with frequent circulation events (cold monomictic, see below) to stable thermal summer stratification due to increasingly warm summer seasons and dimictic behaviour (see below) which finally allowed for accumulation of electrolytes in the deep layers of these lakes.

Längsee, which was first investigated by Findenegg (1935, 1947, 1953) and recognised as a meromictic lake, became a classic case when Frey (1955) tried to identify the time of onset of meromixis and its cause. From his observations, based on sedimentary, pollen and chydorid stratigraphy, he concluded that erosion due to forest

Table 2.5 Limnological data for Längsee, Kleinsee and Lobsigensee.

	A (km^2)	D (km^2)	z_{max} (m)	z_{max}*	z_{max}†
Längsee	0.75	6.2	21	31?	–
Kleinsee	0.012	?	9.3	16	13
Lobsigensee	0.02	?	2.7	17	11

* Earliest lake stage.

† Time of return to holomictic condition.

clearing, some 2000 years ago, caused eutrophication and eventually meromixis. More recent information on geochemistry (Harmsworth 1984), and
on the ostracod fauna (Löffler 1986), however, suggests that meromixis must have occurred much
earlier, about 15,000 years BP. Likewise, Kleinsee
and Lobsigensee became meromictic before the
Bölling. With increasing sediment accumulation,
Kleinsee and Lobsigensee became holomictic
again (Table 2.5) during the late Boreal and the Subatlantic, respectively. Table 2.5 presents the main
limnological features of these three lakes.

In contrast, Lake Bled in Slovenia ($A = 1.44 \, km^2$,
$z_{max} = 30.2 \, m$) did not become meromictic until the
Subatlantic, when a striking decrease of *Fagus* and
Abies pollen indicates contemporaneous human
activities. Meromixis in Lake Bled thus seems to
correspond to Frey's hypothesis expressed for
Längsee.

Among the larger meromictic lakes of
Carinthia, Wörthersee ($A = 19.4 \, km^2$, $z_{max} = 83 \, m$)
exhibits the special feature of three basins of
which the eastern and the western are meromictic. The central basin, however, being the shallow-
est, is holomictic. Analysis of the ostracod fauna
indicates that meromixis in Wörthersee began
its onset only at the beginning of the Holocene
(Pre-Boreal).

A remarkable example of irregular meromictic
and holomictic behaviour is provided by the
Lunzer Obersee in Austria (altitude = 1117 m, $A =$
$0.14 \, km^2$, $z_{max} = 15 \, m$, see also sections 2.5.4 and
2.6.7), which is due to the length of the ice
cover during spring. A late thaw of ice may provide
for only a short and even incomplete circulation period, which does not allow the deep water
with some excess of electrolytes to mix with the
main body of the lake. This creates a meromictic
condition. Such a condition may be missing if early
melting and a long circulation period prevail.
Alternating systems of this type may be more
common than so far realised.

Meromictic lakes are abundant in the northern
temperate and sub-Arctic zones and include the
classic examples of crenogenic and biogenic maars
of the Eifel in Germany (Scharf, 1991), Lago di
Lugano (Italy) and a variety of lakes in Canada and

in USA, for which most of the latter Hutchinson
(1957) gives as biogenic types. In addition, Walker
(1974) lists lakes from the semi-arid regions of
Washington State, of which the majority are ectogenic, but there is one case (Hot Lake) which can be
claimed to be crenogenic in the expanded sense of
Hutchinson (1957, p. 482), including lakes where
waters of high density enter the lake at the surface
and flow into the monimolimnion as density currents. Walker (1974) also gives an account of Hot
Lake's stability.

Further examples of meromictic lakes have
been reported from British Columbia, including
Yellow Lake (Northcote & Halsly 1969) which is
not salinity stratified and at times of isothermy exhibits a stability ostensibly of zero. Obviously a
morphometrically and climatically determined
factor suppresses full circulation there (see final
part of this section).

In the tropics most of the very deep lakes (e.g.
Kivu, Nyos, Tanganyika, Malawi) are meromictic,
often of complex origin. Kivu and Nyos (see Fig.
2.10) are known respectively for their high concentrations of methane and CO_2 in the deep water.

In summarising it should be stressed that simple cases of ectogenic and crenogenic meromixis
are easily recognised. They are less dependant on
morphological and hydrographical features than
the more complex biogenic types. In addition,
some cases of meromixis may come about mainly
because of morphological features (deep lakes with
very small areas, as is often the case in kettles, volcanic and karstic lakes) and only rarely by biological processes. Finally, in lakes disposed eventually
to meromixis, the climatic change at the end of the
Pleistocene (some 15,000–12,000 yr BP) seems to
have been the major trigger. This factor may or
may not be coupled with biogenic processes (according to algal pigments, Lobsigensee was oligotrophic when it entered its meromictic period).
World-wide (including Antarctica), more than 100
meromictic lakes have been described to date, but
only in a few cases is the time of onset known.

Among lakes with transient meromictic conditions, the simplest cases are saline lakes which receive small freshwater inflows. A stable density
gradient may arise from such a condition and may

last as long as the inflows persist and mixing by strong winds is absent. The saline water may attain temperatures far beyond 50°C (see above). A more complex situation arises in lakes in contact with glaciers, which mobilise turbid materials, where abnormal thermal profiles may be observed. An example of this condition is represented by a small shallow lake north of Dingpoche (Lake no. 18 in Löffler 1969) at 5420 m a.s.l. and at a distance of about 10 km from Mt Everest (Fig. 2.20). There, turbid layers cause cooling below 4°C at a depth of 3 m and at distances from the glacier of more than 50 m and up to 100 m. This type of 'meromictic' stratification is always connected with high turbidity and was first described for Andean lakes (Löffler 1960).

2.10   THE CLIMATIC TYPES OF HOLOMICTIC LAKES

Holomictic lakes (see Fig. 2.21) are, by definition, deep enough to allow, in temperate zones, for the development of **summer stratification**. How-

ever, this process is only part of a more complex annual thermal cycle, based on a key physical property of water, which is that its maximum density is reached at about 4°C (actually 3.94°C). In such lakes, as late spring and summer heating proceeds, and the surface of the lake is warmed further beyond 4°C, an upper layer, or **epilimnion**, of relatively freely circulating water and variable temperature develops. This overlies a deep and more or less undisturbed cold layer (or **hypolimnion**) which remains at 4°C or slightly above that value. The plane between these two layers, in which the temperature changes rapidly, is called the **metalimnion**. Within this plane, the **thermocline** may be distinguished as the steepest slope of temperature change. This type of temperature distribution is conveniently labelled as **direct** or **summer stratification**.

During autumn, the epilimnion cools, and there comes a time when the temperature of the entire water column from top to bottom is more or less the same. At this point, stratification breaks down, and a period of lake circulation known as the **fall** or **autumn overturn** takes

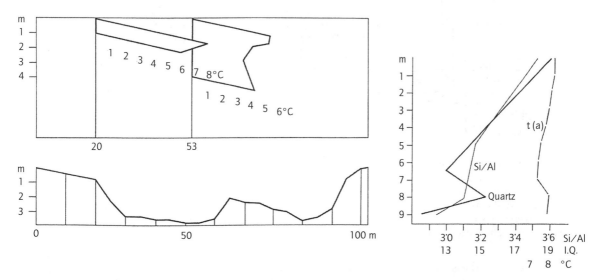

Fig. 2.20 Meromictic symptoms in lakes of the Mt Everest area caused by glacier released turbidity which — distributed within certain depth layers — influences the thermal profiles depending on the distance from the glacier. Lake 18 (left) and Tsola Tso (right). IQ, intensity reading of quartz; Si/Al ratio scale is shown as actual × times; t(a), thermal stratification of profile a.

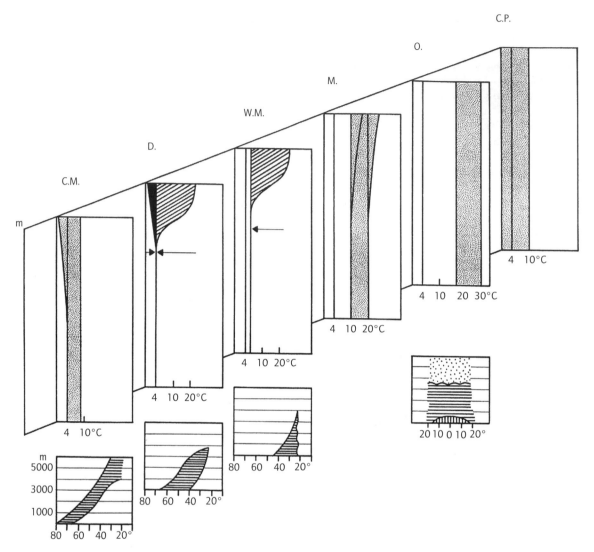

Fig. 2.21 Climatic types of holomictic lakes: C.M., cold monomictic; D., dimictic; W.M., warm monomictic; M., monomictic (tropical regions); O., oligomictic, C.P., cold polymictic. Stippling on main diagram shows the range of actual temperatures possible as a function of lake depth; diagonal hatching in dimictic and warm monomictic lakes corresponds with the summer epilimnion, above the thermocline (arrowed); in dimictic lakes, the upper layer falls below 4°C in winter (shown black), when the stratification is inverse, as it is in the case of cold monomictic lakes. The insets show schematic distributions of each lake type as a function of latitude and altitude.

place. However, if the surface layers of the lake then continue to cool below 4°C, an **inverse** or **winter stratification** develops, usually accompanied by the formation of surface ice. When this begins to melt, there will be a period of **spring** (or **vernal**) **circulation** during which the water

column is again of the same density throughout, and in which circulation requires only minimal forces sufficient to overcome viscosity (Hutchinson 1957). Because of the incidence of two circulation periods annually, such lakes are called **dimictic**.

The idea behind the paper of Hutchinson & Löffler (1956), which presented a proposal for the thermal classification of holomictic lakes (thus excluding shallow lakes), was to examine their mixing behaviour in different climatic zones. It resulted in a system which will be described below in a modified way, owing to more data now being available. Depending on latitude, altitude and climate, the following lake types according to circulation pattern may be distinguished.

Amictic lakes are sealed by ice from any climatic influence. These represent a rare type known from Antarctica, amongst other places, where Lake Vostok, with an area of about 10,000 km^2, is supposed to be covered by an ice sheet estimated to be 4 km thick. Before the recent retreat of glaciers in most parts of the world they were also known to exist in high mountains, such as the Alps, the Andes and Mt Kenya (Curling Pond, Fig. 2.12).

Cold monomictic lakes are characterised by circulation only during summer. This tendency requires that the summer surface temperatures do not exceed 10–12°C. Hutchinson (1957) defines cold monomictic lakes as those where water temperatures never exceed 4°C at any depth, but he admits that such a condition appears to be rather rare. Cold monomictic lakes are confined to certain non-arid sub-polar and extratropical mountain regions. Owing to the incidence of long days during late spring and summer, such lakes may easily attain surface temperatures of about 20°C in dry areas even at a latitude of 70°C. Cold monomictic lakes hardly exist in the Southern Hemisphere.

Dimictic lakes are also mainly restricted to the Northern Hemisphere, and to extratropical mountain regions. They do not occur in New Zealand, or, with the exception of Hokkaido and some mountains, Japan. Owing to strong oceanic influences, they are also absent from most of Chile and Ireland.

Warm monomictic lakes experience circulation during winter at temperatures well above 4°C. They are typical of the Mediterranean climatic zones, of regions strongly influenced by oceans, and some mountains in sub-tropical latitudes (e.g. southern Prealpine lakes, Lake Ohrid, Ireland, New Zealand, most of Japan, southern Chile, etc.).

If warm monomictic lakes are exposed to an abnormally cold winter, the temperature of the hypolimnion may fall to such an extent that circulation during the following years may remain incomplete, which could result in the onset of meromictic conditions.

It should be stressed that between all of these categories of extratropical lakes, transitions may be observed (cold-monomictic–dimictic, dimictic–warm-monomictic), each of which depend on individual lake features and climatic events.

Within the tropics, distinguished by a prevailing diurnal climate, there are two annual maxima of solar radiation. Otherwise the variation in the radiation flux is small, and other factors such as duration and intensity of the rainy season are likely to be of major importance. Two main circulation types can be distinguished. Lakes at great altitudes (well above 4000 m) do not develop persistent thermal stratification. Cooling during the night (beyond a certain altitude, which is different for various tropical mountains, regularly below 0°C) results most often in complete mixing followed by only weak stratification during the day. This kind of lake has been labelled **cold polymictic**. The majority of this type is known from the tropical Andes (Löffler 1960). The rest are located in East Africa (Ruwenzori, Mt Kenya; Löffler 1969) and on the mountain ranges of New Guinea.

In contrast to the statement of Hutchinson & Löffler (1956), it seems that **at lower to low altitudes within the tropics**, holomictic lakes behave mainly as monomictic bodies (but to be distinguished from temperate warm monomictic lakes), with the tendency to mix completely once a year, often during the dry season. The **oligomictic** type suggested by Hutchinson & Löffler (1956), and by Hutchinson (1957), which circulates only at very infrequent irregular intervals when abnormally cold spells occur, seems to be an exception. Lake Nyos in Cameroon with a tendency to meromictic behaviour could be mentioned as an example, whose eventual mixing in 1986 caused the CO_2 catastrophe earlier described (Fig. 2.10). Hutchinson (1957) suggests that Lake Toba (Sumatra) might be another example. Relevant observations are still missing there.

2.11   CLIMATIC INFLUENCE OF LAKES ON THEIR ENVIRONMENT

The effects of lakes on their vicinity may involve influences upon temperature, humidity, rainfall and cloudiness, or fogginess. Such influences are far more frequently observed in extratropical and (semi-)arid regions, and are most often missing under humid tropical conditions with sometimes almost thermostatic characteristics such as the wet altitudinal Paramo zone in the tropical Andes. There are, however, examples of lakes in the temperate regions which in spite of their dimensions do not affect the climate of their surroundings. Lake Musters ($A = 414\,km^2$, $V = 8.3\,km^3$) in southern Argentina (ILEC 1997) has been reported as such a case.

Normally, however, the slow uptake of heat during spring and summer by a moderately sized, or large, lake and its slow heat loss during the autumn compared with the overlying atmosphere lead to a reduction in amplitude of annual temperature variation in its vicinity. When it leads to sup-pression of frost during spring, this influence is often most important for horti- and viticultures. It may also increase the length of the growing season for other crops. When very large lakes are considered, the climatic effect can be quite spectacular. Thus, cold air from the northwest, moving southeast over Lakes Superior and Michigan, is often warmed considerably, and temperature rises of up to 12.5°C have been recorded (Hutchinson 1957).

Apart from the considerable orographic rainfall which occurs on the northern slope of the Elburz Mountains, and which is caused mainly by the presence of the Caspian Sea, the most extraordinary example of thermal influence of a lake on its surrounding region, as seen from the disposition of isotherms for December and for July (Hutchinson 1957; Fig. 2.22), is provided by Lake Baikal. On a modest scale, similar thermal influences have been observed in the vicinity of Neusiedlersee ($A = 300\,km^2$, $V = \sim 0.1\,km^3$), with some cooling effect during summer (Dobesch & Neuwirth 1983; Fig. 2.23). In addition, suppression of frost during spring has been observed.

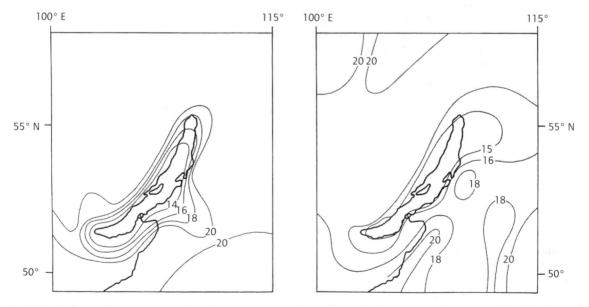

Fig. 2.22   Isotherms for mean ground-level temperatures in December and in July in the Baikal region. (Modified from Hutchinson, 1957.)

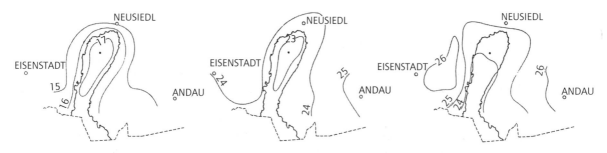

Fig. 2.23 Isotherms for the mean minima in July, for the mean July values at 1400 hours and the mean maxima in July (°C) in the vicinity of Neusiedlersee. (Modified from Dobesch & Neuwirth, 1983.)

The effect of large lakes on humidity and rainfall may also be significant. Along with the Aral catastrophe (see sections 2.5.1 and 2.7), the vicinity of that lake has become significantly drier. A similar effect is likely to be observed in the Lake Chad basin. With permanent low pressure over Lake Victoria, large amounts of rain condensed from rising air masses falling on the open lake should be expected. So far, long-term observations of rainfall on different parts of the open lake are missing. Rainfall around the lake indicates significantly higher values on the western and northern shore, which are most likely due to the wind regime dominated by the southeast trade winds blowing most strongly during May to July. The wind system is, however, complicated by on- and offshore breezes (Crul 1995). Finally, the climatic effect of a lake on its surroundings may produce striking biogeographic consequences. The northward extension of the range of the mistletoe (*Viscum album* L.) along the shore of Lake Mälaren in Sweden constitutes a remarkable example of this phenomenon (Hutchinson 1957).

2.12 ONTOGENY, CHANGES OF SHORELINE CONFIGURATION AND DISPLACEMENT OF LAKES

2.12.1 Ontogeny

Generally, lakes become shallower and decline in area over time because of sedimentation through-

out the body of the lake and near the shore and the inflows. There are, however, exceptions, where erosive events may even deepen basins. Such cases—if tectonic deepening is not considered—are restricted to shallow astatic lakes which are exposed to deflation, or to shallow basins surrounded by emergent vegetation where wind-blown turbid material from the lake's open portion may accumulate. For this, the Neusiedlersee (Austria, Hungary) which is occupied mainly in its west and south by a *Phragmites* belt (about 50% of its total area) presents a paradigmatic case. Since its last desiccation in 1868–1870, the fast expanding vegetated area has received some 150 million m³ of sediment. In addition, in deeper lakes, hyperpycnal inflows, with water more dense than that of the rest of the lake, may erode sub-lacustrine channels as exemplified by Lake Constance.

Most often, however, autochthonous (residues of plankton, physical or biological precipitated $CaCO_3$ and other inorganic material) and allochthonous matter delivered by inflows, surface and subsurface runoff, sub-lacustrine and terrestrial landslides and airborne material accumulate (see Chapter 8). Depending on the properties of the catchment, and its hydrography, autochthonous matter may surpass allochthonous material in importance or vice versa. In addition, floods may increase the allochthonous component in lakes whose sedimentation is otherwise dominated by autochthonous material. Sub-lacustrine landslides are frequent events in basins with littoral and sublittoral inclinations of more than 15°. Sedimentary layers resulting from terrestrial land-slides were described for the first time by Nipkow in 1920 from a section of a core from the Zürichsee (Switzerland). Among airborne material, volcanic ash and scoria, sand and dust (especially in arid zones and in barren landscapes), and pollen may be mentioned. The lakes of the coniferous forest zones receive considerable amounts of this type of pollen during years with heavy blooms. Thus the major fraction of the sediment of the Klopeiner See (Carinthia, Austria) consists of coniferous pollen.

Among autochthonous material, the diatoms deserve special attention. Observations in 1934 from the Lunzer Untersee indicated the presence of 21 million diatoms per cm^2 (Ruttner 1962). Under certain conditions sediments may consist almost exclusively of diatoms. Such lakes (e.g. Myvatn, Iceland) may become important sites of diatomite exploitation. In small, often mesotrophic lakes, high concentrations of shells of Cladocera (above all *Bosmina* and chydorids) are a common feature. Sediments consisting predominantly of shells of snails (less often of mussels) and ostracods are typical of some shallow lakes. The snail *Coxiella* sp. is extremely abundant in the shallow astatic Lake Corangamite (Victoria, Australia, $A = 235\,km^2$, $z_{max} = 4.9\,m$, $V = 0.5\,km^3$) where it also represents a major portion of the littoral substrate.

Owing to the great variety of matter input, sedimentation rates are highly variable both within and between lakes. Moreover, there is often also great variability in sediment accumulation rates through time. The most common annual rates range from 1 to 5 mm, up to 1000-fold greater than the rate of the deep oceans. Values below 1 mm per year have been described by Timms (1992), among others, and are most likely to occur in oligotrophic lakes with extremely small watersheds, such as several volcanic lakes. For Lake Keilambete in Australia, values of $0.4\,mm\,yr^{-1}$ have been reported, and for the playa Lake Buchanan (Queensland, Australia; Timms 1992) a figure even as low as $0.008\,mm\,yr^{-1}$ has been registered. This value comes close to oceanic proportions, and is due to the effects of deflation processes known to be involved, but not measured.

Given such values of annual sedimentation, it appears that most lakes with a depth between 10 and 500 m can be expected to possess a life span between 10^4 and 10^5 years. This is also in accordance with what is known about 'fossil' lakes, such as the freshwater Steinheim Basin (Schwäbische Alb, southern Germany), a meteoritic crater lake which existed for several hundred thousand years (Janz 1993) during the Miocene, about 15 million years BP. Its area was approximately $3.5\,km^2$ and its maximum depth is estimated to have been at least 120 m and perhaps even 180 m. During its early existence, the lake is believed to have passed through meromixis.

In lakes deeper than 500 m, life span will amount to close to one million years, and in lakes of more than 1000 m depth (e.g. Baikal, Tanganyika, etc.), millions of years. More recently, accumulation of sediment in Lake Tanganyika has been assessed by seismic investigations. According to the data obtained, the lake came into existence some 5–10 million years BP and contains a sediment layer about 6000 m thick (Livingstone 1965). With its origin during the Pliocene, Lake Biwa, the largest lake in Japan, overlies sediments more than 1400 m thick (Horie 1991). So far, no data of this kind exist for other Tertiary lakes, especially Lake Baikal, whose age is estimated to be 30–33 million years (Oligocene), which makes it older than any other existing lake (see section 2.5.1). Most recently, cores from Lake Baikal of the order of 200 m in length have been obtained, which obviously represent only a negligible section of a total sediment layer expected to be beyond 10,000 m thick.

In contrast to these examples of long-term changes, rapid filling of transient lakes is typical of basins formed during the late Pleistocene by glaciers. Following withdrawal of such glaciers in large Alpine valleys, short-lived lakes came into existence, dammed by terminal moraines. When eventual rapid retreat of the glaciers caused the power of the rivers to increase rapidly, thus draining the large valleys and their watersheds, the morainic dams were eventually eroded within a rather short time. Afterwards these rivers, with their large load of gravel, filled the lake basins within one or at most a few thousand years. An example of this

kind is the area of Salzburg (Austria), which overlies such a transient lake originally more than 500
m deep. Similarly 'Kühnsdorf Lake' (Carinthia,
Austria) was a short-lived lake dammed by a gravel
barrier across the Drau River which accumulated
via a tributary when the Drau Glacier was rapidly
retreating. Eventually the Drau River became sufficiently powerful and broke the barrier, with the
result that Kühnsdorfer Lake, some 100 m deep,
emptied within a very short time. Only two former
kettle basins (Klopeiner See and Kleinsee, see section 2.9), the deepest parts of the Kühnsdorf Basin,
were left behind.

With respect to allochthonous material, lakes
with active terrigenous input contain different
sediments from those with insignificant inflowing
streams. The first situation exemplified by transient lakes with gravel sediment as we have just
described (and represented by large glacial trough
lakes) implies that coarse material remains restricted to the up-lake, in contrast to the down-
lake where the fine sediment dominates. To a
lesser extent this is frequently the pattern when a
rhithral tributary empties into a lake. Among
many Alpine lakes, the Lunzer Untersee is an example of this kind. Its main inflow has built up a
permanently reshaped gravel barrier at some distance from its mouth, whereas increasingly fine
sediment and littoral marl are typical of the
mid- and down-lake.

In contrast, in maars and other volcanic lakes,
with their limited drainage basins, and less so in
karstic and in dune lakes, one may find exclusively
fine inorganic matter, whereas in peat bog
lakes, often quoted as dystrophic lakes (from **dy**,
Swedish: peaty soil, boggy conditions), organic material prevails. All kinds of transitions between organic and inorganic calcareous and non-calcareous
sediments, the latter often referred to as **gyttja**
(Swedish: mud), occur in lakes depending on the
properties of their catchments. In addition, minerals such as vivianite (iron-phosphate mineral)
and pyrite may form distinct layers or lentil-like
inclusions, which most often are due to bacterial
activities.

Texture may vary greatly along a sediment profile and is characterised by three main types which
can occur in combination. River floods and under-
flow currents may produce discrete sand layers
or sand lenses, whereas turbidite layers contain
distinct sedimentary sequences changing from
coarse to fine material, some of which are due to
laminar flow and others to turbulent water flow
movement.

The third phenomenon concerns various laminations which comprise layers of different texture.
They are of varying origin and may be the result of
seasonal or secular deoxygenation of the overlying
water, or sometimes of discrete sedimentation
events. If the laminations are paired and laid down
in an annual cycle they are conveniently called
varves. They are common in lakes with glacial origin and activities. Often, varves form in lakes
receiving sediment from a melting glacier, the
coarser particles soon settling out in summer, and
the fine particles during winter when the lake is
frozen over and no additional allochthonous sediment is added. Varves are an important tool to estimate the age of lakes if one pair is formed each year.
It was De Geer (1912, 1940) who for the first time
used varved sediment in Sweden for dating of the
period from 12,000 years BP until the present.

Laminated sediments may also come about in
lakes outside of any glacial influence (O'Sullivan
1983). Here, they may be caused by autochthonous
seasonal events, such as the sedimentation of
algae, $CaCO_3$, etc., during summer, whereas iron
precipitation varies during the winter and summer
stagnation periods (as black FeS) when oxygen approaches a minimum. During circulation, when
column oxygen concentrations increase, light
brownish $Fe(OH)_3$ is precipitated. In addition,
rhythmites resulting from a single flood may be
distinguished. Depending on the incidence of
floods, annually or irregularly, sometimes it can be
difficult to distinguish between varves and rhythmites (Sturm 1979; Timms 1992).

2.12.2 Changes of shoreline configuration

Generally, delta formation is the most obvious
change affecting lakes. Often a natural process, it
may, however, also be induced by many human activities. Large sediment supply and little or rare

wave action may produce lobate delta forms as exemplified by Lake Macquarie (southeast Australia; Timms 1992) and to a lesser extent by the discharge of the Rhine into Lake Constance before the river's eastward dislocation and channel-like regulation in 1900 (Fig. 2.24). In contrast to the 'Old Rhine' with its former large flood-plain, the regulated section of the 'New Rhine' lacks any riverine area with sufficient retention capacity which would allow for the deposition of the considerable amounts of gravel and sand transported by the river. Therefore, a growing arcuate delta protrudes into the lake which at present has achieved an area

of more than 1 km². The river flows over its increasing delta without any effect on the lake bottom. This is again in contrast to the 'Old Rhine' which has cut a very distinct sub-lacustrine channel.

If any rhithral tributary enters a lake from a slope it may transfer enough material to cause a narrow passage of the lake, as for example in Wolfgangsee, Austria. Owing to frequent changes of the riverbed of Zinkenbach in its last section, the influent has built up an arcuate delta. Finally a lake may be cut into two as at Interlaken (Switzerland) where the Lutschin Delta separates the Thunersee

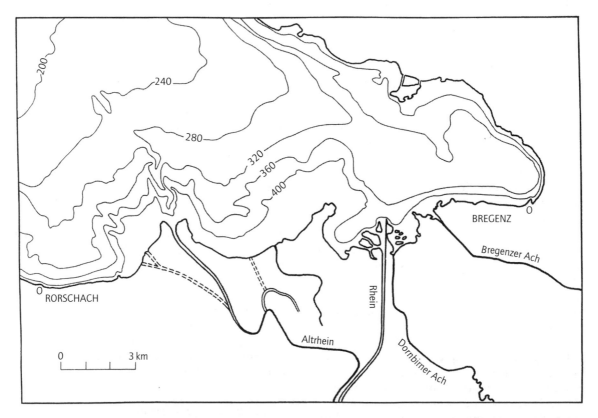

Fig. 2.24 The old (Altrhein) and new (Rhein) lower section of the River Rhine with their respective outlets into Lake Constance. In contrast to the old section which allowed for the deposition of gravel and other bedload material at the outlet, the channelised new section has led to major changes of the relevant coastal section owing to deposition in the lake since 1900. The old section varied greatly (dashed river courses) in its course and resulted in sub-lacustrine channels. (Modified from Müller, 1966.)

and the Brienzersee. Similarly, talus fan growth may encroach on lakes. Timms (1992) refers to Lake Pearson in the Southern New Zealand Alps which is almost cut into two by fans from opposite mountain sides of the lake.

One of the most important agencies of shoreline modification, either in arid or in coastal zones, especially in shallow lakes, is sand dune movement. Many examples of this kind have been described from Australia by Timms (1992). In elongated shallow lakes, sandy substrate segmentation may be brought about if the lake axis corresponds to the dominant wind direction. Under a strong wind regime, triangular (cuspate) spits develop along the shore. If the prevailing winds are monodirectional, the spits do not develop further, and a shoal may emerge to become a barrier and divide the lake (Lees 1989). If winds are bidirectional, the spits grow in response to circulation cells and may eventually join an opposing spit to sub-divide the lake (Zenkovich 1959; from Timms 1992).

Other changes of shoreline configuration are connected with volcanic lakes where lava flows, ash and scoria may contribute to relevant changes. Similarly the retreat (see Lewis Tarn, Fig. 2.12) and protrusion of glaciers, and above all exposure of flood-plains to major floods, contribute to profound changes. Concerning the latter, backwaters and evorsion lakes may even become displaced. Such events will be discussed below.

2.12.3 Spatial displacement of lakes

The most prominent example of a lake's spatial dislocation is represented by Lake Biwa, Japan's largest and oldest lake (Horie 1991). Table 2.6 presents some of the important data on the lake.

The origin of Palaeo-Lake Biwa during the Middle Pliocene implies a continuous lacustrine history, and consequently one can witness the

Table 2.6 Background data of Lake Biwa.

A_E	3848 km^2	z_{max}	104 m
A	685 km^2	z_{mean}	41 m
V	27.5 km^2	Retention time	5.5 yr

presence of numerous endemic species of algae, macrophytes, molluscs, crustaceans and fish. Figure 2.25 shows that a Palaeo-Lake Biwa, much smaller than the modern lake, evolved some 3–2.5 million years BP about 30 km to the south of the southernmost part of the present lake (Yokoyama 1991) and then gradually shifted northward until it finally reached the modern basin 1–0.4 million years BP. Its present extent came about only about 0.3 million years BP. Obviously, both volcanic and tectonic events contributed to this process and to the lake's final shape and location. The whole progress of basin drifting is evidenced by sediment cores but at present the dynamics behind the dislocation of Lake Biwa are still a matter of discussion (Fig. 2.25).

Another remarkable illustration of long-term displacement is Lop-Nor (Sinkiang, China) with an area of about 3000 km^2 and a volume of 5 km^3 (Williams 1996). This lake, situated within the large plain of the Tarim Basin, changes its location over long periods. When Lop-Nor was first mentioned, more than 2000 years ago (during the Han Dynasty), it occupied a northerly part of the basin and was fed by the Kum Darja (= Kuruk Darja) river. During this stage, sedimentation increased in the shallow Lop-Nor, whereas **deflation** took place in the southern part of the Tarim basin. Eventually (in about AD 330), these processes led to the (re-)activation of the southern branch of the Kontche Darja, whilst, as a result of drought, the city of Lou Shan, west of Lop-Nor, became completely abandoned.

The southern branch of the Kontche Dara then flooded the southernmost sections of the Tarim Basin, giving rise to Kara Kushun Lake, located some 50 km southwest of Lop-Nor. In 1876–7 the Russian explorer Przewalski visited this large lake with abundant *Phragmites* stands. Soon afterwards (between 1900 and 1928), a dramatic reverse occurred, whereby, after almost 1600 years, the northern part of the Tarim Basin had been deepened by wind erosion to such an extent that the northern branch of the upper Kontche Darja again became the main river contributing to the lake, leading to the re-emergence of Lop-Nor as a lake, once more occupying the northern part of the

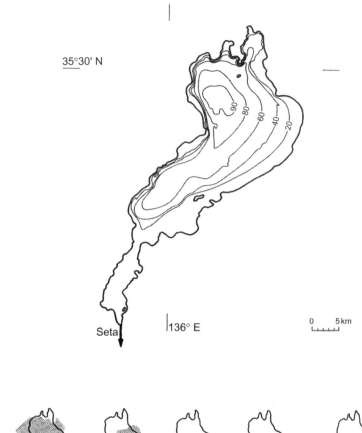

35°30' N

90 80 60 40 20

136° E

Seta

0   5 km

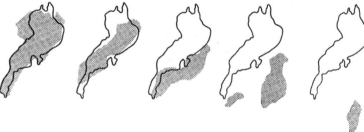

Fig. 2.25 Northward movement of Lake Biwa during the past three million years. Stippled areas represent previous lake positions relative to present; the oldest is to the right and evolution is leftward. (Modified from information provided through the Secretariat of Biwako Prize for Ecology 1996.)

basin. This process was observed and carefully followed by Hedin who devoted several excursions to the study of the area during this critical period (1896–7, 1900–1, 1928 and 1934; see Hedin 1942). Since then, and most likely again for a long but unpredictable period, Lop-Nor will persist as a lake for some centuries if human society does not interfere with the hydrography of the region.

As already mentioned, displacement of lakes on a small scale frequently occurs in flood-plains and in dune areas, where floods or strong persistent winds are involved. Less often volcanic and glacial events may contribute to dislocation. Finally, reference should be made to those once extensive lakes which now are witnessed only by small remnants. The Pleistocene Lakes Lahontan (maximum extent 21,860 km^2) and Bonneville (51,300 km^2) in the USA are remarkable examples of this kind. The relics of Lahontan Lake, such as Pyramid Lake, do not surpass an area of $500\,km^2$ and Lake Bonneville is now reduced to three salt lakes, Great Salt Lake being the most important. Since about 130 years ago it has undergone a water level variation of more than 6 m. Lake Chad, whose area has now shrunk to less than $1000\,km^2$, presumably attained more than $300,000–400,000\,km^2$ some 6000 years ago. With this area, it would have been the second largest lake in the world, after the Caspian Sea. During the mid 1960s it still covered an area of $25,000\,km^2$.

In Asia the precursor of the Dead Sea, Lisan Lake (about 100,000–20,000 yr BP), should be recalled (see section 2.5.1). At this stage it comprised the Sea of Galilee (Lake Kinneret) and attained an area of about three times the present size ($1020\,km^2$). Another example of this kind is Lake Sevan in Armenia ($A = 1416\,km^2$, $V = 58.4\,km^2$) which originated during the Early Quaternary when a Palaeo-Sevan, at least ten times larger than the present lake, came into existence by tectonic formation. Like the Neusiedlersee (see sections 2.1, 2.5.1, 2.5.4, 2.5.7, 2.7, 2.11 and 2.12.1) it is a slightly (sodium) alkaline lake with a salinity of only 0.7‰. With its retention time of about 50 years, it ranks 15th among those of the world's lakes exceeding $500\,km^2$ in area. As a result of water withdrawal for irrigation and hydropower generation, lake level has fallen since the 19th century. Attempts to stabilise the water level were not fully successful, and at present it stands 20 m below the level attained before the economic use of lake water. Plans for major river diversion in order to raise the water level again have so far not been carried out. Moreover, since 1970 the lake has experienced bottom anoxia connected with the presence of H_2S and methane (Meybeck *et al.* 1997).

In conclusion it must be mentioned that there are many more examples of small remnant lakes from once large bodies of water. Many of these are located in China.

REFERENCES

Ammann, B. (1982) Säkulare Seespiegelschwankungen: wo, wie, wann, warum? *Mitteilungen der Naturforschenden Gesellschaft in Bern, neue Folge,* **39**, 97–106.

Bayly, I.A.E. (1967) The general biological classification of aquatic environments with special reference to those of Australia. In: Weatherley, A.H. (ed.), *Australian Inland Waters and their Fauna* (11 studies). Australian National University Press, Canberra, 78–104.

Bayly, I.A.E. & Williams, W.D. (1973) *Inland Waters and their Ecology.* Longmans, Melbourne, 316 pp.

Berger, F. (1971) Zur Morphometrie der Seebecken. *Carinthia,* **II**, Sonderheft, **31**, 29–39.

Bonacci, O. (1987) *Karst Hydrology.* Springer, Heidelberg, 184 pp.

Bonacci, O. (1993) The Vrana Lake hydrology (island of Cres—Croatia). *Water Resources Bulletin,* **19**, 407–14.

Bond, R.M. (1935) Investigations of some Hispaniolan lakes II. Hydrology and hydrography. *Archiv für Hydrobiologie,* **28**, 137–61.

Brunskill, G.J., Ludlam, S.D. & Diment, W.H. (1969) A comparative study of meromixis. *Verhandlungen der Internationalen Vereinigung der Limnologie,* **37**, 137–9.

Cowgill, U.M. & Hutchinson, G.E. (1970) Chemistry and mineralogy of the sediments and their source materials. In: *Ianula: an Account of the History and Development of the Lago di Monterosi, Latium, Italy. Transactions of the American Philosophical Society, New series,* **60**(4), 37–101.

Crul, R.C.M.C. (1995) *Limnology and Hydrology of Lake Victoria.* UNESCO/IHP-IV Project M-5.1, Unesco, Paris, 79 pp.

Davis, W.M. (1882) On the classification of lake basins. *Proceedings, Boston Society of Natural History,* **21**, 315–81.

De Geer, G. (1912) A geochronology of the last 12,000 years. *Compte Rendu Congrès Géologique International Stockholm,* **1910**, 241–53.

De Geer, G. (1940) Geochronologica Suecica Principles. *Kungligar Svenska Vetenskapakademien Handlingar, Stockholm, Series 3,* **18**, 6.

Dobesch, H. & Neuwirth, F. (1983) Das Klima des Raumes Neusiedlersee. *Raumplanung Burgenland,* **1**, 1–110.

Findenegg, I. (1935) Limnologische Untersuchungen im Kärntner Seengebiete. *Internationale Revue für theoretische und angewandte Hydrobiologie,* **32**, 369–423.

Findenegg, I. (1947) Der Längsee. Eine limnologische Untersuchung. *Carinthia,* **136**, 77–93.

Findenegg, I. (1953) Kärntner Seen naturkundlich betrachtet. *Carinthia, Sonderheft*, **15**, 101 pp.

Forel, F.A. (1901) *Handbuch der Seenkunde: allgemeine Limnologie*. J. Engelhorn, Stuttgart, 249 pp.

Frey, D.G. (1955) Längsee: a history of meromixis. *Memorie Istituto Italiano di Idrobiologia Supplemento*, **8**, 141–64.

Goldsmith, E. & Hildyard, N. (1985) *The Social and Environmental Effects of Large Dams. 2 Case Studies*. Wadebridge Ecological Centre, Wadebridge, 327 pp.

Golubev, G.N. (1996) Caspian and Aral Seas: two different paths of environmental degradation. *Verhandlungen der Internationalen Vereinigung für theoretische und angewandte Limnologie*, **26**, 159–66.

Halbfass, W. (1923) *Grundzüge einer vergleichenden Seenkunde*. Borntraeger, Berlin, 354 pp.

Hammer, T.U. (1988) Water resources and their utilisation in Saskatchewan, Canada. *Verhandlungen der Internationalen Vereinigung für theoretische und angewandte Limnologie*, **23**, 228–33.

Harmsworth, R.V. (1984) Längsee: a geochemical history of meromixis. *Hydrobiologia*, **108**, 219–31.

Hedin, S. (1942) *Der wandernde See*. Brockhaus, Leipzig, 295 pp.

Herdendorf, C.E. (1990) Distribution of the world's large lakes. In: Tilzer, M.M. & Serruya, C. (eds), *Large Lakes, Ecological Structure and Function*. Springer, Berlin, Heidelberg, 691 pp.

Horie, S. (1991) *Die Geschichte des Biwasees in Japan*. Universitätsverlag Wagner, Innsbruk, 346 pp.

Hsü, K.J. (1972) When the Mediterranean dried up. *Scientific American*, **227**(6), 27–36.

Hurlbert, S.H. & Chang. C.C.Y. (1984) Ancient ice islands in salt lakes of the Central Andes. *Science*, **224**, 299–302.

Hutchinson, G.E. (1937) A contribution to the limnology of arid regions primarily founded on observations made in the Lahontan Basin. *Transactions of the Connecticut Academy of Arts and Sciences*, **33**, 47–132.

Hutchinson, G.E. (1957) *A Treatise on Limnology*, Vol. 1, *Geography, Physics, Chemistry*. John Wiley, New York, 1015 pp.

Hutchinson, G.E. (1967) *A Treatise on Limnology*, Vol. 2. John Wiley, New York, 1115 pp.

Hutchinson, G.E. & Löffler. H. (1956) The thermal classification of lakes. *Proceedings of the National Academy of Sciences*, **42**, 84–6.

IHP (International Hydrological Programme) (1996). Caspian sea-level rise: an environmental emergency. *UNESCO Newsletter*, **5**, 22–26.

ILEC (International Lake Environment Committee) (1992) Lakes of the world: Lake Baikal (Russia, CIS; Otsu, Shiga, Japan). *Newsletter*, **18**, 6–7.

ILEC (International Lake Environment Committee) (1997) A look at five water bodies in Argentina. *Newsletter*, **29**, 6–7.

Janz, H. (1993) Die Bedeutung der Ostrakoden (Crustacea) des Steinheimer Beckens für die Diskussion um die Gestaltveränderungen der Planorbidae (Gastropoda) im Laufe der Seegeschichte. *Jahreshefte der Gesellschaft für Naturkunde in Württemberg*, **148**, 33–51.

Källquist, T., Lien, L. & Liti, D. (1988) *Lake Turkana Limnological Study 1985–1988*. Norwegian Institute for Water Research, Oslo, 98 pp.

Kjensmo, J. (1997) The influence of road salts on the salinity and the meromictic stability of Lake Svinsjøn, southern Norway. *Hydrobiologia*, **347**, 151–8.

Kosarev, A.N. & Yablonska, E.A. (1994) *The Caspian Sea*. Academic Publishers, New York, 259 pp.

Küpper, H. (1957) *Erläuterungen zur geologischen Karte Mattersburg-Deutschkreuz*. Geologische Bundesanstalt, Wien, 59–67.

Kvet, J., Löffler, H., Gopal, B. & Tundisi, J.G. (1990) Wetland impact assessment. In: Patten, B.C. (ed.), *Wetlands and Shallow Continental Water Bodies 1*. SPB Academic Publishing BV, The Hague, Netherlands, 363–71.

Lees, B. (1989) Lake segmentation and lunette initiation. *Zeitschrift für Geomorphologie*, **33**, 475–84.

Leveque, C. (1976) *Ecology of Lake Chad, a Tropical Shallow Lake (Africa)*. Office de la Recherche Scientifique et Technique outre mer, Paris, 163–87.

Livingstone, D.A. (1965) Sedimentation and history of water level change in Lake Tanganyika. *Limnology and Oceanography*, **10**(4), 607–10.

Löffler, H. (1960) Limnologische Untersuchungen an chilenischen und peruanischen Binnengewässern. *Arkiv för Geofysik*, **3**, 155–254.

Löffler, H. (1968) Die Hochgebirgsseen Ostafrikas. *Hochgebirgsforschung* 1. Universitätsverlag Wagner, Innsbruck—München, 1–65.

Löffler, H. (1969) High altitude lakes in the Mt. Everest region. *Verhandlungen Internationale Vereinigung für Theoretische und Angewandte Limnologie*, **17**, 373–85.

Löffler, H. (1977) 'Fossil' meromixis in Kleinsee (Carinthia) indicated by ostracods. In: Löffler, H. & Danielopol, D. (eds), *Aspects of Ecology and Zoogeography of Recent and Fossil Ostracoda. Proceedings of* the 6th International Symposium on Ostracods. Dr W. Junk, The Hague, 321–25.

Löffler, H. (1981) The winter condition of Lake Niriz in southern Iran. *Verhandlungen der internationalen Vereinigung der Limnologie*, **21**, 528–34.

Löffler, H. (1986) An early meromictic stage in Lobsigensee (Switzerland) as evidenced by ostracods and *Chaoborus*. *Hydrobiologia*, **143**, 309–14.

Löffler, H. (1988) Natural hazards and health risks from lakes. *Water Resources Development*, **4**(4), 276–83.

Löffler, H. (1997) Längsee: a history of meromixis; 40 years later: Homage to Dr D.G. Frey. *Verhandlungen der Internationalen Vereinigung für theoretische und angewandte Limnologie*, **26**, 829–32.

Löffler, H., Berger, F., Brenner, T., Dokulil, M., Schiemer, F. & Schutze, E. (1975) Arbeitsbericht der limnologischen Exkursion Goggausee 1974. *Carinthia II*, **165/85**, 165–96.

Mandych, A.F. (1995) *Enclosed seas and large lakes of Eastern Europe and Middle Asia*. Academic Publishing, Amsterdam, 273 pp.

Margalef, R. (1983) *Limnologia*. Ediciones Omega, Barcelona, 1010 pp.

Margalef, R. (1994) The place of epicontinental waters in global ecology. In: Margalef R. (ed.), *Limnology Now, a Paradigm of Planetary Problems*. Elsevier, Amsterdam, 1–8.

Martonne, E. & Aufrère, L. (1928) L'extension des régions peivées d'ecoulement vers l'océan. *Publications Union Géographiques internationales, Paris*, **3**, 194 pp.

Meybeck, M., Akopian, M. & Andréassian, V. (1997) What happened to Lake Sevan? *SIL News* **23**, 7–10.

Müller, G. (1966) The new Rhine Delta in Lake Constance. In: *Deltas in their Geologic Framework*. Houston Geological Society, Houston, TX, 106–24.

Neev, D. & Emery, K.O. (1967) *The Dead Sea*. Geological Survey Bulletin 41, State of Israel, Ministry of Development, Tel Aviv, 147 pp.

Negrea, S., Botniaruc, N. & Dumont, H. (1999) Phylogeny, evolution and classification of the Branchiopoda (Crustacea). *Hydrobiologia*, **412**, 191–212.

Nipkow, F. (1920) Vorläufige Mitteilungen über Untersuchung des Schlammabsatzes im Zürichsee. *Revue d'Hydrologie*, **1**, 1–27.

Northcote, T.D. & Halsey, T.G. (1969) Seasonal changes in the limnology of some meromictic lakes in southern British Columbia. *Journal of the Fisheries Research Board Canada*, **26**, 1763–87.

Northcote, T.D., Morales, P., Levy, D.A. & Greaven, M.S.

(eds) (1989) *Pollution in Lake Titicaca, Peru: Training, Research and Management*. Westwater Research Centre University of British Columbia, Vancouver, 259 pp.

O'Sullivan, P.E. (1983) Annually-laminated lake sediments and the study of Quaternary environmental changes — a review. *Quaternary Science Reviews*, **1**, 245–313.

Penck, A. (1894) *Morphologie der Erdoberfläche*. Engelhorn, Stuttgart, Vol. 1, 471 pp., Vol. 2, 696 pp.

Richardson, S.J. & Richardson, A.E. (1972) History of an African Rift Lake and its climatic implications. *Ecological Monographs*, **42**, 499–543.

Rögl, F. & Steininger, F.F. (1983) Vom Zerfall der Tethys zu Mediterran und Parathys. *Annalen naturhistorisches Museum Wien*, **85/A**, 135–63.

Ruttner, F. (1931) Hydrographische und hydrochemische Beobachtungen auf Java, Sumatra und Bali. *Archiv für Hydrobiologie, Supplementband*, **8**, 197–460.

Ruttner, F. (1955) Der Lunzer Mittersee, ein Quellsee mit zeitweise meromiktischer Schichtung. *Memorie dell'Istituto Italiano di Idrobiologia*, **8**, 265–80.

Ruttner, F. (1962) *Grundriss der Limnologie, 3. Auflage*. De Gruyter & Co, Berlin.

Scharf, B.W. (1991) *Ektogene, krenogene und biogene Meromixis in Maarseen der Eifel*. Deutsche Gesellschaft für Limnologie ev., Jahrestagung 30.09–6.10.1991 in Mondsee, Frank GmbH, München.

Serruya, C. & Pollingher, U. (1983) *Lakes of the Warm Belt*. Cambridge University Press, Cambridge, 569 pp.

Stumm, W. & Matter-Muller, C. (1984) Wasser — Ein Rohstoff in Gefahr. In: *Vorträge der GDI — Tagung*. Gottlieb Duttweiler Institut, Rüshlikon, Zürich, 1–21.

Thenius, E. (1977) *Meere und Länder im Wechsel der Zeiten*. Springer-Verlag, Berlin, 200 pp.

Thienemann, A. (1913) Physikalische und chemische Untersuchungen in den Maaren der Eifel. *Verhandlungen der naturhistorischen Vereinigung des preussischen Rheinlandes*, **70**, 249–302.

Timms, B.V. (1992) *Lake Geomorphology*. Glenaeagles Publishing, Adelaide, 180 pp.

Troll, C. (1959) Die tropischen Gebirge. *Bonner geographische Abhandlungen*, **25**, 1–93.

Tundisi, J.G. (1994) Tropical South America: Present and perspectives. In: Margalef R. (ed.), *Limnology Now, a Paradigm of Planetary Problems*. Elsevier, Amsterdam, 353–424.

Walker, K.F. (1974) The stability of meromictic lakes in Central Washington. *Limnology and Oceanography*, **19**, 209–22.

Wallace, J.M. & Vogel, S. (1994) *El Niño and Climate Pre-*

diction. Reports to the Nation on our Changing Planet, Office of Global Programs, Washington, DC, 24 pp.

Williams, W.D. (1996) The largest, highest and lowest lakes of the world: saline lakes. *Verhandlungen der Internationalen Vereinigung für theoretische und angewandte Limnologie*, **26**, 61–79.

Yokoyama, T. (1991) Sedimentologie and Tephrastratigraphie des Biwa- See-Beckens. In: Horie S. (ed.), *Die Geschichte des Biwa-See in Japan*. Universitätsverlag Wagner, Innsbruck, 101–8.

Zenkovich, V.P. (1959) On the genesis of cuspate spits along lagoon shores. *Journal of Geology*, **67**, 269–77.

3   The Hydrology of Lakes

THOMAS C. WINTER

3.1   BACKGROUND

3.1.1   Lakes as features of the hydrological system

Lakes are integral features of the global hydrological system. Because they interact directly with atmospheric water, surface water, and groundwater, the hydrology of lakes is greatly influenced both by their physiographic and by their climatic setting. The former affects movement of surface water and groundwater, and the latter, atmospheric-water components, precipitation and evaporation. In addition, the physiography of lake terrain needs to include natural depressions in order to retain lake water, and climate needs to be wet enough in order to provide sources of water to maintain the lake. Because lake distribution is dependent on the combination of physiography and climate, especially of the past 25,000 years, natural lakes are concentrated in specific regions of the world (Hutchinson 1957; Meybeck 1995). Large parts of Earth contain no natural lakes. In order to overcome this uneven natural distribution, and to provide surface storage of water where needed for human use, reservoirs have been constructed in many parts of the world, including those that also contain natural lakes. Reservoirs can be considered a specific type of lake and therefore they are included in this chapter.

Although lakes are intimately connected to all components of the hydrological system, the volume of water they contain world-wide is small. The proportion of water contained in lakes, circulating groundwater, and atmospheric water vapour constitutes about 9% of Earth's inland waters (Hutchinson 1957). Meybeck (1995) indicated that about $82.5 \times 10^3 \, km^3$ is contained in saline lakes (95% of which is in the Caspian Sea), and about $95.14 \times 10^3 \, km^3$ in freshwater lakes. Lakes in tectonic basins contain about 57% and in glacial terrain about 40% of the total volume of water in freshwater lakes. The percentage of lakes in glacial, tectonic and other terrain, grouped according to various size categories, is shown in Fig. 3.1a. The relationship between lake volume and lake area, and lake numbers, is shown in Fig. 3.1b.

Although representing a relatively small percentage of Earth's water, lakes are and always have been of great importance to humanity for water supply, as habitat for food, as a source of power (in the case of reservoirs), for recreation and aesthetic value. Because these are needed in order to manage lakewater volumes, and to determine chemical budgets of lakes, much interest in the hydrology of lakes is centred on water budgets. Furthermore, because this has a great effect on the types and rates of biochemical processes, many biochemical studies are concerned with the residence time of water in lakes. In order to address the need for understanding water budgets and the residence time of lake water, this chapter focuses on developing an understanding of the hydrological processes that affect water budgets and how these influence water residence times of lakes in different hydrogeological and climatic settings.

3.2   HYDROLOGICAL PROCESSES AS RELATED TO LAKE HYDROLOGY

The potential sources of water to lakes are (i)

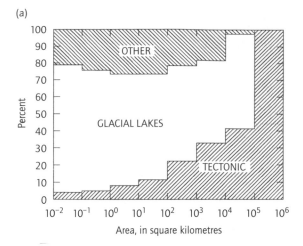

(a)

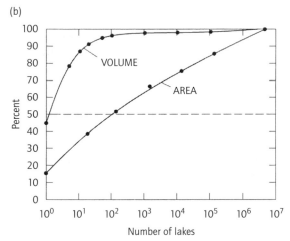

(b)

Fig. 3.1 (a) Percentage of lake types in various size categories. The Caspian Sea represents all of the bar between 10^5 and 10^6 km². (b) Relations of cumulative lake area and volume to number of lakes. In both curves, the Caspian Sea is the first lake plotted. (Modified from Meybeck 1995; reproduced with permission of Springer-Verlag.)

direct precipitation, (ii) inflowing streams, and (iii) groundwater. Losses occur to (iv) evaporation, (v) streams and (vi) groundwater (Fig. 3.2). Although the relative importance of these sources and sinks may vary greatly according to physiographic and climatic settings, a general understanding of processes in all components of the hydrological system is fundamental to understanding the hy-

drology of lakes. In addition to the information presented below, detailed discussions of the hydrology of lakes can be found in Winter (1981), Winter & Woo (1990) and Winter (1995).

3.2.1 The interaction of lakes with atmospheric water

Atmospheric water is the most dynamic component of the hydrological system. Weather patterns are highly variable, resulting in great variability in the distribution of precipitation in time and space. In some physiographic and climatic settings, weather conditions can be relatively broad and uniform, and in such settings, the estimation of precipitation input to lakes can be reasonably accurate. However, if an area is characterised by convective storms, estimation of precipitation may be subject to considerable error. Many regions experience a mix of weather types at different seasons. Although also controlled by weather conditions, evaporation rates are less variable than the distribution of precipitation.

3.2.1.1 Precipitation

Considering the lake as the basic accounting unit, the only aspect of precipitation of major interest to determining water budgets is the volume of water that falls directly on the lake. Few precipitation gauges are, however, positioned directly over a water body. Therefore, precipitation directly on the lake surface must be determined by interpretation of data collected at land-based stations. Even if it were possible to place a precipitation gauge over a lake, its opening would sample only a very small surface area, and interpretation of this point data would still be necessary.

Although measuring precipitation is conceptually simple, determination of the actual volume of precipitation input to a lake can be quite uncertain. The catch efficiency of precipitation gauges may vary considerably depending on factors such as wind velocity during storms. The catching area of a precipitation gauge most accurately represents the equivalent portion of lake surface only when precipitation falls vertically. If, because of wind,

Fig. 3.2 Schematic diagram of the hydrological components associated with the water budget of lakes: P, precipitation; E, evaporation; SWI, surface water in; GWI, groundwater in; SWO, surface water out; GWO, groundwater out.

precipitation approaches the gauge at an angle, its effective catch area is reduced. Furthermore, effective catch area of the gauge orifice becomes smaller as wind velocities increase. Therefore, in order to overcome the problem of catch efficiency, wind shields are used to disrupt the flow of air, so that precipitation falls more vertically. A number of different types have been designed (Allis *et al.* 1963), but the Alter shield seems to be the most commonly used.

The second step in determining precipitation input to a lake is the interpretation (or **regionalisation**) of the data collected at a network of precipitation gauges. The three methods most commonly used are **averaging**, **Theissen polygons** and **isohyets**. When using the averaging method, the data are usually weighted according to the distance of the gauge from the lake. Theissen polygons are used to determine the area represented by a gauge, where the gauge is at the centre of the area. The isohyetal method involves contouring the data, then determining the area between the contours, or isohyets, in order to calculate the quantity of precipitation received. Linsley *et al.* (1975) provided an example where the three methods are compared for determining precipitation to a watershed (Fig. 3.3). The calculated precipitation input differed by

as much as 18% between the three methods. This example is not strictly applicable because gauges are usually not present in the middle of lakes. However, if precipitation gauges are present around the perimeter, the volume of water that falls on the lake's surface can be determined by one of these three methods.

In many studies, precipitation input is determined by regionalisation of data collected at network gauges some distance from the lake of interest. Siegel & Winter (1980) found that precipitation input to a lake is most accurately determined if gauges are placed as close to the lake as possible, rather than using a more distant network.

3.2.1.2 Evaporation

Evaporation is the transformation of water molecules from the liquid to the gaseous state. Therefore, in order to calculate evaporation most accurately, the mechanisms that control the transformation process need to be understood and quantified. In order to use one of the most accurate methods, such as **eddy correlation**, measurements need to be collected of the vertical fluxes of air temperature and water vapour at close time

(a)

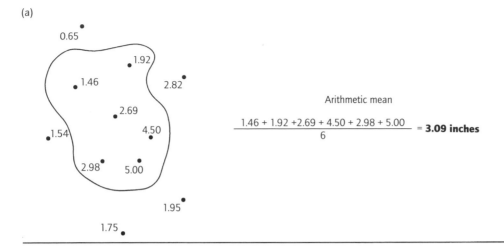

Arithmetic mean

$$\frac{1.46 + 1.92 + 2.69 + 4.50 + 2.98 + 5.00}{6} = \textbf{3.09 inches}$$

(b)

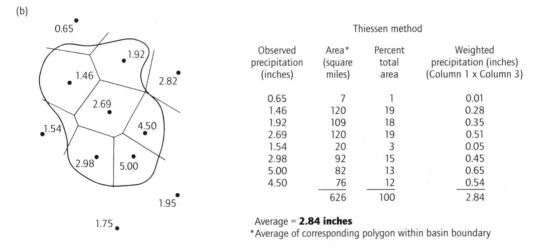

Thiessen method

Observed precipitation (inches)	Area* (square miles)	Percent total area	Weighted precipitation (inches) (Column 1 x Column 3)
0.65	7	1	0.01
1.46	120	19	0.28
1.92	109	18	0.35
2.69	120	19	0.51
1.54	20	3	0.05
2.98	92	15	0.45
5.00	82	13	0.65
4.50	76	12	0.54
	626	100	2.84

Average = **2.84 inches**
*Average of corresponding polygon within basin boundary

(c)

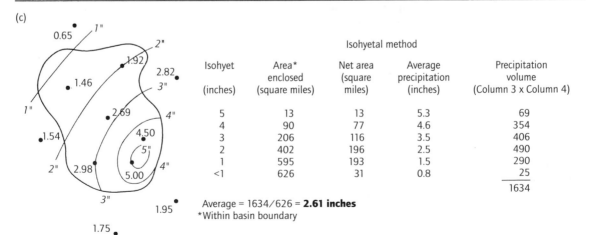

Isohyetal method

Isohyet (inches)	Area* enclosed (square miles)	Net area (square miles)	Average precipitation (inches)	Precipitation volume (Column 3 x Column 4)
5	13	13	5.3	69
4	90	77	4.6	354
3	206	116	3.5	406
2	402	196	2.5	490
1	595	193	1.5	290
<1	626	31	0.8	25
				1634

Average = 1634/626 = **2.61 inches**
*Within basin boundary

Fig. 3.3 Comparison of three methods for estimating the areal distribution of precipitation from point-gauge data. (Modified from Linsley *et al.* 1975; reproduced with permission of the McGraw-Hill Companies.)

intervals. Because of the need for extensive and costly instrumentation, as well as for close, hands-on interaction with the instruments and data collection, the eddy-correlation method generally is not used for evaporation studies lasting more than a few days. Instead the **energy budget method** is one of the most accurate ways of determining evaporation for extended periods of time. This method also requires costly instrumentation to measure radiation, air temperature, water temperature and vapour pressure over the lake surface, but the instruments do not need constant monitoring and attention.

Numerous **methods for estimating evapotranspiration** have also been developed, a number of which have also been used for determining evaporation. Winter *et al.* (1995) evaluated 11 commonly used methods, using the energy budget method as the standard of comparison, for a small lake in the north-central USA. They found that the methods which compared most favourably to the energy budget were those which used data on net radiation and heat stored in the lake. These were the modified Penman (Jensen *et al.* 1974), the modified DeBruin–Keijman (DeBruin & Keijman 1979), and the Priestley–Taylor (Stewart & Rouse 1976) methods. They also found that if data only on air temperature are available, reasonably acceptable values were obtained using the Hamon (1961) method.

Evaporation pans are the instruments most commonly used for estimating evaporation from lakes. Here, the pan is considered to be a surrogate for a lake. However, it was long ago recognised that response of pans to the effects of wind and temperature differ considerably from those of lakes, and need to be accounted for in the estimation of evaporation. Furthermore, in order to make pan data more useful in estimating evaporation from lakes, pan-to-lake coefficients need to be determined. These have not been developed at many sites because the actual evaporation of a lake needs to be known in order to determine the coefficient. Therefore, they have been determined only for lakes which have been studied by more accurate methods of determining evaporation, such as the energy budget method (Harbeck *et al.* 1958). The commonly used pan-to-lake coefficient of 0.7 is valid only for annual estimates of evaporation (Kohler *et al.* 1955). Studies comparing evaporation from pans to that determined by the energy budget method indicate that monthly pan-to-lake coefficients can vary from 0.5 to more than 2.0 (Hounam 1973). If 0.7 is used for monthly values in cases where the actual coefficient is 2.0, the estimate of evaporation could be in error by nearly 200%.

In many studies, pan stations are not established at the lake of interest. Instead, pan data from network stations, sometimes a considerable distance from the lake, are used. In these cases, the atmospheric conditions at the pan site may differ considerably from those at the lake, resulting in large errors.

3.2.2 *The interaction of lakes with streams*

The water budget of lakes with inflowing and outflowing streams is commonly dominated by streamflow. Of the three main components of the hydrological system—atmospheric water, surface runoff and groundwater—surface runoff is the one which can be measured most accurately. Therefore, if it is possible to measure streamflow to and from a lake at the lake's boundary, these will be the most accurately determined components of the lake's water budget. However, if interest is in understanding and predicting streamflow **input**, rather than simply knowing the quantity of streamflow to and from a lake, it is important to understand how the stream interacts with its watershed before entering the lake.

For example, if a watershed contains steep slopes and little subsurface storage, stream inflows will be highly variable. Large volumes of water will flow into the lake during and following storms, but little or none will enter between. Conversely, if a lake's watershed contains gentler slopes and a large volume of groundwater, stream inflows will be relatively constant during most weather conditions. Selection of the type of stream-gauging method to use for these different types of flow conditions is an important consideration when determining the surface-water component of water budgets of lakes as well as rivers (Winter 1981).

In many settings it is not possible to site stream gauges directly at the boundary of a lake. Instead, because of a logistical or other reason, the gauge may need to be placed some distance from the lake. If this is the case upstream of the lake, it may be necessary to estimate the groundwater input to the stream between the stream gauge and the edge of the lake, and to add that quantity to the stream-flow measured at the gauge. If this is the case downstream of the outflow from the lake, the quantity of groundwater inflow between the lake edge and the actual gauging station also needs to be subtracted from the quantity of streamwater measured at the gauge. If this approach is necessary, the hydrological processes of groundwater interaction with lakes, and the techniques of investigation presented in the next section, will also apply to in-flowing and ouflowing streams.

3.2.3   The interaction of lakes with groundwater

Nearly all lakes experience some direct interaction with groundwater. The only exceptions will be those separated from the groundwater system by an unsaturated zone, and even in these cases seepage from the lake may recharge groundwater. In most cases, the presence of an unsaturated zone beneath a lake is a transient phenomenon and is most common for small lakes in groundwater recharge areas. Lakes in groundwater recharge areas are sometimes thought of as being separated from the groundwater system because the levels in adjacent wells may be lower than lake level. However, a downward component of flow is a characteristic of groundwater recharge areas and a water table slope away from the lake-shore is to be expected.

The largest interface between lakes and the three major components of the hydrological system is with groundwater. The proportion of a lake in contact with the atmosphere is much smaller than that of its interface with groundwater, in that a lake's bed is not generally flat, and contact with surface flow occurs only at those points at which

streams enter and leave the lake. Conversely, the entire perimeter of a lake, including the bed, is in direct connection with groundwater, and contiguous uplands are considered to be areas of direct drainage (Fig. 3.4). In most cases, this input takes place through subsurface flow, but where soils are relatively impermeable or when precipitation rates exceed the infiltration capacity of the soil direct drainage across the land surface may occur as overland flow.

Because of their direct connection with groundwater, lakes are greatly influenced by their position within groundwater flow systems. For example, lakes may (i) experience seepage outflow to groundwater, (ii) receive seepage inflow from groundwater, or (iii) both (Born *et al.* 1974; Winter 1976). Lakes which are elevated above the surrounding topography are more likely to contribute to groundwater recharge than those which lie in the lower parts of the landscape. Conversely, low-lying lakes are more likely to receive groundwater input, and to experience little if any seepage loss to groundwater. However, it is also not unusual for lakes which contribute recharge to groundwater to

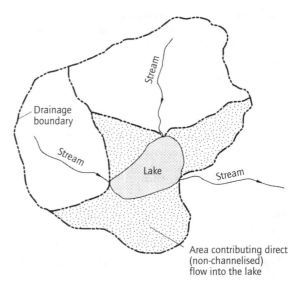

Fig. 3.4   Schematic lake and drainage basin showing area contributing to non-channelised surface and subsurface inflow to a lake.

be present in lowlands, and those which receive groundwater inputs to be found in uplands (Fig. 3.5). Rates of seepage, and the volume of a lake related to its interaction with groundwater, are affected to a large extent by the permeability of its geological substrate and the hydraulic characteristics of its sediments. Furthermore, LaBaugh (1988) has indicated that the chemical characteristics of lakes are strongly related to their position within groundwater flow systems.

In addition, a number of smaller-scale hydrological processes which take place near the edges of lakes also can affect their interactions with groundwater. Two of these (**preferential flow** and **non-linear seepage**) are related to the geology of lake-beds. Two others (**focused recharge** and **transpiration-induced seepage from lakes**) are connected with climate conditions.

Underlying geology exerts a significant influence on the distribution of seepage through a lake-bed. Most lakes contain a profundal zone in which there are limnetic sediments, but also a wave-washed littoral zone which is generally free of such material. The sediments of the deeper parts of lakes commonly consist of silts and clays, overlain by organic deposits. These are generally of very low permeability, permitting very little water to flow through them. The absence of fine-grained limnetic sediments from the littoral zone, however, means that lakes generally interact freely with groundwater over that part of their bed. It is not only groundwater from the surrounding upland which interacts with a lake in the littoral zone. That which encounters the less permeable

profundal sediments is also diverted to this part of the lake-bed.

Preferential flow through lake-beds (commonly termed **springs** if flow is into the lake) is a result of a heterogeneous geological substrate. Flow through a sand lens bounded by silt or clay is faster and more voluminous than that through finer grained deposits.

Non-linear seepage patterns occur as a result of a sloping water table meeting the flat lake surface (Fig. 3.6). This condition causes **non-linear seepage inflow** wherever the water table slope breaks upward where it meets the flat lake surface on the inflow side of lakes. **Non-linear seepage outflow** takes place where the water table slope breaks downward from the lake surface on the outflow side of lakes. These seepage patterns can be modified further, and yet remain non-linear, by anisotropy of the geological substrate.

Anisotropy, the ratio of horizontal to vertical hydraulic conductivity, is the result of orientation of rock grains (rock fabric) in the geological substrate. The anisotropy of a rock fabric composed of flat grains lying on top of one another is greater than that of a fabric characterised by round grains. Anisotropy affects the spacing between the flow lines which intersect a lake-bed (Pfannkuch & Winter 1984). The higher the anisotropy, the greater the distance between given flow lines (Fig. 3.7).

Focused recharge is related to thickness of the unsaturated zone. If precipitation or snowmelt infiltrates the land surface uniformly, it will reach

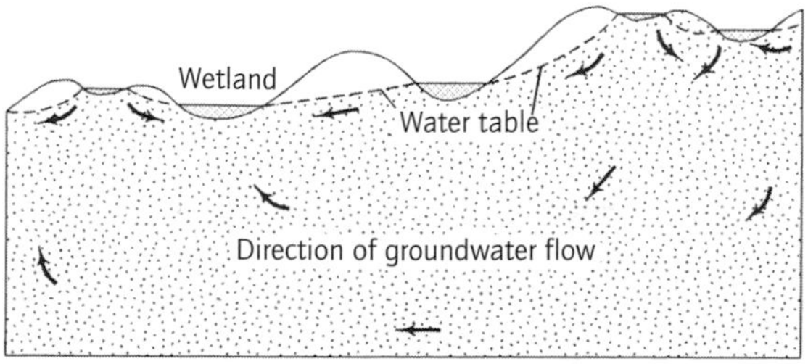

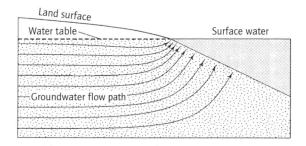

Fig. 3.5 Local and regional groundwater flow associated with lakes in hummocky terrain.

Fig. 3.6 Groundwater inflow to lakes showing decreasing volumes of seepage with distance from shore.

the water table sooner, and will therefore recharge groundwater more rapidly where the distance to the water table is least (Winter 1983). Because the unsaturated zone is very thin where the water table meets the lake, groundwater recharge is commonly concentrated close to the lake-shore (Fig. 3.8). This can cause small groundwater flow systems to discharge into the lake just offshore. Generally these flow systems are highly transient, lasting from hours to weeks, depending on the magnitude and duration of the infiltration event.

Transpiration-induced seepage from lakes also is related to thickness of the unsaturated zone and to climate. Because the water table close to lakes is near the soil surface, upland plants can transpire water directly from groundwater. This can lead to the formation of cones of depression on the water table, similar to the effects of pumping a well, which in turn can cause water to seep from the lake and be transpired by the plants (Fig. 3.9). This process also is transient, but it tends to last longer than focused recharge, often weeks or months.

As with the other components of the hydrological system, water flux can be measured at the interface of the lake with that component, which in the case of groundwater is the lake-bed. Devices such as seepage meters and potentiomanometers (Lee & Cherry 1978; Winter *et al.* 1988) are commonly used in order to measure fluxes across the lake-bed in space and in time. Small-scale flow processes related to geology of the lake-bed (spatial variation) and to climate (temporal variation), discussed above, need to be taken into account when interpreting the data. Furthermore, uncertainties related to the functioning of the devices themselves (Shaw & Prepas 1990; Belanger & Montgomery 1992) need also to be considered.

As well as small-scale effects related to geology of the lake-bed and to climate, field studies of interaction between lakes and groundwater need also to address larger-scale processes related to the position of lakes within groundwater flow systems. The most thorough instrumentation for determining the interaction of lakes with groundwater would use a combination of piezometer nests, water-table wells and devices to measure

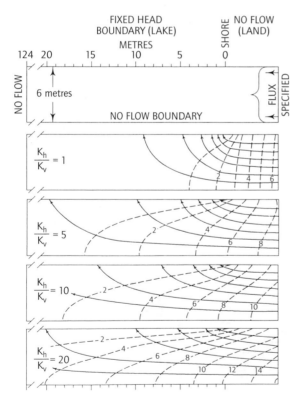

Fig. 3.7 Distribution of seepage inflow to a lake for different values of anisotropy (K_h/K_v). (Modified from Lee *et al.* 1980; reproduced with permission of the American Society of Limnology and Oceanography.)

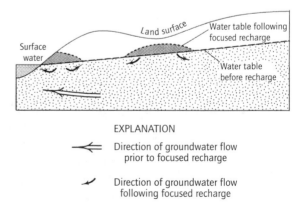

Fig. 3.8 Focused groundwater recharge adjacent to surface water and beneath land-surface depressions.

seepage across the lake-bed (Fig. 3.10). The more extensive instrumentation using permanently installed piezometers and wells provides information on the three-dimensional flow field associated with a lake, and also the opportunity continuously to record data. However, difficulty in characterising the distribution of permeability, and the geological boundaries of the flow system, lead to uncertainties in this more thorough approach to understanding groundwater flow systems.

3.2.4 *Changes in lake volume*

Changes in volume of a lake over a given time integrate the gains and losses of water to and from the

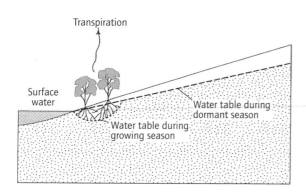

Fig. 3.9 Depressions in the water table caused by transpiration directly from groundwater. This process can induce seepage from the lake toward the water-table depression.

three major components of the hydrological system. Any imbalance between gains and losses leads to a change in the volume of water stored in the lake, and hence its level. For this reason, it is important to gauge the level of lakes as accurately as possible and to prepare accurate bathymetric maps so that stage–volume relationships can be determined. It is particularly critical that the stage–volume relationship for the normal range of water-level fluctuations be accurately known.

3.3 WATER BUDGETS AND HYDRAULIC RESIDENCE TIME OF LAKES

The preceding overview of hydrological processes, as they relate to the hydrology of lakes, provides a background for understanding the relative effects of each component on their water budgets. One way to evaluate the relative effects of hydrological fluxes is to place them in the perspective of **water residence time**. Stated simply, this is the time taken for the water in a lake to be replaced, assuming that fluxes are uniformly replacing lake volume. Residence time is most commonly calculated by dividing lake volume by the rate of outflow, but it also has been calculated by dividing volume by the rate of inflow. Residence time has been used most extensively in studies of eutrophication (Vollenweider 1976). This process is related both to nutrient loading and to the length of time that water remains in lakes.

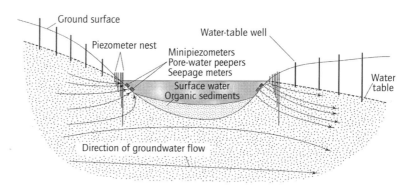

Fig. 3.10 Schematic diagram showing idealised field design for determining the interaction of lakes and groundwater.

Residence time has primarily been determined by using streamflow to or from lakes. However, as discussed in the previous section, water is also lost from lakes by evaporation and seepage to groundwater. Furthermore, many lakes do not possess outflowing streams, which limits the possibilities of calculation of their residence time. In order to determine the actual residence time of such lakes, it is necessary to divide volume by total losses of water by way of streamflow, seepage to groundwater and evaporation.

To place these factors in a unified framework, a trilinear diagram (Fig. 3.11) can be used to classify lakes according to their water losses. This approach follows a more extensive use of trilinear diagrams by Lent *et al.* (1997), who used them to develop hydrological indices for non-tidal wetlands. For calculation of residence time, data on water gains or losses, but not necessarily both, are needed. Losses are used here. The apices of the diagram, stream discharge from the lake, groundwater seepage from the lake and evaporation, indicate that 100% of the water lost from a lake is by way of that component. Most lakes lose water

by way of more than one. Therefore, they fall somewhere within the field of the diagram.

The relationship of the three major components of the hydrological system to the residence time of lake water therefore varies greatly between lakes in different hydrogeological and climatic settings. Therefore, the following discussion is organised according to the major physiographic and climatic landscapes which contain lakes (cf. Chapter 2, section 2.5). The landscapes discussed in the following section are mountainous, riverine, glacial, karst, arid inland basins and those related to major geological structures. Few calculations of residence time reported in the literature are based on total water loss and most are based on stream outflow only. Therefore, the examples given in the following sections were selected not only to illustrate differences in residence time between different types of lakes, but also to compare calculated residence times for a given lake if only one hydrological component is used instead of all three.

Calculation of residence time is particularly tenuous for lakes which experience large changes in volume. This is particularly the case for

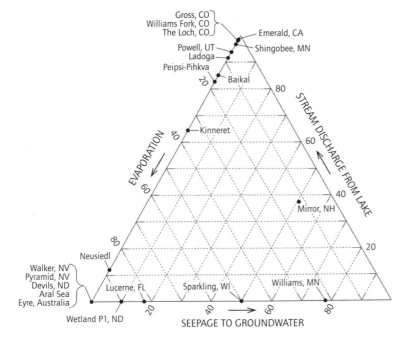

Fig. 3.11 Trilinear diagram for classifying lakes according to water losses.

reservoirs, and for lakes in arid inland basins. Therefore, an attempt has been made to present only those examples where the investigators used long-term averages for volume and water fluxes, or where they provide information which they believe represents close-to-average or pre-development conditions.

Residence time may also be highly seasonal. For example, that of a small high mountain lake during spring snowmelt may represent only hours to days but during late summer may reach several months. Recognising this problem, an attempt has been made to provide a more general estimate of residence time for those types of lakes.

3.3.1 *Lakes in mountainous terrain*

The commonest types of lakes in mountainous terrain are cirque lakes, or lakes dammed by glacial deposits (see Chapter 2, section 2.5.4). These are commonly small, with steep watersheds, and are usually fed by high-gradient streams. Many are underlain by fractured crystalline rocks whose permeability is low. Therefore, the water budget of cirque lakes is generally strongly influenced by streamflow resulting both from snowmelt and from abundant rainfall. Those which are present at high altitudes are ice-covered for most of the year, which minimises losses of water to evaporation. Evaporation is not great, however, when such lakes are ice-free, because their water is cold, and vapour-pressure gradients are not large. The poor permeability of their geological substrates leads to relatively small contributions from (and losses to) groundwater.

The average residence time of water in lakes in mountainous terrain ranges from a matter of days to a few months in the case of cirque lakes at high altitude, to more than a year in lakes lower in the valleys. During high flow, residence times can be less than a day. For example, the average volume of The Loch, a small cirque lake in the Rocky Mountains of Colorado, USA, is about $61 \times 10^3 \, \text{m}^3$. Hydrological fluxes are dominated by streamflow, and residence time is one to three weeks (Baron 1992). Evaporation was not determined specifically for The Loch, but is probably relatively small

because of the lake's high altitude. Similarly, Emerald Lake is a small cirque lake in the Sierra Nevada Mountains of California, USA, with an average volume of $162 \times 10^3 \, \text{m}^3$. Residence times have been estimated to range from 4 to 11 days depending on whether conditions are wet or dry (James Sickman, University of California at Santa Barbara, USA, written communication). On Fig. 3.11, The Loch and Emerald Lake probably would plot as close to the 100% surface-water apex as any type of lake.

In contrast to these two high-mountain lakes, Mirror Lake lies near the lower end of a valley in the White Mountains of New Hampshire, USA. Its volume is about $860 \times 10^3 \, \text{m}^3$. Although its watershed is steep, and although Mirror Lake gains most of its water from streamflow, a much greater volume of groundwater is stored in glacial deposits in its watershed than in those of The Loch or Emerald Lake. This does not lead to groundwater being a major source of water to Mirror Lake, but the greatest loss of water **from** the lake is via seepage to groundwater. This relatively large loss to groundwater is reflected by the position of Mirror Lake on Fig. 3.11. Mirror Lake also experiences higher evaporation rates than the two high-mountain lakes. Data were collected on all components of the hydrological system interacting with Mirror Lake from 1981 to 1990. Based on 10-year averages, the residence time of water in Mirror Lake is about 1.6 years if all losses are considered in the calculation. If stream outflow only is used, its residence time would be about 4 years.

3.3.2 *Lakes in riverine terrain*

Lakes in riverine terrain are primarily reservoirs. Because they are part of rivers, their water balance is usually dominated by streamflow. However, reservoirs which are shallow, and possess a large surface area, may lose a considerable amount of water to evaporation. Furthermore, by raising base level above that which groundwater flow systems had adjusted to before the reservoir was built, many lose water to groundwater, especially in the vicinity of the dam.

A study of two water-supply reservoirs in the

mountains of Colorado, USA (Williams Fork Reservoir, volume = c.119×10^6 m^3, Gross Reservoir, volume = c.51.6×10^6 m^3; LaBaugh & Winter 1984) indicated that streamflow accounted for more than 98% of inflow and outflow for both reservoirs. Evaporation, which was determined by the energy budget method, accounted for about 1% of water lost, and losses to groundwater, based on analysis of the geological setting and a few seepage-meter measurements, were considered to be minimal. On the basis of data collected over a 2-year period, the residence time of water in Williams Fork Reservoir was 0.8 years and in Gross Reservoir it was 0.3 years.

Residence times were also estimated for seven reservoirs constructed on streams in the prairies of Kansas, USA (Marzolf 1984). The results, using stream outflow only for the water loss in the calculation, ranged from 0.17 years to 3.0 years, but for four of the seven the residence time was less than 0.5 years. Marzolf indicated that determinations of residence time could be misleading for reservoirs because of the human control of storage and release. Reservoirs on the Prairies commonly possess large surface areas relative to depth; therefore, their waters can be relatively warm. This factor, combined with the windy conditions characteristic of the Prairies, can lead to large losses via evaporation. For this reason, the estimates should be considered as maximums, and the average values could be much less than indicated. Because of lack of data on evaporation, and on losses to groundwater, these reservoirs are not plotted on Fig. 3.11.

On a different scale, the volume of Lake Powell, a reservoir on the Colorado River in Arizona and Utah, USA, is about 33.3×10^9 m^3. Despite its enormous size, the water residence time of Lake Powell is only about 2.3 years (R.G. Marzolf, United States Geological Survey, pers. comm.) because the volume of water flowing from it also is large. As in the case of the smaller, Kansas reservoirs, the estimate of residence time for Lake Powell also did not take evaporation and groundwater seepage into account. Although it is unlikely that Lake Powell loses much, if any, volume to groundwater, about 135×10^6 m^3 annually may move temporarily into bank storage. This amount of water loss would not

have much effect on the calculation of residence time for the lake. Estimates of annual evaporation range from 0.6 to 1.2×10^9 m^3 (Dawdy 1991). Adding the average of this range of estimates (0.9×10^9 m^3) to the average annual stream outflow (14.8×10^9 m^3) produces a total annual outflow of 15.7×10^9 m^3, which changes the estimate of residence time to 2.12 years. This illustrates the point that the water budgets of reservoirs, even those situated in arid regions, where there are high rates of evaporation, are still largely comprised of surface-water fluxes (Fig. 3.11), and that even in such environments residence times of this kind of water body can be estimated reasonably accurately from lake volume and stream discharge only.

3.3.3   Lakes in glacial terrain

Glacial landscapes contain the largest number of lakes in the world (Wetzel 1983; see also Chapter 2, section 2.5.4). Glacial terrain is geologically young, much of it being formed less than 20,000 years ago, and there has often not been time for integrated drainage networks to develop. Therefore, many lakes in glacial terrain are closed, which means that they do not possess outflowing streams. As a result, because of the tendency, in calculation, to use stream outflow only, residence times of closed lakes have seldom been determined. Perhaps more important, residence time has not been determined for many closed lakes because accurate measurements of evaporation and outflow to groundwater have not been made.

In an attempt to evaluate the effect of residence time on limnological processes in open versus closed lakes in glacial terrain, a paired-lake study was initiated in north-central Minnesota, USA (Averett & Winter 1997). Both lakes lie within the same watershed, which is underlain by more than 120 m of glacial deposits (Winter & Rosenberry 1997). Those directly underlying the lakes are relatively permeable. Precipitation is about equal to evaporation. Extensive instrumentation was established in order to measure all components of the hydrological system interacting with the lakes, including precipitation, evaporation and groundwater (Rosenberry et al. 1997). The water budget of the closed lake (Williams Lake, volume = c.2×10^6 m^3) is made up mainly by groundwater inflow and outflow. Of the total water lost, 75% is to groundwater and 25% by evaporation (Fig. 3.11). Water residence time for Williams Lake is about 3 years. The water budget of the open lake (Shingobee Lake; volume = c.4×10^6 m^3) is composed mainly of stream inflow and outflow. Of the total water lost from Shingobee Lake, 95% is by streamflow and 5% is by evaporation (Fig. 3.11). Water residence time for Shingobee Lake is about 0.5 year.

A similar study is that of Sparkling Lake, which is underlain by about 50 m of a permeable, glacial sand and gravel aquifer in northern Wisconsin, USA (Krabbenhoft *et al.* 1990). This lake has been highly instrumented, in order to determine all influxes and losses of water. That lost from the lake is equally divided between evaporation and groundwater outflow—both constitute about 0.44 $\times 10^6$ m^3. With its volume of 8.8×10^6 m^3, the residence time of Sparkling Lake is about 10 years.

For comparison with these three lakes, a small prairie wetland in North Dakota, USA (Wetland P1) was instrumented and studied in a similar manner. The wetland is underlain by more than 120 m of glacial deposits (Winter & Carr 1980) and, in contrast to the Minnesota lakes, is directly underlain by poorly permeable glacial till. Annual precipitation is about 30 cm less than annual evaporation. The water budget of Wetland P1, with a volume of about 20×10^3 m^3 when water depth is about 1 m, is made up largely of precipitation and evaporation. Of the total water lost from the wetland, about 90% is by evaporation and 10% by way of seepage to groundwater (Fig. 3.11). Water residence time for Wetland P1 is about 0.8 year. This finding was confirmed during 1988 when little precipitation fell during the summer and when the wetland, after being nearly 1 m deep in spring, dried up by autumn (Winter & Rosenberry 1995).

3.3.4 *Lakes in karst terrain*

Lakes in karst terrain may lie directly within carbonate rocks, such as the cenotes of Mexico and Guatemala, and some lakes in Florida, but may also be present where unconsolidated deposits overlie limestone strata (see Chapter 2, section 2.5.5). In this 'mantled karst' terrain, lakes usually occupy depressions in the overlying deposits which are caused by slumpage into sinkholes at depth.

Little detailed hydrological research has been carried out on cenote-type lakes. Most possess no stream outflow, so they interact only with the atmosphere and with groundwater. Flow through the solution openings in carbonate rocks is extremely rapid, therefore it is assumed that karstic lakes directly in contact with carbonate aquifers are composed mainly of groundwater, and that their water residence time is very short.

Detailed hydrological studies have been conducted, however, of several closed lakes in mantled karst in Florida, USA. Each of the studies involved determining evaporation by the energy budget method, and groundwater data were obtained from extensive networks of wells. Lake Lucerne, in Central Florida, lies within porous media which overlie an extensive carbonate aquifer. The lake covers an area of 178×10^3 m^2, and its volume is 463×10^3 m^3. Data evaluated for 1 year of the study indicated that the annual volume of water lost by evaporation, 262×10^3 m^3, represented 82% of total water loss (Fig. 3.11), and that lost to groundwater was 57×10^3 m^3 (Lee & Swancar 1997). On the basis of data from this one year, the water residence time of Lake Lucerne is thought to be 1.45 years.

3.3.5 *Terminal lakes in arid inland basins*

Lakes which occupy the lowest altitude in basins of interior drainage are called 'terminal lakes' (see Chapter 2, section 2.7). Terminal lakes possess no stream outflow, and are commonly assumed to contribute no outflow to groundwater. Therefore, their only loss of water is by evaporation.

Pyramid Lake and Walker Lake are terminal lakes which lie within the Great Basin of Nevada, USA. The geological substrate of this region consists of hundreds of metres of fluvial deposits which have filled structural basins. Annual evaporation, which has been determined from both lakes by the energy budget method, exceeds annual precipitation in this region by about 0.5 to 1.0 m. Pyramid Lake, with a volume of about 35×10^9 m³, experiences an annual evaporation of about 1.25 m (Milne 1987; Hostetler & Benson 1990). Its residence time is about 50 years. The residence time of Walker Lake, with a volume of about 14.4×10^9 m³, and from which annual evaporation is about 1.35 m, is 35 years.

Devils Lake is a terminal lake which lies within glacial terrain in North Dakota, USA. The lake is underlain largely by glacial till, but a linear sand and gravel aquifer forms part of the geology of its basin. Evaporation exceeds precipitation in this area by about 0.3 m. Evaporation from Devils Lake was determined by the energy budget method for more than 10 years. For a relatively normal period of lake volume and climate conditions (1986–1988), Wiche (1992) calculated annual evaporation to be about 172×10^6 m³. Therefore, the residence time of Devils Lake, whose volume at its average elevation is about 966×10^6 m³, is 5.6 years.

Lake Eyre, a terminal lake in arid central Australia, receives intermittent inflows from a huge drainage area of 1.14×10^6 km². Annual evaporation in this region is about 2 m. Lake area is 8430 km², and lake volume 27.7×10^9 km³ (Kotwicki & Clark 1991). Water residence time is 1639 years.

The Aral Sea lies in the semi-arid zone of western Asia along part of the border between Kazakhstan and Uzbekistan. The level of the lake has varied by several metres over the past several centuries (see Chapter 2, sections 2.5.1 and 2.7, also Volume 2, Chapter 7). Shnitnikov (1973) compared the present water balance of the Aral Sea with that of former times, during which it lay at both lower levels and higher levels. Using data provided by this author for present conditions, the volume of the Aral Sea (1023×10^9 m³) divided by annual evaporation (64.5×10^9 m³) gives a residence time of 15.9 years.

3.3.6 Lakes in major geological structures

The majority of the largest lakes in the world are located in major geological structures, which, for the purposes of this chapter, include shield areas of basement rock. Nearly half the largest lakes in the world are located in northern North America, close to, or overlying, the Canadian Shield. The second most common type of geological terrain which contains large lakes are the tectonic belts (Herdendorf 1982; see also Chapter 2, section 2.5.1.)

Estimates of residence time for the North American Great Lakes, which lie near the southern edge of the Canadian Shield, range from 185 years for Lake Superior to 2.6 years for Lake Erie (Vollenweider 1976). Each of these lakes is open, and Vollenweider's calculations of their residence times are based on stream outflow only. Therefore, they are not plotted on Fig. 3.11.

The Neusiedlersee occupies a shallow basin at the western edge of the Small Hungarian Lowland along the border between Austria and Hungary (see also Chapter 2, sections 2.1, 2.5.1, 2.5.4, 2.5.7, 2.7, 2.11 and 2.12). The lake is shallow, with an average depth of 1.2 m. About 60% of its area of 320 km² is occupied by *Phragmites australis* reedswamp. Its volume is 384×10^6 m³. The annual volume of evaporation (230×10^6 m³) represents 88% of total water loss from the lake (Nobilis *et al.* 1991). The remainder is surface outflow to the Hanság Canal (see Chapter 2, sections 2.5.1 and 2.7). Water residence time for the Neusiedlersee is 1.5 years. However, if data on surface outflow only are used, the calculated residence time would be 12.8 years.

Lake Peipsi-Pihkva lies in the Peipsi-Pihkva depression situated between Estonia and Russia. The depression is underlain by a thick sequence of sedimentary rocks. Lake Peipsi-Pihkva covers an area of 3566 km², with an average depth of 7.1 m, and a volume of 25.21×10^9 m³. Water is lost from the lake by flow to the River Narva, and by evaporation. The annual volume of discharge to the Narva (9.6×10^9 m³) represents 82% of the water lost from the lake (Kullus 1973). The remainder is lost by evaporation. Water residence time for Lake Peipsi-Pihkva is 2.2 years, but if surface outflow only is used, the calculated residence of the lake would be 2.6 years.

Lake Ladoga is located in northwest Russia near St Petersburg. The largest lake in Europe, Ladoga covers an area of 18,260 km^2, with an average depth of 51 m. Like several of the Great Lakes of North America, Lake Ladoga lies close to a basement shield (the Baltic Shield) comprised of crystalline rocks, but also is partly underlain by sedimentary rocks and glacial deposits (Zektzer & Kudelin 1966). The volume of Lake Ladoga is 908×10^9 m^3. Water is lost from the lake by streamflow (the River Neva) and by evaporation. The annual flow of the Neva is 73.4×10^9 m^3, which represents about 92% of the total volume of water lost from the lake (Malinina 1966). Water residence time for Lake Ladoga is 11.4 years (Fig. 3.11). If data for surface outflow only are used, the calculated residence time of the lake becomes 12.4 years.

Lake Kinneret (see also Chapter 2, section 2.12.3) lies in the Jordan Valley along the border between Israel and the occupied West Bank to the west and Jordan to the east. The lake covers an area of 167.87×10^6 m^2, with a mean depth of 25.6 m and a volume of 4.3×10^9 m^3. Water is lost from the lake via the River Jordan, and by evaporation. The annual flow of the Jordan discharging from the lake is 546×10^6 m^3, which represents about 65% of the total volume of water lost from the lake (Mero 1978). Water residence time for Lake Kinneret is 5.1 years (Fig. 3.11). If surface outflow only is used, the calculated residence time would be 7.9 years.

Lake Baikal (see also Chapter 2, sections 2.5.1 and 2.12) lies north of Mongolia in Siberian Russia. Globally, the lake ranks eighth in surface area (31,500 km^2) and at 1740 m is the deepest lake in the world. Its volume (23,000×10^9 m^3, see also Chapter 2, Table 2.4) is second only to that of the Caspian Sea (see Chapter 2, sections 2.5.1 and 2.7). Water is lost from the lake by flow to the Angara River, and by evaporation. The annual flow of the outlet to the Angara River (60.8×10^9 m^3) represents about 85% of the total volume of water lost from the lake (Afansyev & Leksakova 1973). Water residence time for Lake Baikal (Fig. 3.11) is 321 years. If surface outflow only is used, the calculated residence time would be 378 years.

3.4 CONCLUSIONS

Lakes interact with all three major components of the hydrological system: atmospheric water, surface water and groundwater. However, according to the examples presented here, in most cases, one of these three major components of water loss is usually much greater than the other two. In the case of open lakes, whether in mountainous, riverine or glacial terrain, or those related to major geological structures, the greatest loss of water is usually by means of surface stream outflow. Because large amounts of water can be lost from such lakes by open-channel flow, water residence time is short for most lakes of this type; the only restriction is the size of the channel. In contrast, water residence time for lakes whose main water losses are via evaporation and by groundwater is generally much longer. This is because the mechanisms for water transfer from the lake by these two processes operate much more slowly than open-channel flow.

Residence times for cirque lakes in mountainous terrain area are also usually short, in that such lakes are generally small, and the volume of surface water flushing through them, especially during snowmelt, is normally relatively large. Open lakes in glacial terrain (such as Shingobee Lake, see section 3.3) commonly also exhibit short residence times because, even though the streams draining them are small, the lakes themselves also usually possess small area and volume.

In general, reservoirs in riverine terrain possess relatively short residence times. This is because even if they are very large, the streams associated with them also are commonly substantial rivers. For example, Lake Powell (section 3.3.1) contains 33 km^3 of water, but because of the great flow of the Colorado River its residence time is only slightly greater than 2 years. Large lakes associated with major geological structures also exhibit relatively short residence times if the rivers draining them possess large fluxes. For example, the residence time of Lake Peipsi-Pihkva, whose volume is about 25.2 km^3, is similar to that of Lake Powell. Only the very largest lakes, such as Lakes Baikal and Superior, which contain enormous amounts of water, possess very long residence times, measured in centuries, even though the streams draining them also are very large.

Including a term for evaporation in the calculation of residence time for lakes whose losses are mainly by streamflow does not change residence time appreciably. For example, water residence time for Lake Powell, whose surface outflow is very large, is reduced by only 0.2 year. As the size of the outflow stream decreases, however, the more significant other losses become. This point can be demonstrated by comparing Lakes Peipsi-Pihkva, Kinneret and the Neusiedlersee. The water residence time for Lake Peipsi-Pihkva is reduced by about half a year, but that of Lake Kinneret, whose stream outflow relative to lake volume is smaller compared with that of Lake Peipsi-Pihkva, declines by about 2.8 years. For the Neusiedlersee, whose surface outflow is very small, water residence time is reduced by about 11.3 years.

Lakes such as Williams and Sparkling Lakes, whose losses are to groundwater and to evaporation only, generally experience residence times measured in years. Terminal lakes, where losses are to evaporation only, generally possess very long residence times measured in decades. This is because such lakes generally are large both in terms of area and in terms of volume, and evaporation rates are relatively slow. Even if closed lakes (such as Devils Lake) are relatively small, residence time is still measured in years. Only the very smallest of lakes, such as Wetland P1, where losses are almost completely to evaporation, possess residence times of less than a year.

REFERENCES

Afansyev, A.N. & Leksakova, V.D. (1973) The water balance of Lake Baikal. *International Association of Hydrological Sciences Publication*, **109**, 170–5.

Allis, J.A., Harris, B. & Sharp, A.L. (1963) A comparison of performance of five rain-gage installations. *Journal of Geophysical Research*, **68**, 4723–29.

Averett, R.C. & Winter, T.C. (1997) *History and present status of the United States Geological Survey's Interdisciplinary Research Initiative in the Shingobee River headwaters area, Minnesota.* Water-resources Investigations Report 96-4215, United States Geological Survey, Washington, DC, 2 pp.

Baron, J. (1992) *Biogeochemistry of a Subalpine Ecosystem—Loch Vale Watershed.* Springer-Verlag, New York, 247 pp.

Belanger, T.V. & Montgomery, M.T. (1992) Seepage meter errors. *Limnology and Oceanography*, **37**, 1787–95.

Born, S.M., Smith, S.A. & Stephenson, D.A. (1974) *The hydrogeologic regime of glacial-terrain lakes, with management and planning applications.* Inland Lake Renewal and Management Demonstration Project, University of Wisconsin Press, Madison, 73 pp.

Dawdy, D.R. (1991) Hydrology of Glen Canyon and the Grand Canyon. *Proceedings of Symposium on Colorado River Ecology and Dam Management.* National Academy Press, Washington, DC, 40–53.

DeBruin, H.A.R. & Keijman, J.Q. (1979) The Priestley–Taylor evaporation model applied to a large, shallow lake in the Netherlands. *Journal of Applied Meteorology*, **18**, 898–903.

Hamon, W.R. (1961) Estimating potential evapotranspiration. *Journal of the Hydraulics Division, American Society of Civil Engineers*, **87**, 107–20.

Harbeck, G.E., Kohler, M.A. & Koberg, G.E. (1958) Water-loss investigations—Lake Mead studies. *United States Geological Survey Professional Paper*, **298**, 100 pp.

Herdendorf, C.E. (1982) Large lakes of the world. *Journal of Great Lakes Research*, **8**, 379–412.

Hostetler, S. & Benson, L.V. (1990) Paleoclimatic implications of the high stand of Lake Lahontan derived from models of evaporation and lake level. *Climate Dynamics*, **4**, 207–17.

Hounam, C.E. (1973) Comparison between pan and lake evaporation. *World Meteorological Organisation Technical Note*, **126**, 52 pp.

Hutchinson, G.E. (1957) *A Treatise on Limnology*, Vol. 1, *Geography, Physics, Chemistry.* John Wiley, New York, 1015 pp.

Jensen, M.E. (ed.) (1974) *Consumptive Use of Water and Irrigation Water Requirements.* American Society of Civil Engineers, New York, 227 pp.

Kohler, M.A., Nordenson, T.J. & Fox, W.E. (1955) Evaporation from pans and lakes. *United States Weather Bureau Research Paper*, **38**, 21 pp.

Kotwicki, V. & Clark, R.D.S. (1991) Aspects of the water balance of three Australian terminal lakes. *International Association of Hydrological Sciences Publication*, **206**, 3–12.

Krabbenhoft, D.P., Bowser, C.J., Anderson, M.P. & Valley, J.W. (1990) Estimating groundwater exchange with lakes 1. The stable isotope mass balance method. *Water Resources Research*, **26**, 2445–53.

Kullus, L.P. (1973) Water balance of Lake Peipsi-Pihkva. *International Association of Hydrological Sciences Publication*, **109**, 158–63.

LaBaugh, J.W. (1988) Relation of hydrogeologic setting to chemical characteristics of selected lakes and wetlands within a climate gradient in the north-central United States. *Verhandlungen International Vereiningen Limnologie*, **23**, 131–7.

LaBaugh, J.W. & Winter, T.C. (1984) The impact of uncertainties in hydrologic measurement on phosphorus budgets and empirical models for two Colorado reservoirs. *Limnology and Oceanography*, **29**, 322–39.

Lee, D.R. & Cherry, J.A. (1978) A field exercise on groundwater flow using seepage meters and mini-piezometers. *Journal of Geological Education*, **27**, 6–10.

Lee, D.R., Cherry, J.A. & Pickens, J.F. (1980) Groundwater transport of a salt tracer through a sandy lakebed. *Limnology and Oceanography*, **25**, 45–61.

Lee, T.M. & Swancar, A. (1997) Influence of evaporation, ground water, and uncertainty in the hydrologic budget of Lake Lucerne, a seepage lake in Polk County, Florida. *United States Geological Survey Water-supply Paper*, **2439**, 61 pp.

Lent, R.M., Weiskel, P.K., Lyford, F.P. & Armstrong, D.S. (1997) Hydrologic indices for nontidal wetlands. *Wetlands*, **17**, 19–30.

Linsley, R.K., Kohler, M.A. & Paulhus, J.L.H. (1975) *Hydrology for Engineers*. McGraw-Hill, New York, 482 pp.

Malinina, T.I. (1966) The water balance of the Ladoga Lake. *International Association of Scientific Hydrology Publication*, **70**, 17–24.

Marzolf, R.G. (1984) Reservoirs in the Great Plains of North America. *Ecosystems of the World 23 — Lakes and reservoirs*. Elsevier, Amsterdam, 291–302.

Mero, F. (1978) The water balance of Lake Kinneret. In: Serruya, C. (ed.), *Lake Kinneret*. Dr W. Junk, The Hague, 501 pp.

Meybeck, M. (1995) Global distribution of lakes. In: Lerman, A., Imboden, D.M. & Gat, J.R. (eds), *Physics and Chemistry of Lakes*. Springer-Verlag, Berlin, 1–35.

Milne, W. (1987) *A comparison of reconstructed lake-level records since the mid-1800s of some Great Basin lakes*. MS thesis, Colorado School of Mines, Golden, CO, 207 pp.

Nobilis, F., Plattner, J. & Pramberger, F. (1991) Investigations of the volume variations of Lake Neusiedl. *International Association of Hydrological Sciences Publication*, **206**, 23–30.

Pfannkuch, H.O. & Winter, T.C. (1984) Effect of anisotropy and ground-water system geometry on seepage through lakebeds, 1. Analog and dimensional analysis. *Journal of Hydrology*, **75**, 213–37.

Rosenberry, D.O., Winter, T.C., Merk, D.A., Leavesley, G.H. & Beaver, L.D. (1997) Hydrology of the Shingobee River headwaters area. *United States Geological Survey Water-resources Investigations Report*, **96-4215**, 19–23.

Shaw, R.D. & Prepas, E.E. (1990) Groundwater-lake interactions: I. Accuracy of seepage meter estimates of lake seepage. *Journal of Hydrology*, **119**, 105–20.

Shnitnikov, A.V. (1973) Water balance variability of Lakes Aral, Balkhash, Issyk-kul and Chany. *International Association of Hydrological Sciences Publication*, **109**, 130–40.

Siegel, D.I. & Winter, T.C. (1980) Hydrologic setting of Williams Lake, Hubbard County, Minnesota. *United States Geological Survey Open-file Report*, **80-403**, 56 pp.

Stewart, R.B. & Rouse, W.R. (1976) A simple equation for determining the evaporation from shallow lakes and ponds. *Water Resources Research*, **12**, 623–8.

Vollenweider, R.A. (1976) Advances in defining critical loading levels for phosphorus in lake eutrophication. *Memorie dell'Istituto Italiano di Idrobiologia*, **33**, 53–83.

Wetzel, R.G. (1983) *Limnology*. Saunders College Publishing, Fort Worth, TX, 767 pp.

Wiche, G.J. (1992) Evaporation computed by energy-budget and mass-transfer methods and water-balance estimates for Devils Lake, North Dakota, 1986–88. *North Dakota State Water Commission Water Resources Investigation*, **11**, 52 pp.

Winter, T.C. (1976) Numerical simulation analysis of the interaction of lakes and ground water. *United States Geological Survey Professional Paper*, **1001**, 45 pp.

Winter, T.C. (1981) Uncertainties in estimating the water balance of lakes. *Water Resources Bulletin*, **17**, 82–115.

Winter, T.C. (1983) The interaction of lakes with variably saturated porous media. *Water Resources Research*, **19**, 1203–18.

Winter, T.C. (1995) Hydrological processes and the water budget of lakes. In: Lerman, A., Imboden, D.M. & Gat, J.R. (eds), *Physics and Chemistry of Lakes*. Springer-Verlag, Berlin, 37–62.

Winter, T.C. & Carr, M.R. (1980) Hydrologic setting of wetlands in the Cottonwood Lake area, Stutsman County, North Dakota. *United States Geological Survey Water-resources Investigations*, **80-99**, 42 pp.

Winter, T.C. & Rosenberry, D.O. (1997) Physiographic and geologic characteristics of the Shingobee River headwaters area. *United States Geological Survey Water-resources Investigations Report*, **96-4215**, 11–17.

Winter, T.C. & Rosenberry, D.O. (1995) The interaction of ground water with prairie pothole wetlands in the Cottonwood Lake area, east-central North Dakota, 1979–1990. *Wetlands*, **15**, 193–211.

Winter, T.C. & Woo, M-K. (1990) Hydrology of lakes and wetlands. In: Wolman, M.G. & Riggs, H.C. (eds), *Surface Water Hydrology*. The Geology of North America, Vol. 0–1. Geological Society of America, Boulder, CO, 159–87.

Winter, T.C., Rosenberry, D.O. & Sturrock, A.M. (1995) Evaluation of eleven equations for determining evaporation for a small lake in the north-central United States. *Water Resources Research*, **31**, 983–93.

Winter, T.C., LaBaugh, J.W. & Rosenberry, D.O. (1988) The design and use of a hydraulic potentiomanometer for direct measurement of differences in hydraulic head between ground water and surface water. *Limnology and Oceanography*, **33**, 1209–14.

Zektzer, I.S. & Kudelin, B.I. (1966) The methods of determining ground water flow to lakes with special reference to Lake Ladoga. *International Association of Scientific Hydrology Publication*, **70**, 31–8.

4 Chemical Processes Regulating the Composition of Lake Waters

WERNER STUMM

4.1 INTRODUCTION: THE ACQUISITION OF SOLUTES

Figure 4.1a (taken from Berner & Berner 1996) summarises inputs and outputs of water to and from lakes. The relative importance of each of the processes involved varies from lake to lake. For example, in arid regions there is often no direct surface or subsurface output, and water is lost only by evaporation (see Chapter 2, section 2.7, **endorheic lakes**). Inputs may also vary considerably. Similarly, Lake Victoria is a shallow lake with only minor stream input, but located in a region of high rainfall (see also Chapter 2, sections 2.5.1, 2.7 and 2.11). As a result, roughly three-quarters of its water input is provided by rainfall directly on to the lake surface (Hutchinson 1957). This is unusually high, and for most larger lakes the major input takes place via rivers and streams, with rainfall accounting only for a small percentage of the total.

In contrast, many very small lakes (**spring-fed lakes**) are totally supplied by underground springs and some, known as **karst lakes**, also lose water by underground seepage. Karst lakes develop in karst terrains, or regions of extensive underground dissolution of limestone (see Chapter 2, section 2.5.5). Because of the permeable nature of the underlying, partly dissolved limestone bedrock, water is able to readily seep into and out of the lake floor. Most lakes, however, do not experience major underground seepage (Berner & Berner 1996).

Figure 4.1b illustrates the various processes which affect the chemical composition of the lake waters. For every substance to be transported into

a lake, we wish to know its source functions, its transport routes, its reservoirs and the chemical and biological processes in which it takes part. The main transport pathways are rivers and the atmosphere. The particulate load of most rivers is about three to five times their dissolved load.

A significant portion of the flux of dissolved and suspended materials transported by the rivers is due to erosion and chemical denudation. These fluxes may recently (during historical times) have substantially increased, owing to changes in land use, and to acidification of rain. The transformation of forest to cropland and grazing areas often increases erosion rates of suspended matter by at least one order of magnitude, and road building, urbanisation or other construction by a further order, or even two, at least until the construction process is complete (Wolman 1967).

Although most solutes and pollutants are contributed to lakes by rivers, industrial outfalls and drainage, for some pollutants transport through the atmosphere must not be underestimated. The atmosphere is a significant and rapid conveyor belt, acting over long distances for various products of fossil fuel combustion (oxides of sulphur and nitrogen, hydrocarbons, heavy metals, acidity), the by-products of steel production (fine particles), emissions of volatile oxides (of arsenic, boron, lead, scandium and zinc) from the production of cement, radioactive substances released from nuclear reactors and weapon tests, and halogenated hydrocarbons (DDT (dichlorodiphenyltrichloroethane), PCBs (polychlorinated biphenyls), hexachlorobenzene, compounds transported mainly in the vapour phase). Some

a)

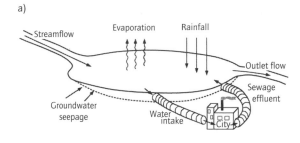

b)

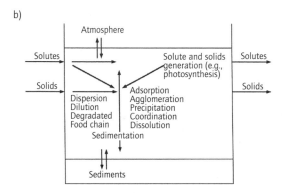

Fig. 4.1 Water balance (a) and factors regulating the concentration of lake components (b).

pollutants (soil components, fertilisers) are also moved into lakes through the action of wind.

4.2   CHEMICAL COMPOSITION

4.2.1   Acid deposition

The steady-state condition of nature, with regard to proton–electron balance, is neutral. Acid atmospheric deposition, therefore, results from the disturbance of the normal cycles which couple atmosphere, land and water. Figure 4.2, adapted from Schnoor & Stumm (1985), shows the various processes which involve atmospheric pollutants and natural components in the atmosphere. The following reactions are of particular importance in the formation of acid precipitation: **oxidative reactions**, either in the gaseous or the aqueous phase, leading to the formation of oxides of carbon, sul-

phur and nitrogen (CO_2, SO_2, SO_3, H_2SO_4, NO, NO_2, HNO_2, HNO_3); **absorption of gases into water** (cloud droplets, falling raindrops, or fog) and interaction of the resulting acids ($SO_2 \cdot H_2O$, H_2SO_4, HNO_3) with ammonia (NH_3) and the carbonates of airborne dust; and the **scavenging and partial dissolution of aerosols into water**. In this case, aerosols are produced from the interaction of vapours and airborne (maritime and dust) particles; they often contain $(NH_4)_2SO_4$ and NH_4NO_3.

4.2.1.1   Deposition

The products of the various chemical and physical reactions are eventually returned to the Earth's surface and hence to the catchments of lakes. Usually, one distinguishes between **wet deposition** (rainout and washout) which includes the flux of all those components which are carried to the Earth's surface by rain or snow, that is, those dissolved and particulate substances contained in rain or snow, and **dry deposition**—the **flux** of particles and gases (especially SO_2, HNO_3 and NH_3) to the receptor surface during the absence of rain or snow. Deposition also may occur via fog aerosols and droplets, which may be deposited on trees, other plants, or on the ground. In forests, approximately half the deposition of SO_4^{2-}, NO_3^- and H^+ occurs as dry deposition (Lindberg *et al.* 1986; Table 4.1).

Evapotranspiration of water from receptor surfaces leads to an increase in the concentration of conservative solutes in the soil solution and in water draining from the root zone. The rainwater shown in Fig. 4.2 contains an excess of strong acids, most of which originate from the oxidation of sulphur during fossil fuel combustion and from the fixation of atmospheric nitrogen to NO and NO_2, for example, during combustion by motor vehicles. It also should be mentioned that there are natural sources of acidity, resulting from volcanic activity, from H_2S from anaerobic sediments, and from dimethyl sulphide and carbonyl sulphide that originate in the ocean. Hydrogen chloride results from the combustion and decomposition of organochlorine compounds such as polyvinyl

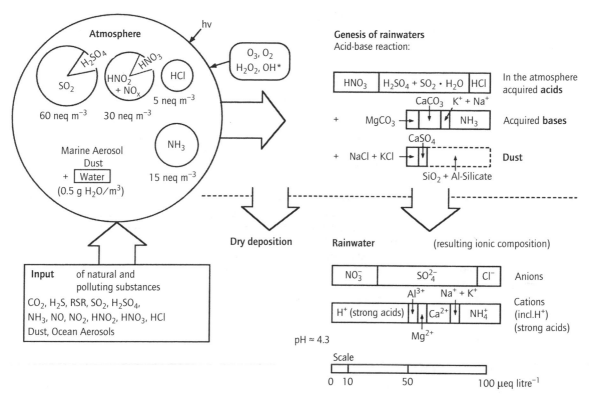

Fig. 4.2 Genesis of acid rain and its deposition. From the oxidation of carbon, sulphur and nitrogen during the combustion of fossil fuels, there is a build-up in the atmosphere (gas phase, aerosol particles, raindrops, snowflakes and fog) of CO_2 and the oxides of sulphur and nitrogen, which leads to acid–base interaction. The importance of absorption of gases into the various phases of gas, aerosol and atmospheric water depends on a number of factors. The genesis of acid rain is shown on the upper right as an acid-base titration. The data given are representative of the situation encountered in Zürich, Switzerland.

chloride. Bases originate in the atmosphere as the carbonate of wind-blown dust and from NH_3, generally of natural origin. The NH_3 comes from NH_4^+ and from the decomposition of urea in soil and agricultural environments.

The reaction rates for oxidation of atmospheric SO_2 (0.05–0.5 day^{-1}) yield a sulphur residence time of several days at the most; this corresponds to a transport distance of several hundred to $1000\,km$. The formation of HNO_3 by oxidation is more rapid and, compared with H_2SO_4, involves much shorter travel distances from the emission source. Sulphuric acid also can react with NH_3 to form aerosols of NH_4HSO_4 or $(NH_4)_2SO_4$. In addition,

NH_4NO_3 aerosols are in equilibrium with $NH_3(g)$ and $HNO_3(g)$.

The foliar canopy receives much of its dry deposition in the form of sulphate, nitrate, and hydrogen ion, which occur primarily as SO_2, HNO_3 and NH_3 vapours. Dry deposition of coarse particles has been shown to be an important source of calcium and potassium ion deposition on deciduous forests in the eastern USA (Table 4.1). In addition to the acid-base components shown in Fig. 4.2, various organic acids are often found. Many of these are by-products of the oxidation of organic matter released into the atmosphere. Of special interest are formic, acetic, oxalic and benzoic acids, which

Table 4.1   Total annual atmospheric deposition of major ions to an oak forest at Walker Branch watershed*

Process	Atmospheric deposition (meq m^{-2} yr^{-1})					
	SO_4^{2-}	NO_3^-	H^+	NH_4^+	Ca^{2+}	K^+
Precipitation	70 ± 5	20 ± 2	69 ± 5	12 ± 1	12 ± 2	0.9 ± 0.1
Dry deposition:						
fine particles	7 ± 2	0.1 ± 0.02	2.0 ± 0.9	3.6 ± 1.3	1.0 ± 0.2	0.1 ± 0.05
coarse particles	19 ± 2	8.3 ± 0.8	0.5 ± 0.2	0.8 ± 0.3	30 ± 3	1.2 ± 0.2
vapours†	62 ± 7	26 ± 4	85 ± 8	1.3	0	0
Total deposition	160 ± 9	54 ± 4	160 ± 9	18 ± 2	43 ± 4	2.2 ± 0.3

*   Values are means ± standard errors for 2 years of data. Numbers of observations range from 15 (HNO_3) to 26 (particles) to 128 (precipitation) to 730 (SO_2). In comparing these deposition rates it must be recalled that any such estimates are subject to considerable uncertainty. The standard errors given provide only a measure of uncertainty in the calculated sample means relative to the population means; hence additional uncertainties in analytical results, hydrological measurements, scaling factors and deposition velocities must be included. The overall uncertainty for wet deposition fluxes is about 20% and that for dry deposition fluxes is approximately 50% for SO_4^{2-}, Ca^{2+}, K^+ and NH_4^+, and approximately 75% for NO_3^- and H^+.

†   Includes SO_2, HNO_3 and NH_3. Complete conversion of deposited SO_2 to H_2SO_4 and of NH_3 to NH_4^+ was assumed in determining the vapour input of H^+. NH_3 deposition was estimated from the literature. (Source: Lindberg *et al.* 1986.)

have been found in rainwater in concentrations occasionally exceeding a few micromoles per litre. Thus, they must be included in the calculated acidity of rainwater, and their presence in larger concentrations influences the pH of rain and fog. The composition of the rain—an 'average' inorganic composition is given in Fig. 4.2—reflects the acid-base titration which occurs in the atmosphere. Total concentrations (the sum of cations and anions) vary typically from 20 to 500 µeq L^{-1}, and pH from 3.5 to 6.

4.2.1.2   Receiving waters

Fresh waters vary in chemical composition, but these variations are at least partially understandable if the environmental history of the water, its pollution, and the rock–water–atmosphere systems are considered. Many constituents of the Earth's crust are thermodynamically unstable in the presence of water and the atmosphere; rocks react primarily with CO_2(g) and H_2O, for example:

$$CaCO_3 + CO_2 + H_2O \rightarrow Ca^{2+} + 2HCO_3^- \quad (4.1)$$
$$\text{calcite}$$

$$NaAlSiO_8(s) + CO_2 + 5.5H_2O \rightarrow Na^+ +$$
$$\text{Na-feldspar}$$
$$HCO_3^- + 2H_4SiO_4 + 0.5Al_2Si_2O_5(OH)_4(s) \quad (4.2)$$
$$\text{kaolinite}$$

$$3Ca_{0.33}Al_{4.67}Si_{7.33}O_{20}(OH)_4 + 2CO_2 + 25H_2O \rightarrow$$
$$\text{montmorillonite}$$
$$Ca^{2+} + 2HCO_3^- + 8H_4SiO_4 + 7Al_2Si_2O_5(OH)_4(s)$$
$$\text{kaolinite}$$
$$(4.3)$$

To a large extent, rock type in the drainage basin of a lake determines the composition of the inorganic fraction of lake water. In Table 4.2 (Garrels & Mackenzie 1971) chemical analyses of river waters whose drainage is dominated by certain rock types are presented. Thus, waters in lakes draining limestone areas, where reactions such as (4.1) predominate, are very much like those of the Danube. In contrast, in areas of crystalline rocks, lake waters are very much like those of the Nile, because their

Table 4.2   Control of inorganic water composition by chemical weathering processes. Composition $(mg\,L^{-1})$ of selected freshwater types. (Source: Garrels & Mackenzie 1971.)

	Limestone terrain: Danube River	Argillaceous terrain: Amazon River at Obidas	Crystalline terrain: Nile River	Salty: Rio Grande River at Laredo, Texas
SiO_2*	5.6	10.6	20.1	30
Ca^{2+}	43.9	5.4	15.8	109
Mg^{2+}	9.9	0.5	8.8	24
Na^+	2.8	1.6	15.6	117
K^+	1.6	1.8	3.9	6.7
HCO_3^-	167.0	17.9	85.8	183.0
SO_4^{2-}	14.7	0.8	4.7	238.0
Cl^-	2.4	2.6	3.4	171.0

*   Dissolved as H_4SiO_4.

chemical composition is regulated more by reactions such as (4.2) and (4.3).

Acquisition of solutes by weathering processes is exemplified by reactions (4.1)–(4.3). The mechanism of silicate dissolution is similar to that of oxides. With decreasing pH, surface protonation of O and OH lattice sites accelerates dissolution. Figure 4.3 gives a brief survey of the dissolution rates of some minerals. A pH effect is observed in all cases. There are dramatic differences in the dissolution rates (at pH = 6, four to five orders of magnitude) between carbonates and silicates. Similarly, the **reductive dissolution** of Fe(III)(hydr)oxides (e.g., by H_2S) is faster than their non-reductive dissolution by orders of magnitude.

4.2.2   Freshwater composition and carbonate equilibrium

In Fig. 4.4 (adapted from Holland 1978), HCO_3^- and Ca^{2+} concentrations of selected rivers of the world are plotted. The drawn-out lines correspond (i) to the charge balance of the most important species in these rivers,

$$2[Ca^{2+}] = [HCO_3^-] \qquad (4.4)$$

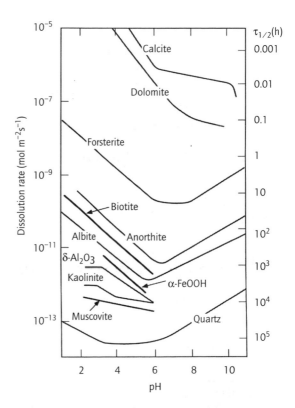

Fig. 4.3   Dissolution rates of different minerals as a function of pH (25°C) on the basis of experimental data by different authors. For calculation of the half-life (right-hand scale) 10 surface groups nm^{-2} were assumed. (Modified from Sigg & Stumm 1994.)

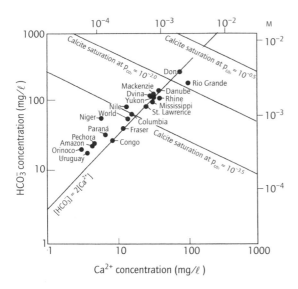

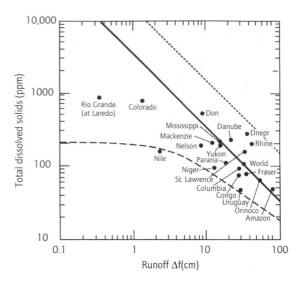

Fig. 4.4 The relationship between the bicarbonate and calcium concentrations in some major rivers. The composition of many rivers is characterised by the charge balance $2[Ca^{2+}] = [HCO_3^-]$ and the saturation of $CaCO_3$ $[CaCO_3(s) + CO_2(g) + H_2O \rightarrow Ca^{2+} + 2HCO_3^-]$.

Fig. 4.5 Dissolved solids of rivers as a function of runoff (major rivers).

and (ii) to the solubility of calcite $(CaCO_3)$,

$$CaCO_3(s) + CO_2(g) + H_2O \rightarrow$$
$$Ca^{2+} + 2HCO_3^-;\ K \quad\quad (4.5)$$

Thus,

$$[Ca^{2+}] + [HCO_3^-]^2 / p_{co_2} = K = 10^{-5.9}(25°C) \quad (4.6)$$

Thus, in a plot of $\log[HCO_3^-]$ versus $\log[Ca^{2+}]$, we obtain a straight line with a slope of

$$d\log[HCO_3^-]/d\log[Ca^{2+}] = -0.5 \quad (4.7)$$

Apparently, many dilute rivers are undersaturated with regard to $CaCO_3$ and many others reach saturation at CO_2 partial pressures between $10^{-3.5}$ and 10^{-2} atm. Because of the organic loading of the waters (respiration of organic material to CO_2) and because of the inflow of groundwaters (characterised by higher p_{CO_2}), the p_{CO_2} of many river waters is higher than that of the atmosphere.

4.2.3 'Dilution' by runoff

The water carried in rivers may be considered as consisting of (i) a fraction made up of subsurface water and groundwater which re-enters (or infiltrates into) surface waters via soils, subsoils and underlying rocks, and (ii) a (direct) surface runoff component which enters the drainage system during and soon after precipitation periods, as overland flow, or lateral throughflow and interflow in the soil profile. The term **runoff** is usually considered to be synonymous with **streamflow**, and is the sum of the surface runoff, and the subsurface and groundwater flow which reaches the stream. Surface runoff therefore equals precipitation minus evaporation, surface retention and infiltration.

The relative proportions of these components, and the concentrations of dissolved species they each contain (as influenced by the interactions of rainwater with minerals and vegetation and by evaporation and transpiration from plants), largely determine the composition of river waters (Fig. 4.5, adapted from Holland 1978). Direct (surface) runoff has experienced only brief contact with soil or vegetation, and is therefore similar in its composition to precipitation. In contrast, reactions below ground are often extensive enough for water which infiltrates into the subsurface zone to impart to baseflow an increased dissolved solid content. Two extreme cases may be considered.

1 Direct runoff (e.g. due to evaporation) is small in comparison with subsurface and groundwater flow; the latter carries a relatively constant solute concentration, for example, by nearly reaching solid–solution equilibria. In this case, solute concentration in the river is relatively independent of river flow.

2 If direct runoff is greater than the flow which has infiltrated to the subsurface, solute concentrations in subsurface waters and groundwater become 'diluted' by the direct runoff. Thus, solute concentrations of river waters tend to be inversely related to river flow. Superimposed on these natural relationships are the influences of pollution and waste disposal.

4.2.4 *Provenance of major constituents of river water*

Holland (1978) has provided a detailed analysis of the variations of the major constituents of the river waters. The most important results are given in Table 4.3. According to these data, nearly two-thirds of the carbon in river HCO_3^- is derived from the atmosphere either directly as gaseous CO_2 or via photosynthesis followed by plant decay. The remaining HCO_3^- is derived largely from weathering of carbonates.

The source of SO_4^{2-} is still somewhat uncertain; roughly 40% is cycled through the atmosphere, and 60% is derived from the weathering of sulphides and the solution of sulphate minerals (gypsum and anhydrite). Atmospheric chloride contributes less than half the average chloride in river water; chloride mostly from evaporitic halite contributes the bulk.

Most of the Ca^{2+} in river waters is derived from solution of carbonates. However, the weathering of silicates contributes about 50% more Mg^{2+} than the solution of carbonates. Most of the Na in rivers is provided by NaCl (from evaporites and from atmospheric recycling); only 35% of the Na^+ present in river water owes its origin to weathering of silicates. The K^+ and dissolved SiO_2 components are largely contributed by silicate weathering. Martin & Meybeck (1979) reassessed the elemental mass balance of material carried by major world rivers. Good examples are given by Drever (1988) and by

Table 4.3 The provenance of solutes in average river water (Modified from Holland 1978.)

Source	Anions (meq kg^{-1})			Cations (meq kg^{-1})				Neutral species (mmol kg^{-1})
	HCO_3^-	SO_4^{2-}	Cl^-	Ca^{2+}	Mg^{2+}	Na^+	K^+	SiO_2
Atmosphere*	0.58†	0.09‡	0.06	0.01	≤0.01	0.05	≤0.01	≤0.01
Weathering or solution of:								
silicates	0	0	0	0.14	0.20	0.10	0.05	0.21
carbonates	0.31	0	0	0.50	0.13	0	0	0
sulphates	0	0.07	0	0.07	0	0	0	0
sulphides	0	0.07	0	0	0	0	0	0
chlorides	0	0	0.16	0.03	≤0.01	0.11	0.01	0
Organic carbon	0.07	0	0	0	0	0	0	0
Sum	0.96	0.23	0.22	0.75	0.35	0.26	0.07	0.22

*   These figures do not include soil-derived material.

†   Largely as atmospheric CO_2.

‡   Much of this is apparently balanced by H^+.

Table 4.4 (a) Chemical composition of average river water. (Source: all river water concentrations and discharge values from Meybeck (1979) except 'actual' concentrations by continent, which were calculated by Berner & Berner (1996) from Meybeck's data.)

By continent	River water concentration* (mg L^{-1})									Water discharge (10^3 km^3 yr^{-1})	Runoff ratio†
	Ca^{2+}	Mg^{2+}	Na^+	K^+	Cl^-	SO_4^{2-}	HCO_3^-	SiO_2	TDS‡		
Africa:											
actual	5.7	2.2	4.4	1.4	4.1	4.2	26.9	12.0	60.5	3.41	0.28
natural	5.3	2.2	3.8	1.4	3.4	3.2	26.7	12.0	57.8		
Asia:											
actual	17.8	4.6	8.7	1.7	10.0	13.3	67.1	11.0	134.6	12.47	0.54
natural	16.6	4.3	6.6	1.6	7.6	9.7	66.2	11.0	123.5		
South America:											
actual	6.3	1.4	3.3	1.0	4.1	3.8	24.4	10.3	54.6	11.04	0.41
natural	6.3	1.4	3.3	1.0	4.1	3.5	24.4	10.3	54.3		
North America:											
actual	21.2	4.9	8.4	1.5	9.2	18.0	72.3	7.2	142.6	5.53	0.38
natural	20.1	4.9	6.5	1.5	7.0	14.9	71.4	7.2	133.5		
Europe:											
actual	31.7	6.7	16.5	1.8	20.0	35.5	86.0	6.8	212.8	2.56	0.42
natural	24.2	5.2	3.2	1.1	4.7	15.1	80.1	6.8	140.3		
Oceania:											
actual	15.2	3.8	7.6	1.1	6.8	7.7	65.6	16.3	125.3	2.40	–
natural	15.0	3.8	7.0	1.1	5.9	6.5	65.1	16.3	120.3		
World average:											
actual	14.7	3.7	7.2	1.4	8.3	11.5	53.0	10.4	110.1	37.4	0.46
natural (unpolluted)	13.4	3.4	5.2	1.3	5.8	8.3	52.0	10.4	99.6	37.4	0.46

*   *Actual* concentrations include pollution; *natural* concentrations are corrected for pollution.
†   Runoff ratio = average runoff per unit area/average rainfall (calculated from Meybeck 1979).
‡   Total dissolved solids.

Berner & Berner (1987). The composition of the world average river water, according to Meybeck (1979), is given in Table 4.4a. Values for natural river water for each continent are also given in this table. Examples of the chemical composition of lakes are given in Table 4.4b.

4.3   WEATHERING AND THE PROTON BALANCE

In order to understand pH regulation, one needs to consider the major H^+-yielding and H^+-consuming processes. Weathering is a major H^+-consuming process and pH-buffering mechanism, not only globally and regionally, but also locally in watersheds, soil processes, nutrient uptake by plants and in reactions in sediments. If neutralisation of acids by weathering is too slow, acid atmospheric deposition causes acidification of waters and soils. Biologically mediated redox processes are important in affecting the H^+ balance. Among the redox processes which exert a major impact on H^+ production and consumption are the synthesis and mineralisation of biomass. Any uncoupling of linkages between photosynthesis and respiration

Table 4.4 (b) Chemical composition of lakes (selected from a tabulation by Meybeck 1995).* Cations and anions are expressed in mol % of their sum.

Category	Lake	$\Sigma+$ μeq L^{-1}	Ca^{2+} %	Mg^{2+} %	Na$^+$ %	K$^+$ %	Cl$^-$ %	SO$_4^{2-}$ %	HCO$_3^-$ %	SiO$_2$ mg L^{-1}
Lakes	Chauvet (F)	240	35	31	28	6	12	16	72	3.0
	Superieur (USA)	910	68	25	5	2	6	7	87	2.4
	Tahoe (Nevada)	985	48	21	27	4	7	7	86	–
	Victoria (Ken)	1040	27	21	43	9	10	4	86	4.2
	Taupo (NZ)	1220	27	15	54	4	23	11	66	14.2
	Baikal	1240	65	20	13	2	1	9	90	2.9
	G.L. Ours (Can)	1577	51	36	12	1	6	20	74	4.6
	Ohrid (Yu)	2580	60	32	8		6	5	89	–
	Geneva	2780	79	17	3	1	3	34	63	2.5
Lakes fed by atmospheric deposition	Waldo (Oregon)	17	24	23	41	12	–	–	–	–
	Godivelle (F)	134	28	36	22	14	30	4	66	0.3
	Tourbières (USA)	182	19	13	6	3	3	8	0	–
	Tourbières (Irl)	899	7	17	66	3	75	9	0	–

Category	Lake	$\Sigma+$ meq L^{-1}	Ca^{2+} %	Mg^{2+} %	Na$^+$ %	K$^+$ %	Cl$^-$ %	SO$_4^{2-}$ %	ΣCO$_2$ %	SiO$_2$ mg L^{-1}
Lakes with hydrothermal inputs	Pavin 70–92 m	3.93	64	21	8	7	6	1	93	50.2
	Tanganyika	7.58	5	45	38	12	8	1	91	–
	Albert	8.75	6	30	45	19	10	8	82	–
	Rotomahana (NZ)	11.5	5	7	83	5	62	6	32	43.4
Saline lakes (water lost by evaporation)	Issyk Kul	94	6	25	69		46	43	11	2.6
	Walker (Nevada)	151	<1	8	90	<1	50	33	17	9.3
	Caspian Sea	215	8	29	63	<1	69	29	2	–
	Van (Turkey)	358	<1	2	94	4	(43)	(14)	(43)	–
	Great Salt Lake	3850	<1	16	81	3	90	10	<1	–
	Magadi (Kenya)	7106	<1	<1	99	1	34	1	65	–

* For references and more examples see Meybeck (1995).

affects acidity and alkalinity in both terrestrial and aquatic ecosystems.

4.3.1 Alkalinity and acidity: neutralising concepts

It is necessary to distinguish between H$^+$ **concentration** (or **activity**) as an intensity factor and the **availability** of H$^+$, the H$^+$-ion reservoir as given by the base neutralising capacity (BNC) or the H-acidity [H-Acy]. The BNC relates to alkalinity [Alk] or acid neutralising capacity (ANC) via:

$$\text{H-Acy} = -[\text{Alk}] \qquad (4.8)$$

base neutralising capacity =
$-$ acid neutralising capacity

The BNC can be defined by a net proton balance with regard to a reference level—the sum of the concentrations of all the species containing protons in deficiency of the proton reference level. For natural waters, a convenient reference level (corresponding to an equivalence point in alkalimetric titrations) includes H$_2$O and H$_2$CO$_3$:

$$[\text{H-Acy}] = [\text{H}^+] - [\text{HCO}_3^-] - 2[\text{CO}_3^{2-}] - [\text{OH}^-] \quad (4.9)$$

The ANC or alkalinity [Alk] is related to [H-Acy] by:

$$-[\text{H-Acy}] = [\text{Alk}] = [\text{HCO}_3^-] + 2[\text{CO}_3^{2-}]$$
$$+ [\text{OH}^-] - [\text{H}^+] \quad (4.10)$$

When we consider a charge balance for typical natural waters

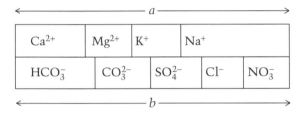

we realise that [Alk] and [H-Acy] can also be expressed by a balance which is the equivalent sum of conservative cations (a), minus the sum of conservative anions (b). Thus, ([Alk] = $a - b$). The conservative cations are those of the strong bases (Ca(OH)$_2$, KOH, etc.), and the conservative anions are the conjugate bases of strong acids (SO$_4^{2-}$, Cl$^-$ and NO$_3^-$). Therefore

$$[\text{Alk}] = [\text{HCO}_3^-] + 2[\text{CO}_3^{2-}] + [\text{OH}^-] - [\text{H}^+]$$
$$= [\text{Na}^+] + [\text{K}^+] + 2[\text{Ca}^{2+}] + 2[\text{Mg}^{2+}]$$
$$- [\text{Cl}^-] - 2[\text{SO}_4^{2-}] - [\text{NO}_3^-] \quad (4.11)$$

The [H-Acy] for these particular waters, obviously negative, is defined ([H-Acy] = $b - a$) as:

$$[\text{H-Acy}] = [\text{Cl}^-] + 2[\text{SO}_4^{2-}] + [\text{NO}_3^-]$$
$$- [\text{Na}^+] - [\text{K}^+] - 2[\text{Ca}^{2+}] - 2[\text{Mg}^{2+}] \quad (4.12)$$

These definitions can be used to interpret interactions of acid precipitation with the environment, in that a simple accounting can be made. Every base cationic charge unit (e.g., Ca^{2+} or K$^+$) removed from waters, by whatever process, is equivalent to the addition of a proton. Every conservative anionic charge unit (from anions of strong acids, e.g. SO$_4^{2-}$, Cl$^-$ and NO$_3^-$) removed, corresponds to removal of a proton, or generally:

$$\text{XA[base cations]} - \text{XA[conservative anions]}$$
$$= +\Delta[\text{Alk}] = -\Delta[\text{H-Acy}] \quad (4.13)$$

If the waters under consideration contain other acid- or base-consuming species, the proton reference level must be extended to include these other components. In addition to the species given, if natural waters contain organic molecules (such as a carboxylic acid; H-Org) or ammonium ions, the reference level is extended to incorporate these species also.

As Fig. 4.6 illustrates, aggrading vegetation (forests and intensive crop production) produces acidity. As more cations are taken up by the plants (trees) H$^+$ is released through the roots. The protons liberated react with weathered minerals to produce some of the cations needed by the plants. The H$^+$ balance in soils (production through the

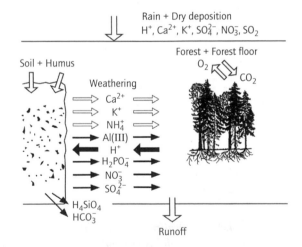

Fig. 4.6 Effect of plants (trees) upon H$^+$ balance in soils. Competition takes place between H$^+$ release by the roots of growing trees (aggrading forest) and H$^+$ ion consumption by chemical weathering of soil minerals. The delicate proton balance can be disturbed by acid deposition.

roots versus consumption by weathering) is delicate and can be disturbed by acid deposition.

If the rate of weathering equals or exceeds the rate of H⁺ release by the biota (such as would be the case in a calcareous soil), the soil will maintain a buffer in base cations and residual alkalinity. On the other hand, in non-calcareous 'acid' soils, the rate of release of H^+ by the biomass may exceed the rate of consumption of H^+ by weathering, and cause progressive acidification of the soil. In some instances, acidic atmospheric deposition may be sufficient to disturb an existing H^+ balance between aggrading vegetation and weathering reactions.

4.3.2 Critical load of acid deposition

The concept of critical load of a substance is based on the dose–response relationship, the critical value being exceeded when the load causes harmful effects to the receptor (the lake or catchment area). With regard to acid deposition, critical load may be defined in terms of the (tolerable) acidity deposition (e.g., in $\mu eq\,m^{-2}\,yr^{-1}$) which must not be exceeded in order to avoid harmful effects. Various methods have been employed in order to establish critical loads, which we will not discuss here. Instead, we refer the reader to publications by Bricker & Rice (1989), Kämäri *et al.* (1992) and Sverdrup *et al.* (1992). The sensitivity of a region to acid precipitation is strongly influenced by bedrock mineralogy. As shown in Fig. 4.3, dissolution rate (the neutralisation rate of acidity) is very rapid in carbonate terrain but very slow in regions of crystalline rocks.

Simple steady-state models have been used in order to determine critical loads for lakes and forest soils. The basic principle is that primary mineral weathering in the watershed is the ultimate supplier of base cations, which are required elements for vegetation and lake water, in order to ensure adequate acid neutralising capacity (see section 4.3.1 for previous definition of ANC as the sum of base cations minus acid anions). If more acid is deposited in a watershed than chemical weathering can neutralise, acidification of soils, and eventually of waters, will occur. This may not

happen immediately, however, because ion-exchange reactions can supply base cations to vegetation and runoff water for a period. Therefore, there may be a delayed response. Eventually, however, the exchange capacity of the soil will become depleted, and acidification will result.

As an example, Fig. 4.7 gives the water composition of four lakes located at the top of the Maggia valley, Switzerland (Giovanoli *et al.* 1988), at an elevation of 2100–2500 m. Their small catchments are characterised by sparse vegetation (no trees), thin soils and steep slopes. Although these lakes are situated less than 10 km apart, they vary markedly in their water composition, as influenced by bedrock.

Lakes Cristallina and Zota, situated within a drainage basin characterised by the preponderance of gneissic rocks, and the absence of calcite and dolomite, exhibit **mineral acidity** (i.e. caused by mineral acids and HNO_3). Their calcium concentrations are 10–$15\,\mu mol\,L^{-1}$ and their pH is less than 5.3. On the other hand, the waters of Lake Val

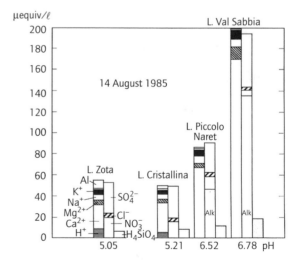

Fig. 4.7 Comparison of water composition of four lakes influenced by different bedrock in their catchments. Drainage areas of Lake Zota and Lake Cristallina contain only gneiss and granitic gneiss; that of Lake Piccolo Naret overlies small amounts of calcareous schist; the catchment of Lake Val Sabbia exhibits a higher proportion of schist.

Sabbia, whose catchment contains dolomite, are characterised by an alkalinity of $130\,\mu mol\,L^{-1}$, and a calcium concentration of $85\,\mu mol\,L^{-1}$. The water of Lake Piccolo Naret is intermediate. Its alkalinity is $<50\,\mu mol\,L^{-1}$, and appears to have been influenced by the presence of some calcite or dolomite in its catchment area.

4.4   HEAVY METALS

Table 4.5 gives representative examples of **dissolved** trace metal concentrations in natural waters. The data show that these waters contain remarkably low concentrations (in some cases as low as $10^{-11}\,M$) of dissolved metal ions. However, many of the data reported in the literature are based on total particulate and dissolved concentrations, and furthermore, analytical procedures have often not been able to discriminate against contamination during sampling and sample processing.

Metal concentrations in fresh waters are a consequence of:

1 geochemistry of rocks in the catchment (metals released by weathering);
2 anthropogenic pollution (by waste inputs and atmospheric deposition);
3 river chemistry (adsorption of metal ions to particles and other surfaces and particle deposition into the sediments).

Windom *et al.* (1991) report that, on average, 62, 40, 90 and 80% respectively of the cadmium, copper, lead and zinc carried by the USA east coast rivers is attached to particles. Similar conclusions were reached earlier by Martin & Whitfield (1983). Such high proportions of particulate metal ions are representative of large rivers, which are often relatively unpolluted, and are characterised by high turbidity and low ionic strength. In small rivers with anthropogenic metal pollution and low turbidity (and often a calcareous drainage area), a significant fraction of metal ions may also be present in dissolved form.

The effect of particles in the residual concentrations is also apparent from the pH dependence. Figure 4.8 (from Sigg *et al.* 1997) shows an example. Similar data on the pH dependence of cadmium and lead were reported by Windom *et al.* (1991). The experimental results (Fig. 4.8) indicate a decrease in [Zn] with increasing pH. Such a dependence is in a first approximation compatible with a reaction of the type

$$S-OH + Me^{2+} \rightarrow S-OMe^+ + H^+ \qquad (4.14)$$

where S–OH is a surface hydroxyl group on an organic or inorganic particle surface.

4.4.1   *Regulation of trace elements in lakes*

When trace elements are introduced into a lake by riverine and atmospheric input, they interact (i) with solutes (in the formation of complexes), and

Table 4.5   Representative concentrations of some dissolved heavy metals in natural waters. (Source: Sigg 1994.)

	Cu (nM)	Zn (nM)	Cd (nM)	Pb (nM)	References
USA East Coast rivers	17	13	0.095	0.11	Windom *et al.* (1991)
Mississipi	23	3	23	–	Shiller & Boyle (1985)
Amazon	24	0.3–0.8	0.06	–	Shiller & Boyle (1985)
Lake Constance	5–20	15–60	0.05–0.1	0.2–0.5	Sigg *et al.* (1983)
Lake Michigan	10	9	0.17	0.25	Shafer & Armstrong (1990)
Lago Cristallina*	5	30	0.5	3	Sigg (1993), personal communication
Pacific Ocean	0.5–5	0.1–10	0.01–1	0.005–0.08	Bruland & Franks (1983)
Rain water†	10–300	80–900	0.4–7	10–200	Sigg (1993), personal communication

*   Alpine Lake at 2200 m altitude in the Southern Swiss Alps. Its dissolved Al(III) concentration at pH 6 is 600 nM.

†   As measured near Zürich, Switzerland.

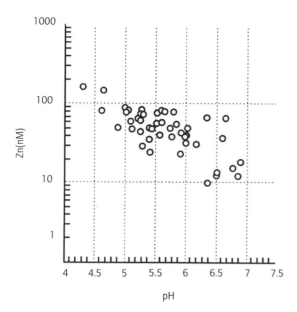

Fig. 4.8 Effect of pH on residual metal concentration. Dissolved zinc is plotted against pH in different mountain lakes in the southern parts of the Swiss Alps. These lakes are less than 10 km apart, so the atmospheric inputs can be considered to be uniform over this scale, but their water composition (pH) is influenced by different bedrock in their catchments. A similar dependence on pH has also been observed for Cd and Pb, but this dependence is less pronounced with Cu(II) when solute complex formation counteracts adsorption (data 1983–1992).

(ii) with inorganic and organic (phytoplanktonic) particles (via adsorption and assimilation). Affinity of reactive elements for particles which settle through the water column essentially determines the relative residence time of these elements, their residual concentrations and their ultimate fate (Fig. 4.9, adapted from Sigg & Stumm 1994). As shown in Table 4.5, concentrations of trace elements in lakewater columns are—despite anthropogenic pollution—extremely small (10^{-11}–10^{-7} M), illustrating the remarkable efficiency of the continuous 'conveyor belt' of the settling, adsorbing, scavenging and assimilating particles. Scavenging of metals such as Zn(II), Pb(II) and Cd(II) is pH dependent (see Fig. 4.8). Thus, the residual con-

centration of these metals tends to decrease with increasing pH. The sedimentary record reflects the accumulation of trace elements in sediments and a profile of concentration versus sediment depth (or age) gives a 'memory record' of past loadings (see Chapter 17).

4.4.2 The role of settling particles

Settling of both biogenic organic (algae, biological debris) and inorganic particles (e.g. manganese and iron oxides) contributes to the binding, assimilation and transport of reactive elements. The photosynthetic production of algae and their sedimentation is a dominant process, especially in eutrophic lakes. Near the sediment–water interface (SWI) anoxic conditions, under which iron and manganese oxides undergo reduction and dissolution, may occur. As discussed before, trace elements are affected by these processes in different ways (via interaction with iron and manganese oxides, and precipitation as, and complexation with, sulphides).

In many lakes, sedimentation rates of 0.1–2 g m^{-2} day^{-1} are typically observed. Still higher values are found in very productive lakes (Sigg & Stumm 1994). The settling material can be collected in sediment traps, and then characterised in terms of its chemical composition, morphology and size distribution of particles (see Chapter 8). The composition of this material is subject to seasonal variations caused primarily by different biological activity during the various seasons (Fig. 4.9).

4.4.2.1 Steady-state models

Simple steady-state models may be used in order to study the quantitative relationship between (i) the mean concentration in the lakewater column and residence time of metal ions and (ii) the removal rate by sedimentation (for a detailed treatment of lake models see Sigg (1994), Schwarzenbach *et al.* (1993) and Stumm & Morgan (1996)). In a simple steady-state model, inputs of material to the lake from its airshed and watershed are deemed equal to its removal by sedimentation and outflow. The

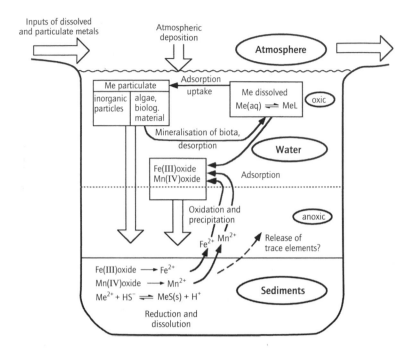

Fig. 4.9 Schematic representation of the cycling of trace elements in a lake. Trace elements are removed to the sediments together with settling material, which consists in large part of biological material.

water column is considered to be fully mixed, and mean concentrations and residence times in the water column may thus be derived from measured sedimentation fluxes. In comparison with the settling process, the binding of metals to particles is rapid.

These quantitative relationships, and an example, are summarised in Table 4.6. Under steady-state assumptions, the residence time of an element (τ_M) being removed from the water column by sedimentation and by outflow is given by

$$\tau_M^{-1} = \tau_W^{-1} + \tau_S^{-1} \qquad (4.15)$$

where τ_W is the residence time of water in the lake and τ_S is the residence time of an element with respect to sedimentation. The removal rate by sedimentation (τ_S^{-1}) can then be expressed as:

$$\tau_S^{-1} = f_p \times k_s \qquad (4.16)$$

where f_p is the fraction of the element in particulate form and k_s (day^{-1}) is the rate constant characterising sedimentation.

The fraction of an element present in the particulate phase relates to its tendency to bind to particles. This depends on the partition coefficient of the element between particulate phase and solution (dependent on the chemical interactions with the particles and in solution), and on the concentrations of particles in the water column.

The removal rate of an element by sedimentation can be calculated from (i) the flux of the element to the sediments and (ii) the total amount present in the water column (Table 4.6). The removal rate by sedimentation is large if both the fraction of the element bound to particles and the sedimentation rate are significant. In this case, the residence time of the element in the water column is much smaller than the water residence time. The mean concentration in the water column can be shown to depend on the removal rate by sedimentation and the water residence time according to equation (iii), Table 4.6. This means that if the removal rate by sedimentation is high, the mean concentration in the water column turns out to be much lower than the input concentration. Low concentrations of metal ions in the water col-

Table 4.6 Removal of metal ions from lake water column by sedimentation. (Source: Sigg 1994.)

Residence time of element M:
$$\tau_M^{-1} = \tau_w^{-1} + \tau_s^{-1} \qquad \text{(i)}$$
where τ_M is the residence time of element M (days), τ_w is the residence time of water (days) and τ_s is the residence time of element with respect to sedimentation (days)

Removal rate by sedimentation (τ_s^{-1}):
$$\tau_s^- = f_p \times k_s = \frac{F_M}{h \times [M]_w} \qquad \text{(ii)}$$

where f_p is the fraction of element bound to particles, F_M is the sedimentation rate of element M (mol m^{-2} day^{-1}), $[M]_w$ is the concentration of element M in water (mol m^{-3}), k_s is the removal rate of particles (day^{-1}) and h is the mean depth of water column (m)

Example: Pb in Lake Zurich, summer

Sedimentation rate of Pb: $\qquad F_{M(Pb)} = 1\ \mu\text{mol m}^{-2}\text{day}^{-1}$

Mean concentration in water column: $\qquad [Pb]_w = 8 \times 10^{-4}\ \mu M = 0.8\ \mu\text{mol m}^{-3}$

Removal rate: $\qquad \dfrac{F_M}{h \times [Pb]_w} = 2.5 \times 10^{-2}\ \text{day}^{-1}$

Residence time of water: $\qquad \tau_w = 400\ \text{days}$

With $\tau_w^{-1} = 2.5 \times 10^{-3}\ \text{day}^{-1}$: residence time of Pb: $\qquad \tau_M = 36\ \text{days}$

Steady-state concentration:
$$[Pb]_w = \frac{\tau_w^{-1}}{\tau_w^{-1} + \tau_s^{-1}} \times c_{in} = 0.09\ c_{in} \qquad \text{(iii)}$$

where c_{in} is the input concentration

umn can thus be expected to occur if a metal ion binds to a significant extent to particles, and if the sedimentation rate is high.

The chemical factors which determine binding to particles (pH, surface ligands, ligands in solution), the tendency of phytoplankton for uptake of trace elements, and the chemical factors which influence coagulation and, in turn, sedimentation rates, affect the distribution of a metal ion between the particulate phase and solution, and thus the fraction found in particulate form (f_p). For a single element under similar chemical conditions in different lakes, residence time and mean concentration depend on sedimentation rates.

Simple steady-state models are only able to predict mean concentrations. Seasonal variations, and concentration depth profiles in a lakewater column, give further insight into the mechanisms governing the removal of metal ions. Nutrient metal ions co-regulate the growth of phytoplankton, but the effect of the phytoplankton community itself on metal ions and their speciation is also pronounced. The bulk of trace elements, adsorbed most efficiently on to biological surfaces in the upper layers, is carried into the profundal zone by settling. As particles fall through the water column, they provide a food source for successive populations of filter feeders, so that the material is repackaged many times *en route* to the sediment (Whitfield & Turner 1987).

4.5 CHEMICAL SPECIATION

Historically, limnologists have been primarily interested in determining collective parameters and elemental composition of lake waters and biotic materials. However, this information alone is often inadequate for identifying those mechanisms which control composition of natural waters, and for understanding their perturbations. To achieve this goal, limnologists also need to know the form and the **species** in which the element is present, in order to gain some insight into the role of the elements in the sedimentary cycles, into the physical chemistry of lake waters and into the nature of pollutant interactions as well as the complexities of the biochemical cycle. (In this context, the term **species** refers to the actual form in which a molecule or ion is present in solution (Stumm & Morgan 1996.) It is especially important to recall that biological availability, and physiological and toxicological effects, depend on the specific structure of the individual substances: e.g. $CuCO_3(aq)$ affects the growth of algae in a different way from $Cu^{2+}(aq)$; organic isomers usually differ in their toxicological effects.

The concentration of **ligands** (which tend to interact with metal ions) in fresh waters is given in Table 4.7. The information on amino acids and the sum of organic acids, including humic acids, is from Buffle (1988). In many waters, anthropogenic ligands such as NTA (nitriloacetic acid) and EDTA (ethylenediaminetetraacetic acid) occur, which form stable complexes with metal ions. For example, in Swiss rivers both NTA and EDTA are found in a concentration range of $10^{-7}–10^{-8}$ M.

4.5.1 Speciation in representative fresh water

We now consider inorganic chemical speciation for a representative freshwater composition. The speciation calculations are illustrated in detail in Sigg & Xue (1994) and Stumm & Morgan (1996). Table 4.8 lists the predominant inorganic species which have been computed for fresh waters (Turner *et al.* 1981; Sigg & Xue 1994)

Table 4.7 Concentration range of some ligands in natural waters (log concentration (M))

Ligand	Fresh water
HCO_3^-	−4 to −2.3
CO_3^{2-}	−6 to −4
Cl^-	−5 to −3
SO_4^{2-}	−5 to −3
F^-	−6 to −4
HS^-/S^{2-} (anoxic conditions)	−6 to −3
Amino acids	−7 to −5
Organic acids	−6 to −4
Particle surface groups	−8 to −4

4.6 REDOX PROCESSES AND REDOX EQUILIBRIA

Only a few elements—carbon, nitrogen, oxygen, sulphur, hydrogen, iron and manganese—are major participants in aquatic redox processes. When comparisons are drawn between calculations for an equilibrium redox state and concentrations in the aquatic dynamic environment, the implicit assumptions involved are (i) that biological mediations are operating essentially in a **reversible** manner at each stage of each process, or (ii) that there is a metastable steady state which approximates to partial equilibrium for the system under consideration. As shown in Fig. 4.9, non-photosynthetic organisms tend to restore equilibrium by catalytically decomposing the unstable products of photosynthesis through energy-yielding redox reactions, thereby obtaining a source of energy for their metabolic needs.

Although conclusions regarding chemical dynamics may not be drawn from thermodynamic considerations, it appears that most relevant redox reactions, except possibly those involving N_2, are biologically mediated in the presence of suitable and generally ubiquitous microorganisms. A redox intensity scale in Fig. 4.9 gives the sequence of reactions observed in an aqueous system at various $p\varepsilon$ values (pH = 7). The various redox combinations are listed in Fig. 4.10a.

Although aqueous solutions do not contain free electrons, it is possible to define a (hypothetical) electron activity, $p\varepsilon = -\log\{e^-\}$; the latter is related (via the Nernst equation) to the redox potential (oxidation intensity), E_H, of the system.

$$p\varepsilon = \frac{E_H}{2.3RT/F} = \frac{E^0}{2.3RT/F} + \frac{1}{n}\log\frac{\Pi\{ox_i\}^{n_i}}{\Pi\{red_i\}^{n_i}} \quad (4.17)$$

For example, for the reaction

$$NO_3^- + 6H^+ + 5e^- \rightleftharpoons 0.5N_2(g) + 3H_2O \quad (4.18)$$

$$E_H = E_H^0 + \frac{2.3RT}{5F}\log\frac{\{NO_3^-\}\{H^+\}^6}{P_{N_2}^{1/2}} \quad (4.19)$$

Table 4.8 Major inorganic metallic or metalloid species in fresh waters

	Element	Major species	Freshwater $[M^{n+}]/M_T$*
Hydrolysed, anionic	B(III)	H_3BO_3, $B(OH)_4^-$	
	V(V)	HVO_4^{2-}, $H_2VO_4^-$	
	Cr(VI)	CrO_4^{2-}	
	As(V)	$HAsO_4^{2-}$	
	Se(VI)	SeO_4^{2-}	
	Mo(VI)	MoO_4^{2-}	
	Si(IV)	$Si(OH)_4$	
Predominantly free	Li	Li^+	1.00
aqueous ions	Na	Na^+	1.00
	Mg	Mg^{2+}	0.94
	K	K^+	1.00
	Ca	Ca^{2+}	0.94
	Sr	Sr^{2+}	0.94
	Cs	Cs^+	1.00
	Ba	Ba^{2+}	0.95
Complexation with	Be(II)	$BeOH^+$, $Be(OH)_2^0$	1.5×10^{-3}
OH^-, CO_3^{2-}, HCO_3^-,	Al(III)	$Al(OH)_3(s)$, $Al(OH)_2^+$, $Al(OH)_4^-$	1×10^{-9}
Cl^-	Ti(IV)	$TiO_2(s)$, $Ti(OH)_4^0$	
	Mn(IV)	$MnO_2(s)$	
	Fe(III)	$Fe(OH)_3(s)$, $Fe(OH)_2^+$, $Fe(OH)_4^-$	2×10^{-11}
	Co(II)	Co^{2+}, $CoCO_3^0$	0.5
	Ni(II)	Ni^{2+}, $NiCO_3^0$	0.4
	Cu(II)	$CuCO_3^0$, $Cu(OH)_2^0$	0.01
	Zn(II)	Zn^{2+}, $ZnCO_3^0$	0.4
	Ag(I)	Ag^+, $AgCl^0$	0.6
	Cd(II)	Cd^{2+}, $CdCO_3^0$	0.5
	La(III)†	$LaCO_3^+$, $La(CO_3)_2^-$	8×10^{-3}
	Hg(II)	$Hg(OH)_2^0$ $(HgCl_4^{2-})$	1×10^{-10}
	Tl(I), (III)	Tl^+, $Tl(OH)_3^0$, $Tl(OH)_4^-$	2×10^{-21}‡
	Pb(II)	$PbCO_3^0$	5×10^{-2}
	Bi(III)	$Bi(OH)_3^0$	7×10^{-16}
	Th(IV)	$Th(OH)_4^0$	
	U(VI)	$UO_2(CO_3)_2^{2-}$, $UO_2(CO_3)_3^{4-}$	1×10^{-7}§

* Data from Sigg & Xue (1994). Freshwater conditions are:
  pH $= 8$ Alk $= 2 \times 10^{-3}$ M
  $[SO_4^{2-}]_T = 3 \times 10^{-4}$ M $[Cl^-]$ $= 2.5 \times 10^{-4}$ M
  $[Ca^{2+}]_T = 10^{-3}$ M $[Mg^{2+}]_T = 0.3 \times 10^{-3}$ M
  $[Na^+]_T = 2.5 \times 10^{-4}$ M
  O_2 at saturation with air, $I = 5 \times 10^{-3}$.
† La(III) is representative of the lanthanides.
‡ Redox state of Tl(I) under natural conditions is uncertain, ratio is for Tl(III).
§ As UO_2^{2+}.

(a)

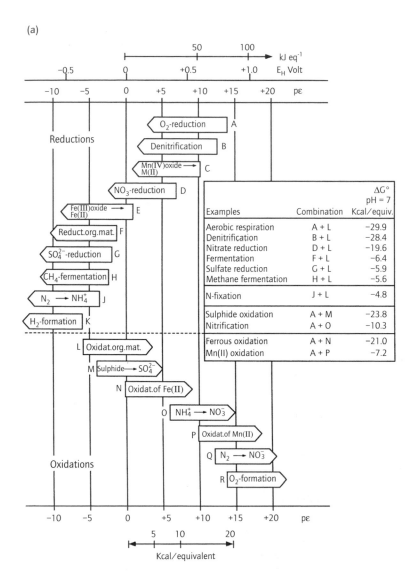

Fig. 4.10 (a) Biologically mediated redox processes calculated for pH = 7. (b) Representative ranges of redox intensity in soil and water. Range 1 is for oxygen-bearing waters. The electron activity, pε, range 2 is representative of many ground- and lake waters where O_2 has been consumed (by degradation of organic matter), but SO_4^{2-} is not yet reduced. In this range soluble Fe(II) and Mn(II) are present; their concentration is redox-buffered because of the presence of solid Fe(III) and Mn(III, IV) oxides. The pε range 3 is characterised by SO_4^{2-}/HS^- or SO_4^{2-}/FeS_2 redox equilibria. pε range 4 occurs in anoxic sediments and sludges.

$$pε = pε^0 + 1/5 \log \frac{\{NO_3^-\}\{H^+\}^6}{P_{N_2}^{1/2}} \qquad (4.20)$$

In an equilibrium system containing a number of redox couples of known total concentration, the activity of each redox species is a function of pε. Since many redox processes are very slow in reaching equilibrium, and do not couple with one anoth-er readily, it is possible for several different apparent redox levels to coexist in the same local environment.

In natural waters, redox reactions tend to occur in order of their thermodynamic possibilities. In a lake, the organic products of photosynthesis set-tling into the deeper waters act as **reductants** (sup-plying electrons), leading to a sequential 'titration' of the various electron acceptors available in the deeper portions of the lake and its sediments (Table 4.9). Incipient reduction of oxygen is followed by NO_3^- and NO_2^-. Reduction of MnO_2 should occur at about the same pε levels as that of NO_3, followed

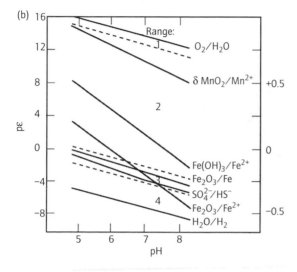

Fig. 4.10 *Continued.*

by FeOOH(s) to Fe^{2+}. When sufficiently negative pε levels have been reached, fermentation reactions and reduction of SO_4^{2-} and CO_2 may occur. Typically, there is mediation (by different enzymes) of redox processes in both directions.

The reaction $NO_3^- \rightarrow N_2$ is an exception, because no enzyme catalyses the reverse reaction, the breaking of the strong bonding of the N_2 molecules. The fact that $N_2(g)$ has not been converted largely into NO_3^- also indicates a lack of efficient biological mediation of the reverse reaction, for the mediating catalysis must operate equally well for reactions in both directions.

Denitrification may occur by an indirect mechanism such as reduction of NO_3^- to NO_2^- followed by reaction of NO_2^- with NH_4^+ to produce N_2 and H_2O. Figure 4.10 gives free energy values (ΔG^0 at pH = 7) for the important biologically mediated redox processes, a proportion of which is exploited by the organisms for synthesis of new cells. A yield proportional to ΔG of the redox reaction is found experimentally

Figure 4.10b, adapted from Drever (1988), gives representative redox intensity ranges of

Table 4.9 Sequence of progressive reduction of redox intensity by organic pollutants

O_2 consumption (respiration)		
$\frac{1}{4}\{CH_2O\} + \frac{1}{4}O_2$	$\rightleftharpoons \frac{1}{4}CO_2 + \frac{1}{4}H_2O$	(1)
Denitrification		
$\frac{1}{4}\{CH_2O\} + \frac{1}{5}NO_3^- + \frac{1}{5}H^+$	$\rightleftharpoons \frac{1}{4}CO_2 + \frac{1}{10}N_2 + \frac{1}{2}H_2O$	(2)
Nitrate reduction		
$\frac{1}{4}\{CH_2O\} + \frac{1}{8}NO_3^- + \frac{1}{4}H^+$	$\rightleftharpoons \frac{1}{4}CO_2 + \frac{1}{8}NH_4^+ + \frac{1}{8}H_2O$	(3)
Production of soluble Mn(II)		
$\frac{1}{4}\{CH_2O\} + \frac{1}{2}MnO_2(s) + H^+$	$\rightleftharpoons \frac{1}{4}CO_2 + \frac{1}{2}Mn^{2+} + \frac{1}{8}H_2O$	(4)
Fermentation		
$\frac{3}{4}\{CH_2O\} + \frac{1}{4}H_2O$	$\rightleftharpoons \frac{1}{4}CO_2 + \frac{1}{2}CH_3OH$	(5)
Production of soluble Fe(II)		
$\frac{1}{4}\{CH_2O\} + FeOOH(s) + 2H^+$	$\rightleftharpoons \frac{1}{4}CO_2 + \frac{7}{4}H_2O + Fe^{2+}$	(6)
Sulphate reduction, production of H_2S		
$\frac{1}{4}\{CH_2O\} + \frac{1}{8}SO_4^{2-} + \frac{1}{8}H^+$	$\rightleftharpoons \frac{1}{8}HS^- + \frac{1}{4}CO_2 + \frac{1}{4}H_2O$	(7)
Methane fermentation		
$\frac{1}{4}\{CH_2O\}$	$\rightleftharpoons \frac{1}{8}CH_4 + \frac{1}{8}CO_2$	(8)

importance in soil, sediments, and surface- and groundwaters. When organic matter is mineralised, alkalinity, $[NO_3^-]$ and $[SO_4^{2-}]$ increase, and Fe(II) and Mn(II) are mobilised. Phosphate, incipiently bound to Fe(III) (hydr)oxides, is released as a consequence of partial reductive dissolution of the Fe(III) solid phases. At lower pε (range 3 in Fig. 4.10), concentrations of Fe(II) and Mn(II) further increase. Sulphate reduction is accompanied by precipitation of FeS and MnS and formation of pyrite. At the oxic–anoxic boundaries, rapid turnover of iron takes place. This oxic–anoxic boundary may occur in deeper layers of the water column of fresh and marine waters, at the SWI, or within the sediments.

During the summer stagnation period, the successive settling of biota into the lower portion of the lake (the **hypolimnion**, see Chapter 2, section 9) causes progressive reduction in redox intensity and can be regarded as a redox 'titration'. The 'titrant' (the sinking biota) can be seen as an 'electron complex'; that is, we can titrate with electrons. Such a calculation is of course only partially

correct, but it shows synoptically how anoxis progresses in such a situation.

Figure 4.11, from Buffle & Stumm (1994), gives the 'titration' of Lake Bret and the progressive development of more reductive conditions. If the lake hypolimnion is considered a fully closed box, except for the input of organic matter, the length of each plateau region represents the maximum oxidising capacity of the corresponding compound, that is, its molar concentration multiplied by the number of electrons exchanged during its reduction.

The total oxidising capacity of the lake bottom is the sum of that of all the individual species; in this particular case it is 2.0×10^{-3} M. Such estimations are simplified, since they do not take into account the diffusion of O_2 and SO_4^{2-} from surface waters and many other factors such as the slow degradation of organic matter. Recent reviews of redox-driven cycling of trace elements have been given by Hamilton-Taylor & Davison (1995), Sigg (1994), Buffle & de Vitre (1994) and Davison (1985).

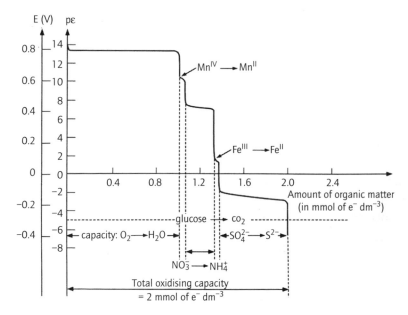

Fig. 4.11 Simplified titration curve of the oxidants with organic matter (decomposing phytoplankton) of Lake Bret (Switzerland). Plateau lengths correspond to the oxidising capacity of each oxidant. Computing does not take into account the input of O_2 by diffusion and the oxidising capacity of Fe(III)OOH and MnO_2 in sediments.

4.7 NUTRIENTS

In a simplified way we may consider a stationary state between photosynthetic production, P, and heterotrophic respiration, R:

$$106\ CO_2 + 16\ NO_3^- + HPO_4^{2-} + 122\ H_2O$$
$$+ 18\ H^+ (+\ \text{trace elements, energy})$$
$$P \downarrow \uparrow R \qquad (4.21)$$
$$C_{106}H_{263}O_{110}N_{16}P_1 + 138\ O_2$$
$$(\text{algal protoplasm })$$

As the stoichiometric equation predicts, during photosynthesis, phosphate and nitrate are eliminated from the water in a fixed ratio. During algal respiration, and as a result of mineralisation of algae, which occurs predominantly in the deeper water layers and in the sediments, phosphate and nitrate are liberated in the same fixed ratio. The influence of photosynthetic and respiratory processes on the composition of lake waters is frequently reflected in a correlation of concentration of soluble phosphate or nitrate and oxygen. A complex series of interrelated biological, geological and physical long-term processes lead to the evolution of lakes with constant proportions of mean annual C:N:P concentrations (Fig. 4.12a modified from Sigg & Stumm 1994; Fig. 4.12b from Ambühl 1975).

4.7.1 Limiting nutrients

For most inland waters, phosphorus is the limiting nutrient determining productivity. In some estuaries and in many marine coastal waters, nitrogen appears to be more limiting to algal growth than phosphorus. A stoichiometry of the nutrients can be observed in the elemental composition of the algae as well as in the concentration depth profiles (see Fig. 4.12).

Because of the complex functional interactions in lake ecosystems, the limiting factor concept needs to be applied with caution. We should distinguish between **rate-determining factors** (individual nutrients, temperature, light, etc.) that determine the rate of biomass production, and a **limiting factor** where a nutrient determines in a stoichiometric sense the maximum possible biomass standing crop.

Schindler (1977) used evidence from whole-lake experiments in order to convincingly demon-

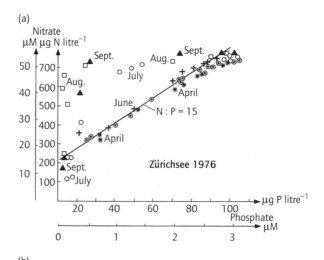

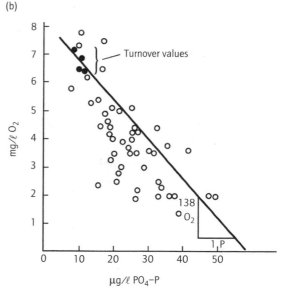

Fig. 4.12 (a) Concentration of nitrate versus phosphate in the Zürichsee. The slope of the drawn out line is ΔP: $\Delta N = 15$. The deviations in July–September are related to the development of a metalimnetic oxygen minimum at c. 20 m depth during this period. (b) Stoichiometric correlation between phosphate and dissolved oxygen (Lake Gersau).

strate phosphorus limitation in lakes. He also showed that natural mechanisms compensate for nitrogen and carbon deficiencies in eutrophicated lakes. In experimentally fertilised lakes, influx of atmospheric carbon dioxide supplied enough carbon to support and maintain phytoplankton populations in proportion to phosphorus concentrations over a wide range of values. There was a strong tendency in every case for lakes to correct carbon deficiencies—obviously, the rate of CO_2 supply from the atmosphere was sufficiently fast—maintaining concentrations of both chlorophyll and carbon that were proportional to the phosphorus concentration (Fig. 4.13, adapted from Schindler 1977).

Schindler (1977) also demonstrated that biological mechanisms were in many cases capable of correcting algal nitrogen deficiencies. While a sudden increase in phosphorus input may cause algae to exhibit symptoms of limitation, either by nitrogen or carbon, or both, long-term processes are at work which appear eventually to correct these deficiencies, once again leaving phytoplankton growth proportional to phosphorus concentration. As Schindler (1977) points out, this 'evolution' of appropriate nutrient ratios in fresh waters involves a complex series of interrelated biological, geological and physical processes, including (i) photosynthesis, (ii) the selection of species of algae which can fix atmospheric nitrogen, (iii) alkalinity, (iv) nutrient supplies and concentrations, (v) rates of water renewal and (vi) turbulence. Various authors have observed shifts in species composition of algal communities with changing N/P ratios. Low ratios appear to favour nitrogen-fixing blue-green algae, whereas high ratios, achieved by controlling phosphorus input by extensive waste treatment, cause a shift from a 'water bloom' consisting of blue-green algae to forms which are less objectionable.

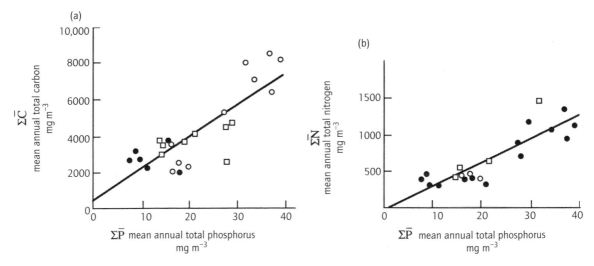

Fig. 4.13 Phosphorus limitations in lakes. In experimentally fertilised lakes of the experimental lakes area of the Freshwater Institute in Winnipeg (Environment Canada), ratios of mean annual concentrations C/P and N/P tend to become constant. In (a) we see that the carbon content increases as a consequence of phosphorus addition to the lakes, while (b) illustrates that the nitrogen content of a lake increases when the phosphorus input is increased, even when little or no nitrogen is added with fertiliser. Each point represents a result for a different lake. Open circles refer to data from enriched or fertilised lakes with a nitrogen deficiency (N : P < 8 by weight); open squares refer to data from enriched or fertilised lakes with a carbon deficiency (C : P < 50 by weight); closed circles refer to natural lakes with natural or higher than natural C : P and N : P ratios.

4.7.2 Consumption resulting from increased productivity

As is also the case in the oceans, the deeper portions of a lake receive phosphorus in two forms: (i) **preformed**, that is phosphorus which enters the lake as such (or adsorbed on clays), and (ii) **biogenic debris**. Most of the latter is oxidised to form phosphates of oxidative origin (P_{ox}). For every phosphorus atom of oxidative origin, a stoichiometric equivalent of 342 atoms of oxygen (or $140\,g\,O_2\,g^{-1}$ P) have been consumed. Accordingly, as indicated schematically in Fig. 4.14, a flux of P_{ox} is paralleled by a flux of O_2 utilisation.

The tolerable phosphorus loading of a lake may be related to the hypolimnetic oxygen consumption as follows. Under the simplifying assumption that all incoming phosphorus eventually becomes phosphorus of oxidative origin (P_{ox}), we may say that during a stagnation period of T_{st} days the annual phosphorus loading per lake surface, L_t (mg P $m^{-2}\,yr^{-1}$), causes an approximate oxygen consumption $\Delta[O_2]$ ($mg\,m^{-3}$) of a hypolimnion assumed to be homogeneously mixed of depth z_H(m) of

$$\Delta O_2 = 140\,\frac{T_{st}\,L_t}{365\,z_H} \qquad (4.22)$$

Correspondingly, a maximum phosphorus loading, L_{max}, for a tolerable oxygen consumption $[O_2]_{max}$ could be estimated as

$$L_{max} = \Delta[O_2]_{max} \times 7 \times 10^{-3} \times \frac{365}{T_{st}}\,z_H \qquad (4.23)$$

The many possible complicating factors, however, must not be overlooked. Simple stoichiometric relations may be too simplified, and may change from lake to lake.

As shown schematically in Fig. 4.14 1 mg of phosphorus (assuming the element to be the limiting factor) can be used to synthesise approximately 0.1 g of algal biomass (dry weight) in a single cycle of the limnological transformation. After settling to the deeper layers, during mineralisation, this biomass exerts a biochemical oxygen demand of approximately 40 mg.

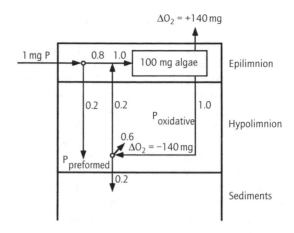

Fig. 4.14 Simplified scheme of typical transformation of phosphorus in a stratified lake. One milligram of phosphorus introduced into a lake during the stagnation period may lead to the synthesis of 100 mg algae (dry mass), which upon mineralisation cause an oxygen consumption in the hypolimnion of $140\,mg\,O_2$. From the organic phosphorus mineralised in the hypolimnion, 0.6 mg is assumed to accumulate during the stagnation period, while 0.2 mg is assumed to be adsorbed (e.g., on iron(III)oxide) and transferred into the sediments. Another 0.2 mg reaches the surface waters as phosphate by eddy diffusion.

From this simple calculation it is obvious that organic material introduced into the lake by domestic wastes (normally some 20–100 mg organic matter per litre) is small in comparison to that which can be biosynthesised from inwashed agricultural fertilisers (3–8 mg P L^{-1}), which can yield 300–800 mg organic matter per litre. Aerobic biological waste treatment with a heterotrophic enrichment culture mineralises substantial fractions of bacterially oxidisable organic substances, but is not capable of eliminating more than 20–50% of nitrogen and phosphate constituents.

4.8 THE SEDIMENT–WATER INTERFACE

Sediments are not just passive depositories for material removed from lake water. The flux of constituents from the sediments into the water, and vice versa, is important in controlling the composition of many lakes. Diagenetic chemical reactions occurring within the sediments consist of abiotic and biogenic reactions. Diagenesis refers to changes that take place within a sediment during and after burial. Because of these reactions, sediments exert a significant effect on the overlying lake waters (Fig. 4.15). Most biogenic reactions depend on the decomposition of organic matter. The sequence of redox reactions occurring in the sediments is the same as that already discussed for the interaction of excess organic matter with O_2, NO_3^-, SO_4^{2-} and HCO_3^-: that is, removal of dissolved O_2; reduction of NO_3^-, SO_4^{2-} and HCO_3^-; and production of CO_2, NH_4^+, phosphate, HS^- and CH_4 (Fig. 4.15a).

The products of these reactions may in turn bring further changes in sediment chemistry, such as the solubilisation of iron and manganese after O_2 has been removed. On the other hand, HS^- resulting from SO_4^{2-} reduction may react with detrital iron minerals to form iron sulphides. Excess HCO_3^- is produced by SO_4^{2-} reduction and NH_4^+ formation; eventually $CaCO_3$ may be precipitated. Build-up of dissolved phosphate may under suitable conditions bring about the precipitation of apatite $[Ca(PO_4)_3OH]$. Precipitation of Mg^{2+} may occur as a result of the removal of clay minerals of iron, which reacts with HS^- to form iron sulphides. Reactions that are not biogenically controlled include the dissolution of silica, $CaCO_3$ and feldspars, and various ion-substitution reactions in sediment minerals and ion exchange processes on clay minerals.

4.8.1 Fluxes of solids, water and solutes

Sedimentation of solid particles, and entrapment of water in the pore spaces, represent two major fluxes of materials across the sediment–water interface (SWI). Further processes also responsible for transport across the SWI are (i) upward flow of (pore) water caused by hydrostatic pressure gradients of groundwater in aquifers or land, (ii) molecular diffusional fluxes in pore-water, and (iii) mixing of sediment and water at the interface (bioturbation and water turbulence). Rates of sediment deposition vary from millimetres per 1000 years in the pelagic ocean up to a centimetre per year in lakes. A sediment can be thought of as being made up of equal volumes of water and solid particles (sediment volume fraction = 0.5) of density $2.5\,\mathrm{g\,cm^{-3}}$ (Lerman 1979).

(a)

(b)

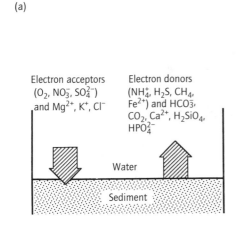

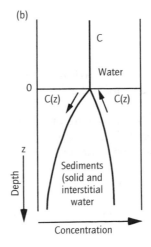

Fig. 4.15 The sediment–water interface. (a) Direction of fluxes expected for dissolved constituents between sediment pore-waters and the overlying waters (oceans and lakes). (b) For sediments and pore-water, the one-dimensional distribution of concentrations is time and depth dependent. Arrows indicate fluxes at the sediment–water interface depending on the concentration gradient in pore-water. The overlying water (lakes) is assumed to be well mixed.

4.8.2 Redox transformation in lake sediments

Figure 4.16, adapted from Wersin *et al.* (1991), gives some pore-water profiles from the Greifensee, Switzerland. The idealised redox sequence generates a series of stratigraphically discrete processes.

Remineralisation of organic nitrogen and phosphorus leads to increases in ammonium and phosphate concentration below the SWI. Increase in alkalinity is also a result of decomposition of organic matter either directly (as organic nitrogen is remineralised to NH_4^+) or indirectly (as calcite dissolves in response to the release of CO_2 associated with remineralisation of organic carbon). In this

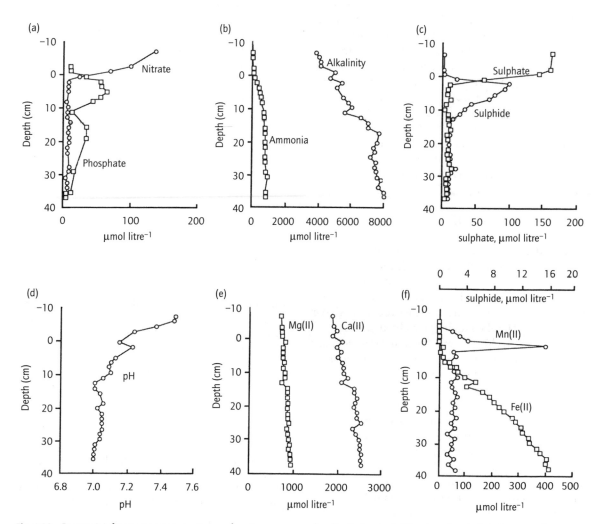

Fig. 4.16 Interstitial water concentrations of major compounds, alkalinity and pH versus depth: (a) nitrate and phosphate, (b) ammonia and alkalinity, (c) sulphate and sulphide, (d) pH, (e) magnesium (II) and calcium (II), and (f) manganese (II) and iron (II).

sediment, the profiles of the redox-active species are not clearly separated.

Depletion of both NO_3^- and SO_4^{2-} and release of Mn^{2+} occurs very near the sediment–water interface; $MnO_2(s)$ is normally found only in the top centimetre of the sediments. The peak in sulphide, however, occurs slightly lower in the sediment profile than the peak in dissolved Mn^{2+}; the successive appearance of these species is in accord with the corresponding redox potentials. The decrease in the dissolved concentrations of Mn(II) and S(-II) with depth are attributable to precipitation of rhodochrosite [$MnCO_3(s)$] and iron sulphides.

Precipitation of phosphate-containing minerals may account for the decrease in pore-water phosphate. Vivianite ($Fe_3(PO_4)_2 \cdot 8H_2O$) was calculated to be supersaturated in the pore-water, although this mineral could not be identified in the sediment. For a recent quantitative evaluation of microbial reactions, chemical speciation and multicomponent diffusion in pore-waters of a eutrophic lake, see Furrer & Wehrli (1996).

4.9 CONCLUDING REMARKS

Every lake is a mirror of its environment. Although the primary influence of human beings is on terrestrial systems, because of the interdependence of terrestrial and aquatic ecosystems, human impact on nature finds its most sensitive response in inland waters, especially in lakes.

Although in this review, the non-biological aspects of limnology are emphasised, lakes must be viewed, and studied, as microcosms. In the 'cosmos' of a lake, solar energy is extracted by the phytoplankton, and used to support the biological community, to organise the ecosystem, and to drive the cycles of nutrients, and of other elements. Therefore, the common ecological goal is to determine how the chemical environment interacts with organisms and with ecosystems, and to assess how they relate to one another.

Furthermore, lakes are excellent models with which to study the interdependence of several biogeochemical cycles, and their interlocking feedback mechanisms. The aquatic biomass in marine and freshwater systems is characterised by a surprising compositional constancy. Rates of cycling of carbon, nitrogen, phosphorus, sulphur, oxygen and some biophile metals, fall within certain limits given by the uptake and release of these elements by living organisms. These rates, in turn, reflect the variation in $C:N:P:S$ ratios in biotic tissues. Often, phosphorus appears capable of controlling major parts of the nitrogen, carbon, sulphur and silicon cycles. Furthermore, the input of phosphorus also appears to enhance trace-metal and metalloid elimination from waters, since the resulting higher productivities and larger particle sedimentation rates increase the efficiency of scavenging.

Understanding ways in which the complex of geochemical cycles in lakes is coupled by organisms may aid in our understanding of the global ecosystem. Especially valuable would be information on the role of the relatively small mass of Earth's biota in regulating and interconnecting the circulation of materials through land, water, biosphere and atmosphere, and the sensitivity of these interacting systems to disturbances caused by civilisation.

REFERENCES

Ambühl, H. (1975) Versuch der Quantifizierung der Beeinflussung des Oekosystems durch chemische Faktoren: stehende Gewässer. *Schweizerische Zeitschrift fuer Hydrologie*, **37**, 35–52.

Berner, E.K. & Berner, R.A. (1987) *The Global Water Cycle: Geochemistry and Environment*. Prentice Hall, Englewood Cliffs, NJ, 397 pp.

Berner, E.K. & Berner, R.A. (1996) *Global Environment: Water, Air, and Geochemical Cycles*. Prentice Hall, Englewood Cliffs, NJ, 376 pp.

Blum, A.E. & Lasaga, A.C. (1988) Role of surface speciation in the low-temperature dissolution of minerals. *Nature*, **331**, 431–3.

Bricker, O. & Rice, K.C. (1989) Acidic deposition in streams. *Environmental Science and Technology*, **23**, 379–85.

Bruland, K.W. & Franks, R.P. (1983) Mn, Ni, Cu, Zn and Cd in the western North Atlantic. In: Wong, S., Boyle, E., Bruland, K.W., Burton, J.D. & Goldberg, E.D. (eds),

Trace Metals in Sea Water. Plenum, New York, 395–410.

Buffle, J. (1988) *Complexation Reactions in Aquatic Systems. An Analytical Approach*. Ellis Horwood, Chichester, 692 pp.

Buffle, J. & de Vitre, R.R. (eds) (1994) *Chemical and Biological Regulation of Aquatic Systems*. Lewis Publishers, Chelsea, MI, 393 pp.

Buffle, J. & Stumm, W. (1994) General chemistry of aquatic systems. In: Buffle, J.D. & de Vitre, R.R. (eds), *Chemical and Biological Regulation of Aquatic Systems*. Lewis Publishers, Chelsea, MI, 1–43.

Davison, W. (1985) Conceptual models for transport at a redox boundary. In: Stumm. W. (ed.), *Chemical Processes in Lakes*. Wiley-Interscience, New York, 31–53.

Drever, J.I. (1988) *The Geochemistry of Natural Waters* (2nd edition). Prentice Hall, New York, 437 pp.

Furrer, G. & Wehrli, B. (1996) Microbial reactions, chemical speciation, and multicomponent diffusion in porewaters of a eutrophical lake. *Geochimica Cosmochimica Acta*, **60**, 2333–46.

Garrels, R.M. & Mackenzie, F.T. (1971) *Evolution of Sedimentary Rocks*. W.W. Norton and Co., New York, 397 pp.

Giovanoli, R., Schnoor, J.L., Sigg, L., Stumm, W. & Zobrist, J. (1988) Chemical weathering of crystalline rocks in the catchment area of acidic Ticino lakes, Switzerland. *Clay and Clay Minerals*, **36**(6), 521–9.

Hamilton-Taylor, J. & Davison, W. (1995) Redox-driven cycling of trace elements in lakes. In: Lerman, A., Imboden, D.M. & Gat, J.R. (eds), *Physics and Chemistry of Lakes*. Springer-Verlag, Berlin, 217–63.

Holland, D.H. (1978) *The Chemistry of the Atmosphere and Oceans*, Wiley-Interscience, New York, 351 pp.

Hutchinson, G.E. (1957) *A Treatise on Limnology*, Vol. 1, *Geography, Physics, Chemistry*. Wiley-Interscience, New York, 1015 pp.

Kämäri, J., Amann, M., Brodin, Y.W., Chadwick, M.J., Henriksen, A., Hettlingh, J.P., Kuylenstierna, J., Posch, M. & Sverdrup, H. (1992) The use of critical loads for the assessment of future alternatives to acidification. *Ambio*, **21**, 377–87.

Lerman, A. (1979) *Geochemical Processes — Water and Sediment Environments*. Wiley-Interscience, New York, 481 pp.

Lindberg, S.E., Lovett, G.M., Richter, D.D. & Johnson, D.W. (1986) Atmospheric deposition and canopy interactions of major ions in a forest. *Science*, **231**, 141–5.

Martin, J.-M. & Meybeck, M. (1979) Elemental mass-balance of material carried by major world rivers. *Marine Chemistry*, **7**, 173–206.

Martin, J.-M. & Whitfield, M. (1983) The significance of river input of chemical elements to the ocean. In: Wong, C.S., Boyle, E., Bruland, K.W., Burton, J.W. & Goldberg, E.D. (eds), *Trace Metals in Sea Water*. Plenum Press, New York, 265–96.

Meybeck, M. (1979) Concentration des eaux fluviales en elements majeurs at apports en solution aux oceans. *Revue de Geographie Physique et de Géologie Dynamique*, **21**(3), 215–46.

Meybeck, P. (1995) Les lacs et leur bassins. In: Pourriot, R. & Maybeck, M. (eds), *Limnologie générale*. Masson, Paris, 6–59.

Schindler, D.W. (1977) Evolution of phosphorus limitation in lakes. *Science*, **195**, 260–2.

Schnoor, J.L. & Stumm, W. (1985) Acidification of aquatic and terrestrial systems. In: Stumm, W. (ed.), *Chemical Processes in Lakes*. Wiley-Interscience, New York, 311–38.

Schwarzenbach, R.P., Imboden, D. & Gschwend, Ph.M. (1993) *Environmental Organic Chemistry*. Wiley-Interscience, New York, 681 pp.

Shafer, M.M. & Armstrong, D.E. (1990) Trace elements cycling in southern Lake Michigan: role of water-column particle components. *Abstracts of Papers, American Chemical Society*, **199**, 22.

Shiller, A.M. & Boyle, E. (1985) Dissolved zinc in rivers. *Nature*, **317**, 49–52.

Sigg, L. (1994) Regulation of trace elements in lakes: the role of sedimentation. In: Buffle, J. & de Vitre, R.R. (eds), *Chemical and Biological Regulation of Aquatic Processes*. Lewis Publishers, Chelsea, MI, 177–97.

Sigg, L. & Stumm, W. (1994) *Aquatische Chemie, Eine Einführung in die Chemie wässriger Lösungen und natürlicher Gewässer* (3rd edition). Teubner, Stuttgart, 498 pp.

Sigg, L. & Xue, H.B. (1994) Metal speciation: concepts, analysis and effects. In: Bidoglio, G. & Stumm, W. eds. *Chemistry of Aquatic Systems: Local and Global Perspectives*. KIuwer Academic, Dordrecht, 153–81.

Sigg, L., Sturm, M., Stumm, W., Mart, L. & Nurnberg, H.W. (1982) Heavy metals in Lake Constance: concentration-regulating mechanisms. *Naturwissenschaften*, **69**, 546–8.

Sigg, L., Schnoor, J.L. & Stumm, W. (1997) Atmosphere–water–rock reactions; as observed in alpine lakes. In: Macalady, D. (ed.), *Perspectives of Environmental Chemistry* Oxford University Press, 456–72.

Stumm, W. & Morgan, J.J. (1996) *Aquatic Chemistry*, 3rd Ed. Wiley-Interscience, New York, 1022 pp.

Sverdrup, H., Warfving, E.P., Frogner, T., Haoya, A.O, Johansson, M. & Andersen, B. (1992) Critical loads for forest soils in the Nordic countries. *Ambio*, **21**, 348–55.

Turner, D.R., Whitfield, M. & Dickson, A.G. (1981) The equilibrium speciation of dissolved components in freshwater and seawater at 25°C and 1 atm pressure. *Geochimica Cosmochimica Acta*, **45**, 855–82.

Wersin, P., Höhener, P., Giovanoli, R. & Stumm, W. (1991) Early diagenetic influences iron transformation in a freshwater lake sediment. *Chemical Geology*, **90**, 223–52.

Windom, H.L., Byrd, T., Smith, R.G. & Huan, F. (1991) Inadequacy of NASQAN data for assessing metal trends in the nation's rivers. *Environmental Science and Technology*, **25**, 1137–42.

Whitfield, M. & Turner, D.R. (1987) The role of particles in regulating the composition of sea water. In: Stumm, W. (ed.), *Aquatic Surface Chemistry*. John Wiley, New York, 457–93.

Wolman, M.G. (1967) A cycle of sedimentation and erosion in urban river channels. *Geografiska Annaler*, **49**, 385–95.

5 Physical Properties of Water Relevant to Limnology and Limnetic Ecology

COLIN S. REYNOLDS

5.1 INTRODUCTION

Despite being relatively abundant and ubiquitous on the planet, and as familiar as it is important to the support of living organisms, there are several quite anomalous physical properties of water that turn out to be crucial to its suitability as a habitat for aquatic organisms. It is, perhaps, important that these are emphasised. The purpose of this short chapter is, therefore, to highlight those special characteristics of water relevant to the subject matter of the book and to furnish some generalised explanations to account for them.

The fact that water is liquid at all, at least over a wide range of normal temperatures, is, for example, crucial to the persistence of a steadily fluid environment. The fact that it is also resistant to rapid temperature change is a function of the high specific heat of water. When it does cool sufficiently to solidify, water freezes first at its surface: this has the curious and unique effect of preserving liquid water beneath a veneer of ice, rather than simply solidifying right through, which is how most other substances behave below their melting points. Surface ice insulates the deeper water from further heat loss to the atmosphere, preserving a relatively equable liquid environment for aquatic organisms in regional climates otherwise at times too hostile to support much terrestrial life.

The density of water is unexpectedly high for what is, ostensibly, a low-molecular-weight substance. Liquid water is approximately 800 times more dense than air. Since the cells of most animals and plants also contain a great deal of water, it follows that they are much more nearly isopycnic with the aquatic medium. This Archimedean effect means that the necessity of mechanically robust supporting structures is much less pressing. Aquatic organisms whose evolution has been confined to water (from *amoebae* to blanket weeds) manage with only enough stiffening to maintain their integrity in turbulence fields.

For a liquid of low molecular weight (it is manifestly more resistant to flow, for example, than petroleum [or gasoline]), water is also relatively viscous, and at about $1 \times 10^{-3}\,\mathrm{N\,s\,m^{-2}}$, is roughly 50 times more viscous than air. Ignoring gravity, an object requires proportionately more effort to progress through water than air. At its edges, where mutual attraction between water molecules acts unilaterally, a powerful surface tension normally exists. However, the air–water interface is not merely an exploitable habitat for neustic plants and animals. For many small invertebrates it is a formidable, sometimes fatal, trap.

These properties arise through curiosities in the molecular structure of water. In addition, its suitability as a medium relates to its specific heat, its transparency and its solvent properties. These topics are addressed below. It is not intended that the article should provide a comprehensive overview of the physical properties. The intention is simply to give a background to the anomalies and to state how important they are to the biology of lakes. More detailed expositions are given in

Hutchinson (1957) and in Lampert & Sommer (1993).

5.2   MOLECULAR STRUCTURE

Many of the physically anomalous properties of water arise as a consequence of the structure and behaviour of water molecules. The entity comprising two atoms of hydrogen and one atom of oxygen, with a molecular weight of c.18 daltons, exists in the gaseous phase (water vapour). The covalent O–H bonds are formed by the sharing of the single electron of either hydrogen with the six in the outer ring of the oxygen atom. The distance from the centre of the hydrogen nucleus to the centre of the oxygen nucleus is 96 pm. The angle formed between the two bonds is approximately 105° and not the 90° predicted from theory. The reason for this is the mutual electrostatic repulsion of their charges. In turn, this leaves the molecule itself with a polarity, on one side (the hydrogen side) a weak positive charge, and on the other (the oxygen side) a weak negative charge. (It is this polarity that is exploited in a microwave oven: an electric field of alternating polarity causes the water molecules to vibrate sufficiently to generate significant fluxes of heat.) The polarity gives to the water molecules a certain attraction one for another: when molecules occur in liquid phase, polarity permits them to polymerise into much larger aggregates, held together by hydrogen bonds, with the general formula $(H_2O)_n$. It is this phenomenon which underpins many of the anomalous physical properties of water.

The incidence of larger molecules raises the freezing and the boiling points. Attraction among molecules increases the viscosity of the liquid phase, and their aggregation raises the density of the liquid. An important feature of the clusters of molecules is that they are dynamic, frequently reforming and breaking apart. The number of molecules in clusters broadly decreases with increasing temperature, but the relationship is probabilistic and, thus, empirically predictable: both the density and the viscosity of pure water (Fig. 5.1) are robust functions of its temperature.

Raising the temperature of a liquid increases the motion of the molecules: accordingly the spaces separating them become, on average, larger, and so the density of the liquid decreases. The behaviour of water is, however, anomalous in this respect. When water is frozen, the molecules are held in a crystalline matrix: the density of ice just below its freezing point is c.916.8 $kg\,m^{-3}$. Upon melting the molecules are able to move more closely together, giving a sharp **rise** in density to c.999.87 $kg\,m^{-3}$. As the temperature of the liquid water is raised a little further, more individual molecules break from the complexes and fall within the matrices: in this way, the same given number of molecules now occupies yet **less** space, and the density of the water is further **increased**. Eventually, however, it is dismemberment of the clusters that predominates: as the temperature is raised further, the main process operating is that of molecules moving further apart. The liquid expands and its density diminishes. Under a pressure of 1 atmosphere, pure water reaches its maximum density of 1000 $kg\,m^{-3}$ at 3.94°C.

This behaviour explains not only why fresh water at sea level achieves its greatest density close to 4°C, but also why, under appropriate conditions, ice forms at the surface of a lake, so insulating the deeper water from further heat-loss. It is also the explanation for the fact that, with every degree above 4°C, the **difference** in density also becomes greater. Later chapters remind us that this latter effect also enhances the mechanical energy required to mix increasingly warmed surface waters with denser layers below. Such considerations govern the whole sequence of cycles of density stratification and mixing in lakes, of all morphometries and at all latitudes, and possess important consequences for the distribution of dissolved nutrients and gases.

The presence of solutes lowers the temperature of greatest density. When the salt content of water, for instance, reaches 25 $g\,kg^{-1}$, the temperature of maximum density and the freezing point coincide (at about −1.3°C). Increased hydrostatic pressure also depresses the temperature of maximum density, by 0.1°C for every 10 bar of pressure. At a depth of 1000 m, the temperature of maximum

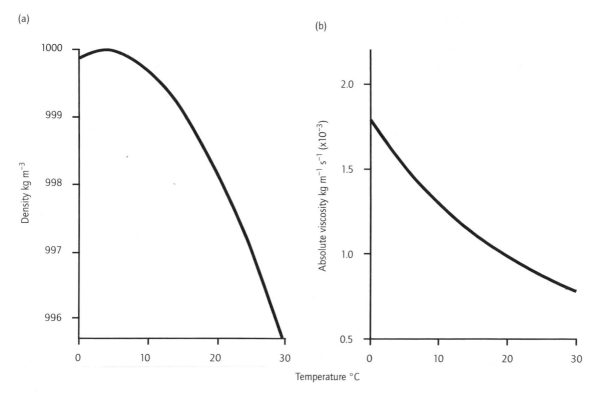

Fig. 5.1 Plots showing (a) the density and (b) the viscosity of water as a function of temperature. (Redrawn from Reynolds, 1997; reproduced with permission of The Ecology Institute.)

density is 2.91°C. At 1700 m, in the world's deepest lake (Ozero Baykal = Lake Baikal), it is c.2.3°C.

The mutual attraction of water molecules resists their free movement and the ability of one layer to slide over one another, if subjected to external forcing. Instead, a temperature-dependent threshold is reached at which the flow is chaotic and turbulent. In turn, momentum is dissipated through a sequence of progressively smaller eddies (the Kolmogorov Spectrum) until viscosity overwhelms the residual motion and re-establishes order. The size of the smallest eddies rather depends upon the strength of the forcing and the physical space available for its dissipation but, in most lakes, the spectrum collapses in the range of 1–4 mm. This means that most of the phytoplankton, though transported freely through turbulent

layers, are not necessarily experiencing any direct turbulence themselves. Their immediate environments are, thus, typically viscous, with all its implications on solute diffusion and locomotory ease. Larger zooplankton, at least, exceed the size of the smallest eddies. Those that they create while feeding greatly increase the rate of encounters with food particles that are themselves entrained (Rothschild & Osborn 1988).

Additionally, molecular attraction generates the powerful surface tension at the air–water interface ($73.5 \times 10^{-3}\,\mathrm{N\,m^{-1}}$ at 15°C). This is higher than that of any other liquid save mercury. For small plants and animals, whether they flourish in it or flounder, surface forces exceed gravitational attraction. Surface tension supports not merely the microorganisms of the neuston but, often, a flux of dry particulate deposit carried in the air. Animals as robust as Gerrid pond-skaters and Gyrinid whirligig beetles are also able to run about on a still water surface as if it were a solid. Powerful surface forces also resist the formation of very small bubbles, save when surfactants are present to reduce the surface tension. This means that gas concentrations in water tend to reach supersaturation of the solubility product before the gas comes out of solution.

Molecular reorganisation associated with state transitions in water is also relevant to the equability of aquatic environments. The so-called 'latent heat of evaporation' of water (that is the energy converted in raising liquid water to the gaseous state) is 2.243 MJ kg^{-1}. Compounding what is already a high specific heat (4.186 kJ kg^{-1} K^{-1}), it is evident that water is consumptive of solar radiative fluxes without commensurate and life-threatening rises in temperature, so long as the surface is open to a dry atmosphere. Under full tropical sunlight (delivering up to 900 W m^{-2}) and with minimal surface reflectance, the potential rate of temperature increase of the top millimetre of a water column approaches 0.2 K s^{-1} (i.e. 0.2°C s^{-1}), nominally sufficient to bring the water to the boil in 5 to 6 minutes.

Several processes prevent this from happening. These include transmission of heat to the water layer below and the air layer above. Water is, however, a poor conductor of heat (<0.006 J cm^{-1} K^{-1} s^{-1}), so dissipation is actually greatly facilitated by turbulence induced by wind- or gravitational forcing. Most important is that the same heat flux consumed in vaporising 0.4 g m^{-2} s^{-1} from the surface (that is, a layer 0.4 μm in thickness every second) would fail to raise the water temperature at all.

Just how much water can be evaporated is a function of the heat flux and the atmospheric saturation. Other factors being equal, the fluxes are accelerated by wind action, which refreshes the air above water and prevents saturation; at the same time the surface area over which heat is lost to the atmosphere is increased.

Differences in air and water temperature are material to the determination of the net direction of heat exchanges. For the limnologist disinterested in the complex quantitative derivation of evaporative and other heat fluxes, it is usually enough to know that, except under conditions of vapour saturation of the atmosphere, generally, there can be net warming of the surface water when the air temperature is greater than that of the water, and net cooling when the water is warmer. Note that condensation of vapour from warm, moist air on to a cool water surface also adds heat to the water, at the rate of 2.243 MJ kg^{-1} of condensate.

5.3   THE TRANSPARENCY OF WATER

Despite being almost colourless, even pure water is severely restrictive to the passage of the photosynthetically active wavelengths of the solar flux (roundly, from 400 to 700 nm, and almost coincidental with 'visible' radiation). Selective absorption at particular wavelengths compounds this effect: it is most powerfully evident to underwater divers, as they see the fading light with increased depth becoming more clearly composed of blue or green wavelengths. In waters moderately to heavily stained with humic acids, or charged with fine particles, the attenuation of light underwater soon becomes prejudicial to maintenance of net plant production and, thus, to the quality and quantity of dependent aquatic communities.

The conceptual complexities of hydrological optics constitute a large and difficult subject area, the theoretical development of which is beyond the scope of this chapter. The key work is still that of Preisendorfer (1976) but the treatment in Kirk (1994) is perhaps the more accessible. The following section is concerned mainly with those features of light penetration in lake waters which affect their biological function.

The properties of the underwater light field are due, in part, to the nature of light and, in part, to inherent properties of the water. 'Light', in fact, refers to the visible wavelengths of the solar electromagnetic flux, specifically those in the radiation band, 400–700 nm. Electromagnetic energy occurs in indivisible units known as **quanta**, or, at least within the visible waveband, as **photons**. They travel, at great velocity (c, approximately $3 \times 10^8 \, \mathrm{m\,s^{-1}}$), along characteristically wave-like pathways, each of distinctive length (λ) and frequency (v), where $\lambda v = c$. The energy, e, thus carried varies with the frequency (and inversely with the wavelength), the relation being:

$$e = hv = hc/\lambda \qquad (5.1)$$

where h is Planck's constant ($6.63 \times 10^{-34} \, \mathrm{J\,s}$). It can be seen that for any given wavelength the quantum flux and its energy equivalent are intercalculable. Note, too, that, within the visible spectrum alone, light at the red end of the spectrum (700 nm) contains little over half the energy of that at the blue extreme (400 nm). The energy content and the flux density of a broad waveband of mixed radiation are not readily interconvertible, owing to the multiplicity of λ values. However, the approximate conversion of $2.77 \times 10^{18} \, \mathrm{quanta\,s^{-1}\,W^{-1}}$ applied to solar radiation within the visible spectrum, and under a wide range of meteorological conditions, is reliable to within a few percentage points (Morel & Smith 1974). Even under the most favourable combination of the maximum intensity of perpendicular solar radiation to the upper atmosphere (averaging $1352 \, \mathrm{W\,m^{-2}}$, the so-called **solar constant**), and the minimal absorption through a dry cloudless atmosphere, no more than c.900 $\mathrm{W\,m^{-2}}$ reaches the surface of the sea or a lake: the visible fraction (nominally, about 45% but, through selective absorption, nearer 50% at ground level) rarely exceeds $450 \, \mathrm{W\,m^{-2}}$, or roundly $1.25 \times 10^{21} \, \mathrm{photons\,m^{-2}\,s^{-1}}$. Dividing through by Avogadro's number (6.023×10^{23}), the derivation ($0.0021 \, \mathrm{mol\,photons\,m^{-2}\,s^{-1}}$) accurately predicts maximum light measurements reported in the literature.

Not all the light reaching the surface of a lake necessarily penetrates the water column. A proportion is reflected, mainly as a function of the angle of incidence of the light rays (c.2% when the sunlight strikes perpendicular to the water surface; <5% at angles >30°, but increasing steeply towards 100% at progressively lower incidences). The passage of light through a surface ruffled by

wind is, in consequence, extremely complex but, when the sun is low in the sky, surface wind action significantly enhances and extends the period of the significant underwater light field.

Photons that do penetrate the water surface are subject to absorption, though not without a degree of prior scattering. Absorption refers to the process by which molecules capture photons passing nearby. The energy of the molecule increases by an amount corresponding to the energy of the photon. Short-wavelength quanta raise the electronic energy of recipient atoms: electrons are lifted from the ground state to a higher, 'excited' one (whence they are potentially shed to constitute chemical reductive power). The energy of longer-wavelength quanta is insufficient to drive more than transitions among the orbits of individual electrons. Excitation is actually very brief (10^{-9} to 10^{-4} s, depending on wavelength). In this way, most of the light energy absorbed by water is only briefly involved in electron excitation and ends up either as heat or as chemical energy. So far as lake ecosystems are concerned, the latter, when incorporated into photosynthate, is the basis of all biological production.

Even in pure water, absorption of photons is not uniform across the spectrum (Fig. 5.2): photons with wavelengths of about 400–480 nm are least likely to be captured by water molecules; those with wavelengths closer to 700 nm are 30 times more likely to be absorbed. For this reason, water is not colourless at all. Selective absorption in the red leaves natural waters of high purity distinctly blue-green in colour. However, the absorption spectrum can be significantly modified by solutes, especially those derived from humic and fulvic substances arising in the oxidation (mainly) of plant material present in, or on, the soils with which water may come in contact (see Chapter 7). Abundance of these substances is influenced by the nature of the catchment, the type of vegetation present and its refractory products, and seasonal variations in hydrology. Low-latitude, acidic forests, in areas of moderate rainfall and slow percolating drainage, as well as the peatlands of higher latitudes, provide good examples of stained waters. Kirk (1994) cites several Australian waters where

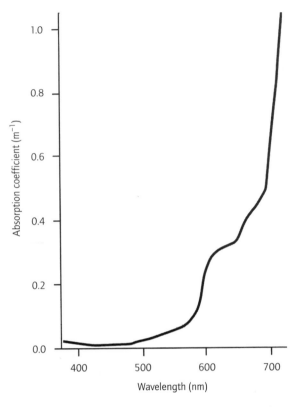

Fig. 5.2 The absorption of visible light (375–725 nm) by pure water. Drawn from data presented by Kirk (1994).

enhanced absorption of blue wavelengths leaves them distinctly yellow in colour.

Although most of the photons entering water are absorbed, many undergo prior scattering: even in pure water, paths of electrons are deflected as they bounce off molecules. Scattering of downwelling photons occurs in all directions, including backwards. Whilst the light is not absorbed any faster, net restriction of the total pathlength to the upper layers before it is absorbed means that scattering does contribute to the enhanced probability that absorption will occur relatively close to the water surface.

The effect is considerably enhanced by suspended particles (**seston**) – inert mineral substances (such as clay, fine silt and other **tripton**), planktonic algae and bacteria and their debris – especially if these are a few orders larger than the wavelength of the downwelling photons. The Secchi Disc, that simple instrument used by limnologists to compare the optical properties of lake waters, both among lakes and, in any given water, through time, is mainly sensitive to scatter. There can be no exact conversion between Secchi Disc measurements and profiles of light penetration, although helpful approximate factors have been proposed many times (but see, especially, Preisendorfer, 1976).

Thus, attenuation of light underwater is roughly the sum of the absorption (which is wavelength-sensitive) and the scattering (which is not). However, it is more usual to consider underwater irradiance – the integral of the unabsorbed light, whether scattered or not, usually referred to as **scalar irradiance** – and the way in which this is diminished with depth. Scalar-irradiance meters may be sensitive to the entire visible spectrum (more or less, the photosynthetically active wavelengths) or, with appropriate filters, to selected wavebands only. Other factors being equal (most, especially, with respect to density and uniformity of suspended particulates), scalar irradiance of a given waveband is diminished exponentially with depth, according to the Beer–Lambert formulation:

$$I_{z2} = I_{z1} \cdot \exp -[\varepsilon\lambda(z_2 - z_1)] \qquad (5.2)$$

where I_{z1} and I_{z2} are the waveband-specific scalar irradiances at two points in the water column, separated vertically by 1 m, and $\varepsilon\lambda$ is the wavelength-specific coefficient of attenuation. Often, the formulation relates an average coefficient to light penetrating the surface (I_0). Thus,

$$I_z = I_0 \cdot \exp -[\varepsilon\lambda z] \qquad (5.3)$$

Owing to differential absorption, attenuation of white light becomes less steep as the fraction penetrating to depth is increasingly made up of wavelengths which are absorbed least rapidly. Nevertheless, it is quite usual to find extinction coefficients quoted in the literature fitted to the near-surface attenuation of scalar irradiance in the full photosynthetically active spectrum. The wavelength specificity is omitted from the equation:

$$I_z = I_0 \cdot \exp^{-\varepsilon z} \qquad (5.4)$$

whence,

$$\varepsilon \approx (\ln I_z - \ln I_0)/z \qquad (5.5)$$

It is recognised that the influence of the different sources of water supplied collectively to natural lakes, on the transparency of the recipient systems, is profound. Reported coefficients of vertical attenuation of photosynthetically active radiation (ε) in the world's clearest lakes varies from 0.06 to 0.18 m^{-1}. At the other extreme, drainage from catchments characterised by slow plant decomposition may be stained brown by humic derivatives (as in several Australian reservoirs; $\varepsilon \approx 1.5$–2.0 m^{-1}), while eroding soils and exposures of unconsolidated deposits may charge the water with clouds of fine yellow or whitish clay (reports place ε between 8 and 20 m^{-1}; resuspension of earlier deposition may contribute substantially to the suspended load at times). The comparison between these influences is scarcely more striking than at the confluence of the ('black water') Rio Negro and the ('white water') Solimões rivers where they meet at Manaus, Brasil, to form the lower Amazon. Where the capacity of chemical nutrients can support it in large concentrations, planktonic biomass itself lends great turbidity to the water: in extremes, absorption by photosynthetic pigments can largely account for values of ε up to c.8 m^{-1}.

In these ways, the hospitability of lake waters to the support of further photoautotrophic biomass can be just as easily constrained by the penetration of light or, at least, by the factors resisting it, as by a more commonly acknowledged scarcity of available nutrients (see Reynolds 1997).

5.4 IMPACTS UPON LAKES AND THEIR LIFE

The behaviour of water masses impounded in lake basins is governed, on the one hand, by the driving environmental variables (solar heat income, mechanical stirring by wind and gravitational forcing and hydrological exchanges; availability of vital solutes, released from the catchment) appropriate to the location (latitude, altitude, aspect, proximity to the ocean, climate and catchment criteria), and, on the other, by the limits to reactivity imposed by the physical properties of water. For instance, the extent, rate and frequency of mixing of lakes represent compromises between the energy of forcing and the inertia incumbent upon high density, high viscosity, poor heat conductance and the buoyant resistance engendered by the coefficient of thermal expansion. These relationships determine if, when and for how long lakes undergo thermal stratification (see Chapters 2 and 6). They interact with the flux of nutrients exported from the catchment or reprocessed in the lake (see Chapter 4) and with the scalar light gradient in determining the proportion of the solar day during which entrained pelagic organisms can be exposed to daylight (see Chapter 9). All the main biota of lakes (see Chapters 10–16) are directly influenced by the Archimedean properties of water, and the stresses incumbent upon its movement; the lives of all of them are touched by the interaction of turbulence and viscosity; all respond to the changes in water temperature and all are sensitive to gas content.

It is easy to become excited by and, in time, to be specialist in certain types of organisms, of microhabitats, or even of selected lakes. Theorists may become fascinated by the behaviour of aquatic ecosystems and their comparability with terrestrial counterparts (I have argued as much elsewhere: see Reynolds, 1997) but it is essential still to recognise that the characteristic constraints of systems themselves frequently relate to those imposed by the physical environment which they inhabit. So far as lakes and limnologists are concerned, the truism is inescapable: the properties of the limnetic environment are substantially governed by the unusual physical properties of water itself.

REFERENCES

Hutchinson, G.E. (1957) *A Treatise on Limnology*, Vol. 1, *Geography, Physics, Chemistry*. Wiley, New York, 1016 pp.

Kirk, J.T.O. (1994) *Light and Photosynthesis in Aquatic Ecosystems.* Cambridge University Press, Cambridge, 528 pp.

Lampert, W. & Sommer, U. (1993) *Limnoökologie.* Georg Thieme Verlag, Stuttgart, 440 pp.

Morel, A. & Smith, R.C. (1974) Relation between total quanta and total energy for photosynthesis. *Limnology and Oceanography*, **19**, 591–600.

Preisendorfer, R.W. (1976) *Hydrological Optics.* United States Department of Commerce, Washington, DC, 382 pp.

Reynolds, C.S. (1997) *Vegetation Processes in the Pelagic: a Model for Ecosystem Theory.* Ecology Institute, Oldendorf, 372 pp.

Rothschild, B.J. & Osborn, T.R (1988). Small-scale turbulence and plankton contact rates. *Journal of Plankton Research*, **10**, 465–474.

6 The Motion of Lake Waters

DIETER M. IMBODEN

6.1 THE ECOLOGICAL RELEVANCE OF FLUID MOTION

Water is the most ubiquitous fluid on Earth. It provided the stage for the development of life, and also symbolises life's central paradigm: constant change. 'Panta rhei' ($\pi\alpha\nu\tau\alpha$ $\rho\varepsilon\iota$)* —'everything flows', as **Heraclites** cites, an unknown Greek philosopher. According to classic theory, a liquid possesses neither shape nor structure. Instead, it is envisioned as an amorphous mass, a continuum, the dynamics of which can be characterised by a few spatial parameters such as density $\rho(\mathbf{r},t)$, temperature $T(\mathbf{r},t)$, pressure $p(\mathbf{r},t)$ and the velocity vector field $\mathbf{u}(\mathbf{r},t)$.† The latter can be interpreted as a combination of three scalar fields, e.g., the three velocity components of the Euclidean coordinate system, $u_x(\mathbf{r},t)$, $u_y(\mathbf{r},t)$ and $u_z(\mathbf{r},t)$.

Although the 'continuum approach' may be sufficient to describe the flow field of a large water body such as an ocean or a large lake, it misses essential features at the microscopic level that are important to chemical and biological processes. Here the molecular nature of matter comes into play. In fact, the liquid phase of matter combines characteristic properties of the solid and the gaseous phase. In the former, the movement and position of atoms are coordinated over large distances such as to bring about the characteristic crystalline structures of solids. In the latter, the interactions of the atoms and molecules are assumed to be confined to the instance of collision between two gas molecules. Molecules in the liquid phase are neither bound to fixed positions (as in the lattice of solid state) nor are they completely free (as in the gaseous phase); they can move rather freely but feel some local interaction. This makes liquids more viscous than gases and fairly incompressible.

It is impossible to understand the motion of water without distinguishing between different time and space scales of motion. At the small end of the spectrum one finds the thermal motion of individual water molecules. Although thermal molecular velocities are very large, the average distance over which a molecule travels before changing direction owing to interaction with other molecules is very small. Thus, the coefficient of molecular diffusivity D in water, roughly the product of mean velocity times mean free path, lies typically between 10^{-6} and 10^{-5} $\mathrm{cm^2\,s^{-1}}$, i.e. is small compared with, e.g., diffusivity in air. In 1827, the botanist Robert Brown observed through the microscope small particles (such as spores) suspended in water or other liquids performing peculiar random movements. It took 80 years before Einstein (1905) and Smoluchowski‡ were able to explain quantitatively so-called Brownian motion as the effect of the thermal motion of the fluid particles, which pushes the macroscopic particle around.

At the other end of the spectrum lie current patterns which extend over the whole lake. In the oceans, these are the large gyres and related currents, such as the Gulf Stream, which keep their structure over decades. Usually, lake-wide currents do not persist over such timescales since most lakes are small compared with the typical at-

---

* The Greek word $\rho\varepsilon\varepsilon\iota\nu$ (to flow) reappears in the word 'rheology' – the theory of liquids.
† $\mathbf{r}$ is the position vector expressed, e.g., by the Cartesian coordinates $r = (x,y,z)$, and t is time. In the following, bold letters will be used for vectors.

‡ Einstein and the Polish physicist Smoluchowksi found this law independently, hence the name 'law of Einstein and Smoluchowski', but only Einstein's publication is available.

mospheric pressure and wind field patterns and thus experience the variability of the driving forces more strongly.

In between the molecular and the 'whole-system' scales exists a continuous size spectrum of motion (see section 6.3). Yet with increasing scale a fundamental difference between the horizontal and vertical axes emerges. At the small end of the spectrum, motion is **isotropic** (i.e. equal in all spatial directions). With increasing size motion turns more and more into the horizontal direction, since lakes are usually much broader than they are deep. Often, vertical density gradients additionally suppress vertical motion relative to horizontal (section 6.4).

Owing to the omnipresence of motion and mixing, lakes and oceans provide aquatic organisms with an environment of remarkable physical and chemical homogeneity. On one hand, 'whole-system' motion is responsible for the transport of nutrients, biomass and other constituents from input areas (e.g. inlets) to consumption areas. On the other hand, small-scale and molecular motion is responsible for the exchange of matter between living cells and the 'bulk' water. Since light intensity strongly decreases with water depth, differentiation of conditions of life occurs mainly along the vertical axis (see also Chapter 5). The special role of the vertical versus the horizontal direction is further amplified by anisotropic mixing.

In comparison, it was necessary for life on land to adapt to a much wider spectrum of spatial and temporal variation (Imboden 1990): the supply of water varies from too much to too little. If nutrients on land become rare, they cannot be replenished as easily as in water, where currents provide for rapid distribution. Competition for light is more sophisticated in terrestrial ecosystems; only large plants can afford the enormous investment in structure which enables a leaf to enjoy the full sunlight high up in the canopy of the tree.

This is not the case in water: here gravity is compensated for by buoyancy. Even the smallest organism can make it to the top. This explains why large plants are rare in aquatic ecosystems.

The group of organisms which makes for the largest proportion of biomass in waters was named **plankton**, the ones which 'are pushed around'. In the ocean, planktonic organisms are responsible for roughly the same total primary production as all terrestrial plants combined, but the total biomass of the former is only about 1‰ of the biomass on land. In fact, the greater homogeneity of aquatic versus terrestrial living conditions, for which the permanent motion of the water is mainly responsible, is also reflected in the relatively small number of aquatic plant species. In Table 6.1 the conditions for plant life in pelagic waters are compared with the conditions on land.

It is impossible to compress the entirety of knowledge about the motion of water in lakes into a single chapter. Thus, the following considerations will intentionally omit many important aspects of lake physics. The selection adopted reflects both the personal experience and preference of the author, as well as an attempt to complement existing textbooks on limnology with more recent findings on the physics of water, and how they refer to the chemistry and biology of lakes. The approach chosen follows the classical physical tradition: the lake is considered as a system under the influence of external forces.* Insight into the **internal structure** and property of the system is gained by analysing its response to the variation of the **external forces**.

In contrast to the field of physical oceanography, there is no integral book on physical limnology, except perhaps for the second volume of F.A. Forel's classic monograph *Le Léman* (Forel, 1895) and the rather voluminous chapters in G.E. Hutchinson's *A Treatise on Limnology* (1957). As a guide for the interested reader, a few articles and books are mentioned which together yield a fairly complete picture of the field: Hutter (1983), Imberger & Patterson (1989), Imboden (1990), Hutter (1993), Imberger (1994), Imboden & Wüest (1995) and Massel (1999).

---

* Here, the word 'force' stands for every external influence. For instance, the chemical composition of an inlet exerts a physical impact on the lake via the density of the inflowing water.

Table 6.1 Conditions for life in pelagic waters and on land. (Adapted from Imboden 1990.)

Factor	Open water	Land
Factors related to the physical properties of water versus air		
Gravity	**Small** or absent (buoyancy)	**Large**: supporting structures needed (trees)
Light	Low transparency of water	High transparency of air
Temperature variation	Moderate	Often large or even extreme
Factors related to water motion		
Control over position	**Limited**. Control by buoyancy or active swimming	**Normal**. By roots, attachments, etc.
Nutrients	Usually growth limiting, but easily distributed within system	Local inhomogeneities coexist (fertile/ barren land)
Spreading of species	Easy, by water currents	Usually difficult: easy by 'tricks' only (seeds transported by animals, through air, etc.)
Light variation for individual	Large and unpredictable (depending on position)	Slow and regular
Ecological key numbers for aquatic flora		
Ecological niches	Few	Many
Number of plant species in total ecosystem*	$\approx 40,000$	$\approx 400,000$
Average standing crop of living biomass†	$2\,g\,C\,m^{-2}$	$6000\,g\,C\,m^{-2}$
Total primary production†	$50 \times 10^{15}\,g\,C\,yr^{-1}$	$40 \times 10^{15}\,g\,C\,yr^{-1}$
Primary production per area†	$150\,g\,C\,m^{-2}\,yr^{-1}$	$250\,g\,C\,m^{-2}\,yr^{-1}$
Productivity per biomass	$75\,yr^{-1}$	$0.04\,yr^{-1}$

* From Flindt (2000).

† Based on values by Bolin & Cook (1983).

6.2 EQUATIONS OF MOTION AND DRIVING FORCES

In this section we address the question of how external forces set water in motion, and why water is so easily moved.* Obviously, internal friction of water (**viscosity**) is extremely small compared with friction between solids. Early in its history, humankind became aware that hauling a floating object through water needs much less energy than dragging the same mass over land. This remained true even after the invention of the wheel, and explains why waterways have played such an important role for transportation of goods (and still do). Yet, even motion in water is not completely without loss, so there must be external forces which drive that motion. It will come as no great surprise to find out that these forces are primarily of solar origin, but the different ways in which solar energy is transformed into water movement can be fairly intricate.

In order to illustrate these mechanisms, a rectangular volume of water with sides of lengths Δx, Δy and Δz is considered (Fig. 6.1). Three kinds of forces are acting on the water volume: (i) body (or volume) forces, (ii) forces perpendicular to the surfaces and (iii) forces tangential to the surfaces. In the first group we find gravitation, but also a pseudo-force called the **Coriolis force**, which originates from the rotation of the Earth. Since water currents are conveniently measured relative to the

* The gas phase is even more 'motion-friendly' than the liquid phase, since energy loss by internal friction (viscosity) is even smaller. The fact that Earth contains two well-developed fluid systems, the atmosphere and the ocean, and that these are rather mobile, plays a key role in understanding the biogeochemical evolution of our planet.

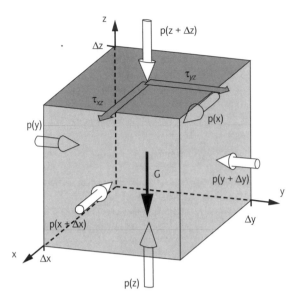

Fig. 6.1 Forces acting on a rectangular water parcel with dimensions Δx, Δy and Δz. Cartesian coordinates x and y are horizontal, and z is vertical (upwards). Three kinds of forces act on the water parcel: (i) gravitation G is a body (or volume) force (black arrow); (ii) pressure p (white arrows) is a surface force that acts perpendicularly to the surface; (iii) shear $\tau_{\alpha\beta}$ (grey arrows) is a tangential surface force. The first subscript of $\tau_{\alpha\beta}$ indicates the direction of the force, the second the orientation of the surface on which it acts, e.g. τ_{xz} acts in the x-direction, on the plane perpendicular to the z-axis.

(rotating) surface of the globe, i.e. relative to a **non-inertial** reference system, a mass on which no (real) force is acting and which thus should move along a straight line seems to be deflected relative to the rotating surface. For an observer on Earth this deflection is interpreted as originating from a hidden force, the Coriolis force. This force is essential for understanding the dynamics of the atmosphere and the ocean, but plays only a minor role in small lakes (see section 6.7).[*]

Pressure is the force acting perpendicular to the surface of the water volume. Equal pressure forces

[*] There may be other body forces (e.g. electromagnetic forces) acting on charged particles, but they are not considered here.

on opposite surfaces compensate for each other, but pressure **gradients** cause an acceleration of the water. As an example it is assumed (Fig. 6.1) that pressure on the surface perpendicular to the x-axis at position x, $p(x)$, is larger than pressure on the opposite surface, $p(x + \Delta x)$. Then the total force on the water cube with mass $m = \rho \cdot \Delta x \cdot \Delta y \cdot \Delta z$ points into the positive x-axis and has the size $F_x = \Delta y \cdot \Delta z \cdot [p(x) - p(x + \Delta x)]$. According to Newton's Law, this causes an acceleration of the water parcel in the positive x-axis of size $\partial u_x / \partial t = F_x / m$, where u_x is the flow velocity in the x-direction. Thus, dividing by $(\Delta x \cdot \Delta y \cdot \Delta z)$ and taking the limit $\Delta x \to 0$ yields:

$$\frac{\partial u_x}{\partial t} = \frac{1}{\rho} \frac{[p(x) - p(x + \Delta x)]}{\Delta x} = -\frac{1}{\rho} \frac{\partial p}{\partial x} \qquad (6.1)$$

Note that the acceleration points in the direction of the negative pressure gradient. Similar expressions hold for the other two coordinate axes. The spatial variation of pressure along the **vertical** axis is approximately equal to the increase of the hydrostatic pressure with depth:

$$\left(\frac{dp}{dz} \right)_{\text{hydrostatic}} = -g\rho \qquad (6.2)$$

The minus sign is a consequence of the orientation of the coordinate system with the vertical axis (z) pointing upwards (Fig. 6.1). For constant water density ρ the solution of eqn. 6.2 is

$$p(z) = p_0 - g\rho z \qquad (6.3)$$

where p_0 is the atmospheric pressure and $z = 0$ at the water surface. If eqn. 6.2 holds, the corresponding vertical acceleration is exactly compensated for by the acceleration of gravity, g, so there is no acceleration in the vertical (hydrostatic approximation).

The third kind of force acting in parallel to the surfaces of the cube is called **shear**. At molecular level, shear results from the viscosity of the water. For instance, if two adjacent, vertically separated water parcels move in the same direction, but with different horizontal velocity u_x, the shear τ_{xz} (expressed as force per unit contact area) is given by

$$\tau_{xz} = \mu \frac{\partial u_x}{\partial z} \qquad (6.4)$$

where μ is the dynamic viscosity of water. The first subscript of τ_{xz} refers to the direction of the force (the x-axis), the second to the orientation of the surface on which it acts (perpendicular to the z-axis). Were it only for molecular viscosity, it would be difficult to propagate momentum, e.g., from the lake surface to deeper parts of the water column. Yet, the small-scale fluctuations of current velocity, which are typical for turbulent flow, lead to an additional and more efficient shear force that can be described by a modified version of eqn 6.4 in which μ is replaced by the so-called eddy viscosity K^m.

The combined effect of all forces leads to the following dynamic equation for the **moving** water volume ($v = \mu/\rho$: kinematic viscosity):

$$\left(\frac{du_\alpha}{dt} \right)_{\text{Lagrange}} = -\delta_{\alpha z} g - \frac{1}{\rho} \frac{\partial p}{\partial x_\alpha}$$

$$+ v \sum_{\beta = x,y,z} \frac{\partial^2 u_\alpha}{\partial x_\beta^2}; \quad \alpha = x, y, z \qquad (6.5)$$

where $\delta_{\alpha z} = 1$ for $\alpha = z$ and $\delta_{\alpha z} = 0$ otherwise. The index 'Lagrange' on the left-hand side points to the particular system of reference, the so-called **Lagrangian** system which moves along with the current. Usually, quantities such as current velocities, water temperature, etc. are recorded at a point which is **fixed** relative to the coordinate system. This is called the **Eulerian** representation of the equation of motion, and the corresponding acceleration is usually written as $\partial u_\alpha / \partial t$. As shown in textbooks on fluid dynamics (for a simple introduction see, e.g., Paterson 1983), the two reference systems are related by the transformation

$$\left(\frac{dG}{dt} \right)_{\text{Lagrange}} = \frac{\partial G}{\partial t} + \sum_{\beta = x,y,z} u_\beta \frac{\partial G}{\partial x_\beta} \qquad (6.6)$$

where G stands for any variable, e.g. for u_α. Thus combining eqns. 6.5 and 6.6 yields the **Navier–Stokes Equations** for viscous, incompressible flow relative to a non-rotating reference system:

$$\frac{\partial u_\alpha}{\partial t} + \sum_{\beta = x,y,z} u_\beta \frac{\partial u_\alpha}{\partial x_\beta}$$

$$= -\delta_{\alpha z} g - \frac{1}{\rho} \frac{\partial p}{\partial x_\alpha} + v \sum_{\beta = x,y,z} \frac{\partial^2 u_\alpha}{\partial x_\beta^2} \qquad (6.7a)$$

$$\sum_{\alpha = x,y,z} \frac{\partial u_\alpha}{\partial x_a} = 0 \qquad (6.7b)$$

Equation 6.7b is the continuity equation for incompressible flow. Additional terms are needed in eqn. 6.7a in order to describe the effect of Earth's rotation (see e.g. Pedlosky 1979).

The fundamental difference between the above equation and Newton's equation of motion for a rigid mass lies in the non-linear terms on the left-hand side of eqn. 6.7a. Essentially, these terms, in combination with system-specific boundary conditions and external forces, are responsible for the great variety of patterns of fluid motion to be observed in nature, from the ripples evoked by a stone thrown into a lake to the enormous transport of water by the Gulf Stream.

Driving forces which keep the water in motion act mainly at the water surface. They comprise the surface shear of the wind field, which produces surface waves but also piles water up at the downwind end of the lake (see Fig. 6.10). Rivers import kinetic energy. If the densities of river and lake water are different (as is usually the case), density gradients are induced at the river mouth which cause pressure gradients, and thus drive water motion (see Fig. 6.6).

A second kind of external force is related to the exchange of thermal energy at the water surface. From an energetic point of view, **heat flux** is much larger than the energy input by wind shear (Table 6.2), but its mechanical efficiency to set water in motion is small. Heat flux gives rise to local density changes in the water. These will be discussed in more detail in section 6.4.

Table 6.2 Typical external energy fluxes and energy content in a lake.

Wind (W_{10})		Weak (1 m s^{-1})	Strong (10 m s^{-1})
Wind stress coefficient C_{10}, eqn. 6.33		4×10^{-3}	1.5×10^{-3}
Wind shear τ_w, eqn. 6.31 (N m^{-2})		5×10^{-3}	0.2
Energy input (W m^{-2}):			
$\quad$10 m above water P_{10}, eqn. 6.34		5×10^{-3}	2
$\quad$to internal seiches P_{seiche}, eqn. 6.38*		$(1-2) \times 10^{-5}$	$(2-4) \times 10^{-2}$
$\quad$to surface waves P_{wave}†		1×10^{-3}	0.4
$\quad$to turbulence P_{turb}‡		5×10^{-5}	2×10^{-2}
Mechanical energy content (J m^{-2}):			
$\quad$internal seiches E_{seiche}		0.1–1	10^{-2}–10^{-3}
$\quad$surface waves E_{wave}		10	10^3
$\quad$turbulent E_{turb}		10^{-3}	1
Energy dissipation (W kg^{-1}):			
$\quad$in wave zone ε		10^{-8}	10^{-5}
$\quad$below wave zone ε		10^{-9}–10^{-8}	10^{-6}–10^{-5}
Kinetic energy input by rivers per unit area§ $P_{kin,River}$ (W m^{-2})	10^{-5}–10^{-3}		
Thermal energy flux at surface H_{net} (W m^{-2}):			
$\quad$annual mean	$\pm(60-120)$		
$\quad$daily maxima	$\pm(150-300)$		
Annual variation of thermal energy content ΔE_{th} (J m^{-2})	$(1-2) \times 10^9$		
Potential energy relative to mixed water column ΔE_{pot} (J m^{-2})	$-(0-10^2)$		
$\quad$in saline lakes	$-(0-10^4)$		
Surface buoyancy flux from	heat flux¶	evaporation**	river††
$\quad J_b$ (W kg^{-1})	10^{-8}–10^{-7}	$<10^{-8}$	$<10^{-7}$
$\quad P_{pot}$ (W m^{-2})	10^{-4}–10^{-3}	$<10^{-4}$	$<10^{-3}$

* 20% of P_{10} valid during initial phase of wind event (duration about one-quarter of period of first internal seiche mode).
† Until wave field is fully developed.
‡ Of the order of 1% of P_{10} (Denman & Miyake 1973).
§ For hydraulic loading q of 10^{-7} to 10^{-5} m s^{-1}, and river flow capacity of c. 0.5 m s^{-1}.
¶ See Table 6.4.
** See Table 6.4, $h_{mix} = 10$ m, salinity 35‰ (sea water), annual evaporation rate = 1 m yr^{-1}.
†† See Table 6.4; larger values in some lakes.

A third source of kinetic energy in a standing body of water* is due to **internal** changes of water density caused, e.g., by the exchange of heat or dissolved substances at the sediment–water interface (geothermal heat flux, redissolution of solutes, etc.), or by internal mixing of water masses of different salinity and temperature. These processes will be discussed in section 6.4.†

* Here, the distinction between a standing water body (lake, ocean) and a body of running water (river) becomes important, since in the latter water which is running continuously downhill gains kinetic energy from potential energy. In most rivers, this is by far the most important motive force.

† In fact, the largest change of water density is due to the compressibility of water, but only its variation with temperature can

Typical orders of magnitude for the various driving forces are summarised in Table 6.2. The basic concepts on which these numbers rely will be presented in the following sections. In most cases, the wind is the most important source of kinetic energy. Above the water surface, the corresponding energy flux is given by P_{10} which grows as the third power of the wind velocity (section 6.5) and lies mostly between 10^{-3} and $1\,\mathrm{W\,m^{-2}}$. Only a small fraction of P_{10} (20% or less) is fed into the different kinds of water motion (P_{seiche}, P_{wave} and P_{turb}; see section 6.5). In some lakes with extremely large hydraulic loading q (input of water per unit lake area and time) the kinetic energy input by inlets (normalised by lake area) may exceed the influence of the wind ($P_{\mathrm{river}}=$ up to $10^{-3}\,\mathrm{W\,m^{-2}}$), but its effect is obviously more localised.

Kinetic energy is primarily stored in organised current patterns such as wave motion (E_{wave}), and large-scale current systems that extend over the whole lake (E_{seiche}). From these currents, a small fraction continuously 'trickles' down to small-scale turbulent motion and, after a typical lifetime of the order of 10^{-3} to $10^{3}\,\mathrm{s}$, is dissipated into heat. When wind shear is large, turbulent kinetic energy (E_{turb}) is also directly produced by breaking waves at the water surface and in the interior of the water column.

Compared with kinetic energy, thermal energy fluxes and contents are larger by several orders of magnitude. Daily mean values of net thermal energy flux reach peak values of $\pm300\,\mathrm{W\,m^{-2}}$. The variation of annual heat energy content is of the order $10^{9}\,\mathrm{J\,m^{-2}}$. Related to seasonal changes of water temperature and salinity is the potential energy stored during stratification of lakes. The corresponding energy variation ΔE_{pot} is typically up to $10^{2}\,\mathrm{J\,m^{-2}}$, i.e. of similar magnitude as the kinetic energy content. In saline lakes it may reach values of the order of $10^{4}\,\mathrm{J\,m^{-2}}$.

Changes of potential energy content affect the kinetic energy balance of lakes. If, in a stratified water column, density is increasing at the surface, the available potential energy is transformed into

trigger water motion, and not the density changes due to compression as such (see section 6.4).

kinetic energy, and water begins to sink spontaneously. The liberation of potential energy is called a (positive) surface buoyancy flux. If expressed as the production of kinetic energy per unit time and **per unit water mass**, the flux is denoted as J_{b}^{0} (section 6.6). If related to **lake area**, the buoyancy flux, denoted as P_{pot}, can be directly compared with the other energy fluxes. During strong cooling in warm lakes, intensive evaporation in saline lakes or under the influence of a large inlet which discharges water with a density exceeding that of the main lake, the buoyancy flux dominates the production of kinetic energy (Table 6.2). In contrast, negative buoyancy (i.e. the upward transport of mass) feeds on kinetic energy, and so decelerates the motion of water. This effect occurs either at the surface (if water density decreases) or in the interior of a stably stratified water column undergoing turbulent mixing. The interplay between the different kinds of energy present in lakes will be further discussed in section 6.6.

6.3 SCALES OF MOTION AND THEIR EFFECTS ON TRANSPORT

Figure 6.2a shows a short section of the temporal variation of the horizontal current speed, $|u_{\mathrm{h}}|=(u_{x}^{2}+u_{y}^{2})^{1/2}$, recorded by two acoustic current meters moored in Alpnachersee, a small basin of Lake Lucerne (Switzerland). One instrument was positioned at 1.6 m depth, the other at 27.3 m, about 2.5 m above the lake bottom. The frequency spectra of the square of the horizontal current are shown in Fig. 6.2b. If multiplied by $(\rho/2)$ the spectra give the kinetic energy density per unit volume. The left portion of the spectra (region I, 10^{-5} to $10^{-3}\,\mathrm{s^{-1}}$) was calculated from a time series of 1 week with 5-minute sampling intervals, and the high frequency portion (region II) from several recordings with 0.25-s sampling intervals each of about 8 min duration.

The spectra exhibit an overlap of several peaks separated by more continuous slopes. Straight lines showing different slopes are added for comparison. At the high-frequency end, the

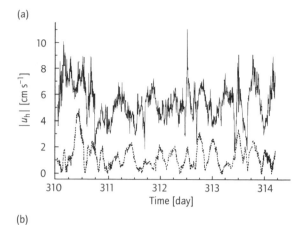

(a)

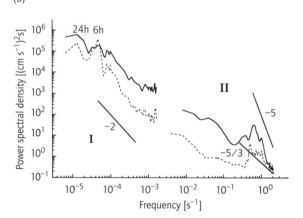

(b)

Fig. 6.2 (a) Section of time series of horizontal current speed, $|u_h| = (u_x^2 + u_y^2)^{1/2}$, recorded in November 1996 by two acoustic current meters moored in the Alpnachersee, a small basin of Lake Lucerne (Switzerland). One instrument (continuous line) was positioned at 1.6 m depth, the other (dashed line) at 27.3 m, about 2.5 m above the lake bottom. (b) Power spectra of u_h^2, calculated from the two current meter recordings. Low-frequency spectra (10^{-5} to $10^{-3}\,\mathrm{s}^{-1}$) were calculated from a record of 1 week, with 5-min sampling interval; high-frequency spectra (up to $2\,\mathrm{s}^{-1}$) form several data segments during which the sampling interval was 0.25 s. Straight lines with slope indicated refer to theoretical concepts which may be relevant for certain parts of the spectrum (see text). (Unpublished data by André Simon, collected for his dissertation (Simon, 1997).)

spectrum of the surface currents reaches a distinct peak at about $0.6\,\mathrm{s}^{-1}$, followed by a sharp fall with slope -5 which is indicative of surface waves (Kitaigorodskii 1983). In the same frequency range, the lower current meter records significantly less energy, and the spectrum falls at a rate of $-5/3$, the value calculated from the Batchelor inertial subrange of isotropic turbulence (Batchelor 1959). Between 10^{-4} and $10^{-3}\,\mathrm{s}^{-1}$, the slope is about -2, the value derived from the spectrum of internal waves (Garrett & Munk 1975).

The spectrum at the low-frequency end records the signal from standing internal waves (internal seiches). Some years earlier, similar peaks were observed by Münnich *et al.* (1992) in the vertical oscillation of the isotherms measured in the same lake (Fig. 6.3b). The first peak from the left corresponds to a second vertical mode.[*] Since its periodicity is close to 24 hours, this otherwise rare mode is excited by resonance with the diurnal wind field (Fig. 6.3a). In contrast, there is no counterpart in the wind field of the period of the first vertical mode, at about 8 hours. Basin-wide motion is further discussed in section 6.7.

The energy content of subsegments of Fig. 6.2b can be estimated by integration. As it turns out, for moderate winds ($W < 4\,\mathrm{m\,s}^{-1}$), most of the energy (about $100\,\mathrm{J\,m}^{-2}$ in the upper spectrum) is concentrated in the high-frequency range ($0.4\,\mathrm{s}^{-1} < f < 2\,\mathrm{s}^{-1}$).[†] This is in accordance with the relative size of the energies listed in Table 6.2 ($E_{\mathrm{wave}} > E_{\mathrm{seiche}}, E_{\mathrm{turb}}$).

The high-frequency velocity variation is a result of small-scale patterns of turbulent motion that pass by the space-fixed current meter with mean advection velocity u_a. According to Taylor's 'frozen turbulence hypothesis', the length scale L of such a structure is approximately

---

[*] Second vertical modes can be visualised by means of three layers, whereby the two faces oscillate against each other such as to produce a thin and a thick intermediate layer.

[†] Note that the logarithmic frequency scale hides the fact that integration in region II extends over a very large frequency segment, so that the integral is larger than in region I, in spite of the greater spectral densities.

$$L = \frac{u_a}{f} \qquad (6.8)$$

Because of internal friction, such structures lose their kinetic energy fairly quickly if their size falls below a critical value called the **Kolmogorov length scale** (L_K; Kolmogorov 1941). The value of L_K is determined by the flux of energy through the system (commonly expressed in terms of the rate of turbulent kinetic energy dissipation ε, section 6.6) and by internal friction (expressed by the kinematic viscosity ν):

$$L_k = \left(\nu^3 / \varepsilon \right)^{1/4} \qquad (6.9)$$

where L_K typically lies between 10^{-3} and 10^{-2} m. According to eqn. 6.8 and with $u_a \approx 0.1\,\mathrm{m\,s^{-1}}$, the Kolomogorov cut-off frequency is of the order 10 to $100\,\mathrm{s^{-1}}$. Note that the smallest turbulent structures are not resolved in Fig. 6.2 since the sampling rate of the current meter was too small.

Figure 6.2b gives the impression that a lake is filled with a complex mixture of current structures, some of them organised as waves, others of pure random nature. How may this complexity be addressed by simple mathematical concepts? Obviously, it is neither possible nor meaningful to describe water motion and transport of solutes in lakes by making **direct** use of the information contained in the current recordings given in Fig. 6.2a. For instance, as a counterpart to the high temporal resolution, one would need the corresponding spatial information from thousands of current meters. But even with all these data, one could not calculate, or even predict, the 'exact' current field, since the non-linearity of the equations of fluid motion (eqn. 6.7) always preserves some unresolved randomness.

Therefore, by the usual method introduced by Reynolds, small-scale variation is treated in a simplified quasi-statistical way by splitting any variable $G(t)$ into a temporal mean $\overline{G}$ and an instantaneous fluctuation G':

$$G(t) = \overline{G}(t) + G'(t) \qquad (6.10a)$$

$$\text{where} \quad \overline{G}(t) = \frac{1}{\tau} \int_{t-\tau/2}^{t+\tau/2} G(s)\,\mathrm{d}s$$

$$\text{and} \quad \overline{G}' = \frac{1}{\tau} \int_{t-\tau/2}^{t+\tau/2} G'(s)\,\mathrm{d}s = 0 \qquad (6.10b)$$

and s is an integration variable. The specific choice of the averaging interval τ determines the boundary between diffusion and advection.

As an example, the one-dimensional transport of a solute of concentration C in a turbulent current field is calculated. The flux of solute per unit area and per unit time, F_x, is given by:

$$F_x = F_x^a + F_x^d = u_x C - D \frac{\partial C}{\partial x} \qquad (6.11)$$

F_x^a describes transport by advection, and F_x^d is the First Fick's Law for diffusive transport, with D the molecular diffusivity of the solute in water. By splitting u_x and C according to eqn. 6.10a and taking the temporal mean, the advective flux F_x^a is:

$$F_x^a = \overline{u}_x \overline{C} - \overline{u_x' C'} \qquad (6.12)$$

According to the standard method the second term on the right-hand side of eqn. 6.12 is approximated by the gradient-flux closure scheme

$$\overline{u_x' C'} = -K_c \frac{\partial \overline{C}}{\partial x} \qquad (6.13)$$

where K_c is turbulent diffusivity of mass. Since eqn. 6.13 has the same mathematical form as F_x^d, and K_c is usually much larger than D, molecular diffusion is usually subsumed under K_c. Then eqn. 6.11 becomes

$$F_x = \overline{u}_x \overline{C} - K_c \frac{\partial \overline{C}}{\partial x} \qquad (6.14)$$

where the two right-hand terms describe transport by large-scale advection and by turbulence, respectively. Remember that the distinction between the two modes of transport implicitly depends on the choice of τ in eqn. 6.10b. As the spectra of Fig. 6.2 demonstrate, there is no distinct frequency which would offer itself as the natural dividing line between large-scale (or advective) and small-scale (or turbulent) motion. In reality, the choice of the proper cut-off time must be made individually for each case, based on the system under consideration, the kind of question to be analysed, and the available data set. Often such a choice is made inadvertently, e.g. by choosing a specific tracer in order to determine mixing characteristics in a lake. For instance, the radioactive isotope ^{222}Rn can be used to quantify vertical turbulent mixing near the lake bottom (Imboden & Joller 1984). Because of its half-life of 3.8 days, ^{222}Rn is only sensitive to current patterns with a maximum timescale of about 1 week. Most of the isotope decays before patterns of longer duration can evolve.

The impact of the two kinds of transport (advection and turbulent diffusion) on the distribution of solutes and suspended particles (e.g., plankton) varies. The distance of advective transport L_a increases linearly with time

$$L_a = u_a t \qquad (6.15)$$

where $u_a = \overline{u}_x$ is the flow velocity. For $u_a = 5\,\mathrm{cm\,s^{-1}}$, L_a is about 4 km after 1 day, provided that the current remains constant and does not change direction.

Diffusion is more complex. Provided (i) that the velocities responsible for the diffusive process are normally distributed around the mean, $\overline{u}_{\mathrm{diff}}$, and (ii) that the mean free path λ of the diffusing objects (e.g. dissolved molecules in water, helium atoms in air, etc.) is small relative to the size of the patch, then a 'point-like' initial distribution is transformed into a normal distribution with variance σ^2 which is increasing with time according to

$$\frac{d\sigma^2}{dt} = \lambda\overline{u}_{\mathrm{diff}} = 2D \qquad (6.16)$$

where

$$D = \frac{1}{2}\lambda\overline{u}_{\mathrm{diff}} \qquad (6.17)$$

is diffusivity. The above conditions are met, for instance, for molecular diffusivity in air, since, in this case, λ is extremely small. Then, assuming that the initial variance $\sigma^2(0)$ is zero, integration of eqn. 6.16 yields the famous relationship of Einstein–Smoluchowski (Einstein 1905):

$$\sigma = L_d = (2Dt)^{1/2} \qquad (6.18)$$

For molecular diffusion, D is of the order $10^{-9}\,\mathrm{m^2\,s^{-1}}$. Thus, for $t = 1$ second, 1 day and 1 year, respectively, L_d is about $4 \times 10^{-3}\,\mathrm{cm}$, 1 cm and 25 cm. This transport is fairly slow, especially for long time periods. For instance, in 1 day the average travelling speed by diffusion—if measured along the line that connects the initial and end points—is just $3 \times 10^{-4}\,\mathrm{cm\,s^{-1}}$. Thus the slightest macroscopic motion of the fluid clearly outruns the effect of molecular diffusion.

The Peclet number, Pe, is twice the ratio of the transport time by diffusion and advection, respectively, for a given distance L.[*] Solving eqns. 6.15 and 6.18 for t and putting $L_a = L_d = L$ yields

$$Pe = 2\frac{t_d}{t_a} = \frac{Lu_a}{D} \qquad (6.19)$$

If $Pe \gg 1$, transport by advection outruns transport by diffusion. Obviously, molecular motion is only relevant for very small distances L or if macroscopic motion is suppressed (u_a is zero or very small). Such conditions are met at boundaries such as the sediment–water or air–water interface, but also at the surface of an algal cell. The transport of nutrient molecules from the water into the cell ultimately occurs by molecular motion. For a spherical cell radius r, the transport time from 'open' water to the cell is proportional to r^2/D (see chapter 19 of Schwarzenbach et al. 2003)—one reason why it may be advantageous to be small!

Turbulent diffusion coefficients are usually much larger than molecular coefficients and can

---

[*] The factor 2 is adopted since, for the definition of non-dimensional numbers such as Pe, numerical factors are usually omitted.

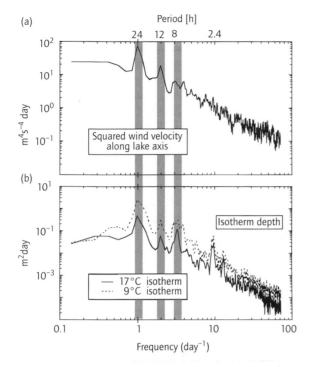

Fig. 6.3 (a) Power spectra of squared wind velocity W^2 (proportional to the wind stress τ_w, see eqn. 6.4; (b) power spectra of vertical displacement of the isotherms calculated from temperature recordings at several depths in the Alpnachersee during summer 1989. The stippled zones indicate the frequency of the standing internal waves (internal seiches). The first peak from the left, a second vertical mode, is close to the 24-hour period of the diurnal wind field. Owing to resonance with the wind, this mode, which is not very common is other lakes, is a distinct feature of the Alpnachersee. (From Münnich *et al.* 1992; reproduced by permission of the American Society of Limnology and Oceanography.)

thus become relevant also for large distances. Yet, as shown in Fig. 6.3, in contrast to molecular diffusion, turbulent diffusion is caused by a mixture of current structures of very different dimension. Thus, if K_c is operationally defined by the growth of a tracer patch size (see eqn. 6.16)

$$K_c = \frac{1}{2}\frac{d\sigma^2}{dt} \qquad (6.20)$$

then we expect K_c to increase with patch size σ^2. The situation is illustrated in Fig. 6.4: the larger the patch, the larger the eddy structure which, rather than moving the patch as a whole, tears it apart. Since the patch size is calculated from the variance relative to the centre of mass, the effect of the larger eddies is interpreted as advection while the smaller structures act as diffusion.

Okubo (1971) has combined results from several tracer diffusion experiments in the ocean into a diagram that relates horizontal diffusivity K_h to length scale L. He found the empirical relation

$$K_h = \alpha L^\eta \qquad (6.21)$$

with η between 1 and 4/3. In the former case, α adopts the meaning of a turbulent velocity $\bar{u}_{diff}$ (eqn. 6.17). The Peclet number would then be independent of scale $(Pe = u_a/\bar{u}_{diff})$. Okubo's diagram suggests that $\bar{u}_{diff} \approx 10^{-3}\,\mathrm{m\,s^{-1}}$.

A slope of $\eta = 4/3$ would be expected for the Batchelor spectrum (Batchelor 1952) describing isotropic turbulence in the inertial subrange. Dimensional considerations for this case yield $\bar{u}_{diff} = $ constant $\varepsilon^{1/3}L^{1/3}$ and $\lambda = $ constant L, thus from eqn. 6.17.

$$K_c = \text{constant } \varepsilon^{1/3}L^{4/3} \qquad (6.22)$$

In the energy spectrum, isotropic turbulence should exhibit a characteristic slope $f^{-5/3}$. In surface waters such slopes are observed only at the high-frequency end of the spectrum—though superimposed by the wave signal (Fig. 6.2b)—since stratification of the water column limits the size of isotropic turbulent structures. Therefore, the origin of the specific K_h/L relationship observed by Okubo (1971), although compatible with eqn. 6.22, must be different. Peeters *et al.* (1996a) showed that the expansion of a tracer cloud is not radially symmetrical (Fig. 6.5). The shear-diffusion model of Carter & Okubo (1965) was found to provide the best explanation for the observations. According to this model the spreading of the cloud mainly occurs as a result of dispersion in the direction of

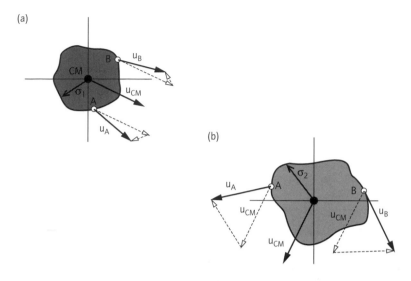

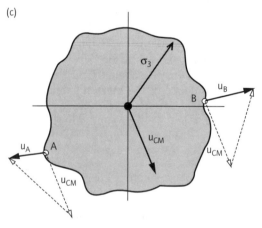

Fig. 6.4 Movement and growth of a tracer patch under the influence of turbulent water currents. For a small patch size (a) the velocities of two water parcels A and B, $\mathbf{u_A}$ and $\mathbf{u_B}$, do not differ much from the velocity of the centre of mass (CM), $\mathbf{u_{CM}}$. At a larger patch size (b) and (c), individual velocities are less closely correlated, and the growth rate of the patch, quantified by $d\sigma^2/dt$, increases, where σ^2 is the variance of the patch around the centre of mass (black dots).

the flow (**longitudinal dispersion**) accompanied by lateral and vertical turbulent diffusion.

In this section it was shown that the motion of lake water is a result of a complex superposition of current structures of different scales. The effect of small scales on transport can be approximated by coefficients of turbulent diffusion. The distinction between advection and turbulent diffusion depends on the time and length scale of observation. Small concentration structures are mainly moved as a whole, i.e. advectively, while at large time and space scales transport acts more randomly, i.e. diffusive.

6.4   THE ROLE OF DENSITY GRADIENTS

Because of vertical density gradients, vertical transport is fundamentally different, especially if the water column is unstable. Under such conditions, water parcels may be transported vertically over large distances. The resulting mass fluxes do not obey Fick's gradient–flux relationship (eqn. 6.11).

Density gradients in the water column are usually extremely small, but are important nonetheless. In combination with pressure

(a)

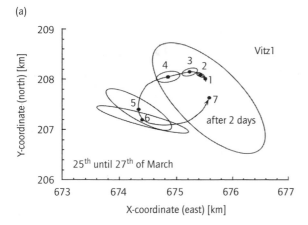

(b)

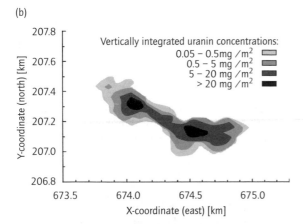

Fig. 6.5 (a) Spatial distribution of artificially added uranin tracer in Lake Lucerne, approximated by ellipses. The black dots indicate the centre of mass as it moves as a result of advection. The temporal change of the cloud size σ_m (defined by the product of the principal axes) obeys the relation $t^{1.4}$ (where t is elapsed time). The exponent is significantly smaller than the t^3-law derived from Bachelor's inertial subrange model (Bachelor 1950). (b) Actual shape of tracer cloud represented in (a) by ellipse number 6. (From Peeters *et al.* 1996a. Copyright 1996 American Geophysical Union. Reproduced by permission of the American Geophysical Union.)

gradients they determine the pattern of currents. Since water density depends on temperature and its chemical composition, the equation of state of water is an important addition to the system

Table 6.3 Density of fresh water and related quantities.

Density of water $(g\,cm^{-3})$ (Chen & Millero 1986)

$$\rho(T,s,p) = \rho^\circ(T,S)\left(1-\frac{p}{K_m}\right)^{-1}$$

$$\rho^0(T,S) = 0.9998395 + 6.7914 \times 10^{-5}T - 9.0894 \times 10^{-6}T^2$$
$$+ 1.0171 \times 10^{-7}T^3 - 1.2846 \times 10^{-9}T^4 + 1.1592 \times$$
$$10^{-11}T^5 - 5.0125 \times 10^{-14}T^6 + \rho^1(T,S)$$
$$\rho^1(T,S) = S(8.181 \times 10^{-4} - 3.85 \times 10^{-6}T + 4.96 \times 10^{-8}T^2)$$

where ρ, ρ^0, ρ^1 $(g\,cm^{-3})$ are contributions to density of water, $T\,(^\circ C)$ is water temperature, $S(\text{‰})$ is salinity, $p(bar)$ is pressure $(1\,bar = 10^5\,Pa = 10^5\,Nm^{-2})$ and

$$K_m^{-1} = \gamma = \frac{1}{\rho}\left(\frac{\partial\rho}{\partial\rho}\right)_{T,S} \quad (bar^{-1})$$

is the compressibility of water (for values see Chen & Millero 1986)

Thermal expansion coefficient

$$\alpha = -\frac{1}{\rho}\left(\frac{\partial\rho}{\partial T}\right)_{S,p} \quad (K^{-1})$$

Coefficient of haline contraction

$$\beta = \frac{1}{\rho}\left(\frac{\partial\rho}{\partial S}\right)_{T,p} \quad (\text{‰}^{-1})$$

For $Ca(HCO_3)_2$: $\beta = 0.807 \times 10^{-3}$ (‰^{-1})
(For other substances see Imboden & Wüest 1995)

Temperature of maximum density $T_{\rho max}$ defined by $\alpha(T_{\rho max}) = 0$
$$T_{\rho max}(S,p) = 3.9839 - 1.9911 \times 10^{-2}p - 5.822 \times 10^{-2}p^2$$
$$- (0.2219 + 1.106 \times 10^{-4}p)S$$

Adiabatic lapse rate

$$\left(\frac{dT}{dz}\right)_{ad} = -\frac{\alpha g T^*}{c_p} = -\Gamma(T,S,p)$$

where $g = 9.81\,m\,s^{-2}$ (acceleration of gravity), $T^* = T + 273.2\,(K)$ (absolute temperature of water) and $c_p = 4.18 \times 10^3\,J\,kg^{-1}\,K^{-1}$ (specific heat of water between $0\,^\circ C$ and $25\,^\circ C$)

of equations which describe currents in lakes (eqn. 6.7).

The density of fresh water is a complex function of temperature T, salinity S and pressure p. Empirical relations of $\rho(T, S, p)$ and related quantities are summarised in Table 6.3. Since the relative chemical composition of seawater is fairly constant, a

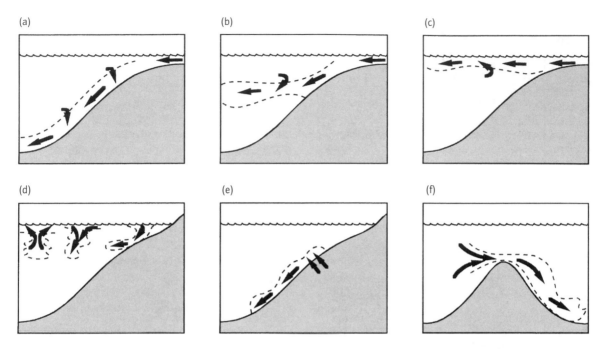

Fig. 6.6   Density-induced lake currents. (a)–(c) The currents at a river mouth depend on the density differences of lake and river water, and on the rate of entrainment of lake water into the river plume. Density currents induced (d) by cooling or heating, (e) by flux of solutes from the sediments, and (f) over sill separating two basins with different temperature and/or salinity.

single number such as the concentration of sodium, or electrical conductivity at a fixed temperature, completely characterises salinity and its influence on water density. This is not the case for lakes; the size of the coefficient of haline contraction β depends on the chemistry of the lake (Wüest *et al.* 1996). Table 6.3 lists a value of β for calcium bicarbonate; other figures are given in Imboden & Wüest (1995).

In many cases, density-driven currents reflect the influence of biogeochemical processes on the physics of lakes (Imboden 1998). This is exemplified by the different situations shown in Fig. 6.6. The effect of all these processes (except Fig. 6.6c) is net mass transport downwards. In section 6.2 we have characterised this by J_b^o, the positive surface buoyancy flux. Analytical expressions of J_b^o for some simple situations are given in Table 6.4.

Figures 6.6a & 6.6b involve the effect of inlets with densities exceeding water density at the lake surface. The maximum potential energy flux per unit lake area, $P_{pot,River}^{max}$, is calculated by assuming that river water 'falls' through the water column to the equilibrium depth $z=-h_{eq}$, where density in the lake equals the density of the incoming river water. The corresponding buoyancy flux, $J_{b,River}^o$, is given by the energy flux divided by the water mass per unit area above $z=-h_{eq}$.

The incoming river plume may eventually reach the deepest point of the lake (or at least of that part of the lake basin into which the river is flowing). On its way down the slope, the river plume entrains water from the lake, thus losing part of its excess buoyancy, a quantitative example of which is given by Wüest *et al.* (1988). River induced energy and buoyancy fluxes may become important in lakes with large hydraulic loading q,

Table 6.4 Energy and surface buoyancy flux caused by inlets and heat exchange at the water surface. Energy flux P is in $W\,m^{-2}$, surface buoyancy flux J_b^0 is in $W\,kg^{-1}$. For definitions see also Table 6.3.

Inlets

Maximum potential energy flux per total lake area without loss by friction

$$P_{pot,River}^{max} = gq \int_{-h_{eq}}^{0} \left[\rho_{River} - \rho(z) \right] dz = gq h_{eq} \overline{\Delta\rho}$$

where $q = Q/A_0$ ($m\,s^{-1}$) (hydraulic loading, i.e. water discharge Q per unit lake area A_0), $\rho(z)$, ρ_{River} ($kg\,m^{-3}$) (density of lake and river water, respectively), $\Delta\rho$ ($kg\,m^{-3}$) (average density excess of river water) and $-h_{eq}$ (m) (equilibrium depth defined by $\rho(-h_{eq}) = \rho_{River}$)

$$J_b^0 = \frac{P_{pot,\,River}^{max}}{\rho h_{eq}}$$

Typical numbers

$q = 10^{-5} - 10^{-7}\,m\,s^{-1}\ (3–300\,m\,yr^{-1})$

$\Delta\rho = 0.5\,kg\,m^{-3}\ [e.g.,\ \rho(4\,°C) - \rho(12\,°C)]$

$h_{eq} = 20\,m$

$\Rightarrow\quad P_{pot,River}^{max} = 10^{-5} - 10^{-3}\,W\,m^{-2}$

$\quad\quad J_{b,River}^0 = 10^{-9} - 10^{-7}\,W\,kg^{-1}$

Note: the largest variation is due to hydraulic loading q. There are river-dominated lakes with $q > 10^4\,m\,yr^{-1}$ and lakes with virtually no throughflow of water ($q \approx 0$).

Heat flux

$$J_{b,\,Heat}^0 = -\frac{\alpha g}{c_p \rho} H_{net} \quad \text{(Imboden \& Wüest 1995)}$$

$$P_{pot,\,Heat} = -\frac{\alpha g h_{mix}}{c_p} H_{net}$$

where $-H_{net}$ ($W\,m^{-2}$) is net heat loss of water and h_{mix}(m) is mixing depth, i.e. thickness of well-mixed surface layer.

Typical numbers

$H_{net} = -100\,W\,m^{-2}$

$h_{mix} = 10\,m$

$\alpha = 10^{-4}\,K^{-1}\ (T \approx 11\,°C)$

$\Rightarrow\quad J_{b,Heat}^0 = 2.4 \times 10^{-8}\,W\,kg^{-1}$

$\quad\quad P_{pot,Heat} = 2.4 \times 10^{-4}\,W\,m^{-2}$

Evaporation from saline water without temperature change

$$J_{b,\,Evap}^0 = \frac{g \beta_S S}{\rho} m_E \quad \text{(Imboden \& Wüest 1995)}$$

$$P_{pot,Evap} = g \beta_S S h_{mix} m_E$$

where $\beta_S = \frac{1}{S}\left(\frac{\partial\rho}{\partial S}\right)_{T,p}$ ($‰^{-1}$) (coefficient of haline contraction), S (‰) is salinity and m_E ($kg\,m^{-2}\,s^{-1}$) is the evaporation rate

Typical numbers

$S = 35‰$

$\beta_S = 0.8 \times 10^{-3}\,(‰^{-1})$

$m_E = 0.3 \times 10^{-4}\,kg\,m^{-2}\,s^{-1}$ (corresponding to $1\,m\,yr^{-1}$)

$h_{mix} = 10\,m$

$\Rightarrow\quad J_{b,Evap}^0 \approx 8 \times 10^{-9}\,W\,kg^{-1}$

$\quad\quad P_{pot,Evap} = 8 \times 10^{-5}\,W\,m^{-2}$

---

and/or in lakes which are too deep to be fully mixed by wind. Deep water exchange in Lake Baikal, the deepest lake on Earth, is partially driven by inlets (Hohmann *et al.* 1977).

Heat exchange and evaporation may also produce a positive buoyancy flux (Fig. 6.6d). If the density of a parcel of water at the surface **increases**, gravitation is no longer fully compensated for by the vertical pressure gradient, and the parcel sinks. This is called **convection**. According to Table 6.4, convection is induced if the product of thermal expansivity α, and the net heat flux H_{net} (defined as positive if the lake gains thermal energy), is negative. This is the case either for a lake with $T > T_{\rho max}$ which loses heat, or for a lake with $T < T_{\rho max}$ which gains heat (where $T_{\rho max}$ = the temperature of maximum density). Evaporation from saline lakes can also induce positive buoyancy fluxes. Typical values for these processes are given in Table 6.4. Buoyancy fluxes may also be internally produced, e.g. by diffusion of salt from the sediments into the overlaying water (Fig. 6.6e).

How large must the density excess $\Delta\rho$ be in order to trigger a convective current? As a simple approach,* we consider a one-dimensional water column containing a layer of thickness H, with excess density $\Delta\rho$. The timescale for the gravitational forces to set the layer in motion is of the order $T_{grav} = (H/g')^{1/2}$, where $g' = (\Delta\rho/\rho)g$ is the reduced gravity. The value for T_{grav} must be compared with the time needed for viscous forces to decay, and for excess density to diffuse out of the layer (see eqn. 6.18). These are, respectively, $T_{visc} = H^2/\nu$ and $T_{diff} = H^2/D$, where ν is kinematic viscosity, and D is molecular diffusivity, either of heat or of salt, depending on which is producing the extra density. A convective current can develop only if both ratios (T_{grav}/T_{visc}) and (T_{grav}/T_{diff}) are much smaller than unity. The product of the two ratios is the so-called Rayleigh number

$$Ra = \frac{g(\Delta\rho/\rho)H^3}{\nu D} = \frac{g\alpha\Delta T H^3}{\nu D} \qquad (6.23)$$

where the expression on the far right-hand side

---

* For a rigorous derivation, see e.g. Malkus (1981).

refers to Rayleigh's original problem in which the excess density was due to temperature variation. Theoretical and experimental information shows that convection is triggered if Ra is larger than 1500–1700 (Reid & Harris 1958). As an example, for $\alpha\Delta T = 10^{-7}$, the layer depth H necessary to trigger convection is only 0.05 m. In freshwater the relative excess density of 10^{-7} would, e.g., be produced by $\Delta T = 0.01$ K at $T = 4.7°C$ and atmospheric pressure.

In contrast, if the density of the water parcel at the surface **decreases**, either a vertical density gradient is built up or an existing gradient is strengthened by the production of negative buoyancy. Such situations occur due to (i) heat exchange [if $(H_{net}\alpha) > 0$], (ii) the effect of an inlet with smaller density (Fig. 6.6c), or (iii) rain on to salty water. Since the vertical density gradient impedes vertical mixing by turbulence, then wind shear and heat flux take over antagonistic roles. Shear acts as a mixing agent, while the negative buoyancy flux produced by the heat exchange extracts energy from the kinetic energy pool and thus slows water movement down. A quantitative discussion of this situation is given in section 6.5.

Figure 6.6f depicts special circumstances in which density-driven currents are created by internally produced positive buoyancy. Compared with convection induced at the water surface, the processes leading to development of internal currents are often the result of extremely small density effects. In order to understand how this process works, the following concept of vertical stability is helpful.

A water column is locally stable if a fluid parcel displaced isentropically† by an infinitesimal vertical distance dz from its initial position experiences a restoring force.‡ This is equivalent to the condition that the density change of the parcel due to its isentropic displacement exceeds that in the water column:

---

† Isentropic: without exchange of heat and mass with the surrounding water.
‡ Global stability means that the water parcel can be displaced over the entire water column and still experiences a restoring force.

$$\left(\frac{d\rho}{dz}\right)_{isen} - \frac{d\rho}{dz} > 0 \qquad (6.24)$$

where z is the vertical coordinate defined as positive upwards.

Equation 6.24 can also be expressed in terms of the Brunt–Väisälä frequency (Gill 1982; Peeters *et al.* 1996b):

$$N^2 = \frac{g}{\rho}\left[\left(\frac{d\rho}{dz}\right)_{isen} - \frac{d\rho}{dz}\right]$$
$$= g\left[\alpha\left(\frac{dT}{dz} + \Gamma\right) - \beta\frac{dS}{dz}\right] \qquad (6.25)$$

where Γ is the adiabatic lapse rate as defined in Table 6.3. Often it is integrated into the gradient of potential temperature θ:

$$\frac{d\theta}{dz} = \left(\frac{dT}{dz} + \Gamma\right) \qquad (6.26)$$

The **equation of state** of water is unique in two ways:

1 the density of its liquid phase is greater than that of ice;

2 water reaches maximum density at a temperature greater than its freezing point (the **density anomaly**).

The consequences of the first property are of the greatest importance ecologically (in that lakes do not turn into solid blocks of ice during the cold season), yet in the context of lake currents the second point is more relevant. According to Table 6.3, the temperature of maximum density $T_{\rho max}$ (which at atmospheric pressure and zero salinity is approximately 4°C) decreases with salinity by about $0.22\,\text{K‰}^{-1}$, and falls below the freezing point (which is also decreasing with S, but less strongly) at $S = 24.7‰$. Thus, in seawater, there is no density anomaly. The value of $T_{\rho max}$ also decreases with pressure, or depth, i.e. by about 0.2 K per 100 m. The thermal expansion coefficient α, as a function of T and p, is shown in Fig. 6.7.

The special form of the equation of state is important for the vertical stability of the water

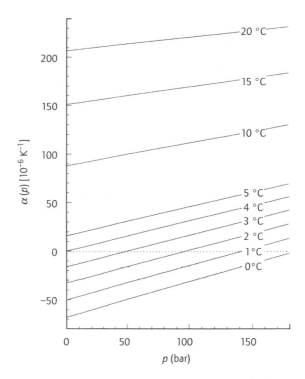

Fig. 6.7 Thermal expansion coefficient $\alpha = -\dfrac{1}{\rho}\left(\dfrac{\partial \rho}{\partial T}\right)_{S,p}$ of fresh water ($S = 0$) as a function of temperature T and pressure p. α changes sign at the temperature of maximum density $T_{\rho max}$.

column. First the effect of salinity will be disregarded. Figure 6.8a shows a stably stratified water body at the surface of a lake. The temperature of the lower layer is equal to $T_{\rho max}$ at atmospheric pressure. It often happens that, owing to wind stress, at the downwind end of the lake a water mass is rapidly displaced downwards. Since the compressibility of the two layers is different, at 100 m depth it is now the temperature of the upper layer which is equal to the local $T_{\rho max}$ of 3.8°C. A positive buoyancy flux is triggered which can create large-scale currents and mixing. This process is called **thermal baricity** (McDougall, 1987a).

Figure 6.8b depicts the so-called **thermal bar phenomenon** which often develops in large lakes during spring or autumn. Since heat exchange is not uniform across the lake, horizontal temperature gradients develop. For instance, during spring, part of the surface water—usually located offshore—may still be colder than 4°C. The line where the 4°C isotherm intercepts the water surface separates two water masses. Exchange of water across this thermal bar is impeded since the mixing water sinks to deeper parts of the lake. Often the thermal bar is clearly visible because of colour or turbidity differences of the water on either side. Thermal bars also develop in the interior of the water column when the isotherm with

temperature T reaches the depth where $T_{\rho max}$ is equal to T.

The thermal bar is a special case of a more general phenomenon called **cabbeling**. Since density ρ is a non-linear function of T, mixing of two water bodies of different temperature always causes the mean density to increase. Figure 6.8c shows an example of two isopycnals characterised by specific combinations of T and S which make density constant. Imagine that due to lateral mixing along the upper isopycnal a mixture is created which consists of 50% of each of the water masses shown on the left and right of the isopycnal. This water mass has $T = 4.75$°C and $S = 0.109$‰. Due to the non-linearity of ρ as a function of T, the density of the mixture is greater than the densities of each of the original water masses (which are equal since they lie on the same isopycnal). Thus, as a result of lateral mixing, the water moves off the isopycnal; a positive buoyancy flux is produced. This effect is common in the ocean from low latitudes to the poles, and is accompanied by a lateral flux of heat and salt (McDougall & You 1990). As discussed below (Fig. 6.9), cabbeling may be important for large-scale vertical mixing. Also note that both phenomena (thermal baricity and cabbeling) may occur in fresh water as well as salt, yet their effects are most spectacular in cold freshwater lakes

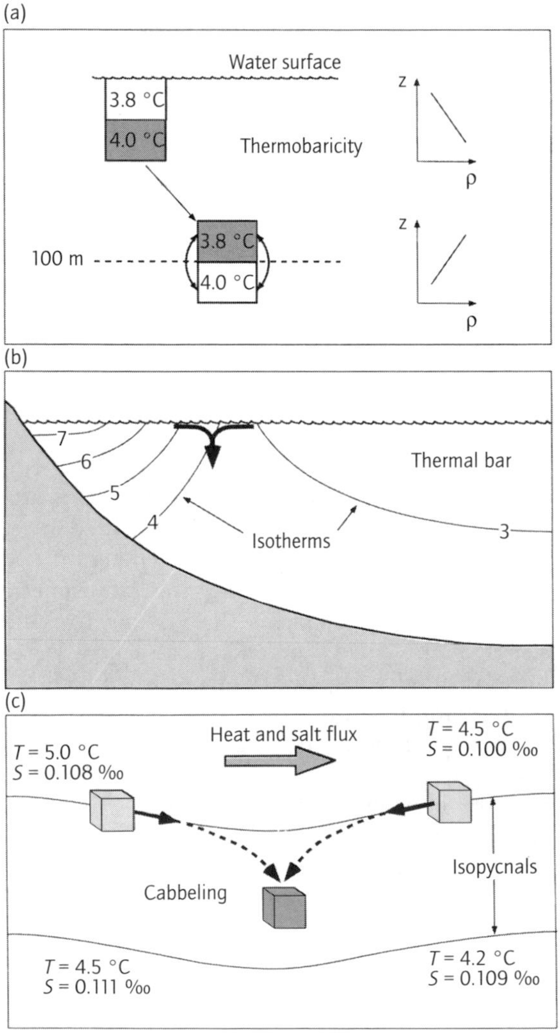

Fig. 6.8 (a) The effect of thermal baricity on vertical mixing: when a stably stratified water column at the surface is displaced downwards to 100 m depth (e.g., as a result of an internal seiche), the water body becomes unstable and a positive buoyancy flux is triggered. Note that at 100 m, $T_{\rho max}$ lies at 3.8°C. (b) Thermal bar: owing to differential heating or cooling in spring and autumn, horizontal temperature gradients frequently develop in large lakes. Where the 4°C isotherm meets the water surface, a distinct front is created by sinking water of maximum density produced by lateral mixing. The thermal bar prevents the surface water from moving across the 4°C isotherm. (c) The thermal bar is a special form of the cabbeling process. Since density ρ is a non-linear function of T, lateral mixing causes the water to sink out of the isopycnal. The T and S numbers exemplify the possible variation along an isopycnal. See text for further explanations. (Adapted from McDougall & You 1990.)

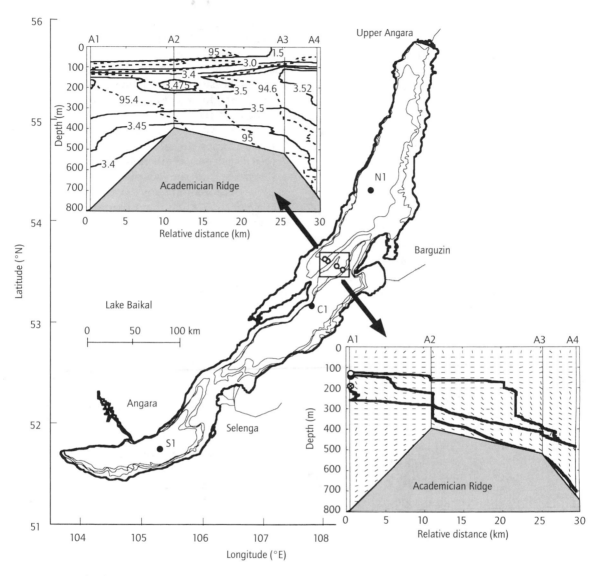

Fig. 6.9 Map of Lake Baikal (Siberia). The upper left inset shows the two-dimensional temperature and salinity variations across the Academician Ridge separating the North (right) and Central Basins (left) of the lake. The full lines are isotherms (temperature in °C), the dashed lines are isohalines (salinity in ‰). Lower right inset: The short lines give the direction of the local neutral surface, the bold solid lines are neutral tracks with their origin marked by different symbols. Note that water from 100 to 200 m depth flows from the more saline Central Basin into the deeper part of the fresher North Basin. (From Hohmann *et al.* 1997; reproduced with permission of the American Society of Limnology and Oceanography.)

where water temperature is close to $T_{\rho max}$. Thus the non-linearity of $\rho(T, p, S)$ is particularly important.

At first sight it seems paradoxical that water

masses do not move along surfaces of constant density. A rather detailed mathematical formalism is needed in order to analyse the buoyancy-free movement of a water parcel in a general three-dimensional (T, p, S)-field (Peeters *et al.* 1996b). The central idea may be summarised as follows.

First, an infinitesimal isentropic displacement of a water parcel, and the forces acting upon it, is considered. In order to do so, eqn. 6.24 must be evaluated for various directions in the x, y, z-space. In a stable water column displacements along the z-axis lead to restoring forces. For an **arbitrary** displacement, the ensemble of all directions $\boldsymbol{n}$ defined by the vector $\boldsymbol{n} = (n_x, n_y, n_z)$, for which

$$\left(\frac{d\rho}{dn}\right)_{\text{isen}} - \frac{d\rho}{dn} = 0 \qquad (6.27)$$

defines a surface through the equilibrium-position of the selected water parcel. McDougall (1984, 1987b) calls this the **neutral surface**. Thus the neutral surface defines all directions of infinitesimal buoyancy-free displacements of a particular water parcel.[*]

As a complementary approach, Peeters *et al.* (1996b) introduced the concept of the **neutral track**. This is the path along which the isentropic transport of a water parcel is buoyancy-free over **arbitrary** distances. Each parcel follows its own neutral track, and those belonging to different parcels may cross (see inset, Fig. 6.9).[†] Although at crossings the density of parcels is equal, their compressibilities vary according to their different combinations of heat and salt. Thus, when moving on, their densities also vary differently. For the special case where salinity is constant (ρ depends on T only), the isotherms (surfaces of constant potential temperature), the neutral surfaces and the neural tracks coincide.

Cabbeling can be important in a cold freshwater lake which exhibits large-scale salinity gradients owing to the different chemical compositions of its inlets. For instance, mixing at the boundary between the North and Central Basin of Lake Baikal (Siberia) produces a mixture of water which can sink to the deeper parts of the North Basin, and which thus may be responsible for the formation of deep water (Fig. 6.9). Here the driving force for the buoyancy flux is the slight salinity difference between the basins (Hohmann *et al.* 1997).

In this section we have shown how the equation of state of water gives rise to the internal production of positive buoyancy as pressure changes, or if water masses with different characteristics are mixed. Concepts were introduced, such as the neutral surface and the neutral track, in order to analyse the potential buoyancy-free movement of water parcels. Yet, it is important to be aware of the limitations of this approach. As discussed in section 6.1, pressure gradients, not density differences, set water in motion (eqns. 6.1 and 6.7). It is true that the latter give rise to pressure gradients (e.g., by making the hydrostatic equilibrium (eqn. 6.2) invalid), but density gradients are by no means the only mechanisms which produce pressure gradients. In fact, the top 200 m of the ocean is driven by wind forcing, either directly by the wind stress τ_w or because of the tilt of both the water surface and isopycnals in the interior of the water body. In contrast, circulation in the interior of the ocean, and the formation of deep water, is strongly influenced by density gradients.

Here one quickly reaches the limits of simple concepts. Ultimately, the study of water currents under the simultaneous action of different forces requires numerical models. These have been developed in a great variety of complexity during the past 20 years (Tee 1987). In order to describe density-driven water currents, non-hydrostatic models are needed in which vertical acceleration is taken into account (e.g., Walker & Watts 1995).

6.5   EXTERNAL MOMENTUM AND ENERGY FLUX

In this section we analyse the influence of the wind on the dynamics of a lake. Owing to the effects of friction, the horizontal velocity $W(z)$ of the wind field decreases from the free atmosphere towards Earth's surface. However, there is a zone immedi-

---

[*]   Note that the neutral surface is not identical with the isopycnal surface.

[†]   Obviously, the neutral track is a theoretical construct. Since in reality water parcels constantly change their composition due to mixing, they do not stay on a given neutral track.

ately above the ground where a constant wind stress τ_w is established. On the basis of dimensional considerations, the vertical velocity profile in the constant stress layer obeys the so-called 'law of the wall'

$$W(z) = \frac{(\tau_w/\rho_{air})^{1/2}}{\kappa} \ln\left(\frac{z}{z_0}\right) \qquad (6.28)$$

where $\kappa = 0.41$ is the von Kármán constant, z is the vertical distance above ground, z_0 is the integration constant which depends on the roughness of the surface, and ρ_{air} is air density. If the wind is blowing over a lake, the stress layer created extends downwards into the water column, thus transferring momentum and energy from the wind field into the lake. Usually, τ_w is expressed by a friction velocity defined by[*] $u_*^{air} = (\tau_w/\rho_{air})^{1/2}$. It is then assumed that the boundary layer in the water is characterised by the same stress τ_w. Thus, the friction velocity in the water is:

$$u_* = (\tau_w/\rho)^{1/2} \qquad (6.29)$$

Equation 6.28 allows us, at least in principle, to determine τ_w from wind measurements recorded at different heights above the water surface, z_i, by solving the following equation for τ_w:

$$W(z_1) - W(z_2) = \frac{(\tau_w/\rho_{air})^{1/2}}{\kappa} \ln\left(\frac{z_1}{z_2}\right) \qquad (6.30)$$

The standard height for reporting wind velocities is $z = 10\,\text{m}$ (W_{10}). According to eqn. 6.28 the wind stress τ_w is linearly related to W_{10}^2. This is expressed as

$$\tau_w = \rho_{air} C_{10} (W_{10})^2 \qquad (6.31)$$

where the non-dimensional wind stress coefficient C_{10} is related to z_0 of eqn. 6.28 by:

$$z_0 = 10[\text{m}] \exp\left(-\frac{\kappa}{\sqrt{C_{10}}}\right) \qquad (6.32)$$

The boundary layer concept has been adopted from atmospheric science, where it was developed in order to describe the wind field above the solid ground. The stress τ_w is assumed to be due to turbulence, and the roughness length z_0 describes the structure of the surface. Therefore the z_0 of grassland should be smaller than that of a forest.

Application of this concept to the air–water interface is more complicated. Not only does the formation of waves change the roughness of the water surface, but waves also add to the stress in the water as well as the air (Janssen 1989). Since the build-up and decay of a wave field takes time, waves are often not in equilibrium with the actual wind velocity. As an example, consider the sudden onset of a wind which then remains constant. While the waves are growing, the effective wave length (and thus the effective phase speed[†] $c_{ph,eff}$) increases. The corresponding rate of change of wave momentum is equal to the wave-induced stress τ_{wave}. Owing to distortion of the water surface by the waves, a 'mirror' pattern of wave-like motion develops in the air which also contributes to the stress in the air-side boundary layer. Simon (1997) showed that the ratio between turbulent and wave-induced stress varies between about 8 at low wind speed to about 0.5 at high speed. Since, at high speeds, waves in lakes are seldom fully developed, the change of this ratio reflects the increasing importance of wave-induced stress for a growing wave field.

The influence of wave age on τ_w explains why experimental information on the wind stress coefficient C_{10} is so ambiguous (Donelan et al. 1993). For strong wind ($W_{10} > 7\,\text{m s}^{-1}$), the values given by Amorocho & deVries (1980) are widely accepted. According to these authors, C_{10} increases with

---

[*] u_*^{air} and u_* quantify the turbulent velocity fluctuations which are ultimately responsible for the momentum transfer from the wind field through the boundary layer to the water. They are not real velocities which could be directly measured with current meters.

[†] The effective phase speed is defined as the weighted mean phase speed of the wave field. See eqn. 6.56 for the relation between wave length and phase speed.

wind velocity, from 10^{-3} at $W_{10} \approx 7\,\mathrm{m\,s^{-1}}$, to 2.5×10^{-3} at $W_{10} \approx 17$ to $20\,\mathrm{m\,s^{-1}}$. These values are based on oceanic data, however; therefore they probably represent rather mature wind fields, and may be too small for lakes.

In contrast to typical oceanic conditions, W_{10} over lakes is often fairly small. For low wind velocities, the surface roughness is influenced by capillary waves which depend on the surface tension of water versus air (see section 6.7). Wu (1994) discussed three different theories to describe the size of C_{10} for small and moderate winds. These models as well as experimental data show C_{10} to increase with decreasing W_{10}. On the basis of Wu's data, Simon (1997) developed an empirical relation for $W_{10} < 5\,\mathrm{m\,s^{-1}}$. In the following equation, this is combined with the relation by Amorocho & deVries (1980) for strong winds:

$$C_{10} = \begin{cases} 4.33 \times 10^{-3} (W_{10})^{-0.872} & \text{for } W_{10} \leq 5\,\mathrm{ms^{-1}} \\ 1 \times 10^{-3} & \text{for } 5\,\mathrm{ms^{-1}} < W_{10} < 7\,\mathrm{ms^{-1}} \\ 1 \times 10^{-3} + 1.5 \times 10^{-4}(W_{10} - 7\,\mathrm{ms^{-1}}) \\ \quad \text{for } W_{10} > 7\,\mathrm{ms^{-1}} \end{cases}$$

$$(6.33)$$

The energy flux through the air–water interface is proportional to the product of the stress τ_w and the wind velocity W. At the 10 m level, the energy flux is:

$$P_{10} = \tau_w W_{10} = \rho_{air} C_{10}(W_{10})^3 \qquad (6.34)$$

Typically P_{10} lies between 10^{-2} and $1\,\mathrm{W\,m^{-2}}$ (Table 6.2). In relation to the specific energy spectrum of lake currents (Fig. 6.2), it is interesting to obtain information on the fraction of this energy transferred to the water, and on its partition among the different spatial scales of the spectrum. For the smallest scale it is found that only about 1–2% of this energy is available for turbulent mixing in the water (Denman & Miyake 1973). The turbulent kinetic energy per unit area, E_{turb}, can be estimated from the friction velocity as $1/2(u_*)^2 \rho h_{turb}$, where h_{turb}, the depth zone containing most of E_{turb}, is of the

order of 10^1 m. As summarised in Table 6.2, E_{turb} usually represents only a minor fraction of the total kinetic energy of the water column.

As discussed above, most of the momentum is transferred to the waves. Simon (1997) developed a model to describe the energy balance of the wave field under the influence of variable wind velocities. Using observations of Lac Neuchâtel (Switzerland), he found that energy transfer into relatively young waves (which are common in lakes) is of the order of 20% of P_{10}. Furthermore, in his analysis, the dissipation rate of wave energy calculated from published theories (e.g., Longuet-Higgins 1969; Thorpe & Humphries 1980) turned out to be too small to explain the measured response of the wave field to decreasing wind speed. He concluded that only a small fraction of the wave energy is dissipated locally and that most wave energy dissappears by horizontal advection. It seems that the overall dissipation timescale is determined by the time taken for waves to reach the shore, where most of the energy is dissipated. Thus, dissipation is expected to be slower in large lakes than in small. Typical values for the **apparent** dissipation time, τ_{wave} (i.e. the time for the wave energy to decay at a fixed location), lie between 10^3 and 10^4 seconds.

At steady-state, the kinetic wave energy per unit area ($E_{wave} = P_{wave}\tau_{wave}$) varies between 10 and $10^3\,\mathrm{J\,m^{-2}}$ (Table 6.2). A similar observation was made for the interior of the water body. Temperature microstructure measurements by Goudsmit & Wüest (1997), in the hypolimnion of a Swiss lake, show that about 90% of the available turbulent kinetic energy is dissipated in the bottom boundary layer.

A small but important fraction of wind energy is incorporated in large-scale motion. Owing to wind stress, water is pushed to the downwind end of the lake, which causes a slight tilt of the water surface. If the wind is blowing at constant velocity for some time*, an equilibrium is reached between the wind stress at the water surface and the pres-

---

* The time needed in order to reach steady state is about one quarter of the period of the first barotropic seiche mode (Spigel & Imberger 1980). Seiches are discussed in section 6.7.

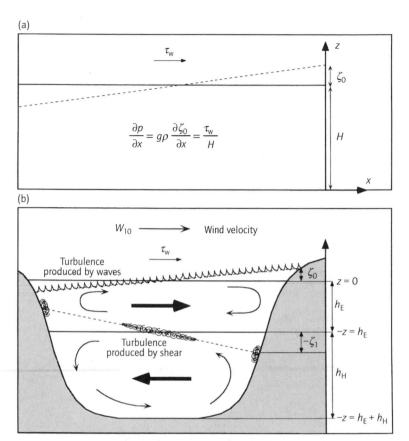

Fig. 6.10 (a) Schematic picture of sloping water surface due to wind stress τ_w. The slope of the water surface $\partial\zeta_0/\partial x$ produces a horizontal pressure gradient which, at steady state, is balanced by the wind stress. For a homogeneous water column, the pressure gradient is called **barotropic**. (b) In a stratified lake, a tilted water surface, with water piling up at the downwind end of the lake, causes the thermocline to tilt in the opposite direction (**baroclinic pressure gradient**). The vertical excursion of the interface $\xi_1(x)$ is much larger than $\xi_0(x)$. The thick black arrows give the initial flow in the water. Once the interfaces reach a steady-state position, the water currents follow the pattern sketched by the thin arrows. Turbulence is produced by breaking surface and internal waves, and by shear at the thermocline.

sure force in the interior of the water column due to the sloping surface (Fig. 6.10a):[*]

$$-H\frac{\partial p}{\partial x} + \tau_w = 0 \quad \text{with} \quad \frac{\partial p}{\partial x} = g\rho\frac{\partial\zeta_0}{\partial x} \qquad (6.35)$$

where H represents total depth and ζ_0 the vertical displacement of the water surface relative to the equilibrium position. From eqn. 6.35 the surface slope is:

$$\frac{\partial\zeta_0}{\partial x} = \frac{\tau_w}{g\rho H} = \frac{u_*^2}{gH} \qquad (6.36)$$

When the wind stops, relaxation of the water

surface produces long surface waves among which **eigenmodes**, i.e. basin-wide standing waves called **seiches**, are particularly important (see section 6.7). If approximated as linear waves, the mean total energy of the seiches, E_{seiche}, is equally partitioned between kinetic and potential energy. The total energy can be calculated from the maximum amplitude at the end of the lake (see Fig. 6.10a), which is

$$A_{\max} = \frac{L}{2}\frac{\partial\zeta_0}{\partial x} = \frac{L}{2}\frac{\tau_w}{g\rho H} \qquad (6.37)$$

The energy of a standing linear wave per unit area is:[†]

---

[*] Note that the pressure force is acting against the pressure gradient (eqn. 6.1), hence the minus sign in eqn. 6.35.

[†] Note the difference between eqn. 6.38 and the energy of a wave field given by eqn. 6.59. For standing waves the energy per unit area is only half the value for the travelling wave field.

$$E_{\text{seiche}} = \frac{1}{4} g\rho A_{\max}^2 = \frac{1}{16} \frac{1}{g\rho} \left(\frac{L}{H}\right)^2 \tau_w^2 \quad [\text{J}\,\text{m}^{-2}] \quad (6.38)$$

If the water is stratified, the surface tilt is accompanied by an inverse motion of the density interface. Figure 6.10b depicts an idealised situation, in which two water layers of depth h_E and h_H are separated by a density jump $\Delta\rho$. The interior interface is tilted in the opposite direction. It can be shown that the above equations remain valid if the following substitutions are applied (e.g., Mortimer 1975):

$$g \rightarrow g' = g\frac{\Delta\rho}{\rho} \quad \text{and}$$

$$H \rightarrow H' = \left(\frac{1}{h_E} + \frac{1}{h_H}\right)^{-1} \quad (6.39)$$

where g' and H' are called reduced gravity and reduced depth. Thus, the slope of the internal interface is

$$\frac{\partial \zeta_1}{\partial x} = \frac{u_\star^2}{g'H'} = Ri^{-1} \quad (6.40)$$

where Ri is the mixed-layer **Richardson number**. This expresses the ratio between stress and gravity forces. If $h_E \ll h_H$, then $H' = h_E$. Typical relative density differences are of the order 10^{-4} to 10^{-3}. By making the corresponding substitutions in eqn. 6.38, and by disregarding the difference between H and H', it can be deduced that, for a given wind stress, the slope of the internal interface and the energy stored in the internal seiche motion exceed the corresponding surface values by $(\Delta\rho/\rho)^{-1}$, i.e. by 10^3 to 10^4. Note that it takes significantly more time for the internal seiche to reach its steady-state amplitude. Furthermore, the linear wave approximation underlying the energy equation (6.38) is no longer applicable once the amplitude of the thermocline at the end of the lake exceeds the depth of the surface layer h_E. According to eqn. 6.40 this happens if the maximum amplitude of the internal seiche $A'_{\max} = \frac{L}{2}\frac{\mathrm{d}\zeta}{\mathrm{d}x}$ obeys:

$$1 > \frac{h_E}{A'_{\max}} = 2\frac{h_E}{L} Ri = 2\,We$$

where

$$We = \left(\frac{h_E}{L}\right) Ri < 0.5 \quad (6.41)$$

was first introduced by Spigel & Imberger (1980), and later named the **Wedderburn number** by Thompson & Imberger (1980). A small Wedderburn number, i.e. a combination of a small Richardson number Ri and a small aspect ratio (h_E/L), favours non-linear motion of the thermocline, accompanied by strong mixing and entrainment between different density layers (Imberger 1985).

Application of the above equations to two lakes of different size, and two wind velocities, is shown in Fig. 6.11. The conclusions drawn from these examples are typical for lakes in general:

1 variation of wind speed between 1 and $10\,\text{m}\,\text{s}^{-1}$ changes energy flux P_{seiche} and energy content E_{seiche} by a factor of the order of 10^3;

2 the energy flux into the surface seiche is only about 0.1 to 1% of the flux into the internal mode;

3 the maximum energy content of the surface seiche is typically 10^{-4} times smaller than the corresponding content of the internal mode;

4 the largest energy input rates are attained for weak stratification and strong winds (of the order of $10^{-2}\,\text{W}\,\text{m}^{-2}$ and greater), but P_{seiche} reaches only a few percent of P_{10}.

If the Wedderburn number We becomes too small, energy is no longer fed into the basin-wide modes but dissipates locally. For the particular examples shown in Fig. 6.11, this occurs for the strong wind, since the maximum internal seiche amplitude for lakes A and B, respectively, is $10\,\text{m}$ and $30\,\text{m}$. In contrast, maximum surface amplitudes do not exceed $10^{-2}\,\text{m}$.

There are two reasons why, in a lake, the energy input of large-scale modes plays a special role. First, energy dissipation of basin-wide eigenmodes is very small, at least in lakes with fairly regular basin topography (Lemmin 1987). It is not uncom-

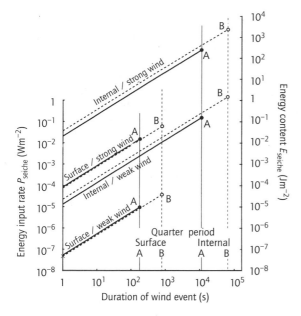

Fig. 6.11 Rate of energy input, P_{seiche}, into surface and internal seiche mode (left end of straight lines measured by left scale) as a function of wind duration, and corresponding maximum energy content E_{seiche} (right scale) after one-quarter of the seiche period (T/4). Values are shown for two wind speeds ($W_{10} = 1$ and $10\,m\,s^{-1}$) and two lakes of different size. Lake A: $L = 5\,km$, $H = 20\,m$, $h_E = 10\,m$, $h_H = 10\,m$, $\Delta\rho/\rho = 10^{-3}$; Lake B: $L = 50\,km$, $H = 100\,m$, $h_E = 20\,m$, $h_H = 80\,m$, $\Delta\rho/\rho = 10^{-3}$. In the double logarithmic scale, the linear approximation of the energy transfer, $E_{seiche} = P_{seiche}t$, appears as a straight line with slope = 1. Seiche periods are calculated with eqn. 6.62. P_{10} for the two wind conditions is 5×10^{-3} and $2\,W\,m^{-2}$, respectively (Table 6.2).

mon for the half-life of this energy to be of the order of several weeks. Second, owing to the non-linearity of the equation of motion of water (eqn. 6.7), energy is continuously transferred to smaller scales.* Thus, once the wind has ceased, basin-wide motion serves as the main source of small-scale and turbulent kinetic energy (Fig. 6.12).

---

* G.E. Hutchinson (1957) cites Richardson with the famous phrase 'Big swirls have little swirls that prey on their velocity, and little swirls have lesser swirls and so on . . . to viscosity.'

Ultimately, most of the energy transferred into the lake is transformed into random turbulent motion and dissipated into heat by viscous forces. According to the 'law of the wall' concept, the rate of turbulent kinetic energy dissipation ε by shear as a function of depth h should follow the equation (see Fig. 6.13a)

$$\varepsilon(h) = \frac{u_*^3}{\kappa h} \qquad (6.42)$$

where u_* is the friction velocity in the water (eqn. 6.29). For the situations given in Table 6.2, typical values of ε calculated for 1 m depth lie between 10^{-9} and $10^{-5}\,W\,kg^{-1}$. Yet, Terray *et al.* (1996) showed that, in the wave zone, ε can exceed the values calculated from eqn. 6.42 by up to one order of magnitude. If we assume that depth of the wave zone (h_{waves}) = 1 m, at steady-state ε would be of the order of $P_{water}/(\rho h_{waves})$, i.e. up to $10^{-4}\,W\,kg^{-1}$. However, only about one-tenth of the energy fed into the wave field is dissipated locally, so that typical wave zone dissipation rates are rather $10^{-5}\,W\,kg^{-1}$, or less. As mentioned above, a significant fraction of energy dissipation occurs at the boundaries of the lake.

The above description of kinetic energy fluxes in lakes is by no means complete. There are further kinds of organised motion, such as other types of waves, and Langmuir cells. Some of those will be discussed in section 6.7.

6.6 TURBULENT KINETIC ENERGY BALANCE

In section 6.3 the simultaneous effect of large-scale (organised) and small-scale (random) water currents was discussed. It was shown that the distribution of solutes and suspended particles is mainly effected by basin-wide currents. In a classic study, Eckart (1948) used the word **stirring** for the redistribution of water by advection. Stirring increases spatial gradients by intertwining water masses of different composition. In contrast, **mixing** reduces spatial gradients. Ultimately, mixing

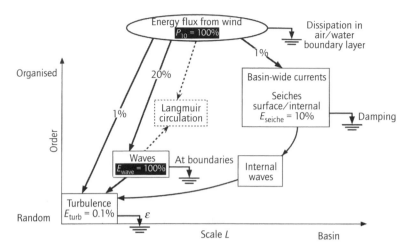

Fig. 6.12 Typical relative sizes of energy fluxes and energy content in a lake. About 20% of the energy input by wind shear 10 m above the water surface (P_{10} taken as 100%) is fed into surface waves, and about 1% into seiches and turbulence each. Most of the kinetic energy is stored in the wave field (E_{wave} taken as 100%), about 10% in the seiche motion and only 0.1% in turbulence. See Table 6.2 for definitions and absolute numbers. The arrows on the right-hand side of the boxes indicate energy dissipation.

occurs by molecular diffusion, but, as an important prior step, turbulence is responsible for 'chopping up' the stirring structures into smaller pieces. Thus, turbulence, located at the small-scale end of the range of water currents is the counterpart to large currents. In fact, the intensity of turbulence is one of the factors that selects for biological species. It is well known that, in rivers, flow velocity is an important ecological parameter (Reynolds 1994). This is also the case for the ocean and for lakes (e.g., Lewis *et al.* 1984; Reynolds 1992).

Table 6.2 summarises the range of energy fluxes and energy reservoirs in lakes. As discussed in section 6.5, the input of energy from the wind is commonly the most important energy source. It was shown (Fig. 6.12) that, ultimately, most kinetic energy ends as turbulent kinetic energy from where it is dissipated into heat. Potential energy is an additional albeit much smaller energy sink. In fact, in a stably stratified water column ($N^2 > 0$, eqn. 6.25) the 1st Fickian Law (see eqn. 6.11), applied to water density, yields an upward transport of mass F_ρ:

$$F_\rho = -K_\rho \left[\frac{d\rho}{dz} - \left(\frac{d\rho}{dz} \right)_{isen} \right]$$

$$= \frac{\rho}{g} K_\rho N^2 > 0 \quad \left[\text{kg m}^{-2} \text{s}^{-1} \right] \quad (6.43)$$

where the term in parentheses corrects for the effect of adiabatic density changes (see eqn. 6.24) and K_ρ is the turbulent diffusion coefficient for water. By analogy with the surface buoyancy flux J_b^o defined in section 6.2, the potential energy change per unit mass related to F_ρ is called the (internal) buoyancy flux J_b^{int}. In the following equation the minus sign means that J_b^{int} is defined as positive if the potential energy is diminished, owing to net downward movement of mass. Then according to eqn. 6.43 kinetic energy is produced:

$$J_b^{int} = -\frac{g}{\rho} F_\rho = -K_\rho N^2 \quad \left[\text{W kg}^{-1} \right] \quad (6.44)$$

The balance of turbulent kinetic energy, E_{turb}, can be expressed by three terms:

$$\frac{\partial E_{turb}}{\partial t} = J_R + J_b - \varepsilon \quad (6.45)$$

with $J_b = J_b^o + J_b^{int}$, and where J_R is the production of E_{turb} due to the so-called Reynolds' stress which, by analogy with eqn. 6.13, may be written in terms of turbulent diffusivity of momentum. For instance, for the vertical velocity gradient of the x-component of the mean current, J_R is:

$$J_R = -\overline{u'_x u'_z} \frac{\partial \overline{u}_x}{\partial z} \quad (6.46)$$

Fig. 6.13 Monin–Obukhov length for two different circumstances: (a) the vertical density profile defines the mixing depth h_{mix}. (b) If $L_M < h_{mix}$, mixing and entrainment at the base of the mixed layer are primarily by convection represented by the depth-constant buoancy flux J_b^o. (c) If $L_M > h_{mix}$, mixing is mainly due to turbulent shear stress represented by a profile which decreases with depth h as h^{-1}. See eqn. 6.42.

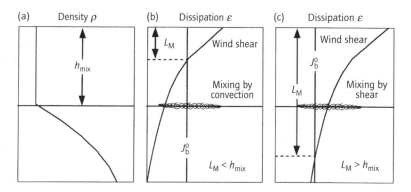

The term J_R includes all of the energy transfer processes, such as energy dissipation from the basin-wide currents, the dissipation of waves, etc., described in section 6.5. In the absence of density gradients, the production of E_{turb} due to the Reynolds' stress must be balanced by energy dissipation, thus $\varepsilon = J_R$. If J_b is positive there is an additional contribution to ε.

As discussed in section 6.4, a positive buoyancy flux is usually triggered at the water surface, either by cooling (if $T > T_{pmax}$), by salt fluxes, or by the effects of inlets (see Table 6.4). Depending on the density structure, the effect of a positive J_b^o may lead to the production of large vertical current patterns called **convection**. This may extend over the entire water column. Both convection and wind mixing act in the same manner. Both contribute to E_{turb}. Yet, whilst within the mixed layer the contribution of the buoyancy flux is roughly constant with depth, the shear-produced dissipation decreases inversely with depth h (eqn. 6.42). At a certain depth both contributions are equal. This depth is called the Monin–Obukhov length L_M, and is given by:

$$L_M = \frac{u_\star^3}{\kappa J_b^o} \qquad (6.47)$$

Depending on the relative size of L_M and h_{mix}, mixing and entrainment at the base of the mixed layer is either produced mainly by convection ($L_M < h_{mix}$) or by wind shear ($L_M > h_{mix}$). Examples are given in Fig. 6.13.

If the water column is stable and the buoyancy flux negative, the Reynolds' stress production term J_R is balanced by two opposing processes, the production of potential energy ($J_b < 0$) and turbulent kinetic energy dissipation ε. The so-called mixing efficiency γ_{mix} is defined by:

$$\gamma_{mix} = -J_b/\varepsilon = K_\rho N^2/\varepsilon \qquad (6.48)$$

Only the vertical velocity component can produce negative buoyancy while dissipation occurs in all three dimensions. Therefore, most of the available energy goes into ε. Experimental data yield values of γ_{mix} between 0.05 and 0.25.

A similar concept is given by the flux Richardson number (Rf), which is defined by

$$Rf = -J_b^{int}/J_R \qquad (6.49)$$

where Rf gives the fraction of available energy that is transformed into potential energy. At steady-state (E_{turb} = constant), the two numbers are related by

$$Rf = \frac{\gamma_{mix}}{1 + \gamma_{mix}} \qquad (6.50)$$

with Rf typically between 0.05 and 0.2. Combining eqns. 6.44 and 6.49 yields:

$$K_\rho = J_R \frac{Rf}{N^2} \qquad (6.51)$$

For constant E_{turb}, this can also be expressed by the energy dissipation rate ε:

$$K_\rho = \gamma_{mix} \frac{\varepsilon}{N^2} \qquad (6.52)$$

In contrast to Rf, ε can be experimentally determined, e.g. by microstructure measurements (Osborn 1980; Gregg 1987, 1991). Experimental data (Lewis & Perkin 1982; Coleman & Armstrong 1987; Wüest *et al.* 1988) confirm that K_ρ indeed increases with decreasing N^2, but often the energy source is stored wind motion, such as internal waves and seiches. Welander (1968) proposed the empirical relation

$$K_\rho = a\left(N^2\right)^{-b} \qquad (6.53)$$

with b lying typically between 0.5 and 1. A more detailed discussion is given by Imboden & Wüest (1995).

6.7 WATER CURRENTS CAUSED BY WAVES AND OTHER ORGANISED MOTION

A great variety of wave types exist in surface waters, among which **surface gravity waves** are the most apparent, but by no means the most important. An overview of important wave types, and the physical forces responsible for their propogation, is given in Table 6.5. Waves are always accompanied by water currents,[*] but do not always produce net transport of water. Instead, they mainly transport and store energy.

As discussed in section 6.5, energy storage by waves is an important process. Transport of energy also means transmission of information. For instance, the El Niño–Southern Oscillation (ENSO) phenomenon involves the action of large-scale Kelvin and Rossby waves in the atmosphere–ocean feedback mechanism, which couple air pressure anomalies over Indonesia with sea-surface temperature (SST) variations on the west coast of Ecuador and Peru (Battisti 1989). The following discussion will be focused on linear gravity waves; an extensive discussion of wave motion goes beyond the scope of this chapter. More information is given in LeBlond & Mysak (1978), Lighthill (1980) and Hutter (1993).

Surface gravity waves along the x-axis are described by

$$\zeta = \zeta_0 \exp[i(kx - \omega t)]$$
$$u = \frac{kg}{\omega} \zeta_0 \frac{\cosh[k(z+H)]}{\cosh(kH)} \exp[i(kx - \omega t)]$$
$$w = \frac{kg}{\omega} \zeta_0 \frac{\sinh[k(z+H)]}{\cosh(kH)} \times$$
$$\exp[i(kx - \omega t) - i\pi/2] \qquad (6.54)$$

where ζ is height of the water surface relative to the equilibrium level, ζ_0 is maximum wave amplitude, H is water depth, u and w are velocity in the x and z directions respectively, and k and ω are angular wave number and frequency. The latter are related to wave length λ and period T by $k = 2p/\lambda$ and $\omega = 2\pi/T$. The vertical coordinate z is positive upwards and zero at the equilibrium water level. If $(kH) \gg 1$, the velocity components are equal in size, but w lags behind u by a quarter period.

Generally, as the wave passes, each water parcel describes an elliptical motion with principal axes that decrease with depth. The relation between ω and k, the **dispersion relation**, is given by:

$$\omega^2 = gk \tanh(kH) \qquad (6.55)$$

Thus, surface gravity waves are generally dispersive, i.e. their phase velocity c_{ph} depends on wave number k or wavelength λ:

$$c_{ph} \equiv \frac{\omega}{k} = \left[\frac{g}{k} \tanh(kH)\right]^{1/2} \qquad (6.56)$$

For short waves (waves with $\lambda \ll H$, i.e. $kH \gg 1$), eqn. 6.56 becomes:

---

[*] Strictly speaking, this is not correct for sound waves, but they are not considered here.

Table 6.5 Important wave types in surface waters. (From Imboden & Wüest 1995.)

Wave	Restoring force	Characteristics of governing wave equation*	Dispersion relation
Capillary	Surface tension		$k^3 = \dfrac{\rho}{\delta}\omega^2$
Surface gravity	Gravity	First class, barotropic	$\tanh(kH) = \omega^2/gk$ long: $k^2 = \omega^2/gH$ short: $k = \omega^2/g$
Internal gravity	Gravity in density-stratified fluid	ρ = variable	$k^2 = \dfrac{\omega^2 - f^2}{N^2 - \omega^2}m^2$
Inertial	Coriolis force		
Poincaré (or Sverdrup)	Gravity and Coriolis force	First class, long gravity waves affected by Coriolis force $f \neq 0$, ρ = constant	$k^2 = \dfrac{\omega^2 - f^2}{gh}$
Kelvin	Like Poincaré, but with boundary conditions	f = constant depth H = constant, $u_{boundary} = 0$	$k = \dfrac{\omega}{\sqrt{gH}}$
Planetary or Rossby	Gravity, Coriolis force; potential vorticity	Large scale, long period, second class, f linear approximation of θ, ρ = constant, H = constant	$k^2 = \dfrac{\beta k}{\omega}$
Topographic Rossby waves	Gravity, Coriolis force; potential vorticity	Like Rossby waves but water depth $H \neq$ constant	

* First class, gravity waves modified by rotation; second class, variation of potential vorticity; ω, wave frequency; ρ, density of water; k, horizontal wave numbers; m, vertical wave number; N, buoyancy frequency; H, water depth; δ, surface tension ($\delta = 0.074\,\mathrm{N\,m^{-1}}$ at $T = 10\,°C$); $f = 2\omega\sin\theta$, inertial frequency ($\omega = 7.27 \times 10^{-5}\,\mathrm{s^{-1}}$); θ, geographical latitude; $\beta = (2\omega/R)\cos\theta$, change of f with latitudinal distance (R = Earth radius).

$$c_{ph} = \left(\frac{g}{k}\right)^{1/2} \quad \text{(short wave approximation)} \quad (6.57)$$

In dispersive systems the so-called **group velocity** c_{gr} is different from the phase velocity c_{ph}. The former describes the velocity of a group of waves, e.g. as triggered by a stone thrown into a pond. It is defined by $c_{gr} = d\omega/dk$. For short waves:

$$c_{gr} = \frac{1}{2}\left(\frac{g}{k}\right)^{1/2}$$
$$= \frac{1}{2}c_{ph} \quad \text{(short wave approximation)} \quad (6.58)$$

For instance, for waves with periods $T = 2$ seconds (see Fig. 6.2b), k is $1\,\mathrm{m^{-1}}$, wavelength $\lambda = 6\,\mathrm{m}$, and the two velocities c_{ph} and c_{gr} are 3 and $1.5\,\mathrm{m\,s^{-1}}$.

It was shown in section 6.5 that surface gravity waves take up the major portion of the wind energy (Table 6.2). The total (kinetic and potential) energy per unit surface area stored in a linear wave described by eqn. 6.54 is:

$$E_{wave} = \frac{1}{2}g\rho\zeta_0^2 \quad (6.59)$$

Thus, the energy $E_{wave} = 10^3\,\mathrm{J\,m^{-2}}$, the largest value listed in Table 6.2, corresponds to a wave amplitude $\zeta_0 \approx 0.5$ m. The corresponding current velocities at the water surface are:

$$u_{max}(z=0) = w_{max}(z=0) = (gk)^{1/2}\zeta_0 \qquad (6.60)$$

At wind speeds of 10 m s^{-1}, significant waves possess periods T of about 2 seconds, and the maximal current velocity owing to wave motion is about 1.5 m s^{-1}. In contrast, at low wind velocity (≈ 1 m s^{-1}), when E_{wave} is of the order of 1 J m^{-2}, waves are not controlled by gravity but by surface tension. These are called **capillary waves**. Their dispersion relation is given in Table 6.5. At a given period they possess a smaller wave length than gravity waves.

When waves move towards the shore, depth H decreases. Since frequency ω remains constant, but (according to eqn. 6.56) phase speed decreases, k increases, i.e. waves become shorter and eventually collapse on the beach. As discussed in section 6.5, wave breaking is an important mechanism of kinetic energy dissipation.

For $kH \ll 1$, eqn. 6.56 becomes:

$$c_{ph} = (gH)^{1/2}$$
$$\text{(shallow water approximation)} \qquad (6.61)$$

Long waves (waves with lengths much larger than water depth) are not dispersive. The longest waves in a lake possess lengths λ of the order of the basin length L. Usually H is much smaller than L, thus such waves are 'shallow waves'. In a lake with depth $H = 100$ m they travel with a speed of about 30 m s^{-1}. Since their phase speed does not change when they reach the shore, long waves are reflected with little energy loss, so they possess the potential to travel back and forth several times across the lake before they decay. In fact, only those waves survive for which the ingoing and outgoing waves are in phase. This is only possible if wavelength is an integral fraction of twice the lake length:

$$\lambda_n = \frac{2}{n}L, \; T_n = \frac{2L}{n}\frac{1}{(gH)^{1/2}} \quad (n=1,2,\ldots) \qquad (6.62)$$

Waves described by the above equation are the eigenmodes of a rectangular basin with length L and constant depth H. As discussed in section 6.5 they are called **seiches**, or more precisely **surface seiches**. Equation 6.62 is actually the well-known Merian formula (Merian 1828) which was introduced into limnology by Forel (1876) in order to explain periodic water level variations observed in Lake Geneva well back in the 18th century.

The pattern produced by waves which obey the relation in eqn. 6.62 can be described by the superposition of two waves of equal amplitude and wavelength which travel in opposite directions. The mathematical description of such waves differs only in the exponential factors, $\exp[i(kx - \omega t)]$ and $\exp[i(kx + \omega t)]$, respectively. Then, from eqn. 6.54, and from the shallow water approximation, there follows:

$$\zeta = 2\zeta_0 \cos(kx)\cos(\omega t)$$
$$u = 2\frac{kg}{\omega}\zeta_0 \sin(kx)\sin(\omega t) \qquad (6.63)$$

Here, vertical velocity w was disregarded since it is very small compared with u. In fact, the water prescribes a highly excentric ellipse with a ratio of principal axes of the order of (H/L). The motion generated by the seiche is nearly horizontal.

Equation 6.63 describes a **standing wave**. The surface elevation remains zero for $kx = (2m-1)(\pi/2)$. Those positions are called **nodes**, and are given by:

$$x_m = L\frac{2m-1}{2n} \quad \text{with } m = 1,\ldots n \qquad (6.64)$$

For the first seiche mode ($n=1$) the only node lies halfway along the lake ($x_1 = L/2$), while the second mode ($n=2$) has two nodes at $x_1 = (1/4)L$ and $x_2 = (3/4)L$. At the nodal points, horizontal velocity reaches its maximum value

$$u_{max}^{seiche} = A_{max}(gk)^{1/2} = A_{max}(g\pi n/L)^{1/2} \qquad (6.65)$$

where $A_{max} = 2\zeta_0$ is the maximum amplitude of the seiche reached, e.g., at the lake end. In turn, u is

Table 6.6 Typical displacements and current velocities due to first-mode $(n = 1)$ surface and internal seiche motion. Lakes A and B as in Fig. 6.11. Seiche amplitudes are assumed which are typical for strong winds. L_R is independent of amplitude, all other results are proportional to A_{max}. Values in parantheses indicate that Wedderburn number We becomes small enough to allow overturn of seiche motion and partial destruction of stratification.

	Surface	Internal
Lake A		
$L = 5$ km; $H = 20$ m, $H' = 5$ m, $\Delta\rho/\rho = 10^{-3}$		
Amplitude A_{max} (assumed)	10^{-3} m	(10 m)
$u_{max,1}$ (eqn. 6.65)	8×10^{-5} m s^{-1}	2.5×10^{-2} m s^{-1}
X_1 (eqn. 6.66)	0.16 m	$(6 \times 10^3$ m)
Rossby radius L_R (eqn. 6.67)	1.4×10^5 m	2.2×10^3 m
Lake B		
$L = 50$ km; $H = 100$ m, $H' = 16$ m, $\Delta\rho/\rho = 10^{-3}$		
Amplitude A_{max} (assumed)	10^{-2} m	(30 m)
$u_{max,1}$ (eqn. 6.65)	2.5×10^{-4} m s^{-1}	2.4×10^{-6} m s^{-1}
X_1 (eqn. 6.66)	3.2 m	$(60 \times 10^3$ m)
Rossby radius L_R (eqn. 6.67 at $\theta = 45°$)	3.2×10^5 m	4×10^3 m

zero at the antinodes defined by $x_m^* = (m/n)L$. Antinodes lie both at the end of the lake and halfway between nodes. At the nodes, the maximum total horizontal displacement for mode n_1, X_n, can be calculated from eqn. 6.63 by setting $\sin(kx_m^*) = 1$, and by integrating u over half a wave period:

$$x_n = \frac{2A_{max}}{\pi n} \frac{L}{H} \qquad (6.66)$$

Note that for a pure wave this displacement is strictly reversible. Numerical examples are given in Table 6.6.

In stratified waters (as most lakes are) waves also develop in the interior of the water column. They are called **internal waves**. As for surface waves, standing internal waves (internal seiches) are especially important since they last long and are able to store a significant amount of the mechanical energy transferred from the wind into the water column. The energy per unit area stored by the first surface mode was given by eqn. 6.38. In section 6.5 it was shown that the internal seiche is more important for storing energy than the surface seiche. For the special case of a sharp interface, i.e. for a density shift between two homogeneous water layers, all the wave equations discussed

above remain valid if g and H are substituted by reduced gravity g' and reduced depth H' (eqn. 6.39). Examples are given in Table 6.6.

The structure of basin-wide waves in real lakes is more complicated than the simple picture shown in Fig. 6.10. Usually the variations of the current velocities and wave amplitudes do not exhibit the sinusoidal form given by eqns. 6.54 and 6.63. As an example, Fig. 6.14 shows the relationship between the internal seiche and water currents measured at the bottom of Baldeggersee, a small but fairly deep lake (area = 5.2 km^2, maximum depth = 66 m) in central Switzerland. The irregular shape of lake basins explains part of the observed complexity. More than 80 years ago A. Defant (1918) and 35 years later F. Defant (1953) developed a one-dimensional procedure for calculating surface seiches in basins of irregular topography. Mortimer (1979) extended this method to internal seiches. Figure 6.15 shows horizontal and vertical displacements of the 1st and 2nd internal seiche modes in Lake Geneva (Switzerland). Owing to the shape of the lake, the amplitude along the lake has lost its sinusoidal shape. At the transition from the deep main basin to the narrower and less deep 'Petit Lac' (Small Lake) at the western end, horizontal displacement suddenly

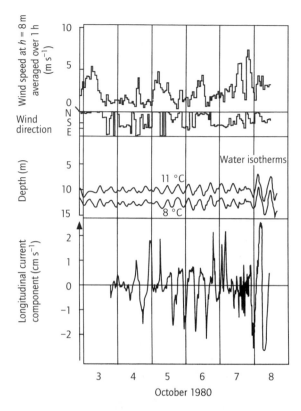

Fig. 6.14 Internal wave dynamics in the Baldeggersee (Switzerland). Wind speed and wind direction (top), vertical oscillation of thermocline (middle), and horizontal water currents measured at the bottom of the lake along its main axis. Frequency spectra of these signals (not shown) give distinct peaks both for isotherm oscillation and currents at the period of the 1st and 2nd horizontal mode, $T_1 = 8.4$ h, $T_2 = 4.1$ h. (From Lemmin & Imboden 1987; reproduced with permission of the American Society of Limnology and Oceanography.)

increases. Furthermore, the maximum amplitude at Geneva (western end) is much larger than at the eastern end of the lake, since the small basin acts like a funnel, taking up the large moving water mass from the main lake. The surface seiche exhibits similar characteristics, which explains why extreme seiche amplitudes have attracted the attention of people in Geneva for many years. According to Forel (1895), Fatio de Dullier, engineer of the city of Geneva, was the first one to use the word

seiches in a written document in which he describes the back and forth flux of the water occasionally leaving the shore at Geneva dry (French sèche).

A second important difference between the idealised wave model and real lakes relates to the continuous, rather than step-like, vertical density variation which is found in the latter. Since isopycnal surfaces are generally horizontal, the basin modes still propogate fairly horizontally, although their periods may differ from those calculated by the Merian formula (eqn. 6.62). However, the structure of short internal waves in continuous density gradients is more complicated. In principle, such waves can travel in all directions, and their reflection at a sloping boundary can lead to special phenomena (see e.g. Ivey & Nokes 1989).

Finally, water currents and thus also waves are affected by Earth's rotation. Since the Coriolis acceleration is acting perpendicular to the current (to the right in the Northern Hemisphere), it changes the direction of the flow without changing its speed. Currents in large fluid systems such as the atmosphere and the ocean cannot be understood without this influence. The flow in these systems is often geostrophic, i.e. determined by the balance between pressure and Coriolis forces. For instance, air does not directly plunge into the centre of a low pressure area to fill the pressure deficit. Instead, it is deflected to the right and then circles counterclockwise around the low such as to attain a dynamic equilibrium between the force from the pressure gradient (directed to the left) and the Coriolis acceleration to the right.

In small lakes, boundaries tend to suppress the influence of the Coriolis acceleration, but in large lakes, currents and waves are modified by the rotation of Earth. Take as an example a surface seiche. Owing to the deflection to the right, the right 'shoulder' of the wave piles up slightly against the right-hand shore. When the wave returns, the hump moves along the opposite shoreline. In essence this looks like a wave attached to the shore which moves counter-clockwise around the lake, and is known as a **Kelvin wave**. Mortimer (1975)

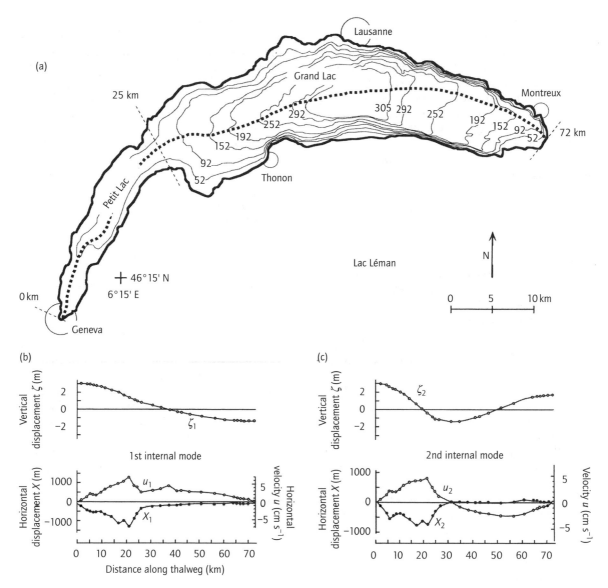

Fig. 6.15 Internal seiches in Lake Geneva (Switzerland/France) calculated by the two-layer Defant procedure (Mortimer 1979). (a) Map of the lake with depth contours, selected cross-sections and thalweg (dotted line, path along deepest point of cross-sections). (b) Maximum amplitude of interface ξ_1 along the thalweg (upper panel), maximum horizontal velocity u_1 and total horizontal displacement X_1 during one half-cycle for first horizontal mode with period $T_1 = 72.6\,h$ (lower panel). (c) As (b) for second internal mode with period $T_2 = 45.7\,h$. The maximum amplitude at 0 km (Geneva) is arbitrarily chosen as 3 m for both modes. (From Lemmin & Mortimer 1986; reproduced by permission of the American Society of Limnology and Oceanography.)

has masterfully analysed and drawn such waves, amongst others, for Lake Geneva and Lake Superior. Another wave type which combines the effects of gravitation and Earth's rotation is the **Poincaré wave**. These waves not only involve longitudinal displacements, but also possess transverse velocity components. Some characteristics of Kelvin and Poincaré waves are given in Table 6.5.

A criterion for the importance of rotational effects on basin waves is given by the **Rossby radius of deformation** L_R defined by

$$L_R = \frac{c_{ph}}{f} = \frac{c_{ph}}{2\pi} T_f \qquad (6.67)$$

Here c_{ph} is the phase velocity of the wave ($c_{ph} = (gH)^{1/2}$ for the surface or barotropic mode, $c_{ph} = (g'H')^{1/2}$ for the internal (baroclinic) mode (eqn. 6.61)), f is the Coriolis parameter (units s^{-1}) and T_f the inertial period defined by

$$T_f = \frac{2\pi}{f} \quad \text{with} \quad f = 2\Omega \sin\phi \qquad (6.68)$$

where $\Omega = 7.29 \times 10^{-5} \, s^{-1}$ is the angular frequency of Earth's rotation, and ϕ is geographical latitude. For mid-latitude lakes, $\phi = 45°$, the inertial period $T_f = 6.1 \times 10^4 \, s$ ($= 16.9 \, h$), and $f = 1.03 \times 10^{-4} \, s$.

Table 6.5 lists a further type of wave called **Rossby waves**. These large-scale waves only exist in rotational systems. They involve the so-called potential vorticity, a quantity that is approximately given by f/H. Such waves develop either because of the meridional variation of the Coriolis parameter f (planetary waves) or owing to the variation of depth H (topographic waves). Both kinds are important in the ocean. In large lakes topographic Rossby waves appear as long-period oscillations of current velocity as observed in Lake Michigan (Saylor et al. 1980). An extensive review of topographic waves is given by Stocker & Hutter (1987).

The appearance of streaks of foam at the water surface aligned in the direction of the wind, and typically 5 to 100 m apart, has intrigued limnologist and oceanographers for many years. Langmuir

(1938) showed that the streaks are a result of convergence of counter-rotating vortices with axes more or less parallel to the wind. Although some aspects of their origin are still only poorly explored, it is clear that they originate from the interaction between the wind and the wave field. A recent review of **Langmuir cells** in the ocean is given by Plueddemann et al. (1996).

In Langmuir cells, the influence of water currents on biological processes is particularly interesting. The cells develop close to the water surface, and occupy the top 5 to 20 m of the water column, where variations of light are strong. Their timescales are of the order of minutes to 1 hour and their impact on processes such as photosynthesis is different from, e.g., that of the motion due to surface waves. Furthermore, the regularly spaced convergence and divergence zones associated with the cells tend to trap floating objects with either positive or negative buoyancy, respectively. Stavn (1971) has studied *Daphnia* distribution in Langmuir cells, and Ledbetter (1979) conducted a theoretical investigation of their influence on plankton patchiness.

In this section, organised current patterns which evolve owing to the operation of different kinds of waves were discussed. Such considerations close the circle from the current spectrum presented in Fig. 6.2 to the mechanisms responsible for the observed structures. The rather extensive treatment of waves reflects the fact that, in closed water bodies, boundary conditions are dominant for the evolution of currents. Since lakes are driven by single and extremely variable weather events, the internal properties of the system such as eigenmodes, seiches, Kelvin and Poincaré waves, but also vertical density stratification, become more important. Therefore, the characteristic spatial displacements and current speeds produced by these modes are a central element in the physics of lakes.

Of course, there is much more lake physics between the very large and the very small processes, such as the dynamics of the diurnal mixed layer (Imberger 1985), the effects of convection, of currents at the bottom of lakes (Gloor et al. 1994; Wüest & Gloor 1998), of turbidity currents triggered by sediment slides or by an inlet. At this point, dear reader, you should be reminded of the beautiful and immense selection of lakes on Earth and of the vast literature in our libraries; both await your visit.

ACKNOWLEDGEMENT

The preparation of this chapter profited greatly from the scientific and technical help of Roland Hohmann who has composed the figures, took care of the references and looked after the many details which are essential for writing such an overwiew.

6.8   REFERENCES

Amorocho, J. & deVries, J.J. (1980) A new evaluation of the wind stress coefficient over water surfaces. *Journal of Geophysical Research*, **C85**, 433–42.

Batchelor, G.K. (1950) The application of the similarity theory of turbulence to atmospheric diffusion. *Quarterly Journal of the Royal Meteorological Society*, **76**, 133–46.

Batchelor, G.K. (1952) Diffusion in a field of homogeneous turbulence II. The relative motion of particles. *Mathematical Proceedings of the Cambridge Philosophical Society*, **48**, 345–62.

Batchelor, G.K. (1959) Small-scale variation of convected quantities like temperature in turbulent fluid. *Journal of Fluid Mechanics*, **5**, 113–39.

Battisti, D.S. (1989) On the role of off-equatorial oceanic Rossby waves during ENSO. *Journal of Physical Oceanography*, **19**, 551–9.

Bolin, B. & Cook R.B. (eds) (1983) *The Major Biogeochemical Cycles and their Interactions*. SCOPE Report No. 21, Wiley, Chichester, 532 pp.

Carter, H.H. & Okubo, A. (1965) *A Study of the Physical Processes of the Movement and Dispersion in the Cape Kennedy Area*. Final Report under the U.S. Atomic Energy Commission, Report NYO-2973-1. Chesapeake Bay Institute, Johns Hopkins University, Baltimore, MD, 164 pp.

Chen, C.T. & Millero, F.J. (1986) Precise thermodynamic properties for natural waters covering only the limnological range. *Limnology and Oceanography*, **31**, 657–62.

Coleman, J.A. & Armstrong, C.E. (1987) Vertical eddy diffusivity determined with radon-222 in the benthic boundary layer of ice covered lakes. *Limnology and Oceanography*, **32**, 577–90.

Defant, A. (1918) Neue Methode zur Ermittlung der Eigenschwingungen (Seiches) von abgeschlossenen Wassermassen (Seen, Buchten, usw.). *Annalen der Hydrographie*, **46**, 78–85.

Defant, F. (1953) Theorie des Seiches des Michigansees und ihre Abwandlung durch Wirkung der Corioliskraft. *Archiv für Meteorologie, Geophysik und Bioklmatologie* Wien A, **6**, 218–41.

Denman, K.L, & Miyake, M. (1973) Upper layer modification at Ocean Station Papa: observations and simulations. *Journal of Physical Oceanography*, **3**, 185–96.

Donelan, M.A., Dobson, F.W., Smith, S.D. & Anderson, R.J. (1993) On the dependence of sea surface roughness on wave development. *Journal of Physical Oceanography*, **23**, 2143–9.

Eckart, C. (1948) An analysis of the stirring and mixing processes in incompressible fluids. *Journal of Marine Research*, **7**, 265–75.

Einstein, A. (1905) Über die von der molekularkinetischen Theorie der Wärme geforderte Bewegung von in ruhenden Flüssigkeiten suspendierten Teilchen. *Annalen der Physik*, **17**, 549–60.

Flindt, R. (2000) *Biologie in Zahlen* (5th edition). Spektrum Akademischer Verlag, Heidelberg, 285 pp.

Forel, F.-A. (1876) La formule des seiches. *Archives des Sciences Physiques et Naturelles*, **57**, 278–92.

Forel, F.-A. (1895) Le Leman. In: Rouge F. (ed.), *Monographie Limnologique*. Librairie de l'Université, Lausanne, Vol. 2, 651 pp.

Garrett, C.J.R. & Munk, W.H. (1975) Space–time scales of internal waves: a progress report. *Journal of Geophysical Research*, **C 80**, 291–7.

Gill, A.E. (1982) *Atmosphere–Ocean Dynamics*. Academic Press, New York, 662 pp.

Gloor, M., Wüest, A. & Münnich, M. (1994) Benthic boundary mixing and resuspension induced by internal seiches. *Hydrobiologia*, **284**, 59–68.

Goudsmit, G.-H. & Wüest, A. (1997) Interior and basin-wide diapycnal mixing in stratified water: a comparison of dissipation and diffusivity. In: Imberger, J. (ed.), *Physical Limnology Coastal and Estuarine Studies*. Clarendon Press, Oxford, 145–63.

Gregg, M.C. (1987) Diapycnal mixing in the thermocline: a review. *Journal of Geophysical Research*, **C 92**, 5249–86.

Gregg, M.C. (1991) The study of mixing in the ocean: a brief history. *Oceanography*, **4**, 39–45.

Hohmann, R., Kipfer, R., Peeters, F., Piepke, G., Imboden, D.M. & Shimaraev, M.N. (1997) Processes of deep-water renewal in Lake Baikal. *Limnology and Oceanography*, **42**, 841–55.

Hutchinson, G.E. (1957) *A Treatise on Limnology*, Vol. 1, *Geography, Physics and Chemistry*. Wiley, New York, 1015 pp.

Hutter, K. (ed.) (1983) *Hydrodynamics of Lakes*. Springer-Verlag, Wien, 341 pp.

Hutter, K. (1993) Waves and oscillations in the ocean and lakes. In: Hutter, K. (ed.), *Continuum Mechanics in Environmental Sciences and Geophysics*. CISM Courses and Lectures No. 337. Springer-Verlag, New York, 79–240.

Imberger, J. (1985) The diurnal mixed layer. *Limnology and Oceanography*, **30**, 737–70.

Imberger, J. & Patterson, J.C. (1989) Physical limnology. In: Hutchinson, J.W. & Wu, T.Y. (eds), *Advances in Applied Mechanics*. Academic Press, Cambridge, 303–475.

Imberger, J. (1994) Transport processes in lakes: a review. In: Margalef, R. (ed.), *Limnology Now: a Paradigm of Planetary Problems*. Elsevier Science, Amsterdam, 99–193.

Imboden, D.M. (1990) Mixing and transport in lakes: mechanisms and ecological relevance. In: Tilzer, M. & Serruya, C. (eds), *Large Lakes: Ecological Structure and Function*. Springer-Verlag, Berlin, 47–80.

Imboden, D.M. (1998) The influence of biogeochemical processes on the physics of lakes. In: Imberger, J. (ed.), *Physical Limnology*. Coastal and Estuarine Studies, American Geophysical Society, Washington, DC, 591–612.

Imboden, D.M. & Joller, T. (1984) Turbulent mixing in the hypolimnion of Baldeggersee (Switzerland) traced by natural radon-222. *Limnology and Oceanography*, **29**, 831–44.

Imboden, D.M. & Wüest A. (1995) Mixing mechanisms in lakes. In: Lerman, A., Imboden, D.M. & Gat, J.R. (eds), *Physics and Chemistry of Lakes*. Springer-Verlag, Berlin, 83–138.

Ivey, G.N. & Nokes, R.I. (1989) Vertical! mixing due to the breaking of critical internal waves on sloping boundaries. *Journal of Fluid Mechanics*, **204**, 479–500.

Janssen, P.A.E.M. (1989) Wave-induced stress and the drag of air flow over sea waves. *Journal of Physical Oceanography*, **19**, 745–54.

Kitaigorodskii, S.A. (1983) On the theory of the equilibrium range in the spectrum of wind-generated waves. *Journal of Physical Oceanography*, **13**, 816–27.

Kolmogorov, A.N. (1941) The local structure of turbulence in an incompressible viscous fluid for very large Reynolds numbers. *Doklad Akademie Nauk USSR*, **30**, 299–303.

Langmuir, I. (1938) Surface motion of water induced by wind. *Science*, **87**, 119–23.

LeBlond, P.H. & Mysak, L.A. (1978) *Waves in the Ocean*. Oceanography Series, No. 20, Elsevier, Amsterdam, 602 pp.

Ledbetter, M. (1979) Langmuir circulations and plankton patchiness. *Ecological Modelling*, **7**, 289–310.

Lemmin, U. (1987) The structure and dynamics of internal waves in Baldeggersee. *Limnology and Oceanography*, **32**, 43–61.

Lemmin, U.P. & Imboden, D.M. (1987) Dynamics of bottom currents in a small lake. *Limnology and Oceanography*, **32**, 62–75.

Lemmin, U. & Mortimer, C.H. (1986) Tests of an extension to internal seiches of Defant's procedure for determination of surface seiche characteristics in real lakes. *Limnology and Oceanography*, **31**, 1207–31.

Lewis, E.L. & Perkin, A.G. (1982) Seasonal mixing processes in a Arctic fjord system. *Journal of Physical Oceanography*, **12**, 74–83.

Lewis, M.R., Horue, E.P.W., Cullen, J.J., Oakey, N.S. & Platt, T. (1984) Turbulent motions may control phytoplankton photosynthesis in the upper ocean. *Nature*, **311**, 49–50.

Lighthill, J. (1980) *Waves in Fluids*. Cambridge University Press, Cambridge, 504 pp.

Longuet-Higgins, M.S. (1969) On wave breaking and the equilibrium range of wind-generated waves. *Proceedings of the Royal Society of London, Series A* **310**, 151–9.

Malkus, W.V.R. (1981) The amplitude of convection. In: Warren, B.A. & Wunsch, C. (eds), *Evolution of Physical Oceanography*. MIT Press, Cambridge, MA, 384–94.

Massel, S.R. (1999) *Fluid Mechanics for Ecologists*. Springer-Verlag, Heidelberg, 566 pp.

McDougall, T.J. (1984) The relative roles of diapycnal and isopycnal mixing on subsurface water mass conversion. *Journal of Physical Oceanography*, **14**, 1577–89.

McDougall, T.J. (1987a) Thermobaricity, cabbeling and

water mass conversion. *Journal of Geophysical Research* **92**, 5448–64.

McDougall, T.J. (1987b) Neutral surface. *Journal of Physical Oceanography*, **17**, 1950–64.

McDougall, T.J. & You, Y. (1990) Implications of the nonlinear equation of state for upwelling in the ocean interior. *Journal of Geophysical Research*, C **95**, 13263–76.

Merian, J.-R. (1828) Über die Bewegung tropfbarer Flüssigkeiten in Gefässen. Abhandlung J-R. Merian, Basel.

Mortimer, C.H. (1975) Substantive corrections to SIL communications (IVL Mitteilung) Numbers 6 and 20. *Verhandlungen Internationale Vereinigung der Limnologie*, **19**, 60–72.

Mortimer, C.H. (1979) Strategies for coupling data collection and analysis with dynamic modelling of lake motion. In: Graf, W.H. & Mortimer, C.H. (eds), *Lake Hydrodynamics*. Elsevier, Amsterdam, 183–277.

Münnich, M., Wüest, A. & Imboden, D.M. (1992) Observations of the second vertical mode of the internal seiche in an alpine lake. *Limnology and Oceanography*, **37**, 1705–19.

Okubo, A. (1971) Oceanic diffusion diagrams. *Deep Sea Research*, **18**, 789–802.

Osborn, T.R. (1980) Estimates of the local rate of vertical diffusion from dissipation measurements. *Journal of Physical Oceanography*, **10**, 83–9.

Paterson, A.R. (1983) *A First Course in Fluid Dynamics*. Cambridge University Press, Cambridge, 528 pp.

Pedlosky, J. (1979) *Geophysical Fluid Dynamics*. Springer-Verlag, Berlin, 624 pp.

Peeters, F., Wüest, A., Piepke, G. & Imboden, D.M. (1996a) Horizontal mixing in lakes. *Journal of Geophysical Research*, C **101**, 18361–75.

Peeters, F., Piepke, G., Kipfer, R., Hohmann, R. & Imboden, D.M. (1996b) Description of stability and neutrally buoyant transport in freshwater lakes. *Limnology and Oceanography*, **41**, 1711–24.

Plueddemann, A.J., Smith, J.A., Farmer, D.M., Weller, R.A., Crawford, W.R., Pinkel, R., Vagle, S. & Gnanadesikan, A. (1996) Structure and variability of Langmuir circulation during the Surface Waves Processes Program. *Journal of Geophysical Research*, C **101**, 3525–43.

Reid, W.H. & Harris, D.L. (1958) Some further results on the Bénard problem. *The Physics of Fluids*, **1**, 102–10.

Reynolds, C.S. (1992) Dynamics, selection and composition of phytoplankton in relation to vertical structure in lakes. *Ergebnisse der Limnologie*, **35**, 13–31.

Reynolds, C.S. (1994) The long, the short and the stalled; on the attributes of phytoplankton selected by physical mixing in lakes and rivers. *Hydrobiologia*, **289**, 9–21.

Saylor, J.H., Huang, J.C.K. & Reid, R.O. (1980) Vortex modes in southern Lake Michigan. *Journal of Physical Oceanography*, **10**, 1814–23.

Schwarzenbach, R.P., Gschwend, P.M. & Imboden, D.M. (2003) *Environmental Organic Chemistry* (2nd edition) Wiley, New York, 1313 pp.

Simon, A. (1997) *Turbulent mixing in the upper layer of lakes*. Dissertation No 12272, Eidgenössische Technische Hochschule, Zürich, 99 pp.

Spigel, R.H. & Imberger, J. (1980) The classification of mixed-layer dynamics in lakes of small to medium size. *Journal of Physical Oceanography*, **10**, 1104–21.

Stavn, R.H. (1971) The horizontal–vertical distribution hypothesis: Langmuir circulation and *Daphnia* distributions. *Limnology and Oceanography*, **16**, 453–66.

Stocker, T. & Hutter, K. (1987) *Topographic Waves in Channels and Lakes on the f-plane*. Lecture Notes on Coastal and Estuarine Studies, No. 4. Springer-Verlag, Berlin, 176 pp.

Tee, K.-T. (1987) Simple models to simulate three-dimensional tidal and residual currents. In: Heaps, N.S. (ed.), *Three-dimensional Ocean Models*. Coastal and Estuarine Sciences, No. 4. American Geophysical Union, Washington, DC, 125–49.

Terray, E.A., Donelan, M.A., Agrawal, Y.C., Drennan, W.M., Kahma, K.K., Williams, A.J. III, Hwang, P.A. & Kitaigorodskii, S.A. (1996) Estimates of kinetic energy dissipation under breaking waves. *Journal of Physical Oceanography*, **26**, 792–807.

Thompson, H.O. & Imberger, J. (1980) Response of a numerical model of a stratified lake to wind stress. In: Carstens, T. & McClimans, T. (eds), *Second International Symposium on Stratified Flows*. Tapir, Trondheim, Vol. 1, 562–70.

Thorpe, S.A. & Humphries, P.N. (1980) Bubbles and breaking waves. *Nature*, **283**, 463–5.

Walker, S.J. & Watts, R.G. (1995) A three-dimensional numerical model of deep ventilation in temperate lakes. *Journal of Geophysical Research*, C **100**, 22711–31.

Welander, P. (1968) Theoretical forms for the vertical

exchange coefficients in a stratified fluid with application to lakes and seas. *Geophys Gothenburg*, **1**, 1–27.

Wu, J. (1994) The sea surface is aerodynamically rough even under light winds. *Boundary-layer Meteorology*, **69**, 149–58.

Wüest, A., Imboden, D.M. & Schurter, M. (1988) Origin and size of hypolimnic mixing in Urnersee, the souther basin of Vierwaldstättersee (Lake Lucerne). *Schweizerische Zeitschrift für Hydrologie*, **50**, 40–70.

Wüest, A., Piepke G. & Halfman, J.D. (1996) Combined effects of dissolved solids and temperature on the density stratification of Lake Malawi. In: Johnson, T.C. & Odada, E.O. (eds), *The Limnology, Climatology and Paleoclimatology of the East African Lakes*. Gordon & Breach, Toronto, 183–202.

Wüest, A. & Gloor, M. (1998) Bottom boundary mixing: the role of near-sediment density stratification. In: Imberger, J. (ed.), *Physical Processes in Lakes and Oceans*. Coastal and Estuarine Studies. American Geophysical Society, Washington, DC, 485–502.

7 Regulatory Impacts of Humic Substances in Lakes

CHRISTIAN E.W. STEINBERG

7.1 INTRODUCTION

The traditional focus of limnoecology is on living organisms and their interactions with the abiotic components of the limnetic medium. Within this appreciation, the most established concept is of a community structure and function regulated by the structure and energy flow through the network of trophic relationships, or food web. This powerful paradigm is dominated by the behaviour of the living components whereas, in reality, aquatic systems are awash with the organic remains of dead organisms which are subject to **humification**. Complex and difficult to characterise though they have been, humic substances should be of primary interest to all limnoecologists, not least because they comprise what is often the largest fraction of the organic carbon to be found in lakes. In many instances, the mass of dead organic matter exceeds that of all the living organisms by, roughly, an order of magnitude (Wetzel 1995), while the fraction of humic substances in solution (dissolved humic matter, or DHM) often makes up the largest part of the dissolved organic matter (DOM) present in inland waters. However, it is only recently that ecologists have begun to devote the attention that these substances deserve. It is now becoming clear that these substances have important regulatory impacts on the limnetic biota, across a range of spatial and temporal scales. Very recently, a comprehensive treatise on the ecology of humic substances has been published (Steinberg 2003).

This chapter addresses some of the key environmental interactions involving DHM. Issues confronted concern the diverse nature and sources of humic solutes, their reactivity with other components of the system and their effects on material cycling within aquatic habitats.

The term 'humic substances' (HS) embraces three principal groups of compounds: **fulvic acids**, **humic acids** and **humin**. They may be separated by a process of fractionation. Treatment of bulk humic material with dilute alkali dissolves the fulvic and humic acids, leaving a residue of insoluble humin (Killops & Killops 1993). Acidification of the alkaline extract precipitates humic acids (HA), leaving the fulvic acids (FA) in solution. The FAs comprise smaller and more oxygenated molecules than HAs and, thus, have been supposed to be products of natural oxidative degeneration of HAs. Both HA and FA are thought to be intermediates in the diagenesis of humin. The reality may be more complicated, because humic substances are not confined to single physical states. For instance, HAs are found in solution in fresh waters, as solids or as colloidal gels in waters emanating from soils and bogs, in streams and lakes, or as dry solids in deposits such as coal (Killops & Killops 1993).

Not all the organic carbon in solution is humic: carbohydrates or their photosynthetic precursors, fatty acids and free amino-acids, are usually present in water supporting active microorganisms. By consensus, 'dissolved organic matter' (DOM) is now defined practically by its filterability through a 0.45-μm pore size, even though filtrate may well contain fine colloids as well as molecules in true solution. The filtrate fraction may even contain fine particles of humin, as well as complexed FAs and (especially) HAs. A graphical representation of 'DOM' is shown in Fig. 7.1. For ease of reference, the abbreviations adopted in this chapter are summarised in Table 7.1, where some other acronyms used are also listed.

Among fresh waters, as much as 80% of the total DOM present in fresh waters may be humic in origin (Steinberg & Münster 1985), of which aromatic carbon compounds make up some 30–40% (Thurman 1985; Kļaviņš 1997). In strongly humic waters, the pool of humic substances (HS) comprises up to 95% of the DOC (Münster *et al.* 1999). In terms of organic carbon, DHM concentrations of 4–8 mg DOC L^{-1} are common among surface waters of lakes and may often exceed 25 mg DOC L^{-1} in the interstitial waters of the organic-rich sediments of wetland systems (Wetzel 1984).

Dissolved humic matter is widely regarded to be refractory and difficult to break down ('recalcitrant'). Certainly, the longevities of terrestrially derived humic substances in water vary widely. There are also appreciable discrepancies between the results of chemical analyses and biological or biochemical estimates. Chemical approaches generally show the age of bulk DOM to range between a few hundred to 6000 years. On the other hand, microbial methods applied to seawaters containing in the order of 1 mg DOC L^{-1}, predominantly of substances of low molecular weight, suggested average longevities of days to weeks (Amon & Benner 1994). Comparable studies on humic fresh waters

are scarce but the examples available reveal even lower ages (Münster & de Haan 1998; Grossart & Plough 2001).

Already, it is obvious that there are differences among the DOM contents of fresh and marine waters, although the general supposition is that oceanic DHM is terrestrially derived. The expectation is that the HS exported, in relatively high con-

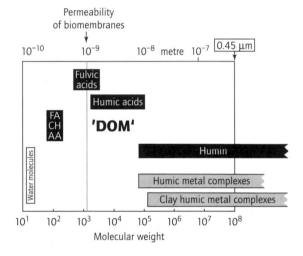

Fig. 7.1 Graphical definition of 'dissolved' organic matter (DOM), including small organic molecules and the three fractions of humic substances: FA, fatty acids; CH, carbohydrates; AA, amino acids. (After Thurman 1985; reproduced with kind permission of Kluwer Academic Publishers.)

Table 7.1 Glossary of abbreviations for organic carbon fractions and other notations used frequently in this chapter.

Organic carbon fractions		Other notations	
DHM	dissolved humic matter	BCF	bioconcentration factor
DOC	dissolved organic carbon	BaP	benzo[a]pyrene
DOM	dissolved organic matter	DNA	deoxyribonucleic acid
FA	fulvic acid	HMW	high molecular weight
HA	humic acid	LMW	low molecular weight
HS	humic substances (in general: HS = FA + HA + humin)	PAH	polycyclic aromatic hydrocarbon
OSC	octanol-soluble carbon	TBA	terbutylazine
TOC	total organic carbon	UV	ultraviolet radiation

centrations through significant river and estuarine flows, would retain signatures of its terrestrial provenance, although there has been no compelling evidence that this is so. Pending such confirmation, supposition of a terrestrial source must be regarded as unsubstantiated. On the other hand, there is no compelling evidence that it originates in the sea either.

We are still some distance from a clear and full understanding of the metabolism of DHM in aquatic systems. Setting aside the danger attaching equal metabolic weightings to all forms of organic matter, several aspects of the regulatory capability of DOC/DHM require further evaluation. However, it is already plain that DHM is not all as refractory as once believed.

As more is learned about the composition and reactivity of the DHM fractions, the significance of the metabolism of dissolved organic carbon is become clearer. This chapter considers some of these reactivities, involving ions, larger associated groups and major cleavages, catalysed by microbial action, by chemical oxidation and the effects of solar radiation. These reactivities interact with nutrient availabilities, metal and contaminant ('xenobiotic') toxicities and with other biopolymers.

In seeking to consolidate a contemporary overview of the role of DHM in the biogeochemical functioning of lakes, it is impossible to escape Wetzel's (1995) deduction that 'the trophic structure of the pelagic zone is not only a particulate world' but one in which 'nearly all of the organic carbon is dissolved or is particulate dead organic carbon'. Moreover, with the ratios DOC to POC (particulate organic carbon) found in lakes and running water typically between 6:1 and 10:1 (Wetzel 1984), it becomes very clear how the size and the fluxes of the energetic pathways founded on organic carbon should dominate the essential metabolism and maintenance of the ecosystem. 'Population fluxes are not representative of the material and energy fluxes of either the composite pelagic region or the lake ecosystem' (Wetzel 1995). Rather, he argued for the dominance of the material and energy fluxes by the metabolism of the particulate and, especially, the dissolved or-

ganic detritus, from many pelagic and autochthonous non-pelagic and allochthonous sources. Because of the large magnitude and relative chemical recalcitrance of these detrital products, their slow metabolism counters the 'ephemeral, volatile' fluctuations at the higher trophic levels, providing a strongly stabilising effect on the ecosystem.

Wetzel's deliberations provide a deep challenge to the adherence to organism-orientated relationships to explain process-driven couplings of aquatic ecosystems. However, they also allow us to evaluate the regulatory capacity of DHM/DOM transformations:

1 over long timescales, changing the chemistry of lakes;
2 over medium to long timescales, being subject to atmospheric and climatic changes;
3 on short time and multi-species scales, through nutritional effects;
4 on short time and single-species scales, affecting fertility, bioconcentration of potentially toxic metals and xenobiotic substances, and the production of oxidant substances.

7.2 LONG-TERM, WHOLE-LAKE EFFECTS OF DISSOLVED HUMIC MATTER

7.2.1 Acidification

Humic substances possess a variety of functional groups, among which carboxylic acids and phenols are prevalent. These impart to humic substances their typically acidic properties. They contribute significantly to the acid-base status of inland waters founded on non-calcareous catchments. Consequently, some authors have considered humic substances to provide a proximal source of acidity in appropriately sensitive inland waters.

Anthropogenic acidification of aquatic ecosystems and their catchments was a major environmental issue during the closing decades of the twentieth century and the processes and the damage caused are now well documented (Steinberg & Wright 1994). However, acidification of soils as a consequence of vegetation development undoubtedly contributes to a lowering of pH in drainage and receiving waters but, in many instances, changes have been complemented or obscured by other influences (see Krug & Frink 1983; Pennington 1984; cf. Johnson *et al.* 1984; Battarbee *et al.* 1985). Nevertheless, before the recent trend towards increased lake acidification was corrected, some authors considered the effects of acid precipitation on particular lakes to have been exaggerated, attributing the balance of the enhanced acidity to changes in the local catchment. In reconstructing the post-glacial development of the Grosser Arbersee, in the Bavarian Forest, Germany, Steinberg (1991) distinguished periods of both natural and anthropogenic acidification. In Fig. 7.2, the fluctuating pH levels in the lake over the last 13,000 years, inferred from the main contemporaneous diatoms deposited in the lake (for principles, see Chapter 18), are compared with the total organic carbon content (TOC) of the sediments from the corresponding depths. During the last part of the glacial period, the deposits (now 4.2 m beneath the sediment surface) indicate that the pH of the Grosser Arbersee was typical for a natural soft-water lake (pH 6). Meltwaters supplied the lake with relatively greater proportions of inorganic, acid-neutralising bases but organic carbon was minimal. Through the post-glacial period of rising TOC accumulation, pH declined gradually to c.5. The correlations, shown in Fig. 7.3, are explained by the leaching of organic acids from the developing soil and vegetation to deeper soil layers (spodosolation). In their passage through the soil, the organic acids condense into humic materials (Larson & Hufnal 1980) and these, in turn, contribute increased acidity to the lake. Positive correlations between acidity and TOC are reported from sediment studies elsewhere (Reinikainen & Hyvärinen 1997; Seppä & Weckström 1999).

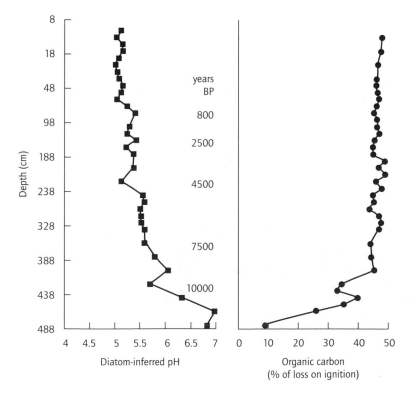

Fig. 7.2 Late and post-glacial development of organic carbon content and diatom-inferred pH in Grosser Arbersee sediments. (From Steinberg 1991. Copyright (1991), with permission from Elsevier Science.)

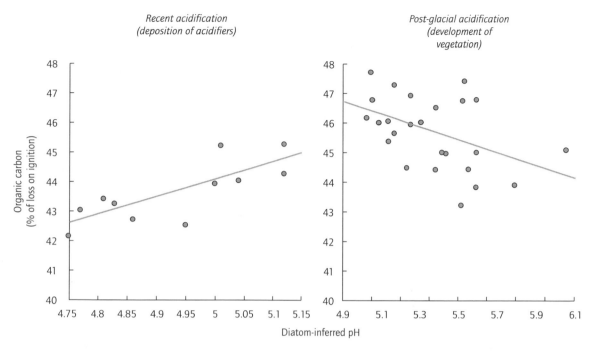

Fig. 7.3 Role of organic carbon in the acidification processes of Grosser Arbersee (Bavarian forest). (From Steinberg 1991. Copyright (1991), with permission from Elsevier Science.)

Acidification often leads to increased transparency of lakes. It has been supposed that acid-mobilised metals, especially aluminium, hydrolyse and co-precipitate with coloured organic matter, while some of the high molecular-weight DOM is cleaved into lighter coloured DOM of lower molecular weight (Dickson 1980; Yan 1983; Steinberg & Kühnel 1987; Schindler *et al.* 1997). Davis *et al.* (1985) showed a loss of organic matter from the water column of each of two acidified Norwegian lakes, with TOC decreasing from c.9 to c.6 mg L^{-1} in either case.

The main drivers of modern acidification are the strong-acid residues from the combustion of fossil hydrocarbon and wastes (Bruckmeier *et al.* 1997). At Grosser Arbersee, the acidity–organic-carbon relationship is quite opposite during the last 150 years, where a decrease in pH has been correlated with a reduction in the sedimentary organic carbon content (see Fig. 7.4). Anthropogenic increases in the acidity of precipitation generally

lead to diminution in the organic carbon content of lake water, while simultaneous increases in the acidity of soils also lead to a decrease in the flux of organic carbon to the lake (Marmorek *et al.* 1987; Schindler *et al.* 1997). They proposed several explanations for this contrasted behaviour:

1 increased deposition of mineral acids results in decreased organic matter mobilisation from soils and wetlands;

2 increases in the concentrations of mineral acids in soils, sediments and surface waters slow the microbial breakdown of organic matter;

3 increased concentrations of mineral acids regulate the dissociation of organic acids, alter the physical structure of dissolved humic substances, and reduce their mobility;

4 increased concentrations of mineral acid increase the rate at which dissolved organic matter is lost from surface waters to the sediments through co-precipitation with metals (Al, Fe) that are themselves mobilised from soils by strong mineral acids.

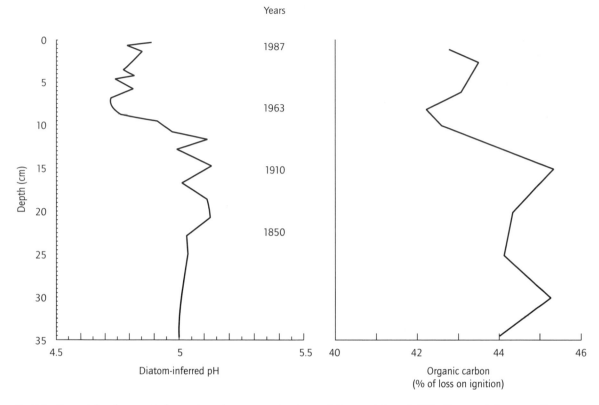

Fig. 7.4 Recent development of organic carbon concentration and diatom-inferred pH in Grosser Arbersee sediments. (From Steinberg 1991. Copyright (1991), with permission from Elsevier Science.)

These hypotheses are not mutually exclusive; they overlap and they may be additive.

The magnitude of DOC loss from waters subject to recent acidification has been estimated by Steinberg (1991), who regressed mass losses on ignition against diatom-inferred pH for a selection of lakes, to vary between c.10 to 50% per one pH unit. The lakes are located at various altitudes with differing subsurface geologies. Although organic content diminishes with lowered pH in each of the four examples (see Fig. 7.5), the effect is relatively more pronounced in the lakes already organically rich prior to the recent anthropogenic acidification (Kleiner Arbersee, Kleiner Bullensee). Thus, there is no unique relationship between the decrease in organic content and the degree of acidification.

7.2.2   Changes in humic matter caused by acidification

Changes in pH and ionic strength of drainage water brought about by catchment acidification also bring about physico-chemical changes in the humic substances present. Gonet & Wegner (1996) showed that the mean molecular weight of HA diminished with lowering pH. Moreover, the solution chemistry of humic substances is influenced by the functional-group heterogeneity and variation in molecule size distributions (Tipping et al. 1990; de Wit et al. 1993; Ephraim et al. 1996).

These, essentially laboratory, findings are backed by the results of the HUMEX Project (Humic Lake Acidification Experiment; for de-

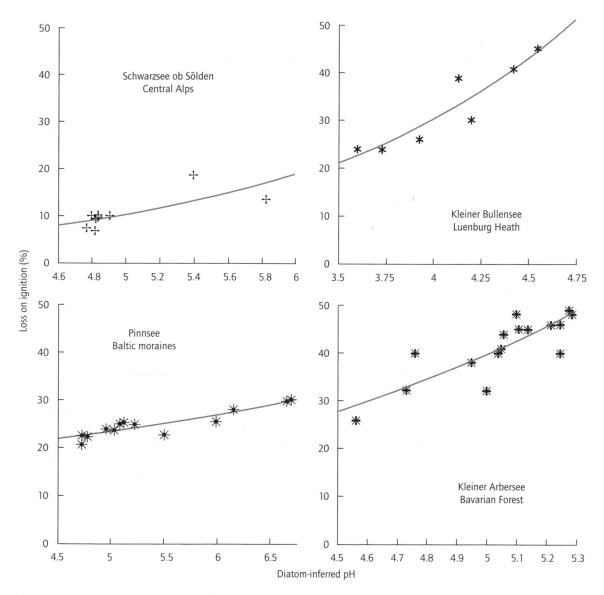

Fig. 7.5 Regressions relating diatom-inferred pH to loss on ignition (surrogate of TOC) in Schwarzsee ob Sölden (Central Alps), Kleiner Bullensee (Lüneburg Heath), Pinnsee (Baltic moraines) and Kleiner Arbersee (Bavarian Forest). (From Steinberg 1991. Copyright (1991), with permission from Elsevier Science.)

tails, see Gjessing 1992) at Skervatjen, Norway, which attempted to elucidate the fate of HS in a whole dystrophic lake subject to artificial acidification of its surface waters. The lake was divided by a plastic curtain, one part being acidified, the other left as a control (Gjessing 1994a). The HS content, approximated from measurements of TOC, colour and UV absorption, increased in both sections but by almost twice as much in the control than in the test. The TOC export from the control section also greatly exceeded that of the acidified test section. Thus, the findings are consistent with the palaeoecological assessment showing HS reduction with acidification. They also concur with the results of studies showing that increased levels of acidity lower the solubility of the hydrophobic acid fraction in spodosolic soils (David et al. 1989; Vance & David 1989).

In the acid-treated section of HUMEX, the soil-to-water transport of octanol-soluble organic carbons (OSCs) showed an average first-year increase (from 46 to $104 \mu g L^{-1}$ OSC; Kullberg et al. 1992). This confirmed that a significant acidification-led change in the physical–chemical properties of the HS had occurred, with some protonation of humic functional groups making them less water-soluble. Whereas the traditional geochemical analysis of DOC revealed only slight decreases in the ratio of hydrophobic to hydrophilic fractions (Gjessing 1994b; Vogt et al. 1994), the use of proton nuclear magnetic resonance (^{1}H-NMR) spectroscopy revealed distinct changes in the composition of the DHM (Malcolm & Hayes 1994): after 1 year of acidification, there had been a 13% decrease in the proton attachment to the N and O atoms of the fulvic acids and a decrease in aromatic protons of between 27 and 31%. Humic acids also showed an 11% decrease in aromatic protonation as a consequence of acidification. Thus, the general trend of FA- and HA-protonation in response to increased acidification appears to be a decrease in the proportion of aromatic attachment in favour of aliphatic protons.

Further evidence of the effects of acidification on the HS content of water comes from palaeolimnological investigations of recent lake sediments in acid-impacted catchments. A case study is provided by stratigraphic analysis of humic substances in short sediment cores corresponding to the acidification of the Grosser Arbersee (Steinberg et al. 2001). The acidification was reconstructed from the subfossil diatom assemblages: over the preceding 130 years, diatom-inferred pH decreased more or less continuously from 5.3 to 5.0. The characteristics of the sedimented HS altered during the acidification, the proportion of HA increasing from 12 to 17%.

Within the FA fraction, carbohydrates increased by c.10%, whereas aliphatic and aromatic compounds decreased, by about 10% and 3% respectively. Acidification accounted roughly for 50% of the observed changes in the constituents.

Although the changes to the HS content of the water are attributable ultimately to enhanced mineral acid deposition, the sensitivity is clearly dominated by changes to the processes affecting the huge reservoir of HS held in the catchment soils. Nevertheless, many chemical and microbial processes are common to lakes and soils: in surface drainage, the open waters of lakes and in their bottom sediments alike, increases in the mineral-acid content depress the rates of microbial breakdown of leaf litter and other major sources of organic matter (Kelly et al. 1984; Schindler 1994). Strong acids seem to affect the enzymatic cleavage of the large polymers, such as cellulose and proteins, rather than the metabolism of compounds of relatively low molecular weight, such as urea (Bewley & Parkinson 1986).

Certainly, the decomposition of both aquatic (Nymphaea) and terrestrial (Betula) leaves is considerably slower in acid water (Kok & van der Velde 1994). Kok et al. (1992) showed that, although the fungi capable of breaking down dead macrophytic tissue are not prevented from developing at pH levels as low as 4.0, their ability to macerate the substrate tissue is severely impaired below pH 5.5. This near-specific inhibition of enzyme action would explain the increased representation of carbohydrate groups in the HS of acidifying systems.

The significant influence on the acid–base relationship in non-calcareous lake catchments that is exerted by the DHM/DOM loads they receive is sketched in Fig. 7.6. It provides the template for the development in this chapter of the impact of internal transformations of the organic load and their impact on other functions.

7.3 DISSOLVED HUMIC MATTER AND MEDIUM-TERM CHANGE

If the environmental threats posed by acidification have lessened, those attributable to global climate

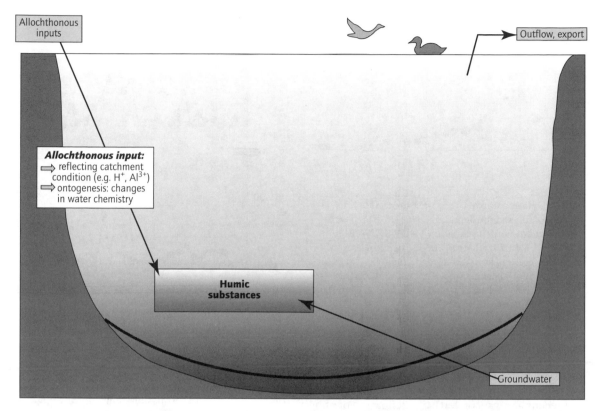

Fig. 7.6 Synoptic view of humic substances as long-term determinant of lakewater chemistry. (Steinberg 2003; reproduced with permission of Spring-Verlag.)

change and, in particular, to the enhanced penetration by ultraviolet (UV) radiation of an ozone-depleted stratosphere have intensified. Like acidification, their effects on lakes are mediated by the dynamics of DHM/DOC. Hence, the main focus of this section is the interaction between DOC and UV absorption in water.

The relative penetration of UV-B through the surface waters of boreal lakes is generally appreciated to be a function of the organic carbon dissolved in the water (Fig. 7.7). By absorbing most of the ultraviolet radiation within the first few decimetres of the water surface, DOC concentrations of the order of several milligrams per litre are sufficient to provide an effective shield for aquatic organisms (Schindler *et al.* 1996). On the other hand, UV-B penetration is greater in lakes of low DOC content. Two factors are mainly responsible

for the relationship: the less is the total DOC pool, the greater is the proportion of colourless DOC produced in the lake, while photobleaching and photodegradation increases as a function of the hydraulic residence time of coloured, mainly allochthonous, DOC in the lake (Schindler *et al.* 1996). The DOC that reaches the lake has had, effectively, to have resisted microbial decomposition in soils; thus, the proportion of coloured DOC in boreal lakes tends to be a function of the catchment/lake ratio (Engstrom 1987).

7.3.1 *Inferred recent changes in dissolved organic matter fluxes and ultraviolet income. Observations from Canada*

The complex interactions between solar radiation and DOC depend predominantly on the relatively

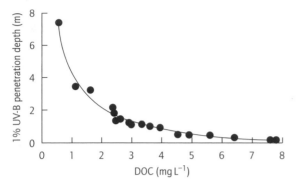

Fig. 7.7 Relationship between measured DOC concentration and the depth of the 1% UV-B isopleth. The line represents the equation 1%UV-B = 5.173 $(DOC)^{-0.706} - 1.029$, where $r^2 = 0.98$, fitted to data from 18 lakes of boreal and northern Canada, including lakes in the Experimental Lakes Area. (From Schindler *et al.* 1996; reprinted with permission from *Nature*. Copyright (1996) Macmillan Magazines Limited.)

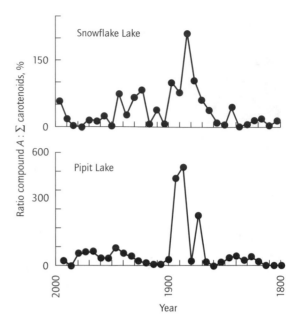

Fig. 7.8 Regional increases in penetration of UV radiation in Snowflake Lake and Pipit Lake following droughts in western Canada, indicated by elevated concentration of UV-radiation-specific pigment (compound *A*) from *c.*1850 to *c.*1900. (From Leavitt *et al.* 1997; reprinted with permission from *Nature*. Copyright (1997) Macmillan Magazines Limited.)

rapid attenuation of visible and ultraviolet wavelengths with increasing subsurface depth (Kirk 1983; Rasmussen *et al.* 1989; Scully & Lean 1994). In a recent analysis of vertical attenuation of photosynthetically active radiation in the boreal lakes of northwestern Ontario, Fee *et al.* (1996) attributed all but 6% of average light absorption to be due to DOC. The magnitude of the variable UV fluxes relative to those in the past are still not well quantified, so prediction of the effects of atmospheric change remain imprecise. Important indications of past changes in the underwater UV environment are provided by pigment residues in lake sediments (Leavitt *et al.* 1997). The recovery of substances resembling scytonemin (a pigment produced in the sheaths of certain Cyanobacteria on exposure to UV; Garcia-Pichel & Castenholz 1991) is a sensitive indicator of high UV penetration. Leavitt *et al.* (1997; see Fig.7.8) found rather elevated levels in sediment profiles from two alpine lakes in the Rockies but which correspond to environments over a century ago, rather than to the recent period of stratospheric oxygen depletion. At that time, droughts are believed to have been prevalent on the prairies and cold tempera-

tures lowered the Albertan tree line (Leavitt *et al.* 1997). Other recent investigations of poorly buffered Shield lakes (Yan *et al.* 1996; Schindler *et al.* 1997) showed that UV penetration increases in droughts because the concentrations of UV absorbing DOC are also reduced. Thus, their results also suggest that droughts may affect much wider land areas than previously supposed, including those of alpine lakes and other locations with naturally low DOC concentrations. Leavitt *et al.* (1997) deduced that biological susceptibility to UV damage may be particularly severe in alpine lakes receiving less photoprotective terrestrial DOC $(0.9 \pm 0.7\,mg\,L^{-1})$ than do subalpine sites $(4.7 \pm 4.4\,mg\,L^{-1})$.

More than 100,000 North American boreal lakes, as well as many mountain lakes, have such low concentrations of DOC $(<2\,mg\,L^{-1})$ that even slight variations in rainfall, acidity and DOC input may significantly affect the penetration of UV. Because the relationship is exponential (Fig. 7.7), drought-induced fluctuations in DOC have a more serious impact on UV penetration than changes in the atmospheric UV flux itself.

Yan *et al.* (1996) revealed another instance of DOC decrease brought on by climatic changes from their 10-year data set on the rainfall and chemistry at Swan Lake, close to the smelting complexes at Sudbury, Ontario. Here drought led to decreased water levels but exposure of the littoral sediments to the atmosphere allowed the oxidation of the abundant deposits of reduced sulphur. Remobilisation of acid decreased the DOC content to the extent that the depth of ultraviolet penetration increased threefold. The relationships are summarised in Fig. 7.9.

The Canadian experiences reveal a common trend of faster evaporation and lowered precipitation. These should lead to lower DOC inputs and longer residence times. Schindler *et al.* (1997) have calculated the DOC loss: with average climatic warming, increased drought and frequency of forest fires through the period 1970–1990, DOC concentrations in the Experimental Lakes Area (ELA) of northwestern Ontario declined by 15–25%. Decreased input of DOC is the primary reason for the

decline, although in-lake removal has accelerated slightly too. The experimental acidification of some of the ELA lakes was marked by accelerated DOC loss to internal removal. In one ELA lake, acidified to pH 4.5 during the 1980s, DOC fell to less than 10% of its pre-acidification level.

7.3.2 *Responses to climatic change. Observations from Scandinavia*

Given regional differences in the pattern of climate changes, it is likely that alternative behaviours will emerge elsewhere. Forsberg (1992) was among the first to argue for the coupling of water colour to climate change. His results emphasise the role of precipitation as the clear intermediary of the flux of humic material from soil to surface waters. His flow diagram of effects of enhanced atmospheric fluxes of greenhouse gases is redrawn here (as Fig. 7.10). It amplifies the broad principles of the changing environmental influences on the water colour in lakes due to DOM. The prevailing paradigm is that the most important effects of enhanced CO_2 are that plants will increase their photosynthesis and productivity of biomass. Net warming should accelerate production and extend growing seasons, leading to higher yields of crops and of forest production. The essential N-

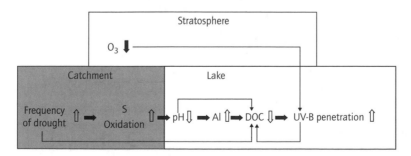

Fig. 7.9 Illustration of linkages among drought, acidification, DOC concentration and UV-B penetration depth in lakes. Black arrows link processes; white arrows indicate changes in frequency, intensity or concentration of constituents or processes. The DOC concentration is affected by several mechanisms including acidification, Al increase and hydrological change (drought), but may also be affected by changes in DOC decomposition rate induced by changes in UV-B penetration depth. (From Yan *et al.* 1996; reprinted with permission from *Nature*. Copyright (1996) Macmillan Magazines Limited.)

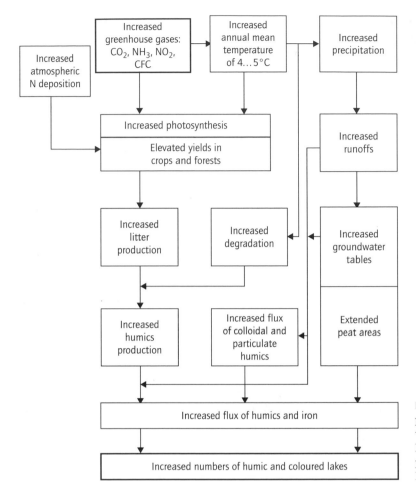

Fig. 7.10 Schematic presentation of possible greenhouse impact on lake water colour in Fennoscandia. (From Forsberg 1992; reproduced with kind permission of Kluwer Academic Publishers.)

limitation of forest growth (Abrahamsen 1980) may already have been alleviated by the elevated atmospheric nitrogen loads of the industrial age (Rodhe & Rood 1985), with present loadings likely to be maintained well into the 21st century. Higher temperatures should permit faster rates of decomposition and mineralisation, leading ultimately to increased fluxes of humic material in drainage.

During sustained warmer temperatures, a larger proportion of precipitation will fall as rain rather than snow. Discharge distribution and frequency will be modified accordingly and possibly favour peatland growth (Boer *et al.* 1990), the type

of vegetation offering the largest fluxes of humic matter to drainage (Kortelainen, 1993). Forsberg's (1992) conclusion is that, if the climate of Fennoscandia develops as hypothesised, then the production of vegetation and humic substances will increase too. If rainfall increases, more of the HS will be transported in drainage waters, with more humic lakes showing more of the physical, chemical and biological symptoms outlined in this chapter. Whether there will be a change in the effectiveness of the DHM-UV shield depends upon region- and basin-specific trends, with drier meaning clearer and wetter meaning more coloured in most instances. Net evapotranspiration may be decisive locally (Clair & Ehrman 1996) and, particularly in alpine and montane districts, even buck the general regional trend (Williams *et al.* 1996).

7.4 MICROBIAL BREAKDOWN OF DISSOLVED HUMIC MATTER

7.4.1 *Dissolved humic matter and dissolved organic matter as energy sources*

Dissolved organic matter provides the main source of energy for microorganisms. However, because of the relative recalcitrance of DHM, the biological importance of the humic components has long been a matter of conjecture (Jørgensen 1976). Experienced intuition supports a common view that humic lakes are rather unproductive ('dystrophic'); there is also an expectation that small organic molecules decompose faster than larger ones, so molecular size (weight) should be critical to microbial DOM-utilisation. Certainly, some low molecular weight (LMW) compounds derived from phytoplankton are decomposed quickly, usually within hours (Münster *et al.* 1999). Compounds of higher molecular weight may require days to weeks while some still larger high molecular weight (HMW) polymers may require the order of months to decompose (Saunders 1976). However, the principle does not apply to all organic matter.

The generally refractory nature of DHM is usually attributed to the diversity and complexity of its composition, featuring minimal repetition of structural units and minimal targeting efficiency by specific microbial enzymes, but even this paradigm is now seriously questioned (Münster *et al.* 1999). Measurements of community respiration and bacterial production have confirmed that heterotrophy in humic waters is comparable to eutrophic autotrophy, suggesting that relevant microorganisms are better adapted to low nutrient concentrations than are phytoplankton. Moreover, whether the autochthonous DOC produced by autotrophic phytoplankton is always sufficient to support the observed bacterial productivity is questionable: whether allochthonous DOC is adequate to support the bacterial activity is crucial (Münster *et al.* 1999).

Tranvik (1988, 1989, 1990, 1992) has been a leading proponent for the importance of DHM oxidation. He tested the ability of ultrafiltered fractions of DOC of differing molecular-weight (MW > 10,000 daltons, or Da; 10,000 to 1000 Da) to support bacterial growth. More bacteria were produced per unit of organic carbon when it was present mainly in larger compounds (MW > 10,000 Da) than as smaller ones. The evidently enhanced availability to bacteria of the HMW carbon decreased with increasing HA content. However, the average bacterial-carrying capacity of the humic brown-water lakes of southern Sweden was higher than the average capacities of comparable clear-water lakes. Tulonen *et al.* (1992) had similar findings from small humic lakes in Finland. Amon & Benner (1994), working in the Gulf of Mexico, also noted that bacterial growth and respiration rates were three to six times greater on HMW than on LMW material.

As neither the chemical structure of the preferred carbon sources nor the precise identities of the bacteria have been resolved in any detail, it is still not possible to substantiate the observations (Münster *et al.* 1999). Natural pools doubtless contain mixtures of compounds with varying susceptibilities and turnover times. The microbial extracellular enzymes ranged against DHM catalyse one or other of two principal pathways (Münster & de Haan 1998): one invokes hydrolytic enzymes, cleaving phospho–ester, lipid–ester–peptide and glucoside bonds; the other, involving phenol oxidases and peroxidases, cleaves the aliphatic carbon bonds and aromatic rings (Münster *et al.* 1998). Neither system seems to have high substrate specificity, meaning that many aquatic microorganisms probably have the capacity to degrade some DHM. The activities of particular enzymes may give more information about the substrate processing than the groups of organisms actually involved (Jackson *et al.* 1995). From the results of Amon & Benner (1994), it does appear that the bulk of the oceanic DOM comprises mainly small molecules but that these cycle only very slowly through being scarcely available to microorganismic enzymes: thus, the reputed recalcitrance of DOM owes to a preponderance of small molecules which, for some reason, are not amenable to microbial breakdown.

In a later study, Amon & Benner (1996a) were able to compare the sources, composition and transience of the varied DOM profiles of rivers, estuaries and the sea. More than 80% of the DOC in samples from the Amazon occurred in HMW fractions whereas most marine DOC is, indeed, in the LMW fraction. In all cases, bacterial growth efficiencies were consistently higher on the LMW substrates but the specific growth rates were highest on HMW carbon sources. Carbon-normalised bacterial DOC utilisation rates were 1.4–4.0 fold greater on HMW fractions and, among all the environments investigated, a greater proportion of HMW DOC was metabolised than from the lower MW fractions. Representing the reactivity of DOM to the diminishing size and increasing diagenetic age explains the observed predominance of refractory, LMW DOC compounds in the 'oldest' water and fits well to the general understanding of the humification of labile plant macromolecules (Hatcher & Spiker 1988). Increasingly oxidised substrates of lower molecular weights and greater solubility become the refractory core of the residual DOM. The model of Amon & Benner clearly explains the existence of more or less refractory HMW and particularly LMW fractions of organic compounds. However, it obviously does not take into account that LMW organic nutrients, like freely dissolved amino acids and carbohydrates, as well as oligopeptides and disaccharides, have very high turnover times. This means that the residual concentrations do not represent the reactivity of these LMW organic compounds (Grossart & Plough 2001).

Several authors have considered the extent of direct and indirect utilisation of DHM by aquatic microorganisms (de Haan 1974; Stabel et al. 1979; Geller 1985a,b, 1986; Tranvik & Höfle 1987; Tranvik 1988; Moran & Hodson 1990). Stabel et al. (1979) observed up to 30% mineralisation of DHM by pure bacterial cultures but, in general, growth rates are generally several times faster on appropri-ate non-humic carbon sources (Steinberg 1977; Moran & Hodson 1990).

Overall, without light, DHM is degraded less fast than some other sources of DOM. That protracted biodegradation may be aggravated by deficiencies in inorganic nutrients (Ryhänen 1968).

7.4.2   Co-factor mediated dissolved humic matter utilisation

Besides direct utilisation of DHM as a source of energy, there exist secondary microbial requirements. Some species in the microbial community produce co-factors, such as growth-stimulants and vitamins. Consumers of HMW DOM which lack the ability to generate their own co-factors nevertheless generate an intracellular state adequate to bring about the oxidation of refractory substrates (Münster et al. 1999). An example has been given by Steinberg & Bach (1996), who compared the effects of DHM on an alga–bacteria system with those on an axenic algal culture. Their experiments were conducted in a groundwater from Fuhrberg, Germany, where the FA content had been well characterised (Alberts et al. 1992). The coccal green alga, Scenedesmus subspicatus, was unable to benefit from the Fuhrberg FA in axenic culture. Moreover, yields diminished with greater DOC loads: this is not surprising for, as field studies confirm, primary production is sensitive to enhanced light attenuation in polyhumic water (Jackson & Hecky 1980; Guildford et al. 1987; del Giorgio & Peters 1994). However, the bacteria-containing Scenedesmus cultures grew remarkably well: with FA-DOC at $20\,mg\,L^{-1}$, the algal yield increased by 23–42% compared with DHM-free controls.

The clear demonstration of algal enhancement of algal productivity by FA mirrors the well-known growth effects of HA mediation. Most of its advocates refer to the enhanced availability of inorganic nutrients and trace metals (Giesy 1976; Steinberg & Münster 1985; Horth et al. 1988; Wojciechowski & Górniak 1990) but this can be one of several contributory mechanisms. As the inorganic nutrient concentrations used by Steinberg & Bach (1996) were far in excess of algal demand, even at the end of the assay, the beneficial effects must owe to microbial metabolism of HS and unidentified growth promoting factors.

Co-factors may also be provided by extra-chromosomal DNA sources such as plasmids. Many have been identified in recent studies (Münster *et al.* 1999).

7.4.3 *Dissolved humic matter input from catchments and alterations during allochthonous transport*

The structure of the catchments may have a very strong influence on the DOC content of lakes and rivers. Peatlands and acidic soils produce especially large amounts of organic carbon. As found, for instance, in a Finnish study (Kortilainen 1993), the leaching of organic carbon was dependent on hydrological conditions: annual leaching was closely related to annual runoff. Fractionation showed that organic carbon in Finnish lakes is dominated typically by organic acids, averaging 84% of DOC, with smaller amounts of neutral and basic compounds. Consequently, the catchment-derived organic acids represent a significant contribution to lakewater acidity. Latitude and the proportion of the catchment covered with peatland are the strongest predictors of TOC and DOC in Finnish lakes (Kortilainen 1993).

Recent studies on DHM residence and water colour in Canadian freshwaters suggest that chromophoric groups are readily lost during the passage through the terrestrial catchment, being accumulated in weakly structured bulk material (Rasmussen *et al.* 1989). In the biogeochemical processing of DHM, the refractory pool increases as it steadily loses those structural subunits and functional groups with counterparts at the enzymatic–microbial levels. Again, the implication is that the interactions and affinity of the microbial degrading enzymes and transport systems for refractory DHM compounds weaken down a gradient of increasing age and diminishing structures, culminating in slower degradation and weaker utilisation. The deduction helps to explain the higher persistence and greater age of DHM in humic waters (Münster *et al.* 1999).

By applying a fingerprint technique based on size exclusion chromatography and simultaneous double detection (UV, DOC), Sachse *et al.* (2001) traced the environmental fate of DOC that originated from an oligotrophic bog and entered a small lake. On its way into the lake, DOC loses molecular weight, whereas upon irradiation and microbial activity in the lake the DOC decreases in both aromaticity and molecular weight.

7.5 DISSOLVED HUMIC MATTER AND SOLAR RADIATION

7.5.1 *Photolysis of dissolved organic matter*

Complementary to the light-absorbing properties of DHM, discussed in section 7.3, is the photolysis of the molecules. As with other aspects of their ecological role, the understanding of the photochemistry of humic substances is hampered by imprecise definition of their structure. Characterisation of the photolytic products is subject to the same limitations. Much of what is known at present applies to integrated data and bulk properties of humic components and end products; only the reactions with small, well-defined and sufficiently long-lived organic species have been determined reliably.

The main photoreaction pathways transfer energy to reactive humic species (RS) and turn them into rapidly reactive radicals. The multifunctional character of HS lies in the wide variation in their ground states and their reactivity to radicals. Molecular rearrangement of the HS and partial degradation to carbon dioxide is frequent (Fig. 7.11); residual HS is available to further photochemical reactions (Frimmel 1994).

The main light-absorbing molecular components are the chromophores. Schwarzenbach *et al.* (1993) estimated that, at the concentrations observed in the epilimnion of Greifensee, Switzerland, individual chromophores would be excited at an average rate of 10 times per hour. The most important acceptor of the excited chromophore is oxygen in its ground state (triplet oxygen, $^3O_2$)

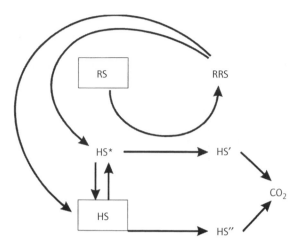

Fig. 7.11 Pathways of photoreactions of humic substances: HS, humic substances; RS, reactive species; RRS, rapidly reactive species. (After Frimmel 1994. Copyright (1994), with permission from Elsevier Science.)

(Zepp *et al.* 1977). Promotion to its first excited state (singlet oxygen, $^1O_2$) requires $94\,kJ\,mol^{-1}$ and most chromophores that absorb visible or UV radiation transfer their energy in this way. Other reactive oxygen species include free peroxy radicals (Mill *et al.* 1980), hydroxyl radicals and organic peroxides (Larson *et al.* 1981). The resultant oxidant is then consumed in reactions with organic molecules (Schwarzenbach *et al.* 1993): although different kinds of DOM react with differing sensitivities and with differing quantum yields of $^1O_2$, there is a clear, near-linear dependence of $^1O_2$ production on the DOC content of the water.

Draper & Crosby (1983) were among the first to show that hydrogen peroxide is frequently the photoxidant: after brief exposure to sunlight, the hydrogen peroxide content of natural waters may rise from below detection limit ($1.5\,\mu M$) to $6.8\,\mu M$, and in highly eutrophic source water to over $30\,\mu M$. Peroxide is also longer lasting in water, likely adding to its ecological significance.

Exposure to natural solar radiation is highly effective in photolysing DOM. It has been estimated that up to 20% of the colour reduction in lakes

is attributable to solar photobleaching (Gjessing & Gjerdahl 1970). The effect of UV wavelengths on HS is of increasing relevance because of fears of enhanced penetration as a consequence of stratospheric ozone depletion, especially at high latitudes. Photooxidation of organic compounds is, in any case, a familiar process. The ultimate product is carbon dioxide, which, in some instances of inorganic carbon deficiency, may be sufficient to enhance the rates of aquatic primary production. At the same time, the release of small quantities of organically bound iron and phosphorus may support higher crops of primary producers. Ultraviolet also promotes breakdown of organic macromolecules (de Haan 1992) and, by increasing the pool of available organic carbon, stimulates the production of bacteria and the phagotrophic food chain (Hessen *et al.* 1994).

7.5.2 *Cleavage and bioavailability*

Lönnerblad (1931, in Lindell & Rai 1994) was one of the first to note the typical epilimnetic oxygen undersaturation of humic lakes. At the time, it was thought to be the consequence of high heterotrophic activity. Once the improved methods for the counting of bacteria and for estimating their biological activity were available, the fallacy of this explanation was quickly recognised (Hutchinson 1957). Later, Miles & Brezonik (1981) showed increased rates of abiotic oxygen consumption followed exposure of the humic content to light.

The photochemical degradation of refractory DOM to smaller and more readily assimilable organic compounds is now well-documented (Backlund 1992; Frimmel 1994). In one of the pioneering studies, Geller (1985a) had shown that the abiotic stability of the macromolecules of recalcitrant, high molecular weight DOM was abruptly weakened when exposed to sunlight in the laboratory. During 6 weeks of subsequent exposure to weak daylight (>300 nm), the mass of HMW substances decreased by 15–25%, with a corresponding increase in substances of LWM. These smaller subunits were found to become biodegraded: photolysis of macromolecular DOM increased the bioavailability of DOM by about 10% (Geller 1986).

Since then, the impact of radiation, especially of UV wavelengths on the character and reactivity of surface-water DOM, has become better understood. The formation of CO_2 when high molecular weight DOM is exposed to UV radiation was shown by Kulovaara & Backlund (1993) and by Salonen & Vähätalo (1994), while the production of various intermediates (carbon monoxide: Conrad & Seiler 1980) and LWM carboxylic acids (Backlund 1992; Allard et al. 1994; Wetzel et al. 1995; Bertilsson & Allard 1996) has been widely reported.

Many of these products of photolytic cleavage are highly bioavailable (Lindell et al. 1995). Kieber et al. (1989) showed that the biological uptake of pyruvate was closely correlated to the rate of its photochemical production in seawater and that its photochemical precursors derive from DOC fractions having molecular weights of about 500 Da. In humic waters having high concentrations of DOM, photochemical production of LMW substances is important for the nutrition of bacteria (Salonen & Vähätalo 1994; Vähätalo & Salonen 1997). The direct coupling of photochemical production and biological utilisation could contribute to the stability of bacterial numbers in water and maintains net heterotrophy in non-eutrophic water bodies (Salonen et al. 1992; see also Weisse, Chapter 13).

In a detailed study, Wetzel et al. (1995) demonstrated a markedly enhanced production of bacteria grown on natural dissolved organic matter exposed to mild UV-B photolysis. Whole leachate and HA and FA fractions of DOM released from senescent littoral aquatic plants were exposed to varying spectra of UV, as well as to natural UV radiation in sunlight over varying periods of time. Examination of the DOM, before and after photolysis, revealed only subtle changes to the bulk DOM but the DOM exposed to natural UV radiation showed immediate stimulation and sustained bacterial growth. Chemical analysis of the photolysed fractions revealed progressively increasing yields of carboxylic acids (acetic, formic, pyruvic and levulinic, among others). Use of radio-labelled

humic substances confirmed that these photolysed products were, indeed, readily metabolised by the bacteria.

In another recent study, Dahlén et al. (1996) investigated the quality and quantity of DOM in humic surface water subjected to continuous, low-intensity (three times lower than natural sunlight on a cloudless summer day in southern Sweden) UV-A irradiation. They found photochemical production of oxalic, malonic, formic and acetic acids (Fig. 7.12). The only organic acid detected in the original samples was lactic acid, which responded differently to the irradiation, though having little to do with the photolysis of DHM, apparently. However, photochemical yields of carboxylic acids of this magnitude seem likely to have a profound influence on surface-water environments, apart from having metal-complexing properties similar to those of FAs, the carboxylic acids make highly available substrates for aerobic heterotrophs. Despite their bioavailability, transformation of organic matter into CO and CO_2 metal chemistry is also rapid. Nevertheless, LMW organic compounds are important intermediates in the photodecomposition of HMW organic matter.

Few studies have sought to calculate the significance of DHM photolysis at the ecosystem scale. Working on the slightly acidic lake Valkea-Kotinen, Vähätalo & Salonen (1997) estimated the contributions of photochemical degradation of DHM as follows: c.30 mg C m^{-2} day^{-1} of chromophoric DOM/DHM degraded photochemically to products of lower absorbance than the original DHM, c.7 mg C m^{-2} day^{-1} was converted photochemically into DIC and c.23 mg C m^{-2} day^{-1} was explained by non-chromophoric degradation DHM products. Comparison of the total (60 mg C m^{-2} day^{-1}) with epilimnetic bacterial respiration (89 mg C m^{-2} day^{-1}) shows the scale of the potential contribution to the pelagic carbon cycle. The experiments of Amon & Benner (1996b) to determine the photoreactivity of biologically refractory DOC in the waters of the Rio Negro (Amazon basin) showed unusually fast rates of photochemical consumption of DOC (c.4.0 μM C h^{-1}) and dissolved oxygen (c.3.6 μM O_2 h^{-1}); the rate of bacterial utili-

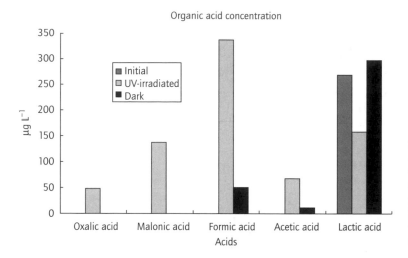

Fig. 7.12 Production of oxalic, malonic, formic, acetic and lactic acids in UV-A-irradiated samples compared with dark controls and initial samples. The values are means of three replicates. (From Dahlén *et al.* 1996. Copyright (1996), with permission from Elsevier Science.)

sation of DOC was about seven times less than the rate of its photochemical consumption, suggesting that the photomineralisation of biologically refractory riverine DOM is the more important mechanism for the removal of terrestrially derived DOC from surface waters generally.

7.5.3 *Photomineralisation*

Photochemical reactions result ultimately in the mineralisation of organic matter (de Haan 1992, 1993; Lindell & Rai 1994; Granéli *et al.* 1996) and, thus, directly contribute to the carbon budget (Salonen & Tulonen 1990). However, studies on the significance on the whole-lake scale are rather scarce. De Haan (1993), recognising that photodegradation of humic substances occurred at depths reached by UV wavelengths (<320 nm), used measured rates of UV degradation of aquatic HS to calculate an annual carbon yield. This was smaller than the consumption needed to sustain the annual primary production of phytoplankton (it was about 8%) or of submerged macrophytes (3%) in the fertile conditions but local photodegradation of aquatic HS on certain cloudless summer days was more rapid than the contemporaneous rates of pelagic photosynthesis (de Haan 1993). Granéli *et al.* (1996), working in lakes of contrasting humic content (3.9–19 mg DOC L^{-1}),

obtained similar results. Measurable yields of DIC were attributable to sunlight action on DOC. Even in sterile water, DIC production was within an order of magnitude of carbon exchange due to planktonic respiration and primary production.

Photooxidative production of DIC from DOC can be significant in the carbon budgets of temperate lakes, especially shallow, well-insolated examples. At lower latitudes, where warmer waters receive more intensive solar radiation, DOC mineralisation is likely to be significant in lacustrine DIC budgets. A summary of the light effects on DHM, incorporating a conceptual model of the epilimnetic carbon pathways, is shown in Fig. 7.13. In Fig. 7.14, the pathways are shown in the developing overview of limnetic humic metabolism.

7.5.4 *Toxic effects of dissolved humic matter irradiation*

So far, the photobleaching effect has been considered as a mechanism for increasing the bioavailability of DOM/DHM, through cleavage of the macromolecules and the release of products of lower molecular weight such as pyruvate and other carboxylates.

There are some negative effects of DOM oxidation, largely to do with the free radicals and oxidants produced and their deleterious effects on

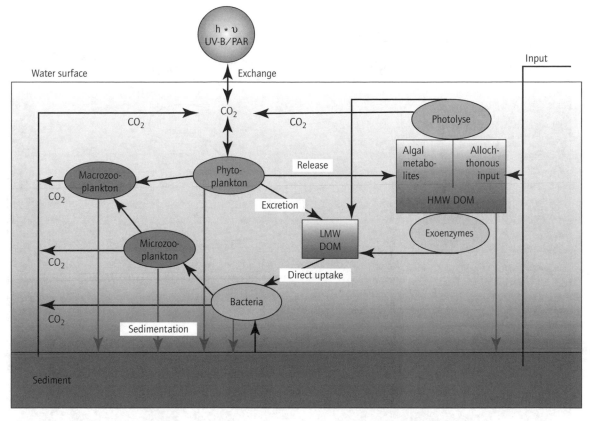

Fig. 7.13   A conceptual model incorporating photolysis of dissolved organic carbon (DOM) into the pelagic carbon cycle. The model includes three sinks of DOM: (i) sedimentation, (ii) reincorporation into organisms by bacterial uptake and (iii) photochemical oxygen consumption with subsequent carbon mono-/dioxide production. The carbon dioxide stimulates primary producers whereas carbon monoxide is a toxicant. Arrows are not quantitative. (Modified after Lindell & Rai 1994; reproduced with permission of E. Schweizerbart'sche Verlagsbuchhandlung.)

biological membranes, proteins and DNA. If the oxidants and their products are short lived (< few hours), they do not accumulate and affect only the top few centimetres of the lake. If they are longer-lived, toxic effects may persist for weeks and toxic compounds may be dispersed though the entire mixed layer.

The longest-lived reactive oxidant is hydrogen peroxide. When UV is absorbed by DHM in aquatic systems, superoxide is formed. This reacts with itself (dismutates) to re-generate hydrogen peroxide. Scully *et al.* (1996) described its production in relation to DOC in a wide range of waters from highly

coloured bogs and dystrophic pools to large clear lakes by the equation. The fitted regression has the equation:

$$[H_2O_2] = 49.65[DOC]^{1.71}(r^2 = 0.94)$$

The biological effects of the photoproducts are generally not well-known. Gjessing & Källqvist (1991) showed that exposure of water containing HS to UV changed the acidity and charge density and also inhibited the growth of the coccal green alga *Ankistrodesmus bibraianum*. The algistatic effect persisted for some weeks. Lund & Hongve

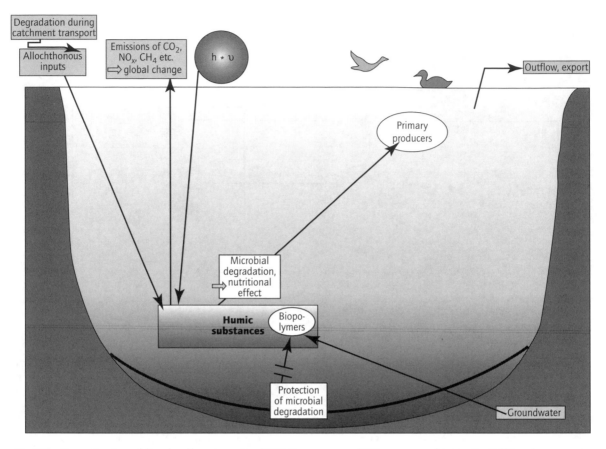

Fig. 7.14 Synoptic view of dissolved humic matter (DHM) as energy and CO_2 source and protection of biopolymers against microbial degradation. (From Steinberg 2003; reproduced with permission of Springer-Verlag.)

(1994) observed a similar effect on heterotrophic bacteria exposed to UV-irradiated humic water, though here, the effects persisted for several days.

A more complex picture, invoking an additional mechanism of toxicity of irradiated humic substances, was described by Hessen & van Donk (1994) They conducted bioassays of UV-irradiated surface waters from two localities They found that phytoplankton production was stimulated by moderate UV doses (1.1–5.4 J cm^{-2} at 312 nm) but subject progressively to inhibition at increasing doses (>10 J cm^{-2}). Inoculation of *Ankistrodesmus bibraianum* after the irradiation showed persistent toxic effects, which they attributed to metals (Al) hitherto complexed by HS. The UV doses were

of no greater intensity than occur naturally but the effects were markedly non-linear in terms of UV dose. No effects on dark respiration were detected—contrary to the results of Lund & Hongve (1994). The short-term effects observed by Hessen & van Donk were presumed to be due to photooxidants, with metal toxicity being responsible for the persistent symptoms. The main conclusion is that natural doses of UV may promote inhibitory effects on primary producers with toxic effects building over a matter of some days.

7.5.5 Photooxidants and xenobiotics

In natural waters rich in DHM, potentially photodegradable xenobiotic (allochthonous and non-natural) compounds may undergo photolysis to potentially harmful or more harmful products.

The processes are similarly initiated by light absorption by the DHM present in the system and they are, therefore, commonly referred to as indirect or sensitised photolysis (Schwarzenbach *et al.* 1993). Despite their adsorption on to seston and subsequent sedimentation, indirect photolysis of xenobiotic compounds may be the major pathway by which a great many of otherwise persistent chemicals are eliminated. Half lives of a few to several hours or days are common (Faust & Hoigné 1987; see also Table 7.2).

With an increasing number of alkyl substituents, the reactivity of a given phenol increases. Comparing the relative reactivities of various 4-alkylphenols, it is apparent that the reactivity decreases with increasing numbers of carbon atoms and increasingly more hydrophobic behaviour of the compound. One explanation offered by Faust & Hoigné (1987) is that, because of hydrophobic interactions of the alkyl group with part of the DHM, the probability of encounters of phenol with the more polar peroxy radicals increases. In a recent study, Canonica *et al.* (1995) showed that the reactive excited triplet states of DHM are important photooxidants in natural waters. The concentration of reactive triplet state in the top metre of Greifensee under midday summer sunshine was estimated to be c.10^{-14} mol L^{-1} but it is sufficient to shorten the half-life of 2,4,6-trimethylphenol from several days to c.7 hours. The authors stress that still uncharacterised photooxidants derived from the DOM are also involved in the phototransformation of the phenols

In many cases, such indirect photolytic consequences are a function of DHM concentration. For instance, the DHM-mediated photolysis of two organoborates (tetraphenylborate and diphenylboric acid) were directly proportional to the concentration of DOM and HA for values <10 mg L^{-1} (Mills & Schwind 1990). At higher concentrations, there was a negative departure from linearity, which suggests scavenging of the transient reactive chemical species by reaction with components of the humic material (Mill *et al.* 1980; Faust & Hoigné 1987). Mills & Schwind (1990) also found that removal of dissolved oxygen increased reaction rates by a factor of two, indicating that the oxygenation mechanism does not involve singlet oxygen.

Kulovaara *et al.* (1995) compared the degradation of DDT (2,2-bis(4-chlorophenyl)-1,1,1-trichloroenthene) under UV radiation at 254 nm and under intense simulated sunlight. The DDT was degraded in both treatments but was substantially faster (minutes rather than hours) in UV radiation. The presence of natural DHM had a retarding effect on the photolytic degradation under UV light but an accelerating effect under artificial sunlight. Different reaction pathways are invoked in these experiments: the first, presumably direct, photolysis of DDT was initiated by UV radiation; the relatively slow, sunlight-initiated direct decomposition is enhanced by an indirect, DHM-sensitised process but, as the DHM absorbs UV as well as visible light, direct photolysis of DDT is accordingly impaired.

Photolysis sensitised by DHM is by no means restricted to xenobiotic chemicals. Biopolymers (e.g., exoenzymes, DNA), allelochemicals (e.g., cyanotoxins) as well as kaironomes (such as those chemicals produced by fish that induce diel vertical migration and helmet spines in *Daphnia*; see also Chapter 14) may also be subject to DHM-mediated photolysis. The information available shows the pigment-sensitised photodegradation of

Table 7.2 Reactivity of alkylphenols relative to phenol and half-lives of *in situ* photodegradation in good sunlight. (Data from Faust & Hoigné 1987 and Schwarzenbach *et al.* 1993).

Substance	Reactivity	Estimated half-life (days)
Phenol	1	200
4-nonylphenol	2.6	78
4-isopropylphenol	5.3	38
2-methylphenol	7.3	27
4-ethylphenol	8.9	22
4-methylphenol	12.4	16
2,6-dimethylphenol	23.8	8
2,4,6-trimethylphenol	40.6	5

microcystin (Tsuji *et al.* 1994). However, the pigment concentrations used in these experiments applied to scum conditions rather than the whole water column. Microbial degradation is efficient in most inocula but is achieved only after a lag phase of several days to weeks (Bourne *et al.* 1996; Welker *et al.* 2001). Microcystin-LR was not degraded by sunlight alone but it was in the presence of fulvic acids and DOC. Compared with solutions of purified fulvic acids, photolysis in natural waters amounted to only one-third for a given optical density. Thus, the natural half-lives of the organic carbon were estimated to be about 90–120 days m^{-1} depth of water. Therefore, a combination of photolytic and microbial degradation seems to be responsible for the weak persistence of microcystins in natural waters (Welker & Steinberg 2000).

7.6   DISSOLVED HUMIC MATTER, NUTRIENTS AND METALS

Dissolved humic matter interacts with the chemistries of a variety of major and trace elements, influencing their bioavailabilities. Owing to their biological importance in lakes, the main attention is directed to phosphorus, nitrogen and iron.

7.6.1   *Dissolved humic matter and phosphorus*

As phosphorus is the principal factor limiting the biomass in many fresh waters, a great many papers have been directed to its interaction with DHM. For instance, the concentrations of DHM may affect directly the rate at which orthophosphate ions can be taken up by the cells of planktic algae and bacteria. Working on lakes of the Canadian Shield, Brassard & Auclair (1984) found that the rates of orthophosphate uptake were mediated directly by DHM in the range 1–10 kDa. Francko (1986) also found that uptake of ^{32}P-labelled orthophosphate by phytoplankton could be repressed or stimulated by small additions of *Typha* DHM, alone or in concert with ferric iron. The degree of stimulation or repression was apparently size- and lake-specific.

Perhaps the best-known interaction is the complexation and sequestration of orthophosphate by DHM and the influence of the presence of iron. Numerous studies have shown that, under conditions of low pH and low redox, DHM, ferric iron and orthophosphate form colloidal aggregates and that the phosphorus thus bound is rendered unavailable to biotic assimilation (Francko & Heath 1982, 1983; Steinberg & Münster 1985; Jones *et al.* 1988, 1993; Shaw 1994), even though a substantial proportion continues to give a positive (blue) reaction to molybdate. Dissolved humic matter is apparently incapable of significant P-binding itself but, in darkly stained, iron-rich waters, most of the orthophosphate may be sorbed on to HMW aggregates in forms which are only partly reactive to acid molybdate reagents. The physico-chemical nature of the metal–P–DHM binding is still not well-understood, and the relationships between DHM–P complexes and the non-DHM colloids (calcium carbonate colloids, algal microfibrils) remain unclear. Jones *et al.* (1993) and Shaw (1994) have shown that, in the absence of DHM, iron and phosphate do not transfer to colloidal material. These studies have also shown that movement of ferric and orthophosphate ions to larger size fractions of greater molecular weight diminishes with decreasing pH. Furthermore, in the presence of DHM, the formation of particulate iron is suppressed at high pH. It is well known that pH influences both the protonation of the functional groups of DHM and the conformation of DHM aggregates. The 'stretched' conformation and low protonation of DHM at high pH appear to favour the colloidal and truly dissolved forms of ferric iron and orthophosphate. This tendency may be attributed to the stabilisation of Fe(III) oxide colloids. At lower pH, with greater protonation and closer, more compact conformation of the DHM aggregates, more of the iron is particulate.

The presence of high concentrations of calcium ions has not been found to have much effect upon the relative quantities of immobilised iron and phosphorus. Calcium carbonate may well have been involved in the formation of iron–phosphate–DHM colloids in Shaw's (1994) studies but the similarity of results to those from experiments using softer waters suggests that the influence of calcium ions on the binding of phosphorus to iron is small. Significant differences in the results with clear-water samples for low or high concentrations of calcium ions indicated a stronger dependence on the ionic composition of the precipitation and the quantities of particulate iron and phosphorus, which may reflect differences of reactivity between iron and calcium.

The most direct evidence for DHM–iron–phosphorus complexation has been presented by de Haan *et al.* (1990) from work on stained waters in southern Finland. Using a double isotope-labelling technique, they showed that radio-labelled iron and orthophosphate added to epilimnetic water samples became simultaneously incorporated into a fraction of nominal molecular weights of between 10 and 20 kDa. Complexation did not occur in the absence of DHM. By disturbing the equilibrium between the isotopic concentrations and DHM, labelled orthophosphate ions could be liberated into solution, presumably by displacement from the colloid. The photosensitivity of these colloids was not investigated.

The importance of DHM–P as both sink and source of bioavailable phosphate has generated considerable interest. A widely-accepted model suggests that orthophosphate can be displaced from HMW bound moieties (de Haan *et al.* 1990; Wetzel 2001). Others suppose that abiotic, photodependent reactions also mediate P release. Steinberg & Baltes (1984) tested the influence of UV radiation and redox-sensitive metals (Fe, Mn) on fulvic-acid linked molybdenum blue-reactive phosphorus (or SRP). In water samples from a raised peat bog, they found a bimodal distribution of phosphorus compounds, HMW–SRP and LMW–SRP. The HMW–SRP was sensitive to UV irradiation. Addition of Fe(II) and of low concentrations of Mn(II) brought about increases in HMW–SRP at the expense of LMW–SRP but higher concentrations of Mn(II) caused an opposite effect, a decrease in HMW–SRP for an increase in LMW–SRP. The effect was attributed to the relative preponderance of catalytic cleavage or displacement within the complexes (see Fig. 7.15).

In another study of the same bog, Francko &

Heath (1982, 1983) showed that the ferric iron was also associated with the UV-sensitive P and that the UV radiation may release phosphorus from its bound, complexed state through the photoreduction of the iron to ferrous. In this study, photoreduction and, thus, the release of phosphorus were freely reversible: both the UV-sensitive P and the ferric state were regenerated when irradiated samples were incubated in the dark. The turnover time for iron reduction and P release was about 1 hour. The plausibility of the mechanism in nature was supported by a diel series of observations on further samples from the same bog: the concentrations of DHM and UV-sensitive P were markedly higher during the night than during daylight. Besides the postulated photoreduction of iron, it is possible that UV radiation disrupts the phenolic groups in the DHM, thus reducing the affinity of iron for orthophosphate (Steinberg & Baltes 1984).

Few studies have been conducted to determine the presence of UV-sensitive P among natural waters, so that the relative importance of this mechanism as a source of phosphorus for planktic primary production cannot be evaluated. The findings of Jones *et al.* (1988), that ^{32}P-labelled orthophosphate added to water from small, iron-rich, humic Finnish lakes became adsorbed into molecules of two distinct size fractions (10–20 and >100 kDa), are consistent with those of Francko & Heath (1983). The bound fractions were similarly unreactive to molybdate and exposure to sunlight resulted in a modest release of SRP.

Cotner & Heath (1990) studied the factors controlling the release of phosphorus from DHM–iron complexes in an acid bog lake. Additions of Fe(III) and UV irradiation increased the concentrations of free orthophosphate ions; oxidation of Fe(II) in lake water proceeded rather more slowly than the photoreduction, which suggests that, in UV light, iron is accumulated as Fe(II). However, Fe(II) concentrations diminished concomitantly with the summer development of phytoplankton in the lake. Thus, biotic uptake of Fe(II) may have depleted the ambient pools of Fe(II) and Fe(III) in summer and so lowered the dependence upon photosensitive reactions.

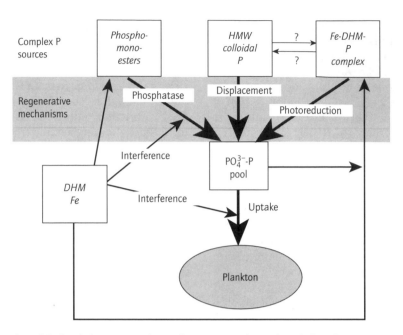

Fig. 7.15 Conceptual model of epilimnetic P cycling. Three groups of complexed phosphorus sources and P-regenerative mechanisms may co-occur in a given lake system. The relative importance of each change in each regenerative machnism is not fixed in time, but rather dependent on dynamic changes in physico-chemical and biotic parameters. Central is that dissolved humic matter (DHM) not only sequesters P into a complexed form but also interacts through unknown mechanisms with alkaline phosphatase activity and ortho-P uptake, as well as with the pool size of phosphomonoesters. (Modified after Francko 1986. Copyright 1986. © John Wiley & Sons Limited. Reproduced with permission.)

Modelling of these many potential influences of DHM on epilimnetic phosphorus cycling is in its infancy. Clearly, there is no single epilimnetic pathway. Francko (1986) envisaged three major sources of complexed P (phosphomonoesters, non-DHM colloidal P and DHM–Fe colloidal P) and several regenerative mechanisms (alkaline phosphatase activity, colloidal displacement and UV reduction). The concentrations of DHM and iron provide important influences on the regenerative mechanisms and thus in regulating P cycling in lakes.

7.6.2 Dissolved humic matter and nitrogen

The fate of the nitrogen bound in aquatic humic substances and its susceptibility to photochemical degradation have not been addressed adequately.

Poor bioavailability of humic-bound nitrogen has made HS seem unlikely sources of nitrogen for marine and microbial food webs. Contrary to a supposition that HS-bound nitrogen is scarcely bioavailable and that the high C:N ratios of HS (averaging 50:1; Thurman 1985) also make it unlikely that bacteria could obtain adequate nitrogen from this source, Bushaw et al. (1996) have shown that simple nitrogen compounds, notably ammonium and amino acids (Tarr et al. 2001), are released from DOM exposed to solar radiation. They are produced most efficiently in UV wavebands.

7.6.3 Dissolved humic matter and iron

In addition to the demonstrable involvement of iron in its relationships with phosphorus, DHM and light impact directly on the Fe cycle, mainly in the kinetics of redox transformations and the reductive solutions of oxides. This applies particularly to open oceanic waters where iron is a major limiting element (Coale *et al.* 1996). Even in the dark, Fe(II) reactivity is affected by the presence of FA (Voelker-Bartschat 1994). She found that, at pH 5, the rate of Fe(II) oxidation increases with increasing concentration of FA and that the principal oxidants are HO_2/O_2. She also showed that FA have the potential to reduce Fe(III). Fe(II) is formed in the light on the surface of particles of fulvate-lepidocrocite (λ-FeOOH) by ligand–metal charge transfer reactions. In the upper, sunlit sea, reduction of Fe(III) with the HO_2/O_2 photoproduct could provide a source of Fe(II): In 0.7 M chloride, and so long as pH > 5.5, Voelker-Bartschat (1994) showed that high HO_2/O_2 concentrations may maintain up to 75% of the iron in solution; she calculated that between 30 and 75% of the dissolved iron in the photic zone during daylight hours is present as Fe(II).

Analogous investigations of DHM- and light-mediated iron cycling in lakes are not available. However, it is known that iron–humic complexes act as oxidants. Luther *et al.* (1996) have investigated the organic Fe(II) and Fe(III) complexes in gel-fractionated porewaters from a saltmarsh, where Fe(III) occurring in the 100–5000 Da fraction was shown to be a significant oxidant of sulphide. Fe(II) was found in fractions of 100–5000 Da but most was located in the <100-Da fraction. Both forms precipitated with humic material at pH < 3. Between pH 3 and pH 5.5, there was a twofold decrease in the <5000-Da fractions but a ninefold increase in the >5000-Da fraction and a tenfold decrease in the concentration of dissolved iron. Such low pH values can occur during drought conditions, when desiccation leads to sulphide oxidation.

7.6.4 Dissolved humic matter and mercury

Several metals (e.g., Hg, Pb) and metalloids (As, Sb) may be incorporated into organic compounds with significantly enhanced bioavailability, and with serious toxicity implications. Mercury is highly toxic; it is encountered in natural ecosystems mainly as a consequence of such anthropogenic activities as burning fossil fuels and industrial applications, sometimes in concentrations causing environmental concern. Sweden has an estimated 83,000 lakes, some 40,000 of which have fish with mercury concentrations in excess of the 0.5 mg Hg kg^{-1} wet weight guideline (Håkanson 1996). In about 10,000 of them, the fish carry more than 1.0 mg kg^{-1}.

Dissolved humic matter affects mercury speciation in two ways, through the reduction of Hg(II) and by the formation of methyl mercury. The first indications of DHM mediation in the reduction of Hg(II) to elemental mercury were found by Alberts *et al.*(1974) and Miller (1975). Since then, it has been suggested that abiotic reduction could account for as much as 70% of the volatile mercury in contaminated streams. Reduction by DHM can be slowed by competing reactions, such as removal of Hg(II) by chloride. Reduction is slower in the presence of air but light enhances it (Allard & Arsenie 1991).

Methyl mercury, Hg $(CH_3)_2$, is more toxic and more bioavailable than inorganic species. The potential for methylation of Hg(II) by methylcobalmin, methyl tin compounds and by humic matter was propounded by Weber (1993). Methylcobalmin is uncommon in the environment and, though all three (mono-, di- and tri-) methyl tin compounds are found in the sea, humic matter is the most omnipresent and relatively abundant methylating agent.

Nevertheless, methylation is not exclusively abiotic as it can be mediated by microorganisms. Matilainen & Verta (1995) considered methylation in aerobic humic waters to be incidental to the activity of bacterial exoenzymes. Anoxia increases the rate of methylation in soils and sediments, though not in water (Porvari & Verta 1995). The proportion of free methyl mercury increases with decreasing pH (Hintelmann *et al.* 1995). However, Porvari & Verta (1995) found a positive correlation between DHM and methyl mercury but a negative one with methylation.

Thus, in lakes, HS may be more carriers of the mercury rather than active methylating agents. Binding to DHM allows mercury to be transported among the various aquatic compartments. Wallschläger *et al.* (1996) showed dispersion of mercury from a point source entirely in HA-bound complexes. Moreover, Mierle & Ingram (1991) deduced that the seasonal pattern of mercury export from lake catchments was closely correlated with the DHM loss rates: thus, DHM may be equally effective in transporting mercury from diffuse, non-point sources.

7.6.5   Dissolved humic matter as a redox catalyst

Lovley *et al.* (1996) showed that, in the absence of oxygen, some kinds of microorganisms in soils and lake sediments use humic substances as electron acceptors for their respiration. The electron transport gives an energy yield sufficient to support their growth. Microbial reduction of humic material enhances the capacity of other microorganisms to reduce other, less accessible electron acceptors, such as Fe(III) oxides.

That microorganisms can donate electrons to DHM has important implications for the relative effectiveness of microbial oxidation of natural and contaminant organic matter in soils and sediments. The model proposed by Lovley *et al.* (1996; see Fig. 7.16) to explain the reactions stimulated by the iron-reducing bacterium *Geobacter metallireducens* illustrates well the mediation of DHM in the redox chemistry. It may also be noted that this method of electron shuttling alleviates the need for iron reducers to contact Fe(III) oxides directly in order to bring about their reduction.

The DHM redox catalysis is inserted into the developing schematic in Fig. 7.17.

7.7   DISSOLVED HUMIC MATTER AND BIOPOLYMERS

Dissolved humic matter has the capacity to bind biopolymers, such as enzymes, toxins, DNA, through hydrogen-, ionic- or covalent bonding. In many instances, the binding protects against microbial degradation, so decreasing the biological activity (Crecchio & Stotzky 1997).

7.7.1   Exoenzymes

Microbial extracellular enzymes cleave biopolymers efficiently (Wetzel 1991; Münster *et al.* 1992a,b; Münster & de Haan 1998). Humic substances have been shown to have a strong influence on the activity of these extracellular enzymes. For instance, Münster *et al.* (1992a,b) found that some 70–90% of the acid phosphomonoesterase (acPME) present in small, humic forest lakes was not particle bound but existed in a free, dissolved state.

This moiety is subject to interactions with HS. They involve dipole–dipole hydrogen bonding between NH_2- and/or carboxylic groups of the enzyme and phenolic -OH groups of the HS (Wetzel 1991). It has been assumed that this bonding of the free enzyme modifies its catalytic properties. It may also shield the enzyme from inhibitors and proteolysis. These assumptions have been veri-

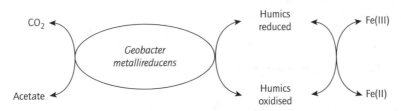

Fig. 7.16   Model for mechanism by which humic acids stimulate Fe(III) reduction by *Geobacter metallireducens*. For details, see text. (From Lovley *et al.* 1996. Reprinted with permission from *Nature*. Copyright (1996) Macmillan Magazines Limited.)

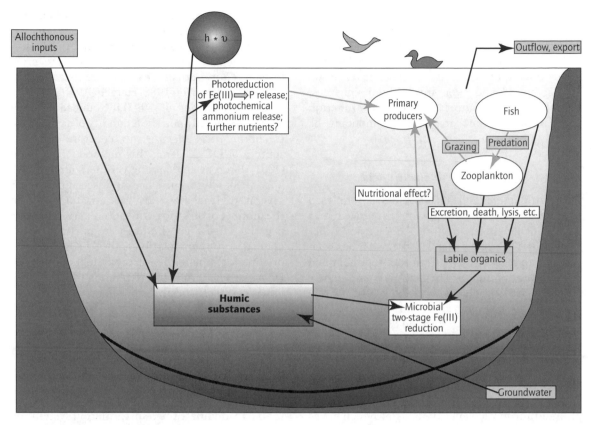

Fig. 7.17 Synoptic view of the interference of humic substances with inorganic nutrient cycles, particularly phosphorus and nitrogen. This sketch includes also the autochthonous production of humic substances, which is not dealt with in the text (but see Steinberg 2003). (From Steinberg 2003; reproduced with permission of Springer-Verlag.)

fied, in part: Wetzel (1991) showed significant impacts of HS on the kinetic characteristics of alkaline phosphatase. Similarly, Münster (1994) observed analogous differences in the activities of free and bound acPME. The addition of polyvinyl pyrolidone (PVP), an effective HS immobiliser, induced a higher acPME activity than was observed in PVP-free controls. Humic substances also modify the catalytic properties of some soil enzymes (review of Schinner & Sonnleitner 1996).

Through its effects on alkaline phosphatase-mediated hydrolysis of phosphomonoester substrates, DHM affects the availability of assimilable P to plankton. Stewart & Wetzel (1982) observed that exogenously applied DHM enhanced the alkaline phosphatase activity of natural bacterial/algal assemblages from Lawrence Lake, especially under low light intensities. Their work indicated that LMW DHM was more stimulatory than HMW moieties. Jones *et al.* (1988) also noted stimulation of phosphatase production by DHM.

The ecophysiological consequences of exoenzyme immobilisation have been considered by Wetzel (1991, 1993) and by Münster & de Haan (1998). It is potentially important in regulating the recycling of nutrients from biomass in lakes with long hydraulic retention times. Though unverified, the view is analogous to the understanding of the role of HS and free enzymes in soil ecosystems, where 'abiotic' enzymes are demonstrably more resistant to degradation and inactivation by (e.g.) heavy metals (Schinner & Sonnleitner 1996). In general, it may be supposed that free and immobilised exoenzymes in humic waters react faster to short pulses of substrate abundance than do surface-bound and even intracellular enzymes. Dissolved humic matter may have a key function in the initiation of biocatalysis and in inducing the cellular biosynthesis of exoenzymes.

7.7.2 *Deoxyribonucleic acid*

Extracellular ('naked') DNA is released from lysed eukaryote cells and by growing bacteria. It has been considered sensitive to DNAase and other nucleases. During the last decade or so it has been possible to recover DNA from soil, sediment and aquatic environments. The DNA is found tightly bound to HS complexes, with up to 10% of the total organic phosphorus in soils being attributable to this one source (Crecchio & Stotzky 1997). These authors showed the adsorption to HS of DNA from the bacterium *Bacillus subtilis*, where resistance to nuclease was increased. They also showed the ability of the bound DNA to transform (acting as 'cryptic genes'), for instance, auxotrophic and antibiotic-sensitive cells, although not as rapidly as free DNA. Protection of free DNA from enzymatic attack and persistence of cryptic genes may be features of aquatic HS as well, though confirmation is awaited.

7.8   EFFECTS OF DISSOLVED HUMIC MATTER ON ORGANISMS

7.8.1 *Direct effects*

While there are many indications in the literature about the part humic materials in the soil play in the growth of plants and microorganisms (e.g., Kononova 1966; Tan & Tantiwiramanond 1983; Lobartini *et al.* 1994; Nardini *et al.* 1994) and about the antiviral and bactericidal uses of humic substances in medicine (e.g., Riede *et al.* 1991; Schneider *et al.* 1994), little is yet known of the direct effects of DHM on aquatic animals or plants. Natural DOM has been argued to benefit the survival and reproductive activity of acid-stressed *Daphnia magna* (Petersen & Persson 1986) and to be similarly beneficial to the macroinvertebrate crustaceans, *Gammarus pulex* and *Asellus aquaticus* (Hargeby & Petersen 1988; Hargeby 1990).

Traunspurger *et al.* (1997) introduced an assay technique, based upon the fecundity of a nematode *Caenorhabditis elegans*, to compare different sources of DOC. Using this method, Höss *et al.* (2001) differentiated among various sources of FA (Fig. 7.18): while some had no discernible effect, several soil leachates and a waste-water increased the number of offspring while one exerted a strong negative effect. The findings are not readily explained, as the composition of the DOCs (aromatic or aliphatic, phenolic or carboxylic) is unknown. However, endocrinic interference is another possibility: consistent oestrogenic effects of humic substances have been reported by Klöcking *et al.* (1992). The hormone-like potential of humic substances towards *C. elegans* has been confirmed recently (Steinberg *et al.* 2002). This potential has been attributed to alkylaromatics structural components that are strongly photostable.

Dissolved humic substances also have the potential to modulate the transformation system and the photosynthetic release of oxygen from macrophytes, aquatic mosses and algae (Pflugmacher *et al.* 1999, 2001). There are indications that this electron trapping effect is due to the quinoid structure of humic substances (see also section 7.6.5).

7.8.2 *Bioconcentration of xenobiotics*

There is a general concern about the potential of natural food webs to bioconcentrate hazardous xenobiotic compounds. Empirical data come mainly from determinations of the uptake of xenobiotics from artificial test media. The latter are formulated without DHM, which, in nature, interacts with pollutants through chemical reaction, binding, adsorption, ion exchange and covalent bonding (Choudhry 1984; Piccolo 1994). Any of these interactions could influence the bioavailability and uptake of the xenobiotic substance by organisms. Thus, solubility of the

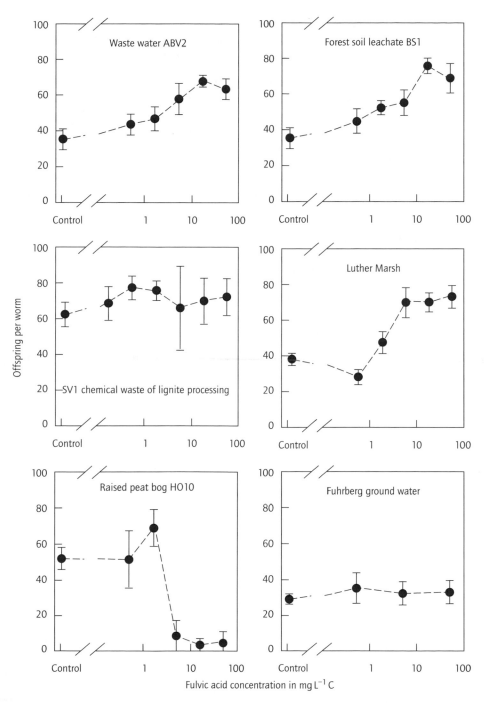

Fig. 7.18 Hormone-like effect of various fulvic acids on the reproductive success of the nematode *Caenorhabditis elegans*. (From Höss *et al.* 2001; reproduced with permission of Blackwell Publishing Ltd.)

substance may be altered (Wershaw *et al.* 1969; Burnison 1994) or the kinetics of its hydrolysis may be affected (Choudhry 1994). Volatilisation (Mackay 1979; Gschwend & Wu 1985), photolytic sensitivity (Zepp *et al.* 1981), assimilability and toxicity (Cary *et al.* 1987; Kukkonen & Oikari 1987; Day 1991; Bollag & Myers 1992; Hodge *et al.* 1993; Kadlec & Benson 1995) are also liable to alteration.

Generally, it is supposed that the reduced bioavailability of xenobiotics is a function of the size of the molecules of the humic substances with which they associate. This has been proved in the case of chlorobenzuron with Aldrich humic substances (Steinberg *et al.* 1993). The diminution in the factor of its bioconcentration (BCF) is a clear function of the DHM content (Fig. 7.19):

$$\mathrm{BCF} = 99.305 e^{-0.0073\mathrm{DHM}} \quad \left(r^2 = 0.9665 \right)$$

The capacity to decrease the extent of bioconcentration differs among DHM sources. This can be seen from the work of Haitzer *et al.* (1998, 1999,

2001), shown in Fig. 7.20, where organic material from Luther Marsh was more effective in its interference with the bioconcentration of benzo[*a*]pyrene (B*a*P) in the nematode *Caenorhabditis elegans* than the other sources tested. Haitzer *et al.* (1998) also sought an overview from the literature of the effect of DOM-mediated changes to the BCF of such common pollutants as polycyclic aromatic hydrocarbons (PAHs). Whereas the BCFs of the various PAHs tested ranged between 2 and 300%, interference by DOC varied between 2% and 98% as a broad correlative of concentration (see Fig. 7.21). In some cases, mild enhancements in bioconcentration were observed but always at the lower concentrations (<10 mg DOC L^{-1}). The mechanism behind this enhancement is still obscure (Haitzer *et al.* 2001). Reduction in BCF was relatively more effective against the hydrophobic compounds (B*a*P, methylchloranthracene) than the less hydrophobic examples (anthracene, naphthalene).

The bioconcentration of some other classes of pollutant may be enhanced by DOC, including

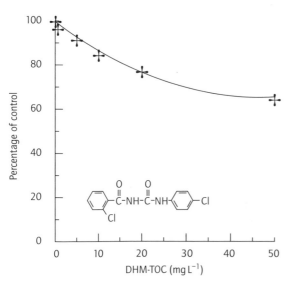

Fig. 7.19 Reduction of *Daphnia magna* bioconcentration of chlorobenzuron (CCU) in the presence of dissolved humic material. (Modified after Steinberg *et al.* 1993.)

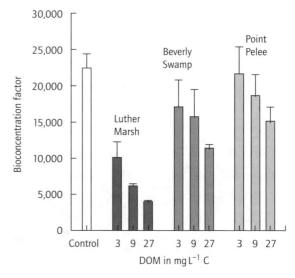

Fig. 7.20 Reduction of bioconcentration of benzo[a]pyrene in *Caenorhabditis elegans* (Nematoda) depending on the dissolved organic matter (DOM) sources. (Modified after Haitzer *et al.* 1999.)

that of pyrethroids by rainbow trout (Muir *et al.* 1994); the bioavailability of BaP and chlorinated dioxins may also be enhanced at low DOC concentrations (Kukkonen *et al.* 1989, 1990; Servos *et al.* 1989; Kukkonen & Oikari 1991). Whereas the depressing influence of DOM on the biotic uptake of hydrophobic compounds may be attributed to their binding to DOM, no mechanism for enhanced transferability has been suggested.

7.8.3 Changes in the toxicity of xenobiotic compounds in the presence of dissolved humic matter

There are several indications in the literature that the presence of humic substances can enhance or suppress the toxicity of xenobiotics. Pollutants can be become bound to HS, through the formation of C—C, C—O, C—N or N—N bonds, and their toxicity is reduced. The bonding may be catalysed biologically by polyphenol oxidases or peroxidase. Bollag & Myers (1992) demonstrated enzymatic catalysis of phenols and aromatic amines and the

formation of their coupling products. These are highly reactive with certain humus constituents. The major bonding mechanism appears to occur through weak interactions among HS and the lipophilic xenobiotics: HS associate with the chemicals, making them unavailable for uptake into organisms and, consequently, reducing the toxic effects. Aromatic properties of the HS are the most powerful predictor of each of these steps (Steinberg *et al.* 2000). The enhanced survivorship of *Daphnia magna* to the insecticide diazinon in the presence (0.5–50 mg TOC L^{-1}) of DHM observed by Lee *et al.* (1993) might be explained by this mechanism.

On the other hand, there is a short-term (c.2 h) enhancement of the acute toxicity to *D. magna* of substituted anilines and phenols in the presence of DHM (Steinberg *et al.* 1992). Significant increases in toxicity to 2,4-dichlorophenol were also observed over longer (c.60 h) periods. The mechanism of enhancement is not known, neither is the influence of contact time explained.

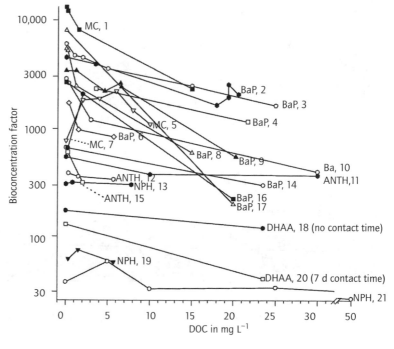

Fig. 7.21 Influence of dissolved organic carbon (DOC) on the bioconcentration of polycyclic aromatic hydrocarbons and deoxyhydroabietic acid: NPH, naphthalene; PaB, benzo[a]pyrene; ANTH, anthracene; MC, methylcholanthracene; BA, benzanthracene; the different numbers at the figures indicate different studies. (From Haitzer *et al.* 1998. Copyright (1998), with permission from Elsevier Science.)

7.8.4   Changes to heavy metal toxicity

There is an extensive literature on the speciation, bioavailability and toxicity of non-methylating heavy metals in the presence of DHM. Best studied is zinc. This is an essential trace nutrient but, like other heavy metals, it is severely toxic to aquatic biota in more than micromolar concentrations, subject to variations in acidity and hardness of the water. In the presence of HS, zinc or zinc ions are absorbed into metallo-organic complexes (Sigg & Stumm 1991). Thus chelated, greater concentrations of zinc are maintained in solution but the toxicity is reduced. Using an embryo-larval assay (Meinelt & Stüber 1991), Meinelt *et al.*

(1995) established that HS have no influence on embryo survival rates at concentrations up to 900 mg TOC L^{-1}. On the other hand, even at the lowest concentrations of zinc they tested (0.5 mg Zn L^{-1}), severe effects on larval survival were apparent. However, in the presence of HS concentrations of 5–20 mg TOC L^{-1}, embryos survived zinc concentrations of up to 2.5 mg L^{-1}. The effect was pronounced in soft water, owing to the weaker competitive interference of calcium and magnesium for binding sites (see also Rashid 1971); although hardness is generally depressive of metal toxicity (Bradley & Sprague 1985; Gundersen *et al.* 1994), the opposite is true when hardness and HS interact. Analogous effects of HS binding on cad-

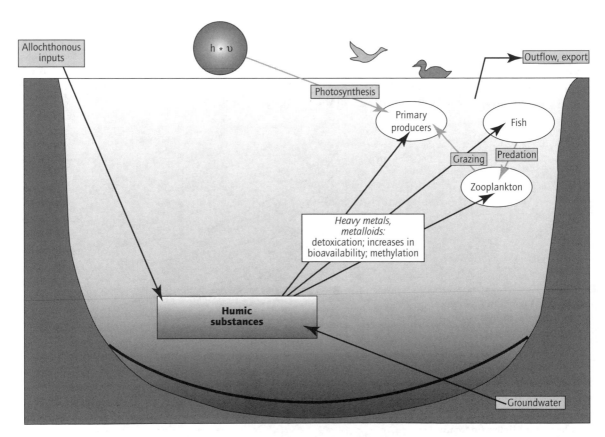

Fig. 7.22   Synoptic view of heavy metal toxicity alteration in the presence of humic substances. Black arrows indicate direct effects and grey ones symbolize indirect effects. (From Steinberg 2003; reproduced with permission of Springer-Verlag.)

mium toxicity in hard water have been noted by Winner (1986).

A synopsis of the probable impacts of HS on heavy metal toxicity in lakes is shown in Fig. 7.22.

7.8.5 *Influence on behaviour of organisms*

Behaviour integrates many cellular processes and is a vital component in the viability of the organism, the population and the community. Thus observations on organismic behaviour link toxicology to ecology and provide a unique ecotoxicological perspective of the consequences of environmental contamination (Little 1990). Changes in the swimming or feeding behaviour of fishes provide certain indications of the sub-lethal levels of contamination, and long before effects on growth or fecundity become apparent (Little & Finger 1990). Such behaviour may now be empiricised: the software BehavioQuant (Spieser & Scholz 1992) automatically records, accumulates and integrates variability in the swimming behaviour (velocity, depth velocity, turn-frequency, shoaling) of captive fishes. Thus, Lorenz *et al.* (1996) have been able to study long-term effects of sub-lethal doses of the triazine herbicide terbutylazine (TBA) on the behaviour of zebra fish (*Danio rerio*), with or without the presence of humic material (Fuhrberg FA; see section 7.4.2). Preferences for light or dark and the inconstancy of swimming were found to be the most sensitive indicators of altered behaviour. Light/dark preferences are summarised in Fig. 7.23 by means of Hasse diagrams (Brüggemann *et al.* 1995) to show how the chemical effects were distinguished. Starting at the bottom of Fig. 7.23, intensification of the light/dark preference to increasing concentrations of TBA, from $1\,\mu g\,L^{-1}$ (no reaction) up to $200\,\mu g\,L^{-1}$ (intense reaction), is tracked vertically upwards. Reactivity is increased, at all TBA concentrations, by the presence of a realistically low fulvic acid concentration ($2\,mg\,DOC\,L^{-1}$). With respect to inconstant swimming, all the TBA treatments were raised to a similar level but were moderated in all the DOC treatments. The apparent contradiction with the light/dark preference emphasises the interpreta-

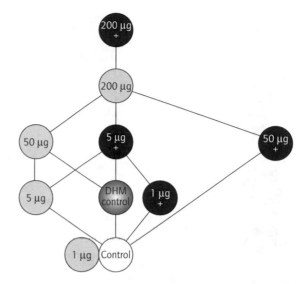

Fig. 7.23 Ranking of the light–dark, preference of *Danio rerio*, when exposed to terbutylazine (grey circles in $\mu g\,L^{-1}$) and to a combination of terbutylazine and Fuhrberg fulvic acid at $2\,mg\,L^{-1}$ dissolved organic carbon (black circles). (From Lorenz *et al.* 1996. Copyright (1996), with permission from Elsevier Science.)

tive care that is required but the combination of less erratic swimming with a cryptic response to minimise predation risk is not an unreasonable one. There is a concentration-related reaction to TBA and it is tempered by the presence of DHM.

As with other toxicological parameters, the extrapolation of behavioural responses to natural populations is fraught with difficulties. The ecological consequences of aberrant swimming behaviour are likely manifest in an impairment to migration, foraging or escape from predators. Hypoactivity, hyperactivity or deviations from natural alternations compound the short-term effects and feed through to the energy allocations to growth and fecundity (Little & Finger 1990).

These effects may also be tempered by the fact that the DHM component is not stable and that its photodegradation may result in a slow return to solution of xenobiotics and previously adsorbed contaminants (Pellissier 1993). Fresh inputs of DHM

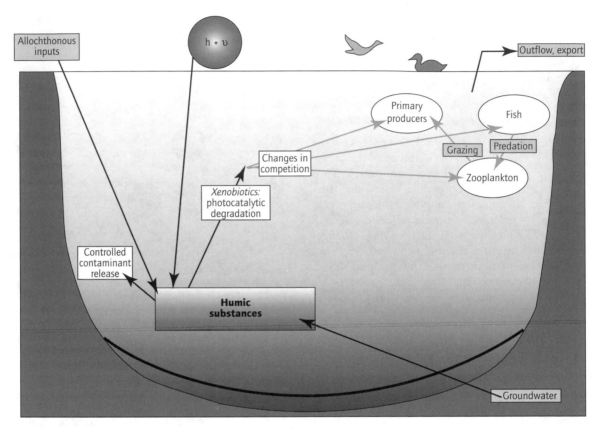

Fig. 7.24 Synoptic view of dissolved humic matter interference in ecotoxic effects of xenobiotic chemicals and potential controlled release of toxicants. Black arrows indicate direct effects and grey ones indirect effects. (From Steinberg 2003; reproduced with permission of Springer-Verlag.)

might scavenge these but, again, the comparative dynamics are not well quantified.

Of course, sublethal effects on lake biota of xenobiotics are already recognised to be potentially profound in the long term and command a considerable research investment. By contrast, the involvement of DHM in damping these effects is less well appreciated but would repay some amplifying study. A generalised scheme of the involvement of humic substances in the ecotoxicity of xenobiotic toxicity is proposed in Fig. 7.24.

7.9 CONCLUSION

Putting together the various pieces of the humic jig-saw puzzle (Figs 7.6, 7.13, 7.14, 7.17, 7.22 and 7.24) the composite Fig. 7.25 proves to be highly complex. There is no aspect of limnology that is untouched by the environmental geochemistry of DHM. Moreover, this summary makes no attempt to separate the behaviours of the autochthonous and allochthonous DOC, or of the dynamic dependence of the latter on its numerous possible sources. This text has made little reference to the work on the provenance of DHM products, for which information the reader is directed to the excellent review of Thomas (1997).

The foregoing chapter has shown that the amounts and the characteristics of the DHM reaching lakes will influence substantially several fundamental features, including their pH and

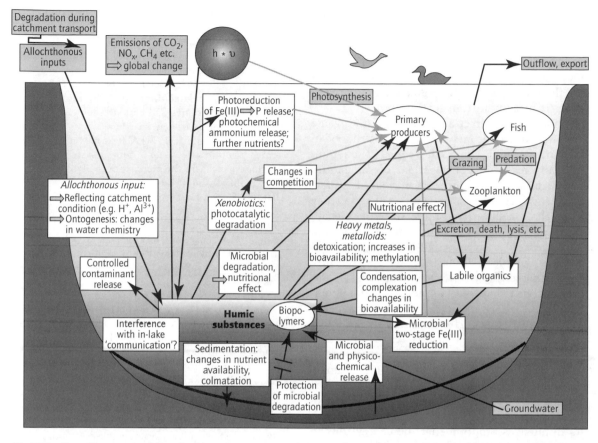

Fig. 7.25   Overall synopsis of dissolved humic matter interference in biotic–biotic and biotic–abiotic interactions within a lake. Black arrows indicate direct effects and grey ones indirect effects. (From Steinberg 2003; reproduced with permission of Springer-Verlag.)

buffering potential. The vegetation that it is possible to support in the catchment also affects the quantities of organic acids and the extent of their condensation. Recent acidification of precipitation by mineral acids has had the effect of diminishing the DHM content of lakes, at once depriving them of an effective protective shield against potentially harmful UV radiation as well as a chemical resistance to toxic substances. Although there has been an encouraging retreat from global acidification, the threats to the DHM content of lakes are now posed by changes in global climate.

With respect to the nutritional value of the various DHM fractions, there has been a major shift in

the paradigm about the metabolism of limnetic ecosystems. There is a consensus that DOM provides an important energy source that is harnessed by microorganisms and that even HMW DHM may be one important source.

Dissolved humic matter may be photolysed and even photomineralised by UV-B and visible radiation, often releasing scarce nutrients (phosphorus, nitrogen and iron) as well as supplementing the inorganic carbon dioxide stock available to photoautotrophic primary producers. Photodegradation of DHM has negative consequences as well, if bonds with toxic chemicals and metals are broken. The longest lived of the photoactive chemicals (hydrogen superoxide) are positively harmful to organisms. Even without light, DHM has the capacity to catalyse redox reactions in the iron cycle.

Dissolved humic matter can adsorb biopolymers, enzymes, toxins and DNA, actually protecting them from degradation. Bound DNA ('cryptic genes') can facilitate transformations.

Short-term interactions of DHM with organic and inorganic toxicants reduces the likelihood of bioconcentration. This applies equally to environmentally realistic concentrations of xenobiotics although sensitive parameters are needed to show this. However, low concentrations of DHM may bring about a short-lived increase in the bioconcentration of xenobiotic compounds. The work with nematodes suggests that DHM acts in an oestrogen-like role. It has been shown very recently that humic substances are weak xenobiotic chemicals and, thus, may structure aquatic guilds (Steinberg 2003.)

This multiplicity of functions may be sufficient to demonstrate the regulatory force of humic substances in limnetic ecosystems. It is not difficult to share Wetzel's (1995) view that 'They dominate the material and energy fluxes'. Neither does Thomas' (1997) deduction seem any longer controversial: 'The central dogma of the foodweb, with its implicit assumption that the energy flow in aquatic ecosystems can be quantified by measuring rates of photosynthesis and the rates of food ingestion by consumer organisms, is invalid'.

This does not mean that energy and material fluxes are exclusively mediated by DOM, or that primary production and trophic linkages are unimportant processes. It is vital to recognise the mutual overlaps and functional interdependencies. Progress depends upon reconciling the classical food-web theory into a deepening knowledge and understanding of the system-regulating functions of dissolved humic matter.

REFERENCES

Abrahamsen, G. (1980) Acid precipitation, plant and forest growth. In: Drabløs, D. & Tollan, A. (eds), *Ecological Impact of Acid Precipitation*. SNSF (Norwegian Council for Scientific and Industrial Research), Ås, 58–63.

Alberts, J.J., Schindler, J.E., Miller, R.W. & Nutter, D.E. (1974) Elemental mercury evolution mediated by humic acid. *Science*, **184**, 895–7.

Alberts, J.J., Filip, Z. & Hertkorn, N. (1992) Fulvic and humic acids isolated from groundwater: compositional characteristics and cation binding. *Journal of Contaminant Hydrology*, **11**, 317–30.

Allard, B. & Arsenie, I. (1991) Abiotic reduction of mercury by humic substances in aquatic systems—an important process for the mercury cycle. *Water, Air and Soil Pollution*, **56**, 457–64.

Allard, B., Borén, H., Pettersson, C. & Zhang, G. (1994) Degradation of humic substances by UV irradiation. *Environment International*, **20**, 97–101.

Amon, R.M.W. & Benner, R. (1994) Rapid cycling of high molecular-weight dissolved organic matter in the ocean. *Nature*, **369**, 549–52.

Amon, R.M.W. & Benner, R. (1996a) Bacterial utilisation of different size classes of dissolved organic matter. *Limnology and Oceanography*, **41**, 41–51.

Amon, R.M.W. & Benner, R. (1996b) Photochemical and microbial consumption of dissolved organic carbon and dissolved oxygen in the Amazon River system. *Geochimica et Cosmochimica Acta*, **60**, 1783–92.

Backlund, P. (1992) Degradation of aquatic humic material by ultraviolet light. *Chemosphere*, **25**, 1869–78.

Battarbee, R.W., Flower, R.J., Stevenson, A.C. & Rippey, B. (1985) Lake acidification in Galloway: a palaeoecological test of competing hypotheses. *Nature*, **314**, 350–2.

Bertilsson, S. & Allard, B. (1996) Sequential photochemical and microbial degradation of refractory dissolved organic matter in a humic freshwater system. *Ergebnisse der Limnologie*, **48**, 133–41.

Bewley, R.J.F. & Parkinson, D. (1986) Sensitivity of certain soil microbial processes to acid precipitation. *Pedobiologia*, **29**, 73–84.

Boer, M.M., Koster, E.A. & Lundberg, H. (1990) Greenhouse impact in Fennoscandia—preliminary findings of a Eurpoean workshop on the effects of climate change. *Ambio*, **19**, 2–10.

Bollag, J.M. & Myers, C. (1992) detoxification of aquatic and terrestrial sites through binding of pollutants to humic substances. *Science of the Total Environment*, **117/118**, 357–66.

Bourne, D.G., Jones, G.J., Blakeley, R.L., Jones, A., Negrie, A.P. & Riddles, P. (1996) Enzymatic pathway for the bacterial degradation of the cyanobacterial cyclic peptide toxin, microcystin LR. *Applied and Environmental Microbiology*, **62**, 1086–94.

Bradley, R.W. & Sprague, J.B. (1985) The influence of pH, water hardness and alkalinity on the acute lethality of zinc to rainbow trout (*Salmo gairdneri*). *Canadian Journal of Fisheries and Aquatic Sciences*, **42**, 731–6.

Brassard, P. & Auclair, J.C. (1984) orthophosphate rate constants are mediated by the 10^3–10^4 molecular weight fraction in Shield lake waters. *Canadian Journal of Fisheries and Aquatic Sciences*, **41**, 166–73.

Bruckmeier, B.F.A., Jüttner, I., Schramm, K.W., Winkler, R., Steinberg, C.E.W. & Kettrup, A. (1997) PCBs and BCDD/Fs in lake sediments at Grosser Arbersee, Bavarian Forest, South Germany. *Environmental Pollution*, **95**, 19–25.

Brüggemann, R., Schwaiger, J. & Negele, R.D. (1995) Applying Hasse diagram technique for evaluation of toxicological fish tests. *Chemosphere*, **30**, 1767–80.

Burnison, K. (1994) Solubility enhancement of fenvalerate by isolated DOC lakewater fractions. In: Senesi, N. & Miano, T. (eds), *Humic Substances in the Global Environment and Implications on Human Health*. Elsevier, Amsterdam, 811–18.

Bushaw, K.L., Zepp, R.G., Tarr, M.A., *et al.* (1996) Photochemical release of biologically available nitrogen from aquatic dissolved organic matter. *Nature*, **381**, 404–7.

Canonica, S., Jans, U., Stemmler, K. & Hoigné, J. (1995) Transformation kinetics of phenols in water: photosensitization by dissolved of natural organic material and aromatic ketones. *Environmental Science and Technology*, **29**, 1822–31.

Cary, G.A., McMahon, J.A. & Kuc, W.J. (1987) The effect of suspended solids on naturally occurring dissolved organics in reducing the acute toxicity of cationic polyelectrolytes to aquatic organisms. *Environmental Toxicology and Chemistry*, **6**, 469–74.

Choudhry, G.G. (1984) Interactions of humic substances with environmental chemicals. In: Hutzinger, O. (ed.), *The Handbook of Environmental Chemistry*, Vol. 2. Springer-Verlag, New York, 103–28.

Clair, T.A. & Ehrman, J.M. (1996) Variations in discharge and dissolved organic carbon and nitrogen from terrestrial basins with changes in climate. A neural-network approach. *Limnology and Oceanography*, **41**, 921–7.

Coale, K.H., Johnson, K.S., Fitzwater, S.E., *et al.* (1996) A massive phytoplankton bloom induced by an ecosystem-scale iron fertilization experiment in the equatorial Pacific Ocean. *Nature*, **383**, 495–501.

Conrad, R. & Seiler, W. (1980) Photooxidative production and microbial consumption of carbon monoxide in seawater. *FEMS Microbiological Letters*, **9**, 61–4.

Cotner, J.B. & Heath, R.T. (1990) Iron redox effects on

photosensitive phosphorus release from dissolved humic materials. *Limnology and Oceanography*, **35**, 1175–81.

Crecchio, C. & Stotzky, G. (1997) Binding of DNA on humic acids: effects on transformation of *Bacillus subtilis*. *Soil Biology and Biochemistry*, **30**, 1061–7.

Dahlén, J., Bertilsson, S. & Pettersen, C. (1996) Effects of UV-A irradiation on dissolved organic matter in humic surface waters. *Environment International*, **22**, 501–6.

David, M.B., Vance, G.F., Rissing, J.M. & Stevenson, F.J. (1989) Organic carbon fraction in extracts of O and B horizons from a New England spodozol. Effects of acid treatment. *Journal of Environmental Quality*, **18**, 212–17.

Davis, R.B., Anderson, D.S. & Berge, F. (1985) Palaolimnological evidence that lake acidification is accompanied by loss of organic matter. *Nature*, **316**: 436–8.

Day, K.E. (1991) Effects of dissolved organic carbon on accumulation and acute toxicity of fenvalerate, deltamethrin and cyhalothrin to *Daphnia magna* (Straus). *Environmental Science and Technology*, **10**, 91–101.

De Haan, H. (1974) Effect of fulvic acid fraction on the growth of a *Pseudomonas* sp. from Tjeukemeer (The Netherlands). *Freshwater Biology*, **4**, 301–10.

De Haan, H. (1992) Impacts of environmental changes on the biogeochemistry of aquatic humic substances. *Hydrobiologia*, **229**, 59–71.

De Haan, H. (1993) Solar UV-light penetration and photodegradation of humic substances in peaty lake water. *Limnology and Oceanography*, **38**, 1072–6.

De Haan, H., Jones, R.I. & Salonen, K. (1990) Abiotic transformations of iron and phosphate in humic lake water revealed by double isotope labeling and gel filtration. *Limnology and Oceanography*, **35**, 491–7.

Dickson, W. (1980) Properties of acidified water. In: Drabløs, D. & Tollan, A. (eds), *Ecological Impact of Acid Precipitation*. SNSF, (Norwegian Council for Scientific and Industrial Research), Ås, 75–83.

Draper, W.M. & Crosby, D.G. (1983) The photochemical generation of hydrogen peroxide in natural waters. *Archives of Environmental Contamination and Toxicology*, **12**, 121–6.

Engstrom, D.R. (1987) Influence of vegetation and hydrology on the humus budgets of Labrador lakes. *Canadian Journal of Fisheries and Aquatic Sciences*, **44**, 1306–10.

Ephraim, J.H., Pettersson, C. & Allard, B. (1996) Correlations between acidity and molecular size distributions of an aquatic fulvic acid. *Environment International*, **22**, 475–83.

Faust, B.C. & Hoigné, J. (1987) Sensitized photooxidation of phenols by fulvic acid in natural waters. *Environmental Science and Technology*, **21**, 957–64.

Fee, E.J., Hecky, R.E., Kasian, S.E. & Cruikshank, D. (1996) Potential size-related effects of climate change on mixing depths in Canadian Shield lakes. *Limnology and Oceanography*, **41**, 912–20.

Forsberg, C. (1992) Will an increased greenhouse effect impact in Fennoscandia give rise to more humic or coloured lakes? *Hydrobiologia*, **229**, 51–8.

Francko, D.A. (1986) Epilimnetic phosphorus cycling: influence of humic materials and iron on coexisting major mechanisms. *Canadian Journal of Fisheries and Aquatic Sciences*, **43**, 302–10.

Francko, D.A. & Heath, R.T. (1982) UV-sensitive complex phosphorus: association with dissolved humic material and iron in a bog lake. *Limnology and Oceanography*, **27**, 564–9.

Francko, D.A. & Heath, R.T. (1983) Abiotic uptake and photodependent release of phosphate from high-molecular weight complexes in a bog lake. In: Christman, R.F. & Gjessing, E.T. (eds), *Aquatic and Terrestrial Humic Materials*. Ann Arbor Science, Ann Arbor, 467–80.

Frimmel, F.H. (1994) Photochemical aspects related to humic substances. *Environment International*, **20**, 373–85.

Garcia-Pichel, F. & Castenholz, R.W. (1991) Characterisation and biological implications of scytonemin, a cyanobacterial sheath pigment. *Journal of Phycology*, **27**, 395–409.

Geller, A. (1985a) Light-induced conversion of refractory, high molecular-weight, lake-water constituents. *Schweizerische Zeitschrift für Hydrologie*, **47**, 21–6.

Geller, A. (1985b) Degradation and formation of refractory DOM by bacteria during simultaneous growth on labile substrates and persistent lake water constituents. *Schweizerische Zeitschrift für Hydrologie*, **47**, 27–44.

Geller, A. (1986) Comparison of mechanisms enhancing biodegradability of refractory lake water constituents. *Limnology and Oceanography*, **31**, 755–64.

Giesy, J.P. (1976) Stimulation of growth in *Scenedesmus obliquus* (Clorophyceae) by humic acids under iron-limited conditions. *Journal of Phycology*, **12**, 172–9.

Giorgio, P.A. del & Peters, R.H. (1994) Patterns in planktonic P : R ratio in lakes: Influence of lake trophy and dissolved organic carbon. *Limnology and Oceanography*, **39**, 772–87.

Gjessing, E.T. (1992) The HUMEX Project: experimental acidification of a catchment and its humic lake. *Environment International*, **18**, 535–43.

Gjessing, E.T. (1994a) HUMEX (Humic Lake Acidification Experiment): chemistry, hydrology and meteorology. *Environment International*, **20**, 267–76.

Gjessing, E.T. (1994b) The role of humic substances in the acidification response of soil and water—results of the humic lake acidification experiment (HUMEX). *Environment International*, **20**, 363–8.

Gjessing, E.T. & Gjerdahl, T.C. (1970) Influence of ultraviolet radiation on aquatic humus. *Vatten*, **26**, 144–5.

Gjessing, E.T. & Källqvist, T. (1991) Algicidal and chemical effects of UV-radiation of water containing humic substances. *Water Research*, **25**, 491–4.

Gonet, S.S. & Wegner, K. (1996) Viscosimetric and chromatographic studies on soil humic acids. *Environment International*, **22**, 485–8.

Granéli, W., Lindell, M. & Tranvik, L. (1996) Photooxidative production of dissolved inorganic carbon in lakes of different humic content. *Limnology and Oceanography*, **41**, 698–706.

Grossart, H.-P. & Plough, H. (2001) Microbial degradation of organic carbon and nitrogen on diatom aggregates. *Limnology and Oceanography*, **46**, 267–77.

Gschwend, P.M. & Wu, S.C. (1985) On the constancy of sediment–water partition coefficients of hydrophobic organic pollutants. *Environmental Science and Technology*, **19**, 90–6.

Guildford, S.J., Healey, F.P. & Hecky, R.E. (1987) Depression of primary production by humic matter and suspended sediment in limnocorral experiments at southern Indian Lake, northern Manitoba. *Canadian Journal of Fisheries and Aquatic Sciences*, **44**, 1408–17.

Gundersen, D.T., Bustaman, S., Seim, W.K. & Curtis, L.R. (1994) pH, hardness and humic acid influence aluminium toxicity to rainbow trout (*Oncorhyncus mykiss*) in weakly alkaline waters. *Canadian Journal of Fisheries and Aquatic Sciences*, **51**, 1345–55.

Haitzer, M., Höss, S., Traunspurger, W. & Steinberg, C. (1998) Effects of dissolved organic matter (DOM) on the bioconcentration of organic chemicals in aquatic organisms—a review. *Chemosphere*, **37**, 1335–62.

Haitzer, M., Burnison, B.K., Hoess, S., Traunspurger, W. & Steinberg, C.E.W. (1999) Effects of quantity, quality and contact time of dissolved organic matter on bioconcentration of benzo[a]pyrene in the nematode, *Caenorhabditis elegans*. *Environmental Toxicology and Chemistry*, **18**, 459–65.

Haitzer, M., Akkanen, J., Steinberg, C. & Kukkonen, J.V. (2001) No enhancement in bioconcentration of organic contaminants by low levels of DOM. *Chemosphere*, **44**, 165–71.

Hargeby, A. (1990) Effects of pH, humic substances and animal interactions on survival and physiological status of *Asellus aquaticus* L. and *Gammarus pulex* (L.)—a field experiment. *Oecologia*, **82**, 348–54.

Hargeby, A. & Petersen, R.C. (1988) Effects of low pH and humus on the survivorship, growth and feeding of *Gammarus pulex* (L.). (Amphipoda). *Freshwater Biology*, **19**, 235–47.

Hatcher, P.G. & Spiker, E.C. (1988) Selective degradation of plant molecules. In: Frimmel, F.H. & Christman, R.F. (eds), *Humic Substances and their Role in the Environment*. Wiley, Chichester, 59–74.

Håkanson, L. (1996) A simple model to predict the duration of the mercury problem in Sweden. *Ecological Modelling*, **93**, 251–62.

Hessen, D.O. & Donk, E. van (1994) Effects of UV-radiation of humic water on primary and secondary production. *Water, Air and Soil Pollution*, **75**, 325–38.

Hessen, D.O., Nygaard, K., Salonen, K. & Vähätalo, A. (1994) The effect of substrate stoichiometry on microbial activity and carbon degradation in humic lakes. *Environment International*, **20**, 67–76.

Hintelmann, H., Welbourn, P.M., Evans, R.D. (1995) Binding of methyl mercury compounds by humic and fulvic acids. *Water, Air and Soil Pollution*, **80**, 1031–4.

Hodge, V.A., Fan, G.T., Solomon, K.R., Kaushik, N.K., Leppard, G.G. & Burnison, K.(1993) Effects of the presence or absence of various fractions of dissolved organic matter on the toxicity of fenvalerate to *Daphnia magna*. *Environmental Technology and Chemistry*, **12**, 167–76.

Horth, H., Frimmel, F.H., Hargitai, L., *et al*. (1988) Environmental reactions and functions. In: Frimmel, C.H. & Christman, R.F. (eds), *Humic Substances and their Role in the Environment*. Wiley, Chichester, 245–56.

Höss, S., Haitzer, M., Traunspurger, W. & Steinberg, C. (2001) Refractory dissolved organic matter can influence the reproduction of *Caenorhabditis elegans* (Nematoda). *Freshwater Biology*, **46**, 1–10.

Hutchinson, G.E. (1957) *A Treatise on Limnology*, Vol. 1, *Geography, Physics, Chemistry*. Wiley, New York, 1016 pp.

Jackson, C.R., Foreman, C.M. & Sinsabaugh, R.L. (1995) Microbial enzyme activities as indicators of organic-matter processing rates in a Lake-Erie coastal wetland. *Freshwater Biology*, **34**, 329–342.

Jackson, T.A. & Hecky, R.E. (1980) depression of primary productivity by humic matter in lake and reservoir waters of the boreal forest zone. *Canadian Journal of Fisheries and Aquatic Sciences*, **37**, 2300–17.

Johnson, N.M., Likens, G.E., Feller, M.C. & Driscoll, C.T. (1984) Acid rain and soil chemistry. *Science*, **225**, 1424–5.

Jørgensen, C.B. (1976) August Pütter, August Krogh and modern ideas on the use of dissolved organic matter in aquatic environments. *Biological Reviews*, **51**, 291–328.

Jones, R.I., Salonen, K. & Haan, H. de (1988) Phosphorus transformations in the epilimnion of humic lakes: abiotic interactions between dissolved humic materials and phosphate. *Freshwater Biology*, **19**, 357–67.

Jones, R.I., Shaw, P.J. & de Haan, H. (1993) Effects of dissolved humic substances on the speciation of iron and phosphate at different pH and ionic strength. *Environmental Science and Technology*, **27**, 1052–9.

Kadlec, M.C. & Benson, W.H. (1995) Relationship of aquatic natural organic material characteristics to the toxicity of selected insecticides. *Ecotoxicology and Environmental Safety*, **31**, 84–97.

Kelly, C.A., Rudd, J.W.M., Furutani, A. & Schindler, D.W. (1984) Effects of lake acidification on rates of organic matter decomposition in sediments. *Limnology and Oceanography*, **29**, 687–94.

Kieber, D.J., McDaniel, J. & Mopper, K. (1989) Photochemical source of biological substrates in sea water: implications for carbon cycling. *Nature*, **341**, 637–9.

Killops, S.D. & Killops, V.J. (1993) *An introduction to organic geochemistry*. Longman, London.

Kirk, J.T.O. (1983) *Light and photosynthesis in Aquatic systems*. Cambridge University Press, Cambridge.

Kļaviņš, M. (1997) *Aquatic Humic Substances: Characterization, Structure and genesis*. Riga University Press, Riga, Latvia, 234 pp.

Klöcking, R., Fernekorn, A. & Stölzner, W. (1992) Nachweis einer östrogenen Aktivität von Huminsäuren und huminsäureähnlch Polmeren. *Telma*, **22**, 187–97.

Kok, C.J. & Velde, G. van der (1994) Decomposition and macroinvertebrate colonization of aquatic and terrestrial leaf material in alkaline and acid still water. *Freshwater Biology*, **31**, 65–75.

Kok, C.J., Haverkamp, W. & Aa, H.A. van der (1992) Influence of pH on the growth and leaf-maceration ability of fungi involved in the decomposition of floating leaves of *Nymphaea alba* in acid water. *Journal of General Microbiology*, **138**, 103–8.

Kononova, M.M. (1966) *Soil Organic Matter*. Pergamon Press, Oxford.

Kortelainen, P. (1993) Contribution of organic acids to

the acidity of Finnish lakes. *Publications of the Water Environment Research Institute*, **13**, 1–48.

Krug, E.C. & Frink, C.R. (1983) Acid rain and acid soil: a new perspective. *Science*, **221**, 520–5.

Kukkonen, J. & Oikari, A. (1987) Effects of aquatic humus on accumulation and acute toxicity of organic micropollutants. *Science of the Total Environment*, **62**, 399–402.

Kukkonen, J. & Oikari, A. (1991) Bioavailability of organic pollutants in boreal waters with varying levels of dissolved organic material. *Water Research*, **25**, 455–63.

Kukkonen, J., Oikari, A., Johnsen, S. & Gjessing, E. (1989) Effects of humus concentrations on benzo[a]pyrene accumulation from water to *Daphnia magna*: comparison of natural waters and standard preparations. *Science of the Total Environment*, **62**, 399–402.

Kukkonen, J., McCarthy, J.F. & Oikari, A. (1990) Effects of XAD-8 fractions of dissolved organic carbon on the sorption and bioavailability of organic micropollutants. *Archives of Environmental Contamination and Toxicology*, **19**, 551–7.

Kullberg, A., Petersen, R.C., Jr., Hargeby, A. & Svensson, A. (1992) Transport of octanol-soluble carbon and dissolved organic carbon through the soil/water interface of the HUMEX experiment. *Environment International*, **18**, 631–6.

Kulovaara, M. & Backlund, P. (1993) Effects of simulated sunlight on aquatic humic matter. *Vatten*, **49**, 100–3.

Kulovaara, M., Backlund, P. & Corin, N. (1995) Light-induced degradation of DDT in humic water. *Science of the Total Environment*, **170**, 185–91.

Larson, R.A. & Hufnal, J.M., Jr. (1980) Oxidative polymerization of dissolved phenols by soluble and insoluble organic species. *Limnology and Oceanography*, **25**, 505–12.

Larson, R.A., Smykowski, K. &, Hunt, L.L. (1981) Occurrence and determination of organic oxidants in rivers and waste waters. *Chemosphere*, **10**, 1335–8.

Leavitt, P.R., Vinebrooke, R.D., Donald, D.B., Smol, J.P. & Schindler, D.W. (1997) Past ultraviolet radiation environments in lakes derived from fossil pigments. *Nature*, **388**, 457–9.

Lee, S.K., Freitag, D., Steinberg, C.E.W., Kettrup, A. & Kim, Y.-H. (1993) Effects of dissolved humic materials on acute toxicity of some organic chemicals to aquatic organisms. *Water Research*, **27**, 199–204.

Lindell, M.J. & Rai, H. (1994) Photochemical oxygen consumption in humic waters *Ergebnisse der Limnologie*, **43**, 145–55.

Lindell, M.J., Graneli, W. & Tranvik, L.J. (1995) Enhanced bacterial growth in response to photochemical transformation of dissolved organic matter. *Limnology and Oceanography*, **40**, 195–9.

Little, E.E. (1990) Behavioural toxicology: stimulating challenges for a growing discipline. *Environmental Toxicology and Chemistry*, **9**, 1–2.

Little, E.E. & Finger, S.E. (1990) Swimming behavior as an indicator of sublethal toxicity in fish. *Environmental Toxicology and Chemistry*, **9**, 13–19.

Lobartini, J.C., Tan, K.H. & Pape, C. (1994) The nature of humic acid–apatite interactions, products and their availability to plant growth. *Communications in Science and Plant Analysis*, **25**, 2355–69.

Lorenz, R., Brüggemann, R., Steinberg, C.E.W. & Spieser, O.H. (1996) Humic material changes effects of terbutylazine on behavior of zebrafish (*Brachydanio rerio*). *Chemosphere*, **33**, 2145–58.

Lovley, D.R., Coates, J.D., Blunt-Harris, E.L., Phillips, E.J.P. & Woodward, J.C. (1996) Humic substances as electron acceptors for microbial respiration. *Nature*, **382**, 445–8.

Lund, V. & Hongve, D. (1994) Ultra-violet irradiated water containing humic substances inhibits bacterial metabolism. *Water Research*, **28**, 111–16.

Luther, G.W., Shellenbarger, P.A. & Brendel, P.J. (1996) Dissolved organic Fe(III) and Fe(II) complexes in salt marsh pore waters. *Geochimica et Cosmochimica Acta*, **60**, 951–60.

Mackay, D. (1979) Finding fugacity feasible. *Environmental Science and Technology*, **13**, 1218–23.

Malcolm, R.L. & Hayes, T. (1994) Organic solute changes with acidification in Lake Skjervatjrn as shown by ^{1}H-NMR spectroscopy. *Environment International*, **20**, 299–305.

Marmorek, D.R., Bernard, D.P., Jones, M.L., Rattie, L.P. & Sullivan, T.J. (1987) *The Effects of Mineral Acid Deposition on Concentrations of Dissolved Organic Acids in Surface Waters*. U.S. Environment Protection Agency, Corvallis.

Matilainen, T. & Verta, M. (1995) Mercury methylation and demethylation in aerobic surface waters. *Canadian Journal of Fisheries and Aquatic Sciences*, **52**, 1597–608.

Meinelt, T. & Stüber, A. (1991) Der Embryo-Larval-Test. Eine subchronische Testmethode für die Fischwirtschaft. *Fischer und Teichwirt*, **42**, 247–9.

Meinelt, T., Staaks, G. & Stüber A. (1995) Veränderung der Zinktoxizität im Wässern geringer und hoher Wasserhärte unter dem Einfluss eines sythetischen Huminstoffes. *Umweltwissenschaften und Schadstoff-Forschung-Zeitschrift für Umweltchemie und Ökotoxicologie*, **3**, 155–8.

Mierle, G. & Ingram, R. (1991) The role of humic substances in the mobilization of mercury from watersheds. *Water, Air and Soil Pollution*, 56, 349–57.

Miles, C.J. & Brezonik, P.L. (1981) Oxygen consumption and humic-coloured waters by a photochemical ferrous-ferric catalytic cycle. *Environmental Science and Technology*, 15, 1089–95.

Mill, M.T., Hendry, D.G. & Richardson, H. (1980) Free-radical oxidants in natural waters. *Science*, 207, 886–8.

Miller, R.W. (1975) The role of humic acids in the uptake and release of mercury by freshwater sediments. *Verhandlungen der internationale Vereinigung für theoretische und angewandte Limnologie*, 19, 2082–6.

Mills, G.L. & Schwind, D. (1990) Photochemical degradation rates of tetraphenylborate and diphenylboric acid sensitized by dissolved organic matter in stream water. *Environmental Toxicology and Chemistry*, 9, 569–74.

Moran, M.A. & Hodson, R.E. (1990) Bacterial production on humic and non-humic components of dissolved organic carbon. *Limnology and Oceanography*, 35, 1744–56.

Muir, D.C.G., Hobden, B.R. & Servos, M.R. (1994) Bioconcentration of pyrethroid insecticides and DTT by rainbow trout: uptake, depuration and effect of dissolved organic carbon. *Aquatic Toxicology*, 29, 230–40.

Münster, U. (1994) Studies on phosphatase activities in humic lakes. *Environment International*, 20, 49–59.

Münster, U. & de Haan, H. (1998) The role of microbial extracellular enzymes (MEE) in the transformation of dissolved organic matter (DOM) in humic waters. In: Hessen, D.O. & Tranvik, L. (eds), *Aquatic Humic Substances, Ecology and Biogeochemistry*. Ecological Studies 133, Springer-Verlag, Heidelberg, 199–257.

Münster, U., Einiö, P., Nurminen, J. & Overbeck, J. (1992a) Extracellular enzymes in a polyhumic lake: important regulators in detritus processing. *Hydrobiologia*, 229, 225–38.

Münster, U., Nurminen, J., Einiö, P. & Overbeck, J. (1992b) Extracellular enzymes in a small polyhumic lake: origin, distribution and activities. *Hydrobiologia*, 243/244, 47–59.

Münster, U., Heikkinen, E., Salonen, K. & de Haan, H. (1998) Tracing of peroxidase activity in humic lake water. *Acta Hydrochimica et Hydrobiologica*, 26, 158–66.

Münster, U., Salonen, K. & Tulonen, T. (1999) Decomposition. In: Keskitalo, J. & Eloranta, P. (eds), *Limnology of Humic Waters*. Backhuys Publishers, Leiden, 225–64.

Nardini, S., Pamuccio, M.R., Abenavolit, M.R. & Muscola, A. (1994) Auxin-like effect of humic substances extracted from the faeces of *Allolobophora caliginosa* and *Antennaria rosea*. *Soil Biology and Biochemistry*, 26, 1341–6.

Pellissier, F. (1993) Allelopathic effect of phenol acids from humic solutions on two spruce mycorrhizal fungi — *Genococcum graniforme* and *Accaria laccata*. *Journal of Chemical Ecology*, 19, 2015–114.

Pennington, W. (1984) Long-term natural acidification of upland sites in Cumbria: evidence from post-glacial lake sediments. *Report of the Freshwater Biological Association*, 52, 28–46.

Petersen, R.C. & Persson, U. (1986) Comparison of the biological effects of humic materials under acidified conditions. *Science of the Total Environment*, 62, 387–98.

Pflugmacher, S., Spangenberg, M. & Steinberg, C.E.W. (1999) Dissolved organic matter (DOM) and effects on the aquatic macrophyte *Ceratophyllum demersum* in relation to photosynthesis, pigment pattern and activity of detoxication enzymes. *Journal of Applied Botany*, 73, 184–90.

Pflugmacher, S., Tidwell, L.F. & Steinberg, C.E.W. (2001) Dissolved humic substances directly affect freshwater organisms. *Acta Hydrochimica et Hydrobiologica*, 29, 34–40.

Piccolo, A. (1994) Interactions between organic pollutants and humic substances in the environment. In: Senesi, N. & Miano, T. (eds), *Humic Substances in the Global Environment and Implications on Human Health*. Elsevier, Amsterdam, 961–79.

Porvari, P. & Verta, M. (1995) Methylmercury production in flooded soils: a laboratory study. *Water, Air and Soil Pollution*, 80, 789–98.

Rashid, M.A. (1971) Role of humic acids of marine origin and their different molecular weight fractions in complexing di- and tri-valent metals. *Soil Science*, 111, 289–306.

Rasmussen, J.B., Godbout, L. & Schallenberg, M. (1989) The humic content of lake water and its relationship to watershed and lake morphometry. *Limnology and Oceanography*, 34, 1336–43.

Reinikainen, J. & Hyvärinen, H. (1997) Humic- and fulvic-acid stratigraphy of the Holocene sediments in Finnish Lappland. *Holocene*, 7, 401–7.

Riede, U.N., Zeck-Knapp, G., Freudenberg, N., Keller, H.K., Seubert, B. (1991) Humate induced activation of human granulocytes. *Virchow's Archives, B: Cell Pathology*, 60, 27–30.

Rodhe, H. & Rood, M.J. (1985) Temporal evaluation of

nitrogen compounds in Swedish precipitation since 1955. *Nature*, **321**, 762–4.

Ryhänen, R. (1968) Die Bedeutung der Humussubstanzen im Stoffhaushalt der Gewässer Finlands. *Mitteilungen der internationalen Vereinigung für theoretische und angewandte Limnologie*, **14**, 168–78.

Sachse, A., Babenzien, D., Ginzel, G., Gelbrecht, J. & Steinberg, C.E.W (2001) Characterization of dissolved organic carbon (DOC) in a dystrophic lake and an adjacent fen. *Biogeochemistry*, **54**, 279–96.

Salonen, K. & Tulonen, T. (1990) Photochemical and biological transformation of dissolved humic substances. *Verhandlungen der internationale Vereinigung für theoretische und angewandte Limnologie*, **24**, 294.

Salonen, K. & Vähätalo, A. (1994) Photochemical mineralisation of dissolved organic matter in Lake Skjervatjern. *Environment International*, **20**, 307–12.

Salonen, K., Kairesalo, T. & Jones, R.I. (1992) Dissolved organic matter in lacustrine ecosystems: energysource and system regulator. *Hydrobiologia*, **229**, vii.

Saunders, G. (1976) Decomposition in fresh water. In: Anderson, J. & Macfadyen, A. (eds), *The Role of Terrestrial and Aquatic Organisms in Decomposition Processes*. Blackwell Scientific Publications, Oxford, 341–74.

Schindler, D.W. (1994) Changes caused by acidification to the biodiversity:productivity and biogeochemical cycles of lakes. In: Steinberg, C.E.W. & Wright, R.F. (eds), *Acidification of Freshwater Ecosystems — Implications for the Future*. Wiley, Chichester, 153–64.

Schindler, D.W., Curtis, P.J., Parker, B.R. & Stainton, M.P. (1996) Consequences of climate warming and lake acidification for UV-B penetration in North American boreal lakes. *Nature*, **379**, 707–8.

Schindler, D.W., Curtis, P.J., Bayley, S.E., Parker, B.R., Beaty, K.G. & Stainton, M.P. (1997) Climate-induced changes in the dissolved organic carbon budgets of boreal lakes. *Biogeochemistry*, **36**, 9–28.

Schinner, F. & Sonnleitner, R. (1996) *Bodenökologie: Mikrobiologie und Bodenenzymatik*. Springer-Verlag, Berlin.

Schneider, J., Werner, A., Weis, R., Männer, C. & Riede, U.N. (1994) HIV-virostatic effects of humic acids. *Pathological Research and Practice*, **190**, 245–8.

Schwarzenbach, R.P., Gschwend, P.M. & Imboden, D.M. (1993) *Environmental Organic Chemistry*. Wiley, Chichester.

Scully, N.M. & Lean, D.R.S. (1994) The attenuation of ultraviolet radiation in temperate lakes. *Ergebnisse der Limnologie*, **43**, 135–144.

Scully, N.M., McQueen, D.J., Lean, D.R.S. & Cooper, W.J. (1996) Hydrogen peroxide formation: the interaction of ultraviolet radiation and dissolved organic carbon in lake waters along a 43–75°N gradient. *Limnology and Oceanography*, **41**, 540–8.

Seppä, H. & Weckström, J. (1999) Holocene vegetational and limnological changes in the Fennoscandian treeline area as documented by pollen and diatom records from Lake Tsuolbmajärvi, Finland. *Ecoscience*, **6**, 621–35.

Servos, M.R., Muir, D.C.G. & Webster, G.R. (1989) The effect of dissolved organic matter on the bioavailability of polychlorinated dibenzo-p-dioxins. *Aquatic Toxicology*, **14**, 169–84.

Shaw, P.J (1994) The effect of pH dissolved humic substances and inic composition on the transfer of iron and phosphate to particulate size fractions in epilimnetic lake water. *Limnology and Oceanography*, **39**, 1734–43.

Sigg, L. & Stumm, W. (1991) *Aquatische Chemie*. Verlag der Fachverine, Zürich.

Spieser, O.H. & Scholz, W. (1992) German Patent, P 4224750.0.

Stabel, H.-H., Moaledj K. & Overbeck, J. (1979) On the degradation of dissolved organic molecules from Plusssee by oligocarbophilic bacteria. *Ergebnisse der Limnologie*, **12**, 95–104.

Steinberg, C. (1977) Schwer abbaubare, stickstoffhaltige gelöste organische Substanzen in Schöhsee und in Algenkulturen. *Archiv für Hydrobiologie (Supplementband)*, **53**, 48–148.

Steinberg, C. (1991) Fate of organic matter during natural and anthropogenic lake acidification. *Water Research*, **25**, 1453–8.

Steinberg, C. (2003) *Ecology of Humic Substances: Determinants from Geochemistry to Ecological Niches*. Springer, Berlin.

Steinberg, C. & Baltes, G.F (1984) Influence of metal compounds on fulvic acid/molybdenum blue reactive phosphate reactions. *Archiv für Hydrobiologie*, **100**, 61–71.

Steinberg, C. & Kühnel, W. (1987) Influence of cation acids on dissolved humic substances under acidified conditions. *Water Research*, **21**, 95–8.

Steinberg, C. & Münster, U. (1985) Geochemistry and ecological role of humic substances in lake water. In: Aiken, G.R., McKnight, D.M., Wershaw, R.L. & MacCarthy, P. (eds), *Humic Substances in Soil, Sediment and Water. Geochmistry, Isolation and Characterisation*. Wiley, New York, 105–45.

Steinberg, C.E.W. & Bach, S. (1996) Growth promotion by a groundwater fulvic acid in a bacteria/algae system. *Acta Hydrochimica et Hydrobiologica*, **24**, 98–100.

Steinberg, C.E.W. & Wright, R.F. (eds) (1994) *Acidification of Freshwater Ecosystems: Implications for the Future.* Wiley, Chichester.

Steinberg, C.E.W., Sturm, A., Kelbel, J., Lee, S.K., Hertkorn, N., Freitag, D. & Kettrup, A. (1992) Changes of acute toxicity of organic chemicals to *Daphnia magna* in the presence of dissolved humic material (DHM). *Acta Hydrochimica et Hydrobiologica*, **20**, 326–32.

Steinberg, C.E.W., Xu, Y., Lee, S.K., Freitag, D. & Kettrup, A. (1993) Effect of dissolved humic material (DHM) on bioavailability of some organic xenobiotics to *Daphnia magna. Chemical Speciation and Bioavailability*, **5**, 1–9.

Steinberg, C.E.W., Haitzer, M., Brüggemann, R., Perminova, I.V., Yashchenko, N.Yu. & Petrosyan, V.S. (2000) Towards a quantitative structure activity relationship (QSAR) of dissolved humic substances as detoxifying agents in freshwaters. *International Revue Hydrobiology*, **85**, 253–66.

Steinberg, C.E.W., Hoppe, A., Jüttner, I., Bruckmeier, B. & Hertkorn, N. (2001) Changes of humic substance constituents in Großer Arbersee during acidification. *Acta Hydrochimica et Hydrobiologica*, **29**, 78–87.

Steinberg, C.E.W., Höss, S. & Brüggemann, R. (2002) Further evidence that humic substances have the potential to modulate the fertility of the nematode, *Caenorhabditis elegans. International Revue of Hydrobiology*, **87**, 121–33.

Stewart, A.J. & Wetzel, R.G. (1982) Influence of dissolved humic materials on carbon assimilation and alkaline phosphatase activity in natural algal-bacterial assemblages. *Freshwater Biology*, **12**, 369–80.

Tan, K.H. & Tantiwiramanond, D. (1983) Effect of humic acid on modulation and dry matter production of soybean, peanut and clover. *Journal of the Soil Science Society of America*, **47**, 1121–4.

Tarr, M.A., Wang, W., Bianchi, T.S. & Engelhaupt, E. (2001) Mechanisms of ammonia and amino-acid production from aquatic humic and colloidal matter. *Water Research*, **35**, 3688–96.

Thomas, J.D. (1997) The role of dissolved organic matter, particularly free amino-acids and humic substances, in freshwater ecosystems. *Freshwater Biology*, **38**, 1–36.

Thurman, E.M. (1985) *Organic Geochemistry of Natural Waters.* Dr W. Junk Publishers, Dordrecht.

Tipping, E., Reddy, M.M. & Hurley, M.A. (1990) Modeling electrostatic and heterogeneity effects on proton dissociation from humic substances. *Environmental Science and Technology*, **24**, 1700–5.

Tranvik, L.J. (1988) Availability of dissolved organic carbon for planktonic bacteria in oligotrophic lakes of differing humic content. *Microbial Ecology*, **16**, 311–22.

Tranvik, L.J. (1989) Bacterioplankton growth, grazing mortality and quantitative relationship to primary production in a humic and a clearwater lake. *Journal of Plankton Research*, **11**, 985–1000.

Tranvik, L.J. (1990) Bacterioplankton growth on fractions of dissolved organic carbon of different molecular weights from humic and clear waters. *Applied and Environmental Microbiology*, **56**, 1672–7.

Tranvik, L.J. (1992) Allochthonous dissolved organic matter as energy source for pelagic bacteria and the concept of the microbial loop. *Hydrobiologia*, **229**, 107–14.

Tranvik, L.J. & Höfle, M.G. (1987) Bacterial growth in mixed cultures on dissolved organic carbon from humic and clear waters. *Applied and Environmental Microbiology*, **53**, 482–8.

Traunspurger, W., Haitzer, M., Höss, S., Beier, S., Ahlf, W. & Steinberg, C. (1997) Ecotoxicological assessment of aquatic sediments with *Caenorhabditis elegans* (Nematoda)—a method for testing liquid medium and whole sediment samples. *Environmental Toxicology and Chemistry*, **16**, 245–50.

Tsuji, K., Naito, S., Kondo, F., Ishikawa, N., Watanabe, M.F., Suzuki, M. & Harada, K.I. (1994) Stability of microcystin from cyanobacteria: effect of light on decomposition and isomerization. *Environmental Science and Technology*, **28**, 173–7.

Tulonen, T., Salonen, K. & Arvola L. (1992) Effects of different molecular weight fractions of dissolved organic matter on the growth of bacteria, algae and protozoa from a highly humic lake. *Hydrobiologia*, **229**, 239–52.

Vance, G.F. & David, M.B. (1989) Forest soil response to acid and salt additions of sulphate. III. Solubilization and composition of dissolved organic carbon. *Soil Science*, **151**, 297–305.

Vähätalo, A. & Salonen, K. (1997) Photochemical degradation of chromophoric dissolved organic matter and its contribution to bacterial respiration in a humic lake. *Nordic Humus News*, **4**, 14.

Voelker-Bartschat, B.M. (1994) *Iron redox cycling in surface waters: effects of humic substances and light.* PhD thesis, Swiss Federal Institute of Environmental Science and Technology, Zürich.

Vogt, R.D., Ranneklev, S.B. & Mykkelbost, T.C. (1994) The impact of acid treatment on soilwater chemistry at the HUMEX site. *Environment International*, **20**, 277–86.

Wallschläger, D., Desai, M.V.M. & Wilken, R.D. (1996) The role of humic substances in the aqueous mobilization of mercury from contaminated floodplain soils. *Water, Air and Soil Pollution*, **90**, 507–20.

Weber, J.H. (1993) Review of possible paths for abiotic methylation of mercury(II) in the aquatic environment. *Chemosphere*, **26**, 2063–77.

Welker, M. & Steinberg, C. (2000) Rates of humic substances photosensitized degradation of microcystin-LR in natural waters. *Environmental Science and Technology*, **34**, 3415–19.

Welker, M., Steinberg, C. & Jones G. (2001) Release and persistence of microcystins in natural waters. In: Chorus, I. (ed.), *Cyanotoxins — Occurrence, Effects, Controlling Factors*. Springer Scientific Publishers, Heidelberg, 83–101.

Wershaw, R.L., Burcar, P.J. & Goldberg, M.C. (1969) Interaction of pesticides with natural organic material. *Environmental Science and Technology*, **3**, 271–3.

Wetzel, R.G. (2001) *Limnology* (3rd edition). Academic Press, San Diego.

Wetzel, R.G. (1984) Detrital dissolved and particulate organic carbon functions in aquatic ecosystems. *Bulletin of Marine Sciences*, **35**, 503–9.

Wetzel, R.G. (1991) Extracellular enzymatic interactions in aquatic ecosystems: storage, redistribution, interspecific communication. In: Chróst, R.J. (ed.), *Microbial enzymes in Aquatic Environments*. Brock-Springer, Madison, 6–28.

Wetzel, R.G. (1993) Humic compounds from wetlands: complexation, inactivation and reactivation of surface-bound and extracellular enzymes. *Verhandlungen der internationale Vereinigung für theoretische und angewandte Limnologie*, **25**, 122–8.

Wetzel, R.G. (1995) Death, detritus and energy flow in aquatic ecosystems. *Freshwater Biology*, **33**, 83–9.

Wetzel, R.G., Hatcher, P.G. & Bianchi, T.S. (1995) Natural photolysis by ultraviolet irradiance of recalcitrant dissolved organic matter to simple substrates for rapid bacterial metabolism. *Limnology and Oceanography*, **40**, 1369–80.

Williams, M.W., Losleben, M., Caine, N. & Greenland, D. (1996) Changes in climate and hydrochemical responses in a high-elevation catchment in the Rocky Mountains, USA. *Limnology and Oceanography*, **41**, 939–46.

Winner, R.W. (1986) Interactive effects of water hardness and humic acid on the chronic toxicity of cadmium to *Daphnia pulex*. *Aquatic Toxicology*, **8**, 281–93.

Wit, J.C.M. de, Riemsdijk, W.H. van & Koopal, L.K. (1993) Proton binding to humic substances. 1. Electrostatic effects. *Environmental Science and Technology*, **27**, 2005–14.

Wojciechowski, I. & Górniak, A. (1990) Influence of the brown humic and fulvic originating from nearby peat bogs on phytoplankton activity in the littoral of two lakes in mid-eastern Poland. *Verhandlungen der internationalen Vereinigung für theoretische und angewandte Limnologie*, **24**, 621–3.

Yan, N.D. (1983) effects of changes in pH on transparency and thermal regimes of Lohi Lake, near Sudbury, Ontario. *Canadian Journal of Fisheries and Aquatic Sciences*, **40**, 621–3.

Yan, N.D., Keller, W., Scully, N.M., Lean, D.R.S. & Dillon, P. (1996) Increased UV-B penetration in a lake owing to drought-induced acidification. *Nature*, **381**, 141–3.

Zepp, R.G., Wolfe, N.L., Baugham, G.L. & Hollis, R.C. (1977) Singlet oxygen in natural waters. *Nature*, **207**, 421–3.

Zepp, R.G., Baughman, G.L. & Schlotzhauer, P.F. (1981) Comparison of photochemical behaviour of various humic substances in water. I. Sunlight-induced reactions of aquatic pollutants photosensitized by humic substances. *Chemosphere*, **10**, 109–17.

8 Sedimentation and Lake Sediment Formation

JÜRG BLOESCH

8.1 INTRODUCTION

The origin and fate of particulate matter substantially influence lake metabolism (Fig. 8.1). This chapter stresses the different sources and composition of particles, the suspension and settling processes, the mechanisms of particle transportation and transformation, and the final burial in bottom sediments. Specific attention is given to lakes in the temperate zone of Europe. Sedimentation of suspended particulate matter (SPM) plays a key role in lake 'perturbation', management and rehabilitation (see Volume 2, Chapter 2).

8.2 THE ORIGIN OF PARTICULATE MATTER IN LAKES

8.2.1 Allochthonous input of suspended particulate matter

Particulate material of allochthonous origin is mainly derived from bedrock and soils and, thus, its composition is dominated by inorganic minerals. Note that instead of 'allochthonous' geologists use the expression 'detrital', which is in contradiction to the limnological definition of detritus as meaning **dead organic matter** (Wetzel 1983). The geology of the catchment, and the human activities (e.g. agriculture) within it, are therefore of crucial importance with respect to allochthonous input, both of solutes (see Chapter 4) and of particulate matter (Prasuhn *et al.* 1996). A key mechanism is the erosive force of streams and rivers, which is documented, for example, by Lewin (1992). Increased erosion, as caused by agricultural

land use, enhances the ontogeny of lakes (see section 8.3.3).

The main point sources of allochthonous particle input are large rivers, e.g. the River Rhone into Lake Geneva (Giovanoli & Lambert 1985) and the River Rhine into Lake Constance (Lambert 1982). Under natural conditions, gravel and sand would be sorted in subsurface delta areas, but most braided delta structures in developed countries have been canalised and destroyed, and the deltas and littoral zones of many rivers excavated by gravel exploitation. The Chiemsee, Germany, and its tributary Tiroler Ache, provide a rare but fine example of intact delta structures in western Europe (Bayerisches Landesamt für Wasserwirtschaft 1987), whereas the River Reuss delta into Lake Lucerne, Switzerland, was destroyed during the previous century but has been partially restored in recent times (Lang & Rutz 1996).

Straightforward riverine inlets promote density underflows by creating jet-like currents, which form canyon-like structures along the subsurface delta zone and are visible from echosoundings or submarine exploration (M. Sturm, pers. comm.). Hence, finer particulates are usually imported into the hypolimnion of lakes, and rarely into the surface (Lambert *et al.* 1976; Wagner & Wagner 1978; Lambert 1982; Wüest 1987; Lambert & Giovanoli 1988). Typically, the coarser particles form a plume, gradually settling out along the subsurface delta; the finer particles may be distributed by eddies and currents over the whole lake area, causing an increase in overall turbidity of lake water. A changed hydraulic river regime, e.g. decreased peak flow and reversed seasonal flow patterns by

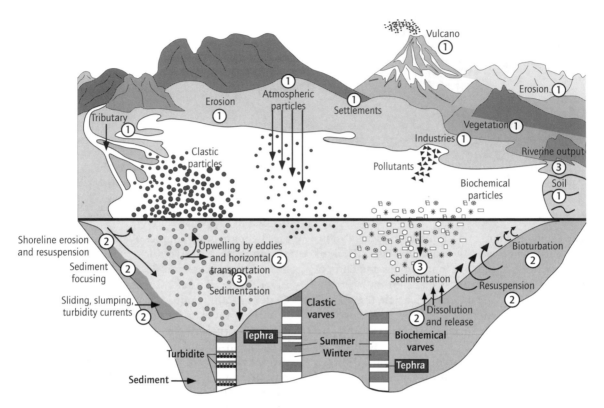

Fig. 8.1 The influence of external and internal environmental factors on suspended sediments, sedimentation and sediment formation in a lake basin: 1, sources of suspended particulate matter (SPM); 2, transportation and transformation of SPM; 3, removal of SPM. (Modified from Sturm & Lotter 1995.)

hydropower use, may add significantly to the increase of lake turbidity (Sturm *et al.* 1996). Aspects of sediment transportation are addressed in section 8.6.

River water flows into the lake in stratified layers with the same density ρ [kg dm^{-3}] (Fig. 8.2). The density of lake water is composed of contributions by temperature (T[°C]), salinity (S[g kg^{-1}]) and SPM concentration (c[mg L^{-1}]) (Sturm *et al.* 1996):

$$\rho(T,S,c) = \rho(T) + \Delta\rho(S) + \Delta\rho(c) \qquad (8.1)$$

where

$$\rho(T) = 0.9998395 + 6.7914 \times 10^{-5} \times T - 9.0894$$
$$\times 10^{-6} \times T^2 + 1.0171 \times 10^{-7} \times T^3 - 1.2846$$
$$\times 10^{-9} \times T^4 + 1.1592 \times 10^{-11} \times T^5 - 5.0125$$
$$\times 10^{-14} \times T^6$$

$$\Delta\rho(S) = \left\{ 8.818 \times 10^{-4} + T \cdot \left(-3.85 \times 10^{-6} + T \right. \right.$$
$$\left. \left. \times 4.96 \times 10^{-8} \right) \right\} \times S$$

$$\Delta\rho(c) = c \times 10^{-6} \times \left(1 - \rho_{(water)} \times \rho_{(SPM)}^{-1} \right)$$

The depth of input varies with lake stratification, river temperature and concentration of dissolved and particulate substances. At times of high flow, density currents may form and progress from the river mouth along the slope into the bottom layers of the lake. Generally, suspensoids are distributed in a counterclockwise current, attributed to the Coriolis force (e.g. Brienzersee; Nydegger 1967).

The general effects of turbidity on lake ecology, and their importance for conservation, have scarcely been addressed. Also, the method of measurement is not easy (see section 8.2.3). Consequently, quantitative data on riverine SPM inputs

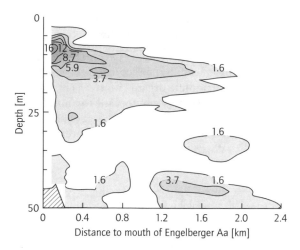

Fig. 8.2 Stratification of waters from the SPM-loaded River Engelberger Aa (mean discharge = 11.5 m³ s⁻¹) into Lake Lucerne, Switzerland, after a major high flow event. The isopleths show SPM concentrations [g m⁻³], measured in the delta area on 1 June 1990. Turbidity due to influx of river water is clearly visible at about 10 m depth. The higher concentrations, about 1.5–2 km in distance from the delta and below 25 m depth, originate from the initial high flow event of 25–26 May 1990. (Data from Frank Peters (Bloesch 1992).)

to lakes are scarce. Sturm *et al.* (1996) measured/calculated catchment-specific SPM loads of 190–392 t km⁻² yr⁻¹ and 435–720 t km⁻² yr⁻¹, respectively, for River Aare and Lütschine into the Brienzersee, in 1994 and 1995. Wagner & Bührer (1989) found similar total SPM loads from all tributaries into the upper part of Lake Constance (231 t km⁻² yr⁻¹). These high loads are typical for mountainous and alpine regions, where erosional forces are strong. In contrast, the total SPM load into the lower part of Lake Constance, from a typical lowland catchment, was only 6 t km⁻² yr⁻¹.

Inputs of carbon (C), phosphorus (P) and nitrogen (N) have been documented mainly in lake restoration studies, since these have dealt with dissolved and particulate nutrients. Catchment-specific loads in various Swiss lakes are in the range of 5–11 × 10³ kg km⁻² yr⁻¹ for total carbon, 1–2 × 10³ kg km⁻² yr⁻¹ for total nitrogen, and 100–500 kg km⁻² yr⁻¹ for total phosphorus, respectively (OECD 1980; Wagner & Bührer 1989;

Bloesch 1992; Prasuhn *et al.* 1996). Whereas for carbon and phosphorus the particulate component was typically dominant (particulate organic carbon POC, 57–83%; particulate phosphorus PP, 44–90%), it was of minor importance for nitrogen (particulate nitrogen PN, 14–19%).

The campaign against eutrophication has led both to significant reductions and to qualitative changes in particulate phosphorus (PP) loads to lakes. Formerly, these originated mainly from untreated wastewater discharged into rivers, and were thus available for phytoplankton that is growth-limited by phosphorus. After installation of wastewater treatment plants, not only were total phosphorus loads drastically reduced, but also the composition of PP changed. It now occurs mostly as apatite, and therefore not in the readily bio-available form (Wagner & Bührer 1989; Bloesch *et al.* 1995a).

Most particulate organic matter (POM) is closely associated with suspended mineral particles, (e.g. adsorbed on to iron oxide surfaces), and is thus transported in this form into lakes (Gu *et al.* 1996; Keil *et al.* 1997). Analysis of flow–load relationships has shown that in most cases >50% of the yearly load is due to few high flow events (Wagner 1976). While the flow–concentration relationship is a dilution-type curve for dissolved substances, for eroded particulate matter it is either an increase-type curve or a combination of the two (Fig. 8.3).

Small streams, and other point and non-point sources, may also contribute significant amounts of allochthonous matter, especially during high flow events and related erosion phenomena. In Hallwilersee, Switzerland, up to 43% of the total phosphorus input was due to nine small streams from an agricultural catchment (Baudepartement des Kantons Aargau 1992). Stadelmann (1988) reported that in the nearby Sempachersee catchment, which is intensively used by pig farms, 78% of the total phosphorus load (or 206 kg km⁻² yr⁻¹) was contributed by these sources; 40% of this was in the particulate form.

In smaller remote lakes in forested areas, plant and wood debris may be washed in larger quantities into the lakes (e.g. Mirror Lake in New Hampshire, USA, whose area is small compared with the large catchment of Hubbard Brook Valley; Likens 1985). In contrast, the particle input via air is normally insignificant for total SPM, except for attached contaminants. For instance, airborne chlorinated compounds were found in high concentrations in the Great Lakes and in Swiss lakes, reflecting the input from industrial areas (Eisenreich *et al.* 1981; Czuczwa *et al.* 1985; see section 8.7).

8.2.2 *Autochthonous input of suspended particulate matter*

Autochthonous suspended particulate matter (SPM) is produced by biological and chemical processes. Particulate organic matter is formed by primary production by phytoplankton and in the subsequent food chain by grazers (zooplankton) and decomposers (bacteria) (see Chapters 10, 13, 14 and 17). When the littoral zone covers a signifi-

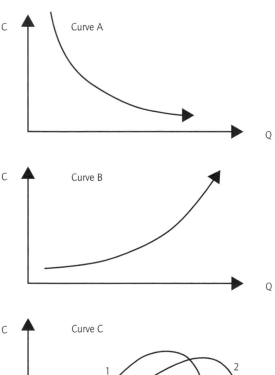

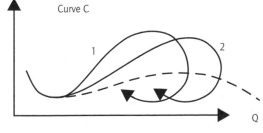

Fig. 8.3   Theoretical flow–concentration $c(Q)$ relationships in flowing waters and lake tributaries: solid lines, real $c(Q)$ curves; dashed line, mean $c(Q)$ curve obtained by modelling. By means of $c(Q)$ relationships the total yearly load of dissolved and particulate matter can be calculated, with a precision of ±30%, by using the following Fourier polynomial regression (Wagner & Bührer 1989):

$$\sum_{i=a}^{b} c_i \cdot Q_i + \sum_{j=d}^{e} c_{j1} \cdot \sin(j\tau) + c_{j2} \cdot \cos(j\tau)$$

where: y is the calculated concentration of dissolved or particulate components [$mg\,m^{-3}$]; c_i represents polynomial coefficients; c_{j1}, c_{j2} are Fourier coefficients; Q is measured flow [$m^3\,s^{-1}$]; i, j are indices; a, b, d and e are constants, which differ for the various formulae shown in the matrix below; $a \le i \le b$ and $d \le j \le e$; and

$$\tau = \text{transformed time} = \frac{(\text{day in the year})}{365} \cdot 2\pi\,[\text{days}]$$

The following is a matrix for calculating eight $c(Q)$ relationships, from which the 'best fit' is chosen, i.e. the curve with the least square deviation, under consideration of the degrees of freedom

Formula number	Constant			
	a	*b*	*d*	*e*
1	0	1	–	–
2	−1	0	–	–
3	−1	1	–	–
4	−1	2	–	–
5	−1	2	1	1
6	−1	1	1	1
7	0	1	1	3
8	−1	3	1	2

a

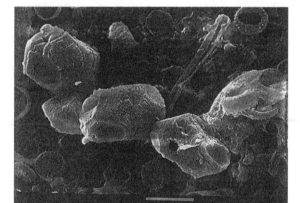

b

Fig. 8.4 SEM pictures of suspended and settling particulate matter (SPM), as typically found in calcareous lakes: (a) SPM entrapped in the Zugersee in an early spring situation, when centric diatoms are blooming (April 1982); (b) SPM entrapped in Lake Constance in the subsequent situation, when biogenic calcium precipitation has been initiated (May 1982). (Photographs from M. Sturm, EAWAG (Swiss Federal Institute for Environmental Science and Technology), with permission.)

cant amount of the lake area (e.g. as in the Neusiedlersee; Löffler 1979, see Chapter 2), littoral plant debris from macrophytes and reeds may also be an important source of SPM.

Chemical and physical processes contribute

significantly to production of autochthonous SPM. Biogenic calcite precipitation is a major source of particulate inorganic matter (PIM) in lakes of the temperate zone (Bloesch *et al.* 1977; Kelts & Hsü 1978; Stabel 1985). Supersaturation of bicarbonate (HCO_3^-) is the prerequisite for calcite production; however, shells of small centric diatoms causing a small zone of CO_2-oversaturation around their cells during intensive production have been identified as nuclei of the calcite crystals and therefore as initiators of calcite precipitation (Fig. 8.4: Sturm *et al.* 1982; Stabel 1985; Niessen & Sturm 1987). Similarly, redox processes in eutrophicated lakes are responsible for particle formation, e.g. in the hypolimnetic manganese and iron cycle, with MnO_2 and $Fe(OH)_3$ precipitation under oxic conditions during turnover, and Mn(II)/Fe(II) dissolution under anoxic conditions during stratification (Bloesch 1974; Hofmann 1996; Schaller 1996; Stumm & Morgan 1996). Also bacterial activity through *Metallogenium* may contribute to the formation of MnO_2 precipitates (Jaquet *et al.* 1982).

Suspended particulate matter transported into, and produced in, lakes is subject to many changes, as discussed in section 8.4. In lakes which are not dominated by large river input, and which are rather productive, autochthonous SPM is more abundant than the allochthonous form. This refers also to the pelagic zone of large lakes with significant river input, e.g. Lake Constance or Lake Geneva, where the effect of allochthonous SPM is mostly restricted to the delta area. In mountainous oligotrophic lakes, such as the Brienzersee, lake-water colour (emerald green) indicates stronger influence of clay mineral suspensoids originating from the erosional force of glaciers ('glacier milk'). However, lake metabolism is governed, in most lakes, by autochthonous production and subsequent processes of particle transformation, decomposition and sedimentation (Bloesch *et al.* 1977; Gries 1995).

8.2.3 *Methodology*

Allochthonous matter input by rivers and streams is best quantified by turbidity measurement or special samplers (Spreafico *et al.* 1987), such as automatic 'ASPEG II' (Sturm *et al.* 1996) for 1-hour samples, or 'QS-3000' (Quantum Science Ltd, Cheltenham, England) for 24-hour samples. Whereas the latter method is obviously advantageous when sampling for dissolved nutrients, it may be biased considerably by oversampling of suspended solids (M. Sturm, personal communication). A rough estimate of lake input–output balance, and a check by reference to accumulation rates estimated using sediment cores, helps to avoid errors in input calculations caused by unrepresentative measurements.

Transmission measurements in delta areas of tributaries (see section 8.3.4) usually do not allow calculation of quantitative mass input balance, since they lack spatial and temporal integration (Sturm *et al.* 1996). Sediment traps provide a much more precise mass balance. An example of methodological approaches to separate allochthonous from autochthonous suspended matter is the use of conservative species derived from rock material, such as apatite (PP: Bloesch *et al.* 1988) and titanium (Stabel 1985).

8.3   SUSPENSION AND SEDIMENTATION OF PARTICLES

8.3.1   *Settling velocity and settling flux of sinking particles*

As their specific weight is greater than that of the surrounding water (i.e., the excess density, $\rho' - \rho$, is positive in eqn. 8.2), most suspended particles are subject to settling. Sedimentation of particulate matter is governed by particle size, shape and specific weight, and by water temperature, viscosity and density. All these factors are integrated in the well-known Stokes' formula, and related theoretical formulae, for spheres and non-spherical particles, respectively (Hutchinson 1967):

$$v_s = 2gr^2(\rho' - \rho)/9\eta \qquad (8.2)$$

where v_s is the sinking velocity of the particle [$m\,s^{-1}$], g is the gravitational acceleration [$m\,s^{-2}$], η is the coefficient of viscosity of the fluid medium

[$kg\,m^{-1}\,s^{-1}$], ρ is its density and ρ' is the density of the particle [$kg\,m^{-3}$] and r is its radius [m]. A comparison of settling velocities of spherical and non-spherical particles is presented in Fig. 8.5.

The mechanisms of phytoplankton buoyancy and particle settling have been thoroughly described (Smayda 1970; Lerman 1979; Reynolds 1984; see also Chapters 4, 6 and 10). Table 8.1 provides a selection of settling velocities for SPM and plankton. Whereas particles smaller than $50\,\mu m$ may sink at a rate of less than $1\,m\,day^{-1}$ (as an order of magnitude), larger objects, such as algal cells and faecal pellets, settle respectively in the range 5–30 $m\,day^{-1}$ and $>100\,m\,day^{-1}$. It should be noted that, owing to the coefficient of form resistance of non-spherical bodies (see table 10 in Reynolds 1984), calculated settling velocities of algae differ from their measured settling velocities by a factor of up to 5.5. Also, senescent and dead phytoplankton sink more rapidly than living cells, most probably because the various devices and mechanisms which in life serve to reduce their density— such as vacuoles, mucilage sheaths and ionic regulation—are absent (Reynolds 1984). Living motile and vacuole-containing algae may sink and float even in daily fluctuations, as demonstrated by *in situ* settling chamber experiments (Burns & Rosa 1980).

Since sinking velocity of particles is not only dependent on particle density, but also on water density (temperature) and viscosity, it is variable, for the same particles, in space and time. Settling velocity is increased in warm waters, i.e. during summer (compared with winter) and in the epilimnion (compared with the hypolimnion). However, the effects of water temperature are usually overridden by other factors such as water turbulence and physiological condition, and form resistance structures of algal cells. Therefore, settling velocities of algae usually increase in the hypolimnion (Bloesch 1974). Finally, particle transformation can change settling velocity considerably, see section 8.4.

Theoretical settling velocities are hardly representative of measured values or *in situ* conditions, as these are for quiescent water, a situation not found in lakes (Smith 1982; Kozerski 1994;

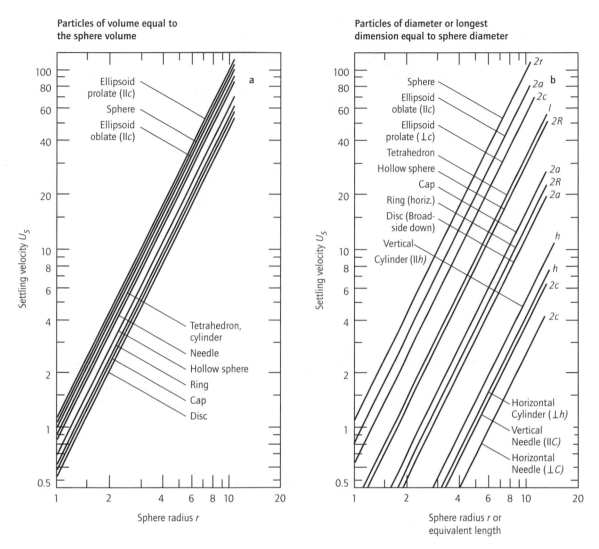

Particles of volume equal to the sphere volume

Ellipsoid prolate (∥c)
Sphere
Ellipsoid oblate (∥c)

a

Tetrahedron, cylinder
Needle
Hollow sphere
Ring
Cap
Disc

Particles of diameter or longest dimension equal to sphere diameter

Sphere
Ellipsoid oblate (∥c)
Ellipsoid prolate (⊥c)
Tetrahedron
Hollow sphere
Cap
Ring (horiz.)
Disc (Broadside down)
Vertical Cylinder (∥h)

b

2r
2a
2c
l
2R

2a
2R
2a

h

h
2c

2c

Horizontal Cylinder (⊥h)
Vertical Needle (∥C)
Horizontal Needle (⊥C)

Fig. 8.5 Settling velocities of spherical and non-spherical particles. (a) Particles of equal volume; the settling velocity of non-spheres vary within a factor of two of the settling velocity of spheres. (b) Particles of equal length or diameter; the spread of settling velocities is considerably greater. U_s is in units of cm s^{-1}, when r is in μm and $B = 1 \times 10^4$ cm^{-1} s^{-1}. (From *Geochemical Processes. Water and Sediment Environments*, A. Lerman, Copyright © (1979). Reprinted by permission of John Wiley & Sons, Inc.)

Imboden & Wüest 1995). Turbulence, vertical mixing and horizontal transportation of water masses influence settling of particles. The role of turbulence on particle suspension and settling is debated and still poorly understood (see discussion in Bloesch & Burns 1980, p. 18). There is controver-

sy as to whether turbulence increases (Jobson & Sayre 1970) or decreases (Murray 1970) the settling velocity of particles.

Epilimnetic Langmuir cells may recycle SPM over longer time periods (Reynolds 1984). Since hydrodynamics and the nature of bulk particles

Table 8.1 Order of magnitude of sinking velocities of inorganic and organic bulk particles, algal cells and zooplankton. Specific differences in sinking velocities according to particle density, form and size, physical algal cell condition (healthy, senescent, dead), space (epilimnion, hypolimnion) and season (summer, winter) are extensively described in Hutchinson (1967), Bloesch (1974), Lerman (1979), and Reynolds (1984).

Particle	Range of mean particle size (μm)	Range of mean sinking velocity* (m day^{-1})	Overall mean sinking velocity (m day^{-1})	Literature
Particulate organic carbon	<50	0.04–3.22 s	0.7	Bloesch & Sturm 1986
Particulate phosphorus		0.11–3.22 s	0.9	Bloesch & Sturm 1986
Centric Diatoms	10–30	0.22–2.39 v	1.3	Reynolds 1984; Gálvez et al. 1993
(*Stephanodiscus astraea*		0.01–9.70 s	1.5	Bloesch 1974; Reynolds 1984
Stephanodiscus hantzschii				
Cyclotella spp.				
Melosira italica subarctica)				
Pennate Diatoms	70–130	0.10–2.4 v	0.6	Reynolds 1984
(*Fragilaria crotonensis*		0.08–36.2 s	4.5	Bloesch 1974; Reynolds 1984
Tabellaria fenestrata				
Asterionella formosa)				
Peridineae				
Peridinium cinctum	20–60	7.74 v	7.7	Gálvez et al. 1993
Rotifers	10^2–10^3			
Brachionus quadridentatus		33.4–43.2 v	38	Hutchinson 1967
Euchlanis triquetra		56.9–74.5 v	64	Hutchinson 1967
Crustaceans	10^3–10^4			
Daphnia galeata		8.64–229 v	107	Hutchinson 1967
Daphnia pulex, magna		389–734 v	610	Hutchinson 1967
Zooplankton faecal pellets	c.250	35–864 v	200	Smayda 1969, 1971; Fowler & Small 1972

* v = *in vitro*; s = *in situ*.

in lake systems are extremely complex, *in situ* settling velocity of SPM cannot be precisely measured (Bloesch & Sturm 1986). The general formula to determine *in situ* sinking velocities is:

$$V = F/C \quad \left[\text{m day}^{-1}\right] \qquad (8.3a)$$

where V is sinking velocity of SPM (particles) [m day^{-1}], F is settling flux of SPM (particles) [mg m^{-2} day^{-1}] and C is mean SPM (particle) concentration [mg m^{-3}].

Various modifications of the above formula do not greatly influence the calculated mean settling velocities of SPM (Bloesch & Sturm 1986). For in-

stance, it is not of crucial importance whether C is depth-averaged over the water column above sediment traps measuring F, or time-averaged at trap depth. Somewhat lower sinking velocity values (by a factor of two to three) are obtained by a formula suggested by S.W. Effler (pers. comm.) in order to account for vertical heterogeneities in SPM concentration:

$$V' = \frac{V \cdot MSD}{D} \quad \left[\text{m day}^{-1}\right] \qquad (8.3b)$$

where D is sediment trap depth [m], MSD is the mean settling depth [m] of SPM

$$= \frac{\sum\limits_{i=1}^{n} C_i H_i}{\sum\limits_{i=1}^{n} C_i}$$

C_i is the average SPM concentration at the mid-point between sampling depths [mg m^{-3}], H_i is the distance from surface to midpoint between sampling depth [m] and n is the number of sampling depths to sediment trap.

Apart from possible methodological bias by flux over/underestimation by sediment traps (outlined in section 8.3.4), the major problem of calculating sinking velocities of algae with eqns. (8.3a) and (8.3b) is to provide a precise value of SPM/algal concentration integrated over time (of trap exposure) and depth (in the water column above the trap). However, these formulae can yield reliable results only if the suspended and entrapped particles are similar in size and shape (e.g. the shells of diatoms). This was demonstrated in short-term *in situ* experiments in waters around Berlin with settling chambers, where the smallest particle size fractions (mean sinking velocities 0.2 m day^{-1}) were high in concentration (4.35 g dry wt m^{-3}) but low in flux (0.9 g dry wt m^{-2} day^{-1}), whereas the largest particles (mean sinking velocities 43 m day^{-1}) were low in concentration (0.35 g dry wt m^{-3}) but high in flux (15.3 g dry wt m^{-2} day^{-1}; Kozerski 1994).

An important issue for understanding the settling behaviour of particles in stagnant waters has been pointed out by Gardner (1977). Since the vertical sinking velocity of settling particles in the range of 1–40 μm is generally one to six orders of magnitude less than that of horizontal water currents, we must dismiss the common concept of a steady 'rain' of detritus or phytoplankton cells sinking downward. In water, particles always possess a small constant component of downward movement. However, they do not appear to sink vertically, or even at a certain angle, but to be carried passively in turbulent eddies (Bloesch & Burns 1980). Large, light aggregates such as 'marine snow' may behave in the same manner as small single particles, whereas large, dense planktonic shells may deviate from this general picture.

It is well documented that algal blooms quantified by water sampling are sinking rapidly, since they are caught simultaneously by vertical series of hypolimnetic sediment traps within the usual deployment period of 2 weeks: there is hardly any time lag observed in traps from, for example, 15 m to 60 m in Horw Bay, Lake Lucerne (Bloesch 1974). An order of magnitude of minimal sinking velocity can be obtained simply by dividing sinking distance (i.e. the depth between maximum algal density and the sediment trap) by the number of trap exposure days. Such estimates yield sinking velocities of the order of several metres per day, which is in general accordance with results obtained by eqns. (8.3a) and (8.3b). The mechanisms behind these findings can be hypothesised, at present, to be that algal blooms are sustained in the bulk of cells by strong eddies, followed by bulk sedimentation after these eddies have weakened.

Such findings are confirmed by measuring short-term settling flux with time-integrating sediment traps (see section 8.3.4). Figure 8.6 shows short-term settling flux in Baldeggersee, Switzerland, with daily SPM peaks, levelling off within 1 week, of three to five times and up to 36 times higher than the average flux of 5.5 g m^{-2} day^{-1} (Sturm & Friedl, unpublished). These peaks, however, do not only consist of algal cells but also consist of calcite precipitated because of algal production.

Primary vertical settling flux of SPM, i.e. SPM sedimenting out of the productive zone, can be calculated, according to eqn. (8.3a), as:

$$F = V \cdot C \quad \left[\mathrm{mg\ m^{-2} day^{-1}}\right] \qquad (8.4a)$$

This flux refers to **export production**, a term first introduced in oceanography (Wassmann 1990). Smith (1975) suggests a formula for turbulent situations, compensating the downward flux component by an upward flux component supplied by eddies:

$$F = V \cdot C = K \cdot \mathrm{d}C_i / \mathrm{d}z \quad \left[\mathrm{mg\ m^{-2} day^{-1}}\right] \qquad (8.4b)$$

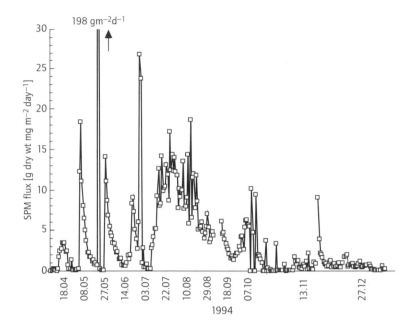

Fig. 8.6 Short-term SPM settling flux measured by time-sequencing sediment traps in 1994 in Baldeggersee, Switzerland. (Unpublished data from M. Sturm, EAWAG, with permission.)

where K is the coefficient of eddy diffusion [m^{-2} day^{-1}] and dC_i/dz is the vertical concentration gradient of particle fraction i [g m^{-4}].

Near lake bottoms, or at any depth of shallow and turbulent lakes, bottom-sediment resuspension, as described in section 8.6, referred to as secondary flux, needs to be taken into account, according to the formula of Partheniades (1965):

$$F = V \cdot C \cdot (1 - \tau_0/\tau_{cr,s}) \left[\text{mg m}^{-2} \text{day}^{-1} \right] \quad (8.5)$$

where τ_0 is the bottom shear stress [N m^{-2}] and $\tau_{cr,s}$ is the critical bottom shear stress for sedimentation of particles [N m^{-2}].

The value of τ can be quantified directly as $\tau = \rho_w \cdot c_1 \cdot u_1^2$ [N m^{-2}] or indirectly by estimating the friction velocity u^* under limiting conditions using the equation $u^* = (\tau/\rho_w)^{1/2}$ [m s^{-1}], where ρ_w is water density [kg m^{-3}], c_1 is drag coefficient above bottom [–], and u_1 is current velocity above bottom [m s^{-1}]. Particles settle down if $\tau_0 < \tau_{cr,s}$. Cohesive material has a very low threshold shear stress ($\tau_{cr,s}$) for sedimentation and a high critical shear stress for resuspension ($\tau_{cr,res}$), while non-cohesive material features a lower critical shear stress (see section

8.6). The shear stress τ_0 lies between these threshold values, i.e. $\tau_{cr,s} < \tau_0 < \tau_{cr,res}$. Thus, most of the organic particles resuspended from bottom sediments remain in suspension.

Theoretical and 'true' settling flux (obtained in experimental calm situations) cannot be exactly measured by sediment traps deployed in *in situ* turbulent conditions. This is because, in most studies, quantification of the terms $K \cdot dC_i/dz$ (from eqn. 8.4b) and $\tau_0/\tau_{cr,s}$ (from eqn. 8.5) has not been achieved (Kozerski 1994).

8.3.2 Spatial and temporal variation in settling flux

Normally, sediment traps used to measure settling flux are deployed at mid-lake stations. The flux determined is then extrapolated to the whole lake area, and used, for instance, as input parameter for lake models. (Models may require either mean settling flux or mean settling velocity of SPM.) There are few studies investigating horizontal sedimentation differences (reviewed in Bloesch & Uehlinger 1986). Increased nearshore sedimentation is usually due to resuspension, whereas increased mid-lake sedimentation is caused by sediment focusing (see section 8.6). Large littoral reed belts act as 'traps' for sedimentation, e.g. in the Neusiedlersee, Austria (Löffler 1979).

The role of seiches (see Chapter 6) in affecting the settling flux measured by moored sediment traps is not yet fully understood. Typical horizontal differences in deep lakes are of the order of two- to threefold, as shown for Hallwilersee (Fig. 8.7a). For shallow lakes with an effective wind fetch, maximal differences of 4.5 (stratification) to 7.7 (turnover) in Lake Erken (Weyhenmeyer 1996), and 6–13 (autumn) in Lake Erie (Bloesch 1982), have been reported. In rare cases, no horizontal sedimentation differences are recorded (Moeller & Likens 1978).

Vertical sediment-flux differences in deep lakes, e.g. between traps located just below the thermocline (epi traps) and those deployed above lake bottom (hypo traps), when compared with horizontal and temporal differences, are rather low. Hypo traps are expected to collect less of the primary flux than epi traps, since decomposition will diminish the amount settling. However, hypo trap flux is usually slightly but consistently increased by sediment focusing (Fig. 8.7c; Bloesch 1993; Bloesch & Ma Xi 1996). Such vertical differences are more distinct in shallow lakes subject to wind fetch and intensive resuspension and, during periods of stratification, may considerably exceed horizontal variations (Weyhenmeyer 1996, 1997).

Temporal variation in sediment flux is mainly governed by the seasonal dynamics of primary production. (Only in shallow lakes are irregular seasonal sediment-flux peaks caused by wind-induced resuspension events; Weyhenmeyer *et al.* 1997.) The settling flux of temperate lakes may, in summer, exceed that of winter by more than an order of magnitude, and usually also exhibits one or two peaks in spring and late summer, respectively (Fig. 8.7b; Bloesch 1974). Basic winter settling flux in temperate lakes is normally $<1\,\mathrm{g}$ dry $\mathrm{wt\,m^{-2}\,day^{-1}}$. Interannual differences in settling flux mainly depend on weather conditions, though they are usually not great. For example, during the years 1984–1987, summer SPM peaks in Sempach-

ersee varied between 12 and 23 g dry $\mathrm{wt\,m^{-2}\,day^{-1}}$ (Gächter *et al.* 1989). Particulate organic carbon and particulate phosphorus fluxes generally exhibited similar seasonal and interannual fluctuation.

Weyhenmeyer & Bloesch (2001) showed that particle flux variability, governed by lake morphology and trophic state, followed a consistent pattern resulting in a significant relationship between minimum, mean and maximum particle flux. The mean particle flux is significantly correlated to the dynamic ratio (square root of area divided by mean depth) of the lake, provided that the lake is shallow (mean lake depth $\leq 9\,\mathrm{m}$), its dynamic ratio is ≥ 0.15 and allochthonous particulate matter input is $\leq 10\%$. In such lakes, found typically in Sweden, the described relationship may be used to predict the variability of particle settling flux without developing complicated models.

8.3.3 Suspended particulate matter concentration, settling flux and sediment accumulation in relation to the trophic state of lakes

The ranges of SPM concentration and settling flux in various types of lakes are compiled in Table 8.2. Key factors are autochthonous primary production and biomass (section 8.2.2), allochthonous SPM input through major tributaries and/or erosion (section 8.2.1) and internal lake-bottom-sediment resuspension (section 8.6). The latter may significantly raise suspensoid concentration and primary settling flux in shallow lakes, but is not accounted for in Table 8.2. Hence, lakes are generally rich in particles and yearly sediment accumulation rates are in the order of a few millimetres up to 1 cm (see section 8.5). Bottom sediment accumulation may be significantly increased in lakes with high allochthonous input (2–3 cm yr^{-1}). This process accelerates lake ontogeny, transforming aquatic ecosystems via swamp, marsh or peat bog stages into terrestrial ecosystems, especially in small and shallow basins (Wetzel 1983).

It should be noted that particle concentrations and fluxes observed in the pelagic zones of lakes are, in general, one to three orders of magnitude greater than those recorded in the open oceans, al-

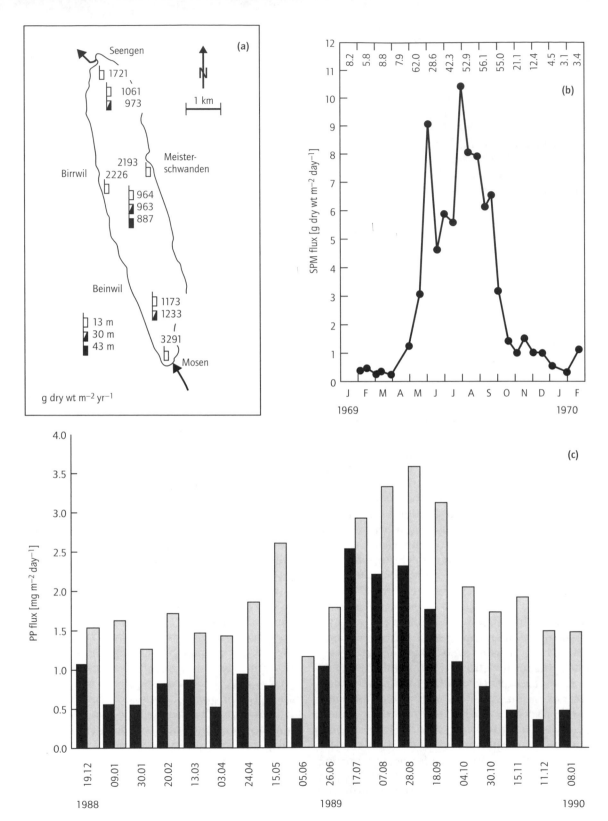

Fig. 8.7 Spatial and temporal sedimentation differences. (a) Horizontal SPM sedimentation differences in Hallwilersee, Switzerland, 1982–1983. (From Bloesch & Uehlinger 1986.) (b) Temporal SPM sedimentation fluctuations in Horw Bay, Lake Lucerne, Switzerland, 1969–1970: figures above the curve correspond to maximum monthly rates of photosynthesis in mg C_{ass} m^{-3} h^{-1}. (From Bloesch 1974.) (c) Vertical PP sedimentation differences in Vitznau Basin, Lake Lucerne, 1989: dark bars, epilimnetic traps; light bars, hypolimnetic traps. (From Bloesch 1995.)

Table 8.2 Primary production, SPM concentration, primary settling flux (export production) and accumulation rates in lakes of the temperate zone, with different trophic state and tributary input. *Note*: The given average figures (yearly means per lake) represent typical order of magnitudes, and specific sites and seasonal patterns may differ significantly. Original data from Bloesch (1974), Kimmel & Goldmann (1976), Moeller & Likens (1978), Wright *et al.* (1980), Lehmann (1983), Stabel (1985), Sturm (1985), Bloesch & Sturm (1986), Bloesch & Uehlinger (1986, 1990), Bloesch (1992), Bloesch *et al.* (1995b), Weyhenmeyer (1996, 1997), Binderheim-Bankay (1998).

Parameter	Allochthonous input to oligotrophic lakes		Allochthonous input to mesotrophic lakes		Allochthonous input to eutrophic lakes	
	Small	Large	Small	Large	Small	Large
Primary production (g C_{ass} m^2 yr^{-1})	<150		150–350		>350–700	
SPM concentration:*						
dry weight (g m^{-3})	1.3–2.3	10–25	2.2	2.9–21	2.5–3.2	>2.5
particulate organic carbon (mg m^{-3})	435–700	230	500	448	1150	3100
particulate nitrogen (mg m^{-3})	32–53	20	–	43	142	536
particulate phosphorus (mg m^{-3})	2–6.6	4	5–12	6.3	18	39–64
Settling flux:*						
dry weight (g m^{-2} day^{-1})	0.1–1.2	7.4–30	0.3–2	2–6.5	2–10	5–31
particulate organic carbon (mg m^{-2} day^{-1})	41–120	250	160–390	268	180–920	380–1300
particulate nitrogen (mg m^{-2} day^{-1})	4.8–12	9	45	20	45–53	131
particulate phosphorus (mg m^{-2} day^{-1})	1.6	13	6–7	5.2	6–13	26
Accumulation rates (mm yr^{-1})	4	>20	3.5–4	4.5–5	3.7–10	13–31

* Data indicating significant resuspension (turbidity, secondary flux) in shallow lakes are excluded.

though yearly primary production differs only by a factor of about five (see table 9 in Rosa *et al.* 1994). Marine fjords and bays are rather more similar to lakes than to open oceans, as they are influenced by tributaries and shoreline erosion and are usually more productive. However, there is a considerable data overlap between lakes and oceans for biomass concentration, planktonic primary production and carbon settling flux (Baines *et al.* 1994). The last is clearly correlated with chlorophyll concentrations in both systems but is lower for a given concentration of chlorophyll in lakes than in the oceans (Fig. 8.8). Also, productivity is positively correlated with increasing biomass (chlorophyll). Because the slopes of the regressions are different in lakes and oceans, the **export ratio** (settling rate/production) in lakes decreases with increasing biomass, whilst it **increases** in the oceans. Possible explanations for these differences may be a more intensive mixing

in the oceans, better adaptation to low nutrient concentrations in marine phytoplankton, higher efficiency of marine grazers and greater nutrient competition between bacteria and phytoplankton in lakes.

Mean SPM concentrations are surprisingly constant over a large variety of lakes and amount—independent of trophic state—to 1–3 mg L^{-1} for lakes dominated by autochthonous particles (Table 8.2). However, the spatial and temporal range may be from 0.1 to 10 mg L^{-1}. In lakes with high allochthonous input, mean SPM concentrations can increase up to 25 mg L^{-1} near deltas, with single peaks of 55 mg L^{-1} (Sturm 1985). Typically, POC, PN and PP concentrations range from 100 to 1500 µg L^{-1}, 10 to 1500 µg L^{-1} and 1 to 100 µg L^{-1}, respectively (Rosa *et al.* 1994). Whilst differences between oligotrophic and mesotrophic lakes are minor, POC, PN and PP concentrations significantly increase in eutrophic lakes. This is caused

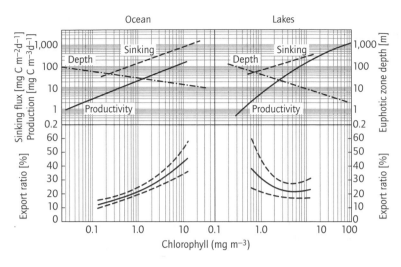

Fig. 8.8 Primary productivity, carbon settling flux and export ratios versus chlorophyll concentration (biomass) in lakes and oceans, according to regression and prediction models of Baines *et al.* (1994). (Modified from Baines *et al.* (1994; reproduced by permission of the American Society of Limnology and Oceanography.)

by high summer maxima induced by intensified phytoplankton production.

In contrast to concentration, a closer relationship between SPM settling flux and trophic state has been found (Table 8.2), which is clearly reflected by POM and PP (Stabel 1985; Bloesch & Uehlinger 1990). Figure 8.9 shows an increase of epilimnetic POC settling flux along a trophic gradient. This is in agreement with a statistical analysis on 15 lakes in the USA performed by Baines & Pace (1994), who found a negative relationship between export ratio and primary production. However, this relationship is not linear: In both oligotrophic as well as eutrophic lakes, POC sedimentation relative to primary production (= export ratio) is higher (21–35%) when compared with mesotrophic bodies (15–18%).

In oligotrophic lakes, usually situated in mountainous regions, allochthonous POC input and sedimentation may be the reason for increased export rates. Both oligotrophic and eutrophic lakes may, in addition, contain an increased fraction of refractory carbon of allochthonous and/or autochthonous origin. The bias from a linear relationship will be further enhanced, when comparing hypolimnetic flux with primary production, by decreased carbon decomposition under anaerobic conditions prevailing in the hypolimnion of eutrophic lakes.

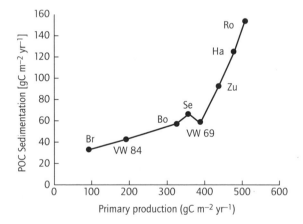

Fig. 8.9 The relationship between particulate organic matter (POC) settling flux and trophic state (primary production). (Modified from Stabel (1985), and including data from Bloesch & Uehlinger (1990).) Br, Brienzersee; VW, Vierwaldstättersee (Lake Lucerne); Bo, Bodensee (Lake Constance); Se, Sempachersee; Zu, Zugersee; Ha, Hallwilersee; Ro, Rotsee.

8.3.4 *Methodology*

Suspended particulate matter in lakes is sampled by water samplers and flow-through centrifuges, or measured by light transmission profiles, as outlined in Murdoch & MacKnight (1994). Calibrated conductivity–temperature–dissolved oxygen (CTD) profiles can be used for relative concentration differences, but cannot yield quantitative information for mass balance calculations (Sturm *et al.* 1996). Principles, methods and applications of particle size analysis have been comprehensively compiled by Syvitski (1991). Special care must be taken for sampling and sample pretreatment, since subsequent particle size analysis may not provide *in situ* particle distribution. Aggregates can decay into single particles during the methodological procedures, and/or filtration and coagulation may bias particle size distribution. Also, flow-through centrifuges may produce artefacts (Rosa *et al.* 1994). In this respect, *in situ* investigations by means of optical and acoustical instruments may provide more realistic results (see literature review in Bloesch 1994). Single particle analysis is dealt with in section 8.4.3.

The settling velocity of particles and algae may be experimentally measured in tubes (Smayda 1974; Bienfang 1981, Gálvez *et al.* 1993), but again the absence of natural turbulence makes extrapolation difficult. As discussed above, *in situ* measurement performed by comparing particle flux with particle concentration can only yield an order of magnitude estimate for bulk particles subject to various transformations (see section 8.4). However, it may yield more precise values for well defined algal cells such as diatoms and Peridineae. Results from *in situ* settling chambers (Burns & Rosa 1980) do not necessarily coincide with sediment trap flux measurements (Rathke *et al.* 1981).

Sediment traps are the only tool with which to study the *in situ* behaviour of settling particles and to measure settling flux in vertical profiles. The trap methodology was first outlined by Bloesch & Burns (1980), Reynolds *et al.* (1980), Blomqvist & Hakanson (1981), and later substantiated by U.S. GOFS Working Group (1989), IOC (1994) and Rosa *et al.* (1994). During the 1980s and 1990s, the method was greatly developed via flume and calibration experiments (see e.g. Gardner 1980a,b; Butman 1986; Baker *et al.* 1988; Buesseler 1991; Buesseler *et al.* 1994; Gust *et al.* 1994), and by testing for preservatives and 'entering swimmers' (see e.g. Karl *et al.* 1988; Lee *et al.* 1988, 1992; Gunder-

sen & Wassmann 1990; Wakeham *et al.* 1993; Hansell & Newton 1994). The present state of the art of trap technology and calibration is given by Gardner (1996) and Bloesch (1996). Unresolved problems of settling flux measurement are (i) the effect of strong currents causing undertrapping of particles (as evident specifically from the oceans; Gardner 1996), (ii) POM decomposition inside the traps leading to the use of preservatives (which may bias the analysis; Rosa *et al.* 1994), and (iii) 'entering swimmers' (Karl & Knauer 1989). In most lakes, cylindrical traps with appropriate dimensions (diameter >5 cm; aspect ratio >5:1) will provide realistic settling flux measurements (with an error range of 5–20%, according to the respective parameter; Bloesch & Burns 1980). Bottle-shaped traps overestimate, and funnel-type traps underestimate, the settling flux.

Relatively new trap designs comprise time-sequencing traps; since these were developed for the oceans, and often consist of funnels, undertrapping effects should be considered. Baffles on top of these funnels can minimise undertrapping (Butman *et al.* 1986). Plate sediment traps with low aspect ratio were recently developed by Kozerski & Leuschner (1999, 2000). They can be used at the bottom and in unidirectional flow, as in slowly flowing rivers, in order to measure net sediment accumulation rates, where cylindrical traps with high aspect ratio measure gross sedimentation rates, i.e. primary and secondary (resuspended) settling flux. For very shallow systems, where measurement of settling flux may be inconsistent (Kozerski 1994), Banas *et al.* (2001) propose lowering the traps into a bucket buried in the sediments and hence to position the trap mouth some centimetres above the bottom.

8.4 PARTICLE TRANSFORMATION DURING SEDIMENTATION

Particle transformation is effected by means of physical, chemical and biological processes. Lake residence time of particles is of crucial importance for transformation, and is usually limited by the rate of downward flux, rather than the loss through river outlets. Processes such as decomposition, mineralisation and dissolution diminish particle size, whereas adsorption, coagulation, aggregation ('marine snow') and precipitation increase it. However, particle size and shape are not the only criteria for sinking velocity, as particle density is of utmost importance (see section 8.3.1).

8.4.1 *Biological particle transformation*

Zooplankton, as a consequence of their grazing activity, produce faecal pellets that may aggregate, thereby increasing general particle size and the rate of their sedimentation (Fig. 8.10b: Ferrante & Parker 1977; Uehlinger & Bloesch 1987; Bloesch &

Bürgi 1989). Bacteria attached to small detrital particles can form polysaccharide fibrils (Seki 1972, Paerl 1973), thus enhancing particle aggregation (Fig. 8.10a). These microorganisms incorporate phosphorus rather than release it (Currie & Kalff 1984; Gächter & Mares 1985; Stöckli 1985), and significantly reduce the C:P ratios of settling particulate material below the Redfield ratio of 106:1 (Gächter & Bloesch 1985). A similar, but less-pronounced, effect was found in C:N ratios of epilimnetic and hypolimnetic SPM. Because C:N ratios of phytoplankton (8–31:1) are higher than those of bacteria (5–8:1), decomposition during settling results in a lowering of C:N ratios as the latter colonise detrital particles and retain or immobilise

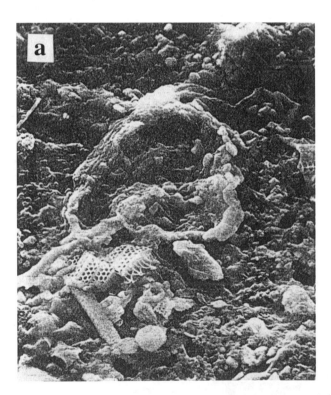 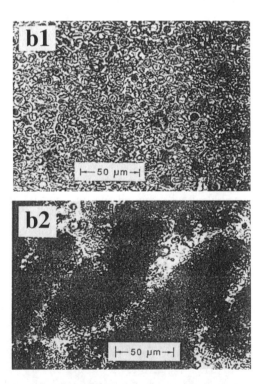

Fig. 8.10 Biological particle transformation in lakes. (a) Particle aggregation through bacteria, an aggregate collected at 4000 m depth in the western North Pacific. (Reprinted with permission from Seki (1982), copyright CRC Press, Boca Raton, Florida.) (b) Faecal pellet production through zooplankton grazing in Lake Lucerne, Switzerland: uniformly distributed phytoplankton (nanoplankton) cells and particles in a 95-μm-filtered limnocorral without crustaceans (b1), flakes of cells and particles and faecal aggregates in a control limnocorral with crustaceans (b2). (From Bloesch & Bürgi 1989.)

much of the mineralised nitrogen (Urban *et al.*, unpublished). Manganese bacteria may significantly influence the manganese cycle, as shown below (section 8.4.2).

8.4.2 *Physical–chemical particle transformation*

Apart from biological mechanisms, physical–chemical processes are likely to play an important role in particle coagulation and, hence, sedimentation, as found in the ocean (Gibbs 1974) and in lakes (Osman-Sigg 1982). Coagulation is mainly driven by the large specific surface area, surface energy and (negative) electrostatic charge of small and colloidal particles (Osman-Sigg 1982). Independent of particle concentration, lake depth and season, particle size distribution follows a general pattern according to the equation

$$\frac{\Delta N(d_{p})}{\Delta d_{p}} = A \cdot d_{p}^{-m} \qquad (8.6)$$

where $\Delta N(d_{p})$ is the number of particles with a diameter between d_{p} and $d_{p} + \Delta d_{p}$ per cm^3, d_{p} is the particle diameter [μm] and A is a constant of particle size distribution.

Phosphorus is not only taken up by bacteria but is also adsorbed on to the surface of settling particles (Stumm & Morgan 1996). During adsorption of organic matter on to iron oxides, strongly-binding organic compounds displace those which are weakly bound (Gu *et al.* 1996). Scavenging of heavy metals and organic compounds by settling particles is addressed in more detail in section 8.7.

In the deep hypolimnia of productive lakes, redox cycles such as Fe/Mn precipitation and redissolution may contribute substantially to the settling particle flux, and to particle formation and transformation (Stumm & Morgan 1996; Hofmann 1996; Friedl *et al.* 1997; see also section 8.6). The behaviours of iron and manganese differ and not only with respect to redox behaviour (iron being more readily adsorbed on to clay minerals as goethite and reduced at lower redox potentials than manganese). Jaquet *et al.* (1982) found that the hypolimnetic manganese cycle in Lake Geneva

is strongly influenced by *Metallogenium*: the amount of Mn released from the anoxic bottom sediments was equivalent to the amount consumed and precipitated as Mn carbonate by the bacteria. While some trace elements (such as Zn, Cd and Pb) followed the same geochemical pattern as manganese, iron behaved in a different manner and was not taken up by *Metallogenium*. Schaller (1996) further elucidated the redox cycling of manganese, and demonstrated a link to focusing from the littoral zone to the hypolimnion (see section 8.6).

Despite the various changes to settling particles which take place, vertical sediment-flux differences are usually exceeded by seasonal sediment-flux differences. This is due to the rather rapid settling velocities (in the range of m day^{-1}) which apply to particles >50 μm.

8.4.3 *Methodology*

Methods of studying particle transformations are not standardised. Limnocorrals may be a valuable tool to study the influence of grazing zooplankton on phytoplankton, and hence on faecal pellet production and sedimentation (Uehlinger & Bloesch 1987; Bloesch & Bürgi 1989). As shown in Fig. 8.4, scanning electron microscopy (SEM) analysis of suspended and settling particles provides a qualitative and semi-quantitative appraisal of mechanisms of particle transformation. However, advances in technology with automated electron probe X-ray microanalysis (EPXMA), and laser ablation–inductively coupled plasma–mass spectrometry (LA–ICP–MS), now provide quantitative data of particle composition (Watmough *et al.* 1997; Jambers & van Grieken 1997).

8.5 BURIAL OF SETTLING PARTICULATE MATTER AND PARTICULATE ORGANIC MATTER

8.5.1 *Nutrient balance of pelagic ecosystems and pelagic–benthic coupling*

Although Karl *et al.* (1996) have recently found a decoupling of particle production and particle flux in the ocean, export production in lakes is generally closely linked to primary production. Material that reaches the lake bottom provides food for bacterial assemblages and benthos living and burrowing in the sediment surface (Cranford & Hargrave 1994). The evident decomposition of POM in the intensive epilimnetic nutrient cycle (Bloesch *et al.* 1977) and during sedimentation is continued at the sediment–water interface (SWI; Bloesch & Wehrli 1995). For example, in the eutrophic Zugursee, Switzerland, only 21% of particulate organic carbon (POC) produced is deposited in the bottom sediments, corresponding to a production : sedimentation (p : s) ratio of 4.4.

Production : sedimentation ratios lie in the range 3.3–6.8 (Bloesch & Uehlinger 1990) and do not seem to vary greatly between oligotrophic and eutrophic lakes. Of POC sedimented in Zugsee, respectively 57% in the uppermost layers and 24% between 0.5 and 15 cm are remineralised by bacteria within 1–2 years (Fig. 8.11). Under oxic conditions, CO_2 is released by fast mineralisation, while CH_4 is generated by the slower anoxic processes. Only 19% of deposited material, or 4% of primary production, are buried over geological timescales as refractory carbon. Other refractory materials such as shells of diatoms, pollen grains, ash of volcanoes and radionuclides, remain in the sediments as 'fingerprints' (see section 8.5.3). Such values for carbon sedimentation and mineralisation are typical for temperate lakes of various trophic states (Bloesch & Uehlinger 1990).

Nitrogen seems to be decomposed to an even greater extent than carbon. In both Lake Lucerne and the Rotsee, about 18–20% of the nitrogen produced settles to the lake bottom (Bloesch 1974), whereas in Lake Biwa this portion is only 5%, with 1% being permanently buried in the bottom sediment (Sakamoto 1980). In contrast, about 36–44% of the phosphorus produced is deposited

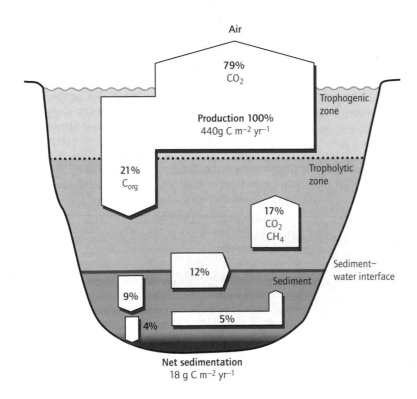

Fig. 8.11 Production, sedimentation and mineralisation of particulate organic carbon (POC) in 200-m-deep Zugersee, Switzerland. The processes in the water column (during settling) and in the upper 15 cm of the bottom sediments (after burial) are represented schematically. The deep hypolimnion of Zugersee is anaerobic during stratification. (From Bloesch & Wehrli 1995.)

in the bottom sediments of Lake Lucerne, the Rotsee and the Bodensee (Lake Constance; Bloesch 1974; Gries 1995). Considering the intensive rates of epilimnetic phosphorus recycling found in such lakes, these high values of settling flux clearly reflect phosphorus uptake by particles during settling through the hypolimnion (see section 8.4.1). In general, both the water column and the sediments are equally important as sources for regeneration of dissolved substances (Carignan & Lean 1991).

Pelagic–benthic coupling is evidenced by the reaction, for example, of oligochaete communities to changes in POM sedimentation, as induced by eutrophication and/or oligotrophication (Lang & Hutter 1981; Robbins *et al.* 1989). Hence, abundance of 'oligotrophic' species is negatively correlated to increase of phosphorus concentration in the water column, organic particle sedimentation and deposition, and occurrence of black, anoxic sediment layers (Lang 1997). Similarly, Sly & Christie (1992) have demonstrated the response of the amphipod *Pontoporeia hoyi* to the seasonal change of downward flux of planktonic detritus in Lake Ontario. In extreme eutrophicated lakes, benthic communities cannot exist in the profundal zone, as both the sediments and the hypolimnia are anoxic. However, if such lakes are restored by aeration, oligochaetes and other benthic animals will recolonise the profundal sediments by invading from the littoral zone (Stössel 1989).

8.5.2 Early diagenesis and sediment–water processes

The sediment–water interface (SWI) of lakes is a very active microbial habitat. Intensive bacterial decomposition and mineralisation of POM within the top few millimetres of bottom sediments leads to very strong redox gradients of dissolved oxidants (e.g., O_2, NO_3^-, SO_4^{2-}) and enhances release of nutrients (e.g., HPO_4^{2-}, NH_4^+) through the SWI (Furrer & Wehrli 1993, 1996; Gächter & Meyer 1993). Recent developments of new methods, involving benthic chambers (Gächter & Meyer

1990; Tengberg *et al.* 1995), dialysis plates (Urban *et al.* 1997), gel samplers (Davison *et al.* 1994) and microelectrodes (Müller *et al.* 1998), have all improved spatial resolution of *in situ* sampling and analysis, and allow precise measurement of diffusion of solutes at the SWI.

Consequently, a series of new models of early diagenesis has emerged, respecting the sequence in use and exhaustion of terminal electron acceptors, with oxygen (as the most powerful oxidant) being consumed first, followed by nitrate and nitrite, manganese oxide, iron oxides, sulphate and finally oxygen-bound organic matter (Furrer & Wehrli 1996; Soetaert *et al.* 1996). In the eutrophic, hardwater Sempachersee, Switzerland, typical release fluxes were 0.05–0.07 mmol m^{-2} day^{-1} for Mn^{2+} and HPO$_4$$^{2-}$, 0.6 mmol m^{-2} day^{-1} for Fe^{2+}, 1.5 mmol m^{-2} day^{-1} for NH$_4^+$ and CH$_4$, 1.8 mmol m^{-2} day^{-1} for Ca^{2+}, and 8.1 mmol m^{-2} day^{-1} for HCO$_3^-$ (Furrer & Wehrli 1996). Determination of stable isotopes (e.g. in $^{15}NO_3^-$) in *in situ* incubation experiments allows detection of specific transformation pathways in microbial metabolism at the sediment surface (Mengis *et al.* 1997).

The basic relation between deep hypolimnetic oxygen concentrations and carbon preservation (Henrichs 1992), denitrification (Rysgaard *et al.* 1994) and phosphorus release could be elucidated by these new methodologies. For example, the release of phosphorus purely by chemical redox (Stumm & Morgan 1996), as described in the 'classic' view of Mortimer (1941, 1942), cannot be substantiated quantitatively, as bacterial action proceeds alongside chemical processes. Long-term data analysis from three Swiss lakes showed that restoration of eutrophic lakes through artificial aeration will not stop internal phosphorus loading (Wehrli & Wüest 1996). The oxidants used penetrate only into the topmost few millimetres of the bottom sediments, thus leaving the deeper layers anoxic, allowing diffusion of reduced solutes along steep gradients.

Early sediment diagenesis is further influenced by the activities of macrobenthos—burrowing worms, insects, amphipods, clams, and other animals, a set of processes cumulatively referred to as **bioturbation**. These change the physical and chemical properties of sediments, and modify the transport characteristics of the SWI (Robbins 1982). Thus, some species of benthic fauna enhance oxygen transport from the sediment surface into layers of sediments 2–3 cm deep (Rippey & Jewson 1982), which—together with transport of easily degradable particulate matter—cause a reduction of phosphorus release from the bottom sediments (Kamp-Nielsen *et al.* 1982). Transport in the opposite direction by other species may increase phosphorus release, as degradable particulate matter is transported to the sediment surface, where microbial activity is greater.

8.5.3 Sediments as archives

As receptors of settling particles, the sediments of lakes act as archives documenting both the history of the lake, and of its airshed and watershed (Sturm & Lotter 1995; see also Chapter 18). Palaeolimnological studies are exemplified, for example, in Windermere, England and in Lake Biwa, Japan (see figs 18-10 and 18-11 in Goldman & Horne 1983). The historical interpretation of environmental changes, e.g. climate changes, eutrophication and other human impact, is possible by evaluating remains of diatoms (Simola *et al.* 1992), ostracods (Geiger 1993), chironomids (Meriläinen & Hamina 1993) and pollen (Lotter 1989). Similarly, allochthonous input events may be interpreted by grain size analysis (Siegenthaler & Sturm 1991).

'Annual laminations' (**varves**; Fig. 8.12) allow direct determination of sediment accumulation rates (mm yr^{-1}), by measuring the thickness of winter and summer layers. Special layers (**markers**) formed by allochthonous input events enable dating of sediment cores. For instance, in the USA Midwest, pollen of ragweed (*Ambrosia*), introduced by European settlers to North America, forms a sedimentary horizon which is dated to 1850, the arrival of plough agriculture in a particular administrative county (Kemp & Harper 1976); layers formed by mountain slides documented in history books can be used for calculating sediment accumulation rates (Lemcke 1992); ^{137}Cs fallout caused by atmospheric explosion of nuclear

Fig. 8.12 Sediment varves from eutrophic Baldeggersee, Switzerland. The light layers are deposited during summer stratification by biogenic calcite precipitation, the blackish layers are deposited during winter turnover. Thus, a set of light and dark layers represent the yearly sediment deposition. (Photograph from M. Sturm, EAWAG, with permission.)

weapons and the Chernobyl accident provide recent sediment markers for the years AD 1959, 1963 and 1986 (Wan *et al.* 1987; Santschi *et al.* 1990); the eruptions of volcanoes, e.g. Katmai, Alaska (in 1913) and Mt St Helens, Washington, USA (in 1980), form layers in lake sediments along the path of the ash plume. From such distinct sediment layers, a mean sediment accumulation rate for a given core (although not for the lake as a whole; Dearing 1983; see also sections 8.3.2 above, and 8.6.1 below) can easily be obtained by dividing the thickness of the sediment above this marker by the number of years which have passed since that specific event.

8.5.4 Methodology

An updated review of sampling lake-bottom sediment, including a variety of grab samplers, gravity and piston corers, is given by Mudroch & MacKnight (1994). It must be stressed that the general sampling technique has been modified by many scientists and manufacturers. One issue of debate is whether the bow-wave produced during free-fall of corers operated from ropes disturbs the surface sediment (Baxter *et al.* 1981). Recently, the freeze-coring technique has been successfully introduced to lake sediment sampling (Saarnisto 1986), preventing gas bubbling caused by depression and thus providing an undisturbed core up to and including the SWI.

In order to measure/calculate annual sedimentation rates ($g m^2 yr^{-1}$) and sediment accumulation rates ($mm yr^{-1}$), various devices and methods such as sediment traps (settling flux, respecting a compaction factor for calculating accumulation rates; Bloesch 1974), ^{210}Pb- and ^{137}Cs-dating techniques, varve counting, or using natural sediment markers, are available. They yield different results according to their strengths and weaknesses (Bloesch & Evans 1982). Sediment accumulation rates ($mm yr^{-1}$) may also be obtained by input–output balance or modelling. In this respect, particular attention should be paid to the effects of sediment resuspension, which produces distortions of results obtained from sediment traps (see Bloesch 1994, and section 8.6).

Diffusion and release rates of solutes within sediments, and across the SWI, can be quantified using dialysis plates (Brandl & Hanselmann 1991), gel samplers (Davison *et al.* 1994) and microelectrodes (Müller *et al.* 1998). *In situ* experiments are best performed by using benthic chambers with remote controlled operation (Gächter & Meyer 1990; Tengberg *et al.* 1995).

8.6 SEDIMENT RESUSPENSION, TRANSPORTATION AND REDEPOSITION

8.6.1 The mechanisms and ecological significance of sediment resuspension and focusing

Lateral particle and sediment transportation is an important lake internal process. Håkanson & Jansson (1983) developed a model for defining erosion, transportation and accumulation at the bottoms of lakes (Fig. 8.13). Hilton (1985) and Hilton *et al.* (1986) have elucidated the mechanisms of sediment resuspension and distribution in lakes, and distinguish between: riverine delta formation, river plume sedimentation, continuous complete mixing, intermittent complete mixing, intermittent epilimnetic complete mixing, peripheral wave attack, random redistribution of sediment, current erosion/redeposition, slumping and sliding on slopes, and organic degradation. In particular, sediment focusing (Davis *et al.* 1985; Bloesch 1995; Weyhenmeyer *et al.* 1997), i.e. lateral sediment transportation, is an effective mechanism of concentrating particulate matter in the deepest parts of deep lakes. Sediment winnowing is clearly related to focusing, as the smallest particles are transported further than coarse objects. These processes cause the horizontal differences in sedimentation patterns outlined in section 8.3.2.

Sediment resuspension is basically induced by wind fetch and wind force, and influenced by lake morphology, size and depth, and by the texture of bottom sediments (Lick 1982). In shallow lakes, resuspension may occur over the total lake area, whereas in deep lakes it is usually restricted to nearshore areas. Shear stress theory has recently been compiled by Bloesch (1995). Shield's diagram (Miller *et al.* 1977), as applied by Gloor *et al.* (1994), where the non-dimensional expressions for relative bottom shear stress θ and particle Reynolds number Re are plotted, indicates the current velocities necessary to resuspend organic and inorganic non-cohesive particles (Fig. 8.14; for cohesive sediments, these critical currents would be somewhat

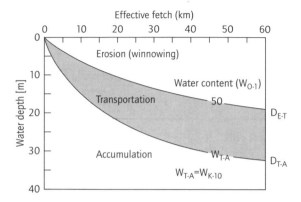

Fig. 8.13 Diagram showing areas of sediment erosion, transportation and accumulation at the bottoms in lakes, as a function of wind fetch and water depth

$$D_{E\text{-}T} = (30.4 \cdot L_f)/(L_f + 34.2)$$
$$D_{T\text{-}A} = (45.7 \cdot L_f)/(L_f + 21.4)$$

where $D_{E\text{-}T}$ and $D_{T\text{-}A}$ are 'critical' water depth [m] and L_f is the effective fetch, i.e. the potential maximum effective fetch [km]. The rough general distinction between erosion and transportation is based upon the water content $W_{0\text{-}1} = 50\%$. The 'critical' limit, between areas of transportation and accumulation, may be given by $W_{T\text{-}A} = W_k - 10$, where W_k is the characteristic water content and $W_{T\text{-}A}$ is the 'critical' water content of surficial lake sediments (0–1 cm). (From Hakanson & Jansson 1983, fig. 7.26; with permission from Springer-Verlag.)

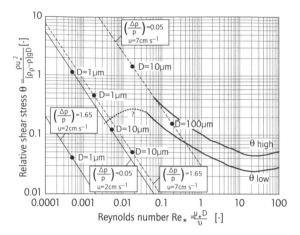

Fig. 8.14 Shields diagram of Miller *et al.* (1977), as applied by Gloor *et al.* (1994), showing the shear stress necessary to resuspend organic and inorganic non-cohesive particles. The solid lines represent the upper (θ_{high}) and lower (θ_{low}) limits of the critical range of $(Re^\star, \theta)$ values for onset of sediment movement. For $(Re^\star, \theta)$ values lying above the upper curve (θ_{high}), particle transport is likely. For $(Re^\star, \theta)$ values lying below the lower curve (θ_{low}), particle transport is unlikely. The dotted line of the θ_{low} curve at $Re^\star < 0.04$ indicates a suggestion for the extrapolation of the critical range for small $Re^\star$ values. The black and dashed lines of slope -1 represent values of $Re^\star$ and θ calculated for $u = 2\,cm\,s^{-1}$ and $u = 7\,cm\,s^{-1}$, respectively, and relative density differences of 1.65 for mineral particles, and 0.05 for organic particles. $Re^\star$ is the particle Reynolds number, θ is the relative bottom shear stress and u is the bottom current. (From Bloesch 1995.)

higher). Bottom currents of $2\,cm\,s^{-1}$, as typically found near the bottom of deep lakes (Lemmin & Imboden 1987), are not sufficient to resuspend even light organic non-cohesive particles with a diameter of $1\,\mu m$.

Burst-like events causing current speeds $>7\,cm\,s^{-1}$ may resuspend organic particles up to $100\,\mu m$ in diameter. However, they are too weak to resuspend mineral clay particles a few micrometres in diameter. Critical friction velocities to resuspend the lightest particles, for unconsolidated sediments in the range of 0.5–$1.7\,cm\,s^{-1}$, and for consolidated sediments of about $5\,cm\,s^{-1}$, have been reported by Bengtsson *et al.* (1990) and Valeur (1994). In shallow unstratified lakes, direct

resuspension is caused by waves/currents (Carper & Bachmann 1984; Dillon *et al.* 1990), while indirect resuspension is induced by seiches in deeper stratified lakes (Münnich *et al.* 1992, Gloor *et al.* 1994).

Recently, Hawley *et al.* (1996) have reported, for the first time, a bottom resuspension event in Lake Ontario at depths below the wave base (63 m) during an unstratified period that has been confirmed by current velocity measurements. Maximum speeds of $20\,cm\,s^{-1}$, equivalent to a critical shear stress between 0.3 and 0.6 dynes m^{-2}, could resuspend bottom material at the site of trap deployment. An unusual high sedimentation peak in November, recorded by a mid-lake epilimnetic trap in the Zugersee, Switzerland, could clearly be attributed to a severe storm event, which led to resuspension and lateral transportation of nearshore bottom sediments (Bloesch & Sturm 1986).

The ecological significance of bottom sediment resuspension is clearly the intense cycling of particulate material and their associated nutrients and contaminants, as outlined specifically in section 8.7 (Lick & Lick 1988; Brassard *et al.* 1994). Accordingly, algal production may be increased by enhanced nutrient cycling, or decreased by increased SPM movement and turbidity. Further, in contradiction to the traditional view that enhanced oxygen input increases phosphorus sedimentation and retention through mixing, intense mixing and subsequent resuspension may actually **reduce** lake phosphorus retention (Wüest & Wehrli 1994).

8.6.2 Methodology

Devices and methods used in order to measure and calculate sediment resuspension are listed in Bloesch (1994). These include optical and acoustic instruments, instantaneous multiple-point water samplers, sediment traps, sediment cores and grabs, radiotracers, mass balance calculations, various modelling approaches, statistical methods (correlation analysis) and laboratory experiments. Recently, an interesting regression model, where the relationship between inorganic settling particles and total settling particles can provide quantitative proportions of planktonic production and sediment resuspension, has been introduced by Weyhenmeyer *et al.* (1995).

8.7 HUMAN IMPACTS ON SEDIMENTS

There are many investigations documenting strong scavenging of heavy metals, radionuclides and organic contaminants on to settling particles originating from the primary flux (e.g. Sigg *et al.* 1987; Kansanen *et al.* 1991; Wieland *et al.* 1991,

1993; Brownawell *et al.* 1997; Weissmahr *et al.* 1997). Moreover, owing to their high surface to volume ratio (and hence their efficient adsorption properties), resuspended particles act as removal agents for heavy metals and other contaminants (Sigg 1985; Allan 1986; Sigg *et al.* 1987; Blomqvist 1992; Blomqvist & Larsson 1992). Specifically, Santschi (1989) showed that in Bielersee, Switzerland, resuspension of rebound particles was more important than river-induced turbidity currents, groundwater input and boundary scavenging in accounting for peaks of natural isotopes at greater depths. Similarly, spatially non-homogeneous distribution of contaminants in sediments is caused by resuspension and focusing (Eadie & Robbins 1987). Contaminant concentration in settling particles may be larger or lower than that in deposited bottom sediments. Apart from physico-chemical adsorption, biological uptake, particularly by bacteria and algae, and allochthonous sediment input, may increase contaminant concentrations in bottom sediments (Hart 1982).

Heavy metal and contaminant concentrations in various lake sediments are compiled in Tables 8.3a & 8.3b. Anthropogenic contamination (i.e. deviations from natural background concentrations) is quantified, for example, by the USA and Canadian guidelines quoted (respectively) in Giesy & Hoke (1990) and Smith *et al.* (1996). While mean heavy metal concentrations in sediments from various Swiss lakes lie generally below tolerance limits, organic contaminant concentrations in the sediments of the Great Lakes is mostly within the range of, or even above, the critical toxic levels given by bioassessment tests.

Contamination of particular lake sediments obviously depends mainly on the sources of pollution in the catchment. Remote lakes may be affected by airborne input of contaminants, but their heavy metal concentration usually is low (Bloesch *et al.* 1995b). High contaminant concentrations clearly reflect point sources along the lake shore and/or in the catchment: For instance, zinc concentrations in the Rotsee (near the Swiss town of Lucerne) are four to five times increased when compared with Baldeggersee (situated in the countryside), as the zinc mainly originates from galvanised piping

Table 8.3 (a) Range of heavy metal concentration in settling SPM and bottom sediments of various Swiss lakes. (From Bloesch *et al.* 1995.)

Heavy metal	Concentration in various Swiss lakes (μg g^{-1} dry wt)	Geochemical standard in clays* (μg g^{-1} dry wt)	Mean concentration in unpolluted soil† (μg g^{-1} dry wt)	EPA–Canadian standards‡	
				No pollution–lowest limit (μg g^{-1} dry wt)	Tolerance limit (μg g^{-1} dry wt)
Zn	81–543	95	1–900	65–110	800
Cu	11–56	45	2–250	15–25	114
Cr	10–51	90	5–1500	22–31	111
Ni	33–76	68	2–750	15–31	90
Cd	0.4–1.7	0.3	0.01–2	0.6–1.0	10.0
Pb	11–90	20	2–300	23–40	250
Hg	0.1–0.3	0.4	0.01–0.5	0.1–1.0	2.0

* From Förstner & Müller (1974); † From Baudo & Muntau (1990); ‡ From Giesy & Hoke (1990).

Table 8.3 (b) Range of organic contaminant concentration in sediments of the Laurentian Great Lakes. (From Papoulias & Buckler 1996.)

Organic contaminant, number of single compounds investigated†	Concentration in various lake and riverine sediments‡	Canadian standards*	
		Minimum–lowest effect level	Severe–toxic effect level
Pesticides, 21 (μg kg^{-1} dry wt)	ND–390	2–10	9–1300
Low molecular PAHs, 8 (μg kg^{-1} dry wt)	ND–151,500	400–560	800–9500
High molecular PAHs, 12 (μg kg^{-1} dry wt)	ND–290,000	320–750	500–14,800
Dioxins, 5 (ng kg^{-1} dry wt)	ND–46,000	–	–
Furans, 5 (ng kg^{-1} dry wt)	ND–22,000	–	–
PCBs, 7 (μg kg^{-1} dry wt)	ND–60,000	70–200	1000–5300

* From Smith *et al.* (1996). † From Papoulias & Buckler (1996). ‡ ND, not detectable.

(Bloesch & Wehrli 1995). Harbour sediments are rich in organotin compounds, e.g. tributyltin used as antifouling paint for boats, with a significant ecotoxicological effect on mussels and fish (Fent 1996).

The question of whether sediments are a sink or source of nutrients and contaminants is important with respect to eutrophication and pollution (Xue *et al.* 1997). The key role of sediments in eutrophication is clearly demonstrated by the countless studies of internal phosphorus loading (e.g. Mortimer 1941, 1942; Thomas 1955; Schindler

1974; Boström *et al.* 1982; Sas 1989). The fact that heavy metals and persistent contaminants are scavenged by settling particles, and thus accumulate in bottom sediments, leads to environmental problems, as these toxic substances are taken up by benthic sediment feeders and may subsequently be concentrated along the food chain. Robbins *et al.* (1977) have demonstrated how the action of deposit-feeding organisms can generalise heavy metal profiles in sediments. Bioaccumulation of toxicants is clearly specific to particular organisms, and strongly dependent on behaviour (such as periodic migration into the water column; Hare *et al.* 1994). Organic contaminants, such as PCBs (polychlorinated biphenyl) and PAHs (polycyclic aromatic hydrocarbon), were reported to deform mouth parts of benthic midge larvae (Canfield *et al.* 1996), and DNA (deoxyribonucleic acid) may be mutated in bacteria (Papoulias & Buckler 1996). Fish contaminant concentrations reflect abundance in the water column (De Vault *et al.* 1996). The biota (and, ultimately, humanity—**although probably not the shareholders of the companies concerned, who can afford to live elsewhere**) will pay the price, in the form of illness and disease.

Hence, contaminated and nutrient-enriched sediments may be removed by dredging (Förstner 1987; Salomons & Förstner 1988; Skei 1992). However, this restoration procedure may not be effective with regard to improvement of the lakewater quality (Ryding 1982). With regard to contaminant removal, this is simply a transfer of materials to special terrestrial solid-waste disposals. Fighting the sources by diminishing contaminant production and emission, i.e., input into rivers and lakes, is a more promising way to solve these problems. Analysis of sediment cores showing peaks of contaminant concentration for the 1960s, and a distinct decrease in recent years, provides evidence of successful protection measures taken in some catchments (Müller 1997) and the success of enforced limits for reducing emissions (Moor *et al.* 1996).

8.8 REFERENCES

Allan, R.J. (1986) *The Role of Particulate Matter in the Fate of Contaminants in Aquatic Ecosystems.* NWRI Report, Scientific Series No. 142, Canada Centre for Inland Waters, Burlington, Ontario, Canada, 128 pp.

Baines, S.B. & Pace, M.L. (1994) Relationships between suspended particulate matter and sinking flux along a trophic gradient and implications for the fate of planktonic primary production. *Canadian Journal of Fisheries and Aquatic Sciences*, **51**, 25–36.

Baines, S.B., Pace, M.L. & Karl, D.M. (1994) Why does the relationship between sinking flux and planktonic primary production differ between lakes and oceans? *Limnology and Oceanography*, **39**, 213–26.

Baker, E.T., Milburn, H.B. & Tennant, D.A. (1988) Field assessment of sediment trap efficiency under varying flow conditions. *Journal of Marine Research*, **46**, 573–92.

Banas, D., Capizzi, S., Masson, G., et al. (2001) New sediment traps for drainable shallow freshwater systems. *Revue des Sciences de l'Eau*, **15**, 263–72.

Baudepartement des Kantons Aargau, Abteilung Umweltschutz (1992) Sanierung des Hallwilersees: Zuflussuntersuchung zur Nährstoffbelastung 1988/90 (Schlussbericht). *Wasser, Energie, Luft*, **84**, 66–7.

Baudo, R. & Muntau, H. (1990) Lesser known in-place pollutants and diffuse source problems. In: Baudo, R., Giesy, J.P. & Muntau, H. (eds), *Sediments: Chemistry and Toxicity of In-place Pollutants.* Lewis Publishers, Boston, pp. 1–14.

Baxter, M.S., Farmer, J.F., McKinley, I.G., Swan, D.S. & Jack, W. (1981) Evidence of the unsuitability of gravity coring for collecting sediment in pollution and sedimentation rate studies. *Environmental Science and Technology*, **15**, 843–6.

Bayerisches Landesamt für Wasserwirtschaft (1987) *Grundzüge der Gewässerpflege, Fliessgewässer.* Schriftenreihe, Heft 21, 112 pp.

Bengtsson, L., Hellström, Th. & Rakoczi, L. (1990) Redistribution of sediments in three Swedish lakes. *Hydrobiologia*, **192**, 167–81.

Bienfang, P.K. (1981) SETCOL—a technologically simple and reliable method for measuring phytoplankton sinking rates. *Canadian Journal of Fisheries and Aquatic Sciences*, **38**, 1289–94.

Binderheim-Bankay, E.A. (1998) *Sanierungsziel für natürlich eutrophe Kleinseen des Schweizer Mittellandes.* PhD thesis, Eidgenössische Technische Hochschule, Zürich, 149 pp.

Bloesch, J. (1974) Sedimentation und Phosphorhaushalt im Vierwaldstättersee (Horwer Bucht) und im Rotsee. *Schweizerische Zeitschrift für Hydrologie*, **36**, 71–186.

Bloesch, J. (1982) Inshore–offshore sedimentation differences resulting from resuspension in the Eastern Basin of Lake Erie. *Canadian Journal of Fisheries and Aquatic Sciences*, **39**, 748–59.

Bloesch, J. (1992) *Studie Gewässerschutz im Einzugsgebiet des Vierwaldstättersees. Zuflussuntersuchung 1989.* EAWAG-Report No. 4752, Swiss Federal Institute for Environmental Science and Technology, Duebendorf, 139 pp.

Bloesch, J. (1993) Phosphorus sedimentation in seven basins of Lake Lucerne, Switzerland. *Hydrobiologia*, **253**, 319.

Bloesch, J. (1994) A review of methods used to measure sediment resuspension. *Hydrobiologia*, **284**, 13–18.

Bloesch, J. (1995) Mechanisms, measurement and importance of sediment resuspension in lakes. *Marine and Freshwater Research*, **46**, 295–304.

Bloesch, J. (1996) Towards a new generation of sediment traps and a better measurement/understanding of settling particle flux in lakes and oceans: a hydrodynamical protocol. *Aquatic Sciences*, **58**, 283–96.

Bloesch, J. & Burns, N.M. (1980) A critical review of sedimentation trap technique. *Schweizerische Zeitschrift für Hydrologie*, **42**, 15–55.

Bloesch, J. & Bürgi, H.R. (1989) Changes in phytoplankton and zooplankton biomass and composition reflected by sedimentation. *Limnology and Oceanography*, **34**, 1048–61.

Bloesch, J. & Evans, R.D. (1982) Lead-210 dating of sediments compared with accumulation rates estimated by natural markers and measured with sediment traps. *Hydrobiologia*, **92**, 579–86.

Bloesch, J. & Ma Xi (1996) 200 meter deep Lake Zug, Switzerland: bottom sediment resuspension or sediment focusing? (Abstract). *7th International Symposium of IASWS, September 22–25*, Baveno, Italy, p. 32.

Bloesch, J. & Sturm, M. (1986) Settling flux and sinking velocities of particulate phosphorus (PP) and particulate organic carbon (POC) in Lake Zug, Switzerland. In: Sly, P.G. (ed.), *Sediments and Water Interaction*. Springer-Verlag, New York, pp. 481–90.

Bloesch, J. & Uehlinger, U. (1986) Horizontal sedimentation differences in a eutrophic Swiss Lake. *Limnology and. Oceanography*, **31**, 1094–109.

Bloesch, J. & Uehlinger, U. (1990) Epilimnetic carbon flux and turnover of different particle size classes in oligo-mesotrophic Lake Lucerne, Switzerland. *Archiv für Hydrobiologie*, **118**, 403–19.

Bloesch, J. & Wehrli, B. (1995) Formation of natural sediment records. *EAWAG News*, **38E**, 10–12.

Bloesch, J., Stadelmann, P. & Bührer, H. (1977) Primary production, mineralisation, and sedimentation in the euphotic zone of two Swiss lakes. *Limnology and Oceanography*, **22**, 511–26.

Bloesch, J., Armengol, J., Giovanoli, F. & Stabel, H.-H. (1988) Phosphorus in suspended and settling particulate matter of lakes. *Ergebnisse der Limnologie*, **30**, 84–90.

Bloesch, J., Bührer, H., Bossard, P., Bürgi, H.-R. & Müller, R. (1995a) Lake oligotrophication due to external phosphorus load reduction in Swiss lakes. *Sixth International Conference on the Conservation and Management of Lakes*, Kasumigaura, Japan, Extended Abstract, pp.738–42.

Bloesch, J., Hohmann, D. & Leemann, A. (1995b) Die Limnologie des Oeschinensees, mit besonderer Berücksichtigung des Planktons, der Sedimentation und der Schwermetallbelastung. *Mitteilungen der Naturforschenden Gesellschaft Bern*, N.F. **52**, 121–45.

Blomqvist, S. (1992) *Geochemistry of coastal Baltic sediments: Processes and sampling procedures*. PhD thesis, University of Stockholm, 137 pp.

Blomqvist, S. & Håkanson, L. (1981) A review on sediment traps in aquatic environments. *Archiv für Hydrobiologie*, **91**, 101–32.

Blomqvist, S. & Larsson, U. (1992) Petrogenic metals as tracers of resuspended and primary settling matter in a coastal area of the Baltic Sea. In: Blomqvist, S., *Geochemistry of Coastal Baltic Sediments: Processes and Sampling Procedures*. PhD thesis, University of Stockholm, 1–37.

Boström, B., Jansson, M. & Forsberg, C. (1982) Phosphorus release from lake sediments. *Ergebnisse der Limnologie*, **18**, 5–59.

Brandl, H. & Hanselmann, K.W. (1991) Evaluation and application of dialysis porewater samplers for microbiological studies at sediment–water interfaces. *Aquatic Sciences*, **53**, 55–73.

Brassard, P., Kramer, J.R., McAndrew, J. & Mueller, E. (1994) Metal–sediment interaction during resuspension. *Hydrobiologia*, **284**, 101–12.

Brownawell, B.J., Chen, H., Zhang, W. & Westall, J.C. (1997) Sorption of nonionic surfactants on sediment materials. *Environmental Science and Technology*, **31**, 1735–41.

Buesseler, K.O. (1991) Do upper-ocean sediment traps provide an accurate record of particle flux? *Nature*, **353**, 420–3.

Buesseler, K.O., Michaels, A.F., Siegel, D.A. & Knap, A.H. (1994) A three dimensional time-dependent approach to calibrating sediment trap fluxes. *Global Biogeochemical Cycles*, **8**, 179–93.

Burns, N.M. & Rosa, F. (1980) *In situ* measurement of the settling velocity of organic carbon particles and 10 species of phytoplankton. *Limnology and Oceanography*, **25**, 855–64.

Butman, C.A. (1986) Sediment trap biases in turbulent flows: results from a laboratory flume study. *Journal of Marine Research*, **44**, 645–93.

Butman, C.A., Grant, W.D. & Stolzenbach, K.D. (1986) Predictions of sediment trap biases in turbulent flows: a theoretical analysis based on observations from the literature. *Journal of Marine Research*, **44**, 601–44.

Canfield, T.J., Dwyer, F.J., Fairchild, J.F., *et al.*(1996) Assessing contamination in Great Lakes sediments using benthic invertebrate communities and the sediment quality triad approach. *Journal of Great Lakes Research*, **22**, 565–83.

Carignan, R. & Lean, D.R.S. (1991) Regeneration of dissolved substances in a seasonally anoxic lake: the relative importance of processes occurring in the water column and in the sediments. *Limnology and Oceanography*, **36**, 683–707.

Carper, G.L. & Bachmann, R.W. (1984) Wind resuspension of sediments in a prairie lake. *Canadian Journal of Fisheries and Aquatic Sciences*, **41**, 1763–7.

Cranford, P.J. & Hargrave, B.T. (1994) *In situ* time-series measurement of ingestion and absorption rates of suspension-feeding bivalves: *Placopecten magellanicus*. *Limnology and Oceanography*, **39**, 730–8.

Currie, D.J. & Kalff, J. (1984) The relative importance of bacterioplankton and phytoplankton in phosphorus uptake in freshwater. *Limnology and Oceanography*, **29**, 311–21.

Czuczwa, J.M., Niessen, F. & Hites, R.A. (1985) Historical record of polychlorinated dibenzo-*p*-dioxins and dibenzofurans in Swiss lake sediments. *Chemosphere*, **14**, 1175–9.

Davis, M.B., Ford, M.S. (Jesse) & Moeller, R.E. (1985) Paleolimnology. In: Likens, G.E. (ed.), *An Ecosystem Approach to Aquatic Ecology. Mirror Lake and its Environment*. Springer-Verlag, New York, pp. 345–429.

Davison, W., Zhang, H. & Grime, G.W. (1994) Performance characteristics of gel probes used for measuring the chemistry of pore waters. *Environmental Science and Technology*, **28**, 1623–32.

Dearing, J.A. (1983) Changing patterns of sediment accumulation in a small in Scania, southern Sweden. *Hydrobiologia*, **103**, 59–64.

De Vault, D.S., Hesselberg, R., Rodgers, P.W. & Feist, T.J. (1996) Contaminant trends in lake trout and walleye from the Laurentian Great Lakes. *Journal of Great Lakes Research*, **22**, 884–95.

Dillon, P.J., Evans, R.D. & Molot, L.A. (1990) Retention and resuspension of phosphorus, nitrogen, and iron in a central Ontario lake. *Canadian Journal of Fisheries and Aquatic Sciences*, **47**, 1269–74.

Eadie, B.J. & Robbins, J.A. (1987) The role of particulate matter in the movement of contaminants in the Great Lakes. In: Hites, R.A. & Eisenreich, S.J. (eds), *Sources and Fates of Aquatic Pollutants*. Advances in Chemistry Series No. 216, American Chemical Society, Washington, pp. 319–64.

Eisenreich, S.J., Looney, B.B. & Thornton, J.D. (1981) Airborne organic contaminants in the Great Lakes ecosystem. *Environmental Science and Technology*, **15**, 30–8.

Fent, K. (1996) Ecotoxicology of organotin compounds. *Critical Reviews in Toxicology*, **26**, 1–117.

Ferrante, J.G. & Parker, J.I. (1977) Transport of diatom frustules by copepod fecal pellets to the sediments of Lake Michigan. *Limnology and Oceanography*, **22**, 92–8.

Fowler, S.W. & Small, L.F. (1972) Sinking rates of euphausiid fecal pellets. *Limnology and Oceanography*, **17**, 293–6.

Förstner, U. (1987) Sediment-associated contamination—an overview of scientific bases for developing remedial options. *Hydrobiologia*, **149**, 221–46.

Förstner, U. & Müller, G. (1974) *Schwermetalle in Flüssen und Seen als Ausdruck der Umweltverschmutzung*. Springer-Verlag, Berlin, 225 pp.

Friedl, G., Wehrli, B. & Manceau, A. (1997) Solid phases in the cycling of manganese in eutrophic lakes: new insights from EXAFS spectroscopy. *Geochimica et Cosmochimica Acta*, **61**, 275–90.

Furrer, G. & Wehrli, B. (1993) Biogeochemical processes at the sediment–water interface: measurement and modelling. *Applied Geochemistry, Supplementary Issue*, **2**, 117–19.

Furrer, G. & Wehrli, B. (1996) Microbial reactions, chemical speciation and multi component diffusion in porewaters of a productive lake. *Geochimica et Cosmochimica Acta*, **60**, 2333–47.

Gächter, R. & Bloesch, J. (1985) Seasonal and vertical variation in the C : P ratio of suspended and settling seston of lakes. *Hydrobiologia*, **128**, 193–200.

Gächter, R. & Mares, A. (1985) Does settling seston release soluble reactive phosphorus in the hypolimnion of lakes? *Limnology and Oceanography*, **30**, 366–73.

Gächter, R. & Meyer, J.S. (1990) Mechanisms controlling fluxes of nutrients across the sediment/water interface in a eutrophic lake. In: Baudo, R., Giesy, J.P. & Muntau, H. (eds), *Sediments and Toxicity of Inplace Pollutants*. Lewis Publishers, Boston, pp. 131–62.

Gächter, R. & Meyer, J.S. (1993) The role of microorganisms in mobilisation and fixation of phosphorus in sediments. *Hydrobiologia*, **253**, 103–21.

Gächter, R., Mares, A., Grieder, E., Zwyssig, A. & Höhener, P. (1989) Auswirkungen der Belüftung und Sauerstoffbegasung auf den P-Haushalt des Sempachersees. *Wasser, Energie, Luft*, **81**, 335–41.

Gálvez, J.A., Niell, F.X. & Lucena, J. (1993) Sinking velocities of principal phytoplankton species in a stratified reservoir: Ecological implications. *Verhandlungen der internationale Vereinigung für theoretische und angewandte Limnologie*, **25**, 1228–31.

Gardner, W.D. (1977) *Fluxes, dynamics and chemistry of particulates in the ocean.* Ph.D. Dissertation, Massachusetts Institute of Technology/Woods Hole Oceanographic Institution Joint Program in Oceanography, Woods Hole, 405 pp.

Gardner, W.D. (1980a) Sediment trap dynamics and calibration. A laboratory evaluation. *Journal of Marine Research*, **38**, 17–39.

Gardner, W.D. (1980b) Field assessment of sediment traps. *Journal of Marine Research*, **38**, 41–52.

Gardner, W.D. (1996) Sediment trap technology and sampling in surface waters. *Report on the JGOFS Symposium in Villefranche sur Mer in May, 1995*. Internet: , 35 pp.

Geiger, W. (1993) *Cytherissa lacustris* (Ostracoda, Crustacea): its use in detecting and reconstructing environmental changes at the sediment–water interface. *Verhandlungen der internationale Vereinigung für theoretische und angewandte Limnologie*, **25**, 1102–7.

Gibbs, R.J. (ed.) (1974) *Suspended Solids in Water.* Plenum Press, New York, 320 pp.

Giovanoli, F. & Lambert, A. (1985) Die Einschichtung der Rhone im Genfersee: Ergebnisse von Strömungsmessungen im August 1983. *Schweizerische Zeitschrift für Hydrologie*, **47**, 159–78.

Giesy, J.P. & Hoke, R.A. (1990) Freshwater sediment quality criteria: toxicity bioassessment. In: Baudo, R., Giesy, J.P. & Muntau, H. (eds), *Sediments: Chemistry and Toxicity of In-place Pollutants.* Lewis Publishers, Boston, pp. 265–348.

Gloor, M., Wüest, A. & Münnich, M. (1994) Benthic boundary mixing and resuspension induced by internal seiches. *Hydrobiologia*, **284**, 59–68.

Goldman, C.J. & Horne, A.J. (1983) *Limnology.* McGraw-Hill, New York, 464 pp.

Gries, T. (1995) Phosphorhaushalt der oberen 20 m des Überlinger Sees (Bodensee) unter besonderer Berücksichtigung der Sedimentation. *Konstanzer Dissertationen*, **488**, 185 pp.

Gu, B., Mehlhorn, T.L., Liang, L. & McCarthy, J.F. (1996) Competitive adsorption, displacement, and transport of organic matter on iron oxide: II. Displacement and transport. *Geochimica et Cosmochimica Acta*, **60**, 2977–92.

Gundersen, K. & Wassmann, P. (1990) Use of chloroform in sediment traps: caution advised. *Marine Ecology — Progress Series*, **64**, 187–95.

Gust, G., Michaels, A.F., Johnson, R., Deuser, W.G. & Bowles, W. (1994) Mooring line motions and sediment trap hydromechanics: *in-situ* intercomparison of three common deployment designs. *Deep-sea Research*, **41**, 831–57.

Håkanson, L. & Jansson, M. (1983) *Principles of Lake Sedimentology.* Springer-Verlag, Berlin, 316 pp.

Hansell, D.A. & Newton, J.A. (1994) Design and evaluation of a 'swimmer'-segregating particle interceptor trap. *Limnology and Oceanography*, **39**, 1487–95.

Hare, L., Carignan, R. & Huerta-Diaz, M.A. (1994) A field study of metal toxicity and accumulation by benthic invertebrates; implications for the acid-volatile sulphide (AVS) model. *Limnology and Oceanography*, **39**, 1653–68.

Hart, B.T. (1982) Uptake of trace metals by sediments and suspended particulates: a review. *Hydrobiologia*, **91**, 299–313.

Hawley, N., Wang, X., Brownawell, B. & Flood, R. (1996) Resuspension of bottom sediments in Lake Ontario during the unstratified period, 1992–1993. *Journal of Great Lakes Research*, **22**, 707–21.

Henrichs, S.M. (1992) Early diagenesis of organic matter in marine sediments: progress and perplexity. *Marine Chemistry*, **39**, 119–49.

Hilton, J. (1985) A conceptual framework for predicting the occurrence of sediment focusing and sediment redistribution in small lakes. *Limnology and Oceanography*, **30**, 1131–43.

Hilton, J., Lishman, J.P. & Allen, P.V. (1986) The dominant processes of sediment distribution and focusing in a small, eutrophic, monomictic lake *Limnology and Oceanography*, **31**, 125–33.

Hofmann, A. (1996) *Caractéristiques géochimiques et processus de transport de la matière particulaire dans le bassin nord du lac de Lugano (Suisse, Italie)*, Vol.8a & b. Thèse No.2853, Institut Forel, Université de Genève, 392 pp.

Hutchinson, G.E. (1967) *A Treatise on Limnology*, Vol. 2, *Introduction to Lake Biology and the Limnoplankton.* Wiley, New York, 1115 pp.

Imboden, D.M. & Wüest, A. (1995) Mixing mechanisms in lakes. In: Lerman, A., Imboden, D.M. & Gat, J.R. (eds), *Physics and Chemistry of Lakes.* Springer-Verlag, Heidelberg, 83–137.

IOC Manual and Guides No. 29 (1994) Protocols for the Joint Global Ocean Flux Study (JGOFS) core measurements. *UNESCO Scientific Committee on Oceanic*

Research, Intergovernmental Oceanographic Commission, Geneva.

Jambers, W. & van Grieken, R. (1997) Single particle characterisation of inorganic suspension in Lake Baikal, Siberia. *Environmental Science and Technology*, **31**, 1525–33.

Jaquet, J.-M., Nembrini, G., Garcia, J. & Vernet, J.-P. (1982) The manganese cycle in Lac Léman, Switzerland: the role of *Metallogenium*. *Hydrobiologia*, **91**, 323–40.

Jobson, H.E. & Sayre, W.W. (1970) Vertical transfer in open channel flow. *Journal of the Hydraulics Division of the American Society of Civil Engineers*, **96**, 703–24.

Kamp-Nielsen, L., Mejer, H. & Jørgensen, S-E. (1982) Modelling the influence of bioturbation on the vertical distribution of sedimentary phosphorus in L. Esrom. *Hydrobiologia*, **91**, 197–206.

Kansanen, P.H., Jaakkola, T., Kulmala, S. & Suutarinen, R. (1991) Sedimentation and distribution of gamma-emitting radionuclides in bottom sediments of southern Lake Päijänne, Finland, after the Chernobyl accident. *Hydrobiologia*, **222**, 121–40.

Karl, D.M. & Knauer, G.A. (1989) Swimmers: a recapitulation of the problems and a potential solution. *Oceanography Magazine*, **2**, 32–5.

Karl, D.M., Knauer, G.A. & Martin, J.H. (1988) Downward flux of particulate organic matter in the ocean: a particle decomposition paradox. *Nature*, **31**, 438–41.

Karl, D.M., Christian, J.R., Dore, J.E., *et al.* (1996) Seasonal and interannual variability in primary production and particle flux at Station ALOHA. *Deep-sea Research*, **43**, 539–68.

Keil, R.G., Mayer, L.M., Quay, P.D., Richey, J.E. & Hedges, J.I. (1997) Loss of organic matter from riverine particles in deltas. *Geochimica et Cosmochimica Acta*, **61**, 1507–11.

Kelts, K. & Hsü, K.J. (1978) Freshwater carbonate sedimentation. In: Lerman, A. (ed.), *Lakes: Chemistry, Geology, Physics*. Springer-Verlag, New York, 295–323.

Kemp, A.L.W. & Harper, N.S. (1976) Sedimentation rates and a sediment budget for Lake Ontario. *Journal of Great Lakes Research*, **2**, 324–40.

Kimmel, B.L. & Goldman, C.R. (1977) Production, sedimentation and accumulation of particulate carbon and nitrogen in a sheltered subalpine lake. In: Golterman, H.L. (ed.), *Interactions between Sediments and Freshwater*. W. Junk, Den Haag, pp. 148–55.

Kozerski, H.-P. (1994) Possibilities and limitations of sediment traps to measure sedimentation and resuspension. *Hydrobiologia*, **284**, 93–100.

Kozerski, H.-P. & Leuschner, K. (1999) Plate sediment traps for slowly moving waters. *Water Research*, **33**, 2913–22.

Kozerski, H.-P. & Leuschner, K. (2000) A new plate sediment trap: design and first experiences. *Verhandlungen der internationalen Vereinigung für theoretische und angewandte Limnologie*, **27**, 242–5.

Lambert, A. (1982) Trübeströme des Rheins am Grund des Bodensees. *Wasserwirtschaft*, **72**, 169–72.

Lambert, A. & Giovanoli, F. (1988) Records of riverborne turbidity currents and indications of slope failures in the Rhone delta of Lake Geneva. *Limnology and Oceanography*, **33**, 458–68.

Lambert, A., Kelts, K. & Marshall, N. (1976) Measurement of density underflows from Walensee, Switzerland. *Sedimentology*, **23**, 87–105.

Lang, C. (1997) Oligochaetes, organic sedimentation, and trophic state: how to assess the biological recovery of sediments in lakes? *Aquatic Sciences*, **59**, 26–33.

Lang, C. & Hutter, P. (1981) Structure, diversity and stability of two oligochaete communities according to sedimentary inputs in Lake Geneva (Switzerland). *Schweizerische Zeitschrift für Hydrologie*, **43**, 265–76.

Lang, O. & Rutz, F. (1996) Die Berücksichtigung der ökologischen Bedingungen, der natürlich-dynamischen Prozesse und der Stellenwert der Ingenieurbiologie in der Gewässerpflege und Entwicklung. *Österreichische Wasser- und Abfallwirtschaft*, **48**, 187–200.

Lee, C., Wakeham, S.G. & Hedges, J.I. (1988) The measurement of oceanic particle flux—are 'swimmers' a problem? *Oceanography*, **2**, 34–6.

Lee, C., Hedges, J.I., Wakeham, S.G. & Zhu, N. (1992) Effectiveness of various treatments in retarding microbial activity in sediment trap material and their effects on the collection of swimmers. *Limnology and Oceanography*, **37**, 117–30.

Lehmann, R. (1983) Untersuchungen zur Sedimentation in einem oligotrophen Alpensee (Königssee) während der sommerlichen Schichtung *Archiv für Hydrobiologie*, **96**, 486–95.

Lemcke, G. (1992) *Ablagerungen aus Extremereignissen als Zeitmarken der Sedimentations geschichte im Becken von Vitznau/Weggis (Vierwaldstättersee, Schweiz)*. Diplomarbeit Universität Göttingen, 154 pp.

Lemmin, U. & Imboden, D.M. (1987) Dynamics of bottom currents in a small lake. *Limnology and Oceanography*, **32**, 62–75.

Lerman, A. (1979) *Geochemical Processes. Water and Sediment Environments.* Wiley-Interscience, New York, 481 pp.

Lewin, J. (1992) Floodplain construction and erosion. In: Calow, P. & Petts, G.E. (eds), *The Rivers Handbook*, Vol.1. *Hydrological and Ecological Principles.* Blackwell Scientific, Oxford, 144–62.

Lick, W. (1982) Entrainment, deposition and transport of fine-grained sediments in lakes. *Hydrobiologia*, **91**, 31–40.

Lick, W. & Lick, J. (1988) Aggregation and disaggregation of fine-grained lake sediments. *Journal of Great Lakes Research*, **14**, 514–23.

Likens, G.E. (ed.) (1985) *An Ecosystem Approach to Aquatic Ecology. Mirror Lake and its Environment.* Springer-Verlag, New York, 516 pp.

Lotter, A.F. (1989) Evidence of annual layering in Holocene sediments of Soppensee, Switzerland. *Aquatic Sciences*, **51**, 19–30.

Löffler, H. (1979) *Neusiedlersee: the Limnology of a Shallow Lake in Central Europe.* Dr. W. Junk, Den Haag, 543 pp.

Mengis, M., Gächter, R., Wehrli, B. & Bernasconi, S. (1997) Nitrogen elimination in two deep eutrophic lakes. *Limnology and Oceanography*, **42**, 1530–43.

Meriläinen, J.J. & Hamina, V. (1993) Changes in biological condition of the profundal area in an unpolluted nutrient-poor lake during the past 400 years. *Verhandlungen der internationalen Vereinigung für theoretische und angewandte Limnologie*, **25**, 1079–81.

Miller, M.C., McCave, I.N. & Komar, P.D. (1977) Threshold of sediment motion under unidirectional currents. *Sedimentology*, **24**, 507–27.

Moeller, R.E. & Likens, G.E. (1978) Seston sedimentation in Mirror Lake, New Hampshire, and its relationship to long-term sediment accumulation. *Verhandlungen der internationalen Vereinigung für theoretische und angewandte Limnologie*, **20**, 525–30.

Moor, H.C., Schaller, T. & Sturm, M. (1996) Recent changes in stable lead isotope ratios in sediments of Lake Zug, Switzerland. *Environmental Science and Technology*, **30**, 2928–33.

Mortimer, C.H. (1941) The exchange of dissolved substances between mud and water in lakes (Parts I and II). *Journal of Ecology*, **29**, 280–329.

Mortimer, C.H. (1942) The exchange of dissolved substances between mud and water in lakes (Parts III, IV, summary and references). *Journal of Ecology*, **30**, 147–201.

Mudroch, A. & MacKnight, S.D. (eds) (1994) *Handbook of Techniques for Aquatic Sediments Sampling* (2nd edition). CRC Press, Boca Raton, FL, 236 pp.

Müller, B., Buis, K., Stierli, R. & Wehrli, B. (1998) High spatial resolution measurements in lake sediments with PVC based liquid membrane ion-selective electrodes. *Limnology and Oceanography*, **43**, 1728–33.

Müller, G. (1997) Nur noch geringer Eintrag anthropogener Schwermetalle in den Bodensee—neue Daten zur Entwicklung der Belastung der Sedimente. *Naturwissenschaften*, **84**, 37–8.

Münnich, M., Wüest, A. & Imboden, D.M. (1992) Observations of the second vertical mode of the internal seiche in an alpine lake. *Limnology and Oceanography*, **37**, 1705–19.

Murray, St. P. (1970) Settling velocities and vertical diffusion of particles in turbulent water. *Journal of Geophysical Research*, **75**, 1647–54.

Niessen, F. & Sturm, M. (1987) Die Sedimente des Baldeggersees (Schweiz): Ablagerungsraum und Eutrophierungsentwicklung der letzten 100 Jahre. *Archiv für Hydrobiologie*, **108**, 365–78.

Nydegger, P. (1967) Untersuchungen über Feinstofftransport in Flüssen und Seen, über Entstehung von Trübungshorizonten und zuflussbedingten Strömungen im Brienzersee und einigen Vergleichsseen. *Beiträge zur Geologie der Schweiz, Hydrologie*, **16**, 92 pp.

OECD (1980) *Eutrophication of Waters: Monitoring, Assessment and Control.* Organisation for Economic Co-operation and Development, Paris, 154 pp.

Osman-Sigg, G.K. (1982) *Kolloidale und suspendierte Teilchen in natürlichen Gewässern; Partikelgrössenverteilung und natürliche Koagulation im Zürichsee.* PhD thesis, Eidgenössische Technische Hochschule, Zürich, 123 pp.

Paerl, H.W. (1973) Detritus in Lake Tahoe: structural modification by attached microflora. *Science*, **180**, 496–8.

Papoulias, D.M. & Buckler, D.R. (1996) Mutagenicity of Great Lakes sediments. *Journal of Great Lakes Research*, **22**, 591–601.

Partheniades, E. (1965) Erosion and deposition of cohesive soils. *Journal of the American Society of Civil Engineers, Hydraulics Division*, **HY1**, 105–39.

Prasuhn, V., Spiess, E. & Braun, M. (1996) Methoden zur Abschätzung der Phosphor- und Stickstoffeinträge aus diffusen Quellen in den Bodensee. *Bericht der internationalen Gewässerschutzkommission Bodensee*, **45**, 113 pp.

Rathke, D.E., Bloesch, J., Burns, N.M. & Rosa, F. (1981) Settling fluxes in Lake Erie (Canada) measured by traps and settling chambers. *Verhandlungen der internationalen Vereinigung für theoretische und angewandte Limnologie*, **21**, 383–8.

Reynolds, C.S. (1984) *The Ecology of Freshwater Phytoplankton*. Cambridge University Press, Cambridge, 384 pp.

Reynolds, C.S., Wiseman, S.W. & Gardner, W.D. (1980) An annotated bibliography of aquatic sediment traps and trapping methods. *Occasional Publications of the Freshwater Biological Association*, **11**, 54 pp.

Rippey, B. & Jewson, D.H. (1982) The rates of sediment–water exchange of oxygen and sediment bioturbation in Lough Neagh, Northern Ireland. *Hydrobiologia*, **92**, 377–82.

Robbins, J.A. (1982) Stratigraphic and dynamic effects of sediment reworking by Great Lakes zoobenthos. *Hydrobiologia*, **92**, 611–22.

Robbins, J.A., Krezoski, J.R. & Mozley, S.C. (1977) Radioactivity in sediments of the Great Lakes: Post-depositional redistribution by deposit-feeding organisms. *Earth Planetary Science Letters*, **36**, 325–33.

Robbins, J.A., Keilty, T., White, D.S. & Eddington, D.N. (1989) Relationship among tubificid abundances, sediment composition and accumulation rates in Lake Erie. *Canadian Journal of Fisheries and Aquatic Sciences*, **46**, 223–31.

Rosa, F., Bloesch, J. & Rathke, D.E. (1994) Sampling the settling and suspended particulate matter (SPM) In: Mudroch, A. & MacKnight, S.D. (eds), *Handbook of Techniques for Aquatic Sediments Sampling* (2nd edition). CRC Press, Boca Raton, Fl, 97–130.

Ryding, S.-O. (1982) Lake Trehörningen restoration project. Changes in water quality after sediment dredging. *Hydrobiologia*, **92**, 549–58.

Rysgaard, S., Risgaard-Petersen, N., Sloth, N.P., Jensen, K. & Nielsen, L.P. (1994) Oxygen regulation of nitrification and denitrification in sediments. *Limnology and Oceanography*, **39**, 1643–52.

Saarnisto, M. (1986) Annually laminated sediments. In: Berglund, B.E. (ed.), *Handbook of Holocene Palaeoecology and Palaeohydrology*. Wiley, Chichester, 343–70.

Sakamoto, M. (1980) Metabolism of organic matter and nutrients in the ecosystem of Lake Biwa. In: Mori, S. (ed.), *An Introduction to Limnology of Lake Biwa*. Twenty-first Societas Internationalis Limnologiae Congress, Kyoto, Japan, 52–9.

Salomons, W. & Förstner, U. (1988) *Environmental Management of Solid Waste. Dredged Material and Mine Tailings*. Springer-Verlag, New York, 396 pp.

Santschi, P.H. (1989) Use of radionuclides in the study of contaminant cycling processes. *Hydrobiologia*, **176/177**, 307–20.

Santschi, P., Bollhalder, S., Zingg, S. & Farrenkothen, K. (1990) The self-cleaning capacity of surface waters after radioactive fallout. Evidence from European waters after Chernobyl, 1986–1988. *Environmental Science and Technology*, **24**, 519–27.

Sas, H. (ed.) (1989) *Lake Restoration by Reduction of Nutrient Loading. Expectations, Experiences, Extrapolations*. Academia Verlag Richarz, Sankt Augustin, 497 pp.

Schaller, T.L. (1996) *Redox-sensitive metals in recent lake sediments—proxy-indicators of deep-water oxygen and climate conditions*. PhD thesis, Eidgenössische Technische Hochschule, Zürich, 115 pp.

Schindler, D.W. (1974) Eutrophication and recovery in experimental lakes: implications for lake management. *Science*, **184**, 897–9.

Seki, H. (1972) The role of microorganisms in the marine food chain with reference to organic aggregates. *Memorie dell' Istituto Italiano Idrobiologia*, **29** (Supplement), 245–59.

Seki, H. (1982) *Organic Materials in Aquatic Ecosystems*. CRC Press, Boca Raton, 201 pp.

Siegenthaler, C. & Sturm, M. (1991) Die Häufigkeit von Ablagerungen extremer Reusshochwasser. Die Sedimentationsgeschichte im Urnersee seit dem Mittelalter. *Mitteilungen des Bundesamtes für Wasserwirtschaft*, **4**, 127–39.

Sigg, L. (1985) Metal transfer mechanisms in lakes; the role of settling particles. In: Stumm, W. (ed.), *Chemical Processes in Lakes*. Wiley, New York, 283–310.

Sigg, L., Sturm, M. & Kistler, D. (1987) Vertical transport of heavy metals by settling particles in Lake Zürich. *Limnology and Oceanography*, **32**, 112–30.

Simola, H., Sandman, O. & Ollikainen, M. (1992) Short-core palaeolimnology of three basins of the Saimaa lake complex. *Publications of the Karelian Institute, University of Joensuu*, **103**, 177–87.

Skei, J.M. (1992) A review of assessment and remediation strategies for hot spot sediments. *Hydrobiologia*, **235/236**, 629–38.

Sly, P.G. & Christie, W.J. (1992) Factors influencing densities and distributions of *Pontoporeia hoyi* in Lake Ontario. *Hydrobiologia*, **235/236**, 321–52.

Smayda, T.J. (1969) Some measurements of the sinking rate of fecal pellets. *Limnology and Oceanography*, **14**, 621–5.

Smayda, T.J. (1970) The suspension and sinking of phytoplankton in the sea. *Oceanography and Marine Biology Annual Review*, **8**, 353–414.

Smayda, T.J. (1971) Normal and accelerated sinking of phytoplankton in the sea. *Marine Geology*, **11**, 105–22.

Smayda, T.J. (1974) Some experiments on the sinking characteristics of two freshwater diatoms. *Limnology and Oceanography*, **19**, 628–35.

Smith, I.R. (1975) Turbulence in lakes and rivers. *Scientific Publications of the Freshwater Biological Association No. 29*, 79 pp.

Smith, I.R. (1982) A simple theory of algal deposition. *Freshwater Biology*, **12**, 445–9.

Smith, S.L., MacDonald, D.D., Keenleyside, K.A., Ingersoll, C.G. & Field, L.J. (1996) A preliminary evaluation of sediment quality assessment values for freshwater ecosystems. *Journal of Great Lakes Research*, **22**, 624–38.

Soetaert, K., Herman, P.M.J. & Middelburg, J.J. (1996) A model of early diagenetic processes from the shelf to abyssal depths. *Geochimica et Cosmochimica Acta*, **60**, 1019–40.

Spreafico, M., Bruschin, J., Geiger, H., *et al.* (1987) *Feststoffbeobachtung in schweizerischen Gewässern: Die mengen-mässige Erfassung von Schwebstoffen und Geschiebefrachten. Erfahrungen und Empfehlungen.* Arbeitsgruppe für operationelle Hydrologie, Mitteilungen No. 2, Bern, 91 pp.

Stabel, H.-H. (1985) Mechanisms controlling the sedimentation sequence of various elements in prealpine lakes. In: Stumm, W. (ed.), *Chemical Processes in Lakes.* Wiley, New York, 143–67.

Stadelmann, P. (1988) Zustand des Sempachersees vor und nach der Inbetriebnahme der see-internen Massnahmen: künstlicher Sauerstoffeintrag und Zwangszirkulation 1980–1987. *Wasser, Energie, Luft*, **80**, 81–96.

Stöckli, A.P. (1985) *Die Rolle der Bakterien bei der Regeneration von Nährstoffen aus Algenexkreten und Autolyseprodukten—Experimente mit gekoppelten, kontinuierlichen Kulturen.* PhD thesis, Eidgenössische Technische Hochschule, Zürich, 183 pp.

Stössel, F. (1989) Die Sanierung des Hallwilersees. Auswirkungen auf die Organismen des Seegrundes nach 2 3/4 Jahren Zirkulationshilfe und Tiefenwasser-Sauerstoffbegasung. *Wasser, Energie, Luft*, **81**, 333–5.

Stumm, W. & Morgan, J.J. (1996) *Aquatic Chemistry. Chemical Equilibria and Rates in Natural Waters* (3rd edition). Wiley, New York, 1022 pp.

Sturm, M. (1985) Schwebstoffe in Seen. *Mitt./Nouv. EAWAG News*, **19**, 9–15.

Sturm, M. & Lotter, A.F. (1995) Lake sediments as environmental archives. A means of distinguishing natural events from human activity. *EAWAG News*, **38E**, 6–9.

Sturm, M., Zeh, U., Müller, J., Sigg, L. & Stabel, H.-H. (1982) Schwebstoffuntersuchungen im Bodensee mit Intervall-Sedimentationsfallen. *Eclogae Geologicae Helveticae*, **75**, 579–88.

Sturm, M., Siegenthaler, C., Suter, H. & Wüest, A. (1996) *Das Verhalten von Schwebstoffen im Brienzersee.* EAWAG Report No. 84109, Swiss Federal Institute for Environmental Science and Technology, Duebendorf, 102 pp.

Syvitsky, J.P.M. (ed.) (1991) *Principles, Methods and Application of Particle Size Analysis.* Cambridge University Press, Cambridge, 368 pp.

Tengberg, A., de Bove, F., Hall, P., *et al.* (1995) Benthic chamber and profiling landers in oceanography—a review of design, technical solutions and functioning. *Progress in Oceanography*, **35**, 253–94.

Thomas, E.A. (1955) Stoffhaushalt und Sedimentation im oligotrophen Ägerisee und im eutrophen Pfäffiker- und Greifensee. *Memorie dell' Istituto Italiano Idrobiologia*, **8**(Supplement), 357–465.

Uehlinger, U. & Bloesch, J. (1987) The influence of crustacean zooplankton on the size structure of algal biomass and suspended and settling seston (biomanipulation in limnocorrals II). *Internationale Revue der Gesamten Hydrobiologie*, **72**, 473–86.

Urban, N.R., Dinkel, C. & Wehrli, B. (1997) Solute transfer across the sediment surface of a eutrophic lake: I. Porewater profiles from dialysis samplers. *Aquatic Sciences*, **59**, 1–25.

U.S.GOFS Working Group (1989) *Sediment Trap Technology and Sampling.* Report No. 10, U.S. Global Ocean Flux Study, August 1989. Woods Hole Oceanographic Institution, Woods Hole, 94 pp.

Valeur, J.R. (1994) Resuspension—mechanisms and measuring methods. In: Floderus, S., Heiskanen, A.-S., Olesen, M. & Wassmann, P. (eds), *Sediment Trap Studies in the Nordic Countries*, No. 3. Nurmi Print Oy, Nurmijärvi, 184–202.

Wagner, G. (1976) Die Untersuchung von Sinkstoffen aus Bodenseezuflüssen. *Schweizerische Zeitschrift für Hydrologie*, **38**, 191–205.

Wagner, G. & Bührer, H. (1989) Die Belastung des Bodensees mit Phosphor- und Stickstoffverbindungen, organisch gebundenem Kohlenstoff und Borat im Abflussjahr 1985/86. *Bericht der internationalen Gewässerschutzkommission Bodensee*, **40**, 52 pp.

Wagner, G. & Wagner, B. (1978) Zur Einschichtung von Flusswasser in den Bodensee-Obersee. *Schweizerische Zeitschrift fuer Hydrologie*, **40**, 231–48.

Wakeham, S.G., Hedges, J.I., Lee, C. & Pease, T.K. (1993) Effects of poisons and preservatives on the composition of organic matter in a sediment trap experiment. *Journal of Marine Research*, **51**, 669–96.

Wan, G.J., Santschi, P.H., Sturm, M., *et al.* (1987) Natural (^{210}Pb, ^{7}Be) and fallout (^{137}Cs, ^{239,240}Pu, ^{90}Sr) radionuclides as geochemical tracers of sedimentation in Greifensee, Switzerland. *Chemical Geology*, **63**, 181–96.

Wassmann, P. (1990) Relationship between primary and export production in the boreal coastal zone of the North Atlantic. *Limnology and Oceanography*, **35**, 464–71.

Watmough, S.A., Hutchinson, T.C. & Evans, R.D. (1997) Application of laser ablation inductively coupled plasma-mass spectrometry in dendrochemical analysis. *Environmental Science and Technolology*, **31**, 114–8.

Wehrli, B. & Wüest, A. (1996) *Zehn Jahre Seenbelüftung: Erfahrungen und Optionen*. EAWAG Report No. 9, Swiss Federal Institute for Environmental Science and Technology, Duebendorf, 157 pp.

Weissmahr, K.W., Haderlein, S.B., Schwarzenbach, R., Hany, R. & Nüesch, R. (1997) *In situ* spectroscopic investigations of adsorption mechanisms of nitroaromatic compounds at clay minerals. *Environmental Science and Technolology*, **31**, 240–7.

Wetzel, R.G. (1983) *Limnology*, (2nd edition). Saunders, Philadelphia, 767 pp.

Weyhenmeyer, G.A. (1996) The influence of stratification on the amount and distribution of different settling particles in Lake Erken. *Canadian Journal of Fisheries and Aquatic Sciences*, **53**, 1254–62.

Weyhenmeyer, G.A. (1997) Quantification of resuspended particles in sedimentation traps. *Verhandlungen der internationalen Vereinigung für theoretische und angewandte Limnologie*, **26**, 271–6.

Weyhenmeyer, G.A. & Bloesch, J. (2001) The pattern of particle flux variability in Swedish and Swiss lakes. *The Science of the Total Environment*, **266**, 69–78.

Weyhenmeyer, G.A., Meili, M. & Pierson, D.C. (1995) A simple method to quantify sources of settling particles in lakes: Resuspension versus new sedimentation of material from planktonic production. *Marine and Freshwater Research*, **46**, 223–31.

Weyhenmeyer, G.A., Håkanson, L. & Meili, M. (1997) A validated model for daily variations in the flux, origin and distribution of settling particles within lakes. *Limnology and Oceanography*, **26**, 1517–29.

Wieland, E., Santschi, P.H. & Beer, J. (1991) A multitracer study of radionuclides in Lake Zürich, Switzerland. 2. Residence times, removal processes, and sediment focusing. *Journal of Geophysical Research*, **96**, 17067–80.

Wieland, E., Santschi, P.H., Höhener, P. & Sturm, M. (1993) Scavenging of Chernobyl ^{137}Cs and natural ^{210}Pb in Lake Sempach, Switzerland. *Geochimica et Cosmochimica Acta*, **57**, 2959–79.

Wright, R.F., Matter, A., Schweingruber, M. & Siegenthaler, U. (1980) Sedimentation in Lake Biel, an eutrophic, hard-water lake in northwestern Switzerland. *Schweizerische Zeitschrift für Hydrologie*, **42**, 101–26.

Wüest, A. (1987) *Ursprung und Grösse von Mischungsprozessen im Hypolimnion natürlicher Seen*. PhD thesis, Eidgenössische Technische Hochschule, Zürich, 144 pp.

Wüest, A. & Wehrli, B. (1994) Mixing and water quality in lakes: Does mixing alter phosphorus retention and productivity? In: Sund, H., Xiaogan, Y. Stabel, H.-H., Kechang, Y., Geller, W. & Fengning, S. (eds), *Environmental Protection and Lake Ecosystems*. China Science and Technology Press, Beijing, 109–28.

Xue Han Bin, Gächter, R. & Sigg, L. (1997) Comparison of Cu and Zn cycling in eutrophic lakes with oxic and anoxic hypolimnion. *Aquatic Sciences*, **59**, 176–89.

9 Organisation and Energetic Partitioning of Limnetic Communities

COLIN S. REYNOLDS

9.1 INTRODUCTION

The essential purpose of this chapter is to provide a general context for those (Chapters 10–16) concerned with the limnetic biota and those which make up the remainder of Volume 1. There are precedents for sequencing the subject matter in any of several particular ways: for instance, it might be arranged by phyletic affinity and in conventional order of ascending complexity; or by sub-habitat (shores, deep water, etc.); or, more usually, according to a functional classification (phytoplankton, zooplankton, macrophytes, etc.). None is wholly superior to any other in a review of the depth and complexity sought in this book. Moreover, the sequence of topics should not convey a false implication of the relative importance of the various biota thus circumscribed, neither should some preferred interdependence of the components be incorporated.

The objective is therefore to review the broad structure and general organisation of limnetic communities and their roles in the energy flow and material processing through limnetic ecosystems. It seeks to emphasise the key linkages among them. So, like the lines of the Homeric epigram reputedly inscribed on the tomb of Midas

A girl in bronze, on Midas' grave I'll go,
As long as water springs and great trees grow
While rising sun and gleaming moon still glow
While rivers run, and waves wash to and fro
Set on this tomb where tears in torrents flow
I'll tell who passes: Midas lies below

the chapters should be readable in any order, without compromise to their theme. The allocation of the subject matter among Chapters 10–16 follows the third orthodoxy but the sequence of chapters is one of editorial whim, not knowingly based upon any precedent. Its logic is, broadly, to follow the pathways along which reduced carbon is passed, from the sites of its fixation time to the point of its ultimate reoxidation.

The sequence relates to the theme of the present chapter on the collective functioning of lake biota. It first considers, empirically wherever possible, the general constraints upon photoautotrophic producers in lakes and how the problems these then pose to the heterotrophic consumers combine to influence the nature and function of the whole limnetic community.

9.2 STRUCTURAL ORGANISATION OF LIMNETIC ENVIRONMENTS

Previous chapters have highlighted the great variety of lakes, in terms of the origins of their basins, their ages and their transience, their hydraulic characters (loading, flushing, delivery of catchment leachates) and their morphometries. Each lake is truly unique in the combination of its distinctive properties. This should never be forgotten when the results of a detailed study of ecosystemic function in one lake are extrapolated to the behaviour of a second. On the other hand, the basic organisation of living systems can be readily reduced to a fairly simple statement about the roles of primary producers, consumers and decomposers in continuously processing carbon and other elements, driven essentially by energy harvested from the sun. Does not this recognition restore a generality of pattern against which we may measure the assembly and functional organisation of the components of limnetic ecosystems?

Of course, the answer must be 'yes', but it is an affirmation which must also be qualified by the condition that both community structure and function are profoundly influenced by the attributes of particular lake environments. Thus, when the powerful effects of the physical properties of water and the solvent properties governing the development of the chemical environment are translated to the lives of aquatic biota, **and** to how much material they can process for the consumption of how much energy, it is essential to embrace an element of scaling and a notion of how limnetic habitats are morphometrically sub-divided. Intuitively, at least, we may separate large from small or deep from shallow, though we may argue about the empiricism of the transitions. It is probably more helpful, however, to adopt a picture of interlocking habitats, albeit differentially represented among individual lakes. The labels applied to these habitat zones—pelagic, bathypelagic, abyssal, benthic, littoral (see Fig. 9.1)—have been borrowed from oceanography, although their original definitions are often quite inappropriate to lakes (the defining bounds of the marine littoral are meaningless in tideless inland waters, while its separation from the truly open waters of the pelagic by the coastal and shelf areas of the neritic zone, where interactions persist between the water and the adjacent land and the bottom sediment, is not recognised in limnology). Except in the world's truly great lakes, the functional isolation of the limnetic pelagic zone from its coasts and bottom sediments is nowhere near so complete as it is in the open ocean.

Unsatisfactory though these borrowed terms may be, they are much too entrenched for replacements to be advocated. In this chapter, the littoral region is considered to be that in which the bottom sediments and its benthos are subject to frequent or continuous shear stress from surface waves or onshore currents. The depth of water will vary with the fetch and exposure but ≤5 m will provide a provisional estimate. Active fluxes of materials and solutes between the water and the substratum ensure that, functionally, they remain mutually inseparable. Fine particles are readily entrained and transported away, to be 'focused' in the unstressed sediments of deeper water (Likens & Davis 1975). Beyond the littoral, the bottom sediment and the benthos are progressively separated physically from the surface water. Well away from the shores (tens of kilometres), we approach the limnetic equivalent of the true pelagic: open water, unprotected from wind, and subject to unrestricted mechanical wind-stirring down to 200 m or so, though, unlike the sea, without the assistance of tidal surge. A bathypelagic layer separates the pelagic from the benthos, though even among deep lakes thermal convection currents often provide a significant transport and gas-exchange mechanism. Most material cycling takes place in the water column.

In between (water of 5–200 m deep and only intermittently or seasonally mixed to the bottom), sediments are more or less protected from shear stress by boundary layer formation. Accretion is by sedimentary recruitment, although the metabolites of diagenesis and inorganic decomposition contribute to internal chemical recycling. In the open water, geochemical and biotic activity involves materials derived mainly externally (from the adjacent land) and, so, is functionally more similar to the neritic zone of the sea.

9.3 PHOTOAUTOTRPHY IN LIMNETIC ENVIRONMENTS

Organisms which increase their mass through the accumulation of the proteins which they manufacture themselves from simple inorganic radicals, rather than by ingesting the assembled proteins of another, are called **autotrophs**. Those which derive the necessary energy from harvested sunlight are described as being **photoautotrophic**. Photoautotrophy is a trait of most pigmented plants and protists. Of the several, essentially anabolic processes that characterise them, the most fundamental is, of course, photosynthesis. Through a complex series of controlled reactions, electrons are first stripped from a donor substance (usually water, which is photochemically oxidised to oxygen). In the second stage, light energy is used to transfer electrons to carbon dioxide. Photosynthate can be subsequently elaborated in to proteins, or it is carefully and progressively reoxidised to provide the energy to drive the mediating biochemical apparatus. Assembly of the cell architecture of the photoautotroph requires up to 20 further elements (or **nutrients**, including especially hydrogen, oxygen, nitrogen, phosphorus and sulphur), all of which must be derived from the adjacent media.

It is an axiom that there are thus four requirements for photoautotrophic organisms: a supply of water, a flux of carbon dioxide, an adequate intensity of light of the appropriate wavelength and a resource base of nutrients to match the demand of carbon-skeleton assembly. While all four requirements are satisfied, the photoautotroph can be expected to increase its mass and at a finite rate of assembly. This is likely to be strongly temperature-dependent, although the growth of aquatic plants is recognised to be otherwise allometrically scaled to organismic size and surface–volume relationships (Nielsen & Sand-Jensen 1990). A corollary, which is not generally recognised, is that where a sufficiency of energy and material resources obtains, we should expect the rates of their collection to be regulated by the rate of growth and not vice versa.

As is well-known, however, light, carbon and nutrients are rarely present in simultaneous excess (we may ignore, in this context, the onset of water deficiency, which can be caused by osmotic gradients). Many aspects of the ecology of aquatic photoautotrophs relate to discontinuities in their distribution and oscillations in their availability (see Chapters 10 and 11). The concern here is how these factors influence the principal life-forms present in limnetic environments, the kinds of special adaptations that may be selected to deal with these constraints and, in turn, the ways in which the responses condition the main avenues of carbon transfer in lakes.

The relationships governing the carrying capacities and replication rates of primary producers are relatively well understood. For instance, in the case of light, the keystone quantity is the **quantum yield of photosynthesis**. 'Light', of course, refers to the visible and photosynthetically-active wavelengths ($\lambda \approx 400$–700 nm) of electromagnetic radiation emanating from the sun, which band accounts for almost half (46–48%) the solar energy reaching Earth. Light beams comprise a continuous flow of indivisible **quanta** (or, in the visible spectrum, **photons**), travelling at a constant velocity, c, of $\approx 3 \times 10^8$ m s^{-1} (differences in wavelength are reciprocal to frequency: $v = c/\lambda$). Thus, light intensity (I) refers to the number of photons arriving at a surface normal to the beam per unit time, or the **photon-flux density** (PFD). Against this flux, the light-harvesting apparatus is highly conserved among all photoautotrophs. Photons are intercepted by specialised membranous organelles, the thylakoids. These occur within the discrete, membrane-bound chloroplasts of eukaryotic algae and all higher plants; in the Cyanobacteria, they are distributed through the body of the cell. Strung out along the thylakoids are the two types of granular structures corresponding to the two associated photochemical systems (PS I and II). Both comprise complexes of up to around 300 molecules of the photosynthetic pigment, chlorophyll, the associated molecules of certain specific proteins, and, usually, some other pigments. Both harvest photons (one, PS II, to strip electrons from water, the other, PS I, to transfer electrons to carbon dioxide) in a light-harvesting core (LHC). The quiescent structure is open to an arriving photon, but the excitation reaction triggered by its reception raises an electron in the LHC chlorophyll from its ground state to its high orbital, whence it is transported down the gradient of reduction–oxidation reactions making up the long-accepted 'Z-scheme' model of photosynthesis (Hill & Bendall 1960). The reaction centre cannot accept another electron until it has been reoxidised and remains closed until the electrons are transferred to the next stage. The time of clearance is temperature dependent (Kolber & Falkowski 1993): around 66 reactions per second may be accommodated at near-freezing temperatures and up to 500 at around

30°C (note, close to one doubling in the rate for every 10°C), if the next electron was always to arrive at the reaction centre the instant that it re-opened.

The molecular biology of photosynthesis and the elaborate mechanisms for its regulation and photoprotection are not within the scope of this chapter (the reader is referred to the excellent reviews of Walker 1992; Demmig-Adams & Adams 1992; Long *et al.* 1994). The foregoing is included to show the productive capacity of the aquatic habitat. The stoichiometry of the Z-scheme requires the harvest of four electrons per photosystem to secure one atom of carbon: hence, it takes 8 mol photons to fix 1 mol carbon; the theoretical quantum yield of photosynthesis is (ϕ) 0.125 mol C (mol photon)$^{-1}$. Attempts to measure ϕ, in a variety of photosynthetic organisms, have come up with values below this theoretical maximum: most determinations for algae fall between 0.07 and 0.09 mol C (mol photon)$^{-1}$ (Bannister & Weidemann 1984; Walker 1992). Nevertheless, at the maximum photon flux density attainable (about 2.3 mmol photons m^{-2} s^{-1} at the equatorial solar zenith, during very dry and cloudless conditions), gross carbon fixation cannot exceed 0.29 mmol C, or 3.45 mg C m^{-2} s^{-1} ($\approx$ 12.4 g C m^{-2} h^{-1}, or a little over 100 g C m^{-2} day^{-1}). To attain anything approaching this yield means, literally, that every photon must be intercepted by an open reaction centre. At 30°C, the requisite (2.3 mmol × 6.023 × 10^{23} photons mol^{-1} $\approx$) 14 × 10^{20} reactions m^{-2} s^{-1} could be achieved by $\approx$ 3 × 10^{18} reaction-centres m^{-2}, involving some 6–9 × 10^{20} molecules of chlorophyll. From the molecular weight of chlorophyll (894 daltons, or $\approx$ 148.4 × 10^{-23} g), we may deduce that an active mass of chlorophyll of 1 g m^{-2} approximates to the self-shaded carrying capacity. Adopting the conventional carbon-chlorophyll conversion supposed by plankton biologists, this capacity approximates to an active (i.e., ignoring mechanical or accessory tissues) biomass of 50 g cell C m^{-2}. We may further deduce a carbon-specific carbon fixation rate of (3.45 mg C m^{-2} s^{-1}/ 50 g cell C m^{-2} =) 0.069 mg C (g cell C)$^{-1}$ s^{-1}. In turn this is equivalent to a theoretical growth rate, uncorrected for respiration, of 69 × 10^{-6} s^{-1}.

The empiricism of this relationship is purely illustrative: the carrying capacity is less at a lower temperature and at the lesser photon-flux densities experienced at other times of day and at other latitudes. Moreover, the differential absorption by the water itself of photons with different wavelengths (see Kirk 1994) and in accordance with Beer–Lambert laws of exponential attenuation with vertical distance further contributes to the variability of the photoautotroph carrying capacities of aquatic environments. A corollary is that the deeper the biomass is situated, the less of it can be supported. Even the clearest lake waters fail to support net photoautotrophic growth at depths much beyond 50 m (Cabrera & Montecino 1984). Thus, a more realistic quantitative estimate of the typical carrying capacity of the open water of otherwise clear lake waters is 0.2–0.6 g chlorophyll m^{-2} (Reynolds 1984: say, 10–30 g C m^{-2}). Even the filling of the capacity available is markedly nonlinear, owing to the consumption of energy in the maintenance of accumulated biomass, to which proportionately more of the intercepted light energy is directed as the capacity is approached (Reynolds 1997). Actual experimental estimates of integral photosynthesis, mostly in the order 0.5–5 g C m^{-2} day^{-1} (see Ahlgren 1970; Kalff & Knoechel 1978), are far below the theoretical maximum (above), even when allowance for temperature is made.

The carbon requirement of submerged photoautotrophs is met primarily from the pool of carbon dioxide dissolved in the adjacent water. This pool is maintained partly by gradient-driven invasion of carbon dioxide across the water surface, partly by the carbon dioxide dissolved in inflows (including bicarbonate ions) and partly by the photolysis and photo-oxidation of dissolved organic matter. Biotic respiration of carbohydrate is a further component. Only macrophytes with aerial leaves and stems are able to derive carbon dioxide directly from the atmosphere. The high solubility of the gas in water compensates, in part, for the low partial pressure exerted by its atmospheric concentration (3 × 10^{-4} atmosphere). In the biologically relevant temperature range (0–35°C), the air-equilibrated concentration in pure water varies between 0.5 and 1.0 g CO_2 m^{-3}. If it is supposed that the carbon-dioxide water supplied to a lake basin is air-equilibrated, its nominal supportive capacity is 0.13–0.27 g active-biomass $C\,m^{-3}$. Not all the carbon dioxide invades across the water surface. Some is derived from the internal biotic metabolism (respiration and decomposition, and from the photolysis of dissolved organic matter). This capacity may be considerably reinforced and regulated by equilibria involving dissolved bicarbonate. Circumneutral equilibrium concentrations of assimilable carbon across a spectrum of alkalinities of from 0.4 to 4.0 meq L^{-1} are in the range 3–35 g $CO_2\,m^{-3}$.

Whatever the size of the pool of dissolved carbon dioxide, however, the response to its photosynthetic withdrawal is that more carbon dioxide invades across the air–water interface. Thus, there is practically no upper limit on the carrying capacity of aquatic environments which is regulated by carbon availability, but the rate of air–water exchange ultimately places an important assembly-rate limitation at high levels of biomass. The flux is not easy to assess, though it is theoretically the product of the solubility of the gas, the difference in partial pressure either side of the interface and the coefficient of gas exchange (roughly the linear migration rate). The outcome is greatly affected by the wind stress on the water surface. Field determinations (for instance, by Frankignoulle (1988) and Watson *et al.* (1991)) confirm that, at the large partial-pressure differences associated with high aquatic carbon demand, the invasion flux approaches $30 \times 10^{-8}\,mol\,m^{-2}\,s^{-1}$ under a wind of $15\,m\,s^{-1}$ but at winds $<3.5\,m\,s^{-1}$ it is an order of magnitude slower. These values correspond to daily fluxes of between 0.03 and 0.3 g $C\,m^{-2}$ day^{-1}. Using the conversion above (C/chlorophyll $=50$), this is theoretically sufficient to sustain a carbon-limited productive increment of between 0.6 and 6 mg chlorophyll $m^{-2}\,day^{-1}$.

The remaining precursor to photoautotrophic accumulation is the nutrient supply. Of the 20 or so nutrients required, in specific and largely invariable relative quantities, the four for which demand most often exceeds supply in limnetic environments (nitrogen and phosphorus; the reputed deficiency of iron in the open ocean occasionally applies to lakes, too, while an inherent skeletal requirement of diatoms and grasses for silicon poly-

mers is capable of exhausting the supply to the system) are considered critical. Moreover, as the silicon requirement is not universal, and as dinitrogen fixers are ostensibly released from limitation by dissolved inorganic species such as nitrate and ammonium, the nutrient-determined carrying capacity of a lake is perceived to be crucially influenced by the availability of phosphorus (Vollenweider 1968; Schindler 1977).

Because, in general, the molecular composition of protoplasm conforms approximately to the Redfield stoichiometry (which supposes 106 carbon atoms for each phosphorus atom), it is simple to deduce that approximately 1.2 g intracellular phosphorus is required in order to comprise the active biomass of a self-limiting population equivalent to $50\,g\,C\,m^{-2}$. However, the natural loads of phosphorus delivered to many lakes from their catchments are often typically $<0.1\,g\,P\,m^{-2}$ per year. In many of these instances, much of what is delivered is bound in calcium salts, clay minerals or occluded in ferric hydroxide flocs, which, pending some chemical dissociation, remain unavailable to plants. Planktonic algae possess very effective scavenging mechanisms, most species probably being able to operate facultatively at concentrations of $<10^{-7}\,mol$. They are quick to respond to an increased availability of phosphorus (or any other nutrient, whilst it is limiting the standing-crop biomass), although the maximal populations sustainable for a given dose of available phosphorus tend not to be in direct proportion (Reynolds 1992).

Availability of nutrients in a lake is also subject to internal processes. Concentrations of non-conservative solutes in the inflowing water are bound to be diluted as they are dispersed into the main circulation. Moreover, nutrients in or quickly transferred to the particulate fraction settle more or less promptly, though such particles are much more likely to be re-entrained from the littoral benthos by wave stress. In this way, there is always likely to exist a declining centripetal gradient, from the shore to open water, in the availability of nutrients (Fig. 9.1). The poverty of photoautotrophy in deep water (see above), the consequent low nutrient demand and a balance in favour of nutrient regenera-

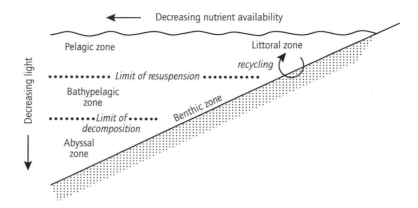

Fig. 9.1 Diagrammatic section through a (very large) lake to show the main sub-habitats (zones) differentiated in the chapter and to show the main gradients of increasing limitation of primary production.

tion by heterotrophs may together contribute to a vertical gradient of increasing availability with depth. Transformation at low redox levels (of such as Fe^{3+} to Fe^{2+}, with the concomitant discharge of any occluded phosphate) may exaggerate this effect. Against this background, the effects of artificial enrichment (lake eutrophication) can be seen to promote a series of positive feedbacks. More deep-water decomposition facilitates greater prevalence of low-redox dissociation, while physical recycling in the littoral zone assists a certain cumulative effect of external nutrient loading (Sas 1989).

Whether bioavailable carbon is supplied to the lake primarily as dissolved carbon dioxide, as degradable organic solutes or in solution as bicarbonate, the regulation of primary production by available carbon is potentially analogous to that by any other nutrient, such as phosphate, but for the direct replenishment through invasion of atmospheric carbon dioxide across the water surface. Photoautotrophic demand may well exceed the capacity of the dissolved pool, the upward flux of gases from the decomposition of biogenic material and, under many circumstances (Watson *et al.* 1991), gaseous invasion across the air–water interface. At least under these conditions, carbon availability is likely to be greater with increasing water depth.

It is nevertheless interesting to deduce that the estimates of annual areal production based on measurements of carbon fixation rate (100–1000 g $C m^{-2} yr^{-1}$; Ahlgren 1970; Kalff & Knoechel 1978)

significantly exceed the maximum biomass supported at any one time. This means either that the turnover of limnetic biomass is extremely dynamic (recruiting 3–30 maximum crops per year) or, as deduced by Forsberg (1985), and Reynolds *et al.* (1985), that large amounts of fixed carbon are never incorporated into new algal tissue at all but are rather metabolised to extracellular products (e.g. glycollate) or photorespired carbon dioxide. Reynolds *et al.* (1985) found that between 18 and 97% of the carbon fixed each day by the phytoplankton community was directed other than to daily cell growth. This carbon is not necessarily all lost to limnetic biomass, as the organic exudates contribute part of the substrate for limnetic bacteria (Bird & Kalff 1984; Vadstein *et al.* 1993).

These various considerations help us to understand the factors governing the scale of primary production encountered in limnetic habitats. The simple section of a lake, from margin to deep-water offshore regions, may be characterised by a gradient of the weakening influence of the land and shallow-water processes, including nutrient inwash and mechanical regeneration cycling and a greater depth of water deprived of light.

9.4 MORPHOLOGICAL ADAPTATIONS OF LIMNETIC AUTOTROPHS

What is the optimum life-form for the limnetic photoautotroph, harvesting light and garnering nutrients in the body of the lake? How are the basic requirements attained against the properties of lakes, not all of which are especially favourable to autotrophs? The answers are not simple but the questions are fundamental and worthy of further development.

Beginning with the more familiar examples provided by terrestrial plants, it is not difficult to recognise those relevant functional and evolutionary constraints which have typically led to the development of the structures we recognise. Efficient light interception obviously requires provision of a physically extensive surface, but one which is simultaneously thin (to facilitate gas and vapour exchanges), elevated (to be higher than those of neighbours), yet is dissected (to lower drag resistance to the passage of the wind), and is arranged to maximise light interception. It will also need to possess an appropriate means of mechanical support, irrigation and nutrient transport to the photosynthetic surface. Before long, the design for living quickly comes to resemble a tree.

For the submerged aquatic macrophyte, the Archimedean support imparted by the density of water reduces reliance upon load-bearing mechanical tissues. On the other hand, water in motion, associated with wind-forcing, wave generation and turbulent eddying, requires plants both to be flexible and tough if, habitually, they are frequently exposed to such stresses. In this case, the photosynthetic surface presented may still share some of the arboreal properties: for example, the architecture of dendritic pondweeds conforms to the generality, although the leaves are frequently filiform rather than blade-like. A rooting system helps to keep the plant at a fixed location; it also offers access to gas and nutrients other than those in the water, though the water-conducting function is less vital than on land. Emergent plants resemble still more closely their terrestrial counterparts: photosynthesis may take place predominantly in the aerial positions which must be self-supporting and, so, are frequently fibrous or woody; provision for ventilation of the submerged parts (especially the roots) is achieved through the linear vacuities of aerenchyma (Sculthorpe 1967). Some form of tidal flow, driven presumably by diel expansion and contraction of

air, must be invoked in order to amplify simple gaseous diffusion.

Emergent and, especially, submerged macrophytes live successfully some distance into the water but the limiting condition imposed by attenuation of downwelling light means that smaller plants receive too little light to grow. One alternative, that of maintaining leaves near, at, or above the water surface, is also quickly obviated because the length of anchoring tissue required in deep water soon becomes prohibitive. A third possibility, to abandon anchorage altogether, seems not to have been widely adopted among plants: mostly, it is a trait of floating plants (variously exemplified by *Lemna*, *Azolla*, *Hydrocharis* and *Eichhornia*, all of which are originally confined to water-bodies (or parts thereof) not normally liable to frequent or excessive wind stress). Supposing unrooted, disconnected structures make teleological sense, why are not pelagic plants like carpets or footballs drifting around beneath the water surface, in the mode of the seaweed, *Sargassum*?

The most likely explanation is that the requirement for skeletal toughening, to provide resistance to the persistent tendency of turbulence to tear, crumple or roll the photosynthetic tissue, is substantially incompatible with its obviation in response to density. Yet it is possible to gain the advantage of the support provided by the density of the water **and** to escape the impact of turbulence, within which the smallest eddies propagated are generally not smaller than 0.5 mm (Reynolds 1994, 1997), before they are overwhelmed by the viscosity of the water. This viscous range of the eddy spectrum is the province of the phytoplankton.

The argument is teleological, but the contiguity of the scales of motion and organisms, in the sea as well as in lakes, is compelling. Many aspects of the suspension, nutrient uptake and even vulnerability to feeders invoke theory that depends upon the viscosity of water. The integrity of the microbial loop relies on the elaboration of structurally intact consortia of microorganisms which, necessarily, must survive (more likely, escape) persistent turbulent dispersion (Azam *et al.* 1995). However, the consequent 'embedding' of phytoplankton in the turbulence field gives rise to another inevitability, which is that it is liable to carriage throughout the turbulent layer—in other words, it goes wherever the water takes it, including beyond the limits of the photic zone and down the river draining the lake. At the same time, small size imparts to planktonic producers the virtue of a potentially much more rapid rate of replication than macrophytes (Nielsen & Sand-Jensen 1990): specific growth rates in the order of 1–3 day^{-1} are not at all unusual). Other factors being equal, phytoplankton is able to exploit growth opportunities (warming and thermal stratification, increasing daylength, nutrient enrichment) quickly and efficiently. The microscopic life-form is appropriate to the pelagic primary producers.

Differences in the communities based upon macrophytic and planktonic producers (effectively those of the littoral and of the open pelagic) may be considered to be self-evident. Nevertheless, it is important to emphasise that the difference in size ranges affects not just potential biomass assembly rates (<0.11 day^{-1}; Nielsen & Sand-Jensen 1990) but the proportion of the standing crop that is eventually mechanical as opposed to functional is much greater. Thus, macrophyte biomass carried per unit area, including the skeletal tissue and the rooting systems, is much greater than the theoretically self-shaded capacity of $50\,g\,C\,m^{-2}$ but it takes months rather than weeks to assemble. Moreover, the size range of macrophytes is such that, for many limnetic species of invertebrates (and even of fish, birds and mammals), stands of littoral plants provide habitat as well as a supply of fixed carbon. The external surfaces of macrophytes are potentially colonisable by epiphytes, such as diatom lawns and gardens of filamentous and dendritic algae. Together with associated microorganisms, sponges, bryozoans and rotifers, the epiphytic flora contributes to the community of the **Aufwuchs**. Like the phytoplankton, epiphytic algae are effectively dependent for their nutrient supplies upon the renewal of solutes in the adjacent water; the rooting systems of macrophytes provide access to an alternative or even primary source of essential nutrients, namely those of the underlying soil—the littoral sediment and its interstitial water.

Considerable attention has been focused on the littoral and immediate sublittoral areas of ponds and lakes, where the respective adaptations of macrophytes and plankton permit them to co-exist; particular interest has been directed towards those factors which influence the variable dominance of the one life-form over the other (see especially Blindow *et al.* 1993; Scheffer *et al.* 1993). The effects on the ecology and perceived quality of the water body are profound. Understanding the complex relationships involved is therefore important to the adoption of appropriate management strategies. As the relationships involve the activities of herbivores, it is helpful first to consider the impact of such disparity of life-forms of alternative primary producers on consumer networks.

9.5 HETEROTROPHY IN LIMNETIC ENVIRONMENTS

The traditional view of food chains in the pelagic zone of lakes is based upon the idea that planktonic herbivores, mainly comprising calanoid and cladoceran crustaceans and rotifers, graze luxuriously upon the planktonic autotrophs, and that the principal reason that they do not eventually eliminate the food supply is that they are effectively controlled by pelagic-feeding planktivorous invertebrates (the decapod, *Mysis,* is one of the examples most cited) and planktivorous and piscivorous fish (generally salmonids and coregonids figure in high-latitude large lakes; clupeids fulfil the planktivorous role in Lake Tanganyika; cichlids in Lake Malawi; see Lowe-McConnell 1975). The same tradition regards the capacity of carbon transfer through the system as a direct function of the carbon-fixation capacity of the primary producers, which, as has been indicated, may sometimes possess the potential to double its mass each day.

The subsequent chapters of this volume reveal how far it has been necessary to revise traditional perceptions in the light of specialist investigations at each level. It is not that the importance of direct consumption of phytoplankton by zooplankton, or of zooplankton by fish, has been invalidated in any way, so much as the fact that it can be placed in a rather wider contextual view of the variety of routes by which carbon is transferred through limnetic networks. It is now clear, for instance that: (i) planktonic primary products are also processed by microbial and protist communities, including those of benthic sediments; (ii) those of macrophytes are exploited largely through decomposer networks; (iii) the contrasted foraging strategies of different components of the zooplankton are decisive in directing the development of pelagic community structure; and (iv) through analogous feed-backs, the food preferences of the main species of fish influence the composition and function of the supportable biota of the entire limnetic system. Moreover, the transport to lakes from their catchments of a variety of fulvic and humic acids may sustain limnetic concentrations of dissolved and particulate organic carbon that exceed, sometimes considerably, the concentrations of live, organismic organic carbon (Wetzel 1995; see also Chapter 7). This allochthonous dissolved organic carbon (v) should also be supposed to contribute to the biological productivity of the open water, though its persistence, even in the presence of substantial populations of microbial heterotrophs, is not indicative of any rapid turnover.

9.6 HETEROTROPHIC OPPORTUNITIES IN THE PELAGIC ENVIRONMENT

Taking the first of these five points, the assertion recognises that the fastest chlorophyll-specific photosynthetic rates of natural phytoplankton drawn from mixed limnetic layers but measured in fixed bottles often exceed the rate at which the carbon thus fixed can be consumed by the demands of growth, even in nutrient-replete environments (Reynolds *et al.* 1985). True, in deep, light-deficient layers or in deep-mixed layers, where the growth rate is demonstrably limited by the aggregate daily light-dose, almost all available fixed carbon will, by definition, be directed to biomass assembly (Talling 1957). Moreover, cells grown in continuous high light actually reduce their specific chlorophyll content, so that the rate of biomass-specific carbon fixation comes into line with the biomass-specific rate of its consumption in growth: the cell adjusts its light harvesting and

photosynthesis in order to match the growth rate and not, as had usually been assumed, the other way around. In a variable environment, such precise matching is not easily achieved; rather, cells use a range of homeostatic mechanisms in order to adjust the internal environment to deal with the accumulation of unstorable photosynthate and of potentially lethal photosynthetically produced oxygen (Demmig-Adams & Adams 1992; Long *et al.* 1994). Especially among green algae (Chlorophyta), some of this organic carbon is reoxidised to carbon dioxide, but many species of plankton counter the accumulation of organic intermediates, or convenient alternative sink metabolites, by voiding them to the exterior. The secretion of such substances, especially glycollate, was recognised, well before its significance was understood, as 'extracellular production' (Fogg 1971). This is particularly evident among nutrient-deficient populations (when little net synthesis of new biomass is possible) or in erstwhile deep-mixed populations trapped near the surface by a sharp relaxation of wind-stirring.

Much of this so-called extracellular production of excess organic carbon becomes the substrate of planktonic bacteria (Cole *et al.* 1982; Chróst 1983; Chróst & Faust 1983; Sell & Overbeck 1992). As is evident from Chapter 13, there is no reason to assume that the microbial community is any less diverse than is the algal plankton, either in its composition or in its metabolic specialisms (see also Jones 1987; DeLong *et al.* 1993). The typical microbial biomass ($\approx 0.01–1\,\mathrm{g\,C\,m^{-3}}$) is also often comparable to that of the phytoplankton ($\approx 0.01–10\,\mathrm{g\,C\,m^{-3}}$), although the potential replication rates of the former are faster. It must be emphasised, however, that there can hardly be more heterotroph carbon assembled than there is autotroph carbon made available. Bacterial production is just as likely to be constrained by nutrient shortage as the phytoplankton; whereas bacterial resource gathering is, in general, at least as efficient as that of planktonic algae (Bratbak & Thingstad 1985), production remains dependent upon the continued production of phytoplankton and its ability to supply sufficient fixed carbon. It is not at all clear that the allochthonous fulvic- and humic-carbon imported from the catchments off-sets the demand for the simpler, low-molecular weight organic intermediates produced *in situ*. The principle underpins the quasi-steady state that is struck among the components of true open-water pelagic communities. Thus, while energy and resource limitation keeps the phytoplankton biomass to $<2\,g\,C\,m^{-2}$, the yield of carbon to bacteria is not likely to sustain the turnover of a bacterial biomass of more than a further 0.2 to 2 g C m^{-2}. Simply, a greater biomass than this would deprive the phytoplankton of nutrients and, in turn, reduce the supply of usable organic carbon. We may deduce that, for a surface layer of some 10 m depth, the average concentration of standing bacterial carbon is equivalent to 0.02 to $0.2\,g\,C\,m^{-3}$, which, according to the much-cited equivalence devised by Lee & Fuhrman (1987), matches the 10^{6}–$10^{7}\,mL^{-1}$ bacterial cells commonly reported in the plankton of oligotrophic lakes (say 10^{3}–10^{4} in a 1-mm^{3} viscous patch).

Consumption of bacteria by micro-zooplankton completes this **microbial loop** in the pelagic plant–animal linkage. The concentration range is hardly guaranteed to satisfy the demands of active filter-feeders, however: the minimum require-ments of obligately filter-feeding Daphniid species, just avoiding starvation, is close to 0.1 µg C ml^{-1} (0.1 g C m^{-3}: Lampert 1977); the foraging strategy that is most favoured involves phagotrophic nanoflagellates and small ciliates which graze on bacteria encountered within the same viscous microcosmic patches. Numbers can reach 10–100 mL^{-1}, more if facultatively mixotrophic chlorophyll-containing nanoplankton cells are included (Riemann *et al.* 1995). Collectively, the microplanktonic ciliates and other protists have extremely varied diets of nanoplanktonic autotrophs, phagotrophs and mixotrophs (see, for instance, Fenchel & Finlay 1994) and their activities carry carbon into the size range of microplanktonic primary producers. With bacterised biogenic particles of comparable dimensions, these are all adequate to become the food of calanoid copepods (such as *Eudiaptomus* and *Limnocalanus*) for which they are adapted to forage. This step in trophic level, at least, conforms to classic food-chain theory, showing the expected fewer but larger sized organisms, with longer generation times, as well as an accumulation of embodied energy ('emergy': Odum 1988). Adult calanoids exceed the size of the smallest scales of pelagic turbulence. Interestingly, this results in an enhancement of energy transfer: it has been demonstrated (Rothschild & Osborn 1988) that, at this scale, the contact rate with potential food particles at a given concentration is greater per unit time in turbulent flow than otherwise. At appropriate food concentrations, however, *Eudiaptomus* is able to switch to filter feeding.

9.7 FORAGING STRATEGIES IN THE ZOOPLANKTON

In oligotrophic lakes, calanoids rarely achieve more than 10–20 adults L^{-1} ($<0.08\,g\,C\,m^{-3}$) and (following Huntley & Lopez 1992) attain such populations slowly, as a consequence of temperature, rather than food limitation. Yet they are always likely to constitute the main pelagic crustaceans while the combined food resource is sustained by a biomass of algal producers and bacteria not exceeding 0.1 g C m^{-3} ($\approx 2\,µg\,L^{-1}$ as chlorophyll). In those slightly more eutrophic lakes where, at least episodically, the capacity to support autotrophic biomass significantly exceeds 0.1 g m^{-3}, and where filter feeding is likely to yield sufficient reward of food for foraging effort, there is a strong possibility that Daphniids will be able to replace Calanoids. Collectively, several species are common world-wide and their distributions are well-studied (see Chapter 14). However, some features are generally shared interspecifically, such as the finding that the filter-feeding is obligate, and the rate of filtration (*sensu* volume per unit time) and the upper size limit of filterable particles are robustly predicted by animal size (Burns 1968, 1969). Moreover, the reproductive capacity of Daphniids is much greater than that of the calanoids (Ferguson *et al.* 1982). Thus, given reasonable ambient temperatures (15–25°C) and adequate resources, Daphniid-dominated populations potentially can increase their filtration to 140-fold in just two generations (less than one calendar month). Over the corresponding period, the loss rate sustained by the food organisms may rise from being negligible (say -0.005 day^{-1}) to a rate (-0.7 day^{-1}) at which the organism will need to self-replicate daily merely to sustain its population. By analogy, 20–25 2-mm *Daphnia* per litre are statistically quite capable of sweeping the entire water volume at c.20°C and, thus effectively, of removing all the relevant-sized particles therefrom, including all the components of a viable microbial loop. The presence of a numerous filter-feeding population in the pelagic should tell us that the resources to support algal growth are not continuously limiting, and that its products are transferred directly to crustacean grazers without resort to the concentrating mechanics of an active microbial loop.

Matters are not always so simple: a third inevitability appears to be that, by themselves, filter-feeders are also capable of removing the resource completely. This process, in fact, largely contributes to the high clarity of lake waters, especially quite eutrophic ones, early in the post-stratification season (Sommer *et al.* 1986). The outcome, desired as it is by lake managers, is, however, quite unsustainable—filter-feeders unable to find food are liable to very substantial mortalities. Filtration rate falls again; microorganisms can grow again; filter-feeders may recover. Indeed, this 'tracking' by filter-feeders of the fluctuating food supplies has been shown to apply in simple, pelagic systems involving herbivores and their algal foods (Ferguson *et al.* 1982). In the real world, however, two sets of mitigating circumstances act to damp the extremes of fluctuation. One is the intervention of higher trophic levels, particularly planktivorous fish (see section 9.10). The other is the development of phytoplankton too large to be easily filtered by the zooplankton (Gliwicz 1990).

One of the events said to be forced by the presence of active filter-feeding populations is the increase in abundance of large or colonial phytoplankton species (including, significantly, large dinoflagellates and the bloom-forming cyanobacteria: see Haney (1987) for discussion). However, this should not be considered inevitable. There is an undoubted successional tendency,

other things being equal, for large species to replace small ones anyway (e.g. Reynolds 1984); thus, it is arguable that grazing merely accelerates a vegetation process. Any such relationship must be tempered by the recognition that filter-feeders may still starve if the only primary food is too large, or if the food is nutritionally poor or even toxic. Moreover, a pre-existing population of filter-feeders seems well equipped to inhibit recruitment of colonial species, while they themselves are still small. Nevertheless, one outcome is predictable: an abundance of large phytoplankton is often accompanied by a zooplankton made up, variously, by small Cladocera (e.g. Chydorids) browsing on the surfaces of the colonies for epiphytes, and Bosminids, together with rotifers, feeding on small algae and bacteria in the water, rather like rabbits in a wood, browsing on herbs rather than the trees. It seems that in the presence of very large phytoplankton, as of the very smallest, the indirect linkages ('loops') assume considerable importance in carbon transfer in the pelagic.

9.8 MACROPHYTE-BASED HETEROTROPHY

Macrophyte communities, involving filamentous and thalloid algae, bryophytes, pteridophytes and, especially, representatives of many Spermatophyte families (see Chapter 11), are confined to the littoral zone and shallows of a lake. Whether the successful plants are free-floating or rooted, submerged, with or without floating plants, or emergent (or whether they are there at all) is influenced by water depth, wave exposure and the nature of the substratum. Climate, shade, level fluctuation, alkalinity, saprobity, aeration and frequency of disturbance each contribute to interspecific selection. Macrophytes in turn provide surfaces for epiphytic algae, which may contribute a substantial fraction of the turnover of carbon in ponds and shallow lake margins. Dense beds of macrophytes acquire and, in part, maintain functional characters of their own, with respect to dissolved gases, pH and nutrient cycles (Frodge *et al.* 1991). The most striking contrast with the pelagic environment is the size and (often) the standing biomass of autotrophs that characterise macrophyte swamps (up to 1–1.5 kg C m^{-2}: Wetzel 1990).

The principal plants are also much larger than most of the aquatic animals. Although there is some herbivory on macrophytes (certainly upon their epiphytes), by gastropods, larval mayflies and rasping cladocerans, the major trophic pathways for macrophytic production are assembled around their death and decomposition. Fungi, and the microorganisms of decay, are central to this process, but the physical fragmentation of plant debris is assisted by the activities of shredding invertebrates (malacostracans, insects). Finer particles are available to gatherers, such as corixids; settled detritus is exploited by burrowing chironomids and oligochaetes, while filtering lamellibranchs exploit the fine end of the detrital spectrum to supply their energy needs; fine detritus that remains in suspension or is periodically resuspended is available to swimming filter-feeders (cladocerans again, this time from such genera as *Simocephalus* and *Ceriodaphnia*).

The time needed in order to achieve total mineralisation of the more robust lignins and sclerenchyma is clearly very protracted. In contrast, the annual turnover of a (supposedly) smaller biomass of epiphytic algae and bacteria seems likely to contribute a significant alternative carbon pathway. To date relatively few attempts have been made to distinguish the respective contributions but available estimates (Jónasson *et al.* 1990; see also Chapter 12; Vadeboncoeur *et al.* 2001) indicate that, on an areal basis, rates of epiphyte turnover can be comparable to, or exceed, those of the plankton (100–300 g C m^{-2} yr^{-1}). An important finding comes from Hecky & Hesslein (1995) whose comparison of stable-isotope analyses of the tissue of littoral detritivores and their invertebrate carnivores shows that their food sources are strongly weighted towards the epiphytic/periphytic sources. The same study also makes it plain that local fish biomass may often be supported mainly by benthivory.

Given the area-specific efficiency and trophic intensity that it is possible to achieve in macrophyte beds, their potential contribution to the overall carbon economy of a lake is expected to bear some relation to the productivity and relative areal extent of the macrophyte beds in the whole system: clearly the contribution is likely to be low in large lakes dominated by their pelagic environments, but in small or shallow lakes where a relatively large area can be dominated by macrophytes the opposite may well be true. The analysis of Jeppesen *et al.* (1997) of a series of Danish lakes found just such an inverse relationship between the relative contribution to secondary production of zoobenthos and the lake depth.

Plainly, the zoobenthos provides an attractive and relatively more concentrated feeding resource to browsing fish than does zooplankton, except where and when they occur at high densities. Despite the physical constraints that the density of macrophytes may confer in terms of providing a refuge from adult fish, this alternative resource availability to fish is the basis of the potentially strong interaction between the food webs of the benthos and of the open water (Schindler *et al.* 1996; Vadeboncoeur *et al.* 2002).

The accumulation of resistant, mainly macrophytic, organic debris may be structurally stabilised by subsequent generations of macrophytes, which may thus contribute to the consolidation and extension of the swamp. The plants may quickly become a steady-state habitat to the invertebrate fauna as well as the source of its foods. They may also become a trap for other detrital material washed in from the adjacent land, or generated in adjacent deep waters. Moreover, fine-particulate organic derivatives and dissolved organic leachates may be transported by convectional currents out into the open-water beyond. In this way, macrophytic communities and their carbon cycles are integrated into those of entire lakes.

9.9   PRODUCT TRANSPORT AND INTERACTIONS

Carbon-processing in lakes is not confined to the sites of the initial carbon fixation. Autotrophic products are interchanged between swamp and open water. The predominant flux is to the water:

in a pond or small lake, the influence of the littoral zone can be clearly very significant, especially where the supply of fixed carbon substitutes for a resource-dependent weakness of primary production in the open water (Wetzel & Allen 1972). The contribution of macrophyte growth and decomposition to the total carbon budget of all but the most hypertrophic small lakes (say, up to 1 km^2 in area) is recognised to be significant, but, to judge by the recovery of periphytic algae in the open water of the Bodensee (Lake of Constance, 540 km^2; Müller 1967; Lehn 1968), littoral products can continue to influence the pelagic environment over substantial distances from the shore.

The pelagic exports material, principally by sedimentation, to the bathypelagic and benthos; in this sense, the pelagic is not an open system. Besides the remnants of dispersed littoral products, biogenic sedimentary flux is dominated by phytoplankton, in varying states of decomposition, of zooplankton and their cadavers, and the faecal remains of the foods ingested by all planktonic consumers. To judge by the evidence reviewed in section 9.3, the possible annual flux is hardly likely to exceed 1000 g C m^{-2}. Jónasson's (1996) data suggest that actual limnogenic fluxes are considerably smaller (<300 g C m^{-2}2 yr^{-1}), especially in oligotrophic lakes (<100 g C m^{-2} yr^{-1}).

The nature of this material, and its role in sustaining the deep water fauna, is the subject of Chapter 12. The specialised bathypelagic heterotrophs are efficient in the exploitation of this sedimenting resource, which, potentially, leads to complete mineralisation of the material and a final output which comprises little more than the remnants of skeletal silica or carbonate. Alternatively, sedimenting material reaches the lake bottom prior to its mineralisation, principally as a function of its sinking rate, relative to the depth and temperature of the water. The effect of a truncated water column is to confine a substantial fraction of potential decomposition to the veneer of the sediment surface. At the same time, the scale of the flux (in g C m^{-2} day^{-1}) sets an oxidative demand upon the reserve of oxygen dissolved in the water column (2.67 g O$_2$ per gram of organic carbon).

In concert, these processes distinguish the con-

trasted environments of the worlds' great lakes (oligotrophic, aerated, largely mineralised deposits), its shallower oligotrophic lakes (in which the renewable oxygen reserves are adequate to meet the oxidation demand of the bottom sediments) and those shallower lakes where the oxidative demands of the exported biogenic products exceed the capacity of the stored oxygen (thus, a concentration of 10 mg O$_2$ L^{-1} in a 6-m deep hypolimnion at the onset of summer stratification would stand to be exhausted by the oxidation of a seasonal flux of 250 g C m^{-2}). Substantially anoxic, organic sediments are accumulated at the base of the water column. As Jónasson (this volume; see also Jónasson 1996) shows, these various traits in lake metabolism are closely matched by differences in the structure and abundance of the heterotroph communities.

These possibilities are not exclusive. Meromictic lakes (such as Tanganyika) may be so poorly ventilated that their vast monimolimnia nevertheless fall devoid of oxygen through cumulative oxidative demand. At the other extreme, wind-generated water movements in shallow lakes entrain and re-entrain limnogenic materials so effectively and so frequently that decomposition takes place mainly in the water column. The main functional consequences of these structural differences are in the sites and rates of carbon recycling and in the diagenesis of the other biological components, including the refractory products involving nitrogen and phosphorus. In this way, the interactions of lake geography, basin morphometry and the supply of materials from the catchment do not merely set the state attributes of the lakes but, through the major pathways of limnetic metabolism, also determine the reprocessing of their resources. On the one hand, very large lakes behave like the sea in recycling nutrient resources during major circulations and seasonal events. On the other, very shallow lakes re-use their resources at much higher frequencies. It is equally clear, however, that there ranges between them many which recycle inefficiently, and which allow little more than a single sedimentary passage of limnogenic products. In passing, it is worth noting that such lakes dominate the data set used by Vollenweider (1968); had this not been the case, the model would never have been so clear!

9.10   THE ROLE OF FISH

The middle to upper levels of aquatic food chains are occupied mainly by fish: collectively, the diversity and mass of fish, together with their variety of dietary habits, contribute key alternative linkages in carbon cycling in lakes (consider the pathways represented by the African cichlids: Ribbink 1994). The complexity of linkages is compounded by the fact that many species of fish are recruited at the scale of plankton, but potentially mature to fresh weights in the order of kilograms, during which their preferred foods and their position in the food chain may change significantly. Thus, to make generalisations about the interactive aspects of carbon flow through limnetic food webs is fraught with difficulties and sometimes contradictory evidence. Chapter 15 by Winfield and Chapter 16 by Sarvala *et al.* convey these problems and bring much needed resolution to the issues. The task of the present section is to continue the theme of the chapter into detecting the main opportunities for harvesting the products of primary production and the order of energetic transfers which are possible in lake systems.

Like those of the sea, the pelagic waters of large lakes rarely provide a hospitable environment for fish: physical variability combined with nutrient paucity militate against a sustained, large resource base of primary products or of their consumers; they offer no shelter or refuge. Almost every function is carried out in open water: the alternative is to migrate inshore. Pelagic fish species tend to share common abilities of rapid swimming and adaptations to an ultimately varied diet which, in many cases, begins with planktivory. As suggested previously, the most prevalent freshwater pelagic fish belong to the order Clupeiformes (which include the salmonids, coregonids and osmerids). In smaller and shallower lakes, or in the inshore areas of larger lakes, benthic production is generally more intense and dispersed among more varied producers offering more and different exploitation opportunities. Consequently, the variety of consumer species is much greater, with, especially, many more species of percids, esocoids (in temperate regions) and cichlids and characinids (at low latitudes). The great diversity of small water bodies and microhabitats is matched by the enormous diversity and distribution of cypriniform species. Eels (Angulliformes), which also occur widely in lakes, are predominantly benthic feeders.

The supply of exploitable foods contributes one of the constraints regulating the abundance and species composition of fish populations in lakes, but the extent to which the structure and abundance of the producer communities are governed by the scale of activities of the fish is still a matter of debate (see Chapter 14). The crucial support for or against the various propositions comes mainly from theory and from studies conducted in relatively very small (<1 km^2), shallow (<10 m) water bodies (Evans 1990). There may still be too few good, long-term databases to validate or otherwise test current ideas on fish community dynamics (Winfield, Chapter 15) and there may still be too few validated, physiologically based models of fish growth and recruitment success (of the type assembled by Kitchell *et al.* (1974), and by Elliott (1975a, b) and Elliott & Hurley (1999)) for the theorising to be translated to the calculation of carrying capacities and resource diminution.

Reynolds (1994) used quantitative data on the food requirements of brown trout (*Salmo trutta*: Elliott 1975a) to determine the size-dependent feeding rates (Burns 1969; Lampert 1977), recruitment and growth of *Daphnia* (Ferguson *et al.* 1982) and the rates of growth of nanoplankton in the size range of particles ingestible by all sizes of *Daphnia* (Reynolds 1984), in order to calculate the stable point of a three-stage (alga–zooplankter–planktivore) food chain at around 18°C. In order to prevent nanoplankton from increasing, despite a replication rate of one doubling every 24 h, the *Daphnia* need to be filtering at least half the volume of the water each day. Under the model circumstances, the population required to achieve this result will possess an aggregate dry mass of about 400–500 mg m^{-3}. The concentration of algae should not be greater than 300–400 mg C m^{-3} ($\approx$ 6–8 mg Chl *a* m^{-3},

$\approx 600\text{--}800\,\mathrm{mg}$ dry mass m^{-3}), or else the individual feeding rates are saturated and the filtering rates slow proportionately and the control on instantaneous algal increase is weakened. In order to prevent an adequately nourished zooplankton from increasing mass and recruiting later generations that, potentially, would raise the demand on the food supply by 20% per day, it is necessary to propose that the equivalent of some $100\,\mathrm{mg}\,\mathrm{m}^{-3}$ dry matter is cropped each day by planktivores. On the supposition that an 11-g trout is satiated by a daily ration of $\geq 300\,\mathrm{mg}$ of food, and a 50-g fish by ≥ 1000 mg, the minimum fresh mass of fish that would be capable of the required level of planktivory is equivalent to $3.5\text{--}5\,\mathrm{g}\,\mathrm{m}^{-3}$. Supposing a mixed water depth of 10 m, this would translate into a stock density of $350\text{--}500\,\mathrm{kg}\,\mathrm{ha}^{-1}$. A closer steady state could be proposed if the resultant increment (0.2% per day) could be simultaneously harvested (perhaps by a piscivorous predator) but expecting a large biomass of fish simply to feed at maintenance levels (about one-third of satiation, according to Elliott 1975b) is unrealistic while food continues to be available.

This last point reminds us that, although relationships among the components are palpably temperature-sensitive, there is very little leeway regarding the quantities present at each trophic level for there to be any other temperature-specific steady state. More fish would simply outstrip the potential food; fewer fish would allow the zooplankton to exceed its resource. This, of course, is the ultimate goal of biomanipulation (see Volume 2), but it should be recognised that, in an exclusively pelagic food chain, the proximal consequences are that the zooplankton starves, with massive mortalities, and the chain is broken: algae are ungrazed and the fish are severely deprived of food. Whereas daphniid numbers might recover after a month or two, with a by-now refreshed food resource, the population responses of fish usually occur over periods of a year or more.

It is clear that, in a wholly pelagic food chain, the aggregate mass of each of the components must, on average, be balanced about some lower ambient level of its respective resource. This means that the fish stock could not prevent the zooplankton from tracking a seasonal increase in the phytoplankton, itself responding sensitively to (say) lengthening days or enhanced nutrient renewal. This is indeed approximately what happens in many of the world's mesotrophic and mildly eutrophic lakes: the spring increase in phytoplankton in temperate regions, and the stimulus to productivity in tropical lakes after seasonal mixing, each typically prompt increases in the zooplankton mass, followed in many instances by the depletion of the edible phytoplankton and a significant clearing of the water (Lewis 1986; Sommer et al. 1986). This lurching relationship between phytoplankton and zooplankton, only periodically representing a primary nutritional source to the fish (nonetheless, one able to influence growth and recruitment success; see Winfield et al. 1998), accommodates all the familiar seasonal cycles of plankton abundance and species composition as well as the annual growth and recruitment cycles of primarily planktivorous and secondarily piscivorous fish.

This model can be developed to explain the control exerted through the food chains of other types of lake. A richer supply of capacity-limiting nutrients permits the seasonal response of phytoplankton to proceed more rapidly, and further, before it is ultimately grazed down by zooplankton. If there is yet nutrient resource to spare, it is likely to become invested in the biomass of the larger algae which, in general, a Daphniid-dominated filter-feeding zooplankton fails to control. The consequent exaggerated fluctuations in the phytoplankton biomass and its late-season dominance by species of *Peridinium*, *Ceratium* or *Microcystis* are arguably the characteristics of very eutrophic systems where the fish populations are quite unable to regulate the lower trophic levels. At the other extreme, there is always insufficient concentration of algae in very oligotrophic, nutrient-deficient lakes to satisfy the minimum requirements of filter-feeding daphniids ($>100\,\mathrm{mg}\,\mathrm{C}$ m^{-3}: section 9.7): the alternative foraging strategies of calanoids are much better suited to feeding in rarified environments. Moreover, calanoids are not, as are obligate filter-feeders, exclusive of the simultaneous maintenance of the microbial food chain, which yields the protists that contribute to the calanoid diet. On the other hand, calanoids are not capable of generating anywhere near the production of daphniids: the best performance of *Eudiaptomus gracilis* in well-fertilised limnetic enclosures (Ferguson *et al.* 1982) was a ten-fold increase in biomass over 4 months (about 2 % day^{-1}) to $20 L^{-1}$ (equivalent to c.350 mg dry matter m^{-3}). The steady-state yield to fish could scarcely exceed 7 mg m^{-3} day^{-1}; hence, the level of sustainable planktivory via calanoids is less than one-tenth that possible by cladocera, and the sustainable fresh biomass of planktivorous fish is, correspondingly, less than 0.5 g m^{-3}.

The important role of fish in the functioning of littoral and benthic habitats should be covered briefly. Trophic pathways built on littoral macrophyte production potentially support a greater intensity and greater variety of small and specialised fish species but, if sufficiently developed, macrophyte beds are effectively denied to pelagic fish. They are supposed to provide an important physical refuge for zooplankton and, although many of the cladoceran and copepod species thus protected are littoral rather than truly planktonic (see section 9.8), their mode of feeding on detritus and bacterial flocs (including, in many instances, by filtration) helps to keep the water in the macrophyte beds clear of suspensoids and is also effective in reducing phytoplankton in the adjacent open water (Canfield *et al.* 1984; Ozimek *et al.* 1990; Schriver *et al.* 1995). Conversely, this additional resource of detrital carbon generated in, or redispersed from, macrophyte beds may be sufficient to secure the survival of planktonic filter-feeders when local concentrations of edible phytoplankton fall below 80 mg C m^{-3}. Macrophyte beds not only exchange carbon to adjacent planktonic food chains, mainly through the browsing activities of populations of adult fish that would otherwise be unsustainable on the basis of pelagic carbon cycling alone, but also serve to dampen the 'plenty-to-famine' oscillations which otherwise characterise plankton cycles in mesotrophic and eutrophic lakes.

Many common limnetic fish species exploit the shallow-water benthos as a main or a secondary

source of food. The harvestable flux of carbon through this pathway cannot, of course, exceed pelagic or littoral inputs. Availability is generally greatest in the shallow areas of productive lakes and, so, is typical of water bodies which are small and eutrophic. Often the most successful fish species in such environments (though not those in Australia and South America), are Cyprinids. Some carp, and their allies, are pivotal in switching the carbon pathways of small lakes between (i) one driven mainly by macrophyte and periphyton production, and (ii) a route by which phytoplankton production is overwhelmingly exclusive, and carbon is passed to the fish via benthic decomposition and burrowing invertebrates (oligochaetes, chironomid larvae; Moss 1983, Blindow *et al.* 1993; Scheffer *et al.* 1993).

9.11 COMMUNITY ASSEMBLY: THE BIOTIC CHARACTER OF WHOLE LAKES

With such structurally contrasted and functionally disparate components potentially contributing to the carbon economy of lakes, it should not be difficult to accept the individuality of their characters. Yet there is still a tendency among researchers who have exhaustively researched the community function in a particular lake (or who have been engaged in structural manipulations of trophic levels in order to achieve better water quality or a more satisfying balance of components) to believe that all lakes work in more or less the same way. Scientific investigation of processes has been and continues to be necessary to the elucidation of interactions among autotrophs, and consuming and decomposer heterotrophs in separate or separated types of habitats; the true limnologist often needs to see a wider picture, to recognise the interactions among adjacent yet distinct habitats in the same lake and the ways in which their relative magnitudes may define many of the differences between individual lakes. In other words, it is often no longer enough to study component processes without recognising their role in the behaviour of the entire system.

This more integrated view of the organisation and energetic partitioning in particular limnetic systems is now emerging, largely in response to the need to understand and to manage the behaviour of whole lakes. This view acknowledges that the carbon cycles, like those of the nutrients, are not independent of the delivery of carbon dioxide and inorganic carbon sources from the catchment, nor of substantial loads of particulate and inorganic carbon. It takes into account the potential of truncated shallow water columns to concentrate the areal fluxes of energy and materials into high-biomass, macrophyte-dominated communities. It accepts the value of such internal stores as macrophyte biomass, detritus and organic deep-water sediments to buffer the rates of carbon turnover in certain lake compartments. Moreover, it recognises the coupling of the biotic components of the open-water, littoral and benthic habitats that is, in part, the inevitable consequence of juxtaposition, horizontal and vertical transport but which is considerably enhanced by the motility of fish.

The variability in the nature and extent of littoral/benthic and pelagic habitats makes for a wide range in the principal carbon flows of individual lakes. Following Fig. 9.1, it can be assumed that the world's larger and deeper lakes are essentially driven by processes in their pelagic and bathypelagic zones, represented to the left of Fig. 9.2. The influence of shores and even inflows may be relatively weak: morphometry will determine that they are oligotrophic (in the modern sense of the term, that is with very diluted biological supporting capacity), or that they are similarly rarified, because the 'dilution' of light energy in the mixed layer produces the analogous effect. A deep photic zone and a reserve of carbon dioxide often means that there is a substantial capacity for phytoplankton photosynthesis: however, of the potential annual areal carbon fixation (approaching $200\,g\,C\,m^{-2}$), most is passed to higher trophic levels, and is conveyed by the microbial–protist pathway to populations of planktivorous and, ultimately, piscivorous pelagic fish, all of which assist in the regeneration and recycling of the resource base.

Contrasting conditions are found with diminishing water-body dimensions and, perhaps, the greater concentrations of nutrient resources in more eutrophic systems (to the right of Fig. 9.2). A greater concentration of phytoplankton can be maintained, but the impact on light attenuation is such that the areal density is not greatly enhanced and the photosynthetic production rarely exceeds $600\,g\,C\,m^{-2}$. The main difference is that a greater part of net production stays in the phytoplankton and is passed to planktonic herbivores or benthic detritivores.

The trend with diminishing size is not just that the pelagic zone becomes smaller in absolute terms, but that the relative influence of the watershed increases: this is manifest in more rapid hydrological renewal, larger nutrient imports and enhanced cycling, as well as the interface with different types of littoral system, alternative primary producers and alternative food webs. As already observed, the size range representing the transition from pelagic to littoral systems is ill-defined, and is subject to several other factors, not the least being the way in which the littoral zone is vegetated: 200–$0.2\,km^2$ is suggested to stretch from the point of detectable influence to dominance over pelagic processes. Similarly, just as depth of water influences the extent to which the sediments act either as a sink for primary products and nutrients or as a source of recycled material, so the types of sediment and the proportions in direct contact with the (seasonally variable) surface mixed layer exert considerable influence on the functioning of the entire system. As in the sea, great depth may permit substantial cadavory and decomposition to occur in the water column, and allow basin-wide circulation to redistribute re-usable raw materials. At the same time, frequent sediment resuspension in very shallow waters may achieve the corresponding effect by renewing opportunities for oxidation of biogenic products and release of nutrients. In the intermediate range, organic products reach the sediment surface, which is sufficiently deep for the deposited material to escape frequent resuspension in applied shear currents, so that oxidation is carried out mainly by benthos, with only infrequent returns of recycled materials to the pelagic food webs. The definition of 'deep/

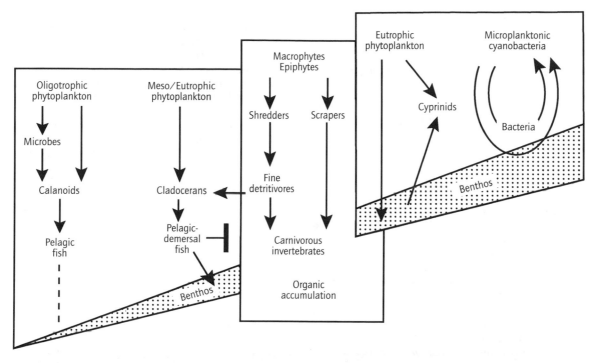

Fig. 9.2 Stylised limnetic food chains, broadly segregated by habitats. The left hand box represents open-water, pelagic, self-sustaining trophic pathways, sub-divided to separate deep-column oligotrophic systems from more eutrophic or benthos-sustaining linkages. The right hand box accommodates two other food chains, the degraded alga-benthos-cyprinid pathway of small, enriched waters and the shallow lake dominated by Oscillatoriales. Between them is a macrophyte-based web which represents a different sort of small-water system and an important interface with open water in larger lakes.

shallow' in this context is no less satisfactory than that for 'large/small', although the frequency of resuspension events, as a function of depth and area or fetch (see Hilton 1985), offers a basis for being able to anticipate the relative importance of sediment resuspension in the production ecology of lakes: most examples >20 m in depth and >6 km² in area will experience transport of sediment from shallow to deep water, in quantities and with frequencies related to the nature of the sedimentary material and the fraction of the total area that is regularly subject to wind shear. In smaller lakes (<6 km²) the critical depth may be less, but larger, shallow lakes (>6 km², wholly <20 m) experience frequent sediment resuspension.

It is this wide range of habitat types and the variety of possibilities for their exploitation that favour broad differences in community composition and organisation. Of the actual species pool present at a given location, those which dominate and most influence the flows of energy and matter through the system generally turn out to be functionally the most appropriate to the blend of microhabitats represented at the site. In turn, the importance of phytoplankton or macrophyte primary production, of zooplankton or zoobenthos, and whether the fish appear primarily to regulate or to respond to the influence of lower trophic levels are not inviolate outcomes of community interactions generally; rather the interactions within the communities are biased towards structural outcomes most favoured by site-specific conditions.

As in the Homeric verse, there is more than one

way to achieve the ecosystemic equivalent of Midas' steady state.

9.12 ACKNOWLEDGEMENT

I am grateful to Charles Sinker for introducing me not only to ecological thought but to the profundity of Homer's verse.

9.13 REFERENCES

Ahlgren, G. (1970) Limnological studies of Lake Norrviken, a eutrophicated Swedish lake. II. Phytoplankton and its production. *Schweizerische Zeitschrift für Hydrologie*, **32**, 353–76.

Azam, F., Smith, D.C., Long, R.A. & Steward, G.F. (1995) Bacteria in oceanic cycling as a molecular problem. In: Joint, I. (ed.), *Molecular Ecology of Aquatic Microbes*. Springer-Verlag, Berlin, 39–54.

Bannister, T.T. & Weidemann, A.D. (1984) The maximum quantum yield of phytoplankton photosynthesis. *Journal of Plankton Research*, **6**, 275–94.

Bird, D.F. & Kalff, J. (1984) Empirical relationships between bacterial abundance and chlorophyll concentrations in fresh and marine waters. *Canadian Journal of Fisheries and Aquatic Sciences*, **41**, 1015–23.

Blindow, I., Andersson, G., Hargeby, A. & Johansson, S. (1993) Long-term pattern of alternative stable states in two eutrophic lakes. *Freshwater Biology*, **30**, 419–32.

Bratbak, G. & Thingstad, T.F. (1985) Phytoplankton–bacteria interactions: an apparent paradox? Analysis of a model system with both competition and commensalism. *Marine Ecology Progress Series*, **25**, 23–30.

Burns, C.W. (1968) The relationship between body size of filter-feeding cladocera and the maximum size of particle ingested. *Limnology and Oceanography*, **13**, 675–8.

Burns, C.W. (1969) Relation between filtering rate, temperature and body size in four species of *Daphnia*. *Limnology and Oceanography*, **14**, 693–700.

Cabrera, S. & Montecino, V. (1984) The meaning of the euphotic chlorophyll *a* measurement. *Verhandlungen Internationale Vereinigung Limnologie*, **22**, 1328–31.

Canfield, D.E., Shireman, J.V., Colle, D.E., Haller, W.T., Watkins, C.E. & Maceina M.J. (1984) Prediction of chlorophyll a concentrations in Florida lakes: importance of macrophytes. *Canadian Journal of Fisheries and Aquatic Sciences*, **41**, 497–501.

Chróst, R.J. (1983) Plankton photosynthesis, extracellular release and bacterial utilization of released dissolved organic carbon in lakes of different trophy. *Acta Microbiologia Polonia*, **32**, 275–87.

Chróst, R.J. & Faust, M.A. (1983) Organic carbon release by phytoplankton: its composition and utilization by bacterioplankton. *Journal of Plankton Research*, **5**, 477–93.

Cole, J.J., Likens, G.E. & Strayer, D.L. (1982) Photosynthetically produced dissolved organic carbon: an important carbon source for planktonic bacteria. *Limnology and Oceanography*, **27**, 1080–90.

DeLong, E.F., Franks D.G. & Alldredge, A.L. (1993) Phylogenetic diversity of aggregate-attached bacteria vs free-living bacterial assemblages. *Limnology and Oceanography*, **38**, 924–34.

Demmig-Adams, B. & Adams W.W. (1992) Photoprotection and other responses of plants to high light stress. *Annual Review of Plant Physiology and Plant Molecular Biology*, **43**, 599–626.

Elliott, J.M. (1975a) The growth rate of brown trout (*Salmo trutta* L.) fed on maximum rations. *Journal of Animal Ecology*, **44**, 805–21.

Elliott, J.M. (1975b) The growth rate of brown trout (*Salmo trutta* L.) fed on reduced rations. *Journal of Animal Ecology*, **44**, 823–42.

Elliott, J.M. & Hurley, M.A. (1999) A new energetics model for brown trout, *Salmo trutta*. *Freshwater Biology*, **42**, 235–46.

Evans, M.S. (1990) Large-lake responses to declines in the abundance of a major fish planktivore—the Lake Michigan example. *Canadian Journal of Fisheries and Aquatic Sciences*, **47**, 1738–54.

Fenchel, T. & Finlay, B.J. (1994) *Ecology and Evolution in Anoxic Worlds*. Oxford University Press, Oxford, 288 pp.

Ferguson, A.J.D., Thompson, J.M. & Reynolds, C.S. (1982) Structure and dynamics of zooplankton communities maintained in closed systems, with special reference to the algal food supply. *Journal of Plankton Research*, **4**, 523–43.

Fogg, G.E. (1971) Extracellular products of algae in freshwater. *Ergebnisse der Limnologie*, **5**, 1–25.

Forsberg, B. (1985) The fate of planktonic primary production. *Limnology and Oceanography*, **30**, 807–19.

Frankignoulle, M. (1988) Field-measurements of air–sea CO_2 exchange. *Limnology and Oceanography*, **33**, 313–22.

Frodge, J.D., Thomas G.L. & Pauley, G.B. (1991) Sediment phosphorus loading beneath dense canopies of aquatic macrophytes. *Lake and Reservoir Management*, 7, 61–71.

Gliwicz, Z.M. (1990) Why do cladocerans fail to control algal blooms? *Hydrobiologia*, **200/201**, 83–97.

Haney, J.F. (1987) Field studies on zooplankton–cyanobacteria interactions. *New Zealand Journal of Marine and Freshwater Research*, **21**, 467–75.

Hecky, R.E. & Hesslein, R.H. (1995) Contributions of benthic algae to lake food webs as revealed by stable isotope analysis. *Journal of the North American Benthological Society*, **14**, 631–53.

Hill, R. & Bendall, F. (1960) Function of two cytochrome components in chloroplasts: a working hypothesis. *Nature*, **186**, 136–7.

Hilton, J.H. (1985) A conceptual framework for predicting the occurrence of sediment focusing and sediment redistribution in lakes. *Limnology and Oceanography*, **30**, 1131–43.

Huntley, M.E. & Lopez, M.D.G. (1992) Temperature-dependent production of marine copepods: a global synthesis. *American Naturalist*, **140**, 201–42.

Jeppesen, E., Jensen, J.P., Sondergaard, M., Lauridsen, T., Pedersen L.J. & Jensen, L. (1997) Top-down control in freshwater lakes: the role of nutrient state, submerged macrophytes and water depth. *Hydrobiologia*, **342/343**, 151–64.

Jónasson, P.M. (1996) Limits for life in the lake ecosystem. *Verhandlungen Internationale Vereinigung Limnologie*, **26**, 1–33.

Jónasson, P.M., Lindgaard, C., Dall P.C., Hamburger, K. & Adalsteinsson, H. (1990) Ecosystem studies on temperate Lake Esrom and the subarctic lakes Myvatn and Thingvallvatn. *Limnologica*, **20**, 259–66.

Jones, J.G. (1987) Diversity in freshwater microbiology In: Fletcher, M., Gray, T.R.G. & Jones, J.G. (eds), *Ecology of Microbial Communities*. Cambridge University Press, Cambridge, pp. 235–59.

Kalff, J. & Knoechel, R. (1978) Phytoplankton and their dynamics in oligotrophic and eutrophic lakes. *Annual Review of Ecology and Systematics*, **9**, 475–95.

Kirk, J.T.O. (1994) *Light and Photosynthesis in Aquatic Ecosystems* (2nd edition). Cambridge University Press, Cambridge, 528 pp.

Kitchell, J.F., Koonce, J.F., O'Neill, R.V., Shugart, H.M., Magnusson, J.J. & Booth, R.S. (1974). Model of fish biomass dynamics. *Transactions of the American Fisheries Society*, **103**, 786–98.

Kolber, Z. & Falkowski, P.G. (1993) Use of active fluorescence to estimate phytoplankton photosynthesis *in situ*. *Limnology and Oceanography*, **38**, 1646–65.

Lampert, W. (1977) Studies on the carbon balance of *Daphnia pulex* De Geer as related to environmental conditions IV. Determination of the threshold concentration as a factor controlling the abundance of zooplankton species. *Archiv für Hydrobiologie*, **48**(Supplement), 361–8.

Lee, S. & Fuhrman, J. (1987) Relationships between biovolume and biomass of naturally derived marine bacterioplankton. *Applied and Environmental Microbiology*, **53**, 1298–303.

Lehn, H. (1968) Litorale Aufwuchsalgen im Pelagial des Bodensees. *Beitrag naturk Forschungen Südwest Deutschland*, **27**, 97–100.

Lewis, W.M. (1986) Phytoplankton succession in Lake Valencia, Venezuela. *Hydrobiologia*, **138**, 189–204.

Likens, G.E. & Davis, M.B. (1975) Post-glacial history of Mirror Lake and its watershed in New Hampshire, U.S.A.: an initial report. *Verhandlungen Internationale Vereinigung Limnologie*, **19**, 982–93.

Long, S.P., Humphries S. & Falkowski, P.G. (1994) Photoinhibition of photosynthesis in nature. *Annual Review of Plant Physiology and Plant Molecular Biology*, **45**, 633–62.

Lowe-McConnell, R.M. (1975) *Fish Communities in Tropical Waters*. Longman, London, 337 pp.

Moss, B. (1983) The Norfolk Broadland; experiments in the restoration of a complex wetland. *Biological Reviews*, **58**, 521–61.

Müller, H. (1967) Eine neue qualitative Bestandsaufnahme des Phytoplanktons des Bodensee-Obersee mit besonderer Berücksichtigung der tychoplanktischen Diatomeen. *Archiv für Hydrobiologie*, **23**(Supplement), 206–36.

Nielsen, S.L. & Sand-Jensen, K (1990) Allometric scaling of maximal photosynthetic growth rate to surface/volume ratio. *Limnology and Oceanography*, **35**, 177–81.

Odum, H.T. (1988) Self-organization, transformity and information. *Science*, **242**, 1132–9.

Ozimek, T., Gulati, R.D. & van. Donk E. (1990) Can macrophytes be useful in the biomanipulation of lakes? The Lake Zwemlust example. *Hydrobiologia*, **200/201**, 83–97.

Reynolds, C.S. (1984) Phytoplankton periodicity: the interaction of form, function and environmental variability. *Freshwater Biology*, **14**, 111–42.

Reynolds, C.S. (1992) Eutrophication and the management of planktonic algae: what Vollenweider couldn't tell us. In: Sutcliffe, D.W. & Jones, J.G. (eds), *Eutrophication: Research and Application to Water Supply*. Freshwater Biological Association, Ambleside, UK, pp. 4–29.

Reynolds, C.S. (1994) The role of fluid motion in the dynamics of phytoplankton in lakes and rivers. In: Giller, P.S., Hildrew, A.G. & Raffaelli, D.G. (eds), *Aquatic Ecology: Scale, Pattern and Process*. Blackwell Scientific Publications, Oxford, pp. 141–87

Reynolds, C.S. (1997) *Vegetation Processes in the Pelagic: a Model for Ecosystem Theory*. Ecological Institute, Oldendorf, 372 pp.

Reynolds, C.S., Harris, G.P. & Gouldney, D.N. (1985) Comparison of carbon-specific growth rates and rates of cellular increase in large limnetic enclosures. *Journal of Plankton Research*, **7**, 791–820.

Ribbink, A.J. (1994) Biodiversity and speciation of freshwater fishes with particular reference to African cichlids. In: Giller, P.S., Hildrew, A.G. & Raffaelli, D.G. (eds), *Aquatic Ecology: Scale, Pattern and Process*, Blackwell Scientific Publications, Oxford, pp. 261–88.

Riemann, B., Havskum, H., Thingstad, F. & Bernard, C. (1995) The role of mixotrophy in pelagic environments. In: Joint, I. (ed.), *Molecular Ecology of Aquatic Microbes*. Springer-Verlag, Berlin, pp. 87–114.

Rothschild, B.J. & Osborn, T.R. (1988) Small-scale turbulence and plankton contact rates. *Journal of Plankton Research*, **10**, 465–74.

Sas, H. (1989) *Lake Restoration by Reduction of Nutrient Loading: Expectations, Experiences, Extrapolations*. Akademia Verlag Richarz, Sankt Augustin, 519 pp.

Scheffer, M., Hosper, S.H., Meijer, M.-L., Moss, B. & Jeppesen, E. (1993) Alternative equilibria in shallow lakes. *Trends in Ecology and Evolution*, **8**, 275–9.

Schindler, D.E., Carpenter, S.R., Cottingham, K.I., *et al.* (1996) Food web structure and littoral zone coupling to trophic cascades. In Polis, G.A. & Winemiller, K.O. (eds), *Food Webs: Integration of Patterns and Dynamics*. Chapman and Hall, New York, 96–105.

Schindler, D.W. (1977) Evolution of phosphorus limitation in lakes. *Science*, **196**, 260–2.

Schriver, P., Bigestrand J., Jeppersen E. (1995) Impact of submerged macrophytes on fish–zooplankton–phytoplankton interactions: large-scale enclosure experiments in a shallow eutrophic lake. *Freshwater Biology*, **33**, 255–70.

Sculthorpe, C.D. (1967) *The Biology of Aquatic Vascular Plants*. Edward Arnold, London, 610 pp.

Sell, A.F. & Overbeck, J (1992) Exudates: phytoplankton–bacterioplankton relationships in Plußsee. *Journal of Plankton Research*, **14**, 1199–215.

Sommer, U., Gliwicz, Z.M., Lampert, W. & Duncan, A. (1986) The PEG-model of seasonal succession of planktonic events in fresh waters. *Archiv für Hydrobiologie*, **106**, 433–71.

Talling, J.F. (1957) The phytoplankton population as a compound photosynthetic system. *New Phytologist*, **56**, 133–49.

Vadeboncoeur, Y., Lodge, D.M. & Carpenter, S.R. (2001) Whole-lake fertilization effects on the distribution of primary production between benthic and pelagic habitats. *Ecology*, **82**, 1065–77.

Vadeboncoeur, Y., Vander Zanden, J. & Lodge, D.M. (2002) Putting the lake back together: reintegrating benthic pathways into lake foodweb models. *BioScience*, **52**, 44–54.

Vadstein, O., Olson, Y., Reinertsen H. & Jensen, A. (1993) The role of planktonic bacteria in lakes — sink and link. *Limnology and Oceanography*, **38**, 539–44.

Vollenweider, R.A. (1968) *Scientific Fundamentals of the Eutrophication of Lakes and Flowing Waters, with Particular Reference to Nitrogen and Phosphorus as Factors in Eutrophication*. Organisation for Economic Co-operation and Development, Paris, 240 pp.

Walker, D. (1992) Excited leaves. *New Phytologist*, **121**, 325–45.

Watson, A.J., Uppstill-Goddard, R.S. & Liss P.S. (1991) Air–sea gas exchange in rough and stormy seas measured by a dual-tracer technique. *Nature*, **349**, 145–7.

Wetzel, R.G. (1990) Detritus, macrophytes and nutrient cycling in lakes. *Memorie dell'Istituto Italiano di Idrobiologia*, **47**, 233–49.

Wetzel, R.G. (1995) Death, detritus and energy flow in aquatic ecosystems. *Freshwater Biology*, **33**, 83–9.

Wetzel, R.G. & Allen H.A. (1972) Functions and interactions of dissolved organic matter and the littoral zone in lake metabolism and eutrophication. In: Kajak, Z. & Hillbricht-Ilkowska, A. (eds), *Productivity Problems in Freshwaters*. PWN Polish Scientific Publications, Warsawa, pp. 333–47.

Winfield I.J., George, D.G., Fletcher, J.M. & Hewitt, D.P. (1998) Environmental factors influencing the recruitment and growth of underyearling perch (*Perca fluviatilis*) in Windermere, U.K., from 1960 to 1990. In: George, D.G., Jones, J.G., Punčochar, P., Reynolds, C.S. & Sutcliffe, D.W. (eds), *Management of Lakes and Reservoirs during Global Change*, Kluwer, Dordrecht, pp. 245–61.

10 Phytoplankton

JUDIT PADISÁK

10.1 INTRODUCTION

The total number of phytoplankton species inhabiting the world's lakes is not known exactly. According to a recent review (Tett & Barton 1995), the number of modern phytoplankton species in the sea is estimated to be some 5000. In contrast to the seas, however, the world's lakes do not form one large, contiguous water body; rather they should be considered sometimes as interconnected (river–lake systems) and sometimes as disconnected (lakes fed exclusively with ground water and precipitation) 'islands' on the terrenum. Although different parts of the world's ocean vary significantly in their salt content, nutrient availability, mixing properties, etc., it is safe to say that the lake environment offers a higher level of habitat diversity. Therefore phytoplankton species number in lakes probably exceeds that in the ocean. These thousands of species differ in their evolutionary adaptations, their ability to grow and their sensitivity to different loss processes, out of which the phytoplankton, characteristically some hundreds of species in any particular lake, is assembled.

Many aspects of the phytoplankton, such as the role of limiting nutrients and their ratios, adaptations to low nutrient concentration or supply ratio, heterotrophic nutrition, overcoming light limitation either by superior shade tolerance or via active mechanisms such as buoyancy regulation or flagellar movement, differential sinking, interference with zooplankton grazing mechanisms, and disturbance sensitivity, etc., have been the subject of exhaustive studies. All of the above factors in almost any lake are also characterised by a permanent variability operating at different spatial and temporal scales, and they induce different biological responses, depending upon the current struc-

ture and the time required for a particular response to become manifest.

For example, even operation of a flash-light (for <1 s) may induce chlorophyll-fluorescence and photosynthesis. Similarly, the day–night cycle occupies 24 h. However, it is usually of no influence on population dynamics, because species are physiologically adapted to it. Conversely, irradiance regimes over intervals of days or weeks may be important, in that they may vary significantly in terms of alternating spells of bright sunshine or permanent cloudiness, which may shift or even reverse successional sequences. Further, the annual solar cycle, and consequent changes in incident light income, exhibit striking differences between the tropics and the polar regions. As a result, lakes in different regions vary considerably in terms of their seasonal patterns of succession and annual gross primary production. Interannual differences (sequences of dry then wet years, anthropogenic eutrophication) may lead to floristic changes. Finally, climatic shifts over intervals of tens to hundreds of years lead to evolutionary adaptation of existing species and evolution of new ones (Reynolds 1990).

Phytoplankton species vary in size, shape, evolutionary and phylogenetic position, energy and nutrient demands, and many other features. However, even if they do not float freely and involuntarily in open water, independent of shores and bottom as the original definition of Hensen (1887) envisaged, they persist in suspension and are liable to passive movement by wind and current (Reynolds 1984a).

This chapter is not intended to discuss in detail any of the above factors and phenomena. Neither is it the aim to provide a review of recently available information. However, it was formulated in accordance with the above basic concepts. Attention is focused in order to provide a basic introduction to phenomena that influence the geographical distribution, annual appearance and seasonal behaviour of phytoplankton species, and to offer a basis for comparisons, especially in tabular form. For a deeper insight into the physiological properties of phytoplankton and their ecology, numerous books (e.g. Morris 1980; Carr & Whitton 1982; Reynolds 1984a, 1997a; Harris 1986; Round 1988; Sandgren 1988a; Sommer 1989) provide detailed information.

10.2 SIZE, SHAPE AND COMPOSITION

10.2.1 Size and shape

The size and shape of phytoplankton species, among other properties, determine their adaptabilities to aquatic environments. Algae obtain the nutrients they require from the surrounding medium via the cell surface, and then transport them to sites of use within the cell. Small size and large surface area therefore assist conversion to active biomass. On the other hand, larger size can deter grazing. Size, shape and composition also greatly influence sinking properties and light harvesting abilities (Reynolds 1984a). Nevertheless, phyto-

planktonic organisms are no bigger than the smallest turbulent eddies generated in dissipating input kinetic energy (Reynolds 1994a; see also Chapter 6).

Size of phytoplankton varies from <1 μm (*Synechococcus*) to >1 mm (*Gloeotrichia* colonies); aggregates of *Microcystis* may even be larger. During the last few decades, numerous terms (such as nanoplankton, μ-algae, ultraplankton, etc.) have been developed in order to categorise phytoplankton species by their size. Confusion can be demonstrated by the point that the upper size of nanoplankton has been considered by different authors to range between 10 and 80 μm (see Reynolds 1984a, p. 3). Rationalisation of the categories proposed by Sommer (1994) is adopted for Table 10.1.

Methods of studying phytoplankton frequently depend on size. The term 'net-plankton' refers to those species which can be collected in plankton nets. These are usually the larger sizes. Picophytoplankton can only be quantitatively studied using microscopes equipped with epifluorescence. The standard technique for studying larger species is inverted microscopy, introduced by Utermöhl (1931). Quantitative methods are described in detail in Padisák *et al.* (1997) and Padisák & Adrian (1999).

The 'size' of individual phytoplankton species can be expressed in different ways. The greatest

Table 10.1 Size classes of the phytoplankton (after Sommer 1994). Note that in the categories femtoplankton (<0.2 μm, viruses and phages) and megaplankton (>2 cm, large zooplankton), no phytoplanktonic organisms are found.

Name	Lower limit	Upper limit	Examples
Picoplankton	0.2 μm	2 μm	*Synechococcus*, *Nanochloris*, *Chlorella*
Nanoplankton	2 μm	20 μm	*Rhodomonas*, many Chlorococcales, small chrysophytes
Microplankton	20 μm	200 μm	*Asterionella*, *Ceratium*, 'Sphaerocystis', *Snowella*, filamentous blue–greens
Mesoplankton	200 μm	2 mm	*Gloeotrichia*, *Aphanizomenon* clusters, *Aulacoseira*, chain-forming Pennales
Macroplankton	2 mm	2 cm	Extremely large *Microcystis* colonies

axial linear dimension (GALD [μm]) comes from direct measurements. Volume (V [μm^3]) and surface area (SA [μm^2]) are usually approximated to the nearest geometric equivalent. Sometimes, for non-spherical forms, size is expressed as the diameter of the spherical equivalent volume (DSE [μm]).

The shapes of most phytoplankton species can be compared with simple geometrical forms such as rods, ellipses, spheres, cones or parallelepipeds, or to combinations of these for compound shapes (*Ceratium, Campylodiscus, Staurastrum*). Spines and other surface structures are common, as is mucilage investment. Morphological properties greatly influence physiological performance (e.g., light harvesting), or the behaviour of the cell in limnetic habitats (Reynolds 1988a).

10.2.2 Chemical composition

The major constituent of live algal cells is water. Dry weight (the air-dried fraction, comprising both organic and inorganic components) to volume relationships of cells are linear on a log–log scale, and can be described by the equation $W_c = 0.47\,V^{0.99}$, where W_c is dry weight [pg] and V is cell volume [μm^3]. The mucilage present in many planktonic algae (a complex polysaccharide, with a high water absorbing capacity) is 99% water (Reynolds 1984a). The ash content of algal cells (the inorganic fraction of the dry weight left after ignition of the organic fraction by heating in air to c.500°C) belonging to different taxonomic groups varies significantly. Nalewajko (1966) found that ash accounted for between 5.3 and 19.9% (mean: 10.2%) of dry weight of planktonic chlorophytes. However, the polymerised silica present in the cell walls of diatoms is heavy (specific gravity c.2.6 g cm^{-3}). Therefore, the ash content of diatoms is much higher than that of other phytoplankton. Silica content accounts for about half the dry weight (range of mean values from different measurements: 26–69%) of diatom species. Alternatively, silica content can be expressed relative to cell volume (typically, 0.2–0.3 pg μm^{-3}; Reynolds 1984a).

The most important components of the ash-free dry weight are carbon, nitrogen and phosphorus. The carbon content of algal cells in optimal culture conditions was found to be 51–56% of dry weight. If light, or inorganic carbon supply, is insufficient, this value can fall to a minimum of 30%. If other elements are limiting, a maximum value of 70% can be reached. Nitrogen accounts for 4–9% of ash-free dry weight, whilst phosphorus content ranges between 0.03 and 0.8% (Reynolds 1984a).

The optimal (Redfield) atomic ratio of C : N : P is approximately 106 : 16 : 1 (about 42 : 7 : 1 by weight). If the characteristic carbon : volume ratio is 0.21–0.24 pg μm^{-3}, then the Redfield ratio predicts 0.04 pg μm^{-3} nitrogen and 0.005 pg μm^{-3} phosphorus (Reynolds 1984a).

10.2.3 Photosynthetic pigments

Apart from its role in photosynthesis, the universal occurrence of chlorophyll a (Chl a) among algae makes it especially useful as an estimator of algal biomass. Chlorophyll a content of algal cells is influenced both by external and internal factors. Low values occur if nutrient limitation prevails, whilst high Chl a content is found under light-limited conditions. Young populations contain relatively more Chl a than senescent ones. The effect of species composition on relative Chl a content is unclear (Vörös & Padisák 1990). A positive, linear relationship between cell volumes of phytoplankton species and their Chl a content has been found ($\log$ Chl $a = 0.984 \log V - 2.072$; where Chl a content is in picograms [pg] and V is in cubic micrometres [μm^3]; Reynolds 1984a). It also can be shown that relative Chl a content is higher if the phytoplankton is made up of smaller rather than larger cells ($\log$ Chl$_{rel}$ $a = -\log 0.205\ V + 1.044$; where Chl$_{rel}$ a is the per unit biomass freshweight [μg g^{-1}] and V is mean cell volume [μm^3]; Vörös & Padisák 1990). Expressed as a function of dry weight, Chl a usually accounts for between 0.5 and 2%, or if corrected for ash-free dry weight, 0.9–3.9% of total mass (Reynolds 1984a). Field measurements, normally carried out using volumetrically estimated phytoplankton fresh-weight, and photometrically measured Chl *a* content, give a range between 0.08 and 1.83%. Mean values are about 0.2–0.3% (Vörös & Padisák 1990, and literature therein).

10.3   PHOTOSYNTHESIS

10.3.1   Photosynthetic oxygen evolution, carbon fixation

The process of photosynthesis can be summarised with the well known equation:

$$6CO_2 + 6H_2O \rightarrow C_6H_{12}O_6 + 6O_2 \quad (10.1)$$

This is a redox system involving a light and a dark reaction. In the first, electrons are stripped from a donor substance (in algae, including prokaryotes, water), and passed through a light-driven transport system. Dioxygen (O_2) is liberated at the initial stage of the photosynthetic electron transport system, whilst the protons stripped are eventually used in order to reduce nicotinamide adenine dinucleotide phosphate (NADP).

Two photochemical systems (PS I and PS II) are involved in the light reaction. Amongst autotrophic bacteria, only the cyanobacteria and the Prochlorophyta possess PS II; therefore they are the only bacteria which liberate oxygen whilst photosynthesising, as in higher plants. Photochemical system II is associated with non-cyclical electron transport derived from water molecules, and PS I is involved in coenzyme reduction. They are linked by a cytochrome carrier system through which adenosine triphosphate (ATP) is generated in an additional cyclic photophosphorylation. Energy and the reduced molecules are developed in the light reaction. Moreover carbon dioxide is used in order to build up organic material during the dark reactions (the Calvin cycle) of photosynthesis.

In both PS II and PS I, photosynthetic pigments trap light energy. Chlorophyll *a* is the main pigment used, but it may be associated with a series of other kinds of pigments (light antennae). Maximal absorption of Chl *a* occurs in two bands centred on 450 and 645 nm, whilst the accessory pigments possess absorption maxima at different wavelengths; therefore these increase the width of the absorption spectrum of the phytoplanktonic light trap. The energy they absorb is transferred to Chl *a*. The accessory pigments are often invoked in algal taxonomy, but they are also of high functional, and therefore great ecological, importance.

Photosynthesis by phytoplankton can be measured by the rate of oxygen evolution (gross production, *BP*). Because photosynthetic organisms not only produce oxygen, but, during respiration, also consume it, respiration (*R*) must also be taken into account. Net production (*NP*) is then

$$NP = BP - R \quad (10.2)$$

In the upper layers (the **euphotic zone**, see Chapter 5), $BP > R$, and *NP* is usually positive. At low light intensities, however, $BP < R$ and the net rate of photosynthesis is negative. The horizon in the lake where $BP = R$ is called the **compensation depth** (of photosynthesis).

10.3.2   Light dependence of photosynthesis

Global radiation is the sum of radiation that originates from the sun and which reaches Earth's surface. It is therefore the sum of residual direct solar radiation and that scattered by the clouds. Because the atmosphere absorbs part of the short wavelengths, radiation reaching the surface is mainly within the waveband 300–3000 nm. This can be divided into three subranges.
1 300–380 nm: ultraviolet (UV) radiation, usually deterious to living organisms.
2 380–750 nm: visible radiation. The 400–700 nm range can be used for photosynthesis, and is therefore usually referred to as photosynthetically active radiation (PAR). It generally represents 46–48% of global radiation reaching Earth's surface.
3 750–3000 nm: infrared radiation (or heat).
The processes of light refraction from the water surface, its penetration into the water column, of light attenuation, penetration and transmission,

together with the appropriate equations and conversion factors between different widely used measures of light, are described in detail in Chapter 5. Concerning vertical light attenuation, it is necessary to note here that the coefficient of attenuation (k) depends, amongst other factors, on algal biomass, according to the equation $k = 0.015\,\text{Chl}$ $(\text{mg m}^{-3}) + 0.27$. From this relationship the value $0.27\,\text{m}^{-1}$ can be understood as **background attenuation**, i.e. that which can be measured with all material in the water except phytoplankton.

The value $0.015\,\text{m}^{-1}\,\text{Chl m}^{-3}$ is the **specific attenuation coefficient** (Sommer 1994). As conventionally defined, the euphotic zone of natural waters extends down to that depth where the intensity of surface radiation reaches 1% — the compensation point (see 10.3.1 above). This is the average limit of net photosynthesis, and is not the same for all species. The **aphotic layers** below the compensation point receive <1% of surface light radiation.

Because of the simplicity of the method, the transparency of surface waters is conventionally measured with Secchi disks. Multiplying Secchi disk transparency (SDT) by a factor of 2.7 is generally assumed to estimate the thickness of the euphotic zone. However, this factor may range from 2.2 to 2.7. The higher multiples apply to waters made turbid by high inorganic loading.

According to Reynolds (1987b), mean irradiance (I^*) in the mixed column can be calculated as

$$\ln I^* = (\ln I_0 + \ln I_\text{m})/2 \qquad (10.3)$$

where I_0 is the surface radiation and I_m is the radiation at the base of the mixed layer.

During summer, and under a clear sky, photon flux densities may reach $2000\,\mu\text{E m}^{-2}\,\text{s}^{-1}$ (common values are about $1600\,\mu\text{E m}^{-2}\,\text{s}^{-1}$), whilst on cloudy days in winter, photon flux density may fall below $100\text{–}200\,\mu\text{E m}^{-2}\,\text{s}^{-1}$. At such low light intensities, photosynthesis becomes continuously and directly limited by light. Dependence of phytoplankton growth upon light intensity is usually described by $P\text{–}I$ curves, with light (I, as PAR [μE $\text{m}^{-2}\,\text{s}^{-1}$]) on the x axis, and some measure of activity (specific photosynthetic rate, P [mg C mg^{-1} Chl h^{-1}]

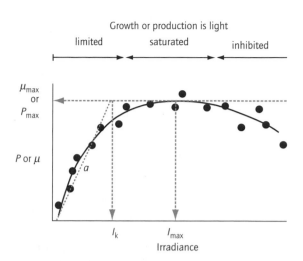

Fig. 10.1 Photosynthetic rate (P) or growth rate (μ) as a function of irradiance intensity: P_max is the light saturated rate of photosynthesis; I_k is the photoadaptational parameter; a is the light-limited curve.

or growth rate, μ [ln units day^{-1}]) on the y axis (Fig. 10.1).

Typical $P\text{–}I$ curves exhibit three regions. In the lowermost part of the curve, growth is directly correlated with light intensity. Within this range, growth is said to be **light limited**. Above this level, increment in growth becomes uncoupled from light, flattens and reaches a plateau. At this point, growth can be said to be **light saturated**: no further increase in light will enhance growth. The light intensity where maximal growth (P_max) occurs is called I_max. The **light saturation parameter** I_k (also referred to as the photoadaptational parameter) is derived from this relationship: it is the light intensity at which the extrapolated light-limited linear section of the $P\text{–}I$ curve intersects the horizontal line drawn by P_max.

Higher light intensities may lead to a decline in photosynthesis; this is the third, **light inhibited**, section of the relationship in Fig. 10.1, and occurs for a multiplicity of reasons. Amongst these are that UV radiation causes photochemical damage and, in some cases, that increased photorespiration occurs. The steepness of light-limited

Table 10.2 Characteristic values of P–I curves of phytoplankton. (From Reynolds 1984a, with some new data.)

Parameter	Usual values	Extremes
I_k (μE m^{-2} s^{-1})	60–100	7–300
I_{max} (μE m^{-2} s^{-1})	200–800	130–1200
P_{max} (mg C mg^{-1} Chl h^{-1})	Up to 7.5	Up to 12
a (mg C mg^{-1} Chl E^{-1} m^2)	6–18	2–37

photosynthesis (a) is of diagnostic importance, with the dimension mg C mg^{-1} Chl μE^{-1} m^2. Typical values of these parameters are given in Table 10.2. It must be noted that the P_{max} and μ_{max} values of different species depend on many factors, for example light–dark cycles and temperature (Kohl & Nicklisch 1988). Optimal temperatures vary in the range 8°C (for cold-adapted species in Antarctica) to 35°C (tropical green and blue-green algae). In the range 0–30°C, light-saturated photosynthetic rates vary by a factor (Q_{10}) of 2.0–2.3 per 10°C rise or fall (Reynolds 1997a).

Numerous functions with which to describe the P–I curve are available in the literature, ranging from the simple empirical relationship described as early as 1936 (Smith 1936), through more complicated equations (e.g. Iwakuma & Yasuno 1983), to dynamic models which take into account both scales of change in the physical environment and characteristic response times of biological processes (Pahl-Wostl 1992). Different kinds of functions used to describe the P–I curve are given in Iwakuma & Yasuno (1983). The value of I_k can be relatively simply calculated by the exponential saturation equation described by Webb et al. (1974) as:

$$P = P_{max}\left(1 - e^{-I/I_k}\right) \qquad (10.4)$$

from which

$$I_k = -I/\ln\left(1 - P/P_{max}\right) \qquad (10.5)$$

where P is rate of photosynthesis (in mg C mg^{-1} Chl h^{-1}), P_{max} is the maximum rate of photosynthesis

[mg C mg Chl^{-1} h^{-1}], I [μE m^{-2} s^{-1}] is light intensity, and I_k [μE m^{-2} s^{-1}] is light intensity at which the initial slope reaches the photosynthetic rate of P_{max}. The values for P and P_{max} can be substituted by daily growth rates μ and μ_{max} [ln units day^{-1}].

10.3.3 Adaptation to low and inhibitory light

Phytoplankton species respond to light fluctuation on different timescales. The so-called 'antenna effect', i.e. distributing pigments across the maximum area of interception which may be achieved in markedly flattened (shortened in one dimension) or attenuated shapes (shortened in two dimensions; Reynolds 1989) requires evolutionary time. Increasing the concentration of photosynthetic pigments per unit biomass (Foy & Gibson 1982a) raises the cell-specific potential for light harvesting, and may be effective within the lifetime of individuals. In photoadapting natural populations of Asterionella formosa, it has been observed that, during one cell-doubling, Chl a concentration per cell may increase by a factor of 1.8. A similar increase (1.5×) was observed for Cryptomonas cf. ovata, and in a Planktothrix mougeotii population a striking nine-fold difference has been demonstrated (Reynolds 1997a).

Species with the ability to adjust their vertical position in the water column (via buoyancy or swimming) can actively select that part which is optimal for their own photosynthesis. For example, in both shallow and deep lakes, and in laboratory cultures, Ceratium hirundinella select for water layers receiving 125–440 μE m^{-2} s^{-1} PAR (Harris et al. 1979; Heaney & Furnass 1980; Padisák 1985). Another means to increase the spectral width of light absorption (**chromatic adaptation**) is achieved by raising the concentrations of accessory pigments, especially phycobiliproteins (cyanoprokaryota and cryptophytes) and xanthophylls (Kirk 1983; Raven & Richardson 1984).

Phytoplankton species are variously adapted in order to utilise low light intensities (see examples in Table 10.3) and to tolerate high intensities. Characteristic values of the P–I curve are 'species-specific', and in some cases can explain develop-

Table 10.3 Characteristic values of *P–I* curves of some selected species growing under optimal conditions. Except those with asterisks, all data are from laboratory cultures.

Species	Temperature (°C)	I_k ($\mu E\,m^{-2}\,s^{-1}$)	I_{max} ($\mu E\,m^{-2}\,s^{-1}$)	k_{max} (day^{-1}, ln units)	Reference
Freshwater phytoplankton					
Synechococcus BGS-171	28–30	52	150–200	2.27	Mastala *et al.* (1996)
Picoplankton fraction	–	7*	20**	–	Petersen (1991)
Microcystis aeruginosa	–	–	550	–	Visser (1989)
	15–30	47–103	–	–	Nicklisch *et al.* (1983)
	–	73–119	–	0.12–0.36	Kromkamp *et al.* (1989)
Anabaena cylindrica	32	50	100	0.7	Dauta (1982)
Anabaena circinalis	20	198–317	–	–	Foy & Gibson (1982a)
Anabaena flos-aquae	20	131–190	–	–	Foy & Gibson (1982a)
Anabaena solitaria	20	160–211	–	–	Foy & Gibson (1982a)
	20	65–80	–	1.1	Schlangstedt *et al.* 1987
Anabaena spiroides	20	157–269	–	–	Foy & Gibson (1982a)
Anabaena variabilis	10	70	–	–	Collins & Boylen (1982)
	20	150	–	–	Collins & Boylen (1982)
	30	360	–	–	Collins & Boylen (1982)
	40	570	–	–	Collins & Boylen (1982)
Aphanizomenon flos-aquae	20	162–231	–	–	Collins & Boylen (1982)
	20	46	–	–	Zevenboom & Mur (1980)
Cylindrospermopsis raciborskii	–	20	121	0.95	Vörös (1995)
Cylindrospermopsis raciborskii 90% dominance	–	30–60*	70–240*	–	Dokulil & Mayer (1996)
Planktothrix agardhii	–	20	>180	0.4	Van Liere & Mur (1980)
	28	20	>180	0.56	Van Liere & Mur (1980)
	20	10	185	0.44	Zevenboom & Mur (1980)
	20	47–214	–	–	Foy & Gibson (1982a)
Planktothrix rubescens	–	10	130	–	Mur & Bejsdorf (1978)
	–	–	180*	–	Konopka (1982)
	20	90–173	–	–	Foy & Gibson (1982a)
'Oscillatoria' limnetica	–	73–90	–	–	Rijkeboer *et al.* (1992)
	20	95–168	–	–	Foy & Gibson (1982a)
Limnothrix redekei	8	80	–	0.49	Nicklisch *et al.* (1981)
	12	67	–	0.74	Nicklisch *et al.* (1981)
	17	50	–	0.83	Nicklisch *et al.* (1981)
	20	55	–	1.13	Nicklisch *et al.* (1981)
	24	102	–	2.05	Nicklisch *et al.* (1981)
	20	78–139	–	–	Foy & Gibson (1982b)
Prochlorothrix hollandica	–	80–140	–	–	Rijkeboer *et al.* (1992)
Chlorella vulgaris	–	98	300	–	Grobbelaar (1994)
Chlorella sp.	18	200	350	–	Flik (1985)
Dunaliella salina	–	–	220	–	Peng *et al.* (1994)
	–	73	345	–	Grobbelaar (1994)
Scenedesmus protuberan	20	28	154	0.5	Zevenboom & Mur (1980)
Scenedesmus crassus	30	85	270	1.6	Dauta (1982)
Coelastrum microporum	32	85	270	1.7	Dauta (1982)
Dictyosphaerium pulchellum	32	280	430	1.65	Dauta (1982)

Table 10.3 *Continued*

Species	Temperature (°C)	I_k ($\mu E\,m^{-2}\,s^{-1}$)	I_{max} ($\mu E\,m^{-2}\,s^{-1}$)	k_{max} (ln units day^{-1})	Reference
Scenedesmus quadricauda	32	210	300	1.65	Dauta (1982)
Scenedesmus obliquus	–	185–510	–	–	Kirk (1983)
Pediastrum boryanum	32	100	260	1.2	Dauta (1982)
Monoraphidium minutum	35	200	380	1.9	Dauta (1982)
Closterium acutum var. *variabile*	20	80–150			Coesel (1993)
Staurastrum chaetoceras	20	20	70	0.5	Coesel & Wardenaar (1994)
Cosmarium abbreviatum	20	30	70	0.35	Coesel & Wardenaar (1994)
Fragilaria bidens	20	125	285	1.1	Coesel & Wardenaar (1994)
Thalassiosira fluviatilis	–	70	–	1.1	Laws & Bannister (1980)
Skeletonema costatum	20	90	–	–	Kirk (1983)
	7	34	–	–	Kirk (1983)
	2	16	–	–	Kirk (1983)
Stephanodiscus minutulus	20	8–18	–	0.4–1.2	Kohl & Geisdorf (1991)
Nitzschia acicularis	20	10–32	–	0.32–1.67	Kohl & Geisdorf (1991)
Nitzschia palea	–	37–61	300	–	Kirk (1983)
Glenodinium sp.	–	65–130	–	–	Kirk (1983)
Peridinium gatunense	20	50–200	–	0.11	Berman & Dubinsky (1985)
	20	130	2000	–	Lindstrom (1984)
Ceratium hirundinella	–	140*	–	–	Harris *et al.* (1979)
Choromonas sp.	–	60–80	–	–	Humphrey (1979)
Other freshwater photosynthetic organisms					
Purple bacteria	–	25–70	–	–	Pfennig (1978)
Periphyton					
spring	–	250	–	–	Meulemans & Heinis (1983)
autumn	–	100–200	–	–	Meulemans & Heinis (1983)
Chilocyphus rivularis (freshwater bryophyte)	13	25	70	–	Farmer *et al.* (1988)
Marine species					
Oscillatoria thiebautii	–	500	650	–	Li *et al.* (1980)
Synechococcus sp.	–	44*	144*	–	Platt *et al.* (1983)

ment of community composition, or its vertical structure. The initial slope of the curve (a) in Fig. 10.1 for shade tolerant species is steep. Therefore I_k is attained at low light intensities. These taxa, such as *Planktothrix rubescens*, or picoplanktonic *Synechococcus*, are found in deep-layer maxima receiving 1% or less of surface incident radiation. Interestingly, the deep-layer freshwater bryophyte *Chilocyphus rivularis* exhibits similar characteristics.

Another group of shade-tolerant species is that which forms dense populations, but which still generally remains buoyant. Classic examples are *Planktothrix agardhii* and *Limnothrix redekei*. Recent investigations have shown that the nitrogen fixer *Cylindrospermopsis raciborskii* must be placed in this group (Padisák 1997). Photosynthesis of the above taxa becomes light-inhibited at relatively low light intensities ($<200\,\mu E\,m^{-2}\,s^{-1}$). Heterocytic, bloom-forming cyanoprokaryota (e.g. *Anabaena, Aphanizomenon*) usually have relatively high I_k, requirements, and lose their buoyancy at higher light intensities. *Microcystis*, which often forms dense surface blooms, seems to be especially adapted to high intensities: its I_k values fall within higher ranges (Reynolds' (1990) model calculated an I_k of 753 μE m^{-2} s^{-1} at 20°C). However, inhibitory intensities are very high. In Visser's (1989) experiments, no growth reduction was experienced at 550 μE m^{-2} s^{-1}, and in natural populations optimal photosynthetic rates were found even at 1000 2600 μE m^{-2} s^{-1} (Paerl *et al.* 1985; Zohary & Robarts 1989). Green algae commonly possess higher I_k than most cyanoprokaryotes and diatoms, consequently their particular photosynthesis becomes inhibited only at high light intensities.

10.4   NUTRIENTS AND NUTRIENT UPTAKE

The dry matter (DM) of phytoplankton cells is comprised of up to 90% carbon, oxygen and hydrogen. In some groups, the substance of which the cell wall is composed, such as silica (in diatoms) or calcium carbonate (in *Phacotus*), may contribute significantly to DM content. Macronutrients (N, P, S, K, Mg, Ca, Na, Cl) typically constitute >0.1% of dry weight, whilst micronutrients (Fe, Mn, Cu, Zn, Mo, Co, B, V) occur in traces (<<1%). Some micronutrients, for example Cu and Zn, are toxic if present at higher concentrations.

Nutrients are obtained from the external medium, but can only be taken up if they occur as soluble compounds present as diffusible ions or, sometimes, as non-dissociated small molecules such as silicic acid. Concentrations of these compounds are usually higher inside the cell than in the surrounding medium. Therefore uptake by passive diffusion rarely occurs. Under natural conditions, transport into the cell will thus proceed against an (electro)chemical concentration gradient, and must be mediated by energy-consuming, enzyme-like, transport systems (ion pumps).

In nature, unlimited growth of organisms, including phytoplankton species, does not occur.

When growing, a population will, sooner or later, exhaust at least one of the resources necessary for growth. Supply of this substance will then limit further increase of biomass (Liebig's 'Law of the Minimum'). It is well known that resources do not occur in constant concentrations: their amounts and availability change both in time and space. In more dynamic terms, that resource whose supply fails to satisfy the rate of its consumption by the organisms will becoming limiting.

10.4.1   Kinetics of nutrient uptake

Various empirical models have been developed to describe uptake kinetics and nutrient-related growth of phytoplankton. Michaelis–Menten kinetics are invoked in order to relate nutrient uptake rates to external concentrations (Dugdale 1967) as:

$$v = v_{max}S/(k_m + S) \qquad (10.6)$$

Description of the symbols used in the above, and in the following uptake equations, is summarised in Table 10.4. The variables v and v_{max} can be substituted by specific biomass increase rates in the Monod equation, to describe nutrient impact of growth as:

$$\mu = \mu_{max}S/(k_s + S) \qquad (10.7)$$

At low ambient nutrient concentrations, species with low k_s can, in theory, outcompete others. However, combinations of kinetic constants (μ_{max} and k_s) have led to the idea that there are several basic growth strategies (Sommer 1984). These are (i) **affinity adapted species** (high μ_{max}/k_s), (ii) **velocity adapted species** (high k_{max} is associated with high k_s) and (c) **storage adapted species** (where relatively slow growth is based on nutrients previously stored). Common and extreme values for phosphorus, nitrogen and silica uptake, and some specific half-saturation constants for phosphorus and silicon, are given in Tables 10.5–10.7.

Numerous deviations from the above empirical relationships have been described. The Monod model is based on the assumption that the possible

Table 10.4 Symbols of nutrient uptake equations used in section 10.4.

v	nutrient uptake rate (the amount of nutrient that is taken up in unit time per unit amount of phytoplankton)
v_{max}	maximal uptake rate
S	external nutrient concentration
S_{ma}	external concentration of the nutrient when growth is saturated
k_m	half-saturation constant; the nutrient concentration at $v_{max}/2$
μ	growth rate (in ln units day^{-1})
μ_{max}	maximal growth rate growth rate (in ln units day^{-1}) that occurs when growth is not limited by the nutrient
k_s	half-saturation constant; the nutrient concentration at $\mu_{max}/2$
μ'_{max}	theoretical maximum of growth rate
q	cell quota; amount of nutrient available in the cell
q_0	minimal cell quota; the minimal intracellular amount of the nutrient that is necessary for minimal functioning
q_{max}	maximum cell quota
v'_{max}	nutrient uptake rate at q_{max}
v''_{max}	nutrient uptake rate at q_0
S_t	the amount of a given nutrient capable of being accumulated in the cell
S_0	nutrient concentration in the cell at time t
$S_¥$	hypothetical nutrient concentration in the cell provided continued and unlimited capable of being accumulated in the cell
K_i	the inhibition constant for nutrient uptake
B_{max}	maximal achievable biomass
S_{tot}	total amount of limiting nutrient

Table 10.5 Common and extreme values of k_s, q_0 and q_{max} for phosphorus, nitrogen and silicon. (From Sommer 1994.)

Element	Symbol	Common values	Extremes
Phosphorus	k_s (μmol L^{-1})	0.02–0.2	0.003–1.83
	q_0 (mol P mol^{-1} C)	0.0008–0.002	0.0003–0.008
	q_{max} (mol P mol^{-1} C)	0.01	0.008–0.04
Nitrogen	k_s (μmol L^{-1})	0.3–3.0	0.036–11.6
	q_0 (mol N mol^{-1} C)	0.02–0.05	0.014–0.18
	q_{max} (mol N mol^{-1} C)	0.15	0.09–0.28
Silicon	k_s (μmol L^{-1})	2–5	0.88–19.3
	q_0 (mol Si mol^{-1} C)		0.05–0.16
	q_{max} (mol Si mol^{-1} C)		0.12–0.8

can be demonstrated that maximum nutrient uptake rate is a linear function of cell quota. Starving cells will take up nutrients at a greater rate than those with maximal storage (Morel 1987):

$$v_{max} = v''_{max} = (v''_{max} - v'_{max})$$
$$(q - q_0)/(q_{max} - q_0) \quad (10.9)$$

The amount of a given nutrient capable of being accumulated in the cell can be described (Kromkamp *et al.* 1989) as

$$S_t = (S_0 - S_¥)\exp(-K_i t) + S_¥ \quad (10.10)$$

Some specific K_i values are given in Table 10.8.

If maximum nutrient uptake rate is high, and the inhibition constant is low, the cell should be able to store a large surplus of the nutrient. This is called **storage strategy**. The lower the inhibition constant, the longer nutrient uptake will continue. Temporarily favourable nutrient conditions (**nutrient pulses**) may be utilised in order to build up an internal store of nutrient (**luxury uptake**). However, this strategy can only be successful if the time span between nutrient pulses is not too prolonged.

For example, *Microcystis aeruginosa* may be a significant organism in waters with fluctuating phosphorus input. The success of this species

threshold for uptake is negligible. This model can be applied if, for example, external concentration of the nutrient can be kept constant (e.g. in chemostat cultures) or for steady-state conditions for which nutrients are not stored (e.g. silicon in diatoms). However, in nature this condition is rarely fulfilled. In the Droop model (Droop 1973), growth is related to the intracellular concentration of the nutrient:

$$\mu = \mu'_{max}(1 - q/q_0) \quad (10.8)$$

Whilst, in the Monod equation, μ_{max} is a realistic, achievable maximum rate of growth, μ'_{max} is a theoretical value achieved at 'unlimited' cell quota. It

Table 10.6 Phosphorus half-saturation constants established in laboratory cultures.

Species	Temperature (°C)	μ_{max} (day^{-1})	k_s(P) (μg P L^{-1})	Reference
Microcystis wesenbergii	20	0.4–0.5	0.5	Ahlgren (1985)
Synechococcus BGS-171	28–30	2.27	1.2	Mastala *et al.* (1996)
Anabaena flos-aquae			49.3	Kromkamp *et al.* (1989)
Aphanizomenon flos-aquae			7.4	Kromkamp *et al.* (1989)
Cylindrospermopsis raciborskii	26.5	1.6	1.5–2.5	Istvánovics (1997)
Planktothrix agardhii	15	0.4	1.0	Ahlgren (1988)
Prochlorothrix hollandica			6.7	Kromkamp *et al.* (1989)
Cyclotella meneghiniana			8	Tilman & Kilham (1976)
Stephanodiscus hantzschii	10	0.5	4.0	Kilham (1984)
Asterionella formosa			0.64–1.28	Tilman & Kilham (1976)
Synedra filiformis	20	0.65	0.1	Tilman *et al.* (1982)
Peridinium gatunense	20–25		5.6–9.5	Berman & Dubinsky (1985)
Chlorella pyrenoidosa			21.8	Nyholm (1977)
Chlorella minutissima	20	2.15	7.0	Sommer (1986a)
Scenedesmus spinosus		2.78	7.2	Shafik (1991)
Cosmarium abbreviatum	20	0.5	0.35	Spijkerman & Coesel (1996)
Staurastrum pingue	20	1.03	1.22	Spijkerman & Coesel (1996)
Volvox globator	20	0.43	58.6	Tilman *et al.* (1982)

Table 10.7 Parameters of the Monod equation: μ_{max} [ln units day^{-1}], $K_{s,Si}$ and threshold concentrations (Si$_0$) for silicon-limited growth of some diatom species.

Species	Temperature (°C)	$K_{s,Si}$ (μmol L^{-1})	Si$_0$ (μmol L^{-1})	μ_{max} (day^{-1})	Reference
Synedra filiformis	20	17.7		1.1	Tilman (1977)
Synedra ulna	8	4.9	0.2	0.16	Tilman *et al.* (1982)
	13	3.8	0.3	0.71	Tilman *et al.* (1982)
	20	4.0	0.6	0.65	Tilman *et al.* (1982)
	24	4.4	0.3	0.78	Tilman *et al.* (1982)
Asterionella formosa	4	1.3	0.6	0.35	Tilman *et al.* (1982)
	8	1.6	0.2	0.52	Tilman *et al.* (1982)
	13	2.5	0.2	0.79	Tilman *et al.* (1982)
	20	3.7	0.1		Tilman *et al.* (1982)
Fragilaria crotonensis	20	1.5		0.62	Tilman (1982)
Tabellaria flocculosa	20	19.0		0.74	Tilman (1982)
Diatoma tenuis	20	0.85		1.51	Tilman (1982)
Cyclotella meneghiniana	20	1.44		0.92	Tilman & Kilham (1976)
Stephanodiscus minutulus	10	0.31		0.88	Kilham (1984)
	20	0.88		0.15	Kilham (1984)

Table 10.8 Some examples of inhibition constant (K_i) for phosphorus uptake. (Data from Kromkamp *et al.* 1989.)

Species	K_i ($\mu mol\, P\, L^{-1}\, h^{-1}$)
Aphanizomenon flos-aquae	0.72
Microcystis aeruginosa	0.82
Planktothrix agardhii	1.14
Synechococcus 6301	2.22
Prochlorothrix hollandica	0.78

under such circumstances is due to its high v_{max} relative to its low K_i. It is able to accumulate c. 110 $\mu g\, P\, mg$ protein^{-1}, whereas *Planktothrix agardhii* (whose v_{max} is 60% lower, and K_i, 40% higher) may accumulate only 30 μg. Using this stored phosphorus, *M. aeruginosa* can then produce 18 new cells (4.2 doublings), whilst *P. agardhii* can form only three cells (1.6 doublings). Combined with excellent buoyancy regulation ability, this strategy may be highly successful in nutrient-rich waters with strong vertical differentiation of the euphotic zone (Olrik 1994).

10.4.2 Phosphorus

The phosphorus available for phytoplankton uptake occurs (in the inorganic form) as orthophosphate ions, and as organic, largely biogenic, low molecular weight, dissolved phosphorus compounds, in the concentration range 0.1–1000 $\mu g\, P\, L^{-1}$. The Monod equation can be applied if external phosphorus concentrations are relatively high, but at the lower concentrations which frequently occur in nature, other considerations must be taken into account. According to the Monod function, the concentration dependence of the specific growth rate passes through the origin, i.e. at $S = 0$. Thus, any increase in external phosphate concentration above zero should lead to renewed growth. For kinetic and energetic reasons, however, phosphate uptake at very low ambient concentrations is impaired, since gross uptake rates are compensated by leakage or efflux rates. Therefore, growth will cease before external concentration declines

to zero. Concentrations of substrates below which cells cannot acquire nutrients are called **threshold concentrations**. Phosphorus thresholds for phosphorus-deficient algae usually fall into the nanomolar range. Using a linear flow–force relationship, Falkner & Falkner (1989) have shown that the threshold value can be extrapolated from a plot of steady-state uptake rates versus the logarithm of external phosphate concentration.

Recent studies of phosphorus uptake and phosphorus leakage rates in the range of the threshold concentrations introduce some new and important considerations. **Affinity adaptation** (i.e. reduction of the threshold to the lowest possible value), operating on an energetic basis (Falkner *et al.* 1995), may lead to reduction of leakage. As a result, regulation of leakage, or efflux of phosphate, may be of evolutionary importance for the adaptive adjustment of the threshold value of a population. While small algae and bacteria, owing to more intense affinity adaptation, seem to be better competitors for phosphorus (Currie & Kalff 1984a, b), no similar size relationship is apparent in leakage. In experiments by Lean & Nalewajko (1976), *Chlorella pyrenoidosa* was the most conservative in leakage reduction, followed by *Scenedesmus quadricauda*, *Navicula pelliculosa* and *Anabaena flos-aquae*. Efficient leakage reduction may provide a competitive advantage to larger and more complex organisms, which possess a greater number of intracellular compartments. Lower leakage will increase the efficiency with which organisms use acquired phosphorus for growth (Istvánovics & Herodek 1995).

10.4.3 Nitrogen

The main sources of nitrogen for phytoplankton growth include inorganic compounds (nitrate, nitrite, ammonia), organic molecules (urea, free amino-acids, peptides) and, for cyanobacteria, atmospheric dinitrogen (N_2). If nitrates (or, less significantly, nitrites) serve as a nitrogen source, they must first be reduced within the cell before they enter into its metabolism. This requires the availability both of NADPH$_2$ and those enzymes (nitrate reductase and nitrite reductase) which utilise (respectively) NADPH and reduced ferredoxin. The energy requirement of this process roughly equals that needed in order to break the highly stable triple bond of dinitrogen during atmospheric fixation (Postgate & Cannon 1981). In the case of ammonia or organic nitrogen consumption, there is no such energy demand.

Nitrate may be very abundant in polluted waters (in the range of $mg\,L^{-1}$); elsewhere, concentrations lie typically within the range $10–1000\,\mu g\,L^{-1}$. Low nitrate concentrations occur in the upper 50–100 m of the world's oceans, and in many tropical lakes. Many high-altitude lakes are also nitrate-deficient. In temperate lakes, nitrate concentrations may exhibit marked seasonality: high concentrations occur during full circulation, whilst low epilimnetic concentrations are found during stratification (see Chapter 2). In the anoxic hypolimnion, nitrate is reduced to ammonia. In unpolluted waters, ammonia concentrations rarely exceed $150\,\mu g\,L^{-1}$. Low ammonia concentrations do not necessarily imply ammonium deficiency, since in aquatic ecosystems the ion is rapidly processed.

Atmospheric dinitrogen is, theoretically, an unlimited source. However, the ability to carry out direct fixation is limited to certain bacteria and cyanobacteria. The process can be described by the following bulk equation (Stewart 1974):

$$N_2 + 4NADPH + 6H^+ + 12ATP \Leftrightarrow$$

$$2NH_4^+ + H_2 + 12ADP + 4NADP^+ + 12P \quad (10.11)$$

Nitrogen fixation is energy consumptive, and requires a high light income, as well as the presence of micronutrients (Mo, Fe). Thus, a low dissolved nitrogen concentration is far from being a guarantee that nitrogen fixers will thrive (Reynolds 1987b). The process is driven by the nitrogenase enzyme system which is rapidly destroyed in the presence of oxygen (irreversibly in cell-free extracts but reversibly under most *in vivo* conditions). Numerous morphological and physiological adaptations have been evolved in order to overcome this biochemical paradox, for example:

1 During differentiation of the heterocyte, a change in colour, from bright green to yellowish,

takes place. This signals changes in the light trapping apparatus (PS I and PS II; see section 10.3.1.). 'Yellowing' of heterocytes can be attributed to loss of phycobiliproteins (especially the blue-coloured phycocyanin), whilst Chl *a* and carotenoids remain. Thus PS I activity is retained, whilst PS II ceases to function. Elimination of PS II negates photolysis of water, which then leads to oxygen evolution. Remaining PS I activity needs an alternative electron donor. The source appears to be the neighbouring vegetative cells which supply PS I in the heterocytes with reduced carbon compounds (Wolk 1982). Thus, PS I function is able to fulfil the intermediary electron acceptor demand of nitrogen fixation. As a consequence of elimination of PS II, intracellular oxygen evolution is reduced.

2 The thickened wall of heterocytes reduces physical diffusion from the surrounding, sometimes hypersaturated, medium. Parallel mucilage production also retards oxygen diffusion.

3 Association with heterotrophic bacteria leads to localised oxygen consumption (Paerl 1988).

4 Among other mechanisms, physiological oxygen removal near the site of nitrogen fixation and photorespiration involve the production of enzymes (superoxide dismutate and catalase) which convert oxygen to hydrogen peroxide (H_2O_2), and finally to water.

Although nitrogen fixation is typical of the Nostocales, it is well established that it can be induced in non-heterocytic genera of cyanobacteria. Such genera occur in temporarily oxygen-deficient environments. A variety of adaptations provide oxygen-deficient cellular environments in non-planktonic algae (Paerl 1988). In respect of phytoplankton ecology in lakes, nitrogen fixation by non-heterocytic genera may become important if it occurs at night in phosphorus-rich lakes with dense algal populations, when photosynthesis cannot take place and when respiration of the biota, including the algae themselves, reduces dissolved oxygen concentrations significantly (Olrik 1994).

10.4.4 *Heterotrophic nutrition*

The carbon requirement of algal growth is usually accounted for by photosynthetic assimilation.

Many species of phytoplankton possess an obligate requirement for externally produced organic carbon, in the form of vitamins (Gaines & Elbrächter 1987), or may utilise organic carbon compounds in the dark. Such 'heterotrophy' was for a long time thought to be of negligible quantitative importance (Sanders & Porter 1987). Therefore, it was surprising to discover that some common freshwater genera (*Dinobryon, Uroglena, Chrysosphaerella, Ochromonas*) may depend more on ingested prey than on photosynthesis, at least under low light conditions (Bird & Kalff 1986, 1987).

However, it would be misleading to discuss such phagotrophy as a deviant mode of nutrition. Since the photosynthetic function of each eukaryote can be traced to one or more endosymbiotic events involving ingestion and retention of unicellular photosynthetic organisms (e.g., the mitochondria: Larkham 1999), phagotrophy has been a crucial element in the evolution of photosynthetic eukaryotes (Raven 1997). From the ecological point of view, algal phagotrophy (particle digestion in phagocytic vacuoles) appears in some cases to be more than a simple facultative supplement. Kimura & Ishida (1986) showed that *Uroglena americana* can only grow in the presence of bacterial prey. Other taxa are facultative mixotrophs. For example, *Dinobryon* can be maintained axenically (Lehman 1976), but can also rely on particulate prey. *Dinobryon sertularia* collected from a thin, metalimnetic layer of Lake Gilbert, Quebec, incorporated carbon from radio-labelled phagocytosed bacterial prey with an efficiency of 54% (range 40–79%; Bird & Kalff 1989).

In a strongly humic lake investigated by Salonen & Jokinen (1988), *Ochromonas* sp. and *Chromulina* sp. ingested 75–203% of their body carbon per day from bacteria. During a chrysophyte bloom, Porter (1988) found that phytoflagellate grazing exceeded heterotrophic grazing, and constituted 55% of the total bactivory both by microflagellates and by zooplankton. She demonstrated an inverse relationship between the rates of ingestion and light, whilst primary production was correlated positively with light.

Mixotrophy may help phytoplankton to escape either carbon limitation or light limitation, espe-

cially in deep layers. Many of the species involved can only take up carbon dioxide as their inorganic carbon source. Phagotrophy occurs in the Cryptophyceae (*Cryptomonas*), Dinophyceae (*Amphidinium, Ceratium, Gymnodinium, Peridinium*), Chrysophyceae (*Chromulina, Chrysococcus, Chrysosphaerella, Chrysamoeba, Dinobryon, Mallomonas Paraphysomonas, Poterioochromonas, Ochromonas, Uroglena*), the Prymnesiophyceae (*Chrysochromulina*), and the Chlorophyceae (*Chlamydomonas, Chlorogonium*).

10.4.5 Eutrophication and oligotrophication

Eutrophication is defined as the enrichment of water with plant nutrients, primarily phosphorus and nitrogen. This leads to enhanced plant growth, which produces visible algal blooms, floating algal and/or macrophyte mats, and benthic algal and submerged macrophyte agglomerations. On decaying, this plant material causes depletion of oxygen, leading to an array of secondary problems such as fish mortality, liberation of corrosive gases, and other undesirable and toxic substances (Vollenweider 1989). Each of the above seriously jeopardises almost every kind of water use. Following eutrophication, the fauna and flora of lakes are often observed to change radically.

Characterisation of lakes according to trophic states is usually achieved using **trophic scales**, of which numerous examples are available. The most widely used system (Table 10.9) was developed by the OECD (1982). Recent studies have shown that picophytoplankton are a very sensitive estimator of trophic status, at least in the range ultra-oligotrophic to moderately eutrophic. Therefore these variables are also included in Table 10.9.

Recently, it has become clear that what leads fundamentally to eutrophication is an increase in one or more of the resources essential for plant growth. Although each may be critical (e.g. in caves frequented by tourists, illumination has been shown to be a crucial factor), increased phosphorus loading was recognised very early to be the commonest reason for eutrophication. In contrast, in fresh waters, nitrogen enrichment alone only

Table 10.9 Boundary value criteria for trophic categories according to [1]OECD (1982), [2]Lampert & Sommer (1993) and [3]Stockner (1991).

Component†	Ultra-oligotrophic	Oligotrophic	Mesotrophic	Eutrophic	Hypertrophic
[1]TP_a	<4	4–10	10–35	35–100	>100
[2]TP_{sot}	<5	5–10	10–30	30–100	>100
[1]Chl_m	<1	1–2.5	2.5–8	8–25	>25
[1]Chl_{max}	<2.5	2.5–8	8–25	25–75	>75
[1]Sec_m*	>12	12–6	6–3	3–1.5	<1.5
[1]Sec_{min}*	>6	6–3	3–1.5	1.5–0.7	<0.7
[1]O_{sat}**	90%	80%	90–40%	40–0%	10–0%
[3]N_{pico}***	$<2 \times 10^4$	2×10^4–2×10^5	2×10^5–2×10^6	?	Up to 10^8
[3]C_{pico}***	>80%	80–55%	55–25%	>25%	>25%

* Only for lakes whose turbidity is organic; ** only for stratifying lakes; *** only for non-acidic lakes.

† TP_a, annual mean total phosphorus [µg L^{-1}]; TP_{sot}, total phosphorus during the spring overturn [µg L^{-1}]; Chl_m, annual mean chlorophyll [µg L^{-1}]; Chl_{max}, annual maximum of chlorophyll [µg L^{-1}]; Sec_m, annual average of Secchi disc transparency [m]; Sec_{min}, minimum of Secchi disc transparency [m]; O_{sat}, percentage O_2 saturation over ground; N_{pico}, annual average of cell number of picophytoplankton [ind mL^{-1}]; C_{pico}, percentage contribution of picophytoplankton to annual average of total phytoplankton biomass and production.

exceptionally produces increased phytoplankton growth. However, in subtropical and tropical regions, and especially at high altitudes, nitrogen limitation seems commonly to occur. Other kinds of phosphorus-rich and nitrogen-deficient systems include very young waters in craters formed by vulcanism (Jolly & Brown 1975), and those which almost continuously receive groundwater inputs (Boucher *et al.* 1984).

Using data from many lakes, Vollenweider & Kerekes (1980) developed an empirical relationship between phosphorus loading and Chl *a* concentration which has proved to be widely applicable. However, experience has shown that individual patterns of eutrophication and oligotrophication may significantly vary from its actual predictions. Subsequent research, without undermining the validity of Vollenweider's work, has shown that discrepancies between the model and observational data may be attributed to several reasons (most notably that the Vollenweider model describes only the 'linear' part of a non-linear relationship, whilst perturbations occur mostly at the 'margins').

It is often observed that a response of the phytoplankton biomass to decreased phosphorus loads is delayed (Stage 1: no response stage; Fig. 10.2). This is partly due to **community resilience**. The phytoplankton, as with any other organised community, is able to absorb fluctuation of external resources within a certain range. Another factor is that the inorganic components of seston often possess a phosphorus-binding capacity which restricts the amount of the element readily available. Because the sediment : water-volume ratio in shallow bodies is much higher than that in deep lakes, accumulation of phosphorus in sediments of shallow lakes is relatively more important. Both community resilience and binding capacity are finite, so that, ultimately, increase in external phosphorus load leads to enhanced availability, and a rise in the capacity of the lake to support phytoplankton biomass.

This increase, in the long term, is roughly linear and can be described by the Vollenweider equation (Stage 2: biomass increase; Reynolds 1992b). Phytoplankton can absorb phosphorus enrichment so long as other factors, usually either nitrogen or light, do not limit growth. The lake's nitrogen-determined capacity may lie either below or above its light-determined capacity. If the former is reached sooner, heterocytic blue-green algae,

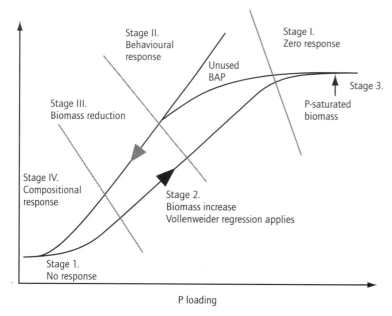

Fig. 10.2   Phytoplankton biomass as a function of increasing phosphorus load: 1–3 indicate stages of eutrophication; I–IV indicate stages of phytoplankton response during restoration (decreasing phosphorus load); BAP, biologically available phosphorus.

which are able to fix their own atmospheric nitrogen, may well begin to increase. Therefore phytoplankton biomass replete with nitrogen and phosphorus should then grow further and finally reach the light-determined capacity of the lake (Stage 3: resource saturated biomass). If the development of nitrogen-fixing blue-green algae is impaired for any reason, maximal biomass capacity will be limited by available nitrogen.

When phosphorus loading is reduced in order to improve water quality, phytoplankton biomass does not respond, so long as the lake remains phosphorus-saturated (Stage I: zero response) and declines only slightly whilst phosphorus is available (Stage II: behavioural response). In shallow lakes, these phases are prolonged, because sedimentary-bound phosphorus previously accumulated (the sedimentary **internal load**) remains significant. Further load-reduction and exhaustion of internal phosphorus sources leads to a roughly linear decrease in phytoplankton biomass capacity (Stage III: biomass reduction response). Finally a phytoplankton composition characteristic of a more oligotrophic lake (Stage IV: compositional response) may be re-established.

As demonstrated in many case studies (Sas 1989), lakes behave rather individually, both during eutrophication and oligotrophication. During the first decade of oligotrophication of the large, deep Lake Constance (the Bodensee), spring total phosphorus (TP) declined from c.130 to $50\,\mu g\,L^{-1}$. This decrease was mirrored by an approximately proportional reduction in summer phytoplankton biomass, although spring biomass seemed unresponsive. At genus level, the majority of trends for the eutrophication and oligotrophication periods are found to be symmetrical (Sommer et al. 1993). In accordance with the oligotrophication model (Fig. 10.2), the response corresponds to Stages III and IV.

Lake Balaton is a shallow lake of similar size to the Bodensee, which, as a consequence of increased phosphorus loading, underwent rapid eutrophication during the 1960s–1970s. Phytoplankton, especially summer biomass maximum, increased. However, most of the external phosphorus load found its way into the sediments. Restoration management since 1984 has led to an approximate 50% reduction in external biologically available phosphorus load (Istvánovics &

Herodek 1994), but response of the phytoplankton has been disappointingly insignificant. Increase in summer maximum biomass continued until 1992. During subsequent years, both biomass maximum and species composition were highly variable. During some years (1992, 1994) heavy blooms occurred, whilst during others (e.g., 1993) both maximum biomass and species composition characteristic of a mesotrophic state developed. It was also shown that development of phytoplankton in individual years is highly dependent upon stochasticity of the weather (Padisák 1993, 1998). The phytoplankton of the lake has probably been in an insufficiently defined 'behavioural response' stage (Padisák & Reynolds 1998).

10.4.6 *Toxicity of Cyanoprokaryota*

Since eutrophication often leads to an increase of cyanoprokaryotes, many of which are at least potentially toxic, the problem of toxicity in fresh waters is closely related to that of eutrophication. Algal toxicity has also long been associated with marine flagellates, but toxic effects of freshwater cyanobacteria have been repeatedly reported since the 19th century (Francis 1878). Poisoning cases involve sickness and death of livestock, pets and wildlife following ingestion of water containing toxic algae or toxins released by senescent cells. No acute lethal poisoning of humans has been confirmed but, nevertheless, sublethal effects should not be ignored. Because world-wide eutrophication of inland waters facilitates dense cyanoprokaryote blooms, an increase in 'toxic cases' is predictable.

Cyanotoxins are secondary metabolites, i.e. they are not used by the producing organism for its own primary metabolism. They include compounds that act as hormones, antibiotics, allelochemicals and toxins. By definition, toxins are secondary compounds which harmfully affect other tissues, cells or organisms (Carmichael 1992). By their chemical structure and metabolic effect, toxins are either neurotoxic, hepatotoxic or dermatic lipopolysaccharides. Little is known about the newly described cylindrospermopsin;

it may, probably, establish a new group of cyanotoxins.

The first cyanotoxin recognised, and chemically defined, was anatoxin-a, which belongs to the group of neurotoxins. These are alkaloids with relatively low molecular weight, which are potent postsynaptic cholinegic nicotinic antagonists, causing a depolarising neuromuscular blockade. Signs of toxicosis in field cases of wild and domestic animals include staggering, muscle fasciculations, gasping, convulsions and opistothonos (birds). Death is most probably due to respiratory arrest, and occurs within minutes or a few hours. Numerous freshwater planktonic cyanoprokaryotes produce various kinds of neurotoxins, the most important are *Anabaena flos-aquae*, *A. spiroides*, *A. circinalis*, *Aphanizomenon flosaquae* and *Planktothrix rubescens*.

The best known hepatototoxin is microcystin, which is a cyclic heptapeptide. The biochemical basis of the toxic effect is that, using an irreversible covalent bond, it blocks proteinphosphateses 1 and 2a, which are basic metabolic enzymes in every eukaryotic cell. Clinical signs of hepatotoxicosis have been observed in field poisonings of cattle, sheep, horses, pigs, ducks and other domestic and wild animals. In acute toxicosis, death due to intrahepatic haemorrhage occurs in some hours to some days. Subacute exposure to hepatotoxins leads to an increased death rate related to insufficient liver function and liver cancer. Most human intoxications fall into the subacute range. Besides the various structural varieties of microcystin, nodularin (produced by *Nodularia spumigena*) is the best known hepatotoxin. Chemically undefined microcystins have been found in numerous other planktonic genera, including *Anabaena*, *Oscillatoria*, *Nostoc*, *Aphanizomenon*, *Gloeotrichia* and *Coelosphaerium*.

Cylindrospermopsin is produced by *Cylindrospermopsis raciborskii*. Its toxic effect is manifested via proteinsynthesis blocking. Symptoms of field toxicosis are somewhat similar to those of hepatotoxins, but neurotoxic effects on isolated snail neurons may also be detected (Kiss *et al.* 1997). Interestingly, it considerably suppresses some of the ontogenetic phases in higher plants (among them

Table 10.10 Surveys of toxic cyanoprokaryotes.
(From Chorus 1997.)

Region/country	Number of samples	Percentage found to be toxic
England	78	70
Scandinavia	51	59
Finland	188	44
Denmark	96	72
Baltic See	25	72
Wisconsin, USA	102	27
The Netherlands I	10	90
The Netherlands II	29	79
Hungary	35	82
Germany I	6	67
Germany II	12	83

Phragmites), but in experiments with algae either a stimulating or a neutral effect was found (Borbély *et al.* 1997, Ördög 1997).

According to present knowledge, the 'one species–one toxin' assumption cannot be sustained: a single species may produce toxins of different kinds. Surveys of the frequency of occurrence of toxic cyanoprokaryota have led to a surprisingly high record: more than half the samples proved to be toxic in most countries (Table 10.10). Although it is tempting to characterise toxicity as a potentially adaptive and competitive feature, the ecological and physiological traits related to the environmental signals leading to toxin production remain unclear.

10.5 POPULATION GROWTH

The number of specimens in a phytoplankton population increases via a series of cell divisions. In the case of increase at a constant rate, each doubling of cell number may be accomplished over similar time intervals, so that increase in cell numbers is exponential. Each generation of cells attains approximately the dimensions of the mother cell. Therefore, the increase in the total cell volume of the population as a whole will re-

main proportional to the increase in the number of cells. If no loss takes place, growth (= gross increase) and observed biovolume increase (= net increase) will be equivalent. Expressed as rates, they will be $k = \mu$, where μ is the rate of gross increase and k is the rate of net increase.

In natural populations, however, as a consequence of loss of cells through washout, grazing, sinking, etc., net increase is lower than gross. Then,

$$k = \mu - \lambda \qquad (10.12)$$

where λ is the rate of loss.

Both gross and net increase per unit time are non-linear functions of time. Net rate of increase per unit time can be estimated from the logarithms of biomass (cell number) data when continuous increase occurs as

$$k = (\ln N_2 - \ln N_1)/(t_2 - t_1) \qquad (10.13)$$

where N_1 is the biomass at time t_1 and N_2 is the biomass at time t_2. In phytoplankton ecology, 1 day is usually considered as the unit of time. Then the unit of k is day^{-1}.

Numerous data on maximal growth rates observed in laboratory cultures have been published. The fastest rate (7.97 day^{-1}, corresponding to a t_g (see below) of 2.09 h) was observed in a strain of *Synechococcus* (Kratz & Myers 1955). Because growth conditions in experimental systems are optimised for light, nutrients, temperature, etc., and loss is reduced (no grazing, no sinking, etc.), significantly higher rates are measured than in field populations. Examples in Table 10.11 reflect that in some cases (*Ceratium, Cryptomonas*) field populations approach maximal rates observed in laboratory cultures. However, most phytoplankton populations increase at a much lower rate in nature.

For example, similarly high rates of increase (1.6 day^{-1} and 1.7 day^{-1}) were established by two different laboratories (Vörös 1995: Hawkins 1996) for *Cylindrospermopsis raciborskii*. A net increase rate of 0.23 day^{-1} was found in two exponentially increasing field populations; in both cases

Table 10.11 Net growth rates (k [day^{-1}]) of some species observed in laboratory cultures (k_{lab}) and in natural populations (k_{field}).

Species	k_{lab}	Reference	k_{field}	Reference
Microcystis aeruginosa	0.48*	Reynolds (1984a)	0.37	Padisák, unpublished
Synechococcus sp.	1.05	Fahnenstiel *et al.* (1991)	0.42	Fahnenstiel *et al.* (1991)
Anabaena flos-aquae	0.78	Reynolds (1984a)	0.41	Sommer (1981)
Aphanizomenon flos-aquae	0.98	Reynolds (1984a)	0.43	Sommer (1981)
Cylindrospermopsis raciborskii	1.6	Vörös (1995)	0.23	Padisák (1994)
	1.7	Hawkins (1996)	0.23	Fabbro & Duivervoorden (1996)
Cryptomonas ovata	0.81	Reynolds (1984a)	0.89	Sommer (1981)
Cryptomonas erosa	0.83	Reynolds (1984a)	0.48	Sommer (1981)
Ceratium hirundinella	0.21–0.27	Reynolds (1984a)	0.14–0.24	Padisák (1985)
Peridinium cinctum	0.22	Reynolds (1984a)	0.1	Sommer (1981)
Stephanodiscus hantzschii	1.18	Reynolds (1984a)	0.59	Sommer (1991)
Asterionella formosa	0.62–1.74	Reynolds (1984a)	0.36	Sommer (1991)
Chlamydomonas spp.	2.29–2.91	Reynolds (1984a)	0.6	Sommer (1991)

* Colony culture.

Cylindrospermopsis reached a bloom density and replaced all the other species (Padisák 1994; Fabbro & Duiverndoornen 1996). Field populations rarely achieve their maximal growth capacity because conditions are commonly far from optimal, and loss processes take place. In nature, phytoplankton populations rarely grow faster than one doubling in every second to third day.

A further measure of population increase is **doubling time**, which expresses the time taken for the population to double. Doubling time (t_d, unit is day if k was expressed as day^{-1}) can be derived from the daily net rate of increase as

$$t_d = \ln 2/k \cong 0.693/k \qquad (10.14)$$

It is necessary to note that doubling time is often confused with generation time (t_g: the time that elapses until the next generation develops). But $t_d = t_g$ only for those species in which mother cells divide into two (*Ceratium*, *Synechococcus*, diatoms, etc.), whereas many species of planktonic algae produce more than two daughter cells during one division cycle. For example, mother cells of many coccal green algae divide into four. Therefore generation time increases according to exponentials of

four. In such species generation time and doubling time are not the same. For comparative usage, t_d became widespread.

For some species, it is possible to estimate the parameters of population dynamics based on frequency of dividing cells or by estimation of division time (see examples and mathematical formulations in Heller 1977; Pollingher & Serruya 1976; Ramberg 1980; Frempong 1982; Braunwarth & Sommer 1985; Elser & Smith 1985; McDuff & Chisholm 1985; Alvarez-Cobelas *et al.* 1988; Fahnenstiel *et al.* 1991). In organisms reproducing by binary fission, μ can be calculated from the fraction of cells undergoing mitosis (p_D) per unit time (usually day) as

$$\mu = \ln(1 + p_D) \qquad (10.15)$$

In coccal, non-motile freshwater phytoplankton species, the highest observed net increase rates correlate negatively with the logarithm of the cell volume. Filamentous species (both green algae and cyanoprokaryotes) usually exhibit lower k than expected from their volumes, and only a weak trend was found for flagellates (Sommer 1981).

Maximum growth rates (r) of algae in

continuously light- and nutrient-saturated cultures at 20°C, increase with increasing surface/volume (S/V) ratio by the regression equation $r = 1.142(S/V)^{0.325}$. With the exception of cold-water and thermophilic species, most planktonic algae achieve their maximal rate of growth at temperatures between 25 and 35°C. However, the sensitivity of different species to incremental change of temperature is different, and its slope decreases as a function of increased S/V (Reynolds 1997a).

10.6 LOSS PROCESSES

The loss rate (λ) includes several components, each of which may be of greater or lesser importance:

$$\lambda = \gamma + \sigma + \delta + \pi + \omega \qquad (10.16)$$

where γ is the grazing rate, σ is the rate of sedimentation, δ is physiological mortality, π expresses parasitism and θ is flushing rate. In most cases some of the above factors are negligible.

10.6.1 Leakage

The above equation considers losses from assembled populations of complete cells, although it is well known that the rate of carbon fixation exceeds, often considerably, the rate of its incorporation into complete cells. This is partly, but not exclusively, due to organic matter release by the producers themselves. This process is certainly a serious loss from the point of view of cells that release this material, but not from that of the whole ecosystem, including not only bacteria but algae as well. It has been shown in a recent study that whilst picoplankton (bacteria and small unicellular coccal picoprokaryotes) take up inorganic phosphorus at a much higher efficiency than larger sized phytoplankton, in return, the latter utilise organically bound phosphorus (glucose-6-P, nucleotide-P) at a higher rate. In this way, small, leaky cells may transform phosphorus to a form more amenable to classic phytoplankton. Organic substances, released by zooplankton and other animals, contribute significantly to this material.

Zooplankton production, however, can never exceed the rate of carbon fixation, at least not for long.

10.6.2 Consumption by heterotrophs

Consumption by heterotrophs is certainly one of the main causes of phytoplankton loss. Being a biotic interaction, it cannot be considered as a mechanical process, as, for example, with hydraulic washout. In the extreme, some members of the phytoplankton may act as heterotrophic consumers of other species of phytoplankton within the same community! Many phytoplankton are able to graze others, as described in section 10.4.4.

Filter feeding approaches the traditional view of the planktonic food web. Efficiency is determined by the size of populations involved, and the physical sizes (and other properties, such as digestibility) of phytoplankton particles as compared with the abilities of grazers to take up the given size-spectrum. There are numerous data in the literature (see also Chapter 14) concerning the upper size limit of grazeable particles for different species of grazers. A particular, and in many cases contradictory, issue is the ability of filter feeders to graze on filamentous algae. The lower limit of grazeable particles is usually estimated as 1–2 μm.

According to Reynolds *et al.* (1982), grazing rate may be expressed from the proportion of the water volume filtered by zooplankton per day (G) and the coefficient of selectivity (w) for the given species as $\gamma = Gw$. In the optimal case, $w = 1$ (all specimens of the given species are removed from the volume filtered and ingested). However, in most cases, $w < 1$. In the Bodensee, in the presence of filter feeders, small *Stephanodiscus hantzschii* possessed relatively high (>0.5) selectivity indices, whilst large forms (*Aulacoseira granulata*, *Stephanodiscus binderanus*, *Asterionella formosa*) did not (>0.1). Raptorial feeders (e.g. *Cyclops*) prefer larger particles (Sommer 1988b). For more details see Chapter 14.

10.6.3 Sinking and resuspension

Assuming laminar flow past the particle, the terminal sinking velocity (V_t [m s^{-1}]) of a settling spherical particle depends on its radius (r [m]), the density of the medium (ρ [kg m^{-3}]), the density of the particle (ρ' [kg m^{-3}]) and the viscosity of the medium (η [kg m^{-1} s^{-1}]) according to the Stokes equation:

$$V_t = 2gr^2(\rho' - \rho)/9\eta \qquad (10.17)$$

where g (9.8 m s^{-2}) is the gravitational acceleration (see also Chapter 8).

It follows from the above equation that the denser phytoplankton cells (e.g. diatoms) sink faster than lighter cells, as do larger cells than smaller. If we assume that the density of non-siliceous algae is 1.04 g cm^{-3}, then a diatom of density >1.08 g cm^{-3}, with similar geometrical properties, will sink in fresh water (density 1 g cm^{-3}) at least twice as fast. Most diatom species sink 3 to 5.5 times faster, and the heaviest ones 16 times faster, than non-siliceous algae. Densities for *Asterionella formosa* range between 1.15 and 1.26 g cm^{-3}, 1.08 g cm^{-3} for *Stephanodiscus binderanus*, 1.13 g cm^{-3} for *Fragilaria crotonensis* and *Aulacoseira granulata* and 1.16 for *Diatoma elongatum*. The silica content per unit cell volume of some species is given in Table 10.12. Non-siliceous algae range from 1.019 (*Pandorina morum*) to 1.067 (*Cosmarium* sp.). If $\rho' < \rho$ (gas-vacuolated cyanoprokaryota), V_t is negative, and the particle is positively buoyant, i.e. moves towards the water surface. Peripheral mucilage reduces average density, but this advantage is readily overcome by increased particle size. For details see Walsby & Reynolds (1980).

For non-spherical particles, a dimensionless factor for form resistance (ϕ), which expresses how much faster or slower the given form is sinking than an equivalent sphere of the same volume, must be used, as in:

$$V_t = 2gr^2(\rho' - v)/9\eta\phi \qquad (10.18)$$

Most geometric forms, except teardrops, sink more slowly than the equivalent sphere (Walsby & Reynolds 1980). It has been demonstrated by many authors that the nutritional and physiological sta-

Table 10.12   Silica content of natural diatom populations expressed per unit cell volume. (Data from Sommer 1988b.)

Species	pg Si μm^{-3}
Asterionella formosa	0.14–0.19
	0.16
	0.11
	0.15–0.27
Fragilaria crotonensis	0.01–0.11
	0.09–0.14
Tabellaria fenestrata	0.14–0.19
Synedra acus	0.14–0.19
	0.36
Aulacoseira granulata	0.14
	0.09–0.11
Aulacoseira italica	0.19
Cyclotella glomerata	0.26–0.28
Cyclotella radiosa	0.46
Stephanodiscus neoastraea	0.16
Stephanodiscus binderanus	0.11–0.13
Stephanodiscus hantzschii	0.10–0.17

tus of cells influences their sinking: living cells may reduce their sinking velocity ('vital factor') considerably (some two- to eight-fold) but this phenomenon is still unexplained (Reynolds 1984a, 1997a).

As turbulence counteracts sinking, and tends to resuspend cells (see Chapter 6), the velocity at which they sink through lake water is not identical to V. If mixing is uniform and continuous, the loss of cells from the mixed layer can be calculated as

$$N_t = N_0 e^{V_t/z_m} \qquad (10.19)$$

where N_0 is the initial number of cells, N_t is their number after time t [day], V_t is the terminal sinking velocity [m day^{-1}] and z_m (m) is the thickness of the mixed layer. If we consider that daytime heating leads to the formation of temperature gradients within the epilimnion, and nocturnal cooling leads to homothermy, we can assume one mixing event per day. Then

$$N_t = N_0(1 - V_t/z_m)^t \qquad (10.20)$$

The higher its V_t, the more a species depends on deep mixing for continued suspension. Accordingly, large and heavy cells require relatively deep mixing depths for persistence in the epilimnion. It must be noted that sinking is not necessarily disadvantageous. If nutrients in the epilimnion are exhausted, viable hypolimnetic populations can be viewed as unspecialised survival forms.

In cyanoprokaryota, buoyancy is regulated by variation of gas-vacuole formation and adjustment of cytoplasmatic composition. Both mechanisms are regulated through synthesis and breakdown of photosynthetic products. If photosynthesis is reduced (e.g., in deep layers with insufficient light), osmotic pressure of newly produced sugar is small, moreover carbohydrates and secondarily produced ballast materials are not produced at a high rate. Therefore cell density (which will increase sinking) does not increase. Under such conditions, gas-vacuoles may be produced at a high rate, as a consequence of which the buoyancy of cells with relatively low photosynthetic rates tends to increase. Conversely, if cell osmotic potential is high (high sugar production, increased amount of secondary photosynthetic products) cells maintain relatively few gas vacuoles. Thus, cells with poor photosynthetic histories often tend to exhibit maximum buoyancy, whereas those with optimal histories become negatively buoyant (Paerl 1988).

Vertical migration rates may well exceed maximal swimming speeds of flagellar eukaryotic phytoplankton (see section 10.7). Persistent photosynthetic production can lead to accumulation of surface scums, in extreme cases to thick and dense hyperscums stable enough to support ducks (Zohary 1985). Cells near the surface are exposed to very high irradiation. Bloom-forming species usually possess higher photoadaptational parameters (I_k) than others (see section 10.3) and also effective cellular adaptations protecting against photooxidation; moreover, scums effectively shade other populations in the subsurface layers. In this way, buoyancy regulation of cyanobacteria may be considered both as an effective means of optimal use of resources and a highly competitive mechanism.

10.6.4 Physiological mortality, pathogens

The rate of physiological mortality of phytoplankton can be estimated only if recognisable remains (empty cell walls, frustules) survive after the cells die. Information on pathogens of phytoplankton species is scarce (Canter 1979; Canter & Jaworski 1980, 1981; Canter & Heaney 1984), and their effects are difficult to quantify, for at least two reasons. One is that in most cases it is not known whether the parasite attack acted on perfectly healthy populations, or whether the population physiologically affected was weak and therefore more susceptible to infection. Another problem is that so long as the time between infection and death of the cells remains unknown, it is impossible to calculate rate of mortality due to parasitism (Sommer 1988b). In a population dynamic model developed by Brüning et al. (1992) to assess the impact of pathogens and related fungal parasites, four parameters (prevalence of infection, development time of sporangia of the parasite, specific growth rate of the uninfected host, and the difference between loss rates of infected and uninfected hosts) were needed in order to describe the parasite effect on the host population. If the above variables are unknown, it is safer to treat death due to parasitism as a fraction of physiological mortality.

10.6.5 Hydraulic washout

The daily loss rate of phytoplankton due to washout can be calculated by dividing daily discharge at the lake outlet by the volume of the mixed layer. In large lakes, loss due to washout is usually negligible, whilst in small, highly flushed lakes not only loss via the outlet but also the dilution effect of the inflow must be considered. If theoretical lake retention time is shorter than 30–40 days, the effect of washout-dilution on species selection and community organisation is significant (Reynolds 1997a) in that it interferes with the time required for establishment of successional equilibrium states.

10.7 TURBULENT MIXING

If a force (commonly, directional wind) is applied to a water surface (in an idealised case with infinite depth and extension, and without density gradients), its energy will dissipate within the water column through a cascade of smaller and smaller downwelling eddies (laminar dissipation is possible but not considered here). The consequent wind-mixed layer rarely extends to a depth >200–250 m, even in the open ocean, which is the environment closest to an ideal case. It can be shown (see details and mathematical formulation in Reynolds 1994b, 1997a) that the most important feature of the environment of a planktonic organism is determined not by input energy but by the rate of its dissipation, and the sizes of the smallest eddies. The smallest sizes in oceans, and in deeper lakes, are rarely smaller than 1.3 mm; in highly kinetic environments, such as a tidally mixed estuary, they fall into the range 200–400 μm. As all but the largest planktonic algae are smaller than this by one or two orders of magnitude, it follows that they live most of their lives in a predominantly viscous environment which is, itself, liable to be transported far and rapidly through the turbulence field (see also Chapter 6).

10.7.1 Stir-up effect of wave action

One of the main selective features for lake phytoplankton is the number of mixing events per unit time, usually per year (i.e. how many times a year does complete mixing of the entire water column take place?). In shallow lakes, particles at the sediment surface are stirred up during storms. Such events affect most biological interactions, both in the sediments themselves and in the open water.

The wind speed at the required height can be calculated from the so-called Logarithmic Law (Simiu & Scanlan 1986) as

$$V/V_1 = \ln(H/z_0)/\ln(H_1/z_0) \qquad (10.21)$$

where V is the wind speed at the required height (H) above the surface, V_1 is the measured wind speed at H_1 and z_0 [cm] is the value of the surface roughness

Table 10.13 Values of surface roughness (z_0).

Type of surface	z_0 (m)
Sand/water	0.0001–0.001
Snow surface	0.001–0.006
Mown grass (0.01 m)	0.001–0.01
Low grass, steppe	0.01–0.04
Fallow field	0.02–0.04
Tall grass	0.04–0.1
Palmetto	0.1–0.3
Pine forest (15 m high, 1 tree, 10 m^{-2})	0.9–1
Sparsely built-up settlements	0.2–0.4
Densely built-up settlements	0.8–1.2
Centres of large cities	2–3

(Table 10.13). The relationship between wind speed over land and over water at the same height can be expressed after Ford & Stefan (1980) as

$$V_w = V_{10}\left[\ln(H_{10}/z_2)\cdot\ln(z_b/z_1)\right]/ \\ \left[\ln(H_{10}/z_1)\cdot\ln(z_b/z_2)\right] \qquad (10.22)$$

where V_w is wind speed over water at 10 m, V_{10} is wind speed over land at 10 m, z_2 is the surface roughness of water (0.0001 m; Table 10.13), z_1 is the surface roughness of the land (m; from Table 10.13), and z_b, the equivalent boundary layer over water, can be calculated (Elliot 1958), provided that each variable is expressed in metres, as

$$z_b = 0.86Fz_1^{0.23} \qquad (10.23)$$

where F is fetch of wind over the water surface [m].

The period and wavelength of waves formed by wind action can be calculated (Carper & Bachmann 1984) as

$$L = gT^2/2\pi \qquad (10.24)$$

where L is the wavelength [m], g is the gravitational acceleration (9.8 m s^{-2}), T is the wave period and $\pi = 3.1416$. Wave period is related to the wind velocity (V) and fetch (F) by the following equation:

$$V(gT/2\pi) = 1.2\tanh x \qquad (10.25)$$

where

$$x = 0.077\left(gF/V^2\right)^{0.25} \qquad (10.26)$$

and the definition of the hyperbolic tangent function (tanh) is

$$\tanh x = \left(e^x - e^{-x}\right)/\left(e^x + e^{-x}\right) \qquad (10.27)$$

which can be calculated either from tables or from the above relationship.

The **wavebase** (i.e. the depth to which the wave effect extends) is taken as being one-half the wavelength (Smith & Sinclair 1972; Carper & Bachmann 1984). If the wavebase reaches the lake bottom complete mixing is guaranteed.

Using the above equations it can be calculated whether, for a given wind speed, a lake of a particular size and depth is liable to be completely mixed. An application of this relationship to Lake Zeekoevlei (Cape Town, South Africa: dimensions 2.36 × 1.56 km, mean depth 1.9 m, max. depth 5 m, partially divided into two basins by a peninsula, moderate surface roughness; Harding 1996, 1997), is given in Fig. 10.3.

The extent of vertical mixing in a lake will depend on (i) its location in regard to surrounding environment (z values), (ii) its shape and extension with regard to wind direction (fetch), (iii) wind velocity, (iv) basin morphology (wave generation) and

(v) the depth of eddy propagation in relation to the vertical density (temperature) gradient in the water column. Phytoplankton environments depend on many other features, such as the light field, chemical gradients, inorganic dissolved or particulate components, etc. However, the importance of vertical extent of mixing is best demonstrated by the point that in a deep lake with continuous mixing no autotrophic production is possible. This example is not theoretical: Lake Mantano in Sulawesi is 540 m deep, yet completely isothermal (Haffner *et al.* 2001).

In view of the above considerations, it is clear that shallowness itself does not imply frequent or continuous mixing. The temperature gradient in tropical deep lakes is often less than 5°C! Small, sheltered lakes may well also be physically stable, even if a classic thermocline cannot develop. Extended macrophyte stands may significantly contribute to their physical stability.

10.7.2 Vertical structure of phytoplankton assemblages

Numerous environmental variables which are of importance for phytoplankton vary along vertical gradients. For example, light becomes attenuated, and its spectral composition is changed. Nutrients usually become more concentrated with depth. Vertical migration of zooplankton species may

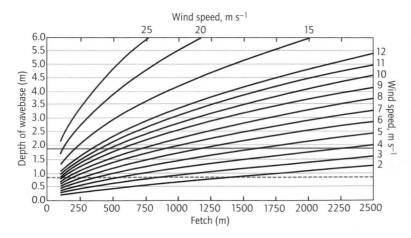

Fig. 10.3 Relationship of wind speed (right-hand and top ordinates) and wind fetch (bottom ordinate) to wave depth (left-hand ordinate), calculated for Lake Zeekoevlei, South Africa. Thick horizontal line represents the mean water depth of Zeekoevlei and the dotted line the depth of the euphotic zone. The shaded region illustrates the range of wind speeds typical for the Zeekoevlei region. (From Harding 1997; reproduced with kind permission of Kluwer Academic Publishers.)

lead to multiple influences on phytoplankton standing crops, etc. (Lampert 1992). Therefore, the potential is there for a large number of species each to seek and attain their preferred position, and thus for a tightly structured community of co-existing species to develop.

That few planktonic communities conform to this ideal may be easily explained. Vertical rates of advection of planktonic algae (10^2–10^4 s: consequential upon convectional, rotational and wind-driven mixing processes) are generally orders of magnitude less than times required for phytoplankton cells to replicate (10^5–10^6 s), and much less than that required for development of a stable, equilibrated community (10^6–10^7 s). In contrast to a supposed potential vertical structure, many planktonic assemblages experience randomised dispersion through vertical gradients. This does not offer an infinite range of vertical niches, but rather a single one, characterised by frequent alterations (Reynolds 1992a).

Phytoplankton usually occupy the mid-ground between these extremes, depending on the stability, and variability in stability, of vertical structures (for details see Reynolds 1992a), and the abilities of individual species to attain and to remain at preferred depths. Once a relatively stable vertical structure in the water column develops, availability and regeneration rates of resources will vary between the different water layers.

In photosynthesis, carbon dioxide is usually taken up from dissolved sources at a faster rate than that at which the gas diffuses from the atmosphere to the water. Without a bicarbonate reserve (in solution), rapid consumption of carbon dioxide will be followed by carbon shortage, and a rise in pH. Under such conditions, cyanobacteria are better adapted than most eukaryotic species (Shapiro 1984), among which chrysophytes will receive a particular setback (Reynolds 1986). Nutrients (nitrogen, phosphorus and silica) are taken up from the upper euphotic layers, as a consequence of which their concentrations fall to an extent sufficient to limit the further growth of algae.

Although nitrogen can be replenished from the atmosphere by nitrogen-fixing blue-green algae, and although short-term recycling of nutrients in the upper layers allows further growth, the lower strata rapidly become more rich in nutrients than the surface layers. Oxygen status of the lower layers, whether oxic or anoxic, profoundly affect the form, quantity and availability of nutrients. Onset of stratification leads to separation of a well-illuminated but increasingly nutrient deficient surface layer, from a dark, nutrient-rich, sometimes anoxic bottom layer. Under such circumstances, nutrient assimilation in the deeper strata, coupled with photosynthesis in the upper layers, offers a superior strategy, provided that algae are periodically able to change their vertical position, either by manipulation of gas-vacuoles or by flagellar movement. The ability of species to gain benefit from such environmental circumstances depends both on their ability to migrate between strata and the steepness of environmental gradients.

The maximum recorded amplitude of diel vertical migration of different phytoplankton species varies between 2.5 and 18 m (Table 10.14). Comparison between maximal amplitude (A [μm]) and

Table 10.14 Maximum recorded amplitude of diel vertical migrations and size expressed as diameter of spherical equivalent (DSE) of some freshwater flagellates. (Data from Sommer 1988a.)

Species	Amplitude (m)	DSE (μm)
Peridinium cinctum	10	49
Ceratium hirundinella	5	46
Peridinium aciculiferum	8	34
Gymnodinium uberrimum	5.5	33
Peridinium penardii	5	31
Gymnodinium lacustre	2.5	13
Synura petersenii	2.5	53
Mallomonas tonsurata	2.8	13
Mallomonas sp.	7.5	8.4
Dinobryon sertularia	7.5	16
Cryptomonas ovata	5	19
Cryptomonas erosa	3	19.5
Cryptomonas marssonii	5.5	12
Rhodomonas minuta	2.5	5.4
Rhodomonas lens	4	7.8
Chroomonas acuta	3	8.3
Volvox sp.	18	108

size (expressed as diameter of spherical equivalents, DSE [μm]) leads to the relationship $A = 1.51 \times 10^6 DSE^{0.4}$ (with a correlation coefficient of 0.61), showing that larger flagellates are able to migrate greater distances than small ones. If the trend of the above equation is extrapolated to very small flagellates (>5 μm), it can be shown that vertical migration would be profitable for these taxa only in water bodies with steep environmental gradients, where relevant differences occur at a vertical distance of less than 2 m (Sommer 1988a).

Such steep gradients can exist in lakes where stratification is very stable. For example, in the meromictic Lake Cisó, Spain, a vertical distance of as little as 45 cm is enough to change the environment completely (Gasol et al. 1991). The migratory ability of chrysophyte flagellates is weaker than that of species belonging to other groups. For example, diel oscillations of 0.5 to 2 m have been reported (Sandgren 1988b) for variably sized chrysophytes, including a variety of sizes (Mallomonas, Dinobryon, Synura, Chrysococcus, Ochromonas, Synura, Chrysosphaerella). Swimming velocities of marine phytoplankton vary in a range of 50–1500 μm s^{-1}, velocities of 100–400 μm s^{-1} being characteristic. Some data on vertical migration velocities of freshwater species are given in Table 10.15. Metabolic costs of flagellar movement have been calculated to be extremely low (Raven & Richardson 1984).

Table 10.15 Vertical migration velocities of some phytoplankton species. If ranges are given, the smaller value refers to the epilimnion, the larger to the hypolimnion.

Species	Velocity (μm s^{-1})	Reference
Anabaena circinalis***	56	Reynolds (1972)
Anabaena flos-aquae***	105	Reynolds et al. (1982)
Aphanizomenon flos-aquae***	111–764	Paerl & Ustach (1982), Paerl (1982)
	70	Reynolds et al. (1982)
Aphanizomenon flos-aquae*	0.04	Sommer (1988b)
Microcystis aeruginosa***	830	Ganf 1975
	95–160	Reynolds (1992a)
Planktothrix agardhii***	11	Reynolds et al. (1982)
Planktothrix agardhii*	4.6	Walsby et al. (1983)
Stephanodiscus astraea*	25	Reynolds (1984a)
Fragilaria crotonensis*	19–118	Sommer (1988b)
Asterionella formosa*	13–81	Sommer (1988b)
Aulacoseira granulata*	9–80	Sommer (1988b)
Stephanodiscus binderanus*	7–83	Sommer (1988b)
Stephanodiscus hantzschii*	0.03–52	Sommer (1988b)
Ceratium hirundinella**	195–278	Heaney & Eppley (1981)
	116	Reynolds et al. (1982)
Peridinium umbonatum**	250	Heaney & Eppley (1981)
Peridinium cinctum**	200	Metzner (1929)
Pandorina morum*	0.75–7	Sommer (1988b)
Volvox sp.***	1160	Reynolds (1992a)
Staurastrum cingulum*	5–56	Sommer (1988b)
Staurastrum sp.*	3.5	Reynolds et al. (1982)
Mougeotia thylespora*	1–15	Sommer (1988b)

* Sinking, ** upward migration of cyanoprokaryotes, *** swimming speed.

10.7.3 Deep-layer chlorophyll maxima (DCM)

Another feature of phytoplankton vertical distribution is persistent development of dense populations in deeper strata. These are often referred to as **deep-layer chlorophyll maxima** (DCM). A selection of species described in the DCM of freshwater lakes (Table 10.16) exhibits a high diversity, and implies that reasons for the development of DCM are also diverse. Deep-layer chlorophyll maxima are well known from the open ocean (Cullen 1982), where they occur at depths rather greater than the thermocline, frequently at the nitricline. The commonest contributors are species of autotrophic picoplankton.

In the oligotrophic Stechlinsee (Germany), the characteristics of the DCM are very close to those in the ocean. Main growth of the picoplanktonic *Cyanobium* (formerly *Synechococcus*) occurs during spring isothermal conditions. When the lake stratifies (during late May), the *Cyanobium* population forms a thin layer below the thermocline in the upper hypolimnion (15–17 m; temperature: 6–8°C), which receives a variable amount of surface photosynthetically active radiation. Light

Table 10.16 Species reported from freshwater deep chlorophyll maxima (DCM).

Species	Group*	Reference
Planktothrix agardhii	Cya	Lindholm (1992)
Plantothrix rubescens	Cya	Konopka (1982)
Aphanizomenon flos–aquae	Cya	Konopka (1989)
Anabaena sp.	Cya	Konopka (1989)
Synechococcus	Cya	Steenbergen & Korthals (1982)
	Cya	Craig (1987)
	Cya	Padisák *et al.* (1999)
Cryptomonas paseolus	Cry	Pedrós-Alió *et al.* (1987)
		Miracle *et al.* (1992)
Cryptomonas erosa	Cry	Miracle *et al.* (1992)
Cryptomonas obovata	Cry	Miracle *et al.* (1992)
Trachelomonas hispida	Eug	Miracle *et al.* (1992)
Euglena spp.	Eug	Miracle *et al.* (1992)
Ceratium hirundinella	Pyrrho	Gálvez *et al.* (1988)
	Pyrrho	Konopka (1989)
Dinobryon sertularia	Chryso	Bird & Kalff (1989)
Ochromonas sp.	Chryso	Bird & Kalff (1989)
Chrysosphaerella longispina	Chryso	Pick *et al.* (1984)
Mallomonas caudata	Chryso	Steenbergen & Korthals (1982)
Tribonema sp.	Xantho	Larson *et al.* (1987)
Scourfieldia caeca	Prymnesio	Croome & Tyler (1984)
Urosolenia eriensis	Centr	Jackson *et al.* (1989)
Cyclotella spp.	Centr	Jackson *et al.* (1989)
Stephanodiscus hantzschii	Centr	Larson *et al.* (1987)
Astasia sp.	Chloro	Miracle *et al.* (1992)
Hyaloraphidium contortum	Chloro	Gasol & Pedrós-Alió (1991)
Crucigenia tetrapedia	Chloro	Gasol & Pedrós-Alió (1991)
Selenastrum capricornutum	Chloro	Gasol & Pedrós-Alió (1991)

* Cya, cyanobacteria; Cry, Cryptomonadae; Eug, Euglenidae; Pyrrho, Pyrrhophycaeae; Chryso, Chrysophycaeae; Xantho, Xanthophycaeae; Prymnesio, Prymnesiophyceae; Centr, Centrales (Bacilllariophycaeae); Chloro, Chlorophycaeae.

intensity at the depth of the DCM at noon varies from $65 \mu Em^{-2}s^{-1}$ during sunny weather to only $3.6 \mu Em^{-2}s^{-1}$ on cloudy days (Gervais et al. 1997; Padisák et al. 1998). The DCM persists during summer, and is associated with the nitricline. After breakdown of the spring Cyclotella maximum, sinking diatoms temporarily contribute to the DCM.

The deepest DCM reported from freshwater environments occur in the clearest ultra-oligotrophic lakes. In Lake Tahoe (California), summer chlorophyll maxima occur at 100 m, owing partly to sinking of diatoms and partly to in situ growth. This DCM was not found to be constant either in time or space (Abbott et al. 1984; Coon et al. 1987). In Crater Lake (Oregon), Chl a and primary production reach a maximum at 80–140 m depth. Tribonema sp. occupies the 80–120 m layers, whilst Stephanodiscus is found in the lowermost 160–200 m (Larson et al. 1987).

Apart from the above cases, freshwater DCM are usually related to the thermocline; the commonest occurrences are found in the metalimnion. The best known metalimnetic maxima in temperate lakes are associated with the occurrence of Planktothrix rubescens (e.g. Konopka 1982; Feulliade et al. 1984).

A possible reason for DCM formation is behavioural aggregation of dinoflagellates. In the La Concepción Reservoir, southern Spain, Ceratium hirundinella was found to be abundant in the DCM that occurs in the metalimnion, where it receives 1–3% of surface incident light (Galvez et al. 1988). Duration of the DCM was limited to 2–3 weeks. Mass excystment of the chrysophyte Chrysosphaerella longispina was found to cause a short-lasting DCM in Jacks Lake, Ontario (Pick et al. 1984). In the DCM observed by Bird & Kalff (1989) in Lake Gilbert, Quebec, phagotrophic chrysoflagellates (Dinobryon sertularia, Ochromonas sp.) were abundant, and carbon uptake was based on bacteriovory at 40–79%.

Development of DCM is especially prevalent in meromictic lakes. Their longevity, stability and position depend on the resilience and location of the thermocline, and moreover on its relative position to the chemoclines. Detailed case studies are available from Lake Cisó, Spain (Pedrós-Alió et al. 1987; Gasol & Pedrós-Alió 1991; Gasol et al. 1991), Laguna de la Cruz, Spain (Miracle et al. 1992) and various Tasmanian lakes (Croome & Tyler 1984).

There are also numerous observations of high density of phytoplankton under ice. Although considerable water movement takes place via convection, conditions under ice in temperate regions are fairly similar to those in the deep strata of stratifying lakes during summer. Light is often very scarce but nutrient conditions are usually favourable, and the environment is relatively stable (Lindholm 1992).

10.8 BASIC ADAPTATIONS AND SURVIVAL STRATEGIES OF FRESHWATER PHYTOPLANKTON

Very few ecological adaptations are restricted to particular taxonomic groups of algae. Phylogeny is therefore not at all a good predictor of ecology. Small and large forms, with all the consequences for nutrient uptake, planktonic food webs, etc., exist in each group, as do motile and non-motile forms (except the cyanoprokaryota).

10.8.1 Phylogeny-related properties

Despite the above, several features of ecological relevance are largely phylogenetically based (see a detailed set of examples in Sandgren 1988a, and the individual contributions in that volume). The best-known examples are the obligate silica demand of diatoms, and of silica-scaled flagellates. Similarly well known is the finding that cyanoprokaryota avoid acidic environments, and that they exclusively possess gas-vacuoles which enable them rapidly to adjust their position in a water column. Fixation of atmospheric nitrogen is also exclusive to cyanoprokaryota. The cells of cyanoprokaryota and chrysophytes are, in general, more leaky than those of other groups.

In response to phosphorus limitation, or in the presence of organophosphates, some species produce an extracellular organophosphate-degrading enzyme (acidic phosphatase). In alkaline conditions, however, alkaline phosphatases are absent, or are not secreted in significant amounts. In many taxa (e.g. chrysophytes), the inability to take up carbon in the form of free carbon dioxide (Saxby-Rouen *et al.* 1996) may explain their observed preference for neutral or slightly acidic waters. Bacteriovory may be of major importance (see section 10.3.3.), especially in permanently low-light environments.

In Sandgren's (1986) germination experiments, neither nutrient concentrations nor temperature regime consistently affected germination of chrysophyte cysts. According to his conclusions, small numbers of stratospores may be excysting at any time, and establishment of vegetative populations in the plankton therefore depends on the germinating cell encountering conditions favourable for vegetative growth. Such a recruitment strategy is extremely opportunistic, in that it provides a mechanism by which chrysophytes can persist by capitalising on brief windows of the time during which competitive or grazing stress is temporarily reduced.

The cysts of dinoflagellates need an endogenous period of dormancy. Encystment, and especially excystment, occur under specific environmental conditions in which temperature seems to be highly relevant. Akinete germination of cyanoprokaryota is also triggered by environmental factors (temperature, light, nitrogen deficiency) but, at least in some species, without an obligate requirement of a dormancy period.

In diatoms, formation of the zygote (the auxospore) is a mechanism designed to restore initial cell size after continued cell divisions have led to progressive size reduction (McDonald–Pfitzer rule). The process is very costly from a demographic point of view: four consecutive division events produce only one or two cells from the two mother cells, whereas four consecutive divisions of vegetative cells would have led to the production of 32 cells (Sommer 1988b). Only in rapidly dividing populations may cell sizes decrease to a lower limit necessary for auxosporulation of the diatom zygote, which, in contrast to many other algal groups, is not suitable as a resting stage. Many planktonic diatoms have circumvented size decrease and sex-

uality; among members of Fragilariaceae sexuality is absent in pelagic populations (Lewis 1984). Among freshwater diatoms, only the predominantly marine taxa *Rhizosolenia*, *Chaetoceros* and *Aulacoseira italica* possess resting spores.

Possession of siliceous frustules renders diatoms heavier than most other algae of comparable size. Therefore they sink more rapidly than other phytoplankton species. In accordance with the Stokes equation (see section 10.6, and also Chapters 6 and 8), the larger the cell the faster the sinking. In general, diatoms need greater mixing depths for persistence within the near-surface layers than other algae of similar size and form (Reynolds *et al.* 1984). Accumulation of oil droplets reduces specific density, and consequently counteracts sinking. Non-flagellar forms of green algae are also subjected to significant sinking loss.

Compared with other groups of phytoplankton, diatoms are dependent on an additional nutrient (silica), which is less readily recycled than either nitrogen or phosphorus. Many species are affinity-adapted in regard to phosphorus-uptake strategy, which enables them better to survive if phosphorus limitation prevails. Two hypotheses explain this characteristic. One is that diatoms are less susceptible to development of depleted microhabitats around the cells, and the other is that silica may act as an effective adsorbing agent for dissolved substances at low concentrations (Sommer 1988b).

10.7.2 *Complex phytoplankton strategies*

Independent of taxonomic affiliation, there are several principal ecological strategies of phytoplankton species. The distinction between r- and K-selected species (that the former are able to grow rapidly when environmental resources are unlimited, and that K-selected species grow more slowly, but can better exploit scarce resources), originally developed by MacArthur & Wilson (1967), was shown to apply to phytoplankton by Sommer (1981).

Following Grime (1979), Reynolds (1988a) described phytoplankton strategies in relation to intensity of environmental stress and disturbance. If both are intense, no pelagic life is possible. If both disturbance and stress are low, rich resources (practically, light and nutrients) favour species with high intrinsic growth rates, these are therefore **C-strategists**. If nutrients or light energy are scarce (high stress), but disturbances are insignificant, only the effective competitors can survive.

To be an effective competitor a species must possess some special adaptation: for example, nutrient storage, low threshold and/or half-saturation constants for nutrient uptake, low I_k, effective migration, ability to utilise organic resources, etc. These species are called **specialised, stress tolerant S-species**. The third type, **R-strategy**, includes disturbance tolerant, ruderal species, which are able to tolerate, for example, frequent transport through the light gradient.

C-strategists are small, and possess high S/V ratios and relatively high metabolic activity over a wide range of temperatures. The higher the S/V ratio, the greater the light-harvesting capacity, the more successful the nutrient uptake capacity and the subsequent growth rate. The negative aspect of this strategy is sensitivity to high temperature. This is because temperature-regulated processes, especially photosynthesis and respiration, run too fast in relation to the rate of exhaustion of resources from the surrounding environment. Furthermore, these forms often possess relatively high I_k and k_s, and are sensitive to zooplankton grazing. Consequently, they benefit from high concentrations of nutrients and low grazing pressure. Typical C-strategists are small green algae (*Chlorella, Ankyra, Monoraphidium, Scenedesmus, Chlamydomonas*), cryptophytes and centric diatoms.

R-strategists are of medium to large size, but with morphologies which maintain high S/V ratio. These species are often filamentous, or maintain a high projected area per unit biomass. They are quite resistant to grazing. R-selected species are specifically adapted to conditions characterised by high frequency physical variability, reduced light, low temperatures and plentiful nutrients. Typical R-strategists are diatoms such as *Aulacoseira, Melosira, Fragilaria, Asterionella, Acanthoceras*, larger centric diatoms, and green algae such as *Staurastrum* and *Closterium*, which possess protuberances.

S-strategists are large, with low S/V ratios and hence lower rates of growth and respiration. They are less sensitive to strong light, because their capacity for light harvesting per unit volume is lower than that of small or attenuated forms and, moreover, because they possess special features designed to avoid inhibitory doses of light. Their reduced growth rate allows them to prefer high temperatures and to tolerate low nutrient concentrations. Many S-strategists possess properties (large nutrient storage capacity, ability to migrate between nutrient rich and well-illuminated strata) designed to resist nutrient stress. The disadvantage of the S-strategy is reduced tolerance to: (i) low temperatures—because these species can easily become temperature-limited in their nutrient uptake, (ii) low light and (iii) physical disturbances. Typical S-strategists are cyanobacteria—both colonial (*Microcystis, Aphanothece*) and filamentous (*Anabaena, Aphanizomenon, Gloeotrichia, Cylindrspermopsis*) forms, dinoflagellates (*Ceratium, Peridinium*), some chrysophytes (*Uroglena, Dinobryon*) and large colonial green algae (*Volvox, Pandorina*).

A rich set of examples, explaining phytoplankton sequences based on the C–S–R concept, is found in Olrik (1994). These strategies are not wholly exclusive and individual species may possess intermediate characteristics. Examples are given in Tables 10.17 & 10.18.

10.8.3 Picophytoplankton

By the original definition (Sieburth *et al.* 1978), picoplankton are free-living single cells in the size class (longest dimension) 0.2–2 μm. Thus, colony-forming species with colonies >2 μm, comprised of cells <2 μm, are not true picoplankton. Minute cells in aggregated colonies represent a different life-form and life strategy, which is consequential upon, for example, grazing.

Colonial forms of medium size (up to 32–64 cells) may take part in the classic food web, whilst unicells smaller than the lower size limit for grazeable particles (see Chapter 13) enter the higher

Table 10.17 Morphometric and physiological characteristics of planktonic C-, R- and S-strategists. (From Reynolds 1988a.)

Properties	C	R	S
Unit biovolume [μm^3]	5–5000	500–10^5	10^4–10^7
SA/V [μm^{-1}]	0.3–3.0	0.3–2.0	0.03–3.0
GALD [μm]	3–80	10–300	30–500
Photosynthetic efficiency [mg C mg^{-1} Chl a E^{-1} m^2]	18	5–12	<6
Maximum growth rate at 20°C, μ_{max} [day^{-1}]	0.8–1.8	0.8–1.8	0.2–0.9
Cellular P uptake relative to μ_{max} [μM P 10^{-9} cell^{-1} day^{-1}]	0.2–0.5	0.1–0.7	0.4–0.9
Temperature sensitivity of μ (Q_{10})	2.2–3.2	1.9–3.4	2.4–4.4
Threshold dose for exposure to saturated light intensity [h day^{-1}]	3–8	3–6	>5
Motility	Variable	–(mostly)	+
Minimum sinking rate [m day^{-1}]	Generally ≪0.6	Generally 0.2–1	0
Susceptibility to grazing	c.1	<0.6	<0.3

Table 10.18 Environmental factors which promote (+) or inhibit (–) occurrence of planktonic C-, R- and S-strategists. (From Reynolds 1988a.)

Variable	C	R	S
Low temperature	+	+	–
Stability of water column	++	–	++
Shallow mixing depth (z_m < 3 m)	++	–	++
High turbidity in the mixed zone (z_m/z_{eu} > 1.5)	–	+	–
P deficiency (PO$_4$–P < 0.3 μg L^{-1})	–	–	+
Grazing rate >0.6 day^{-1}	–	+	++

trophic levels via the microbial loop. Restriction of the term 'picoplankton' to solitary living unicells is therefore necessary both from an ecological and a taxonomic viewpoint. However, such colonies often disintegrate, or lose single cells. Therefore they may potentially belong to the picoplankton. Some apparent colonies are often only aggregates of unicells (Stockner 1991) but 'true' colonies with distinguishable morphologies (*Aphanocapsa*, *Aphanothece*, *Cyanodyction*, *Merismopedia*, *Cyanonephron*, *Coelosphaerium*, *Lemmermanniella*) also exist.

The ecological term 'picoplankton' covers several phylogenetic groups. The most prominent are chroococcal cyanoprokaryotes (*Cyanobium*, *Synechococcus*), and coccal green algae usually referred to as '*Chlorella*-like' cells. Occurrence of picoplanktonic *Prochlorophyta* in fresh waters has not yet been confirmed (Stockner 1991). Picoplanktonic cyanoprokaryotic organisms are morphologically very simple and their identification on a classic morphological basis alone is often impossible.

A combined approach, during which separate strains of unicellular forms were studied with different methods (morphology, ecophysiology, isozyme composition, pigment composition, ultrastructure, division type, ecological limits, nucleoid structure) led to recognition of much higher diversity than expected. Within the simplest cyanoprokaryotes living in solitary cells, and dividing cross-wise in one plane only, the material investigated yielded four clearly delimited genera, *Cyanobium*, *Cyanothece*, *Synechococcus* and another whose nomenclature is not yet resolved (Komárek 1996).

Picoalgae are found in almost every kind of lake. In acidic, dystrophic bodies, the picoplankton is made up of green algae, whereas the importance of blue-green algae increases sharply in neutral to alkaline lakes. In most lakes, picophytoplankton abundance exhibits a summer maximum (Stockner 1991; Malinsky-Rushansky *et al.* 1995). Exceptionally, they grow under isothermal conditions and provide a spring maximum (Padisák *et al.* 1997). Concerning trophic status, importance of picophytoplankton (cyanoprokaryota) declines from ultra-oligotrophic to moderately eutrophic lakes (Stockner 1991). There is an apparent scarcity of picophytoplankton in eutrophic to moderately hypertrophic systems, but their importance sharply increases in very hypertrophic lakes.

Recent studies have demonstrated how the high affinity of *Cyanobium* for inorganic phosphorus uptake may explain the high importance of picophytoplankton in phosphorus-deficient lakes. Increasing evidence substantiates the view that cyanoprokaryotic picoplankton can grow under very poor light conditions, for example, in deep-layer maxima of oligotrophic lakes (Padisák *et al.* 1999), in inorganically turbid environments (Carrick *et al.* 1993; Padisák & Dokulil 1994), or hypertrophic lakes (Vörös *et al.* 1991).

According to recent knowledge, lakes with abundant picoprokaryotes are either extremely phosphorus-deficient or very turbid. Therefore, either low phosphorus concentration or low light levels favour the existence of blue-green picoplankton in non-acidic environments, which explains their irregular incidence along the trophic scale. At low total phosphorus concentration they are favoured by low availability of inorganic phosphorus, whilst at very high total phosphorus concentration, low light levels favour picoplankton. Certainly, different species are involved.

10.9 PHYTOGEOGRAPHY, DISTRIBUTION AND DISPERSAL OF PHYTOPLANKTON SPECIES

10.9.1 *Species concepts in phytoplankton*

For most phycologists, the species concept is based on morphological characteristics. Hence the term 'species' means morphospecies. On the other hand, for evolutionary biologists, a species, e.g. according to Mayr (1982), can be defined as a reproductively isolated community of populations which occupies a specific niche in nature. Recent

authors (McCourt & Horshaw 1990) argue that it would be unproductive and inconvenient to restrict the term 'species' exclusively to one meaning or the other.

Most freshwater phytoplankton species appear to be cosmopolitan: if studied extensively by specialists they may be found all over the world. In terms of evolutionary biology, this means either that they possess highly efficient means of dispersal, or that their morphological characteristics have been constant through long evolutionary times (Ichimura 1996). Some species ('good species') are indeed very constant in their morphological characters, whilst others apparently vary within a wide range.

Former taxonomic concepts gave weightings to minor morphological differences by giving them a taxonomic, usually intraspecific, rank; this practice allowed Lange-Bertalot & Simonsen (1978, p. 63) to conclude 'that the traditional definition of species, because of lack of a sufficient species concept, must lead to infinite separation . . . finally to individuals'.

Recent investigators tend to regard species formerly considered separate as synonymous. Without a deeper discussion of the species concepts applicable to phytoplankton (see some considerations in Kristiansen 1996a), it is necessary to realise that phytogeographical qualifications (subtropical, polar, endemic, etc.) are ultimately dependent upon the species concept applied.

10.9.2 *Dispersal of phytoplankton species*

The occurrence of so many common freshwater algae throughout the world is certainly a reflection of ease of dispersal; yet for the majority of species there is little information on their mechanisms of transport (Round 1981). Among factors shaping the geographical distribution of phytoplankton species, temperature is usually considered to be of prime importance. If, however, it was the only factor, there would only be a limited number of distribution types (such as pantropical, temperate and circumpolar). Therefore we must assume that other factors are also important and that the role of migratory barriers (seas, arid areas, mountain chains) must be considered. However, it must also be stated that owing to incomplete taxonomy, and a general lack of understanding of the autecology, distribution and speciation of freshwater phytoplankton, there are serious obstacles to detailed phytogeographical analyses: our knowledge is therefore incomplete and fragmentary.

10.9.2.1 Active versus passive mechanisms

Active mechanisms in phytoplankton dispersal (e.g. the species spreads via the efforts of individuals) can be ruled out. Nevertheless, it is interesting to note that epilithic algae may be able to move from one habitat to another along wet rock surfaces (Hustedt 1943). Passive dispersal agents involve water, air, different animals and human agency.

10.9.2.2 Water, river courses

Travel along river courses is the most obvious means of dispersal. The phytoplankton flora of rivers, however, does not, in general, contain any species that would not be found in lakes, and it is similarly the case that phytoplankton species are adapted to survive lotic conditions. In the Northern Hemisphere, unicellular centrics, chlorococcalean green algae are the best fitted to live in the main courses. In South American rivers, chain-forming centrics and desmids are often found, whilst blue-green algae are common in Australian rivers. Riverine dead zones offer a chance for less well-adapted species to survive (Stoyneva 1994).

The shorelines of most large rivers, especially in the industrialised world, are highly regulated. As a consequence, they are reduced to a geometrically simple flow channel. In the past, river channels were more natural, offering a much higher proportion of dead zones, and therefore a wider habitat diversity for many species to survive. The recent and very successful dispersion of the cyanoprokaryote *Cylindrospermopsis raciborskii* may be largely attributed to its ability to tolerate prolonged travel along river courses (Padisák 1997). Inocula, by which newly constructed reservoirs are colonised, derive primarily from the catchment area; species not occurring in the catchment take a much longer time to appear in the newly created body (Atkinson 1988).

10.9.2.3 Animals

The gut contents of aquatic animals, from filter feeders to fish, often contain dozens of planktonic algae which may pass undamaged through the intestines. Dispersal from one lake to another requires overland transport, during which desiccation is the most immediate danger. Water beetles and water-living mammals may carry algae from one pond to another; dragonflies (Odonata) may be more effective transporters and waterfowl are of prime importance.

In judging the efficiency of transport on the feet or the feathers of birds, the ability of algae to survive desiccation must be compared with the travel distance achievable within a given time span. In Schlichting's (1960) experiments, most algae transported externally survived for 4 hours. Some were still alive after 8 hours, but following that hardly any survived. Atkinson (1980, and other works cited therein) has shown that transport via the digestive tract of birds may be more efficient, because desiccation is ruled out. Specialised life-cycle forms (spores, cysts, akinetes), as well as thick, resistant cell walls, will better survive longer transport, either externally or internally, than vegetative forms. For example, Talling (1951) found a correlation between widespread distribution and more resistant resting stages in desmids. In Antarctica, all dispersal must have been carried out by birds, since the atmosphere itself is almost sterile (Schlichtling *et al.* 1978).

10.9.2.4 Airborne algae

Airborne dispersal of algae has long been a focus of interest. Ehrenberg (1849) found 18 species of diatoms in atmospheric dust collected on HMS Beagle by Darwin (1839), 300 km from the nearest coast, but he did not test for viability. During his transatlantic flight in 1933, Lindberg also collected atmospheric samples (Gislén 1948). Later, Overeem (1937) filtered air up to an altitude of

2000 m, and found a maximum concentration of algae at 500 m. The samples contained aerophilic green algae such as *Chlorococcum*, *Stichococcus*, *Pleurococcus*, *Hormidium*, and some cyanoprokaryotes. Later investigations reviewed by Kristiansen (1996b) have demonstrated the presence of planktonic algae (*Chlorella*, *Chalmydomonas*, *Nostoc*, *Anabaena*, *Planctonema*) in aerosols.

10.9.2.5   Human agency

The different kinds of human activity which lead to the introduction of species to lakes where they had not been present originally is difficult to list. A summary would certainly involve large- and small-scale commerce (shipments of large tropical and ornamental fish), the overall increasing tourism that ranges from the transport of large ships to the small plastic toys of children, and, sad enough to say, the activities of energetic field naturalists carrying algae on their plankton nets from one lake to another, as described by Talling (1951).

10.9.2.6   Dispersal distances

The probability of successful dispersal depends on the effectiveness (with respect to velocity, distance) of the carrier, and the ability of algae to tolerate the transport conditions (see Kristiansen 1996b, and literature cited therein). Water beetles and aquatic mammals may be effective over a distance of a few kilometres, whereas dragonflies have been shown to transport viable algal material almost 1000 km. In Atkinson's (1980) experiments, *Asterionella* remained viable in the faeces of waterfowl for a maximum of 20 hours, perhaps corresponding to a flight of 220 km. Many ducks migrate up to 4800 km, but with many stops. Nonstop flight distances of the order of 3200 km in 48 hours have been noted for the white-fronted goose (*Anser albifrons*: Owen & Black 1990). Long-distance north–south transequatorial transport can be effected by migrating birds such as the south polar skua (*Catharacta antarctica*) and the Antarctic tern (*Sterna vittata*). For airborne transport, it is difficult to give maximal distances. Since viable

Melosira were found at 3000 m altitude, a considerable distance may be possible. Dust containing diatom frustules can certainly be blown across the Atlantic Ocean (Kristiansen 1996b).

10.9.2.7   Dispersal times

As the time which has elapsed since phytoplankton data were first collected is quite short compared with evolutionary time, even for such rapidly multiplying organisms, it is difficult to say how long it might be before a species with a non-uniform distribution populates many waters of the world. Expansion of some southern species (*Aphanizomenon aphanizomenoides*, *A. issatschenkoii*, *Cylindospermopsis raciborskii*; Krienitz et al. 1996; Krienitz & Hegewald 1996; Padisák & Kovács 1997) has been recorded during the past 30–50 years. In the case of *C. raciborskii*, less than 100 years has elapsed since this originally pantropical species has spread into the Northern Hemisphere as far north as latitude 53° (Padisák 1997).

10.9.3   *Biogeography and speciation*

In biogeography it is customary to consider the region richest in species or in morphological diversity as the area of evolutionary origin of a given taxonomic group. The genus *Cylindrospermopsis* and desmids in general have their evolutionary origin in the tropics. Almost nothing is known about the time taken for new phytoplankton species to evolve. For example, blue-green algae are a very ancient group, and even 3.5 billion (10^9) years ago forms occurred which are morphologically similar to modern taxa (Knell & Golubic 1992). Other work, based on molecular genetics, or phytogeography, however, argues for the relatively recent evolutionary origin of certain blue-green algal species (Castenholz 1983; Stulp & Stam 1984).

10.9.4   *Types of geographical distributions*

Although the ease and long distance of dispersal of phytoplankton theoretically allow the majority of species to become cosmopolitan, this idea has seldom been tested, and is unlikely, especially in term of genetics, to be the case. The absence of certain species from whole continents cannot generally be explained by the lack of suitable habitats. In this respect, K-selected species, which make higher demands upon their environments, are more likely to occur less frequently or less widely than r-strategists (MacArthur 1972). The establishment of a cosmopolitan, pantropical, circumpolar, regional, etc., flora (see later) depends on an efficient distribution in relation to the rate of evolution. If the dispersal rate of new genotypes is faster than the rate of evolution, then a pantropical, holarctic or cosmopolitan distribution will result. In the opposite case (the rate of dispersal of genotypes is slower than the rate of evolution), floristic regionality will develop, which explains the existence of local, regional or endemic species (Hoffmann 1996).

Owing to the paucity of sampling of many regions of the world, the question of whether there is non-uniform distribution of taxa is generally answerable only on a spatial scale of thousands of kilometres. The chance of finding new or endemic species in remote locations is always there.

10.9.4.1 Subcosmopolitan taxa

It is reasonable to replace the term 'cosmopolitan' (meaning, strictly speaking, 'occurring almost anywhere and in many kinds of lakes') by the term 'subcosmopolitan', which applies to species occurring throughout the world, but always in specialised environments. The clearest examples are the algae of hotsprings, which flourish within a narrow range of environmental conditions and occur more or less wherever those are fulfilled. The wide distribution of such species is, therefore, mosaic-like, reflecting the corresponding distribution of habitats.

Among planktonic organisms, species such as *Microcystis aeruginosa, M. wesenbergii, Planktothrix agardhii* and, indeed, a majority of planktonic algae are widely distributed geographically. The relative proportion of subcosmopolitan taxa in phytoplankton is certainly higher than in any other group of freshwater non-planktonic algae. Comparison of typical summer assemblages in

temperate lakes (Reynolds 1984a) indicates a great overlap with year-round associations in the tropics. Many species which are restricted to the warmest seasons in the temperate zone (e.g., *Urosolenia* spp., *Aulacoseira granulata* var. *angustissima, Cylindrospermopsis raciborskii*) may be perennial in the tropics.

Tropical lowland lakes are rich in cosmopolitan and pantropical taxa. On the other hand, a detailed survey of the Indo-Malayan phycogeographical region (Vyverman 1996) has shown that altitudinal gradients do exist. Between 1700 and 2500 m a.s.l., tropical taxa are gradually replaced by those characteristic of cooler climates. For example, the typical temperate species *Ceratium hirundinella* has been reported from the Indo-Malayan region, but only from high mountain lakes.

10.9.4.2 Pantropical, temperate and circumpolar (holarctic) taxa

Many phytoplankton species are globally distributed in large latitudinal bands. Amongst these, species found only in the area located roughly between the two tropics are called **pantropical**. Numerous algae are found in this group (e.g. all species of *Cylindrospermopsis* except *C. raciborskii, Anabaena fuellebornii, A. iyengarii, A. leonardii, A. oblonga, A. recta, Anabaenopsis tanganyikae, Aulacoseira agassizii, A. ikapoensis, Schroederia indica, Mallomonas bangladeshica, M. bronchoartiana, M. tropica, Synura australiensis, Ceratium bracyceros* and *Peridinium gutwinskii*). Several well known species (*Planktothrix rubescens, Limnothrix redekei, Anabaena solitaria, A. flos-aquae, A. lemmermannii, Anabaenopsis arnoldii, A. milleri, Aphanizomenon flos-aquae, A. issatschenkoi*) are restricted to the temperate zones (Hoffmann 1996; Vyverman 1996). Some are found at quite high latitudes, others (e.g. *A. issatschenkoi*) only in warmer temperate regions.

Some rare algal species occur in a wide latitudinal range. However, there are also several phycogeographical regions where they do not. For example, *Asterionella formosa* has never been reported from the Indo-Malayan region, even though there are many lakes where, theoretically, it could occur (Vyverman 1996). Kristiansen & Vigna (1996) found a marked bipolarity in geographical distribution of silica-scaled chrysoflagellates: a high degree of similarity is apparent between the floras of the climatically comparable regions of the Northern and Southern Hemispheres. Typical taxa are: *Mallomonas parvula, M. paxillata, M. transsylvanica, M. cristata, M. alata* and *M. alveolata.*

A more restricted distributional type characterises the occurrence of *Cyclotella tripartita,* which is confined to latitudes greater than 50°, but only in the Northern Hemisphere (Scheffler & Padisák 1997). Such species are **northern circumpolar taxa**. Desmids provide another example, but in contrast to that of the Arctic the desmid flora of Antarctica is extremely poor (Coesel 1996).

10.9.4.3 Regional taxa, endemic species

Endemism may originate in two quite different ways. The purest form is where a species evolves uniquely in a given location, and remains exclusive to that location. Endemism can also occur, however, as a result of habitat fragmentation or destruction, and subsequent extinction from all localities except one. Distinction of the two types from each other is difficult, because, in most cases, no fossils remain.

Quite a number of species (**local endemic species**) have, for a long time, been known only from the type locality. Others (**regional endemic species**) are found to be restricted to a certain region. Tyler (1996) differentiates between 'fragile endemics' (with restricted distribution) and 'robust endemics', which are more widely distributed. For example, both *Aphanizomenon manguinii* and *Trichormus subtropicus* have so far only been recorded on several islands in the Caribbean region (Komárek 1985). Similarly, *Cyclotella tasmanica* occurs exclusively in Tasmania, but is widely distributed there (Haworth & Tyler 1993). Australia, and especially Tasmania, contain a considerable number of endemic species and genera, among them planktonic algae called 'flagship' taxa by Tyler (1996). Such species are, for

example, *Tessellaria volvocina, Dinobryon ungentariforme, Chrysonephele palustris, Prorocentrum playfairi, P. foveolata* and *Thecadiniopsis tasmanica.*

10.10   INTERACTIONS BETWEEN PHYTOPLANKTON POPULATIONS

10.10.1   Competition for resources

Several types of interactions between phytoplankton species occur. Competition (in some rare cases, **allelopathy**) is of major importance.

10.10.1.1   Steady-state competition

According to Hardin's competitive exclusion principle (1960), the number of species which can co-exist in any particular environment is the same as the number of limiting factors present. In the case of phytoplankton assemblages (assuming nitrogen, phosphorus, silicon, one trace element and light as being potentially limiting), the principle predicts the equilibrium co-existence of no more than five species. It was demonstrated in Tilman's competition models and experiments (1982) that, if several species are competing for one resource, the species with the highest affinity (lowest k_s and threshold) will outcompete all the others, regardless of initial intrinsic growth rates (rate of loss is constant). If two species are competing for two resources (supposing that the threshold of species X is low for resource A and high for resource B, whilst that of species Y is high for resource A and low for resource B), there may be three outcomes, as follows.

1 If concentration of resource A falls below the threshold of species Y but remains above that of species X, then X will outcompete Y.

2 If concentration of resource B falls below the threshold of species X but remains above that of species Y, then Y will outcompete X.

3 If concentration of resource A is close to, but not below, the threshold of species Y, and the concentration of resource B is close to, but not below, the threshold of species X, then the two species can steadily co-exist: growth of species A is limited by resource Y and that of species B by resource X.

Tilman (1977), in chemostat experiments, convincingly substantiated the above scenarios, using *Asterionella* (relatively low k_{sP} for phosphorus and high k_{sSi} for silicon) and *Cyclotella* (high k_{sP} and low k_{sSi}).

10.10.1.2 Competition in fluctuating environments

The above experiment is only reproducible, however, if the resources required are supplied at a rate necessary to keep a constant concentration. Sommer (1985), using natural phytoplankton from the Bodensee (Lake Constance), demonstrated that in chemostats receiving a continuous supply of nutrients (throughflow rate 0.3 day^{-1}, with Si:P ratio 20:1), *Synedra–Ankyra* coexistence developed. All the other species were eliminated. In parallel chemostats receiving pulsed nutrients (throughflow rate 0.3 day^{-1} but without nutrients; Si and P at ratio of 20:1 supplied only once a week), a multispecies assemblage (*Mougeotia, Staurastrum, Aphanizomenon, Pediastrum, Scenedesmus, Chlorella, Synedra, Nitzschia, Asterionella*) developed.

10.11 SCALES OF PHYTOPLANKTON-RELATED ENVIRONMENTAL CHANGE

Steady-state conditions develop very rarely in nature and, even if they do, they are far from long-lasting. Concentrations of potential limiting factors (temperature, light, loss rates, etc.) all fluctuate at different frequencies, but their effects are 'species-specific'. Crucial environmental factors fluctuate on different timescales.

10.11.1 Multi-annual cycles

In many temperate regions, dry and wet years oscillate with a circa-decadal period of some 5–10 years, and cycles are arranged in a longer oscillation with a frequency of centuries. According to

Wigley & Jones (1987), global mean temperatures have fluctuated throughout recorded history with a mode of 200–300 years. The influence of such changes on shallow, polymictic, astatic lakes can be considerable.

For example, central European lakes, even those as large as the Neusiedlersee (surface area 300 km^2; see Chapter 2), dry out completely at a frequency of 100 years. Quantitative phytoplankton data sets of comparable length are not available, but the similarity between Grunow's (1860, 1862, 1863) diatom data from the period prior to the last complete drying out, and those of Pantocsek (1912), from the period soon after the lake refilled, show that even such a profound change as this is tolerable at a floristic level. All of the characteristic (and rare) species of diatom found by Grunow were also recorded by Pantocsek. It was also demonstrated, in a recent quantitative phytoplankton survey of the lake, that a higher frequency (5–10 years), wet–dry periodicity induces similar oscillations in annual crop of the most abundant species.

Abundance of tropical species in temperate waters, among them such a strong invader as *Cylindrospermopsis raciborskii*, accords with a multi-annual scale: they reach high biomass only in exceptionally warm summers (Padisák 1998). Fluctuations in the position of the Pacific intertropical convergence and its associated air circulation and rainfall vary with a periodicity of 7 years. The cycle affecting June temperature in the English Lake District exhibits a 9–10-year periodicity (George & Harris 1985).

10.11.2 Annual cycles, the PEG model of plankton succession

In temperate latitudes, the solar, hydrological and temperature cycles vary on an annual timescale, and shape the basic stability and chemical background of phytoplankton succession. These annual changes shape the best known cyclicity of phytoplankton, their **seasonal succession**. The first coherent model of seasonal plankton succession, the so-called PEG (Plankton Ecology Group) model (Sommer *et al.* 1986), was produced as a result of several years of discussions among about 30 plankton ecologists, and is based on successional sequences observed in 24 lakes. The assumptions upon which the model is based are:

1 resource limitation and competition occur;

2 algae will always reproduce at the maximum rate sustainable under given physical and chemical conditions;

3 small phytoplankton species are edible for zooplankton;

4 zooplankton is temporarily able to overexploit phytoplankton populations (clear-water phase);

5 during the clear-water phase, nutrient concentrations will regenerate;

6 after the clear-water phase, a further phase characterised by faster growing algae with larger cell sizes (*Cryptomonas*, green algae) begins;

7 phosphorus-limitation, and competition for phosphorus, combined with high silicon concentrations, favours a transition to diatoms (provided that resuspension events allow them to survive in the euphotic zone);

8 silicon limitation, in combination with phosphorus limitation, leads to increase of dinoflagellates—additional depletion of nitrogen favours nitrogen-fixing cyanoprokaryotes;

9 the sequence of summer species abundance observed is the result of a combination of limitational pattern and selective grazing.

Typical successional events observed during a plankton year are indicated in 24 consecutive steps below (in parentheses), summarised as follows.

• Towards the end of winter, nutrient availability, and increased light, permit growth of phytoplankton (diatoms, cryptophytes) (1).

• The spring maximum is then grazed by herbivorous zooplankton (first the faster-, then the slower-growing species), which soon become abundant as a consequence of high fecundity induced by the large biomass of edible algae. This continues until zooplankton community filtration rate and biomass exceed the reproduction rate of phytoplankton (2–4).

• As a result of zooplankton grazing, phytoplankton biomass declines rapidly, to very low amounts. There then follows the **clear-water phase**, which persists until inedible algal species appear in suffi-

cient numbers. Nutrients become remineralised by the grazing process, and may accumulate during the clear-water phase (5).

• Herbivorous zooplankton become food-limited, and fish predation accelerates the decline of herbivorous zooplankton populations (6, 7).

• Under conditions of reduced grazing pressure, the summer phytoplankton begins to increase. The composition of the phytoplankton becomes complex owing both to increase in species richness and to functional diversification into small, 'undergrowth' species (available as food for filter-feeders) and inedible 'canopy' species (8).

• Nutrient depletion prevents explosive growth of edible algae. Phytoplankton composition shifts towards taxa which can utilise high hypolimnetic phosphorus concentrations (*Ceratium*), and which can fix atmospheric nitrogen (heterocytic cyanoprokaryota), whilst diatoms are limited by low silica concentrations (9–13).

• Larger species of herbivorous crustaceans are replaced by smaller species, and by rotifers, which are less vulnerable to fish predation and are less affected by the interference with their food-collecting apparatus which can be caused by some forms of inedible algae (14, 16).

• The period of autogenic succession is terminated by factors related to physical changes, including increased mixing depths resulting in nutrient replenishment and deterioration of the effective underwater light climate (17).

• Large inedible forms and diatoms then appear, accompanied by small edible phytoplankton species. Together with a decrease in fish predator pressure this leads to an autumn maximum of zooplankton, including larger forms and species (18–20).

• Further reduction of light energy leads to low or negative net primary production, and an imbalance. Consequently, phytoplankton biomass declines to the winter minimum, which is followed by a decline in herbivorous zooplankton biomass (21, 22).

• Some zooplankton species produce resting stages. Cyclopoid copepods awake from their diapause and contribute to overwintering populations in the zooplankton (23, 24).

Since this original model was published, numerous cases have been put forward which show that it is not universally applicable. For example, shallow lakes seldom experience a clear-water phase, so much as a detritus phase (Olrik 1994). In large, deep, temperate lakes, where successional events are closest to the sequence described above, a number of other occurrences are also possible. Similarly, the reason for termination of the spring diatom maximum may well be exhaustion of silica, nitrogen and/or phosphorus. Decline in numbers is not necessarily due to overgrazing, because the effect of onset of stratification may be similar (heavy diatoms simply sink to the hypolimnion; Padisák *et al.* 1998).

Another insufficiency of the model is that physical destabilisation (a subject which will be discussed later) is largely neglected. Also, whilst the activity of the microbial loop may be essential in nutrient regeneration, here it is not considered at all. Nevertheless, the model provides a coherent conceptual framework for interpretation of plankton succession in relation to nutrient resources, competition and loss processes (Olrik 1994).

Towards lower latitudes and, most prominently, in equatorial regions, the intra-annual range of variations in light and temperature declines, and a high degree of environmental stability can develop. This may allow a given phytoplankton assemblage to be sustained over periods longer than a year (Melack 1979; Zohary *et al.* 1996). However, repetitive seasonal successional patterns are also frequently observed in the tropics, as a consequence of regular changes in stream discharge, and the subsequent cyclicity of physically stable and unstable periods. In view of the great differences observed in the short term, as well as in annual nutrient supply rates, in mixing regimes, in residence times and the morphological differences in tropical lakes (Kalff & Watson 1986), it is difficult to reach a general conclusion regarding seasonal successional sequences in these lakes.

10.11.3 *Variability on a scale of weeks to months*

Oscillations in herbivorous zooplankton density,

and consequently in its grazing pressure on phytoplankton populations, vary with a periodicity of 30 to 50 days. In moderately deep lakes with unstable stratification, only very strong mixing events can erode the thermocline. Therefore, such an effect operates at monthly intervals. Meteorological conditions in temperate regions typically vary on a timescale of 5 to 15 days. Decreases in incident heat income, and therefore temperature decline, may well lead to deepening of the thermocline. Therefore, more nutrient-rich, deeper water replenishes the nutrient-depleted euphotic zones.

In shallow lakes, such meteorological changes can lead to complete mixing, and the entrainment of nutrient-rich sedimentary pore-water, reducing the penetration of light. Frequent resuspensions in shallow, turbid lakes may lead to almost continuous persistence of 'merophytoplankton', comprised of diatoms with large individual volumes (*Aulacoseira*, *Fragilaria* chains, *Camplylodiscus*, *Surirella*). These algae ultimately depend on resuspension, because, even if they are evolutionarily adapted, their photosynthesis would be stongly reduced on the aphotic sediment surface to which they descend during calm periods (Carrick *et al.* 1993; Padisák & Dokulil 1994).

10.11.4 *Diurnal (diel) cycles*

Diurnal cycles of zooplankton (Chapter 14) lead, on the one hand, to spatially variable grazing pressure at different depths, and, on the other, to variations in zooplankton excretion. Therefore, nutrient regeneration is also greater at depths characterised by denser zooplankton populations. Circadian rhythms of phytoplankton populations can be considered as a specific adaptational behaviour. Practically, diurnal changes can be regarded as the shortest which can influence populations, and therefore community dynamics. Shorter fluctuations, for example day–night , also occur, and may influence, for example, photosynthetic rates. However, most species are physiologically adapted to such fluctuations, and so far no influences on community dynamics have been reported.

10.12 DIVERSITY AND SUCCESSION OF PHYTOPLANKTON COMMUNITIES

10.12.1 Species number

The simplest measure of diversity is the number of species in a given area, in this case, that of the lake. The frequently-used term **biodiversity** refers most commonly to species number. Numerous attempts have been made to compare this variable between different lakes, but no clear relationships have been obtained.

In general, species number is higher in large lakes than in small, because they offer a greater degree of habitat diversity. However, in contrast to zooplankton (Dodson 1992), no clear evidence is generally accepted for the phytoplankton. Findings have been contradictory. For example, in small hypertrophic ponds, numbers of chlorococcalean taxa alone may well exceed total species number in a larger lake.

Species lists compiled by different workers are difficult to compare, partly because their taxonomic expertise and interests are different, and because the effort required to lengthen the list also varies from site to site. In this way, species lists accumulated during year-long quantitative studies, with similar sample numbers, are becoming acceptable for such comparisons provided that sufficient background effort is given to correct species identification. Such studies usually result in lists numbering 80–130 species.

The assumption that species number (and diversity) is lower in eutrophic lakes than in oligotrophic bodies (Margalef 1964) has not yet been substantiated. Nutrient scarcity may select for species with high affinity for nutrient uptake, or any other kind of specialised, nutrient-related strategy, whilst extreme nutrient richness favours species with high intrinsic growth rates, perhaps combined with substantial shade tolerance. Species number is maximised in the oligo-mesotrophic to moderately eutrophic portion of the trophic gradient. However, this statement needs to be tested in the future.

Another early belief was that temperate lakes are richer in species than lakes in tropical regions. Several studies (e.g. Lewis 1978) support this view. The observational basis of this opinion is quite clear: as a consequence of generally higher environmental stability, development of equilibrium states is more complete when (and in such states) overwhelming abundance of one or two species is common. As more and more data on seasonal successional sequences of tropical phytoplankton became available, this view also changed. In their comparative study, Kalff & Watson (1986) found no fundamental difference between tropical and temperate lakes. Similar conclusions can be derived from the study of Lake Titicaca by Carney et al. (1987).

In general, lakes with 'average' properties (no extreme pH, conductivity, ionic composition, trophy, turbidity, mixing stress, or any other kind of variable relevant for phytoplankton growth) provide conditions favourable to hundreds of phytoplankton species. If any of the variables mentioned approximates an extreme, all species unable to tolerate this change are excluded, and species number begins to decline. For example, in reservoirs in Sicily, Calvo et al. (1993) found a close negative correlation between species number and conductivity, in the normal to extremely high conductivity range. Similarly, in oligotrophic high-mountain lakes in Spain, Morales-Baquero et al. (1992) found a positive relationship in the conductivity range from very low to 'normal'. This example demonstrates that either extremely low or extremely high conductivities delimit species number, which maximises at 'average' values.

10.12.2 Evenness, the Shannon–Weaver function

Community diversity is most frequently estimated by considering not only number of species, but also the evenness: how evenly is total biomass (numbers) distributed among species? Such estimates need to be quantified. Numerous functions, each with different sensitivity profiles, are described in the literature. However, the Shannon–Weaver index of diversity (H''; Shannon

& Weaver 1949), originating from information theory, has become the most widely used:

$$H'' = -\sum_{i=1}^{s} N_i/N \log_2 N_i/N \qquad (10.28)$$

where N_i is the density of the ith species; s is the number of species in the sample; and

$$N = \sum_{i=1}^{s} N_i \qquad (10.29)$$

As follows from the mathematical properties of the Shannon function, the theoretical maximum (H''_{max}) of diversity in any sample may be expressed as

$$H''_{max} = \log_2 s \qquad (10.30)$$

The more equally distributed the total density among the species, the higher the diversity. **Equitability** (or evenness, E) can be expressed as

$$E = H''/H''_{max} \qquad (10.31)$$

or as a percentage

$$E\% = 100 H''/H''_{max} \qquad (10.32)$$

The various diversity functions described in the literature (see review by Washington 1984) allocate different weights to the number of species and their evenness. In some cases, species number, and in others, evenness, is weighted. The Shannon–Weaver function offers a compromise between these two extremes, in that it gives weight both to evenness and to species number. However, in special cases, other functions may be more applicable.

Although it is not exceptional to calculate diversity values based on cell numbers or on individual numbers, it is becoming more and more customary to use biomass (biovolume). This is not just because, for example, one single cell of *Ceratium* (volume: $53,000\,\mu m^3$) would accommodate approximately 10^5 picophytoplankton cells with diameter of $1\,\mu m$. The main reason is that many concepts of community development rely on re-

source partitioning and are therefore closely related to biomass attained.

Comparisons of values of diversity functions from different lakes provide no clear relationships to background variables: this finding initiated a scepticism about the usefulness of diversity indices in general. For example, Kalff & Knoechel (1978, p. 477) urged 'algal ecologists to stop wasting their time on such mathematically simple but ecologically obscure indexes'. Indeed, the mechanical usage of function values does not provide applicable information, but it is also correct that the diversity approach to understanding community development remains of profound importance.

10.12.3 Community change rate, measurement of successional rates, similarity, dissimilarity

Although diversity functions are the most common means used in order to describe organisational changes within phytoplankton communities, several other functions are used to analyse such properties. According to Reynolds (1997a) the rate of community change (δ_S) can be estimated as

$$\sigma_s = \sum_{I=1}^{s} \{[b_i(t_1)/B(t_1)] - [b_i(t_2)/B(t_2)]\}/(t_1 - t_2)$$

$$(10.33)$$

where $b_i(t_1)$ is the biomass of the ith species and $B(t)$ is the total biomass, each measured on separate dates, t_1 and t_2; s is the total species number. The shorter the interval $(t_1 - t_2)$, the greater the sensitivity of the index to change in community structure.

Without attempting to discuss different similarity indices available in the literature (see Washington 1984, and the literature cited therein), only two indices, representing two basic types, are mentioned here. Jaccard's (1908) index is one of the simplest and oldest, and calculates similarity (S_J) based on percentage of species in common as

$$S_J = 100[n_c/(n_i + n_j)] \qquad (10.34)$$

where n_c is the number of species found in both samples, n_i is the number of species found exclusively in the ith sample, and n_j is the number of species found exclusively in the jth sample. As a measure of community structure it only takes into account species number and not abundance.

Among indices taking species abundance into account, Czekanowski's (1909) index is quite common and expresses similarity (S_{CZ}) as

$$S_{CZ} = \frac{2\sum_{i=1}^{s} \min(x_{ki}, x_{kj})}{\sum_{i=1}^{s} (x_{ki} + x_{kj})} \qquad (10.35)$$

where $\min(x_{ki}, x_{kj})$ is the biomass of the kth species in that sample of the sample-pair in which its biomass is smaller, x_{ki} is the biomass of the kth species in sample I, and x_{kj} is the biomass of the kth species in sample j. If the two samples compared contain similar species at similar biomass, then $S_{CZ} = 1$. If no species is shared it resolves to 0.

According to Geviatkin (1979), successional index (I_s) can be derived from S_{CZ} as

$$I_s = 100[(1 - S_{CZ})/t] \qquad (10.36)$$

where t is the time [days] elapsed between the two samplings.

10.12.4 Plant succession, equilibrium model of community development

The question of diversity remains one of the most controversial issues in succession theory in general. According to the classic view (e.g. Clements 1916; Odum 1969), diversity of communities increases towards the late phases of succession, whilst the competitive exclusion theory (Hardin 1960; see above, section 10.9) predicts that succession should tend towards establishment of low-diversity equilibrium.

One of the most basic ecological concepts, that of **niche**, dates back to the 1920s (Vandermeer 1972), and states that, in nature, in an equilibrium state, niches (which are multidimensional both in time and space) are filled with species, and that each must use the environment differently, in

order for them to coexist (Gause's (1964) principle of exclusion). Moreover, exploitation of environmental resources can be measured from the number of coexisting species depending on **niche partitioning**; in other words, the relationship between niche overlap and width (MacArthur 1972). Hardin's theory also depends on equilibrium concepts. As shown in section 10.10, it can be evaluated both by theoretical models and by chemostat experiments with lake phytoplankton.

Therefore, both the Clementsian and the niche-partitioning based concepts of diversity rely on the supposition that ecological equilibrium exists. The first, however, predicts increasing diversity along a successional series, the second, decreasing.

10.12.5 Hutchinson's Paradox of the Plankton, non-equilibrium models

The competitive exclusion principle predicts that assemblages will consist of a maximum of four to five species, each limited by a different resource (further resources are not likely to limit growth under natural conditions). This clearly contradicts the experience that it is not unusual to find dozens, sometimes more than 100 species, of phytoplankton in one single plankton sample. This contradiction between theory and experience has been formulated by Hutchinson (1961), raising the question of how so many species can coexist in such a supposedly homogeneous environment as the open water. In his original paper, Hutchinson suggested several possible explanations for his 'paradox', among them, that the boundary conditions of competition change frequently enough to disrupt competitive hierarchies before exclusion occurs.

This is a typical non-equilibrium explanation of species coexistence: there is no time for competitive exclusion to complete. This view had been expressed by many authors in referring to many kinds of ecosystem, from plankton to tropical rain forests (Richerson et al. 1970; Wiens 1976, 1977; Connell 1978; Shugart & West 1981; Harris 1983; Gaedeke & Sommer 1986; Wilson 1990). In fact, in nature, equilibrium is more the exception than the rule (Reynolds 1997a).

10.12.6 *Connell's intermediate disturbance hypothesis*

Connell's (1978) Intermediate Disturbance Hypothesis (IDH) states that:

1 In the absence of disturbance (eternal steady state), competitive exclusion will reduce the number of species surviving to minimal levels.

2 Under very intense disturbance, only a few populations of pioneer species can re-establish themselves after each disturbance event. This would also maintain low diversity.

3 If disturbances are of intermediate frequency and/or intensity there will be repeated opportunities for the re-establishment of pioneer populations which would otherwise be outcompeted. Populations of successful competitors can withstand disturbance without completely taking over the community. Thus, a peak of diversity should be found at intermediate frequencies and intensities of disturbance.

According to Reynolds (1988b), at frequencies of the order of a few hours (less than one generation time) responses are physiological. Low-frequency pulses, separated by intervals of 10 days or more, can initiate a successional sequence; and progressively smaller intermediate scales (200–20 h) interact with growth rates of phytoplankton species, and tend to preserve high species diversity. Repeated disturbances achieve a kind of 'pseudoclimax' ('plagioclimax' in Reynolds 1984b).

In order to test the applicability of the IDH to phytoplankton, it is essential to define both disturbance and equilibrium.

10.12.6.1 Definition, origin and nature of disturbance

According to the definition adopted by the Eighth Workshop of the International Association of Phytoplankton Taxonomy and Ecology, 'disturbances are primarily non-biotic, stochastic events that result in distinct and abrupt changes in the composition and which interfere with internally-driven progress towards self-organisation and ecological equilibrium; such events are understood to operate through the medium of (e.g.) weather and the frequency scale of algal generation times' (Reynolds *et al.* 1993, p. 187).

It is necessary to make it clear that disturbance *per se* does not exist, and that, what is more, it is impossible to define without involving the entity affected by the disturbance itself. In his original paper, Connell conceived of disturbances as primarily originating from internal processes (e.g. treefall gaps caused by the death of senescent trees). However, there is no a priori reason why the consequences for species diversity of disturbances of external origin should be different. Obviously, there are instances when the contribution of internal and external factors cannot be separated easily. But such distinctions are less important than the occurrence of disturbance *per se*.

It is also necessary to mention here that planktonologists tend to perceive sustained thermal stratification as the 'undisturbed state' of the lake. However, permanent circulation can also represent low disturbance status, in which case, sudden stratification then qualifies as disturbance (a terrestrial analogue may be to consider regular livestock grazing of a grassland, or mowing the garden lawn, as undisturbed states). For this reason, Chorus & Schlag (1993) found it helpful to refer to 'Intermediate Quiescence'. This concept shows clearly that perhaps the best way to understand disturbance is as a more or less abrupt shift in the existing *status quo*. In plankton (or in any other community in which disturbances of external origin are significant), the term 'frequency of environmental change' could well replace the term 'disturbance'.

10.12.6.2 Identification of equilibrium

As, in nature, equilibrated communities rarely occur, there is no widely accepted definition or description of equilibrium. In natural phytoplankton communities, it is often difficult to determine whether a given 'phase' in a seasonal sequence can be considered to be in an equilibrium state or not, either because of a lack of chemical data or of sufficient sampling frequency, or some other cause. For practical purposes, provided that (i) one, two or three species of algae contribute more than 80% of total biomass, (ii) their existence or coexistence persists for long enough (more than 1–2 weeks) and (iii) during that period the total biomass does not increase significantly, then a phase is sufficiently established for us to treat it as being an equilibrated state.

10.12.6.3 Applicability of IDH in phytoplankton

Individual analyses of phytoplankton structure versus disturbance frequency (see case studies in Padisák et al. 1993) support Connell's theory. Diversity is low at a high frequency of disturbance, because species number is very limited, whilst at low frequencies, low equitability diminishes diversity. Development of a near-equilibrium state may require some 12–16 generations, perhaps occupying 35–60 days in a temperate summer. Diversity is maximal somewhere between the second and the fourth generation (i.e. 5–15 days).

Diversity is high, or increases, when species replacement rates are rapid, as, for instance, in warm water. These changes are neither necessarily nor exclusively driven by disturbances, although communities repeatedly destabilised at frequencies of the order of three generation times are likely to support a high diversity of species. Diversity is promoted when fast-growing (i.e. usually small-sized) algae spread through the residual structure.

Diversity is low, or declines, in advanced successions, where a large biomass is made up of a single, generally 'large' (>200 mm) motile algal species. Diversity is rarely high in strongly selective environments, such as highly flushed systems or in lakes characterised by extremes of acidity, alkalinity, turbidity, oligitrophy or physical constancy (save by spatial niche differentiation). Without disturbance of the status quo, equilibrium dynamics predict an eventual total suppression of diversity (Reynolds et al. 1993).

10.12.6.4 Strengths and weaknesses of the IDH concept

Two outstanding merits of the IDH are obvious. One is the ease of its comprehension. As a word

model it is readily understood that, whereas natural communities under unchanged external conditions tend to become uniform, their progress towards the anticipated, equilibrated, competitively excluded outcome can, at any time, be shifted, reversed, slowed, interrupted or overridden by forces emanating from outside, which are beyond the capability of the existing community to absorb. The second is that it provides an explanation for the uncomfortable gulf between what we expect to happen on the basis of controlled experimentation, and what we generally observe. Immaturity becomes a recognisable attribute of systems; equally, disturbance becomes a valid factor contributing to their productivity, species structure, biomass and organisation. Ranged against the deceptive simplicity of the IDH are some quite fundamental questions of logic and/or practicality. Until these are adequately addressed, its utility remains in doubt.

The principal difficulty concerns the recognition and measurement of disturbance. Although it is becoming increasingly straightforward to measure a force imposed on an ecosystem, it is quite another matter to relate quantitatively the extent of the impact, if any, of that force. Although the effect upon the biomass, species composition and productivity of a respondent system, of a sudden storm, or a flash flood attributable to sharply enhanced summer rains would be anticipated to be immediate, it is in no way 'guaranteed'. Imposition of the same external forcing may well invoke quite different community responses, depending upon the resilience shown by the developmental stage achieved.

A more fundamental difficulty with the IDH is that it seems inescapable that disturbance is a phenomenon recognised and measured only as an effect. The stimulus may generally be external, but its nature is judged exclusively by the reaction which it engenders (Reynolds et al. 1993). It is a future task of plankton ecologists to overcome the above indicated weaknesses.

10.13 ORGANISATION OF THE PHYTOPLANKTON; A COMPARISON WITH TERRESTRIAL PLANT COMMUNITIES

The profound differences between terrestrial and planktonic vegetation are evident. Phytoplankton species are small (generally $10^{-2} - 10^{-7}$ m) and live short lives ($10^4 - 10^7$ s). Disregarding bacteria and viruses, they are the smallest organisms in the pelagic food webs. Primary producers on land, especially in forests, are the largest and longest living life-forms inhabiting their ecosystems. Plotting generation time against body size of characteristic groups (plants, herbivores, other invertebrates, vertebrates), Cohen (1994) clearly demonstrated the 'upside-down' structure of marine food webs compared with those of terrestrial ecosystems.

Both aquatic and terrestrial ecosystems can be very rich in species. Such high diversity can be explained by reference to internal and external mechanisms, and their heterogeneity in space or time. Traditionally, internal mechanisms are associated with terrestrial ecosystems, whilst community changes in the open water match external forces. These differences are justified via the way in which the scales of space and time in ocean physics and biology overlap, compared with their inherent separation in most atmospheric systems (Steele 1991).

Spatial scales, from eddy sizes to deep circulations, are generally smaller in inland waters (lakes) than in the open sea. Nevertheless, this 'shortening' is not of sufficient magnitude to challenge the basic assumption that terrestrial populations exist in a 'white noise' environment, whereas aquatic populations are embedded in 'red noise' (Steele & Henderson 1994). Indeed, the differences between water and land are much easier to perceive than the similarities.

Elton (1927, pp. 59–60) was probably the first person in modern ecology who assumed that uniform principles operate in the sea and on land. Present ecologists also feel it necessary to stress that they 'assume that the laws of physics, chemistry and biology are the same in oceans and on continents' (Cohen 1994, p. 47); or that '. . . "Ecology"

does not alter simply because its factual basis is assembled from observations on different systems' (Reynolds 1997b, p. 196).

Whilst succession theories (for terrestrial habitats), most notably the 'superorganismic' approach of Clements (1916), and the individualistic concept of Gleason (1917, 1927), date from the beginning of the 20th century, for a long time there were no similar theories for phytoplankton. Early attempts to classify algal associations (e.g. Pankin 1941, 1945) have never received wide acceptance (see the review by Symoens *et al.* 1988). Nevertheless, there is no doubt that phytoplankton associations exist, even if phytosociological terminology (usage of the '-etum' suffix after characteristic genera; see Braun-Blanquet 1964) has not been applied. Sommer (1986b) found a high level of similarity in species composition and seasonal sequences in alpine lakes. Reynolds (1980, 1997a) applied the classic phytosociological approach to a long series of phytoplankton data from lakes in northwest England and differentiated numerous reproducible species associations (Table 10.19).

By 1753, when Linnaeus published the binominal nomenclature in his famous work, *Species Plantarum* (Linnaeus 1753), thousands of higher plant species had already been described in medieval botanical monographs. Scientific description of phytoplankton, however, did not begin in earnest until the 19th century. Quantitative counting methods (the inverted microscope technique) were not elaborated until the 20th century (Utermöhl 1931). Quantitative phytoplankton time-series data began to appear only during the 1930s–1950s, and studies in which the former were coupled with zooplankton and water chemical data only during the 1960s (e.g., Nauwerck 1963). Utilising dozens of case studies, the first sufficiently general concept of plankton succession, the so-called PEG model, was developed as recently as the mid-1980s (Sommer *et al.* 1986) as a joint effort of an international team (Plankton Ecology Group) of contemporary plankton ecologists.

Meanwhile, classic succession theory underwent significant development. I refer here to Watt's (1947) 'pattern and process' concept, and the work of Drury & Nisbet (1973), who demon-

Table 10.19 Phytoplankton associations with different morphological and physiological characteristics and the subsequent adaptation strategies. (From Reynolds 1984b, 1987a,b. Partly modified.)

Representative species	Size range (μm^3)	SA/V (μm^{-1})	μ_{max} 20°C (day^{-1}) (* 30°C)	Q_{10} of μ_{max} at 10–20°C (* at 20–30°C)	Motility	Sensitivity to grazing	Adaptation strategy
Synechococcus/Chlorella/Monorpahidium	10^0–10^3	1–4	Often >2.0	<2.5	Slow sinking	High	C
Chlamydomonas/Chrysochromulina, Rhodomonas/Glenodinium	10^1–10^3	1–3	Often >1.4	<2.5	Weak swimming	High	C
Cyclotella/Rhizosolenia	10^2–10^4	0.5–1.5	>0.8	<2.5	Sinking	Low/moderate	C–R
Diatoma/Stephanodiscus hantzschii	10^2–10^4	0.5–1.5	1.0–1.8	<2.5	Quick sinking	Low/moderate	R
Tabellaria/Cosmarium/Staurastrum	10^3–10^4	0.3–1.0	0.7–1.4	2.5–3.0	Slow sinking	Low	R–S
Asterionella/Aulacoseira italica Cyclotella	10^3–10^5	0.5–1.5	0.8–1.8	<2.5	Relatively fast sinking	Low/moderate	C–R
Asterionella/Stephanodiscus rotula	10^3–10^5	0.3–1.5	0.8–1.8	<2.5	Relatively fast sinking	Low/moderate	R
Cryptomonas	10^3–10^4	0.3–1.0	0.8	>2.5	Moderate swimming	Moderate/high	C–S–R
Fragilaria/Aulacoseira granulata Staurastrum/Closterium	10^3–10^5	0.3–1.5	0.7–1.4	2.2–3.0	Relatively quick sinking	Low	R
Planktothrix rubescens	10^5–10^7	<0.5	<0.7	3.5	Slow sinking/floating	Low	R–S
Dinobryon/Synura/Mallomonas	10^3–10^5	0.5–1.0	?	?2.5	Weak/moderate Swimming	Low/moderate	S–R
Sphaerocystis/Gemellocystis Coenococcus	10^4–10^6	<0.3	0.6	>2.5?	Slow sinking	Moderate/low	C–S
Eudorina/Pandorina/Volvox	10^4–10^6	<0.3	0.6	>2.5	Moderate swimming	Moderate/low	C–S
Anabaena/Aphanizomenon Gloeotrichia	10^4–10^6	<0.3	0.8–1.0	2.5–3.0	Adaptation of floating abilities to moderate/quick movements	Generally low	S
Pediastrum/Coelastrum/Oocystis	10^3–10^6	0.3–1.2	?	?	Slow sinking	Moderate/low	R–C
Aphanothece/Aphanocapsa	10^3–10^5	<0.5	?	?	Slow sinking	Moderate/low	R–S
Ceratium/Peridinium	10^4–10^6	<0.3	<0.4	>4.0	Effective swimming	Low	S
Planktothrix agardhii/Limnothrix redekei	10^3–10^4	<0.5	<0.9	3.0	Slow sinking or floating	Low	R–S
Cylindrospermopsis raciborskii	10^2–10^3	1.5	1.7*	4*	Slow sinking or floating	Low	S
Uroglena	10^4–10^8	<0.3	<0.6?	>0.3?	Moderate/fast swimming	Low	S
Microcystis/Ceratium	10^4–10^8	<0.3	<0.5	>4	Effective swimming or adaptation to floating	Low	S

strated that succession, like all biological processes operating at the supra-individual level, is a stochastic phenomenon in which the transitions can only be described by their probabilities. This consideration is readily perceptible among phytoplankton ecologists, who cannot avoid the stochasticity of the communities which they study, on a daily basis.

Recent investigations (Czárán & Bartha 1992) demonstrated that, in early stages of plant succession, transitions are far from being as clearly directional than was formerly supposed. The Mosaic Cycle Theory of plant succession (Remmert 1991) states that each plant community is a mosaic of its different successional stages, and questions the self-sustainability of the final state (because existing individuals inhibit development of conspecific juveniles). Consequently, it also casts doubt upon the existence of a true climactic state of terrestrial plant communities. This statement is very important from the point of view of making connections between terrestrial and planktonic successions, since it was almost dogmatically believed for a long time that succession in plankton is impossible because climax vegetation can never develop (see, e.g., Whittaker 1974). The Mosaic Cycle Theory is also notable for our purposes in another context: it was the first time that a concept was published which considered its implications for plankton as well as terrestrial systems (Sommer 1991).

From the above considerations, we may deduce that:

1 Phytoplankton communities typically undergo significant changes within individual calendar years. These changes have been termed 'seasonal succession' by plankton ecologists.

2 Seasonal succession of phytoplankton is more analogous to long-term succession of terrestrial vegetation than to seasonal phenomena in the terrestrial environment.

3 Many phytoplankton generations, characteristically about 40, are involved in the process of succession. This is a much higher number than, for example, the number of tree generations in a terrestrial succession.

4 Seasonal succession of phytoplankton exhibits several quite predictable and distinct phases; succession is directional and its outcome is predictable within quite clearly defined limits of probability. Under certain physical and chemical conditions, the intrinsically transient nature of early and mid-successional phases can be demonstrated. For example, the terminator of spring diatom development is often a decline in silica concentration. In other words, because populations themselves have modified their environment to the degree that no further growth is possible, a new phase must begin.

5 Under favourable physical conditions, self-sustainability of the final stages of seasonal succession can be demonstrated. The clearest examples are provided by tropical lakes, in which a given association can persist over a number of years. Sommer (1991) describes a case study from a deep temperate lake when, in a meteorologically exceptional year, stratification was established very early and produced a very stable thermocline. The main *Ceratium* population, therefore, developed earlier than in other years, and persisted until the autumn destratification, when it encysted and was lost from the mixed layer. This example allows us to deduce that the final (climax) stages of phytoplankton succession, if achieved at all, are highly self-sustainable.

6 In experimental systems, phytoplankton succession is convergent. Physical and chemical conditions, and the physiological abilities and requirements of the various species present, but not their initial abundance, determine the equilibrium state. Superimposing a higher trophic level may increase the time of convergence. Convergences in nature are implicit in the observation that successional sequences in a given lake are quite similar in different plankton years.

7 'Good' terrestrial communities (e.g., *Quercetum petreae-cerris*), characterised by similar species composition and abundance, develop in many forms in geographically isolated locations. The greater the number and fidelity of such forms, the higher the level of coordination (*sensu* Juhász-Nagy & Vida 1978). Such templates are readily applicable to phytoplankton succession (Reynolds 1980; 1997a).

8 On the scale of generation times, periods of several months in plankton succession are equivalent to decades in a grassland and centuries in a forest. It is only the external cycle of climatic and hydrological conditions which forces plankton succession to begin again each year (Sommer 1991). In this way, the 'plankton year' is somewhat analogous to the glacial–interglacial cycle.

10.14 ACKNOWLEDGEMENTS

I thank Dr Gernot Falkner, Dr Renate Falkner and Dr Colin Reynolds for their constructive comments on the manuscript.

10.15 REFERENCES

Abbott, M.R., Denmann, K.L., Powell, T.M., Richerson, P.J., Richards, R.C. & Goldman, C.R. (1984) Mixing and the dynamics of a deep chlorophyll maximum in Lake Tahoe. *Limnology and Oceanography*, **29**, 862–78.

Ahlgren, G. (1985) Growth of *Microcystis wesenbergii* in batch and chemostat cultures. *Verhandlungen der Internationale Vereinigung für Limnologie*, **22**, 2813–20.

Ahlgren, G. (1988) Comparison of algal C14-uptake and growth rate in situ and in vitro. *Verhandlungen der Internationale Vereinigung für Limnologie*, **23**, 1898–907.

Alvarez-Cobelas, M., Velasco, J.L., Rubio, A. & Brook, A.J. (1988) Phased cell division by a field population of *Staurastrum longiradiatum* (Conjugatophyceae, Desmidiaceae). *Archiv für Hydrobiologie*, **112**, 1–20.

Atkinson, K.M. (1980) Experiments in dispersal of phytoplankton by ducks. *British Phycological Journal*, **15**, 49–58.

Atkinson, K.M. (1988) The initial development of net phytoplankton in Cow Green Reservoir (Upper Teesdale), a new impoundment in northern England. In: Round, F.E. (ed.), *Algae and the Aquatic Environment*. Biopress, Bristol, 30–40.

Berman, T. & Dubinsky, Z. (1985) The autecology of *Peridinium cinctum* fo. *westii* from Lake Kinneret. *Verhandlungen der Internationale Vereinigung für Limnologie*, **22**, 2850–4.

Bird, D.F. & Kalff, J. (1986) Bacterial grazing by planktonic algae. *Science*, **231**, 493–5.

Bird, D.F. & Kalff, J. (1987) Algal phagotrophy: regulating factors and importance relative to photosynthesis in *Dinobryon* (Chrysophyceae). *Limnology and Oceanography*, **32**, 277–84.

Bird, D.F. & Kalff, J. (1989) Phagotrophic sustenance of a metalimnetic phytoplankton peak. *Limnology and Oceanography*, **34**, 155–62.

Borbély, G., Máthé, C., Hamvas, M.M. & Kós, P (1997) A növények és a cianotoxinok interakciója [Cyanotoxin and plant interaction]. *Hidrológiai Közlöny*, **77**, 24–8. [In Hungarian with English summary.]

Boucher, P., Blinn, D.W. & Johnson, D.B. (1984) Phytoplankton ecology in an unusually stable environment (Montezuma Well, Arizona, USA). *Hydrobiologia*, **119**, 149–60.

Braun-Blanquet, J. (1964) *Pflanzensoziologie*. Springer-Verlag, Wien, 330 pp.

Braunwarth, C. & Sommer, U. (1985) Analysis of the *in situ* growth rates of Cryptophyceae by use of mitotic index technique. *Limnology and Oceanography*, **30**, 893–7.

Brüning, K., Lingeman, N. & Ringelberg, J. (1992) Estimating the impact of fungal parasites on phytoplankton. *Limnology and Oceanography*, **37**, 252–60.

Calvo, S., Barone, R., Naselli Flores, L., Fradà Orestano, C., Dongarrà, G. & Genchi, G. (1993) Limnological studies on lakes and reservoirs of Sicily. *Il Naturista Siciliano*, **27**(Supplement), 292 pp.

Canter, H.M. (1979) Fungal and protozoan parasites and their importance in the ecology of phytoplankton. *Reports of the Freshwater Biological Association*, **47**, 43–50.

Canter, H.M. & Heaney, S.I. (1984) Observations on zoosporic fungi of *Ceratium* spp. in lakes of the English Lake District; importance for phytoplankton and population dynamics. *New Phytologist*, **97**, 601–12.

Canter, H.M. & Jaworski, G.H.M. (1980) Some observations on zoospores of the chytrid *Rhizophydium planktonicum* Canter emend. *New Phytologist*, **84**, 515–31.

Canter, H.M. & Jaworski, G.H.M. (1981) The effect of light and darkness upon infection of *Asterionella formosa* Hassall by the chytrid *Rhizophydium planktonicum* Canter emend. *Annals of Botany*, **47**, 13–30.

Carmichael, W.W. (1992) Cyanobacteria secondary metabolites—the cyanotoxins. *Journal of Applied Bacteriology*, **72**, 445–59.

Carney, H.J., Richerson, P.J. & Eloranta, P. (1987) Lake Titicaca (Peru/Bolivia) phytoplankton: species composition and structural comparison with other tropical and temperate lakes. *Archiv für Hydrobiologie*, **110**, 365–85.

Carper, G.L. & Bachmann, R.W. (1984) Wind resuspension of sediments in a prairie lake. *Canadian Journal of Fisheries and Aquatic Sciences*, **41**, 1763–7.

Carr, N.G. & Whitton, B.A. (1982) *The Biology of the Cyanobacteria*. Botanical Monographs 9, Blackwell Scientific Publications, Oxford, 688 pp.

Carrick, H.J.F., Aldridge, J. & Schelske, C.L. (1993) Wind influences phytoplankton biomass and composition in a shallow, productive lake. *Limnology and Oceanography*, **38**, 1179–92.

Castenholz, R.W. (1983) Ecology of blue-green algae in hotsprings. In: Carr, N.G. & Whitton, B.A. (eds), *The Biology of the Cyanobacteria*. Botanical Monographs 9, Blackwell Scientific Publications, Oxford, 379–414.

Chorus, I. (1997) Cyanobakterietoxine: Kenntnisstand und Forshchungsprogramme. *Mitteilungen der Deutschen Gesellschaft für Limnologie, Jahrestagung Berlin 1995*.

Chorus, I. & Schlag, G. (1993) Importance of intermediate disturbances for the species composition and diversity in two very different Berlin lakes. *Hydrobiologia*, **249**, 67–92.

Clements, F.E. (1916) *Plant Succession: an Analysis of the Development of Vegetation*. Publication 242, Carnegie Institute, Washington, DC, 517 pp.

Coesel, P.F.M. (1993) Poor physiological adaptation to alkaline culture conditions in *Closterium acutum* var. *variabile*, a planktonic desmid from eutrophic waters. *European Journal of Phycology*, **28**, 53–7.

Coesel, P.F.M. (1996) Biogeography of desmids. *Hydrobiologia*, **336**, 41–53.

Coesel, P.F.M. & Wardenaar, K. (1994) Light limited growth and photosynthetic characteristics of two planktonic desmid species. *Freshwater Biology*, **31**, 221–6.

Cohen, J.E. (1994) Marine and continental food webs: three paradoxes? *Philosophical Transactions of the Royal Society London, Series B*, **343**, 57–69.

Collins, C.D. & Boylen, C.W. (1982) Physiological responses of *Anabaena variabilis* (Cyanophyceae) to instantaneous exposure to various combinations of light intensity and temperature. *Journal of Phycology*, **18**, 206–11.

Connell, J.H. (1978) Diversity in tropical rainforests and coral reefs. *Science*, **199**, 1302–10.

Coon, T.G., Lopez, M.M., Richerson, P.J., Powell, T.M. & Goldman, C.R. (1987) Summer dynamics of the deep chlorophyll maximum in Lake Tahoe. *Journal of Plankton Research*, **9**, 327–44.

Craig, S.R. (1987) The distribution and contribution of picoplankton to deep photosynthetic layers in some meromictic lakes. *Acta Academica Aboensis*, **B 47**, 55–81.

Croome, R.L. & Tyler, P.A. (1984) Microbial microstratification and crepuscular photosynthesis in meromictic Tasmanian lakes. *Verhandlungen der Internationale Vereinigung für Limnologie*, **22**, 1216–3.

Cullen, J.J. (1982) The deep chlorophyll maximum: comparing vertical profiles of chlorophyll *a*. *Canadian Journal of Fisheries and Aquatic Sciences*, **39**, 791–803.

Currie, D.J. & Kalff, J. (1984a) A comparison of the abilities of freshwater algae and bacteria to acquire and retain phosphorus. *Limnology and Oceanography*, **29**, 298–310.

Currie, D.J. & Kalff, J. (1984b) The relative importance of bacterioplankton and phytoplankton in phosphorus uptake in freshwater. *Limnology and Oceanography*, **29**, 311–21.

Czárán, T. & Bartha, S. (1992) Spatiotemporal dynamic models of plant populations and communities. *Trends in Ecology and Evolution*, **7**, 38–42.

Czekanowski, J. (1909) Differentialdiagnose der Neanderthalgruppe. *Kosses-publication deutsche Antropologische Gesellschaft*, **40**, 44–7.

Darwin, C. (1839) *Journal of Researches in the Geology and Natural History of the Various Countries Visited by H. M. S. Beagle*. Ward, Lock & Bowden, London, 615 pp.

Dauta, A. (1982) Conditions de developpement du phytoplancton etude comparative du comportement de huit especes en culture. I. Detarmination des parameters de croissance en fonction de la lumiere et de la temperature. *Annales de Limnologie*, **18**, 217–62.

Dodson, S. (1992) Predicting zooplankton species richness. *Limnology and Oceanography*, **37**, 848–56.

Dokulil, M. & Mayer, J. (1996) Population dynamics and photosynthetic rates of a *Cylindrospermopsis-Limnothrix* association in a highly eutrophic urban lake, Alte Donau, Vienna, Austria. *Algological Studies*, **83**, 179–95.

Droop, M.R. (1973) Some thoughts on nutrient limitation in algae. *Journal of Phycology*, **9**, 264–72.

Drury, W.H. & Nisbet, I.C.T. (1973) Succession. *Journal of the Arnold Arboretum*, **54**, 331–68.

Dugdale, R.C. (1967) Nutrient limitation in the sea: dynamics, identification and significance. *Limnology and Oceanography*, **12**, 685–95.

Ehrenberg, C.G. (1849) Passatstaub und Blutregen. *Abhandlungen der Königlich Akademie der Wissenschaften zu Berlin*, **1847**, 269–460.

Elliot, W.P. (1958) The growth of atmospheric internal boundary layer. *Transactions of the American Geophysical Union*, **39**, 1048–54.

Elser, M.M. & Smith, W.O. (1985) Phased cell division and growth rate of a planktonic dinoflagellate, *Ceratium hirundinella*, in relation to environmental variables. *Archiv für Hydrobiologie*, **104**, 477–91.

Elton, C. (1927) *Animal Ecology*. Sedgwick and Jackson, London, 207 pp.

Fabbro, L.D. & Duiverndoornen, L.J. (1996) Profile of a bloom of the cyanobacterium *Cylindrospermopsis raciborskii* (Woloszynska) Seeneya and Subba Raju in the Fitzroy River in tropical Central Queensland. *Marine and Freshwater Research*, **47**, 685–94.

Fahnenstiel, G.L., Patton, T.R., Carrick, H.J. & McCormick, M.J. (1991) Diel division cycle and growth rates of *Synechococcus* in lakes Huron and Michigan. *Internationale Revue der gesamten Hydrobiologie*, **76**, 657–64.

Falkner, R & Falkner, G. (1989) Phosphate uptake by eukaryotic algae in cultures and in mixed phytoplankton population in a lake: An analysis by a force–flow relationship. *Botanica Acta*, **102**, 283–6.

Falkner, G., Wagner, F., Small, J.V. & Falkner, R. (1995) Influence of fluctuating phosphate supply on the regulation of phosphate uptake by the blue-green alga *Anacystis nidulans*. *Journal of Phycology*, **31**, 745–53.

Farmer, A.M., Boston, H.L. & Adams, M.S. (1988) Photosynthetic characters of a deepwater bryophyte from a clear, oligotrophic lake in Wisconsin, U.S.A. *Verhandlungen der Internationale Vereinigung für Limnologie*, **23**, 1912–5.

Feuillade, J., Davies, A. & Feuillade, J. (1984) Seasonal variations and long-term trends in phytoplankton pigments. *Ergebnisse in Limnologie*, **41**, 95–111.

Flik, B.J.G. (1985) Measurements of time dependent inhibition and recovery of a laboratory culture of *Chlorella* sp. at various light intensities. *Verhandlungen der Internationale Vereinigung für Limnologie*, **22**, 2888–92.

Ford, D.E. & Stefan, H. (1980) Thermal predictions using integral energy model. *Journal of Hydrology Division, American Society of Chemical Engineers*, **106**, 39–55.

Foy, R.H. & Gibson, C.E. (1982a) Photosynthetic characteristics of planktonic blue-green algae: The response of twenty strains grown under high and low light. *British Phycological Journal*, **17**, 169–82.

Foy, R.H. & Gibson, C.E. (1982b) Photosynthetic characteristics of planktonic blue-green algae: changes in photosynthetic capacity and pigmentation of *Oscillatoria redekei* Van Goor under high and low light. *Phycological Journal*, **17**, 183–93.

Francis, G. (1878) Poisonous Australian lakes. *Nature*, **18**, 11–12.

Frempong, E. (1982) The space–time resolution of phased cell division in natural populations of the freshwater dinoflagellate *Ceratium hirundinella*. *Internationale Revue der Gesamten Hydrobiologie*, **67**, 323–40.

Gaedeke, A. & Sommer, U. (1986) The influence of the frequency of periodic disturbances on the maintenance of phytoplankton diversity. *Oecologia*, **71**, 25–8.

Gaines, G. & Elbrächter, M. (1987) Heterotrophic nutrition. In: Taylor, F.J.R. (ed.), *The Biology of Dinoflagellates*. Botanical Monographs 22, Blackwell Scientific Publications, Oxford, 224–68.

Galvez, J.A., Niell, F.X. & Lucena, J. (1988) Description and mechanism of formation of a deep chlorophyll maximum due to *Ceratium hirundinella* (O.F. Müller) Bergh. *Archiv für Hydrobiologie*, **112**, 143–55.

Ganf, G.G. (1975) Photosynthetic production and irradiance–photosynthesis relationships of the phytoplankton from a shallow, equatorial lake (Lake George, Uganda). *Oecologia*, **18**, 165–83.

Gasol, J.M. & Pedrós-Alió, C. (1991) On the origin of deep algal maxima: the case of lake Cisó. *Verhandlungen der Internationale Vereinigung für Limnologie*, **24**, 1024–8.

Gasol, J.M., García-Cantizano, J., Massana, R., Peters, F., Guerrero, R. & Pedrós-Alió, C. (1991) Diel changes in the microstratification of the metalimnetic community in Lake Cisó. *Hydrobiologia*, **211**, 227–40.

Gause, G.F. (1964) *The Struggle for Existence*. William and Wilkins, Baltimore, 160 pp.

George, D.G. & Harris, G.P. (1985) The effect of climate on long-term changes in the crustacean zooplankton biomass in Lake Windermere, UK. *Nature*, **316**, 536–9.

Gervais, F., Padisák, J. & Koschel, R. (1997). Do light quality and low nutrient concentration favour picocyanobacteria below the thermocline of the oligotrophic Lake Stechlin? *Journal of Plankton Research*, **19**, 771–81.

Geviatkin, V.G. (1979) Dinamika razvitiya algaflori obrastaniy v Ribinskom Vodohranilise. V. Flora Rastitelnosti Vodoyamov Basseyna Verhney Volgi. *Trudy Instituta Biologii Vodokhranilishch*, **42–45**, 78–108.

Gislén, T. (1948) Aerial plankton and its conditions of life. *Biological Reviews*, **23**, 109–26.

Gleason, H. (1917) The structure and development of plant association. *Bulletin of the Torrey Botanical Club*, **44**, 463–81.

Gleason, H. (1927) Further views on the succession-concept. *Ecology*, **8**, 299–326.

Grime, J.P. (1979) *Plant Strategies and Vegetation Processes*. Wiley, New York, 222 pp.

Grobbelaar, J.U. (1994) The role of turbulence and light/dark cycles on the photosynthetic rates of phytoplankton. *Verhandlungen der Internationale Vereinigung für Limnologie*, **25**, 2242–4.

Grunow, A. (1860) Über neue oder ungenügend gekannte Algen. *Verhandlungen der Zoologisch-Botanischen Gesellschaft in Wien*, **10**, 503–82.

Grunow, A. (1862) Die österreichischen Diatomaceen nebst Anschluß einiger neuer Arten von anderen Lokalitäten und einer kritischen Übersicht der bisher bekannten Gattungen. *Verhandlungen der Zoologisch-Botanischen Gesellschaft in Wien*, **12**, 315–472, 545–88.

Grunow, A. (1863) Über einige neue und ungenügend bekannte Arten und Gattungen von Diatomaceen. *Verhandlungen der Zoologisch-Botanischen Gesellschaft in Wien*, **13**, 137–62.

Haffner, D.G., Hehanussa, P.E. & Hartoto, D.I. (2001) The biology and physical processes of large lakes of Indonesia: Lakes Matano and Towuti. In: Munawar, M. & Hecky, R.E. (eds), *Great Lakes of the World, Food-web, Health and Integrity*. Backhuys Publishers, Leiden, 183–92.

Hardin, G. (1960) The competitive exclusion hypothesis. *Science*, **131**, 1292–7.

Harding, W.R. (1996) The *phytoplankton ecology of a hypertrophic, shallow lake, with particular reference to primary production, periodicity and diversity*. PhD thesis, University of Cape Town.

Harding, W.R. (1997) Phytoplankton primary production in a shallow, well-mixed, hypertrophic South African Lake. *Hydrobiologia*, **344**, 87–102.

Harris, G.P. (1983) Mixed layer physics and phytoplankton populations; studies in equilibrium and non-equilibrium ecology. In: Round, F.E. & Chapman, D. (eds), *Progress in Phycological Research* 2, Elsevier, Amsterdam, 1–52.

Harris, G.P. (1986) *Phytoplankton Ecology. Structure, Function and Fluctuation*. Chapman and Hall, London, 384 pp.

Harris, G.P., Heaney, S.I. & Talling, J.F. (1979) Physiological and environmental constraints in the ecology of the planktonic dinoflagellate *Ceratium hirundinella*. *Freshwater Biology*, **9**, 413–28.

Hawkins, P.R. (1996) Factors which influence the development of blooms of *Cylindrospermopsis*. *Cylindrospermopsis*—a new toxic algal bloom challenge for Australia. *Symposium Abstracts, Brisbane City Hall, 24 October*.

Haworth, E.Y. & Tyler, P.A. (1993) Morphology and taxonomy of *Cyclotella tasmanica* spec. nov., a newly described diatom from Tasmanian lakes. *Hydrobiologia*, **269/270**, 49–56.

Heaney, S.I. & Eppley, R.W. (1981) Light, temperature and nitrogen as interacting factors affecting diel vertical migration in the dinoflagellate *Ceratium hirundinella*. *Freshwater Biology*, **10**, 163–70.

Heaney, S.I. & Furnass, T.I. (1980) Laboratory models of diel vertical migration in the dinoflagellate *Ceratium hirundinella*. *Freshwater Biology*, **10**, 163–70.

Heller, H.D. (1977) The phased division of the freshwater dinoflagellate *Ceratium hirundinella* and its use as a method of assessing growth in natural populations. *Freshwater Biology*, **7**, 527–33.

Hensen, V. (1887) Über die Bestimmung des Planktons oder des im Meere treibenden Materials an Pflanzen und Tieren. *Bericht des deutschen wissenschafltichen Kommission für Meerenforschung*, **5**, 1–109.

Hoffmann, L. (1996) Geographic distribution of freshwater blue-green algae. *Hydrobiologia*, **336**, 33–40.

Humphrey, G.F. (1979) Photosynthetic characteristics of algae grown under constant illumination and light–dark regimes. *Journal of Experimental Marine Biology and Ecology*, **40**, 63–70.

Hustedt, F. (1943) Die Diatomeenflora einiger Hochgebirgsseen der Landschaft Davos in der Schweizer Alpen. *Internationale Revue der gesamten Hydrobiologie*, **43**, 225–80.

Hutchinson, G.E. (1961) The Paradox of Plankton. *American Naturalist*, **95**, 137–45.

Ichimura, T. (1996) Genome rearrangement and speciation in freshwater algae. *Hydrobiologia*, **336**, 1–17.

Istvánovics, V. (1997) *A balatoni fitoplankton foszfor felvételi és növekedési stratégiái a belsö foszfor terhelés és annak csökkentése függvényében. [Growth and P-uptake strategies of Lake Balaton phytoplankton in dependence of internal P-load and its reduction]*. Research Report, Balaton Limnological Institute, Hungary, 37 pp.

Istvánovics, V. & Herodek, S. (1994) Principles of eutrophication and the case of Lake Balaton. In: Salánki, J. & Bíró, P. (eds), *Limnological Bases of Lake Management*. Proceedings of the ILEC/UNEP International Training Course, Tihany, Hungary, 11–23 October, 1993. International Lakes Environment Committee, Shiga, 112–134.

Istvánovics, V. & Herodek, S. (1995) Estimation of net uptake and leakage rates of orthophosphate from ^{32}P-uptake kinetics by a linear force–flow model. *Limnology and Oceanography*, **40**, 17–32.

Iwakuma, T. & Yasuno, M. (1983) A comparison of several mathematical equations describing photosynthesis-light curve for natural phytoplankton populations. *Archiv für Hydrobiologie*, **97**, 208–26.

Jaccard, P. (1908) Nouvelles recherches sur la distribution florale. *Bulletin de la Société Vaudoise des Sciences Naturelles*, **44**, 163, 223–69.

Jackson, L.J., Stockner, J.D. & Harrison, J.P. (1989) Contribution of *Rhizosolenia ereinsis* and *Cyclotella* spp. to the deep chlorophyll maximum of Sproat Lake, British Columbia, Canada. *Canadian Journal of Fisheries and Aquatic Sciences*, **46**, 128–35.

Jolly, V.H. & Brown, J.M.A. (1975) *New Zealand Lakes*. Auckland University Press, Auckland, 388 pp.

Juhász-Nagy, P. & Vida, G. (1978) Szupraindividuális organizáció [Supraindividual organization]. In: Csaba, G. (ed.), *A biológiai szabályozás [Biological Regulation]*, Medicina, Budapest, 337–406. [In Hungarian.]

Kalff, J. & Knoechel, R. (1978) Phytoplankton and their dynamics in oligotrophic and eutrophic lakes. *Annual Reviews in Ecology and Systematics*, **9**, 475–95.

Kalff, J. & Watson, S. (1986) Phytoplankton and its dynamics in two tropical lakes: a tropical and temperate zone comparison. *Hydrobiologia*, **138**, 161–76.

Kilham, S.S. (1984) Silicon and phosphorus growth kinetics and comparative interactions between *Stephanodiscus minutus* and *Synedra* sp. *Verhandlungen der Internationale Vereinigung für Limnologie*, **22**, 435–9.

Kimura, B. & Ishida, Y.B. (1986) Possible phagotrophic feeding of bacteria in a freshwater red tide Chrysophyceae *Uroglena americana*. *Bulletin of the Japanese Society of Scientific Fishery*, **52**, 697–701.

Kirk, J.T.O. (1983) *Light and Photosynthesis in Aquatic Ecosystems*. Cambridge University Press, Cambridge, 401 pp.

Kiss, T., Vehovszky, Á., Hiripi, L. & Vörös, L. (1997) Cyanotoxinok hatása az idegsejtek müködésére [Effect of cyanotoxins on the activity of gastropod nerve cells]. *Hidrológiai Közlöny*, **77**, 69–70. [In Hungarian with English summary.]

Knell, A.H. & Golubic, S. (1992) Proterozoic and living cyanobacteria. In: Schidlowski, M., Golubic, S., Kimberley, M.M., McKirdy, D.M. & Trudinger, P.A. (eds), *Early Organic Evolution: Implications for Mineral and Energy Resources*. Springer-Verlag, Heidelberg, 450–62.

Kohl, J.-G. & Geisdorf, K. (1991) Competition ability of two planktonic diatoms under different vertical light gradients, mixing depth and frequencies: an experimental approach. *Verhandlungen der Internationale Vereinigung für Limnologie*, **24**, 2652–6.

Kohl, J.-G. & Nicklisch, A. (1988) *Ökophysiologie der Algen. Wachstum und Ressourcennutzung*. Akademie-Verlag, Berlin, 253 pp.

Komárek, J. (1985) Do all cyanophytes have a cosmopolitan distribution? Survey of freshwater cyanophyte flora of Cuba. *Algological Studies*, **75**, 11–29.

Komárek, J. (1996) Towards a combined approach for taxonomy and species delimitation of picoplanktonic cyanoprokayotes. *Algological Studies*, **83**, 377–401.

Konopka, A. (1982) Physiological ecology of a metalimnetic *Oscillatoria rubescens* population. *Limnology and Oceanography*, **27**, 1154–61.

Konopka, A. (1989) Metalimnetic cyanobacteria in hard-water lakes: Buoyancy regulation and physiological state. *Limnology and Oceanography*, **34**, 1174–84.

Kratz, W.A. & Myers, J. (1955) Nutrition and growth of several blue-green algae. *American Journal of Botany*, **42**, 282–7.

Krienitz, L., Kasprzak, P. & Koschel, R. (1996) Long term study on the influence of eutrophication, restoration and biomanipulation on the structure and development of phytoplankton communities in Feldberger Haussee (Baltic Lake District, Germany). *Hydrobiologia*, **330**, 89–110.

Krienitz, L. & Hegewald, E. (1996) Über das Vorkommen von wärmliebenden Blauaglgenarten in einem norddeutschen See. *Lauterbornia*, **26**, 55–64.

Kristiansen, J. (ed.) (1996a) *Biogeography of Freshwater Algae*. Developments in Hydrobiology 118, Kluwer Academic Publishers, Dordrecht, 161 pp.

Kristiansen, J. (1996b) Dispersal of freshwater algae—a review. *Hydrobiologia*, **336**, 151–7.

Kristiansen, J. & Vigna, M.S. (1996) Bipolarity in the distribution of silica-scaled chrysophytes. *Hydrobiologia*, **336**, 121–36.

Kromkamp, J.C., van der Heuvel, A. & Mur, L.R. (1989) Phosphorus uptake and photosynthesis by phosphate limited cultures of *Microcystis aeruginosa*. *British Phycological Journal*, **24**, 347–55.

Lampert, W. (1992) Zooplankton vertical migrations: Implications for phytoplankton-zooplankton interactions. *Archiv für Hydrobiologie Beiheft Ergebnisse der Limnologie*, **35**, 69–78.

Lampert, W. & Sommer, U. (1993) *Limnoökologie*. Georg Thieme Verlag, Stuttgart, 440 pp.

Lange-Bertalot, H. & Simonsen, R. (1978) A taxonomic revision of Nitzschiae lanceolatae Grunow 2. European and related extra-European fresh water and brackish water taxa. *Bacillaria*, **1**, 11–111.

Larkham, A.W.D. (1999) The evolution of the algae. In: Sekcbach, J. (ed.), *Enigmatic Microrganisms and Life in Extreme Environments*, Kluwer Academic Publishers, Dordrecht, 31–49.

Larson, D.W., Dahm, C.N. & Geiger, N.S. (1987) Vertical partitioning of the phytoplankton assemblage in the ultraoligotrophic Crater Lake, Oregon, U.S.A. *Freshwater Biology*, **18**, 429–42.

Laws, E.A. & Bannister, T.T. (1980) Nutrient- and light-limited growth of *Thalassiosira fluviatilis* in continuous culture, with implications for phytoplankton growth in the ocean. *Limnology and Oceanography*, **25**, 457–73.

Lean, D.R.S. & Nalewajko, C. (1976) Phosphate exchange and organic phosphorus excretion by freshwater algae. *Journal of Fisheries Research Board of Canada*, **33**, 1312–23.

Lehman, J.T. (1976) Ecological and nutritional studies on *Dinobryon* Ehrenb.: seasonal periodicity and the phosphate toxicity problem. *Limnology and Oceanography*, **21**, 646–58.

Lewis, W.M., Jr. (1978) A compositional, phytogeographical and elementary structural analysis of the phytoplankton in a tropical lake: Lake Lanao, Philippines. *Journal of Ecology*, **66**, 213–26.

Lewis, W.M., Jr. (1984) The diatom sexuality and its evolutionary significance. *American Naturalist*, **123**, 73–80.

Li, W.K.W., Glover, H.E. & Morris, I. (1980) Physiology of carbon photoassimilation by *Oscillatoria thiebautii* in the Caribbean Sea. *Limnology and Oceanography*, **25**, 447–56.

Lindholm, T. (1992) Ecological role of depth maxima of phytoplankton. *Archiv für Hydrobiologie Beiheft Ergebnisse der Limnologie*, **35**, 33–45.

Lindstrom, K. (1984) Effect of temperature, light and pH on growth, photosynthesis and respiration of the dinoflagellate *Peridinium cinctum* fa. *westii* in laboratory cultures. *Journal of Phycology*, **20**, 212–20.

Linné, K. (1753) *Species Plantarum*. Salvii, Laurentii, Stockholm, 1200 pp.

MacArthur, R.H. (1972) *Geographical Ecology*. Harper and Row, New York, 269 pp.

MacArthur, R.H. & Wilson, E.O. (1967) *The Theory of Island Biogeography*. Princeton University Press, Princeton, 203 pp.

Malinsky-Rushansky, N., Berman, T. & Dubinsky, Z. (1995) Seasonal dynamics of picophytoplankton in Lake Kinneret, Israel. *Freshwater Biology*, **34**, 241–54.

Margalef, R. (1964) Correspondence between classic types of lakes and the structural and dynamic properties of their populations. *Verhandlungen der Internationale Vereinigung für Limnologie*, **15**, 548–59.

Mastala, Z., Herodek, S., V.-Balogh, K., Borbély, G., Shafik, H.M. & Vörös L. (1996) Nutrient requirement and growth of a *Synechococcus* species isolated from Lake Balaton. *Internationale Revue der gesamten Hydrobiologie*, **81**, 503–12.

Mayr, E. (1982) *The Growth of Biological Thought: Diversity, Evolution and Inheritance*. Harvard University Press, Cambridge (MA), 974 pp.

McCourt, R.M. & Hoshaw, R.W. (1990) Noncorrespondence of breeding groups, morphology, and monophyletic groups in *Spirogyra* (Zygnemataceae: Chlorophyta) and the application of species concepts. *Systematic Botany* **15**, 69–78.

McDuff, R.E. & Chisholm, S.W. (1985) The calculation of *in situ* growth rates of phytoplankton populations from fractions of cells undergoing mitosis: A clarification. *Limnology and Oceanography*, **27**, 783–8.

Melack, J.M. (1979) Temporal variability of phytoplankton in tropical lakes. *Oecologia*, **44**, 1–7.

Metzner, P. (1929) Bewegungstudien in Peridiniien. *Zeitschrift für Botanik*, **22**, 225–65.

Meulemans, J.T. & Heinis, F. (1983) Biomass and production of periphyton attached to dead read stems in Lake Maarseveen. In: Wetzel, R.G. (ed.), *Periphyton in Freshwater Environments*. Dr W Junk, The Hague, 169–73.

Miracle, M.R., Vicente, E. & Pedrós-Alió, C. (1992) Biological studies of Spanish meromictic and stratified karstic lakes. *Limnetica*, **8**, 59–77.

Morales-Baquero, R., Carillo, P., Cruz-Pizarro, L. & Sanchez-Castillo, P. (1992) Southernmost high mountain lakes in Europe (Sierra Nevada) as reference sites for pollution and climate change monitoring. *Limnologica*, **8**, 39–47.

Morel, M.M. (1987) Kinetics of nutrient uptake and growth in phytoplankton. *Journal of Phycology*, **23**, 137–50.

Morris, I. (1980) *The Physiological Ecology of Phytoplankton*. Blackwell Scientific Publications, Oxford, 625 pp.

Mur, L. & Bejsdorf, R.O. (1978) A model of the succession from green to blue-green algae based on light limitation. *Verhandlungen der Internationale Vereinigung für Limnologie*, **20**, 2314–2321.

Nalewajko, C. (1966) Dry weight, ash and volume data for some freshwater planktonic algae. *Journal of the Fisheries Research Board of Canada*, **23**, 1285–8.

Nauwerck, A. (1963) Die Beziehungen zwischen Zooplankton und Phytoplankton im See Erken. *Symbolae Botanicae Upsaliensis*, **12/5**, 1–163.

Nicklisch, A., Conrad, B. & Kohl, J-G. (1981) Growth kinetics of *Oscillatoria redekei* van Goor as a basis to understand its mass development in eutrophic lakes. *Verhandlungen der Internationale Vereinigung für Limnologie*, **21**, 1427–31.

Nicklisch, A., Conrad, B. & Kohl, J-G. (1983) Growth kinetics of *Microcystis aeruginosa* (Kütz.) Kütz as a basis for modelling its population dynamics. *Internationale Revue der gesamten Hydrobiologie*, **68**, 317–26.

Nyholm, N. (1977) Kinetics of phosphate limited algal growth. *Biotechnology and Bioengineering*, **19**, 467–92.

Odum, E.P. (1969) The strategy of ecosystem development. *Science*, **164**, 262–70.

OECD (1982) *Eutrophication of Waters. Monitoring, Assessment and Control.* Organization for Economic Cooperation and Development, Paris, 152 pp.

Olrik, K. (1994) *Phytoplankton Ecology.* Miljøprojekt 251, Danish Environmental Protection Agency, Copenhagen, 183 pp.

Ördög, V. (1997) Growth stimulation/inhibition of some cyanobacteria and algae caused by *Cylindrospermopsis raciborskii* (Wolosz.) Seenaya et Subba Raju. 8. *Hungarian Algological Meeting, Abstracts*, **17**.

Overeem, M.A. (1937) On green organisms occurring in the lower troposphere. *Travaux Botaniques Neerlands*, **34**, 388–442.

Owen, M. & Black, J.M. (1990) *Waterfowl Ecology.* Blackie, Glasgow, 203 pp.

Padisák, J. (1985) Population dynamics of the dinoflagellate *Ceratium hirundinella* in the largest shallow lake of Central Europe, Lake Balaton, Hungary. *Freshwater Biology*, **15**, 43–52.

Padisák, J. (1993) The influence of different timescale disturbances on the species richness, diversity and equitability of phytoplankton in shallow lakes. *Hydrobiologia*, **249**, 135–56.

Padisák, J. (1994) Relationships between short-term and long-term responses of phytoplankton to eutrophication of the largest shallow lake in Central Europe (Balaton, Hungary). In: Sund, H., Stabel, H-H., Geller, W., Xiaogan, Y., Kechang, Y. & Fengning, S. (eds), *Environmental Protection and Lake Ecosystems.* China Science and Technology Press, Beijing, 419–37.

Padisák, J. (1997) *Cylindrospermopsis raciborskii* (Woloszynska) Seenayya et Subba Raju, an expanding, highly adaptive blue-green algal species: geographic distribution and autecology. *Archiv für Hydrobiologie*, **107**(Supplementband), 563–593.

Padisák, J. (1998) Sudden and gradual responses of phytoplankton to global change: case studies from two large, shallow lakes (Balaton, Hungary; Neusiedlersee Austria/Hungary). In, George D.G., Jones, J.G., Puncochar, P. & Reynolds, C.S. (eds), *Management of Lakes and Reservoirs during Global Change.* Proceedings of the NATO Advanced Research Workshop, Prague. Kluwer, Dordrecht, 111–125.

Padisák, J. & Adrian, R. (1999) Biomassebestimmung. In: von Tümpling, W. & Friedrich, G. (eds), *Methoden der biologischen Gewässeruntersuchung.* Gustav Fischer Verlag, Jena, 334–367.

Padisák, J. & Dokulil, M. (1994) Meroplankton dynamics in a saline, turbulent, turbid shallow lake (Neusiedlersee, Austria and Hungary). *Hydrobiologia*, **289**, 23–42.

Padisák J. & Kovács, A. (1997) *Anabaena compacta* (Nygaard) Hickel—új kékalga faj a Balaton üledékében és planktonjában. [*Anabaena compacta* (Nygaard) Hickel—a new blue-green algal species in the sediments and plankton of lake Balaton.]. *Hidrológiai Közlöny*, **77**, 29–32.

Padisák, J. & Reynolds, C.S. (1998). Selection of phytoplankton associations in lake Balaton, Hungary, in response to eutrophication and restoration measures, with special reference to cyanoprokaryotes. *Hydrobiologia*, **384**, 41–53.

Padisák, J., Reynolds, C.S. & Sommer, U. (1993) *Intermediate Disturbance Hypothesis in Phytoplankton Ecology.* Developments in Hydrobiology 81, Kluwer Academic Publishers, Dordrecht, 199 pp.

Padisák, J., Krienitz, L., Koschel, R. & Nedoma, J. (1997) Deep-layer autotrophic picoplankton maximum in the oligotrophic Lake Stechlin, Germany: origin, activity, development and erosion. *European Journal of Phycology*, **32**, 403–16.

Padisák, J., Krienitz, L., Scheffler, W., Kristiansen, J. & Grigorszky, I. (1998) Phytoplankton succession in the oligotrophic Lake Stechlin, Germany. *Hydrobiologia*, **370**, 178–97.

Padisák, J., Krienitz, L. & Scheffler, W. (1999) Phytoplankton. In: von Tümpling, W. & Friedrich, G. (eds), *Methoden der biologischen Gewässeruntersuchung.* Gustav Fischer Verlag, Jena, 35–52.

Paerl, H.W. (1982) Interactions with bacteria. In: Carr, N.G. & Whitton, B.A. (eds), *The Biology of Blue-green Algae.* Blackwell Scientific Publications, Oxford, 441–61.

Paerl, H.W. (1988) Growth and reproductive strategies of freshwater blue-green algae (Cyanobacteria). In: Sandgren, C.G. (ed.), *Growth and Reproductive Strategies of Freshwater Phytoplankton.* Cambridge University Press, Cambridge, 261–315.

Paerl, H.W. & Ustach, J.F. (1982) Blue-green algal scums: an explanation for their occurrence during freshwater blooms. *Limnology and Oceanography*, **21**, 212–7.

Paerl, H.W., Bland, P.T., Bowles, N.D. & Haibach, M.E. (1985) Adaptation to high intensity, low wavelength light among surface blooms of the cyanobacterium *Microcystis aeruginosa*. *Applied Environmental Microbiology*, **49**, 1046–52.

Pahl-Wostl, C. (1992) Dynamic versus static models for photosynthesis. *Hydrobiologia*, **238**, 189–96.

Pankin, W. (1941) Die Vegetation einiger Seen und Umgebung von Joachimsthal. *Bibliotheca Botanica*, **119**, 1–162.

Pankin, W. (1945) Zur Entwicklungsgeschichte der Algensoziologie und zum Problem der 'echten' und 'zugehörigen' Algengesellschaften. *Archiv für Hydrobiologie*, **41**, 92–111.

Pantocsek, J. (1912) *Bacillariae lacus Peisonis*. Veda Tech-Dejimach Slovakie, Bratislava, 1–43.

Pedros-Alio, C., Gasol, J.M. & Guerrero, R. (1987) On the ecology of a *Cosmarium phaseolus* population forming a metalimnetic bloom in Lake Cisó, Spain: annual distribution and loss factor. *Limnology and Oceanography*, **32**, 285–98.

Peng, C., Lin, Z., Sun, G. & Shi, D. (1994) Comparative study of photosynthetic property of two species of *Dunaliella*. *Oceanologica and Limnologica Sinica*, **25**, 606–11.

Petersen, R. (1991) Carbon-14 uptake by picoplankton and total phytoplankton in eight New Zealand lakes. *Internationale Revue der gesamten Hydrobiologie*, **76**, 631–41.

Pfennig, N. (1978) General physiology and ecology of photosynthetic bacteria. In: Clayton, R.K. & Sistrom, W.R. (eds), *The Photosynthetic Bacteria*. Plenum, New York, 3–18.

Pick, F.R., Nalewajko, C. & Lean, D.R.S. (1984) The origin of a metalimnetic chrysophyte peak. *Limnology and Oceanography*, **29**, 125–34.

Platt, T., Subba Rao, D.V. & Irwin, B. (1983) Photosynthesis of picoplankton in the oligotrophic ocean. *Nature*, **301**, 702–4.

Pollingher, U. & Serruya, C. (1976) Phased division of *Peridinium cinctum* f. *westii* (Dinophyceae) and development of Lake Kinneret (Israel) bloom. *Journal of Phycology*, **12**, 162–70.

Porter, K.G. (1988) Phagotrophic phytoflagellates in microbial food webs. *Hydrobiologia*, **159**, 89–97.

Postgate, J.R. & Cannon, F.C. (1981) The molecular and genetic manipulation of nitrogen fixation. *Philosophical Transactions of the Royal Society, London, Series B*, **292**, 589–99.

Ramberg, L. (1980) A population dynamics model for *Oocystis parva* (Chlorophyceae) *Archiv für Hydrobiologie*, **89**, 118–34.

Raven, J.A. (1997) Phagotrophy in phototrophs. *Limnology and Oceanography*, **42**, 198–205.

Raven, J.A. & Richardson, K. (1984) Dinophyte flagella: a cost benefit analysis. *New Phytologist*, **98**, 259–76.

Remmert, H. (1991) The mosaic-cycle concept of ecosystems—an overview. In: Remmert, H. (ed.), *The Mosaic-cycle Concept of Ecosystems*. Springer-Verlag, Berlin, 1–21.

Reynolds, C.S. (1972) Growth, gas-vacuolation and buoyancy in a natural population of a blue-green alga. *Freshwater Biology*, **2**, 87–106.

Reynolds, C.S. (1980) Phytoplankton assemblages and their periodicity in stratifying lake systems. *Holarctic Ecology* **3**, 141–59.

Reynolds, C.S. (1984a) *The Ecology of Freshwater Phytoplankton*. Cambridge University Press, Cambridge, 384 pp.

Reynolds, C.S. (1984b) Phytoplankton periodicity: the interactions of form, function and environmental variability. *Freshwater Biology*, **14**, 111–42.

Reynolds, C.S. (1986) Experimental manipulation of phytoplankton periodicity in large limnetic enclosures in Blelham Tarn, English Lake District. *Hydrobiologia*, **138**, 43–64.

Reynolds, C.S. (1987a) Community organisation in freshwater plankton. In: Gee, J.H.R. & Giller, P.S. (eds), *Organization of Communities*. Blackwell Scientific Publications, Oxford, 297–325.

Reynolds, C.S. (1987b) The response of phytoplankton communities to changing lake environments. *Swiss Journal of Hydrology* **49**, 220–36.

Reynolds, C.S. (1988a) Functional morphology and the adaptive strategies of freshwater phytoplankton. In: Sandgren, C.D. (ed.), *Growth and Reproductive Strategies of Freshwater Phytoplankton*. Cambridge University Press, Cambridge, 388–433.

Reynolds, C.S. (1988b) The concept of biological succession applied to seasonal periodicity of phytoplankton. *Verhandlungen der Internationale Vereinigung für Limnologie*, **23**, 683–9.

Reynolds, C.S. (1989) Physical determinants of phytoplankton succession. In: Sommer, U. (ed.), *Plankton Ecology*. Springer-Verlag, Berlin, 9–56.

Reynolds, C.S. (1990) Temporal scales of variability in pelagic environments and the response of phytoplankton. *Freshwater Biology*, **23**, 25–53.

Reynolds, C.S. (1992a) Dynamics, selection and composition of phytoplankton in relation to vertical structure in lakes. *Archiv für Hydrobiologie Beiheft Ergebnisse der Limnologie*, **35**, 13–31.

Reynolds, C.S. (1992b) Eutrophication and the management of planktonic algae: What Vollenweider couldn't tell us. In: Sutcliffe, D.W. & Jones, J.G. (eds), *Eutrophication: Research and Application to Water Supply*. Freshwater Biological Association, Ambleside, Cumbria, UK, 4–29.

Reynolds, C.S. (1994) The long, the short and the stalled: on the attributes of phytoplankton selected by physical mixing in lakes and rivers. *Hydrobiologia*, **289**, 9–21.

Reynolds, C.S. (1994b) The role of fluid motion in the dynamics of phytoplankton in lakes and rivers. In: Giller P.S., Hildrew, A.G. & Raffelli, D.G. (eds), *Aquatic Ecology: Scale, Pattern and Process*. Blackwell Scientific Publications, Oxford, 141–87.

Reynolds, C.S. (1997a) *Vegetation Processes in the Pelagic*. Ecology Institute, Oldendorf/Luhe, Germany, 370 pp.

Reynolds, C.S. (1997b) Successional development, energetics and diversity in planktonic communities. In: Abe, T., Levin, S.A. & Higashi, M. (eds), *Biodiversity— an Ecological Perspective*. Springer-Verlag, Berlin, 167–202.

Reynolds, C.S., Thompson, J.M., Ferguson, A.J.D. & Wiseman, S.W. (1982) Loss processes in the population dynamics of phytoplankton maintained in closed systems. *Journal of Plankton Research*, **4**, 561–600.

Reynolds, C.S., Wiseman, S.W. & Clarke, M.J.O. (1984) Growth and loss rate responses of phytoplankton to intermittent artificial mixing and their potential application to the control of planktonic algal biomass. *Journal of Applied Ecology*, **21**, 11–39.

Reynolds, C.S., Padisák, J. & Sommer, U. (1993) Intermediate disturbance in the ecology of phytoplankton and the maintenance of species diversity: a synthesis. *Hydrobiologia*, **249**, 183–8.

Richerson, P., Armstrong, R. & Goldman, C.R. (1970) Contemporaneous disequilibrium, a new hypothesis to explain the ' paradox of plankton'. *Proceedings of the National Academy of Science*, **67**, 1710–4.

Rijkeboer, M., De Kloet, W.A. & Gons, K.H. (1992) Interspecific variation in pigmentation: implications for production estimates for shallow eutrophic lakes using an incubator. *Hydrobiologia*, **238**, 197–202.

Round, F.E. (1981) *The Ecology of Algae*. Cambridge University Press, Cambridge, 653 pp.

Round, F.E. (1988) *Algae and the Aquatic Environment*. Biopress, Bristol, 460 pp.

Salonen, K. & Jokinen, S. (1988) Flagellate grazing on bacteria in a small dystrophic lake. *Hydrobiologia*, **161**, 203–9.

Sanders, R.W. & Porter, K.G. (1987) Phagotrophic phytoflagellates. *Advances in Microbial Ecology*, **10**, 167–92.

Sandgren, C.D. (1986) Effects of environmental temperature on the vegetative growth and sexual life history of *Dinobryon cylindricum* Imhof. In: Kristiansen, J. & Andersen, R.A. (eds), *Chrysophytes: Aspects and Problems*. Cambridge University Press, Cambridge, 207–25.

Sandgren, C.D. (1988a) *Growth and Reproductive Strategies of Freshwater Phytoplankton*. Cambridge University Press, Cambridge, 442 pp.

Sandgren, C.D. (1988b) The ecology of chrysophyte flagellates: their growth and perennation as freshwater phytoplankton. In: Sandgren, C.D. (ed.), *Growth and Reproductive Strategies of Freshwater Phytoplankton*. Cambridge University Press, Cambridge, 9–104.

Sas, H. (1989) *Lake Restoration by Reduction of Nutrient Loading: Expectations, Experiences, Extrapolations*. Academia Verlag Richarz, St Augustin, 497 pp.

Saxby-Rouen, K.J., Leadbeater, B.S.C. & Reynolds, C.S. (1996) Ecophysiological studies on *Synura petersenii* (Synurophyceae). *Nova Hedwigia Beiheft*, **114**, 111–24.

Scheffler, W. & Padisák, J. (1997) *Cyclotella tripartita* Håkansson (Bacillariophyceae), a dominant diatom species in the oligotrophic Stechlinsee (Germany). *Nova Hedwigia*, **65**, 221–232.

Schlangstedt, M., Bisen, P.S., Dudel, G. & Kohl, J-G. (1987) Interaction of combined nitrogen availability and light in the regulation of growth, heterocyst differentiation and dinitrogen fixation of the planktic blue-green alga *Anabaena solitaria* Kleb. *Archiv für Protistenkunde*, **134**, 389–96.

Schlichtling, H.E. (1960) The role of waterfowl in dispersal of algae. *Transactions of the American Microscopical Society*, **79**, 160–6.

Schlichtling, H.E., Speziale, B.J. & Zink, R.M. (1978) Dispersal of algae and Protozoa by Antarctic flying birds. *Antarctic Journal of the United States*, **13**, 1147–9.

Shafik, H.M. (1991) *Growth, nutrient uptake and competition of algae of Lake Balaton in flow-through cultures*. PhD thesis, Balaton Limnological Institute, Tihany.

Shannon, C.E. & Weaver, W. (1949) *The Mathematical Theory of Communication*. University of Illinois Press, Urbana, IL 117 pp.

Shapiro, J. (1984) Blue-green dominance in lakes: the role and management significance of pH and CO_2. *Inter-

nationale Revue der gesamten Hydrobiologie, **69**, 765–80.

Shugart, H.H. & West, D.C. (1981) Long-term dynamics of forest ecosystems. *American Scientist*, **69**, 647–52.

Sieburth, J.M.N., Smetacek, V. & Lenz, J. (1978) Pelagic ecosystem structure: heterotrophic compartments of the plankton and their relationship to plankton size fractionation. *Limnology and Oceanography*, **23**, 1256–63.

Simiu, E. & Scanlan, R.H. (1986) *Wind Effect on Structures*. John Wiley and Sons, New York, 589 pp.

Smith, E.L. (1936) Photosynthesis in relation to light and carbon dioxide. *Proceedings of the National Academy of Science, Washington*, **22**, 504.

Smith, I.R. & Sinclair, I.J. (1972) Deep water waves in lakes. *Freshwater Biology*, **2**, 387–99.

Sommer, U. (1981) The role of r- and K-selection in the succession of phytoplankton in Lake Constance. *Acta Oecologica Oecologia, Generale*, **2**, 327–42.

Sommer, U. (1984) The paradox of plankton: fluctuations in phosphorus availability maintain diversity in flow-through cultures. *Limnology and Oceanography*, **29**, 633–6.

Sommer, U. (1985) Comparison between steady-state and non steady-state competition: experiments with natural phytoplankton. *Limnology and Oceanography*, **30**, 335–46.

Sommer, U. (1986a) Phytoplankton competition along a gradient of dilution rates. *Oecologia*, **68**, 503–6.

Sommer, U. (1986b) The periodicity of phytoplankton in Lake Constance (Bodensee) in comparison to other deep lakes of central Europe. *Hydrobiologia*, **138**, 1–7.

Sommer, U. (1988a) Some size relationships of phytoflagellate motility. *Hydrobiologia*, **161**, 125–31.

Sommer, U. (1988b) Growth and survival strategies of planktonic diatoms. In: Sandgren, C.G. (ed.), *Growth and Reproductive Strategies of Freshwater Phytoplankton*. Cambridge University Press, Cambridge, 227–60.

Sommer, U. (1989) *Plankton Ecology. Succession in Plankton Communities*. Brock/Springer-Verlag, Berlin, 369 pp.

Sommer, U. (1991) Phytoplankton: directional succession and forced cycles. In: Remmert, H. (ed.), *The Mosaic-cycle Concept of Ecosystems*. Springer-Verlag, Berlin, 132–46.

Sommer, U. (1994) *Planktologie*. Springer-Verlag, Berlin, 274 pp.

Sommer, U., Gaedke, U. & Schweizer, A. (1993) The first decade of oligotrophication of Lake Constance. II. The

response of phytoplankton taxonomic composition. *Oecologia*, **93**, 276–84.

Sommer, U., Gliwicz, M., Lampert, W. & Duncan, A. (1986) The PEG model of seasonal succession of planktonic events in fresh waters. *Archiv für Hydrobiologie*, **106**, 433–71.

Spijkerman, E. & Coesel, P.F.M. (1996) Phosphorrus uptake and growth kinetics of two planktonic desmid species. *European Journal of Phycology*, **31**, 53–60.

Steele, J.H. (1991) Can ecological theory cross the land–sea boundary? *Journal of Theoretical Biology*, **153**, 425–36.

Steele, J.H. & Henderson, E.W. (1994) Coupling between physical and biological scales. *Philosophical Transactions of the Royal Society, London, Series B*, **343**, 5–9.

Steenbergen, C.L.M. & Korthals, H.J. (1982) Distribution of phototrophic microorganisms in the anaerobic and microaerophilic strata of Lake Vachten (The Netherlands). Pigment analysis and role in primary production. *Limnology and Oceanography*, **27**, 883–95.

Stewart, W.D.P. (1974) *Algal Physiology and Biochemistry*. Blackwell Scientific Publications, Oxford, 989 pp.

Stockner, J.G. (1991) Autotrophic picoplankton in freshwater ecosystems: the view from the summit. *Internationale Revue der gesamten Hydrobiologie*, **76**, 483–92.

Stoyneva, M.P. (1994) Shallows of the lower Danube as additional sources of potamoplankton. *Hydrobiologia*, **289**, 171–8.

Stulp, B.K. & Stam, W.T. (1984) Genotypic relationships between strains of *Anabaena* (Cyanophyceae) and their correlations with morphological affinities. *British Phycological Journal*, **19**, 287–301.

Symoens, J-J., Kusel-Fetzmann, E. & Descy, J-P. (1988) Algal communities of continental waters. In: Symoens, J-J. (ed.), *Vegetation of Inland Waters*. Kluwer Academic Publishers, Dordrecht, 183–221.

Talling, J.F. (1951) The element of chance in pond populations. *The Naturalist* **1951**, 157–70.

Tett, P. & Barton, E.D. (1995) Why are there about 5000 species of phytoplankton in the sea? *Journal of Plankton Research*, **17**, 1693–704.

Tilman, D. (1977) Resource competition between planktonic algae: An experimental and theoretical approach. *Ecology*, **58**, 338–48.

Tilman, D. (1982) *Resource Competition and Community Structure*. Princeton University Press, Princeton, 296 pp.

Tilman, D. & Kilham, S.S. (1976) Phosphate and silicate uptake and growth kinetics of the diatoms *Asterionel-*

la formosa and *Cyclotella meneghiniana* in batch and semi-continuous culture. *Journal of Phycology*, **12**, 375–83.

Tilman, D., Kilham, S.S. & Kilham, P. (1982) Phytoplankton community ecology: the role of limiting nutrients. *Annual Review of Ecology and Systematics*, **13**, 349–72.

Tyler, P.A. (1996) Endemism in freshwater algae. With special reference to the Australian region. *Hydrobiologia*, **336**, 127–35.

Utermöhl, H. (1931) Neue Wege in der quantitativen Erfassung des Planktons. *Verhandlungen der Internationale Vereinigung für Limnologie*, **5**, 567–96.

Vandermeer, J.H. (1972) Niche theory. *Annual Review of Ecology and Systematics*, **3**, 107–32.

Van Liere, L. & Mur, L.R. (1980) Occurrence of *Oscillatoria agardhii* and some related species, a survey. *Developments in Hydrobiology*, **2**, 67–77.

Visser, P.M. (1989) *Experimentele studie naar de invloed van een wisselend lichtklimat op het groenwier Scenedesmus protuberans en de cyanobakterie Microcystis aeruginosa*. Project Report, Laboratory for Microbiology, University of Amsterdam, 22 pp.

Vollenweider, R.A. (1989) Global problems of eutrophication and its control. In: Salánki, J. & Herodek, S. (eds), *Conservation and Management of Lakes*. Akadémiai Kiadó, Budapest, 19–41.

Vollenweider, R.A. & Kerekes, J. (1980) The loading concept as basis for controlling eutrophication philosophy and preliminary results of the OECD programme on eutrophication. *Progress in Water Technology*, **12**, 5–38.

Vörös, L. (1995) *A kékalgák elszaporodását befolyásoló tényezők és toxicitásuk kutatása [Factors affecting growth and toxicity of cyanobacteria]*. Ministry for Environment and Regional Affairs, Budapest, Research Report, 204 pp.

Vörös, L. & Padisák, J. (1990) Phytoplankton biomass and chlorophyll-a in some shallow lakes in central Europe. *Hydrobiologia*, **215**, 111–9.

Vörös, L., Gulyás, P. & Németh, J. (1991) Occurrence, dynamics and production of picoplankton in Hungarian shallow lakes. *Internationale Revue der gesamten Hydrobiologie*, **76**, 617–29.

Vyverman, W. (1996) The Indo-Malaysian North-Australian phycogeographical region revised. *Hydrobiologia*, **336**, 107–20.

Walsby, A.E. & Reynolds, C.S.R. (1980) Sinking and floating. In: Morris, I. (ed.), *The Physiological Ecology of Phytoplankton*. University of California Press, Berkeley & Los Angeles, 371–412.

Walsby, A.E., Utkilen, H.C. & Johnsen, I.J. (1983) Buoyancy changes of a red coloured Oscillatoria agardhii in Lake Gjersjøen, Norway. *Archiv für Hydrobiologie*, **97**, 18–38.

Washington, H.G. (1984) Diversity, biotic and similarity indices. A review with special relevance to aquatic ecosystems. *Water Research*, **18**, 653–94.

Watt, A.S. (1947) Pattern and process in the plant community. *Journal of Ecology*, **35**, 1–22.

Webb, W.L., Newton, N. & Starr, D. (1974) Carbon dioxide exchange of *Alnus rubra*: a mathematical model. *Oecologia*, **20**, 419–25.

Whittaker, R.H. (1974) Climax concepts and recognition. In: Knapp, R. (ed.), *Handbook of Vegetation Science* 8. Dr W. Junk, The Hague, 138–54.

Wiens, J.A. (1976) Population responses to patchy environments. *Annual Review of Ecology and Systematics*, **7**, 81–120.

Wiens, J.A. (1977) On competition and variable environments. *American Scientist*, **65**, 590–7.

Wigley, T.M.L. & Jones, P.D. (1987) England and Wales precipitation: a discussion of recent changes in variability and an update to 1985. *Journal of Climatology*, **7**, 231–46.

Wilson, J.B. (1990) Mechanisms of species coexistence: twelve explanations for Hutchinson's 'Paradox of the Plankton': evidence from New Zealand plant communities. *New Zealand Journal of Ecology*, **13**, 17–42.

Wolk, C.P. (1982) Heterocysts. In: Carr, N.G. & Whitton, B.A. (eds), *The Biology of Blue-green Algae*. Blackwell Scientific Publications, Oxford, 359–86.

Zevenboom, W. & Mur, L.C. (1980) N-fixing cyanobacteria: why they do not become dominant in Dutch hypertrophic lakes. *Developments in Hydrobiology*, **2**, 123–30.

Zohary, T. (1985) Hyperscums of the cyanobacterium *Microcystis aeruginosa* in a hypertrophic lake (Hartbeespoort Dam, South Africa). *Journal of Plankton Research*, **7**, 399–409.

Zohary, T. & Robarts, R.D. (1989) Diurnal mixed layers and the long-term dominance of *Microcystis aeruginosa*. *Journal of Plankton Research*, **11**, 25–48.

Zohary, T., Pais-Madeira, A.M., Robarts, R. & Hambright, K.D. (1996) Interannual phytoplankton dynamics of a hypertrophic African lake. *Archiv für Hydrobiologie*, **136**, 105–26.

11 Aquatic Plants and Lake Ecosystms

JAN POKORNÝ AND JAN KVĚT

11.1 INTRODUCTION

Aquatic **macrophytes** are those water plants which can be viewed without a microscope. They comprise vascular plants, mosses and the larger filamentous algae. The term 'freshwater macrophytes' is to be interpreted as including the Charophyta (stoneworts), Bryophyta (mosses and liverworts), Pteridophyta (ferns and their allies) and some Spermatophyta (seed-bearing plants), whose photosynthetically active parts are permanently, or for some months of each year, submerged in freshwater, or float on the water surface or emerge above it. The term is also taken to include certain macrophytic taxa of Chlorophyta (green algae), which therefore are known as **macroalgae**.

All aquatic vascular plants are descended from terrestrial ancestors and have returned secondarily to the water. Many of them exhibit some reduction in their ancestral features, often possessing thin cuticles, functionless stomata and weak lignification of the xylem elements. They have not, however, reverted to fertilisation by motile sperms: most aquatic angiosperms still present their flowers above the water surface, to be pollinated aerially by insects or by wind. A few species have developed submerged flowers, but fertilisation continues to involve gamete and pollen-tube structures analogous to those of terrestrial plants (Sculthorpe 1985).

Aquatic macrophytes are able to colonise standing and flowing waters in all climatic zones. Although most are rooted, some species float freely in the water and a few are epiphytic. There is some considerable plasticity in somatic organisation, which often creates problems for the taxonomist, particularly when diagnostic flowers are absent. Indeed, many macrophytes are able to reproduce and spread rapidly by vegetative means. For example, *Elodea canadensis* Michx. was first recorded in Europe in 1836; by the end of the century it had become well-established in Scandinavia, central Europe and east to Russia and Hungary. During the 20th century it has become equally common throughout eastern and southeastern Europe and western Siberia. The remarkable fact is that this invasion has been accomplished exclusively by vegetative propagules (Sculthorpe 1985).

The sweet flag (*Acorus calamus* L.) is also a neophyte in Europe. It is probably derived from East Asia, and it is very likely that, at least in southern Europe, the present populations are descended from a single ancestral rhizome. The plant was introduced to Europe at an early but unknown time (Hendrych 2003), ostensibly for its medicinal value (it has been used in the treatment of ophthalmic and stomach disorders, hysteria, epilepsy and chronic rheumatism; its fragrance has led to its use in perfumery and brewing). Sweet flag flowers infrequently, and fertile seeds are unknown, but it has spread throughout central Europe, Belgium, France, Germany and, later, to the British Isles (Wein 1939; Casper & Krausch 1980; Weber & Brändle 1996). In tropical and subtropical regions, the South American *Salvinia molesta* (giant salvinia) has invaded large water bodies in Africa, Asia and Australia and caused serious management problems. By 1962, shortly after the creation of the lake, *Salvinia* covered 100,000 ha of Lake

Kariba (Zambia/Zimbabwe). The invasion has since been successfully contained, thanks to biological control (Mitchell 1976).

On the other hand, stands of macrophytes may persist for centuries, with unchanged species composition. This is especially so for plant associations in lakes of low trophic status, such as those of the Bohemian Forest, Czech Republic, with their enduring stands of *Isoëtes lacustris* L. and *I. echinospora* Durieu.

Macrophytes respond to changes in water quality, to water-level fluctuations and to other environmental factors. They are a natural component of lake ecosystems, where they provide habitat and food for animals, concentrate nutrients and release oxygen during photosynthesis. In general, the influence of macrophytes on lake ecosystems is stabilising but, with increasing nutrient status, their development sometimes causes its own problems. Thus, they sometimes also earn the name of **aquatic weeds**.

11.2   LIFE-FORMS OF AQUATIC MACROPHYTES

The morphology, anatomy, life-forms and physiological features of water macrophytes are described in several well known monographs (Gessner 1955, 1959; Hutchinson 1977; Sculthorpe 1985). Detailed descriptions of individual macrophyte species, including their biology, distribution (especially in relation to water chemistry and human impact) and their indicator value, are widely available. A relevant bibliography is that by Hejný & Sytnik (1993). A widespread feature of water macrophytes is the presence of aerenchyma, a loose tissue with large air spaces, or lacunae. The lacunal system within the plants facilitates the internal transport of gases (air, carbon dioxide, etc.) to roots, stems and leaves.

Aquatic macrophytes are usefully classified by their life-forms (Fig. 11.1), which distinguishes **emergent macrophytes** (such as *Phragmites, Typha*), **surface-floating macrophytes** (such as *Eichhornia, Lemna*), the **euhydrophytes with floating leaves** (such as *Nymphaea, Potamogeton*

natans) and the fully **submersed euhydrophytes** (such as *Najas, Ceratophyllum, Lemna trisulca*; see Denny 1987). More elaborate systems of classifying the life- and growth-forms of aquatic macrophytes have been devised (see, e.g., Tansley 1939; Hejný 1957, 1960; den Hartog & Segal 1964; Spence 1964; Hejný *et al.* 1998).

Among cold and temperate water bodies, where factors such as littoral topography, bottom conditions, exposure to wind and waves, light penetration in the water and presence of grazers variously favour development of a constellation of life-forms, the distribution of aquatic macrophytes is essentially littoral, conforming to the schematic illustration in Fig. 11.2. The littoral zone is divisible, however, into a series of sub-zones based on the occurrence of different life-forms. The zonation system is useful for the characterisation of the littoral vegetation and its various adaptive traits. The groups of macrophytic life-forms comprise:

1 The hyperhydates (helophytes)—emergent plants such as the graminoids *Phragmites* and *Typha*, and the herbids *Alisma and Cicuta*.

2 The ephydates—floating-leaved (**natant**) plants represented by spirodelids (*Spirodela, Lemna*) and nymphaeids such as *Nymphaea* and the species of *Potamogeton* with floating leaves.

3 The hyphydates—the group of submerged plants including riccids (*Riccia*), elodeids (i.e. plants with long shoots such as *Elodea, Myriophyllum* and *Potamogeton* species without floating leaves), isoëtids (submerged plants with short shoots, exemplified by *Isoëtes* and *Littorella*) and muscids (submerged mosses).

The uppermost part of the littoral zone, the **eulittoral**, is delimited by the extreme highest (generally spring) and lowest (usually summer) water levels. Differentiation of permanently submerged sublittoral zones is based on the vertical distributions of the hyperhydate, nymphaeid, elodeid and isoetid plant types. Thus, the lakeward limit of the upper sublittoral zone coincides with the extent of the emergent hyperhydates (helophytes), and that of the lower sublittoral zone with the deepest occurrence of nymphaeids. In the same way, the upper elittoral zone ends with the deepest elodeids, and the lower elittoral zone with that of

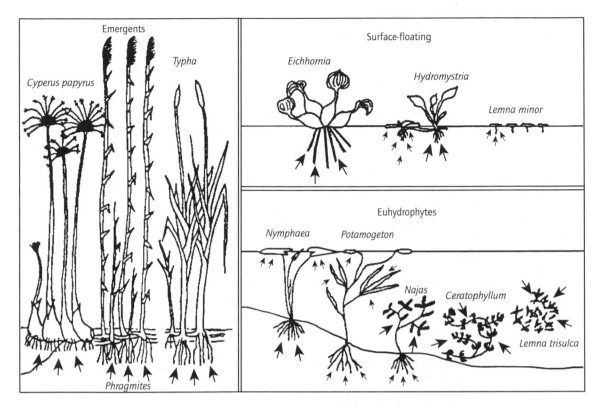

Fig. 11.1 Wetland plant life-forms: emergent, surface-floating and euhydrophytes. (From Denny 1987.)

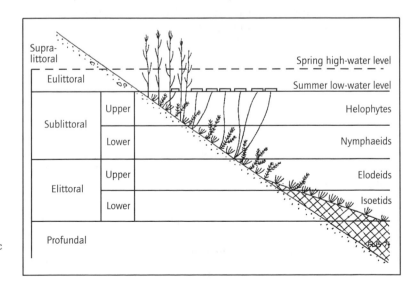

Fig. 11.2 Schematic illustration of the littoral zonation of macrophytic vegetation in an oligotrophic lake. (From Pokorný 1994.)

isoetids and mosses. Beyond the littoral zones is the profundal zone, which is typically outside the range in which rooted photoautotrophs are able to function.

Depending on environmental conditions, the types of plants present in lake systems can vary tremendously. Similarly, there is a considerable range of depth distributions. Thus, in the clear-water lakes of southern Scandinavia, macrophytes occur down to a depth of c.5 m. In some of the Eifel maar lakes of Germany, *Nitella flexilis* grows at a depth of more than 20 m (Melzer 1992), whilst in Bavarian lakes, meadows of *Chara* have been observed at depths of around 15 m (Melzer 1976). The most diverse vegetation is found in the upper littoral sub-zones, where there is typically a mixture of life-forms. With increasing water depth, first the hyperhydates, then the nymphaeids, the elodeids and the isoetids are excluded sequentially, as their respective ecological limits are exceeded.

The full spectrum of macrophyte life-forms, as represented in Fig. 11.2, may be encountered in clear, oligotrophic lakes ('*Lobelia* lakes'), in which dense eulittoral carpets of isoetids are able to develop. In humic, brown-water lakes, the sub-zones are compressed and displaced upwards. With increased turbidity, including that attributable to the abundance of phytoplankton (stimulated by greater availability of dissolved nutrients, or with enhanced overgrowth of periphytic algae), first the isoetids and then the elodeids, are excluded: i.e. turbid lakes lose their lower and upper elittoral zones. At the same time, the biomass and density of emergent and floating-leaved plants usually increases.

With advanced nutrient enrichment ecosystems are increasingly characterised by instability, being subject to abrupt changes and population collapses. Such systems are exemplified by the heavily fertilised fishponds of Bohemia (Czech Republic; see below).

11.3   PRACTICAL ASPECTS

11.3.1   Species identification

The identification of aquatic macrophytes is particularly difficult. Many are quite plastic in their fea-

tures, and easily modified in response to environmental conditions. They are frequently found without reproductive structures. As the present system of classification, and most keys to the identity of aquatic macrophytes, are based largely on reproductive anatomy, correct identification requires considerable practical experience. Besides, many macrophytes are very mobile, appearing at or disappearing from new locations, apparently quite spontaneously, so there is a risk that local floras quickly become out of date. Moreover, many botanists are apparently reluctant to get their feet wet; thus the state of knowledge on aquatic plants is weak in comparison to that for terrestrial species.

It is recommended that, as part of any site survey, specimens of macrophytes are collected and their identities checked against herbarium material. Each specimen needs a correct label. The following manuals and keys for determination of water macrophytes are recommended: Hotchkiss (1972), Haslam *et al.* (1975), Casper & Krausch (1980), Clapham *et al.* (1981), Spencer-Jones & Wade (1986), Sainty & Jacobs (1994) and Rothmaler (1995).

11.3.2   Biomass determination

For assessing the role of macrophytes in aquatic and wetland ecosystems, it is often necessary to know their aggregate biomass, net production and chemical composition. It is valuable to be able to express the chemical content of plants in terms both of **concentration per unit plant dry mass** and of standing stock, i.e. **amount per unit area**. The precision required depends upon the level of the assessment. Such assessments can be made either directly, by harvesting plants from a certain area, or indirectly, by calculating biomass from estimates of stand density and measurements of those morphological characters which are correlated with the dry mass of individual plants or shoots.

11.3.3   Sampling aquatic macrophytes

Examples of devices for the direct quantitative sampling of aquatic macrophytes are illustrated in Fig. 11.3. Metallic frames, enclosing a square or circular area (usually) of 0.25 m^2, are used to sample

Fig. 11.3 Photograph of various samplers, especially of a rotary sampler.

stands in relatively deep water. A diver places the frame on the vegetated bottom and all plants rooted within the frame are collected. Such devices are suitable for sampling rooted macrophytes which are short and not too dense. For sampling in shallow water, a wooden or metallic cage with wire-mesh sides (up to 0.8 m in height) and enclosing a known area (between 0.25 and 1 m^2) is more appropriate. All plants rooted or floating within the cage are sampled by hand. A sieve may be used to sample small free-floating plants (*Azolla*, *Lemna*, etc.). Even dense aquatic vegetation can be sampled in this way.

The rotary sampler described by Howard-Williams & Longman (1976) cuts plants at their base, and then removes them from a circular area determined by the length of the rotating knife blade. A blade of 35.68 cm in diameter harvests plants from an area of 0.1 m^2. Extension poles enable sampling from depths between 1 and 4 m. The apparatus is suitable for sampling loose to medium-density rooted aquatic vegetation. Losses of sampled plants must be checked, for over-estimates of biomass are possible in dense stands.

It is usual to collect several replicate samples in order to obtain reliable average data. Random variation in macrophyte biomass occurs even under identical habitat conditions. Systematic variations in macrophyte biomass may correlate with the presence of environmental gradients (e.g. of water depth). The statistical validity of any deductions based upon biomass data needs to be borne in mind in the initial sampling design.

Sampling of the aerial parts of emergent vegetation (helophytes) growing in very shallow water (to about 0.5 m depth), or in wet mud, is accomplished in a manner similar to that for terrestrial vegetation, i.e., by cutting off the shoots at ground (bottom) level. Underground biomass may be estimated by direct coring (where a large number of samples may be needed in order to detect non-random variation) or from correlations between above-ground and underground biomass (see, e.g., Hejný *et al.* 1981; Květ *et al.* 1998).

11.4 PRODUCTIVITY

As a general rule, net primary production (P) over a certain time interval ($t_2 - t_1$) is estimated from the equation:

$$P = W_2 - W_1 + L - T \qquad (11.1)$$

where $W_{1/2}$ is biomass assessed at times $t_{1/2}$, respectively, and L denotes loss of biomass due to mortality or consumption of plants or plant parts over the interval $(t_2 - t_1)$. The facultative term T is used only when partial (above-ground or under-ground) production is considered: T then denotes transport of assimilates to or from the plant parts considered.

The assumptions that $P = W_2 - W_1$ and that L may be neglected are valid only over very short time intervals, or over longer intervals if the growth in the plant stands is well synchronised. For this reason, seasonal maximum biomass, W_{max}, approaches above-ground production in macrophyte communities growing in seasonal climates (e.g., temperate, subarctic). The **biomass turnover factor** used for estimating annual net production of such communities from W_{max} usually varies between 1.05 and 1.5 in well synchronised stands of rooted macrophytes. The turnover factor, calculated with respect to average biomass recorded, cannot be ignored and acquires much higher values over longer time intervals in communities of short-lived plants (e.g. *Lemna*, *Azolla*), or in communities where plant growth is unsynchronised. This is the case in most tropical and subtropical macrophyte communities. A correct assessment of biomass turnover is necessary for making reliable estimates of annual net primary production of macrophyte stands from their average biomass (W), i.e. for estimating the P/W ratio. The most reliable estimates of biomass turnover are those based on observation of growth, production, development and mortality in carefully selected (either individually or by cohorts, i.e., sets of much the same age) plants, shoots or otherwise defined population or community elements (**ramets** of clonal plants). For correct estimation of the organic matter production of macrophytes, the ash content must be deducted from the dry weight. This is particularly important in many euhydrophytes, in which the ash content can be high (up to some 50% of dry mass in some Charophytes).

In temperate zones, the annual net production of submerged water macrophytes usually reaches not more than 0.5 kg of ash-free dry mass m^{-2}. The annual above-ground production of emergent macrophytes can reach 1 to 2 kg m^{-2}. When an optimised harvest regime is applied, annual net production by macrophytes may reach as much as 3 kg m^{-2} in eutrophic habitats.

The relationship between production and dry biomass of a submersed plant stand is shown in Fig. 11.4. The continuous line describes the seasonal course of biomass in a stand of *Elodea canadensis* (Pokorný *et al.* 1984). The course of cumulative net production in the same stand is calculated by adding the cumulative biomass losses due to death or consumption of certain plants or plant parts (as well as the increments recorded after the seasonal peak was reached) to the biomass course. Thus, the dashed line describes seasonal cumulative net production. Total annual net production is represented by the ordinate of this curve, when biomass has fallen to its minimum value, at the end of the growing season.

11.5   DECOMPOSITION

The assessment of macrophyte decomposition is important for an evaluation of their role in aquatic or wetland ecosystems, particularly with respect to the budgets for carbon, oxygen, nitrogen and

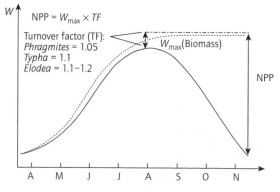

Estimating net primary production from macrophyte biomass

Fig. 11.4   Fitted growth curve of the total stand biomass (W), net primary production (NPP) and turnover factor (TF). (According to data from Pokorný *et al.* 1984.)

other nutrients in these systems. The rate of release of nutrients contained in the macrophytes and their detritus is proportional to the decomposition rate. As a rule, algal macrophytes decompose most readily, followed by submerged and floating-leaved vascular plants (especially their leaves and stems). The half-lives of their decomposition (the time needed for the decomposition of 50% of the initial biomass) usually range between several days and a few weeks. On the other hand, emergent macrophytes decompose more slowly, with half-lives ranging from a few months to a few years, depending on the nature of the plant material and the climatic and hydrological conditions. Decomposition of detritus may cease at the stage of humic substance formation: the result is an accumulation of peat (nutrient-poor in **bogs** and nutrient-rich in **fens**). In this way, peat layers several metres thick are formed in bogs under humid and cool climatic conditions.

Rapid decomposition rates often occur in eutrophic waters, where a surplus of nutrients stimulates the development of microbes. Detritus accumulates at the lake bottom, and decomposes with an equivalent consumption of oxygen. Consequently, anaerobic conditions and low redox potential are created. Black sediments release methane, hydrogen sulphide and also phosphorus. Usually, organic matter does not simply accumulate *in situ*.

11.6 NUTRIENT CONTENT

Aquatic macrophytes take up nutrients both through their leaves and via their roots. The old controversy over the role of leaves and roots in nutrient uptake by submersed macrophytes has been ended by experimental studies (see literature reviewed by Bristow 1975; Hutchinson 1975; Denny 1980; Dykyjová & Úlehlová 1998). Helophytes such as *Phragmites, Typha, Scirpus* and *Carex* take up mineral nutrients not only via their **edaphic roots** (which penetrate the bottom mud or the subsoil), but also via their **accessory roots**, which grow directly into the water, or into detritus. Rooted plants do not need to compete with non-rooted

macrophytes or phytoplankton for nutrients dissolved in water. Generally, concentrations of both major and minor elements are higher in the biomass than in the environment.

The three different kinds of vascular water plants (emergent, floating and submerged) differ in their ash and mineral contents. Submersed plants contain a higher water and ash content than the emergent vegetation, whilst floating species fall between the two (Westlake 1965; Dykyjová & Úlehlová 1998). Concentration factors (concentration of an element in dried plant/concentration of the same element in water) mostly exceed 1000 and over 10,000 for P, K and Mn (Boyd 1969; Dykyjová & Úlehlová 1998).

The accumulation of minerals in individual organs varies with metabolic activity. The greatest nutrient content is generally found in the stems and leaves; phosphorus and magnesium often accumulate in flowers and fruits. In the aerial parts of the plant, concentrations of most nutrients other than calcium tend to decline during the growing season. Those aerial and submerged organs which decompose each season release their nutrients into the environment. The large underground storage organs of perennial macrophytes (rhizomes, roots and tubers) retain a substantial portion of the nutrients accumulated by the parent plants, and do so for several years. Return of nutrients to the bottom subsoil occurs mainly only after decomposition. Vymazal (1995) gives a large literature review on nutrient and heavy metal contents of the biomass of algae and water macrophytes. The data include information on nutrient concentrations and standing stocks, rates of daily production, daily nutrient uptake and decomposition rates of macrophytes. Figure 11.5 presents an example of the seasonal course of nutrient standing stock in a submersed macrophyte stand (*Elodea canadensis*).

Besides the natural sources of variability in chemical composition of aquatic plants, many discrepancies exist in the analytical techniques used. Even sets of data on water content are often not comparable with each other, as drying in ovens may be carried out at 60, 70, 80 or 105°C. Sometimes, only air drying is carried out. For a rough estimation of amounts of nutrients bound in the

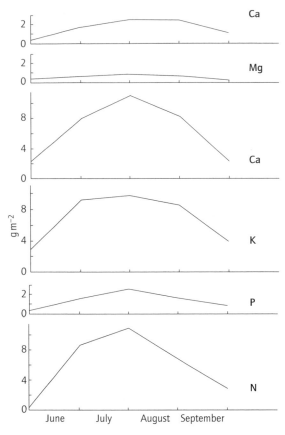

Fig. 11.5 Seasonal course of the mineral nutrient contents in *Elodea canadensis* biomass per unit area of the stand (standing stock [g m^{-2} dry mass]). (From Pokorný *et al.* 1984.)

biomass of macrophytes (the **standing stock**) the following approximate ranges of concentrations (% dry mass) may be used (as a rule of thumb): nitrogen 1–4%, phosphorus 0.1–0.5%, sulphur 0.1–0.7%, potassium 1–5%, magnesium 0.1–0.6%, calcium 0.1–4% (variability caused by precipitated calcium carbonate), iron 0.02–5%, manganese 0.01–0.5% in submersed macrophytes and 0.01–0.1 for helophytes.

11.7   THE ECOPHYSIOLOGICAL ROLE OF MACROPHYTES IN LAKE ECOSYSTEMS

The lives of macrophytes in lakewater or on flooded soil are conditioned by the solubility and slow rate of movement of gases. Concentrations of gases are different in air and soil. Moreover, as a result of the metabolic activities of plants and other biota, concentrations of oxygen and carbon dioxide in water may change markedly during the course of the day, whereas atmospheric concentrations are effectively constant. In this section, we consider the impact of gases exchanged in photosynthesis and respiration.

11.7.1   Oxygen in water and in flooded soil, and its estimation

The solubility of oxygen in water is weak: at 20°C air-equilibrated water contains over thirty times less oxygen than the same volume of air (9 mg versus 285 mg). Its solubility is also temperature sensitive, saturation at 30°C being reached at half the concentration (7.4 mg O$_2$ L^{-1}) at which it occurs at 10°C (14.2 mg O$_2$ L^{-1}). Diffusion of gases is also slower in water than in air (coefficients of molecular diffusion in the order $\approx$ 10^{-9} m^2 s^{-1}) by about an order of magnitude. Unreplenished oxygen demand in water quickly leads to exhaustion.

The greater the plant biomass, the higher the potential photosynthetic release of oxygen and the greater the net respiratory oxygen consumption in the dark. Large diel amplitudes in the concentration of oxygen are thus typical of dense, active macrophyte stands. Dissolved oxygen concentrations may be restored by daytime production and release of the gas during plant photosynthesis. Oxygen content of the water increases around the leaves of submersed macrophytes, and may well rise above the value of air saturation. Under sunny conditions, water with abundant macrophytes may well become considerably supersaturated with oxygen (concentrations corresponding to 150–200% of air saturation are not unknown). In contrast, the oxygen content of clear, unpolluted, plant-free water varies little from the air-saturation concentration.

Accumulation and decomposition of dead macrophyte biomass places extra consumptive demand on oxygen exchanges, to the extent of local exhaustion, both in deep water and in flooded soils. Environments in which the oxygen concentrations are too low to be detected by standard methods (Winkler, electrochemical Clark type oxygen sensor) are described as being **anoxic**, or **anaerobic**. In practice, the label applies to water where the concentration of dissolved oxygen is $<0.1\,mg\,L^{-1}$ ($<10^{-5}\,mol\,L^{-1}$). At such concentrations, metabolism is necessarily anaerobic: sugars cannot be respired to carbon dioxide and water, and fermentation products of anaerobic metabolism, such as ethanol and organic acids, accumulate instead. The energy released in anaerobic metabolism is small, 1 mol glucose yielding only 2 mol adenosine triphosphate (ATP), rather than the 32 mol released by aerobic respiration. Anaerobic, reducing conditions also see nitrate (NO_3^-) converted to the ammonium ion (NH_4^+), sulphate (SO_4^{2-}) to sulphide (S^{2-}) and carbon dioxide (CO_2) to methane (CH_4). The ratio between oxidised and reduced forms (SO_4^{2-}/S^{2-}, Fe^{3+}/Fe^{2+}) thus indicates the extent of redox conditions in wetland soil or in water.

Redox potential is a useful environmental measure which describes the degree of anaerobiosis in conditions under which concentration of oxygen is conventionally unmeasurable. It is determined with a platinum electrode, and related to a standard hydrogen or calomel reference electrode.

11.7.2 The oxygen regime in aquatic environments

A general scheme depicting the movements of oxygen which take place within aquatic media is shown in Fig. 11.6. Concentrations change through time in response to variations in source (photosynthesis), sinks (respiration, oxidation processes) and physical connections (diffusion, exchange with the atmosphere, advection). Were all fluxes to be equilibrated, a steady state would be achieved. In fact, when vegetation is present in water, the system is highly variable. Gaseous exchanges at the water surface are generally slower than biotic processes, and often fail to compensate for sub- or supersaturation of the water with oxygen. Nevertheless, when oxygen concentration is sufficiently high in water, and the abundance of vegetation is low, the system can approach the equilibrium state with respect to the atmosphere (i.e. oxygen concentration remains almost constant within a time period, for example, of 1 day).

In dense stands of submersed macrophytes,

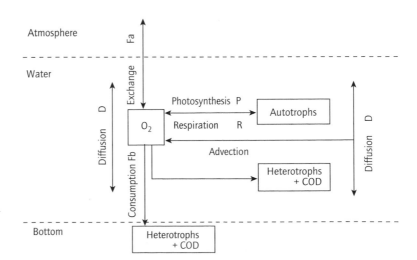

Fig. 11.6 General scheme of oxygen regime in a shallow lake with macrophytes: COD, chemical oxygen demand.

however, the oxygen regime represents a system disturbed by biotic processes in which equilibrium is steadily re-established by physical processes of aeration and diffusion (possibly also by advection). Periodicity in system behaviour appears to be a result of dependence of biotic processes, which are themselves periodic, on solar radiation input. The amplitude of diurnal (diel) changes in oxygen concentration shows how strongly biotic activity affects the dynamics of oxygen regime in the habitat. By way of an illustration, Fig. 11.7 tracks diurnal changes of oxygen concentration in a sublittoral stand of *Utricularia vulgaris*. The difference between the maximum oxygen concentration (at about 1800 hours) and the minimum (at about 0500 hours) was ≈ 11 mg L^{-1}. High daytime irradiance (photosynthetically active radiation (PAR) of 350 W m^{-2} at midday) was favourable for photosynthesis.

A further feature of the oxygen regime in stands of submersed macrophytes, illustrated in Fig. 11.7, is the vertical stratification of oxygen concentration. This may be generated via the occurrence of differing rates of photosynthetic production and respiratory consumption of oxygen at different depths in the stand. As the plant biomass is not distributed uniformly, each layer may contain con-

trasting amounts of biomass and receive different fluxes of PAR. Thus gross rates of photosynthesis may vary considerably. Usually, oxygen is produced at a faster rate in the near-surface layers (which support a dense plant biomass) than in shaded or deeper stand layers. Characteristic profiles of oxygen concentration may develop within the submersed stands.

Another source of variation in oxygen concentration is the flux of the element into the bottom sediments, where it is consumed by the respiration of benthic organisms, and by chemical oxidation processes. In phytoplankton communities, stratification of oxygen is smoothed out by the relatively homogeneous biomass distribution of the phytoplankton and by turbulent diffusion, which compensates for stratification of oxygen concentration more effectively than in stands of submersed macrophytes where the turbulent diffusion may be weak (see Chapter 10). However, stratification of oxygen near the lake bottom is similar in phytoplankton- and in macrophyte-dominated systems.

Differentiation of the water column in terms of dissolved oxygen concentration mirrors the typical course of diurnal density stratification, stabilising during the morning, when irradiance is

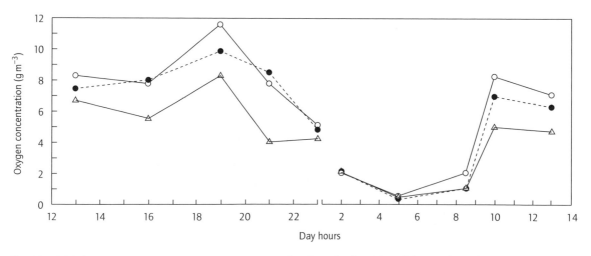

Fig. 11.7 Diel changes in oxygen concentrations measured at three depths: 0.05 m (o), 0.2 m (•) and 0.4 m (Δ) in a stand of *Utricularia vulgaris* (Neusiedlersee Austria). (Redrawn from Pokorný & Ondok 1991.)

increasing in intensity, and weakening through the afternoon. During the night, when depletion of oxygen is proportional to the density of the respiring biomass, convection and turbulent diffusivity override some of the differences in oxygen concentration between various individual layers. In shallow waters, stratification may almost disappear, except in a thin layer just above the lake bottom, where the flux of oxygen to the sediments permanently diminishes its concentration.

Other oxygen sinks, such as bacterial respiration, and consumption in oxidation processes, generally play a proportionately lesser role in the oxygen balance of stands of submersed macrophytes than they do in planktonic communities. Because separate measurement of individual components of the oxygen balance is difficult, they are usually determined as BOD (biological oxygen demand) and COD (chemical oxygen demand). In the model of oxygen dynamics of submerged vegetation of Ondok & Pokorný (1987) they are treated as a single sink and simply compounded with the other main sinks identified.

11.7.3 Carbon dioxide, carbonate equilibrium and photosynthesis of water plants

Carbon dioxide (CO_2) is more soluble in water than oxygen. Although air contains several orders of magnitude less of carbon dioxide (about 0.03%) than of oxygen (21%), water may be supersaturated with carbon dioxide above the air-equilibrated concentration in pure water (0.5–1.0 mg CO_2 L^{-1}). Aqueous concentrations of carbon dioxide are also influenced by equilibria involving various forms of inorganic carbon, C_t found in water, which include: (i) free carbon dioxide (CO_2), (ii) the bicarbonate (HCO_3^-) and carbonate (CO_3^{2-}) ions and (iii) carbonic acid (H_2CO_3). Concentrations (C) of each form tend towards equilibrium, as described in the following equations for total inorganic (C_t) and free CO_2:

$$C_t = \left(TA - \left[OH^-\right] + \left[H^+\right]\right) \frac{\left[H^+\right]^2 + K_1\left[H^+\right] + K_1K_2}{K_1\left[H^+\right] + 2K_1K_2} \quad (11.2)$$

$$\text{Free } CO_2 = \frac{\left[H^+\right]^2 C_1}{\left[H^+\right]^2 + K_1\left[H^+\right] + K_1K_2} \quad (11.3)$$

$$\ln K_1 = -14554.21T^{-1} + 290.9097 - 45.0575 \ln T \quad (11.4)$$

$$\ln K_2 = -11843.79T^{-1} + 207.6548 - 33.6484 \ln T \quad (11.5)$$

$$\ln K_w = -13847.26T^{-1} + 148.9802 - 23.6521 \ln T \quad (11.6)$$

where TA is total alkalinity (see below) and $K1$, $K2$, . . . are the reaction constants given.

The carbonate equilibrium system regulates the pH of the oceans, of most freshwaters, of most soils and even of vertebrate blood. This makes it one of the most ubiquitous, and one of the most important, buffering agents on the planet. When free carbon dioxide (CO_2) is added to water (e.g. when released during respiration), pH falls because CO_2 forms a weak carbonic acid with water ($CO_2 + H_2O = H_2CO_3$). When free carbon dioxide is removed from water (e.g. taken up during photosynthesis), then pH rises. The well-known relationship shown in Fig. 11.8 plots the distribution of free carbon dioxide, bicarbonate and carbonate as a function of pH. It is evident that at pH 4.3, all inorganic carbon is present as free CO_2. At pH 8.3, bicarbonate (HCO_3^-) predominates, whilst at pH 12.3 all inorganic carbon is in the form of carbonate ions (CO_3^{2-}).

Whilst photosynthesising, aquatic plants take up carbon dioxide from the water around them. When respiration exceeds photosynthesis, there is net release of carbon dioxide, and pH falls. The range of pH variation depends on the amount of carbon dioxide released or consumed, and on the buffering capacity of the water. The latter is determined by its alkalinity, which is the sum of $HCO_3^- + 2CO_2 + OH^- - H^+$ concentrations. The total alkalinity of water is relatively easily determined: a given volume of the test water is titrated with hydrochloric acid (HCl) of known strength to pH 4.3, at which point, the HCl consumed is equivalent to the original alkalinity of the water. At pH 4.3, all

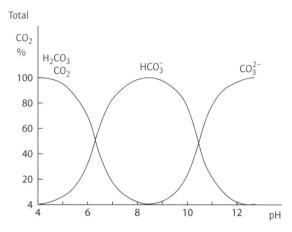

Fig. 11.8 Relationship between pH and distribution of inorganic carbon forms (free CO_2, HCO_3^-, CO_3^{2-}) expressed as percentages of total inorganic carbon (Golterman & Cymo 1969).

carbonate and bicarbonate ions have been converted to free carbon dioxide. At very low total alkalinity, a correction of end-point should be made owing to interaction with carbon dioxide from air (the Gran titration: see Talling 1973).

The alkalinity of water emanating from nutrient-poor soil is low (typically $<0.5\,mmol\,L^{-1}$). Alkalinity of lakes in granite mountains is also low, and close to the zero values found in distilled water or in rainwater. On the other hand, water from limestone areas, or percolating from eutrophic (i.e. nutrient-rich) soils, possesses a high alkalinity of up to $10\,mmol\,L^{-1}$. This water is well buffered, and contains more inorganic carbon as bicarbonate and carbonate than water of low alkalinity. Waters with high alkalinity provide more inorganic carbon for photosynthesis of water plants than do those with low alkalinity.

In general, water plants are exposed to a wider range of inorganic carbon concentrations than terrestrial ones. Photosynthetic uptake of free carbon dioxide and bicarbonate often leads to precipitation of calcium carbonate on the surface of water macrophytes (including filamentous algae). In a simple system, calcium carbonate is precipitated according to the equation of its solubility. In a lake ecosystem and, in particular, in the plankton com-munity, this equation is inadequate to describe precipitation of calcium carbonate, which does not form on the surfaces of microscopic algae. Although it complies with the solubility equation $[Ca^{2+}] \times [CO_3^{2-}] = K_s$, precipitation is avoided, possibly because calcium cations are bound into chelates of organic acids and alginate.

As several species of submerged macrophytes can take up both free carbon dioxide and bicarbonate, we may assume that these compounds are the most important components in the carbon dynamics of submersed vegetation. A simplified scheme of the inorganic carbon system, where these two components are included, is presented in Fig. 11.9. A particular example of changes in pH linked with changes in concentrations of the two most important forms of total inorganic carbon, free CO_2 and HCO_3^-, is shown in Fig. 11.10.

11.7.4 Solar energy and its dissipation

Solar energy arrives at Earth's surface in a full spectrum of electromagnetic wavelengths (ultraviolet, visible light, infrared and longwave thermal radiation; see also Chapters 5 and 6). On a summer day in the temperate zones, global solar irradiance reaches about $1000\,W\,m^{-2}$ at midday, and over a whole day 4–6 kWh (≈ 14.5–22 MJ) of solar energy reach each square metre of Earth's surface. This is a relatively large amount, equivalent to the energy content of about 1 kg coal or dry plant biomass. Solar energy flux (irradiance) can be expressed in energy units (W), in quantum terms (Einstein) or as illuminance (lux). For rough calculations, the following conversion factors may be used:

$$1\,W\,m^{-2}\ (irradiance)$$
$$= 5\,\mu E\,m^{-2}\,s^{-1}\ (quantum\ irradiance)$$
$$= 250\ lux\ (illuminance)$$

The energy realised in the net annual production of biomass in eutrophic waters in the temperate zone (expressed in dry mass) averages 1–2 kg m^{-2} for emergent plants, and up to 0.5 kg m^{-2} for submerged examples (Westlake 1975; Hejný et al. 1981). Thus, the energy content of the biomass produced on $1\,m^2$ during a whole growing season is

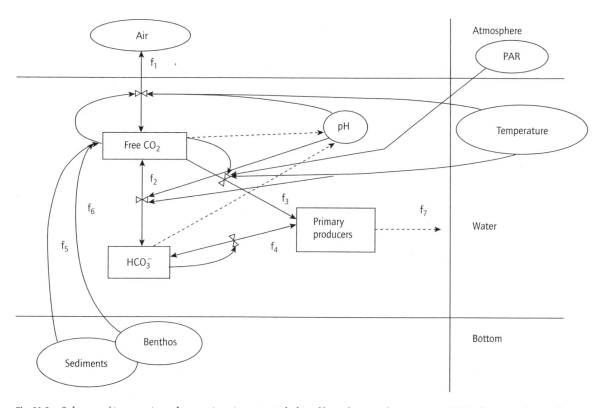

Fig. 11.9 Scheme of inorganic carbon regime in water inhabited by submersed vegetation: PAR, photosynthetically active radiation; f_1–f_7, reaction functions.

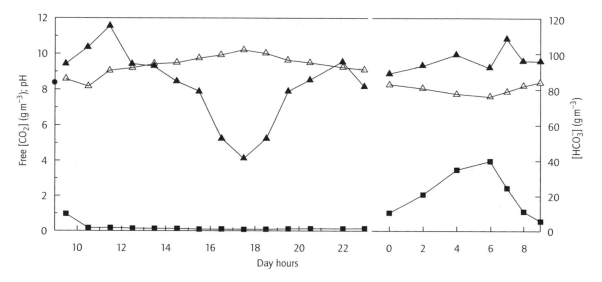

Fig. 11.10 Diurnal changes in free CO_2 (■) and HCO_3^- (▲) concentrations, and pH (△) in a stand of *Elodea canadensis*, calculated from measured pH and alkalinity.

roughly equivalent to the amount of solar energy reaching 1 m² on a single clear day. The total annual solar energy flux amounts to 1000–1200 kWh m⁻² (≈ 3600–4300 MJ m⁻²), while the amount of solar energy bound in 1 kg of dry biomass is about 5 kWh (≈ 18 MJ), i.e. less than 0.5% of annual incident solar energy. One may well enquire as to the fate of the more than 99% of incoming solar energy not realised in photoautotrophy. In fact, plants which are well supplied with water are engaged in the dissipation of solar energy in other important ways, which are relevant to the local and regional climate.

Most of the solar energy incident on the water surface, and of the irrigated surfaces of aquatic plants, is consumed in evaporation. The amount needed for evaporation of 1 L of water is 2.5 MJ = 0.7 kWh. This is a relatively large amount, which is invested in water vapour and only released again (as the latent heat of evaporation) when it condenses, in clouds or as dew. Evaporation from plants is called **transpiration**, therefore evaporation of water from plant stands (plants and soil or water surface) is called **evapotranspiration** (see also Chapter 3). This process damps the potential local temperature fluctuations in space (cooler and hotter places) and time (day and night). By draining land, wetlands, lakes and their littoral zones, steppe-like conditions are created. If plants are unable to dissipate heat by vaporising water, solar energy is converted directly to heat (see Fig. 11.11).

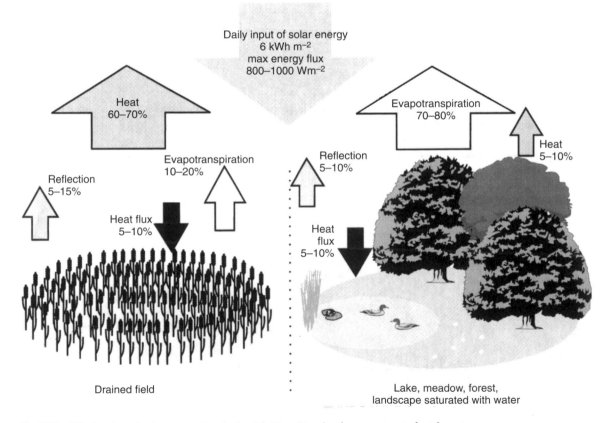

Fig. 11.11   Dissipation of solar energy in a drained field and in a landscape saturated with water.

11.7.5 Solar radiation in water

Solar radiation reaching the water surface is partly reflected (Fig. 11.11). Reflection (**albedo**) increases with decreasing declination (or angle of incidence), and is also increased by surface wave formation. Ultraviolet radiation and infrared wavelength are more easily absorbed than visible light, of which the blue wavelengths naturally penetrate the furthest in pure water, though in lakes they are subject to absorbance by dissolved substances. Relative penetration of light determines the transparency of the water, which is found from the ratio between the irradiance observed at a given depth and that recorded at the water surface. Transparency is commonly estimated using a Secchi disc, a disc about 25 cm in diameter, divided into two white and two black equal quarters. This is lowered into the water to the depth where the black and white sectors become indistinguishable. This is the 'Secchi-disc depth', which is a reasonable guide to the 'euphotic depth' that marks the lower bounds of the layer in which net photosynthetic production is possible and where the growth of water plants becomes limited by lack of light (see Chapter 10). Roughly, it corresponds with about 1% of full daylight. Euphotic depth is usually supposed to approximate to $2.0\text{--}2.5 \times$ Secchi-disc depth (see Chapter 5). In the most transparent, oligotrophic lakes, the euphotic zone may extend to 30 m depth or more—50 m in extreme cases—and certain macrophytes (e.g. *Isoëtes lacustris*, Characeae) may grow in such lakes to depths of 15–20 m. In contrast, transparency in dense macrophyte stands may be a few decimetres only.

Transparency is roughly reciprocal to the underwater attenuation of photosynthetically active radiation (PAR). Light energy penetrating the water surface, I_0, is progressively extinguished with depth; I_z, plots of which generally conform to an exponential equation (the Lambert–Beer law) of the form,

$$I_z = I_0 \cdot \exp- \varepsilon z \qquad (11.7a)$$

or

$$\ln I_z - \ln I_0 = -\varepsilon z \qquad (11.7b)$$

where ε is the coefficient of vertical light attenuation. Typical light attenuation plots in macrophyte beds conform to this general equation, but the attenuation coefficients vary with plant density. The plots shown in Fig. 11.12 have been standardised to a common base of percentage penetration, $(I_z/I_0)\,100\%$.

There are significant differences between the spectral composition of light reaching submersed vegetation and that received by terrestrial stands. For instance, almost all infrared radiation is absorbed during the first 20 mm of its passage through water; thus the radiant energy penetrating into water is diminished by approximately 50%. On the other hand, the reflectance, or albedo, of the water surface, defined as the percentage of incident global radiation reflected, is significantly less (4–10%) than that of terrestrial vegetation (20–30%). Reflectance of light by water varies with the angle of incidence, and is modified by wave ruffling and by turbulence (Bykovskij 1980). Moreover, spectral composition of radiation penetrating natural waters varies more rapidly with depth than does that of light penetrating into

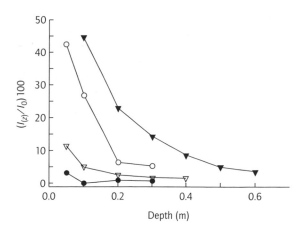

Fig. 11.12 Examples of vertical profiles of photosynthetically active radiation (PAR) extinction in stands of *Potamogeton lucens* (▲), *Ceratophyllum demersum* (o), *Nuphar lutea* (Δ) and in a mixed stand of *Ceratophyllum demersum* and *Lemna* sp. (●): I_0, I_z, PAR irradiance above water surface and at the depth z, respectively.

terrestrial vegetation. This is principally due to preferential absorption by water of certain spectral bands, but behaviour attributable to pure water is modified by dissolved material (especially humic and fulvic acids), and by selective dispersal by organic and inorganic particles suspended in water (Wetzel 1975). In phytoplankton, the fraction of the radiation thus dispersed may be c.20% (Ondok 1977), whereas in emergent vegetation the value is usually less than 5%.

Radiation not absorbed or dispersed in water is available to plants, although the amount which they absorb is related to the density of the submersed vegetation and to the extent of its own self-shading. Many species of macrophytes, especially those of eutrophic habitats, accumulate biomass in the upper layers where radiation is most favourable for photosynthesis. Vertical biomass distribution is then heterogeneous, and the biomass accumulated in the upper layers causes rapid radiation extinction in water. In such cases, light extinction with depth through dense plant stands is distorted with respect to the ordered Lambert–Beer expression (Monsi & Saeki 1953; Anderson 1971; Pokorný & Ondok 1991). In shallow lakes, macrophytes occupy the entire water column, but in deeper lakes with clear water they may grow within a depth range of 2–6 m or more. The upper parts of the stand may then be within a metre or so of the water surface: this was the case in a stand of *Potamogeton lucens* rooted in Stechlinsee (Germany) at a depth of 2 m below the water surface but with a stand height of about 1 m, for which a representative light-extinction profile is shown in Fig. 11.13. Ikushima (1970) suggested that for such cases an extinction formula that separated the two coefficients be used, as in:

$$I_z/I_0 = \exp(-\alpha z - h) \qquad (11.8)$$

where α and h are the respective extinction coefficients in the upper zone (void of submersed vegetation), and z expresses the distance from 0 to z_0 and h that from z_0 to h_0 (see Fig. 11.13). If the stand biomass is homogeneously distributed the formula describes the actual extinction profile.

In shallow ponds with macrophyte biomass dis-

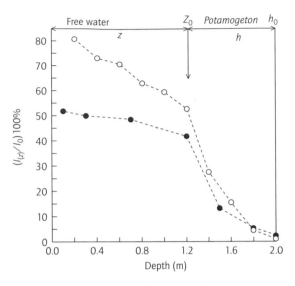

Fig. 11.13 Photosynthetically active radiation extinction profile above and inside the stand of *Potamogeton lucens*, measured under clear (o) and overcast (•) sky. In the upper part, extinction is due mainly to absorption by water and phytoplankton; in the deeper part, extinction is mainly effected by *Potamogeton* (Stechlinsee, Germany). (Redrawn from Pokorný & Ondok 1991.)

tributed unevenly in the water column, the fit of extinction data to the Lambert–Beer law is usually not satisfactory. Ondok & Pokorný (1982) introduced a further modification to the formulation, to which actual PAR profiles fit well:

$$I_z = I_0 \exp(-\varepsilon z^\alpha) \qquad (11.9)$$

where I_z and I_0 are PAR irradiances at the depth z [m] below the water surface and above the water, respectively, ε is the extinction coefficient and α is a parameter expressing the degree of heterogeneity in the spatial distribution of plant biomass. Figure 11.14 shows an example of fitted PAR extinction profiles in a stand of *Elodea canadensis* on two separate occasions. The coefficients ε and α were estimated by a standard linear regression technique. Their values were approximated to be, respectively, 4.01 and 0.73 for the June profile, and 8.31 and 0.57 for August.

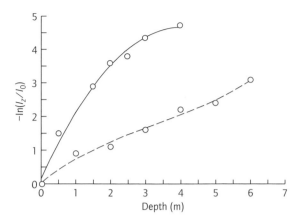

Fig. 11.14 Examples of photosynthetically active radiation extinction profiles in a stand of *Elodea canadensis* fitted by the extinction formula for two dates: 15 June 1976 (dashed line), dry mass = $120 \, \mathrm{g \, m^{-2}}$, $k = 4.01$, $\alpha = 0.73$; August 3 (solid line), dry mass = $420 \, \mathrm{g \, m^{-2}}$, $k = 8.31$, $\alpha = 0.57$.

Specific radiation regimes are found in submersed macrophyte stands growing in water under floating-leaved or free-floating vegetation (e.g., *Nuphar lutea, Nymphaea candida, Lemna* spp.) which may significantly shade the plots beneath. Light penetrates the water only through gaps between the floating structures, and the resultant mosaic of light may be very uneven. It may be that, in such cases, barely submersed plants may receive less light than those at a greater depth.

11.7.6 Water temperature

The influence of macrophytic vegetation on the daily temperature cycle in ponds and shallow lake margins is rather greater than that of phytoplankton. For example, temperature profiles in stands of submersed macrophytes exhibit stronger gradients and higher temperatures in their upper layers. In stands of submersed and floating-leaved or free-floating macrophytes, water temperature is higher during the day and falls during the night. This is explained by a greater heating rate of the plants at the water surface, where a fraction of the heat is conducted from the biomass to the surrounding water.

A steep gradient of water temperature towards the water surface thus develops. In contrast, during the night, plant biomass loses heat more rapidly than water. Development of a steeper temperature gradient in stands of submersed macrophytes is also favoured by weaker turbulent advection in plant stands than in open water. An example of the daily course of water temperature in a macrophyte stand is compared with the corresponding plot for the adjacent open water in Fig. 11.15.

11.8 PHOTOSYNTHETIC PRODUCTION IN SUBMERSED MACROPHYTES

11.8.1 Photosynthesis

Photosynthesis of water plants is measured either as a rate of oxygen yield or of carbon dioxide consumption. Various designs of closed chamber are available for measurement of gas exchange under controlled conditions of irradiance and temperature. Relevant comparisons of photosynthesis and respiration in macrophytes can be obtained by the combination of measurements in dark and light bottles. Photosynthetic rates in the field may be approximated from *in situ* measurements of irradiance, pH and oxygen concentration in macrophyte stands (e.g. Pokorný *et al.* 1987a).

As established above, submersed macrophytes are exposed to lower concentrations of oxygen and carbon dioxide, and to weaker PAR than terrestrial plants, and their photosynthetic rates are commonly lower. Westlake (1980) estimated the net photosynthetic activity of submersed macrophytes to produce between 6 and 40 g of oxygen per kilogram dry mass per hour. He suggested a value of $10 \, \mathrm{g \, O_2 \, kg^{-1}}$ dry mass $\mathrm{h^{-1}}$ to be a reasonable expression of normal photosynthetic activity. His review of the literature relating to the photosynthetic productivity of various species of submerged macrophytes (Westlake 1980) also notes that the typical rate of dark respiration is about 10% of maximum photosynthetic rate.

Generally, four factors are considered to exert the main influence on photosynthesis of macro-

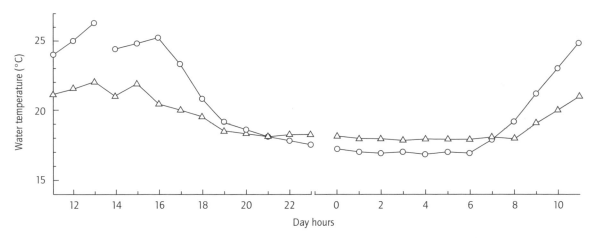

Fig. 11.15 Diurnal course (August 2) of water temperature at 0.2 m depth in a stand of *Nuphar lutea* (L.) J.E. Smith in Sibthorp et J.E. Smith (o) and in surrounding water (∆) (Přibáň *et al.* 1986).

phytes: irradiance, temperature, concentration of inorganic carbon and oxygen concentration. All of these factors mutually interact; further complexity arises through photosynthetic adaptation processes, and the development and senescence of the plants themselves. Analytical formulae are available which relate photosynthetic responses to various driving components. Sometimes, empirical formulae are used in default of adequate relevant measurements. Guides to the calculation and interpretation of photosynthetic behaviour in plants are given by Tenhunen *et al.* (1976a, b, 1977), Weber *et al.* (1981), Ondok & Pokorný (1987) and Pokorný & Ondok (1991).

11.8.2 *Photosynthetic responses to light*

Submersed water plants are generally adapted to low irradiances. Typically, they possess very thin leaves, with chloroplasts present in their epidermal cells. These are **shade plants**, well able to achieve net photosynthesis at quite low levels of PAR (1–3 W m^{-2}; <15 µE m^{-2} s^{-1}). In contrast, **helophytes** (including emergent macrophytes) that are adapted to direct sunshine usually require rather higher irradiances in order to compensate for their respiratory requirement (c.10 W m^{-2}). Photosyn-

thetic rate increases linearly with irradiance up to **light saturation**, which is generally at about 100 W m^{-2}. At levels greater than this value, small increases may be detected as plants adapt to the high irradiances.

Photoinhibitory effects of high irradiance on photosynthesis, of the type commonly observed in phytoplankton communities, is only rarely found in submersed macrophytes; presumably most are able to invoke photorespiration of the excess fixed carbon. The general difference in the photosynthetic behaviour of sun- and shade plants is shown in Fig. 11.16. In submersed leaves of *Myriophyllum brasiliense*, photosynthesis is saturated at about 20% of full sun radiation, whereas in its aerial leaves, photosynthesis remains unsaturated up to full irradiance.

11.8.3 *Photosynthetic responses to temperature*

Photosynthetic rate is understood to increase with increasing temperature up to some optimum, and then to decrease with further temperature rise. Because aquatic plants adapt to ambient temperature shifts throughout the growing season, determination of temperature dependence and of the opti-

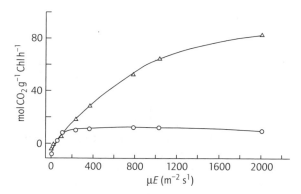

Fig. 11.16 Photosynthetic response to irradiance of submersed (o) and aerial (Δ) leaves of *Myriophyllum brasiliense*. (After Salvucci & Bowes 1982; from Pokorný & Ondok 1991.)

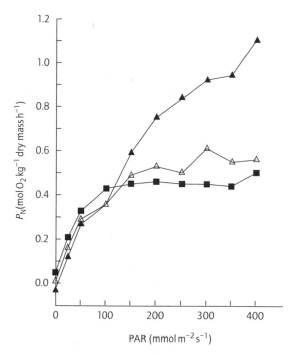

Fig. 11.17 Photosynthetic response of *Myriophyllum salsugineum* to photosynthetically active radiation (PAR) irradiance at constant water temperature: 15°C (■), 25°C (△) and 35°C (▲). (Data from Pokorný *et al.* 1989.)

mum for photosynthesis is not easy. In experimental investigations of temperature dependence of photosynthesis in submersed macrophytes, the effects of abrupt and gradual changes in temperature must also be distinguished (Kirk 1983). Rapid responses to controlled temperature increase conform to a temperature-driven reaction, up to an optimum. Respiration rate also increases with rising temperature: it seems probable that the supraoptimal decline in net photosynthesis owes more to enhanced respiration than to thermal damage to the photosynthetic apparatus. As expected, the temperature rise elicits very little response under subsaturating light levels.

These various trends are evident in the measurements of photosynthetic response in *Myriophyllum salsugineum* (Fig. 11.17): photosynthesis is hardly temperature sensitive at all up to about 100 μmol m^{-2} s^{-1}. Above this value, mass-specific photosynthesis reflects higher temperatures. The temperature optimum for photosynthesis is high, as shown also for *Vallisneria americana* and *Myriophyllum spicatum*. Reported optimal temperatures for photosynthesis among submerged macrophytes vary within a broad range (Stanley & Naylor 1972; Fornwall & Glime 1982; Pokorný & Ondok 1991).

11.8.4 Photosynthetic responses to inorganic carbon

Michaelis–Menten kinetics have been invoked in order to describe the photosynthetic responses of submersed macrophytes to inorganic carbon (Adams *et al.* 1978; Salvucci & Bowes 1982; Titus & Stone 1982; and others). The parameters P_m (maximum photosynthetic rate) and K_c (the carbon half-saturation constant) estimated for various species of submersed macrophytes vary strongly with pH. Photosynthetic response to total carbon dioxide is, in fact, a combined response to free carbon dioxide and bicarbonate. Either source of carbon exerts its own photosynthetic dependence, each with different 'species-specific' P_m and K_c values, and with separate rate-saturating concentrations.

In formulating the photosynthetic response to total inorganic carbon the parameters P_m and K_c should be introduced as pH-sensitive functions. The compensating and saturating concentrations are, equally, pH-dependent, not least because of the relationship of pH to the bicarbonate–carbon dioxide system (Fig. 11.8). Thus, the pH-dependence of photosynthesis invokes the ratio between free carbon dioxide and bicarbonate uptake. Photosynthetic rate is preferentially faster with free carbon dioxide than with bicarbonate; thus rapid photosynthesis eventually leads to diminution of free CO_2, elevation of pH, and a decrease in photosynthetic rate. This is correct, even when total inorganic carbon is held constant. The effect has been reported by several authors (see, e.g., Lucas 1975; Van *et al.* 1976; Titus & Stone 1982; Sand-Jensen 1983).

Although this depression of photosynthetic rate is commonly found at pH > 8, discrepancies exist among the results obtained in photosynthesis measurements at pH < 7. Some data indicate that photosynthetic rate is approximately constant for pH values < 6 (Van *et al.* 1976; Kadono 1980). However, other authors have found that photosynthetic rate decreases more or less sharply whenever ambient pH levels vary above or below the optimum. The pH values favouring maximal photosynthetic rate may well be different for various species of submersed macrophytes, though they usually lie between pH 6 and 8. Decreases in photosynthetic rate at low pH have been reported by Weber *et al.* (1981) for *Elodea densa*, and by Allen & Spence (1981) for *Elodea canadensis* and *Fontinalis antipyretica*. Photosynthetic responses of *Elodea canadensis* to the concentration of total inorganic carbon at selected pH values exhibit sensitivity both to low and to high concentrations of free carbon dioxide (above c.0.5 mmol L^{-1}; Pokorný *et al.* 1985).

As a result of photosynthetic carbon uptake with increasing pH, the proportion of bicarbonate rises, and at pH > 8.3 almost no free carbon dioxide is available. At pH > 8.3, bicarbonate reacts with hydroxyl ions to form carbonate. Above pH 10, there is more carbonate than bicarbonate (see Fig. 11.8). The carbonates of calcium and magnesium are much less soluble than their bicarbonates. Thus, rapid photosynthetic uptake of carbon dioxide by water plants in calcareous waters may eventually lead to precipitation of carbonate (marl deposition) on the surfaces of the plants.

Whereas all water plants readily utilise freely assimilable carbon dioxide, the ability to use bicarbonate is not universal, and none are able to take up carbon from carbonate ions. The ability of water plants to use bicarbonate, and to photosynthesise at high pH, is tested by enclosing plants in glass bottles of lakewater, and exposing them to light until a constant pH is reached. The higher the final pH, the greater the affinity of the plants for inorganic carbon. Macrophytes adapted to utilise bicarbonate are able to raise pH to 10, a value which a handful of planktonic algae and cyanobacteria are also able to reach. Typical bicarbonate users include small *Potamogeton* species, *Elodea* and *Najas*. In contrast, most mosses, *Utricularia*, and the water plants of acidic or peaty waters (e.g. *Hottonia*) are unable to use bicarbonate, and their photosynthesis ceases at pH 8.5–9.

11.8.5 Photosynthetic adaptation to high pH and high oxygen

It may be deduced that photosynthesis in water plants involves two negative feedbacks: the suppression caused by pH increase and the effect of inorganic carbon depletion. In addition, the excess of oxygen produced is also eventually inhibitory to photosynthetic carbon reduction. Water plants, however, possess an elegant strategy for minimising inhibition at high pH. Broad leaves (for example, those of *Potamogeton lucens*) take up inorganic carbon (carbon dioxide and bicarbonate) through their lower (**abaxial**) surfaces and release hydroxyl ions at the upper (**adaxial**) surface. In this way, pH on the lower side of the leaf increases less than it does on the upper side (Raven 1984). Acidic and alkaline zones have also been found in plants of *Chara* sp., indicating areas of preferential bicarbonate uptake and hydroxyl release, distinguished by differences in pH of up to two units (Lucas 1975). As a result of this polarity, calcium carbonate is precipitated only in regions of hydroxyl release.

Some macrophytes (e.g. *Isoëtes*) have developed the adaptation of crassulacean acid metabolism (CAM). They take up carbon dioxide during the night, when its concentration is relatively high, through the action of phosphoenol pyruvate (PEP) carboxylase. The product malate is accumulated in the cells. By day, malate is metabolised to release a dedicated supply of carbon dioxide for conventional fixation by ribulose biphosphate carboxylase (Keeley & Morton 1982). These plants can also use carbon dioxide produced by root respiration, which passes to the assimilatory tissues through the lacunal system in the aerenchyma.

The amount of calcium carbonate precipitated on leaves of water plants, and on filamentous algae, can reach values of tens of grams per square metre, i.e. several hundred kilograms per hectare. In this way, the concentration of dissociated calcium is reduced, phosphorus can be bound in the precipitate, and total alkalinity of water decreases owing to precipitation of carbonate. The pH of the water in macrophyte stands rich in calcium carbonate precipitate is high and stable, even during the night, because carbon dioxide released at that time reacts with calcium carbonate. The pH does not change but total alkalinity increases owing to the release of bicarbonate.

The problem with excess oxygen is manifest in an effect on the compensation level of carbon uptake. When oxygen concentration increases from 8 to 40% air saturation, the carbon compensation point in low-pH-adapted plants of *Myriophyllum salsugineum* rises from 5.3 to $8.9\,\mu mol\,L^{-1}$, but from 1.8 to $2.8\,\mu mol\,L^{-1}$ in *M. salsugineum* adapted to high pH. However, some aquatic macrophytes exhibit some of the features of C4 plants, in that they possess low photorespiration levels and low CO_2 compensation points, which are relatively insensitive to oxygen concentration (Bowes & Salvucci 1989; Lambers *et al*. 1998). These features may be attributed to high PEP carboxylase activity, which has been demonstrated in *Hydrilla verticillata* adapted to a high temperature and low free carbon dioxide concentration (Holaday *et al*. 1983). Internal accumulation of inorganic carbon in ex-

cess of the concentration in the external medium suppresses photorespiration in most submersed macrophytes with high ribulose biphosphate carboxylase activity. Most are able to acclimatise to summer conditions of high temperature, high oxygen and low carbon dioxide concentrations.

11.8.6 Dark respiration and photorespiration

The ratio of dark respiration rate to net photosynthesis in submerged macrophytes has been reported as being three to five times greater than in terrestrial plants (e.g., Van *et al*. 1976; Søndergaard & Wetzel 1980; Westlake 1980). Van *et al*. (1976) reported a dark respiration rate in three species of submersed macrophytes (*Myriophyllum spicatum*, *Ceratophyllum demersum* and *Hydrilla verticillata*) which constituted over 50% of net photosynthesis. In terrestrial plants, dark respiration is typically only 5–10% of the net photosynthesis. Pokorný & Ondok (1991) measured the dark respiration rate of several species of submerged macrophytes, which ranged between 0.6 and $1.8\,g\,O_2\,kg^{-1}$ dry matter h^{-1}. In *Utricularia vulgaris* this corresponds to a range of 6–20% of maximum net photosynthesis. Dark respiration in *Elodea densa* was 7–9% of the maximum photosynthesis; *Myriophyllum salsugineum*, *Lagarosiphon major*, *Cabomba caroliniana*, *Egeria densa* and *Nitella* sp. possessed an even smaller ratio. The most important factor controlling this ratio is plant age and periphyton density on the plant surface. Plants from eutrophic water bodies overgrown by periphytic organisms exhibit the highest respiration rate at the end of the growing season. This also leads to oxygen deficits in stands of water macrophytes.

According to several authors (Hough & Wetzel 1972; Prins & Wolf 1974; Hough 1979) photorespiration occurring in submersed macrophytes is less intense than in terrestrial plants. Salvucci & Bowes (1982) found photorespiration to be about 15% of the net photosynthetic rate in *Myriophyllum brasiliense*. Jana & Choudhuri (1971) studied photorespiration in *Hydrilla verticillata*, *Valisneria spiralis* and *Potamogeton pectinatus*. Both dark respiration and photorespiration were highest in *P. pectinatus*. Photorespiration of the above three species was about 50% of net photosynthesis. In contrast to the results of Hough (1979), photorespiration exceeded dark respiration: the ratio of light/dark respiration was 1.62, 1.6 and 1.23, respectively, for *Valisneria spiralis*, *Potamogeton pectinatus* and *Hydrilla verticillata*. The ratios of light/dark respiration measured by Søndergaard & Wetzel (1980) in *Elodea canadensis*, *Lobelia dortmana*, *Fontinalis antipyretica* and *Littorella uniflora* were in agreement with the data reported by Hough (1979), ranging between 0.07 and 0.64, depending upon the oxygen concentration. The ratio increased when oxygen concentration rose. The ratio of photosynthesis to photorespiration is similar to that found in C3 plants, but it is recognised that precise practical measurement of the refixation of carbon dioxide from the internal lacunal system is difficult to achieve.

11.9   MACROPHYTES AND ASSOCIATED BIOTA

11.9.1   *Invertebrates associated with aerial parts of emergent and floating vegetation*

Grazing invertebrates directly impinge on the development of water macrophytes by damaging their stems and meristems. Dvořák *et al.* (1998) reviewed the results of the International Biological Programme and other findings on the role of animals and animal communities in wetlands, with particular emphasis on the role of macroinvertebrates. They distinguished between those groups of invertebrates which use the emergent vegetation only as a refuge or shelter (adults of *Chironomidae*, *Culicidae*, *Ephemeroptera*, *Trichoptera*, etc.), and several groups of organisms which are plant feeders, or which have adopted special feeding mechanisms prejudicial to the plants, many of which figure prominently in the energy budget of the ecosystem. Gaevskaya (1966, 1969) showed that, when they were at their most abundant, adult populations of two coleopteran species, *Donacia dentata* and *Hydronomus alismatis*, together con-

sumed up to 6% of leaf biomass of *Sagittaria sagittifolia* each day. *Donacia dentata* and *D. crassipes* grazing on *Nymphaea candida* were found to consume 2.6% of the biomass per day. The feeding intensity of individual invertebrates depends on their age, and on the suitability of the plant species as a food. In some species, daily consumption by larvae is equivalent to 50% of body mass. The older and larger the larvae, however, the smaller the mean daily consumption per unit body weight.

Such consumption data indicate intensity of feeding on plant biomass at selected times, but give little clue as to annual food consumption, or the loss of macrophyte biomass sustained. These quantities are determined by a complex of factors, including the seasonal dynamics of recruitment and growth of consumers, seasonal changes in the quality and abundance of the plant parts consumed, and temperature. Smirnov (1961) estimated the consumption of *Alisma plantago-aquatica* and *Oenanthe aquatica* by *Donacia dentata* to be, respectively, 3 and 6.1% of their annual production. Imhof (1973) expressed the plant biomass consumed by all phytophagous insects present in a stand of *Phragmites* in terms of energy intake — the total of which, 170–250 kJ m^{-2} yr^{-1}, represented some 0.3–0.4% of annual net production.

The complex impacts of invertebrates on macrophyte growth in terms of biomass consumed or potential energy sequestered cannot always be gauged. For instance, the stem-boring lepidopteran, *Phragmataecia*, takes up 6.3 kJ per *Phragmites* stem per year without visibly affecting its growth, whereas the gall-forming fly, *Lipara*, taking only 0.25 kJ, inhibits stem growth and prevents flowering (Dvořák *et al.* 1998).

11.9.2   *Invertebrate macrofauna associated with the submerged parts of the vegetation*

Studies of invertebrates associated with macrophyte stands are somewhat inconsistent, some concentrating on benthos but not plankton, others on submerged plants, but ignoring free-living bottom fauna. Sampling equipment is non-standard, and sieves are of differing mesh sizes (see also Chapter 12). Few studies include an attempt to quantify plant biomass. Thus, comparison is limited mainly to generalisations. Dvořák *et al.* (1998) reviewed data on invertebrate biomass found in different types of vegetation from various northern temperate and tropical wetlands, within the context of a derived unit of vegetated area. The highest non-benthic macrofaunal biomass concentrations (3.2–10 g m^{-2} dry mass) were found among *Lemna trisulca*, *Potamogeton crispus*, *Schoenoplectus lacustris* and on *Ranunculus aquatilis* and *Elodea canadensis*. Other plant species supported 0.2–1.9 g m^{-2}. There are often no apparent differences between colonisation of submerged plants and emergent species.

It generally may be assumed that detritus and periphyton are the prime trophic links governing the productivity of macrophyte–invertebrate communities. Although these invertebrates are often called **phytophilous**, they survive mostly as consumers of detritus or of periphyton (**aufwuchs**), for which the living plants provide suitable environments for accumulation, namely shelter and substrate. Some invertebrate species directly consume the living tissues of submerged plants (*Acentropus niveus*, *Elophila nymphaeata*, *Phryganea grandis*; Dvořák *et al.* 1998) but these are generally distinguished by low biomass and by temporal variability, their feeding periods being restricted to certain, often relatively short, periods. In contrast, soft, decaying and bacteria-rich plant tissues are consumed for longer periods. Entrapment of foods of allochthonous origin, such as terrestrial leaves and detritus, can contribute an important component of available resources.

11.9.3 Role of filamentous algae in development of macrophyte stands

Owing to their fast rates of growth, filamentous algae are able rapidly to overwhelm vascular plants, shading them and lowering the availability of carbon dioxide, sometimes below their compensation points. In this case, vascular plants become little more than substrata for filamentous algae.

The development of filamentous algae is affected by concentrations of phosphorus and nitrogen in water. It is at high concentrations of nutrients, and low feeding pressures, that filamentous algae grow most rapidly. When phosphorus concentration in the algal dry mass exceeds 0.25%, growth of *Cladophora* is considered not to be phosphorus limited (Auer & Canale 1982). Phosphorus concentrations in *Cladophora* from the fishponds studied in the Třeboň Basin Biosphere Reserve exceeded this value, and the nitrogen content ranged between 1.3 and 4.3% of dry mass (Eiseltová & Pokorný 1994). Ettl *et al.* (1973) estimated the maximum rate of biomass doubling of *Cladophora* to be about 20 hours: maximum oxygen production reached 400% saturation and pH rose to 11.6.

These observations of the conditions that develop in dense algal mats exceed those catalogued by Hillebrand (1983). The expectation is that depletion of carbon dioxide and high pH restrict the growth of most other autotrophs. Simultaneous release of free ammonia is toxic to other organisms, causing, *inter alia*, gill necrosis in fish. High pH also slows microbial activity, including decomposition. Decay of *Cladophora fracta* can be as slow (90% dry-weight loss in 95 days; Pieczyńska 1972) as reported for *Cladophora albida* in tidal estuaries (Birch *et al.* 1983).

Light sensitivity of photosynthesis in *Cladophora fracta* is demonstrable (authors' data): faster rates, equivalent to 6.5 mg O$_2$ g^{-1} h^{-1}, have been measured at the water surface (PAR irradiance >100 W m^{-2}) than at the lake bottom (40 W m^{-2}). The estimated compensating light intensity, about 6 W m^{-2}, does not suggest that there is much adaptation to low irradiance in this alga. The photosynthetic activity of filamentous algae may lead to precipitation of calcium carbonate on filaments. For example, 32 mg of CaCO$_3$ were deposited on 1 g of *Cladophora fracta* dry mass (i.e., 134 kg CaCO$_3$ ha^{-1} at biomass of 420 g dry matter m^{-2}).

11.10 DEVELOPMENT OF MACROPHYTE STANDS IN RELATION TO ANTHROPOGENIC IMPACTS

The various responses of limnetic macrophytic vegetation to nutrient load, and to management of land fertility, drainage, fish stocking, application of lime and organic fertiliser (manure), are demonstrable among the fishponds of southern Bohemia. During the early decades of the twentieth century, fishponds were mostly slightly acidic (pH 6–6.5), with alkalinity generally <1 mmol⁻¹. Nutrient concentrations in these small, shallow lakes were generally low, most being oligotrophic or mesotrophic in character. From the 1930s fish production in many of these ponds was intensified by liming, fertilisation and the addition of grain and food pellets. Since that time it has increased by about one order of magnitude. Nowadays, production (mostly of the common carp, *Cyprinus carpio* L.) yields between 500 and 1000 kg fresh weight ha⁻¹ yr⁻¹ (50–100 g m⁻²). In the process, the ponds have advanced to hypertrophy.

With increased fertility, stands of aquatic plants within the fishponds have become more dense, with increased biomass per unit area and altered vertical distribution. Carbon shortage and light constraints are now more crucial to development and succession of submersed vegetation than is any shortage of nutrients.

These broad impacts of eutrophication on the composition and distribution of submersed vegetation and their relation to light extinction, pH and dissolved oxygen are shown graphically in Fig. 11.18A–F. This scheme is based on the works of DeNie (1987), Pokorný *et al.* (1990, 1994), Pokorný & Ondok (1991) and Pokorný *et al.* (1999). The stages depicted in the figures are now briefly reviewed.

11.10.1   Oligotrophic stage

In oligotrophic water bodies (Fig. 11.18A), growth of macrophytes and phytoplankton is typically limited by chronic lack of one or more nutrients in the water. Plants able to take up nutrients through their roots possess an additional resource opportunity. Transparency remains high, so plant biomass can develop and thrive at the lake bottom, to some depth. Net production remains low. Concentrations of carbon dioxide and oxygen, and pH, equilibrate with the air and are stable on a diel basis. Such conditions describe '**Lobelia** lakes' where

species of *Littorella*, *Lobelia* and *Isoëtes* clothe the marginal substrate.

11.10.2   Oligotrophic–mesotrophic stage

In somewhat more fertile water bodies (Fig. 11.18B), aquatic vegetation may be comprised of a greater variety of species. Nutrient availability remains a key constraint for most of the year, especially for phytoplankton and epiphytes. Potentially, rooted plant biomass is evenly distributed between the surface and a depth of ≥ 2 m. The tips of submersed macrophyte shoots only rarely reach the water surface, except for those of species which develop floating leaves. The concentration of dissolved oxygen remains close to air saturation, with no marked day–night fluctuations. Carbon dioxide exchanges only slightly affect pH. Growth of epiphytes is muted, via the grazing activity of molluscs, insect larvae and fish. Water bodies in this stage of development are now rather rare in agricultural landscapes, as a consequence of high nutrient inputs.

11.10.3   Early eutrophic stage

With progressively higher nutrient loading (whether as a consequence of deliberate manuring and fertiliser application, or unintended run-off from agricultural land) more vigorous growth of macrophytes occurs, leading to greater standing biomass (Fig. 11.18C). Plants grow fast, and biomass accumulates at the water surface. The young, green parts of macrophytes shade the deeper water. Transparency decreases and, even in water bodies of <1 m depth, irradiance at the bottom may fail to compensate respiration (PAR < 2 W m⁻²). While photosynthesis takes place at the water surface, respiration prevails at the bottom: if there is no overriding wind mixing, steep gradients of oxygen concentrations develop during daylight hours.

As nutrient loading continues, species diversity decreases, even though stand biomass increases. This stage is also characterised by mass development of periphyton, which suppresses the growth of some macrophytes. Respiration rate of the whole community exceeds that of previous stages,

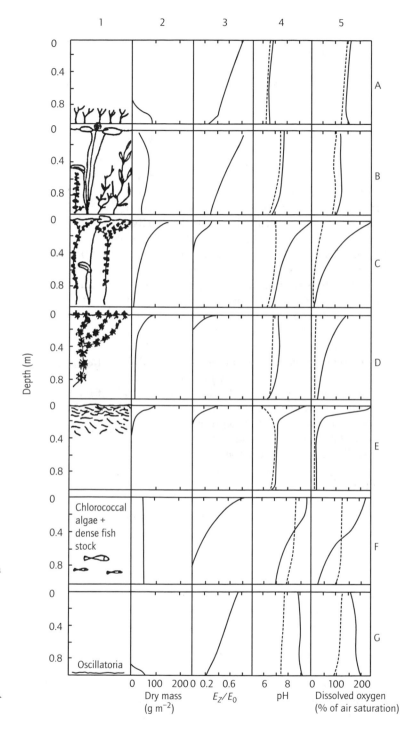

Fig. 11.18 Development of aquatic vegetation in ponds of increasing trophic status (1) from oligotrophic (A) over mildly eutrophic (B), eutrophic (C, D) to hypertrophic (E, F, G) stages; (2) biomass distribution in vertical profile; (3) extinction profile, E_0 = photosynthetically active radiation (PAR) irradiance at water surface, E_Z = PAR irradiance at the depth z; (4) pH; (5) oxygen concentrations in the vertical profile. Dashed line, night minimum values; solid line, day maximum values. (From Pokorný & Ondok 1991.)

and an oxygen deficit often develops near the lake bottom. Here, decomposition of organic matter leads to low oxygen concentration, and release of phosphorus, ammonium and ferrous iron may occur. Internal nutrient cycling therefore now becomes important.

11.10.4   The progression to hypertrophy

In the stages represented by Fig. 11.18D–F, nutrients have ceased to be a controlling factor in community structure. Instead, the fish stock itself may exert a crucial role in the extent of the submersed vegetation. At a low fish stock (seasonal mean live biomass of fish $<$c.350 kg ha^{-1}), large *Daphnia* (filter-feeders of phytoplankton) are not wholly consumed by the fish, and so continue to regulate recruitment of planktonic algae. Transparency remains relatively high, in spite of high nutrient loading. This combination of circumstances is not stable, however, and may lead **either** to expansion of certain macrophytes (Fig. 11.18D) **or** to the development of blue-green algal blooms (Fig. 11.18E).

These alternative outcomes, and their determining factors, are of considerable interest. At high nutrient concentrations, macrophytes are very often suppressed by filamentous algae (namely *Cladophora* sp.). These use inorganic carbon effectively even at low concentrations, grow fast and develop dense mats throughout the water column. When the filamentous algae reach the water surface (Fig. 11.18E) they shade the complete water column. Now, pH may increase to >11 (Eiseltová & Pokorný 1994), and the macrophytes die back. Development of blue-green algae is particularly tolerant of a low availability of inorganic carbon, but is favoured by the rapid cycling of phosphorus and ammonium nitrogen (Pokorný *et al.* 1999). Furthermore, blue-greens are less intensely consumed by zooplanktonic filter-feeders than chlorococcal algae.

At high fish-stocking densities (seasonal mean biomass of fish $\geq$400 kg ha^{-1}), large *Daphnia* spp. are likely to be eliminated. This situation allows fast-growing chlorococcal algae to develop dense populations of several hundred µg L^{-1} of chlorophyll *a*. The dense algal populations (Fig. 11.18F)

create low transparency (less than 0.4 m), which prevents the development of blue-green algae such as *Aphanizomenon* and *Microcystis*. Other, low-light tolerant blue-greens (*Limnothrix*, *Planktothrix*) may occur in the phytoplankton and persist (see Chapter 10). Dense stocking of planktivorous fish thus prevents the development of macrophytes in very eutrophicated water bodies. A factor leading to positive feedback is that large fish searching for food in the bottom sediments interfere with establishment of rooted macrophytes.

11.10.5   Development of water plants on bottom sediments in spring and on fish-kills

Spring development of filamentous blue-green algae (such as *Oscillatoria* spp., Fig. 11.18G) at the sediment surface leads to high pH in the whole water column. High pH in the entire water column may also be caused by *Chara* sp., and by other species of fast growing macrophytes. However, the subsequent conversion of ammonium ions to ammonia gas is lethal to fish. Fish-kills are frequent in fishponds at times when the water is cold, respiration is slow and net photosynthesis is strong. Generation of ammonia is represented thus:

$$NH_4^+ + OH^- \Leftrightarrow NH_3 + H_2O \qquad (11.10)$$

$$\frac{[NH_4^+][OH^-]}{[NH_3]} = K_1 = 10^{-4.8} \qquad (11.11)$$

$$\frac{[NH_4^+]}{[NH_3]} = \frac{10^{-4.8}}{[OH^-]} = \frac{10^{-4.8}}{(10^{-14})(10^{pH})} = \frac{10^{9.2}}{10^{pH}} \qquad (11.12)$$

Concentrations of NH_3 in excess of 0.01 mg L^{-1} are toxic to most fish species, especially their fry (Schäperclaus 1979). As a result, stocking of fish in ammonium-rich fishponds is delayed during periods of high pH.

11.11   AQUATIC WEEDS

'Macrophytes' become 'weeds' when their vigorous development causes problems by:

• interference with flow in irrigation systems and in flood mitigation drains, overgrowing of water bodies, etc.;

• promotion of certain diseases, such as those borne by mosquitoes, or molluscs living in association with aquatic macrophytes (bilharsiosis), etc.;

• interference with recreation;

• interference with fish production, the dense stands of aquatic weeds preventing movements of fish. Anaerobic conditions develop at the bottom and high assimilatory pH at the water surface. Soft macrophyte biomass may rapidly decompose and consume oxygen;

• aesthetic degradation of the environment;

Management of aquatic weeds involves a number of control techniques. Physical removal, by hand or with a machine, is effective and immediate but slow and expensive, and requires sites for disposal of the vegetation removed. In principle, composting or fermentation is the best way of dealing with the material, as its nutrient content may be recovered. Chemical controls are still widely favoured, often without due attention to their polluting effects on the water (killing of non-target, sometimes remote organisms) or to deleterious secondary effects when the killed material decomposes, including nutrient release and oxygen consumption. Acquired resistance may make herbicide application selectively favourable to target species. Many herbicides indirectly stimulate growth of algae, through release of nutrients from dead macrophytes.

11.11.1   Application of methods of aquatic weed control: biological methods

Tree shading is an important means of controlling macrophytes at the water's edge, stunting but not eliminating submersed plants. Tree litter in the water (fallen leaves, shed branches), increasing detrital material at the bottom and releasing humic and other organic substances may be counter-productive.

Herbivorous, weed-eating fish (such as the Chinese grass carp *Ctenopharyngodon idella*) can effectively control growth and propagation of macrophytes, or even eliminate them. Enhanced

nutrient turnover is again the drawback, risking the development of dense phytoplankton, perhaps cyanobacteria. Selective effects are also apparent in that grass carp prefers soft submersed vegetation (such as *Elodea* sp.) to hard macrophytes such as *Najas marina* or *Myriophyllum* sp. Grass carp does not reproduce naturally in cooler regions of the temperate zone, where the species has been used to control macrophytes without risk to the existing fish fauna.

Insects or other invertebrates can also be used to control 'aquatic weeds'. Examples of successful application of this technique are provided by the use of the curculionid beetles *Neochetlina eichhorniae* and *Cytobagous salviniae*, against *Eichhornia crassipes* and *Salvinia molesta*, respectively. The control effect of insects or other animals can be augmented by that of fungal plant pathogens, which may be inoculated on the animal-damaged plants (e.g., *Cercospora rodmanii* on *Eichhornia* in the USA; Pieterse & Murphy 1990).

11.11.2   Mechanical methods

Laborious hand cutting or chain scything techniques have been all but finally superseded by specialised mechanical cutters: flails, weed-cutting boats and dredges, of which there are many different types available. In order to avoid release of nutrients when decomposing, plants are harvested and removed from the shores or other sites from which nutrients might leach back into the water body. Removal of the harvested material also avoids oxygen deficit which may develop when plants are rapidly decomposing in a relatively warm water rich in nutrients. The harvested plants can be used as animal feed or can be composted, provided that they are substantially free (or contain only permitted concentrations) of toxic substances such as heavy metals, pesticide residues or PCBs.

11.11.3   Chemical methods

There is now available a sufficient range of aquatic herbicides to permit chemical control of almost all aquatic plants. Each should be subjected to exhaustive toxicological and environmental testing and, as a rule, only those compounds which precisely satisfy the registration authorities should be available for commercial use. Nevertheless, attention must always be paid to the manufacturers' instructions, and warnings about the harmful effects of over-application Even then, the fate of the herbicide in the food web should be understood if deleterious consequences of use are to be avoided.

11.12   COMMERCIAL USES OF AQUATIC PLANTS

There are also positive commercial benefits of aquatic macrophyte growth (National Academy of Sciences 1976). Among the economic uses to which aquatic plants are put are:
• energy generation: fermentation to yield methane gas, direct burning after drying (especially emergent life-forms—with the ash, if it is not toxically contaminated, providing fertiliser).
• waste-water treatment and 'nutrient stripping': living natural and artificial macrophyte beds are used to treat waste waters by absorbing and incorporating dissolved compounds of nitrogen and phosphorus into their biomass. Emergent macrophytes (reed, cattail, etc.) are now widely used in constructed wetlands to treat waste waters from small to medium rural settlements (artificial reed bed treatment systems; RBTS). Some plants, for example, the water hyacinth, also accumulate phenols, heavy metals or other toxic substances, thereby reducing the concentrations of noxious chemicals in the water. For this treatment to be persistently effective, macrophytes must continue to be harvested and removed.
• pulp, paper and fibre: fibrous, reed-like plants can be processed in cellulose and paper manufacture.
• food: paddy rice (*Oryza sativa*), Chinese water chestnut (*Eleocharis dulcis*), watercress (*Nasturtium*), water spinach (*Ipomea aquatica*), wild rice (*Zizania aquatica*), lotus (*Nelumbo nucifera*), taro (*Colocasia*), swamp taro (*Cyrtosperma*), arrow-

head (*Sagittaria trifolia*) are among the macrophytes raised deliberately for their yields of edible products.
• animal feed: this applies to a great many species of macrophytes. Quite often, however, the fresh plants must be thoroughly dried before they are fed to domestic animals as a precaution against parasites or toxicity of the fresh plants.

11.13   CONCLUDING REMARKS

This chapter has explored the enormous phylogenetic and adaptive diversity among water plants, as well as their ecology and their contribution to ecological functioning and the metabolism of lake ecosystems. They are prominent producers and processors of carbon in shallow lakes, and around the margins of larger bodies. Their direct impacts on lake-wide carbon cycling, not least through generating the habitat of an important part of its animal biomass, are clearly significant, continuing to contribute to and influence carbon cycling into the very largest lakes on the planet. Macrophytes also influence our perception of lake beauty, and impinge upon the societal judgement of their value: interest in their well-being is not solely the concern of the limnetic ecologist.

11.14   REFERENCES

Adams, M., Guillizoni, P. & Adams, S. (1978) Relationship of dissolved inorganic carbon to macrophyte photosynthesis in some Italian lakes. *Limnology and Oceanography*, **23**, 912–9.

Allen, E.D. & Spence, D.H.N. (1981) The differential ability of aquatic plants to utilise the inorganic carbon supply in freshwaters. *New Phytologist*, **87**, 269–83.

Anderson, M.C. (1971) Radiation and crop structure. In: Šesták, Z., Čatský, J. & Jarvis, P.G. (eds), *Plant Photosynthetic Production. Manual of Methods*. Dr. W. Junk, The Hague, 412–66.

Auer, M.T. & Canale, R.P. (1982) Ecological studies and mathematical modelling of *Cladophora* in Lake Huron: 3. The dependence of growth rates on internal phosphorus pool size. *Journal of Great Lakes Research*, **8**, 93–9.

Birch, P.B., Gabrielson, J.O. & Hamel, K.S. (1983) Decomposition of Cladophora. I. Field Studies in the Peel–Harvey Estuarine System, Western Australia. *Botanica Marina*, **26**, 165–71.

Bowes, G. & Salvucci, M.E. (1989) Plasticity in the photosynthetic carbon metabolism of submersed aquatic macrophytes. *Aquatic Botany*, **34**, 233–86.

Boyd, C.E. (1969) Production, mineral nutrient absorption and biochemical assimilation by *Justicia americana* and *Alternanthera philoxeroides*. *Archiv für Hydrobiologie*, **66**, 139–60.

Bristow, J.M. (1975) The structure and function of roots in aquatic vascular plants. In: Torrey, J.G. & Clarkson, D.T. (eds), *The Development and Function of Roots*. Academic Press, London, 231–3.

Bykovskij, V.J. (1980) Vlianie dvizhenia vody na osveschennost' jejo tolshchi (Effect of water movement on irradiance in water layers.) *Gidrobiologicheskii Zhurnal*, **16**, 58–61.

Casper, S.J. & Krausch, H.D. (1980) *Pteridophyta and Anthophyta 1. Teil: Lycopodiaceae bis Orchidaceae*. Die Süsswasserflora von Mitteleuropa Band 23. Gustav Fischer Verlag, Jena, 403 pp.

Casper, S.J. & Krausch, H.D. (1981) *Pteridophyta and Anthophyta 2. Teil Saururaceae bis Asteraceae*. Die Süsswasserflora von Mitteleuropa Band 24. Gustav Fischer Verlag, Jena, 539 pp.

Clapham, A.R., Tutin, T.G. & Warburg, E.F. (1981) *Excursion Flora of the British Isles*. Cambridge University Press, Cambridge, 499 pp.

De Nie, H.W (1987) *The decrease in aquatic vegetation in Europe and its consequences for fish populations*. EIFAC/CECPI Occasional paper No. 19, Food and Agriculture Organization, Rome, 52 pp.

Den Hartog, C. & Segal, S. (1964) A new classification of the water-plant communities. *Acta Botanica Neerlandica*, **13**, 367–93.

Denny, P. (1980) Solute movement in submerged angiosperms. *Biological Reviews*, **55**, 65–92.

Denny, P. (1987) Mineral cycling by wetland plants—a review. *Archiv für Hydrobiologie Beiheft Ergebnisse der Limnologie*, **27**, 1–25.

Denny, P. (1998) *Tropical Wetlands and their Management*. Lecture course handout for the UNESCO Course in Limnology, International Hydraulics Laboratory, Delft, The Netherlands, 122 pp.

Dvořák, J. & Best, E.P.H. (1982) Macro-invertebrate communities associated with macrophytes of Lake Wechten: Structural and functional relationships. *Hydrobiologia*, **339**, 27–36.

Dvořák, J., Imhof, G., Day J.W., Jr., *et al.* (1998) The role of

animals and animal communities in wetlands. In: Westlake, D.F., Květ, J. & Szczepański, A. (eds), *The Production Ecology of Wetlands. The IBP Synthesis*. Cambridge University Press, Cambridge, 211–318.

Dykyjová, D. & Úlehlová, B. (1998) Mineral economy and cycling of minerals in wetlands. In: Westlake, D.F., Květ, J. & Szczepański, A. (eds), *The Production Ecology of Wetlands. The IBP Synthesis*. Cambridge University Press, Cambridge, 319–66.

Eiseltová, M. & Pokorný, J. (1994) Filamentous algae in fish ponds of the Třeboň Biosphere Reserve—ecophysiological study. *Vegetatio*, **113**, 155–70.

Ettl, H., Březina, V. & Marvan, P. (1973) Methodical notes on assessment of productivity in littoral algae. In: Květ, J. (ed.), *Littoral of the Nesyt Fishpond*. Studie 73/15, ČSAV (Czechoslovakian Academy of Science), Praha, 111–15.

Fornwall, M.D. & Glime, J.M. (1982) Cold and warm-adapted phases in *Fontinalis duriali* Schimp. as evidenced by net assimilatory and respiratory responses to temperatures. *Aquatic Botany*, **13**, 165–77.

Gaevskaya, N.S. (1966/1969) *Rol' Vysshikh Vodnykh Rastenii v Pitanii Zhivotnykh Presnykh Vodoemov*. Nauka, Moskva, 327 pp. [*The Role of Higher Aquatic Plants in the Nutrition of the Animals of Freshwater Basins*, Vol. 1, 1–231, Vol. 2, 232–417, Vol. 3, 418–629. Translated by D.G. Maitland-Muller, edited by K.H. Mann, 1969. National Lending Library for Science and Technology, Boston Spa.]

Gessner, F. (1955) *Hydrobotanik: die physiologischen Grundlagen der Pflanzenverbreitung im Wasser. I. Energiehaushalt*. Deutscher Verlag der Wissenschaften, Berlin, 517 pp.

Gessner, F. (1959) *Hydrobotanik: die physiologischen Grundlagen der Pflanzenverbreitung im Wasser. II. Stoffhaushalt*. Deutscher Verlag der Wisseschaften, Berlin, 701 pp.

Golterman, H.L. & Clymo, R.S. (eds) (1969) *Methods for Chemical Analysis of Fresh Waters* (International Biological Programme Handbook 8). Blackwell Scientific Publishers, Oxford, 172 pp.

Haslam, S.M., Sinker, C.A. & Wolseley, P.A. (1975) British water plants. *Field Studies*, **4**, 243–351.

Hejný, S. (1957) Ein Beitrag zur Ökologischen Gliederung der Makrophyten der tschechoslowakischen Niederungsgewässer. *Preslia*, **29**, 349–68.

Hejný, S. (1960) *Ökologische Charakteristik der Wasser- und Sumpfpflanzen in den slowakischen Tiefebenen (Donau- und Theissgebiet)*. Vydavatel'stvo Slovenskej Akádemie Vied., Bratislava, 492 pp.

Hejný, S. & Sytnik, K. M. (1993) *Makrofity—indikatory*

izmenenii prirodnoi sredy. [Macrophytes—Indicators of Changes of the Natural Environment]. Institute of Botany, Ukrainian Academy of Sciences, Kiev, Ukraine, 430 pp. (In Russian.)

Hejný, S., Květ, J. & Dykyjová, D. (1981) Survey of biomass and net production of higher plant communities in fishponds. *Folia Geobotanica Phytotaxonomica,* **16**, 73–94.

Hejný, S., Segal, S. & Raspopov, I.M. (1998) General ecology of wetlands. In: Westlake, D.F., Květ, J. & Szczepański, A. *The Production Ecology of Wetlands. The IBP Synthesis.* Cambridge University Press, Cambridge, 1–77.

Hendrych, R. (2003) Herkunft und vorkommen von *Acorus calamus* in unseren Länderen. *Zprávy Česke botanicke Společnosti, Praha,* **38**, 95–109. (In Czech with German and English summaries.)

Hillebrand, H. (1983) Development and dynamics of floating cultures of filamentous algae. In: Wetzel, R. G. (ed.), *Periphyton of Freshwater Ecosystems.* Dr W. Junk, The Hague, 31–9.

Holaday, A.S., Salvucci, M.E. & Bowes, G. (1983) Variable photosynthesis/photorespiration ratios in *Hydrilla* and other submersed macrophyte species. *Canadian Journal of Botany,* **61**, 229–36.

Hotchkiss, N. (1972) *Common Marsh, Underwater and Floating-leaved Plants of the United States and Canada.* Dover Publications, New York, 223 pp.

Hough, R. A. (1979) Photosynthesis, respiration and inorganic carbon release in *Elodea canadensis* Michx. *Aquatic Botany,* **7**, 1–11.

Hough, R. A. & Wetzel, R. G. (1972) A ^{14}C-assay for photorespiration in aquatic plants. *Plant Physiology,* **49**, 987–90.

Howard-Williams, C. & Longman, T.G. (1976) A quantitative sampler for submerged aquatic macrophytes. *Journal of the Limnology Society of South Africa,* **2**, 31–3.

Hutchinson, G.E. (1975) *A Treatise on Limnology.* III. *Limnological Botany.* Wiley, New York, 660 pp.

Ikushima, I. (1970) Ecological studies on the productivity of aquatic plant communities. IV. Light conditions and community photosynthetic production. *Botanical Magazine, Tokyo,* **83**, 330–41.

Imhof, G. (1973) Aspects of energy flow by different food chains in a reed bed. A review. *Polskie Archiwum Hydrobiologii,* **20**, 165–8.

Jana, S. & Choudhury, M.A. (1971) Photosynthetic and photorespiratory behaviour of three submersed angiosperms. *Aquatic Botany,* **7**, 13–20.

Kadono, A. (1980) Photosynthetic carbon sources in some *Potamogeton* species. *Botanical Magazine, Tokyo,* **93**, 185–94.

Keely, J.E. & Morton, B.A. (1982) Distribution of diurnal acid metabolism in submersed aquatic plants outside the genus *Isöetes. Photosynthetica,* **16**, 546–53.

Kirk, J.T.O. (1983) *Light and Photosynthesis in Aquatic Ecosystems.* Cambridge University Press, Cambridge, 401 pp.

Květ, J. (1973) Mineral nutrients in shoots of reed (*Phragmites communis* Trin.). *Polskie Archiwum Hydrobiologii,* **20**, 137–47.

Květ, J., Westlake, D.F., Dykyjová, D., Marshall, E.J.P. & Ondok, J.P. (1998) Primary production in wetlands. In: Westlake, D.F., Květ, J. & Szczepański, A. (eds), *The Production Ecology of Wetlands. The IBP Synthesis.* Cambridge University Press, Cambridge, 78–139.

Lambers, H., Chapin, F.S., III & Pons, T.L. 1990. *Plant Physiological Ecology.* Springer-Verlag, New York, 540 pp.

Lucas, W.J. (1975) Photosynthetic fixation of ^{14}carbon dioxide by internodial cells of *Chara coralina. Journal of Experimental Botany,* **26**, 331–46

Lucas, W.J. & Berry, J.A. (1985) *Inorganic Carbon Uptake by Aquatic Photosynthetic Organisms.* University of California, Davis, Rockville, 483 pp.

Melzer, A. (1976) Makrophytische Wasserpflanzen als Indikatoren des Gewässerzustandes oberbayerischer Seen. *Dissertationes Botanicae,* Vol. 34. J. Crammer, Vaduz, 1–195.

Melzer, A. (1992) Submersed macrophytes. In: Scharf, B.W. & Björk, S. (eds),. *Limnology of Eifel Maar Lakes. Ergebnisse der Limnologie,* **38**, 223–37.

Mitchell, D. S. (1976) The growth and management of *Eichhornia crassipes* and *Salvinia* spp. in their native environment and in alien situations. In: Varshney, C.K. & Rzóska, J. (eds), *Aquatic Weeds in South-east Asia.* Dr W. Junk, The Hague, 167–76.

Monsi, M. & Saeki, T. (1953) Über den Lichtfaktor in den Pflanzengesellschaften und seine Bedeutung für die Stoffproduktion. *Japanese Journal of Botany,* **14**, 22–52.

National Academy of Sciences, USA. (1976) *Making Aquatic Weeds Useful: some Perspectives for Developing Countries.* National Academy of Sciences, Washington, 175 pp.

Ondok, J.P. (1977) *Regime of Global and Photosynthetically Active Radiation in Helophyte Stands.* Studie 77/15, ČSAV (Czechoslovakian Academy of Science), Praha, 1–112.

Ondok, J.P. & Pokorný, J. (1982) Model of diurnal regime of O_2 and CO_2 in stands of submerged aquatic vegetation. *Ekológia* (ČSSR) **1**, 381–94.

Ondok, J.P. & Pokorný, J. (1987) Modelling photosynthesis of submersed macrophyte stands in habitats with limiting inorganic carbon. 2. Application to a stand of *Elodea canadensis* Michx. *Photosynthetica*, **21**, 543–54.

Pieczyńska, E. (1972) Production and decomposition in the eulittoral zone of lakes. In: Kajak, Z. & Hillbricht-Ilkowska, A. (eds), *Productivity Problems of Freshwaters. Proceedings of the IBP–UNESCO Symposium on Productivity Problems of Freshwater, Kazimierz Dolny, May 1970.* Polish Science Publishers, Warszawa, 615–35.

Pieterse, A.H. & Murphy, K.J. (1990) *Aquatic Weeds. The Ecology and Management of Nuisance Aquatic Vegetation.* Oxford Science Publications, Oxford, 594 pp.

Pokorný, J. (1994) Development of aquatic macrophytes in shallow lakes and ponds. In: Eiseltová, M. (ed.), *Restoration of Lake Ecosystems—a Holistic Approach.* International Wildlife and Wetlands Research Bureau Publications, Slimbridge, 36–43.

Pokorný, J. & Ondok, J.P. (1991) *Macrophyte Photosynthesis and Aquatic Environment.* Rozpravy Československé Akademie Věd, řada matematických a přírodních věd, Academia, Praha, 1–117.

Pokorný, J., Květ, J., Ondok, J.P., Toul, Z. & Ostrý, I. (1984) Production—ecological analysis of a plant community dominated by *Elodea canadensis* Michx. *Aquatic Botany*, **19**, 263–92.

Pokorný, J., Ondok, J.P. & Končalová, H. (1985) Photosynthetic response to inorganic carbon in *Elodea densa* (Planchon) Caspary. *Photosynthetica*, **19**, 366–72.

Pokorný, J., Hammer L., & Ondok, J.P. (1987a) Oxygen budget in the reed belt and open water of shallow lake. *Archiv für Hydrobiologie Beiheft Ergebnisse der Limnologie*, **27**, 185–201.

Pokorný, J., Lhotsky, O., Denny, P. & Turner, E.G. (eds) (1987b) Water plants and wetland processes. *Archiv für Hydrobiologie Beiheft Ergebnisse der Limnologie*, **27**, 1–265.

Pokorný, J., Orr, P.T., Ondok, J.P., Denny, P. (1989) Photosynthetic quotients of some aquatic macrophyte species. *Photosynthetica*, **23**, 494–506.

Pokorný, J., Květ, J. & Ondok, J.P. (1990) Functioning of the plant component in densely stocked fishponds. *Bull. Ecol. Brunoy*, **21**, 44–8.

Pokorný, J., Schlott, G., Schlott, K., Pechar, L. & Koutníkova, J. (1994) Monitoring of changes in fishpond ecosystems. In: Aubrecht, G., Dick, G. & Prentice, C. (eds), *Monitoring of Ecological Change in Wetlands of Middle Europe. Proceedings of International Workshop, Linz, Austria, October 1993.* Linz-Dornach, Linz, 37–46.

Pokorný, J., Fleischer, S., Pechar, L. & Pansar, J. (1999) Nitrogen distribution in hypertrophic fishponds and composition of gas produced in sediment. In: Vymazal, J. (ed.), *Nutrient Cycling and Retention in Natural and Constructed Wetlands.* Backhuys Publishers, Leiden, 111–20.

Přibáň, K., Ondok, J.P., Šmíd, P. (1986) Impact of vegetation on physical factors in the aquatic and wetland environment. In: Hejný, S., Raspopov, I.M. & Květ, J. (eds), *Studies on Shallow Lakes and Ponds.* Academia, Praha, 185–91.

Prins, H.B.A. & Wolf, W. (1974) Photorespiration in leaves of *Valisneria spiralis*. The effect of oxygen on the carbon dioxide compensation point. *Proceedings, Koninklijke Nederlandse Akademie van Wetenschappen, Series C*, **77**, 171–245.

Raven, J.A. (1984) *Energetics and Transport in Aquatic Plants.* MBL Lectures in Biology. Allan R. Liss, New York, 587 pp.

Rothmaler, W. (1995) *Exkursionsflora von Deutschland, Gefässpflanzen,* III *Atlasband.* Gustav Fischer Verlag, Stuttgart, 753 pp.

Salvucci, M.E. & Bowes, G. (1982) Photosynthetic and respiratory responses of the aerial and submerged leaves of *Myriophyllum brasiliense*. *Aquatic Botany*, **13**, 147–64.

Sainty, G.R. & Jacobs, S.W.L. (1994) *Waterplants in Australia. A Field Guide.* Sainty and Associates, Sydney, 327 pp.

Sand-Jensen, K. (1983) Environmental variables and their effect on photosynthesis of aquatic communities. *Aquatic Botany*, **34**, 5–25.

Schäperclaus, W. (1979) *Fischkrankheiten. Teil 2.* Akademic-Verlag, Berlin, 511–1089.

Sculthorpe, C.D. (1985) *The Biology of Aquatic Plants*, Edward Arnold Ltd. London, 610 pp. [Reprint by Koeltz Scientific Books, D-6240 Königstein, Germany.]

Smirnov, N. N. (1961) Consumption of emergent plants by insects. *Verhandlungen der Internationalen Vereinigung Limnologie*, **14**, 421–9.

Søndergaard, M. & Wetzel, R.G. (1980) Photorespiration and internal recycling of CO_2 in the submersed angiosperm *Scirpus subterminalis* Torr. *Canadian Journal of Botany*, **58**, 591–7.

Spence, D.H.N. (1964) The macrophytic vegetation of freshwater lochs, swamps and associated fens. In: Burnett, J.H. (ed.), *The Vegetation of Scotland*. Oliver and Boyd, Edinburgh, 306–425.

Spencer-Jones, D. & Wade, M. (1986) *Aquatic Plants — a Guide to Recognition*. ICI Professional Products, Farnham, 169 pp.

Stanley, RA. & Naylor, A.V. (1972) Photosynthesis in Eurasian watermilfoil (*Myriophyllum spicatum* L.). *Plant Physiology*, **50**, 149–51.

Talling, J.F. (1973) The application of some electrochemical methods to the measurement of photosynthesis and respiration in fresh waters. *Freshwater Biology*, **3** 335–62.

Tansley, A.G. (1939) *The British Islands and their Vegetation*. Cambridge University Press, Cambridge, 930 pp.

Tenhunen, J.D., Yocum, C.S. & Gates, D.M. (1976a) Development of a photosynthesis model with an emphasis on ecological applications. I. Theory. *Oecologia*, **26**, 89–100.

Tenhunen, J.D., Weber, J.A., Yocum, C.S. & Gates, D. (1976b) Development of a photosynthesis model with an emphasis on ecological applications. II. Analysis of data set describing the Pm surface. *Oecologia*, **26**, 101–19.

Tenhunen, J.D., Weber, J.A., Filipek, L.M., & Gates, D.M. (1977) Development of a photosynthesis model with an emphasis on ecological applications. III. Carbon dioxide and oxygen dependencies. *Oecologia*, **30**, 189–207.

Titus, J.E. & Stone, W.M. (1982) Photosynthetic response of two submersed macrophytes to dissolved inorganic carbon concentration and pH. *Limnology and Oceanography*, **27**, 151–60.

Van, K.T., Haller, W.T. & Bowes, G. (1976) Comparison of the photosynthetic characteristics of three submersed aquatic plants. *Plant Physiology*, **58**, 761–8.

Vymazal, J. (1995) *Algae and Element Cycling in Wetlands*. Lewis Publishers, Boca Raton, FL, 689 pp.

Weber, M. & Brändle, R. (1996) Some aspects of the extreme anoxia tolerance of the sweet flag, *Acorus calamus* L. In: Brändle R, Čížková, H. & Pokorný J. (eds), *Adaptation Strategies in Wetland Plants: Link between Ecology and Physiology. Folia Geobotanica and Phytotaxonomica*, **31**, 43–52.

Weber, J.A., Tenhunen, J.D., Westrin, S.S., Yocum, C.S. & Gates, M.D. (1981) An analytical model of photosynthetic response of aquatic plants to inorganic carbon and pH. *Ecology*, **62**, 697–705.

Wein, K. (1939) Die älteste Einführungs- und Ausbreitungsgeschichte von *Acorus calamus* (erster Teil). *Hercynia*, **1**(3), 367–450.

Westlake, D.F. (1965) Some basic data for investigation of the productivity of aquatic macrophytes. *Memorie dell' Instituto Italiano Idrobiologia*, **18** (Supplement), 229–48.

Westlake, D.F. (1975) Primary production of freshwater macrophytes. In: Cooper, J.P. (ed.), *Photosynthesis and Productivity in Different Environments*. International Biological Programme Publication No. 3, Cambridge University Press, Cambridge, 189–206.

Westlake, D.F. (1980) Primary production. In: LeCren, E.D. & Lowe-McConnell, R.H. (eds), *The Functioning of Freshwater Ecosystems*. International Biological Programme Publication No. 22, Cambridge University Press, Cambridge, 141–6.

Westlake, D.F., Květ, J. & Szczepański, A. (eds) (1998) *The Production Ecology of Wetlands. The IBP Synthesis*. Cambridge University Press, Cambridge, 568 pp.

Wetzel, R.G. (1975) *Limnology*. W.B. Saunders Company, Philadelphia, PA, 743 pp.

12 Benthic Invertebrates

PÉTUR M. JÓNASSON

12.1 INTRODUCTION

The zoobenthos of lakes is among the commonest and most widespread of freshwater faunas. It encompasses all those species, mainly invertebrates, inhabiting the solid–liquid interface at the bottom of all kinds of freshwater habitats from ponds to lakes. It comprises a large number of animal phyla with enormous anatomical variation. Therefore, the possible adaptations to the limnological environment are both fascinating and almost unlimited. The result is an unbelievable wealth of life-forms, and the zoobenthos thus provides fundamental information about freshwater habitats.

Zoobenthic organisms often occur in enormous numbers and form important links in the food web of lakes. Many are microphagous in their feeding habits, using either phytoplankton, other algae, periphyton, submerged macrophytes, bacteria or organic mud constituents as their food. In their turn, they are widely eaten by many aquatic carnivores and are an important food item for the top carnivores, the fish. Fundamentally, the zoobenthos changes fine organic matter into food for larger carnivores.

Phytoplankton and periphyton, together with macrophytes in the euphotic zone, are important determinants of the oxygen regime of the lake, and of various other chemical factors, especially their temporal and vertical variations. It follows that the zoobenthos is affected in quality, quantity and distribution. Above the thermocline, species diversity is high and the fauna is abundant, with a high oxygen demand; below, a few specialists exist in large numbers tolerating the lower oxygen tensions or the reduced food supply. The main aim of the present chapter is to quantify the role of zoobenthos in the lake ecosystem, and especially

to trace its utilisation of the organic matter produced. Limnological parameters, however, greatly affect the pathways of energy flow in the lake ecosystem. Therefore, another main theme is devoted to the influence of such parameters on the ecology of benthos. These factors are clearly illustrated in the bathymetric changes of community structure of lakes.

The most important of these are:
1 physical factors such as wind exposure and wind-induced currents responsible for food and oxygen circulation;
2 chemical factors such as concentration of oxygen, and carbon dioxide (CO_2), bicarbonate $[HCO_3]^-$ and carbonate (CO_3^{2-}) of the carbon dioxide system;
3 physiological adaptations of species to maintain growth at, for example, low levels of oxygen concentration;
4 food quality and quantity and the ability of individual species to utilise the food available;
5 the type of substrate;
6 (more difficult to evaluate) intra- and interspecific competition (Jónasson 1978, 1996).

An early attempt, perhaps the first, to link quantitatively the detritivorous zoobenthos of lakes to their primary food source, and other limnological parameters, was made by Jónasson (1964). The outburst of experiments on the feeding of aquatic invertebrates from the early 1960s and onwards adds much to our knowledge. Better sieving techniques have produced more realistic estimates of densities of zoobenthos (Jónasson 1955, 1958) and fundamentally changed our view on their food requirements and the role of zoobenthos in the ecosystem. The profundal of Lake Esrom (Denmark), with its natural monocultures of species, has been a useful laboratory.

The aim of this chapter is to consider the environmental requirements of zoobenthos in dimictic lakes in the light of species ecology, and to attempt a synthesis of a holistic ecosystem picture. Complete ecosystem studies on various lake types have recently given a more adequate understanding of the function of zoobenthos in the ecosystems of Lake Esrom, Mývatn and Thingvallavatn (Iceland) (Jónasson 1972, 1979a, 1992a).

12.2   HISTORY OF STUDIES ON BENTHIC INVERTEBRATES

Before the invention of the microscope, in 1590, benthic freshwater invertebrates were largely unknown. During the 17th century, the Dutch microscopist A. van Leeuwenhoek (1632–1723) succeeded with his technical skill in improving this instrument to a magnification of 270 times. He was able to send descriptions of wheel-animalicules and Infusoria to The Royal Society in London, where his discoveries were later published. During the Enlightenment period of the 18th century, a series of monographs on invertebrates was produced. At first their content was rather sweeping, but gradually they became more specialised. Most famous is probably *Biblia Naturae* by the Dutch physician and anatomist Jan Swammerdam (1637–1680). It was not published until more than half a century after his death (1738), but his correct descriptions were epoch-making. The German engraver and painter Roesel von Rosenhof (1705–1759), an amateur, gave outstanding and correct descriptions of the biology of insects in his *Insektenbelustigung* (1755). The hand-coloured paintings are of great beauty. The classic *Memoires pour servir à l'histoire des Insectes*, Vols 1–6 (1734–1742) by R. Réaumur (1683–1757) is still valid today.

The German Lutheran pastor I.C. Schaeffer published numerous valuable original descriptions (1754, 1755a, b, 1756, 1762) of phyllopods, daphnids, rotifers and hydroids from ponds around Regensburg. The accompanying hand-coloured paintings were outstandingly correct. Another monograph of a high standard is *Mémoires pour servir à l'histoire d'un genre de Polypes d'eau*

douce, à bras en forme de cornes (1744) by A. Trembley.

Probably the most comprehensive 18th century writer on benthic invertebrates was O.F. Müller (1730–1784) of Denmark. He was a contemporary of Carl von Linné, and the two corresponded. Using the best available microscopes he explored a new field in freshwater fauna: the small invertebrates. In spite of his limited period in active science (1762–1784) he attained fundamental biological and systematic results on freshwater invertebrate faunas. His monographs were published in German and Latin, the scientific languages of that time, and therefore they became well-known all over Europe. Another reason for the great propagation and use of his monographs is due to the detailed descriptions and beautifully coloured plates designed by his brother.

The first of O.F. Müller's papers was on Odonata (1767). The following, *Von Würmern des süssen und salzigen Wasser*s (1771), is an outstanding monograph on vegetative reproduction. It is probably the first paper where graphic presentation of results was used. In *Vermium terrestrium et fluviatilium* (1773) he treated the Hydra, Oligochaeta, Nematomorpha, Nematoda, Hirudinea, Trematoda and Mollusca. In 1778 he published on the Acanthocephala, in 1781 a famous monograph on the water mites, Hydracarina, and in 1782 a paper on *Schistocephalus*. Two volumes were published after O.F. Müller's death in 1784. In 1785 came the basic work on Crustacea: *Entomostraca seu Insecta testacea*, with 21 coloured plates, where the names Entomostraca, *Daphnia*, *Cyclops* and many others still valid were created. Finally, in 1786, appeared the largest freshwater monograph: *Animaliculi Infusoria fluviatilia et marina*, with 50 coloured plates.

During the 19th and 20th centuries numerous monographs on benthic invertebrates were published, especially from central and northern Europe, and mainly in German. The father of limnology, F.-A. Forel, began his investigations on zoobenthos in Lake Geneva, in c.1870. His classic work *Le Léman*, in three volumes (1892–1904), which treated all facets of lake limnology, was similar to the *Bathymetric Survey of the Freshwater Lochs of Scotland* in six volumes by J. Murray and L. Pullar (1910): both milestones in limnological research. Both initiated new studies and monographs on entire zoobenthos communities, e.g., *Die Tierwelt der Hochgebirgsseen* (1910) and *Die Tiefseefauna der Seen Mitteleuropas* (1911) by F. Zschokke.

Fundamental also are the studies of S. Ekman, the inventor of the Ekman sampler: *Die Bodenfauna des Vätterns* (1915). During the first half of last century, Lundbeck (1926, 1936), Berg (1938) and Berg & Petersen (1956) carried out quantitative benthological studies on north European lakes. Jónasson (1972, 1979a, 1992a) studied production processes of lakes at all ecological levels, and was able to measure the share of zoobenthos in the total energy budget of the lake. A unique investigation is the ecosystem approach to Mirror Lake and its environment (Likens 1985), and the study of its zoobenthos and micrometazoans (Strayer 1985). Brinkhurst *et al.* (1974) provided a review of lake typology and produced an approach to various aspects of zoobenthos of lakes.

Decades separate the publication of handbooks on benthic invertebrates. The only complete textbooks ever published on zoobenthos are those of Wesenberg-Lund (*Biologie der Süsswassertiere*, i.e. 'Biology of Freshwater Invertebrates', 1939), covering all systematic groups except insects. In 1943 the continuation (*Biologie der Süsswasserinsekten*) appeared. Altogether, these two volumes constitute 1500 pages, but unfortunately both were published only in German, the limnological language of that time, even though Wesenberg-Lund wrote his other monographs in English. As a background he had published 30 monographs and numerous shorter papers on the various groups of freshwater fauna before he wrote his two volumes of synthesis. The Wesenberg-Lund books are still valuable, thanks to their enthusiastic and fascinating description of the biology of freshwater animals.

In the meantime, half a century has passed, and limnological research has progressed tremendously. Hutchinson's textbook on *Zoobenthos* (1993) was, therefore, extremely welcome, but

covers only part of the zoobenthos, because many animal classes (e.g. Turbellaria, Trematoda, Cestoidea, Nematoda, Nematomorpha, Oligochaeta, Hirudinea, groups of Crustacea, Hydracarina, Lamellibranchiata, and, among the insects, Trichoptera, Neuroptera, Lepidoptera and Diptera) are omitted from his treatment. The Hutchinsonian approach is deductive, focusing on **details** of the biology of zoobenthos, and giving a critical examination of all facets of their ecology. His coverage of papers from central and northern Europe published in German is weak.

12.3 EFFICIENCY OF SIEVING AND SAMPLING METHODS

Valid sampling and sieving methods are of fundamental importance when estimating the true abundance of bottom faunas. It therefore follows that the description of life-histories, population dynamics, production and vertical distribution of benthic invertebrates are all affected by the techniques of sampling used. When quantitative investigations of the bottom fauna are carried out, a series of factors affect the reliability of the results. A knowledge of the methods used, and their accuracy, is therefore necessary in order to evaluate the results.

The equipment used in investigations of this kind comprises various types of bottom samplers, and sieves with varying mesh gauges. Furthermore, there are varying techniques for separating the animals from the sieve residue. Therefore, the use of the bottom sampler, and the processes of sieving and sorting adopted, largely determine the degree of accuracy of results obtained. The suitability of bottom samplers for varying types of substratum at different depths has often been considered (e.g. Berg 1938). Several modifications of the Ekman bottom sampler, the apparatus most frequently used in freshwater, have been described (Ekman 1911; Birge 1922; Lenz 1931, 1932; Rawson 1947), and a quite new apparatus has been introduced (Mortimer 1941–1942, fig. 2; Kajak *et al.* 1965). Sampling via scuba diving has improved the efficiency of the sampler on hard bottoms. By increasing the height of the sampler itself from 15 to 40 cm so that it would penetrate deeper into the mud, Berg (1938) observed an apparent increase of 15% in the abundance of bottom fauna living on soft substrata.

A survivorship curve for *Chironomus anthracinus* Zett., sampled and sorted by hand at 2.5× magnification, from the profundal of Lake Esrom (Fig. 12.1), changes shape according to mesh size. Efficiency of the various mesh gauges appears to be closely correlated with the width of the head capsule. Successively varying the mesh gauge from the previous standard of 600 μm to 510, 260 and 200 μm gave the following results.

A sieve with mesh gauge 200 μm captured larvae quantitatively from July onwards. At this time (during the stratification period, when natural mortality is low) the number of individuals was 16,000 m⁻², decreasing to 14,500 m⁻². The 210-μm size group was seen to be of short duration, however. By 24 July the greater part of the population belonged to the 400-μm group. This larval stage lasted until November, when the final instar, with a head capsule width of 690 μm, became the most abundant.

Towards the end of the observation period, conditions changed greatly. From 22 September to 20 November the population was reduced to 11,000 m⁻² by fish predation (a mortality rate almost twice as large as that prevailing during 9 August to 22 September). During the period October to November, length of larvae increased from 6 to 13 mm. During winter and spring, the number of individuals remained at 10,000–11,000 m⁻² owing to low temperatures, and the consequent light fish predation. Following emergence in May, the number decreased to about half.

Whilst head capsule width was 210 μm, the 260-μm sieve captured 50% of the larval population. In contrast, the results of using the 510-μm sieve were as follows. With head capsule width of 210 μm, this sieve only captured 200 individuals out of a population of 16,000 m⁻². With head capsule width of 400 μm, 4000 individuals, or only 25% of the population, were recovered. Not until larvae had reached the last instar in November were they captured quantitatively. Sieving with a

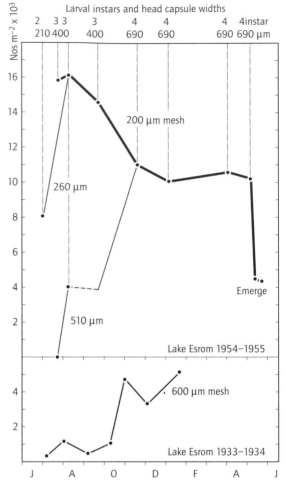

Fig. 12.1   Population size and life-histories of *Chironomus anthracinus* larvae in the profundal zone of Lake Esrom obtained by using different mesh gauges. Curves are based on use of mesh gauges 200, 260 and 510 μm for sampling different instars. (Modified from Jónasson 1955.)

600-μm mesh gave results similar to those of 1933–34 (Berg 1938), indicating the same population size.

A 100-μm mesh sieve captured the first instar in June, with a head capsule width of 113 μm. However, first instars are difficult to sort out from large amounts of residue, and an apparent negative mortality would result (see Jónasson 1972, fig. 55).

It is thus seen that, with most of the techniques previously used in freshwater zoobenthos research, capture results do not provide a reliable indication of true population. **The spring maximum of zoobenthos reported by most authors is simply explained by the observation that the mesh gauges used usually only allow efficient capture of the later instars.** It also simultaneously explains the small densities of zoobenthos, or even their complete absence, which is observed during the following seasons, as these are the periods when the smaller first instars occur. Adequate estimates of abundance can only be obtained when mesh gauge is related to the size of the animals. This provides the true life-history. When appropriate sieves were used, estimates of the total abundance of the profundal fauna of Lake Esrom increased by 640% (Jónasson 1955, 1958).

The above results shed an important light upon the seasonal migration of *Chironomus* larvae postulated by various authors. Young larval instars are present in very high numbers in the profundal zone during summer stratification, but were not captured by previous investigators, because the mesh gauge used was too coarse. Therefore, the theory that young larvae escape hypolimnetic oxygen deficits by migrating into shallower water cannot be substantiated.

Previous investigations (Lundbeck 1926; Berg 1938) also suggested the incidence of two peaks of abundance of zoobenthos with depth—one in the littoral macrophyte zone, another in the sublittoral zone. However, this also appears to be an artefact caused by sampling inefficiency. When adequate methods are used (as shown for a 200 µm mesh sieve in Lake Esrom; Fig. 12.2), numbers of littoral and profundal benthos become very high (Jónasson 1972; Dall *et al.* 1984a). Earlier reviews of bathymetric distribution therefore need revision (e.g. Brinkhurst *et al.* 1974).

12.4 THE EFFECT OF LAKE MORPHOMETRY ON ZOOBENTHOS

Figure 12.3 shows the basins of three lakes—the shallow, eutrophic 22-m-deep Lake Esrom (Denmark), the oligotrophic, 370-m-deep Lago Maggiore (Italy), and the oligotrophic, 114-m-deep Thingvallavatn (Iceland). Environmental and other differences between these lakes are briefly described by Jónasson (1996). A prominent feature

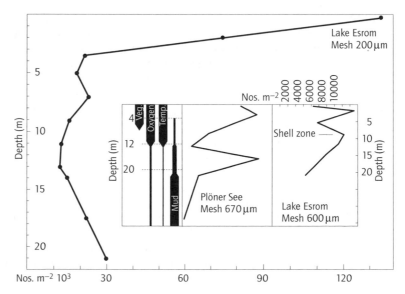

Fig. 12.2 Quantitative vertical distribution of benthos in typical eutrophic lakes in relation to depth distribution of vegetation, thickness of mud layer, temperature and oxygen content of lakewater. Previous investigations in Plöner See and Lake Esrom are compared with recent investigations in Lake Esrom using a mesh gauge of 200 µm. (Modified from Lundbeck 1926; Berg 1938; Jónasson 1972; Dall *et al.* 1984a.)

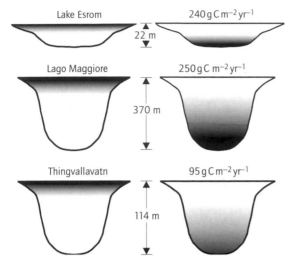

Fig. 12.3 Diagram showing eutrophic Lake Esrom, oligotrophic Lake Maggiore and oligotrophic Lake Thingvallavatn. The internal loading of organic matter, presented as annual primary production, is shown as dots in the epilimnion. The sedimentation of organic matter into the hypolimnion is illustrated by distributing the relevant number of dots over the entire hypolimnion (Jónasson 1996; idea from Thienemann 1928a).

of the data is that whilst average annual phytoplankton production in Lakes Esrom and Maggiore is of the same order $(250\,g\,C\,m^{-2}\,yr^{-1})$, in the former, internal loading of organic matter is diluted into only a small hypolimnion volume, but, in the latter, into a large volume. Consequently, the hypolimnion of the Lago Maggiore remains oxygen rich, but in Lake Esrom a steep gradient of hypolimnetic oxygen concentration occurs. In the subarctic, oligotrophic Thingvallavatn, production is considerably lower $(95\,g\,C\,m^{-2}\,yr^{-1})$ and the number of species is limited (Fig. 12.3).

What are the long-term consequences for life in the epilimnion and hypolimnion of these two lake types? Fixation of organic matter has been measured by the ^{14}C method since the mid-1950s and the results are well-suited to answer the question. Oddly enough we find 'bottle necks' in the epilimnion and the hypolimnion of both oligotrophic

and eutrophic lakes. Lago Maggiore is now phosphorus limited, with a unimodal phytoplankton production curve (de Bernardi *et al.* 1984), whereas the nitrogen-limited Thingvallavatn and Lake Esrom exhibit bimodal curves.

Nitrogen-limitation of production offers the possibility of separating production sequences related to various groups of algae. Therefore, Thingvallavatn, where no nitrogen fixation occurs, and where there is a large spring peak but only a small one at the end of stratification (Jónasson 1996, fig. 3), and Lake Esrom, where nitrogen fixation does take place, and where there is both a spring and a summer peak (see Fig. 12.5b), were selected as sites for study of this problem (Jónasson *et al.* 1992a; Jónasson 1996). Furthermore, planktonic food chains may be traced, via the zoobenthos, through to sedimentation in the hypolimnion of both lakes. However, for detailed zoobenthos studies Lake Esrom, with its steep oxygen gradients, and typical status as a eutrophic lake (Jónasson 1972; 1993; 1996), was chosen. Thingvallavatn was quantified at all ecological levels (Jónasson 1992b).

During the first two decades of the 20th century, August Thienemann (1913a, b, 1915a, b, 1918a, b, 1921), through his studies in *Maaren der Eifel* and northern Germany, elaborated a causal lake-type system. He discovered that as the fertility of their respective catchment areas was highly different, neighbouring lakes may possess an oxygen-rich or an oxygen-poor hypolimnion. The most interesting discovery was also that the profundal chironomid fauna differed between lakes. *Tanytarsus* larvae were abundant in oxygen-rich lakes and *Chironomus* larvae, red owing to their high haemoglobin content, were abundant in the hypolimnia of oxygen-poor bodies.

In 1921, Thienemann discovered that his oxygen-rich and oxygen-poor lake types were identical to the oligotrophic and eutrophic lake types of Naumann, which were based on plankton studies in southern Sweden (Naumann 1917). The Naumann nomenclature became accepted and is still in world-wide use. The zoobenthos system was later differentiated on a global basis (Brundin 1956). Thus, the classification of phytoplankton and of profundal zoobenthos came together.

12.5   ZONATION OF LAKES AND THE FOOD RESOURCES OF ZOOBENTHOS

Primary producers are the most important food source for zoobenthos. Particulate organic matter (POM), and possibly also dissolved organic matter (DOM), contribute to the food of zoobenthos (Pütter 1909; Wetzel *et al.* 1972). However, our interests here are concerned more with the food value for the zoobenthos of the various fractions of this organic matter and its seasonal and spatial variations in the different zones of a lake. The littoral zone is defined as extending from the water edge to the limit of rooted vegetation. It can, with advantage, be divided into the littoral surf zone down to approximately 0.5–2.0 m and the littoral macrophyte zone which continues the euphotic zone down the slope until light limits its existence (see also Chapter 11).

In the surf zone, the phytobenthos consists of epilithic algae with an overgrowth of diatoms, and of epipsammic diatoms. Macrophytes with an overgrowth of epiphytic algae characterise the lower parts of the euphotic zone. In both zones, mobile epipelic algae may cover the bottom. In the upper littoral zone, elodeids, such as the angiosperms *Potamogeton* and *Myriophyllum*, with rich epiphytic growth are abundant, but in the lower part, Characeans are more common and, owing to allelopathy, they are almost devoid of epiphytes. Finally, the sublittoral zone extends from the littoral to the depth limit of living molluscs, with the profundal zone below. The zoobenthic inhabitants of these two zones rely entirely upon organic matter produced in the littoral and the pelagic (Fig. 12.4) zones.

The thickness of the layer of phytoplankton primary production and the seasonal sequence of algal species influence the feeding of zoobenthos throughout the year (see Chapter 10). Figure 12.5a shows the sequence of algal production in the naturally eutrophic lake type. During late winter, under a rather thick layer of ice, considerable pro-

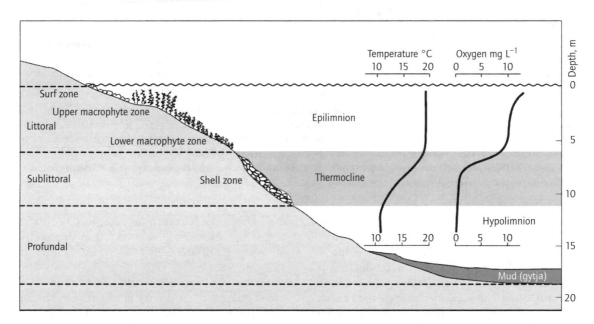

Fig. 12.4   Schematic transect of a dimictic eutrophic lake (Lake Esrom) showing zonation of the substrate and the relevant zoobenthos habitats, together with temperature and oxygen profiles and lake stratification. A late summer situation is illustrated.

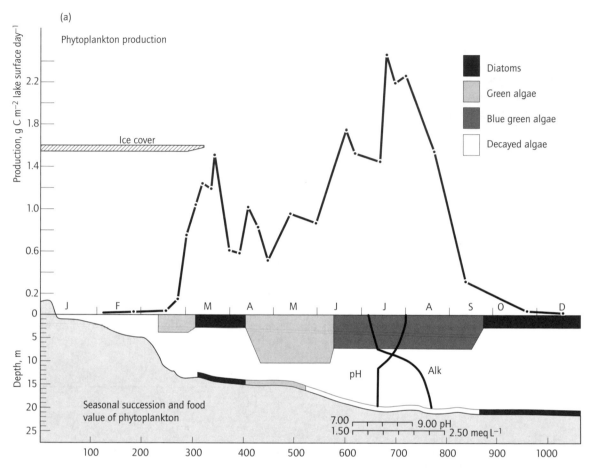

Fig. 12.5 (a) Top: seasonal variation in phytoplankton production per square metre of lake surface. Bottom: main seasonal occurrence of phytoplankton, approximate thickness of the primary producing layer in various seasons and the condition of algae arriving at the bottom. Vertical variation of pH and alkalinity at summer stagnation peak. Different shading represents prominent forms: Bacillariophyceae (diatoms), Chlorophyceae (green algae) and Cyanophyceae (blue-green algae). Algae arriving fresh at the bottom are shown by the density of the relevant shading. Decayed phytoplankton is without shading (Jónasson 1978).

duction by green algae takes place in a layer 4–5 m thick. Immediately after ice-break, a coffee-brown diatom plankton occurs. Transparency is low, so that during the spring maximum, production occurs in a thin layer only 2–3 m thick. Then, during May and June green algae produce less food but in a layer up to 10 m thick. During summer (July to September) the plankton consists of blue-green algae able to fix nitrogen. They build up the summer maximum and occupy an upper layer of 4–5 m

thickness. During autumn, diatoms proliferate again, but biomass and production decline (cf. Jónasson & Kristiansen 1967, table 4; Jónasson 1972, fig. 9).

Figures 12.4 and 12.5a also illustrate the combined effects of (i) primary production at the surface, (ii) vertical temperature gradient and (iii) oxygen concentration upon the condition of organic matter as it reaches the zoobenthos. Both during spring and autumn overturn, and during winter in

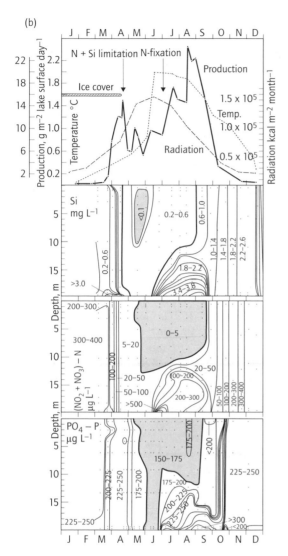

Fig. 12.5 (b) Seasonal variation of primary production m^{-2} surface of Lake Esrom in relation to temperature, monthly means of radiation and isopleths representing silicon, nitrogen and phosphorus content. Please note the seasonal changes in stratification (Data from Jónasson *et al.* 1974).

an ice-free lake, diatoms are rapidly distributed to the bottom, in a fresh condition. Then, as stratification proceeds, green algae are conveyed to the bottom, with increasing retardation. Blue-green algae circulate within the epilimnion, float to the surface at death and then sink slowly through the metalimnion. In the hypolimnion they decay and reach the zoobenthos in that condition (Likens & Hasler 1962; Lund *et al.* 1963; Likens & Ragotzkie 1966; Jónasson & Kristiansen 1967; Lastein 1976; Goedkoop & Johnson 1996; Kamp-Nielsen 1997). Thus, the value of organic matter as food for zoobenthos varies with season and lake rhythm.

A further facet of the production curve in Fig. 12.5a is explained in Fig. 12.5b. This shows that in the temperate, dimictic Lake Esrom, seasonal succession of production depends on light, temperature and the replenishment of nutrients (see Chapter 10). In contrast to the unimodal light and temperature curves, primary production forms a bimodal curve. Primary production is light-limited during winter, but during spring, because nutrients are now in excess and because algae are more efficient at low light intensities, it increases relatively more than light, despite the low temperature. The spring peak of diatoms occurs immediately after ice-break, but lasts only for a short while, until depletion of nitrogen and silicon limits production (Fig. 12.5b). Summer is characterised by a period of low production of green algae during June (owing to lack of nutrients), and a subsequent peak of blue-green algae. The summer maximum of primary production coincides with a rise in insolation and temperature. Calm, warm and sunny summers provide the most regular and highest maxima.

It is well known that blue-green algae, with few exceptions, occur only at the high temperatures recorded during summer and autumn. In Lake Esrom, during this season, owing to their ability to fix nitrogen, they constitute an important component of the standing crop (Jónasson & Kristiansen 1967), and form a broad peak of production which lasts from July to October. After autumn overturn, when nitrogen and silicon become available again, reduction in incoming light limits production. Note that PO$_4$-P is found in excess all through summer. Comparable observations were recorded in Windermere and Esthwaite Water, UK. For the period 1945 to 1960 a rapid decline in numbers of *Asterionella formosa* Hass. was linked to a decline in the concentration of SiO$_2$ to 500 μg L^{-1} (i.e. the equivalent of $230\,\mu g\,Si\,L^{-1}$ or below; Lund 1950, 1964; Lund *et al.* 1963), whilst nitrate remained abundant.

The concept of limiting nutrients is complex (see Chapter 4). Basically, the theoretical ratio of carbon:nitrogen:phosphorus in phytoplankton **by atomic weight** is equal to 100:14:2. That means that the N:P ratio (the **Redfield ratio**) is equal to 7:1. All three elements with an addition of micronutrients are a necessary condition for algal growth. If these conditions are not fulfilled, production curves show that algae are in fact not able to produce organic matter when the nitrogen concentration in the epilimnion is limited (Fig. 12.5b; Jónasson 1996, fig. 3).

The bimodal phytoplankton production curve, together with environmental parameters mentioned below, is thus an excellent tool with which to study interactions between the timing of growth and the survival of zoobenthos (i.e. pelagic–benthic coupling; see sections 12.10–12.12). In contrast, both in the surf zone and in the macrophyte littoral zones, phytobenthos production peaks at the height of summer stagnation (see Fig. 12.22). In the surface layers pH increases because of photosynthesis, and alkalinity declines because of precipitation of $CaCO_3$. In the hypolimnion, conditions are the reverse: pH is low and alkalinity high (Fig. 12.5a). These changes, together with low oxygen and high carbon dioxide concentration, critically affect respiration and survival of zoobenthos (see sections 12.11 and 12.17; Jónasson *et al.* 1974, figs 3–12; Walshe 1950, fig. 4).

12.6 HYDRODYNAMIC DISTRIBUTION OF FOOD AND OXYGEN – THE PELAGIC–BENTHIC COUPLING

Organic matter produced higher in the water column is transported by currents and sedimented along the sides of the basin, to the benefit of local zoobenthos. The occurrence and extent of stratification are very variable, and vitally influence lake benthos. During former times, the thermocline was considered a more or less static feature, but the

outstanding experiments of Mortimer, combined with his simultaneous field measurements of temperature profiles in Windermere, UK (Mortimer 1952, 1953, 1954, 1961, 1974), demonstrated the mechanisms of motion in thermoclines. During calm summers the thermocline is located close to the surface, but during windy seasons it lies much deeper (see Fig. 12.11). One of the biological effects of this contrast is to prolong the residence time of material in the epilimnion, including phytoplankton, particulate organic matter (POM) and bacteria (Lund *et al.* 1963; Ambühl 1969; Wetzel *et al.* 1972; Overbeck 1974, 1993).

The water movements demonstrated by Mortimer influence sedimentation and transport of material down the sides of the basin. Sedimentation of organic matter is much less in the epilimnion, increases markedly in the metalimnion and peaks in the hypolimnion (Lastein 1976; Kamp-Nielsen 1997). Comparing sedimentation rates with organic matter content of sediment indicates that very little organic matter accumulates in those parts of the lake shallower than 15 m, but that much is deposited below this depth (Fig. 12.6). These data indicate net removal of sediment, especially from the littoral, the sublittoral and the

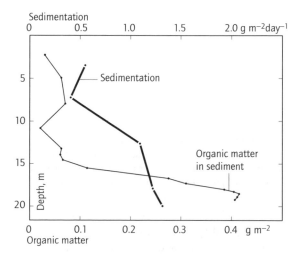

Fig. 12.6 Distribution of sedimentation of organic matter and organic matter in the sediment in Lake Esrom (Jónasson 1978).

upper profundal zones (Kamp-Nielsen 1997; cf. Berg 1938, p. 15). Organic matter which collects below 17 m in Lake Esrom contains approximately twice the amount of carbon and nitrogen of that in deposits above 15 m (Hargrave & Kamp-Nielsen 1977). This confirms Ohle's idea (1960) that a lake acts as a funnel, where organic deposits increase in thickness with depth.

In the profundal zone, sedimentation of suspended organic matter and phytoplankton during the year is also bimodal (Fig. 12.7). However, whilst the spring peak corresponds to that of phytoplankton production, the autumn peak is shown to be due to resuspension of superficial bottom material at autumn overturn, a period when phytoplankton production is declining. In addition to sedimented material, the zoobenthos is also resuspended during autumn (Fig. 12.7). Jónasson (1972, figs 39–42) showed that during winter the tubes of chironomids increase in size and become more solid, thus fixing the sediment and preventing resuspension during spring. This has been confirmed by underwater television (Ohle 1960; Lindegaard *et al.* 1997).

Internal waves generated in a three-layered Mortimer model after wind stress are illustrated in the upper part of Fig. 12.8. Below is shown the effect of the three-layered model applied to the survival and bathymetric distribution of zoobenthos of Lake Esrom. In the three-layer basin, the current on the downwind side of lakes transports large quantities of surface water rich in organic matter and oxygen down the slope. On the upwind side, upwelling metalimnic water occasionally mixes with surface water, adding nutrients and causing higher production (Thomas 1950; Jónasson & Mathiesen 1959).

The consequences for zoobenthos are twofold. Owing to downward transportation, a large accumulation of relatively fresh food and oxygen occurs in the sublittoral zone, and zoobenthos accumulates here in order to feed upon it. This is illustrated very clearly by the mollusc *Dreissena polymorpha*, the pelagic larvae of which (see Fig. 12.24) spread across the basin according to wave movements as shown by Mortimer. Downwind, the settled population is larger and extends to a

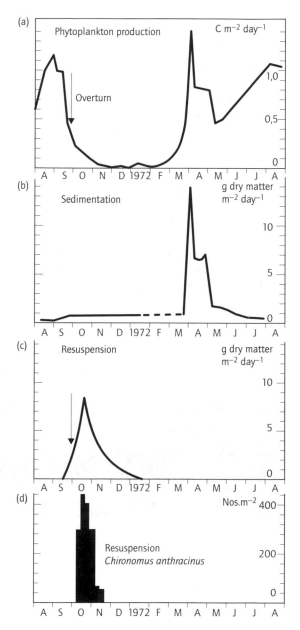

Fig. 12.7 Primary production and sedimentation in Lake Esrom. The solid curve in (a) connects dots representing measurements of phytoplankton primary production [g C m^{-2} day^{-1}]. True sedimentation [dry weight in g m^{-2} day^{-1}] is shown in (b), collected in sediment traps suspended 1.5 m above the bottom in central Lake Esrom. Resuspension [g dry weight m^{-2} day^{-1}] is shown in (c) as a solid curve. Sampling during the period of ice cover failed. Resuspension of *C. anthracinus* larvae [numbers m^{-2}] is given in (d). (Modified from Lastein 1976.)

greater depth, owing to distribution of food, oxygen and temperature. Upwind, the population is smaller and lies nearer to the surface, owing to the influence of hypolimnetic water (Fig. 12.8). Oxygen saturation in the sediment matrix of a downwind station at 0.4 m depth in Lake Esrom was 85% during maximum in winter, but only 40% at the upwind side. For summer, the corresponding values are 45 and 5% (Fig. 12.9). The consequence for the zoobenthos is a much higher abundance at the downwind side (Dall *et al.* 1990). Lundbeck (1928) observed similar distributions for north German lakes.

The conclusion of this section is that the sedimentation rate in the littoral zone is low and that the same is true for the sublittoral. However, owing to wave action, and the tilting of the thermocline, transport of organic matter takes place

down the sides of the basin in the sublittoral zone. The main bulk of sedimentary material sinks into the profundal zone after a prolonged stay in the metalimnion, and the deeper the lake, the more decomposed it becomes. In the absence of stratification, organic matter is evenly distributed throughout the lake volume, but not over the lake bottom (cf. Goedkoop & Johnson 1996).

12.7   LAKE RHYTHM AND ITS EFFECT ON THE SURVIVAL OF ZOOBENTHOS IN EUTROPHIC LAKES

In the preceding sections an analysis was conducted of the single factors which are well-known to be of importance for the life of zoobenthos in

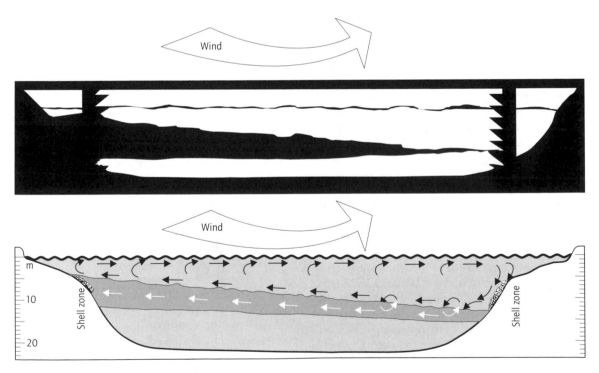

Fig. 12.8   Top: internal waves, after wind stress, in a model lake with three layers of differing density. (From Mortimer 1952.) Below: vertical distribution of *Dreissena polymorpha* in a three-layered basin with wind-induced currents (Lake Esrom; Jónasson 1978). Please note the wind direction at the lake surface. The metalimnion is shown by dark shading.

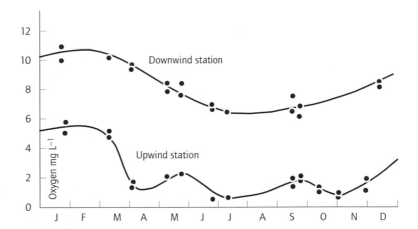

Fig. 12.9 Oxygen concentration at 5 and 10 cm depth in the sediment matrix at two sampling sites at 0.4 m depth in Lake Esrom. Top: the downwind station. Below: the upwind station (Dall *et al.* 1990).

lakes. Vertical variations in temperature, oxygen, pH and alkalinity are shown in Figs 12.4 and 12.5a (and see Fig. 12.11). However, a better understanding of zoobenthos ecology must be based on a synthesis of all data available. This section thus represents a necessary introduction to the community structure, and the bathymetric survival of zoobenthos, in a eutrophic lake.

Figure 12.10 illustrates the seasonal pattern of temperature at 0, 10, 15 and 20 m depth in Lake Esrom, Denmark, during a warm calm summer. The ecological implications of this pattern are clear: the fauna of the epilimnion experiences rising temperatures during spring and is favoured by high temperatures during the whole summer until the autumn decline. The annual amplitude of temperature is 0.4–22.2°C. Quite opposite conditions prevail for the fauna of the hypolimnion. A steep increase in temperature in spring coincides with that of the surface, but during summer the temperature remains low and peak values do not occur until the autumn overturn, when (owing to mixing) surface and bottom values coincide. Thus, the epilimnetic fauna lives under conditions of continuous high temperature during summer, whilst for the hypolimnetic fauna the period is divided into: (i) a rapid increase during spring, (ii) a more stable summer temperature at a low level and (iii) a short autumn peak.

The fauna of the hypolimnion is thus forced to survive the summer at temperatures lower by 10–12°C than those observed at the surface (amplitude about 0.4–16.5°C). It is as if the hypolimnetic fauna spends its life in a refrigerator. Figure 12.11 (bottom) clearly shows the adverse temperature conditions to which the bottom community in the deepest part of the lake is exposed. Paradoxically, during a cool, windy summer, this fauna is favoured by a higher temperature, while in contrast, a fine, warm, calm summer gives rise to lower temperatures in the hypolimnion (Jónasson 1972).

Figure 12.10 illustrates simultaneous seasonal variations in oxygen concentration of the profundal bottom layer during the ice-free period in two successive years, with a large difference in phytoplankton primary production. Oxygen concentration is high during spring but declines gradually to zero during summer stratification, provided that a stable thermocline persists. The higher the production, the earlier oxygen deficit occurs. During autumn overturn, oxygen concentration rises again and autumn values are high even though the water is not fully saturated (Jónasson 1972).

A combination of temperature and oxygen measurements also shows the following results for a transverse section of Lake Esrom (Figs 12.10 & 12.11). In general, the diagrams show that, at first, the thermocline forms a thick stratum. Wind and wave action then gradually force its upper limit

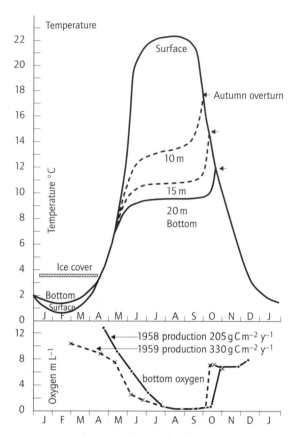

Fig. 12.10 Top: schematic diagram to illustrate temperature regime during a warm, sunny calm summer, based on actual values in Lake Esrom. Uniform heating occurs in spring, and at about 7.5–8°C stratification begins and conditions change. Once a vertical gradient becomes established, the upper layers exhibit increasing temperature, whilst the bottom layers stabilise. Surface temperature is about 22°C for almost 3 months, whilst the bottom remains at 9–10°C until November, i.e. for nearly 5 months. For several months the temperature of the bottom waters is thus c.10° lower than that of the surface. During such a summer, the epilimnion is thin, and the temperature of the intermediate layers (10 and 15 m) closely follows that of the lake bottom, with late mixing. Thus autumn overturn occurs late (October–November), and at a low temperature (12°C). Bottom: Seasonal variation in oxygen content of the profundal bottom layers of Lake Esrom influenced by differences in primary production (Jónasson 1972).

down to greater depth. This process continues throughout the summer, until the autumn overturn: when the thermocline recedes to greater depth and the volume of the epilimnion increases, whilst the volume of the metalimnion and the hypolimnion decrease. As a consequence, an increasing proportion of the lake basin warms. Autumn overturn then occurs at various times ranging from September to November. For a subsequent short period, bottom temperatures reach peak values, followed by a gradual decline. The annual cycle of surface and bottom temperatures shown in Fig. 12.10 illustrates these events clearly. Furthermore, the diagrams emphasise the slower increase in bottom temperature which takes place during summer, and the relatively short period of highest temperature during autumn, but with large differences between years.

If we compare temperature and oxygen measurements with the morphometry of the basin it is clear that for the months of June to November temperature over extensive areas of the lake bottom (70% of lake area) is lower than that at the surface, by 12°C. During the same period, oxygen concentrations in the sublittoral and in the profundal bottom layers (Fig. 12.11, top) are so low that only few animal species are able to survive (Jónasson 1972, figs 33–34). However, during windy years conditions are quite different, and the percentage of lake area exposed to warm, oxygenated water (Fig. 12.11, bottom) is materially higher.

The conditions described above give rise to other differences between the epi- and the hypolimnion, as shown for photosynthesis, alkalinity and pH. No doubt other contrasts (e.g. in H_2S) could be included. Two systems (those for oxygen and for carbon dioxide) are intimately linked and may largely determine the distribution of life in lakes. In a eutrophic alkaline body such as Lake Esrom, these processes are subject to regular fluctuations.

The functioning of essential biological processes linked to photosynthesis and the carbon dioxide cycle of the lake may thus be deduced from primary production data and environmental parameters. In the epilimnion, oxygen is produced in

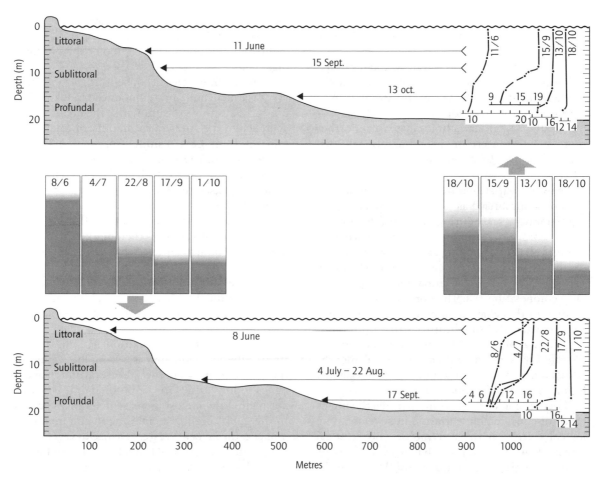

Fig. 12.11 Transect of Lake Esrom during a warm calm summer (top) and a windy cool one (bottom). To the right, bathymetric variation in temperature during summer is inserted. For oxygen values see Jónasson (1972). During the warm calm summer (top), oxygen concentration in the bottom layers is very low, or nil (from early July to November). Arrows show that by July only the water of the littoral zone, or the upper 5 m, is warm and well oxygenated, whilst the deeper layers are cold and oxygen deficient. By mid-September, warm, well oxygenated water only extends to 9 m depth, but this increases to 15 m a month later. During late October, mixing of layers accelerates, and by the end of the month extends to the lower profundal zone. In the bottom transect, wind and wave action caused early mixing of the upper layers. Large areas of the lake were exposed to warm and oxygenated water at least 2 months earlier than in the warm, calm summer (Jónasson 1972).

the euphotic zone during photosynthesis by phytoplankton and by macrophytic vegetation. Oxygen is respired by plants and the lake fauna, and is also used in decomposition of organic matter. Carbon dioxide is taken up during photosynthesis in the euphotic zone, and produced by respiration and oxidation processes. Aquatic plants are able to utilise half the carbon dioxide in the bicarbonate component (see Chapter 11), whilst at the same time calcium carbonate is formed as whitish coatings on macrophytic vegetation (**Seekreide**) and as a thick layer in the littoral zone. Curves of pH and alkalinity (Fig. 12.5a; Jónasson 1972, fig. 31) illustrate the inverse vertical relationship of oxygen and carbon dioxide.

In the hypolimnion, respiratory and decomposition processes predominate and carbon dioxide is regenerated. The $CaCO_3$ returns to solution according to the reaction $CaCO_3 + CO_2 + H_2O \rightarrow Ca^{2+} + 2[HCO_3]^-$ (i.e. alkalinity increases). Furthermore, pH declines from approximately 9.0–9.2 in the epilimnion to 7.4 in the hypolimnion (Fig. 12.5a).

Seasonal variations in the hypolimnion may thus be summarised as follows: temperature and food supply increase during spring and decline during autumn. An adequate concentration of oxygen is found all the year round, except during summer. At this time oxygen deficiency seriously affects survival of the bottom fauna and the processes of decomposition. Depth profiles discussed previously in this chapter show that life conditions vary enormously from the littoral to the profundal; e.g., during one year the upper profundal may be completely without oxygen for several months, whilst during another an abundance of oxygen will be found most of the time (Fig. 12.11).

This section thus has demonstrated the importance of considerable variation in several limnological factors, both bathymetric and seasonal. The zoological section which follows will provide many examples of the influence of the factors discussed above on survival, growth, feeding habits, life-histories, population dynamics and community structure of zoobenthos in a eutrophic lake.

12.8 BATHYMETRIC CHANGES IN COMMUNITY STRUCTURE IN A EUTROPHIC LAKE

In the ecosystem approach, benthic animals are divided into herbivores and detritivores. However, plant growth in temperate climates lasts only for 4–5 months, and, since they rely on macrophytes, the life-cycle of epiphytes is even shorter. How does the zoobenthos survive, unless they are able to change their feeding habits during the year? A single species must clearly rely upon different food items from season to season. Cummins (1973) has expressed similar views for running waters, and defined many subdivisions of food habits of lotic organisms.

At least 500 species of zoobenthic metazoa survive on the organic matter produced in lakes. Figure 12.12 attempts to summarise the quantitative approach to the study of species distribution, in a transect from Lake Esrom. Approximately 200 species live in the stony surf zone of eutrophic lakes (Ehrenberg 1957; Dall et al. 1984), a number which increases to 300 in the macrophyte zone where the habitat is more diverse and ecological separation is easier to carry out (Müller-Liebenau 1956). Species diversity is then markedly reduced in the sublittoral (to 50 species), and in the profundal to fewer than 20. Numbers per unit surface area peak in the littoral surf zone, and to a lesser extent in the profundal owing to a complete change in feeding, community structure and physiological adaptation.

However, vertical distribution of biomass is entirely different. This increases from low values in the surf zone, peaks in the sublittoral and then declines to low values in the profundal. It is worth noting that the largest increase in biomass occurs in the sublittoral, coinciding with the metalimnion, i.e. the layer of greatest change in vegetation, temperature, oxygen, pH, alkalinity and sedimentation. Figure 12.12 also shows that, in littoral vegetation, diel variation in oxygen saturation occurs. Littoral water is supersaturated during the afternoon (125%), but oxygen concentration then decreases gradually, and just before sun-rise in the surf zone saturation may be only 80% (Dall et al. 1984).

The following community and habitat approach attempts to relate the main feeding types of animals to the depth zones of the lake. Description at species level is reviewed by Macan (1974).

12.8.1 The littoral surf zone community

This habitat is exposed to marked seasonal variation. During the growing season, the stones are successively overgrown by diatoms, encrusting blue-green algae and Cladophora. During autumn

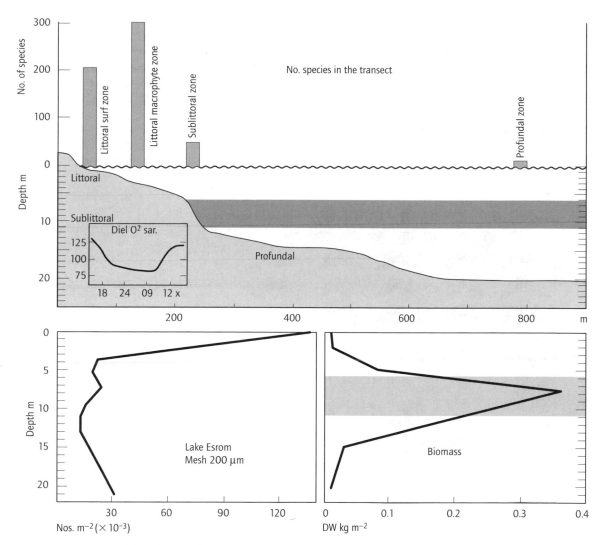

Fig. 12.12 Top: vertical distribution of species abundance in the transect of Lake Esrom, and diel variation of oxygen in the littoral surf zone. Bottom: quantitative vertical distribution of zoobenthos in the Lake Esrom sublittoral zone is shaded. (Data from Berg 1938; Jónasson 1972; Dall *et al.* 1984a).

these algae gradually decay. During winter, the habitat is ice-covered and the algal crust on the stones is removed by abrasion.

Strong wave action produces a special environment, with permanently high oxygen concentration, not only in the water column but also the underlying sedimentary matrix (Fig. 12.9; Dall *et al.* 1984, 1990). The water of the littoral surf zone is thus considered to be 'physiologically oxygen rich', in the same sense as running water, a property which minimises microgradients around the organisms. The zoobenthos of the surf zone community exhibit respiratory adaptation to this habitat. A linear relationship between oxygen uptake and oxygen concentration occurs (see section 12.11). Broadly this is the case for *Theodoxus fluviatilis*, *Gammarus lacustris*, *Erpobdella octoculata* and *Erpobdella testacea* (Walshe 1956; Lumbye 1958; Mann 1961).

Wesenberg-Lund (1908) was the first to draw attention to the interesting community of the surf zone, and its ecological characteristics on a stony shore (Fig. 12.13). Why are the animals of running-water and the surf-zone flat? The pioneer work of

Ambühl (1959) on current as an ecological factor explained the existence and the significance of the **boundary layer**, the morphological and physiological adaptation of animals to current. This community was quantified later by a number of authors (e.g. Ehrenberg 1957; Whiteside & Lindegaard 1982; Dall *et al.* 1984; Lindegaard & Dall 1988).

Ehrenberg (1957) expressed the view that food is the most important of all environmental factors. A

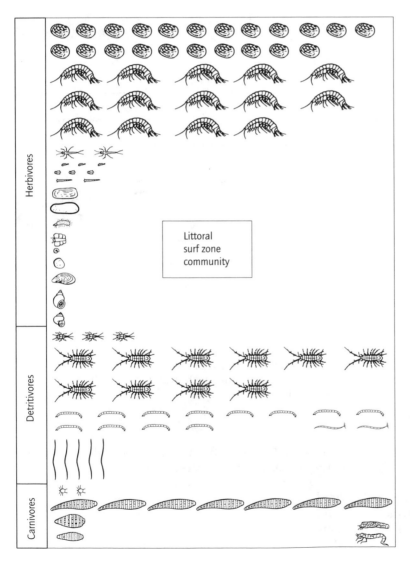

Fig. 12.13 The littoral surf zone community in Lake Esrom (nos 0.02 m^{-2}). Herbivores: *Theodoxus fluviatilis*, *Gammarus lacustris*, *Nemoura avicularis*, *Oulimnius tuberculatus*, *Micronecta poweri poweri*, *Haliplus* sp., *Agraylea multipunctata*, *Hydroptila tineoides*, *Tinodes waeneri*, *Goera pilosa*, *Valvata cristata*, *Sphaerium corneum*, *Dreissena polymorpha*, *Bithynia tentaculata*, *Valvata piscinalis*. Detritivores: *Caenis moesta*, *Asellus aquaticus*, Chironomidae, Ceratopogonidae, Oligochaeta. Carnivores: Hydracarina, *Erpobdella octoculata* and *Erpobdella testacea*, *Glossiphonia complanata*, *Helobdella stagnalis*, Leptoceridae, Polycentropidae (Jónasson 1978; data from Berg 1938).

Within the figure: Herbivores / Detritivores / Carnivores (left labels); Littoral surf zone community (central box).

number of herbivorous specialists graze the littoral surf zone, assisted by ingenious adaptive devices (such as suckers, flattening of the body or heavy shelters made of stones, big claws, cylindrical bodies or small size) for withstanding the surf. Figure 12.13 shows the surf zone community in Lake Esrom at 0.2 m depth. Species diversity is high, and Ehrenberg (1957) was able to show that, of 155 species found in this zone in north German lakes, 85 (or 55%) were also characteristic of lotic habitats. Herbivores make up the major part of the community and increase in numbers during summer as vegetative growth proceeds, but are then succeeded by carnivores. The emergence period for the main types of insects is typically late summer. Figure 12.18 (below) shows that the lower limit for many of the typical herbivorous species which graze in the surf zone is extremely sharp, and lies above 2 m depth.

12.8.2 The littoral macrophyte zone community

Plants in this zone overwinter as buds, which germinate at the beginning of May. By mid-June they are full-grown, but by the end of August their decay begins, and is completed by late November. Large elodeids (such as *Potamogeton* species) provide substrata for epiphytes which are, in many cases, the same as those found on the stones in the surf zone; some species of zoobenthos are also the same. Slight wave action occurs, and a few lotic species are also found here. Two habitats (the macrophytes and the mud between them) are utilised most. Diel variation in oxygen concentration also occurs.

Figure 12.14 shows that community structure in the littoral macrophyte zone in Lake Esrom differs markedly from that of the surf zone. The numerous algal grazers of the surf zone are considerably reduced, and their role is taken over by many detritivores living in the rich but thin mud layer. This statement is consistent with the earlier observations that organic matter settles in the littoral, and that continuous transport of fresh detritus provides food for a large community in an area with optimum facilities for habitat selection.

Müller-Liebenau (1956) listed 270 species of Metazoa in this zone, living mainly on *Potamogeton*. Of these, 41 are also found in the surf zone. It is worth noting that the grazers of this community, mainly snails (see Fig. 12.18), are distributed over a much greater depth range than those of the surf zone.

The richness of the fauna is emphasised by the presence of many predators, and the species of this zone exhibit the most diverse ecological strategies of any zoobenthos. An outstanding example is the prosobranch snail *Bithynia tentaculata*, which in the macrophyte zone is both a filtrator and a detritus feeder, but in the sublittoral is a detritus feeder only. Before the invasion of northwest Europe by the zebra mussel (*D. polymorpha*) from the Black Sea, together with *Valvata piscinalis*, it occupied the shell zone of eutrophic lakes in central and northern Europe (cf. Figs 12.14, 12.15 and 12.18; Tsikon-Lukanina 1965; Monakov 1972). Its respiratory adaptation to oxygen occurs in two steps which is characteristic of several species of the macrophyte habitat, e.g. *Helobdella stagnalis* and *Glossiphonia complanata* (see Mann 1961, fig. 20). In contrast, the habitat of the closely related filter feeder *Bithynia leachi* (see Fig. 12.18) occurs only in a narrow strip in the macrophyte belt and, interestingly, its respiration is more than twice that of *B. tentaculata* (Berg & Ockelmann 1959).

12.8.3 The sublittoral community

The sublittoral habitat is characterised by a shell zone with a thin mud layer. Owing to motion in the thermocline, especially during circulation, this is characterised by a rather high renewal rate. The food resources of the animals of the sublittoral, which are detrital feeders, are produced in the euphotic zone. For example, the mussel *D. polymorpha* is ideally suited for successful exploitation of suspended organic matter (Fig. 12.15), and the isopod *Asellus aquaticus* also utilises decaying organic matter. Of 50 species permanently living here, only five or six detritivores are of quantitative importance. Some of these (e.g. *Caenis moesta*, *B. tentaculata* and *V. piscinalis*) also occur

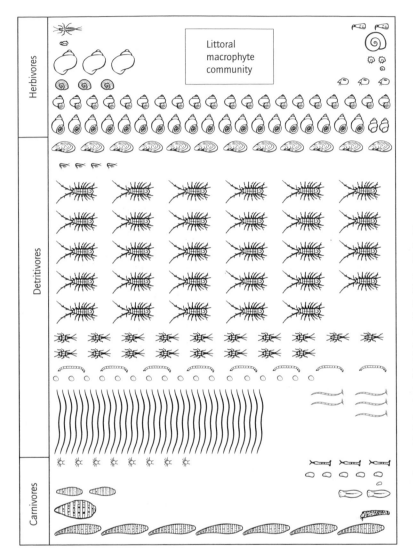

Fig. 12.14 The littoral macrophyte community in Lake Esrom (nos 0.02 m^{-2}). Herbivores: *Centroptilum luteolum*, *Micronecta poweri poweri*, *Lymnaea pereger*, *Gyraulus albus*, *Valvata piscinalis*, *Bithynia tentaculata*, *Oxyethira flavicornis*, *Planorbis planorbis*, *Gyraulus crista*, and *Eurycercus lamellatus*. Detritivores: *Dreissena polymorpha*, *Canthocamptus staphylinus*, *Asellus aquaticus*, *Caenis moesta*, Chironomidae, *Pisidium* sp., Oligochaeta and Ceratopogonidae. Carnivores: Hydracarina, *Helobdella stagnalis*, *Glossiphonia complanata*, *Erpobdella octoculata*, *Erpobdella testacea*, Tanypodinae, *Candona candida*, *Candona neglecta*, *Polycelis tenuis*, Leptoceridae (Jónasson 1978; data from Berg 1938).

in fresh detritus at 2 m but decline with increasing depth (Fig. 12.18), whilst numbers of *A. aquaticus* peak again in the sublittoral. *Dreissena polymorpha* is also abundant here, as is the amphipod *Pallasea quadrispinosa*, a well known detritivore from deep Scandinavian lakes.

Carnivorous invertebrates occur over a wide depth range, i.e. 0–15 m (see the species list in Figs 12.13 & 12.14). Apparently, the limiting factor is

low oxygen saturation during stratification periods. Only two predators, the migrating *Chaoborus flavicans* and the active swimmer *Procladius pectinatus*, occur deeper than this; the latter is present only during the circulation period.

12.8.4 The profundal community

The profundal environment is characterised by

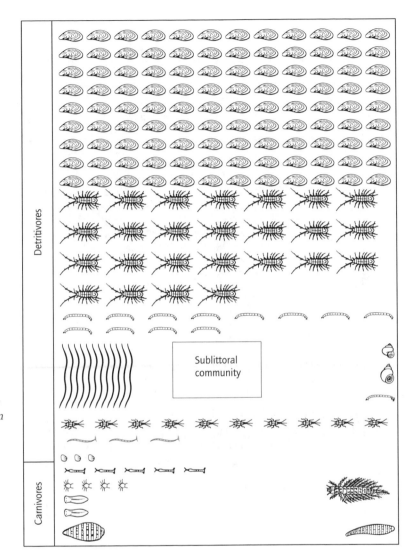

Fig. 12.15 The sublittoral community in Lake Esrom (nos 0.02 m^{-2}). Detritivores: *Dreissena polymorpha*, *Asellus aquaticus*, Chironomidae, Tubificidae, *Caenis moesta*, Ceratopogonidae, *Pisidium* sp. *Valvata piscinalis*, *Bithynia tentaculata*, and *Chironomus anthracinus*. Carnivores: Tanypodinae, *Hydracarina*, *Planaria torva*, *Polycelis tenuis*, *Glossiphonia complanata*, *Sialis lutaria*, *Erpobdella octoculata* (Jónasson 1978; data from Berg 1938).

sedimentation of degraded material during stratification periods and fresh material, including algae, during circulation. Annual temperature range (Fig. 12.10) is narrow. Pronounced yearly variation of oxygen saturation occurs, and the sediment of the profundal is always oxygen-depleted (see Fig. 12.28; Jónasson 1972, fig. 85). Of the 20 species recorded, three detritivores feed on mud deposits (Fig. 12.16). *Chironomus anthracinus* feeds from the surface and builds various types of tubes (see Figs 12.28 & 12.38). The tubificid *Potamothrix hammoniensis* is a subsurface feeder and transports mud to the surface from below (see Fig. 12.32). The bivalves *Pisidium casertanum* and *Pisidium subtruncatum* feed parallel to the surface and are apparently bacterial feeders (see Fig. 12.35; Jónasson 1972; Holopainen 1987; Meier-Brook 1969).

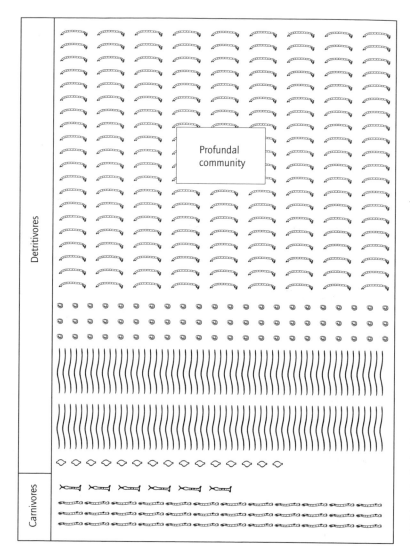

Profundal
community

Detritivores

Carnivores

Fig. 12.16   The profundal
community in Lake Esrom (nos
0.02 m^{-2}). Detritivores: *Chironomus
anthracinus*, *Pisidium casertanum*
and *Pisidium subtruncatum*,
Potamothrix hammoniensis and
cocoons. Carnivores: *Procladius
pectinatus*, *Chaoborus flavicans*
(Jónasson 1978; data from Jónasson
1972).

12.9   BATHYMETRIC DISTRIBUTION OF PHYTO- AND ZOOBENTHOS IN AN OLIGOTROPHIC LAKE

In a recent study, zonation of phytobenthos and the bathymetric distribution of zoobenthos were investigated together in order to connect food-web relationships in an oligotrophic lake (Fig. 12.17). Zonation of phytobenthos is comparable to that

found in the euphotic zone of eutrophic lakes, but the species present are entirely different. For zoobenthos, results from deeper parts of the lake are also different.

12.9.1   *Community structure of phytobenthos*

The benthic community of Thingvallavatn (Iceland) is a highly diverse system, with many species on each trophic level. The phytobenthic community comprises more than 150 species, of which 137 are diatoms. As to the most abundant macroscopic algae, the community is divided into four zones: the *Ulothrix* zone (from 0 to 1 m depth), the *Nostoc* zone (c.1 to c.2 m), the *Cladophora* zone (2–10 m) and the *Nitella* zone, from 10 m to approximately 25 m depth (G. St. Jónsson 1987; Kairesalo *et al.* 1992). In total, the phytobenthos covers c.38% of the area of the lake bottom (Fig. 12.17).

Growth and production of phytobenthos in spring is later than that of phytoplankton, and peaks during June–July. In autumn it continues after the phytoplankton peak, since the plants involved are adapted to low light intensities (Jónsson 1992; Kairesalo *et al.* 1992). Jónsson (1992) found that maximum net production (118 g C m^{-2} yr^{-1}) occurs at 1 m. Both above and below this depth (0.4 and 2 m) the value is c.85 g C m^{-2} year^{-1}. Lower production at 0.4 m is due to the presence of smaller biomass during spring, owing to abrasive action of waves and ice on the stones of the surf zone during winter (see section 12.8.1). The value for 2 m depth and below is attributed to lower irradiance. At 6 m production is about 55 g C m^{-2} year^{-1}. This value is

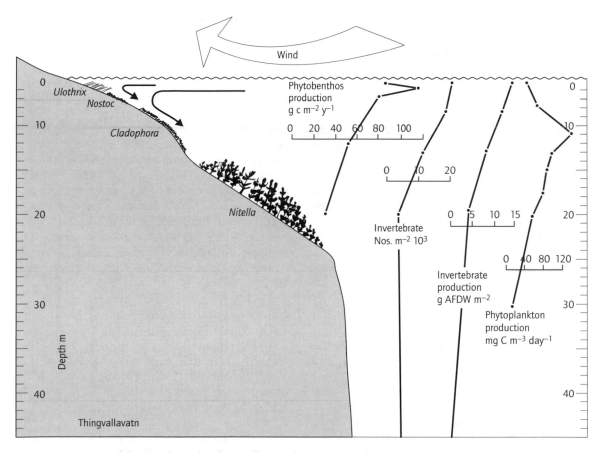

Fig. 12.17 Transect of oligotrophic Lake Thingvallavatn showing vertical distribution of *Ulothrix*, *Nostoc*, *Cladophora* and *Nitella* and the profundal zone. Bathymetric changes in production of phytobenthos and numbers and production of benthic invertebrates are also given. Vertical distribution of primary production of phytoplankton is shown on an occasion in May. (Data from Jónsson 1992; Lindegaard 1992a; Jónasson *et al.* 1992a.)

assumed to be valid throughout the lower stony littoral zone, down to a depth of 10 m.

Net production of *Nitella opaca* from 10 to 25 m is as low as 25–30 g C m^{-2} year^{-1} (Kairesalo *et al.* 1992). However, according to Kairesalo *et al.* (1989), planktonic algae contribute nearly 20% and epiphytic algae 40–50% of this value. Epipelic algae on the sediments were included as epiphytes, but were not quantified separately. It was not explicitly stated whether they originated from planktonic algae.

12.9.2 Community structure of zoobenthos

The macrozoobenthos is composed of about 60 taxa, including 24 Chironomidae and 16 species of Oligochaeta. The remaining groups are represented by a single species, or only a few. Zoobenthos exhibit a distinctive vertical distribution, and five communities are defined (see Fig. 12.19; Lindegaard 1992a, figs 30–33).

The major zoobenthic species present in the upper surf zone community of Thingvallavatn nearly all occur exclusively at lower latitudes as running-water species. A large variety of unicellular and filamentous epilithic algae provides the invertebrates of the surf zone with an excess of high-quality food, and in this habitat herbivorous grazers predominate (cf. Jónasson 1978). Abrasive action of waves diminishes the number of species, and accordingly interspecific competition.

Most interesting is a highly productive upper surf zone community of midges, extending from 0 to 1 m. *Thienemanniella* sp. cf. *morosa* (Edwards), *Rheocricotopus effusus* (Walker), *Eukiefferiella minor* (Edwards) and *Euorthocladius frigidus* (Zetterstedt) are all known from running-water localities as herbivorous and rheophilic species, and produce two generations per year. Other species of the upper surf-zone community are the trichopteran *Limnephilus affinis* Curtis and the Hydracarina *Sperchon glandulosa* Koenike. Eleven taxa reach high densities, evenly distributed throughout the upper 6 m.

The main species in this group is *Lymnaea pereger* L., which is a grazer, feeding mainly on the dense algal carpet on the stones. This snail possesses a 2-year life-cycle in Thingvallavatn, and constitutes the main food source for the bottom-dwelling Arctic char (*Salvelinus alpinus* (L.); Sandlund *et al.* 1992). Another nine species are associated with the stony littoral zone from 2–10 m depth. Enchytraeidae are the most prominent taxon in this group and are also one of those most obviously restricted to the depth range 2–6 m.

The enormous *Nitella opaca* belt serves as a sieve for sedimentary particles, and here in Thingvallavatn, between 10 and 25 m, *Chironomus islandicus* Kieffer lives as a filtrator within tubes 10 cm long, requiring 2 years for its development. Its life-cycle is the same in Mývatn (Iceland), but in this shallow lake it occurs all across the lake bottom, seston being readily available throughout owing to continuous resuspension and sedimentation.

The profundal zone of Thingvallavatn extends from 20 to 30 m, depending on the bottom contour, to a depth of 114 m. The sediment here consists partly of siliceous diatom exoskeletons with low organic content (Lastein 1983). Wind-blown eroded matter from surrounding areas and volcanic ash also contribute (Haflidason *et al.* 1992). Owing to the low energy content of the sediment, the profundal fauna is largely limited to five species of detritivorous tubificids, one of which (*Tubifex tubifex* Müller) is by far the most abundant. The predatory turbellarian *Otomesostoma auditivum* (Forest et Du Plessis) also occurs (Lindegaard 1992a).

Zoobenthos production is as much as 14.2 g ash-free dry weight (AFDW) m^{-2} year^{-1} in the surf zone, and 11.5 g AFDW m^{-2} year^{-1} in the zone between 2 and 6 m. It then declines to 7.9 g AFDW at 6–10 m and 3.6 g AFDW m^{-2} year^{-1} in the *Nitella* zone (Fig. 12.17; Lindegaard 1992a). On the basis of abundance, diversity of zoobenthos in Thingvallavatn is high. Nevertheless, three main species, the snail *L. pereger*, the oligochaete *T. tubifex* and the chironomid *Pseudodiamesa nivosa* (Goetghebuer), account for more than 50% of macrozoobenthos production, totalling 3.4 g AFDW m^{-2} year^{-1}. These calculations do not include the complete micro- and meiofauna elements, as sieves with mesh size 150 μm were used (see section 12.3), and sorting was carried out with 2.5 times magnification only. Therefore, Lindegaard (1992a) suggested (on the basis of literature) that calculated invertebrate production for Thingvallavatn should be increased by approximately 25% in order to allow for total zoobenthic fauna. Mean overall zoobenthos production is then estimated to be c.100 kJ m^{-2} year^{-1}.

12.10 FEEDING TYPES OF ZOOBENTHOS AND THEIR BATHYMETRIC DISTRIBUTION

This brief description of zoobenthos communities has shown us that in lakes it is mainly a few species with large populations that are responsible for degrading the organic matter produced. Direct utilisers of fresh green plants comprise relatively few, and very specialised, species. Some gastropods possess a cellulase and are known for their ability to break down the cellulose of fresh plants. Few other species (e.g. *Gammarus pulex*, *Gammarus lacustris* and the trichopteran *Neuroclipsis*) exhibit the same ability (Bjarnov 1972; Monk 1976). Jacobsen (1993) gives a list of the most important herbivorous invertebrates on macrophytes, which belong to various groups of insects, Malacostraca and snails. In most cases though, decomposition must be via bacteria, and detritus is ingested and utilised again and again (e.g. by *C. anthracinus* and *A. aquaticus*).

Figures 12.13–12.16 portray the community structure of zoobenthos in Lake Esrom, which develops from a community with high species diversity in the littoral, to highly specialised and dense communities in the sublittoral and profundal. Figures 12.18 & 12.19 summarise the vertical bathymetric distribution of selected species (primary consumers only) for Lake Esrom and Thingvallavatn. In both lakes there is a strikingly sharp delimitation in bathymetric distribution. Each species possesses its favourite bathymetric space. In accordance with their ecological role as hunters of prey, carnivores exhibit wider bathymetric distribution (Jónasson 1978).

The surf zone communities in both lakes are very limited in their bathymetric distribution, but species composition is entirely different. Interesting to note is the adaptation of small **bivoltine** chironomids (see section 12.18.5) and the stonefly *Capnia vidua* Klapalek to the stony surf zone, which is covered with the high-energy green alga *Ulothrix*. It seems that, in the deeper parts of the littoral, various species (e.g. the high numbers of the May-fly larva *Caenis* in the macrophytic zone) are specialised and feed on different types of detritus (Fig. 12.18).

Also of interest is the difference between the high concentrations of highly specialised detritivorous filtrators in the sublittoral zone of Lake Esrom (Fig. 12.18) and the corresponding zone of filtrators in Thingvallavatn. Here, the sublittoral corresponds to the *Nitella* belt, a habitat which constitutes a 'forest' of *Nitella*, with inhabitants such as the abundant filtrating *C. islandicus* and many species of sessile daphnids. The tubificid worm *Spirosperma ferox* also finds its main habitat here (Fig. 12.19).

Quantitatively, the *Nitella* community accounts for little (cf. Fig. 12.17), but ecologically it is important as the habitat of the most common fish, the three-spined stickleback *Gasterosteus aculeatus* L., which build their nests there and feed on the filtrators. The sediment of the profundal of oligotrophic lakes contains little energy and therefore supports only tubificids with low metabolism (see Fig. 12.43). In contrast, the profundal sediments of eutrophic lakes are high in energy content and support a highly specialised zoobenthos (e.g. *C. anthracinus*, *Pisidium hammoniensis* and *P. casertanum*), which can withstand the low oxygen tensions.

The conclusion of this section is that in Lake Esrom feeding types vary according to depth (Table 12.1). Herbivores are abundant in the surf zone, but in the macrophyte zone their importance is markedly reduced. Species diversity is high in the littoral, but in the sublittoral and the profundal the reverse applies, detritivores predominate entirely and species diversity is reduced. However, it should be remembered that during circulation detritivores receive fresh food.

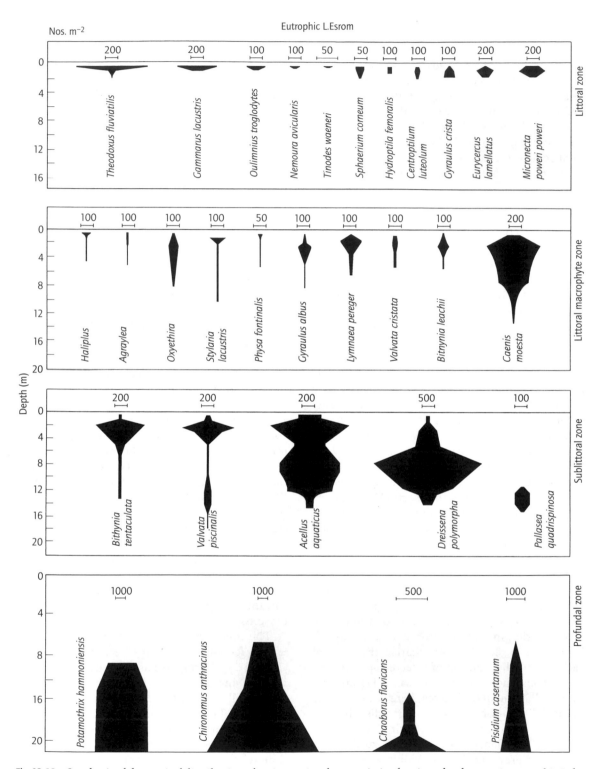

Fig. 12.18 Synthesis of the vertical distribution of main species characteristic of various depth zones in eutrophic Lake Esrom: herbivorous surf zone grazers, macrophyte and periphyton grazers as well as detritus feeders (*Caenis*), main filter feeders in the sublittoral zone, and major deposit feeders in the profundal zone (nos m^{-2}). (Modified from Jónasson 1978. Data from Berg (1938) and Jónasson (1972, 1996).)

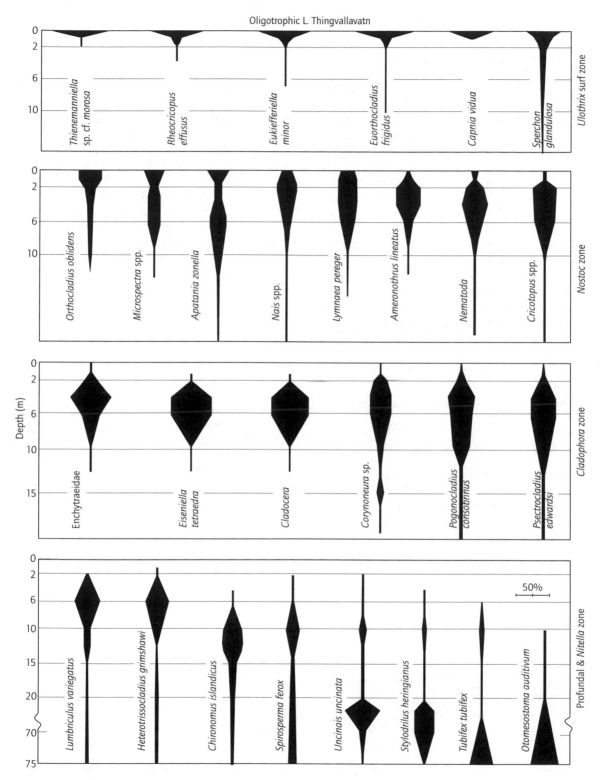

Fig. 12.19 Synthesis of the vertical distribution of main species characteristic of various depth zones in oligotrophic Lake Thingvallavatn: herbivorous surf zone grazers in the *Ulothrix* belt, herbivorous grazers in the *Nostoc* and surf zones, detritus feeders and filtrators in the *Cladophora* belt, filtrators and deposit feeders in the *Nitella* and profundal zones (nos m⁻²). (Data from Lindegaard 1992a.)

Carnivores constitute approximately 10% of the total fauna, except in the sublittoral (1%). Table 12.1 shows that it is profitable to be a filtrator, especially in the sublittoral, but also in the littoral. This is probably related to the factors mentioned previously (that currents continuously transport organic matter all year round into these zones, and organic matter is trapped in the sublittoral). In the profundal this takes place only during circulation periods, and therefore deposit feeding by detritivores is the main zoobenthic feeding process.

In Thingvallavatn, variation of species with depth is not the same as that of temperate eutrophic lakes (e.g. Lake Esrom; Jónasson 1978). However, in terms of **feeding types**, variations are much the same. Thus, as a percentage of biomass, feeding types vary from roughly equal amounts of grazers and filtrators in the littoral to almost 100% filtrators in the sublittoral, and about 75–90% detritivorous deposit feeders in the profundal (Lindegaard 1992a).

12.11 RESPIRATORY ADAPTATION OF ZOOBENTHOS

Physiological aspects of oxygen uptake by various species of zoobenthos are undoubtedly among the most effective factors determining concentrations

of animals in a lake transect. The main idea of this section is therefore to relate respiratory adaptation of a few species characteristic of the Lake Esrom transect to actual oxygen concentrations of the lake and to their choice of habitat and feeding. Table 12.2 illustrates this adaptation and its critical limits. Different responses of various species to low oxygen concentration may account for their depth distribution, and it is possible to relate such differences to the habitat of the animals. Figures 12.20 & 12.21 illustrate the adaptation of some characteristic and quantitatively important species to oxygen availability.

The distribution of *T. fluviatilis*, a herbivorous prosobranch grazer characteristic of the surf zone, follows an approximately linear relation between oxygen uptake and oxygen concentration of lakewater, and declines as oxygen supply decreases

Table 12.1 Feeding types in relation to depth as a percentage of total weight in Lake Esrom, Denmark (Jónasson 1978).

	Herbivorous grazers	Filtrators	Detritivorous deposit feeders
Littoral	40	46	7
Sublittoral		99	
Profundal		16	72

Table 12.2 Examples of respiratory adaptations in littoral, sublittoral and profundal zones. The only conformer is *Theodoxus fluviatilis* living in its surf zone habitat. Other species are regulators with varying degrees of regulation. (Modified from Jónasson 1996.)

Zone	Species	Critical limits (% air saturation)	Bathymetric distribution depth (m)
Littoral	*Theodoxus fluviatilis*	Dependent	0–2
	Helobdella stagnalis	25–65	1.5–4.5
	Valvata piscinalis	15–75	2 and 10–14
Sublittoral	*Chironomus plumosus*	8	10–15
	Dreissena polymorpha	36	0–12
	Psammoryctides barbatus	19	
Profundal	*Pisidium casertanum*	21.5–29.0	8–22
	Chironomus anthracinus	25	9–22
	Potamothrix hammoniensis	9.5	0–22

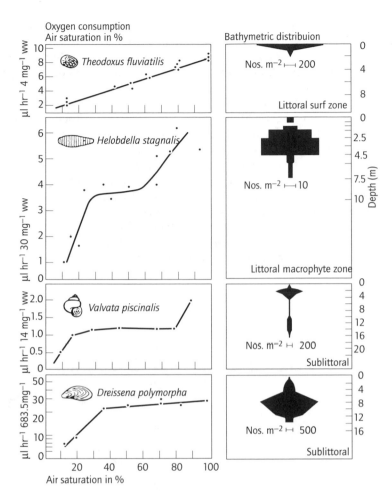

Fig. 12.20 Oxygen consumption in relation to oxygen content for species living in the surf zone, the littoral macrophyte zone and the sublittoral zone of Lake Esrom, together with their bathymetric distribution (Jónasson 1996; data from Berg 1938; Lumbye 1958; Mann 1956; 1961; Berg & Ockelmann 1959; Bennike 1943; Jónasson 1978).

(Fig. 12.20). Comparison with the diel oxygen saturation in the surf zone of Lake Esrom illustrates a variation from 80% saturation at night to 125% during the afternoon (Fig. 12.12). *Theodoxus fluviatilis* thus is a **conformer**, which lives in an oxygen saturated environment with no need to adapt to low oxygen.

In contrast, *H. stagnalis*, the most abundant leech of the littoral macrophyte zone, lives in eutrophic lakes between 1.5 and 4.5 m depth, where diel oxygen concentration may be low. *Helobdella stagnalis* is therefore adapted to maintain a constant rate of oxygen consumption over a broader range, from 25 to 65% saturation (Fig. 12.20). Ac-

cording to data calculated from observations of an *Elodea canadensis* vegetation mat (Buscemi, 1958), lower diel oxygen saturation at this depth is assumed to be c.30 to 50%.

However, the prosobranch *V. piscinalis* exhibits **two** modes of nutrition, and correspondingly two maxima of bathymetric distribution. At 2 m it occurs as a scraper of epiphytes on littoral macrophytic vegetation, but it is also able to extend its habitat to the sublittoral zone at 10–15 m depth, where it acts as a filtrator (Fig. 12.20). Prior to the invasion of northern Europe by the zebra mussel *D. polymorpha*, this species was the main component of the shell zone in north European lakes. The respiratory adaptation of *Valvata* suits its biology, because it possesses two critical limits, the first at 75% air saturation (which corresponds to its habitat at 2 m depth) and another at 15% air saturation (which fits its sublittoral habitat). Between these two limits, respiration is independent of oxygen concentration (Fig. 12.20).

This pattern (of an initial decline in oxygen consumption near full air saturation, followed by an oxygen range where respiration is independent, and then a final decline at lower concentrations) seems to be a general feature of zoobenthos living in the lower littoral zone. Species with this type of respiratory adaptation can withstand diel oxygen variation, but are not able to survive below the littoral zone (Berg 1952; Lumbye 1958; Berg &

Ockelmann 1959; Mann 1961, 1962; Toman & Dall 1998).

In the sublittoral or profundal zones, variation in oxygen concentration is only seasonal. In contrast to the littoral zone, no diel variation occurs. Species living in sublittoral and profundal habitats therefore possess a still wider range of independent respiration, until a critical limit is reached. For example, *D. polymorpha* is highly adapted to its habitat, since its respiration declines only moderately (26%) from full saturation to its critical limit at 36% air saturation, and then falls rapidly (Fig. 12.20).

Figure 12.21 compares the respiratory characteristics of two chironomid (*Chironomus plumosus* L. and *C. anthracinus* Zett.) and two tubificid

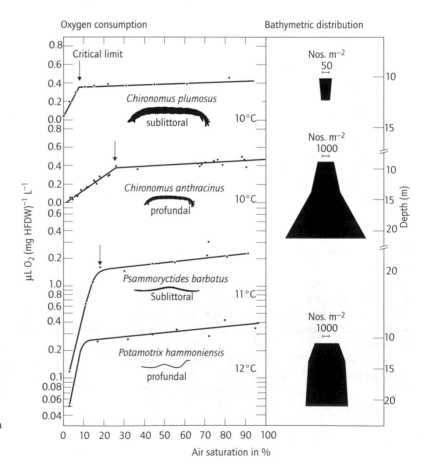

Fig. 12.21 Oxygen consumption in relation to oxygen content for closely related sublittoral and profundal species in Lake Esrom, together with their bathymetric distribution. Arrows indicate critical limits of respiration. (Data from Berg *et al.* 1962; Jónasson 1972; Hamburger *et al.* 1994.)

species (*Psammoryctides barbatus* and *P. hammoniensis*). Bathymetric distribution is shown to the right. Critical limits of respiration for *C. anthracinus* and *P. hammoniensis* have recently been verified (Hamburger *et al.* 1994, 1997, 1998) and for *C. plumosus* new unpublished data have been provided (courtesy of the late P.C. Dall). The respiration rate of *C. plumosus*, which lives in the sublittoral, is almost unchanged from full air saturation down to 8%, when the critical limit is reached and respiration declines. The range of the profundal *C. anthracinus* is nearly independent until air saturation reaches 25%, but then declines. The critical limits of the sublittoral tubificid *P. barbatus* and the profundal *P. hammoniensis* are, respectively, 19 and 9.5%. *Chironomus plumosus* is only found between 10 and 15 m depth, whilst the limits of *P. barbatus* are unknown. It was shown above that profundal species are able to maintain high respiratory rates, even at low oxygen tensions in the profundal. Respiration declined by c.10–15% from full air saturation, down to their critical limits.

12.12 GROWTH OF ZOOBENTHOS IN RELATION TO BATHYMETRIC DISTRIBUTION

12.12.1 Growth and life-histories in the littoral surf zone

The littoral phytobenthos is divided into the psammolittoral and litholittoral. The fraction of lake area covered by these communities is usually small (e.g. 10% in Lake Esrom). The production curve (Fig. 12.22) is unimodal, with a peak in July–August. Epilithic littoral production in the surf zone (June–October) amounts to 437 g C m^{-2} year^{-1}, which implies an annual production of at least c.500 g C m^{-2} year^{-1} (Dall *et al.* 1984a). Psammolittoral surf-zone production, also unimodal, constitutes 143 g C m^{-2} year^{-1}, or about one-third of epilithic periphyton production (Hunding 1971). Surf-zone production is limited by low light intensities, ice-cover and low temperature during winter. Figure 12.12 describes the diel oxygen

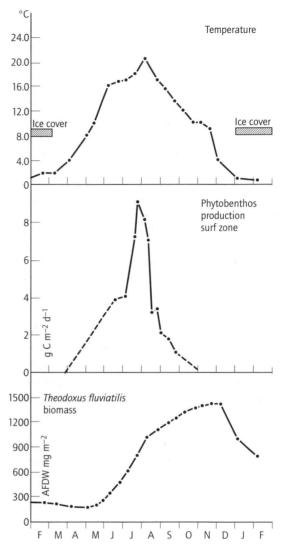

Fig. 12.22 Seasonal variation in littoral surf-zone temperature, primary production of epilithic periphyton from June to Sept, and seasonal accumulation of *Theodoxus fluviatilis* biomass in Lake Esrom. (Modified from Kirkegaard (1980) and Dall *et al.* (1984a).)

regime (Dall *et al.* 1984a). Measurements of macrophytic production are still inadequate but, compared with phytoplankton, production values are low, especially in relation to lake area (Jónasson *et al.* 1990; Kairesalo *et al.* 1992). Light is suboptimal and is thus a limiting factor.

The period of plant growth in temperate zones extends over 5–6 months, and growth in littoral organisms proceeds at high temperatures during high primary production. This involves high feeding rates and the possibility of optimum growth during summer. Figure 12.22 clearly illustrates the relationship between temperature, production and seasonal variation in biomass of the herbivorous grazer *T. fluviatilis*. Increase in biomass proceeds from June to December. It then declines, and is delayed in relation to production of phytobenthos. Growth ceases during winter, when light is limiting for production, and the lake may be covered by ice. The result is that *T. fluviatilis* exhibits slow growth, with high winter mortality, and a 4-year life-cycle (as described by its survivorship curve; Fig. 12.23).

12.12.2 Growth and life-histories in the sublittoral zone

Interaction between environment, food and growth become more complicated for detritivores in the sublittoral than for herbivores in the littoral. The predominant animal of the sublittoral of Lake Esrom is the zebra mussel *D. polymorpha* Pallas, which accounts for 99% of the fauna by weight. It is a filtrator, and its life-cycle contains both a planktonic and a benthic phase. The interesting example of its planktonic habit, and selection of food, is shown in Fig. 12.24. Its pelagic, veliger larvae are produced in July and settle in late August. They are filter feeders and feed on phytoplankton during the period between spring and autumn diatom maxima (Fig. 12.24; Jørgensen 1971; Jónasson *et al.* 1974). The excellent fit between temperature and oxygen stratification and phytoplankton production in the epilimnion controls the spatial distribution of larvae well above the thermocline in their search for food.

On hatching, larvae settle out all over the lake bottom, but only survive in the littoral and the sublittoral. They then aggregate in the sublittoral, where they represent the main filtrators present at 8 m depth. In former times, they used to migrate towards deeper water during winter (Berg 1938, Fig. 179). However, increasing hypolimnetic oxygen deficit in this lake has, in the long term, changed the depth distribution of this species (see Figs 12.20, 12.45 & 12.46). During 1933–1934 it mainly occurred at 8 m depth, but it was also found down to 14 m (Berg 1938). By 1971 (Olsen 1972) its main zone of occurrence had moved to 6.5 m, and its lowest abundance was at 12 m owing to asphyxiation of individuals between 12 and 14 m (Fig. 12.25). Figure 12.26 shows that growth of *D. polymorpha* is strictly temperature dependent, and that it takes the population 4 years to reach a length of 22 mm (Dall & Hamburger 1996).

12.12.3 Growth and life-histories in the profundal zone

Life-histories and population dynamics of animals become more complicated the deeper in the lake they live. This section describes adaptations of detritivorous surface or subsurface feeders to the solid–liquid interface of profundal mud, their interactions with their environment, their food resources and their growth.

The organic matter present in lake sediment (**gyttja**) is a loose material in which lives the profundal bottom fauna. At the surface, approximately 95% of this sediment consists of water and 5% of dry matter. Deeper in the sediment, water content

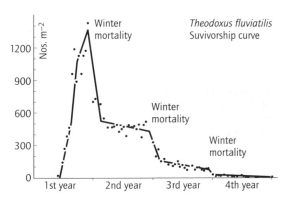

Fig. 12.23 Survivorship curve for *Theodoxus fluviatilis* through 4 years in Lake Esrom. (Modified from Kirkegaard 1980.)

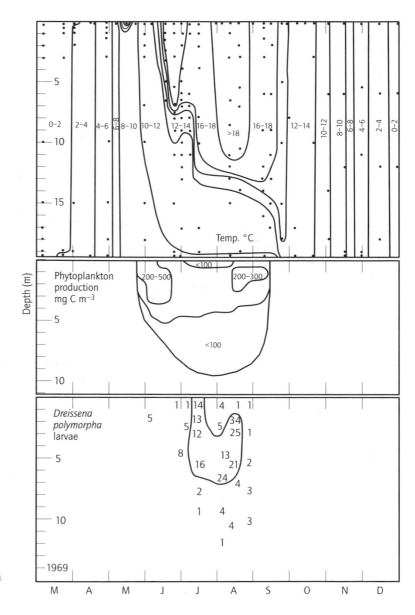

Fig. 12.24 Temperature, phytoplankton production (mg C m^{-3} day^{-1}) and spatial distribution of veliger larvae of *Dreissena polymorpha* in Lake Esrom (in numbers per litre) (Jónasson 1996; data from Jørgensen 1971; Jónasson *et al.* 1974).

declines. At 40 cm in Lake Esrom, no further decomposition seems to occur, and the dry matter is mainly composed of **marl**. Percentage composition is now 80% water and 20% dry matter. Figure 12.27 shows that dry matter content increases from 5% at the surface to 15% at 15 cm depth. Loss on ignition (approximately equivalent to organic matter) varies inversely with depth in the upper 15 cm, with dry matter decreasing from 37 to 25%, to remain nearly constant below. Nitrogen concentration is greatest at the surface and declines slightly with depth.

Animals burrow into the mud and build their cases and tube systems within it. They feed either

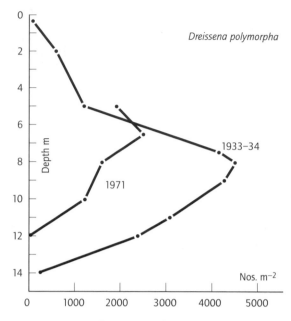

Fig. 12.25 Upward migration of *Dreissena polymorpha* in Lake Esrom from 1933–34 to 1971 (Jónasson 1996).

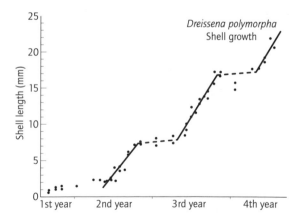

Fig. 12.26 Growth of the *Dreissena polymorpha* population at 0.4 m depth in Lake Esrom. Four year-classes are shown (Dall & Hamburger 1996).

from the surface (e.g. Chironomidae) or on subsurface bacteria (e.g. Tubificidae and *Pisidium*). In doing so they mix the sediment, an activity which takes place mainly in the upper 15 cm (Fig. 12.27).

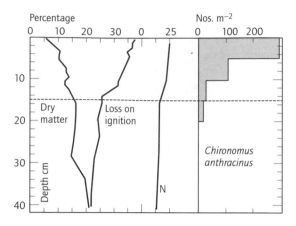

Fig. 12.27 Left: vertical variation of dry matter, loss on ignition and nitrogen concentration of dry sediment in Lake Esrom. Right: vertical distribution of *Chironomus anthracinus* in the profundal sediment. (Modified from Berg 1938; Jónasson 1972, 1978.)

In this layer animals feed on and break down organic material. Their faeces are often water soluble and thus revert to the liquid phase. The effect of the fauna is shown in the change in dry matter and loss on ignition in the uppermost 15 cm of the sediment just described.

12.12.3.1 The surface deposit feeder *Chironomus anthracinus*

The detritivore *C. anthracinus* feeds at the mud surface (Jónasson 1972, figs 39–42), living in vertical, conical tubes which open at the surface (Fig. 12.28). Three millimetres below the surface the oxygen content is zero. Thus, any oxygen used by the larvae must be brought into the sediment by irrigation currents. These are easily observed in experiments. Growth of *C. anthracinus* is limited to two very short periods, one during the spring phytoplankton maximum when the water is oxygen rich, the other after the autumn overturn when oxygen is again available but food production is declining (Fig. 12.29). Decline in oxygen concentration in the summer hypolimnion halts growth (limit $<1 \, \text{mg} \, O_2 \, L^{-1}$), which continues in an ice-free lake during winter but ceases under ice cover.

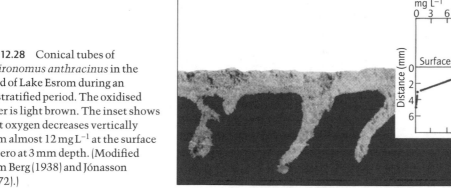

Fig. 12.28 Conical tubes of *Chironomus anthracinus* in the mud of Lake Esrom during an unstratified period. The oxidised layer is light brown. The inset shows that oxygen decreases vertically from almost 12 mg L^{-1} at the surface to zero at 3 mm depth. (Modified from Berg (1938) and Jónasson (1972).)

Oxygen and food are therefore the factors governing growth. Seasonal variations in water, nitrogen, fat and glycogen content of *C. anthracinus* larvae also follow environmental changes, and allow an estimate of the accumulation of organic matter during the larval cycle to be made (Fig. 12.30). The first three instars of *C. anthracinus* contain very high water content (95%) during summer stagnation and for the first fortnight after autumn overturn (Jónasson 1972, table 16). Figure 12.30 also shows that the fourth instar begins with 95% water content, but that this falls to 91% during late autumn and to 89% from December to March. Increase of organic matter during the spring maximum further reduces water content to 81%. During summer stagnation of the second year of the larval cycle, water content again increases, to reach a peak of 87% during the following autumn. A gradual and final decrease to 83% during winter, and particularly during spring, repeats the events of the preceding year.

The protein content of very young larvae is 75% of dry weight (Fig. 12.30). Consequently, they consist of 98.5% protein and water. Decrease in water and protein content means a relative increase in other components. At the spring maximum of phytoplankton, water content is reduced to 82.6% and protein to 46% of dry weight. For comparison, we are only able to account for about 46% of dry weight instead of 75% at the beginning of the larval life-cycle. A study of these events in relation to primary production shows that, during the spring maximum of primary production, water content declines from 90 to 83% and protein from 63 to 46% of dry weight, while fat increases from 10 to 19% (Fig. 12.30; Jónasson 1972, tables 16 & 24).

Glycogen plays an important role during the ontogeny of first-year larvae of *C. anthracinus* (Fig. 12.30), and large fluctuations occur. During periods of spring growth, glycogen concentrations rise to a maximum of 25% of dry weight prior to stratification. During microxic conditions most metabolism is aerobic, but glycogen is used during periods of anaerobic metabolism without traces of oxygen. Concentrations fall to about 5% at autumn overturn. In an ice-free lake a slight increase of glycogen occurs during winter, but during the spring phytoplankton maximum it accumulates again, and prior to pupation the concentration reaches 28% (Hamburger *et al.* 1995, 1996). Protein, fat and glycogen then account for 93–95% of dry weight.

Periods with high primary production, especially spring, are therefore important for metamorphosis and emergence, and influence the population dynamics of *C. anthracinus*. Further, it is clear that glycogen is the main energy source during anaerobic metabolism (Hamburger *et al.*

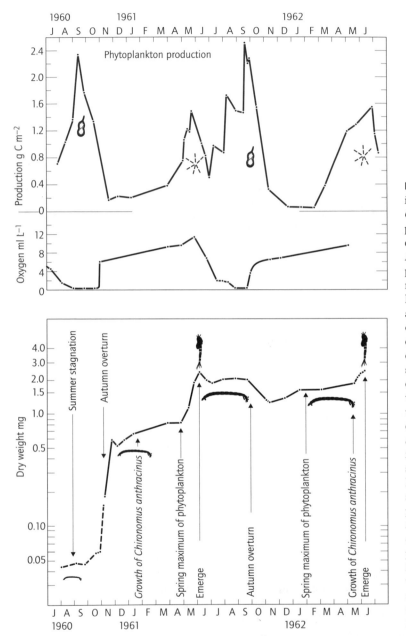

Fig. 12.29 Top: seasonal changes in factors controlling growth of *Chironomus anthracinus*, phytoplankton production and oxygen concentration in Lake Esrom. A spring (April) and summer (August) peak is easily recognised. Decline is rapid during autumn, and values are low in winter. Oxygen concentration at the bottom decreases to zero during summer stratification. In other periods, there is sufficient oxygen, even below ice. The silhouettes show the main types of algae. Bottom: variation in dry weight of *C. anthracinus* larvae through a 2-year period in relation to oxygen and production. *Chironomus anthracinus* occurs in four larval instars, and fourth instar weights are shown. Many larvae exhibit a 2-year life-cycle, but part of the population show a 1-year cycle. Growth takes place in limited periods after the autumn overturn of the first year, and during spring maximum of phytoplankton, prior to emergence in May. Second-year larvae increase in weight in June, lose weight after autumn overturn, and regain weight during spring maximum (Jónasson 1978).

1996). Timing of periods of growth, alternating with intervals with or without negative growth, crucially affects the population dynamics of *C. anthracinus* in Lake Esrom. For second-year larvae these fluctuations are less conspicuous.

At 21 m depth, a 2-year life-cycle is usually observed in populations of *C. anthracinus*, but some emerge after 1 year. In contrast, at 11 m and at 14 m, periods of growth and of exposure to predation are of long duration. This leads to the

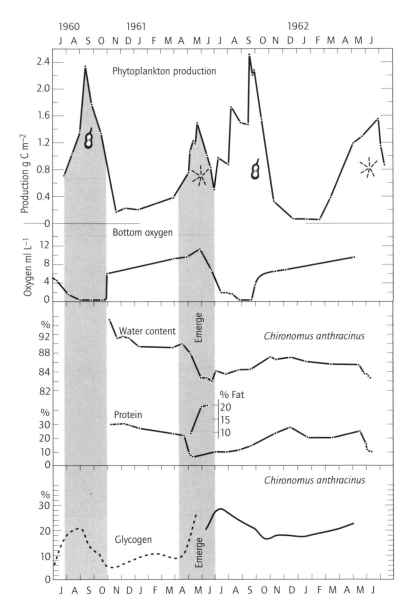

Fig. 12.30 Seasonal succession of primary production $(gCm^{-2}day^{-1})$, seasonal variation of bottom oxygen $(mgO_2 L^{-1})$ and water content (% wet wt), protein, fat and glycogen (% dry wt) of the surface feeding *Chironomus anthracinus* in Lake Esrom. The shaded areas represent spring maximum of diatoms and summer maximum of nitrogen fixing blue-green algae (Jónasson 1972, 1978; Hamburger *et al.* 1995).

presence of lower numbers of larvae per square metre and, as a consequence, higher larval weight (Jónasson 1972, figs 52–57). At 11 m, larvae are consistently able to complete their life-cycle in 1 year, owing to better access to oxygen and thus longer periods of feeding (Fig. 12.31). The transitional zone between 1- and 2-year life-cycles lies between 14 and 17 m, since at the latter depth recruitment of larvae is not successful every year. At 14 m, larvae succeed in yearly recruitment in some but not all years. Thus, at 17 and at 20 m depth a 2-year life-cycle predominates. Population size also increases with depth, primarily as a result of reduced predation.

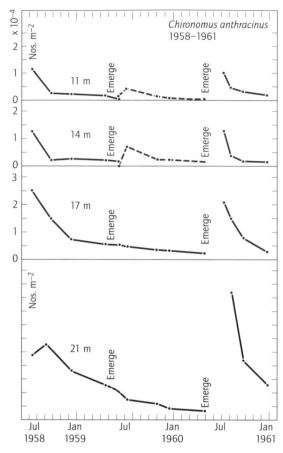

Fig. 12.31   Seasonal fluctuations in numbers of *Chironomus anthracinus* at different depths along a Lake Esrom transect during the period 1958–1961. One-year life-cycles occur in shallow water, at 11 m and 14 m, and 2-year cycles at 17 and 21 m, beginning in even numbered years. At 11 and 14 m, the life-cycle varies in length owing to changes in population size and oxygen supply at different depths (Jónasson 1972).

12.12.3.2   The subsurface deposit feeders Potamothrix hammoniensis and Pisidium casertanum

Tubificid worms (*P. hammoniensis* Michaelsen) occur in great numbers in the profundal of Lake Esrom (Figs 12.32–12.34). In effect they form a tubificid monoculture and transport degraded organic material from 5–10 cm depth within the sediment to the sediment surface, through vertical tubes (Fig. 12.32). They thus convert considerable amounts of organic matter to living material. Further, they are the most numerous and important bacterial feeders of the profundal, apparently exploiting bacterial plates within the sediment (see section 12.19).

Figure 12.33 shows that in *P. hammoniensis* glycogen accumulates during spring when it is needed for reproduction (Hamburger *et al.* 2000). Production of cocoons is limited to periods of high glycogen content, and glycogen is fully used during anaerobiosis in summer. This seasonal succession is the same as that described for the surface feeder *C. anthracinus* in Fig. 12.30, but, in contrast, in *P. hammoniensis* accumulation of organic matter in spring is delayed. The spring maximum of phytoplankton occurs in April–May, but the increase in glycogen is delayed to May–June, and the production of cocoons peaks in July (Fig. 12.33). Correspondingly, water content declines, with accumulation of organic matter from 90% in April to 83% in July (Jónasson 1972, table 27). Growth of *P. hammoniensis* is slow (Fig. 12.34). Seasonal effects on growth are difficult to demonstrate, but larvae certainly grow rapidly after the autumn overturn and in spring (Jónasson 1972, fig. 62).

The life-cycle of *P. hammoniensis* occupies 4–6 years (Jónasson & Thorhauge 1972, 1976a, table 6). Life-cycles begin in even numbered years, as do those of *C. anthracinus*. No doubt the explanation is a predator–prey relationship, in that *Chironomus* feeds extensively on oligochaete cocoons on the mud surface (Jónasson & Thorhauge 1972, 1976a; Thorhauge 1976). In the profundal environment cocoons occur from May to October, and are probably laid from May to July, with a peak in June (Fig. 12.34). During even-numbered years the number of surviving cocoons is high, coinciding with complete emergence of the adult midge stage of the *C. anthracinus* population (i.e. the mud surface is now 'empty' of predators). During odd-numbered years the number of cocoons is low, and full-grown *C. anthracinus* larvae are abundant in the benthos. However, since cocoons of *P. hammoniensis* are laid deep within the sediment some survive preda-

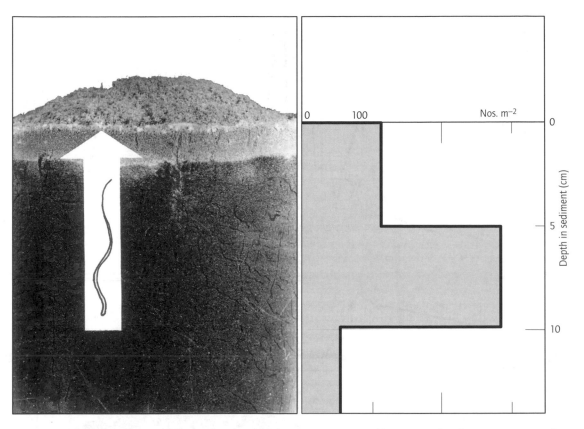

Fig. 12.32 Conical faeces heaps on the mud surface of Lake Esrom transported by *Potamothrix hammoniensis*. Left: the animal lives in vertical tubes reaching 10 cm depth, with head oriented downwards. The white arrow illustrates the transport of organic matter up to the surface. *Chironomus anthracinus* was removed in the experiment. Right: the highest accumulation of animals is between 5 and 10 cm depth in the profundal sediment (Berg 1938; Jónasson 1972).

tion by *C. anthracinus*, but numbers of embryos per cocoon are reduced (Jónasson & Thorhauge 1972, figs 1A and 2; Thorhauge 1975) apparently as a result of interspecific competition (see section 12.13).

The state of the *C. anthracinus* population thus greatly influences the beginning of a *P. hammoniensis* cycle. Large populations of *P. hammoniensis* (40,000–60,000 m⁻²) originated during even-numbered years (i.e. 1952, 1954, 1956, 1958) as did those of *C. anthracinus*. The total life-span of *P. hammoniensis*, as described by survivorship curves (numbers) and by weight (Fig. 12.34; Jónasson & Thorhauge 1976a, fig. 5), lasts up to 6 years. At high densities of *C. anthracinus*, growth and fecundity of *P. hammoniensis* decline and breeding is delayed until worms are 4 years old. In contrast, low densities of *C. anthracinus* lead to high densities of *P. hammoniensis*, which breed at 3 years of age (Jónasson & Thorhauge 1976a, table 6). Inter- and intraspecific competition is discussed in more detail in section 12.13, immediately below.

Pisidium casertanum (Poli) lives in horizontal tubes parallel with and very close to the surface (Fig. 12.35). It is a bacterial feeder which practises interstitial feeding in the mud (Meier-Brook 1969; Jónasson 1972; Lopez & Holopainen 1987). Aver-

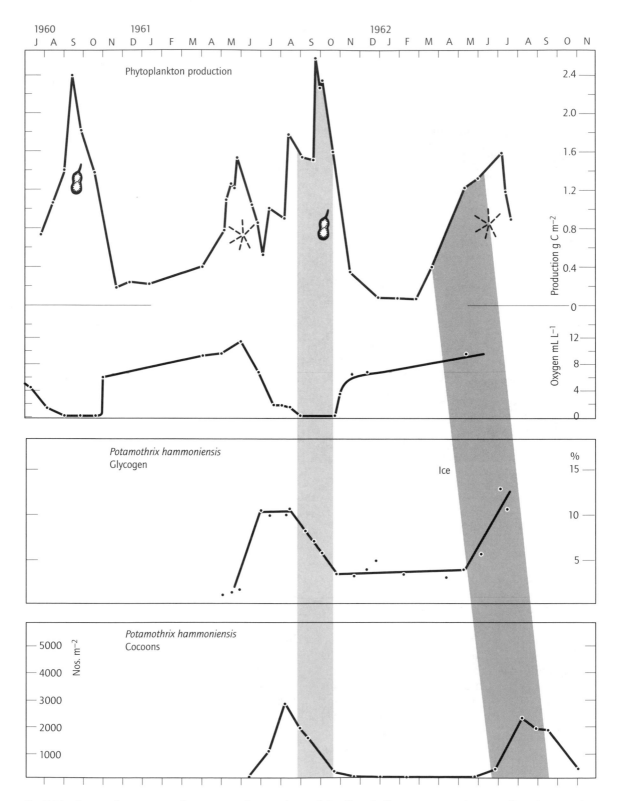

Fig. 12.33 Seasonal succession of primary production $(g\,C\,m^{-2}\,day^{-1})$, and of bottom oxygen $(mg\,O_2\,L^{-1})$, and seasonal variation in glycogen and occurrence of *Potamothrix hammoniensis* cocoons in Lake Esrom. Note that the effect of the spring diatom maximum is delayed in this subsurface feeder.

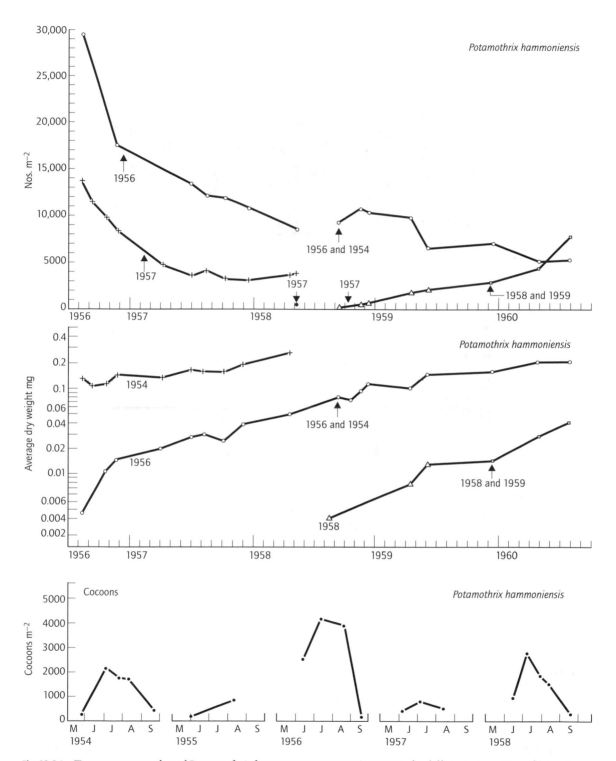

Fig. 12.34 Top: average number of *Potamothrix hammoniensis* per square metre for different generations during 5 years in Lake Esrom. The initial number of the 1956 generation was calculated from the mean number of embryos per cocoon. Middle: average dry weight for the same generations (mg individual^{-1}). Bottom: number of cocoons per square metre in different years. (Modified from Jónasson 1978.)

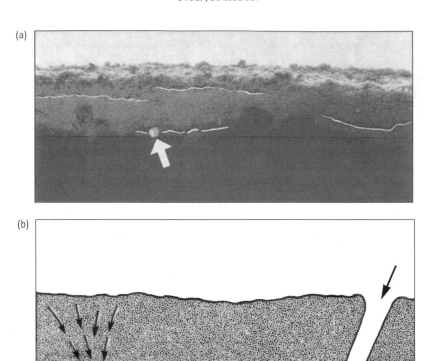

Fig. 12.35   (a) Burrows of *Pisidium casertanum* in Lake Esrom. The upper burrows were empty at the time of photographing, but new burrows on the borderline between the oxygenated (light) and deoxygenated (dark) mud are still made and there *Pisidium* spends its life (white arrow). Burrow openings to the surface are indicated by a small restricted elevation on the surface (black arrow) (Jónasson 1972). (b) Position of filtering *Pisidium* in the tube; the burrow contains faeces. (Modified from Meier-Brook 1969.)

age water content varies between 53 and 60% of dry weight (Jónasson 1972, table 28) and glycogen constitutes c.4% in adult *P. casertanum*, except after autumn overturn when it declines to 3% (the period when viviparous mussels give birth to young). However, in juvenile *P. casertanum* the glycogen content declines gradually from 5% in spring, throughout the summer stratification, to 2% of dry weight in autumn (Hamburger *et al.* 2000).

Pisidium casertanum exhibits a respiration trend slightly different from that of other major profundal species (Fig. 12.36). Its critical limit of

respiration (if one exists) lies at 21.5% air saturation, but respiration at this point is reduced to 50% of initial respiration at air saturation. In spite of this large decline, bathymetric distribution of *P. casertanum* shows that it survives severe oxygen depletion in the profundal (Fig. 12.36). *Pisidium casertanum* follows the same growth pattern as *P. hammoniensis* (Fig. 12.37). Stratification halts breeding but interestingly this begins again 2 months after the autumn overturn. At 4–5 years old, periods of seasonally intermittent growth are well separated (Fig. 12.37). Growth ceases during winter (especially during ice-cover) and during

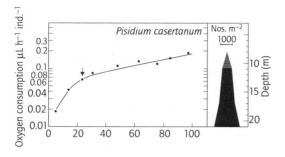

Fig. 12.36 Oxygen consumption in relation to oxygen content together with bathymetric distribution of *Pisidium casertanum* in Lake Esrom. Arrow indicates critical limit of respiration. (Modified from Jónasson (1972) and Holopainen & Jónasson (1989a).)

summer stratification, when oxygen is scarce. Embryonic growth also halts during summer stratification and lasts for 6 months (Holopainen & Jónasson 1983). Thus the life-style, growth, age and population dynamics of *Pisidium* species are quite similar to those of *P. hammoniensis*.

12.13   INTER- AND INTRASPECIFIC COMPETITION

Competition in lakes is both difficult to investigate and to demonstrate, and therefore only a few well-established cases exist. Reynoldson & Bellamy (1971, p. 283) base their definition of

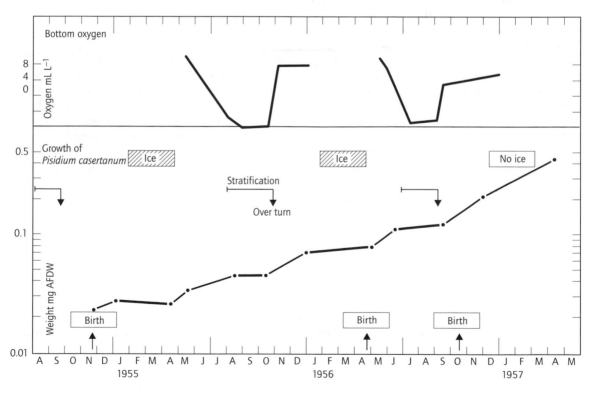

Fig. 12.37 Weight increase of the 1954 cohort of *Pisidium casertanum* in relation to oxygen content in Lake Esrom. The arrows on the abscissa indicate the three main birth periods. (Modified from Holopainen & Jónasson 1983.)

wider generality on the concept of energy, namely that 'competition is occurring when an organism uses more energy to obtain (e.g. food) or maintain (e.g. living space) a unit of resource due to the presence of another individual than it would otherwise do'. Such a definition applies equally well to intra- and to interspecific competition. They give five criteria for identification of competition, of which the most important in our context is: 'It is necessary to show that the competing species are utilising a common resource which may provide the basis of competition' (Reynoldson & Bellamy 1971, p. 283).

12.13.1 Interspecific competition in littoral populations

Throughout four decades Reynoldson (1983) investigated competition (or **niche specification**) between predatory triclad Turbellaria. He found that a more specific partitioning of prey occurred among the four common species. Although individual species feed on a wide and overlapping spectrum of prey, each one also feeds more on one particular invertebrate taxon than on others (Reynoldson & Davies 1970), i.e. **partial competition**. Thus *Dendrocoelum lacteum* took 63% of its meals on Crustacea (mainly *Asellus* sp.), *Dugesia polychroa* 57% on gastropods and the two *Polycelis* spp. 57–68% on oligochaetes (Table 12.3). Such contrasts seem to be present at all seasons.

Competitive interactions have also been thoroughly examined in another group of predators, namely leeches. Interspecific competition for

food among Glossiphoniidae and Erpobdellidae is described in Young (1981) and Young & Spelling (1989). Resource partitioning between littoral populations of *E. octoculata* (L.) and *E. testacea* Sav. was described in Dall (1983). It appears that, in as much as these species swallow most of their prey whole, food niches, especially in the Erpobdellids, are a matter of predator biomass. This was recently confirmed in a study of lotic populations of *E. octoculata* (Toman & Dall 1997).

12.13.2 Competition and coexistence in the profundal — cannibalism as a factor of competition

The profundal of Lake Esrom provides an 'experimental field' of 10 km² in which to study the interplay of population dynamics between three prominent species, i.e. *C. anthracinus*, *P. hammoniensis* and *P. casertanum*. *Chironomus anthracinus* exhibits intraspecific competition by cannibalism. It lives in conical tubes in the rich, soft mud and is a deposit feeder sweeping the mud in circles around the opening of the tubes (Fig. 12.38). A survivorship curve shows the life-history of *C. anthracinus* larvae over a 2-year period (Fig. 12.39). It is obvious that part of the population is **univoltine** and another part **hemivoltine** (see section 12.18.5). What is the explanation?

Chironomus anthracinus produces 600 eggs per female. Figure 12.39 describes the fate of a population through 2 years. Arrows indicate external factors and the curve is the response of the larval population. Initial numbers are very high (up to 70,000 m⁻²) and the large early decline is no doubt

Table 12.3   The proportion of meals taken by the four common lake species of triclad on each prey item expressed as a percentage of total meals per 500 triclads per day. Food refuges are shown in bold type (Reynoldson & Davies 1970).

	Asellus	Gammarus	Oligochaeta	Gastropoda	Total meals
Polycelis tenuis	25	8	**57**	10	156
Polycelis nigra	8	7	**68**	17	155
Dugesia polychroa	16	4	22	**57**	176
Dendrocoelum lacteum	**63**	23	14	0	157

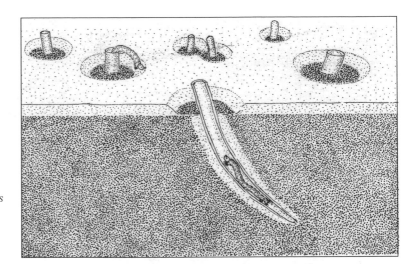

Fig. 12.38 *Chironomus anthracinus* larvae feed from the oxidised (light) substrate in Lake Esrom, and leave the (dark) subsurface mud exposed. (Bjarnov, from Jónasson 1972.)

density dependent (Jónasson 1972, fig. 56). Larval mortality is low during summer stratification, when the population is protected against fish predation by hypolimnetic oxygen scarcity. The low death rate which occurs is also due to other less tangible causes which can be referred to as 'natural mortality'. Autumn overturn produces higher temperatures and oxygenated water, larval activity increases and fish predation begins. The large decline in larval numbers during this period (by c.4000 individuals m^{-2}) is mainly due to fish predation. During winter the activity of poikilothermic animals is low (e.g. eels burrow into the mud), fish predation ceases and larval numbers remain constant. It is thus possible to separate mortality from fish predation and that from other forms of mortality.

Another large decline in numbers (of c.2500 individuals m^{-2}) occurs during emergence in May. It is rather surprising that in 1955 (Fig. 12.39) no new generation appeared at 20 m depth, and that, in general, a new generation is produced only every second year. In fact, alternating generations should not be expected. The most likely explanation is that those eggs of the emerging population which reach the deeper profundal are eaten by remaining *C. anthracinus* larvae. During the spring of 1955 this population amounted to 7000–7500 m^{-2}, and their feeding on the mud surface was sufficiently effective to clear it of eggs and small larvae.

The explanation is intraspecific competition, which leads to cannibalism. Feeding *Chironomus* larvae sweep the surface (Fig. 12.38) and measurements show that larval bodies extend three-quarters of their body length out of their tubes. If we assume this distance to delimit the radius of the feeding area, then burrows of newborn and young larvae cover the surface of the mud completely, with no overlap. After the first autumn overturn, and the resulting larval growth, feeding areas overlap to such an extent that they amount to three times the actual area of the mud surface. Because of increased larval size, areas of feeding continue to overlap, despite reduction in numbers in autumn and partial emergence in spring. Reduction in feeding area only occurs when the population declines during the second autumn of a 2-year cycle. After emergence in May 1956, no larvae were left to feed (Fig. 12.39).

On the basis of the above data, during their feeding activities a total of 2000 m^{-2} of fully grown larvae would completely cover the sediment surface. If the population exceeds this size, neither eggs nor

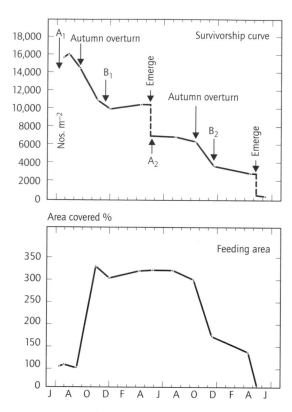

Fig. 12.39 Top: seasonal fluctuation (nos m^{-2}) during a 2-year larval life-cycle of the population of *Chironomus anthracinus* (1954–1956) in Lake Esrom, in relation to external factors. A_1 and A_2 indicate the beginning of summer stratification period in first and second year of life-cycle, B_1 and B_2 the end of fish predation caused by temperature decline (Jónasson 1972). Bottom: seasonal variation in the area covered by the substrate feeding *C. anthracinus* (Jónasson 1972).

young larvae could survive on the surface sediment. If the population is below this number, a new generation of larvae is able to settle and a mixed population of 1-year old and newly hatched larvae occurs. This mechanism therefore determines whether or not a new population is established each year (Fig. 12.39).

A major characteristic of this population is the 2-year life-cycle, each beginning in an even-numbered year (Fig. 12.40). A general feature of

these life-cycles is high population density, especially high initial numbers. The large number of larvae remaining after the first year's emergence prevents a new generation from becoming established. If these conditions are no longer present and population density after the first year's emergence is low, a new generation is able to occupy the vacant space and become established every year. This accounts for the successful recruitment of larvae at 2-year intervals from 1952 to 1960.

If the larval population is very small (e.g. at 11 m and 14 m depth), young larvae are recruited to the population every year (Fig. 12.31). Interspecific competition for space between *C. anthracinus* and *P. hammoniensis*, which results in predation, is in fact described in Figs 12.33 & 12.34. In odd-numbered years almost all cocoons were preyed upon by *C. anthracinus*. Therefore, *P. hammoniensis* populations also begin in even-numbered years, when they succeeded in settling into an 'empty' mud surface.

The lifestyle, growth and population dynamics of *P. casertanum* are quite similar to those of *P. hammoniensis*. Interspecific competition with *C. anthracinus* is likely, but not yet firmly established.

From oligochaete studies in lakes in the English Lake District, Reynoldson (1990) discussed those species likely to act as predators and concluded that, on the basis of present data, the question of whether competition, intra- or interspecifically, plays a significant role in determining abundance could not be answered. On the basis of migration studies by Berg (1937), he hypothesised that high numbers of *Chaoborus* in Blelham Tarn (70,000 m^{-2}) have reduced the number of oligochaetes in the lake but not in experimental tubes (Reynoldson 1990), to which *Chaoborus* had no access. That this might be the case is supported by observations by Jónasson (1955), which showed that in experimental jars *C. flavicans* predated heavily on *P. hammoniensis*, reducing its abundance by c.90% over 6 days. Simultaneously, the transparent *Chaoborus* larvae became pink in colour, like those of *Potamothrix*. This may well happen in nature from autumn to spring, when zooplankton biomass is low (Jónasson 1972).

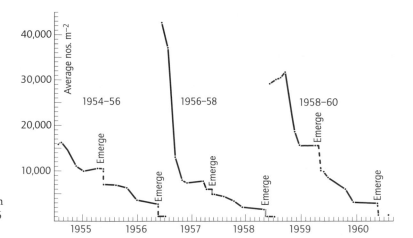

Fig. 12.40 Seasonal fluctuations in numbers of *Chironomus anthracinus* at 20 m depth during 1954–1960 in Lake Esrom. Two-year life-cycles predominate, beginning in even-numbered years (i.e. 1954, 1956 and 1958). (From Jónasson 1972.)

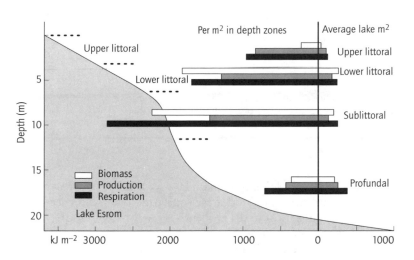

Fig. 12.41 Annual average biomass, annual net production and annual respiration of zoobenthos in their respective depth zones in Lake Esrom, calculated m⁻² and average lake m⁻². (Data from Berg 1938; Berg *et al.* 1962; Jónasson 1972; Dall *et al.* 1984a; Jónasson *et al.* 1990; Hamburger *et al.* 1994.)

12.14 PRODUCTION OF ZOOBENTHOS IN RELATION TO DEPTH IN LAKES ESROM AND THINGVALLAVATN

Because it combines individual growth and population survivorship, and is essential when attempting to quantify energy flow pathways in ecosystem analyses, secondary production of invertebrate populations is an important variable for the structure and functioning of ecosystems. In this section an attempt is made to visualise changes in energy budget with depth for lakes Esrom and Thingvallavatn, at community as well as species level.

Figure 12.41 and Table 12.4 illustrate the presence of low biomass per square metre in the upper littoral benthos, but production is high. In contrast, in the lower littoral and in the sublittoral both biomass **and** production are high. However, both decline materially in the profundal. Figure 12.41 then shows the same values calculated per average lake per square metre, and the bathymetric distribution becomes entirely different. Per aver-

Table 12.4 Annual average biomass (B), annual net production (P), annual respiration (R), P/B, R/P and net production efficiency (NPE) coefficients for zoobenthos in their respective depth zones in Lake Esrom, Denmark (Jónasson et al. 1990).

	Depth (m)	B (kJ m^{-2} yr^{-1})	P (kJ m^{-2} yr^{-1})	R (kJ m^{-2} yr^{-1})	P/B	R/P	NPE
Shallow littoral	0–3	230	840	960	3.64	1.15	0.47
Lower littoral	3–6	1850	1300	1720	0.71	1.32	0.43
Sublittoral	6–12	2260	1470	2830	0.65	1.93	0.34
Profundal	12–22	350	422	673	1.21	1.59	0.39

Table 12.5 Annual net production (P), mortality (M), emergence (E), respiration (R), P/B (biomass) and net production efficiency (NPE) of profundal zoobenthos at the species level in Lake Esrom, Denmark (Jónasson 1972; Jónasson et al. 1990).

	P (kJ m^{-2} yr^{-1})	M (kJ m^{-2} yr^{-1})	E (kJ m^{-2} yr^{-1})	R (kJ m^{-2} yr^{-1})	P/B	NPE
Chironomus anthracinus	316	152	165	378	2.1	0.45
Chaoborus flavicans	60	7	48	215	1.7	0.22
Procladius pectinatus	12	6	7	30	1.9	0.29
Potamothrix hammoniensis	25			30	0.7	0.46
Pisidium spp.	9			20	1.0	0.31
Total	422	166	220	673		

age square metre of lake, the profundal—with its large area but lowest production per square metre site—contributes most to the zoobenthos production in Lake Esrom.

Community respiration data are highly interesting because they express the costs of living. Respiration per square metre is very high in the sublittoral, owing to the abundance of *D. polymorpha* (cf. Hamburger *et al.* 1990), but the same does not apply to the profundal owing to the respiratory adaptation of its zoobenthos. Thus, until the profundal is reached, cost of living increases with depth (Table 12.4 & Fig. 12.41). This is also illustrated by the quotient respiration/production. The littoral therefore turns out to be energetically the most efficient community. The same ranking order applies to net production efficiency (NPE; Table 12.5). Turnover ratios of organic matter for the various communities show, as expected, the

greatest values for the shallow littoral but, quite unexpectedly, a ratio of 1.2 for the profundal community. This occurs in spite of the regular incidence of long periods of serious oxygen depletion, which delay growth.

A more penetrating analysis is given, at species level, for the detritivores and carnivores of the profundal (Table 12.5). The main species present is the surface feeding detritivore *C. anthracinus* Zett., which contributes the highest net production within the profundal community, and whose respiration is 20% higher than its production. Turnover rates of biomass (annual net production/biomass (P/B)) are also the highest, 3.8 for the first year generation and 0.4 for the second (average 2.1, Table 12.5). The subsurface feeders *P. hammoniensis* (Michaelsen) and *Pisidium* sp. exhibit low production, low respiration and low P/B. The invertebrate carnivores *C. flavicans* Meigen and *P.*

pectinatus K. account for approximately 10% of the community by numbers. Their respiration is approximately three times their production. Quite unexpectedly, their NPE as well as their turnover ratio of biomass (P/B) are intermediate to those of the surface and subsurface feeders (Jónasson 1972, 1978; Holopainen & Jónasson 1983).

Production and biomass values for zoobenthos in Thingvallavatn are low compared with those of Lake Esrom (Fig. 12.42). In contrast to Lake Esrom, where peak values are observed in the sublittoral, bathymetric changes in biomass and production decline continuously from the littoral surf zone to the profundal (Fig. 12.41). This means that the *Nitella* filtering belt of Thingvallavatn is much

less efficient than the sublittoral of Lake Esrom. However, calculations for the average square metre of either lake produce almost equal values, owing to the extensive area of the *Nitella* belt, and of the profundal zone (Fig. 12.42). Respiration was not determined in Thingvallavatn.

12.15 PRODUCTION AND COMMUNITY STRUCTURE OF PROFUNDAL ZOOBENTHOS IN RELATION TO ENERGY CONTENT OF SEDIMENT

Production of organic matter is the subject of sev-

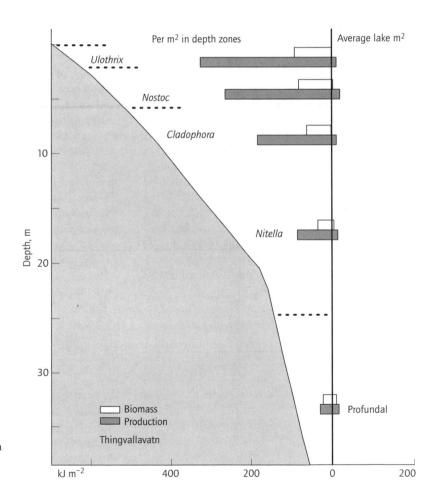

Fig. 12.42 Annual average biomass and annual net production of zoobenthos in their respective depth zones of Thingvallavatn, calculated m^{-2} and average lake m^{-2}. (Data from Jónsson 1992; Kairesalo *et al.* 1992; Lindegaard 1992a.)

eral papers from various lake types (e.g. Jónasson 1972; Jónasson & Adalsteinsson 1979; Lindegaard & Jónasson 1979; Jónasson 1992a; Jónsson 1992; Kairesalo *et al.* 1992). In this section we discuss the decomposition of organic matter in the water column and the influence of calorific content of organic detritus on the community structure of the profundal benthos community.

In sedimentation experiments carried out at 80 m depth (close to the station at 75 m depth) in the south basin of Thingvallavatn (Lastein 1983), it was found that the theoretical atomic ratio of $C:N:P$ recorded in phytoplankton is different from that determined in sedimenting seston, and in sediment (Table 12.6). Material was collected from traps at 20, 40, 60 and 80 m depth, over 22 and 42 days, respectively, values which embrace the sinking times of phytoplankton. The figures represent dry matter analyses of trap material and the top 10 cm of sediment.

The data show that only 20% of carbon produced in the water column is sedimented, and only 30% of the nitrogen, whilst the proportion of phosphorus remains constant. Thus 70–80% of the phytoplankton is decomposed during the sinking process. Table 12.6 further shows that the $C:N$ ratio remains almost constant during sedimentation in the traps (originally 7:1, and then 6.5:1 after 22 days and 7:1 after 42 days), but that in the sediment itself it declines to 4.5:1.

The calorific value of sediment varies between lakes, depending on production and depth. Figure 12.43 illustrates the calorific content of the sediment in Mývatn at 4 m depth, in Lake Esrom at 22

m and in Thingvallavatn at 75 m. Energy content is greater in Lake Esrom (6–7 kJ g^{-1} dry weight) and in Mývatn, and these two lakes exhibit by far the greatest abundance of benthic fauna (an average of 75,500 and 29,100 individuals m^{-2}, respectively), with corresponding net productions of 861 and 431 kJ m^{-2}. The fauna in both lakes is largely made up by chironomids, but in Lake Esrom other groups also occur (Fig. 12.43). The reason why Lake Esrom data apparently represent a rather low aggregate calorific value is that its profundal is deoxygenated during summer, and so, despite the fact that the fauna is adapted to low oxygen tensions, it does not feed or grow or process organic matter through long periods of time.

In the shallow Mývatn the turnover of zoobenthos is higher, with an annual P/B of 4, whilst those of the profundal fauna of Lake Esrom and of Thingvallavatn are 1.8 and 1.4, respectively. This contrast may be related to differences in oxygen content of the hypolimnion and food energy content.

The sediment in the oligotrophic Thingvallavatn is of poor nutritional value for the profundal zoobenthos. Its calorific value is lower (e.g. at 75 m it is 1 kJ g^{-1} dry weight). At this depth, abundance of the profundal fauna is reduced to 9800 individuals m^{-2}, but, instead of chironomids, Thingvallavatn supports tubificids (Lindegaard 1992a). From an energetic point of view the explanation is clear. The energy demand per unit mass of *T. tubifex* is lower than that of chironomids. At 110 m depth the fauna still consists of *T. tubifex*, with a considerable number of the turbellarian predator *O. auditivum*, but numbers are now reduced to 4200 individuals m^{-2} (Fig. 12.43).

Figure 12.44 shows that respiration of *C. anthracinus* in oxygen saturated water from the sublittoral of Lake Esrom is five times higher than that of *P. hammoniensis* from the profundal, and twice that of *T. tubifex* from a nearby river. Thus the energy resource in the profundal sediment of Thingvallavatn seems able to support a tubificid fauna only. The explanation may well lie in respective feeding behaviours, since chironomids feed from the top of the sediment surface, where the energy content is higher, whilst tubificids feed on the

Table 12.6 Atomic ratio of sedimenting phytoplankton (Lastein 1983).

	Atomic ratio		
	C	N	P
Phytoplankton	50.0	7.0	1.0
Trap material, 22 days	22.7	3.5	1.0
Trap material, 42 days	19.7	2.8	1.0
Sediment	10.1	2.2	1.0

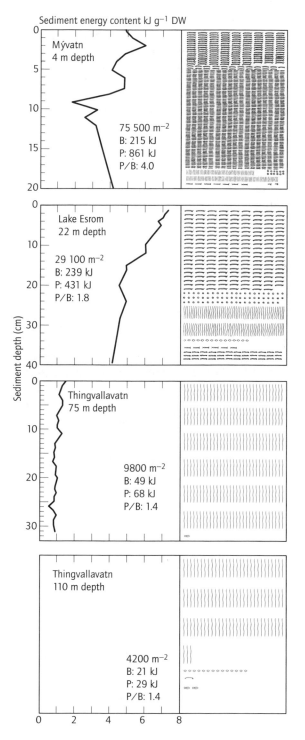

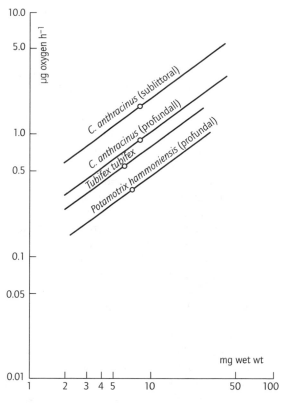

Fig. 12.44 Oxygen uptake ($\mu g\,O_2\,h^{-1}$) in relation to wet weight of two species of tubificid worms and larvae of *Chironomus anthracinus* from Lake Esrom. Respiration of *C. anthracinus* is up to five times greater, and indicates the much higher energy demand in chironomids than in tubificids. Experiments were carried out at 11°C, and at oxygen saturation. (Data from Berg *et al.* 1962.)

Fig. 12.43 Qualitative and quantitative changes of zoobenthos in three lakes (Mývatn, Lake Esrom and Thingvallavatn) in relation to declining energy content of the sediment. Quantitative changes expressed as average abundance, average biomass, average net production and average annual production/biomass (P/B). (Modified from Jónasson 1992b.)

low-energy layer below the surface (cf. Figs 12.32 & 12.38).

12.16   THE ROLE OF ZOOBENTHOS IN THE ENERGY BUDGET OF LAKES

Some lakes contain abundant zoobenthos, others not. A fundamental requirement is a habitat which is stable, where benthic invertebrates are able to survive on the food available. In shallow lakes, where continuous resuspension of bottom material occurs, abundances are low. The aim of this chapter is to calculate the share of benthic invertebrates in the lake budget. Complete or partial energy flow diagrams are available for Lake Esrom, Thingvallavatn and Mývatn (Jónasson 1972, 1979b, 1992b; Lindegaard 1992b).

A balance sheet for the energy flow of an average square metre of Lake Esrom is given in Table 12.7. Photosynthetic fixation of energy accounts for a gross input of 12,410 kJ for an average square metre of lake per year. Phytoplankton and benthic algae account for 92% of production. It is clear that respiratory processes are the main expenditure. Photorespiration is calculated (as in previous papers) to be 25% of gross primary production. Bacterial respiration was obtained by difference. Together they account for half the primary gross production (Table 12.7). Respiration of secondary producers contributes 25%. Respiration in the profundal equals 7.2% of that of secondary producers. Including microbial and chemical contributions, profundal respiration accounts for 15% (Fig. 12.41 & Table 12.7).

Hargrave (1973) found a sediment oxygen uptake of 32% year^{-1} in Lake Esrom. This value is consistent with previous data from north German lakes (Ohle 1962) and Lake Washington (Edmondson 1966), where 70% of production is oxidised in the water column before reaching the bottom. In view of its history of eutrophication, sedimentation in Lake Esrom is highly interesting. Annual sedimentation was measured at 215 g AFDW m^{-2} year^{-1} (Lastein 1972, 1976). Recalculated as per square metre of lake per year, an average phytoplankton production of 240 g C m^{-2} year^{-1} amounts to 4106 kJ, or 33% of total primary production in the lake. Such material is utilised by the profundal zoobenthos and respired in the sediment, and only 3% actually accumulates in this sediment. This conclusion is based on present-day sedimentation rates in a lake under eutrophication, and is thus higher than the previous value of

Table 12.7   Energy budget of an average square metre of Lake Esrom, Denmark. (Modified from Jónasson *et al.* 1990.)

	Average square metre of lake (kJ yr^{-1})		%
	Input	Output	
Gross primary production	12,410		
Photorespiration		3025	24.4
Bacterial respiration in water		3326	26.8
Secondary respiration		3084	24.9
Microbial respiration in sediment		560	4.5
Chemical oxidation in sediment		430	3.4
Secondary production		1470	11.9
Emergence		165	1.4
Accumulation in sediment		350	2.8
Total	12,410	12,410	100.0

1% based on 10,000 years' sedimentation (Whiteside 1970; Hargrave 1973).

In a lake the bathymetric shape and depth of Lake Esrom, plankton communities are bound to be the main primary and secondary producers. Zooplankton accounts for 53% of secondary production and consumes approximately 30% of gross phytoplankton production (Hamburger 1986). The other half is effected by benthic invertebrates. Altogether, secondary production accounts for 12% of primary gross production, and 1% is lost through emergence (Table 12.7).

Dynamics of the profundal zoobenthos indicate close ecological relations to the fish. Mortality of zoobenthos accounts for $166\,kJ\,m^{-2}$ year^{-1} (Table 12.5), the main part being attributed to fish predation. Another source of energy is emerging aquatic insects. Production of fish in Lake Esrom accounts for $18\,kJ\,m^{-2}$ year^{-1}, calculated from the catch of commercial fisheries. However, an outline of energy pathways in Lake Esrom is bound to be inaccurate. Apparently, the weak links are the magnitude of photorespiration and of bacterial respiration. These require detailed investigation in the future.

Lake Esrom is a medium deep lake, where the relative proportions of zooplankton and of zoobenthos biomass are each approximately 50%. This is entirely different from shallow, zoobenthos-rich lakes such as Mývatn and Hjarbæk Fjord, where benthos accounts for the main energy flow with

the high proportion (86%) of secondary production (Table 12.8). In contrast, in the 114 m deep, subarctic Thingvallavatn, zooplankton are the main secondary producers, with a total of 68% of secondary production.

A new aspect to the study of zoobenthos is given for Mirror Lake, USA, where Strayer & Likens (1986) included the meiofauna and constructed a model of energy flow through the zoobenthos of this oligotrophic lake. Alimov (1983) gives reviews of the productivity of zoobenthos communities, their metabolism, energy flow and ecological efficiencies, based on the many Russian (then Soviet) International Biological Programme (IBP) projects. The most recent and penetrating review, covering all 25 known projects, world-wide, is given by Lindegaard (1992a).

12.17 LIMITS FOR ANIMAL LIFE IN THE HYPOLIMNION

12.17.1 *What do critical limits of respiration mean?*

The influence of oxygen depletion on vital parameters of population dynamics (e.g. metabolism, reproduction, growth, mortality and duration of life-cycle) is fundamental (Jónasson 1972, 1984b). Secondary production decreases or becomes negligible owing to environmental stress. What do critical limits mean?

Table 12.8 Trophic efficiency level and percentage contribution of zooplankton (ZP) and zoobenthos (ZB) to total annual net production of primary producers (PC) compared with mean depth and net primary production in four lakes. (Data for Lake Esrom and Thingvallavatn are from Jónasson *et al.* (1990), Jónasson (1992b) and Lindegaard (1992a,b).)

Lake	Mean depth (m)	Net primary production $(kJ\,m^{-2}\,yr^{-1})$	Net production PC $(kJ\,m^{-2}\,yr^{-1})$	ZP (% of PC)	ZB (% of PC)
Hjarbaek Fjord	1.9	10,500	1400	14	86
Mývatn	2.3	13,800	1258	14	86
Esrom	13.5	9400	1470	53	47
Thingvallavatn	34.1	4500	310	68	32

Figure 12.21 (above) showed that the lower the critical limit, the better the adaptation to low oxygen concentrations. Consequently, well adapted animals (e.g. *C. anthracinus*) are able to maintain a high respiratory rate and a high energy metabolism under very severe oxygen stress. *Chironomus anthracinus* hardly shows any decline in respiration between full air saturation and 24%, whereas *P. hammoniensis* exhibits a 30% decline from full saturation to 9.5%.

The best-adapted profundal species possess haemoglobin as a blood pigment. This property lengthens the period of active life for *C. anthracinus* and without doubt also for *P. hammoniensis*. The significance of haemoglobin in *Chironomus* is threefold:

1 it acts in oxygen transport at very low oxygen concentrations, thereby making continued respiration possible;

2 it enables the larva to maintain feeding activity when relatively little oxygen is present;

3 it greatly increases the rate of recovery from periods of oxygen scarcity, and makes such recovery possible even under adverse respiratory conditions (Walshe 1950).

Below the critical limit, animals gradually shift to anaerobic respiration (Berg *et al.* 1962; Jónasson 1972, 1978; Hamburger *et al.* 1997, 1998), but this is highly inefficient (see also Chapter 11). The polysaccharide glycogen is used as an energy source and is degraded to alcohol or other intermediary metabolites. Therefore the animals accumulate glycogen for use under severe oxygen stress (see also section 12.12).

12.17.2   Reduction of hypolimnetic oxygen content

Figure 12.45 illustrates 16 years' primary production data for Lake Esrom, covering a period of 27 years from 1955 to 1982. Despite wide fluctuations, no progressive trend due to eutrophication is indicated. It is reasonable to assume, in spite of the high phosphorus content of local groundwater, that the original background production of phytoplankton in this lake was $100\,g\,C\,m^{-2}$ $year^{-1}$ (Riemann & Mathiesen 1977). Analysis of two

cores from the lake supports this hypothesis. Lake ontogeny (developmental history) is described from the Pleistocene to the present (Hansen 1968; Whiteside 1970; Brodersen *et al.* 2001), and a considerable increase in sedimentation during the past 200 years is observed (Berg 1938). During the 20th century primary production seems gradually to have increased to reach $240\,g\,C\,m^{-2}$ $year^{-1}$ by the middle of the century.

Average internal loading per year of organic matter of phytoplankton production in Lake Esrom is c.10,500 t, including c.3400 t sedimented in the hypolimnion over the period 1955 to 1982. Approximately 70% of total organic production is decomposed in the photic layer. The hypolimnion accounts for 60% of the lake area, but only one-third of lake volume. Decomposition of sedimenting algae requires twice the amount of oxygen available in the hypolimnion at the beginning of stratification, the average duration of which is 135 days, from 15 May to 1 October. The regression line (Fig. 12.45) shows the number of days which pass after stratification in spring, until oxygen concentration in the lower hypolimnion falls below $1\,mg\,O_2\,L^{-1}$, a limit which is critical for many animals. The result is an increasing length of the oxygen-free period during the 20th century, in spite of the fact that since the middle of the century primary production was stable, with an average of $240\,g\,C\,m^{-2}$ $year^{-1}$.

In 1908 there was no oxygen depletion, but by 1933 the period between the beginning of stratification and the onset of oxygen depletion ($<1\,mg\,O_2\,L^{-1}$) had reached 110 days. Finally, by 1992, only 25 days elapsed between stratification and oxygen depletion. Figure 12.46 shows the increasing length of the ecologically important period of extremely low oxygen content (i.e. $<0.5\,mg\,L^{-1}$) during summer stratification for the 20th century. Comparison of data from 1908, 1933 and 1955–1992, at depths just above the lake bottom, shows that the concentrations of dissolved oxygen declined and the length of the period of oxygen depletion in the hypolimnion increased (Fig. 12.46). On 11 August 1908 oxygen saturation of the hypolimnion at 18 m depth was 65%, or about $7.4\,mg\,L^{-1}$ at 10°C. In 1933, measurements

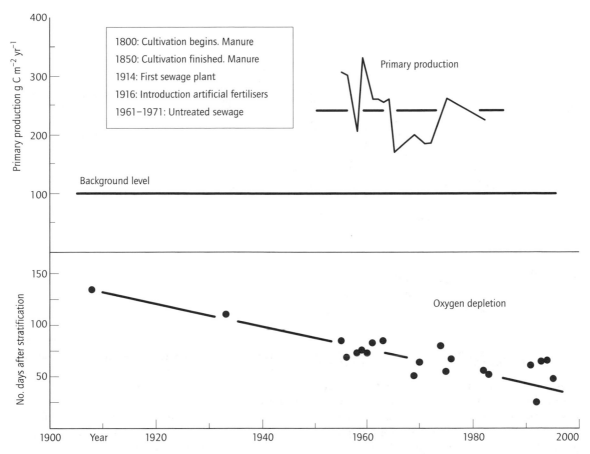

Fig. 12.45 Top: primary production from 1955 to 1982 (inset: recorded changes in the catchment of Lake Esrom relevant to the eutrophication history of the lake). Thick horizontal lines represent average primary production from 1955 to 1982 ($240\,\mathrm{g\,C\,m^{-2}\,yr^{-1}}$, top) and the expected average during the last century ($100\,\mathrm{g\,C\,m^{-2}\,yr^{-1}}$, below). Bottom: oxygen depletion. The solid line shows the number of days from the onset of stratification until decline of oxygen to $1\,\mathrm{mg\,O_2\,L^{-1}}$ in the bottom layers, for the 20th century (Jónasson 1993).

were recorded at 19 m depth and both values were lower (i.e., on 1 August, $3.26\,\mathrm{mg\,L^{-1}}$ or 27% saturation, and on 6 September, $1.42\,\mathrm{mg\,L^{-1}}$ or 12%). By 6 October, oxygen content was $0.24\,\mathrm{mg\,L^{-1}}$ and saturation 2%. These data show that oxygen depletion was still moderate in 1933, but considerably higher than in 1908. From 1955 to 1960 the critical limit occurred in the first half of August, but in 1992 by mid-June. Thus, in spite of decreasing sewage loading, the length of the critical period has

increased from 1 week in 1933 to 120 days in 1992 (Fig. 12.46). There is little doubt that the process accelerated during the second half of the 20th century.

The conclusion is that oxygen-free periods in Lake Esrom are still increasing in length, in spite of constant primary production. Apparently, 100–150 years were needed in order to increase production (i.e. from 100 to $240\,\mathrm{g\,C\,m^{-2}\,yr^{-1}}$) and another 90 years were needed in order to produce a

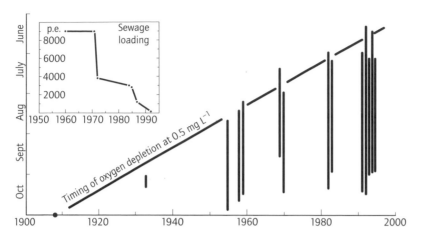

Fig. 12.46 The advancing timing of oxygen depletion at the 0.5 mg $O_2 L^{-1}$ level in the bottom layers of Lake Esrom from 1908 in relation to declining sewage loading 1972–1992. Vertical bars show the duration of the anoxic period during respective years (Jónasson 1993); p.e., person equivalents.

critically low oxygen regime in the hypolimnion. This process now endangers the existence of the rich profundal invertebrate community in Lake Esrom.

12.17.3 Survival of profundal zoobenthos

Most eutrophication studies focus on nutrients, chlorophyll a (Chl a) and primary production. In Lake Esrom, however, eutrophication data are available also from six decades of sampling of profundal benthos. They are now discussed in this section.

Abundance and population dynamics of zoobenthos in relation to declining oxygen content of the profundal were investigated from 1953 to 1995, i.e. through c.40 years. Further, the zoobenthos was also investigated during 1932–34 (Berg 1938) and back calculations are available, thus extending the observation period to six decades. In this context, we will consider long-term fluctuations of three taxa, i.e. the midge *C. anthracinus*, the oligochaete *P. hammoniensis*, and three species of *Pisidium* mussels. From physiological experiments we know that *C. anthracinus* exhibits almost no decline in respiration from 100 to 25% air saturation. This critical limit equals $2.9 \, mg \, O_2 \, L^{-1}$. The *Pisidium* species are highly affected by oxygen decline, because their respiration declines rapidly to a critical limit of

21.5% air saturation (Berg *et al.* 1962; Jónasson 1972, 1984a, b). But recent unpublished data indicate a very low critical level fitting perfectly to its habitat (Dall, unpublished, 2001).

Potamothrix hammoniensis is extremely well adapted to low oxygen tensions, with a critical limit at 9.5% air saturation, in spite of a decline of 25% in oxygen uptake between 100 and 9.5% saturation. Fluctuations in population size appear natural and not influenced by oxygen depletion, since during 1990–1995 the population seems unaffected by oxygen scarcity.

The main conclusion is that, up until 1961, the zoobenthos community in Lake Esrom remained almost unaffected by changing oxygen availability (Fig. 12.47). Because *Pisidium* seems more sensitive to lack of oxygen its decline began in 1958, and it has not recovered since. *Chironomus anthracinus* was affected later, in 1961, owing to its greater physiological adaptation. From 1954 to 1962, *C. anthracinus* underwent 2-year cycles, with high average density. After introduction of untreated sewage into the lake, population densities declined and shifted to alternating 1- and 2-year life-cycles. Owing to the high number of eggs per female (500–700), it is easy for the population to recover quickly, but high reproduction during the years 1965 and 1972 shows that mortality of young larvae in the profundal is very high under insufficient oxygen availability. In fact, compared with

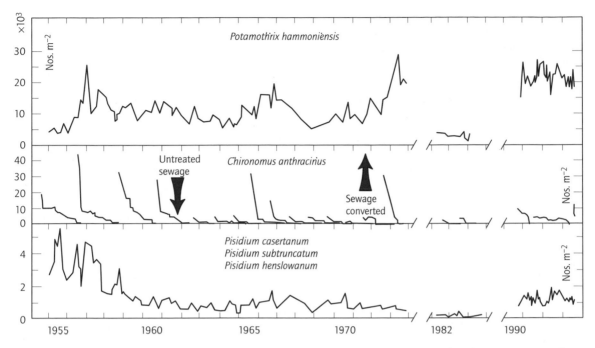

Fig. 12.47 Changes in abundance and annual cycles of *Chironomus anthracinus*, *Potamothrix hammoniensis* and three species of Pisidium (*Pisidium casertanum*, *Pisidium subtruncatum* and *Pisidium henslowanum*) during the period 1953–1994 in Lake Esrom (Jónasson 1993; Lindegaard *et al.* 1997).

earlier periods (Jónasson 1972, 1984b; Lindegaard *et al.* 1992), the profundal of the lake now resembles a desert. However, *P. hammoniensis* seems unaffected, apparently not only owing to efficient adaptation, but also to the absence of interspecific competition from the small population of *C. anthracinus*. Thus *P. hammoniensis* has become the most numerous species, with a population of 20,000 individuals m^{-2} (Lindegaard *et al.* 1997).

12.18 ZOOBENTHOS AND THE ECOSYSTEM

Numerous monographs describe the biology and ecology of zoobenthos of lakes from various viewpoints especially for the last century, and a touch of history was given in section 12.2 above. Wesenberg-Lund (1939, 1943), in two textbooks, gave a fascinating interpretation of the biology of freshwater invertebrates, and Hutchinson (1993) presented a detailed ecology of selected groups of freshwater invertebrates. Benthic invertebrates exhibit innumerable modes of adaptation to variations in lake environment, to limitations of food resources and to interaction between species. The aim of this section is to integrate benthic invertebrates into the various trophic levels in a lake, to follow their ecological pathways and provide a synthesis which ties together previous sections.

Linking of ecological processes into an ecosystem framework to obtain an overall working system is a complex task. In some cases links will break, and the ecosystem will travel in unforeseen directions. However, new methods to quantify ecological processes have been developed during recent decades, and have greatly improved our ability to study causal relations. Tracer techniques (e.g. ^{14}C for measuring primary production and ^{210}Pb for sedimentation, serological and tracer techniques applied to feeding experiments in invertebrates, micro-electrodes for respiration and redox measurements, and many others) have been very helpful.

12.18.1   Energy equivalents of food resources

In section 12.16 it was shown how the energy content of lake sediment influences the community structure of the profundal zoobenthos. Sediment energy content ranges from 1 to 7 kJ g^{-1} dry weight. If we consider the energy content of benthic invertebrates as per gram AFDW, it amounts to 20–25 kJ. Therefore, for herbivores and detritivores to transform their food (the primary producers of lakes) into invertebrate flesh demands hard work. The energy content of primary producers ranges from 5 kJ g^{-1} dry weight in the common green alga *Cladophora* to 25 kJ g^{-1} dry weight for the green alga *Ulothrix*. The Charophyta and the diatoms fall into intermediate values. Particulate organic matter ranges from 2 to 25 kJ, the last being from running waters (Cummins & Wuycheck 1971). Benthic invertebrates use ingenious methods to break down the thick cell walls of plants, but apparently most of the process is carried out by bacteria, which break down the organic detritus. Bacteria, in their turn, act as food for invertebrates. Figures 12.18 & 12.19 support the idea that the detritivores are the main zoobenthos below the littoral surf zone.

In deep lakes, phytoplankton are the main primary producers. A comparison between their energy content (10–25 kJ g^{-1} dry weight) and that of the sediment in an oligotrophic lake (1 kJ g^{-1} dry weight) illustrates the opportunity for benthic life in deep lakes. In shallower, small lakes the macrophytic vegetation is the basic energy source, but its coarse texture renders it difficult for benthic invertebrates to degrade this material. Only a few larvae, such as those of the aquatic moth *Acentropus niveus* (Oliv) and the beetle larvae of Donaciinae, live in the stems of macrophytes. In deeper lakes, POM is transported along the bottom, or is directly sedimented (cf. sections 12.5 & 12.6). Here, macrophytic vegetation is of great value as habitat for a variety of species (cf. Fig. 12.12 & section 12.8), but it is of limited value in the total energy balance (Figs 12.41 & 12.42 & section 12.14).

12.18.2   Tracing energy pathways to zoobenthos

Two lines of approach have been followed in order to investigate the energy pathways of lakes. First, nitrogen limitation of phytoplankton encouraged the idea of quantification of processes, from production of phytoplankton, through sedimentation, to zoobenthos in the hypolimnion. A second approach was offered by oxygen gradients in the hypolimnion of eutrophic lakes, and the linking of the environmental regime to the animals surviving in the hypolimnion. This includes their feeding, respiratory adaptations and population dynamics. Both approaches have proved profitable.

Extensive research has been carried out on phosphorus limitation in lakes, and its consequences for pelagic life, in the Canadian Experimental Lakes Area (ELA) project (Schindler 1988). Pioneering research was carried out on the *A. formosa* Hassall population of Windermere, where the main focus was on silica (Lund 1950, 1964; Lund *et al.* 1963). Few nitrogen-limited lakes have been investigated. Best known, however, is the long-term investigation on nitrogen limitation and eutrophication of Lake Tahoe (Goldman 1988, 1993). Thingvallavatn and Mývatn, Iceland, also belong to this group (Jónasson & Adalsteinsson 1979; Ólafsson 1979, 1992; Jónasson 1992b, 1993, 1996). Further, Lake Esrom, Denmark is also nitrogen limited, in spite of being partly situated in a farmland area (Jónasson *et al.* 1974; Jónasson 1996).

Nitrogen-limited lakes exhibit very distinct peaks of phytoplankton production, which are suddenly interrupted by nitrogen limitation. This bimodal trend is especially clear after the spring maximum (Fig. 12.5). What happens when nitrogen limitation stops phytoplankton growth? These cases have mostly been studied during the spring maximum. Algae with high nitrogen and silicon requirements are replaced by other species with lower demands. In Windermere, owing to the decline of silica, *A. formosa* Hassall is succeeded by *Fragilaria crotonensis* Kitton and *Tabellaria flocculosa* var. *asterionelloides* (Lund 1950, 1964).

In Lake Esrom, *A. formosa, Stephanodiscus astrea* (Ehrenb.) Grun., *Stephanodiscus hantzschhii* Grun. and *Nitzschia avicularis* Smith are replaced by Chlorophyceae and Cryptophyceae (Jónasson & Kristiansen 1967). In Thingvallavatn, *Melosira islandica* and *Melosira italica* are followed by small diatoms, Chlorophyceae, Pyrrhophyta and Chrysophyceae (Jónasson *et al.* 1992a; Jónasson 1996; Kristiansen 1996). The result is that transparency increases from 5 m in spring to 15 m at the height of summer, and the colour of the lake changes from pale green to deep blue.

Such shifts in algal succession, from large to small species, meet physiological demands to increase their surface : volume ratio, and thereby better utilise lower nutrient concentration (see Chapter 10). This reduction in size means a fundamental change for zooplankton grazers and for the pelagic food chain of the epilimnion, but is also of great importance for small benthic herbivores. In fact, by reducing its size, the phytoplankton makes itself **more** available as food for zoobenthos.

The coordination between abundance of the pelagic larvae of the zebra mussel *D. polymorpha* and their phytoplankton food source is fascinating. The larvae feed on algae and are concentrated wherever and whenever primary production peaks. Furthermore, they occur only during stratification and at high temperatures (Fig. 12.24). They settle all over the lake, but most successfully in the sublittoral (Figs 12.12, 12.15 & 12.8). The billions of larvae which settle in the profundal perish. Although carnivorous leeches, thanks to their suckers, are especially fitted to live in the surf zone, *H. stagnalis* finds its niche on the leaves of macrophytes, especially in the lower littoral, where its prey lives (Fig. 12.20).

12.18.3 *Timing of energy accumulation in profundal zoobenthos*

In the hypolimnion conditions are entirely different. Food produced in the epilimnion is utilised by zoobenthos in the hypolimnion (cf. Jónasson 1978). Most importantly we have been able to trace phytoplankton maxima of bimodal production curves to the profundal, and to document accumulation of organic matter in the profundal zoobenthos. Figures 12.7, 12.30 & 12.33 show how the spring maximum of fresh phytoplankton may be traced to the lake bottom, and how it is further incorporated into the bodies of *C. anthracinus* and *P. hammoniensis*, as fat and as glycogen. *Chironomus anthracinus* uses this organic matter for metamorphosis, and for emergence and flight during the swarming season (Figs 12.29 & 12.31). *Potamothrix hammoniensis* uses it for growth and for production of cocoons (Figs 12.33 & 12.34; Jónasson 1972). *Pisidium amnicum* (O.F. Müller) and *Sphaerium corneum* (L.) have also been shown to accumulate glycogen (Holopainen 1987).

The profundal zoobenthos therefore uses the spring maximum of phytoplankton as a source of energy for reproduction, but a nitrogen-limited spring maximum is very limited in duration. It is a matter of surprise that the profundal zoobenthos *C. anthracinus, P. hammoniensis* and *P. casertanum* succeed in accumulating their glycogen reserves during this short span of time.

Data confirm that diatoms are sedimented more rapidly during the spring maximum than during late summer maximum and stratification (Jónasson *et al.* 1974, fig. 9). Again, this is a confirmation that detritivores survive on degraded organic matter, but grow and reproduce during periodic access to fresh food. Sedimenting diatoms are nutrient rich, in that 50% of the soft parts consists of fat, and they are easily digestible food for the detritivores, i.e. they are the 'cream on the detritus cake' (Fig. 12.7; see Jónasson 1964, 1965, 1972, 1978; Kajak & Warda 1968; Morton 1969; Lastein 1976).

Food is abundant in the profundal zone of eutrophic lakes during the summer stratification, but cannot be utilised by animals owing to low oxygen concentrations. Instead, the profundal zoobenthos ceases feeding, reduces its energy costs and turns to anaerobic respiration, utilising its reserves of glycogen. A sharp distinction must be made between surface and subsurface feeders. In the profundal, subsurface feeders utilise bacteria and organic compounds which are likely to have passed through the intestines of the surface feeders. *Chironomus anthracinus* thus feeds on high-energy food at the mud surface whereas *P. hammoniensis* and *P. casertanum* subsist on sub-surface, low-energy food. This point is demonstrated convincingly by Figs 12.27, 12.28 & 12.32 (see also Jónasson 1972, figs 39–42). The sediment characteristics in terms of energy are shown in Fig. 12.43.

Experiments with *C. anthracinus* show that when fed at 13°C on Lake Esrom mud enriched with dry fish food, animals grew to full size and emerged within 3 weeks. Larvae offered mud from below 2 cm depth starved to death. In contrast, *P. hammoniensis* feeds and grows on this mud as well as on mixed mud. There is competition for food if population densities are high (Jónasson 1978). Another consequence of subsurface feeding is that glycogen accumulation in *P. hammoniensis* is delayed by 2 months from the spring maximum of phytoplankton (Fig. 12.33). Food resources in terms of phytoplankton, their sedimentation and intrinsic food value, and oxygen concentration at the mud surface are all factors which are linked together and thus capable of regulating growth, timing and quantity of metamorphosis in *C. anthracinus*, and timing of growth and reproduction in *P. hammoniensis*.

12.18.4   Metabolism and growth

Metabolic rates vary drastically with depth in the sediment. Table 12.9 shows that the respiration rate of the surface feeding *C. anthracinus* at air saturation is twice that of the subsurface feeding *P. hammoniensis* and *P. casertanum*, and this is even more pronounced at 5% air saturation. Recent data confirm this difference in metabolism at air saturation between *C. anthracinus* and *P. hammoniensis* (Hamburger *et al.* 1994, 1998).

The mechanism of tilting of the thermocline, and of the resulting periodical supply of oxygen to the profundal zoobenthos, is described in Figs 12.8, 12.10 & 12.11. Field measurements in Lake Esrom demonstrated the existence of a gradient of declining oxygen just above the mud surface (Berg & Jónasson 1965), whilst laboratory experiments emphasised the permanent lack of oxygen in the sediment itself (Fig. 12.28).

Respiratory metabolism becomes entirely different above and below the thermocline, owing to contrasting limits of oxygen uptake, and consequently of growth, amongst various species. Table 12.10 summarises respiratory metabolism for sublittoral and profundal species at air saturation. It shows that the metabolic rate of *C. anthracinus* in the sublittoral is almost twice that of the same species in the profundal (Berg *et al.* 1962, fig. 16). However, Hamburger *et al.* (1994) were not able to

Table 12.9   Production, production/biomass (P/B) and respiration for surface and subsurface feeding profundal species.

	Production ($kJ\,m^{-2}\,yr^{-1}$)	P/B (yr^{-1})	Respiration ($\mu L\,O_2\,(g\,wet\,wt)^{-1}\,h^{-1}$)	
			Air saturation	5% saturation
Surface feeding				
Chironomus anthracinus	315.5	2.1–3.8	29.9	18.0
Subsurface feeding				
Potamothrix hammoniensis	25.0	0.8	13.1	3.3
Pisidium casertanum	3.6	1.0	16.8	1.8

Table 12.10 Oxygen consumption for sublittoral and profundal species at air saturation (Jónasson 1978).

Species	Respiration ($\mu L O_2 g^{-1}$ wet wt h^{-1} at 11 °C)	
	Sublittoral	Profundal
Psammoryctides barbatus	23.1	
Potamothrix hammoniensis		13.1
Chironomus anthracinus	51.6	29.9

Table 12.11 Biomass of *Chironomus anthracinus* larvae at various depths during late September, at the end of stratification in Lake Esrom (Berg 1938).

Depth (m)	Wet weight (mg individual^{-1})
11	7.4
14	7.6
17	3.9
20	1.6

confirm this result, perhaps because the former set of respiration measurements were related to wet weight and the latter ones to dry.

In terms of growth, this point is best illustrated by the biomass of *C. anthracinus* larvae at various depths during late September, at the end of stratification (Table 12.11; Berg 1938). The explanation is clear, animals that have been in touch with oxygen-rich epilimnetic water grow faster than those solely exposed to hypolimnetic water. The limit at which growth ceases during stratification is at approximately $0.5–1 \, mg O_2 \, L^{-1}$ (Jónasson 1972, figs 53A–53B). At 10°C and under varying oxygen regimes, the oxygen consumption and carbon dioxide production of fourth instar larvae from saturation to about $3 \, mg O_2 \, L^{-1}$ are almost unaltered, but decline steeply below this value (Fig. 12.21; Hamburger *et al.* 1994).

The respiratory quotient increases from 0.82 at saturation to about 3.4 at oxygen concentrations close to $0.5 \, mg O_2 \, L^{-1}$. This change implies a shift from aerobic to partially anaerobic metabolism. At $0.5 \, mg O_2 \, L^{-1}$ the total metabolism of fourth instar larvae equalled 20% of the rate at saturation, and more than one-third was accounted for by anaerobic degradation of glycogen. This corresponds to a daily loss of $12 \, \mu g \, mg \, AFDW^{-1}$, or approximately 5% of body reserves. At an unchanged metabolic rate the glycogen store would last for 3 weeks, but long-term oxygen deficiency causes further suppression of energy metabolism in *C. anthracinus* (Hamburger *et al.* 1994). Different metabolic rates in the sublittoral and the profundal thus funda-

mentally influence the population dynamics of *C. anthracinus*. The organism changes from a univoltine species at 11–14 m depth to partially hemivoltine at 17–21 m (Fig. 12.31).

A further striking example is the contrasting metabolism and vertical distribution of *Psammoryctides barbatus* and *Potamothrix hammoniensis* in Lake Esrom. The metabolic rate of the former (a large species which occurs only in the littoral and sublittoral) is twice that of the latter (Table 12.10 & Fig. 12.21). However, *P. hammoniensis* not only occurs in the littoral (with 3500–8000 individuals m^{-2}) and the sublittoral (where it is common), but is also found as a monoculture in the profundal, with an average of 10,000–24,000 individuals m^{-2} (Jónasson & Thorhauge 1972; Lindegaard & Dall 1988). The explanation is that its critical limit of respiration lies at 9.5% air saturation, versus 19% for *Psammoryctides barbatus* (Fig. 12.21). *Potamothrix hammoniensis* is thus especially adapted to grow at very low oxygen saturations, and becomes a typical inhabitant of the profundal of eutrophic lakes.

In the oxygen-rich hypolimnion of deep, oligotrophic lakes, ecological conditions are entirely different. The example of the 114 m deep Thingvallavatn demonstrates well the sedimentation and degradation of organic matter in the oxygen rich hypolimnion. The low energy content of the sediment supports only those zoobenthos capable of existing on low energy food, i.e. almost a monoculture of the oligochaete *T. tubifex* (Müller) (Fig. 12.43). Furthermore, enchytraeids occupy the upper 10 m, whilst *T. tubifex* occur throughout the depth range 10–114 m (Lindegaard 1992a).

The explanation may well be that the higher energy content of littoral detritus, as opposed to that of the profundal, favours enchytraeids, because their metabolic rate is twice that of *T. tubifex* (cf. Lindegaard *et al.* 1994; section 12.15). The 370 m deep Lake Maggiore supports only a few species of oligochaete, and (if oxygen concentration is above 6.5 mg L⁻¹) the amphipod *Niphargus foreli* (Bonomi 1967). Certain ice-age relict species of central and northern Europe (*Pontoporeia affinis* Lundström, *Mysis relicta* Lóvén) cannot exist below c.4–6 mg O₂ L⁻¹, whilst *P. quadrispinosa* G.O. Sars survives at 3 mg L⁻¹. This is the general scheme for the profundal of oligotrophic lakes (Thienemann 1925b, 1928b).

The respiratory activity of various invertebrate phyla may be illustrated by Fig. 12.48. Generally, Malacostraca possess the highest oxygen uptake and are thus best fitted for life in the epilimnion. Oxygen uptake of molluscs, oligochaetes and insects is lower and profundal species survive in that environment owing to special adaptations to low oxygen availability (Hamburger & Dall 1990, cf. sections 12.11 & 12.12).

In summary: in the profundal of eutrophic lakes, where ample organic matter is available and substrate facilities are favourable, we may find a large population of chironomids with high metabolic demands and with a subtle adaptation to oxygen uptake. Only a few species (e.g., *C. anthracinus*, *P. hammoniensis*, especially, and *P. casertanum*) are able to force the barrier of the metalimnion and adapt to low anaerobic metabolism (Fig. 12.21). Momentary dwellers, such as *C. flavicans*, migrate daily. In deep oligotrophic lakes, where organic matter is degraded, we find a population of tubificids with low metabolic demands (see section 12.12, and Zschokke 1911; Ekman 1915; Wiederholm 1974b; Lindegaard & Jónasson 1979; Lindegaard 1992a; Jónasson 1996).

12.18.5 Voltinism, changes in life-histories and turnover ratios of organic matter in the lake transect

From previous sections it may be concluded that the influence of the distribution of organic matter in a lake basin on the species diversity of zoobenthos, types of feeding, growth, accumulation of biomass, population dynamics and production as well as turnover (P/B) ratios of materials is profound. Faster growth in the littoral often allows the production of more generations per unit time. Thus, the number of generations that a species is able to produce within 1 year (**voltinism**) is a significant expression of life conditions (Tokeshi 1995).

Lindegaard (1992a) described a number of multivoltine benthic invertebrates from various phyla with asexual reproduction and high values of P/B. Bivoltine chironomids from the littoral of Thingvallavatn exhibited very high values, ranging from 10.2 to 11.8. They therefore almost reach the high metabolic turnover rates of zooplankton (see Chapter 14). Lindegaard (1992a) also gave data for 41 species of zoobenthos from Thingvallavatn. In contrast to chironomids, he found 3–7 year lifecycles for the oligochaetes *S. ferox* and the profundal *T. tubifex*, with P/B ratios of 0.5 and 1.4, respectively.

In the littoral zone of Lake Esrom, at 2 m depth, E. Jónsson (1985, 1987) examined life-cycles of 11

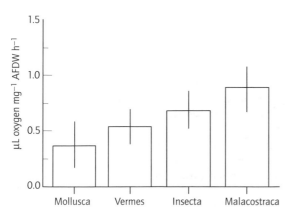

Fig. 12.48 Respiratory activity of various invertebrate phyla (at 10°C and O₂ saturation). (Modified from Hamburger & Dall 1990.)

species of chironomids and found three types of voltinism: **uni**-, **bi**- and **trivoltine** species, but a fourth type could be either uni- or bivoltine. The main type of life-cycle present was bivoltine. Trichoptera are mostly univoltine (Dall *et al.* 1984a; E. Jónsson 1987; Wermer 1994), but those of the surf zone (e.g. *Tinodes waeneri*) may be bivoltine (Dall *et al.* 1984b). In contrast, the profundal-living *C. anthracinus* adopts both uni- and hemivoltine life-cycles, depending partly on depth (Fig. 12.31), whilst *C. flavicans* and *P. pectinatus* are univoltine, owing to seasonal migration to other habitats.

For the profundal, a summary of surface- and subsurface-living species was given in Table 12.5, and in Figs 12.29, 12.31, 12.34 & 12.37. Growth rate of the surface feeder *C. anthracinus* is much higher, and its life-cycle much shorter (1–2 yr versus 4–6 yr), than those of *P. hammoniensis* and *P. casertanum*. Production rates are at least ten times higher. Turnover rates (P/B) are two to five times higher, ranging from 0.7 to 3.8 (Jónasson 1972, 1975, 1978; Jónasson & Thorhauge 1976a, b; Holopainen & Jónasson 1983, 1989a, b).

In the littoral of Lake Esrom, P/B ratios of molluscs are low — 1.3 for the prosobranch snail *T. fluviatilis*, which follows a 4-year life-cycle living in the stony surf zone, in some years surviving long periods of ice-cover (Fig. 12.23). The detritus feeding snail *Potamopyrgus antipodarum* (Gray) exhibits a 1-year life-cycle and a P/B ratio in the range of 2.7–4.8. In Thingvallavatn, the P/B of the common hemivoltine snail *L. pereger* is 2.8 (Kirkegaard 1980; Lindegaard 1992a; Winther & Dahl 1998). Proceeding to the sublittoral zone, the life-span of *Dreissena* is long (5–6 yr). Although accumulation of biomass is very high, production and P/B ratios are as low as 0.57 (Morton 1969; Stanczykowska 1976, Dall & Hamburger 1996). Habitat is important, since the P/B ratio of the pulmonate *L. pereger* in the River Thames is 11.5 (Mann 1964, Mann *et al.* 1972).

Metabolism, growth and voltinism thus exhibit considerable bathymetric changes in a lake transect. Chironomids change from a bi- or trivoltine cycle in the littoral to a hemivoltine cycle in the profundal. Other benthic invertebrates change from multivoltine life-cycles in the littoral to 3–7 year cycles in the profundal.

12.19 ZOOBENTHOS AND THE OLIGOTROPHIC/EUTROPHIC LAKE TYPE

Each lake is different but we have attempted to analyse and to synthesise some relations between zoobenthos and lakes. A lake is a microcosm, its own world, where the ecological processes are closely tied together (Forbes 1887). Yet, it is possible to distinguish lake types, and within those of the same type ecological processes proceed in a similar way. The theme in this section is restricted to 'biological lake types' and zoobenthos of temperate lakes.

In a previous section the importance of lake morphometry and the general terms oligo- and eutrophic were introduced (Fig. 12.3). The visionary limnologist August Thienemann (1918a, b; 1925a, 1928a) was the first to realise the importance of lake basin shape for characterisation of profundal zoobenthos. Is it relevant, and is there a real basis for discussing these two lake types in the light of present research? Yes, in fact these types are so different that they have become an abstract concept. The Thienemann system was causal. His approach was to restrict the system — the *Tanytarsus* and *Chironomus* lakes — to two lake types, and two genera of profundal chironomids and the oxygen content of their habitat, which was measurable.

Thienemann (1922) first discovered that the tolerance of lack of oxygen of *C. plumosus* was greater than that of *C. anthracinus*, and that it survived oxygen scarcity down to zero in the profundal of central and north European eutrophic lakes. This resistance to oxygen scarcity could not be explained through the measurements by Walshe (1950), who found a critical limit of respiration at 56% air saturation. However, new data on respiratory adaptation show the critical limit of *C. plumosus* to be lower than that of *C. anthracinus* (8 versus 25% air saturation). They therefore finally confirm Thienemann's theses (Fig. 12.21;

Hamburger *et al.* 1994; Dall & Hamburger unpublished data, personal communication, 2001).

12.19.1 The trophic approach

For almost 100 years activity has been devoted to the study of chironomids in the profundal of different lake types. Lundbeck (1926, 1936) established a new system based on the nutrient status of the sediment. He further divided the Thienemann system into three trophic types:

1 oligotrophic lakes with *Orthocladius* and *Tanytarsus*;

2 mesotrophic lakes with *Stictochironomus* and *Sergentia*;

3 eutrophic lakes with *C. plumosus* and *C. anthracinus*.

This empirical system was also causal, and based on nutrient uptake and requirements of chironomids. However, in contrast to Thienemann's oxygen-based system the requirements were not known. In fact, the nutrient value of lake sediment was not measured until recently (Fig. 12.43; Jónasson 1992b).

The system became accepted in central and northern Europe (Lenz 1925, 1927), and Brundin (1949, 1956, 1958) further added the ultra-oligotrophic lake type, with *Heterotrissocladius subpilosus*. This empirical system then became accepted for Europe, central Asia, Japan and South America (Table 12.12), and adapted by Sæther (1975) to North America. Sæther (1979) then further developed it to comprise 15 profundal communities of chironomids and their relation to total phosphorus or Chl *a* and mean lake depth. The system has been tested by various authors (Gerstmaier 1989; Aagaard 1986; Lindegaard 1992a; Lindegaard & Mæhl 1993).

Fundamental ecological contributions were made by Wiederholm (1980) during his studies of the Great Lakes of central Sweden. He introduced a **benthic quality index** (BQI) of chironomids for seven indicator taxa, and also an oligochaete/oligochaete plus chironomid ratio (O/O+C), based on numbers of individuals. Both indexes are related to total phosphorus, Chl *a* and to mean lake depth when used for describing pollution status.

Numerous qualitative and quantitative investigations of oligochaetes, the other important profundal group, and their vertical distribution have appeared (cf. Figs 12.18 & 12.19; Dall *et al.* 1984a; Lindegaard & Dall 1988; Lindegaard 1992a). Much has been published on the bacterial feeding of Oligochaeta (Ivlev 1939; Sorokin 1966; Coler *et al.* 1967; Wavre & Brinkhurst 1971; Brinkhurst *et al.* 1972). Recent work has confirmed the ability of oligochaetes to select their food (Poddubnaja 1961; Brinkhurst & Austin 1979). Sorokin (1966) demonstrated with ^{14}C experiments that *Limnodrilus hoffmeisteri* foraged between 1.8 and 2.7 cm below the mud surface, whilst Krezoski *et al.* (1978), using ^{137}Cs, showed *Stylodrilus heringianus* to feed between 3 and 6 cm depth.

From his extensive oligochaete studies in the English Lake District, Reynoldson (1990, p. 333)

Table 12.12 Indicator species of chironomids and oligochaetes for various lake types (C. Lindegaard, pers. comm.).

Trophic state	Indicator chironomids	Indicator oligochaetes
Ultraoligotrophic	*Heterotrissocladius subpilosus*	?
Oligotrophic	*Tanytarsus lugens*	*Stylodrilus heringianus*
	Heterotrissocladius grimshawi	*Spirosperma ferox*
	Heterotrissocladius scutellatus	(*Tubifex tubifex*)
Mesotrophic	*Stictochironomus rosenschoeldi*	*Psammoryctides barbatus*
	Sergentia coracina	
Eutrophic	*Chironomus anthracinus*	*Potamothrix hammoniensis*
Strongly eutrophic	*Chironomus plumosus*	

concluded that: 'it is hypothesized that the major factors determining the profundal oligochaete community structure in the English Lakes are temperature, dissolved oxygen and primary productivity and the interaction of these factors determines the relative dominance and distribution of oligochaete species'. These factors coincide with those used for classification of oligotrophic and eutrophic profundal zoobenthos (cf. Fig. 12.43), but improved quantification and physiological evidence is needed. In a sample of 187 Estonian lakes, *T. tubifex* was present in 23, and dominant in a few, whilst *P. hammoniensis* occurred as the only species in 136 eutrophic lakes (Timm 1996).

Milbrink (1980) classified 28 species of oligochaetes in relation to degree of pollution. According to present evidence, Lindegaard selected five indicator species of oligochaetes which fit into the chironomid system (Table 12.11). In an ultra-oligotrophic lake in Greenland, no oligochaetes were found (Lindegaard & Mæhl 1993). In Thingvallavatn, the three characteristic species of the oligotrophic lake profundal, *S. heringianus*, *S. ferox* and *T. tubifex*, were all found (Lindegaard 1992a). The ecology of *P. hammoniensis* has been treated in several papers, which describe its respiratory physiology, utilisation of glycogen and general ecology (Figs 12.21 & 12.32–12.34). The respiratory adaptation of *Psammoryctides barbatus*, with its critical limit at 19% air saturation, was described in Fig. 12.21 (above).

An empirical classification system, based on a few key species of profundal zoobenthos, has been proposed for characterisation of lake types (Table 12.12), but it has been criticised from various points of view (Alsterberg 1930, 1931; Berg 1938; Nygaard 1938; Wesenberg-Lund 1943). Because of the difficulties of obtaining unambiguous quantification of the relationship to environmental factors (Macan 1970; Brodersen *et al.* 1998), the littoral zoobenthos, though species-rich, would not be any easier to use than the profundal fauna.

12.19.2   The dynamic approach

The concept of production was tied to lake typology at the very beginning of limnology, but merely as a phrase. This was because, at that time, production was not measurable. This section aims at a dynamic approach to zoobenthos and lake typology, and especially to the links between primary and secondary producers. The profundal zoobenthos involves specialists with fascinating ecology and special characteristics. Recent research has quantified many ecological processes in lakes, but only few attempts have been made to measure production at all ecological levels.

Of fundamental importance for herbivorous zoobenthos is the existence of extensive grazing fields of algae, but the production of these phytobenthos has only been measured in a few cases (Alexander & Barsdate 1971; Hunding 1971; Welch & Kalff 1974; Dall *et al.* 1984a; Jónsson 1992; Kairesalo *et al.* 1992). Actual values of primary and secondary net production may be presented for Thingvallavatn and for Lake Esrom because we possess representative and comparable data from these lakes. The comparison is as follows. In Fig. 12.17 it was shown that, in the oligotrophic Thingvallavatn, production of both phytobenthos and zoobenthos decline in parallel with depth. This indicates that secondary production is directly related to net phytobenthos production. Also of fundamental importance for zoobenthic grazers is that the biomass of the phytobenthos should be ten times that of the phytoplankton (500 versus 50 kJ m^{-2}; Jónasson 1992b, fig. 20), but that the turnover rate P/B should only be 2.6. In contrast, the turnover rate of phytoplankton may be as high as 64.

Secondary production of zoobenthos in the oligotrophic lake is thus directly dependent on the high biomass of phytobenthos, whilst the low phytoplankton biomass is distributed over a large volume of water. The result is an oxygen-rich hypolimnion, with an orthograde oxygen profile and profundal sediments low in energy content (cf. Fig. 12.43). In contrast, in dimictic, eutrophic lakes, phytobenthos and phytoplankton produce large amounts of organic matter, which when transported to a small hypolimnion volume produces a clinograde oxygen curve.

For example, in the eutrophic, dimictic Lake Esrom, phytobenthos production in the surf zone is five times greater (20,000–25,000 kJ m^{-2} yr^{-1}) than that of the same zone in the oligotrophic Thingvallavatn (4000 kJ m^{-2} yr^{-1}.). Along with a phytoplankton production which is 2.5 greater in Lake Esrom (Fig. 12.3), this leads to an entirely different vertical distribution and production of zoobenthos (Figs 12.12 & 12.49, & Table 12.4). Secondary production increases from the surf zone, to peak in the sublittoral. In spite of high primary phytobenthos production in the surf zone, secondary production in the sublittoral is almost twice that of the surf zone. This indicates that transport of organic matter produced by phytoplankton and phytobenthos is very important in the lower littoral and sublittoral zones.

In fact, Fig. 12.49 illustrates the fundamental difference between the zoobenthos of oligotrophic and eutrophic lakes. Secondary production in the littoral of Lake Esrom is 2.5–5 times greater than in Thingvallavatn, but, on comparing the sublittoral and profundal of Lake Esrom to the *Nitella opaca*

(Charophyta) belt and the profundal of Thingvallavatn, production is **fifteen** times higher. This finding confirms previous assumptions that oligotrophic lakes may provide a sublittoral belt of Charophyta as an important ecological niche for the sparse population of filtrating zoobenthos (cf. Fig. 12.17).

In oligotrophic lakes, organic matter is the limiting factor for zoobenthos, whereas in the eutrophic lake it is oxygen. This may well be one of the reasons why filtrating zoobenthos in the Charophyta belt of oligotrophic lakes never become as numerous as those in eutrophic lakes. In the oligotrophic lake, P/B ratios of the zoobenthos decline gradually from the littoral surf zone to the profundal, whilst in the eutrophic lake this decline begins in the littoral macrophyte zone (Fig. 12.49). This figure also shows that in the oligotrophic lake the main zoobenthos community consists of the herbivorous grazers of the littoral, whilst in the eutrophic lake the main zoobenthos are the filtering

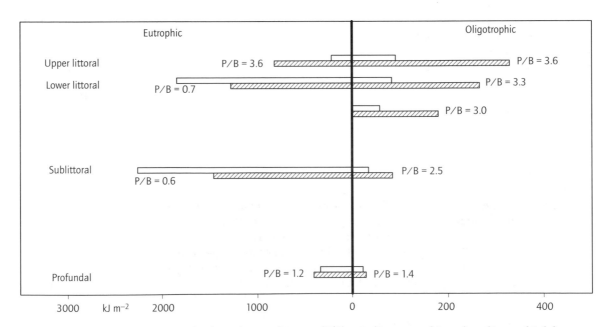

Fig. 12.49 Biomass and production (with production/biomass (P/B) ratios) in a eutrophic and an oligotrophic lake shown as a vertical transect through the littoral, sublittoral and profundal zones. Please note the different scales of the abscissa (modified from Figs 12.41 & 12.42).

community of the sublittoral and (if oxygen conditions permit) the sediment-feeding community of the profundal (cf. Table 12.1).

It is likely that production in European eutrophic lakes would be less without *D. polymorpha*, although other mollusc species are numerous (Fig. 12.15). It will therefore be interesting to follow the establishment of *D. polymorpha* in the oligotrophic regions of the Great Lakes of North America (Fraleigh *et al.* 1993, Coakley *et al.* 1997) in order to see whether this will also be the case on that continent.

12.19.3 *Ecological characterisation of the profundal key species*

Investigations have shown that *C. plumosus* prefers rather shallow, warm, highly eutrophic lakes where it acts as a filtrator of fresh, high-energy plankton. Simultaneously, it irrigates its body and the tube in which it lives with oxygen, in an oxygen-free sediment (Walshe 1947, 1951; Frank 1982, 1983; Lastein, personal communication, 1997). *Chironomus plumosus* also occurs in deep, eutrophic lakes, but then often in the sublittoral, in warm metalimnetic water, with a higher concentration of organic matter available for filtration (Wiederholm 1974a, fig. 21). In contrast, *C. anthracinus* occurs as a deposit feeder in the cool profundal of temperate, eutrophic lakes. In eutrophic, dimictic lakes it accumulates fresh high-energy food as fatty acids and glycogen, but only during the spring and autumn diatom blooms (Fig. 12.30; Jónasson 1972).

Thus, *C. plumosus* utilises high-energy food at high temperatures and at a low critical limit of respiration, which seems to explain its habitat selection. The dietary differences between it and *C. anthracinus* were recently confirmed, in that *C. plumosus* utilises the fatty acids of the algal spring bloom immediately for growth, whilst utilisation by *C. anthracinus* is delayed (Goedkoop *et al.* 1998). In the subsurface feeder *P. hammoniensis* this response is further delayed by 2 months, in relation to *C. anthracinus* (cf. Figs 12.30 & 12.33). Further, similar resource partitioning is exhibited by *C. islandicus*. This species covers the profundal

of the shallow, eutrophic Mývatn with an average of 7000 individuals m^{-2} (compared with 275 m^{-2} in the sublittoral *Nitella* belt of the oligotrophic Thingvallavatn).

This section has emphasised new aspects of the study of zoobenthos (e.g. the close relation between survival and habitat selection, and feeding and the timing of energy requirements). It is to be expected that, in eutrophic lakes, respiratory adaptation is a crucial factor in distributing benthic invertebrates according to depth. This factor operates via various critical limits of respiration, which also affect the level of metabolism. Anaerobic respiration, which contributes only a fraction compared with aerobic metabolism, fundamentally influences growth and life-cycles of zoobenthos in the profundal.

The ability of *C. anthracinus* and *P. hammoniensis* to survive long-term oxygen deficiency depends on their ability to save energy through an effectively suppressed energy metabolism. Whereas aerobic metabolism appears constant with time, after a few days of oxygen deficiency anaerobic metabolism declines, and constitutes an insignificant part of total energy production. Thus these species survive long microxic periods (Berg & Jónasson 1965; Hamburger *et al.* 1994).

Glycogen was found to be an important energy source in *C. anthracinus* during periods such as ecdyses and metamorphosis, with high internal energy requirements, and also during the non-feeding adult life-stage, as well as periods with microxic conditions (cf. Fig. 12.30; Augenfeld 1967; Jónasson 1972, 1996; Hamburger *et al.* 1995, 1996). A similar mechanism seems to apply to *P. hammoniensis*, which increases in weight, matures and produces cocoons during periods of high glycogen content (cf. Figs 12.33 & 12.34; Jónasson 1972, 1996; Hamburger *et al.* 1998).

The third important taxon living in the profundal consists of the genus *Pisidium*. The critical limit of respiration of *P. casertanum* lies at 21.5% air saturation (Fig. 12.36), which seems too high for a species living permanently in the profundal. In fact, various levels of respiration for this species have been described, and both *P. amnicum* and *Sphaerium corneum* survive anoxic periods using glycogen resources (Holopainen 1987; Holopainen & Penttinen 1993). Preliminary measurement on single specimens indicates a lower critical limit of respiration and a more favourable adaptation (Dall, personal communication, 1997).

12.20 ACKNOWLEDGEMENTS

I am much indebted to Professors T.B. Reynoldson D.Sc. and the late C. Overgaard Nielsen Dr Phil., to Drs P. C. Dall, K. Hamburger, E. Lastein and C. Lindegaard for valuable criticism during preparation of this paper, and to H. Heegaard, K. Therkildsen, L. Andersen, F. Pedersen, F. Brundam and E. Leenders for technical assistance.

12.21 REFERENCES

Aagaard, K. (1986) The chironomid fauna of north Norwegian lakes, with a discussion on methods of community classification. *Holarctic Ecology*, **9**, 1–12.

Alexander, V. & Barsdate, R.J. (1971) Physical limnology, chemistry and plant productivity of a Taiga lake. *Internationale Revue der Gesamten Hydrobiologie*, **56**, 825–72.

Alimov, A.A. (1983) Energy flows in populations and communities of aquatic animals. *Internationale Revue der Gesamten Hydrobiologie*, **68**, 1–12.

Alsterberg, G. (1930) Die thermischen und chemischen Ausgleiche in den Seen zwischen Boden- und Wasserkontakt sowie ihre biologische Bedeutung. *Internationale Revue der Gesamten Hydrobiologie*, **24**, 290–327.

Alsterberg, G. (1931) Die Ausgleichströme in den Seen im Sommerhalbjahr bei Abwesenheit der Windwirkung. *Internationale Revue der Gesamten Hydrobiologie*, **25**, 1–32.

Ambühl, H. (1959) Die Bedeutung der Strömung als ökologischer Faktor Physikalische, biologische und physiologische Untersuchungen über Wesen und Wirkung der Strömung im Fliessgewässer. *Schweizerische Zeitschrift für Hydrologie*, **21**, 133–264.

Ambühl, H. (1969) Die neueste Entwicklung des Vierwaldstättersees (Lake of Lucerne). *Verhandlungen der Internationalen Vereinigung Limnologie*, **17**, 219–30.

Augenfeld, J.M. (1967) Effects of oxygen deprivation on aquatic midge larvae under natural and laboratory conditions. *Physiological Zoology*, **40**, 149–58.

Bennike, S.A. (1943) Contributions to the ecology and biology of the Danish freshwater leeches (Hirudinea). *Folia Limnologica Scandinavia*, **2**, 1–109.

Berg, K. (1937) Contributions to the biology of *Corethra* Meigen (*Chaoborus* Lichtenstein). *Kongelige Danske Videnskabernes Selskab, Biologiske Meddelelser*, **13** (11), 1–101.

Berg, K. (1938) Studies on the bottom animals of Esrom Lake. *Kongelige Danske Videnskabernes Selskab Skrifter, Naturvidemskabelig og Mathematisch Afdeling*, **9**(8), 1–255.

Berg, K. (1952) On the oxygen consumption of Ancylidae (Gastropoda) from an ecological point of view. *Hydrobiologia*, **4**, 225–67.

Berg, K. & Jónasson, P.M. (1965) Oxygen consumption of profundal lake animals at low oxygen content of the water. *Hydrobiologia*, **26**, 131–43.

Berg, K. & Ockelmann, K.W. (1959) The respiration of freshwater snails. *Journal of Experimental Biology*, **36**, 690–708.

Berg, K. & Petersen, I.C. (1956) Studies on the humic acid Lake Gribsø. *Folia Limnologica Scandinavia*, **8**, 1–264.

Berg, K., Jónasson, P.M. & Ockelmann, K.W. (1962) The respiration of some animals from the profundal zone of a lake. *Hydrobiologia*, **19**, 1–40.

Birge, E.A. (1922) A second report on limnological apparatus. *Transactions of the Wisconsin Academy of Science, Arts and Letters*, **20**, 533–53.

Bjarnov, N. (1972) Carbohydrases in *Chironomus*, *Gammarus* and Trichoptera of various trophic levels. *Oikos*, **23**, 261–3.

Bonomi, G. (1967) L'evoluzione recente del Lago Maggiore rivelata dalle cospicue profondo. *Memorie dell' Istuto Italiano di Idrobiologia*, **21**, 197–212.

Brinkhurst, R.O. & Austin M.J. (1979). Assimilation by aquatic Oligochaeta. *Internationale Revue der Gesamten Hydrobiologie*, **63**, 863–68.

Brinkhurst, R.O., Chua, K.E. & Kaushik, N.K. (1972) Interspecific interactions and selective feeding by tubificid oligochaetes. *Limnology and Oceanography*, **17**, 122–33.

Brinkhurst, R.O., Boltt, R.E., Johnson, M.G., Mozley, S. & Tyler, A.V. (1974). *The Benthos of Lakes*. Macmillan, 190 pp.

Brodersen, K.P., Dall, P.C. & Lindegaard, C. (1998) The surf zone fauna of Danish lakes: macroinvertebrates as trophic indicators. *Freshwater Biology*, **39**, 577–92.

Brodersen, K.P., Anderson, N.J. & Odgaard, B.V. (2001) Long-term trends in the profundal chironomid-fauna in nitrogen-limited Lake Esrom, Denmark: a combined palaeological/historical approach. *Archiv für Hydrobiologie*, **150**, 393–409.

Brundin, L. (1949) Chironomiden und andere Bodentiere der Südschwedischen Urgebirgseen. *Report of the Institute of Freshwater Research, Drottningholm*, **30**, 1–915.

Brundin, L. (1956) Die bodenfaunistischen Seetypen und ihre Anwendbarkeit auf die Südhalbkugel. Zugleich eine Theorie der produktions-biologischen Bedeutung der glazialen Erosion. *Report of the Institute of Freshwater Research, Drottningholm*, **37**, 186–235.

Brundin, L. (1958) The bottom faunistical lake type system and its application to the southern hemisphere. Moreover a theory of glacial erosion as a factor of productivity in lakes and oceans. *Verhandlungen der Internationalen Vereinigung Limnologie*, **13**, 288–97.

Buscemi, P.A. (1958) Littoral oxygen depletion produced by a cover of *Elodea canadensis*. *Oikos*, **9**, 239–45.

Coakley, J.P., Brown, G.R., Ioannou, S.E. & Charlton, M.N. (1997) Colonization patterns and densities of zebra mussel *Dreissena* in muddy offshore sediments of Western Lake Erie, Canada. *Water, Air and Soil Pollution*, **99**, 623–32.

Coler, R.A., Gunner, H.B. & Zuckerman, B.M. (1967) Selective feeding of tubificids on bacteria. *Nature*, **216**, 1143–4.

Cummins, K.W. (1973) Trophic relations of aquatic insects. *Annual Review of Entomology*, **18**, 183–206.

Cummins, K.W. & Wuycheck J.C. (1971) Caloric equivalents for investigations in ecological energetics. *Mitteilungen der Internationale Vereinigung für Limnologie*, **18**, 1–158.

Dall, P.C. (1983) The natural feeding and resource partitioning of *Erpobdella octoculata* L., and *Erpobdella testacea* SAV. in Lake Esrom, Denmark. *Internationale Revue der Gesamten Hydrobiologie*, **68**, 473–500.

Dall, P.C. & Hamburger, K. (1996) Recruitment and growth of *Dreissena polymorpha* in Lake Esrom, Denmark. *Limnologica*, **26**, 27–37.

Dall, P.C., Lindegaard, C., Jónsson, E., Jónsson, G. & Jónasson, P.M. (1984a) Invertebrate communities and their environment in the exposed littoral zone of Lake Esrom, Denmark. *Archiv für Hydrobiologie*, **69** (Supplement), 477–524.

Dall, P.C., Heegaard, H. & Fullerton, A.F. (1984b) Life-history strategies and production of *Tinodes waeneri* (L.) (Trichoptera) in Lake Esrom, Denmark. *Hydrobiologia*, **112**, 93–104.

Dall, P.C., Lindegaard C. & Jónasson, P.M. (1990) In-lake

variations in the composition of zoobenthos in the littoral of Lake Esrom, Denmark. *Verhandlungen der Internationalen Vereinigung Limnologie*, **24**, 613–20.

De Bernardi, R., Giussani, G. & Grimaldi, E. (1984) Lago Maggiore. In: Taub, F. (ed.) *Ecosystems of The World*, Vol. 23. Elsevier, Amsterdam, 247–66.

Edmondson, W.T. (1966) Changes in the oxygen deficit of Lake Washington. *Verhandlungen der Internationalen Vereinigung Limnologie*, **16**, 153–8.

Ehrenberg, H. (1957) Die Steinfauna der Brandungsufer ostholsteinischer Seen. *Archiv für Hydrobiologie*, **53**, 87–159.

Ekman, S. (1911) Neue Apparate zur qualitativen und quantitativen Erforschung der Bodenfauna. *Internationale Revue der Gesamten Hydrobiologie*, **3**, 553–61.

Ekman, S. (1915) Die Bodenfauna des Vätterns, qualitativ und quantitativ untersucht. *Internationale Revue der Gesamten Hydrobiologie*, **7**, 146–205, 275–426.

Forbes, S.A. (1887) The lake as a microcosm. *Bulletin of the Peoria Scientific Association* (3rd edition). *Bulletin of the Natural History Survey, State of Illinois, 1925*, **15**: 537–50.

Forel, F-A. (1892–1904). Le Léman 1–3. *Monographie Limnologique*. Lausanne, 1925 pp.

Fraleigh, P.C., Klerks P.L., Gubanich, G., Matisoff, G. & Stevenson, R.C. (1993) Abundance and settling of zebra mussel (*Dreissena polymorpha*) veligers in western and central Lake Erie In: Nalepa, T.F. & Schloesser, D.W. (eds), *Zebra Mussels. Biology, Impacts, and Control*. Lewis Publishers, Ann Arbor, MI, 129–42.

Frank, C. (1982) Ecology, production and anaerobic metabolism of *Chironomus plumosus* L. larvae in a shallow lake. I. Ecology and production. *Archiv für Hydrobiologie*, **94**, 460–91.

Frank, C. (1983) Ecology, production and anaerobic metabolism of *Chironomus plumosus* L. larvae in a shallow lake. II. Anaerobic metabolism. *Archiv für Hydrobiologie*, **96**, 354–62.

Gerstmeier, F. (1989) Lake typology and indicator organisms in application to the profundal chironomid fauna of Starnberger See (Diptera, Chironomidae). *Archiv für Hydrobiologie*, **116**, 227–34.

Goedkop, W. & Johnson, R. (1996) Pelagic–benthic coupling: profundal benthic community response to spring diatom deposition in mesotrophic Lake Erken. *Limnology and Oceanography*, **41**, 636–47.

Goedkop, W., Ahlgren, C., Sonesten, L. & Markensten, H. (1998) Fatty acid biomarkers show dietary differences between dominant chironomid taxa in Lake Erken. *Freshwater Biology*, **39**, 101–9.

Goldman, C.R. (1988) Primary productivity, nutrients, and transparency during the early onset of eutrophication in ultra-oligotrophic Lake Tahoe, California–Nevada. *Limnology and Oceanography*, **33**, 1321–33.

Goldman, C.R. (1993) Failures, successes and problems in controlling eutrophication. *Memorie dell' Istuto Italiano di Idrobiologia*, **52**, 79–87.

Haflidason, H., Larsen, G. & Ólafsson, G. (1992) The recent sedimentation history of Thingvallavatn, Iceland. *Oikos*, **64**, 80–95.

Hamburger, K. (1986) Energy flow in the population of *Eudiaptomus graciloides* and *Daphnia galeata* in Lake Esrom. *Archiv für Hydrobiologie*, **105**, 517–30.

Hamburger, K. & Dall, P.C. (1990) The respiration of common benthic invertebrate species from the shallow littoral zone of Lake Esrom, Denmark. *Hydrobiologia*, **199**, 117–30.

Hamburger, K., Dall, P.C. & Lindegaard, C. (1994) Energy metabolism of *Chironomus anthracinus* (Diptera: Chironomidae) from the profundal zone of Lake Esrom, Denmark, as a function of body size, temperature and oxygen concentration. *Hydrobiologia*, **294**, 43–50.

Hamburger, K., Dall, P.C. & Lindegaard, C. (1995) Effects of oxygen deficiency on survival and glycogen content of *Chironomus anthracinus* (Diptera, Chironomidae) under laboratory and field conditions. *Hydrobiologia*, **297**, 187–200.

Hamburger, K., Lindegaard, C. & Dall, P.C. (1996) The role of glycogen during the ontogenesis of *Chironomus anthracinus* (Chironomidae, Diptera). *Hydrobiologia*, **318**, 51–9.

Hamburger, K., Lindegaard, C. & Dall, P.C. (1997) Metabolism and survival of benthic animals short of oxygen. In: ed. Sand Jensen, K. & Pedersen, O. (eds), *Freshwater Biology: Priorities and Development in Danish Research*. Gad. Copenhagen, 183–96.

Hamburger, K., Lindegaard, C., Dall, P.C. & Nilsson, I. (1998) Strategies of respiration and glycogen metabolism in oligochaetes and chironomids from habitats exposed to different oxygen deficits. *Verhandlungen der Internationalen Vereinigung Limnologie*, **26**, 2070–5.

Hamburger, K., Dall, P.C., Lindegaard, C., Nilson, I.B. (2000) Survival and energy metabolism in an oxygen deficient environment. Field and laboratory studies on the bottom fauna from the profundal zone of Lake Esrom, Denmark. *Hydrobiologia*, **432**, 173–88.

Hansen, K. (1968) En boring i Esrom sø. *Meddelelser fra Dansk geologisk Forening* **18**, 244–6.

Hargrave, B.T. (1973) A comparison of sediment oxygen uptake, hypolimnetic oxygen deficit and primary production in Lake Esrom, Denmark. *Verhandlungen der Internationalen Vereinigung Limnologie*, **18**, 134–9.

Hargrave, B.T. & Kamp-Nielsen L. (1977) Accumulation of sedimentary organic matter at the base of steep bottom gradients. In: Golterman, H. (ed.), *Interactions between Sediments and Fresh Water:* Dr W. Junk, The Hague, 168–73.

Holopainen, I.J. (1987) Seasonal variation of survival time in anoxic water and the glycogen content of *Sphaerium corneum* and *Pisidium amnicum* (Bivalvia, Pisidiidae). *American Malacology Bulletin*, **5**, 41–8.

Holopainen, I.J. & Jónasson, P.M. (1983) Long-term population dynamics and production of *Pisidium* (Bivalvia) in the profundal of Lake Esrom, Denmark. *Oikos*, **41**, 99–117.

Holopainen, I.J. & Jónasson, P.M. (1989a) Bathymetric distribution and abundance of *Pisidium* (Bivalvia, Sphaeriidae) in Lake Esrom from 1954 to 1988. *Oikos*, **55**, 324–34.

Holopainen, I.J. & Jónasson, P.M. (1989b) Reproduction of *Pisidium* (Bivalvia, Sphaeriidae) at different depths in Lake Esrom, Denmark. *Archiv für Hydrobiologie*, **116**, 85–95.

Holopainen, I.J. & Penttinen, O-P. (1993) Normoxic and anoxic heat output of the freshwater bivalves *Pisidium* and *Sphaerium*. *Oecologia*, **93**, 215–23.

Hunding, C. (1971) Production of benthic microalgae in the littoral zone of a eutrophic lake. *Oikos*, **22**, 389–97.

Hutchinson, G.E. (1993) *A Treatise on Limnology*. IV. *The Zoobenthos* (Edmondson, Y.H. (ed.)). Wiley, New York, 944 pp.

Ivlev, V.S. (1939) Transformation of energy by aquatic animals. Coefficient of energy consumption by *Tubifex tubifex* Oligochaeta). *Internationale Revue der Gesamten Hydrobiologie*, **38**, 449–59.

Jacobsen, D. (1993) *Herbivori of invertebrater på submerse makrofyter i ferskvand (Herbivory of invertebrates on submerse macrophytes in freshwater)*. PhD thesis, University of Copenhagen, 101 pp.

Jónasson, P.M. (1955) The efficiency of sieving techniques for sampling freshwater bottom fauna. *Oikos*, **6**, 183–208.

Jónasson, P.M. (1958) The mesh factor in sieving techniques. *Verhandlungen der Internationalen Vereinigung Limnologie*, **13**, 860–66.

Jónasson, P.M. (1964) The relationship between primary production and profundal bottom invertebrates in a

Danish eutrophic lake. *Verhandlungen der Internationalen Vereinigung Limnologie*, **15**, 471–9.

Jónasson, P.M. (1965) Factors determining population size of *Chironomus anthracinus* in Lake Esrom. *Mitteilungen der Internationale Vereinigung für Limnologie*, **13**, 139–62.

Jónasson, P.M. (1972) Ecology and production of the profundal benthos in relation to phytoplankton in Lake Esrom. *Oikos*, **14**(Supplement), 1–148.

Jónasson, P.M. (1975) Population ecology and production of benthic detritivores. *Verhandlungen der Internationalen Vereinigung Limnologie*, **19**, 1066–72.

Jónasson, P.M. (1978) Zoobenthos of lakes. (Edgardo Baldi Memorial Lecture). *Verhandlungen der Internationalen Vereinigung Limnologie*, **20**, 13–37.

Jónasson, P.M. (ed.) (1979a) Ecology of eutrophic, subarctic Lake Mývatn. *Oikos*, **32**, 1–308.

Jónasson, P.M. (1979b) The Lake Mývatn ecosystem, Iceland. *Oikos*, **32**, 289–305.

Jónasson, P.M. (1984a) Oxygen demand and long-term changes of profundal zoobenthos. *Hydrobiologia*, **115**, 121–6.

Jónasson, P.M. (1984b) Decline of zoobenthos through five decades of eutrophication in Lake Esrom. *Verhandlungen der Internationalen Vereinigung Limnologie*, **22**, 800–4.

Jónasson, P.M. (ed.) (1992a) Ecology of oligotrophic, subarctic Thingvallavatn. *Oikos*, **64**, 1–439.

Jónasson, P.M. (1992b) The ecosystem of Thingvallavatn: a synthesis. *Oikos*, **64**, 405–34.

Jónasson, P.M. (1993) Lakes as a basic resource for development: the role of limnology. *Memorie dell' Istuto Italiano di Idrobiologia*, **52**, 9–26.

Jónasson, P.M. (1996) Limits for life in the lake ecosystem. Presidential Address. *Verhandlungen der Internationalen Vereinigung Limnologie*, **26**, 1–33.

Jónasson, P.M. & Adalsteinsson, H. (1979) Phytoplankton production in shallow eutrophic Lake Mývatn, Iceland. *Oikos*, **32**, 113–38.

Jónasson, P.M. & Kristiansen, J. (1967) Primary and secondary production in Lake Esrom. Growth of *Chironomus anthracinus* in relation to seasonal cycles of phytoplankton and dissolved oxygen. *Internationale Revue der Gesamten Hydrobiologie*, **52**, 163–217.

Jónasson, P.M. & Mathiesen, H. (1959) Measurements of primary production in two Danish eutrophic lakes, Esrom sø and Furesø. *Oikos*, **10**, 137–67.

Jónasson, P.M. & Thorhauge, F. (1972) Life cycle of *Potamothrix hammoniensis* (Tubificidae) in the profundal of a eutrophic lake. *Oikos*, **23**, 151–8.

Jónasson, P.M. & Thorhauge, F. (1976a) Population dynamics of *Potamothrix hammoniensis* in the profundal of Lake Esrom, with special reference to environmental and competitive factors. *Oikos*, **27**, 193–203.

Jónasson, P.M. & Thorhauge, F. (1976b) Production of *Potamothrix hammoniensis* in the profundal of eutrophic Lake Esrom. *Oikos*, **27**, 204–9.

Jónasson, P.M., Lastein, E. & Rebsdorf, A. (1974) Production, insolation, and nutrient budget of eutrophic Lake Esrom. *Oikos*, **25**, 255–77.

Jónasson, P.M., Lindegaard, C. & Hamburger, K. (1990) Energy budget of Lake Esrom, Denmark. *Verhandlungen der Internationalen Vereinigung Limnologie*, **24**, 632–40.

Jónasson, P.M., Adalsteinsson, H. & Jónsson, G. St. (1992a) Production and nutrient supply of phytoplankton in subarctic, dimictic Thingvallavatn, Iceland. *Oikos*, **64**, 162–87.

Jónasson, P.M., Lindegaard, C., Dall, P.C., Hamburger, K. & Adalsteinsson, H. (1992b) Ecosystem studies on temperate lake Esrom and the subarctic lakes, Mývatn and Thingvallavatn. *Limnologica* **20**, 259–66.

Jónsson, E. (1985) Population dynamics and production of Chironomidae (Diptera) at 2 m depth in Lake Esrom, Denmark. *Archiv für Hydrobiologie*, **70** (Supplement), 239–78.

Jónsson, E. (1987) Flight periods of aquatic insects at Lake Esrom, Denmark. *Archiv für Hydrobiologie*, **110**, 259–74.

Jónsson, G.St. (1987) The depth-distribution and biomass of epilithic periphyton in Lake Thingvallavatn, Iceland. *Archiv für Hydrobiologie*, **108**, 531–47.

Jónsson, G.St. (1992) Photosynthesis and production of epilithic algal communities in Thingvallavatn. *Oikos*, **64**, 222–40.

Jørgensen, J. (1971) *En kvantitativ undersøgelse af zooplanktonet i Esrom sø med særlig henblik på Rotatoria. (A quantitative investigation of zooplankton in Lake Esrom with special reference to Rotatoria)*. MS thesis, University of Copenhagen, Denmark, 22 pp.

Kairesalo, T., Jónsson, G.St., Gunnarsson, K. & Jónasson, P.M. (1989) Macro- and microalgal production within an *Nitella opaca* bed in Lake Thingvallavatn, Iceland. *Journal of Ecology*, **77**, 332–42.

Kairesalo T., Jónsson, G.St., Gunnarsson, K., Lindegaard, C. & Jónasson, P.M. (1992) Metabolism and community dynamics within *Nitella opaca* (Charophyceae) beds in Thingvallavatn. *Oikos*, **64**, 241–56.

Kajak, Z. & Warda, J. (1968) Feeding of benthic non-predatory Chironomidae in lakes. *Annales Zoologici Fennici*, **5**, 57–64.

Kajak, Z., Kacprzak, K. & Polkowski, R. (1965) Tubular bottom sampler. *Ekologia Polska*, **B11**, 159–65.

Kamp-Nielsen, L. (1997) Nutrient dynamics and modelling in lakes and coastal waters. In: Sand Jensen, K. & Pedersen, O. (eds), *Freshwater Biology: Priorities and Development in Danish research*. Gad, Copenhagen, 116–38.

Kirkegaard, J. (1980) *Livscyclus, vækst og produktion hos Theodoxus fluviatilis L. i Esrom sø. (Life cycle, growth and production of Theodoxus fluviatilis in Lake Esrom)*. MS thesis, University of Copenhagen, 66 pp.

Krezoski, J.R., Mozley, S.C. & Robbins, J.A. (1978) Influence of benthic macroinvertebrates on mixing of profundal sediments in southeastern Lake Huron, Canada. *Limnology and Oceanography*, **23**, 1011–6.

Kristiansen, J. (1996) Silica-scaled chrysophytes from Lake Thingvallavatn, Iceland. *Algological Studies*, **79**, 67–76.

Lastein, E. (1972) *Sedimentation i Esrom sø. (Sedimentation in Lake Esrom)*. MS thesis, University of Copenhagen, 42 pp.

Lastein, E. (1976) Recent sedimentation and resuspension of organic matter in eutrophic Lake Esrom, Denmark. *Oikos*, **27**, 44–9.

Lastein, E. (1983) Decomposition and sedimentation processes in oligotrophic, subarctic Lake Thingvalla, Iceland. *Oikos*, **40**, 103–12.

Lenz, F. (1925) Chironomiden und Seetypenlehre. *Die Naturwissenschaften*, **13**, 5–10.

Lenz, F. (1927) Chironomiden aus norwegischen Hochgebirgsseen. Zugleich ein Beitrag zur Seetypenfrage. *Nyt Magazin for Naturvidenskaberne* **88**, 111–92.

Lenz, F. (1931) Untersuchungen über die Vertikalverteilung der Bodenfauna mit Zerteilungsvorrichtung. *Verhandlungen der Internationalen Vereinigung Limnologie*, **5**, 232–61.

Lenz, F. (1932) Zur Methodik der quantitativen Bodenfauna-Untersuchung. Der Stockhalter, ein neues Hilfsgerät zum Bodengreifer. *Archiv für Hydrobiologie*, **23**, 375–80.

Likens, G.E. (ed.) (1985) *An Ecosystem Approach to Aquatic Ecology*. Springer-Verlag, Berlin, 516 pp.

Likens, G.E. & Hasler, A.D. (1962) Movements of radiosodium (Na^{24}) within an ice covered lake. *Limnology and Oceanography*, **7**, 48–56.

Likens, G.E. & Ragotzkie, R.A. (1966) Water movements in an ice-covered lake. *Verhandlungen der Internationalen Vereinigung Limnologie*, **16**, 126–33.

Lindegaard, C. (1992a) Zoobenthos ecology of Thing-

vallavatn: vertical distribution, abundance, population dynamics and production. *Oikos*, **64**, 257–304.

Lindegaard, C. (1992b) The role of zoobenthos in energy flow in deep oligotrophic Lake Thingvallavatn, Iceland. *Hydrobiologie*, **243–4**, 185–95.

Lindegaard, C. & Dall, P.C. (1988) Abundance and distribution of Oligochaeta in the exposed littoral zone of Lake Esrom, Denmark. *Archiv für Hydrobiologie*, **81**(Supplement), 533–62.

Lindegaard, C. & Jónasson, P.M. (1979) Abundance, population dynamics and production of zoobenthos in Lake Mývatn, Iceland. *Oikos*, **32**, 202–28.

Lindegaard, C. & Mæhl, P. (1993) Abundance, population dynamics and production of Chironomidae (Diptera) of an ultraoligotrophic lake in South Greenland. *Netherlands Journal of Aquatic Ecology*, **26**, 297–308.

Lindegaard, C., Dall, P.C. & Hansen, S.B. (1992) Natural and imposed variability in the profundal fauna of Lake Esrom, Denmark. *Verhandlungen der Internationalen Vereinigung Limnologie*, **25**, 576–82.

Lindegaard, C., Hamburger, K. & Dall, P.C. (1994) Population dynamics and energy budget of *Marionina southerni* (Cernosvitov) (Enchytraeidae, Oligochaeta) in the shallow littoral of Lake Esrom, Denmark. *Hydrobiologie*, **278**, 291–301.

Lindegaard, C., Dall, P.C. & Jónasson, P.M. (1997) Long-term patterns of the profundal fauna in Lake Esrom. In: Sand Jensen, K. & Pedersen, O. *Freshwater Biology Priorities and Development in Danish Research*. Gad, Copenhagen, 39–54.

Lopez, G.R. & Holopainen, I.J. (1987) Interstitial suspension-feeding by *Pisidium* spp. (Pisidiidae: Bivalvia): a new guild in the lentic benthos? *American Malacology Bulletin*, **5**, 21–30.

Lumbye, J. (1958) The oxygen consumption of *Theodoxus fluviatilis* (L.) and *Potamopyrgus jensinki* (Smith) in brackish and fresh water. *Hydrobiologia*, **10**, 245–62.

Lund, J.W.G. (1950) Studies on *Asterionella formosa* Hass. II. Nutrient depletion and the spring maximum. *Journal of Ecology*, **38**, 1–35.

Lund, J.W.G. (1964) Primary production and periodicity of phytoplankton (Edgardo Baldi Memorial Lecture). *Verhandlungen der Internationalen Vereinigung Limnologie*, **15**, 37–56.

Lund, J.W.G., Mackereth, F.J.H. & Mortimer, C.H. (1963) Changes in depth and time of certain chemical and physical conditions of the standing crop of *Asterionella formosa* Hass. in the North Basin of Windermere in 1947. *Philosophical Transactions of the Royal Society, London, Series B*, **246**, 255–90.

Lundbeck, J. (1926) Die Bodentierwelt norddeutscher Seen. *Archiv für Hydrobiologie*, **7**(Supplement), 1–471.

Lundbeck, J. (1928) Die 'Schalenzone' der norddeutschen Seen. *Jahrbuch Preussischen Geologischen Landesanstalt*, **49**, 1127–51.

Lundbeck, J. (1936) Untersuchungen über die Bodenbesiedelung der Alpenrandseen. *Archiv für Hydrobiologie*, **10**(Supplement), 207–358.

Macan, T.T. (1970) *Biological Studies of the English Lakes*. Longman, Harlow, 260 pp.

Macan, T.T. (1974) *Freshwater Ecology* (2nd edition). Longman, Harlow, 338 pp.

Mann, K.H. (1956) A study of the oxygen consumption of five species of leech. *Journal of Experimental Biology*, **33**, 615–26.

Mann, K.H. (1961) The oxygen requirements of leeches considered in relation to their habitats. *Verhandlungen der Internationalen Vereinigung Limnologie*, **14**, 1009–13.

Mann, K.H. (1962) *Leeches (Hirudinea). Their Structure, Physiology, Ecology and Embryology*. International Series, Monographs in Pure and Applied Biology, 11. Pergamon, Oxford, 201 pp.

Mann, K.H. (1964) The pattern of energy flow in the fish and invertebrate fauna of the River Thames. *Verhandlungen der Internationalen Vereinigung Limnologie*, **15**, 485–95.

Mann, K.H., Britton, R.H., Kowalczewski, A., Lack, T.J., Mathews, C.P. & McDonald, I. (1972) Productivity and energy flow at all trophic levels in the River Thames, England. In: Kajak, Z. & Hillbricht-Ilkowska, A. (eds), *Productivity Problems of Freshwaters*. PWN Polish Scientific Publisher, Warsaw, 579–97.

Meier-Brook, C. (1969) Substrate relations in some *Pisidium* species (Eulamellibranchiata: Sphaeriidae). *Malacologia*, **9**, 121–5.

Milbrink, G. (1980) Oligochaete communities in pollution biology: The European situation with special reference to lakes in Scandinavia. In: Brinkhurst, R.O. & Cook, D.G. (eds), *Aquatic Oligochaete Biology*. Plenum Press, New York, 433–57.

Monakov, A.V. (1972) Review of studies on feeding of aquatic invertebrates conducted at the Institute of Biology of Inland Waters, Academy of Science, USSR. *Journal of the Fisheries Research Board of Canada*, **29**, 363–83.

Monk, D.C. (1976) The distribution of cellulase in freshwater invertebrates of different feeding habits. *Freshwater Biology*, **6**, 471–5.

Mortimer, C.H. (1941–1942) The exchange of dissolved substances between mud and water in lakes. *Journal of Ecology*, **29**, 280–329; **30**, 147–201.

Mortimer, C.H. (1952) Water movements in lakes during summer stratification; evidence from the distribution of temperature in Windermere. *Philosophical Transactions of the Royal Society, London, Series B*, **236**, 355–404.

Mortimer, C.H. (1953) The resonant response of stratified lakes to mud. *Schweizerische Zeitschrift für Hydrologie*, **15**, 94–151.

Mortimer, C.H. (1954) Models of flow-patterns in lakes. *Weather*, **9**, 177–84.

Mortimer, C.H. (1961) Motion in thermoclines. *Verhandlungen der Internationalen Vereinigung Limnologie*, **14**, 79–83.

Mortimer, C.H. (1974) Lake hydrodynamics. *Mitteilungen der Internationale Vereinigung für Limnologie*, **20**, 124–97.

Morton, B.S. (1969) Studies on the biology of *Dreissena polymorpha* Pall. *Proceedings of the Malacological Society, London*, **38**, 471–82.

Müller, O.F. (1767) *Enumeratio ac Descriptio Libellularum agri Fridrichsdalensis*. Nova Acta Academiæ Leop. Carol. NaturæCuriosorum **III**, 122–31.

Müller, O.F. (1771) *Von Würmen des süssen und salzigen Wassers*. Kopenhagen, 200 pp.

Müller, O.F. (1773) *Vermium terrestrium et fluviatilium, seu animalium infusorium, helmintorum et testaceorum, non marinorum succincta historia*, 72+214+ XXV pp. Index. Hafniae & Lipsiae.

Müller, O.F. (1778) Von Thieren in den Eingeweiden der Thiere, insonderheit vom Kratzer im Hecht. *Der Naturforscher*, **12**, 178–96.

Müller, O.F. (1781) *Hydrachnae quas in aqvis Daniae Palustribus detexit, descripsit pingi et Tabulis XI Aeneis Incidi*. Lipsiae, 82 pp.

Müller, O.F. (1782) Vom Bandwurme des Stichlings und vom milchigten Plattwurme. *Der Naturforscher*, **18**, 21–37.

Müller, O.F. (1785) *Entomostraca seu insecta, Testacea, quae ni aqvis Daniae et Norvegiae reperit descripsit, et i comibus illustrant*. Havniae, 135 pp.

Müller, O.F. (1786) *Animalculi infusoria fluviatiliae et marina*. Hafniae, 367 pp.

Müller-Liebenau, I. (1956) Die Besiedlung der *Potamogeton*-Zone ostholsteinischer Seen. *Archiv für Hydrobiologie*, **52**, 470–606.

Murray, J. & Pullar, L. (1910) *Bathymetric Survey of the Scottish Fresh-water Lochs*. The Challenger Office, Edinburgh, 1499 pp.

Naumann, E. (1917) Undersökninger öfver fytoplankton

och under den pelagiska regionen försiggående gyttje- og dybildningar inom vissa syd- och mellensvenska urbergsvatten. *Kungliga Svenska Vetenskapsakademiens Handlinger*, **Band 56**(6), 1–165.

Nygaard, G. (1938) Hydrobiologische Studien über dänische Teiche und Seen. *I. Teil. Archiv für Hydrobiologie*, **32**, 523–692.

Ohle, W. (1960) Fernsehen, Photographie und Schallortung der Sedimentoberfläche in Seen. *Archiv für Hydrobiologie*, **57**, 135–60.

Ohle, W. (1962) Der Stoffhaushalt der Seen als Grundlage einer allgeneinen Stoffwechseldynamik der Gewässer. *Kieler Meeresforschungen*, **18**, 107–20.

Ólafsson, J. (1979) The chemistry of Lake Mývatn and River Laxá. *Oikos*, **32**, 82–112.

Ólafsson, J. (1992) Chemical characteristics and trace elements of Thingvallavatn. *Oikos*, **64**, 151–61.

Olsen, H.S. (1972) *Populationsanalyse af vandremuslingen, Dreissena polymorpha (Pallas) in relation til miljøfaktorer med særlig henblik på forekomsten i Esrom Sø(Population dynamics of D. polymorpha in relation to environmental factors in Lake Esrom)*. MS thesis, University of Copenhagen, 50 pp.

Overbeck, J. (1974) Microbiology and biochemistry. *Mitteilungen der Internationale Vereinigung für Limnologie*, **20**, 198–228.

Overbeck, J. (1993) Saprophytic and oligotrophic bacteria in Plussee. In: Overbeck, J. & Chrost, R. J. (eds), *Microbial Ecology of Lake Plussee*. Springer-Verlag, Berlin, 175–91.

Poddubnaja, T.L. (1961) Materials on the nutrition of mass species of tubificid worms in the Rybinsk Reservoir. *Transactions, Institute of Biology of Inland Waters, Academy of Sciences USSR* **4**, 219–31 (in Russian).

Pütter, A. (1909) *Die Ernährung der Wassertiere und der Stoffhaushalt der Gewässer*. Gustav Fischer, Jena, 168 pp.

Rawson, D.S. (1947) An automatic-closing Ekman dredge and other equipment for use in extremely deep water. *Limnological Society of America Special Publication*, **18**, 1–8.

Réaumur, R. (1734–1742) *Memoires pour servir à l'histoire des Insects*, 1–6. Paris, 3679 pp, 267 pl.

Reynoldson, T.B. (1983) The population biology of the Turbellaria with special reference to the freshwater Triclads of the British Isles. *Advances in Ecological Research*, **13**, 235–326.

Reynoldson, T.B. (1990) Distribution patterns of oligochaetes in the English Lake District. *Archiv für Hydrobiologie*, **118**, 303–39.

Reynoldson, T.B. & Bellamy L.S. (1971) Interspecific competition in lake-dwelling triclads. *Oikos* **22**, 315–28.

Reynoldson, T.B. & Davies R.W. (1970) Food niche and co-existence in lake-dwelling triclads. *Journal of Animal Ecology*, **39**, 599–617.

Riemann, B. & Mathiesen, H. (1977) Danish research into phytoplankton primary production. *Folia Limnologica Scandinavia*, **17**, 49–54.

Roesel von Rosenhof, A.J. (1755) *Insektenbelustigungen*. Nürnberg, Dutch ed. (undated). De naturlyke Historie der Insecten, Amsterdam. 2996 pp, 287 pl.

Sandlund, O.T., Jónasson, P.M., Jonsson, B., Malmquist, H.J., Skúlason, S. & Snorrason, S.S. (1992) Threespine stickleback *Gasterosteus aculeatus* in Thingvallavatn: habitat and food in a lake dominated by arctic charr *Salvelinus alpinus*. *Oikos*, **64**, 365–70.

Sæther, O.A. (1975) Nearctic chironomids as indicators of lake typology. *Verhandlungen der internationaalen vereiniging für theoretische und angewandte Limnologie*, **19**, 3127–33.

Sæther, O.A. (1979) Chironomid communities as water quality indicators. *Holarctic Ecology*, **2**, 65–74.

Schaeffer, J.C. (1754) *De Armpolypen in den süssen Wasern um Regensburg*. Regensburg, 84 pp.

Schaeffer, J.C. (1755a) *Die grünen Armpolypen, Die geschwänzten und ungeschwänzten Wasserflöte, Wasserade etc.* Regensburg, 94 pp.

Schaeffer, J.C. (1755b) *Die Blumenpolypen der süssen Wasser. Blumenpolypen der süssen Wasser. Blumenpolypen der salzigen Wasser.* Regensburg, 54 pp.

Schaeffer, J.C. (1756) *Der krebsartige Kieferfuss mit kurzer und langer Schwanzkappe.* Regensburg, 142 pp.

Schaeffer, J.C. (1762) *Der fischförmige Kieferfuss in den stehenden Wassern um Regensburg.* 22 pp.

Schindler, D.W. (1988) Experimental studies of chemical stressors on whole lake ecosystems. (Edgardo Baldi Memorial Lecture). *Verhandlungen der Internationalen Vereinigung Limnologie*, **23**, 11–41.

Sorokin, J.L. (1966) Carbon 14 method in the study of the nutrition of aquatic animals. *Internationale Revue der Gesamten Hydrobiologie*, **51**, 209–24.

Stanczykowska, A. (1976) Biomass and production of *Dreissena polymorpha* (Pall.) in some Masurian lakes. *Ekologia Polska*, **24**, 103–12.

Strayer, D. (1985) The benthic micrometazoans of Mirror Lake, New Hampshire. *Archiv für Hydrobiologie*, **72**(Supplement), 287–426.

Strayer, D. & Likens, G.E. (1986) An energy budget for the zoobenthos of Mirror Lake, New Hampshire. *Ecology*, **67**, 303–13.

Swammerdam, J. (1738) *Biblia Naturae 1–2.* Leyden, 910 pp.

Thienemann, A. (1913a) Der Zusammenhang zwischen dem Sauerstoffgehalt des Tiefenwassers und der Zusammensetzung der Tiefenfauna unserer Seen. Vorläufige Mitteilung *Internationale Revue der Gesamten Hydrobiologie*, **6**, 243–9.

Thienemann, A. (1913b) Physikalische und chemische Untersuchungen in dem Maaren der Eifel. I. *Verhandlungen des naturhistorischen Vereins der Prüßischen Rheinlande, Westfalen und des Regierungsbezirks Osnabrück* **70**, 249–302.

Thienemann, A. (1915a) Physikalische und chemische Untersuchungen in den Maaren der Eifel. II. *Verhandlungen des naturhistorischen Vereins der Prüßischen Rheinlande, Westfalen und des Regierungsbezirks Osnabrück* **71**, 273–389.

Thienemann, A. (1915b) Die Chironomidenfauna der Eifelmaare. *Verhandlungen des naturhistorischen Vereins der Prüßischen Rheinlande, Westfalen und des Regierungsbezirks Osnabrück* **72**, 1–58.

Thienemann, A. (1918a) Untersuchungen über die Beziehungen zwischem dem Sauerstoffgehalt des Wassers und der Zusammensetzung der Fauna in norddeutschen Seen. *Archiv für Hydrobiologie*, **12**, 1–65.

Thienemann, A. (1918b) Untersuchungen über die Beziehungen zwischen dem Sauerstoffgehalt des Wassers und der Zusammensetzung der Fauna in norddeutschen Seen. 2. Mitteilung. *Zeitschrift für wissenschaftliche Insektenbiologie*, **14**, 209–17.

Thienemann, A. (1921) Seetypen. *Die Naturwissenschaften*, **3**, Heft 18, 3 pp.

Thienemann, A. (1922) Die beiden *Chironomus*arten der Tiefenfauna der norddeutschen Seen. Ein hydrobiologisches Problem. *Archiv für Hydrobiologie*, **13**, 609–46.

Thienemann, A. (1925a) Die Binnengewässer Mitteleuropas. Eine limnologische Einführung. *Die Binnengewässer*, **1**, 1–255.

Thienemann, A. (1925b) *Mysis relicta. Zeitschrift für Morphologie und Ökologie*, **3**, 389–440.

Thienemann, A. (1928a) Der Sauerstoff im eutrophen und oligotrophen See. Ein Beitrag zur Seetypenlehre. *Die Binnengewässer*, **4**, 1–174.

Thienemann, A. (1928b) Die Reliktenkrebse *Mysis relicta, Pontoporeia affinis, Pallasea quadrispinosa* und die von ihnen bewohnten norddeutschen Seen. *Archiv für Hydrobiologie*, **19**, 521–82.

Thomas, E.A. (1950) Auffällige biologische Folgen von Sprungschichtneigungen im Zürichsee. *Schweizerische Zeitschrift für Hydrologie*, **12**, 1–24.

Thorhauge, F. (1975) Reproduction of *Potamothrix hammoniensis* (Tubificidae, Oligochaeta) in Lake Esrom, Denmark. A field and laboratory study. *Archiv für Hydrobiologie*, **76**, 449–74.

Thorhauge, F. (1976) Growth and life cycle of *Potamothrix hammoniensis* (Tubificidae, Oligochaeta) in the profundal of eutrophic Lake Esrom. A field and laboratory study. *Archiv für Hydrobiologie*, **78**, 71–86.

Timm, T. (1996) *Tubifex tubifex* (Müller, 1774) (Oligochaeta, Tubificidae) in the profundal of Estonian Lakes. *Internationale Revue der Gesamten Hydrobiologie*, **81**, 589–96.

Tokeshi, M. (1995) Life cycles and population dynamics. In: Armitage, P.D., Cranston, P.S. & Pinder, L.C.V. (eds), *The Chironomidae. Biology and Ecology of Non-biting Midges.* Chapman & Hall, London, 225–68.

Toman, M.J. & Dall, P.C. (1997) The diet of *Erpobdella octoculata* (Hirudinea: Erpohdellidae) in two Danish lowland streams. *Archiv für Hydrobiologie*, **140**, 549–63.

Toman, M.J. & Dall P.C. (1998) Respiratory levels and adaptations in four freshwater species of *Gammarus* spp. (Crustacea: Amphipoda). *Internationale Revue der Gesamten Hydrobiologie*, **83**, 251–63.

Trembley, A. (1744) *Mémoires pour servir à l'histoire d'un genre de Polypes d'eau douce, a bras en forme de corne.* Leiden, 324 pp.

Tsikhon-Lukanina, E.A. (1965) The nutrition and growth of freshwater gastropods. *Transactions, Institute of Biology of Inland Waters, Academy of Sciences USSR* **9**, 191–209.

Walshe, B.M. (1947) Feeding mechanisms of *Chironomus* larvae. *Nature*, **160**, 474–6.

Walshe, B.M. (1950) The function of haemoglobin in *Chironomus plumosus* under natural conditions. *Journal of Experimental Biology*, **27**, 73–95.

Walshe, B.M. (1951) The feeding habits of certain chironomid larvae (subfamily Tendipedinae). *Proceedings of the Zoological Society of London*, **121**, 63–79.

Walshe, B.M. (1956) Controle respiratoire et métabolisme chex les crustacés. *Vie et Milieu*, **7**, 523–41.

Wavre, M. & Brinkhurst, R.O. (1971) Interactions between some tubificid oligochaetes and bacteria found in the sediment of Toronto Harbour, Ontario. *Journal of the Fisheries Research Board of Canada*, **28**, 335–41.

Welch, H.E. & Kalff, J. (1974) Benthic photosynthesis and respiration in Char Lake. *Journal of the Fisheries Research Board of Canada*, **31**, 609–20.

Wermer, M. (1994) *Økologien af Polycentropidae i Esrom Sø. (Ecology of Polycentropidae in Lake Esrom).* MSc thesis, University of Copenhagen, 77 pp.

Wesenberg-Lund, C. (1908) Die littoralen Tierge-
sellschaften unserer grösseren Seen. *Internationale
Revue der Gesamten Hydrobiologie*, **1**, 574–609.

Wesenberg-Lund, C. (1939) *Biologie der Süsswassertiere,
Wirbellose Tiere*. Springer-Verlag, Berlin, 817 pp.

Wesenberg-Lund, C. (1943) *Die Biologie der Süsswasserinsekten*. Gyldendal, Copenhagen, 682 pp.

Wetzel, R.G., Rich, P.H., Miller, M.C. & Allen, H.L.
(1972) Metabolism of dissolved and particulate detrital
carbon in a temperate hard-water lake. *Memorie dell'
Istuto Italiano di Idrobiologia*, **29**, 185–243.

Whiteside, M.C. (1970) Danish chydorid Cladocera:
Modern ecology and core studies. *Ecological Monographs*, **40**, 79–118.

Whiteside, M.C. & Lindegaard, C. (1982) Summer distribution of zoobenthos in Grane Langsø, Denmark.
Freshwater Invertebrate Biology, **1**, 1–16.

Wiederholm, T. (1974a) Studier av bottenfaunan i
Mälaren. *Statens Naturvårdsverk*, **71**, 1–77.

Wiederholm, T. (1974b) Studier av bottenfaunan i Vät-
tern. *Statens Naturvårdsverk*, **72**, 1–63.

Wiederholm, T. (1980) Use of benthos in lake monitoring
Journal of the Water Pollution Control Federation, **52**,
537–47.

Winther, L.B. & Dahl, A. (1998) Respiration and production of the prosobranch snail *Potamopyrgus antipodarum* (GRAY) [= *P. jenkinsi* (SMITH)] in Lake Esrom,
Denmark. *Limnologica*, **28**, 285–92.

Young, J.O. (1981) A comparative study of the food niches
of lake-dwelling triclads and leeches. *Hydrobiologia*,
84, 91–102.

Young, J.O. & Spelling, S.M. (1989) Food utilisation and
niche overlap in three species of lake dwelling leeches
(Hirudinea). *Journal of Zoology*, **219**, 231–43.

Zschokke, F. (1900) Die Tierwelt in den Hochgebirgseen.
*Neue Denkschriften Allgemeinen schweizerischen
Gesellschaft für die gesamten Naturwissenschaften,
Zürich* **6**, 400 pp.

Zschokke, F. (1911) Die Tiefseefauna der Seen Mitteleuropas. *Monographien des internationalen Revue der
Hydrobiologie und Hydrographie* **4**, 246 pp.

13 Pelagic Microbes – Protozoa and the Microbial Food Web

THOMAS WEISSE

13.1 INTRODUCTION

Within aquatic research, microbial ecology is probably the discipline which is progressing most rapidly at present. During the 1970s and early 1980s, the advent of novel methods led to a change in the conceptual paradigm. Now, the increasingly widespread application of molecular techniques is currently revolutionising the field once more. It has thus become almost impossible for a single researcher to summarise the entire breadth of aquatic microbial ecology, giving fair account of such diverse subjects ranging from viruses to ciliates. I will therefore focus on the trophic aspects and interactions among microbes within pelagic microbial food webs. Wherever appropriate, I will refer to competent reviews of various aspects of microbial ecology that have recently been published. This also holds for methods and techniques, which I will not cover in any detail in this chapter.

I will, however, liberally include information originating from marine research, because, in many aspects, microbial ecology has been more innovative in salt than in freshwater during the past 25 years. Still, in order to avoid frustrating some potential readers, I need to announce that I will further restrict my review to the aerobic portion of lakes. I will not cover some very intriguing subjects, such as phototrophic bacteria (consult, for review, Schlegel & Bowien 1989; Oberhäuser-Nehls et al. 1993; Fenchel & Finlay 1995; Garrity 2001, and references therein), nor anaerobic and microaerophilic ciliates, which are prominent in hy-

pertrophic ponds and lakes (Fenchel & Finlay 1990; Finlay 1990; Finlay & Fenchel 1993, 1996).

13.2 THE CONCEPTUAL FRAMEWORK

The structure and functioning of microbial food webs has been the focus of aquatic microbiology during the last two decades. Together with the 'classic' grazer food chain (e.g., Fenchel 1988), microbial food webs are an integral part of any plankton community. However, because all plankton organisms are linked by manifold and varying interactions (Fig. 13.1) these categories are arbitrary, and there is no firm borderline between the grazer food chain and the microbial food web. Such interactions include numerous feedback and indirect effects. Protozoan bacterivory may indirectly favour phytoplankton because bacterial competition for nutrients is reduced, and the pool of inorganic nutrients is increased via protozoan excretion (Sherr et al. 1988; see also section 13.4.3). Feeding-related processes are also more important in replenishing the pool of dissolved organic matter (DOM) than leakage from phytoplankton (Lampert 1978; Jumars et al. 1989; see also Chapters 10 and 14).

In addition to internal sources, the pelagic DOM pool also receives inputs of external origin, such as from the littoral zone and the drainage basin (del Giorgio & Peters 1993, 1994; Coveney & Wetzel 1995, and references therein; see also

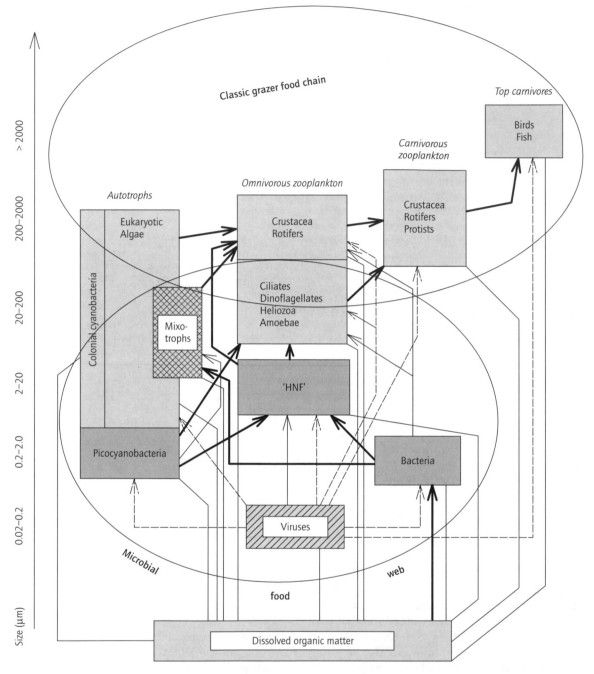

Fig. 13.1 Structure of the planktonic food web in lakes. Algal production is transferred along the the 'classic' food chain (light grey boxes) via omnivorous and carnivorous zooplankton to top predators (above). Microorganisms form the microbial food web (below), which in part overlaps with the grazer food chain. Within the microbial food web, a 'microbial loop' (dark grey boxes) is operating between picoplankton and the taxonomically diverse heterotrophic nanoflagellates ('HNF'). Feeding interactions and bacterial substrate uptake are indicated by solid lines and arrow heads. Viruses attack several components of the planktonic food web (broken lines and arrows). The pool of dissolved organic matter is replenished by various release processes (excretion, exudation, cell lysis, 'sloppy' feeding) and used as substrate by bacteria. The size categories from femto- (0.02–0.2 μm) to pico- (0.2–2 μm), nano- (2–20 μm), micro- (20–200 μm), meso- (200–2000 μm) and macroplankton (>2000 μm) are shown on the left.

Chapter 7). Like Metazoa, bacteria and Protozoa have developed various strategies designed to resist or to escape grazing (e.g., Kuhlmann & Heckmann 1985; Kusch 1993, 1995; Gilbert 1994; Jürgens & Güde 1994; Jürgens *et al.* 1994a; Pernthaler *et al.* 1997; Kuhlmann *et al.* 1999; Weisse & Frahm 2001). Depending on their different ecological functions, the same organisms may become part of the classic grazer food chain or the microbial food web. Mixotrophic algae, for instance, belong to the former when they are photosynthetically active, and are being consumed by crustacea, but they are part of the microbial food web when they graze bacteria. Most mixotrophs will fulfil both functions, varying in space and time. Mixotrophy represents a spectrum of nutritional strategies (reviewed by Sanders & Porter 1988; Sanders 1991a) which cannot easily be classified in the major categories shown in Fig. 13.1.

Both in the ocean and in lentic freshwater systems, microbial food webs are composed of heterotrophic bacteria, autotrophic picoplankton (0.2–2 μm), and auto- and heterotrophic protists in the size range of nanoplankton (2–20 μm) and microplankton (20–200 μm). Some microorganisms such as large ciliates and dinoflagellates belong to the mesozooplankton (200–2000 μm). Viruses, although they are not organisms, functionally belong to planktonic webs because they interact with various microbial components such as bacteria (Torella & Morita 1979; Bergh *et al.* 1989; Bratbak *et al.* 1990; Hennes & Simon 1995; Weinbauer & Suttle 1996), cyanobacteria (Safferman & Morris 1963; Proctor & Fuhrman 1990; Suttle & Chan 1993, 1994), phytoplankton (Suttle *et al.* 1990; Cottrell & Suttle 1991; Suttle *et al.* 1991), and heterotrophic protists (Nagasaki *et al.* 1993, 1994; Garza & Suttle 1995). Most viruses belong to the smallest size class (femtoplankton; 0.02–0.2 μm). The logarithmic size classification proposed by Sieburth *et al.* (1978) has proved operationally useful, but does not set strict limits between functional groups. The large majority of *Synechococcus*-like chroococcoid cyanobacteria, for example, fall into the picoplankton, whilst in some areas a considerable part of the *Synechococcus* population is larger than 2 μm (Jochem 1988).

In contrast to the microbial food web, the existence of the grazer food chain has been known since the beginning of the last century (Lohmann 1908, 1911; Stiasny 1913). From their work in the western Baltic Sea, Lohmann and co-workers concluded that primary production of phytoplankton such as diatoms and dinoflagellates is cropped by herbivorous zooplankton, which in turn are preyed upon by carnivorous zooplankton. The latter serve as food for small fish, which may also feed directly on herbivorous zooplankton. Note that because the 'herbivorous' zooplankton also ingests small heterotrophic protists, and to some extent bacteria, the respective box in Fig. 13.1 has been named *omnivorous zooplankton*. At one end of this marine food chain (see Fenchel 1988 for review) are piscivorous fish, birds, and mammals. It was soon recognised that, because pelagic organisms are connected by more than two trophic relationships (one to their food or substrate, one to their predators) in a web-like structure (Hardy 1924), a linear food chain is usually an oversimplification. This concept of a food web fuelled by primary production was adopted by limnologists, and remained valid until the mid-1970s (Steele 1974; reviewed by Banse 1990, 1995).

Like the classic grazer food chain, the conceptual framework describing the structure and function of microbial food webs was formulated first by marine researchers. This took place during the 1970s and early 1980s (Pomeroy 1974; Williams 1981; Azam *et al.* 1983). The new concept recognised that a substantial fraction of primary production is not consumed directly by herbivorous metazooplankton, but is channelled through a pool of DOM before it is made available to phagotrophs via bacterial production. Protozoa, whose importance had been overlooked in previous studies, were identified as the primary bacterivores.

Conceptual changes in the understanding of the significance of pelagic microorganisms have been so drastic that in 1974 Pomeroy wrote of a changing paradigm with respect to the ocean's food web. Williams (1981) further pointed to the great significance of bacteria and Protozoa for marine carbon flow and nutrient cycling. In 1983, Azam and co-

workers coined the term *microbial loop* for the incorporation of the microorganisms and their metabolic activities into the pelagic food web. Pelagic
microorganisms form a loop within the pelagic
food web because (i) part of the photosynthetically
produced carbon and some of the organically
bound nutrients are transferred to the pelagic bacteria and thus withdrawn from the classic grazing
food chain, whereas on the other hand (ii) part of
the bacterial production and some of the nutrients
are returned to the planktonic food web via protozoan and metazoan grazing and by bacterial and
protozoan remineralisation.

Since 1983, the microbial loop hypothesis has
received several modifications (Pomeroy & Wiebe
1988; Sherr & Sherr 1988, 1994) and expansions
(Porter *et al.* 1988; Sherr *et al.* 1988; Jumars *et al.*
1989; Christoffersen *et al.* 1990; Weisse 1991b,
1993). Porter and co-authors (1988) stated that the
term microbial loop is misleading, in that it implies the existence of an additional system which
can be separated from the classic planktonic food
web. Coupling or uncoupling of the loop from the
grazing food chain has led to the 'sink or link' controversy (Banse 1984; Ducklow *et al.* 1986; Sherr *et
al.* 1987; Fenchel 1988). The fraction of bacterial
production which reaches the grazer food chain depends mainly on how many trophic steps are involved between bacteria and metazoa. A major
part of the organically bound energy is respired at
each trophic level, and only 5–40% of the production of a given trophic level is transferred to the
next (Banse 1990; Wylie & Currie 1991; Gaedke &
Straile 1994b; Gaedke *et al.* 1995). Accordingly, if
bacteria are ingested primarily by small heterotrophic nanoflagellates (HNF), which are eaten
by larger protists and these then fall victim to crustacea, only a small amount of bacterial production
will reach metazoa.

Some authors (Ducklow *et al.* 1986, 1987;
Lyche *et al.* 1996) therefore argue that, in the pelagic carbon flow, the microbial loop primarily acts as
a sink. Others (Sherr & Sherr 1987) consider bacteria to be an additional food source for metazoans,
which is made available via the protozoan link.
Sherr & Sherr (1988) conclude that because the microbial loop is an integral part of a larger microbial

food web, which cannot be separated from the classic planktonic food web (Porter *et al.* 1988; Weisse
1989), the 'sink or link' controversy is a non-issue.
This does not preclude uncoupling of the microbial loop from the grazer food chain by experimental manipulation (Weisse & Scheffel-Möser 1991).
The extent of trophic transfer from bacteria to
metazoa depends on the food-web structure
(Ducklow 1991; Stone *et al.* 1993). In mesotrophic
and eutrophic lakes, transfer efficiency from bacteria to omnivorous zooplankton may vary greatly
during different seasonal phases (Stone *et al.* 1993;
Gaedke & Straile 1994b; Gaedke *et al.* 1995).

The 'sink or link' debate focuses on the carbon
flow in pelagic systems. With respect to functioning of the system, it is important that there is an
inverse relationship between inefficient carbon
transfer and efficient nutrient recycling. Whenever the microbial food web accounts for the main
part of pelagic carbon flow, most nutrients will be
regenerated within the system. Efficient nutrient
recycling is a fundamental function of microbial
food webs (see section 13.5.1). Bacteria, which are
rich in phosphorus and nitrogen, may therefore
simultaneously act as a sink for nutrients and as
a link in terms of carbon (Vadstein *et al.* 1993).

Integration of the microbial loop into the larger
microbial food web possesses an important conceptual disadvantage. The microbial loop is driven
by dead organic matter, which is metabolised by
bacteria. In this respect, the microbial loop resembles benthic and terrestrial detritus food chains
(Fenchel 1988). If all phytoplankton is considered
part of the microbial food web (Sherr & Sherr
1988), the principal difference between the
primary-production-based grazing food chain and
the detritus food chain, is blurred. As a consequence, the microbial loop is well-defined (Azam
et al. 1983), whilst the microbial food web is not
(Lenz 1992).

13.3   THE CLASSIC FOOD CHAIN
REVERSED—AN EVOLUTIONARY
PERSPECTIVE

With respect to evolution, it appears safe to conclude that primordial aquatic food webs were exclusively microbial, benthic and anaerobic. Hausmann & Hülsmann (1996) assume that the first microbial ecosystems arose four billion years ago. Anaerobic prokaryotes, such as the ancestors of archaebacteria and eubacteria (including pico-cyanobacteria), existed for a long time before eukaryotes evolved (Woese 1987; Cavalier-Smith 1991, 1993). Stromatolites and other benthic microfossils are at least 3.5 billion years old (Schopf 1977; Walter 1983). The advent of rRNA sequencing for analysis of phylogenetic relationships has revealed, however, that eukaryotes are much older than previously believed (Woese 1987). The first amitochondriate eukaryotes (the Archezoa) evolved during the Proterozoic Era, several hundred million years before the first metazoans appeared (Cavalier-Smith 1991, 1993). Radiation of the eukaryotes occurred during the interval between oxidation of the hydrosphere, about 1.5 billion years ago, and the beginning of the Palaeozoic Era, about 570 million years ago (Sogin *et al.* 1986). Seven hundred million years ago, the major phyla among both Protozoa and Animalia had already appeared (Cavalier-Smith 1991, 1993, 1995; Adoutte 1994, and references therein). Note that in the phylogenetic classification system advocated by Cavalier-Smith, and adopted by Corliss (1994), Protozoa are given the taxonomic rank of a kingdom, whilst metazoans are no longer restricted to a single kingdom and therefore possess no taxonomic category.

In the context of microbial food webs, it is important to note that the (mostly) free-living Percolozoa, which include the marine bacterivorous *Percolomonas* (Fenchel & Patterson 1986) and the Euglenozoa (with two ecologically important classes of free-living flagellates, Euglenoidea and Kinetoplastidea), are phylogenetically distinctly older than choanoflagellates (Choanozoa), dinoflagellates (Dinozoa) and ciliates (Ciliophora) within the kingdom Protozoa, heterokont heterotrophic flagellates and most 'algae' within the kingdom Chromista, and all Animalia (Sogin *et al.* 1986; Perasso *et al.* 1989; Cavalier-Smith 1991, 1993, 1995; Corliss 1994). There is a tremendous evolutionary distance between heterotrophic protists in

general, and among HNF in particular. Ciliates, dinoflagellates, most 'algal' groups, metazoa and metaphytes all belong to the poorly resolved 'crown' at the top of the evolutionary tree (Adoutte 1994).

Although the timing of formation of the first aerobic planktonic food chains, and their structure, remains unknown (Lenz 1992), they were almost certainly composed of single celled organisms. Simple microbial food chains may have evolved into more complex microbial food webs. The most advanced heterotrophic protists, the ciliates and dinoflagellates, are often prominent in extant microbial food webs, and have co-evolved with the algal classes and the animal phyla mainly in the Palaeozoic Era. It is likely that biological interactions among, and between, these taxa have developed over the course of hundreds of millions of years. It is thus not surprising that, in lakes in particular, many ciliates and dinoflagellates compete with rotifers, copepods and cladocerans for the same algal food (Weisse & Müller 1998).

In contrast to previous illustrations of the microbial loop (Azam *et al.* 1983; Fenchel 1988), I therefore group the majority of ciliates and dinoflagellates into the grazer food chain (Fig. 13.1). Except for an evolutionary and functional perspective, which will be discussed in more detail in section 13.5, this grouping is also justified in terms of semantics, because some Protozoa (such as dinoflagellates and tintinnids) were among the organisms caught and quantified during the first planktological studies (Hensen 1887; Brandt 1906; Lohmann 1908; summarised by Zeitzschel 1967; Smetacek 1985; Florey 1995; Zissler 1995).

From an evolutionary perspective, contemporary microbial food webs (which are present in all aquatic environments) may be regarded as successors of primordial food chains or webs. The classic grazer food chain seems to be relatively modern, and is dependent on, and, therefore, more or less superimposed on to, the microbial food web. Paradoxically, research grew in the opposite direction. What was named the 'classic' food chain is in fact, in evolutionary terms, a young, 'modern' food web. It should be noted that this reasoning is not necessarily correct for lakes, which are young compared with the ancient ocean. The great structural similarities between marine and freshwater microbial food webs (Cole *et al.* 1988; Weisse 1991b; Riemann & Christoffersen 1993) suggest that, in lakes, microbial food webs also represent the original form of trophic interactions.

13.4   COMPONENTS OF MICROBIAL FOOD WEBS IN LAKES

Competent reviews of the taxonomic composition and ecology of each component of the microbial food web have recently been published (consult Cole *et al.* (1988), Ducklow & Carlson (1992) for bacteria; Fogg (1986), Stockner & Antia (1986), Weisse (1993) for picocyanobacteria; Fenchel (1986), Patterson & Larsen (1991) and Leadbeater & Green (2000) for heterotrophic flagellates; Caron & Swanberg (1990) and Arndt (1993) for amoebae and heliozoa; Lessard (1991) for dinoflagellates; and Finlay & Fenchel (1996) for ciliates). Phytoplankton and zooplankton are dealt with in Chapters 10 and 14 of this volume. In the following section I will focus on those components of microbial food webs that have not been discussed in detail elsewhere in this volume.

13.4.1   *Viruses and the 'viral loop'*

Bacteriophages and other viruses have long been known to infect bacteria, cyanobacteria, and eukaryotic algae (Spencer 1955; Safferman & Morris 1963; Adams 1966; Padan & Shilo 1973). Until recently, because their numbers seemed to be generally too low, and below the minimum bacterial abundance level necessary for bacteriophage replication (Wiggins & Alexander 1985), they were considered to be ecologically unimportant. A report by Torella & Morita (1979), who provided minimum estimates of viral abundances of >10^4 mL^{-1} in coastal Oregon waters, received little attention on the part of microbial ecologists for a decade. Research on aquatic viruses was enormously spurred when Bergh *et al.* (1989) reported high viral abundances up to 10^8 mL^{-1} from direct counting using transmission electron microscopy. Their highest

record (2.5×10^8 viral particles per millilitre) originated from the eutrophic lake Plußsee in northern Germany. These high numbers, which exceeded earlier estimates derived from using specific hosts and plaque assays by several orders of magnitude (Paul *et al.* 1991; Bratbak & Heldal 1993; Witzel *et al.* 1994), have been confirmed in the Plußsee (Demuth *et al.* 1993) and other lakes (Klut & Stockner 1990; Hennes & Suttle 1995; Witzel *et al.* 1994). Viral numbers in surface fresh and marine waters range from <10^4 mL^{-1} to >10^8 mL^{-1} (Bratbak *et al.* 1994; Witzel *et al.* 1994; Maranger & Bird 1995).

Common methods used in order to estimate viral abundance in natural waters have been reviewed by Bratbak & Heldal (1993), Suttle (1993) and Marie *et al.* (1999). There is at present no generally accepted standard technique. Of the five methods used for enumerating viruses (plaque assays and most-probable-number assays, transmission electron microscopy (TEM), epifluorescence microscopy (Suttle 1993) and flow cytometry (Marie *et al.* 1999)), direct counting by TEM is currently most widespread. Recent evidence suggests, however, that TEM may considerably underestimate viral abundance (Hennes & Suttle 1995; Weinbauer & Suttle 1997). Using a new DNA dye (Yo-Pro-1) and epifluorescence microscopy, these authors found viral numbers 1.2 to 7.1 times greater than with TEM. The introduction of a new generation of nucleic acid-specific stains that can be excited with argon lasers enabled detection and quantification of aquatic viruses by flow cytometry (Marie *et al.* 1999). Using flow cytometry and the dye SYBR Green-1, Marie *et al.* (1999) detected at least two different viral populations in natural seawater samples.

There is a positive general trend between trophic status and viral abundance (Paul *et al.* 1991; Paul 1993; Weinbauer *et al.* 1993; Witzel *et al.* 1994; Maranger & Bird 1995). From their analysis of marine environments Cochlan *et al.* (1993) found that bacterial abundance alone explained 69% of spatial variability in viral numbers, whilst chlorophyll *a* (Chl *a*) explained only 45%. The opposite conclusion was reached by Maranger & Bird (1995) from a cross-system comparison. In their regression models, Chl *a* concentration explained 77% of variation in viral abundance in aquatic environments as opposed to bacterial abundance explaining only 72%. The virus-to-bacteria ratio was weakly positively related to Chl *a* concentrations and significantly higher in freshwater.

Maranger & Bird (1995) concluded that cyanobacteria and algae are relatively more important in fresh water than they are in the ocean, so that lakes support a higher number of cyanobacterial and algal viruses. In the 22 Québec lakes they studied, viral abundance was also positively related to bacterial production and total phosphorus concentration, and there were more bacteria per unit of chlorophyll as compared with marine sites. The authors presume that this indicates an increased dependence of freshwater bacteria on allochthonous material relative to marine systems. This supports the conclusion of an earlier analysis (Simon *et al.* 1992), which was recently confirmed for picophytoplankton (Bell & Kalff 2001).

In aquatic food webs, viruses fulfil more than one ecological functional role. The function which has received the most attention in recent years is phage-induced mortality of their hosts. Shortly after Bergh *et al.* (1989), Proctor & Fuhrman (1990) estimated that, in the ocean, up to 30% of total mortality in cyanobacteria, and up to 60% in heterotrophic bacteria, are caused by viral infection. Several studies conducted mainly in marine waters have shown that phage-induced bacterial mortality is highly variable, ranging from 0.1 to 100%, and is positively related to trophic state (Hennes & Suttle 1995; Weinbauer & Suttle 1996, and references therein).

The estimate of bacterial mortality caused by viruses is strongly dependent on a number of assumptions and conversion factors which are still debatable (Proctor *et al.* 1993). Accordingly, current estimates of viral mortality of pelagic microbes differ widely. Fuhrman & Noble (1995) reported that, in coastal seawater, phage-induced bacterial mortality is similar to bacterial mortality by grazing. In the Bodensee (Lake Constance), bacteriophages are responsible for only 1–24% of total bacterial mortality (Hennes & Simon 1995; Simon *et al.* 1998a). From his literature review, Suttle

(1994) concluded that, on average, 15% of heterotrophic bacteria, and approximately 3% of picocyanobacteria, are lysed daily by viruses, and that, in the absence of viruses, production by eukaryotic algae would be approximately 2% higher. This latter estimate is much lower than the initial assumption that as much as 78% of primary productivity may be reduced owing to infection of phytoplankton by viruses (Suttle *et al.* 1990).

The potential significance of viruses for controlling production of eukaryotic algae was also shown by Bratbak *et al.* (1990), who assessed the concentration of viruses during the spring phytoplankton bloom in Raunefjorden, Norway. The maximum number of free viruses occurred 1 week after the peak of the diatom bloom. Similarly, Hennes & Simon (1995) observed a temporal shift between bacterial abundance and peaks in the number of infected bacteria and free bacteriophages in the prealpine Lake Constance (the Bodensee; Fig. 13.2).

A second potential ecological function of viruses related to virus-induced mortality, and regulating the microbial community structure, has been difficult to demonstrate. It is clear that viral–host interactions are highly dynamic. Because viral infection is host-specific, and lysogenic hosts will develop resistance to the virus they carry, viral parasitism may lead to structural shifts in the bacterial community (Hennes *et al.* 1995; Simon *et al.* 1998a). It is important to note that over 90% of all known phages are thought to be temperate (Maloy *et al.* 1994), i.e., after infection, the viral genome is inserted into the host DNA and replicated together with the bacterial DNA in non-infectious form for a variable number of generations (the lysogenic cycle, e.g., Luria & Darnell 1978; Maloy *et al.* 1994). Laboratory evidence suggests that environmental factors such as UV radiation or vigorous bacterial growth induce the onset of a lytic cycle (i.e., the production of new viruses), whilst the significance of these factors under natural conditions currently remains unknown (Thingstad *et al.* 1993; Bratbak *et al.* 1994).

Third, viruses may also fulfil a positive trophic function. González & Suttle (1993) demonstrated that nanoflagellates are able to graze upon viruses.

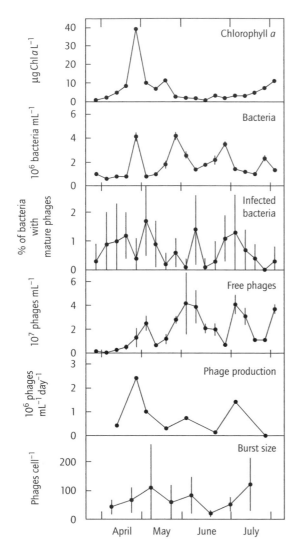

Fig. 13.2 Relationship between phytoplankton biomass (Chl *a*), bacteria, and viruses in Lake Constance. From top to bottom: Chl *a*, bacterial abundance, bacteria with intracellular mature phages, free phages <100 μm in diameter, phage production and burst size during different seasons between late March and August 1992. Error bars denote standard deviations. (Redrawn from Hennes & Simon 1995; reproduced with permission of the American Society for Microbiology Journals Division.)

These authors concluded that for flagellates, when grazing on viruses, nutrient uptake may be as important as carbon uptake. Utilisation of colloidal macromolecules by heterotrophic flagellates was also demonstrated by Tranvik *et al.* (1993). Phagotrophy by HNF seems to contribute very little to total viral loss rates in seawater (González & Suttle 1993). Viral infection of those grazers which consume them may offset the nutritional significance of viruses for heterotrophic nanoflagellates (Garza & Suttle 1995).

In the context of microbial food-web dynamics, a fourth ecological function of viruses may also be significant. Coupled with viral lysogeny is the release of nutrients and DOM. Virus-induced cell lysis may be a major pathway for nutrient release, which is important for the trophic significance of bacterial production (Bratbak *et al.* 1990). The 'viral loop' could result in bacterial production which is not available for the food web, but which is directly channelled back into the DOM pool (Fig. 13.3). This 'short circuit' of bacterial production may lead to enhanced bacterial production, which theoretically could even exceed primary production (Bratbak *et al.* 1990; see also Strayer 1988). Nucleic acids and organic matter possessed by the phages may also be released into the DOM pool after they have been digested by extracellular enzymes (Bratbak *et al.* 1990) or from viral decay caused by abiotic factors (Noble & Fuhrman 1997). Recent evidence suggests, however, that viral contribution to the pool of dissolved DNA is small (Jiang & Paul 1995).

Viral lysogeny could thus significantly alter the ratio between the particulate and the dissolved organic carbon pools (Thingstad *et al.* 1993; Bratbak *et al.* 1994). Because bacteria, cyanobacteria and phytoplankton compete for nutrients, and phytoplankton are the superior competitors at elevated nutrient levels (see section 13.4.3), bacteriophage-induced nutrient release may favour algae at the expense of picoplankton. As a consequence, the viral loop could indirectly favour the classic grazer food chain relative to the microbial food web.

Finally, with the discovery of the high viral numbers in natural waters, it was speculated that the major function of viruses might be to enhance

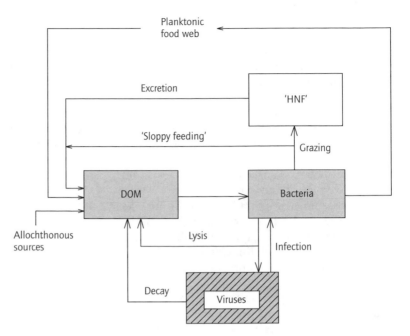

Fig. 13.3 The 'viral loop': by lysing infected bacteria and as a result of viral decay, viruses replenish the pool of dissolved organic matter (DOM) that contains bacterial substrates. This part of bacterial production that is destroyed by viral activity is not available for heterotrophic nanoflagellates ('HNF') and other bacterial grazers. With respect to the food web, the viral loop therefore acts as a short circuit of bacterial production. (Modified after Bratbak *et al.* 1990; Thingstad *et al.* 1993.)

gene transfer between different bacteria (Bergh *et al.* 1989; Sherr 1989; Proctor & Fuhrman 1990). The extent to which lysogenic bacteria may provide potential hosts with foreign DNA in the environment is still unknown (Witzel *et al.* 1994).

13.4.2 *Autotrophs: picocyanobacteria*

Phytoplankton composition and ecophysiology are discussed in detail in Chapter 10. Here, I will therefore consider prokaryotic phototrophs only. The tremendous significance for carbon flow in the ocean and in lakes of the smallest size class, the autotrophic picoplankton (APP), was recognised during the 1980s (reviewed by Fogg 1986; Stockner & Antia 1986; Stockner 1988; Weisse 1993), after Waterbury *et al.* (1979) had shown that coccoid cyanobacteria of the genus *Synechococcus* occur in the central parts of the ocean in concentrations of up to >10^5 cells mL^{-1}. Historical accounts of APP research in marine and fresh waters are given by Fogg (1986) and Stockner (1988, 1991).

The APP consists of single-celled cyanobacteria (including prochlorophytes) and minute

eukaryotic algae (Stockner & Antia 1986; Stockner 1988). Coccoid prochlorophytes, which are abundant in seawater (Chisholm *et al.* 1988), and which may contribute substantially to total primary production at nutrient-depleted oceanic stations (Goericke & Welschmeyer 1993; Li 1994), have not yet been detected in freshwater plankton. The recent discovery (by flow cytometry) of *Prochlorococcus*-like cells in a eutrophic Spanish reservoir (Corzo *et al.* 1999) needs to be verified by alternative methods (i.e., high-performance liquid chromotography (HPLC) analysis and electron microscopy). The lack of planktonic prochlorophytes in fresh water is somewhat surprising, in that a filamentous species, *Prochlorothrix hollandica*, is abundant in shallow eutrophic Dutch lakes (Burger-Wiersma *et al.* 1986; Morden & Golden 1989). The significance of cyanobacteria for primary production in lakes, and the trophic role of picocyanobacteria and colonial cyanobacteria, will be discussed in section 13.5.

Sensitivity of the autotrophic picoplankton (picocyanobacteria in particular) to heavy metal and organic pollution has been demonstrated by

Munawar and colleagues in the Laurentian Great Lakes (Munawar & Munawar 1982; Munawar *et al.* 1987; Severn *et al.* 1989). They propose using the APP as an 'early warning indicator' for anthropogenic lake contamination (Munawar & Weisse 1989; Munawar *et al.* 1994). The consequences of eutrophication and contaminant enrichment on the functioning of microbial food webs have been discussed earlier (Weisse 1991b), and will not be repeated here.

Recent evidence suggests that species and/or clonal diversity among the APP is greater than hitherto assumed, especially in freshwater systems (Stockner 1991). A wealth of information arises from novel (polymerase chain reaction) PCR-based and immunofluorescence techniques. Molecular investigations of 16S ribosomal RNA (ssu rRNA) genes of picocyanobacteria were first applied to marine species. The analyses indicated that a variety of sequences were present which did not match known sequences from cultivated cyanobacteria isolated from similar habitats (Giovannoni *et al.* 1990; Olsen 1990). High genetic diversity among marine *Synechococcus* was confirmed by Wood & Townsend (1990), on the basis of restriction fragment length polymorphism (RFLP) analyses. Sites of sequence variation among clones from the marine *Synechococcus* cluster were identified by Britschgi & Giovannoni (1991).

High genetic diversity has also been found in morphologically similar freshwater picoplankton (Ward *et al.* 1990; Ernst *et al.* 1995; Robertson *et al.* 2001; Semenova *et al.* 2001). Isolates from Lake Constance (the Bodensee) revealed a great variety of morphologically, physiologically and genetically different *Synechococcus*-like cyanobacteria (Ernst 1991; Ernst *et al.* 1992, 1995, 1996). This finding is in contrast to earlier observations that, in this lake, picocyanobacteria are largely made up of phycoerythrin-rich forms, which appear relatively uniform under epifluorescence microscopy (Weisse 1988; Weisse & Kenter 1991; Weisse & Schweizer 1991). Although culture conditions probably favoured growth of phycocyanin-rich strains (Ernst 1991; Ernst *et al.* 1995), and genetic diversity *in situ* has not yet been assessed, seasonal succession of several strains with distinct physio-

logical properties is likely (Ernst *et al.* 1995, 2000; Postius *et al.* 1996; Postius & Ernst 1999). *In situ* hybridisation using strain-specific oligonucleotide probes (discussed below) is a promising tool for providing answers to this question. It seems clear now that freshwater *Synechococcus* are genetically diverse, polyphyletic in origin and should be divided into several different genera (Robertson *et al.* 2001; Ernst *et al.* 2003).

The earlier notion that picocyanobacterial numbers in lakes are primarily controlled by grazing of heterotrophic flagellates and ciliates (Weisse 1988; Wehr 1991) has been confirmed experimentally in a number of recent studies (Šimek *et al.* 1995, 1996, 1997a; Müller 1996; Pernthaler *et al.* 1996). Results yielded major differences with respect to palatability of picocyanobacteria for different species of ciliates and flagellates, as well as selective feeding of protists on various cyanobacterial strains.

13.4.3 Heterotrophic bacteria

The advent of novel methods has revolutionised microbial ecology in general, and our understanding of the ecological role of planktonic bacteria in particular, during the last 25 years. Until the early 1970s, bacteria were regarded as relatively specialised remineralisers, which contributed very little to carbon flow in the open waters. Furthermore, the majority of pelagic bacteria were thought to be dormant (Stevenson 1978). Size-fractionated measurements of respiration (Pomeroy & Johannes 1968), and of heterotrophic utilisation of dissolved organic material (Williams 1970), indicated that the bulk of heterotrophic activity in the sea could not be attributed to larger (>20 μm) net phytoplankton and zooplankton, but was apparently due to pelagic microorganisms in the picoplankton and nanoplankton size range. Direct cell counts using epifluorescence microscopy (Francisco *et al.* 1973) and polycarbonate filters (Zimmermann & Meyer-Reil 1974; Hobbie *et al.* 1977) revealed that bacterial abundance is one or two orders of magnitude higher than data formerly derived from using plate counts, serial dilutions or phase-contrast microscopy. Bacterial numbers in natural aquatic ecosystems generally range from 10^4 to 10^8 mL^{-1}. In most surface waters, variation is restricted to the range 10^5 to 10^7 cells mL^{-1}. Finally, with development of the **frequency of dividing cells** (FDC) technique for the assessment of bacterial production (Hagström *et al.* 1979), and especially the ^{3}H-thymidine-incorporation method (Fuhrman & Azam 1980, 1982), it became apparent that bacteria process a large fraction of the pelagic carbon flow.

From a comparison of marine and freshwater ecosystems, Cole *et al.* (1988) concluded that annual bacterial net production equals about 30% of primary production per unit area (i.e., averaged over the water column). Since bacterial growth efficiency is about 50%, bacterial carbon demand would amount to 60% of net primary production (Cole *et al.* 1988). Variation of these ratios in relation to the trophic state of lakes is discussed in section 13.5. Recent advances in measuring bacterial biomass and production have been reviewed by Riemann & Bell (1990), Ducklow & Carlson (1992), Bratbak (1993), Kirchman (1993) and Kemp (1994). Most of the bacterial biomass and production is provided by free-living, suspended bacteria (Azam *et al.* 1983; Azam *et al.* 1990), although larger attached bacteria may be important at times or specific localities (Simon 1987). Sedimentation of bacteria associated with 'lake snow' particles contributes significantly to total losses of bacterial production from the pelagic (Grosshart & Simon 1993, 1998; Grosshart *et al.* 1997; Simon *et al.* 1998a).

Understanding of the ecological role of bacteria has not only changed regarding their quantitative importance as carbon pools and sources. In contrast to earlier conjecture, bacteria are now considered effective competitors with phytoplankton for mineral nutrients. Owing to their higher substrate affinities, bacteria may out-compete phytoplankton at low nutrient concentrations both in the sea (Bratbak & Thingstad 1985) and in lakes (Currie & Kalff 1984a,b,c; Ammermann & Azam 1985; Jürgens & Güde 1990). A recent study suggests that bacterial nutrient limitation is more severe in fresh water than in marine ecosystems (Elser *et al.* 1995). Remineralisation is instead mediated via

the bacterial grazing activity of protists and their subsequent excretion (Caron & Goldman 1988; Caron *et al.* 1988; Bloem *et al.* 1989; Caron 1991; Vadstein *et al.* 1993). Rothhaupt (1992) could demonstrate that in phosphorus-limited experiments algae were able to grow only when bacterivorous flagellates reduced bacterial numbers and made bacterial phosphorus available to the phytoplankton.

Bacterioplankton dynamics in lakes is controlled by substrate supply, temperature and losses such as grazing, parasitism and sedimentation. Few lakes have been studied sufficiently well to draw firm conclusions regarding the relative impact of the environmental factors on bacterial population size and production (Ochs *et al.* 1995). The role of various substrates in controlling bacterial production is still poorly understood. Using mainly HPLC analyses, Simon and colleagues showed that bacteria in Lake Constance utilise different carbon sources at various times of the year (Simon & Rosenstock 1992; Rosenstock & Simon 1993; Hanisch *et al.* 1996; Simon *et al.* 1998a). In contrast to earlier observations in marine environments (Pomeroy & Deibel 1986; White *et al.* 1991; Shia & Ducklow 1994) and in other lakes (White *et al.* 1991; Coveney & Wetzel 1995; Ochs *et al.* 1995; Felip *et al.* 1996), temperature affected bacterial growth rates relatively little (Simon *et al.* 1998a).

During most of the growing season, bacterial growth was co-limited by DOM and soluble reactive phosphorus (Simon *et al.* 1998a). This finding supports earlier results from a small, acidic, oligotrophic Swedish Lake (Bell *et al.* 1993). Among bacterial grazing losses in Lake Constance, HNF contributed the most at 52–68%, followed by ciliates with 14–19% and daphnids at 9–12%. These estimates agree with earlier measurements of bacterial grazing by HNF in the lake (Weisse 1990) and results from mass balanced carbon flow models (Straile 1995). Viral mortality accounted for an estimated 1–24% of total bacterial mortality (Hennes & Simon 1995; Simon *et al.* 1998a). The factors regulating bacterial abundance and production in Lake Constance and, most likely, causing taxonomic shifts in the bacterioplankton assemblage are summarised in Fig. 13.4.

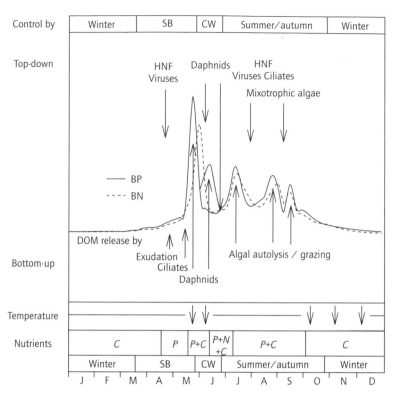

Fig. 13.4 Idealised pattern of dynamics and control of bacterial production and abundance in the pelagial of Lake Constance in the course of the year: SB, phytoplankton 'spring bloom'; CW, clear-water phase; BP, bacterial production; BN, bacterial numbers; HNF, heterotrophic nanoflagellates. The direction of arrows indicates positive (upwards) or negative (downward) effects. The two lines of BP and BN in the CW phase indicate the differing impact of lower and higher abundance of daphnids. (From Simon et al. 1998a. © E.Schweizerbart'sche. Reproduced with permission of the publisher.)

Until recently, identification of bacterial species was carried out by cultivation. Owing to selectivity of the medium, and/or the physiological state of the bacteria, only a very small percentage of the natural bacteria are amenable to cultivation (see Fuhrman et al. 1994; Amann 1995; Amann et al. 1995 for review). Research conducted during the last 10 years, using novel molecular techniques, provides increasing evidence that the bacterioplankton in lakes is a heterogeneous assemblage, and that seasonal succession analogous to that of phytoplankton is common. Immunofluorescence has been applied in order to identify various types of aquatic bacteria (Ward & Perry 1980; Ward 1982; Ward et al. 1982) and cyanobacteria (Fliermans & Schmidt 1977; Campbell & Carpenter 1987; Campbell & Iturriaga 1988).

The advent of nucleic acid-based molecular techniques, particularly rRNA sequence analysis and group-specific rRNA-targeted oligonucleotide probes (in situ hybridisation), demonstrated that bacterial consortia vary considerably both between and within different aquatic habitats (e.g. Giovannoni et al. 1990; Lee & Fuhrman 1990; Hicks et al. 1992; Alfreider et al. 1996). It now appears that there are bacterial species or groups of species that are indigenous to fresh water (Zwart et al. 2002). Research is currently progressing from using kingdom-specific to genus-specific or species-specific probes (summarised by Paul 1993; Fuhrmann et al. 1994; Ammann 1995). In spite of current sensitivity problems, and the general need of optimisation (reviewed by Amann et al. 1995), I anticipate that application of these techniques will greatly improve perception of bacterioplankton dynamics and their controlling mechanisms in the near future (section 13.7).

13.4.4 Protozoa

The ecophysiology of protozoans has been summarised in an excellent book by Fenchel (1987).

Major aspects of protozoan plankton ecology have also been treated comprehensively by Laybourn-Parry (1992) and Anderson (1988). Reviews on individual taxa have been quoted at the beginning of section 13.4.

Free-living HNF are a heterogeneous assemblage consisting of zooflagellates *sensu stricto*, choanoflagellates and kinetoplastids, and of heterotrophic species which have evolved independently in various algal classes (Cavalier-Smith 1991, 1993; Patterson & Zölffel 1991; Corliss 1994). Total HNF abundance in aquatic ecosystems ranges from 10^2 to 10^5 cells mL^{-1} and is thus somewhat less variable than bacterial cell numbers.

Abundance of HNF is positively related to bacterial concentrations. Below a bacterial concentration of approximately 10^6 cells mL^{-1}, the relative increase of HNF is higher than that of bacteria in oligotrophic Canadian lakes (Weisse & MacIsaac 2000). This may imply a bacterial threshold abundance (summarised by Weisse 1989) above which HNF numbers rapidly increase and their grazing becomes effective. A different pattern emerges, however, if the ratio between bacteria and HNF is compared over a wide range of lake trophy.

In contrast to initial analyses (Berninger *et al.* 1991a, 1991b), the slope of the logarithmic regression between HNF and bacterial numbers was found to be significantly below unity (Sanders *et al.* 1992; Gasol & Vaqué 1993; Gasol *et al.* 1995). Therefore, HNF increase less rapidly than bacteria along the trophic gradient leading to eutrophy. Bacteria and HNF are therefore much less coupled in eutrophic lakes than in oligotrophic bodies, and the ratio of HNF to bacteria declines in eutrophic waters (Sanders *et al.* 1992; Gasol & Vaqué 1993). Similarly, bacterial grazing loss rate is related positively to total HNF abundance (Weisse 1991b) with a slope of 0.84 (Sanders *et al.* 1992). As will be discussed in section 13.5, a plausible explanation is that in mesoeutrophic lakes HNF (more than bacteria) are 'top-down' controlled by grazing ciliates and mesozooplankton, namely *Daphnia* (Pace *et al.* 1990; Sanders & Porter 1990; Weisse 1991a; Sanders & Wickham 1993; Jürgens 1994; Jürgens *et al.* 1994b, 1996; Cleven 1996). Strong grazing pressure by HNF may lead to great structural

(Güde 1989; Jürgens & Güde 1994; Jürgens *et al.* 1994a) and genotypic (Pernthaler *et al.* 1997) shifts in the bacterial community (reviewed by Hahn & Höfle 2001).

Investigation of the taxonomic composition of natural HNF communities and their seasonal succession is still in its infancy (Laybourn-Parry 1992, 1994) and has recently been reviewed by Arndt *et al.* (2000). The lack of taxonomic resolution is mainly because HNF cannot be identified to species level with common epifluorescence techniques, and there is no comprehensive key available for the identification of HNF species (Arndt *et al.* 2000). The novel advent of rRNA-based fluorescent oligonucleotide probes may help to identify small HNF species unequivocally without cultivation (Lim *et al.* 1993, 1996; Rice *et al.* 1997). Some trends have been deduced from the mesotrophic lakes Mondsee and Bodensee (Constance), where seasonal succession of HNF has been studied in broad taxonomic categories (Salbrechter & Arndt 1994; Cleven & Weisse 2001).

For example, Salbrechter & Arndt (1994), in their annual investigation of the protozooplankton in the Mondsee, Austria, using a live-counting technique, differentiated between five groups of heterotrophic nanoflagellates (kinetoplastids, choanoflagellates, bicoecids, other small heterokont HNF such as *Spumella*, and *Kathablepharis*). At least 15 HNF species from seven orders were identified in routine samples from Lake Constance (summarised in Weisse & Müller 1998) using a combination of epifluorescence microscopy and live observations. In both lakes, small chrysomonads such as *Spumella* spp. were abundant thoughout the year. On annual average, these small HNF contributed 50–60% of the total HNF biomass averaged over the euphotic zone.

The relatively large flagellate *Kathablepharis* sp., of uncertain taxonomic position (Patterson & Zölffel 1991; Clay & Kugrens 1999), accounted for at least one-fifth of total HNF biomass. *Kathablepharis* exhibits well defined peaks in spring, and smaller maxima in autumn. The regular presence of eukaryotic algae in its food vacuoles suggests that *Kathablepharis* feeds upon algae and is a superior competitor during phytoplankton blooms (Weisse 1997; Weisse & Müller 1998). *Kathable-*

pharis sp. does complement its nutrition by feeding upon bacteria, particularly in summer/autumn (Cleven & Weisse 2001), but in contrast to the smaller *Spumella* sp. it cannot satisfy its carbon demand solely from this source. Bacterial abundance was as important as temperature for *in situ* growth rates of *Spumella* sp., whereas bacteria had no effect on growth rates of *Kathablepharis* sp. (Weisse 1997).

Peak production of the obligate bacterivore *Spumella* sp. and the omnivore *Kathablepharis* sp. alternate seasonally in Lake Constance (Weisse 1997). Choanoflagellates and bicoecids are prominent in summer and late summer, and are generally found attached to algae or other substrates. Choanoflagellates seem, on average, to possess higher bacterial ingestion rates than the small heterokont species (Vaqué & Pace 1992). The preponderance of small heterokont flagellates, in particular *Spumella* (syn. *Monas*), *Ochromonas* and *Paraphysomonas*, has also been noted from the eutrophic lakes Müggelsee, Germany (Arndt & Mathes 1991), and Oglethorpe, Georgia, USA (Sanders *et al.* 1989; Bennett *et al.* 1990).

Mixotrophy is defined as the combination of autotrophic and heterotrophic nutrition, i.e., phagotrophy by photosynthetic organisms (e.g. Riemann *et al.* 1995). Mixotrophic flagellates are found in lakes which vary in origin, size, morphometry and trophic state (see Sanders & Porter 1988; Sanders 1991a,b; Jones 2000 for review). Mixotrophic chrysophytes (Bird & Kalff 1987) and dinoflagellates (Gaines & Elbrächter 1987) can contribute the major proportion of total bacterivory in some lakes (Bird & Kalff 1986; Sanders *et al.* 1989). In Lake Constance, mixotrophic algae of the genus *Dinobryon* became relatively more important during its recent oligotrophication (Sommer *et al.* 1993; Gaedke 1998). Many aspects of nutrient uptake, nutrient release and carbon utilization in mixotrophic algae have been studied experimentally with cultured species (Caron *et al.* 1990; Sanders *et al.* 1990; Rothhaupt 1996a,b, 1997). It is clear that mixotrophy has evolved polyphyletically and serves several ecological functions in lakes.

The majority of bacterivorous HNF species are smaller than 8 µm in size (Bennett *et al.* 1990; Weisse 1991a; Cleven & Weisse 2001). Large heterotrophic flagellates (LHF), mainly dinoflagellates and chrysomonads in the nanoplankton and microplankton size range, seem to be more important for pelagic carbon flow in lakes than previously thought (Arndt & Mathes 1991; Arndt & Nixdorf 1991; Mathes & Arndt 1994). In the Mondsee and Lake Constance, HNF and LHF contribute approximately equal amounts to total protozoan biomass (Salbrechter & Arndt 1994; Weisse & Müller 1998). Maximum growth rates of the heterotrophic dinoflagellate *Gymnodinium helveticum* measured in Lake Michigan (0.47 day^{-1}; Carrick *et al.* 1992) and Lake Constance (0.46 day^{-1}; Weisse & Müller 1998) are distinctly lower than maximum growth rates of HNF and ciliates recorded from the same lakes. Accordingly, *G. helveticum* in mesotrophic lakes seems to be much less important in terms of production than biomass (Fig. 13.5). This may not be the case in highly eutrophic lakes, where LHF growth rates up to 2 day^{-1} have been measured (Arndt & Nixdorf 1991).

The trophic role and ecological significance of LHF in lakes is poorly understood (Arndt *et al.* 2000). It appears that the larger genera amongst HNF (e.g. *Kathablepharis*, *Paraphysomonas*), and most LHF, are omnivorous, but prefer algae and cyanobacteria over the smaller heterotrophic bacteria (Weisse & Müller 1998). *Paraphysomonas* sp. has been observed to feed upon the much larger diatom *Synedra* sp. in an oligotrophic lake in British Columbia (Suttle *et al.* 1986). Uptake of diatoms by the heterotrophic dinoflagellate *Gymnodinium helveticum* f. *achroum* has also been documented (Popovsky 1982). Heterotrophic dinoflagellates have developed very diverse strategies for phagotrophic food uptake, including engulfment of entire food particles, myzocytosis and pallium feeding (reviewed by Gaines & Elbrächter 1987; Elbrächter 1991). Myzocytotic species, which use a special feeding organelle (the peduncle; Schnepf & Deichgräber 1984), and pallium feeders (Jacobson & Anderson 1986) prey upon a large array of algae and other heterotrophic protists (Jacobson & Anderson 1986; Goldman *et al.* 1989; Hansen

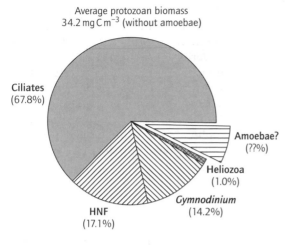

Average protozoan biomass
34.2 mg C m^{-3} (without amoebae)

Ciliates
(67.8%)

Amoebae?
(??%)

Heliozoa
(1.0%)

Gymnodinium
(14.2%)

HNF
(17.1%)

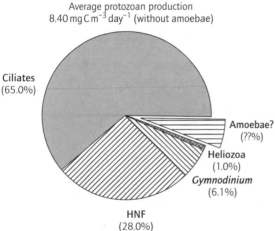

Average protozoan production
8.40 mg C m^{-3} day^{-1} (without amoebae)

Ciliates
(65.0%)

Amoebae?
(??%)

Heliozoa
(1.0%)

Gymnodinium
(6.1%)

HNF
(28.0%)

Fig. 13.5 Average annual protozoan biomass (top) and production (bottom) in Lake Constance (0–20 m): HNF, heterotrophic nanoflagellates. *Gymnodinium helveticum* is the only dominant species among large heterotrophic flagellates. Biomass of naked amoebae was measured in 2–4 m water depth only. The contribution of amoebae to total protozoan biomass and production integrated over the upper 20 m of the water column therefore remains speculative. (Modified from Weisse & Müller 1998.)

Moestrup 1997; Weisse & Kirchhoff 1997). Ciliates, copepods and daphnids prey upon LHF (Arndt *et al.* 2000); the fate of LHF production in lakes remains, however, largely unknown.

Compared with flagellates, a wealth of information is available on the ecology and seasonal succession of ciliates in lakes (reviewed by Laybourn-Parry 1992, 1994). This is because most ciliates are more conspicuous and easier to identify with conventional microscopy than HNF. In general, various types of aquatic habitats support different ciliate communities (Finlay & Fenchel 1996). Ciliates are often the quantitatively most important heterotrophic protists in microbial food webs (Fig. 13.5). The taxonomic composition of ciliates in larger lakes appears to be largely composed of oligotrichs (now grouped into the Subclass Spirotrichia; Corliss 1994), peritrichs (Subclass Peritrichia), haptorids (Class Litostomatea), and scuticociliates (Subclass Scuticociliatia; Beaver & Crisman 1989; Laybourn-Parry 1992, 1994). Yet, recent findings suggest that small prostomatids (Class Prostomatea), which may have been overlooked in previous studies, account for the majority of the ciliate population and a large part of ciliate production in lakes (Müller 1989, 1991; Müller *et al.* 1991; Sommaruga & Psenner 1993; Salbrechter & Arndt 1994; Weisse & Müller 1998).

Ciliates are generally less important than flagellates as bacterial consumers in natural aquatic environments. Fenchel (1980) argued that, as their volume-specific clearance rate is only about one-tenth of that either of bacterivorous HNF or of ciliates feeding on larger food items, bacterivorous ciliates play only a minor role in planktonic food webs. This conclusion is supported by some empirical evidence from lakes (Beaver & Crisman 1989). Using a model of suspension feeding, Fenchel (1984) further elaborated on energetic grounds that organisms in the HNF size range are more effective in extracting small suspended particles such as bacteria than larger specimens.

At particular times and localities, ciliate bacterivory may contribute substantially to bacterial losses in fresh water (Beaver & Crisman 1989; Šimek *et al.* 1990a,b; Müller *et al.* 1991; Simon

1991, 1992). The majority of these investigations were conducted with marine species, but recent evidence suggests that the ecological niches of freshwater dinoflagellates are similar (Calado &

et al. 1998b) and salt water (Rivier *et al.* 1985; Sherr *et al.* 1986; Albright *et al.* 1987; Sherr & Sherr 1987; Weisse 1999). Beaver & Crisman (1982, 1989) suggested that large algivorous ciliates, predominantly oligotrichs, are progressively replaced by small bacterivorous species, namely scuticociliates. In meso- to eutrophic lakes and reservoirs which are characterised by large seasonal changes, facultative or obligate bacterivorous ciliates tend to become the main bacterivores in summer when bacterial concentrations are relatively high (Šimek *et al.* 1990a,b, 1995, 1996, 1998, 2000; Šimek & Straskrabova 1992; Müller *et al.* 1991). As previously mentioned (Section 13.4.3), bacterivory has been studied in great detail in Lake Constance using various methods. Results from using a dilution technique (Weisse 1990), fluorescently labelled bacteria (FLB; Cleven 1995), ^{14}C-labelled bacteria (Simon *et al.* 1998a) and from modelling (Straile 1995, 1998) consistently showed that total ciliate bacterivory is, on annual average, four to five times less than bacterial grazing by HNF.

Ciliates play a major role in planktonic food webs as consumers of algae. Since the preferred size-feeding spectrum of many ciliate species is about 3–20 μm, it overlaps with that of many rotifers and crustacea (reviewed by Sterner 1989). It follows that ciliates compete with metazoa for the same algal food (Weisse & Frahm 2002). In seasonally variable lakes, they are the major herbivores during the early stages of planktonic seasonal succession (Weisse *et al.* 1990). This is because ciliates, owing to their higher intrinsic growth rates, are better able than metazooplankton to respond rapidly to developing phytoplankton blooms (Smetacek 1981). Results from mass balanced carbon flow models averaged over five consecutive years indicate that ciliates are responsible for as much as 43% of total phytoplankton grazing in Lake Constance (Straile 1995). This is almost double algal ingestion by the 'classic' herbivorous crustacean zooplankton. The consequences of this finding will be discussed in section 13.5. An example of pronounced seasonal succession of ciliates in a large lake is shown in Fig. 13.6.

Ciliates may also efficiently control heterotrophic nanoflagellates (Weisse 1990, 1991a;

Cleven 1996; Jürgens *et al.* 1996), although HNF contribute relatively little to total food uptake by the ciliate community (Müller *et al.* 1991; Straile 1995, 1998). Verity & Villareal (1986) and Verity (1991) demonstrated that, for marine ciliate species, the nutritional value of different nano-sized algae and heterotrophic flagellates is 'species-specific', and that ingestion of similar-sized autotrophic and heterotrophic nanoplankton may lead to comparable growth rates and growth efficiencies of ciliates. However, species-specific and even intraspecific differences in growth, grazing and production rates have recently been documented for freshwater ciliates (Weisse & Montagnes 1998; Montagnes & Weisse 2000; Weisse *et al.* 2001). Ciliates with algal symbionts which are functionally autotrophic are found in fresh (Kahl 1932; Beaver *et al.* 1988; Beaver & Crisman 1989; Müller 1989; Müller *et al.* 1991), brackish (Lindholm 1981, 1985) and salt waters (Stoecker *et al.* 1987; Stoecker 1991).

Comparatively little is known about the significance and ecological role of other protozoan taxa. Naked and testate amoebae ranging from about five to several thousand micrometres in size (Caron & Swanberg 1990; Sleigh 1991) are common in marine and freshwater plankton (Caron & Swanberg 1990; Arndt 1993). Small, naked amoeba (5–20 μm), which may have been overlooked in conventional studies owing to methodological shortcomings, seem to prevail in lakes (Arndt 1993; Mathes & Arndt 1994; Salbrechter & Arndt 1994). Evidence based mainly on direct counts of live samples suggests that these small cells occur in abundances several orders of magnitude higher than hitherto believed. Average abundance appears to increase along a trophic gradient ranging from <0.2 cells mL^{-1} in oligotrophic lakes up to >100 cells mL^{-1} in (hyper)eutrophic environments (Arndt 1993). For example, in the highly eutrophic Müggelsee, up to 380 naked amoeba per millilitre were found (Arndt 1993).

In Lake Constance, the near-surface abundance of unidentified naked lobose species varies seasonally from <1 cell mL^{-1} to 118 cells mL^{-1} (Weisse & Müller 1998). Peaks of amoebae occur for a short time during phytoplankton maxima, which cor-

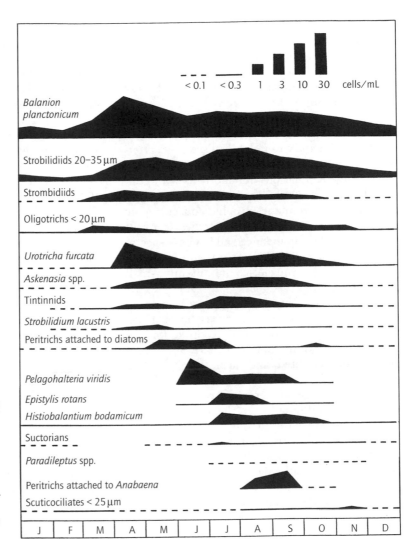

Fig. 13.6 Seasonal succession of pelagic ciliates in the epilimnion of Lake Constance (0–8 m). Monthly mean values were averaged from three consecutive years (1987–1989); peaks are therefore less obvious than in individual years. (From Weisse & Müller 1998. © E.Schweizerbart'sche. Reproduced with permission of the publisher.)

roborates observations from other lakes. However, in contrast to other (limited) records mentioned above, the amoebae in this lake are larger, 50–90 μm in size. Schweizer (1994) found a significant increase in epilimnetic cell numbers of naked amoebae along an inshore-offshore transect in Lake Constance. This may challenge the common assumption that free-living amoebae are generally associated with solid surfaces or substrates such as macroaggregates (Rogerson & Laybourn-Parry 1992; Arndt 1993; Murzov & Caron 1996).

Averaged over the season, naked amoebae seem to contribute very little, normally <1%, to total protozoan biomass in lakes (Arndt 1993). High growth rates with doubling times ranging from 0.74 to 1.74 day^{-1} measured in the Müggelsee (Arndt 1993) indicate that amoebae may become major members of the microbial food web during peaks in their populations. Even less information than on naked amoebae in lakes is available for testate species. Typical numbers are lower than those of naked species, ranging from several up to several hundred cells per litre (reviewed by Arndt 1993). The nutritional ecology of amoebae in lakes, as well as their fate in the food web, awaits further research.

Planktonic heliozoans have received very little attention in limnological studies (Arndt 1993; Zimmermann *et al.* 1996, and references therein). Knowledge of the ecology of heliozoa is restricted mainly to individual species (Patterson & Hausmann 1981; Hausmann & Patterson 1982). Large populations of heliozoans with peak abundances exceeding 50 cells mL^{-1} have recently been recorded from mesotrophic and eutrophic freshwater lakes (Arndt 1993; Mathes & Arndt 1994, 1995). The abundance and biomass spectra of heliozoa and naked amoebae were virtually identical in the Müggelsee, for mainly small species <30 μm.

Community composition and seasonal dynamics of planktonic heliozoans in a large, mesotrophic lake have been reported for the first time by Zimmermann *et al.* (1996). Heliozoans were counted and identified to genus in live samples from two stations at different depths. Seven genera were observed at both stations from mid-June, after the *Daphnia* peak, through to late autumn (Fig. 13.7). Heliozoan numbers fell below the limit of detection during the phytoplankton spring bloom. The small genus *Heterophrys* (mean diameter 15.6 μm, mean cell volume 2065 μm^3) was numerically most important; in terms of biomass, *Heterophrys* and the two largest genera *Actinophrys* (42.9 μm, 47,000 μm^3; volume) and *Raphidocystis* (25.0 μm, 9500 μm^3 volume) dominated. In Fig. 13.7, seasonal succession of the various heliozoan genera is clear.

Zimmermann *et al.* (1996) also measured heliozoan growth rates *in situ* on two occasions during summer and autumn. Growth rates of individual genera ranged from 0.06 day^{-1} in *Raphidocystis* to 0.59 day^{-1} in *Choanocystis*, equivalent to a minimum generation time of 1.2 days. In spite of these relatively high growth rates, heliozoans contribute little to total protozoan biomass and production in Lake Constance (cf. Fig. 13.5).

13.5 THE SIGNIFICANCE OF MICROBIAL FOOD WEBS IN LAKES

After the term 'microbial loop' was coined, discussion of the significance of microbial food webs during the 1980s focused on two general issues.
1 Does the microbial loop act as 'sink' or 'link' for the classic grazer food chain (see section 13.2)?
2 Are individual components (and is the microbial food web as a whole) primarily controlled by 'bottom-up' or 'top-down' forces (McQueen *et al.* 1989)?
The latter debate can be traced back to the 1960s and 1970s (Brooks & Dodson 1965; McQueen *et al.* 1986; Threlkeld 1988; Banse 1995, and references therein). 'Bottom-up' forces include substrates or food, nutrients and abiotic factors such as temperature. 'Top-down' forces are exerted 'from above' by grazers and predators. I will show that these two questions are related, and will address the problem of the relative significance of the microbial food web from different perspectives.

13.5.1 Theoretical attempts to quantify the strength of the microbial food web

Stimulated by accumulating empirical data, several attempts have been made to quantify the significance of microbial production relative to the grazer food chain. Porter *et al.* (1988) proposed to calculate combined autotrophic and heterotrophic picoplankton production as a fraction of total plankton production. The result should indicate the **measure of microbial strength** (MOMS). The authors postulated that the MOMS value should decline along a gradient leading to eutrophic and hypertrophic systems (Fig. 13.8). This is because both autotrophic and heterotrophic picoplankton are better able to sequester nutrients and DOM at low ambient concentrations than are larger algae. The former also possess the advantage that they remain constantly in the water column and are virtually never sedimented, therefore retaining nutrients in the euphotic zone. Picoplankton production is largely based on regenerated nutrients, whilst larger algae are favoured by input of new material leading to elevated nutrient concentra-

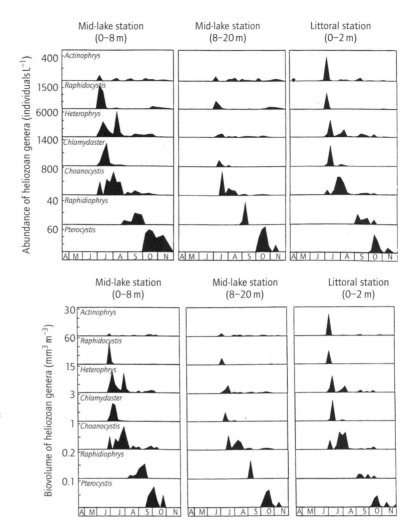

Fig. 13.7 Seasonal course of pelagic heliozoan genera at two different sampling stations and three different depths strata in Lake Constance during 1993. Abundance is shown in the upper panel and biomass (biovolume of live cells) in the lower panel. Note that the scale is different for the individual taxa. (Modified after Zimmermann *et al.* 1996; reproduced with permission of Inter-Research.)

tions. Although there is some evidence that the MOMS hypothesis is correct, it has not yet been rigidly tested. This is mainly because it requires comprehensive data sets on all plankton components from lakes across a trophic gradient. Except for a few cases, there are not enough data.

Weisse (1993) elaborated on the MOMS concept. From empirical data, and on theoretical grounds, he concluded that the ratio between eukaryotic and prokaryotic autotrophic picoplankton production is indicative of lake trophy, and the relative significance of the microbial food web (Fig. 13.8). This ratio is relatively easy to measure, and several studies have demonstrated that the significance of eukaryotic picoplankton is positively related to nutrient supply in lakes (summarised in Weisse 1993). Similarly, recent analyses have revealed that the relative contribution of autotrophic picoplankton to total phytoplankton biomass is strongly reduced in hypertrophic systems (Sommaruga & Robarts 1997; Agawin *et al.* 2000; Bell & Kalff 2001).

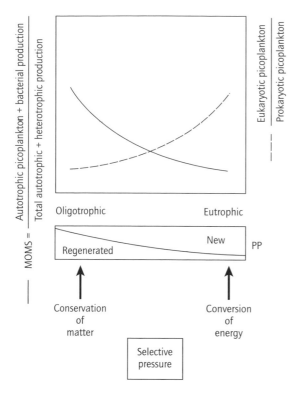

Fig. 13.8 The 'measure of microbial strength' (MOMS) concept (upper panel, modified after Porter *et al.* 1988) in the system context. Prokaryotic picoplankton production dominates in oligotrophic waters where total primary production (PP) is largely based on regenerated nutrients and in which, therefore, organisms have been selected for the conservation of matter (lower panel). In eutrophic systems, in contrast, larger eukaryotic organisms prevail which efficiently use the ample substrate supply converting physical energy into energy-rich organic compounds. Under these conditions, the significance of eukaryotic algae among the picoplankton is also highest. (Modified after Weisse 1993.)

Since this concept explains the prevalence of prokaryotes and the microbial food web in oligotrophic systems as adaptive traits which conserve scarce matter, it may be called the **conservation of matter (COM) hypothesis**. The basic difference between the COM hypothesis and the MOMS model is that the former states that the

simple internal level of organisation of prokaryotic picoplankton is the primary cause of their prevalence in oligotrophic systems rather than cell size. This simple cellular organisation reflects the phylogentic age of heterotrophic bacteria and cyanobacteria. As with the MOMS concept, an experimental test of the COM hypothesis is still unavailable.

Ducklow (1991) introduced the term **microbial food web efficiency** (MFWE), which represents the combined HNF and ciliate production consumed by rotifers and crustaceans, as a fraction of the total ingestion by HNF and ciliates. The MFWE hypothesis was tested in Lake Constance, where the most complete data set on all major pelagic components from bacteria to fish is available (Geller *et al.* 1991; Gaedke & Straile 1994a; Gaedke *et al.* 1995; Straile 1995, 1998). In this seasonally variable lake, the MFWE concept was found to be misleading, because it was low in spring and summer when Protozoa represented the main channel of planktonic carbon flow (Gaedke & Straile 1994a; Straile 1995, 1998). The MFWE hypothesis disregards the point that HNF are primarily bacterivores, whilst the bulk of ciliates are omnivorous, feeding on algae and heterotrophic flagellates. Ciliates thus belong to the same trophic level as rotifers and herbivorous crustacea (Straile 1995).

13.5.2 *Empirical evidence*

The structure and functioning of the microbial food web depend on lake trophy and the general structure of the pelagic community. In pristine oligotrophic waters, pelagic carbon production is largely based on picoplankton, i.e. bacteria and picocyanobacteria. Bacterial biomass reaches or exceeds phytoplankton biomass (Simon *et al.* 1992), whilst picocyanobacteria contribute >70% of total primary production in (ultra)oligotrophic lakes (Stockner 1991). In the previous section I pointed out that the significance of picocyanobacteria relative to eukaryotic algae **decreases** in more eutrophic environments.

On a logarithmic scale, both bacterial abundance and biomass are linearly related to Chl *a* (Bird & Kalff 1984; Cole *et al.* 1988; Currie 1990;

White *et al.* 1991; Cole & Caraco 1993) and phytoplankton carbon (Simon *et al.* 1992). A similar positive empirical relationship was found between bacterial production and primary production (Cole *et al.* 1988). The slope of all of the corresponding regression equations lies typically within the range 0.6–0.8, but in the case of various bacterial to phytoplankton biomass ratios it is well below 0.5 (Simon *et al.* 1992). These allometric ratios thus indicate unequivocally that, as with picocyanobacteria, heterotrophic bacteria are, relative to eukaryotic phytoplankton, much more important in oligotrophic than in eutrophic ecosystems. In the former, bacteria and cyanobacteria are superior competitors at low nutrient concentrations, whereas in nutrient-rich waters, because of their superior nutrient uptake capacities at elevated nutrient concentrations, the contribution of algae to the pelagic carbon flow is higher (section 13.4.3). In other words, nutrient supply is a key factor determining relative significance of bacteria versus algae.

Because both bacteria and autotrophic picoplankton are consumed primarily by fast growing HNF and ciliates, combined annual production of Protozoa in oligotrophic to mesoeutrophic lakes is similar to, or may even exceed, metazoan production (Fig. 13.9). Despite some inconsistency in the methods used to calculate zooplankton production, data presented in Fig. 13.9 suggest that the relative protozoan production is highest in oligotrophic lakes, and decreases in mesoeutrophic systems.

A recent study conducted across a trophic gradient in Lake Erie supported this conclusion. Relative protozoan biomass increased from eutrophic coastal stations towards oligo-mesotrophic offshore sites (Hwang & Heath 1997). Offshore, HNF and ciliates contributed 20–80% of total plankton biomass. Owing to the much shorter generation times of Protozoa compared with larger metazoans, average protozoan production should considerably exceed metazoan production at the oligo-mesotrophic sites in Lake Erie. Results presented in Fig. 13.9, and by Hwang & Heath (1997), sharply contrast with the carbon budget established for the oligotrophic Mirror Lake, New

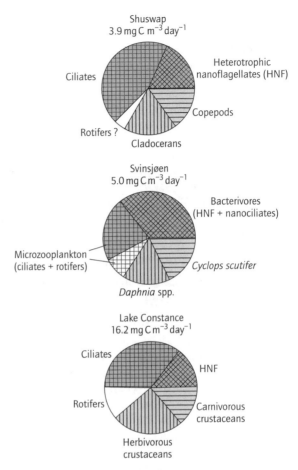

Fig. 13.9 Zooplankton production in oligotrophic–mesoeutrophic lakes. (Top) Average seasonal (March–November) production in the epilimnion of oligotrophic Lake Shuswap, British Columbia, Canada, estimated from measured plankton biomasses and growth rates (protozoa) and from production/biomass ratios (crustacea). Rotifer production is small but remained unmeasured (T. Weisse, original). (Middle) Production estimates from enclosure experiments conducted in the epilimnion of mesotrophic Svinsjøen, southeastern Norway, during a 10-day period in September. Values are based on measurements and estimates from a carbon flow network model (modified after Lyche *et al.* 1996). Microzooplankton represents ciliates > 20 µm and rotifers (almost exclusively *Synchaeta* cf. *kitina*). (Lower) Zooplankton production in Lake Constance averaged over 0–20 m and four seasonal phases during April to August 1987, when the lake was mesoeutrophic. Results stem from a mass-balanced carbon flow model based upon measured values (data from Gaedke & Straile 1994a).

Hampshire. By measuring biomasses and bacterial respiration, and assuming theoretical respiration rates for Protozoa and metazooplankton, Cole and co-workers estimate that, there, metazooplankton respiration is roughly tenfold greater than protozoan respiration (Cole *et al.* 1989).

Both in Svinsjøen, Norway, and in Lake Constance, protozoan and metazoan respiration are about equal (Straile 1995; Lyche *et al.* 1996). Bacterial production in Mirror Lake was equivalent to 16% of primary production, and thus lower than that predicted by a general empirical regression between bacterial and primary production calculated by the same authors (Cole *et al.* 1988, see above). Bacterial respiration was equal to 19% of net primary production and lower than algal respiration in Mirror Lake. Recent evidence has revealed that, in oligotrophic lakes, bacterial respiration generally exceeds both phytoplankton net production and respiration manifold (del Giorgio & Peters 1993, 1994; Coveney & Wetzel 1995; Lyche *et al.* 1996). The estimated zooplankton demand for algal carbon exceeded measured primary production in Mirror Lake. In order to bring their budget closer to balance, Cole & Caraco (1993) speculated that part of bacterial production was consumed by predatory bacteria such as *Bdellovibrio*.

An alternative explanation seems much more likely. Cole and co-workers measured heterotrophic protists using epifluorescence microscopy after filtering on 0.8-μm Nuclepore filters. This technique is adequate for small HNF, but drastically underestimates biomasses of both LHF and ciliates, which in oligotrophic lakes should be counted from settling chambers of relatively large volumes (100 mL). In Mirror Lake, zooplankton was assessed from 45-μm mesh filtered samples. As a consequence of inadequate sampling of LHF and ciliates, and probably an unrealistically high assumption of HNF growth efficiency (60%, see Straile 1995), protozoan respiration was grossly underestimated and the carbon budget inaccurately assessed. I use this example as a warning to the reader: the general trends outlined in this section are strongly dependent on the methods used, and there are at present no standard techniques available with which to measure protozoan production

and grazing rates (Landry 1994; Weisse 1997, and references therein). Accordingly, all carbon budgets for lakes published so far suffer from methodological shortcomings or compromises.

As a consequence of the general trends outlined above, several authors support the 'MOMS' concept that the relative significance of the microbial loop and the microbial food web declines from oligotrophic to eutrophic systems (Stockner & Porter 1988; Weisse 1991b; Simon *et al.* 1992; Weisse & Stockner 1993). It may be surprising that others advocate the opposite (Riemann *et al.* 1986; Søndergaard & Riemann 1986; Riemann & Christoffersen 1993, Mathes & Arndt 1994). The former authors based on their work on shallow Danish lakes (Riemann & Søndergaard 1986; Christoffersen *et al.* 1990; Christoffersen *et al.* 1993) state that 'the importance of bacteria and the microbial loop increases along a productivity gradient relative to the grazer food chain' (Riemann & Christoffersen 1993, p. 91). The answer to this seeming paradox is that the 'MOMS' and 'COM' hypotheses stress bottom-up effects, whilst these Danish researchers strongly favour top-down effects originating from the feeding impact of planktivorous fish.

Changes in nutrient supply are accompanied by taxonomic shifts in the phytoplankton and the zooplankton community (see Chapters 10 and 14). In oligotrophic lakes in British Columbia, the metazooplankton, which support only low fish production, is composed mainly of rotifers, small bodied cladocera such as *Eubosmina* and cyclopoid copepods (Stockner & Shortreed 1989). The microbial food web is slightly affected by these metazoans (Weisse & Stockner 1993). It appears that the relative significance of cladocerans increases at the expense of calanoid copepods along a eutrophication gradient (Richman *et al.* 1990, Riemann & Christoffersen 1993). Calanoid copepods are better adapted to oligotrophic conditions, because their food thresholds are lower than those of daphnids (Lehman 1988; DeMott 1989).

Major contrasts between calanoid copepods and daphnids are that copepods exercise more distinct size preferences, use chemical cues and exhibit lower clearance rates (reviewed by Sterner 1989). A recent field study performed in a mesotrophic lake yielded a surprising result; neither of the two zooplankton groups significantly reduced phytoplankton biomass (Sommer *et al.* 2001). However, cladocerans suppressed small phytoplankton, while copepods suppressed large phytoplankton. *Daphnia* is the main genus in most mesotrophic and eutrophic lakes (see Chapter 14), where coupling between phytoplankton and zooplankton is strongest (Carney & Elser 1990). Raptorial cyclopoid copepods may become more abundant in the most productive lakes, which are often characterised by blooms of colonial cyanobacteria (Riemann & Christoffersen 1993). In hypertrophic lakes, smaller species of *Bosmina* replace *Daphnia*, particularly when fish predation is high (Jeppesen *et al.* 1992).

The consequences of these taxonomic trends in metazooplankton composition in relation to lake trophic status are important for the significance of the microbial food web relative to the zooplankton grazer food chain. *Bosmina*, *Eubosmina* and most calanoid copepods do not feed on picoplankton or the smallest HNF (reviewed by Sterner 1989) and therefore,'unlike *Daphnia*', cannot 'break' the microbial loop (Pace *et al.* 1990; Riemann & Christoffersen 1993). Some experiments (Pace *et al.* 1990; Wylie & Currie 1991) and various modelling approaches (Wylie & Currie 1991; Gaedke & Straile 1994a,b; Gaedke *et al.* 1995) have demonstrated that overall transfer efficiency from bacteria to crustacea is markedly increased if *Daphnia* are abundant in lakes. Many *Daphnia* species can consume a broad size spectrum of prey (c.0.5–50 μm), so that they can affect nearly every component of the microbial food web (Porter *et al.* 1988; Stockner & Porter 1988; Pace *et al.* 1990; Riemann & Christoffersen 1993; Weisse & Stockner 1993). In mesotrophic and slightly eutrophic lakes, *Daphnia* may act as 'keystone' predators, which structure the whole planktonic food web via their feeding activity (Stockner & Porter 1988; Sanders & Porter 1990; Pace *et al.* 1990; Weisse 1991a; Weisse & Stockner 1993; Jürgens 1994; Jürgens *et al.* 1994b).

Bacteria and HNF are much less coupled and their numbers more variable in eutrophic lakes than in oligotrophic waters. The ratio of bacteria to HNF is higher in eutrophic waters, which may be the result of the impact of *Daphnia* on the system (Sanders *et al.* 1992; Gasol & Vaqué 1993). Bacteria apparently possess a twofold advantage in lakes where *Daphnia* are moderately abundant:

1 bacteria enjoy some refuge from grazing when HNF are suppressed by their predators;

2 *Daphnia* considerably fuel the DOM pool via their 'sloppy feeding' (Lampert 1978), and subsequent excretion.

It is important to note that the high abundance of *Daphnia* is mostly restricted to short periods. Averaging over the season will therefore lead to very different conclusions with respect to the structure and function of the microbial food web, as compared with the period when *Daphnia* numbers peak (Gaedke & Straile 1994a; Straile 1995). The large potential significance of ciliates and large heterotrophic flagellates as consumers of algae and competitors of crustacea and rotifers in oligo-mesoeutrophic lakes needs to be considered in conceptual as well as in mathematical models. A similar inference was made by Sherr & Sherr (1994) for marine ecosystems. The impact of *Daphnia* on all components of the pelagic food web is so apparent that some lakes have been characterised as typical '*Daphnia* lakes' although herbivory by protists has not been measured. The example from Lake Constance where phytoplankton grazing by all zooplankton compartments has been estimated (section 13.4.4) shows that this may be premature because herbivory by protists may, on annual average, exceed algal ingestion of crustacea.

Since daphnids are also excellent food for fish and invertebrate planktivores, fish production is enhanced in lakes where *Daphnia* biomass is high (Porter *et al.* 1988; Stockner & Shortreed 1989). Heavy grazing pressure by planktivorous fish may control the abundance of *Daphnia* and lead to a 'revival' of the microbial food web (Christoffersen *et al.* 1993; Riemann & Christoffersen 1993; Weisse & Stockner 1993). Figure 13.10 shows schematically how strong feeding pressure by planktivorous fish may 'cascade' down to the

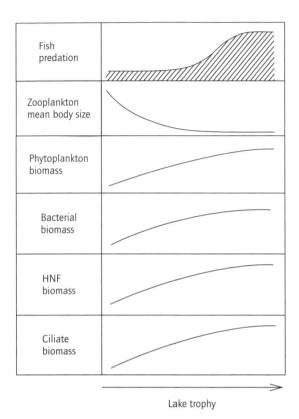

Lake trophy

Fig. 13.10 Idealised scheme illustrating effects of fish predation (top panel) in eutrophic, temperate lakes on community/species composition of the meso- and macrozooplankton (second panel from top); lower panels show the subsequent consequences for phytoplankton, bacteria, heterotrophic nanoflagellates (HNF) and ciliates. (Modified after Riemann & Christoffersen 1993; reproduced with permission from Gauthier-Villars Editeur.)

lower food levels (Carpenter *et al.* 1985; McQueen *et al.* 1989).

Increase in fish predation is often accompanied by a change in lake trophic status in the mesoeutrophic range. This concept seems to support the inference by Riemann and colleagues (above) of increased importance of the microbial food web in more nutrient-rich lakes. If the impact of fish is removed, however, the Danish lakes which they studied do not violate the opposite trend, which is

obvious from comparison across a wider trophic range. The same conclusion is apparent from a study by the same Danish research team, conducted in the oligotrophic Lake Almind, mid-Jutland, where grazing pressure from planktivorous fish is low (Søndergaard *et al.* 1988). In this lake, and in experimental enclosures which received no nutrient additions, bacteria accounted for 75% and 66% of the daily carbon fixation. In enclosures where nutrients were added, bacterial carbon fixation was reduced to only 33%. Concomitantly, phytoplankton size distribution shifted from picoplankton and nanoplankton (<8 μm) at low nutrient concentrations towards larger algae (>40 μm) at high nutrient status. Addition of perch (*Perca fluviatilis*) produced no measurable effect on the algae, probably because of low abundance of mesozooplankton (>140 μm) in the lake. Mesozooplankton contributed little to total bacterial losses, and the authors therefore assumed that (unmeasured) grazing by protists, including the mixotrophic *Dinobryon*, was mainly responsible for bacterial mortality. Søndergaard and colleagues concluded that the detrital food web via the bacteria (i.e., the microbial loop) constituted the main route of organic carbon flow in oligotrophic Lake Almind, and became less important when nutrients stimulated larger algae. Clearly, these observations support the general trends of the decreasing importance of the microbial loop/microbial food web with increasing nutrient status, and the significance of 'bottom-up' control in oligotrophic systems.

A similar effect of secondarily increased significance of the microbial food web may be reached in hypertrophic systems, owing to inedible and/or toxic effects of colony-forming cyanobacteria. This is because *Daphnia* are more vulnerable to cyanobacterial blooms than *Bosmina* and cyclopoid copepods (Lampert 1981; Paerl 1988; Hawkins & Lampert 1989). Cyanobacteria play a pivotal role in planktonic food web structure at both extremes of the trophic gradient: at the (ultra)oligotrophic end, growth dynamics of single-celled species largely determine total biological production; at the (hyper)eutrophic end, potential toxicity and susceptibility (or resistance) to grazing of colonial species exert a major impact on the success of other biota (Weisse 1994). For various reasons outlined above, microbial food webs benefit from the abundance of both types of cyanobacteria.

In conclusion, the various substrate (nutrient) uptake kinetics of bacteria/cyanobacteria and eukaryotic algae are primarily responsible for an increase in algae relative to picoplankton. Therefore, the classic grazer food chain increases in importance relative to the microbial food web, when comparing oligotrophic to eutrophic ecosystems. Top-down effects are generally more important in eutrophic than in oligotrophic systems, and may become most significant in the mesoeutrophic range. Top-down control, and inedible colonial cyanobacteria, may lead to a (temporal) secondary enhancement of the microbial food web in eutrophic-hypertrophic systems, which offsets the general trend apparent from a cross-system comparison.

13.6 FROM ARTIFICIAL COMMUNITIES TO 'SPECIES-SPECIFIC' INTERACTIONS

The above discussion reflects the 'bottom-up versus top-down' controversy in aquatic ecology from the 1980s and early 1990s. Owing to insufficiently discriminatory techniques, and in order to arrive at general conclusions, many microorganisms which are only remotely related, if at all, were lumped together in presumed functional guilds such as 'heterotrophic bacteria' or 'HNF'. There is increasing evidence that such oversimplifications are not productive in furthering our conceptual understanding of the functioning of microbial food webs. One reason is that, when relatively simple carbon budgets based upon functional guilds were established, feedback effects (see section 13.2) were often neglected. More important, broad trophic categories do not adequately describe the complexity and diversity of the ecological niches of aquatic protists (Arndt & Berninger 1995; Weisse 2002).

As discussed in section 13.4.4, the nutritional ecology of HNF, dinoflagellates, and ciliates is di-

verse, and in many cases unknown. If the assumption that the majority of ciliates feed upon nanoplanktonic phytoplankton is correct, the question is, in which parameters do their ecological niches differ? The closely related sympatric prostomatid ciliates *Balanion planctonicum* and *Urotricha furcata*, which dominate the ciliate community in many lakes (Weisse & Müller 1998), vary with respect to their feeding behaviour, temperature response, and interactions with rotifer species of the genus *Keratella* (Weisse *et al.* 2001; Weisse & Frahm 2001, 2002). Relationships between the two prostomatids and the two *Keratella* species were 'species-specific' complex, and pointed to the existence of a chemical defense in *Urotricha* (Weisse & Frahm 2001; see also Adrian *et al.* 2001). Relative to direct rotifer feeding on the ciliates, exploitative competition (for review see DeMott 1989) was less important (Weisse & Frahm 2002). Such direct and indirect interactions among Protozoa, and between protists and metazooplankton, need to be investigated in more detail. For a zooplanktologist it is indisputable that exploitative competition can be as important as direct feeding for the fitness and the ecological success of two or more coexisting species (see Chapter 14). The significance of behavioural features as defence mechanisms of ciliates against predatory rotifers was presented earlier (Gilbert 1994).

Heterotrophic nanoflagellates and ciliates possess chemosensory capabilities (e.g. Spero 1985; Sibbald *et al.* 1987; Bennett *et al.* 1988; Verity 1988), which they use in order to select different food items (Pace & Bailiff 1987; Nygaard *et al.* 1988; Landry *et al.* 1991). Jürgens & De Mott (1995) reported that prey selection among the flagellates *Bodo saltans* and *Spumella* sp. is dependent upon the concentration and quality of food particles, and follows optimal diet models (Stephens & Krebs 1986; Hughes 1993). Although the mechanisms responsible for food selection are not yet well understood (Boenigk & Arndt 2000b), a suite of processes seems to be involved during the uptake and handling of prey particles. Heterotrophic nanoflagellates and ciliates may select their prey based upon size, motility, electrical charge, surface hydrophobicity and chemical cues (reviewed by Weisse 2002). Apparently, bacteria have developed diverse strategies by which to resist flagellate grazing (Güde 1989; Pernthaler *et al.* 1997; Hahn & Höfle 2001). Intense grazing by flagellates may cause structural and genetic shifts in the bacterial assemblage (Jürgens & Güde 1994; Jürgens *et al.* 1994a; Pernthaler *et al.* 1997; Šimek *et al.* 1997b, 1999; Hahn & Höfle 1999 2001; Posch *et al.* 1999).

The role of chemical communication in planktonic food webs is still poorly understood. Kairomones (i.e. chemical substances released by a predator, which are perceived by its potential prey) have been known for a long time to play a major role in inducing morphological and behavioural defence mechanisms among rotifers and daphnids (e.g. Gilbert 1966; Grant & Bayly 1981; Krueger & Dodson 1981; Hebert & Grewe 1985; Parejko & Dodson 1990; Loose *et al.* 1993; Tollrian & Dodson 1999). Although direct confirmation is not yet available, allelopathic interactions which inhibit competing species chemically seem also likely to occur amongst rotifers (Halbach 1969) and daphnids (Seitz 1984).

Amongst Protozoa, the ciliate *Euplotes octocarinatus*, and some other species, respond to factors released from several of their predators by morphological and behavioural changes (Kuhlmann & Heckmann 1985; Kusch 1993, 1995; Wicklow 1997, Görtz *et al.* 1999). Since the phenomenon was first observed in the hymenostome *Lembadion lucens*, the substance was named the 'Lembadion-factor' (Kusch & Heckmann 1992). The chemical cues released from *Lembadion* and other predators have been isolated and characterised as proteins of various sizes (Peters-Regehr *et al.* 1997; Görtz *et al.* 1999). *Euplotes* produces giant cells which cannot be ingested by *Lembadion* or other predators when exposed to the *Lembadion*-factor. The degree of defensive morphology by *E. daidaleos* is both predator-specific and dependent on the abundance of the predator (Kusch 1995). The biochemical and morphological changes induced by the *Lembadion*-factor are associated with energetic costs which reduce reproduction rate in *Euplotes* (Kusch & Kuhlmann 1994). Apparently, there is a trade-off between increased resistance to grazing and lowered reproductive success. This may explain why morphological changes in *Euplotes* are inducible to varying degrees (Kusch 1995).

I use these examples in order to illustrate the urgent need in aquatic microbial ecology to study 'species-specific' interactions which go beyond direct feeding relationships. Fuhrman *et al.* (1994) concluded that ignorance of bacterial species and their distributions is responsible for missing important ecological interactions such as competition. This does not imply that I am advocating autecological studies in their classic sense. Rather, in order to study processes and phenomena of general relevance for the functioning of planktonic food webs, microbial ecologists should make more use of the specific advantages that the organisms which they study offer, relative to the larger zooplankton. Microbes possess much shorter generation times, and are often more accessible and easier to culture, than zooplankton. Furthermore, most protists thrive asexually for many generations and can be cloned relatively easily. Significant intraspecific (i.e. clonal) and ontogenetic differences in ecologically relevant features, such as growth (Pérez-Uz 1995; Weisse & Montagnes 1998; Montagnes & Weisse 2000) and grazing rates (Pfister & Arndt 1998; Weisse *et al.* 2001), have recently been demonstrated for several planktonic ciliates. Ciliates have been used as model organisms in many general physiological and biochemical studies. In ecological, process-oriented laboratory investigations with pelagic microbes, the importance of the study organisms *in situ* should be an important criterion of choice.

13.7   THE APPLICATION OF NOVEL TECHNIQUES — POPULATION DYNAMICS *SENSU STRICTO*

In this review I have repeatedly stressed that the advent of novel techniques offers a tremendous potential for future studies at species level. Viruses, bacteria, cyanobacteria and protists can now be identified without cultivation by various molecular techniques (summarised in Paul 1993; Fuhrman *et al.* 1994). Although the development of 'species-specific' probes for prokaryotes and, in particular, for Protozoa, is still in its infancy (Paul 1993; Caron 1996; Pace 1996; Rice *et al.* 1997), it seems realistic to predict that we will soon overcome the current major problems of unknown probe specificity and selectivity. Several molecular techniques (cited in Paul 1993) can also be used in order to assess the metabolic activity among natural microbes, which is at present an issue of some controversy.

Another important step is assessing variation among a given population. Variation among individuals is an important biological phenomenon, and reporting only population mean values loses potentially relevant information (Gerritsen *et al.* 1987), in particular with respect to the adaptive potential of a population or species to changing environmental conditions. Previous research was too much orientated towards mean values, which were used to characterise populations, species, or even 'functional guilds'. In many grazing studies, for instance, variation around the mean was neglected or completely unknown if the sample size was too small. 'Standard' parameters such as threshold concentrations, or half-saturation constants, were extensively used by modellers. From a conceptual point of view, this approach is unsatisfactory, because natural selection does not select for the average. Those individuals which are best adapted to a given factor will be favoured so long as the environmental conditions do not change; those which are poorly adapted will die out first. Disregard of this fundamental biological principle may in part explain why the predictive value of most models of the pelagic carbon flow, or food web structure, is rather low.

Tools for measuring variation among bacterial and protist populations are already at hand. Image analysis, in combination with epifluorescence microscopy (Psenner 1990, 1993; Sieracki & Webb 1991; Verity & Sieracki 1993, and references therein), video microscopy (Boenigk & Arndt 2000a,b), and flow cytometry (for review consult Davey & Kell 1996) are especially promising techniques for measuring various physical and biochemical properties of individual cells. Recent microscopic observations from video microscopy demonstrated large individual variability among a given HNF species, which was apparent even within flagellate populations of comparable nutritional status and which could not be reduced by enlarging the sample size (Boenigk & Arndt 2000a,b). Flow cytometry is increasingly being applied in marine studies in order to discriminate among picocyanobacteria, prochlorophytes and eukaryotic algae, which cannot be differentiated adequately with conventional microscopic techniques (Yentsch *et al.* 1983; Burkill 1987; Burkill & Mantoura 1990). Flow cytometric cell sorting can be used to measure cell-specific rates of primary production (Yentsch & Campbell 1991; Li 1994). Flow cytometry has been combined with immunofluorescence (reviewed by Ward 1990; Vrieling & Anderson 1996) in order to detect toxic marine dinoflagellates *in situ* (Vrieling *et al.* 1996, and references therein). Similarly, flow cytometry in combination with *in situ* hybridisation has been used to identify marine flagellates and ciliates according to their fluorescence and size characteristics (Lim *et al.* 1993; Rice *et al.* 1997). Due to more adequate nucleic acid stains, which are excitable by argon lasers at 488 nm, it is now possible to measure cultured ciliates by commercially available flow cytometers (Lindström *et al.* 2002).

Although most, if not all, of the above techniques will need considerable refinement before they can be applied in routine field investigations, collectively they have opened a new door to aquatic microbial ecology. In principle, we possess the tools to measure 'what is out there, and what they are doing'. With the aid of 'species-specific' immunofluorescent or oligonucleotide probes, it should be possible to analyse population dynamics of the vast majority of unculturable microbes. For protists, which can be identified with conventional techniques, the time has come to study population dynamics properly, i.e., to give up describing changes occurring within a population in inadequate terms (mean values). Instead, the effect of selected environmental parameters should be analysed on a cellular basis, with reliable statistics.

13.8 CONCLUSIONS

With some time delay, the implications of Pomeroy's (1974) seminal paper on the oceanic food web became evident to limnologists. The significance of the microbial food web has been established in lakes of varying trophic states, and major processes regulating microbial biomass and production have been identified. There is a need for more comprehensive process-oriented studies, including all planktonic components in lakes. Only sporadic evidence exists on some major fluxes such as the carbon demand and production of heterotrophic protists.

In this review, the microbial food web has been portrayed (in evolutionary terms) as the original form of trophic interaction, and one which is most effective in pristine oligotrophic lakes. Implications of widely differing phylogenetic age and, thus, structural complexity, amongst the components of the microbial food web, have not yet been fully explored. The most advanced heterotrophic protists compete with metazooplankton for algal food. Interactions with crustacean zooplankton generally increase with lake trophy, and peak in mesotrophic-eutrophic lakes. Direct and indirect interactions between the various players of the microbial food web and the grazer food chain await further study at the species level. This is essential in order to characterise the ecological niches of the various pelagic microbes more properly.

The advent of molecular techniques and optical methods will help to identify unculturable microbes. Finally, in order to make predictions of potential consequences of future environmental changes more reliable, more attention should be paid to the significance of biological variation among a given microbial population.

13.9 ACKNOWLEDGEMENTS

Thanks to M. Hahn and K. Jürgens who commented on an earlier version of this manuscript, and Nancy Zehrbach for providing linguistic corrections.

13.10 REFERENCES

Adams, M.H. (1966) *Bacteriophages* (3rd edition). Interscience, New York, 592 pp.

Adoutte, A. (1994) Molecular perspectives on evolution. In: Hausmann, K. & Hülsmann, N. (eds), *Progress in Protozoology*. Proceedings of the IX International Congress of Protozoology, Berlin 1993. Gustav Fischer, Stuttgart, 91–6.

Adrian, R., Wickham, A.A. & Butler, N.M. (2001) Trophic interactions between zooplankton and the microbial community in contrasting food webs: the epilimnion and the deep chlorophyll maximum of a mesotrophic lake. *Aquatic Microbial Ecology*, **24**, 83–97.

Agawin, N.S.R., Duarte, C.M. & Agustí, S. (2000) Nutrient and temperature control of the contribution of picoplankton to phytoplankton biomass and production. *Limnology and Oceanography*, **45**, 591–600.

Albright, L.J., Sherr, E.B., Sherr, B.F. & Fallon, R.D. (1987) Grazing of ciliated protozoa on free and particle-attached bacteria. *Marine Ecology Progress Series*, **38**, 125–9.

Alfreider, A., Pernthaler, J., Amann, R., *et al.* (1996) Community analysis of the bacterial assemblages in the winter cover and pelagic layers of a high mountain lake by *in situ* hybridization. *Applied and Environmental Microbiology*, **62**, 2138–44.

Amann, R.I. (1995) Fluorescently labelled, rRNA-targeted oligonucleotide probes in the study of microbial ecology. *Molecular Ecology*, **4**, 543–53.

Amann, R.I., Ludwig, W. & Schleifer, K-H. (1995) Phylogenetic identification and in situ detection of individual microbial cells without cultivation. *Microbiological Reviews*, **59**, 143–69.

Ammermann, J.W. & Azam, F. (1985) Bacterial 5′-nucleotidase in aquatic ecosystems: A novel mechanisms of phosphorus regeneration. *Science*, **227**, 1338–40.

Anderson, O.R. (1988) *Comparative Protoozoology*. Springer-Verlag, Berlin, 482 pp.

Arndt, H. (1993) A critical review of the importance of rhizopods (naked and testate amoebae) and actinopods (helizoa) in lake plankton. *Marine Microbial Food Webs*, **7**, 3–29.

Arndt, H. & Berninger, U-G. (1995) Protists in aquatic food webs—complex interactions. In: Brugerolle, G. & Mignot, J-P. (eds), *Protistological Actualities* (Proceedings of the Second European Congress of Protistology, Clermont-Ferrand 1995), Université Blaise Pascal de Clermont-Ferrand, Clermont-Ferrand, 224–32.

Arndt, H. & Mathes, J. (1991) Large heterotrophic flagellates form a significant part of protozooplankton biomass in lakes and rivers. *Ophelia*, **33**, 225–34.

Arndt, H. & Nixdorf, B. (1991) Spring-clearwater phase on a eutrophic lake: control by herbivorous zooplankton enhanced by grazing on components of the microbial web. *Verhandlungen der Internationalen Vereinigung Limnologie*, **24**, 879–83.

Arndt, H., Dietrich, D., Auer, B., *et al.* (2000) Functional diversity of heterotrophic flagellates in aquatic ecosystems. In: Leadbeater, B.S.C. & Green, J.C. (eds), *The Flagellates — Unity, Diversity and Evolution*. Taylor & Francis, London, 240–68.

Azam, F., Fenchel, T., Field, J.G., Gray, J.S., Meyer-Reil, L.A. & Thingstad, F. (1983) The ecological role of water-column microbes in the sea. *Marine Ecology Progress Series*, **10**, 257–63.

Azam, F., Cho, B.C., Smith, D.C. & Simon, M. (1990) Bacterial cycling of matter in the pelagic zone of aquatic ecosystems. In: Tilzer, M.M. & Serruya, C. (eds), *Large Lakes — Ecological Structure and Function*, Springer-Verlag, Berlin, 477–88.

Banse, K. (1984) Review of Bougis, P. (ed.), *Marine Pelagic Protozoa and Microzooplankton Ecology. Limnology and Oceanography*, **29**, 445–6.

Banse, K. (1990) On pelagic food web interactions in large water bodies. In: Tilzer, M.M. & Serruya, C. (eds), *Large Lakes — Ecological Structure and Function*. Springer-Verlag, Berlin, 556–79.

Banse, K. (1995) Science and organization in open-sea research: the plankton. *Helgoländer Meeresuntersuchungen*, **49**, 3–18.

Beaver, J.R. & Crisman, T.L. (1982) The trophic response of ciliated protozoans in freshwater lakes. *Limnology and Oceanography*, **27**, 246–53.

Beaver, J.R. & Crisman, T.L. (1989) The role of ciliated protozoa in pelagic freshwater ecosystems. *Microbial Ecology*, **17**, 111–36.

Beaver, J.R., Crisman, T.L. & Bienert, R.W. (1988) Distribution of planktonic ciliates in highly coloured subtropical lakes: comparison with clearwater ciliate communities and the contribution of mixotrophic taxa to total autotrophic biomass. *Limnology and Oceanography*, **20**, 51–60.

Bell, R.T., Vrede, K., Stensdotter-Blomberg, U. & Blomqvist, P. (1993) Stimulation of the microbial food web in an oligotrophic, slightly acidified lake. *Limnology and Oceanography*, **38**, 1532–8.

Bell, T. & Kalff, J. (2001) The contribution of picophytoplankton in marine and freshwater systems of different trophic status and depth. *Limnology and Oceanography*, **46**, 1243–8.

Bennett, S.J., Sanders, R.W. & Porter, K.G. (1988) Chemosensory responses of heterotrophic and mixotrophic flagellates to potential food sources. *Bulletin of Marine Science*, **43**, 764–71.

Bennett, S.J., Sanders, R.W. & Porter, K.G. (1990) Heterotrophic, autotrophic, and mixotrophic nanoflagellates: seasonal abundances and bacterivory in a eutrophic lake. *Limnology and Oceanography*, **35**, 1821–32.

Bergh, O., Børsheim, K.Y., Bratbak, G. & Heldal, M. (1989) High abundances of viruses found in aquatic environments. *Nature*, **340**, 467–8.

Berninger, U-G., Finlay, B.J. & Kuuppo-Leinikki, P. (1991a) Protozoan control of bacterial abundances in freshwater. *Limnology and Oceanography*, **36**, 139–47.

Berninger, U-G., Caron, D.A., Sanders, R.W. & Finlay, B.J. (1991b) Heterotrophic flagellates of planktonic communities, their characteristics and methods of study. In: Patterson, D.J. & Larsen, J (eds), *The Biology of Free-living Heterotrophic Flagellates*. Systematics Association Special Volume 45, Clarendon Press, Oxford, 39–56.

Bird, D.F. & Kalff, J. (1984) Empirical relationships between bacterial abundance and chlorophyll concentration in fresh and marine waters. *Canadian Journal of Fisheries and Aquatic Sciences*, **41**, 1015–23.

Bird, D.F. & Kalff, J. (1986) Bacterial grazing by planktonic lake algae. *Science*, **231**, 493–5.

Bird, D.F. & Kalff, J. (1987) Algal phagotrophy: regulating factors and importance relative to photosynthesis in *Dinobryon* (Chrysophyceae). *Limnology and Oceanography*, **32**, 277–84.

Bloem, J., Albert, C., Bär-Gilissen, M-JB., Berman, T. & Cappenberg, T.E. (1989) Nutrient cycling through phytoplankton, bacteria and protozoa, in selectively filtered Lake Vechten water. *Journal of Plankton Research*, **11**, 119–31.

Boenigk, J. & Arndt, H. (2000a) Comparative studies on the feeding behaviour of two heterotrophic nanoflagellates: the filter-feeding choanoflagellate *Monosiga ovata* and the raptorial-feeding kinetoplastid *Rhynchomonas nasuta*. *Aquatic Microbial Ecology*, **22**, 243–9.

Boenigk, J. & Arndt, H. (2000b) Particle handling during interception feeding by four species of heterotrophic nanoflagellates. *Journal of Eukaryotic Microbiology*, **47**, 350–8.

Brandt, K. (1906) Die Tintinnideen der Plankton-Expedition. Tafelerklärungen nebst kurzer Diagnose der neuen Arten. In: Hensen, V. (ed.), *Ergebnisse der*

Plankton-Expedition der Humboldt-Stiftung, Lipsius & Tischler, Kiel, 1–33.

Bratbak, G. (1993) Microscopic methods for measuring bacterial biovolume: epifluorescence microscopy, scanning electron microscopy, and transmission electron microscopy. In: Kemp, P.F., Sherr, B.F., Sherr, E.B. & Cole, J.J (eds), *Handbook of Methods in Aquatic Microbial Ecology*. Lewis, Boca Raton, FL, 309–17.

Bratbak, G. & Heldal, M. (1993) Total count of viruses in aquatic environments. In: Kemp, P.F., Sherr, B.F., Sherr, E.B. & Cole, J. J (eds), *Handbook of Methods in Aquatic Microbial Ecology*. Lewis, Boca Raton, FL, 135–8.

Bratbak, G. & Thingstad, F. (1985) Phytoplankton–bacteria interactions: an apparent paradox? Analysis of a model system with both competition and commensalism. *Marine Ecology Progress Series*, **25**, 23–30.

Bratbak, G., Heldal, M., Norland, S. & Thingstad, F. (1990) Viruses as partners in spring bloom microbial trophodynamics. *Applied and Environmental Microbiology*, **56**, 1400–5.

Bratbak, G., Thingstad, F. & Heldal, M. (1994) Viruses and the microbial loop. *Microbial Ecology*, **28**, 209–21.

Britschgi, T.B. & Giovannoni, S.J. (1991) Phylogenetic analysis of a natural marine bacterioplankton population by rRNA gene cloning and sequencing. *Applied and Environmental Microbiology*, **57**, 1707–13.

Brooks, J.L. & Dodson, S.I. (1965) Predation, body size and composition of plankton. *Science*, **150**, 28–35.

Burger-Wiersma, T., Veenhuis, M., Korthals, H.J., Van de Wiel, C.C.M. & Mur, L.R. (1986) A new prokaryote containing chlorophylls *a* and *b*. *Nature*, **320**, 262–64.

Burkill, P.H. (1987) Analytical flow cytometry and its application to marine microbial ecology. In: Sleigh, M.A. (ed.), *Microbes in The Sea*. Ellis Horwood, Chichester, 139–66.

Burkill, P.H. & Mantoura, R.F.C. (1990) The rapid analysis of single marine cells by flow cytometry. *Philosophical Transactions of the Royal Society London*, **333**, 99–112.

Calado, A.J. & Moestrup, Ø. (1997) Feeding in *Peridiniopsis berolinensis* (Dinophyceae): new observations on tube feeding by an omnivorous, heterotrophic dinoflagellate. *Phycologia*, **36**, 47–59.

Campbell, L. & Carpenter, E.J. (1987) Characterization of phycoerythrin-containing *Synechococcus* spp. populations by immunofluorescence. *Journal of Plankton Research*, **9**, 1167–81.

Campbell, L. & Iturriaga, R. (1988) Identification of *Synechococcus* sp. in the Sargasso Sea by immunofluorescence and fluorescence excitation spectroscopy

performed on individual cells. *Limnology and Oceanography*, **33**, 1196–201.

Carney, H.J. & Elser, J.J. (1990) Strength of zooplankton–phytoplankton coupling in relation to lake trophic state. In: Tilzer, M.M., Serruya, C. (eds), *Large Lakes — Ecological Structure and Function*. Springer-Verlag, Berlin, 615–31.

Caron, D.A. (1991) Evolving role of protozoa in aquatic nutrient cycles. In: Reid, P.C., Turley, C.M. & Burkill, P.H. (eds), *Protozoa and their Role in Marine Processes*. Springer-Verlag, Berlin, 387–415.

Caron, D.A. (1996) Symposium introductory remarks: 'Protistan Molecular Ecology, and systematics'. *Journal of Eukaryotic Microbiology*, **43**, 87–8.

Caron, D.A. & Goldman, J.C. (1988) Dynamics of protistan carbon and nutrient cycling. *Journal of Protozoology*, **35**, 247–9.

Caron, D.A. & Swanberg, N.R. (1990) The ecology of planktonic sarcodines. *Reviews in Aquatic Sciences*, **3**, 147–80.

Caron, D.A., Goldman, J.C. & Dennett, M.R. (1988) Experimental demonstration of the role of bacteria and bacterivorous protozoa in plankton nutrient cycles. *Hydrobiologia*, **159**, 27–40.

Caron, D.A., Porter, K.G. & Sanders, R.W. (1990) Carbon, nitrogen, and phosphorus budgets for the mixotrophic phytoflagellate *Poterioochromonas malhamensis* (Chrysophyceae) during bacterial ingestion. *Limnology and Oceanography*, **35**, 433–43.

Carpenter, S.R., Kitchell, J.F. & Hodgson, J.R. (1985) Cascading trophic interactions and lake productivity. *BioScience*, **35**, 634–9.

Carrick, H.J., Fahnenstiel, G.A. & Taylor, W.D. (1992) Growth and production of planktonic protozoa in Lake Michigan: *In situ* versus *in vitro* comparisons and importance to food web dynamics. *Limnology and Oceanography*, **37**, 1221–35.

Cavalier-Smith, T. (1991) Cell diversification in heterotrophic flagellates. In: Patterson, D.J., Larsen, J. (eds), *The Biology of Free-Living Heterotrophic Flagellates*. Systematics Association Special Volume 45, Clarendon Press, Oxford, 113–31.

Cavalier-Smith, T. (1993) Kingdom Protozoa and its eighteen phyla. *Microbiological Reviews*, **57**, 653–994.

Cavalier-Smith, T. (1995) Evolutionary protistology comes to age: biodiversity and molecular cell biology. *Archiv für Protistenkunde*, **145**, 145–54.

Chisholm, S.W., Olson, R.J., Zettler, E.R., Goericke, R., Waterbury, J.B. & Welschmeyer, N.A. (1988) A novel free-living prochlorophyte abundant in the oceanic euphotic zone. *Nature*, **334**, 340–3.

Christoffersen, K., Riemann, B., Klysner, A. & Sønder-gaard, M. (1993) Potential role of fish predation and natural populations of zooplankton in structuring a plankton community in eutrophic lake water. *Limnology and Oceanography*, **38**, 561–73.

Christoffersen, K., Riemann, B., Hansen, L.R., Klysner, A. & Sorensen, H.B. (1990) Qualitative importance of the microbial loop and plankton community structure in a eutrophic lake during a bloom of cyanobacteria. *Microbial Ecology*, **20**, 253–72.

Clay, B. & Kugrens, P. (1999) Systematics of the enigmatic kathablepharids, including EM characterization of the type species, *Kathablepharis phoenikoston*, and new observations on *K. remigera* comb. nov. *Protist*, **150**, 43–59.

Cleven, E-J. (1996) Indirectly fluorescently labelled flagellates (IFLF): a tool to estimate the predation on free-living heterotrophic flagellates. *Journal of Plankton Research*, **18**, 429–42.

Cleven, E-J. & Weisse, T. (2001) Seasonal succession and taxon-specific bacterial grazing rates of heterotrophic nanoflagellates in Lake Constance. *Aquatic Microbial Ecology*, **23**, 147–61.

Cochlan, W.P., Wikner, J., Stewart, G.F., Smith, D.C. & Azam, F. (1993) Spatial distribution of viruses, bacteria and chlorophyll a in neritic, oceanic and estuarine environments. *Marine Ecology Progress Series*, **92**, 77–87.

Cole, J.J. & Caraco, N.F. (1993) The pelagic microbial food web of oligotrophic lakes. In: Ford, T.E. (ed.), *Aquatic Microbiology*. Blackwell Scientific Publications, Oxford, 101–11.

Cole, J.J., Findlay, S. & Pace, M.L. (1988) Bacterial production in fresh and saltwater ecosystems: a cross-system overview. *Marine Ecology Progress Series*, **43**, 1–10.

Cole, J.J., Caraco, N.F., Strayer, D.L., Ochs, C. & Nolan, S. (1989) A detailed carbon budget as an ecosystem-level calibration of bacterial respiration in an oligotrophic lake during midsummer. *Limnology and Oceanography*, **34**, 286–96.

Corliss, J.O. (1994) An interim utilitarian ('user-friendly') hierarchical classification and characterization of the protists. *Acta Protozoologica*, **33**, 1–51.

Corzo, A., Jiménez-Gómez, F., Gordillo, F.J.L., García-Ruíz, R. & Niell, F.X. (1999) *Synechococcus* and *Prochlorococcus*-like populations detected by flow cytometry in a eutrophic reservoir in summer. *Journal of Plankton Research*, **21**, 1575–81.

Cottrell, M.T. & Suttle, C.A. (1991) Wide-spread occurrence and clonal variations in viruses which cause lysis of a cosmopolitan, eukaryotic marine phyto-plankter, *Micromonas pusilla*. *Marine Ecology Progress Series*, **78**, 1–9.

Coveney, M.F. & Wetzel, R.G. (1995) Biomass, production, and specific growth rate of bacterioplankton and coupling to phytoplankton in an oligotrophic lake. *Limnology and Oceanography*, **40**, 1187–200.

Currie, D.J. (1990) Large-scale variability and interactions among phytoplankton, bacterioplankton, and phosphorus. *Limnology and Oceanography*, **35**, 1437–55.

Currie, D.J. & Kalff, J. (1984a) Can bacteria outcompete phytoplankton for phosphorus? A chemostat test. *Microbial Ecology*, **10**, 205–16.

Currie, D.J. & Kalff, J. (1984b) A comparison of the abilities of freshwater algae and bacteria to acquire and retain phosphorus. *Limnology and Oceanography*, **29**, 298–310.

Currie, D.J & Kalff, J. (1984c) The relative importance of bacterioplankton and phytoplankton in phosphorus uptake in freshwater. *Limnology and Oceanography*, **29**, 311–21.

Davey, H.M. & Kell, D.B. (1996) Flow cytometry and cell sorting of heterogeneous microbial populations: the importance of single-cell analysis. *Microbiological Reviews*, **60**, 641–96.

Del Giorgio, P.A. & Peters, R.H. (1993) Balance between phytoplankton production and plankton respiration in lakes. *Canadian Journal of Fisheries and Aquatic Sciences*, **50**, 282–89.

Del Giorgio, P.A & Peters, R.H. (1994) Patterns in planktonic $P:R$ ratios in lakes: Influence of lake trophy and dissolved organic carbon. *Limnology and Oceanography*, **39**, 772–87.

DeMott, W.R. (1989) The role of competition in zooplankton succession. In: Sommer, U. (ed.), *Plankton, Ecology, Succession in Plankton Communities*. Springer-Verlag, Berlin, 195–252.

Demuth, J., Neve, H. & Witzel, K.P. (1993) Morphological diversity of bacteriophage populations in lake Plußsee, studied by direct electron microscopy. *Applied and Environmental Microbiology*, **59**, 3378–84.

Ducklow, H.W. (1991) The passage of carbon through microbial foodwebs: results from flow network models. *Marine Microbial Food Webs*, **5**, 129–44.

Ducklow, H.W. & Carlson, C.A. (1992) Oceanic bacterial production. In: Marshall, K.C. (ed.), *Advances in Microbial Ecology*. Plenum Press, New York, 113–81.

Ducklow, H.W., Purdie, D.A., Williams, P.J.l. & Davies, J.M. (1986) Bacterioplankton: a sink for carbon in a coastal marine plankton community. *Science*, **232**, 865–67.

Ducklow, H.W., Purdie, D.A., Williams, P.J.l. & Davies, J.M. (1987) Response to 'Bacteria: link or sink?'. *Science*, **235**, 88–9.

Elbrächter, M. (1991) Food uptake mechanisms in phagotrophic dinoflagellates and classification. In: Patterson, D.J. & Larsen, J. (eds), *The Biology of Free-living Heterotrophic Flagellates*. Systematics Association Special Volume 45, Clarendon Press, Oxford, 303–12.

Elser, J.J., Stabler, L.B. & Hasset, R.P. (1995) Nutrient limitation of bacterial growth and rates of bacterivory in lakes and oceans: a comparative study. *Aquatic Microbial Ecology*, **9**, 105–10.

Ernst, A. (1991) Cyanobacterial picoplankton from Lake Constance: I. Isolation by fluorescence characteristics. *Journal of Plankton Research*, **13**, 1307–12.

Ernst, A., Sandmann, G., Postius, C., Brass, S., Kenter, U. & Böger, P. (1992) Cyanobacterial picoplankton from Lake Constance. II. Classification of isolates by cell morphology and pigment composition. *Botanica Acta*, **105**, 161–7.

Ernst, A., Marschall, P. & Postius, C. (1995) Genetic diversity among *Synechococcus* spp. (cyanobacteria) isolated from the pelagial of Lake Constance. *FEMS Microbiology Ecology*, **17**, 197–204.

Ernst, A., Postius, C. & Böger, P. (1996) Glycosylated surface proteins reflect genetic diversity among *Synechococcus* species of Lake Constance. *Archiv für Hydrobiologie Special issues: Ergebnisse der Limnologie*, **48**, 1–6.

Ernst, A., Becker, S., Hennes, K. & Postius, C. (2000) Is there a succession in the autotrophic picoplankton of temperate zone lakes? In: Bell, C.R., Brylinski, M. & Johnson-Green, P. (eds), *Microbial Biosystems: New Frontiers*. Proceedings of the 8th International Symposium on Microbial Ecology. Atlantic Canada Society for Microbial Ecology, Halifax, Nova Scotia, 623–9.

Ernst, A., Becker, S., Hennes, K. & Postious, C. (2003) Ecosystem-dependent adaptive radiations of pycocyanobacteria inferred from 16S rRNA and ITS-1 sequence analysis. *Microbiology*, **149**, 217–28.

Felip, M., Pace, M.L. & Cole, J.J. (1996) Regulation of planktonic bacterial growth rates: the effects of temperature and resources. *Microbial Ecology*, **31**, 15–28.

Fenchel, T. (1980) Relation between particle size selection and clearance in suspension-feeding ciliates. *Limnology and Oceanography*, **25**, 733–8.

Fenchel, T. (1984) Suspended marine bacteria as a food source. In: Fasham, M.J.R (ed.), *Flows of Energy and Materials in Marine Ecosystems*. Plenum, New York, 301–15.

Fenchel, T. (1986) The ecology of heterotrophic microflagellates. In: Marshall, K.C. (ed.), *Advances in Miocrobial Ecology*. Plenum Press, New York, 57–97.

Fenchel, T. (1987) *Ecology of Protozoa. The Biology of Free-living Phagotrophic Protists*. Springer-Verlag, Berlin, 197 pp.

Fenchel, T. (1988) Marine plankton food chains. *Annual Reviews of Ecology and Systematics*, **19**, 19–38.

Fenchel, T. & Finlay, B.J. (1990) Anaerobic free-living protozoa: growth efficiencies and the structure of anaerobic communities. *FEMS Microbiology Ecology*, **74**, 269–76.

Fenchel, T. & Finlay, B.J. (1995) *Ecology and Evolution in Anoxic Worlds*. Oxford University Press, Oxford, 288 pp.

Fenchel, T. & Patterson, D.J. (1986) *Percolomonas cosmopolitus* (Ruinen) n. gen., a new type of filter feeding flagellate from marine plankton. *Journal of the Marine Biological Association of the United Kingdom*, **66**, 465–87.

Finlay, B.J. (1990) Physiological ecology of free-living protozoa. In: Marshall, K.C. (ed.), *Advances in Microbial Ecology*, Vol. 11. Plenum Press, New York, London, 1–35

Finlay, B.J. & Fenchel, T. (1993) Methanogens and other bacteria as symbionts of free-living anaerobic ciliates. *Symbiosis*, **14**, 375–90.

Finlay, B.J. & Fenchel, T. (1996) Ecology: role of ciliates in the natural environment. In: Hausmann, K. & Bradbury, P.C. (ed.), *Ciliates: Cells as Organisms*. Fischer-Verlag, Berlin, 417–40.

Fliermans, C.B. & Schmidt, E.L. (1977) Immunofluorescence for autecological study of a unicellular bluegreen alga. *Journal of Phycology*, **13**, 364–8.

Florey, E. (1995) Highlights and sidelights of early Biology on Heligoland. *Helgoländer Meeresuntersuchungen*, **49**, 77–101.

Fogg, G.E. (1986) Picoplankton. *Proceedings of the Royal Society of London*, **228**, 1–30.

Francisco, D.E., Mah, R.A. & Rabin, A.C. (1973) Acridine orange-epifluorescence technique for counting bacteria in natural waters. *Transactions of the American Microscopical Society*, **92**, 416–21.

Fuhrman, J.A & Azam, F. (1980) Bacterioplankton secondary production estimates for coastal waters of British Columbia, Antarctica, and California. *Applied and Environmental Microbiology*, **39**: 1985–95.

Fuhrman, J.A. & Azam, F. (1982) Thymidine incorporation as a measure of heterotrophic bacterioplankton in marine surface waters: evaluation and field results. *Marine Biology*, **66**, 109–20.

Fuhrman, J.A. & Noble, R.T. (1995) Viruses and protists cause similar bacterial mortality in coastal seawater. *Limnology and Oceanography*, **40**, 1236–42.

Fuhrman, J.A., Lee, S.H., Masuchi, Y., Davis, A.A. & Wilcox, R.M. (1994) Characterization of marine prokaryotic communities via DNA and RNA. *Microbial Ecology*, **28**, 133–45.

Gaedke, U. (1998) Functional and taxonomical properties of the phytoplankton community of large and deep Lake Constance: interannual variability and response to re-oligotrophication (1979–93). *Archiv für Hydrobiologie Special Issues Advancances in Limnology*, **53**, 119–41.

Gaedke, U. & Straile, D. (1994a) Seasonal changes of the quantitative importance of protozoans in a large lake. An ecosystem approach using mass–balanced carbon flow diagrams. *Marine Microbial Food Webs*, **8**, 163–88.

Gaedke, U. & Straile, D. (1994b) Seasonal changes of trophic transfer efficiencies in a plankton food web derived from biomass size distributions and network analysis. *Ecological Modelling*, **75/76**, 435–45.

Gaedke, U., Straile, D. & Pahl-Wostl, C. (1995) Trophic structure and carbon flow dynamics in the pelagic community of a large lake. In: Polis, G. & Winemiller, K. (eds), *Food Webs: Integration of Pattern and Dynamics*, Chapman & Hall, New York, 60–71.

Gaines, G. & Elbrächter, M. (1987) Heterotrophic nutrition. In: Taylor, F.J.R. (ed.), *The Biology of Dinoflagellates*. Blackwell, Oxford, 224–68.

Garrity, G. (2001) *Bergey's Manual of Systematic Bacteriology*, No. 1 (2nd edition). Springer-Verlag, New York, 750 pp.

Garza, R.D. & Suttle, C.A. (1995) Large double-stranded DNA viruses which cause the lysis of a marine heterotrophic nanoflagellate (*Bodo* sp.) occur in natural marine viral communities. *Aquatic Microbial Ecology*, **9**, 203–10.

Gasol, J.M. & Vaqué, D. (1993) Lack of coupling between heterotrophic nanoflagellates and bacteria: A general phenomenon across aquatic systems? *Limnology and Oceanography*, **38**, 657–65.

Gasol, J.M., Simons, A.M. & Kalff, J. (1995) Patterns in the top-down versus bottom-up regulation of heterotrophic nanoflagellates in temperate lakes. *Journal of Plankton Research*, **17**, 1879–903.

Geller, W., Berberovic, U., Gaedke, U., Müller, H, Pauli, H-R., Tilzer, M.M. & Weisse, T. (1991) Relations among the components of autotrophic and heterotrophic plankton during the seasonal cycle 1987 in Lake Constance. *Verhandlungen der Internationalen Vereinigung Limnologie*, **24**, 831–6.

Gerritsen, J., Sanders, R.W., Bradley, S.W. & Porter, K.G. (1987) Individual feeding variability of protozoan and crustacean zooplankton analyzed with flow cytometry. *Limnology and Oceanography*, **32**, 691–9.

Gilbert, J.J. (1966) Rotifer ecology and embryological indiction. *Science*, **151**, 1234–7.

Gilbert, J.J. (1994) Jumping behavior in the oligotrich ciliates *Strobilidium velox* and *Halteria grandinella* and its significance as a defense against rotifers. *Microbial Ecology*, **27**, 189–200.

Giovannoni, S.J., Britschgi, T.B., Moyer, C.L. & Field, K.G. (1990) Genetic diversity in Sargassso Sea bacterioplankton. *Nature*, **345**, 60–3.

Goericke, R. & Welschmeyer, N.A. (1993) The marine prochlorophyte *Prochlorococcus* contributes significantly to phytoplankton biomass and primary production in the Sargasso Sea. *Deep-Sea Research*, **40**, 2283–94.

Görtz, H-D., Kuhlmann, H-W., Möllenbeck, M., *et al.* (1999) Intra- and intercellular communication systems in ciliates. *Naturwissenschaften*, **86**, 422–34.

Goldman, J.C., Dennett, M.R. & Gordin, H. (1989) Dynamics of herbivorous grazing by the heterotrophic dinoflagellate *Oxyrrhis marina*. *Journal of Plankton Research*, **11**, 391–407.

González, J.M. & Suttle, C.A. (1993) Grazing by marine nanoflagellates on viruses and virus-sized particles: Ingestion and digestion. *Marine Ecology Progress Series*, **94**, 1–10.

Grant, J.W.C. & Bayly, I.A.E. (1981) Predator indiction of crests in morphs of the *Daphnia carinata* King complex. *Limnology and Oceanography*, **26**, 201–18.

Grosshart, H-P. & Simon, M. (1993) Limnetic macroscopic organic aggregates (lake snow): Occurrence, characteristics, and microbial dynamics in Lake Constance. *Limnology and Oceanography*, **38**, 532–46.

Grosshart, H-P. & Simon, M. (1998) The significance of lake snow aggregates for the sinking flux of particulate organic matter in a large lake. *Aquatic Microbial Ecology*, **15**, 115–25.

Grossart, H.P., Simon, M. & Logan, B.E. (1997) Formation of macroscopic organic aggregates (lake snow) in a large lake: The significance of transparent exopolymer particles, phytoplankton, and zooplankton. *Limnology and Oceanography*, **42**, 1651–9.

Güde, H. (1989) The role of grazing on bacteria in plankton sucession. In: Sommer, U. (ed.), *Plankton Ecology: Sucession in Plankton Communities*. Brock/Springer-Verlag, Berlin, 337–64.

Hagström, A., Larsson, U., Hörstedt, P. & Normark, S. (1979) Frequency of dividing cells, a new approach to the determination of bacterial growth rates in aquatic environments. *Applied and Environmental Microbiology*, **37**, 805–12.

Halbach, U. (1969) Das Zusammenwirken von Konkurrenz und Räuber-Beute-Beziehungen bei Rädertieren. *Zoologischer Anzeiger (Supplementband)*, **33**, 72–9.

Hahn, M.W. & Höfle, M.G. (1999) Flagellate predation on a bacterial model community: Interplay of size-selective grazing, specific bacterial cell size, and bacterial community composition. *Applied and Environmental Microbiology*, **65**, 4863–72.

Hahn, M.W. & Höfle, M.G. (2001) Grazing of protozoa and its effect on populations of aquatic bacteria. *FEMS Microbiology Ecology*, **35**, 113–21.

Hanisch, K., Schweitzer, B. & Simon, M. (1996) Utilization of dissolved carbohydrates by planktonic bacteria in a mesotrophic lake. *Microbial Ecology*, **31**, 41–55.

Hansen, P.J. (1991) Quantitative importance and trophic role of heterotrophic dinoflagellates in a coastal pelagial food web. *Marine Ecology Progress Series*, **73**, 253–61.

Hansen, P.J. (1992) Prey size selection, feeding rates and growth dynamics of heterotrophic dinoflagellates with special emphasis on *Gyrodinium spirale*. *Marine Biology*, **114**, 327–34.

Hardy, A.C. (1924) The herring in relation to its animate environment. Part II. Report on trials with the plankton indicator. *Fishery Investigations (London)*, *Series II*, **7**, 1–53.

Hausmann, K. & Hülsmann, N. (1996) *Protozoology*. Georg Thieme, Stuttgart, 338 pp.

Hausmann, K. & Patterson, D.J. (1982) Pseudopod formation and membrane production during prey capture by a heliozoan (feeding by *Actinophrys*, II). *Cell Motility*, **2**, 9–24.

Hawkins, P. & Lampert, W. (1989) The effect of *Daphnia* body size on filtering rate inhibition in the presence of a filamentous cyanobacterium. *Limnology and Oceanography*, **34**, 1084–89.

Hebert, P.D.N. & Grewe, P.M. (1985) *Chaoborus*-induced shifts in the morphology of *Daphnia ambigua*. *Limnology and Oceanography*, **30**, 1291–7.

Hennes, K.P. & Simon, M. (1995) Significance of bacteriophages for controlling bacterioplankton growth in a mesotrophic lake. *Applied and Environmental Microbiology*, **61**, 333–40.

Hennes, K.P. & Suttle, C.A. (1995) Direct counts of viruses in natural waters and laboratory cultures by epifluo-rescence microscopy. *Limnology and Oceanography*, **40**, 1050–5.

Hennes, K.P., Suttle, C.A. & Chan, A.M. (1995) Fluorescently labeled virus probes show that natural virus populations can control the structure of marine microbial communities. *Applied and Environmental Microbiology*, **61**, 3623–7.

Hensen, V. (1887) Über die Bestimmung des Planktons oder des im Meere treibenden Materials an Pflanzen und Tieren. *Berichte der deutschen wissenschaftlichen Kommission für Meeresforschung*, **5**, 1–108.

Hicks, R., Amann, R.I. & Stahl, D.A. (1992) Dual staining of natural bacterioplankton with 4′,6-diamidino-2-phenylindole and fluorescent oligonucleotide probes targeting kingdom-level 16S rRNA sequences. *Applied and Environmental Microbiology*, **58**, 2158–63.

Hobbie, J.E., Daley, R.J. & Jasper, S. (1977) Use of Nuclepore filters for counting bacteria by fluorescence microscopy. *Applied and Environmental Microbiology*, **33**, 1225–8.

Hughes, R. (ed.) (1993) *Diet Selection: an Interdisciplinary Approach to Foraging Behaviour*. Blackwell Science, Oxford, 232 pp.

Hwang, S-J. & Heath, R.T. (1997) The distribution of protozoa across a trophic gradient, factors controlling their abundance and importance in the plankton food web. *Journal of Plankton Research*, **19**, 491–518.

Jacobson, D.M. & Anderson, D.M. (1986) Thecate heterotrophic dinoflagellates: feeding behavior and mechanisms. *Journal of Phycology*, **22**, 249–58.

Jeppesen, E., Sortkjaer, O., Søndergaard, M. & Erlandsen, M. (1992) Impact of a trophic cascade on heterotrophic bacterioplankton production in two shallow fish-manipulated lakes. *Archiv für Hydrobiologie Beihefte Ergebnisse der Limnologie*, **37**, 219–31.

Jiang, S.C. & Paul, J.H. (1995) Viral contribution to dissolved DNA in the marine environment as determined by differential centrifugation and kingdom probing. *Applied and Environmental Microbiology*, **61**, 317–25.

Jochem, F. (1988) On the distribution and importance of picocyanobacteria in a boreal inshore area (Kiel Bight, western Baltic). *Journal of Plankton Research*, **10**, 1009–22.

Jones, R.I. (2000) Mixotrophy in planktonic protists: an overview. *Freshwater Biology*, **45**, 219–26.

Jumars, P.A., Penry, D.L., Baross, J.A., Perry, M.J. & Frost, B.W. (1989) Closing the microbial loop: dissolved carbon pathway to heterotrophic bacteria from incomplete ingestion, digestion and absorption in animals. *Deep-Sea Research*, **36**, 483–95.

Jürgens, K. (1994) Impact of *Daphnia* on planktonic microbial food webs—A review. *Marine Microbial Food Webs*, **8**, 295–324.

Jürgens, K. & De Mott, W.R. (1995) Behavioral flexibility in prey selection by bacterivorous nanoflagellates. *Limnology and Oceanography*, **40**, 1503–7.

Jürgens, K. & Güde, H. (1990) Incorporation and release of phosphorus by planktonic bacteria and phagotrophic flagellates. *Marine Ecology Progress Series*, **59**, 271–84.

Jürgens, K. & Güde, H. (1994) The potential importance of grazing-resistant bacteria in planktonic systems. *Marine Ecology Progress Series*, **112**, 169–88.

Jürgens, K., Arndt, H. & Rothhaupt, K-O. (1994a) Zooplankton-mediated changes of bacterial community structure. *Microbial Ecology*, **27**, 27–42.

Jürgens, K., Gasol, J.M., Massana, R. & Pedros-Alió, C. (1994b) Control of heterotrophic bacteria and protozoans by *Daphnia pulex* in the epilimnion of Lake Ciso. *Archiv für Hydrobiologie*, **131**, 55–78.

Jürgens, K., Wickham, S.A., Rothhaupt, K.O. & Santer, B. (1996) Feeding rates of macro- and microzooplankton on heterotrophic nanoflagellates. *Limnology and Oceanography*, **41**, 1833–9.

Kahl, A. (1932) *Urtiere oder Protozoa. I: Wimpertiere oder Ciliata (Infusoria). 3. Spritricha.* Die Tierwelt Deutschlands, Vol. 25, Gustav Fischer, Jena, 399–486.

Kemp, P.F. (1994) A philosophy of methods development: the assimilation of new methods and information into aquatic microbial ecology. *Microbial Ecology*, **28**, 159–66.

Kirchman, D.L. (1993) Leucine incorporation as a measure of biomass production by heterotrophic bacteria. In: Kemp, P.F., Sherr, B.F., Sherr, E.B., Cole, J.J. (eds), *Handbook of Methods in Aquatic Microbial Ecology*, Lewis, Boca Raton, FL, 509–12.

Klut, M.E. & Stockner, J.G. (1990) Virus-like particles in an ultra-oligotrophic lake on Vancouver Island, British Columbia. *Canadian Journal of Fisheries and Aquatic Sciences*, **47**, 725–30.

Krueger, D.A. & Dodson, S.I. (1981) Embryological induction and predation ecology in *Daphnia pulex*. *Limnology and Oceanography*, **26**, 219–23.

Kuhlmann, H-W. & Heckmann, K. (1985) Interspecific morphogens regulating prey-predator relationships in protozoa. *Science*, **227**, 1347–9.

Kuhlmann, H-W., Kusch, J. & Heckmann, K. (1999) Predator-induced defenses in ciliated protozoa. In: Tollrian, R. & Harvell, C.D. (eds), *The Ecology and Evolution of Inducible Defenses.* Princeton University Press, Princeton, NJ, 142–59.

Kusch, J. (1993) Behavioural and morphological changes in ciliates induced by the predator *Amoeba proteus*. *Oecologia*, **96**, 354–9.

Kusch, J. (1995) Adaptation of inducible defense in *Euplotes daidaleos* (Ciliophora) to predation risks by various predators. *Microbial Ecology*, **30**, 79–88.

Kusch, J. & Heckmann, K. (1992) Isolation of the *Lembadion* factor, a morphologenetically active signal, that induces *Euplotes* cells to change from their ovoid form into a larger lateral winged form. *Developments in Genetics*, **13**, 241–6.

Kusch, J. & Kuhlmann, H-W. (1994) Cost of *Stenostomum*-induced morphological defense in the ciliate *Euplotes octocarinatus*. *Archiv für Hydrobiologie*, **130**, 257–67.

Lampert, W. (1978) Release of dissolved organic carbon by grazing zooplankton. *Limnology and Oceanography*, **23**, 831–4.

Lampert, W. (1981) Inhibitory and toxic effects of bluegreen algae on *Daphnia*. *Internationale Revue der gesamten Hydrobiologie*, **66**, 285–98.

Landry, M.R. (1994) Methods and controls for measuring the grazing impact of planktonic protist. *Marine Microbial Food Webs*, **8**, 37–57.

Landry, M.R., Lehner-Fournier, J.M., Sundstrom, J.A., Fagerness, L. & Selph, K.E. (1991) Discrimination between living and heat-killed prey by a marine zooflagellate, *Paraphysomonas vestita* (Stokes). *Journal of Experimental Marine Biology and Ecology*, **146**, 139–51.

Laybourn-Parry, J. (1992) *Protozoan Plankton Ecology.* Chapman & Hall, London, 301 pp.

Laybourn-Parry, J. (1994) Seasonal successions of protozooplankton in freshwater ecosystems of different latitudes. *Marine Microbial Food Webs*, **8**, 145–62.

Leadbeater, B.S.C. & Green, J.C. (eds), (2000) *The Flagellates—Unity, Diversity and Evolution.* Taylor & Francis, London, 401 pp.

Lee, S. & Fuhrman, J.A. (1990) DNA hybridization to compare species compositions of natural bacterioplankton assemblages. *Applied and Environmental Microbiology*, **56**, 739–46.

Lehman, J.T. (1988) Ecological principles affecting community structure and secondary production by zooplankton in marine and freshwater ecosystems. *Limnology and Oceanography*, **33**, 931–45.

Lenz, J. (1992) Microbial loop, microbial food web and classical food chain: their significance in pelagic marine ecosystems. *Archiv für Hydrobiologie Bei-hefte Ergebnisse der Limnologie*, **37**, 265–78.

Lessard, E.J. (1991) The role of heterotrophic dinoflagellates in diverse environments. *Marine Microbial Food Webs*, **5**, 49–58.

Li, W.K.W. (1994) Primary production of prochlorophytes, cyanobacteria, and eucaryotic ultraphytoplankton: Measurements from flow cytometric sorting. *Limnology and Oceanography*, **39**, 169–75.

Lim, E.L., Amaral, L.A., Caron, D.A. & DeLong, E.F. (1993) Application of rRNA-based probes for observing marine nanoplanktonic protists. *Applied and Environmental Microbiology*, **59**, 1647–55.

Lim, E.L., Caron, D.A. & DeLong, E.F. (1996) Development and field application of a quantitative method for examining natural assemblages of protists with oligonucleotide probes. *Applied and Environmental Microbiology*, **62**, 1416–23.

Lindholm, T. (1981) On the ecology of *Mesodinium rubrum* (Lohmann) (Ciliata) in a stagnant brackish basin on Åland, SW Finland. *Kieler Meeresforschungen* Sonderheft, **5**, 117–23.

Lindholm, T. (1985) *Mesodinium rubrum* — a unique photosynthetic ciliate. *Advances in Aquatic Microbiology*, **3**, 1–48.

Lindström, E.S., Stadler, P. & Weisse, T. (2002) Enumeration of small ciliates in culture by flow cytometry and nucleic acid staining. *Journal of Microbiological Methods*, **49**, 173–82.

Lohmann, H. (1908) Untersuchungen zur Feststellung des vollständigen Gehaltes des Meeres and Plankton. *Wissenschaftliche. Meeresuntersuchungen, Abteilung Kiel, Neue Folge*, **10**, 131–370.

Lohmann, H. (1911) Über das Nannoplankton und die Zentrifugierung kleinster Wasserproben zur Gewinnung desselben im lebenden Zustande. *Internationale Revue der gesamten Hydrobiologie*, **4**, 1–38.

Loose, C.J., von Elert, E. & Dawidowicz, P. (1993) Chemically-induced diel vertical migration in *Daphnia*: a new bioassay for kairomones exuded by fish. *Archiv für Hydrobiologie*, **126**, 329–37.

Luria, S.E. & Darnell, J.E.J. (1978) *General Virology* (3rd edition). Wiley, New York, 578 pp.

Lyche, A., Andersen, T., Christoffersen, K., Hessen, D.O., Berger Hansen, P.H. & Klysner, A. (1996) Mesocosm tracer studies. 2. The fate of primary production and the role of consumers in the pelagic carbon cycle of a mesotrophic lake. *Limnology and Oceanography*, **21**, 475–87.

Maloy, S.R., Cronan, J.Jr. & Freifelder, D. (1994) *Micro-*

bial Genetics (2nd edition). Jones and Bartlett, Boston, 512 pp.

Marie, D., Brussard, C.P.D., Thyrhaug, R., Bratbak, G. & Vaulot, D. (1999) Enumeration of marine viruses in culture and natural samples by flow cytometry. *Applied and Environmental Microbiology*, **65**, 45–52.

Maranger, R. & Bird, D.F. (1995) Viral abundance in aquatic systems: a comparison between marine and fresh waters. *Marine Ecology Progress Series*, **121**, 217–26.

Mathes, J. & Arndt, H. (1994) Biomass and composition of protozooplankton in relation to lake trophy in north German lakes. *Marine Microbial Food Webs*, **8**, 357–75.

Mathes, J. & Arndt, H. (1995) Annual cycle of protozooplankton (ciliates, flagellates and sarcodines) in relation to phyto- and metazooplankton in Lake Neumühler See (Mecklenburg, Germany). *Archiv für Hydrobiologie*, **134**, 337–58.

McQueen, D.L., Johannes, M.R.S., Post, J.R., Stewart, T.J. & Lean, D.R.S. (1989) Bottom-up and top-down impacts on freshwater pelagic community structure. *Ecological Monographs*, **59**, 289–309.

McQueen, D.L., Post, J.R. & Mills, E.L. (1986) Trophic relationships in freshwater pelagic ecosystems. *Canadian Journal of Fisheries and Aquatic Sciences*, **43**, 1571–81.

Montagnes, D.J.S. & Weisse, T. (2000) Fluctuating temperatures affect growth and production rates of planktonic ciliates. *Aquatic Microbial Ecology*, **21**, 97–102.

Morden, C.W. & Golden, S.S. (1989) psbA genes indicate common ancestry of prochlorophytes and chloroplasts. *Nature*, **337**, 382–4.

Müller, H. (1989) The relative importance of different ciliate taxa in the pelagic food web of Lake Constance. *Microbial Ecology*, **18**, 261–73.

Müller, H. (1991) *Pseudobalanion planctonicum* (Ciliophora, Prostomatida): ecological significance of an algivorous nanociliate in a deep meso-eutrophic lake. *Journal of Plankton Research*, **13**, 247–62.

Müller, H. (1996) Selective feeding of a freshwater chrysomonad, *Paraphysomonas* sp., on chroococcoid cyanobacteria and nanoflagellates. *Archiv für Hydrobiologie Special Issues Advances in Limnology* **48**, 63–71.

Müller, H., Schöne, A., Pinto-Coelho, R.M., Schweizer, A. & Weisse, T. (1991) Seasonal succession of ciliates in Lake Constance. *Microbial Ecology*, **21**, 119–38.

Munawar, M. & Munawar, I.F. (1982) Phycological studies in Lakes Ontario, Erie, Huron and Superior. *Canadian Journal of Botany*, **60**, 1837–58.

Munawar, M. & Weisse, T. (1989) Is the 'microbial loop' an early warning indicator of anthropogenic stress? *Hydrobiologia*, **188/189**, 163–74.

Munawar, M., Munawar, I.F., Norwood, W.P. & Mayfield, C.I. (1987) Significance of autotrophic picoplankton in the Great Lakes and their use as early indicators of contaminant stress. *Archiv für Hydrobiologie Beihefte Ergebnisse der Limnologie*, **25**, 141–55.

Munawar, M., Munawar, I.F., Weisse, T., Leppard, G.G. & Legner, M. (1994) The significance and future potential of using microbes for assessing ecosystem health: The Great Lakes example. *Journal of Aquatic Ecosystem Health*, **3**, 295–30.

Murzov, S.A. & Caron, D.A. (1996) Sporadic high abundances of naked amoebae in the Black Sea plankton. *Aquatic Microbial Ecology*, **11**, 161–9.

Nagasaki, K., Ando, M., Imai, I., Itakura, S. & Ishida, Y. (1993) Virus-like particles in an apochlorotic flagellate in Hiroshima Bay, Japan. *Marine Ecology Progress Series*, **96**, 307–10.

Nagasaki, K., Ando, M., Itakura, S., Imai, I. & Ishida, Y. (1994) Viral mortality in the final stages of *Heterosigma akashiwo* (Raphidiophyceae) red tide. *Journal of Plankton Research*, **16**, 1595–99.

Noble, R.T. & Fuhrman, J.A. (1997) Virus decay and its causes in coastal waters. *Applied and Environmental Microbiology*, **63**, 77–83.

Nygaard, K., Borsheim, K.Y. & Thingstad, T.F. (1988) Grazing rates on bacteria by marine heterotrophic microflagellates compared to uptake rates of bacterial-sized monodisperse fluorescent latex beads. *Marine Ecology Progress Series*, **44**, 159–65.

Oberhäuser-Nehls, R., Anagnostidis, K. & Overbeck, J. (1993) Phototrophic bacteria in the Plußsee: ecology of *Sulfuretum*. In: Overbeck, J. & Chróst, R.J. (eds), *Microbial Ecology, of Lake Plußsee*. Springer-Verlag, New York, 287–325.

Ochs, C.A., Cole, J.J. & Likens, G.E. (1995) Population dynamics of bacterioplankton in an oligotrophic lake. *Journal of Plankton Research*, **17**, 365–91.

Olsen, G.J. (1990) Variation among the masses. *Nature*, **345**, 20–1.

Pace, M.L. & Bailiff, M.D. (1987) An evaluation of the fluorescent microsphere technique for measuring grazing rates of phagotrophic organisms. *Marine Ecology Progress Series*, **40**, 185–93.

Pace, M.L., McManus, G.B. & Findlay, E.G. (1990) Plankton community structure determines the fate of bacterial production in a temperate lake. *Limnology and Oceanography*, **35**, 795–808.

Pace, N.R. (1996) New perspectives on the natural microbial world: molecular microbial ecology. *ASM News* **62**, 463–70.

Padan, E. & Shilo, M. (1973) Cyanophages—viruses attacking blue-green algae. *Bacteriological Review*, **37**, 343–70.

Paerl, H. (1988) Nuisance phytoplankton blooms in coastal, estuarine, and inland waters. *Limnology and Oceanography*, **33**, 823–47.

Parejko, K. & Dodson, S.I. (1990) Progress towards characterization of a predator/prey kairomone: *Daphnia pulex* and *Chaoborus americanus*. *Hydrobiologia*, **198**, 51–9.

Patterson, D.J. & Hausmann, K. (1981) Feeding by *Actinophyrs sol* (Protista, Heliozoa): I. Light microscopy. *Microbios*, **31**, 39–55.

Patterson, D.J. & Larsen, J. (1991) *The Biology of Free-living Heterotrophic Flagellates*. Systematics Association Special Volume 45, Clarendon Press, Oxford, 502 pp.

Patterson, D.J. & Zölffel, M. (1991) Heterotrophic flagellates of uncertain taxonomic position. In: Patterson, D.J. & Larsen, J. (eds), *The Biology of Free-living Heterotrophic Flagellates*. Systematics Association Special Volume 45, Clarendon Press, Oxford, 427–76.

Paul, J.H. (1993) The advances and limitations of methodology. In: Ford, T.E. (ed.), *Aquatic Microbiology*. Blackwell Scientific Publications, Oxford, 15–46.

Paul, J.H., Jiang, S.C. & Rose, J.B. (1991) Concentration of viruses and dissolved DNA from aquatic environments by vortex flow filtration. *Applied and Environmental Microbiology*, **57**, 2197–204.

Perasso, R., Baroin, A., Hu Qu, L., Bachellerie, J.P. & Adoutte, A. (1989) Origin of the algae. *Nature*, **339**, 142–4.

Pérez-Uz, B. (1995) Growth rate variability in geographically diverse clones of *Uronema* (Ciliophora: Scuticociliatida). *FEMS Microbiology Ecology*, **16**, 193–204.

Pernthaler, J., Posch, T., Šimek, K., Vrba, J., Amann, R. & Psenner, R. (1997) Contrasting bacterial strategies to coexist with a flagellate predator in an experimental microbial assemblage. *Applied and Environmental Microbiology*, **63**, 596–601.

Pernthaler, J., Šimek, K., Sattler, B., Schwarzenbacher, A., Bobková, J. & Psenner, R. (1996) Short-term changes of protozoan control on autotrophic picoplankton in an oligo-mesotrophic lake. *Journal of Plankton Research*, **18**, 443–62.

Peters-Regehr, T., Kusch, J. & Heckmann, K. (1997) Primary structure and origin of a predator released protein that induces defensive morphological changes in *Euplotes*. *European Journal of Protistology*, **33**, 389–95.

Pfister, G. & Arndt, H. (1998) Food selectivity and feeding

behaviour in omnivorous filter-feeding ciliates: a case study for *Stylonychia*. *European Journal of Protistology*, **34**, 446–57.

Pomeroy, L.R. (1974) The ocean's food web: a changing paradigm. *BioScience*, **24**, 499–504.

Pomeroy, L.R. & Deibel, D. (1986) Temperature regulation of bacterial activity during the spring bloom in Newfoundland coastal waters. *Science*, **233**, 359–61.

Pomeroy, L.R. & Johannes, R.E. (1968) Occurrence and respiration of ultraplankton in the upper 500 meters of the ocean. *Deep-Sea Research*, **15**, 381–91.

Pomeroy, L.R. & Wiebe, W.J. (1988) Energetics of microbial food webs. *Hydrobiologia*, **159**, 7–18.

Popovsky, J. (1982) Another case of phagotrophy by *Gymnodinium helveticum* f. *achroum* Skuja. *Archiv für Protistenkunde*, **125**, 73–8.

Porter, K.G., Pearl, H., Hodson, R., *et al.* (1988) Microbial interactions in lake foodwebs. In: Carpenter, S.R. (ed.), *Complex Interactions in Lake Communities*. Springer-Verlag, Berlin, 234–55.

Posch, T., Šimek, K., Vrba, J., *et al.* (1999) Predator-induced changes of bacterial size-structure and productivity studied on an experimental microbial community. *Aquatic Microbial Ecology*, **18**, 235–46.

Postius, C. & Ernst, A. (1999) Mechanisms of dominance: coexistence of picocyanobacterial genotypes in a freshwater ecosystem. *Archives of Microbiology*, **172**, 69–75.

Postius, C., Ernst, A., Kenter, U. & Böger, P. (1996) Persistence and genetic diversity among strains of phycoerythrin-rich cyanobacteria from the picoplankton of Lake Constance. *Journal of Plankton Research*, **18**, 1159–66.

Proctor, L.M. & Fuhrman, J.A. (1990) Viral mortality of marine bacteria and cyanobacteria. *Nature*, **343**, 60–2.

Proctor, L.M., Okubo, A. & Fuhrman, J.A. (1993) Calibrating estimates of phage-induced mortality in marine bacteria: ultrastructural studies of marine bacteriophage development from one-step growth experiments. *Microbial Ecology*, **25**, 161–82.

Psenner, R. (1990) From image analysis to chemical analysis of bacteria: A long-term study? *Limnology and Oceanography*, **35**, 234–7.

Psenner, R. (1993) Determination of size and morphology of aquatic bacteria by automated image analysis. In: Kemp, P.F., Sherr, B.F., Sherr, E.B. & Cole, J.J. (eds), *Handbook of Methods in Aquatic Microbial Ecology*. Lewis Publishers, Boca Raton, FL, 339–45.

Rice, J., Sleigh, M.A., Burkill, P.H., Tarran, G.A., O'Connor, C.D. & Zubkov, M.V. (1997) Flow cytometric analysis of characteristics of hybridization of species-specific fluorescent oligonucleotide probes to rRNA of marine nanoflagellates. *Applied and Environmental Microbiology*, **63**, 938–44.

Richman, S., Branstrator, D.K. & Huber-Villegas, M. (1990) Impact of zooplankton grazing on phytoplankton along a eutrophication gradient. In: Tilzer, M.M. & Serruya, C. (eds), *Large Lakes. Ecological Structure and Function*. Springer-Verlag, Berlin, 592–614.

Riemann, B. & Bell, R.T. (1990) Advances in estimating bacterial biomass and growth in aquatic systems. *Archiv für Hydrobiologie*, **118**, 385–402.

Riemann, B. & Christoffersen, K. (1993) Microbial trophodynamics in temperate lakes. *Marine Microbial Food Webs*, **7**, 69–100.

Riemann, B. & Søndergaard, M. (1986) *Carbon Dynamics in Eutrophic, Temperate Lakes*. Elsevier, Amsterdam, 284 pp.

Riemann, B., Søndergaard, M., Persson, L. & Johansson, L. (1986) Carbon metabolism and community regulation in eutrophic, temperate lakes. In: Riemann, B. & Søndergaard, M. (eds), *Carbon dynamics in eutrophic, temperate lakes*. Elsevier, Amsterdam, 267–80.

Riemann, B., Havskum, H., Thingstad, F. & Bernard, C. (1995) The role of mixotrophy in pelagic environments. In: Joint, I. (ed.), *Molecular Ecology of Aquatic Microbes*. Springer-Verlag, Berlin, 87–114.

Rivier, A., Brownlee, D.C., Sheldon, R.W. & Rassoulzadegan, F. (1985) Growth of microzooplankton: a comparative study of bacterivorous zooflagellates and ciliates. *Marine Microbial Food Webs*, **1**, 51–60.

Robertson, B.R., Tezuda, N. & Watanabe, M.M. (2001) Phylogenetic analyses of *Synechococcus* strains (cyanobacteria) using sequences of 16S rDNA and part of the phycocyanin operon reveal multiple evolutionary lines and reflect phycobilin content. *International Journal of Systematic and Evolutionary Microbiology*, **51**, 861–71.

Rogerson, A. & Laybourn-Parry, J. (1992) The abundance of marine naked amoebae in the water column of the Clyde Estuary. *Estuarine and Coastal Shelf Science*, **34**, 187–96.

Rosenstock, B. & Simon, M. (1993) Utilization of dissolved combined and free amino acids by planktonic bacteria in Lake Constance. *Limnology and Oceanography*, **38**, 1521–31.

Rothhaupt, K.O. (1992) Simulation of phosphorus-limited phytoplankton by bacteriovorous flagellates in laboratory experiments. *Limnology and Oceanography*, **37**, 750–9.

Rothhaupt, K.O. (1996a) Laboratory experiments with a

mixotrophic chrysophyte and obligately phagotrophic and phototrophic competitors. *Ecology*, 77, 716–24.

Rothhaupt, K.O. (1996b) Utilization of substitutable carbon and phosphorus sources by the mixotrophic chrysophyte *Ochromonas* sp. *Ecology*, 77, 706–15.

Rothhaupt, K.O. (1997) Nutrient turnover by freshwater bacterivorous flagellates: differences between a heterotrophic and a mixotrophic chrysophyte. *Aquatic Microbial Ecology*, 12, 65–70.

Safferman, R.S. & Morris, M.E. (1963) Algal virus: isolation. *Science*, 140, 679–80.

Salbrechter, M. & Arndt, H. (1994) The annual cycle of protozooplankton in the alpine, mesotrophic Lake Mondsee (Austria). *Marine Microbial Food Webs*, 8, 217–34.

Sanders, R.W. (1991a) Mixotrophic protists in marine and freshwater ecosystems. *Journal of Protozoology*, 38, 76–81.

Sanders, R.W. (1991b) Trophic strategies among heterotrophic flagellates. In: Patterson, D.J. & Larsen, J. (eds), *The Biology of Free-living Heterotrophic Flagellates.* Systematics Association Special Volume 45, Clarendon Press, Oxford, 21–38.

Sanders, R.W. & Porter, K.G. (1988) Phagotrophic phytoflagellates. *Advances in Microbial Ecology*, 10, 167–92.

Sanders, R.W. & Porter, K.G. (1990) Bacterivorous flagellates as food resources for the freshwater crustacean zooplankter *Daphnia ambigua*. *Limnology and Oceanography*, 35, 188–91.

Sanders, R.W. & Wickham, S.A. (1993) Planktonic protozoa and metazoa: predation, food quality and population control. *Marine Microbial Food Webs*, 7, 197–223.

Sanders, R.W., Porter, K.G., Bennett, S.J. & DeBiase, A.E. (1989) Seasonal patterns of bacterivory by flagellatges, cilliates, rotifers, and cladocerans in a freshwater plankton community. *Limnology and Oceanography*, 34, 673–87.

Sanders, R.W., Porter, K.G. & Caron, D.A. (1990) Relationship between phototrophy and phagotrophy in the mixotrophic chrysophyte *Poterioochromonas malhamensis*. *Microbial Ecology*, 19, 97–109.

Sanders, R.W., Caron, D.A. & Berninger, U-G. (1992) Relationships between bacteria and heterotrophic nanoplankton in marine and fresh waters: an intersystem comparison. *Marine Ecology Progress Series*, 86, 1–14.

Schlegel, H.G. & Bowien, B. (eds) (1989) *Autotrophic Bacteria.* Science and Technology, Madison, WI, 528 pp.

Schnepf, E. & Deichgräber, G. (1984) 'Myzocystosis', a kind of endocytosis with implications to compartmentation in endosymbiosis: observations in *Paulsenella* (Dinophyta). *Naturwissenschaften*, 71, 218–19.

Schopf, J.W. (1977) Biostratigraphic usefulness of stromatolitic Precambrian microbiotas: a preliminary analysis. *Precambrian Research*, 5, 143–73.

Schweizer, A. (1994) Seasonal dynamics of planktonic Ciliophora along a depth transect in Lake Constance. *Marine Microbial Food Webs*, 8, 283–93.

Seitz, A. (1984) Are there allelopathic interactions in zooplankton? Laboratory experiments with *Daphnia*. *Oecologia*, 62, 94–6.

Semenova, E.A., Kuznedelov, K.D. & Grachev, M.A. (2001) Nucleotide sequences of fragments of 16S rRNA of the Baikal natural populations and laboratory cultures of cyanobacteria. *Molecular Biology*, 35, 405–10.

Severn, S.R.T., Munawar, M. & Mayfield, C.I. (1989) Measurements of sediment toxicity of autotrophic and heterotrophic picoplankton by epifluorescence microscopy. *Hydrobiologia*, 176/177, 525–30.

Sherr, B.F., Sherr, E.B. & Albright, L.J. (1987) Bacteria: link or sink? *Science*, 235, 88.

Sherr, B.F., Sherr, E.B. & Hopkinson, C.S. (1988) Trophic interactions within pelagic microbial communities: indications of feedback regulation of carbon flow. *Hydrobiologia*, 159, 19–26.

Sherr, E.B. (1989) And now, small is plentiful. *Nature*, 340, 429.

Sherr, E.B. & Sherr, B.F. (1987) High rates of consumption of bacteria by pelagic ciliates. *Nature*, 325, 710–11.

Sherr, E.B. & Sherr, B.F. (1988) Role of microbes in pelagic food webs: a revised concept. *Limnology and Oceanography*, 33, 1225–7.

Sherr, E.B. & Sherr, B.F. (1994) Bacterivory and herbivory: key roles of phagotrophic protists in pelagic food webs. *Microbial Ecology*, 28, 223–35.

Sherr, B.F., Sherr, E.B., Andrew, T.L., Andrew, L., Fallon, R.D. & Newell, S.Y. (1986) Trophic interactions between heterotrophic protozoa and bacterioplankton in estuarine water analyzed with selective metabolic inhibitors. *Marine Ecology Progress Series*, 32, 169–79.

Shia, F.K. & Ducklow, H.W. (1994) Temperature and substrate regulation of bacterial abundance, production and specific growth rate in Chesapeake Bay, USA. *Marine Ecology Progress Series*, 103, 297–308.

Sibbald, M.J., Albright, L.J. & Sibbald, P.R. (1987) Chemosensory response of a heterotrophic flagellate to bacteria and several nitrogen compounds. *Marine Ecology Progress Series*, 36, 201–4.

Sieburth, J.M., Smetacek, V. & Lenz, J. (1978) Pelagic ecosystem structure: heterotrophic compartments of the plankton and their relationship to plankton size fractions. *Limnology and Oceanography*, **23**, 1256–63.

Sieracki, M.E. & Webb, K.L. (1991) The application of image analysed fluorescence microsopy for characterising planktonic bacteria and protists. In: Reid, P.C., Turley, C.M. & Burkill, P.H. (eds), *Protozoa and their Role in Marine Processes*. Springer-Verlag, Berlin, 77–100.

Šimek, K. & Straskrabova, V. (1992) Bacterioplankton production and protozoan bacterivory in a mesotrophic reservoir. *Journal of Plankton Research*, **14**, 773–87.

Šimek, K., Macek, M. & Vyhnalek, V. (1990a) Uptake of bacteria-sized fluorescent particles by natural protozoan assemblage in a reservoir. *Archiv für Hydrobiologie Beihefte Ergebnisse der Limnologie*, **34**, 275–81.

Šimek, K., Macek, M., Seda, J. & Vyhnalek, V. (1990b). Possible food chain relationships between bacterioplankton, protozoans, and cladocerans in a reservoir. *Internationale Revue der gesamten Hydrobiologie*, **75**, 583–96.

Šimek, K., Bobková, J., Macek, M., Nedoma, J. & Psenner, R. (1995) Ciliate grazing on picoplankton in a eutrophic reservoir during the summer phytoplankton maximum: a study at the species and community level. *Limnology and Oceanography*, **40**, 1077–90.

Šimek, K., Macek, M., Pernthaler, J., Straskrabová, V. & Psenner, R. (1996) Can freshwater planktonic ciliates survive on a diet of picoplankton? *Journal of Plankton Research*, **18**, 597–613.

Šimek, K., Hartman, P., Nedoma, J., *et al.* (1997a) Community structure, picoplankton grazing and zooplankton control of heterotrophic nanoflagellates in a eutrophic reservoir during the summer phytoplankton maximum. *Aquatic Microbial Ecology*, **12**, 49–63.

Šimek, K., Vraba, J., Pernthaler, J., *et al.* (1997b) Morphological and compositional shifts in an experimental bacterial community influenced by protists with contrasting feeding modes. *Applied and Environmental Microbiology*, **63**, 587–95.

Šimek, K., Armengol, J., Comerma, M., *et al.* (1998) Characteristics of protistan control of bacterial production in three reservoirs of different trophy. *Internationale Revue der gesamten Hydrobiologie*, **83**, 485–94.

Šimek, K., Kojecká, P., Nedoma, J., Hartman, P., Vrba, J. & Dolan, J.R. (1999) Shifts in bacterial community composition associated with different microzooplankton size fractions in a eutrophic reservoir. *Limnology and Oceanography*, **44**, 1634–44.

Šimek, K., Jürgens, K., Comerma, M., Armengol, J. &

Nedoma, J. (2000) Ecological role and bacterial grazing of *Halteria* spp.: small freshwater oligotrichs as dominant pelagic ciliate bacterivores. *Aquatic Microbial Ecology*, **22**, 43–56.

Simon, M. (1987) Biomass and production of small and large free-living and attached bacteria in Lake Constance. *Limnology and Oceanography*, **32**, 591–607.

Simon, M. & Rosenstock, B. (1992) Carbon and nitrogen sources of planktonic bacteria in Lake Constance studied by the composition and isotope dilution of intracellular amino acids. *Limnology and Oceanography*, **37**, 1496–511.

Simon, M., Cho, B.C. & Azam, F. (1992) Significance of bacterial biomass in lakes and the ocean: comparison to phytoplankton biomass and biogeochemical implications. *Marine Ecology Progress Series*, **86**, 103–10.

Simon, M., Bunte, C., Schulz, M., Weiss, M. & Wünsch, C. (1998a) Bacterioplankton dynamics in Lake Constance (Bodensee): Substrate utilization, growth control, and long-term trends. *Archiv für Hydrobiologie Special Issues Advancances in Limnology*, **53**, 195–221.

Simon, M., Tilzer, M.M. & Müller, H. (1998b) Bacterioplankton dynamics in a large mesotrophic lake: I. abundance, production and growth control. *Archiv für Hydrobiologie*, **143**, 385–407.

Sleigh, M.A. (1991) A taxonomic review of heterotrophic protists important in marine ecology. In: Reid, P.C., Turley, C.M., Burkill, P.H. (eds), *Protozoa and their Role in Marine Processes*. Springer-Verlag, Berlin, 9–38.

Smetacek, V. (1981) The annual cycle of protozooplankton in Kiel Bight. *Marine Biology*, **63**, 1–11.

Smetacek, V. (1985) The annual cycle of Kiel Bight plankton: a long-term analysis. *Estuaries* **8**, 145–57.

Sogin, M.L., Elwood, H.J. & Gunderson, J.H. (1986) Evolutionary diversity of eukaryotic small subunit rRNA genes. *Proceedings of the National Academy of Sciences of the U.S.A.*, **83**, 1383–7.

Sommaruga, R. & Psenner, R. (1993) Nanociliates of the order Proostomatida: Their relevance in the microbial food web of a mesotrophic lake. *Aquatic Sciences*, **55**, 179–87.

Sommaruga, R. & Robarts, R.D. (1997) The significance of autotrophic and heterotrophic picoplankton in hypertrophic ecosystems. *FEMS Microbiology Ecology*, **24**, 187–200.

Sommer, U., Gaedke, U. & Schweizer, A. (1993) The first decade of oligotrophication of Lake Constance. II. The response of phytoplankton taxonomic composition. *Oecologia*, **93**, 276–84.

Sommer, U., Sommer, F., Santer, B., *et al.* (2001) Complementary impact of copepods and cladocerans on phytoplankton. *Ecology Letters* **4**, 545–50.

Søndergaard, M. & Riemann, B. (1986) Epilogue. In: Riemann, B. & Søndergaard, M. (eds), *Carbon Dynamics in Eutrophic, Temperate Lakes.* Elsevier, Amsterdam, 281–3.

Søndergaard, M., Riemann, B., Møller Jensen, L., *et al.* (1988) Pelagic food web processes in an oligotrophic lake. *Hydrobiologia*, **164**, 271–86.

Spencer, R. (1955) A marine bacteriophage. *Nature*, **175**, 690.

Spero, H.J. (1985) Chemosensory capabilities in the phagotrophic dinoflagellate *Gymnodinium fungiforme*. *Journal of Phycology*, **21**, 181–4.

Steele, J.H. (1974) *The Structure of Marine Ecosystems.* Harvard University Press, Cambridge, MA.

Stephens, D.W. & Krebs, J.R. (1986) *Foraging Theory.* Princeton University Press, Princeton, NJ, 247 pp.

Sterner, R.W. (1989) The role of grazers in phytoplankton succession. In: Sommer, U. (ed.), *Plankton Ecology.* Springer-Verlag, Berlin, 337–64.

Stevenson, L. (1978) A case for bacterial dormancy in aquatic systems. *Microbial Ecology*, **4**, 127–33.

Stiasny, G. (1913) *Das Plankton des Meeres.* Sammlung Göschen, Berlin, 356 pp.

Stockner, J.G. (1988) Phototrophic picoplankton: an overview from marine and freshwater ecosystems. *Limnology and Oceanography*, **33**, 765–75.

Stockner, J.G. (1991) Autotrophic picoplankton in freshwater ecosystems: the view from the summit. *Internationale Revue der gesamten Hydrobiologie*, **76**, 483–92.

Stockner, J.G. & Antia, N.J. (1986) Algal picoplankton from marine and freshwater ecosytems: a multidisciplinary perspective. *Canadian Journal of Fisheries and Aquatic Sciences*, **43**, 2472–503.

Stockner, J.G. & Porter, K.G. (1988) Microbial food webs in freshwater planktonic ecosystems. In: Carpenter, S.R. (ed.), *Complex Interactions in Lake Communities.* Springer-Verlag, Berlin, 69–83.

Stockner, J.G. & Shortreed, K.S. (1989) Algal picoplankton production and contribution to food-webs in oligotrophic British Columbia lakes. *Hydrobiologia*, **173**, 151–66.

Stoecker, D.K. (1991) Mixotrophy in marine planktonic ciliates: physiological and ecological aspects of plastid retention by oligotrichs. In: Reid, P.C., Turley, C.M. &

Burkill, P.H. (eds), *Protozoa and their Role in Marine Processes.* Springer-Verlag, Berlin, 161–79.

Stoecker, D.K., Michaelis, A.E. & Davis, L.H. (1987) Large proportion of marine planktonic ciliates found to contain functional chloroplasts. *Nature*, **326**, 790–2.

Stone, L., Berman, T., Bonner, R., Barry, S. & Weeks, S.W. (1993) Lake Kinneret: a seasonal model for carbon flux through the planktonic biota. *Limnology and Oceanography*, **38**, 1680–95.

Straile, T. (1995) *Die saisonale Entwicklung des Kohlenstoffkreislaufes im pelagischen Nahrungsnetz des Bodensees.—Eine Analyse von massenbilanzierten Flußdiagrammen mit Hilfe der Netzwerktheorie.* Hartung-Gorre Verlag, Konstanz, 143 pp.

Straile, T. (1998) Biomass allocation and carbon flow in the pelagic food web of Lake Constance. *Archiv für Hydrobiologie Special Issues Advancances in Limnology*, **53**, 545–63.

Strayer, D. (1988) On the limits to secondary production. *Limnology and Oceanography*, **33**, 1217–20.

Suttle, C.A. (1993) Enumeration and isolation of viruses. In: Kemp, P.F., Sherr, B.F., Sherr, E.B. & Cole, J.J. (eds), *Handbook of Methods in Aquatic Microbial Ecology.* Lewis, Boca Raton, FL, 121–34.

Suttle, C.A. (1994) The significance of viruses to mortality in aquatic microbial communities. *Microbial Ecology*, **28**, 237–43.

Suttle, C.A. & Chan, A.M. (1993) Marine cyanophages infecting oceanic and coastal strains of *Synechococcus*: abundance, morphology, cross-infectivity and growth characteristics. *Marine Ecology Progress Series*, **92**, 99–109.

Suttle, C.A. & Chan, A.M. (1994) Dynamics and distribution of cyanophages and their effect on marine *Synechococcus* spp. *Applied and Environmental Microbiology*, **60**, 3167–74.

Suttle, C.A., Chan, A.M., Taylor, W.D. & Harrison, P.J. (1986) Grazing of planktonic diatoms by microflagellates. *Journal of Plankton Research*, **8**, 393–8.

Suttle, C.A., Chan, A.M. & Cottrell, M.T. (1990) Infection of phytoplankton by viruses and reduction of primary productivity. *Nature*, **347**, 467–9.

Suttle, C.A., Chan, A.M. & Cottrell, M.T. (1991) Use of ultrafiltration to isolate viruses from seawater which are the pathogens of marine phytplankton. *Applied and Environmental Microbiology*, **57**, 721–6.

Thingstad, F., Heldal, M., Bratbak, G. & Dundas, I. (1993) Are viruses important partners in pelagic food webs? *Trends in Ecology and Evolution*, **8**, 209–13.

Threlkeld, S.T. (1988) Planktivory and planktivore biomass effects on zooplankton, phytoplankton, and the trophic cascade. *Limnology and Oceanography*, **33**, 1362–75.

Tollrian, R. & Dodson, S.I. (1999) Inducible defenses in cladocera: constraints, costs, and multipredator environments. In: Tollrian, R. & Harvell, C.D. (eds), *The ecology and evolution of inducible defenses*. Princeton University Press, Princeton, NJ, 177–202.

Torella, F. & Morita, R.Y. (1979) Evidence for a high incidence of bacteriophage particles in the waters of Yaquina Bay, Oregon: ecological and taxonomical implications. *Applied and Environmental Microbiology*, **37**, 774–8.

Tranvik, L.J., Sherr, E.B. & Sherr, B.F. (1993) Uptake and utilization of 'colloidal DOM' by heterotrophic flagellates in seawater. *Marine Ecology Progress Series*, **92**, 301–9.

Vaqué, D. & Pace, M.L. (1992) Grazing on bacteria by flagellates and cladocerans in lakes of contrasting food-web structure. *Journal of Plankton Research*, **14**, 307–21.

Vadstein, O., Olsen, Y., Reinertsen, H. & Jensen, A. (1993) The role of planktonic bacteria in phosphorus cycling in lakes—sink and link. *Limnology and Oceanography*, **38**, 1539–44.

Verity, P.G. (1988) Chemosensory behavior in marine planktonic ciliates. *Bulletin of Marine Science*, **43**, 772–82.

Verity, P.G. (1991) Measurement and simulation of prey uptake by marine planktonic cilliates fed plastidic and aplastidic nanoplankton. *Limnology and Oceanography*, **36**, 729–50.

Verity, P.G. & Sieracki, M.E. (1993) Use of color image analysis and epifluorescence microscopy to measure plankton biomass. In: Kemp, P.F., Sherr, B.F., Sherr, E.B. & Cole, J.J. (eds), *Handbook of Methods in Aquatic Microbial Ecology*. Lewis Publishers, Boca Raton, FL, 327–38.

Verity, P.G. & Villareal, T.A. (1986) The relative food value of diatoms, dinoflagellates, flagellates, and cyanobacteria for tintinnid ciliates. *Archiv für Protistenkunde*, **131**, 71–84.

Vrieling, E.G. & Anderson, D.M. (1996) Immunofluorescence phytoplankton research: application and potential. *Journal of Phycology*, **32**, 1–16.

Vrieling, E.G., Vriezekolk, G., Gieskes, W.W., Veenhuis, M. & Harder, W. (1996) Immuno-flow cytometric identification and enumeration of the ichthyotoxic dinoflagellate *Gyrodinium aureolum* Hulburt in artifically mixed algal populations. *Journal of Plankton Research*, **18**, 1503–12.

Walter, M.R. (1983) Archean stromatolites: evidence of

the Earth's earliest benthos. In: Schopf, J.W. (ed.), *Earth's Earliest Biosphere*. Princeton University Press, Princeton, NJ, 187–213.

Ward, B.B. (1982) Oceanic distribution of ammonium-oxidizing bacteria determined by immunofluorescence assay. *Journal of Marine Research*, **40**, 1155–72.

Ward, B.B. (1990) Immunology in biological oceanography and marine ecology. *Oceanography*, **3**, 30–5.

Ward, B.B. & Perry, M.J. (1980) Immunofluorescent assay for the marine ammonium-oxidizing bacterium *Nitrococcus oceanus*. *Applied and Environmental Microbiology*, **39**, 913–18.

Ward, B.B., Olson, R.J. & Perry, M.J. (1982) Microbial nitrification rates in the primary nitrite maximum off southern California. *Deep-Sea Research*, **29**, 247–55.

Ward, D.M., Weller, R. & Bateson, M.M. (1990) 16S rRNA sequences reveal numerous uncultured microorganisms in a natural community. *Nature*, **345**, 63–5.

Waterbury, J.B., Watson, S.W., Guillard, R.R.L. & Brand, L.E. (1979) Widespread occurrence of a unicellular, marine, planktonic, cyanobacterium. *Nature*, **277**, 293–4.

Wehr, J.D. (1991) Nutrient and grazer-mediated effects on picoplankton and size structure in phytoplankton communities. *Internationale Revue der gesamten Hydrobiologie*, **76**, 643–56.

Weinbauer, M.G. & Suttle, C.A. (1996) Potential significance of lysogeny to bacteriophage production and bacterial mortality in coastal waters of the Gulf of Mexico. *Applied and Environmental Microbiology*, **62**, 4374–80.

Weinbauer, M.G. & Suttle, C.A. (1997) Comparison of epifluorescence and transmission electron microscopy for counting viruses in natural marine waters. *Aquatic Microbial Ecology*, **13**, 225–32.

Weinbauer, M.G., Fuks, D. & Peduzzi, P. (1993) Distribution of viruses and dissolved DNA along a coastal trophic gradient in the northern Adriatic Sea. *Applied and Environmental Microbiology*, **59**, 4047–82.

Weisse, T. (1988) Dynamics of autotrophic picoplankton in Lake Constance. *Journal of Plankton Research*, **10**, 1179–88.

Weisse, T. (1989) The microbial loop in the Red Sea: dynamics of pelagic bacteria and heterotrophic nanoflagellates. *Marine Ecology Progress Series*, **55**, 241–50.

Weisse, T. (1990) Trophic interactions among heterotrophic microplankton, nanoplankton and bacteria in Lake Constance. *Hydrobiologia*, **191**, 111–22.

Weisse, T. (1991a) The annual cycle of heterotrophic freshwater nanoflagellates: role of bottom-up versus top-down control. *Journal of Plankton Research*, **13**, 167–85.

Weisse, T. (1991b) The microbial loop and its sensitivity to eutrophication and contaminant enrichment: a cross-system overview. *Internationale Revue der gesamten Hydrobiologie*, **76**, 327–37.

Weisse, T. (1993) Dynamics of autotrophic picoplankton in marine and freshwater ecosystems. In: Jones, J.G. (ed.), *Advances in Microbial Ecology*, Vol. 13. Plenum Press, New York, 327–70.

Weisse, T. (1994) Structure of microbial food webs in relation to the trophic status of lakes and fish grazing pressure: a key role of cyanobacteria? In: Pinto-Coelho, R.M., Giani, A. & von Sperling, E. (eds), *Ecology and Human Impact on Lakes and Reservoirs in Minas Gerais with Special Reference to Future Development and Management Strategies*. Sociedad Editoria e Gráfica de Ação Communitária, Belo Horizonte, Brazil, 55–70.

Weisse, T. (1997) Growth and production of heterotrophic nanoflagellates in a meso-eutrophic lake. *Journal of Plankton Research*, **19**, 703–22.

Weisse, T. (1999) Bacterivory in the northwestern Indian Ocean during the intermonsoon–northeast monsoon period. *Deep-Sea Research*, **46**, 795–814.

Weisse, T. (2002) The significance of inter- and intraspecific variation in bacterivorous and herbivorous protists. *Antonie van Leeuwenhoek*, **81**, 327–41.

Weisse, T. & Frahm, A. (2001) Species-specific interactions between small planktonic ciliates (*Urotricha* spp.) and rotifers (*Keratella* spp.). *Journal of Plankton Research*, **23**, 1329–38.

Weisse, T. & Frahm, A. (2002) Direct and indirect impact of two common rotifer species (*Keratella* spp.) on two abundant ciliate species (*Urotricha furcata, Balanion planctonicum*). *Freshwater Biology*, **47**, 53–64.

Weisse, T. & Kenter, U. (1991) Ecological characteristics of autotrophic picoplankton in a prealpine lake. *Internationale Revue der gesamten Hydrobiologie*, **76**, 493–504.

Weisse, T. & Kirchhoff, B. (1997) Feeding of the heterotrophic freshwater dinoflagellate *Peridiinopsis berolinense* on cryptophytes: analyses by flow cytometry and electronic particle counting. *Aquatic Microbial Ecology*, **12**, 153–64.

Weisse, T. & MacIsaac, E.A. (2000) Significance and fate of bacterial production in oligotrophic lakes in British Columbia. *Canadian Journal of Fisheries and Aquatic Sciences*, **57**, 96–105.

Weisse, T. & Montagnes, D.J.S. (1998) Effect of temperature on inter- and intraspecific isolates of *Urotricha* (Prostomatida, Ciliophora). *Aquatic Microbial Ecology*, **15**, 285–91.

Weisse, T. & Müller, H. (1998) Planktonic protozoa and the microbial food web in Lake Constance. *Archiv für Hydrobiologie Special Issues Advancances in Limnology*, **53**, 223–54.

Weisse, T. & Scheffel-Möser, U. (1991) Uncoupling the microbial loop: growth and grazing loss rates of bacteria and heterotrophic nanoflagellates in the North Atlantic. *Marine Ecology Progress Series*, **71**, 195–205.

Weisse, T. & Schweizer, A. (1991) Seasonal and interannual variation of autotrophic picoplankton in a large prealpine lake (Lake Constance). *Verhandlungen der Internationalen Vereinigung für theoretische und angewandte Limnologie*, **24**, 821–5.

Weisse, T. & Stockner, J.G. (1993) Eutrophication: the role of microbial food webs. *Memorie dell'Istituto Italiano di Idrobiologia*, **52**, 133–50.

Weisse, T., Müller, H., Pinto-Coelho, R.M., Schweizer, A., Springmann, D. & Baldringer, G. (1990) Response of the microbial loop to the phytoplankton spring bloom in a large prealpine lake. *Limnology and Oceanography*, **35**, 781–94.

Weisse, T., Karstens, N., Meyer, V.C.M., Janke, L., Lettner, S. & Teichgräber, K. (2001) Niche separation in common prostome freshwater ciliates: the effect of food and temperature. *Aquatic Microbial Ecology*, **26**, 167–79.

White, P.A., Kalff, J., Rasmussen, B. & Gasol, J.M. (1991) The effect of temperature and algal biomass on bacterial production and specific growth rate in freshwater and marine habitats. *Microbial Ecology*, **21**, 99–118.

Wicklow, B.J. (1997) Signal-induced defensive phenotypic changes in ciliated protists: morphological and ecological implications for predator and prey. *Journal of Eukaryotic Microbiology*, **44**, 176–88.

Wiggins, B.A. & Alexander, M. (1985) Minimum bacterial density for bacteriophage replication: Implications for significance of bacteriophages in natural ecosystems. *Applied and Environmental Microbiology*, **49**, 19–23.

Williams, P.J.Le B. (1970) Heterotrophic utilization of dissolved organic compounds in the Sea. I. Size distribution of population and relationship between respiration and incorporation of growth substances. *Journal of the Marine Biological Association of the United Kingdom*, **50**, 859–70.

Williams, P.J.Le B. (1981) Incorporation of micro-heterotrophic processes into the classical paradigm of the planktonic food web. *Kieler Meeresforschungen Sonderheft*, **1**, 1–28.

Witzel, K-P., Demuth, J. & Schütt, C. (1994) Viruses. In: Overbeck, J. & Chróst, R.J. (eds), *Microbial Ecology, of Lake Plußsee*. Springer-Verlag, New York, 270–86.

Woese, C.R. (1987) Bacterial evolution. *Microbiological Reviews*, **51**, 221–71.

Wood, A.M. & Townsend, D. (1990) DNA polymorphism within the WH7803 serogroup of marine *Synechococcus* spp. (cyanobacteria). *Journal of Phycology*, **26**, 576–85.

Wylie, J.L. & Currie, D.J. (1991) The relative importance of bacteria and algae as food sources for crustacean zooplankton. *Limnology and Oceanography*, **36**, 708–28.

Yentsch, C.M. & Campbell, J.W. (1991) Phytoplankton growth: perspectives gained by flow cytometry. *Journal of Plankton Research, Supplement*, **13**, 83–108.

Yentsch, C.M., Horan, P.K., Muirhead, K., *et al.* (1983) Flow cytometry and cell sorting: a technique for analysis and sorting of aquatic particles. *Limnology and Oceanography*, **28**, 1275–80.

Zeitzschel, B. (1967) Die Bedeutung der Tintinnen als Glied der Nahrungskette. *Helgoländer wissenschaftliche Meeresuntersuchungen*, **15**, 589–601.

Zimmermann, R. & Meyer-Reil, L.A. (1974) A new method for fluorescence staining of bacterial populations on membrane filters. *Kieler Meeresforschungen*, **30**, 24–7.

Zimmermann, U., Müller, H. & Weisse, T. (1996) Seasonal and spatial variability of planktonic heliozoa in Lake Constance. *Aquatic Microbial Ecology*, **11**, 21–9.

Zissler, D. (1995) Five scientists on excursion—a picture of Marine Biology, on Helgoland before 1892. *Helgoländer Meeresuntersuchungen*, **49**, 103–12.

Zwart, G., Crump, B.C., Kamst-van-Agterveld, J., *et al.* (2002) Typical freshwater bacteria: an analysis of available 16s rRNA gene sequences from plankton of lakes and rivers. *Aquatic Microbial Ecology*, **28**, 141–55.

14 Zooplankton

Z. MACIEJ GLIWICZ

14.1 INTRODUCTION

The nature of zooplankton is remarkably similar in freshwater and marine habitats. In lakes, much as in the sea, the zooplankton is composed of tiny animals suspended in the mass of water that are not strong enough to oppose water currents, and therefore are compelled to rely for food on even tinier plants, which they must sieve out from the viscous liquid medium. The unique nature of zooplankton stems from the peculiarities of the habitat, which force animals to lead a nomadic life. Short life-spans, a parthenogenetic way of reproduction, small body size, lucidity of body structures, the type of motion and the distinctive means of food collection all arise from the simplicity and instability of the physical structures of the water column, as much as from the nature of food resources, the phytoplankton, bacteria and detritus dispersed in the lake water as a dilute food suspension.

From the time it hatches from an egg, until the time it produces a brood of eggs itself, a planktonic animal is lost in an endless three-dimensional space, without any possibility of learning the topography of its surroundings, except which is the way up, towards the abundant food resources of the euphotic layer and the high temperature of the epilimnion, and which is the way down, towards the safe depths of the dark hypolimnion, where the danger of falling prey to a visually orientated predator is much reduced or absent. Within this space, however, there are vertical gradients of light, temperature, water density and viscosity, food particle concentration, dissolved oxygen, and concentra-tion of different ions and assorted inorganic and organic compounds, many of which play important roles in providing information on the current food situation, and on the possible risk to predation, all yielding knowledge on where is the top and where is the bottom, and whether it is a time when it would pay to ascend or descend.

There are, however, no permanent structures that would allow for horizontal orientation. There are no places to remember and recall as more dangerous or less, or more profitable or less, in terms of food collection, no surfaces on which information for allies or contenders could be left, no places to remember for a rendezvous with a partner, and no locations where a piece of surplus food could be left over for later consumption. Finally, there are no shelters in which to hide, and no specific sites where a planktonic animal could find protection at times when it is neither hungry nor compelled to look for food or seek a partner. There is no chance of a hole, a burrow or a nest, at least for a lair, which an animal in a terrestrial habitat would possess. There is no prospect of mining into stem tissue, or for hiding in the cover provided by lush periphytic vegetation which a crustacean or rotifer would find as a littoral dweller.

Protection is not even available at those times when a planktonic animal becomes hopelessly vulnerable. In crustaceans, this happens during sexual intercourse, or when moulting from instar to instar, with the locomotory system remaining entangled within the multiple tubes and pouches of the chitinous exoskeleton of the old, cast moult. Vulnerability may also increase after mating is over, or at the time of asexual reproduction, when a female carries a clutch of eggs in an egg-sack or in the brood cavity, the yolk-loaded eggs making her more conspicuous to visually orientated predators owing to egg opaqueness, and more attractive to an optimally foraging predator such as a planktivorous fish. The possibility of danger is always there, there is no time for leisure, and no time for a rest, except in a resting stage or a resting egg, when an animal would decide to rest by entering a diapause for a time, or to produce oviparous (diapausing) eggs, thus sending its siblings to such a 'time refuge' in the bottom sediments.

Diapause may be an efficient way to evade increased predation risk as well as to elude times of harsh climate or low food levels. Low food availability and low temperature often come together, and many zooplankton species diapause in winter. Low food levels may arise at any time in the season, however, either as a result of reduced photosynthetic activity in the phytoplankton owing to a sudden nutrient deficiency, or as a consequence of overgrazing owing to the high density of zooplankton and hence low densities of the edible fraction of the phytoplankton. The latter can happen in the absence of effective planktivores (for instance in fishless lakes), or early in the season when most planktivorous fish are at spawning grounds and the new year class is not yet ready to prey on zooplankton.

Despite small body sizes, planktonic animals are capable of withstanding periods of starvation, many species being able to grow at quite low food availabilities. This is enabled by remarkably effective filter-feeding systems, which are used to concentrate dispersed food resources: small algae, bacteria and particles of organic material suspended in the lake water. In the majority of planktonic animals, the system of food collection is coupled with that of gas exchange, and the machinery of motion, so that the same energy investment is allocated to all three functions, each needing to work continuously, without pause, keeping an animal in constant motion, pumping water with a fresh oxygen supply, and at the same time forcing that water through the system of filters which sieve out food particles, or into rotary currents from which food particles are centrifuged.

14.2   THE TAXONOMY AND DIVERSITY OF ZOOPLANKTON

In lakes, as in the sea, the overwhelming majority of zooplankton taxa are cosmopolitan and can be found from the Arctic to Antarctica, or from small lakes in the Scandinavian tundra to Lake Malawi in southeast Africa, and Lake Washington on the west coast of North America. The difference between lakes and the sea is that the freshwater zooplankton is phylogenetically much less diverse than the marine, and that unlike calanoid copepods (the most important group of marine zooplankton), which depend solely on sexual reproduction, its most important taxa (cladocerans and rotifers) are capable of asexual reproduction by parthenogenesis.

The most common, and most abundant truly planktonic species of freshwater zooplankton are free-living Protista, mainly holotrich and spirotrich ciliates (most common genera: *Dileptus, Bursaria, Strombidium* of 0.02–0.20 mm cell length), Rotifera, exclusively Monogononta (most common genera: *Keratella, Polyarthra, Trichocerca, Asplanchna* of 0.06–0.40 mm body length), and the entomostracan Crustacea, mainly subclass Branchiopoda and order Copepoda (subclass Maxillopoda). Branchiopoda are represented by two distinctly different groups, each important in a different way. The suborder Cladocera (order Phyllopoda) are the chief herbivores in the majority of offshore habitats, typically with several species coexisting in the same habitat (most common genera: *Daphnia, Bosmina, Diaphanosoma, Chydorus* of 0.40–3.00 mm adult body length). The order Anostraca is restricted to small temporary water bodies (fairy shrimps), saline lakes (brine shrimps) and other fish-free habitats of high arctic and high mountains, where one of the species often monopolises resources as the sole herbivore (see section 14.9; most common genera: *Branchinecta, Chirocephalus, Artemia* of 0.50–9.00 mm adult body length). Copepoda are also represented by two different groups (suborders): Calanoida with most species being typical herbivores throughout their whole life span (most common genera: *Eudiaptomus, Eurytemora, Boeckella* of 0.80–3.00 mm adult body length), and Cyclopoida, with most species being predaceous as adults (most common genera: *Cyclops*, *Eucyclops*, *Mesocyclops* of 0.50–3.00 mm adult body length).

The representatives of two more groups of entomostracan Crustacea can be found sporadically in plankton offshore, when brought by currents along with resuspended bottom sediments. These are the Harpacticoida (yet another suborder of Copepoda) and the Ostracoda (a separate subclass). Representatives of several orders of malacostracan crustacea are also sometimes recorded in plankton; most of them are typically brackish-water species, with Mysidacea being the only order containing typically lacustrine species such as *Mysis relicta*, which may be an important constituent of zooplankton in some particular locations.

Many different species of other taxonomic groups can also be found sporadically in freshwater zooplankton. Their absence from plankton samples collected from most offshore habitats is either because they are simply rare, even though truly planktonic (euplanktonic, i.e. completing their whole life-cycles as components of the zooplankton), or because they only rarely enter the water column, despite being abundant in one of the neighbouring habitats, in sediments or among littoral vegetation. The best-known genera are among the coelenterates (e.g. the limnomeduse *Craspedacusta*), flatworm (e.g. the turbellarian *Mesostoma*), gastrotrichs and lammellibranchs (larval *Dreissena*), water mites (e.g. *Piona*) and different orders of aquatic insects, some living in aquatic habitat all their lives (e.g. *Trichocorixa*) and some only in the larval stages (*Chaoborus*). The quantitative significance of these animals is generally ambiguous, but there are several exceptions, with phantom midge *Chaoborus* larvae being the best known.

Like other aquatic dipteran families, the Chaoboridae have a long-lasting larval stage. The larvae grow on aquatic resources for nearly a year, and emerge after a short pupation period to find partners, mate and reproduce. Eggs are deposited at the water surface and freshly hatched larvae begin to feed on the aquatic resources again. They spend most of the time in the sediments, but enter the water column every evening in order to feed on planktonic ciliates, rotifers and crustaceans during the night. Phantom midge larvae include the best-known *Chaoborus flavicans* and *Chaoborus punctipennis*, which are both very common and abundant enough to be regarded as important interplanktonic predators.

Numerous monographs on the biology and ecology of particular zooplankton groups have been produced, as well as many guides to the identification of species within each of the major taxa mentioned above. Many of these were edited within such series as Elster and Ohle's 'Die Binnengewässer' (e.g. Kiefer & Fryer 1978). The most recent editorial venture coordinated by H.J.F. Dumont is a series of 'Guides to the Identification of the Microinvertebrates of the Continental Waters of the World', which contains both textbooks on general biology, ecology and systematics (e.g. Nogrady *et al.* 1993; Dussart & Defaye 1995) and practical guides to the identification of species from different families of rotifer, copepod and cladoceran (e.g., respectively, Reddy 1994; Segers 1995; Smirnov 1992). There are also valuable proceedings from international symposia, many of them in one of the two important series, 'Advances in Limnology', and 'Developments in Hydrobiology', and each devoted to one of the major zooplankton taxa such as rotifers (Ejsmont-Karabin & Pontin 1995) or cladocerans (Larsson & Weider 1995), or to one of the general questions of zooplankton ecology such as diel vertical migrations (Ringelberg 1993) or the diapause (Alekseev & Fryer 1996). However, there is no book, so far, which can even match, let alone supersede, Hutchinson's 'Introduction to Lake Biology and the Limnoplankton' (Hutchinson 1967), in which the basic knowledge on the distribution and ecology of each of the major groups of freshwater zooplankton was carefully compiled. This remarkable book was the most competent and valuable aid in zooplankton studies for several decades. It is still useful, especially when bolstered by several more recent and important multi-author books on zooplankton (Kerfoot 1980; Lampert 1985; Peters & De Bernardi 1987) and on freshwater plankton in general (Sommer 1989).

The typical composition of a lake zooplankton community from an offshore station would be as follows: different stages of cyclopoid and calanoid copepods from such genera as *Thermocyclops* and *Eudiaptomus* (nauplii, copepodites, adult males and females, some carrying egg pouches), various instars of parthenogenetically reproducing cladoceran females representing several species of such genera as *Daphnia*, *Bosmina*, *Diaphanosoma* and *Chydorus* (including neonates freshly hatched from eggs in brood cavities), several species of rotifers, most likely of such genera as *Keratella* or *Kellicottia* (some of them such as *Conochilus* or *Synchaeta* are soft-bodied, and impossible to identify in fixed samples), and a few species of ciliates of such genera as *Strombidium* or *Dileptus*, most impossible to distinguish from contracted rotifers and many epibiontic (e.g. *Vorticella*, which is often attached to the exterior surface of a cladoceran's carapace). This is what can be seen when a column of lake water is sieved by a plankton tow-net made of a dense silk or nylon gauze of 50–60 μm mesh size, and the converged residue of a vertical or horizontal haul flushed to a vial, fixed with a 2–4% formaldehyde solution (with sugar added in order to prevent cladoceran eggs from falling out from brood cavities), and a subsample transferred into a dish and inspected under a dissecting microscope (Fig. 14.1). Such a sample would only allow for depiction of the zooplankton community when supported by:

1 a fast examination of live zooplankters in an unfixed sample to assess the quantitative importance of soft-bodied rotifers and free-living protozoans;
2 an assessment of the importance of organisms small or soft enough to pass the gauze of the plankton net using a sample of unfiltered water fixed with Lugol solution and viewed under a high magnification inverted microscope after concentrating all plankton organisms and seston particles by sedimentation on the bottom of an Utermohl chamber.

If taken from a eutrophic (more fertile) lake, such a sample would also contain some 'net phytoplankton': algae and cyanobacteria, larger single-cell hard-covered desmids or peridinians, colonial greens and cyanobacteria in gelatinous sheaths, colonial diatoms, and long trichome of filamentous cyanobacteria (Fig. 14.1, right). Unlike that from less fertile lakes (Fig. 14.2, left), the zooplankton community of eutrophic lakes would be composed of small-bodied species, with a greater proportion of rotifers and a smaller fraction of large-bodied cladocerans (Fig. 14.2, right). The

Fig. 14.1   Two examples of plankton sieved on 60 μm mesh net, from a mesotrophic lake (left) and a eutrophic lake (right). In each the content of 0.5 L of lake water is shown as it can be seen under a dissecting microscope.

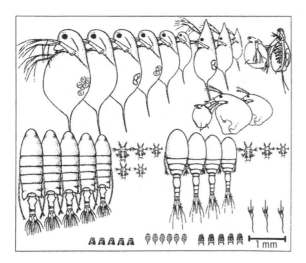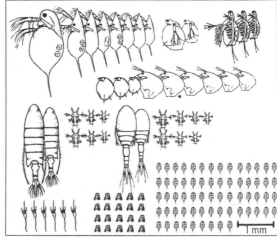

Fig. 14.2 Two examples of plankton from Fig. 14.1, from a mesotrophic lake (left) and a eutrophic lake (right), with algae excluded, and animals arranged according to taxa and body sizes. Cladocerans (*Daphnia hyalina, Daphnia cucullata, Ceriodaphnia reticulata, Diaphanosoma brachyurum, Chydorus sphaericus, Bosmina coregoni*) are at the top, adult and naupliar copepods (*Eudiaptomus graciloides, Thermocyclops leuckarti*) in the middle, and rotifers (*Kellicottia longispina, Polyarthra vulgaris, Keratella cochlearis, Keratella quadrata*) at the bottom.

complex reasons why the zooplankton of more- or less-eutrophic lakes differ are discussed in section 14.9.

Interestingly, a similar contrast is revealed when the spring and the late-summer zooplankton is compared, from the same eutrophic lake. A stronger contribution of large-bodied species is more typical for the spring-time, whereas greater proportions of small-bodied cladocerans and rotifers are characteristic of midsummer. Spring is frequently also a time of low algal densities (Fig. 14.1, left) whereas high summer is often a period of great abundance of algae, mostly large colonial forms, frequently in a 'summer bloom' of a single algal or cyanobacterial species (Fig. 14.1, right; see also Chapter 10).

In neither phytoplankton nor zooplankton is it very common for a single species to monopolise resources (exceptions are discussed in section 14.9). On the contrary, despite the limited number of freshwater taxa, the diversity of lake zooplankton is often astounding. Many rotifer, calanoid copepod and cladoceran species, which seek the same resources, coexist with each other in the

same habitat, contrary to what would be expected from Hardin's (1960) principle of competitive exclusion, also known as Gause's 'one niche–one species' hypothesis. This was first pondered by Hutchinson (1961) as 'The Paradox of the Plankton' (see also Chapter 10) with regard to phytoplankton communities, but it is also considered to be correct in respect of lake zooplankton as a whole community, as well as to its particular constituent taxa. High diversity is often revealed within single genera. In many lakes several species of a single rotifer or cladoceran genus coexist in all seasons, the best known examples being those of *Keratella* (*K. cochlearis* and *K. quadrata*) or *Daphnia* (*D. cucullata, D. cristata, D. galeata* and *D. hyalina*).

14.3 HERBIVORY AND PREDATION: FOOD NICHES AMONG PLANKTONIC ANIMALS

Although the theory from Pütter (1909) on the importance of dissolved organic matter in zooplankton nutrition has never been entirely rejected, and some evidence shows that dissolved compounds can be used as a source of carbon by protozoans (see Chapter 13), it is commonly accepted that the majority of zooplankton taxa are not able to live on dissolved organic matter as their only source of nutrition, needing also to rely on particulate resources. The most abundant source of food in the water column comprises small seston particles (**nanoseston**), made up of unicellular algae, bacteria and detritus suspended in the water, and scattered more or less homogeneously through the pelagic mixed layer. The only way to live on these resources is to develop an ability to concentrate them by sieving lake water through dense filters, or by generating rotary currents from which particles can be centrifuged on to oral surfaces such as the peristome of ciliates or the corona of rotifers. The mode of feeding by concentrating small seston particles does not allow for much selection other than that based on particle size. This is most arduous in typically filter-feeding cladocerans such as different *Daphnia* species. Food choice based on taste is troublesome in these animals when the particles retained on filters are passed to the mouth area as a densely packed mixture of edible and inedible pieces, and when rejection means the loss of desired particles, together with those which they dislike.

The alternative source of food in the water column comprises relatively larger but much scarcer objects, other animals (especially small protozoans, rotifers and crustacean juveniles) or large algal cells and colonies (the 'net phytoplankton'). These require an entirely different mode of feeding, where each food item is dealt with individually from the first step of detecting its presence, through its pursuit and seizure, to its ingestion. Owing to much possible variation in the way of resolving each of the above problems, this raptorial mode of feeding differs from group to group, and even from species to species. Feeding on individually seized algae or animals can be much more selective than filter-feeding, allowing for the tasting of each food item before it is eventually ingested (or rejected).

Animals which use this raptorial mode of feeding may either be typical predators, such as the rotifer *Asplanchna priodonta* and the cladoceran *Leptodora kindti*, or typical herbivores, such as the rotifer *Synchaeta pectinata* and the cyclopoid copepod *Eucyclops macruroides*. Many raptorial feeders are herbivores when juvenile, and predatory carnivores when adults. This is the case for the majority of cyclopoid copepods, with some harvesting algae on the juvenile naupliar stages, then frequently feeding on cladoceran eggs in brood cavities when becoming copepodites, before shifting to their typical prey, rotifers and small-bodied cladocerans, in the late copepodite or adult stages.

14.3.1   *Filter-feeding in cladocerans and copepods*

The mechanism of filter-feeding is most thoroughly studied, and best understood, in *Daphnia*, with the morphology and function of the filtering appendages having been described by Storch (1924), and broad knowledge on *Daphnia* feeding summarised by Fryer (1987) and Lampert (1987). The mechanism and quantitative aspects of filter-feeding in *Daphnia* are best-known in large-bodied species of the genus such as *D. magna*, *D. pulex* or *D. pulicaria*, which are capable of clearing large volumes of water containing seston particles over a wide range of sizes (1–50 μm diameter). Rates may be as much as $4\,mL\,h^{-1}$ per animal at 20°C in *D. magna* late instars of 3 mm body length (McMahon & Rigler 1965), making these the most efficient herbivores at controlling densities of algal populations (see section 14.4).

The filtering machinery of *Daphnia* is regarded as one of the most conclusive (definite) systems of food collection, and has evolved as a proficient apparatus for effective sieving of food particles from the water medium without loss of time for tasting and selecting those which are more nutritive. Most of the immense knowledge available today on *Daphnia* feeding, and *Daphnia* grazing effects on the phytoplankton, has been acquired by bearing in mind this view. It has, however, frequently been contested by those who believe that the

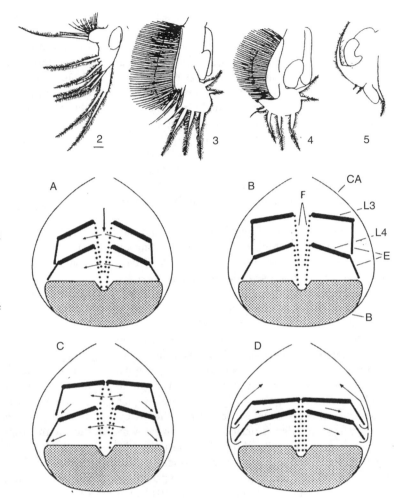

Fig. 14.3 Filter-feeding in *Daphnia*: thoracic limbs (2nd, 3rd, 4th and 5th), of which the 3rd and 4th carry filter screens (top), and the sequence of the filtration process shown in simplified oblique cross-section at its four different phases A, B, C and D (bottom). Arrows indicate the flow of water, which is first sucked into the filter chamber as filtering limbs move forward and sideways (A–B), then squeezed out into the space between the limbs (C–D). Particles are retained on the inner side of the screens and pushed down into the food groove by the gradual closure of the screens. F, filter screens; CA, carapace valves; L3 and L4, 3rd and 4th limbs; E, 3rd and 4th limb exopodites; B, animal body. (After Storch 1924; from Lampert 1987; reproduced with permission of the Istituto Italiano di Idrobiologia.)

Daphnia feeding-mechanism is not restricted to a mechanical suction-and-pressure pump, pressing lake water across the setular meshes of filtering screens, and that some attachment and mucus may be necessary for the ingestion of ultrafine particles such as bacteria (Gerritsen *et al.* 1988; Hartmann & Kunkel 1991).

The most important parts of the filtering machinery of *Daphnia* are the five pairs of thoracic limbs, of which the third and fourth each carry large-area filter screens on which food particles are retained when the water with the food suspension is being pumped across them (Fig. 14.3). The filter-

ing screens separate the medial space of the filter chamber from that between the limbs and the ventral body wall. Water is pumped first by the force of suction as filtering limbs move forward and sideways (Fig. 14.3A & B), and then by the pressure, as water is squeezed out into the space between the limbs (Fig. 14.3C & D). Particles retained on the inner side of the screens are gently pushed down into the food groove by gradual closure of the screens, and pushed forward by the tips of the filters' setae towards the mandibles, which push the concentrated food mass into the mouth. When the quantity of food collected in the groove becomes too large, both it, and the inner sides of the filtering chamber, may be cleaned by postabdominal claws. Rejection movements of the postabdomen combing the filter chamber become quite frequent when high densities of unsuitable particles such as cotton wool, or long cyanobacterial filaments, are suspended in *Daphnia*'s feeding medium and carried into the filtering chamber by feeding currents (Burns 1968; Gliwicz & Siedlar 1980).

A similar mechanism of food collection is utilised by cladocerans of other genera. Morphological details vary from species to species (e.g. all five pairs of thoracic appendages can be equipped with filter screens, as in *Sida cristalina*), and they are often more complex, especially in the littoral-dwelling species which specialise in feeding on alternative sources of nutrition, such as detrital particles stirred from sediments, periphytic algae or large algal colonies. The diversity of morphology and function is most astounding in Chydoridae and Macrothricidae (Fryer 1968, 1974).

The system is also rather more complex, and less understood, in those truly planktonic cladocerans which are capable of supplementing typical filter-feeding with the seizing of larger individual food items, for instance the Bosminidae (e.g. *Bosmina coregoni*: De Mott 1982). The filter-feeding system also functions in a similar way in other groups of Brachiopoda, including the quantitatively most important Anostraca (such as brine shrimps of the genus *Artemia*, known for their very high clearance rates of up to 10 mL per animal per hour in grown adults of *A. salina* (Reeve 1963)).

A more sophisticated mechanism of food collection is used by Calanoid copepods. Its morphology, function and selectivity have been described for marine *Calanus* species as well as for freshwater *Eudiaptomus* and *Eurytemora* species (see reviews by Koehl (1984) and Price (1988) respectively), as the result of two semi-independent mechanisms. One, involving a basket made of maxillules and maxillae comparable to the filter chamber of *Daphnia*, is capable of the passive sieving of fine particles which are retained on the inner sides of the screens. The other, dependent on maxillipeds, chemoreceptors and mechanoreceptors, is capable of capturing individual particles. The first retains small particles, the second captures larger particles, with the selected size being changed in accordance with the animal's choice and in relation to both the nutritional quality of particles and their relative abundance. Neither of the two systems can be as effective as that of *Daphnia*, so the overall maximum feeding rate can not be as high in calanoid copepods as in *Daphnia* of comparable body size (Lampert & Muck 1985).

14.3.2 Sedimentation in rotifers and ciliates

Since the pioneering observations by Naumann, Beauchamp and Pourriot, all summarised by Hutchinson (1967), much new information on rotifer feeding has been collected and generalised (Gilbert & Bogdan 1984; Nogrady *et al.* 1993). The most common and abundant planktonic rotifer species are typical suspension feeders, which collect food particles by settling them, or 'centrifuging' them into the buccal area around the mouth from the feeding eddies produced by the motion of coronal cilia. These feeding eddies also help in gas exchange, and are at the same time the swimming currents for the rotifer, so three different functions are fulfilled by the constant motion of the coronal cilia: locomotion, feeding and gas exchange.

Food items settling in the buccal area can thus be ingested more or less selectively. The first selection may take place before particles are driven to the mouth, and further into the mastax, by the movement of buccal cilia (Fig. 14.4). If not retained and rejected by the pseudotrochal cirri, which may screen the mouth opening, they may still be submitted to the next step of selection in the mastax cavity, and rejected by moving them back. This mechanism allows some genera (e.g. *Keratella* and *Kellicottia*) to select for particles of bacterial size or slightly larger, and others (e.g. *Polyarthra* and *Synchaeta*) to exert a preference for even larger particles of *Cryptomonas* size.

The process of food collection is quite similar in ciliates, even though the fate of food items which have been settled in the peristome by the effect of gravity in the cilia-generated eddy is entirely different. The feeding current, and manner in which a food particle is settled, look nearly the same in the sessile ciliate *Vorticella* as in the sessile rotifer *Brachionus rubens*, both often being attached side by side to a *Daphnia* carapace. Also probably quite similar is the process of food collection in the planktonic larvae of other invertebrate taxa, for instance in the veligers of such bivalves as *Dreissena polymorpha* which at times are very abundant in the water column (see Chapter 12).

14.3.3 *Capturing individual objects*

The raptorial type of feeding can be found in the majority of freshwater zooplankton taxa, but only rarely becomes as important as suspension feeding. Small food particles are simply much more abundant in the water column, and suspension feeding gives higher returns for energy invested in food collection than foraging for larger but less numerous individual food objects. On the other hand, capture of individual objects can be very selective, allowing for higher efficiency in food assimilation. In terms of animal behaviour, as well as from the

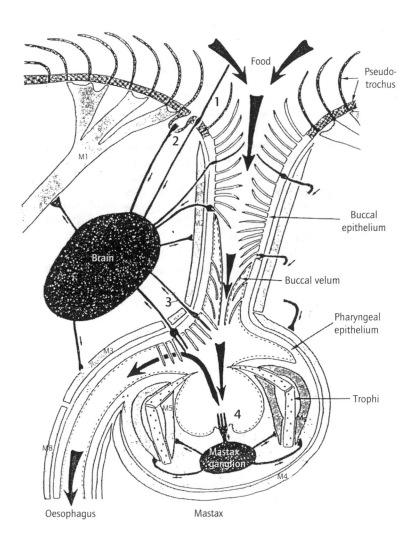

Fig. 14.4 Food collection and ingestion in rotifers: an example of the mouth and mastax interconnected by buccal area in Collothecidae, with anterior mechanoreceptors (1), chemoreceptors (2) and sensory mastax receptors (3 and 4) allowing for determination of which food particles should be directed to the oesophagus. (After Clement *et al.* 1983; from Nogrady *et al.* 1993; reproduced with permission from Backhuys Publishers.)

point of view of optimal foraging, this way of feeding is certainly much closer to typical predation
than suspension feeding. Unlike typical suspension feeding, such as that of *Daphnia*, the way of
feeding by capturing individual objects is also
closer to the principles of the Eltonian pyramids of
numbers where individual body size at the higher
trophic level would be expected to be only an order
of magnitude greater than at the level below.

Diverse modes of raptorial feeding are used by
planktonic herbivores, predatory carnivores and
omnivores. Some mechanisms utilised by ciliates
and rotifers do not differ drastically from suspension feeding, and sometimes it is difficult to distinguish one mode of feeding from the other. Even in
those ciliates and rotifers known as important
predators which ingest large food items whole, the
feeding currents produced by cilia are important in
seizing prey. This can be observed most easily in
such rotifers as *A. priodonta*, when either another
rotifer (e.g. *Keratella cochlearis* or *Brachionus
calyciflorus*), or a large algal cell or algal colony, a
peridinian (*Ceratium hirundinella*) or a diatom
(e.g. *Fragilaria*) is first hauled by the current to the
corona and then seized with the trophi of its mastax. There are exceptions, however, and examples
of specialisation for a narrow food niche can be
found in ciliates (Fiałkowska & Pajdak-Stós 1997),
as well as in rotifers (Fig. 14.5).

The difference between the two modes of feeding is more distinct in crustaceans, and most obvious in cladocerans, which use more specialised
systems of food collection than copepods. The
best-known, frequently abundant cladocerans
that specialise in capturing individual food
objects are *L. kindtii*, *Bythotrephes longimanus*
and *Polyphemus pediculus*, all three of which are
important invertebrate predators. Their thoracic
limbs are developed as mobile parts of a basket
which can be closed around a small prey item such
as a ciliate, a rotifer, a copepod nauplius or a juvenile cladoceran. A similar feature can be formed
of maxillae and maxillipedia in cyclopoid copepods, in many species known as typical predators,
as well as in those which apply a similar system to
seize individual algal cells and colonies. The same
approach is also used by those few calanoid cope-

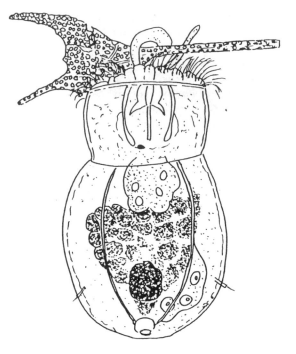

Fig. 14.5 An example of narrow food specialisation in
planktonic rotifers: *Ascomorpha ecaudis*
(Gastropodidae) holding *Ceratium hirundinella*
captured to suck the cytoplasm from its cell. (From
Nogrady *et al.* 1993; reproduced with permission from
Backhuys Publishers.)

pods which are predominantly carnivorous (e.g.
members of the genera *Heterocope* and *Epischura*).
Less is known of the mode of feeding of important
carnivores representing other crustacean taxa, for
instance *M. relicta*.

Capturing individual food objects requires an
ability to sense and locate prey. Since the introduction of sophisticated high-speed photography
equipment into *in vivo* microscopic observations
(see Kerfoot *et al.* 1980), it has become evident that
there are only a very few exceptions, such as the
large-eyed *P. pediculus*, in which vision is important in the process of locating and pursuing prey.
Instead, free-swimming invertebrate predators
usually detect prey by sensing the water turbulence produced by its locomotory or feeding motion. This has been most thoroughly described in

cyclopoid copepods, which are equipped with sensitive mechanoreceptors allowing them to distinguish prey species, and making possible a long chase in the quest for a rotifer or a cladoceran.

Similar mechanoreceptors are also used by phantom midge larvae such as *C. flavicans* and *C. punctipennis*, which are the most common, and whose feeding habits are best known. Both are typical 'ambush predators', or 'sit-and-wait strategists', which enter the water column every evening in order to feed on planktonic ciliates, rotifers and crustaceans during the night. They drift slowly through the water column without causing turbulence, and strike only when the prey swims within a few millimetres range of a successful attack (Pastorok 1980). They are also known for a clear ontogenetic shift in their diet, the upper size of successfully captured and ingested prey increasing from the first to the last larval instars (Swift 1992).

14.3.4 Herbivores and predators

The overwhelming majority of planktonic animals should be regarded as herbivores, even though none should be expected to live on a pure plant diet. Even in the least-selective suspension feeders, filter-feeding crustaceans as well as sedimentatory rotifers, algae would rarely be the sole food, unless the animal is deliberately transferred to a laboratory food suspension made of a clonal culture of *Chlorella vulgaris* or *Scenedesmus acutus* or another single alga species. The natural food suspension in the field would always contain some free bacteria, fine detrital particles and some mineral grains with organic compounds bound to their surfaces. Often the proportion of live algae might be minor. Such a natural food suspension in a lake would also probably contain some small heterotrophic organisms, flagellates, ciliates and soft-bodied rotifers.

On the other hand, even the most resolute herbivores with a raptorial mode of feeding would not always confine themselves to algal food, but will also take in live or dead heterotrophic organisms of comparable size whenever encountered. The only exceptions to this rule are extreme specialists adapted to a single food source, such as *Ascomorpha ecaudis* (Fig. 14.5). However, such specialists are neither common nor abundant.

Narrow food specialisation is more common among predators, but many would certainly be able to shift to algal or detrital food in the absence of prey. Many were probably classified as omnivorous when sampled at a time of low prey density, or observed in the laboratory with no appropriate animal food offered. The most determined carnivores, which usually play an important role as invertebrate predators in lakes offshore, are phantom midge larvae (e.g. *C. flavicans*, *C. punctipennis*, *C. trivittatus*), the adults of large copepods (mostly cyclopoid copepods, e.g. *Cyclops*, *Acanthocyclops*), predatory cladocerans (most frequently *L. kindtii*), and also, on a smaller scale, predatory rotifers (mainly *A. priodonta*). On unique occasions, other taxa may also become important as planktonic predators offshore: mysids (*M. relicta* — Threlkeld *et al.* 1980), water mites (*Piona limnetica* — Gliwicz & Biesiadka 1975), and corixids (*Trichocorixa verticalis* — Wurtsbaugh 1992). Unlike their effects on zooplankton behaviour and life histories, mortality exerted by invertebrate predators on zooplankton populations are only rarely comparable to the influence of predation by planktivorous fishes (see section 14.6).

14.4 DIRECT AND INDIRECT EFFECTS ON PHYTOPLANKTON: GRAZING AND NUTRIENT REGENERATION

14.4.1 Food selection

More important to phytoplankton than herbivores with a raptorial mode of feeding, or small rotifers sedimenting food particles in a much less efficient way, are suspension feeders, or large-bodied filter-feeding cladocerans. For this reason, the ecology of food selection has been studied more thoroughly in large-bodied cladocerans than in copepods or rotifers.

Even though some role may be played by the many different properties of a food particle (such as shape, taste, texture and the electrostatic charge of its surface, or even the presence of flagella), it is a particle's size which is the most important factor. This has been demonstrated experimentally by many authors (reviewed recently by Lampert 1987; De Mott 1989; Sterner 1989; Hartmann & Kunkel 1991), mainly using *Daphnia* as a model. Rather than the particle's longest linear dimension, the most reasonable expression of particle size would be particle volume, or the diameter of the spherical equivalent of real particle volume. This approach, however, would solve neither the problem of flat food items such as colonial, practically two-dimensional green algae (e.g. *Pediastrum duplex*) and pennate diatoms (e.g. *Tabellaria flocculosa*), nor that of elongated shapes, for instance the long rods of centric diatom colonies (e.g. *Aulacoseira granulata*), or the even longer filamentous trichome of cyanobacteria (e.g. *Planktothrix agardhii*).

As compared with more spherical algae, where food selectivity is concerned, filaments are particularly arduous (Fig. 14.6). They are not only difficult for a filter-feeding cladoceran to manipulate, but also disturb the process of food collection, by obstructing movements of thoracic limbs and rendering tight closure of the filtering chamber impossible. Together with other inconvenient particles (such as large, hard-covered peridinians, e.g. *C. hirundinella*), they are usually rejected from the median chamber, or from the food groove, by combined movements of the postabdominal claws.

High densities of such algae in the food suspension may provoke frequent combing of the filtering chamber, and the ensuing loss of most of the food already collected in the food groove, including those particles which are the most convenient and valuable source of nutrition. Large-bodied cladocerans, larger species and later instars are more vulnerable to interference from long filaments than small-bodied cladocerans (see review by Gliwicz 1990a). However, owing to their ability to divide long filaments into short fragments, and those which are more easy to manipulate, large-bodied cladocerans are also more proficient than small-bodied zooplankton at controlling the abundance of filamentous cyanobacteria, and thus preventing thick blooms in natural habitats (Dawidowicz 1990).

Compared with other suspension feeders, cladocerans, especially daphnids, are relatively unselective across quite a large range of food particle sizes. Depending on body size, this may be narrower or broader, but it is generally accepted that particles in the range 3–20 μm are sieved with similar efficiency by each instar in each filter-feeding cladoceran species. The lower size limit is set by the distance between the setules in the filter screens (Fig. 14.7), whilst the upper limit depends on the width of the gap between the ventral edges of the thoracic limbs in the phase when they are moving up and sideways, sucking food particles retained in the median chamber into the space between the limbs and the body's ventral surface (Fig. 14.3). Very large particles and more-elongated filaments can be prevented from entering the median chamber by narrowing the gap between the edges of the valves of the carapace. This may reduce the frequency of the postabdominal combing movements necessary to clean the filtering machinery of filaments entangling the limbs' filters, and large elements obstructing the closure of the median chamber.

The 'mesh size' of the filter screens (the open space between the setules of the thoracic limbs' setae; Fig. 14.7) is quite uniform, and should assign a sharp lower size limit for food particles. In fact, this limit is not as distinct as would result from measurements of the intersetular space (from 0.3 to 1.8 μm in ten different *Daphnia* species; Lampert 1987), because there is always some leakage between the tips of the seta of the filter screens and the bottom of the food groove, and many smaller particles may eventually be lost from the median chamber. This is reflected in the observation that the efficiency of retention of the smallest algal or bacterial cells within the size range of food particles retained (1–3 μm diameter) is often significantly lower than that for slightly larger cells of 5–10 μm. Nevertheless, the lower size limit of grazed particles is remarkably similar in most cladoceran species, regardless of whether it is generated by measurements of filter screen mesh size,

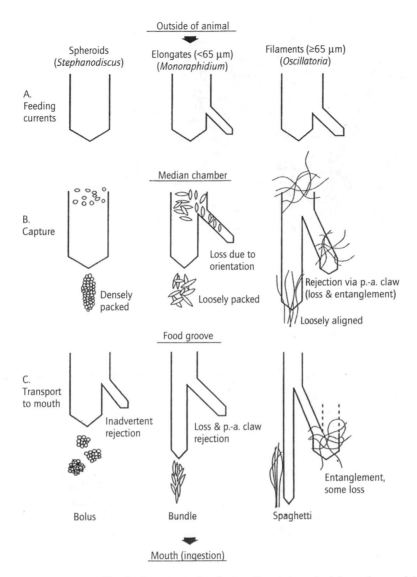

Fig. 14.6 Schematic representation of food selectivity in *Daphnia pulicaria* in each of three phases of the feeding process: (A) encounter and pursuit when particles are sucked into the filter chamber, (B) capture when particles are pushed into the food groove, and (C) ingestion when particles are pushed forward to the mouth. Vertical arrows indicate particles retained; side arrows indicate those lost or rejected; arrow thickness reflects relative proportions of food particle volumes; arrow length indicates handling time relative to spheroid particles; p.a., post-abdominal. (From Hartmann & Kunkel 1991, fig. 26; with kind permission of Kluwer Academic Publishers.)

or estimated from experimental tests of filtering rates for the same algal species. According to Geller & Müller (1981), who compared mesh sizes in neonates and adults of 11 cladoceran species, the mesh sizes in adults range from 0.2–0.4 μm in *Diaphanosoma brachyurum, Chydorus sphaericus* and small-bodied daphnids (*Ceriodaphnia quadrangula* and *Daphnia cucullata*) to 2.0–4.2 μm in

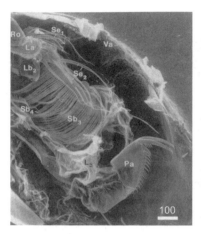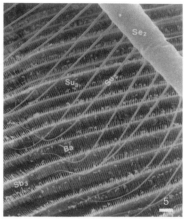

Fig. 14.7 Filter feeding machinery of *Daphnia* in scanning electron microscope photographs: (left) interior of filter chamber after right-half of carapace valve and right limbs have been removed, (centre) filter screen of third thoracic limb at 15 times higher magnification, and (right) details of filter screen under magnification further increased. Different magnifications are indicated by horizontal bars showing 100, 5 and 4 μm, respectively. The two rows of setules projecting from each setum at different angles are typically hooked with each other at the tips. Left and centre in *Daphnia pulicaria* (from Hartmann & Kunkel 1991, figs 1 & 3; with kind permission of Kluwer Academic Publishers), right in *Daphnia magna* (after Brandelberger, from Lampert 1987).

Holopedium gibberum and *Sida crystalina*, and increase ontogenetically in each species, e.g. from 0.4 μm in neonates to 2 μm in fully grown adults in *Daphnia hyalina* from Lake Constance (the Bodensee).

The upper size limit of grazed particles is never as distinct as the lower limit, and may change in seconds, depending on the gap between the edges of the carapace valves and the distance between the edges of the left and right thoracic limbs. This value determines the largest size to be allowed into the median chamber, and can also change in seconds. The upper size limit may also depend on the next steps in food collection, including the passing of the food bolus or food bundle to the mouth by the mandibles, during which some particles may be selectively rejected (Fig. 14.6). Nevertheless, the upper size limit is distinctly different in various species, and depends on body size, with larger species being able to collect and ingest larger food items than smaller ones.

This has been shown for feeding animals of various species on the basis of spherical plastic beads allowing for determination of maximum sizes of particles ingested by cladocerans of different body sizes (Burns 1968), and characterisation of the efficiency of particle clearance across the size spectrum (Gliwicz 1969). In small cladocerans such as *D. brachyurum*, *C. sphaericus*, *C. quadrangula* and *D. cucullata*, the upper size would be around 10 μm diameter of the spherical volume equivalent, and between 20 and 50 μm for larger *Daphnia* species. The complexity of food particle shape, texture and taste makes the problem of food size selection more complex, however, and grazing experiments in the field show that specific selectivities often depend on algal features other than size (for an example see Knisely & Geller 1986).

Lower and upper particle size limits are both more variable in calanoid copepods. They vary from species to species, as well as between early naupliar stages and grown adults. Owing to the alternate mode of raptorial feeding by individual handling of larger food particles, the limits are also more dependent on properties other than size. Although more is known of food collection in marine than freshwater calanoid copepods, there are freshwater species which have attracted considerable attention. The pattern of food selectivity has been best described in several species of the two major genera (*Diaptomus* and *Eudiaptomus*), with some species such as *Diaptomus sicilis* being considered quite consistent in their size preferences for particles of about 15 μm (Vanderploeg *et al.* 1984).

Other species have, however, been described as extremely variant filter feeders, which vary preferred particle size in response to a change in the particle spectrum. Examples of such variant selectivity have been described in *Diaptomus minutus* (Chow-Fraser 1986) and *Eudiaptomus japonicus* (Okamoto 1984). Such complex optimal foraging behaviour makes it difficult to foresee which phytoplankton taxa would or would not be affected by calanoid copepods.

In rotifers too, predictability is rather low in regard to which algae are or are not subject to grazing. Although rotifers are much smaller than crustaceans, they are often much more abundant, and possess higher metabolic rates per unit volume. Their overall clearance rate may thus be as high as that exhibited by crustacean herbivores in the same location. Their impact, however, may be highly selective depending on which species are most abundant.

The lower limit of particle size grazed cannot be found as easily in rotifers as in cladocerans. Neither the lower nor upper size limits are related to body size in rotifers, and nor are they predictable from taxonomy, with some species, even those as small as *K. cochlearis*, being known to feed effectively on bacteria, small algae and large flagellates (Bogdan & Gilbert 1987). Usually, it is simply assumed either that rotifers are unimportant, or that they graze mostly on small particles from 1 to 15 μm, with a preference for those larger than bacteria (especially when species from such genera as *Keratella*, *Kellicottia*, *Conochilus* are abundant) and larger algae when such taxa as *Polyarthra*, *Synchaeta* and *Trichocerca* occur in similar numbers.

14.4.2 *The role of taste*

There is already some evidence to show that herbivorous zooplankton can also remotely detect the quality of food particles by chemoreception (Larsson & Dodson 1993). However, the ability to discriminate between high- and low-quality food particles is probably more often based on taste, which can only be used after particles have been retained and passed into the mouth region. This may explain why the degree of taste discrimination varies among zooplankton taxa, and why it is more common among raptorial than filter-feeding species. DeMott (1986) used *Chlamydomonas*-flavoured polystyrene beads of the same size as natural *Chlamydomonas* cells (6 μm) to show:

1 that naupli and copepodites of cyclopoid and calanoid copepods were more selective in favour of flavoured beads than cladocerans and rotifers;

2 that some rotifers such as *Filinia terminalis* were more selective then others such as *B. calyciflorus*;

3 that some cladocerans (such as *Bosmina longirostris* and *B. coregoni*) were more selective than others regarded as true filter-feeders using the pump-and-filter system as the only means of food collection (e.g. *Daphnia magna* and *D. galeata*, *C. quadrangula*, *Simocephalus vetulus*, *C. sphaericus* and *Diaphanosoma birgei*). Even though they seem capable of discriminating between food properties, they only rarely reject food accumulated in the food groove, unless this is crammed or the filters are entangled by filaments.

Rejections are costly, because all food particles which have been accumulated in the groove, even those of the best nutritional quality, are lost when postabdominal claws comb the filtering machinery. This is probably why the gut content of a typical filter feeding cladoceran is very seldom packed with algae which can be easily identified. Most often, it consists of a mass of shapeless detrital debris mixed with small grains of mineral particles, which sometimes seem to be the exclusive food. Such mineral content of intestines is quite common in *Daphnia* or *Diaphanosoma* from fertile clay-loaded lakes and reservoirs, where phytoplankton is extremely scarce owing to high light extinction, and where a large fraction of the organic matter is bound to micron-scale montmorillonite or kaolinite particles (Arruda *et al.* 1983; Gliwicz 1986a).

The high cost of taste-based rejections is also a probable reason why cladocerans such as *Daphnia* do not even try to repulse toxic algae or cyanobacteria when they are mixed with other non-toxic food particles. On the contrary, these animals ingest, digest and assimilate them, albeit at reduced ingestion and assimilation rates, and if the proportion of toxic food is too high, they die. This takes place even though the animals are capable of discriminating toxic cells or colonies of such a cyanobacterium as *Microcystis aeruginosa* by taste: a finding evident from the reduced filtering and feeding rates which are noted as soon as toxic *Microcystis* is added to the food suspension, and the immediate cessation of the inhibition after this toxic cyanobacterium is replaced again by high quality food (Lampert 1982). The high cost of rejection is also evident from the finding that algae resistant to digestion are ingested by filter feeding zooplankton as readily as those which can be utilised as a valuable source of nutrition.

14.4.3   Resistance to digestion

Many green algae, and a number of cyanobacteria, are resistant to digestion by herbivorous zooplankton. Green algal genera such as *Sphaerocystis* or *Elakatothrix*, and cyanobacteria such as *Chroococcus*, are protected from the digestive systems of filter-feeding herbivores by gelatinous sheaths, and are not only capable of viable passage through a *Daphnia* intestine, but are also able to use the opportunity of being amongst high nutrient concentrations during gut passage, to take up limiting nutrients, store them and to use them effectively after being released from the intestines (Porter 1976). Although resistance to digestion is rarely absolute, and some gelatinous cells or colonies may be killed during gut passage, the profit of gaining extra nutrients must be much higher than the loss, because populations of gelatinous-sheathed species often expand when the density of filter-feeding zooplankton is experimentally increased (Vanni 1987; Sterner 1989).

14.4.4   Filtering rate and food particle size

The **filtering rate** (filtration rate, clearance rate,

grazing rate), or the volume of water swept clear of food particles by an individual in a time unit (e.g. mL individual^{-1} h^{-1}), reflects the ability of a single animal to obtain food resources. A high individual filtering rate would point to competitive superiority of an animal over another with a low filtering rate. On the other hand, filtering rate also demonstrates the potential of herbivores to control the densities of particles suspended in their medium, algae and bacteria included. Especially when integrated for all the filter-feeding animals in a given water volume, such an overall community filtering rate may be used as an accurate measure of the effect of zooplankton grazing on algal and bacterial populations.

The integrated filtering rate for an entire zooplankton community depends on the number of animals and their individual sizes, as individual filtering rates are higher for larger than for smaller animals: for the later instars of the same species as well as for larger bodied species. Such an integrated filtering rate can be expressed as the percentage of the volume of lake water swept clear per hour or per day. When zooplankton abundance and summer temperatures are both high, the daily community-filtering rate may exceed 200%, which shows that the whole volume of a lake or a lake's epilimnion is filtered more than twice a day.

However, filtering rate is not the same for all kinds of particles, or for all the algal or cyanobacterial species present in the lake water. There are particles of small size that are only rarely retained on filters, and filtering rates for these will be extremely low compared with those for particles only a little larger, which are retained with high efficiency. There are also particles too large to be taken by the feeding currents into the filtering basket or filtering chamber, and these will only rarely be retained, leaving their filtering rate also very low. Only for particles of the optimum size will the filtering rate be equal, so that a curve to show change in filtering rates with gradual increase in particle size will be bell-shaped, and skewed at greater size.

Its shape will also vary from species to species, and from instar to instar, but the curve which might show the overall filtering rate across the particle size spectrum for the entire zooplankton community would be of a similar shape, and would vary from lake to lake depending on the relative proportions of cladocerans, copepods and rotifers, and on the abundance of small- and large-bodied animals in the community (Fig. 14.8). In general, the latter possess higher maximum filtering rates and graze a broader size spectrum of particles than the former, suggesting that larger animals are superior utilisers of limiting resources, being able to sieve larger volumes of water of a broader range of food sizes and shapes than smaller animals. The competitive inferiority of smaller animals, smaller species and earlier instars of the same species result from many different factors, but one of the most important is that their inclusive food niches completely overlap the lower particle size fraction of those of larger animals (see section 14.5).

14.4.5 Filtering rates and ingestion rates at different food concentrations

The **ingestion rate** (feeding rate), or the amount of food ingested per animal per unit time (e.g., mL individual^{-1} h^{-1}, mg dry wt individual^{-1} h^{-1}, mg organic carbon individual^{-1} h^{-1}), will match perfectly the **filtering rate** (filtration rate, clearance rate, grazing rate) if all food particles swept from the food medium and retained on the filters are passed to the mouth and enter the gut. For example, if the filtering rate is 1 mL individual^{-1} h^{-1}, and the concentration of food particles is 1 mg organic carbon L^{-1}, the ingestion rate will be 0.01 mg organic carbon individual^{-1} h^{-1}. But this may occur only when the process of food collection is 100% effective and no food particles are lost or deliberately rejected. Otherwise, smaller numbers of

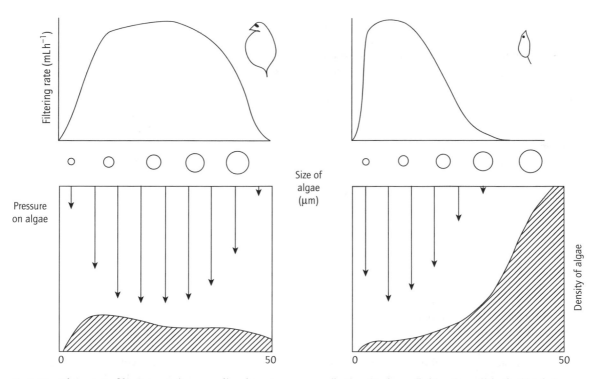

Fig. 14.8   Changes in filtering rate (top panel) and grazing pressure (bottom) in large- (left) and small-bodied (right) *Daphnia* along food particle size spectrum. Size distribution of phytoplankton (hatched area) in the presence of abundant large- or small-bodied zooplankton of oligotrophic or eutrophic lakes, respectively, is a result of the combination of the superiority of small algae in competition for resources, and the superiority of large algae in resistance to grazing. (After Dawidowicz & Gliwicz 1987; reproduced with permission of the Instytut Ekologii PAN.)

food particles will be involved in respect of the ingestion rate than the filtering rate.

When the efficiency of feeding is depressed by high food concentrations, or by high abundance of algae interfering with food collection, the difference is very great. The ingestion rate seems to be more relevant from the animal's point of view, whereas from the algal or bacterial viewpoint the filtering rate seems more important, because the majority of algal or bacterial cells or colonies retained on filters and eventually rejected are either killed or severely damaged by the process of filtration.

Divergence between the ingestion rate and the feeding rate will be smallest at low food availability, and when all the food particles are well within the size range most convenient to an animal (i.e. neither too close to the lower limit, at which a large proportion of particles are too small to be retained and eventually pass the filters, nor too close to the upper size limit, at which a proportion of particles are too large to be allowed into the filtering chamber or a filtering basket). This is often the case in laboratory studies when a food suspension is made of a highly-edible algal monoculture of a clone of a species occurring in single cells of proper size, texture and taste.

Such an ideal source of food will allow for accurate estimation of the functional responses of filtering and ingestion rates to food concentration. At low concentration, a filter-feeding animal eager to obtain as much food as possible will filter at maximum speed, so that the filtering rate will remain constant over a broad spectrum of food concentrations (Fig. 14.9a). A linear increase in food concentration will allow for a linear increase in the ingestion rate, with each millilitre of the medium swept clear of a linearly increasing number of food particles, e.g. *S. acutus* cells (Fig. 14.9b).

A constant increase in food concentration will eventually lead to saturation with food when an upper limit to ingestion is reached, and an animal is finally satiated. A continued increase in food concentration will not allow for any further rise in the ingestion rate, which will remain constant owing to a gradual decrease in the filtering rate. The food concentration at which this saturation is

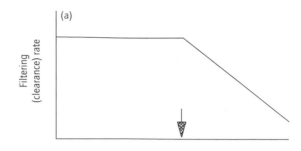

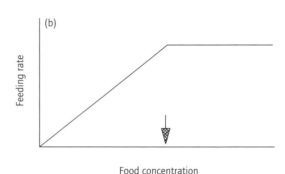

Fig. 14.9 Functional responses of filtering rate and feeding rate to changes in food concentration, with the incipient limiting concentration shown by an arrow. (After McMahon & Rigler 1963.)

reached has been named by McMahon & Rigler (1965) the **incipient limiting food concentration**. This rectilinear model of zooplankton feeding, identical to a Holling type 1 functional response, is most frequently used as it is the best fit for most data on copepod, cladoceran and rotifer feeding, the maximum filtering rate and the incipient limiting concentration being both species- and age-specific (for a review see Lampert 1987).

The decrease in the filtering rate above the incipient limiting concentration may result either from a decrease in the volume of water cleared or from an increase in the intensity of food rejection. In *Daphnia*, at food levels only a little higher than the incipient limiting level, this decrease is mostly due to a decline in the rate at which the filtering appendages beat. However, at higher food levels, it is mostly due to an increase in the rejection rate, when filtering appendages must keep up a minimum beat rate in order to allow for gas exchange.

The simple rectilinear model in Fig. 14.9 does not reflect the functional responses of filtering and ingestion rates to the change in food concentration at extremely low food availabilities. It is obvious that no filter-feeding animal will be able to pace its maximum filtering rate at food levels below a threshold food concentration (i.e. a minimum food level necessary to equal respiration with food assimilation). Below this level, the ingestion rate will be simply too low to allow assimilation at a rate high enough to compensate for the high respiration rate, much of it resulting from the great filtering effort at the maximum filtering rate. Therefore, at very low food levels, a shift to a more economic value for filtering rate should be expected. However, this is not easy to test, for the precision of filtering rate estimates becomes very low at low food levels.

14.4.6 Food digestion and the efficiency of food assimilation

Food assimilation efficiency in herbivorous zooplankton varies widely, depending on the composition and abundance of seston particles suspended in the lake water. Where there is a high proportion of mineral particles (e.g. in the silt-loaded waters of glacial lakes and arid-zone reservoirs), or a high relative abundance of poorly digestible organic particles (e.g. in humic lakes), or an abundance of detrital particles resuspended from sediments (e.g. in shallow lakes); assimilation efficiency may be as low as 1%, with the mass of food being constantly collected, ingested and passed ineffectively through intestines. Food assimilation will also be very limited at high relative abundances of algae protected from digestion by gelatinous sheaths or hard cell walls. Efficiency of food assimilation will be highest of all at low food abilities, when the food suspension is composed of highly digestible algae. Assimilation efficiencies differ from one algal species to another. For instance, in *Daphnia* they may range from 10 to 100% of food ingested for different green algae, or from 0.1 to 7% of *Daphnia*

body carbon per hour for a range of algae differing in size, shape and texture (Fig. 14.10).

14.4.7 Nutrient regeneration as an effect of grazing

The higher the overall community filtering rate, and the more efficient the food digestion and assimilation, the faster the nutrient release by the excretion of waste products from the zooplankton. Waste metabolites most important to planktonic algae are orthophosphate and ammonia, which are the main forms of phosphorus and nitrogen in the excretion products of zooplankton. These compounds are also present in faecal material, but, because they are densely packed into more or less tightly enveloped packages, which usually sink rapidly, taking nutrients down to the sediments, access to them is limited and short-lived. In consequence, of much greater importance are those portions of phosphorus and nitrogen which leach from live animals, or which are deliberately released as useless end-products which are highly soluble in water (for a review see Peters 1987). Some excretion is achieved by a variety of specialised glands which are specific to different taxa, but the major part of excreted soluble orthophosphate and ammonia is released through the body surface, especially through those regions which are also used in osmoregulation and gas exchange. In crustaceans, these are the epipodites of the thoracic appendages modified in a variety of ways, and in cladocerans such as *Daphnia* they are developed as branchial sacs.

A number of experiments have recorded high and persistent release of orthophosphate and ammonia by various zooplankton taxa (Peters 1987). A large number of experiments have also shown that an experimental increase in grazing pressure (by an increase in herbivore zooplankton density) will not only cause a decline in algal densities but will also lead to elevated photosynthetic activity among phytoplankton, which can only be attributed to the indirect effects of grazing by nutrient regeneration (Sterner 1989).

Part of the stimulation of algae into a higher photosynthetic rate may stem from the process by

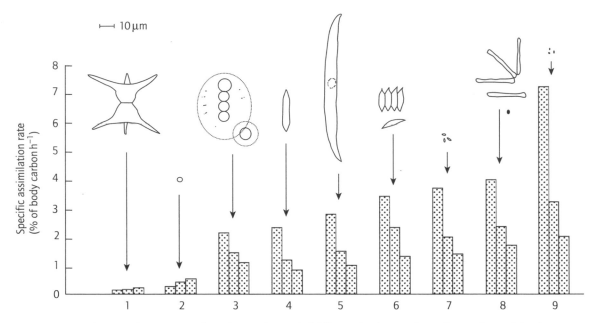

Fig. 14.10 Specific assimilation rates of *Daphnia pulicaria* fed different food types shown for *Daphnia* of the three different body sizes 1, 2 and 3 µm in length from left to right, respectively. Numbers identify food types: 1, *Staurastrum*; 2, *Microcystis*; 3, *Anabaena*; 4, *Nitzschia*; 5, *Closterium*; 6, *Scenedesmus*; 7, *Stichococcus*; 8, *Asterionella*; 9, *Synechococcus*. (After Lampert 1977; from Lampert 1987.)

which zooplankton secrete acid or alkaline phosphatases, the enzymes which may accelerate orthophosphate release from dissolved organic compounds. Although the phenomenon was even considered a kind of 'gardening' of algae by *Daphnia*, through support for nutrient production (Boavida & Heath 1984), it seems more likely merely to be a by-product of the process by which the enzymes are released from algal cells damaged during filtration, or partly digested in the course of gut passage. There is no evidence of the 'deliberate' secretion of 'exoenzymes', a phenomenon which would require a group-selection interpretation.

The rate of nutrient regeneration per unit body mass has been found to depend strongly on body size, with the increase from large to small-bodied animals being attributed to the rising ratio of body surface area to volume. This has been found in various rotifer and cladoceran taxa, and shown to

be the case for both phosphorus and nitrogen (Ejsmont-Karabin 1984; Peters 1987).

14.4.8 Coupling the direct and indirect effects of zooplankton feeding

Both of the two main effects of zooplankton feeding (namely top-down mortality effects due to grazing, and bottom-up effects of nutrient release) are highly selective. Each will disadvantage those algae which are highly edible and easy to digest, and favour those which are difficult to handle, too large to be ingested and too reluctant to be digested. These are also likely to be inferior competitors for nutrients, and thus to require higher nutrient concentrations. The combined effect of selective grazing and nutrient release will eventually cause shifts in the phytoplankton community, from a typical spring assemblage composed of highly edible species, to a typical midsummer community composed of larger, less edible, colonial, filamentous and hard-covered forms (see section 14.9 and Chapter 10).

Such a process of phytoplankton community transformation will be less likely in a habitat with a zooplankton composed of large-bodied herbivores such as *Daphnia pulicaria* or *D. hyalina* (Fig. 14.1, left), and will be more likely to occur, and to develop much more rapidly, in a habitat with zooplankton composed mostly of rotifers and small-bodied cladocerans such as *B. coregoni*, *C. sphaericus*, *C. quadrangula* or *D. cucullata* (Fig. 14.1, right). These small-bodied species will be much less effective at controlling large forms of phytoplankton, and also much more effective in fertilising the habitat through more intense nutrient release, than large-bodied species (Fig. 14.11).

14.5 FOOD LIMITATION, COMPETITION FOR LIMITING RESOURCES, AND VULNERABILITY TO STARVATION

The clearest evidence for strong food limitation in most field zooplankton populations is the number of eggs per clutch which adult females of most species carry in egg-sacs or brood chambers, prior to hatching. In the majority of field populations, clutch size is only a fraction of the maximum that can be observed in conspecific animals of equal size grown in conditions with unlimited food and similar temperatures in the laboratory. Although a reduced number of eggs per clutch may also result from predation (by selective removal of females with larger clutches, or removal of a fraction of a clutch from the brood cavity, see section 14.6), clutch size distribution over different body sizes will eventually give a reliable estimate of the degree of food limitation.

More accurate measures of food limitation in the field will be those based on body lipid reserves (Tessier & Goulden 1982) and on body weight or body carbon regressions on body length (Duncan 1985). Such measures can also be applied to juveniles, males and adult females during the time

which females withhold from egg production (see section 14.11), and they are quite precise. For instance, a *D. pulicaria* of 1.5 mm body length may weigh 30 μg, or 2 mg dry wt (15 μg, or 1 μg body carbon, respectively), depending on whether it has grown with high food availability, or has starved for several days. Even though they are more precise, there are two reasons why body lipid, or body carbon content, may perhaps also be affected by predation. First, predators may select for well-fed prey, which is more conspicuous and which gives higher returns for energy invested in foraging activity. Second, animals often eat less than the maximum possible amount out of fear of being exposed to predation risk (see sections 14.8 & 14.9).

14.5.1 *Threshold food concentration*

There is a critical minimum food concentration at which an animal is able to ingest food in a quantity just great enough to equal its metabolic losses with assimilation. When the food level is below this threshold concentration, an animal loses weight because its respiration rate is higher than that of assimilation. When the food level is above this threshold concentration, the assimilation rate may be higher, and an animal's weight can increase (Fig. 14.12). **Threshold food concentration**, defined as the concentration at which assimilation equals respiration or body growth equals zero, can be regarded as a sound measure of an individual's ability to use resources effectively. It was first estimated by Lampert (1977) for *Daphnia pulex*, and found to be dependent on animal body size, as well as on the type of food and the temperature. There is also a **critical minimum food concentration**, at which growth and reproduction compensate for losses in a population (zero net population growth rate in the absence of predation). This **population threshold** must be significantly higher than the **individual threshold**, since the food level sufficient for minimum body growth may not be adequate to allow maturation and egg production (Lampert & Schober 1980).

Each of the two critical food levels (the threshold for the individual at which assimilation equals respiration, and the threshold for a population at

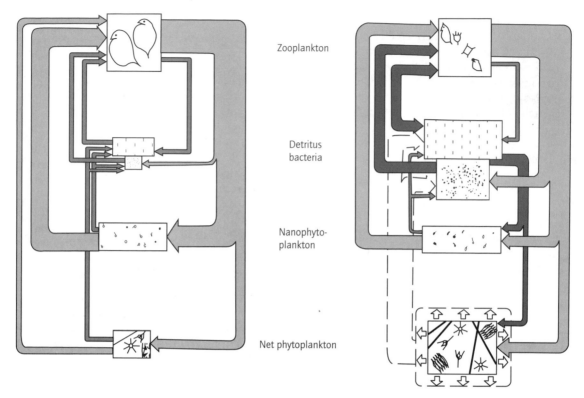

Zooplankton

Detritus
bacteria

Nanophyto-
plankton

Net phytoplankton

Fig. 14.11 Divergence in the ability of herbivore zooplankton to control phytoplankton standing crop in (left) less fertile and (right) more fertile lakes (from Dawidowicz & Gliwicz 1987). In less-fertile lakes (left), the low phytoplankton standing crop remains under the effective control of zooplankton composed of large-bodied filter-feeding cladocerans. Regenerated nutrients are incorporated in both fractions of phytoplankton in amounts proportional to their standing crops and rates of nutrient uptake, and then recycled by herbivore feeding activity (light-shaded arrows). Nutrient flow through detritus and bacteria (dark-shaded arrows) is insignificant. In more-fertile habitats (right), net phytoplankton is not effectively controlled by zooplankton composed of small-bodied herbivores, owing to their restricted food particle size spectra (Fig. 14.8). Regenerated nutrients flow to both net- and nanoplankton, but can only be recycled effectively from the grazeable nanoplanktonic fraction (light-shaded arrows). Thus, zooplankton act as a pump, transferring mobile nutrients from nanoplankton to net phytoplankton. This allows for a consistent increase in net phytoplankton standing crop, even though initial nutrient concentration is low. Nutrients accumulated in net phytoplankton can eventually be recycled with a delay through detritus and bacteria available to herbivores (unshaded arrows). The role of bacteria in nutrient recycling increases when bacteria become an important source of nutrition to zooplankton (dark-shaded arrows again) as the finest food particles (see Fig. 14.8 right). Nutrients incorporated in detritus can be utilised by phytoplankton (dark-shaded arrows) owing to the increased role of phosphatase excreted by algae and bacteria.

which reproduction compensates for mortality) could be used as an accurate determinant of competitive ability in filter-feeding zooplankton. Each of the two is set by physiological constraints, and may be expected to be 'species-specific'. Species with lower food threshold concentrations should be superior competitors for limited resources, compared with those with higher thresholds. When two filter-feeding herbivore species compete for a single food resource under steady-state

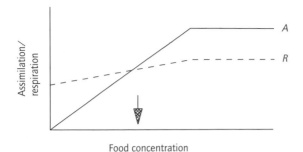

Fig. 14.12 Functional response of assimilation rate (A) and respiration rate (R) to a change in food concentration. The threshold food concentration at which $A = R$, resulting in zero body growth rate, is shown by an arrow (according to Lampert 1977).

conditions and in the absence of predation, the species with the lower threshold should always succeed, regardless of its maximum intrinsic reproductive potential. In an environment with high predation and large fluctuations in food levels, a high intrinsic rate of population increase and the ability to withstand long periods of starvation may also be important. Under steady-state and predation-free conditions, however, the species which can grow and reproduce at the lowest food concentration may maintain the resource level below that necessary for other species long enough to cause their exclusion.

Threshold levels are usually expressed in terms of the energetic value of food, as **food dry weight** or **food organic carbon concentration** (both dry weight and carbon being closely related to energy content). No account is taken of the nutrient content of food particles. As far as the maintenance metabolism of an animal is concerned, the principal value of food is energy. Food particles with low mineral content, for instance senescent algae with low nitrogen and phosphorus, are not inferior compared with those with a high mineral content, for instance phosphorous-rich algae with high growth rate. However, for animal growth and reproduction, energy alone does not appear to be an adequate measure in all circumstances, with food quality also needing to be taken into account (Sterner & Robinson 1994).

14.5.2 Body size and superiority in competition for limiting resources

There are three different reasons why large body size has been intuitively assumed to be superior in overcoming food limitation. First, larger animals, large-bodied species and later instars of the same species should be superior competitors for resources, being able to grow at lower food levels than smaller animals. Second, larger animals, with their lower weight-specific metabolic rates (lower costs of maintaining unit body weight), should be more resistant to starvation than smaller animals. Third, larger animals should exhibit broader spectra for food particle sizes, since they are able to ingest particles of larger maximum size. This is why the food niche of smaller species would simply be included within those of larger species, the latter being able to feed well even after the common source of nutrition has been jointly overexploited. This was already known to Hrbáček (1962), who noted that, in the absence of fish predation, one of the large-bodied *Daphnia* species would always monopolise the resources. This notion was later pondered by Brooks & Dodson (1965) (who believed that, with increasing body size, rates of food ingestion and assimilation should increase more rapidly than respiration rate) as the **size efficiency hypothesis** (for the full importance of the hypothesis, see section 14.9).

The cornerstone of the size–efficiency hypothesis is the notion that large-bodied animals are superior competitors for limiting resources, being able to grow and reproduce at lower food levels. This, however, has frequently been questioned, and there is much contradictory evidence to show that both large and small-bodied species may be superior in overcoming food limitation, depending on the taxon and on whether the population- or individual-threshold values are determined. In rotifers, small body size has been advanced as competitively superior: the threshold food concentration for a population has been shown to increase over an 18-fold range of body mass from small to large bodied species, with two congeneric species of *Keratella* (the small *K. cochlearis* and large *K. crassa*), with extreme values of 0.06 and

1.00 mg dry wt L^{-1} (Stemberger & Gilbert 1985). In cladocerans, large body size has been found to be competitively superior, with threshold food concentrations for individual growth shown to decrease over a tenfold range in body mass, from the small-bodied *Ceriodaphnia reticulata* to the large-bodied *D. magna* (threshold values of 0.016 and 0.040 mg C L^{-1}, respectively, see Fig. 14.13). Thus, threshold values are low for small rotifer species and high for small cladocerans, with much overlap between the two. There is still no evidence as to whether cladoceran or rotifer species are superior in competition for resources in general. Small rotifer species are known to coexist with large-bodied *D. pulicaria* in habitats with extremely low food levels, from which smaller cladoceran species have been excluded (see section 14.9).

Even though, in its notion about the competitive superiority of large-bodied cladocerans and small-bodied rotifers, the size–efficiency hypothesis may be quite correct, its predictive value is low. This is due to the complex nature of natural sources of nutrition, to fluctuating food levels in the field, to the overlap in body size among species, to ontogenetic shifts in body size and to phenotypic plasticity in the morphology of the filtering appendages in such filter-feeders as *Daphnia*, which are capable of changing the area of filter screens in the next instar, in accordance with the experienced food level (building broader filters when grown at low food concentrations; Koza & Kořinek 1985; Lampert 1994).

The complexity of competitive relations may be further increased by the possibility of resource partitioning between different species, as well as by the fact that they usually compete for many resources. Such concepts as the threshold food concentration should really be considered in respect of each of these factors (for a review, see Rothhaupt 1990). The Tilman model of steady-state competition for two or more resources, however, seems much less important in the understanding of interspecific competition in zooplankton than in grasping the sense of resource competition in phytoplankton (see Chapter 10). Unlike planktonic algae, which search for different resources at different places, and at different times (e.g. dissolved

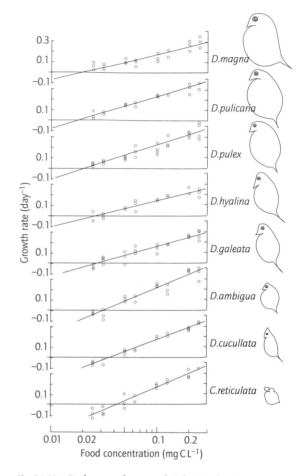

Fig. 14.13 Body growth rates of eight *Daphnia* species at different concentrations of food (*Scenedesmus acutus*) from 0.025 to 0.25 mg C L^{-1}. Each growth rate is estimated as the difference between 2-day and 6-day-old animals. The intercept of each species regression line with zero growth rate indicates specific threshold food concentration, which shifts to higher food levels with a decrease in body length, from 16 and 17 μg C L^{-1} in large-bodied *D. magna* and *D. pulicaria*, to 36 and 41 μg C L^{-1} in the smallest *D. cucullata* and *Ceriodaphnia reticulata*. (From Gliwicz 1990b; reprinted with permission from *Nature*. Copyright (1990) Macmillan Magazines Limited.)

phosphorus in deep strata at night, but radiation close to the surface by day), all of the nutrient and energy requirements of planktonic animals are bound together into live cells or detrital particles, which accumulate in certain places and at certain times.

Among the many possible reasons why either large or small-bodied species are found to be superior in competition for limiting resources, one has been shown to be important specifically to cladocerans. Owing to their lower threshold food concentrations, large-bodied cladocerans (both large-bodied species and later instars of the same species) are superior competitors compared with small-bodied ones. At the same time, however, large-bodied cladocerans are also more susceptible to interference from large algal cells and colonies, and from cyanobacterial filaments, which may cause severe reduction in ingestion rates.

At such times, superiority of large body size can become inferiority. Filamentous cyanobacterial particles interfere with the process of food collection in all filter-feeding cladocerans, but their detrimental effects are greater in large-bodied species than in small. Large body and high upper limit for the spectrum of food particle size make an animal more vulnerable, allowing more filaments to enter the median chamber, forcing it to clean its food groove more frequently (see section 14.4).

When three *Daphnia* species of different body sizes were grown on a readily available single-food resource, the largest, *D. pulicaria*, exhibited the lowest food threshold concentration, and the smallest, *D. cucullata*, the highest. The threshold for the intermediate-sized *D. hyalina* lay between the two. In the presence of high densities of a filamentous cyanobacterium, however, the sequence was reversed, showing that the largest bodied *D. pulicaria* is indeed superior in competition for resources, but only when the filaments interfering with food collection are absent (a situation typical for its natural habitat of oligotrophic lakes). The competitively inferior, small-sized *D. cucullata* may become superior when interfering algae are abundant (as is typical for its natural habitat of highly eutrophic lakes in Europe; Gliwicz & Lampert 1990).

The ability to grow and reproduce at low food levels is only the most important determinant of competitive superiority under the kind of steady-state and single-resource conditions which are neither likely to be encountered in the field nor frequently created in the laboratory. Under conditions of wide food-level fluctuations, specific reproductive potential and the ability to withstand long periods of starvation may become equally important in determining which species outperform others.

14.5.3 *Reproductive potential*

Population increase in zooplankton is usually limited by low levels of the resources exploited by many competing species, some of which, at least, are abundant enough to keep resources at a low level, usually well below the incipient limiting concentration of most coexisting herbivore species. However, there are periods when resource limitation is replaced by **time limitation**. This most often happens to the zooplankton of temperate-zone lakes during spring, when increased solar radiation allows for high phytoplankton production, whilst low temperature does not yet permit fast growth and reproduction among herbivores. A time-limited population of a planktonic herbivore increases with its maximum intrinsic rate of increase, r_{max}. All individuals grow at the physiologically maximal rate (the maximum rate of body mass production), to reach reproductive stage in the shortest time and to produce maximum-size clutches of eggs every adult instar.

Species with short generation times build high-density populations in a shorter time than those with slower maturation, and hence become more abundant. This would suggest that they are superior competitors, but in fact they should be regarded merely as faster growers. As soon as slower growing species are able to build abundant populations, their time is over, especially if one of them (with the lowest threshold food concentration) can monopolise resources by holding the resource level well below the threshold value for other species.

Being a 'superior competitor' at a high food level merely means being faster to build up a population when resources are non-limiting. According to the *r–K* dichotomy, such a capacity for rapid population increase is associated with an opportunistic strategy of taking quick advantage of abundant resources, but of being a poor competitor when resources are scarce. Nevertheless, this strategy allows for exploitation of those short-lasting pulses of food abundance which may be produced by wide fluctuations in the overall density of herbivorous zooplankton.

14.5.4   Resistance to starvation

The other way in which to cope with fluctuations in food levels is to be resistant to long periods of starvation. In habitats with wide fluctuations in food resources, periods with low food availabilities may be frequent and long-lasting. Food scarcity usually stems from a pulse of high herbivore zooplankton density, and lasts longer when animals are resistant to starvation, and/or able to cope with extremely low food levels well below threshold food concentration by using energy reserves and keeping hunger at bay. This is often observed in fish-free habitats, where a superior competitor has monopolised resources, and its population remains composed of old, large-bodied adults which keep food resources well below the threshold level necessary for the growth of juveniles (see sections 14.7 & 14.9).

As the weight-specific metabolic rates of large-bodied animals are lower, they are expected to be more resistant to starvation than small-bodied animals. This is certainly the case when different ontogenetic stages of the same species are compared. In many cladoceran and copepod species, juveniles have been found to be more vulnerable to starvation than adults. However, this is not always the case when different species are compared. Rotifers are certainly more vulnerable to starvation than cladocerans and copepods, but some small cladoceran species can withstand longer periods of starvation than cladocerans of greater body size. For instance, *D. hyalina* exposed to starvation (grown in membrane-filtered lake water containing less than $0.005\,mg\,C\,L^{-1}$) from 2 days old survived up to 10 days, with body lipid reserves still visible after 7 days of starvation. In contrast, 2-day-old *D. pulicaria*, larger in body size, were

able to survive in the same conditions for 5 days only, their body lipid reserves being exhausted after 2 days of starvation (Gliwicz 1991).

Varying vulnerability to starvation in different cladoceran species may stem from differences in access to lipid energy reserves already allocated for reproduction. The ability to withstand a longer or shorter period of starvation may depend on whether these reserves can, or cannot, be taken from the ovaries and used for the maintenance metabolic requirements of a starving animal. This may also be the reason why small-bodied cladocerans are sometimes more successful than large-bodied species in habitats with wide resource variability (Romanovsky & Feniova 1985; see also review by DeMott 1989).

Another way in which to survive long periods of starvation in habitats with wide fluctuations in food resources is to escape to a time refuge by entering diapause, by producing resting eggs or resting stages. It has been suggested that, in many *Daphnia* species, diapausing eggs in ephippial envelopes are produced in response to overcrowding and starvation (Hutchinson 1967), while diapausing copepodites of cyclopoid copepods, or resting stages, are considered to be entered into at times of food shortages (Papińska 1984). It seems that diapause may be used as an effective way to avoid low food levels, even though production of diapausing eggs may rather have been selected for as a forceful means to colonise new habitats, and recolonise old ones, after catastrophic events, even though it can also be used as an avoidance mechanism where and when predation risk is high (see section 14.6).

14.5.5   Clutch size, egg size and food level

Low food levels restrict reproductive effort by reducing the frequency of production of egg clutches (through slower growth) and the number of eggs in each clutch. The latter is not only a direct effect of the smaller amount of resources which can be allocated to eggs (fewer eggs with fewer resources) but also of the need to make larger eggs when the food level has not only been assessed as low, but is also predicted to stay low. The vulnerability of offspring to starvation is reduced by allocating more reserves to a single egg. Whilst increasing egg size is advantageous to offspring survival, there must always be a compromise between investment per offspring and total number of offspring produced. In *Daphnia*, this is an optimal decision between the mean size of a single egg and the number of eggs in a clutch.

The notion that per-egg investment should be higher at low food availability than at high and non-fluctuating food levels has been commonplace in the ecological literature, including zooplankton studies, for decades (Wesenberg-Lund 1908; Hutchinson 1967). However, data which demonstrate the trade-off between egg size and egg number in a clutch are contradictory, with some experimental studies showing that eggs of mothers grown at low food availabilities were smaller than those reared at higher food levels (e.g. Lynch 1989), and others showing the opposite. For instance, in *D. pulicaria* and *D. hyalina*, mothers grown under conditions of abundant food produced large clutches of smaller eggs, leaving offspring unable to survive under starvation conditions. In contrast, mothers grown at low food availabilities produced small clutches of larger eggs, and their offspring, albeit few in number, were able to survive long periods of starvation (Gliwicz & Guisande 1992). The reason why this is not always the case may be that the trade-off between offspring size and offspring number is not the only compromise in which the size of an egg and the size of a freshly hatched neonate are involved (see section 14.6).

14.6 VULNERABILITY TO FISH AND INVERTEBRATE PREDATION— ANTI-PREDATOR DEFENCES

The most apparent symptom of the overwhelming importance of predation to planktonic animals is the low density of zooplankton populations in the field, despite the high reproductive effort evident at the majority of locations, the large proportion of egg-bearing females in each population and the large number of eggs in each egg-bearing female (see section 14.7). Less ostensible, but equally evident, symptoms of the importance of predation can easily be traced in the individual behaviour, life-history and morphological structures of planktonic animals which have developed as inducible defences against different visually and tactilely orientated predators.

14.6.1 The cost of defences

Behavioural, life-history related and morphological defences are costly, so that they are rarely displayed throughout the whole life-span, or whole season, or at every location. On the contrary, owing to wide phenotypic flexibility of genotypes, and thanks to individual aptitude in assessing predation risk, planktonic animals display antipredatory defences only when and where the risk is appraised as real and high enough to pay, first for a defence's induction, and second for its display. Each stage will bear costs relating to the synthesis of extra body tissue, or of a defensive chemical, and to the burden imposed by living with a defensive structure, or in a suboptimal habitat.

14.6.2 Constitutive and inducible defences

Many behavioural, life-history related and morphological defences once believed to be constitutive, or genetically fixed, with their display in field populations being attributed to genotypic succession of different clones, have recently been recognised as inducible responses (Pijanowska 1993) mediated by chemical cues through which the prey becomes aware of the presence of a predator (predator odour), or of predator foraging activity, resulting in the injury of sister prey (an alarm substance). Reviews by Havel (1987), Kalinowska & Pijanowska (1987), Harvell (1990), Adler & Harvell (1990) and Larsson & Dodson (1993) provide many examples of inducible defences in planktonic rotifers, cladocerans and copepods, against fish predation as well as invertebrate predation.

Even though prey body size seems to be the most important determinant of vulnerability to each of the two distinct kinds of predation, there are also other characteristics which may at times play a quite important role in making prey more or less susceptible. In the face of an invertebrate predator which depends on tactile information, prey motility may determine the predator's ability to locate, pursue and attack it, while various small morphological structures, such as the 'neckteeth' of *D. pulex*, may be crucial in determining prey vulnerability in the phase of capture and handling. In the face of a visually orientated vertebrate predator, however, prey body shape and opaqueness will be of far greater importance.

14.6.3 Body size

The difference between the impact of predation by fish and that by most invertebrate predators (such as midge larvae, copepods, predaceous cladocerans and rotifers) stems from their entirely different feeding modes, and whether or not vision is used in foraging for prey. Planktivorous fish are joined by some waterbirds, salamanders, a few other vertebrates and aquatic insects such as *Notonecta*, in being visual predators selecting for more conspicuous prey of large body size. This is not only because such prey is easier to spot, but also because, to an optimally foraging predator, feeding on larger prey would bring a higher return for the energy invested in location, pursuit, attack, capture, handling and ingestion (see section 14.15). Other invertebrate predators rely on tactile information rather than vision in order to detect and capture their prey, and usually being small in size, and gape-limited, they select for small body size. These two different patterns of prey-size selection, originally proposed by Zaret (1978, 1980), are most important implications in respect of the vulnerability of zooplankton to predation by fish or invertebrate predators (Fig. 14.14).

As invertebrate predators are large-bodied compared with herbivorous rotifers, cladocerans and copepods, they are extremely vulnerable to predation by fish, and other visually orientated planktivores. This is why their densities are higher, and their impact greater, in habitats with low fish populations. However, in the majority of lakes, the two opposing impacts of predation are concurrent, as invertebrate predators successfully coexist with an abundance of planktivorous fish.

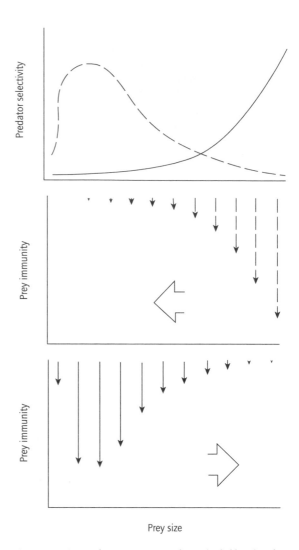

Fig. 14.14 Size selectivity in vertebrate (solid line) and invertebrate (dashed line) predators, according to Zaret 1978 (top), and the resulting size distribution of the immunity of prey to predation by visually oriented vertebrates such as planktivorous fish (middle), and to tactile invertebrate predators such as phantom midge larvae or cyclopoid copepods (bottom). Predation risk is reduced either by a decrease or an increase in body size, depending on which of the two kinds of predation is more important.

14.6.4 *Morphological defences*

Small body size provides an effective escape from the danger of falling prey to a planktivorous fish or another visually orientated predator, but it makes an animal vulnerable to gape-limited invertebrate predators, such as phantom midge larvae, cyclopoid copepods and predatory cladocerans. This is probably why morphological defences are more evident in small-bodied species, which periodically develop protrusive exo-skeletal spines, horns and teeth, or produce body-surrounding gelatinous sheaths in order to reduce vulnerability when being captured and handled by a gape-limited predator (Fig. 14.15).

Such defences are known in many rotifer and cladoceran species, and have been shown to be induced by a chemical cue on the risk to predation. An example is provided by the postero-lateral spines produced in *B. calyciflorus* in the presence of *Asplanchna* as an efficient predator, when the spines are detained (Gilbert 1966), or by the exo-skeletal tooth on the neck of *D. pulex* in the presence of a cue on predation risk concerning phantom midge larvae (Tollrian 1995). Broad inventories of morphological defences in planktonic rotifers and crustaceans can be found in the reviews by Stemberger & Gilbert (1987) and by Havel (1987).

Morphological defences are often displayed at various magnitudes in different seasons. Such a pattern of annual change in body size and shape has long been known in many cladoceran species, especially in *Bosmina* and *Daphnia*, as **cyclomorphosis** (for review, see Jacobs 1987). The most prominent cyclomorphotic changes, involving the shape of the head helmet and the length of the tail spine, and attributable to seasonally changing levels of invertebrate predation, have been demonstrated in *D. cucullata* (Lampert & Wolf 1986; Pijanowska 1990, 1992).

14.6.5 *Pigmentation and nutritional status*

The susceptibility of a planktonic animal to a visual predator will increase with each extra component of its body substance, extra tissue, extra body

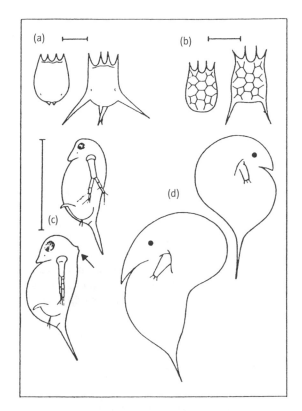

Fig. 14.15 Examples of inducible morphological antipredator defences in two rotifer and two cladoceran species: postero-lateral spines in *Brachionus calyciflorus* and in *Keratella testudo* induced by a chemical cue on the risk to predation by the predacious rotifer *Asplanchna* (a & b, respectively), exo-skeletal neck tooth in *Daphnia pulex* induced by a cue on the risk to predation by phantom midge larvae *Chaoborus* (c), and helmet in *Daphnia carinata* increased in response to a cue on predation by the corixid *Anisops* (d). (After, respectively, Halbach 1969; Stemberger 1988; Havel & Dodson 1984; Grant & Bayly 1981; from Lampert & Sommer 1993; reproduced with permission of Georg Thieme Verlag.)

reserves, extra pigmentation and an extra load of food in its gut or intestine. Dark-pigmented species of *Daphnia* will be more vulnerable to visual predators than those with low pigmentation. This is probably a reason why intensely pigmented cladocerans and copepods inhabit Alpine and shallow Arctic lakes, where the absence of fish allows for the high pigmentation, protecting animals from strong ultraviolet radiation. This is very evident in the biogeography of various *Daphnia* species (for a review, see Haney & Buchanan 1987). The degree of pigmentation can be either an inducible defence, as in many high-altitude and Arctic *Daphnia* species, or a constitutive defence that varies in the polymorphic clones of the same species, such as that suggested by Zaret (1972) in relation to two morphs of *Ceriodaphnia cornuta* with different sizes of the deeply pigmented eye.

Difference in susceptibility to visual predators may also stem from different contents of an animal's body. A *D. pulicaria* of 1.5 mm body length which has grown at high food availability, storing large amounts of body lipid reserves in its ovaries, and weighing 30 mg dry weight, will be much more conspicuous to a hungry smelt or roach than her sister of the same body length who has starved for several days, using all body reserves for basic metabolic needs. In contrast, she will be a ghost of only 1 mg dry wt, and nearly invisible even to a human eye armed with a high-quality dissecting microscope (see section 14.5).

Difference in susceptibility may also be quite clear, if the intestine of one of two equally sized individuals is filled with a nutritious food and that of the other is empty. This may easily be tested by feeding *Daphnia* with dark ink suspended in the water, as has been shown in *C. cornuta* by Zaret (1972). Such differences in susceptibility of hungry and satiated animals to visual predation may be behind the extraordinary differences in feeding rates which have been observed in a natural *D. pulex* population. Night-time filtering rates may be up to an order of magnitude higher than those of daytime, the difference being greatest in the largest individuals, which are those most vulnerable to predation by fish (Haney 1985, see section 14.9).

14.6.6   Clutch size

The difference in susceptibility to visually orientated planktivores may be even sharper between two sisters of the same body length if one carries a large clutch of eggs in its brood chamber and the other does not. This is how fish predation may change reproductive effort in a cladoceran population, by selective removal of females with larger clutches of eggs. These are more vulnerable than the rest, because they are larger as well as more conspicuous, owing to the load of eggs (see section 14.7).

When a new batch of eggs is deposited in the egg sacs, or in the brood chamber, the number in the clutch reflects (i) the food availability in the habitat, (ii) the amount of stored resources allocated for reproduction, as well as (iii) the mother's aptitude for correctly assessing the food abundance for her offspring at the time of hatching (see section 14.5). Later, during the time that the eggs remain in the sacs, or in the brood cavities (in cladocerans nearly 2 days, at 20°C), the mean number of eggs in a clutch becomes more or less reduced owing to the impact of fish, which selectively remove ovigerous females. In cladocerans such as *Daphnia*, clutch size may also be reduced by juvenile copepods which enter the brood cavities in order to feed on eggs and lipid reserves, which they suck out from the ovaries. Mean clutch size may be thus further reduced, with the selective impact of the two types of predation being very difficult to distinguish (Fig. 14.16).

14.6.7   Life-histories: egg size, neonate size and size at maturation

Even though the size of an egg is critical, from the point of view of the food level which a freshly hatched juvenile must face (see section 14.5), it is also critical in respect of its vulnerability to invertebrate and fish predation. Large eggs, and large neonates will be less vulnerable to invertebrate predators such as cyclopoid copepods, predatory cladocerans and phantom midge larvae, whilst small ones individually will be less visible to visually orientated and optimally foraging planktivorous fish selecting for larger prey. This notion was first pondered by Kerfoot (1974), who observed the production of small eggs in summer and large eggs in winter in *B. longirostris*, and suggested that this was a strategy designed to avoid fish predation during summer and invertebrate predation in winter.

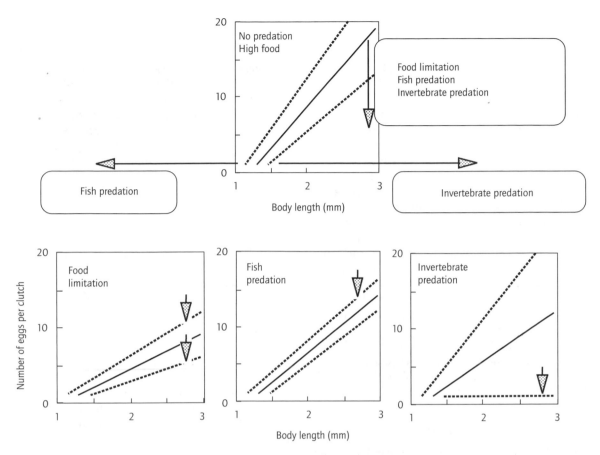

Fig. 14.16 Effects of food limitation and predation on size at first reproduction and number of eggs per clutch in *Daphnia* (top), and hypothetical differences in the mechanisms of clutch-size reduction owing to food limitation, fish predation and invertebrate predation on eggs in brood cavities (bottom). Solid lines, mean clutch size on body length; dotted lines, clutch size upper and lower limits reflecting equal coefficient of variation for different body lengths. Although each of the three factors leads to a decline in the mean number of eggs per clutch in every class of body size, the nature of the decrease is different. Food limitation (left) leads to a symmetrical shift of all clutches in the population, fish predation (middle) to a decline of the clutch upper limit owing to selective removal of females with a greater number of eggs, and invertebrate predation (right) in a decrease of clutch lower limit owing to clutch-size reduction in a fraction of randomly infested females. Fish predation should also produce shortening of the regression line as an effect of decrease in maximum body length, owing to selective removal of the largest individuals from the population. (From Gliwicz & Boavida 1996; reproduced with permission of Oxford University Press.)

Small, summer eggs develop into small adults, which mature before reaching the size to attract fish, whilst the large, winter eggs grow into large adults to become invulnerable to invertebrate predators in the shortest time possible. Similar results, reported by other authors, led Lynch (1980)

to the conclusion that selection pressure imposed by predators was of primary importance for the evolution of cladoceran life-histories. Under conditions of fish predation, small offspring that mature early in the life-span at a small body size should be produced, whereas large offspring and fast somatic growth to postponed maturity and reproduction should be favoured under conditions of invertebrate predation.

This notion has been supported by numerous data from field and experimental studies, many of them showing that each life-history trait (the size of the egg, the size of the neonate at the time of hatching, the time and the number of instars necessary to reach maturity, and the size at first reproduction) can be shifted phenotypically in response to the degree of predation risk assessed by an animal able to detect a chemical cue on either fish or vertebrate predation. Such shifts, induced experimentally by adding fish kairomones as chemical cues on predation risk, are best known in various species of *Daphnia*. Separate clones of the same *Daphnia* species may differ in the norm of reaction to fish kairomones, depending on whether they live in a habitat with or without planktivorous fishes. Such differences have been shown between the two clones of *D. magna* (Weider & Pijanowska 1993). Animals from clones inexperienced with fish are much more vulnerable to predation than those from lakes with abundant planktivorous fish (Pijanowska *et al.* 1993).

14.6.8 *Diapause*

Chemical cues on fish predation also have been used to show that diapause may be an effective mechanism of predator avoidance in zooplankton. In order to elude a temporal overlap with high predator densities, many rotifer, cladoceran and copepod species enter dormant stages at various stages of ontogenesis. Resting eggs (or **cysts**) are known to be produced in many rotifer species as overwintering eggs, but little is known of the role of dormancy as a predation-avoidance mechanism. The diapause and, in many *Daphnia* species especially, the ephippia were thought rather to represent a way to escape either harsh physical conditions or long periods of low food availability (see section 14.5).

Resting stages are observed in many cyclopoid copepods which have reached the late copepodite stages (Nilssen 1977). Resting eggs are reported both in calanoid copepods (Hairston 1987) and in cladocerans, the production of cladoceran diapausing eggs in ephippial envelopes having already been proved to be induced by a fish kairomone as a chemical cue on the risk to fish predation (Ślusarczyk 1995; Pijanowska & Stolpe 1996). Experimental induction of the shift in mode of reproduction in *D. magna*, from parthenogenetic to ephippial eggs, which must be preceded by production of males, shows that risk of predation is one of many possible factors ultimately responsible for the mode of reproduction in zooplankton. This does not, however, preclude other environmental factors such as food limitation from being equally important.

Therefore, diapause or dormancy should still be traditionally considered as an escape into a time refuge in order to avoid harsh environmental conditions of any kind, from the low temperatures and low radiation intensities of a harsh winter, to the low food availability, high predation risks and hypolimnetic anoxia of a warm summer. Diapause associated with a short life-span allows for rapid colonisation of temporary habitats, and should also be considered as an alternative means of existence to that of longer life-span, nondiapausing organisms from permanent waters (see reviews by Fryer (1996) and Hairston & Cáceres (1996)).

14.6.9 *Depth selection and diel vertical migration*

A planktonic animal may greatly reduce its vulnerability to predation by selecting a safer habitat, where it will be less conspicuous, or where predator hunting efficiency will be drastically limited. Even the simplest eye should be able to evaluate the amount of illumination, and allow habitat adjustment by moving towards a dense stand of aquatic vegetation, or by descending deeper into the water column, where the light intensity is not high enough for visually orientated predators successfully to locate and capture their prey. The level of illumination necessary for a given visual predator to see prey depends on prey size and opaqueness. This is why large-bodied and heavily pigmented animals select greater depths than

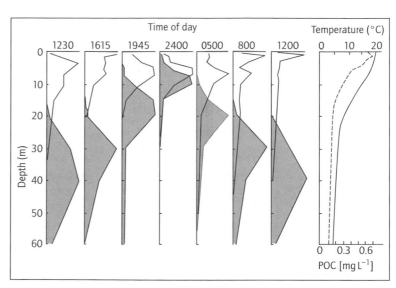

Fig. 14.17 Diel changes in depth distribution of small-bodied *Daphnia galeata* (unshaded) and large-bodied *Daphnia hyalina* (shaded) in Lake Constance during August, with profiles of temperature (°C) and food concentration (mg C L^{-1}) in the right panel: POC, particulate organic carbon. (After Stich & Lampert 1981; from Lampert & Sommer 1993; reproduced with permission of Georg Thieme Verlag.)

small-bodied and transparent species. Later ontogenetic stages, and ovigerous females of the same species, are found to reside deeper in the water column than earlier ontogenetic stages, and females without eggs. Risks gradually diminish during evening, with daylight fading away, and even very conspicuous animals may ascend close to the surface under the darkness of the night (Fig. 14.17).

Such a **habitat shift** (a diel vertical migration between daytime occupancy of deep strata and night-time residence close to the surface) is best-known and most apparent in large-bodied cladocerans, copepods and phantom midge larvae. Besides such normal or nocturnal movements, **reverse migration**, when animals reside deeper in the water column during the night than during the day, can also be distinguished. Reverse migration is less common and best known in smaller bodied species, which are more vulnerable to invertebrate predators such as phantom midge larvae, than to predation by planktivorous fish. Such behaviour may be adaptive for small-bodied animals when phantom midge larvae are abundant, and they are forced to display normal diel vertical migration in order to avoid the risk from visually orientated planktivorous fish.

Diel vertical migration is costly, because descent into deeper strata for long hours of the day entails a decrease in feeding rate owing to scarcer food, and a decrease in individual growth rate owing to lower temperatures in the hypolimnion (see section 14.8). Despite its high costs, however, diel vertical migration is a commonplace phenomenon, and may be regarded as an inducible mechanism of predation risk avoidance. This was first shown in experimental studies by Dawidowicz *et al.* (1990).

14.6.10 *Motion*

Vulnerability to predation may also be reduced by changing the manner and intensity of motion. Planktonic animals can be located by a predator from a greater distance when moving than when stationary. 'Dead men' behaviour and passive sinking is a strategy frequently used in order to minimise the risk of being located and attacked by an approaching visual or tactile predator. The reverse may also be the case. For instance a fast looping movement (a kind of somersaulting) is often displayed by *Daphnia* in the presence of a chemical cue on fish predation. This may confuse the predator and cause an increase in the ability of the prey to escape the predator's mouth.

Such an increase in alertness may be detected by the random capture of animals with a pipette, and by comparing the time needed to pick up the same number of experimental animals in the presence or the absence of the cue on predation (De Meester & Pijanowska 1997). This simple pipette method has allowed us to show that copepods are better than cladocerans in the escape performance, and that large antennae make *Diaphanosoma* a more effective evader than *Daphnia*. A wide assortment of specific methods of predator avoidance and escape responses in planktonic crustaceans has been reviewed by O'Brien (1987), and in planktonic rotifers by Stemberger & Gilbert (1987).

14.6.11   Aggregating

Vulnerability to predation may also be reduced by aggregation into swarms, thereby confusing visual predators and causing a drastic decrease in capture efficiency of such predators as planktivorous fish (Folt 1987). Swarming has been shown, by means of experiments comparing the tendency to aggregate in the presence or absence of a fish kairomone used as a chemical cue on fish predation in *Daphnia*, to be an inducible defence against visual predation (Pijanowska 1994; Pijanowska & Kowalczewski 1997). Aggregations are observed during the day rather than by night, when the risk of visual predation is reduced (Jakobsen & Johnsen 1988). For the same reasons as those relevant in the case of diel habitat shifts, they also entail high costs and are therefore only displayed when predation risk is high enough for a decrease in feeding rate to pay off (see section 14.8).

14.7   BETWEEN FOOD LIMITATION AND PREDATION: THE DEMOGRAPHY OF A POPULATION

Large population sizes in zooplankton do not necessarily reflect the most favourable conditions of the current environment. On the contrary,

conditions may already have deteriorated and the habitat been spoiled by the multitude of animals present via decreased pH, partial expenditure of dissolved oxygen and depletion of food resources. Densely packed animals may attract predators and facilitate the spread of parasites and diseases. The population is also likely to collapse, owing to high mortality and reduced reproduction. It is more correct to assume, therefore, that high population density reflects past optimal environmental conditions, rather than present.

14.7.1   Birth rate, death rate and the instantaneous rate of population growth

High population density therefore reveals that conditions have been favourable for some time, allowing for maximum rate of population increase, by granting the maximum instantaneous per capita rate of population growth ($r = b - d$), owing to the highest possible instantaneous birth rate (b) and low instantaneous death rate (d) amongst the population. This notion was first considered by Edmondson (1960), who introduced the **egg ratio method** for a direct estimate of b and an indirect estimate of d in rotifer populations in the field.

In this simple method, the number of eggs per female in field samples is combined with developmental times for eggs at different temperatures previously assessed in laboratory experiments. The egg number divided by the duration in days of egg development at a given temperature gives daily egg production per capita, which can be taken as the instantaneous per capita birth rate (b), allowing for indirect calculation of the instantaneous per capita death rate (d) as the difference between the b and r values, of which the latter is taken from two neighbouring population density estimates. Such estimates, combined with data on individual body mass, or body carbon, can further be used in the evaluation of the population production.

There are two reasons why, out of many factors which may be important in controlling population growth in any zooplankton species, two biotic factors (food and predation) seem to be of overwhelming importance. First, they are both density-dependent and, second, being a functional response of the increasing intensity of food limitation or predation, each translates directly either into birth rate or death rate. Food level, or the intensity of food limitation, sets limits for the pace of individual body growth to maturity and the intensity of reproduction, thus being responsible for the instantaneous per capita birth rate (b). Intensity of predation will be directly responsible for mortality in the population, thus setting the level of the instantaneous per capita death rate (d). Contrary to what is usually observed in terrestrial and other aquatic animals, mortality in planktonic animals is rarely caused by reasons other than predation. Very few individuals in a planktonic population live to the maximum life-span often displayed in the laboratory, under predation-free conditions.

14.7.2 Temperature

Both birth rate and death rate are also dependent on ambient temperature. All planktonic animals are eurythermal, and will respond to a rise in environmental temperature by increasing their metabolic rate. This increase will continue until an upper 'species-specific' critical temperature level is reached (at 35–45°C), with metabolic rates (respiration, food ingestion and assimilation, excretion and body growth) doubling on average with each 10°C rise ($Q_{10} = 2$).

The actual Q_{10} range is 1.3–3.2 for *Daphnia* (Peters 1987), and even wider for other taxa. The limits for intense reproduction are narrower than those for basic metabolic activity, and laboratory cultures of the majority of planktonic animals do best in the range 15–25°C, although they can easily be maintained at any temperature from 2 to 35°C. In many rotifer and crustacean species, temperatures below 2°C enhance diapause, and resting eggs will be produced, or resting stages entered, even in the absence of predation and at a high food availability (see section 14.6). The rate of individual body growth, the time needed for maturation, the time that resources are being allocated to reproduction, the duration of egg development, and

thus also the rapidity of offspring production or the number of offspring which can be produced and released by an individual per unit time all eventually depend not only on food level, food concentration and food quality (see section 14.5), but also on the temperature of the habitat.

14.7.3 Time-limited and resource-limited populations

There are times, however, when both individual and population growth will be limited by scarce food resources, low temperature of the habitat, or by the long time needed for organisms to convert resources into their own organic compounds, and to allocate them for somatic growth and reproduction. The concept of resource-limited and time-limited populations, which is usually used in order to compare different species, or conspecific populations in different habitats (Schoener 1973), can also be used in order to compare the same population from the same habitat during the course of the seasons. During early spring there are abundant resources, but low temperature prevents fast individual growth and maturation, whilst during midsummer, temperature is high but fast individual growth and maturation are precluded by low food levels, resulting from high zooplankton densities (see section 14.9; compare also Chapters 10 and 12). The 'temperature-or-resource-limitation' dilemma may also become important when comparing two different species, two distinct clones, or two groups of individuals of the same population, one living in the warm strata of the epilimnion, the other in the cool hypolimnion, both by day and by night, or during daytime only (see section 14.8).

14.7.4 Population density—food limitation or predation?

There are three good reasons why population density changes among planktonic animals are only rarely attributed to one of the two major alternative forces that shape the demography of zooplankton, by governing birth rates and death rates (Fig. 14.18). First, it is often an illusion that the impact

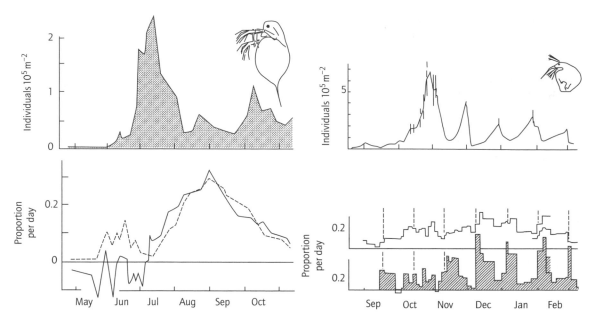

Fig. 14.18 Two different examples of seasonal changes in population density (top), and birth and death rates (*b* and *d*, bottom), in two cladoceran species: *Daphnia longispina*, from Schöhsee, Germany (left, from Lampert & Sommer 1993; reproduced with permission of Georg Thieme Verlag), and *Bosmina longirostris*, from Lake Cabora Bassa, Mozambique (right, from Gliwicz 1986b). Most rapid population increase can be seen when *b* is much higher than *d* (e.g. during June in the temperate lake), just before food becomes depleted owing to heavy grazing ('clear water' phase), and *Daphnia* become exposed to predation by planktivorous fish (*b*, dashed line; *d*, solid line). Most rapid decreases can be seen when *b* is much lower than *d* (e.g. at full moon in the tropical lake), when *Bosmina* is heavily cropped by planktivorous fish by the light of the sudden rising moon (*b*, unshaded; *d*, shaded area).

of food limitation can easily be distinguished from that of predation. A small clutch of eggs in a small-bodied *Daphnia* will be expected to indicate a low food level, even though it might be due to selective removal of females with large clutches by a visually orientated planktivorous fish. Food limitation may also be expected to be the cause of a small clutch of eggs in a large-bodied *Daphnia*, whereas it may also stem from the activity of copepods, feeding on eggs in the *Daphnia* brood cavity (see section 14.6).

Second, food limitation and predation often interact with each other at community and population levels, especially when mortality is caused by a vertebrate predator such as a planktivorous fish. Feeding activity of fish may enhance primary production, giving an increase in food level for zooplankton, both by the release of nutrients (which can then be utilised by algae) and by restricting planktonic herbivores to low population densities (thus relaxing pressure on the phytoplankton). The effect of increased mortality is thus partly balanced by increased reproduction. Population age distribution and reproductive potential change, however, as younger individuals enter reproduction at increased food levels (increased fecundity in small-bodied individuals). Larger individuals, with larger clutches of eggs, are more intensively removed if the impact of size-selective predation is higher (decreased fecundity in large-bodied individuals; Fig. 14.19).

Third, the two effects may also interact at individual level, and often it cannot be judged whether decreased reproduction is a direct effect of low food level, or is rather an indirect effect of predation. High predation risk either prevents prey animals

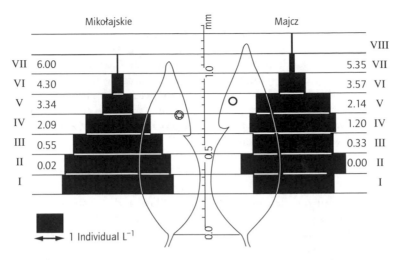

Fig. 14.19 Mean body size, size distribution and mean number of eggs per clutch in subsequent size classes from I to VIII, in two populations of *Daphnia cucullata* from two neighbouring lakes of different trophic state, the more- and less-eutrophic lakes Mikolajskie and Majcz (left and right, respectively). The stronger impact of fish predation in Lake Mikolajskie generates smaller body size and earlier reproduction by removal of large individuals (direct effect) and by fertilising the habitat and thus allowing food level increase (indirect effect). (From Gliwicz *et al.* 1981, fig. 7; reproduced with kind permission of Kluwer Academic Publishers.)

from entering a food-profitable habitat, or provokes them into a high level of defensive activity, which may be very expensive energetically (see section 14.8).

The relative importance of food limitation and predation varies distinctly during the course of ontogenic change. In cladocerans, when predation by fish is more important than that by invertebrates, later instars become more effective at food assimilation and thus superior in competition for resources but, at the same time, more vulnerable to predation. A time of low food levels, and severe food limitation, therefore seems to be more detrimental to juveniles, whereas periods of high danger from visual predators seem to be more precarious for adult animals (see section 14.9 for examples of intraspecific resource competition).

14.7.5 Population renewal time

The life-span of all planktonic protozoans, rotifers and crustaceans is short, and their generation times even shorter. At high food level, and high temperature (20°C), the time needed for a freshly hatched neonate to produce its own offspring lies between a few hours in ciliates, 1 day in many rotifer species, 6 days in most cladocerans and as much as 2 weeks in copepods, whose ontogeny is more complex. All planktonic animals also possess great reproductive potential, which stems either from short generation times (as in ciliate protozoans or rotifers) or from an ability to produce large clutches of eggs in later instars (as in large-bodied cladocerans and copepods). These features allow for extremely high instantaneous rates of population growth, and, in the absence of intense predation, enable a rapid increase in density.

Such features may be displayed also by a population subject to heavy predation if there is fulfilment of the condition that individuals do not react to predation risk by exhibiting costly anti-predator defences (see section 14.8). At a high food level, high mortality generated by predators such as a planktivorous fish can be compensated for by intense reproduction, with death rate (d) equalled by birth rate (*b*), and population density remaining the same (rate of population growth *r* = 0). Coexisting populations of two closely related species may possess exactly the same stable population density, or even the same constant standing crop of population biomass, but differ dramatically in instantaneous rates of population growth. One population may exhibit very low *d*, with very low *b*, and the other very high *d*, with equally high *b*. Renewal times are very long in the former, and very short in the latter (low or high 'renewal coefficients'; Elster 1954).

This may happen when one population is limited by low reproduction and a second is constrained by high mortality. In the majority of lake habitats, high mortality will ensue from the strong impact of predation, whilst low reproduction may be due either to (i) low food levels in resource-limited populations or (ii) low temperature in time-limited populations. An example is that of two constant-density, co-occurring populations, the density of one controlled by low reproduction due to 'time-limitation', and that of the other by high mortality resulting from the great impact of predation (Fig. 14.17).

14.7.6 Population distribution and its diel changes — diel vertical migrations

The strong impact of predation, combined with the ability to compensate via fast responses of greatly increased reproduction, makes population density in planktonic animals extremely vulnerable to sudden changes in predation risk, and the related variations in food levels. This is why it is rather difficult to judge whether patterns of population distribution with respect to depth, or along an inshore-offshore gradient, are the result of deliberate movements of individual animals, or the results of different mortality rates.

Examples of horizontal distribution can be found in a number of publications, some of them summarised in Hutchinson's *Treatise* (1967). Many exhibit distinct gradients of density change along inshore–offshore transects, with the highest densities offshore and the lowest inshore. Such distribution was long regarded as a side-effect of

vertical migration, or a result of animals' ability to orientate themselves with regard to underwater angular light distribution, by moving away from shady near-shore areas during the day and ascending towards better illuminated offshore areas by night (Ringelberg 1969; Siebeck 1969).

Such inshore-offshore population density variations may also be explained as gradients created and sustained on a night-to-night basis by mortality rates induced by the impact of predation by planktivorous fish (mostly cyprinids), which tend to aggregate in littoral vegetation during the day and to surge offshore every evening (Gliwicz & Rykowska 1992; Węgleńska *et al.* 1997; see also Chapter 15). When beginning to move out of their daytime littoral refuge, in the fading light of dusk, they crop zooplankton, selecting for more conspicuous individuals of greater body size, as well as those carrying larger clutches of eggs. They probably cannot be so efficient later on, in the dark, and they eventually reach the most distant patch of large-bodied zooplankton, in the centre of the lake, where many more adults are left alive and able to continue reproduction (Fig. 14.20).

Alternative explanations of depth selection behaviour and diel vertical migrations (one of the most-explored and best-understood phenomena in zooplankton ecology) exist. Many examples are summarised by Hutchinson (1967), and many more are reviewed in Ringelberg (1993) and De Meester *et al.* (1999). Much field and laboratory evidence has recently been collected which shows that selection of the low-food-level, low-temperature strata of the deep hypolimnion as the habitat of permanent or daytime residence is one of the most effective, albeit costly, inducible behavioural defences against predation (see section 14.8). It is also now quite apparent that: small-bodied zooplankton taxa, which are less vulnerable to predation by selectively feeding planktivorous fish, do not migrate, or migrate within a much smaller range than large-bodied taxa (Fig. 14.17); later ontogenetic stages display a wider range of migration than earlier instars; and the range of migration is wider in ovigerous than in non-ovigerous females.

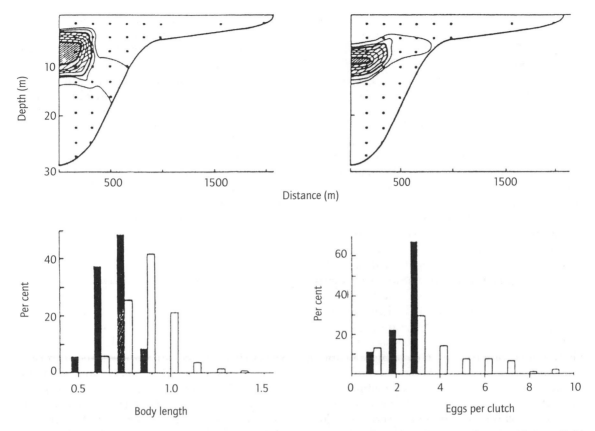

Fig. 14.20 Examples of *Daphnia galeata* distribution along an offshore–inshore transect on two dates in July 1986 (left) and 1987 (right) in the mesotrophic Lake Ros, Poland (top panels). Difference between offshore (unshaded bars) and inshore (shaded) fractions of the population in respect of body size (left bottom panel) and clutch-size distribution during 1986 (right bottom panel). The greatest population density on both dates is >16 individuals L^{-1}, each isopleth indicating a decrease in density of 2 individuals L^{-1}. Smaller body size and fewer eggs per clutch in the inshore fraction indicate that the impact of predation by planktivorous fish is stronger inshore than offshore. (From Gliwicz & Rykowska 1992; reproduced with permission of Oxford University Press.)

Field data are not always as clear as experimental results, for even the most distinct differences in depth distribution of population density, mean body size or clutch size may equally be considered (i) an effect of animals' ability to orientate themselves with regard to changes in underwater light intensity with depth (with more conspicuous individuals deliberately selecting for greater depth as more negatively photoresponsive), or (ii) as a gradient in mortality rates induced as a predation impact of planktivorous fish (with this impact weakening at increasing depth owing to fading illumination). The difference between the two is greater and thus easier to note when wide-ranging diel vertical migration is displayed. However, such migrations are not as common as might appear from a review of the vast number of publications on zooplankton depth distribution, because, when the anticipated migrations are not observed, papers on migration are just not written. The reason why migrations are not common may lie in the ability of planktonic animals to choose between alternative defences against predation (see section 14.8).

14.8 BETWEEN FOOD LIMITATION AND PREDATION: BODY SIZE AND MORPHOLOGY, LIFE-HISTORIES AND BEHAVIOUR. COST OF DEFENCES AND NON-LETHAL EFFECTS OF PREDATION

To a planktonic animal, availability of food and risk of predation are the two most important and never-ending concerns, each critical to individual fitness: the prospects for survival and growth to the maturity needed to produce as large a number of offspring as possible. Each is highly selective, especially with regard to animal body size and sensitivity to light. Animals which are less sensitive to radiation tend to stay close to the surface where they can grow more rapidly, with abundant food and a high temperature, but where they must also face higher predation risks. Each of the two components is also density-dependent, with both food limitation and predation becoming more intense as population density increases. Each can play an important role as a potent force of natural selection in the evolution of different traits of zooplankton behaviour, life-histories and morphology. These two forces of natural selection oppose each other, as the need to feed effectively at low food concentration promotes large body size, whilst the requirement to avoid predation risk fosters small body size. This is the case in planktonic cladocerans, and in lakes where predation by visually orientated fish is much more important than that by invertebrates.

Each of two conflicting demands (the imperative to feed, grow and reproduce, and to avoid predation and stay alive) must be satisfied by fulfilling completely different requirements. A compromise solution entails high costs for high probability of survival, as well as for a high rate of individual growth, which must be paid. Increase in the chance of survival to maturity must also be compensated for by slower growth and reduced reproduction, whilst increase in the rate of individual growth and reproduction must be paid for by an increase in predation risk. The kind of compromise reached will depend on the relative importance of food limitation and predation. In habitats with abundant food, animals are more reluctant to suffer increased predation risk than in those with low food levels, even though this is what may be needed in order to increase rates of assimilation, growth and reproduction. In habitats of low predation risk, animals are more reluctant to move to less food-profitable areas, even when this is what is required in order to increase chances of survival.

The trade-off between the need to grow fast and the requirement to survive is most frequently considered in terms of the costs an animal must pay in order to reduce the risk of being killed before maturation. These are paid more or less indirectly in a number of ways, and are difficult to assess in any other way unless eventually they appear as a decrease in reproduction.

The most indirect way of paying for increased safety, where the cost of decreased predation risk is covered by those body reserves which could otherwise have been allocated to reproduction, involves investment in morphological defences. These are slow in being induced, and their costs are most difficult to assess (Tollrian 1995). This also applies to the costs entailed in life-history shifts, such as those in age and body size at first reproduction, and in egg-and-offspring size. However, these costs can also be evaluated as a decrease in reproduction, for instance as a result of earlier maturation at smaller body size, which does not allow for large clutches (Taylor & Gabriel 1992).

Behavioural defences are much faster in being induced, and their costs can be measured in a more direct way, as energy or body weight (body carbon) units, so that they may be easily compared with the energy or carbon assimilated by an animal in the same unit of time. Thus, the cost of a defence may be expressed as a decrease in individual growth rate, or in individual production $(P = A - R)$, being reflected either as increased respiration (R) or decreased assimilation (A).

A decrease in individual growth rate owing to an increase in respiration will be caused by any increase in evasion activity by an animal. This may be reflected as an increase in its alertness, and seen as fast swimming or intense looping (see section 14.6). A fall in individual growth rate due to a decrease in assimilation may be generated by the immobility of an animal seeking not to attract a predator's attention by conspicuous feeding movements, or by its filled gut, which would be more visible than an empty intestine against diffused light.

A decrease in individual growth rate due to reduced assimilation would also be the expected price to pay for the reduction in predation risk which results from swarming. Food often becomes depleted in dense aggregations of planktonic animals such as swarms of *Bosmina longispina* (Jakobsen & Johnsen 1988) or *H. gibberum* (Tessier 1983). It would also be the price of reduced predation risk in animals which shift to a safer habitat, a refuge among littoral vegetation or in the deep strata below the euphotic layer, where both food quantity and quality are usually much lower.

Sometimes, the costs of reduced predation risk can be clearly seen in drastically reduced individual growth rates, without knowing whether they stem from increased respiration or decreased assimilation. An experimental example shows that different instars of *Daphnia* species of varying body sizes, grown in the presence either of the cyclopoid copepod *Acanthocyclops robustus*, or a chemical cue on its predation of sister *Daphnia*, displayed growth retarded to 40% of that of control animals grown at the same food level in a medium free of the predator and of the chemical cue concerning its activity (Fig. 14.21). Retarded growth was probably caused by a combination of increased respiration due to *Daphnia* evasion activity, and decreased assimilation as a consequence of narrowing the gaps between the edges of the carapace, to prevent copepods from exploring *Daphnia* brood chambers. The impact of the predator and the effect of the chemical cue were both much stronger in small-bodied species and in earlier instars (which are more vulnerable to invertebrate predation than large-bodied prey), as would be predicted from Fig. 14.14. Mortality induced by the predator was less intense in larger than in smaller *Daphnia*.

The predatory effects of planktivorous fish on *Daphnia* are more difficult to evaluate than those of invertebrate predators. The latter are often unsuccessful in their assaults, and usually encounter their prey more than once, leaving the prey alive and anticipating a subsequent threat. Since death is the usual fate of planktonic prey exposed to fish, non-lethal effects on prey growth are more likely to be observed (and assessed) experimentally for invertebrate predators than for vertebrates. This may be why the overwhelming majority of examples of non-lethal effects of predation by planktivorous fish come from indirect field observations, and from experimental work on the effects of chemical cues on predation risk, rather than from experiments on the effects of predation itself (for a review see Larsson & Dodson 1993).

Increase in respiration or decrease in assimilation are not the only possible reasons for retarded individual growth as the cost of decreased predation risk. Another reason, quite unique and specific to offshore habitats in stratifying lakes, is the low temperature of the hypolimnion. Animals which try to avoid predation by visually orientated planktivorous fish in the euphotic strata of the warm and food-rich epilimnion do so either by living permanently in the dark and cool hypolimnion, or by descending there for long hours of the day (12h in tropical lakes, but up to 20h in lakes of the temperate zone). In this way, they pay a high cost of growth rate reduced due to low temperature, which may be more dramatic than a reduction resulting from increased respiration or decreased assimilation.

This subject was first considered by Dawidowicz & Loose (1992a,b) and Dawidowicz (1993, 1994), and experimentally assessed in *D. magna* by Loose & Dawidowicz (1994). Individual growth rate (in individuals forced to perform diel vertical migration by a fish kairomone used as a chemical cue on predation risk) is reduced to 25% of its reference value in individuals which are not forced to migrate, and can stay in the warm epilimnion day and night (Fig. 14.22). Growth rate (to 25%) was reduced more severely at the low food level of $0.1 \, \text{mg C L}^{-1}$ than at $2 \, \text{mg C L}^{-1}$ (to 35% of the reference value). No eggs were produced in *Daphnia* displaying the behaviour of migration in the presence of the fish kairomone, whereas large numbers of eggs were produced in the reference

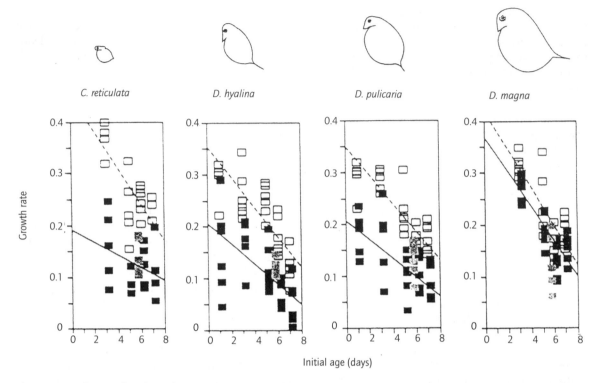

Fig. 14.21 Body growth rate per day in various instars of four daphnids (one *Ceriodaphnia* and three *Daphnia* species, 1, 3, 5, 6 and 7 days old at the beginning of the experiment) grown for 2 or 4 days in the absence (empty squares and dashed regression lines) or presence (dark-shaded squares and solid regression lines) of adults of the cyclopoid copepod *Acanthocyclops robustus* as an invertebrate predator. Data on 6-day-old cladocerans grown in the absence of copepods, but in a continuous flow of water from copepods fed *Daphnia* neonates, are also given as light-shaded squares. (From Gliwicz 1994, figs 1 & 2; reproduced with permission of Springer-Verlag.)

animals, with mean clutches of eight and sixteen eggs being observed respectively at low and high food levels.

The notion that low hypolimnetic temperature may be more important in causing retarded growth and delayed reproduction than low food level of the hypolimnion corresponds with field observations on feeding rates reported to be dramatically lower during the day than at night (Haney 1985, see section 14.6). In the temperate zone, animals may deliberately cease feeding in order to become less conspicuous, and to remain in the warm epilimnion, rather than descending to the hypolimnion for the long hours of the summer day.

As they make a choice from a broad inventory of structural and inducible defences, ranging from morphological adjustments and life-history modifications to shifts in behaviour and habitat exploitation, the high costs resulting from a display of each anti-predator strategy provides zooplankton prey with a dilemma. It has recently been suggested that the costs of decreased predation risk can be reduced when anti-predator strategies are adopted in order to be displayed alternatively to each other, when different clones may display separate traits in structural or inducible defences. For instance, one of the hybrid clones of *D. hyalina* and *D. galeata* may display life-history modification (smaller size at first re-

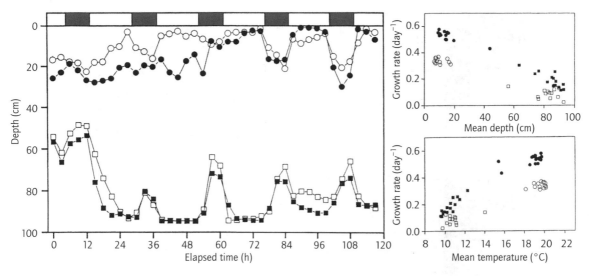

Fig. 14.22 Depth selection by *Daphnia magna* in experimental chambers in the absence (circles) or presence (squares) of a fish kairomone as a cue on predation (left), each at the two different food levels of $0.1\,mg\,C\,L^{-1}$ (empty symbols) and $2\,mg\,C\,L^{-1}$ (filled symbols), and individual body growth rates versus mean depth (top right) and versus the mean temperature (bottom right) that the animals experienced during the 5 days of the experiment. (From Loose & Dawidowicz 1994.)

production in the presence of fish kairomone), whereas another may possess traits for inducible behavioural defences (wide-amplitude diel vertical migration in the presence of fish kairomone; De Meester *et al.* 1995).

This notion has also been tested in the field. Whereas large-bodied *D. pulicaria* from the crystal clear waters of alpine lakes display (in their life-histories and their behaviour) strong anti-predator defences (Fig. 14.23), smaller *Daphnia hyalina* from less-transparent lowland lakes exhibit only one of two alternative strategies. In lakes where migration to deep strata is prevented by hypolimnetic anoxis, they exhibit small size at first maturation, whereas in lakes with high hypolimnetic dissolved oxygen concentration there is a tendency for broad-amplitude diel vertical migrations (Fig. 14.24).

Without appropriate enzyme identification, whether such a difference in habitat use or life-history observed in the field stems from separate traits in different clones with different genotypes, or from a wide reaction norm in the behaviour or

life-history of a single clone, cannot be known. Wide variability in behavioural and life-history anti-predator defences within a single genotype may be the equivalent of an evolutionarily stable strategy, which, in an abundant clone of a planktonic animal such as *Daphnia*, may be exhibited at the same time, with some individuals playing doves and some hawks, and with no need for switching between different behaviours necessary in a genetically unique single individual of a sexually reproducing species.

14.9 BETWEEN FOOD LIMITATION AND PREDATION: ZOOPLANKTON COMMUNITIES

The ability to compete for limiting resources, combined with vulnerability to body-size-selective predation by planktivorous fish, will eventually determine the prospects for each zooplanktonic species or clone (i) to be present in, or absent from, the community in a given habitat

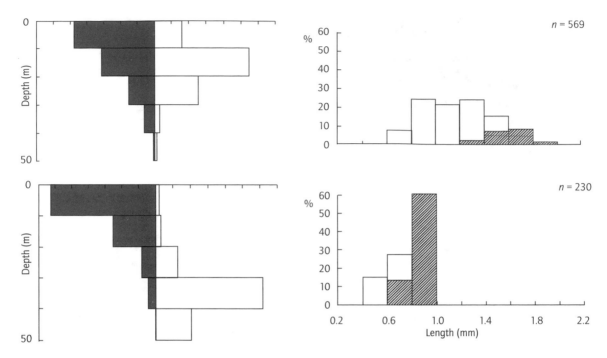

Fig. 14.23 Midnight (shaded) and midday (unshaded) depth (left) and body length distribution (right, ovigerous females – shaded) in *Daphnia pulicaria* in two neighbouring ultraoligotrophic alpine lakes: the fishless Czarny Lake (top) and Morskie Oko which contains natural fish populations (bottom). The comparison shows that, in the absence of planktivorous fish (top), neither of two possible anti-predator defences is displayed, whilst both are exhibited in the presence of fish (bottom): *Daphnia* demonstrate diel vertical migration, and begin reproduction at smaller body size. (From Sakwińska 1997.)

at a given time, and (ii) to be able (or not) to overcome the problem of limiting resources, and to secure enough food for fast growth and intensive reproduction (in order to offset the high mortality rates caused by predation). Both the ability to grow and reproduce at a low food level, and vulnerability to predation by fish, have long been related to body size, at least in crustacean zooplankton, with the two notions becoming cornerstones of the **size–efficiency hypothesis**.

14.9.1 The size–efficiency hypothesis

According to this hypothesis, proposed by Brooks & Dodson (1965), body size is one of the most important determinants of relative abundance of different species in a zooplankton community. In the

absence of predation by planktivorous fish, large-bodied species will monopolise resources, small-bodied species eventually being excluded as unable to grow at reduced food levels. In the presence of planktivorous fish, however, large-bodied species, which are more vulnerable to predation, will be exterminated. Large populations of small-bodied species will eventually build up at food levels increased by the absence of more effective large-bodied herbivores.

The relative abundance of large and small-bodied species will thus depend on the intensity of predation by planktivorous fish, on fish densities, and on fish feeding activity. This was first suggested by Hrbáček (1962) who was able to demonstrate a dramatic difference between the zooplankton of stocked and unstocked fish-ponds. In the absence

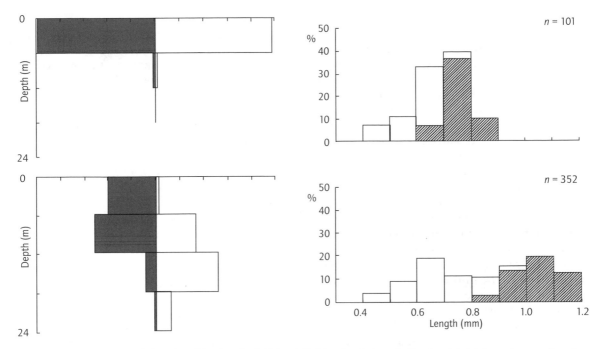

Fig. 14.24 Midnight (shaded) and midday (unshaded) depth (left) and body length distribution (right, ovigerous females – shaded) in *Daphnia hyalina* in two neighbouring eutrophic lakes differing in availability of hypolimnetic refugia: one with an anoxic hypolimnion (top), the other with high hypolimnetic dissolved oxygen concentrations (bottom), both with abundant planktivorous fishes. The comparison shows that only one of the two feasible anti-predatory defences (see Fig. 14.23) is displayed in each population. In the absence of the hypolimnetic refuge (anoxic hypolimnion – top), there is no diel vertical migration, but a clear shift in life-history to early reproduction before growing to as large a body size as in the other lake. Here, with high hypolimnetic dissolved oxygen concentrations (bottom), relaxation of fish predation owing to diel vertical migration allows for larger body size at first reproduction. (From Sakwińska 1997.)

of fish, large-bodied *D. pulicaria* was overwhelmingly abundant, whilst in fish-ponds well stocked with fish a diverse small-bodied zooplankton occurred. Hrbáček's observations were soon confirmed by multiple examples of similar divergence in zooplankton owing to varying levels of fish predation in separate lakes, different seasons, or different years. One of the first datasets to demonstrate a dramatic change in the zooplankton community of a single lake was that of Brooks & Dodson (1965), who compared species composition and body size distribution before and after introduction of an effective planktivorous fish to Crystal Lake, Connecticut.

14.9.2 The zooplankton community throughout the seasons

Intensity of predation depends not only on abundance of fish but also on fish feeding rates, which in turn are related to fish community composition and fish body size, as well as water transparency, intensity of incident radiation and duration of daylight, all of which are not only lake-specific but also strongly dependent on the season. This is why, in the majority of lakes, a clear sequence of seasonal successional events can be seen (see Chapter 10). This sequence is highly predictable, and is most apparent in stratifying lakes of the temperate

zone, where distinct annual variations in solar radiation foster gradual changes in temperature,
stratification patterns, primary production, transparency, and intensity and daily duration of underwater light. The sequence is less predictable in
shallow lakes, where it may depend on the extent
of wind-induced mixing. Wind and rainfall periodicity are responsible for annual changes in tropical
lakes and reservoirs, and lunar periodicity may
also be significant, as the intensity of moonlight
may influence intensity of predation by planktivorous fish (Fig. 14.18).

In a moderately eutrophic, stratifying lake of
the temperate zone, early spring is the time when
ciliates, rotifers and small-bodied crustaceans are
the first zooplanktonic species to become abundant. Low temperature renders all planktonic
populations time-limited, and small-bodied
species with short generation times possess the advantage of faster population growth. They are soon
replaced, however, by large-bodied species, often
large *Daphnia*, which are competitively superior.
High densities of large-bodied herbivores then
cause a drastic decline in the phytoplankton. Resource-limited populations of large-bodied cladocerans and copepods persist in an equilibrium
with low phytoplankton numbers (the 'clear-
water phase', see Chapter 10), until their birth
rates can no longer compensate for increasing
death rates, the latter increase owing to rising
intensity of predation caused by increased fish
feeding activity at higher temperature, and by
recruitment of young-of-the-year fish.

Late spring or early summer is a time when
large-bodied crustaceans are gradually replaced by
small-bodied crustaceans and rotifers, which are
less vulnerable to fish predation, and less affected
by large, interfering algae, which have already become abundant as small algae are heavily grazed.
During summer time, large-bodied species are
sometimes entirely exterminated (the 'midsummer decline'), a phenomenon which is most common in lakes with abundant planktivorous fish
and hypolimnetic anoxia, which therefore does
not allow for wide-ranging diel vertical migration
as an effective anti-predator defence (see section

14.8). Small-bodied and diverse zooplankton
persist in an equilibrium with an abundant
phytoplankton composed mostly of inedible or in-
digestible greens, peridinians and cyanobacteria,
until algal and cyanobacterial populations collapse at the end of summer.

This pattern of seasonal succession is based on
data from many different lakes, and is described in
detail by Sommer *et al.* (1986; see also Chapter 10).
The mechanisms of interspecific competition
for resources and size-selective predation which
underlie these successional events have been
reviewed by DeMott (1989) and by Gliwicz &
Pijanowska (1989). The two contrasting situations
(spring clear-water phase and midsummer) are
similar to those already known from lakes of different trophic status (Fig. 14.1).

14.9.3 Zooplankton communities in
different lakes

Contrasts between zooplankton communities
from different lakes, such as those shown in Fig.
14.1, do not necessarily stem solely from different
levels of predation by planktivorous fish. They
may also arise from the varying vulnerability of
different species to interfering algae, or from their
different abilities to cope with low food levels.
They may also arise from different food levels
themselves, which in turn stem from differences
in fertility allowing varying rates of primary
production. It seems that single-factor explanations are rarely correct, unless the factor is manipulated in lakes whose zooplankton communities
did not differ beforehand. One such example is a series of similar lakes stocked with different densities of planktivorous fish, resulting in drastic
variations in body-size distributions (Fig. 14.25).
Many similar examples have been demonstrated
from a large number of field studies, which first
aimed (during the 1960s and 1970s) to test the
Brooks & Dodson (1965) hypothesis on the role of
size-selective predation by planktivorous fish.
Later (during the 1980s), they sought to test the
concept of **biomanipulation** (for reviews see Gulati
et al. 1990).

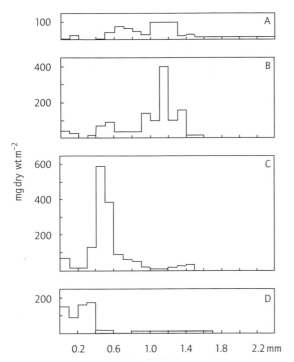

Fig. 14.25 Body size distributions of zooplankton communities from four Norwegian lakes with different stocks of planktivorous fish: (A) fishless lake, (B) lake with small *Salmo trutta* population, (C) lake with abundant *Salvelinus alpinus*, and (D) lake with abundant populations of *S. alpinus* and *Gasterosteus aculeatus*. (After Langeland 1982; from Gliwicz & Pijanowska 1989, fig 7.5; reproduced with permission of Springer-Verlag.)

14.9.4 Biomanipulation

The notion that relaxing the impact of fish predation will promote high densities of large-bodied zooplankton is the cornerstone of the concept of biomanipulation promoted since 1980 by Joe Shapiro (Shapiro & Wright 1984). The key concept of biomanipulation is the reduction of planktivorous fish densities by proper fishery management, and also by promoting numbers of piscivores. Release of zooplankton from the impact of fish predation will lead to growth of abundant populations of large-bodied cladocerans, which will in turn hold

phytoplankton standing crops at a low level, despite high nutrient load: a situation comparable to the spring clear-water phase (see section 14.9.2 and Chapter 10).

This notion attracted much attention, and produced much valuable field experimental evidence which shows that zooplankton released from the pressure of fish predation will be transformed from a diverse community of small-bodied crustaceans and rotifers with short generation-times (which are therefore just able to compensate for high mortalities by high reproduction) into a simpler community, with a few large-bodied *Daphnia* species that merely match minimum body growth to metabolic requirements by exploiting food resources at a low level, which does not allow small-bodied species to persist (see DeMott 1989, and Gliwicz & Pijanowska 1989).

14.9.5 The zooplankton of fishless lakes

Because most bodies of water are interconnected (see Chapter 1) and because humans cannot resist the whim of stocking every available water hole, fishless lakes are rare. Yet, there are still lakes, high in the mountains, from which fish have always been precluded by high cascades at their outflows, and by indigent trails which discourage people from climbing up with buckets filled with fish fry. Whether in the Alps, the Tatra, or the Estrela of Europe, or the Uintas and the Colorado Rockies of North America, the zooplankton community in such fishless, alpine lakes is very simple, with large-bodied *Daphnia* such as *D. pulicaria* as the exclusive crustacean herbivores, one cyclopoid copepod, *Cyclops* or *Tropocyclops*, which preys on *Daphnia* juveniles and eats eggs from brood cavities, and a few species of rotifer from two or three genera (*Keratella*, *Polyarthra* and *Asplanchna*; Fig. 14.26; see also Gliwicz *et al.* 2001).

Daphnia, the only filter-feeding crustacean in the habitat, seems to have monopolised resources, in very much the same way as it does in Hrbáček's (1962) fishponds. During summer, *Daphnia* is found as a cohort of grown individuals, mostly

 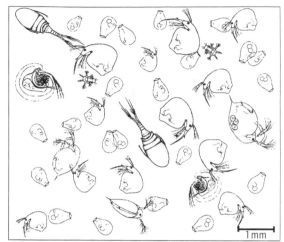

Fig. 14.26 Zooplankton communities of two ultraoligotrophic alpine lakes in the Tatra Mountains, one fishless (Czarny Lake, left), the other containing natural trout (*Salmo trutta fario*) populations (Morskie Oko, right). A simple community, with a single copepod *Cyclops abyssorum tatricus*, a single cladoceran and the large-bodied *Daphnia pulicaria* monopolising resources in the absence of fish (left), is replaced by a multi-species community with the three cladocerans *Bosmina longirostris*, *Holopedium gibberum* and *D. pulicaria*, cyclopoid copepods and three rotifer species, *Asplanchna priodonta* being most abundant (according to Orłowska 1997).

adults, many with eggs in brood chambers. There are no juveniles, or only a few, since these are unable to cope with the low food level (and with *Cyclops* predation) and usually die before being able to grow and reproduce (Fig. 14.27). Small-bodied species had been excluded a long time ago. Interspecific competition has been replaced by intraspecific competition, with grown-up individuals persisting. This is why, in the absence of fish, there is often a single cohort of synchronised individuals (Fig. 14.28), which hatched at the right time: (i) not too early to starve before spring overturn allows some primary production, and (b) not too late to starve with food already depleted by the majority of juveniles hatched at the proper time.

A similar condition is observed where another filter-feeding branchiopod monopolises resources in the absence of fish predation. This may happen even in very fertile habitats, and was probably commonplace before the mid-Mesozoic appearance of fish (Kerfoot & Lynch 1987). Fortunately, a few examples can still be found today. Some quite eutrophic lakes are fish-free because of extremely high salinity, one of which is the Great Salt Lake, traditionally viewed as 'the world's simplest ecosystem', composed of a single producer (the green alga *Dunaliella viridis*) and a single herbivore, the brine shrimp *Artemia franciscana*. The *Artemia* population is a single cohort which hatched at an appropriate time and keeps the algae at the extremely low level of 1 mg chlorophyll L^{-1}. This is enough for slow growth in already large individuals, but much too low for juveniles, which therefore die of starvation. *Dunaliella* density increases dramatically as soon as *Artemia* becomes absent, and reaches 50 mg chlorophyll L^{-1} in winter, when *Artemia* dies off, as is the case in experimental tanks from which *Artemia* has been removed (Gliwicz *et al.* 1995).

Both in winter time and when *Artemia* has been experimentally removed, in order to allow for an increase in phytoplankton standing crop the water of the Great Salt Lake becomes colonised by abundant and diverse ciliates (Gliwicz *et al.* 1995). During exceptionally high rainfall, the Lake's salinity falls from >100 to 50 g L^{-1}, allowing for

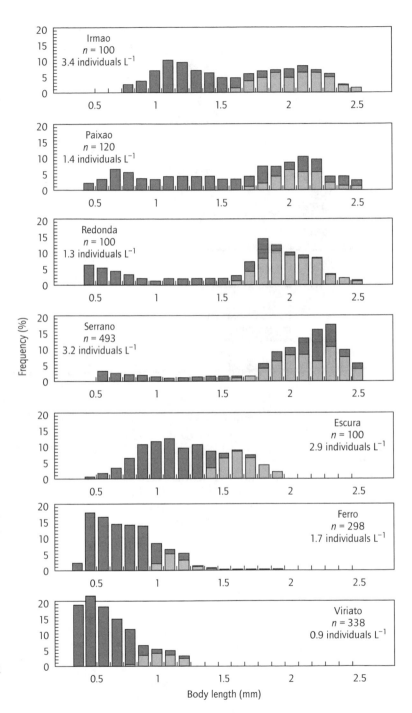

Fig. 14.27 Body length distributions in *Daphnia pulicaria* in seven alpine lakes in Estrela, Portugal, four (Irmao, Paixao, Redonda and Serrano) free of fish, and three stocked with low (Escura) or high (Ferro and Viriato) densities of rainbow trout. Light shaded – egg-bearing females. Sample size (*n*) and population density (individuals L⁻¹) given below lakes' names. (From Gliwicz & Boavida 1996; reproduced with permission of Oxford University Press.)

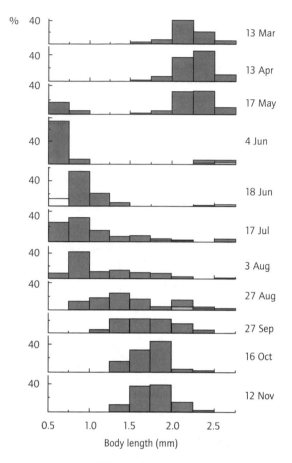

Fig. 14.28 Seasonal changes in body length distributions of *Daphnia pulicaria* in the fishless Czarny Lake in the Tatra Mountains (see also Figs 14.23 & 14.26). Light shading indicates the fraction of egg-bearing females. Note that, in the absence of predation, food limitation becomes the main selecting force and allows only for survival in those individuals that have hatched in June and have not attempted to reproduce in the same season. Maturity can be reached by individuals of the new generation, and there are already females with eggs in brood chambers at the end of August, but recruitment is unsuccessful. No increase occurs in the smallest body-size class (according to Gliwicz *et al.* 2001).

intrusion by an aquatic insect, the corixid *T. verticalis*, an effective invertebrate predator. The single-species zooplankton 'community' of the Great Salt Lake is then replaced by a more diverse community consisting of *Artemia*, rotifers (*Brachionus*) and copepods (*Diaptomus* and *Cletocampus*; Wurtsbaugh 1992).

14.9.6   The paradox of the plankton revisited

What can be learned from the example of the Great Salt Lake is that the 'paradox of the plankton' (see Chapter 10) does not really apply to zooplankton. The competitive exclusion principle of Hardin (1960), or Gause's 'one niche—one species' rule (see section 14.2), does fit the real zooplankton world nicely, however, and succinctly explains the phenomenon of a single species monopolising resources in the absence of predation. It is true that, in the presence of predation by fish, or by an effective invertebrate, diversity of the zooplankton community increases greatly (Fig. 14.26). There is, however, a question as to whether this should be treated as the 'paradox of the plankton', or merely as diversity increased by predation, with the predator allowing inferior competitors to coexist by keeping superior competitors at a level below the carrying capacity of the habitat, an idea known since the 'Food web complexity and species diversity' article by Paine (1966). What we are left with, once again, is evidence as to how important predation is to a planktonic animal, to a population of a zooplankton species and to a community of zooplankton.

14.10   REFERENCES

Adler, F.R. & Harvell, C.D. (1990) Inducible defenses, phenotypic variability and biotic environments. *Trends in Ecology and Evolution*, **5**, 407–10.

Alekseev, V.R. & Fryer, G. (eds) (1996) Diapause in the Crustacea. *Developments in Hydrobiology*, No. 114. *Hydrobiologia*, **320**, 241 pp.

Arruda, J.A., Marzolf, G.R. & Faulk, R.T. (1983) The role of suspended sediments in the nutrition of zooplankton in turbid reservoirs. *Ecology*, **64**, 1225–35.

Burns, CW. (1968) Direct observation of mechanisms regulating feeding behavior of Daphnia in lake water. *Internationale Revue des gesamten Hydrobiologie*, **53**, 83–100.

Boavida, M.J. & Heath, R.T. (1984) Are the phosphatases

released by *Daphnia magna* components of its food? *Limnology and Oceanography*, **29**, 641–5.

Bogdan, K.G. & Gilbert, J.J. (1987) Quantitative comparison of food niches in some freshwater zooplankton. A multi-tracer cell approach. *Oecologia*, **72**, 331–40.

Brooks, J.L. & Dodson, S.I. (1965) Predation, body size and composition of plankton. *Science*, **150**, 28–35.

Chow-Frazer, P. (1986) An empirical model to predict *in situ* grazing rates of *Diaptomus minutus* Lillieborg on small algal particles. *Canadian Journal of Fisheries and Aquatic Sciences*, **43**, 1065–70.

Clement, P., Wurdak, E. & Amsellem, J. (1983) Behavior and structure of sensory organs in rotifers. *Hydrobiologia*, **104**, 89–130.

Dawidowicz, P. (1990) Effectiveness of phytoplankton control by large-bodied and small-bodied zooplankton. *Hydrobiologia*, **200/201**, 43–7.

Dawidowicz, P. (1993) Diel vertical migrations in *Chaoborus flavicans*: population patterns vs individual tracks. *Ergebnisse der Limnologie*, **39**, 19–28.

Dawidowicz, P. (1994) Which is the most costly component in diel vertical migration of zooplankton? *Verhandlungen des internationale Vereinigung für theoretische und angewandte Limnologie*, **25**, 2396–9.

Dawidowicz, P. & Gliwicz, Z.M. (1987) Biomanipulation. III. The role of direct and indirect relationship between phytoplankton and zooplankton. *Wiadomosci Ekologiczne*, **33**, 259–77. (In Polish with English summary.)

Dawidowicz, P. & Loose, C.J. (1992a) Cost of swimming by *Daphnia* during diel vertical migration. *Limnology and Oceanography*, **37**, 665–9.

Dawidowicz, P. & Loose, C.J. (1992b) Metabolic costs during predator-induced diel vertical migration of *Daphnia*. *Limnology and Oceanography*, **37**, 1589–95.

Dawidowicz, P., Pijanowska, J. & Ciechomski, K. (1990) Vertical migration of *Chaoborus* larvae is induced by the presence of fish. *Limnology and Oceanography*, **35**, 1631–7.

De Mott, W.R. (1982) Feeding selectivities and relative ingestion rates of *Daphnia* and *Bosmina*. *Limnology and Oceanography*, **27**, 518–27.

De Mott, W.R. (1986) The role of taste in food selection by freshwater zooplankton. *Oecologia*, **69**, 334–40.

De Mott, W.R. (1989) The role of competition in zooplankton succession. In: Sommer, U. (ed.), *Plankton Ecology. Succession in Plankton Communities.* Springer-Verlag, Berlin, 195–252.

De Meester, L. & Pijanowska, J. (1997) On the trait-specificity of the response of *Daphnia* genotypes to the chemical presence of a predator. In: Lenz, P.H.,

Hartline, D.H., Purcell, J.E. & Macmillan, D.L. (eds), *Zooplankton: Sensory Ecology and Physiology.* Gordon and Breach, Amsterdam, 407–17.

De Meester, L., Dawidowicz, P., van Gool, E. & Loose, C.J. (1999) Ecology and evolution of predator-induced behavior of zooplankton: depth selection behavior and diel vertical migration. In: Tollrian, R. & Harvell, C.D. (eds), *The Evolution of Inducible Defenses.* Princeton University Press, Princeton, NJ, 160–76.

De Meester, L., Weider, L.J. & Tollrian, R. (1995) Alternative antipredator defences and genetic polymorphism in a pelagic predator–prey system. *Nature*, **378**, 483–5.

Duncan, A. (1985) Body carbon in daphnids as an indicator of the food concentration available in the field. *Ergebnisse der Limnologie*, **21**, 81–90.

Dussart, B.H. & Defaye, D. (1995) Copepoda, Introduction to Copepoda. In: Dumont, H.J.F. (ed.), *Guides to the Identification of the Microinvertebrates of the Continental Waters of the World*, No. 7. SPB Academic Publishing, Amsterdam, 277 pp.

Edmondson, W.T. (1960) Reproductive rates of rotifers in natural populations. *Memorie dell'Istituto Italiano di Idrobiologia*, **12**, 21–77.

Ejsmont-Karabin, J. (1984) Phosphorus and nitrogen excretion by lake zooplankton (rotifers and crustaceans) in relationship to individual body weights of the animal, ambient temperature and presence or absence of food. *Ekologia Polska*, **32**, 3–42.

Ejsmont-Karabin, J. & Pontin, R.M. (1995) Rotifera. *Developments in Hydrobiology*, No. 107, *Hydrobiologia*, **313–14**, 405 pp.

Elster, H.J. (1954) Über die Populationsdynamik von *Eudiaptomus gracilis* Sars und *Heterocope borealis* Fischer im Bodensee-Obersee. *Archiv für Hydrobiologie Supplement*, **20**, 546–614.

Fiałkowska, E. & Pajdak-Stós, A. (1997) Inducible defence against a ciliate grazer *Pseudomicrothorax dubius* in two strains of *Phormidium* (Cyanobacteria). *Proceedings of the Royal Society of London, Series B*, **264**, 937–41.

Folt, C.L. (1987) An experimental analysis of costs and benefits of zooplankton aggregation. In: Kerfoot, W.C. & Sih, A. (eds), *Predation. Direct and Indirect Impacts on Aquatic Communities.* University Press of New England, Hanover, New Hampshire, 300–14.

Fryer, G. (1968) Evolution and adaptive radiation in the Chydoridae (Crustacea, Cladocera): a study in comparative functional morphology and ecology. *Philosophical Transactions of the Royal Society of London, Series B*, **254**, 221–385.

Fryer, G. (1974) Evolution and adaptive radiation in the Macrothricidae (Crustacea, Cladocera): a study in comparative functional morphology and ecology. *Philosophical Transactions of the Royal Society of London, Series B*, **269**, 137–274.

Fryer, G. (1987) Morphology and the classification of the so-called Cladocera. *Hydrobiologia*, **145**, 19–28.

Fryer, G. (1996) Diapause, a potent force in the evolution of freshwater crustaceans. *Hydrobiologia*, **320**, 1–14.

Geller, W. & Müller, H. (1981) The filtration apparatus of cladocera: filter mesh-sizes and their implications on food selectivity. *Oecologia*, **49**, 316–21.

Gerritsen, J., Porter, K.G. & Strickler, J.R. (1988) Not by sieving alone: observation of suspension feeding in *Daphnia*. *Bulletin of Marine Science*, **43**, 366–7.

Gilbert, J.J. (1966) Rotifer ecology and embryological induction. *Science*, **151**, 1234–7.

Gilbert, J.J. & Bogdan, K.G. (1984) Rotifer grazing: *in situ* studies on selectivity and rates. In: Meyers, D.G. & Strickler, J.R. (eds), *Trophic Interactions within Aquatic Ecosystems*. Symposium Series 85, American Association for the Advancement of Science, Westview, Boulder, CO, 97–133.

Gliwicz, Z.M. (1969) Studies on the feeding of pelagic zooplankton in lakes with varying trophy. *Ekologia Polska A*, **17**, 663–708.

Gliwicz, Z.M. (1986a) Suspended clay concentration controlled by filter-feeding zooplankton in a tropical reservoir. *Nature*, **321**, 330–2.

Gliwicz, Z.M. (1986b). A lunar cycle in zooplankton. *Ecology*, **67**, 882–97.

Gliwicz, Z.M. (1990a) Why do cladocerans fail to control algal blooms? *Hydrobiologia*, **200/201**, 83–97.

Gliwicz, Z.M. (1990b) Food thresholds and body size in cladocerans. *Nature*, **343**, 638–40.

Gliwicz, Z.M. (1991) Food thresholds, resistance to starvation, and cladoceran body size. *Verhandlungen des internationale Vereinigung für theoretische und angewandte Limnologie*, **24**, 2795–8.

Gliwicz, Z.M. (1994). Retarded growth of cladoceran zooplankton in the presence of a copepod predator. *Oecologia*, **97**, 458–61.

Gliwicz, Z.M. & Biesiadka, E. (1975) Pelagic water mites (Hydracarina) and their effect on the plankton community in a neotropical man-made lake. *Archiv für Hydrobiologie*, **76**, 65–88.

Gliwicz, Z.M. & Boavida, M.J. (1996) Clutch size and body size at first reproduction in *Daphnia pulicaria* at different levels of food and predation. *Journal of Plankton Research*, **18**, 863–80.

Gliwicz, Z.M. & Guisande, C. (1992) Family planning in *Daphnia*: resistance to starvation in offspring born to mothers grown at different food levels. *Oecologia*, **91**, 463–7.

Gliwicz, Z.M. & Lampert, W. (1990) Food thresholds in *Daphnia* species in the absence and presence of blue-green filaments. *Ecology*, **71**, 691–702.

Gliwicz, Z.M. & Pijanowska, J. (1989) The role of predation in zooplankton succession. In: Sommer, U. (ed.), *Plankton Ecology. Succession in Plankton Communities*. Springer-Verlag, Berlin, 253–96.

Gliwicz, Z.M. & Rykowska, A. (1992) 'Shore avoidance' in zooplankton: a predator-induced behavior or predator-induced mortality? *Journal of Plankton Research*, **14**, 1331–42.

Gliwicz, Z.M. & Siedlar, E. (1980) Food size limitation and algae interfering with food collection in *Daphnia*. *Archiv für Hydrobiologie*, **88**, 155–77.

Gliwicz, Z.M., Ghilarov, A.M. & Pijanowska, J. (1981) Food and predation as major factors limiting two natural populations of *Daphnia cucullata* Sars. *Hydrobiologia*, **80**, 205–18.

Gliwicz, Z.M., Wurtsbaugh, W.A. & Ward, A. (1995) *Brine Shrimp Ecology in the Great Salt Lake, Utah*. June 1994–May 1995 Performance Report to the Utah Division of Wildlife Resources, Salt Lake City, Utah, 1–83.

Gliwicz, Z.M., Ślusarczyk, M. & Ślusarczyk, A. (2001) Life-history synchronization in a long-lifespan single-cohort *Daphnia* population of an alpine lake free of fish. *Oecologia*, **128**, 368–78.

Grant, J.W.G. & Bayly, I.A.E. (1981) Predator induction of crests in morphs of the *Daphnia carinita* King complex. *Limnology and Oceanography*, **26**, 201–18.

Gulati, R.D., Lammens, E.H.R.R., Meier, M.L. & van Donk E. (eds) (1990) Biomanipulation. Tool for water management. *Developments in Hydrobiology*, No. 61. *Hydrobiologia*, **200/201**, 628 pp.

Halbach, U. (1969) Das Zusammenwirken von Konkurrenz und Räuber bei Rädertieren. *Zoologiwscher Anzeiger (Supplementband)*, **33**, 72–91.

Hairston, N.G., Jr. (1987) Diapause as a predator avoidance adaptation. In: Kerfoot, W.C. & Sih, A. (eds), *Predation. Direct and Indirect Impacts on Aquatic Communities*. University Press of New England, Hanover, New Hampshire, 281–90.

Hairston, N.G., Jr. & Cáceres, C.E. (1996) Distribution of crustacean diapause: micro- and macroevolutionary pattern and process. *Hydrobiologia*, **320**, 27–44.

Haney, J.F. (1985) Regulation of cladoceran filtering rates in nature by body size, food concentration, and diel feeding patterns. *Limnology and Oceanography*, **30**, 397–411.

Haney, J.F. & Buchanan, C. (1987) Distribution and biogeography of *Daphnia* in the arctic. *Memorie dell'Istituto Italiano di Idrobiologia*, **45**, 77–105.

Hardin, G. (1960) The competitive exclusion principle. *Science*, **131**, 1292–7.

Hartmann, H.J. & Kunkel, D.D. (1991) Mechanisms of food selection in *Daphnia*. *Hydrobiologia*, **225**, 129–54.

Harvell, C.D. (1990) The ecology and evolution of inducible defenses. *The Quarterly Review of Biology*, **65**, 323–40.

Havel, J.E. (1987) Predator-induced defenses: a review. In: Kerfoot, W.C. & Sih, A. (eds), *Predation. Direct and Indirect Impacts on Aquatic Communities*. University Press of New England, Hanover, New Hampshire, 263–78.

Havel, J.E. & Dodson, S.I. (1984) *Chaoborus* predation on typical and spined morphs on *Daphnia pulex*: behavioral observations. *Limnology and Oceanography*, **29**, 487–94.

Hrbáček, J. (1962) Species composition and the amount of zooplankton in relation to the fish stock. *Rozpravy Československé Akademie Véd, Rada Matematickych a Přirodonich Véd*, **72**, 1–114.

Hutchinson, G.E. (1961) The paradox of the plankton. *American Naturalist*, **95**, 137–46.

Hutchinson, G.E. (1967) *A Treatise in Limnology*, Vol. II, *Introduction to Lake Biology and the Limnoplankton*. Wiley, New York, 1115 pp.

Jacobs, J. (1987) Cyclomorphosis in *Daphnia*. *Memorie dell'Istituto Italiano di Idrobiologia*, **45**, 325–52.

Jakobsen, P.J. & Johnsen, G.H. (1988) The influence of food limitation on swarming behaviour in the waterflea *Bosmina longispina*. *Animal Behaviour*, **36**, 991–5.

Kalinowska, A. & Pijanowska, J. (1987) How not to be eaten? Plant, prey and host on the defensive. *Wiadomości Ekologiczne*, **33**, 3–20. (In Polish with English summary.)

Kerfoot, W.C. (1974) Egg-size cycle of a cladoceran. *Ecology*, **55**, 1259–70.

Kerfoot, W.C. (ed.) (1980) *Evolution and Ecology of Zooplankton Communities*. University Press of New England, Hanover, New Hampshire, 793 pp.

Kerfoot, W.C. & Lynch, M. (1987) Branchiopod communities: associations with planktivorous fish in space and time. In: Kerfoot, W.C. & Sih, A. (eds), *Predation.*

Direct and Indirect Impacts on Aquatic Communities. University Press of New England, Hanover, New Hampshire, 367–78.

Kerfoot, W.C., Kellogg, D.L. Jr. & Strickler, J.R. (1980) Visual observations of live zooplankters: evasion, escape, and chemical defenses. In: Kerfoot, W.C. (ed.), *Evolution and Ecology of Zooplankton Communities*. University Press of New England, Hanover, New Hampshire, 10–27.

Kiefer, F. & Fryer, G. (1978) Das Zooplankton der Binnengewässer. *Die Binnengewässer* 2 Teil, 26. E. Schweizerbart'sche Verlagsbuchhandlung, Stuttgart, 380 pp.

Knisely, K. & Geller, W. (1986) Selective feeding of four zooplankton species on natural lake phytoplankton. *Oecologia*, **69**, 86–94.

Koehl, M.A.R. (1984) Mechanisms of particle capture by copepods at low Reynolds numbers: possible modes of selective feeding. In: Meyers, D.G. & Strickler, J.R. (eds), *Trophic Interactions within Aquatic Ecosystems*. Symposium Series 85, American Association for the Advancement of Science, Westview, Boulder, CO, 135–66.

Koza, V. & Kořinek, V. (1985) Adaptability of the filter screen in *Daphnia*: Another answer to the selective pressure of the environment. *Ergebnisse der Limnologie*, **21**, 193–8.

Lampert, W. (1977) Studies on the carbon balance of *Daphnia pulex* De Geer as related to environmental conditions. II. Determination of the 'threshold' concentration as a factor controlling the abundance of zooplankton species. *Archiv für Hydrobiologie, Supplement*, **48**, 361–8.

Lampert, W. (1982) Further studies on the inhibitory effects of the toxic blue-green *Microcystis aeruginosa* on the filtering rate of zooplankton. *Archiv für Hydrobiologie*, **95**, 207–20.

Lampert, W. (ed.) (1985) Food limitation and the structure of zooplankton communities. *Ergebnisse der Limnologie*, **21**, 497 pp.

Lampert, W. (1987) Feeding and nutrition in *Daphnia*. *Memorie dell'Istituto Italiano di Idrobiologia*, **45**, 143–92.

Lampert, W. (1994) Phenotypic plasticity of the filter screens in *Daphnia*: Adaptation to a low-food environment. *Limnology and Oceanography*, **39**, 997–1006.

Lampert, W. & Muck, P. (1985) Multiple aspects of food limitation in zooplankton communities: the *Daphnia-Eudiaptomus* example. *Ergebnisse der Limnologie*, **21**, 311–22.

Lampert, W. & Schober, U. (1980) The importance of 'threshold' food concentrations. In: Kerfoot, W.C. (ed.), *Evolution and Ecology of Zooplankton Communities.* University Press of New England, Hanover, New Hampshire, 264–7.

Lampert, W. & Sommer, U. (1993) *Limnoökologie.* Georg Thieme Verlag, Stuttgart, 440 pp.

Lampert, W. & Wolf, H.G. (1986) Cyclomorphosis in *Daphnia cucullata*: morphometric and population genetics analyses. *Journal of Plankton Research*, **8**, 289–303.

Langeland, A. (1982) Interactions between zooplankton and fish in a fertile lake. *Holarctic Ecology*, **5**, 273–310.

Larsson, P. & Dodson, S.I. (1993) Chemical communication in planktonic animals. *Archiv für Hydrobiologie*, **129**, 129–55.

Larsson, P. & Waider, L.J. (eds) (1995) Cladocera as model organisms in biology. *Developments in Hydrobiology*, No. 107. *Hydrobiologia*, **307**, 307 pp.

Loose, C.J. & Dawidowicz, P. (1994) Trade-offs in diel vertical migration by zooplankton: the costs of predator avoidance. *Ecology*, **75**, 2255–63.

Lynch, M. (1980) *Aphanizomenon* blooms: alternate control and cultivation by *Daphnia pulex*. In: Kerfoot, W.C. (ed.), *Evolution and Ecology of Zooplankton Communities.* University Press of New England, Hanover, New Hampshire, 299–304.

Lynch, M. (1989) The life history consequences of resource depression in *Daphnia pulex*. *Ecology*, **70**, 246–56.

McMahon, J.W. & Rigler, F.H. (1963) Mechanisms regulating the feeding rate of *Daphnia magna* Straus. *Canadian Journal of Zoology*, **41**, 321–32.

McMahon, J.W. & Rigler, F.H. (1965) Feeding rate of *Daphnia magna* Straus in different foods labeled with radioactive phosphorus. *Limnology and Oceanography*, **10**, 105–13.

Nilssen, J.P. (1977) Cryptic predation and the demographic strategy of two limnetic cyclopoid copepods. *Memorie dell'Istituto Italiano di Idrobiologia*, **34**, 187–96.

Nogrady, T., Wallace, R.L. & Snell, T.W. (1993) Rotifera, Volume 1: Biology, Ecology and Systematics. In: H.J.F. Dumont (ed.), *Guides to the Identification of the Microinvertebrates of the Continental Waters of the World, No. 4.* SPB Academic Publishing, Amsterdam, 142 pp.

O'Brien, W.J. (1987) Planktivory by freshwater fish: thrust and parry in the pelagia. In: Kerfoot, W.C. & Sih, A. (eds), *Predation. Direct and Indirect Impacts on Aquatic Communities.* University Press of New England, Hanover, New Hampshire, 3–16.

Okamoto, K. (1984) Size-selective feeding of *Daphnia longispina hyalina* and *Eudiaptomus japonicus* on a natural phytoplankton assemblage with the fractionizing method. *Memoirs of the Faculty of Science, Kyoto University, Series of Biology*, **9**, 23–40.

Orłowska, A. (1997) *Demography of Daphnia pulicaria in two alpine ultra-oligotrophic lakes of the Tatra.* MSc thesis, University of Warsaw, 28 pp.

Paine, R.T. (1966) Food web complexity and species diversity. *American Naturalist*, **110**, 65–75.

Papińska, K. (1984) The life cycle and the zone of occurrence of *Mesocyclops leuckarti* Claus (Cyclopoida, Copepoda). *Ekologia Polska*, **32**, 493–531.

Pastorok, R.A. (1980) Selection of prey by *Chaoborus* larvae: a review and new evidence of behavioral flexibility. In: Kerfoot, W.C. (ed.), *Evolution and Ecology of Zooplankton Communities.* University Press of New England, Hanover, New Hampshire, 538–54.

Peters, R.H. (1987) Metabolism in *Daphnia. Memorie dell'Istituto Italiano di Idrobiologia*, **45**, 193–243.

Peters, R.H. & De Bernardi, R. (eds) (1987) Daphnia. *Memorie dell'Istituto Italiano di Idrobiologia*, **45**, 502 pp.

Pijanowska, J. (1990) Cyclomorphosis in *Daphnia*: an adaptation to avoid invertebrate predation. *Hydrobiologia*, **198**, 41–50.

Pijanowska, J. (1992) Anti-predator defense in three *Daphnia* species. *Internationale Revue des gesamten Hydrobiologie*, **77**, 153–63.

Pijanowska, J. (1993) Diel vertical migration in zooplankton: fixed or inducible behavior? *Ergebnisse der Limnologie*, **39**, 89–97.

Pijanowska, J. (1994) Fish-enhanced patchiness in *Daphnia* distribution. *Verhandlungen des internationale Vereinigung für theoretische und angewandte Limnologie*, **25**, 2366–8.

Pijanowska, J. & Kowalczewski, A. (1997) Predators can induce swarming behaviour and locomotory responses in *Daphnia. Freshwater Biology*, **37**: 649–56.

Pijanowska, J. & Stolpe, G. (1996) A cue from a fish predator can induce summer diapause in *Daphnia. Journal of Plankton Research*, **18**, 1407–12.

Pijanowska, J., Weider, L.W. & Lampert, W. (1993) Predator-mediated genotypic shifts in a prey population: experimental evidence. *Oecologia*, **96**, 40–2.

Porter, K.G. (1976) Enhancement of algal growth and productivity by grazing zooplankton. *Science*, **192**, 1332–4.

Price, H.J. (1988) Feeding mechanisms in marine and freshwater zooplankton. *Bulletin of Marine Science*, **43**, 327–43.

Pütter, A. (1909) *Die Ernährung der Wassertiere und der Stoffhaushalt der Gewässer.* Fischer, Jena, 168 pp.

Reddy, Y.R. (1994) Copepoda: Calanoida: Diaptomidae, Key to the genera *Heliodiaptomus, Allodiaptomus, Neodiaptomus, Phyllodiaptomus, Eodiaptomus, Arctodiaptomus and Sinodiaptomus.* In: Dumont, H.J.F. (ed.), *Guides to the Identification of the Microinvertebrates of the Continental Waters of the World No. 5.* SPB Academic Publishing, Amsterdam, 221 pp.

Reeve, M.R. (1963) The filter-feeding of *Artemia.* I. In pure cultures of plant cells. *Journal of Experimental Biology,* **40**, 195–205.

Ringelberg, J. (1969) Spatial orientation of planktonic crustaceans. 2. The swimming behavior in a vertical plane. *Verhandlungen des internationale Vereinigung für theoretische und angewandte Limnologie,* **17**, 841–7.

Ringelberg, J. (ed.) (1993) Diel vertical migration of zooplankton. *Ergebnisse der Limnologie,* **39**, 222 pp.

Romanovsky, Y.E. & Feniova, I.Y. (1985) Competition among cladocera: effects of different levels of food supply. *Oikos,* **44**, 243–52.

Rothhaupt, K.O. (1990) Resource competition of herbivorous zooplankton: a review of approaches and perspectives. *Archiv für Hydrobiologie,* **118**, 1–29.

Sakwińska, O. (1997) *Life history and habitat shifts as Daphnia's anti-predator defences in fifteen Polish lakes.* MSc thesis, University of Warsaw, 22 pp.

Schoener, T.W. (1973) Population growth regulated by intraspecific competition for energy or time: some simple representations. *Theoretical Population Biology,* **4**, 56–84.

Segers, H. (1995) Rotifera, Volume 2: The Lecanidae (Monogononta). In: Dumont, H.J.F. (ed.), *Guides to the Identification of the Microinvertebrates of the Continental Waters of the World No. 6.* SPB Academic Publishing, Amsterdam, 223 pp.

Siebeck, O. (1969) Spatial orientation of planktonic crustaceans. 1. The swimming behavior in a horizontal plane. *Verhandlungen des internationale Vereinigung für theoretische und angewandte Limnologie,* **17**, 831–40.

Shapiro, J. & Wright, D.I. (1984) Lake restoration by biomanipulation: Round Lake, Minnesota, the first two years. *Freshwater Biology,* **14**, 371–83.

Ślusarczyk, M. (1995) Predator-induced diapause in *Daphnia. Ecology,* **76**, 1008–13.

Smirnov, N.N. (1992) The Macrothricidae of the World. In: Dumont, H.J.F. (ed.), *Guides to the Identification of the Microinvertebrates of the Continental Waters*

of the World, No. 1. SPB Academic Publishing, Amsterdam, 43 pp.

Sommer, U. (ed.) (1989) *Plankton Ecology. Succession in Plankton Communities.* Springer-Verlag, Berlin, 369 pp.

Sommer, U., Gliwicz, Z.M., Lampert, W. & Duncan, A. (1986) The PEG-model of seasonal succession of planktonic events in fresh waters. *Archiv für Hydrobiologie,* **106**, 433–71.

Stemberger, R.S. (1988) Reproductive costs and hydrodynamic benefits of chemically induced defenses in *Karatella testudo. Limnology and Oceanography,* **29**, 487–94.

Stemberger, R.S. & Gilbert, J.J. (1985) Assessment of threshold food levels and population growth in planktonic rotifers. *Ergebnisse der Limnologie,* **31**, 269–76.

Stemberger, R.S. & Gilbert, J.J. (1987) Defense of planktonic rotifers against predators. In: Kerfoot, W.C. & Sih, A. (eds), *Predation. Direct and Indirect Impacts on Aquatic Communities.* University Press of New England, Hanover, New Hampshire, 227–39.

Sterner, R.W. (1989) The role of grazers in phytoplankton succession. In: Sommer, U. (ed.), *Plankton Ecology. Succession in Plankton Communities.* Springer-Verlag, Berlin, 107–70.

Sterner, R.W. & Robinson, J.L. (1994) Threshold for growth in *Daphnia magna* with high and low phosphorus diets. *Limnology and Oceanography,* **39**, 1228–32.

Stich, J. & Lampert, W. (1981) Predator evasion as an explanation for vertical migration by zooplankton. *Nature,* **293**, 396–8.

Storch, O. (1924) Der Phyllopoden-Fangapparat. 1 Theil. *Internationale Revue des gesamten Hydrobiologie,* **12**, 369–91.

Swift, M.C. (1992) Prey capture by the four larval instars of *Chaoborus crystallinus. Limnology and Oceanography,* **37**, 14–24.

Taylor, B.E. & Gabriel, W. (1992) To grow or not to grow: optimal resource allocation for *Daphnia. American Naturalist,* **139**, 248–66.

Tessier, A.J. (1983) Coherence and horizontal movements of patches of *Holopedium gibberum* (Cladocera). *Oecologia,* **60**, 71–5.

Tessier, A.J. & Goulden, C.E. (1982) Estimating food limitation in cladoceran population. *Limnology and Oceanography,* **27**, 707–17.

Threlkeld, S.T., Rybock, J.T., Morgan, M.D., Folt, C.L. & Goldman, C.R. (1980) The effects of an introduced invertebrate predator and food resource variation on zooplankton dynamics in an ultraoligotrophic lake. In: Kerfoot, W.C. (ed.), *Evolution and Ecology of*

Zooplankton Communities. University Press of New England, Hanover, New Hampshire, 555–68.

Tollrian, R. (1995) Predator-induced morphological defenses: costs, life history shifts, and maternal effects in *Daphnia pulex*. *Ecology*, **76**, 1691–705.

Vanderploeg, H.A., Scavia, D. & Liebig, J.R. (1984) Feeding rate of *Diaptomus sicilis* and its relation to selectivity and effective food concentration in algal mixtures in Lake Michigan. *Journal of Plankton Research*, **6**, 919–41.

Vanni, M.J. (1987) Effects of nutrients and zooplankton size on the structure of a phytoplanktonic community. *Ecology*, **68**, 624–35.

Wesenberg-Lund, C. (1908) *Plankton Investigations of the Danish Lakes*. Gyldendalske Boghandel, Nordisk Vorlag, Copenhagen, 389 pp.

Węgleńska, T., Ejsmont-Karabin, J. & Rybak, J.I. (1997) Biotic interactions of the zooplankton community of a shallow, humic lake. *Hydrobiologia*, **342/343**, 185–95.

Weider, L.J. & Pijanowska, J. (1993) Plasticity of *Daphnia* life histories in response to chemical cues from predators. *Oikos*, **67**, 385–92.

Wurtsbaugh, W.A. (1992) Food-dash web modification by an invertebrate predator in the Great Salt Lake (USA). *Oecologia*, **89**, 168–75.

Zaret, T.M. (1972) Predators, invisible prey, and the nature of polymorphism in the Cladocera (class Crustacea). *Limnology and Oceanography*, **17**, 171–84.

Zaret, T.M. (1978) A predation model of zooplankton community structure. *Verhandlungen des internationale Vereinigung für theoretische und angewandte Limnologie*, **20**, 2496–500.

Zaret, T.M. (1980) *Predation and Freshwater Communities*. Yale University Press, 180 pp.

15 Fish Population Ecology

IAN J. WINFIELD

15.1 INTRODUCTION

This chapter is written not for the lake fish ecologist but primarily for limnologists with other expertise. Whilst literature relevant to the former audience will be indicated, the primary objective is to introduce limnologists in general to the population ecology of lake fish, with an emphasis where possible on interactions with other aspects of lake ecosystems. Interactions between and among fish populations **of different species** are not addressed here, but are considered in Chapter 16.

In order to achieve the above objective, and indeed better to understand that chapter, some appreciation of a few key aspects of fish biology is essential, and so these will first briefly be described. This will be followed by consideration of the fundamental aspects of lake fish population ecology as they have been developed by fish biologists, with a historical emphasis on fish populations themselves, but also considering how this subject has more recently been widened by addressing environmental influences.

The ways in which the ecology of fish populations transforms individual feeding behaviour into the ecological roles of fish in lake ecosystems will then be considered. This complex subject, which cannot be thoroughly reviewed within the confines of the present chapter, nevertheless possesses strong links with other chapters (e.g. Chapter 16 and Volume 2, Chapter 18) and so warrants at least limited consideration here. Finally, some personal thoughts will be given in the form of concluding remarks which will address aspects of lake fish population ecology as it stands today, together with some suggestions concerning its possible development in the foreseeable future.

15.2 SOME KEY ASPECTS OF FISH BIOLOGY

15.2.1 Introduction

Those aspects of lake fish biology which are essential for the appreciation and understanding of fish population ecology include species diversity, body size, movement, sensory systems, behaviour and the manifestation of the cumulative effects of these features in the form of ontogenetic niche shifts. For the enquiring limnologist, a much more comprehensive treatment of basic fish biology may be found in Bone *et al.* (1995).

Before considering the above topics, it is worth emphasising that these and other aspects of fish biology are typically portrayed against a background of relatively great longevity, which itself constitutes a major difference between fish and most other lake fauna. As a result, the population ecology of lake fish must be both studied and interpreted on a timescale of years rather than months, weeks or days. In addition to increasing considerably the theoretical complexity of fish population dynamics, longevity also complicates their practical study, through a combination of substantial time lags between cause and effect, and sampling problems arising from the immense size and morphological changes exhibited by individual fish during their lifetime.

15.2.2 Species diversity

The word 'fish' is viewed with misgivings by many biologists. This is because it is a very broad term which encompasses a great variety of vertebrate animals, many of which share few biological, ecological or even evolutionary features. It is not surprising therefore that these occur in habitats as diverse as small mountain head waters and oceanic depths, and together probably constitute up to 28,000 species (Nelson 1994). Of these, 41% occur in fresh waters, even though such habitats cover only 1% of Earth's surface, in contrast to the 70% occupied by salt water (Moyle & Cech 1996). On a global basis, the three most important orders of freshwater fish are the Cypriniformes (c.3000 species), the Siluriformes (c.1950 species), and the Perciformes (c.950 species; Moss 1988), although in the temperate parts of the world, where, historically, most fish population ecology has been carried out, the Siluriformes are replaced in importance by the Salmoniformes.

Because of their commercial importance in terms of food or recreational fisheries, the diversity of fish has been relatively well studied when compared with that of other vertebrates, although coverage has been somewhat selective. Major multi-authored texts have been published on lake fish families such as the Cichlidae (Keenleyside 1991) and the Cyprinidae (Winfield & Nelson 1991), and even largely on single genera such as *Esox* (Craig 1996). In addition, genera of outstanding commercial importance, examples of which include *Salvelinus* (e.g. Magnan *et al.* 2002), *Perca* (e.g. Rask *et al.* 1996) and *Coregonus* (e.g. Eckmann *et al.* 1998), are the subjects of repeated international symposia. Consequently, lake fish population ecology is best understood for these particular taxa, all of which fall within the three major Orders listed above and which together will represent the main examples given in the present chapter.

15.2.3 Body size

As well as their great diversity, perhaps the other most distinguishing feature of fish to the limnologist is their relatively large body size. In most lakes, fish are by far the largest organisms present in the water column, often by orders of magnitude. However, the vast majority of species begin life at a size smaller than that of some invertebrates. Thus, as it grows from a newly hatched larva into an organism which can out-swim and out-manoeuvre the most agile human, great changes in the biology and ecology of an individual fish occur (Fig. 15.1). Such changes usually involve significant ontogenetic niche shifts, which will be briefly considered later.

In addition to the numerous advantages which accrue from large body size, two major disadvantages relate to the basic physics of life in water. As the medium is c.800 times more dense than air, fish must first overcome considerable resistance to movement. Most species spend some time as larvae, experiencing an intermediate range of Reynolds numbers between 30 and 200, where viscous and inertial forces are equally important (as they are for many other aquatic taxa). The growth of fish, however, means that inertial forces eventually become predominant (Fuiman & Webb 1988). In such a context, movement can only be achieved by expending considerable amounts of energy.

The second problem is that the required energy production must be achieved even though oxygen in water is usually a far scarcer commodity than it is in air, so that it can easily become limiting to a

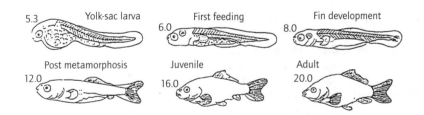

Fig. 15.1   Changes in the morphology of common carp (*Cyprinus carpio* L.) with increasing age and size. Lengths are given in millimetres beside each figure, but note that the adult stage may attain a length of 1000 mm. (Redrawn from Bone *et al.* 1995; reproduced with kind permission of Kluwer Academic & Lippincott-Raven Publishers incorporating Chapman & Hall and Rapid Science.)

large-bodied animal with an inherently low surface area to volume ratio. Fish usually overcome this difficulty via the use of gills, in which an immense effective surface area acts to allow efficient acquisition of oxygen from environmental densities as little as c.1 mg L^{-1}, depending on the species involved. Nevertheless, particularly low concentrations of dissolved oxygen commonly restrict the distribution and movement of lake fish, despite the ability of some species to make brief trips into apparently lethal hypoxic environments (Rahel & Nutzman 1994).

15.2.4 Movement

With large body size comes an ability to overcome almost all water movements encountered, and thus to leave the plankton and to travel in three dimensions to specific habitats around a lake, or even beyond. Such movements, which are usually very complex and specific, even within a relatively simple assemblage of lake species (Helfman 1981), may be for the purposes of (i) feeding (e.g. bluegill sunfish (*Lepomis macrochirus*); Werner *et al.* 1983a), (ii) predator avoidance (e.g. bluegill sunfish; Werner *et al.* 1983b), or (iii) spawning (e.g. whitefish (*Coregonus lavaretus*); Eckmann 1991). It is particularly important to appreciate that most fish species utilise various habitats within a lake for different purposes, even if some uses (such as laying eggs on just 1 day on very localised spawning grounds) are only for very restricted periods.

In addition to facilitating deliberate movements around the lake, the relatively large body size of most fish species allows some of them to leave such water bodies entirely, by entering inflowing and outflowing tributaries. Such journeys may be undertaken for the same reasons as within-lake movements, and may even involve migrations to and from the marine environment (as shown, for example, by some populations of Arctic charr (*Salvelinus alpinus*) for feeding purposes (Berg 1995), and by all populations of the European freshwater eel (*Anguilla anguilla*) for spawning (Svedäng *et al.* 1996)). The population ecology of such species may therefore actually be influenced by factors operating far away from the lake in which they spend the greater part of their lives.

15.2.5 Sensory systems

Large body size also allows development and accommodation of sensory systems with which fish can perceive many subtle aspects of their environment, including their own location, and that of their prey, predators, competitors, peers and potential mates. Vision, olfaction and the acoustico-lateralis system are three of the most important senses used by most lake fish.

Whilst some form of vision is used by many lake animals, even if only for detecting the direction of a light source, this sense is at its most elaborate and sensitive in fish. Colour and binocular vision are used extensively by many species (Guthrie & Muntz 1993), and some are capable of feeding using vision, even when light is limited by depth or the time of day. In particular, members of the Cypriniformes are particularly tolerant of low light intensities when foraging on zooplankton and other prey, and can even feed efficiently under light levels associated with starlight (Townsend & Winfield 1985).

Olfactory abilities are also extremely well developed in many lake fish species and are used to find food, mates and spawning grounds, to avoid predators and (in some species) to navigate during extensive migrations. A review of this important sense is provided by Hara (1993). Finally, the acoustico-lateralis system is most obviously visible in many fish species as a distinct lateral line, although closer examination usually reveals that other structures are also present elsewhere on the body surface. In most lake fish species, this system is composed chiefly of mechanoreceptive neuromasts, although specialised electroreceptors stimulated by weak electric fields are sometimes also present (Bleckmann 1993). The lateral line is essentially an organ of 'distant touch' and allows fish to detect potential prey, predators and other organisms, and in addition plays a role in orientation and on social behaviour such as schooling.

In summary, the sensory systems of lake fish are very highly developed and allow them to detect and to react rapidly to many aspects of their environment. For the limnologist, an important point to note is that, whilst many species will preferentially forage most actively and efficiently under the brightly lit conditions of the day (at least when not under intense predation pressure themselves), many are equally active either at night or in otherwise darkened habitats.

15.2.6 Behaviour

Large body size, high movement capability and elaborate sensory systems combine to facilitate complex patterns of behaviour, including learning, in many lake fish. Foraging, predator avoidance, mate selection and many other aspects of fish behaviour have been extensively researched and a number of excellent summaries are presented in Pitcher (1993). Further summary is inappropriate here, but the limnologist is reminded that behavioural issues have been found to be of crucial importance for the understanding both of lake fish population dynamics (e.g. Carpenter & Kitchell 1996) and of interactions with other components of lake biota (e.g. Jacobsen *et al.* 1997).

15.2.7 Ontogenetic niche shifts

Relatively large body size is usually only attained by lake fish over the course of several years. Another major difference between fish and most other lake fauna is therefore considerable longevity. During attainment of these considerable sizes and ages, an individual fish may go through a series of marked ontogenetic shifts in its ecology which are comparable in extent to the interspecific differences commonly observed among invertebrate fauna. Thus, environmental tolerance ranges may change markedly over the lifetime of a fish, and what may be suitable habitat for juvenile stages may be completely unsuitable for adults. Such niche shifts have been found to be of significance for population and community ecology both in North American lake systems (e.g. Werner & Gilliam 1984) and European lakes (e.g. Persson 1988).

15.3 FUNDAMENTAL ASPECTS OF AND ENVIRONMENTAL INFLUENCES ON FISH POPULATION ECOLOGY

15.3.1 Introduction

Aquisition of much present knowledge about fish population ecology has been driven by the pragmatic requirements of fisheries science. For the limnologist, there have been three important consequences. First, the greatest research effort has been directed at those species which are of greatest economic importance. On a global basis, this has led to concentration on marine taxa, which undeniably support the world's most important fisheries. Even in fresh waters, the study of lake fish populations has generally received less attention than that of their near-universally exploited riverine counterparts. Thus, the population ecology of lake fishes is still very much a developing science.

Second, most effort has also been centred on the larger size classes of fish populations, which are of greatest direct importance to fisheries. Such classes are not, however, necessarily those of greatest importance to the limnologist. Third, even amongst the research on lake fish population ecology which has been undertaken, most has been carried out at fisheries institutes rather than at limnological or more general ecological laboratories. Thus, historically, there has been a gulf between most freshwater fish research and that on other components of lake ecosystems. Often, it is only the latter which is thought of as true limnology.

The net result of these restraints is that the most abundant components of lake fish populations, i.e. the youngest and thus smallest size groups, are amongst the most poorly studied of the world's piscine fauna. This is further compounded by the fact that lake fish generally, and small fish in particular, are often extremely difficult to sample in a scientific and quantitative manner (see papers in Mehner & Winfield 1997). Nevertheless, much progress in the field of lake fish population ecology has taken place, and the topics which will be covered in this review include distribution and abundance, mortality, growth, reproduction, and determination of year-class strength. Particular emphasis will be placed on the ways in which these aspects of fish population ecology are influenced by environmental factors.

15.3.2 *Distribution and abundance*

Given their locomotory abilities, patterns of spatial distribution should always be acknowledged in any examination of lake fish population abundance. Movement in terms of general spatial distributions and specific avoidance reactions to sampling equipment remain major problems both in studies of absolute and of relative population abundance (Cowx 1991, 1996). Studies of fish population dynamics are usually conducted on a group of individuals which may be referred to as a **unit stock**. This is defined as a self-perpetuating discrete group, with a shared gene pool, and little connection with adjacent populations (King 1995). Clearly, in determining the degree of any such connections which may exist among groups of fish within a lake, or even between those in different lakes (via riverine migrations), and thus the number of unit stocks to be considered, knowledge of fish movements is invaluable.

General spatial distribution patterns and accompanying movements also contribute to the complex sampling demands of fish population studies and, together with avoidance movements, mean that the study of both pelagic populations (e.g. Post *et al.* 1994) and their littoral counterparts (e.g. Weaver & Magnuson 1993) requires a flexible approach. Put simply, there is no single methodology which can produce an unbiased sample of a lake fish population for all species under all field conditions. Moreover, even the most appropriate sampling method for a given set of circumstances may produce data of low accuracy (e.g. gill net sampling for adults; Kelso & Shuter 1989) or precision (e.g. buoyant net sampling for larvae; Cryer *et al.* 1986). Consequently, it is recommended that limnologists should always be wary of lake fish population abundance data.

Given the above methodological problems, and the additional point that the great longevity of fish requires that population studies be of long duration, it is not surprising that there have been relatively few studies of lake fish population dynamics of appropriate timescales. Moreover, most of those which have been conducted owe their origins to, or are closely allied to, commercial fisheries (e.g. for yellow perch (*Perca flavescens*) in North America (Bronte *et al.* 1993) and vendace (*Coregonus albula*) in Europe (Helminen & Sarvala 1994)) and so are not equally distributed between lake fish taxa.

The two preceding examples were purposely selected because they constitute specific instances of lake fish population dynamics driven primarily by natural mechanisms, rather than by fishing mortality. They therefore form examples of lake fish population ecology, rather than fisheries ecology. Most of the remaining examples probably involve considerable effects of fishing.

It is self-evident that to study fundamental population ecology in fish which are subject to significant effects originating from fisheries is not an easy task. Moreover, the population dynamics of many species, including even those in unexploited systems, are a function not simply of their own autecology but of the ecology of interactions within the fish communities in which they reside. Such forms of population ecology are properly considered within the context of community ecology and so are covered in Chapter 16, although one example is given here in Fig. 15.2. This illustrates changes in abundance (measured in the relative term of **catch-per-unit-effort**) of perch (*Perca fluviatilis*), pike (*Esox lucius*) and Arctic charr in Windermere, UK. Note the considerable short-term variation in relative abundance of perch following a dramatic reduction in numbers by disease in 1976, a long-term increase in pike attributable to decreased fishing pressure, and a long-term increase and subsequent decrease in Arctic charr with the latter feature probably attributable at least in part to increased predation pressure from pike. All of these features are only apparent because the sampling programmes for these three species are of appropriate lengths for animals with generation times of several years. Further details of earlier parts of the Windermere long-term fish

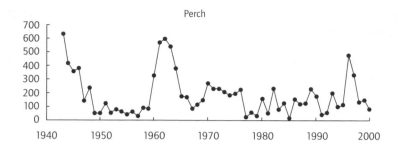

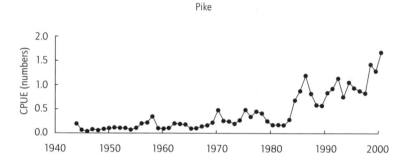

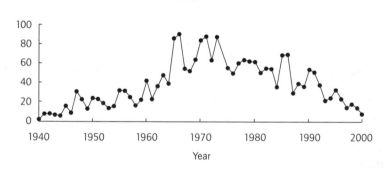

Fig. 15.2 Changes in catch-per-unit-effort (CPUE) of perch (*Perca fluviatilis*), pike (*Esox lucius*) and Arctic charr (*Salvelinus alpinus*) in the north basin of Windermere, UK, from 1940 to 2000. (Data are from long-term monitoring programmes of CEH Windermere and the Freshwater Biological Association, UK.)

population monitoring, and some investigation of their interactions, are given in Le Cren (2001) and references therein.

There are, however, two groups of studies of fish populations in which fishing effects, and interspecific interactions, are not important. They therefore serve as examples of relatively simple lake fish population ecology, and will be referred to a number of times in this review. These are the studies of vendace in Pyhäjärvi, south west Finland (see also chapter 16), and roach (*Rutilus rutilus*) in Alderfen Broad, UK.

Although the vendace population of Pyhäjärvi has been exploited by a commercial fishery for many years, analysis of population data from 1971 to 1990 reveals a 2-year cycle in abundance, driven by changes in recruitment rather than the effects of fishery (Helminen & Sarvala 1994). The mechanisms producing such variations in year-class strength will be returned to later, although it is pertinent to note here that, whilst environmental factors (including predator abundance and the timing of lake warming) were shown by multiple regressions to be significant, the basic reason is asymmetrical food competition within the vendace population. This same mechanism was also found to be responsible for a 2-year cycle of fish abundance in a very different species and lake (i.e. unexploited roach in Alderfen Broad, UK). The cycle was empirically evident in an analysis of roach abundance from 1978 to 1985 (Perrow *et al.* 1990) and its mechanism will be considered later.

All of the above examples of lake fish population dynamics share at least one key feature (i.e. they all display a noticeable absence of stability in population abundance). Moreover, such variation is only likely to increase for fish populations in which community interactions are more important. As a consequence, it is unlikely that any lake fish population will be modelled in the near future with any degree of success approaching that attained by Elliott (1994) for brown trout (*Salmo trutta*) inhabiting an essentially monospecific system in a small stream. In addition, as noted by Reynolds (1997), variability in lake fish population abundance operates on a temporal scale much longer than that exhibited by lower trophic levels in lakes. The consequences of such asynchrony for the organisation of pelagic ecosystems are likely to be considerable.

15.3.3 *Mortality*

Quantitative studies of mortality in lake fish populations are extremely rare, largely because of the difficulty of making the necessary serial estimates of population abundance. It is notable that none of the examples used by Wootton (1990) to show patterns of mortality in fish populations originated in lakes. Studies of mortality of lake fish have also usually been of the early larval or juvenile stages, before individual growth and development has rendered quantitative sampling extremely difficult in all but the most tractable of situations. Thus, Rice & Crowder (1985) have shown that

year-to-year variation in survival of the bloater (*Coregonus hoyi*) in Lake Michigan, USA, is due to mortality from predation in the first month after hatching, whilst in a laboratory study Papoulias & Minckley (1990) found evidence that mortality from starvation between 8 and 19 days after hatching is similarly important for razorback sucker (*Xyrauchen texanus*). Many more examples could be given of the magnitude of mortality during this early period of the first few days, or even the first summer of life, including the effects of environmental degradation such as acidification (Mohr *et al.* 1990). In addition, further significant mortality may occur over the following first winter. The latter may be size-selective for smaller individuals with lower energy reserves (e.g. for roach in Lough Neagh, UK; Griffiths & Kirkwood 1995), or may not (e.g. for largemouth bass (*Micropterus salmoides*) in Illinois reservoirs, USA; Kohler *et al.* 1993).

Estimates of mortality rates for older fish are less frequent in the literature, certainly in the context of their importance for population dynamics, but two examples are given here. First, Mooij *et al.* (1996), in a study of bream (*Abramis brama*) in Tjeukemeer, The Netherlands, between 1976 and 1986, found that year-class strength was determined by mortality during the second summer, even though first summer mortality was also high. This later mortality was characterised by instantaneous rates varying from 0.93 to 6.17, and predation by zander (pikeperch; *Stizostedion lucioperca*) was suggested to be its source. The second example is again of the roach population of Alderfen Broad, UK, for which Townsend *et al.* (1990) investigated the basis of the observed 2-year cycle by using a population model based on a Leslie matrix with variable survival rates. It was found that when annual survival rate in the population was higher than c.50% (as it is in most roach populations) cycling did not occur. In Alderfen Broad, survival rate from 1979 to 1987 was estimated by annual surveys at 19% and so the cycling dynamics of this population corresponded with the model's predictions.

In view of the paucity of appropriate and robust data, it is difficult to draw any general conclusions about mortality in the role of lake fish population ecology. It is regrettable that no lake fish population has yet been subjected to a k-factor analysis, in which the mortality rate of each life-stage is determined, as performed by Elliott (1994) for a stream-dwelling population of brown trout. Unfortunately, these circumstances are likely to persist until quantitative estimation of lake fish population abundance at all life-stages becomes far more accurate and precise. This is unlikely to occur in the foreseeable future.

15.3.4   Growth

For the limnologist, the most important features of fish growth are that (i) it is great in absolute terms, (ii) highly variable and (iii), in population terms, is concentrated in the young life stages. As shown in Fig. 15.1, during its lifetime, an individual fish may grow from a few millimetres to perhaps 1000 mm in length, with a corresponding increase in weight from c.1 g to several tens of kilograms. However, in addition to the great interspecific differences in such growth rates which might be expected amongst any vertebrate taxa, fish are remarkable for the plasticity of their intraspecific growth patterns. This variation is illustrated in Fig. 15.3, which depicts growth curves determined for roach from seven lakes and rivers within its natural Eurasian distribution range. Although several populations exhibit little variation in length of individuals 1 year old, differences thereafter develop rapidly, so that, for example, a 10 year old fish may be between c.100 and 500 mm in length. In addition to differences in the ultimate length, the rate at which this size is attained may also vary.

Individual longevity, and the common restriction of reproduction to a limited period each year, combine in many lake fish populations to produce a multi-modal length frequency distribution, with peaks in abundance representing individuals originating from successive years. An example for an unexploited population of largemouth bass in Paul Lake, Wisconsin, USA, from 1984 to 1990, is given in Fig. 15.4a. Variations in numbers of the smallest size group of individuals, which set the basis for the abundance of larger fish in subsequent years,

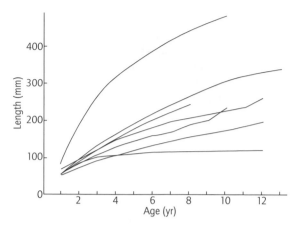

Fig. 15.3   Growth curves of seven roach (*Rutilus rutilus*) populations. (Redrawn from Mann 1991; reproduced with kind permission of Kluwer Academic & Lippincott-Raven Publishers incorporating Chapman & Hall and Rapid Science.)

arise from inter-annual differences in recruitment, or year-class strength (a topic returned to below). Figure 15.4b shows similar data for largemouth bass from nearby Peter Lake, from which most individuals were experimentally removed during 1985 (Hodgson *et al.* 1996). At the same time, a very strong year class was produced which then progressed as a peak in the population length frequency distribution of the following years, being joined by further year classes in 1988, 1989 and 1990. Although this example is particularly extreme (because it is drawn from an experimental manipulation), it serves to illustrate the occasional natural production of a strong year class which has been observed in many lake fish populations. Occurrence of such strongly size-structured populations means that the average size of fish in a given lake is an almost meaningless parameter. Unfortunately for the limnologist, most fish populations in fact comprise a series of size classes, each potentially with its own ecology.

Given this complexity of individual growth, and the importance of individual size as a key issue in any fishery, it is not surprising that the literature on the growth of lake fish is immense. For example, a search of *Aquatic Sciences and Fisheries*

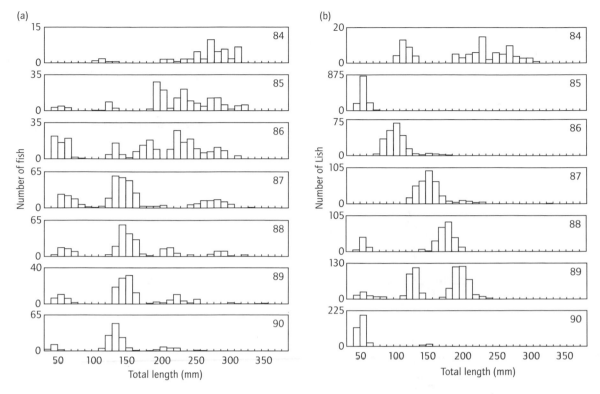

Fig. 15.4 Length-frequency histograms of largemouth bass (*Micropterus salmoides*) in (a) Paul Lake and (b) Peter Lake, Wisconsin, USA, from 1984 to 1990. (Redrawn from Hodgson *et al.* 1996; reproduced with kind permission of Cambridge University Press.)

Abstracts from 1978 to April 2002, using the search terms 'fish', 'lake or reservoir', and 'growth', produced 3728 references. Review of this entire field is both inappropriate and impossible in the present context, but the interested reader may instead consult an excellent and extensive review of the biology of fish growth produced by Weatherley & Gill (1987). In contrast, attention will be focused on aspects of fish growth of particular relevance to the limnologist.

As for most aquatic biota, increasing temperature generally exerts a strong positive influence on the growth of fish, although upper limits do of course exist and vary markedly between species. In general, members of the Salmoniformes and the Cypriniformes may be respectively considered as cool- and warm-water species, with Perciformes occupying an intermediate position. Temperature clearly varies between water bodies and is no doubt one of the main factors involved in producing the contrasting growth curves of roach shown in Fig. 15.3, but it may also exhibit inter-annual variation within a given lake, which will be superimposed on any seasonal patterns. Among many potential examples, Fig. 15.5a shows that such long-term variation is associated with differences in the mean length achieved by underyearling perch in Windermere, UK. Fish growth may also be liable to significant influence from global warming, although such effects are unlikely to be simple (Magnuson & Destasio 1997) and may contain a negative component (Trippel *et al.* 1991). In addition to direct physiological effects, temperature differences probably also generate significant indi-

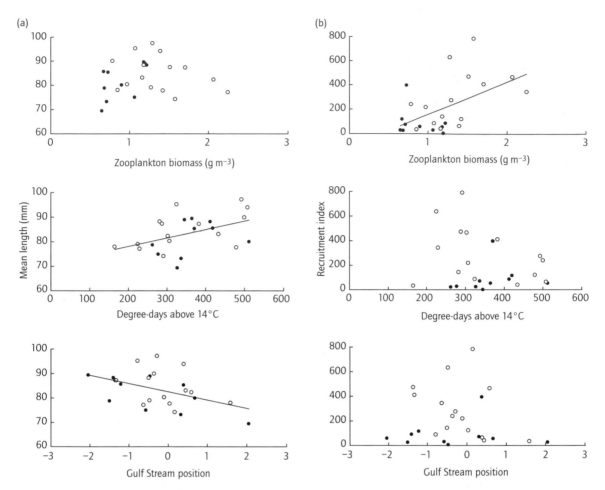

Fig. 15.5 Relationships between (a) mean body length for underyearling perch (*Perca fluviatilis*) in the north basin of Windermere, UK, and mean summer zooplankton biomass, the annual numbers of degree-days above 14°C, and the position of the Gulf Stream, for the period 1966–1990, and (b) between a recruitment index for underyearling perch from the same location and period and the same environmental variables. Data from before (1966–1975) and after (1976–1990) an outbreak of disease in the perch population are shown as closed and open circles, respectively. (Redrawn from Winfield *et al.* 1998; reproduced with kind permission of Kluwer Academic Publishers.)

rect influences by impinging on the abundance of prey populations.

Abundance of prey, and indeed any environmental factor which affects fish feeding opportunities, also widely influences fish growth. Examples of the effects of intraspecific competition for food on the abundance of vendace (Helminen & Sarvala 1994) and roach (Perrow *et al.* 1990) have already been given. In both cases, such competition was associated with decreased growth rates of adults. Interspecific competition for food, and interspecific interactions in the form of predation pressure, may also reduce growth rates, but these issues are beyond the remit of the present chapter. Although the effects of food availability on fish growth are common, they are not universal. For example, they have not been identified in underyearling perch in Windermere (Fig. 15.5a), nor in underyearling smelt (*Osmerus eperlanus*), perch, ruffe (*Gymnocephalus cernuus*), bream or roach in Tjeukemeer, The Netherlands (Mooij *et al.* 1994).

Emphasis on the growth of young fish is appropriate because, in many populations, it is at such early life-stages that growth and production are probably at their greatest and most variable. While sampling difficulties associated with assessing the growth of fish throughout their life restrict the number of lacustrine studies, relevant investigations of growth and production of roach and bleak (*Alburnus alburnus*) in the River Thames, UK, showed that 69% of population production occurs during the first year of life (Mann 1965). Similarly, Craig (1980) found that production of perch in Windermere, UK, during their first 2 years accounts for 60–80% of the total production of a year class, whilst Helminen *et al.* (1990) recorded that during summer highly productive underyearling vendace in Pyhäjärvi, Finland, were responsible for 83% of food consumption by the total vendace population.

Although the above discussion only considers immediate causal mechanisms in the form of temperature and food availability, these factors may themselves be a function of agencies operating elsewhere. A familiar example is lake temperature and global warming, but a less obvious relationship is illustrated in Fig. 15.5a in which the mean length achieved by underyearling perch in Windermere, UK, exhibits a significant relationship with north-south movements of the Gulf Stream in the Atlantic Ocean, with smaller mean lengths being associated with northerly displacements. Although the mechanism of Gulf Stream shift remains to be determined, it probably involves induction of local weather patterns, particularly wind speed, which influences the timing of lake stratification and so development of phytoplankton and zooplankton populations during the spring and summer (Winfield *et al.* 1998). The latter are the major prey of underyearling perch in Windermere (Craig 1978).

15.3.5 Reproduction and the determination of year-class strength

Like growth, fish reproduction has been extensively studied for many years, and a comprehensive review here would again be inappropriate. Instead, the interested reader is directed towards Bone *et al.* (1995), which provides extensive treatments of life-histories, fecundity, maturation and parental care, including many freshwater examples. In this section, attention will be focused on that aspect of reproduction arguably most relevant to population ecology and limnology, i.e. the formation of **year-class strength** (defined as the relative number of young fish produced by a population in a given year). Examples of the variability of this feature of lake fish reproduction have already been implied within examples of population abundance and individual growth given in earlier sections, and so examples here will instead illustrate the causes of such variations.

Causal factors may themselves be grouped as abiotic or biotic in origin and this dichotomy affords a useful structure with which to assemble the following examples, although it must be appreciated that these two groups are not completely independent of each other.

Abiotic factors which have been shown to influence the year-class strength of lake fish are overwhelmingly dominated by temperature, although other associated factors, such as other components of the local climate, may also be involved. By far the most commonly reported example of a temperature effect producing a strong year class is a warm summer. For example, such an effect of this factor, whether measured as mean temperature, accumulated number of degree-days or some other parameter, has been found in two cool-water species (alewife (*Alosa pseudoharengus*) (Henderson & Brown 1985) and rainbow trout (*Oncorhynchus mykiss*) (Donald & Alger 1986)), the intermediate species perch (Mills & Hurley 1990) and zander (Buijse & Houthuijzen 1992), and the warm-water species roach (Goldspink 1978) and bream (Goldspink 1981).

Given this background, it is surprising that Winfield *et al.* (1998), in an analysis of perch recruitment in Windermere, UK, over a 25 year period (Fig. 15.5b), did not find a similar effect, although these were detected in this population over a longer timescale (Mills & Hurley 1990). In addition, temperature effects are not confined to the summer alone. Although most lake fish species are very inactive over winter, owing to low temperature, members of the Salmoniformes typically spawn immediately before or during this time of the year and the prevailing temperature can have a significant effect on egg survival and so year-class strength. Although such winter studies are rare in the literature, Freeberg et al. (1990) reported such an effect for lake whitefish (Coregonus clupeaformis) in Lake Michigan, USA.

More general climatic effects, of which temperature is a component, have also been identified as being important for several species, among which (in an analysis of 46 years of commercial catch data from Minnesota, USA; Smith 1977) are yellow perch and walleye (Stizostedion vitreum). In a particularly detailed multiple linear regression analysis of data from 1962 to 1982, Eckmann et al. (1988) found that climatic factors were of overriding importance in determining year-class strength in a whitefish population in Lake Constance, central Europe. Moreover, and partly as a consequence of the quantity and quality of the suite of limnological data available for analysis, it was found that meteorological conditions that lead to early thermal stratification in spring were of greatest importance, with early stratification giving rise to higher year-class strength.

Biotic factors that have been shown to determine year-class strength may be grouped into (i) those which determine the number of eggs from which the year class originates, and (ii) those which determine the subsequent fate of those eggs and resulting young fish. The number of eggs from which a year class originates is influenced by the number of spawning adults, their fecundity and the quality of their eggs. Whilst at low population densities there is typically a positive relationship between the number of adults and the number of successfully hatching eggs, at higher densities this may become negative, and lead to an overall dome-shaped curve, as found by Mills & Hurley (1990) for

perch in Windermere, UK, by Salojärvi (1991) for vendace in Oulujärvi, Finland, and by Bronte et al. (1993) for yellow perch in western Lake Superior, North America.

Given the variations in growth discussed above, it is not surprising that fecundity in fish also varies greatly. For example, the changes in population abundance of roach (Perrow et al. 1990) and vendace (Helminen & Sarvala 1994) cited earlier are driven by variations in adult fecundity caused by intraspecific competition with juveniles for food resources. Food availability has also been suggested as influencing the fecundity of walleye (Ritchie & Colby 1988) and whitefish (Hartmann et al. 1995). Finally, egg quality itself may vary, and so influence numbers of successfully hatching eggs. Such variation may include differences between populations and between age classes within a population, as reported, for example, by Kamler et al. (1982) for Finnish and Polish vendace populations.

Once eggs have hatched, biotic factors which may influence the resulting year-class strength include food availability, competition and predation. Note that interspecific examples of the last two fall beyond the scope of the present chapter and so will not be considered fully here. Furthermore, limited food availability may often itself be a function of the feeding activities of fish, and so in such cases any effects should be more appropriately classified as competition. However, an apparent example of the effects of low food availability not involving competition is shown in Fig. 15.5b, in which a recruitment index of perch in Windermere, UK, is positively correlated with zooplankton abundance, indicating 'bottom-up' control of this part of the lake food chain (Winfield et al. 1998).

Intraspecific competition for food among larvae has been found to be the main factor determining year-class strength of lake whitefish (Freeberg et al. 1990) and vendace (Auvinen et al. 1992), whilst Sandlund et al. (1991) have suggested that the same mechanism operates between juvenile and adult vendace. In the latter case, the authors argued that juveniles were outcompeted by the adults, leading to an increase in juvenile mortality, rather than a decrease in adult fecundity (as found for the vendace population of Pyhäjärvi, Finland; Helminen & Sarvala 1994).

Competition for space probably rarely influences year-class strength in lake fish, owing to the absence of territorial behaviour in most species, but Borgstrom *et al.* (1993) have documented one such example for a brown trout population in an alpine lake in Norway. In this lake, strong year classes of brown trout were observed at regular time intervals, and were attributed to competition for limited littoral cobble areas. Such habitat is used by fish from their first autumn until they attain 3 years of age. Once such areas become inhabited by a strong year class, fish hatched in subsequent years are excluded by aggression, and so suffer high mortality. The next strong year class is thus not produced until the established year class has vacated the littoral cobble areas following migration to other parts of the lake, at the age of 3 years.

Predation is generally considered to be an important structuring force in lacustrine communities, and so a role in determining the year-class strength of many fish populations is to be expected. Even though the subject of interspecific interactions is beyond the scope of the present chapter, the occurrence of significant predation on young fish by other fish species is so common that to make no mention of it at all would be misleading. One such example which has already been given is the determination of year-class strength of bream in Tjeukemeer, The Netherlands, by predation by zander (Mooij *et al.* 1996), to which may be added a similar relationship between yellow perch and walleye, respectively, in Lake Erie (Hartman & Margraf 1993). Predation originating from older conspecifics, in the form of cannibalism, has been shown to be significant for year-class strength in several species, including pike (Kipling 1983), walleye (Ritchie & Colby 1988) and perch (Mills & Hurley 1990).

Finally, on the issue of the production of young individuals by fish populations, it is pertinent to consider some conclusions drawn by Cushing (1996) after a working lifetime of research on recruitment in extensively researched marine species. He suggests that the stock–recruitment relationship, i.e. that between numbers of spawning adults and those of adults or near-adults in the next generation, will only be understood by examination of life in the larval stages. Moreover, whilst he accepts that for marine fish species such recruitment can now at least be measured, he believes that there is little understanding of the mechanisms by which magnitudes of variation are generated and that 'The study of recruitment, in spite of considerable advances, has remained a science in its infancy' (Cushing 1996, p. 1). It is the present author's belief that the study of recruitment in lake fish populations is even less developed, and that the limnologist should be wary of any fish ecologist claiming otherwise.

15.4 FISH POPULATION ECOLOGY, FEEDING BEHAVIOUR AND ECOLOGICAL ROLES

15.4.1 *Introduction*

As far as the limnologist is concerned, the primary mechanism via which fish populations manifest their ecological role is their feeding behaviour. Although fish feeding ecology is also addressed in other chapters of this text, the ways in which population ecology interacts with feeding behaviour in order to precipitate ecological roles will be briefly considered here. As explained in the introduction to the present chapter, this complex area cannot be thoroughly reviewed, but its importance to other chapters (Chapter 16 and Volume 2, Chapter 18) warrants that it is given a limited consideration. After some key features of fish population ecology are reprised, this section will consider relevant aspects of their feeding behaviour, before offering some comment on the resulting ecological roles of fish in lake ecosystems.

15.4.2 *Key features of fish population ecology*

Fish population ecology is of course in large part a function of their biology. In this context, it is pertinent to recall that fish are diverse, of large body size, highly mobile, possess acute sensory systems and are capable of complex behaviour patterns including learning. Moreover, their typically considerable longevity and great growth result in the frequent occurrence of ontogenetic niche shifts.

The dynamics of lake fish populations are characterised by great variability in terms of population abundance, although, because of sampling difficulties, studies of absolute population densities are rare. For the same reason, patterns of mortality are also infrequently reported, but are probably at their highest during early life-stages. Growth has been much more extensively described and has been shown to be extremely variable within a species with respect to population and time, being influenced both by abiotic factors (e.g. temperature) and biotic factors (e.g. intraspecific competition). Again, it is usually greatest during early life.

As a consequence, biomass and food consumption of a population may be dominated by those components arising from young fish. Growth patterns, longevity and a typical seasonality of reproduction all combine in order to make many lake fish populations highly size-structured, with each size class potentially exhibiting its own ecology. Finally, reproduction, as summarised by year-class strength, is extremely variable, with inter-annual differences often being of the order of several magnitudes. This recruitment of young individuals has been shown to be influenced by an array of abiotic and biotic factors in different populations, although this extremely important aspect of lake fish population ecology remains generally poorly understood.

15.4.3   Key features of fish feeding behaviour

Scientific observations of lake fish feeding behaviour have been collected since at least the early part of last century (e.g. Triplett 1901), and now form an immense literature which is fortunately subject to frequent review. In the present context, the reviews of feeding behaviour by Hart (1993), and of predation risk and feeding behaviour by Milinski (1993), are recommended for the limnologist seeking more information than can be presented here. Given the importance of planktivory in the ecology of lake fishes and their interactions (see below), the review by Lazzaro (1987) is also relevant. In general terms, it may be noted that the size, mobility and sensory abilities of fish enable them to exploit a diversity of feeding opportunities.

A very obvious feature of fish feeding behaviour is that they are usually extremely selective in their choice of prey, and that this choice varies with environmental conditions, including local prey abundance. An early example of this aspect of feeding behaviour was shown by Werner & Hall (1974) in a laboratory study of bluegill sunfish predating different size classes of *Daphnia*. In the more complex situation of the field, the process of learning may become important as different prey types are encountered (Winfield *et al.* 1983). Feeding in a more complex environment also requires that fish decide where to feed in terms both of within and between habitats. Examples of studies addressing these issues include assessment of patch use by goldfish (*Carassius auratus*) within a laboratory 'habitat' (Pitcher *et al.* 1982), and of feeding movements between habitats by bluegill sunfish in experimental ponds (Werner *et al.* 1983a).

Although the limnologist may rightly consider that fish typically occupy the higher trophic levels in most lakes, it is important to realise that many species, or at least the younger and smaller components of their populations, are themselves also preyed upon. As a result, much fish feeding behaviour is a compromise between the goals of eating and avoiding being eaten, the effects of which impinge on prey choice (e.g. Milinski & Heller 1978), foraging location (e.g. Werner *et al.* 1983b) and foraging time (e.g. Bohl 1980).

15.4.4   Ecological roles

As implied above, the primary ecological role of fish in lakes, as far as the limnologist is concerned, is probably that of predator, albeit one often compromised by the need to avoid being predated. Moreover, the fact that the young of almost all lake fish pass through a zooplanktivorous stage lasting from days to months means that the impact of fish populations is usually greatest on the zooplankton.

Research on fish–zooplankton interactions began in earnest with the pioneering limnological studies of Hrbáček *et al.* (1961) and Brooks & Dodson (1965), whilst during recent decades it has also been a focus for studies of fish community ecology (see Chapter 16). Further impetus arrived during the mid-1980s, with the advent of the science of biomanipulation, in which the fish-zooplankton-algae axis is manipulated with the aim of improving water quality (see Volume 2, Chapter 18). Given such coverage elsewhere in this text, further review will not be attempted here, except to note that in addition to direct mortality effects arising from actual predation it is now becoming apparent that lake fish populations may also influence zooplankton populations by altering their behaviour (e.g. Lauridsen & Lodge 1996; Chapter 14).

The ecological roles of lake fish in terms of their effects on other lake biota have received far less attention. However, direct or indirect effects on aquatic macrophytes, both positive (e.g. Hansson *et al.* 1987; Brönmark 1994) and negative (e.g. Crivelli 1983; Giles 1994), have been documented, whilst complex effects on benthic (e.g. Brönmark 1994) and macrophyte-associated (e.g. Chilton & Margraf 1990) macroinvertebrates have also been found.

Finally, interactions of lake fish populations with aquatic bird populations are the subject of developing interest. In such situations, fish may act as prey or as competitors. For example, Fig. 15.6a illustrates the positive relationship between abundance of roach in Lough Neagh, UK, and numbers of piscivorous great crested grebes (*Podiceps cristatus*), but a negative relationship with those of benthivorous tufted duck (*Aythya fuligula*). Winfield & Winfield (1994) suggested that the latter relationship results from competition for molluscan prey, an interpretation which is supported by an analysis which shows that roach and tufted duck cluster together in terms of diet similarity (Fig. 15.6b). Negative correlations have also been recorded for fish and aquatic bird abundances by Eriksson (1979) and by Andersson (1981), whilst experimental evidence of competitive interactions has been produced by Pehrsson (1984) and Giles (1994).

15.5 CONCLUDING REMARKS

The preceding sections have illustrated many of the successes of lake fish population ecology as it stands today, and have drawn attention to areas where progress has been slower or even effectively absent. This section will offer further personal comments on a number of these important topics, together with some suggestions concerning possible developments in the foreseeable future.

Undoubtedly, a major constriction both on present understanding and future development of lake fish population ecology arises simply from sampling difficulties. The multitude of techniques now available to the lake fish ecologist is presented in Murphy & Willis (1996), but many of them remain extremely expensive in terms of time or finance, and so are unlikely to see extensive use outside of their contribution to management of commercial fisheries. However, one subject which is presently developing rapidly and which holds promise of extensive future application is quantitative echo sounding, or **hydroacoustics**. Technological advances in this field are lowering costs and opening up new applications in many areas of fish ecology, including population ecology in lakes. Whilst echo sounding is best developed in marine applications, where its state-of-the-art and likely future developments were recently reviewed by Misund (1997), an introduction to this sampling technique and its application in lakes is provided by Winfield & Bean (1995).

A second major restriction under which lake fish population ecology is practised is its requirement for long-term studies, ideally of duration of decades. In addition to the arguments implicit in the earlier sections of this chapter, the requirement for extensive investigations in fish population ecology is addressed by Elliott (1994), whilst Jones (1990) provides a limnological perspective.

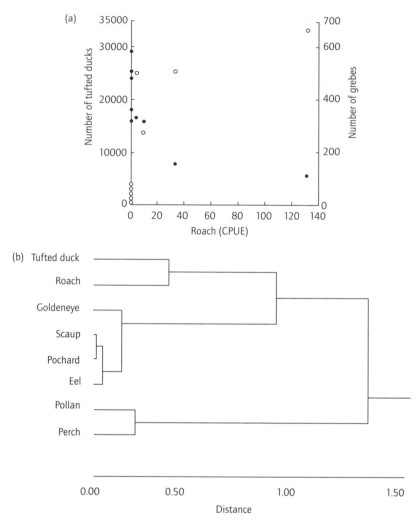

Fig. 15.6 (a) Relationships between mean catch-per-unit-effort (CPUE) of roach (*Rutilus rutilus*) in a given summer, and the mean numbers of tufted duck (*Aythya fuligula*; closed circles) and great crested grebes (*Podiceps cristatus*; open circles) present during the following winter at Lough Neagh, UK, for 9 years within the period 1964–1988, and (b) a dendrogram showing the diet similarities among the major fish and diving duck populations of this lake. Other scientific names are as follows: goldeneye (*Bucephela clangula*), scaup (*Aythya marila*), pochard (*Aythya ferina*), eel (*Anguilla anguilla*), pollan (*Coregonus autumnalis*) and perch (*Perca fluviatilis*). (Redrawn from Winfield *et al.* 1992; reproduced with kind permission of Kluwer Academic Publishers, and from Winfield & Winfield 1994; reproduced with kind permission of Blackwell Science Ltd, respectively.)

Even though the scientific case for studies of long duration is unquestionable, they unfortunately remain rare. In a time of increasing concern over long-term environmental changes, it is essential that the few long-term studies which do exist are maintained in future.

One aspect of lake fish population ecology which has virtually been ignored in the present review is the effect of parasites and diseases. Whilst some examples of the effects of extensive infestations or infections of parasites, e.g. of *Ligula intestinalis* on roach in Slapton Ley, UK (Wyatt 1988), or diseases, e.g. furunculosis on perch in Windermere, UK (Bucke *et al.* 1979), have been recorded, this field remains little explored and is worthy of further study.

The subject of population stability has already been considered, but it is such an important issue that further comment here is justified. The variability in lake fish population abundance commonly observed operates on a temporal scale much longer than those exhibited by lower trophic levels in lakes, and so the dynamics of fish such as zooplanktivorous underyearlings are a function not only of contemporary zooplankton populations, but also of earlier feeding conditions for the previous generation, and perhaps even climatic conditions. The consequent destabilising influence of this asynchrony for the organisation of pelagic ecosystems is considered to be significant by Kerfoot & DeAngelis (1989), and it is suggested here that stability is probably a rare feature of lake fish population ecology. If so, this has profound implications for applied issues such as fisheries management and biomanipulation.

In partial continuation of the above theme, a future goal of lake fish population ecology must be its continued integration into mainstream limnology. Although this requires that the historical gulf in communication between fish ecologists and investigators of other lake components is bridged, the consequent benefits to both parties are likely to be considerable, as exemplified by the wideranging studies of Eckmann *et al.* (1988) at Lake Constance, central Europe, and Winfield *et al.* (1998) at Windermere, UK. If more evidence is needed, an eloquent argument for the continued assimilation of fish ecology, fisheries ecology and limnology is presented by Magnuson (1991).

Finally, some remark on the implications for lake fish population ecology of developments in the rapidly moving field of information technology is appropriate. As in most other areas of science, in recent years there has been a great expansion in the volume of scientific literature published on fish, and so the tool of the literature search is ra-

pidly becoming as indispensable to the fish population ecologist as the plotting of length frequency distributions has been for several generations of scientists. Recent years have seen a number of initiatives towards making published fish data readily accessible without the need for repeated literature searches, among which FishBase 2000 (Froese & Pauly 2000) has become clearly established as the forerunner and is available on CD-ROM and online at *www.fishbase.org*. Following several years of development, this interactive database now contains information on 26,550 fish species, each of which is referenced to its origin in the primary literature. In addition to its extensive use by the lake fish population ecologist, the establishment of this data source will prove invaluable to the limnologist.

15.6   ACKNOWLEDGEMENTS

I am deeply grateful to Colin Reynolds for his invitation to write this chapter, and for his patience, encouragement and understanding so generously given during its gestation. The approaches and views expressed in this work were developed in part during discussions held over many years with innumerable colleagues near and far, and my great appreciation is extended to them all.

15.7   REFERENCES

Andersson, G. (1981) Fiskars inverkan på sjöfågel och fågelsjör. (Influence of fish on waterfowl and lakes. *Anser*, **20**, 21–34. (In Swedish with English Summary).

Auvinen, H., Vuorimies, O. & Huusko, A. (1992) Forecasting and management of vendace (*Coregonus albula* L.) stocks in Lake Lentua, Northern Finland. *Polskie Archiwum Hydrobiologii*, **39**, 789–95.

Berg, O.K. (1995) Downstream migration of anadromous Arctic char (*Salvelinus alpinus* (L.)) in the Vardnes River, northern Norway. *Nordic Journal of Freshwater Research*, **71**, 157–62.

Bleckmann, H. (1993) Role of the lateral line in fish behaviour. In: Pitcher, T.J. (ed.), *Behaviour of Teleost Fishes*. Chapman & Hall, London, 201–46.

Bohl, E. (1980) Diel pattern of pelagic distribution and feeding in planktivorous fish. *Oecologia*, **44**, 368–75.

Bone, Q., Marshall, N.B. & Blaxter. J.H.S. (1995) *Biology of Fishes* (2nd edition). Chapman & Hall, London, 332 pp.

Borgstrom, R., Heggenes, J. & Northcote, T.G. (1993) Regular, cyclic oscillations in cohort strength in an allopatric population of brown trout, *Salmo trutta* L. *Ecology of Freshwater Fish*, **2**, 8–15.

Bronte, C.R., Selgeby, J.H. & Swedberg, D.V. (1993) Dynamics of a yellow perch population in western Lake Superior. *North American Journal of Fisheries Management*, **13**, 511–23.

Brooks, J.L. & Dodson, S.I. (1965) Predation, body size and composition of plankton. *Science*, **150**, 28–35.

Brönmark, C. (1994) Effects of tench and perch on interactions in a freshwater, benthic food chain. *Ecology*, **75**, 1818–28.

Bucke, D., Cawley, G.D., Craig, J.F., Pickering, A.D. & Willoughby, L.G. (1979) Further studies on an epizootic of perch, *Perca fluviatilis* L., of uncertain aetiology. *Journal of Fish Diseases*, **2**, 297–311.

Buijse, A.D. & Houthuijzen, R.P. (1992) Piscivory, growth, and size-selective mortality of age 0 pikeperch (*Stizostedion lucioperca*). *Canadian Journal of Fisheries and Aquatic Sciences*, **49**, 894–902.

Carpenter, S.R. & Kitchell, J.F. (1996) *The Trophic Cascade in Lakes*. Cambridge University Press, Cambridge, 385 pp.

Chilton, E.W. & Margraf, F.J. (1990) Effects of fish predation on invertebrates associated with a macrophyte in Lake Onalaska, Wisconsin. *Journal of Freshwater Ecology*, **5**, 289–96.

Cowx, I.G. (ed.) (1991) *Catch Effort Sampling Strategies: their Application in Freshwater Fisheries Management*. Fishing News Books, Oxford, 420 pp.

Cowx, I.G. (ed.) (1996) *Stock Assessment in Inland Fisheries*. Fishing News Books, Oxford, 513 pp.

Craig, J.F. (1978) A study of the food and feeding of perch, *Perca fluviatilis* L., in Windermere. *Freshwater Biology*, **8**, 59–68.

Craig, J.F. (1980) Growth and production of the 1955–1972 cohorts of perch, *Perca fluviatilis* L., in Windermere. *Journal of Animal Ecology*, **49**, 291–315.

Craig, J.F. (ed.) (1996) *Pike: Biology and Exploitation*. Chapman & Hall, London, 298 pp.

Crivelli, A.J. (1983) The destruction of aquatic vegetation by carp. *Hydrobiolgia*, **106**, 37–41.

Cryer, M., Peirson, G. & Townsend, C.R. (1986) Reciprocal interactions between roach, *Rutilus rutilus*, and zooplankton in a small lake: prey dynamics and fish growth and recruitment. *Limnology and Oceanography*, **31**, 1022–38.

Cushing, D.H. (1996) *Towards a Science of Recruitment in Fish Populations*. Ecology Institute, Oldendorf/ Luhe, 175 pp.

Donald, D.B. & Alger, D.J. (1986) Dynamics of unexploited and lightly exploited populations of rainbow trout (*Salmo gairdneri*) from coastal, montane, and subalpine lakes in western Canada. *Canadian Journal of Fisheries and Aquatic Sciences*, **43**, 1733–41.

Eckmann, R. (1991) A hydroacoustic study of the pelagic spawning behavior of whitefish (*Coregonus lavaretus*) in Lake Constance. *Canadian Journal of Fisheries and Aquatic Sciences*, **48**, 995–1002.

Eckmann, R., Appenzeller, A. & Rösch, R. (eds) (1998) *Biology and Management of Coregonid Fishes – 1996*. *Advances in Limnology*, **50**, 553 pp.

Eckmann, R., Gaedke, U. & Wetzlar, H.J. (1988) Effects of climatic and density-dependent factors on year-class strength of *Coregonus lavaretus* in Lake Constance. *Canadian Journal of Fisheries and Aquatic Sciences*, **45**, 1088–93.

Elliott, J.M. (1994) *Quantitative Ecology and the Brown Trout*. Oxford University Press, Oxford, 286 pp.

Eriksson, M.O.G. (1979) Competition between freshwater fish and goldeneyes *Bucephala clangula* (L.) for common prey. *Oecologia*, **41**, 99–107.

Freeberg, M.H., Taylor, W.W. & Brown, R.W. (1990) Effect of egg and larval survival on year-class strength of lake whitefish in Grand Traverse Bay, Lake Michigan. *Transactions of the American Fisheries Society*, **119**, 92–100.

Froese, R. & Pauly, D. (eds) (2000) *FishBase 2000: Concepts, Design and Data Sources*. International Center for Living Aquatic Resources Management, Los Baños, Laguna, Philipines, 344 pp.

Fuiman, L.A. & Webb, P.A. (1988) Ontogeny of routine swimming activity and performance in zebra danios (Teleostei: Cyprinidae). *Animal Behavior*, **36**, 250–61.

Giles, N. (1994). Tufted duck (*Aythya fuligula*) habitat use and brood survival increases after fish removal from gravel pit lakes. *Hydrobiologia*, **279/280**, 387–92.

Goldspink, C.R. (1978) Comparative observations on the growth rate and year class strength of roach *Rutilus rutilus* L. in two Cheshire lakes, England. *Journal of Fish Biology*, **12**, 421–33.

Goldspink, C.R. (1981) A note on the growth-rate and year-class strength of bream, *Abramis brama* (L.), in three eutrophic lakes, England. *Journal of Fish Biology*, **19**, 665–73.

Griffiths, D. & Kirkwood, R.C. (1995) Seasonal variation in growth, mortality and fat stores of roach and perch in Lough Neagh, Northern Ireland. *Journal of Fish Biology*, **47**, 537–54.

Guthrie, D.M. & Muntz, W.R.A. (1993) Role of vision in fish behaviour. In: Pitcher, T.J. (ed.), *Behaviour of Teleost Fishes*. Chapman & Hall, London, 89–128.

Hansson, L.A., Johansson, L. & Persson, L. (1987) Effects of fish grazing on nutrient release and succession of primary producers. *Limnology and Oceanography*, **32**, 723–9.

Hara, T.J. (1993) Role of olfaction in fish behaviour. In: Pitcher, T.J. (ed.), *Behaviour of Teleost Fishes*. Chapman & Hall, London, 171–200.

Hart, P.J.B. (1993) Teleost foraging: facts and theories. In: Pitcher, T.J. (ed.), *Behaviour of Teleost Fishes*. Chapman & Hall, London, 253–84.

Hartman, K.J. & Margraf, F.J. (1993) Evidence of predatory control of yellow perch (*Perca flavescens*) recruitment in Lake Erie, U.S.A. *Journal of Fish Biology*, **43**, 109–19.

Hartmann, J., Quoss, H. & Knopfler, G. (1995) Response of the fishes to eu- and oligotrophication of Lake Constance. *Osterreichs Fischerei*, **48**, 231–36.

Helfman, G.S. (1981) Twilight activities and temporal structure in a freshwater fish community. *Canadian Journal of Fisheries and Aquatic Sciences*, **38**, 1405–20.

Helminen, H. & Sarvala, J. (1994). Population regulation of vendace (*Coregonus albula*) in Lake Pyhäjärvi, southwest Finland. *Journal of Fish Biology*, **45**, 387–400.

Helminen, H., Sarvala, J. & Hirvonen, A. (1990). Growth and food consumption of vendace (*Coregonus albula* (L.)) in Lake Pyhäjärvi, SW Finland: a bioenergetics modeling analysis. *Hydrobiologia*, **200/201**, 511–22.

Henderson, B.A. & Brown, E.H. (1985) Effects of abundance and water temperature on recruitment and growth of alewife (*Alosa pseudoharengus*) in South Bay, Lake Huron, 1954–82. *Canadian Journal of Fisheries and Aquatic Sciences*, **42**, 1608–13.

Hodgson, J.R., He, X. & Kitchell, J.F. (1996) The fish populations. In: Carpenter, S.R. & Kitchell, J.F. (eds), *The Trophic Cascade in Lakes*. Cambridge University Press, Cambridge, 43–68.

Hrbáček, J., Dvoráková, M., Korínek, V. & Procházková, L. (1961) Demonstration of the effect of the fish stock on the species composition of zooplankton and the intensity of metabolism of the whole plankton association. *Verhandlungen der Internationalen Vereinigung fur Theoretische und Angewandte Limnologie*, **14**, 192–5.

Jacobsen, L., Perrow, M.R., Landkildehus, F., Hojrne, M.,

Lauridsen, T.L. & Berg, S. (1997) Interactions between piscivores, zooplanktivores and zooplankton in submerged macrophytes: preliminary observations from enclosure and pond experiments. *Hydrobiologia*, **342/343**, 197–205.

Jones, J.G. (1990). Long-term research and the future: Resumé. *Freshwater Biology*, **23**, 161–4.

Kamler, E.K., Zuromska, H. & Nissinen, T. (1982) Bioenergetical evaluation of environmental and physiological factors determining egg quality and growth in *Coregonus albula* (L.). *Polish Archives of Hydrobiology*, **29**, 72–121.

Keenleyside, M.H.A. (ed.) (1991) *Cichlid Fishes: Behaviour, Ecology and Evolution*. Chapman & Hall, London, 378 pp.

Kelso, J.R.M. & Shuter, B.J. (1989) Validity of the removal method for fish population estimation in a small lake. *North American Journal of Fisheries Management*, **9**, 471–76.

Kerfoot, W.C. & DeAngelis, D.L. (1989) Scale-dependent dynamics: zooplankton and the stability of freshwater food webs. *Trends in Ecology and Evolution*, **4**, 167–71.

King, M. (1995) *Fisheries Biology, Assessment and Management*. Fishing News Books, Oxford, 341 pp.

Kipling, C. (1983) Changes in the population of pike (*Esox lucius*) in Windermere from 1944 to 1981. *Journal of Animal Ecology*, **52**, 989–99.

Kohler, C.C., Sheehan, R.J. & Sweatman, J.J. (1993) Largemouth bass hatching success and first-winter survival in two Illinois reservoirs. *North American Journal of Fisheries Management*, **13**, 125–33.

Lauridsen, T.L. & Lodge, D.M. (1996) Avoidance by *Daphnia magna* of fish and macrophytes: chemical cues and predator-mediated use of macrophyte habitat. *Limnology and Oceanography*, **41**, 794–8.

Lazzaro, X. (1987) A review of planktivorous fishes: their evolution, feeding behaviours, selectivities and impacts. *Hydrobiologia*, **146**, 97–167.

Le Cren, D. (2001) The Windermere perch and pike project: an historical review. *Freshwater Forum*, **15**, 3–34.

Magnan, P., Audet, C., Glémet, H., Legault, M., Rodríguez, M.A. & Taylor, E.B. (2002) *Ecology, Behavior and Conservation of the Charrs, Genus Salvelinus*. Developments in Environmental Biology of Fishes, No. 22, Kluwer Academic Publishers, Dordrecht, 360 pp.

Magnuson, J.J. (1991) Fish and fisheries ecology. *Ecological Applications*, **1**, 13–26.

Magnuson, J.J. & Destasio, B.T. (1997) Thermal niche of fishes and global warming. In: Wood, C.M. & McDonald, D.G. (eds), *Global Warming: Implications for Freshwater and Marine Fish*. Society for Experimental Biology Seminar Series, Vol. 61. Cambridge University Press, Cambridge, 377–408.

Mann, K.H. (1965) Energy transformations by a population of fish in the River Thames. *Journal of Animal Ecology*, **34**, 253–75.

Mann, R.H.K. (1991) Growth and production. In: Winfield, I.J. & Nelson, J.S. (eds), *Cyprinid Fishes: Systematics, Biology and Exploitation*. Chapman & Hall, London, 456–82.

Mehner, T. & Winfield, I.J. (eds) (1997) Trophic interactions of age-0 fish and zooplankton in temperate waters. *Advances in Limnology*, **49**, 152 pp.

Milinski, M. (1993) Predation risk and feeding. In: Pitcher T.J. (ed.), *Behaviour of Teleost Fishes*. Chapman & Hall, London, 285–305.

Milinski, M. & Heller, R. (1978) Influence of a predator on the optimal foraging behaviour of sticklebacks (*Gasterosteus aculeatus* L.). *Nature*, **275**, 642–4.

Mills, C.A. & Hurley, M.A. (1990) Long-term studies on the Windermere populations of perch (*Perca fluviatilis*), pike (*Esox lucius*) and Arctic charr (*Salvelinus alpinus*). *Freshwater Biology*, **23**, 119–36.

Misund, O.A. (1997) Underwater acoustics in marine fisheries and fisheries research. *Reviews in Fish Biology and Fisheries*, **7**, 1–34.

Mohr, L.C., Mills, K.H. & Klaverkamp, J.F. (1990) Survival and development of lake trout (*Salvelinus namaycush*) embryos in an acidified lake in northwestern Ontario. *Canadian Journal of Fisheries and Aquatic Sciences*, **47**, 236–43.

Mooij, W.M., Lammens, E.H.R.R. & Van Densen, W.L.T. (1994) Growth rate of 0+ fish in relation to temperature, body size, and food in shallow eutrophic Lake Tjeukemeer. *Canadian Journal of Fisheries and Aquatic Sciences*, **51**, 516–26.

Mooij, W.M., Van Densen, W.L.T. & Lammens, E.H.R.R. (1996) Formation of year-class strength in the bream population in the shallow eutrophic lake Tjeukemeer. *Journal of Fish Biology*, **48**, 30–9.

Moss, B. (1988) *Ecology of Fresh Waters: Man and Medium* (2nd edition). Blackwell Scientific Publications, Oxford, 417 pp.

Moyle, P.B. & Cech, J.J. (1996) *Fishes: an Introduction to Ichthyology* (3rd edition). Prentice Hall, New Jersey, 590 pp.

Murphy, B.R. & Willis, D.W. (1996) *Fisheries Techniques* (2nd edition). American Fisheries Society, Bethesda, MD, 732 pp.

Nelson, J.S. (1994) *Fishes of the World* (3rd edition). Wiley, New York, 600 pp.

Papoulias, D. & Minckley, W.L. (1990). Food limited survival of razorback sucker, *Xyrauchen texanus*, in the laboratory. *Environmental Biology of Fishes*, **29**, 73–8.

Pehrsson, P. (1984) Relationships of food to spatial and temporal breeding strategies of mallard in Sweden. *Journal of Wildlife Management*, **48**, 322–39.

Perrow, M.R., Peirson, G. & Townsend, C.R. (1990) The dynamics of a population of roach (*Rutilus rutilus* (L.)) in a shallow lake: is there a two-year cycle in recruitment? *Hydrobiologia*, **191**, 67–73.

Persson, L. (1988) Asymmetries in competitive and predatory interactions in fish populations. In: Ebenman, B. & Persson, L (eds), *Size-structured Populations – Ecology and Evolution*. Springer-Verlag, Berlin, 203–18.

Pitcher, T.J. (1993) *The Behaviour of Teleost Fishes* (2nd edition). Chapman & Hall, London, 715 pp.

Pitcher, T.J., Magurran, A.E. & Winfield, I.J. (1982) Fish in larger shoals find food faster. *Behavioural Ecology and Sociobiology*, **10**, 149–51.

Post, J.R., Rudstam, L.G & Schael, D.M. (1994) Temporal and spatial distribution of pelagic age-0 fish in Lake Mendota, Wisconsin. *Transactions of the American Fisheries Society*, **124**, 84–93.

Rahel, F.J. & Nutzman, J.W. (1994) Foraging in a lethal environment: fish predation in hypoxic waters of a stratified lake. *Ecology*, **75**, 1246–53.

Rask, M., van Densen, W., Lehtonen, H. & Rutherford, E. (eds) (1996) PERCIS II, Second International Percid Fish Symposium, Vaasa, Finland, 21–25 August 1995. *Annales Zoologici Fennici*, **33** (3/4), 303–723.

Reynolds, C.S. (1997) *Vegetation Processes in the Pelagic: a Model for Ecosystem Theory*. Ecology Institute, Oldendorf/Luhe, 371 pp.

Rice, J.A. & Crowder, L.B. (1985) Mechanisms regulating survival of larval bloater in Lake Michigan. *Programs and Abstracts of the 28th Conference on Great Lakes Research*, University of Wisconsin, Milwaukee, 60.

Ritchie, B.J. & Colby, P.J. (1988) Even-odd year differences in walleye year-class strength related to mayfly production. *North American Journal of Fisheries Management*, **8**, 210–5.

Salojärvi, K. (1991) Stock-recruitment relationships in the vendace (*Coregonus albula* (L.)) in Oulujärvi, northern Finland. *Aqua-Fennica*, **21**, 153–61.

Sandlund, O.T., Jonsson, B., Naesje, T.F. & Aass, P. (1991). Year-class fluctuations in vendace, *Coregonus albula* (Linnaeus): who's got the upper hand in intraspecific competition? *Journal of Fish Biology*, **38**, 873–85.

Smith, L.L. (1977) Walleye (*Stizostedion vitreum vitreum*) and yellow perch (*Perca flavescens*) populations and fisheries of the Red Lakes, Minnesota, 1930–75. *Journal of the Fisheries Research Board of Canada*, **34**, 1774–83.

Svedäng, H., Neuman, E. & Wickström, H. (1996) Maturation patterns in female European eel: age and size at the silver eel stage. *Journal of Fish Biology*, **48**, 342–51.

Townsend, C.R. & Winfield, I.J. (1985) The application of optimal foraging theory to feeding behaviour in fish. In: Tytler, P. & Calow, P. (eds), *Fish Energetics: New Perspectives*. Croom Helm, London, 67–98.

Townsend, C.R., Sutherland, W.J. & Perrow, M.R. (1990) A modelling investigation of population cycles in the fish *Rutilus rutilus*. *Journal of Animal Ecology*, **59**, 469–85.

Triplett, N.B. (1901) The educability of the perch. *American Journal of Psychology*, **12**, 354.

Trippel, E.A., Eckmann, R. & Hartmann, J. (1991) Potential effects of global warming on whitefish in Lake Constance, Germany. *Ambio*, **20**, 226–31.

Weatherley, A.H. & Gill, H.S. (eds) (1987) *The Biology of Fish Growth*. Academic Press, London, 443 pp.

Weaver, M.J. & Magnuson, J.J. (1993) Analyses for differentiating littoral fish assemblages with catch data from multiple sampling gears. *Transactions of the American Fisheries Society*, **122**, 1111–9.

Werner, E.E. & Gilliam, J. (1984). The ontogenetic niche and species interactions in size-structured populations. *Annual Review of Ecology and Systematics*, **15**, 393–425.

Werner, E.E. & Hall, D.J. (1974) Optimal foraging and size selection of prey by the bluegill sunfish (*Lepomis macrochirus*). *Ecology*, **55**, 1042–52.

Werner, E.E., Mittelbach, G., Hall, D.J., Gilliam, J.F. (1983a) Experimental tests of optimal habitat use in fish: the role of relative habitat profitability. *Ecology*, **64**, 1525–39.

Werner, E.E., Gilliam, J.F., Hall, D.J. & Mittelbach, G.G. (1983b) An experimental test of the effects of predation risk on habitat use in fish. *Ecology*, **64**, 1540–48.

Winfield, I.J. & Bean, C.W. (1995) The applications and limitations of new developments in echo-sounding technology in studies of lake fish populations. *Proceedings of the Institute of Fisheries Management Annual Study Course*, University of Lancaster, 13–15 September, 1994. Institute of Fisheries management, Nottingham, pp. 227–42.

Winfield, I.J. & Nelson, J.S. (eds) (1991) *Cyprinid Fishes: Systematics, Biology and Exploitation*. Chapman & Hall, London, 667 pp.

Winfield, I.J. & Winfield, D.K. (1994). Feeding ecology of the diving ducks pochard (*Aythya ferina*), tufted duck (*A. fuligula*), scaup (*A. marila*) and goldeneye (*Bucephala clangula*) overwintering on Lough Neagh, Northern Ireland. *Freshwater Biology*, **32**, 467–77.

Winfield, I.J., Peirson, G., Cryer, M. & Townsend, C.R. (1983) The behavioural basis of prey selection by underyearling bream (*Abramis brama* (L.)) and roach (*Rutilus rutilus* (L.)). *Freshwater Biology*, **13**, 139–49.

Winfield, I.J., Winfield, D.K. & Tobin, C.M. (1992). Interactions between the roach, *Rutilus rutilus*, and waterfowl populations of Lough Neagh, Northern Ireland. *Environmental Biology of Fishes*, **33**, 207–14.

Winfield, I.J., George, D.G., Fletcher, J.M. & Hewitt, D.P. (1998) Environmental factors influencing the recruitment and growth of underyearling perch (*Perca fluviatilis*) in Windermere North Basin, UK, from 1966 to 1990. In: George, D.G., Jones, J.G., Punčochář, P., Reynolds, C.S. & Sutcliffe, D.W. (eds), *Proceedings of NATO Advanced Research Workshop on Management of Lakes and Reservoirs during Global Change*. Kluwer Academic Publishers, Dordrecht, 245–61.

Wootton, R.J. (1990) *Ecology of Teleost Fishes*. Chapman & Hall, London, 404 pp.

Wyatt, R.J. (1988) The cause of extreme year class variation in a population of roach, *Rutilus rutilus* L., from a eutrophic lake in southern England. *Journal of Fish Biology*, **32**, 409–21.

16 Fish Community Ecology

JOUKO SARVALA, MARTTI RASK AND
JUHA KARJALAINEN

16.1 THE STRUCTURE OF LAKE FISH COMMUNITIES

16.1.1 *Factors determining the composition of fish communities*

Fish assemblages in lakes are the product of several sets of processes, past, recent and present (Fig. 16.1). On a global scale, only speciation produces new diversity. Regional fish species pools are moulded by historical speciation, colonisation and extinction events, each influenced by long-term geological and climatic factors (habitat age, changing distribution barriers, climatic differences and geomorphological limitations). Within the limits set by the regional species pool, composition of the local fish community is dependent on colonisations and extinctions taking place in short-term ecological time, and influenced by such contemporary lake characteristics as size, isolation (or lack of it), habitat complexity, major abiotic factors (such as oxygen and temperature regimes) and trophic status, as well as by various species interactions.

From the point of view of the presence or absence of a single species of fish, the first question is whether it has ever had an opportunity to colonise the lake. Successful colonisation is only possible if (i) the regional species pool still contains taxa new to the habitat, (ii) if suitable immigration routes exist and (iii) if the ecological context allows invasion and establishment of a species additional to the assemblage. Geographical distribution barriers and the suitability of climatic conditions are important here. Most fish are only able to disperse via continuous waterways, and, even then, their movement upstream may effectively be barred by steep gradients or high waterfalls. Low temperatures prevent dispersal of warm-water fishes, and high temperatures that of cold-water species.

For successful establishment of a new population, various habitats must be available over time which fulfil the demands of the life-cycle of the species, from spawning to adult fish. Success is finally determined by the characteristics of the biotic environment, such as production and availability of suitable food, the possibilities of avoiding predation, or to find prey, etc. The balance between probability of extinction and re-colonisation also pertains to local communities. Local extinctions may result from large-scale natural disasters (e.g. volcanic activity, as in Lake Kivu), or more trivial adverse events such as temporary hypoxia, but they may also depend on interactions within the fish community (e.g. competition or predation, the latter including fishing by humans).

16.1.2. *Global patterns of lacustrine fish species richness*

Roughly 40% (>8500) of all living fish species (>20,000) inhabit freshwaters, but only about 10% of those are endemic and restricted to lake habitats (Fernando 1994). Likewise, amongst the 445 families of fish, only three are exclusively lacustrine. Representatives of 12 more families inhabit lakes, compared with the 157 riverine fish families (Fernando 1994). The small number of truly lacustrine fish forms is understandable, in that most lakes are young compared with rivers (Fernando & Holcik 1989).

On a global scale, freshwater fish diversity (including both riverine and lacustrine species) is greatest in tropical regions, and decreases with increasing latitude and altitude. The greatest numbers of species are known from South America

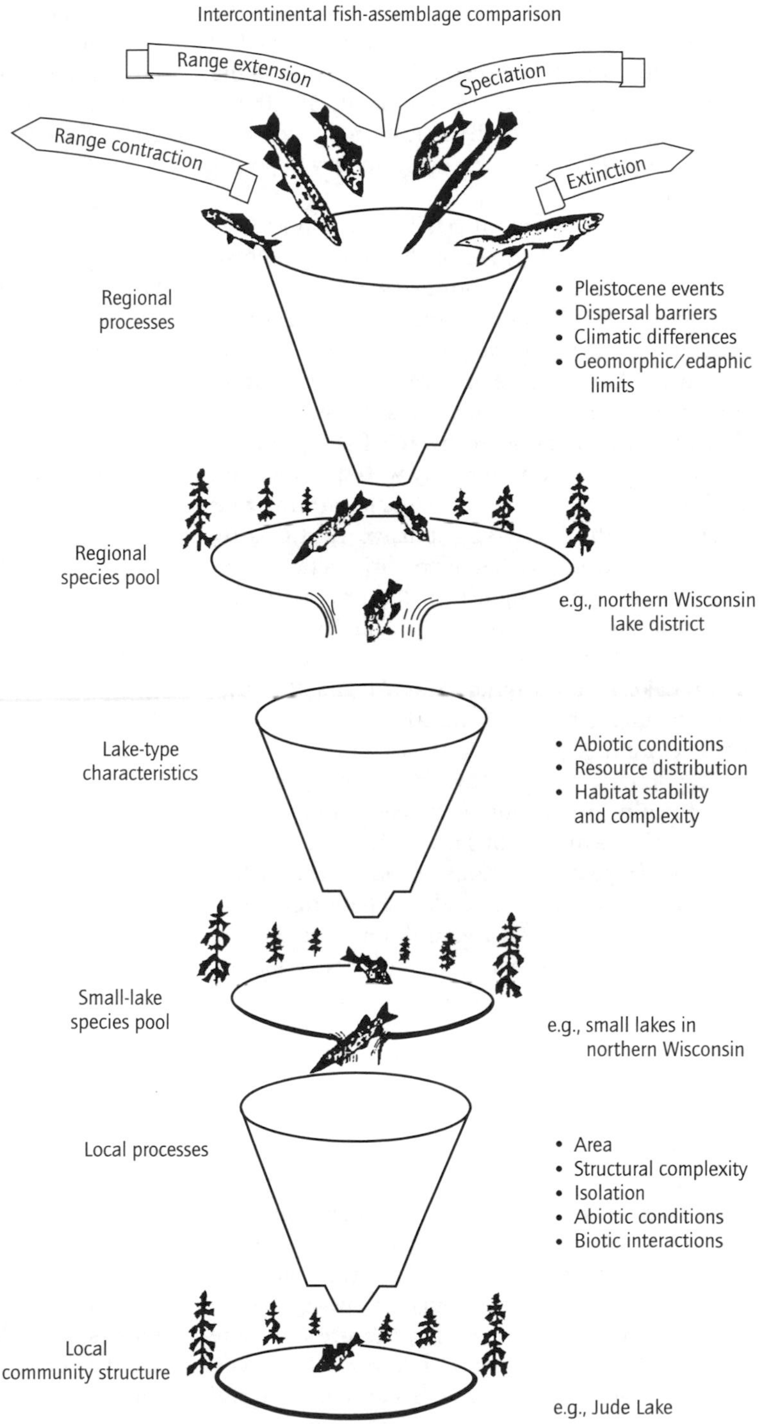

Fig. 16.1 An overview of processes affecting fish species assemblages on different spatial and temporal scales. (Redrawn from Tonn *et al.* 1990; reproduced with permission of the University of Chicago Press.)

(2800) and from Africa (2200 species), whilst the North American Great Lakes area harbours 160 species, Alaska 34, Japan 160, Europe 192 and the whole Palearctic region 450 (Fernando 1994). On this continental scale, diversity differences to a large extent reflect varying evolutionary opportunities in each area.

The number of extant freshwater fish species rises with increasing area of the biogeographical province examined (Welcomme 1985; Rosenzweig & Sandlin 1997). Greater degree of isolation within larger areas enhances the probability of allopatric speciation. Larger areas are also likely to contain more varied habitats, with the same effect. The high total numbers of freshwater fish species in Africa and South America may well depend upon the huge area of these continents (assuming that diversity in the likewise extensive Arctic is restricted by extremely harsh conditions (low temperature, short growing season, low productivity)). But the differences in lacustrine fish diversity are not as simple to explain: the three East African Great Lakes (Tanganyika, Malawi and Victoria) contain more fish species than the whole Holarctic region combined (Fernando 1994).

The length of time available for speciation is certainly important for lake faunas: Lake Tanganyika, which is at least 9–12 (and possibly 20) million years old, contains around 300 fish species, 250 of which are endemic (Coulter 1994); Lake Malawi is at least two million years old and supports >550 fish species, most of them endemic (Ribbink 1994a). Many non-African ancient lakes also contain endemic species flocks (e.g. Baikal 30 (Fernando 1994); Titicaca 24, or 79% of its fish species (Dejoux 1994); Lanao 30 (Fernando 1994); Biwa 8, or 14% of its fish species (Mori & Miura 1980)). On the other hand, although the age of the basin of the African Lake Kivu may be counted in millions of years, the existing lake contains a depauperate fauna, with only 28 fish species (Snoeks et al. 1997). This finding may not repudiate the evolutionary age argument, in that, owing to relatively recent volcanic activity, the present Lake Kivu may be no older than 15,000 years. However, there is now evidence from Lake Victoria, and some other African lakes (e.g. Lake Nabugabo with

five endemic cichlids, and only 4000 years old), that the extraordinary diversity of their fish populations has arisen very recently indeed (see section 16.1.5), so that evolutionary age is not the only decisive factor behind diversity.

16.1.3 Lacustrine fish assemblages in Eurasia and North America

Separate fish families occur in different geographical areas. The fish assemblages of Eurasia and North America are characterised by Salmoniformes (Salmonidae, Coregonidae), Perciformes (Percidae: *Perca*; Centrarchidae: *Lepomis*) and Cypriniformes (Cyprinidae: *Rutilus*, *Phoxinus*), but these groups are not distributed uniformly among different lake types. For a variety of reasons, Salmoniformes are typically found in oligotrophic waters, whilst Cypriniformes usually comprise the major part of the fish communities of eutrophic waters. The Perciformes are commonly found in both types of lake, and in addition are often numerous in North American lakes, in the form of *Lepomis* sunfishes.

Salmoniformes are particularly characteristic of sub-Arctic and Arctic lakes, which in extreme cases may contain only one to three coexisting species, whilst cyprinids in Eurasia, and cyprinids and centrarchids in North America, characterise the more speciose lakes of warm temperate regions. The extant fish fauna of North America is more diverse than that of Europe, probably because of differential survival during and after the Pleistocene glaciations. In North America, small species are more important in assemblages at low latitudes (Griffiths 1997). Fish communities in China and Japan are characterised by cyprinids; typical of China are several species of phyto- and zooplanktivorous carp, which migrate between rivers and lakes, and which have been widely introduced elsewhere for fish culture (Qizhe & Qiuling 1994).

In spite of taxonomic differences between fish faunas from distant areas, it may be possible to identify ecologically equivalent assemblage types which are associated with certain combinations of physical and chemical properties of lakes (e.g.

Tonn & Magnuson 1982; Tonn *et al.* 1983, 1990). Ecologically similar groups of fish species—**guilds** (Austen *et al.* 1994)—would be expected to occur under comparable environmental conditions. Lake area, depth, habitat heterogeneity, isolation, winter oxygen conditions, pH and general productivity have been shown to be key factors distinguishing different assemblages (Tonn & Magnuson 1982; Rahel 1984; Peterson & Martin-Robichaud 1988). Biological interactions such as predation (Robinson & Tonn 1989; Hinch *et al.* 1991) and competition (Persson 1994) also play an important role in structuring the fish assemblages. The fish assemblage concept is especially well suited to relatively small lakes with more or less uniform conditions, whereas large lakes thousands of square kilometres in area may be inhabited by almost the entire regional fish species pool. In such cases, different parts of the lake may provide, in many ways, contrasting ecological circumstances, and exhibit corresponding differences in fish community structure (e.g., from mesotrophic or eutrophic littoral areas, to more oligotrophic pelagial waters (Svärdson 1976)).

A few studies (Mahon 1984; Tonn *et al.* 1990) have examined whether corresponding fish assemblage types would appear in similar present-day conditions in different continents. Mahon (1984) compared the fish communities of two river systems, one in Ontario (Canada) and the other in Poland. However, the hypothesis of ecological convergence was rejected and the main reasons for differences in fish assemblages were suggested to be: (i) differential patterns of colonisation and isolation during the last glaciation; and (ii) the occurrence of entire families of fish in North America which do not belong to the fauna of Europe.

Tonn *et al.* (1990) compared fish communities in small lakes of Wisconsin (USA) and Finland. In this study, the hypothesis of ecological convergence was accepted, and three comparable and predictable fish assemblage types, made up largely of unrelated species, were identified from environmentally similar lakes by multivariate analyses. In the Wisconsin lakes, the three assemblage types were mudminnow (*Umbra limi*), pike (*Esox lucius*) and bass (*Micropterus salmoides*), charac-

terised by different combinations of environmental (low pH, low winter oxygen) and biotic severity (especially predation), and the presence or absence of refuges from these conditions (Tonn *et al.* 1983). Mudminnow assemblages are found in small, shallow, isolated lakes, which frequently experience low oxygen concentrations during the winter (Tonn & Magnuson 1982). These assemblages lack specialised piscivores, and are characterised by small, soft-rayed species of cyprinids, such as central mudminnow (*U. limi*). Bass assemblages also occur in small, isolated lakes, but these are deeper and possess higher oxygen concentrations in winter. In addition to largemouth bass, these assemblages consist primarily of large, spiny-rayed species. Pike lakes are larger and more productive than mudminnow- or bass lakes, and although they are shallow and contain low oxygen concentrations during winter they are connected to streams which provide refuges from seasonally harsh abiotic conditions. Like bass assemblages, pike assemblages consist mainly of large, spiny-rayed species (Tonn & Magnuson 1982).

Fish communities in Ontario (Canada) lakes can likewise be arranged into three groups: salmonid–percid, brook trout (*Salvelinus fontinalis*) and centrarchid communities. There are no exact parallels to the first two in Wisconsin, but the last corresponds roughly to the pike and bass assemblages. Key species in the percid lakes are walleye (*Stizostedion vitreum*), northern pike, white sucker (*Catostomus commersoni*) and yellow perch (*Perca flavescens*), with lake whitefish (*Coregonus clupeaformis*), lake herring (*Coregonus artedii*), spottail shiner (*Notropis hudsonius*) and burbot (*Lota lota*) as additional components (Ryder & Kerr 1990). A further analysis of Ontario fish communities (Jackson & Harvey 1993) identified significant associations with lake morphological characteristics, but no correlations with water chemistry. Large, deep lakes contained richer faunas than shallow lakes, owing to the presence of additional cold-water taxa. Centrarchid species occurred more frequently in small, shallow lakes.

In a community analysis of Finnish lakes (Tonn *et al.* 1990), no distinct fish assemblage types could

be defined on the basis of species presence or absence. Rather, the data exhibited a hierarchical
continuum of species richness across several abiotic variables (lake area, isolation, conductivity,
pH). Environmentally induced extinction, augmented by recolonisation in less isolated lakes,
was proposed as the primary ecological process behind the observed presence–absence continuum.
However, analyses based on relative biomasses of
fish species distinguished three groups of lakes,
each characterised by a different major species
(perch (*Perca fluviatilis*), roach (*Rutilus rutilus*),
and crucian carp (*Carassius carassius*)). These assemblage types possess relatively distinct environmental characteristics, crucian carp lakes being
the most distinct, whilst perch and roach lakes
seem to form more of a continuum. Crucian carp
lakes are the smallest, the most shallow and the
most isolated, and also exhibit the lowest conductivity. Roach lakes are larger and less isolated, and
possess the highest conductivity.

The Finnish assemblage types are therefore
very similar to those defined for Wisconsin (Tonn
et al. 1990). Those that are ecologically the most
similar occur in the most extreme conditions,
mudminnow lakes in Wisconsin and crucian carp
lakes in Finland. Both types are based on the preponderance of predation-sensitive cyprinids in
lakes with low winter oxygen concentrations, or in
ponds that lack predatory fishes. The assemblages
of bass lakes in Wisconsin and perch lakes in
Finland are rather more dissimilar ecologically.
The main abiotic characteristics of these lakes
are low pH and high isolation, both of which limit
the occurrence of cyprinid fishes. In Wisconsin,
predation probably also plays a role. Pike lakes in
Wisconsin and roach lakes in Finland are ecologically the most different, being also the largest, the
least isolated, the least acidic and probably the
most productive of the lakes studied. The main
biological interaction in Wisconsin lakes is
supposed to be predation, but in Finnish lakes it is
competition.

Regional species richness in northern
Wisconsin lakes (65) is almost double that of
southern Finland (37; Tonn *et al.* 1990), yet there is
little difference between numbers of fish species

found in small forest lakes (23 versus 20, respectively), and no significant difference in average
species richness in individual lakes (4.4 versus
3.7). Similar ranges of single-lake species richness
have been reported from lakes of similar size in
Russia, Sweden and Ontario (Tonn *et al.* 1990).

Variability in species composition is lower in
Finland, suggesting that a large proportion of its
fish fauna can maintain populations across a broad
range of environmental conditions, whereas the
fauna of Wisconsin is composed of more specialist
species. In North American lakes in general, piscivores are able to exclude many small prey species,
such as central mudminnow and northern redbelly
dace (*Phoxinus eos*; Tonn *et al.* 1990). In contrast,
most soft-rayed species in northern Europe, including cyprinids, resemble white suckers in
Wisconsin, in that they are able to grow large
enough to attain a size refuge from predation (Tonn
et al. 1989). In this context, interspecific interactions modify relative abundances of species, rather
than their presence or absence.

16.1.4 *Fish communities in tropical lakes*

The fish families common in tropical lakes are
the Cyprinidae, Gobiidae, Characidae, Cichlidae,
Labridae, Losicariidae and Serranidae (Lowe-
McConnell 1987). Primary colonisers of young
lakes are riverine fish, which return to rivers to
breed (Fernando 1994). True lacustrine communities are composed largely of secondary freshwater
fishes: cichlids in the littoral zone, and clupeids
and centropomids in the pelagic. In Africa alone,
there are at least 20 species of pelagic freshwater
clupeids (Marshall 1984; in Fernando 1994). In
tropical lakes, the fish fauna is characterised by
high numbers of specialised species, whereas in
northern latitudes the number of species is lower,
but their ecological flexibility greater.

The Neotropical fish fauna is the most diversified and richest freshwater fish fauna in the world,
with more than 2800 species described (Lowe-
McConnell 1987; Fernando 1994). It is derived
from fewer basic fish stocks than the African
fauna, and lacks the endemic families and cypri-
noids of that continent, but is rich in characoid and

siluroid fish, which developed during isolation of South America during the Tertiary. The neotropical fish fauna can be divided into three groups:

1 a few representatives of widely distributed taxa such as the lungfish *Lepidosiren*, the osteoglossids *Arapaima* and *Osteoglossum*, two small nandids, numerous cyprinodontiform killifishes (including the endemic Poeciliidae, Anablepidae and the genus *Orestis*, which have speciated in Lake Titicaca and other High Andean lakes);

2 about 50 representatives of predominantly marine groups such as stingrays (*Potamotrygon*), clupeids, engraulids, sciaenids, achirine flatfish, belonid needlefish, plus endemic freshwater species of six other families;

3 fishes of 'Gondwanaland', or otophysan characoids and siluroids (catfishes) and percomorph cichlids (Lowe-McConnell 1987).

In South America, cichlids are prominent in the lateral lakes of the river systems and are, in general, more differentiated than in Africa. There are no cichlid species flocks as in Africa, although there are more than 70 species of *Cichlasoma* in Central America, and cichlids are well represented in Nicaraguan lakes. Characoids, which are very diversified in South America (more than 1100 species in 11–13 families, most of them endemic), are predominantly small riverine fish (Lowe-McConnell 1987).

South America lacks large deep lakes (except Titicaca), but extensive temporary lake systems form along flooding rivers. In the equatorial Amazon, at Manaus, water level fluctuations may approach 15 m. Chemical differences between nutrient-rich, silt-laden whitewaters and the black or clear waters from other tributaries are very large. Fishes in these river systems are very mobile, moving from rivers in and out of the lateral lakes (flooded forest and tributary streams). Fish communities are therefore very dynamic, their species composition and relative abundances continually changing (Lowe-McConnell 1987), although the seasonal changes may be similar from year to year (Rodríguez & Lewis 1994). Local diversity is often high, up to 50 species. All feeding niches are normally in use, with some overlaps, often with closely related species sharing the same

resources. Temporal resource partitioning is marked, both seasonally and diurnally.

Fish faunas in **varzea** lakes, formed along western whitewaters, are different from those in clear or blackwater **terra firme** lakes. In varzea lakes, most fish are either dependent on terrestrial food or are piscivores. Varzea lakes are often deficient in oxygen for long periods, and thus many of their fish species are able to live at very low concentrations. Some are air-breathing, whilst others have developed special adaptations for coping with oxygen deficiency. Such lateral lakes regularly dry up, either every year, or at longer intervals, and fish use them mainly as juvenile feeding habitats; spawning occurs in the river. Food webs are very complex even in small lakes. There are many zooplankton-feeding juvenile fishes (e.g. *Colossoma*), but no phytoplankton-feeding fishes. Fish here are important components linking the different habitats and facilitating nutrient recycling and retention. Blackwater rivers are very acid, and lack aquatic insects; only a few fish can live here, on allochthonous forest foods. Clearwater rivers are lake-like, especially along their lower stretches; such river-lakes are characteristic of the Lower and Middle Amazon. The slower the current, the more their fish communities acquire lacustrine characteristics, with a high proportion of characoids and cichlids, and fewer large catfish (Lowe-McConnell 1987).

Lagoons along Venezuelan highways contain numerous specially adapted fish species, including those which bury themselves in sandy bottom (stingrays, a gymnotoid and a characoid), crevice-dwellers (nocturnal electric eels and other gymnotoids and catfishes), fish associated with floating vegetation (most often cichlids, small characoids and catfishes), and annual fishes, which withstand dry periods as resistant eggs (cyprinodonts *Austrofundulus* and *Pterolebias*). Some fish are known for their ability to move overland, others for their sound production (doradid catfishes) or mimicry (Lowe-McConnell 1987).

Lake Titicaca in the Bolivian-Peruvian Altiplano of the Andes, at an altitude of 3809 m, is an ancient lake, some three million years old. Accordingly, its fish fauna exhibits a high degree of endemism: 24 of 25 species of *Orestes* (or 79% of the total number of native fish species) are endemic. Introduction of two exotic predatory fishes, the North American rainbow trout (*Oncorhynchus mykiss*) and the South American *Basilichthys bonariensis*, has led to declining catches of the native fish species (Dejoux 1994).

There are also few lakes in Southeast Asia, but reservoirs are numerous, some of them up to 4000 years old (Lowe-McConnell 1987). Their total extent is comparable to that of natural and floodplain lakes, but the area of ricefields is almost 30 times larger. Fish culture has long been important in Asia (e.g. Sarvala 1993), so that it is difficult to know which are the indigenous species. In ancient times, large parts of the Far East belonged to the same river system. The tropical Asian fish fauna is composed mainly of cyprinids, all of which forage on plant and invertebrate food. Siluroids are also abundant, including nine endemic families and four families (Ariidae, Bagridae, Schilbeidae, Clariidae) common with Africa. As in all tropical fish communities, numbers of predatory species are high (Lowe-McConnell 1987). In Indonesia, there are planktivorous, or vegetarian, anabantids which possess accessory respiratory organs, enabling them to live in deoxygenated water—again a case of convergent evolution.

Malaysia possesses few lacustrine fish species, and the ricefield fauna is also impoverished. Many cyprinids (*Rasbora*) live in acid and unproductive blackwaters. The 10,000 year old Lake Lanao in the Philippines once contained 18 endemic cyprinids, most of which are now extinct owing to introductions of exotic species (Kornfield & Carpenter 1984). Now the lake is inhabited by a mixture of species from many geographical areas—a feature also typical of many other lakes, even in the temperate zone.

The major zoogeographical boundary between the Oriental and Australian biogeographical regions (Wallace's line) is visible even in the fish faunas. Even during the Mesozoic, there was no dry land connection between Asia and Australia. On the Oriental side, Borneo contains >300 freshwater fish species in 17 families, and Java 100 species in 12 families, but on the Australian side,

Sulawesi contains only two species, both probably introduced, and Lombok five, three of which have probably been introduced. Only three relics of the Mesozoic freshwater fish fauna remain: an osteoglossid (*Scleropages formosus*) in Sundaland, another (*S. leichardti*) in Queensland and New Guinea, and the lungfish (*Neoceratodus forsteri*) in Queensland (Lowe-McConnell 1987). The remainder of the freshwater fish fauna of New Guinea and Australia is of marine origin, and the same applies to New Zealand. The latter harbours 27 species, 23 of which are endemic, most of them diadromous species (McDowall 1987a). In addition, the present fauna includes 18 introduced exotic species, many of them predators (McDowall 1987b).

The African freshwater fish fauna is almost as rich as that of South America, but includes a much larger group of lacustrine species. As a relic from their early connections, South America, Africa and Australia share an ancient group of freshwater fish, the lungfishes (Dipnoi). Some cyprinid genera (*Barbus*, *Labeo* and *Clarias*) attest to later exchanges between the African and the Asian faunas. Many fish genera are, however, unique to Africa, having evolved from marine ancestors during long isolation. Distribution of the modern African fish fauna can be divided into several main zones (Worthington & Lowe-McConnell 1994): the **Victorian**, including Lakes Victoria, Kyoga, Edward and George; the **Nilotic**, including Lakes Albert, Turkana, Chad, and the Nasser and Kainji reservoirs; the **Zairean**, including Lakes Tanganyika, Bangweulu and Mweru; the **Zambezian**, including Lake Malawi and the Kariba and Cahora Bassa reservoirs. Other zones include the **Ethiopian** Lake Tana and the **Cameroonian** crater lakes. Ancient connections extend the Nilotic zone westwards beyond the present drainage area of the River Nile.

The fish communities of East African lakes are characterised by very high diversity in general, especially in the cichlids, which have developed rich species flocks in several lakes. The number of cichlid species in African rivers is relatively low (around 100), compared with the >460 species of riverine cyprinids (Goldschmidt & Witte 1992). This distribution is reversed in the African Great

Lakes, however, where over 1000 species of cichlids occur, whilst the number of cyprinid species is less than that found in rivers. In the six major African Great Lakes, the number of fish families varies from eight to 15, and the approximate number of species from 47 in Lake Turkana to either 545 (Worthington & Lowe-McConnell 1994) or 700–1000 in Lake Malawi (Turner 1999; note, not all species have been formally described yet).

Lakes in the Nilotic and Ethiopian regions exhibit only slightly modified faunas, with low endemicity and few cichlid species (Lowe-McConnell 1987; Craig 1992), although in Lake Tana there has been trophic radiation of large cyprinid (*Barbus*) species (Worthington & Lowe-McConnell 1994). For example, in Lake Chad there is little distinction between riverine and lacustrine fish communities, and many of the non-cichlids return to the rivers to spawn. The Nilotic region is characterised by the piscivorous Nile Perch (*Lates niloticus*), which grows to very large size. Lakes Victoria, Edward, Tanganyika and Malawi exhibit high levels of endemism and an abundance of cichlids. Numerous ecologically equivalent cichlid endemic taxa occur in Lakes Malawi, Tanganyika and Victoria. In a few lakes (Victoria, Malawi, Turkana) there are native endemic pelagic cyprinids, and Lakes Victoria and Malawi also contain pelagic cichlids which are absent from Lake Tanganyika.

Even in these cases, pelagic fish communities remain extremely simple, in contrast to the very diverse and complex communities of the littoral zone (Craig 1992). In Lake Tanganyika, the pelagic fish community consists of two endemic clupeids (*Stolothrissa tanganicae* and *Limnothrissa miodon*) and four endemic piscivorous species of *Lates* (family Centropomidae). The clupeids are mainly zooplanktivorous, although *Limnothrissa* also feeds upon shrimps and small fish. *Lates mariae* lives in the benthic zone, whilst *L. angustifrons* favours rocky inshore areas, and *L. microlepis* and *L. stappersi* occupy the offshore pelagic zone. The latter also feeds extensively on shrimps (Coulter 1991). Such large piscivores were absent from Lake Malawi, and (before introductions) from lakes of the Victorian region.

The steep and often rocky shores of Lake Tanganyika harbour a most diverse fish fauna: 12 families with >200 species, mostly cichlids (lamprologines abundant). From 40 m depth to the limits of dissolved oxygen, the benthic and bathypelagic zones contain over 80 species from seven families. Scuba diver censuses of a 20×20 m quadrat in the shallow rocky littoral revealed about 7000 fish belonging to 38 species (Hori *et al.* 1993). Species number remained approximately constant over 10 years. Twelve feeding guilds were present, and resource partitioning within guilds was accomplished via different foraging methods. Keen competition takes place for breeding sites, which are a critical resource (Gashagaza 1991).

Until recently, Lake Victoria contained at least 500 species of haplochromine cichlids (Seehausen *et al.* 1997). Although only moderately diversified morphologically, they were ecologically so diverse that they utilised almost all of the resources available to freshwater fishes in general (Goldschmidt & Witte 1992). In contrast to the deeper Rift Valley lakes with their deep, anoxic waters, this species flock also included benthic species, which inhabited the offshore waters, in addition to the ooze and detritus bottoms of the littoral zone. Similarly, in Lake Victoria, although there are few patches of rocky shore, its rocky shore cichlid fauna has been found to be astonishingly diverse (Seehausen 1996). The pelagic zone once contained numerous zooplanktivorous and piscivorous cichlid species, as well as a pelagic cyprinid, *Rastrineobola argentea*. Coexistence of such a large number of closely related species was possible through extreme trophic diversification; 15 such groups have been defined (Witte & van Oijen 1990). Species of a particular trophic group exhibited differences in body size, diet and horizontal or vertical distributions. For zooplanktivores, niche partitioning by habitat was the predominant method (Goldschmidt & Witte 1992). Many of these species have vanished within two decades, owing to predation by introduced exotic Nile perch and eutrophication (see below).

Lake Malawi, the southernmost of the African Rift Valley lakes, contains more species of fish than any other lake in the world. The lake is deep, steep-sided, warm and permanently stratified, and resembles Lake Tanganyika in many respects. Its hypolimnetic waters below about 200 m depth are anoxic, and lake age has been estimated at about two million years (Ribbink 1994a). Lake level has varied in both lakes, although only Tanganyika has been divided into separate basins. Nearly all of its fish species are endemic, indicating that speciation took place in the lake. As in the other African Great Lakes, the cichlid species flocks are numerically most abundant, the estimated species number being 700–1000 (Turner 1994, 1999). According to genetic studies, all but six of these belong to one major lineage, which, together with the cichlids of Lake Victoria, is descended from a single lineage found in East African rivers (Meyer et al. 1994; Turner 1999). Mouthbrooding cichlids living in the rocky littoral zone of Malawi form a group of ten genera with 200 species. Sandy shores, which cover two-thirds of Malawi's coastline, are inhabited by another group of haplochromine cichlids of 38 genera and 200 species, and more species are known but not yet described (Lowe-McConnell 1996). These two main cichlid groups are also genetically distinct, although derived from the same ancestral stock (Meyer 1993). Although the sandy and muddy bottoms offer fewer opportunities for resource partitioning than the rocky shores, they also contain large numbers of species. Many species exhibit bathymetric segregation. Cichlids in the offshore pelagic zone include numerous zooplanktivores and piscivores. Malawi also possesses a small species flock of tilapias, living pelagically and over sediment bottoms (Lowe-McConnell 1996), and there is one pelagic cyprinid species, *Engraulicypris sardella*.

16.1.5 Origins of the high fish diversity in East African lakes

The origin of the extraordinary cichlid diversity in the large East African Rift Valley lakes has been the subject of much speculation (reviewed by Turner 1999). These lakes have been isolated for many millions of years, and have each developed a distinct endemic fauna. The very high diversity of Lake Tanganyika derives partly from the Zaire River, which contains the most diverse riverine fish fauna in Africa. Tanganyika itself has also been a centre of evolution for a very long time, as evidenced by the high numbers of endemic species even in non-cichlid fishes: there are at least 185 cichlid (180 endemic) and 145 non-cichlid species (61 endemic; Coulter 1994; Snoeks et al. 1994). In Tanganyika, the cichlid group consists of 12 main lineages, whilst only two are found in Lake Malawi. However, the entire cichlid diversity of Lake Victoria (see above) comes from a single lineage. Genetic studies indicate that divergence of haplochromine cichlids in Lake Malawi and especially in Lake Victoria is extremely recent, whilst the genetic distances between the cichlids of Lake Tanganyika suggest a much longer evolutionary history (Meyer 1993).

It should be noted that the total number of cichlid species in these lakes is still unknown. One reason is that detailed taxonomic studies have been areally restricted. Moreover, although recent molecular and behavioural studies and crossing experiments confirm that reported within-site diversity estimates are robust, the overall estimates of cichlid species richness are heavily dependent on the assignation of species status to allopatric populations differing in male colour (Turner et al. 2001). Species delimitation has been more conservative in Lake Tanganyika than in the lakes Malawi and Victoria, and it is possible that the true numbers of species may differ less between these lakes than previously indicated (Turner et al. 2001).

On the basis of allozyme and mitochondrial DNA data, the cichlid lineages of Lake Tanganyika seem to be around five million years old (Meyer et al. 1990; Nishida 1991), whilst the haplochromine species flock in Lake Malawi may be around 700,000 years old (Meyer 1993). Astoundingly, the age of the Victorian haplochromine species flock is <200,000 years, and possibly only 14,000 years (Meyer 1993; Meyer et al. 1994). When compared with the geological ages of these lakes these ages suggest that, at least in Lakes Tanganyika and Malawi, speciation has been intralacustrine.

The lower number of cichlid species in Lake Tanganyika (in spite of its greatest age) may be due to the presence of more efficient native piscivores. Allopatric speciation has probably been important in the evolution of the Tanganyikan fauna. During its geological history, Lake Tanganyika has experienced multiple periods of much lower water level, dividing the lake into three separate basins, and many species may owe their origin to these periods of isolation. Indeed, mitochondrial DNA patterns in several cichlid species still follow the outlines of these ancient lakes (Meyer 1993). About 30% of Tanganyikan cichlids are substratum-spawners, which are absent from other East African lakes.

As to Lake Victoria, the most astonishing finding is that recent geological research shows that although there was probably an ancient lake in this area during the Miocene (22 million years ago), from 17,300 to 12,400 years ago the present lake basin was completely dry (Johnson *et al.* 1996). This explains why certain fish taxa present in the palaeolake (e.g. *Lates*, air-breathing *Polypterus*) are absent from the modern Lake Victoria. Essentially, it also means that all of the more than 300 endemic haplochromine species in the lake, belonging to 21 separate genera, have diverged from each other since this very recent date. It also makes it understandable why all of the cichlids in Lake Victoria seem to be genetically practically identical, because there has not yet been time for any genetic divergence (Meyer *et al.* 1990), and also explains why no haplochromine fossils have been found.

Seventeen of the lake's 40 non-cichlid fish species are likewise endemic. Most of these still ascend the rivers for spawning, in accordance with the idea of their very recent and ultimately riverine origin. Relatively late interconnections with Lake Edward, and between the latter and Lake Kivu, are thought to be reflected in the close phylogenetic relationship of the cichlid species flocks in these lakes (Greenwood 1994), but if indeed Lake Victoria was dry during the very recent past, these conjectures must be reassessed. At present, the fish fauna of Lake Victoria is undergoing a new era of very rapid change, owing to introduction of exotic species and eutrophication (see below). Even if the real number of species is lower, and the actual age of the species flock greater (molecular data have suggested an age of <200,000 to 14,000 years for the species flock; Meyer 1993), the haplochromine cichlids of Lake Victoria still represent one of the most dramatic examples of the importance of rapid speciation in creating biotic diversity in the entire world.

Indications of very rapid speciation also exist from Lake Malawi. In this lake, between AD 1500 and 1850, the level lay at least 120 m below that of the present, implying that, at that time, the shallower southern arm of the lake was dry. However, the presence of numerous endemic rocky-shore cichlid species in that very area suggests that they may have originated in less than 200 years (Owen *et al.* 1990).

Part of cichlid diversity is thought to be due to the great evolutionary flexibility of their versatile pharyngeal jaw apparatus, which allows ready adaptation for various functions (Liem 1973). Cichlid fishes are thus, in a way, preadapted for rapid evolutionary radiation. However, it may be noted that teleost fishes in general are both very speciose compared with other vertebrates, and characterised by a particularly large number of loosely connected bony elements in their heads (Galis & Metz 1998). The striking diversity of trophic groups amongst Lake Victoria cichlids (Witte & van Oijen 1990) suggests rapid specialisation for different feeding niches. Sibling species are always characterised by small differences in feeding behaviour (Galis & Metz 1998). However, structural flexibility alone does not yet explain how reproductive isolation has repeatedly evolved (Goldschmidt & Witte 1992; Turner 1999). The fact that species flock formation in the African Great Lakes has occurred in several other taxa demonstrates that the factors in operation are not unique to cichlids (Goldschmidt & Witte 1992).

Low dispersal capacity, low fecundity and protection of offspring seem to be features connected with strong speciation. The territorial breeding habits of these fish also enhance diversification, via microallopatric speciation. Most cichlids are bound to specific localities, and, in a patchy environment, are effectively isolated by stretches of unsuitable habitat. Recent studies on microsatellite DNA have demonstrated the existence of genetic differences between cichlid populations of neighbouring rocky headlands of Lake Malawi, separated only by a 700-m-wide sandy bay (van Oppen *et al.* 1997). Sandy beaches likewise act as distribution barriers for rock-dwelling cichlids in Lake Tanganyika (Sturmbauer & Meyer 1992; Kohda *et al.* 1996). This shows that, in these lakes, populations remain so local that within them hundreds of geographically isolated populations exist, all of which are potentially able to diverge into new species.

An old hypothesis concerning the haplochromines of Lake Victoria was that speciation began during dry periods, and in temporary geographical isolation in satellite lakes around a reduced Lake Victoria (Greenwood 1994). However, such satellite lakes contain only a very few species of cichlids; moreover, there are no rocky islands in the satellite lakes, whilst every group of rocky islands in Lake Victoria possesses its own species assemblage. Moreover, this explanation is not applicable to Lakes Malawi and Tanganyika.

Recent observational and experimental studies (Seehausen *et al.* 1997) provide a simple explanation for the ease and rapidity of cichlid speciation in Lake Victoria and other East African lakes: they show that sexual selection may be the major mechanism. All of the endemic cichlids of Victoria and Malawi, and those belonging to some of the most diverse groups in Tanganyika, are maternal mouthbrooders. Although the cichlid fish species of Lake Victoria are able to interbreed without loss of fertility, they are sexually isolated by mate choice. In these species, males, which are larger and brighter than the females, and possess longer fins, play no part in parental care. Males of sympatric, closely related species always differ in colour, one species consisting of blue individuals, the other red or yellow (females are usually inconspicuously coloured). The genetic importance of colour hues has been shown experimentally. The main male colours (blue versus red or yellow) match the two absorbance peaks of the retinal pigments. When light conditions are favourable, females exert strong preference for males of a particular colour. Females of a sympatric red/blue sibling species pair (*Pundamilla nyererei*, *P. pundamilla*, the

'zebra nyererei') preferred conspecific over heterospecific males under broad-spectrum illumination, but mated indiscriminately under monochromatic light where colour differences were masked. Disruptive selection on feeding and other specialisations promotes the divergence of these incipient species, and the resulting niche shifts promote their coexistence (Galis & Metz 1998).

There are thus two selection processes acting together: a process of species divergence due to sexual selection, and one of adaptive radiation, owing to disruptive selection. Speciation by sexual selection can explain the large numbers of cichlid species of each ecological and anatomical type in Lake Victoria. Other mating barriers develop more slowly, and coexistence of several hundred species relies on visual mate choice. This model explains allopatric speciation, but it allows also wholly sympatric speciation, for which there is now strong evidence from endemic cichlid species flocks in small Cameroon crater lakes (Schliewen *et al.* 1994). Spatial heterogeneity also contributes to sympatric speciation. Allopatric speciation of cichlids will certainly have played a role in Lake Victoria, because of its size and diversity of habitats, but for many of the haplochromine cichlids of Lake Malawi an allopatric origin is difficult to visualise (Turner 1994, 1999).

Seehausen *et al.* (1997) also showed that, by curbing the impact of sexual selection, increasing turbidity of the waters of Lake Victoria is at least partially responsible for the recent decline in cichlid diversity. Eutrophication decreases light penetration and narrows the light spectrum, so that the waters of the seven African Great Lakes that do possess haplochromine cichlids forming endemic species flocks contain distinctly clearer waters than the five which do not. This relationship remains significant even when the largest lakes are excluded. In this way, human activities that increase turbidity destroy both the mechanism of diversification and that which maintains diversity. Water transparency in Lake Victoria has declined rapidly, from 5.5–8 m during the 1920s to 1.3–3 m in the 1990s. By constraining colour vision, turbidity interferes with mate choice, relaxes sexual selection, and blocks the mechanisms of reproductive isolation. Colour morph diversity and species diversity of rock-dwelling cichlids around 13 islands in southern Lake Victoria correlated positively with water transparency and breadth of light transmission spectrum. Dull fish colouration, few colour morphs and low species diversity are found in areas which have become turbid as a result of recent eutrophication. Females can no longer distinguish males of sibling species from their own when visibility is poor, and so hybridise with males from other species.

Ribbink (1994b) emphasised the role of environmental characteristics, together with the biological characteristics of the organisms, in moulding evolutionary trends. He noted that, in African fresh waters, diversification is greater in the seasonally stable, uniform environments of large lakes than in the seasonally fluctuating, mainly heterogeneous environments of most rivers. On the other hand, diversification of stenotopic organisms appears to be more rapid than that of eurytopic taxa (the terms **stenotopic** and **eurytopic** here referring to habitat tolerance range throughout the whole life-history). Stenotopic species possess narrow habitat tolerance ranges and distribution, are K-selected and sedentary, exist in small populations and small body size, and exhibit parental care. Stable environment allows the evolution of stenotopy, which in turn provides for increasingly rapid speciation. In Lake Malawi, those species flocks with the greatest number of stenotopic members are larger than those which comprise eurytopic species. Within cichlids, the tilapines are eurytopic and are represented in the African Great Lakes by about ten species; in contrast, many haplochromines are stenotopic, and, in these lakes, the group contains more than 1000 species (Ribbink 1994b). Turner (1994) has noted, however, that many of the haplochromines in Lake Malawi are not stenotopic.

The fish diversity of Lake Kivu is unexpectedly low in relation to the age of the lake. Probably because of the volcanic instability of the area, all 15 haplochromine species present are distributed throughout the entire lake, and are eurytopic (Snoeks 1994). The high diversity and endemism of fish (compared with other African rivers) in the Zaire River, which, in evolutionary terms, resembles the large African lakes, attests to the significance of environmental stability for evolution. The reason for the high diversity of cichlids may be that most of them are stenotopic, whereas most other families are eurytopic. Those small non-cichlid species flocks which have formed are the deep-water catfishes (Clariidae) of Lake Malawi, and the Mastacembeloidei of Lake Tanganyika, most of which are cryptic rock-dwelling species (Ribbink 1994b). An important prerequisite for speciation is the presence of ecological opportunity; e.g. the only African instance of cyprinid radiation (Lake Tana; Worthington & Lowe-McConnell 1994) may have been possible owing to lack of strong competitors in this isolated lake.

16.1.6 Determinants of local fish species richness in lakes

Understanding spatial and temporal patterns in species richness is a central goal of community ecology (Oberdorff et al. 1995). Northern lakes, and those at high altitude, are usually inhabited by only two to four species (mostly salmonids), whereas lakes in the European plains possess up to 50 species (with cyprinids most abundant; Morgan 1980), As noted above, in large tropical lakes, several hundred fish species (mostly cichlids) have been identified.

Lake size is the main abiotic variable affecting the number of fish species in lakes, as consistently documented in several studies (Barbour & Brown 1974; Eadie et al. 1986; Rahel 1986; Matuszek & Beggs 1988; Tonn et al. 1990; Eckmann 1995; Griffiths 1997). In a sample of 70 lakes from all over the world, fish species richness was explained by (i) latitude and (ii) lake area (Barbour & Brown 1974). However, in northern North America, latitude explained a larger fraction of total variation in species richness (64%) than lake area (25%), whilst in Africa, latitude was insignificant, and lake area explained 49% of total variation. In 12 lakes in New York State (USA), lake area accounted for 72% of variation of fish species numbers (Browne 1981). Fish species richness in 82 lakes in Ontario,

Canada, was significantly correlated with surface area (80%), whilst latitude explained only 5% of species numbers (Eadie *et al.* 1986). Rivers support more species per surface area than lakes, but the slopes of species–area regressions of lakes and rivers do not differ significantly, suggesting that species are added to both of these habitats at similar rates.

Although these results are consistent with the theory of island biogeography (MacArthur & Wilson 1967), Eadie *et al.* (1986) suggest that patterns of fish species richness are more likely to be a function of habitat diversity (Eadie & Keast 1984) than an equilibrium balance between immigration and extinction. In northern temperate regions (e.g. Ontario, Finland), many lakes are remnants of large glacial waters, and contemporary colonisation rates for most of them may approach zero (Eadie *et al.* 1986). New species are probably added to large lake systems mainly by speciation, or by human agency (Barbour & Brown 1974). In Wisconsin lakes, vegetation diversity and substrate/vegetation diversity are related to fish species richness, and are both correlated with lake area (Tonn & Magnuson 1982). Likewise, large differences in fish community diversity between the pelagic and littoral zones in tropical African lakes are parallel to, and may be causally linked to, those of habitat structural complexity (Lowe-McConnell 1987).

In a very large survey of 2931 lakes in Ontario (Matuszek & Beggs 1988), in which 19 physical and chemical variables were measured, lake area and pH were the principal factors determining fish diversity. In a subset of 992 lakes, lake area explained 18% of variation in species number (slope 0.20), and total aluminium (correlated with pH), latitude, dissolved organic carbon and altitude together an additional 16%. In a larger subset of 2798 lakes, for which area and pH were available, lake area alone explained 28% of variation in fish species number.

Species–area regressions for fish in small forest lakes did not differ between Wisconsin (exponent = 0.32, R^2 = 0.41) and southern Finland (exponent 0.34, R^2 = 0.40; Tonn *et al.* 1990), but were in both cases clearly steeper than those observed during

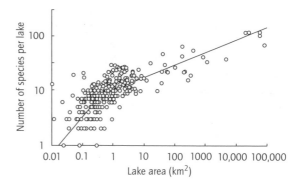

Fig. 16.2   Species–area plot for lacustrine fish in the North American Great Lakes region. (Redrawn from Griffiths 1997.)

earlier studies in North America (0.16–0.24). Griffiths (1997) showed that these differences may be related to lake size. In a reanalysis of data from 651 lakes from the North American Great Lakes region, species–area relationship was found to be steeper for small lakes than for large: for lakes smaller than 1.5 km² the (reduced major axis) slope was 0.38, and for larger lakes, 0.21 (Fig. 16.2; Griffiths 1997). These slopes did not differ between regions, i.e. species numbers changed with lake size in a similar way in different regions.

In the combined data, lake area explained 53% of species richness in small lakes, and 64% in large, whereas lake altitude and maximum depth had little effect. Slopes of species–area regressions previously reported for different data sets were negatively correlated with geometric mean lake areas. The rapid decline in species richness in small lakes may be due to a lack of suitable habitats for certain open-water fish taxa such as coregonids and salmonids, or to extinction of large species in small water bodies (Rahel 1986; Griffiths 1997). Indeed, the percentage of large fish increased with lake size in two sets of small lakes from Ontario and Wisconsin, and was still higher in the North American Great Lakes themselves.

Productivity has rarely been considered as a factor behind fish species diversity, although it may be a candidate for being the major factor explaining global diversity patterns in many other groups

(Wright 1983; Rosenzweig & Sandlin 1997; Rohde 1998). Littoral fish diversity was positively correlated with known phosphorus and primary production gradients in Lake Memphremagog (Canada; Nakashima *et al.* 1977). In a set of 22 south Finnish lakes, after accounting for the effect of lake area, species number was likewise positively correlated with total phosphorus concentration (Helminen *et al.* 2000), whilst species richness in German lowland lakes was correlated with total dissolved solids (Eckmann 1995). Riverine fish diversity patterns can be predicted on a global scale from factors such as size of drainage area, energy availability, habitat heterogeneity and continental species pool (Oberdorff *et al.* 1995; Guégan *et al.* 1998). However, high productivity is not considered a likely reason for high fish diversity in tropical regions (Lowe-McConnell 1987), and recent evidence from Lake Victoria shows that increasing productivity is reducing cichlid species diversity (Seehausen *et al.* 1997).

Rohde (1998) suggested that, in determining global gradients in species diversity, **effective evolutionary time** is even more important than productivity or habitat complexity, taking into account evolutionary speed and habitat age. Thus, diversity should be greater in the tropics than in the temperate zone, in that evolutionary speed is faster at high temperatures, and because tropical habitats have existed undisturbed (by glaciation) for a much longer time than those of temperate latitudes.

16.1.7 *Community saturation*

Relations between local and regional species richness may be used to infer the relative importance of local versus regional processes for community organisation (Ricklefs 1987; Cornell & Lawton 1992). If the relation is asymptotic (i.e. if local richness levels off with increasing regional richness), local communities are **species-saturated** (Fig. 16.3). In a saturated community, all niches are in use and community structure is largely determined via biological interactions between species (e.g. competition, predation). In contrast, if local (log–log-) richness increases linearly with

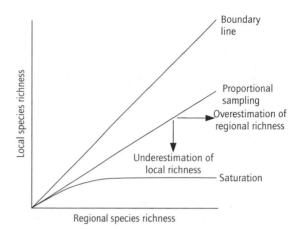

Fig. 16.3 Theoretical relations between local and regional species richness. Linear relation corresponds to unsaturated communities, whilst curvilinear, asymptotic relation implies community saturation. The arrows indicate two ways in which sampling bias can lead to apparent saturation when the community is actually unsaturated. (Redrawn from Griffiths 1997.)

increasing regional richness over the entire range of regional diversities, local communities are **unsaturated** (i.e. there are vacant niches, which may be invaded by new species). In such a case, interspecific interactions may not be important for community structure, which is more dependent on regional and historical processes determining the regional species pool (Cornell & Lawton 1992). However, a variety of community–organisation models predict that local species richness can either be saturated or unsaturated (Cornell & Lawton 1992). Moreover, various types of sampling bias (e.g. insufficient sample sizes relative to regional diversity) may also lead to an asymptotic relation between local and regional richness (Caley & Schluter 1997). Therefore it is not possible to distinguish between different models solely on the basis of observed patterns in this relationship.

There have been few attempts to assess saturation of fish communities in lakes. The constant local diversity of North American (Wisconsin) and North European (Finland) forest lake fish assemblages, in spite of different regional species pools, suggests species saturation of local communities

(Tonn *et al.* 1990). Although this observation was based on two data points only, it is given more weight by the information that the Finnish lake district was, in postglacial (Holocene) times, covered by a single, large, glacial lake (the *Ancylus* Lake, Eronen *et al.* 2001; see Chapter 2). Thus, the present fish assemblages of this region are the result of repeated losses of species from an initial much larger species pool, after separation of the *Ancylus* Lake into smaller and smaller basins. The importance of extinction factors in moulding the fish assemblages of small forest lakes in Finland and Wisconsin was confirmed by Magnuson *et al.* (1998). Minns (1989) also suggested that, in Ontario lakes, mean lake species richness levelled off with increasing watershed richness, although Griffiths (1997) challenged this interpretation.

In contrast, Griffiths (1997), combining several published North American lake data sets, concluded that lacustrine fish communities are unsaturated. The data are consistent with proportional sampling, exhibiting a log–log-linear relationship between regional species pool and local species richness, adjusted to constant lake area (Fig. 16.4). He also considered the possible reasons which

might lead to **pseudosaturation**. Undersampling of small fish species, coupled with increased importance of small fish at low latitudes, would underestimate local richness in the more species-rich southern regions. The more localised distributions of these smaller, southern species are compatible with higher turnover diversity and pseudosaturation. Introduced species tended to be more restricted in their distribution than native fish, so that their inclusion in the calculation would increase the regional species pool but have little effect on average local species richness. However, in lakes in Minnesota (USA), fish stocking tends to increase local diversity, at the same time homogenising fish assemblages (Radomski & Goeman 1995). Presence of localised endemic species might also lead to bias in comparison between local and regional diversity, but this is not a problem in North American lakes, where endemism is rare.

Several studies (Hugueny & Paugy 1995; Oberdorff *et al.* 1998) have concluded that river fish assemblages are also unsaturated, although recent evidence for saturation of the native stream fish community also exists (Angermeier & Winston 1998). Unsaturation was found in a comparison between continents of local and regional diversity of all freshwater fishes in regions and localities of standardised size (Caley & Schluter 1997). Unsaturation of freshwater fish communities thus seems to be the rule, which means that invasions/introductions of new species are often likely to be successful.

In recently created habitats, such as new reservoirs, open niches can be filled by introduced species; saturation is related to the productivity and stability of the system (Milbrink & Holmgren 1981). As fisheries managers well know, especially heavy predation by resident species is sometimes able to prevent establishment of new fish species in a lake. However, there are also numerous examples of successful introduction of new species into lakes, although it may be difficult to judge whether these introductions have in the long term really resulted in increased species richness. Evidence from Lake Malawi shows that, even in the most species-rich communities, invasions of new species may be possible: aquarium fish dealers

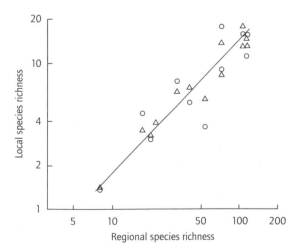

Fig. 16.4 Average local fish species richness in lakes (adjusted to standard size) plotted against regional native species richness (triangles, large lakes; circles, small lakes). (Redrawn from Griffiths 1997.)

have illegally transferred some species of cichlids from their native sites to other parts of the lake, and they have established new populations among the original rocky shore assemblage (Trendall 1988; Hert 1990), and this has happened in a community where practically all niches were already occupied. This may be typical for fish communities: the great plasticity shown by fishes in their resource use could mean that full niche utilisation occurs without communities being fully species-saturated (Griffiths 1997). Examples of such plasticity are found in some northern lakes: in extreme cases there may be only one species, e.g. perch (*P. fluviatilis*) or arctic charr (*Salvelinus alpinus*), occupying several feeding niches, from planktivory and benthivory, to piscivory (Sandlund *et al.* 1992).

16.1.8 Community stability

Long-term data on lake fish communities exhibit relative stability if environmental changes are limited (Inskip & Magnuson 1983). On the other hand, as evidenced by the history of the North American and African Great Lakes (Witte *et al.* 1992; Kitchell *et al.* 1994), rapid community changes also take place as a consequence of species introductions. In highly productive lakes, massive species kills, related to anoxia, epidemic diseases or overpopulation, may occur intermittently and lead to dramatic shifts in fish species composition. For example, in Lake Mendota (Wisconsin, USA) die-offs of cisco (*C. artedii*) are known to have occurred in 1884, 1932, 1940 and 1987; the yellow perch population also collapsed in 1884, and white bass (*Morone chrysops*) in 1884 and 1976 (Magnuson & Lathrop 1992). In between these episodes, community composition remained more or less stable, white bass being abundant during periods of cisco scarcity (Johnson & Kitchell 1996). Species richness in lowland German lakes has not changed during recent decades (Eckmann 1995).

Most fish communities are also vulnerable to change owing to exploitation by fisheries, especially if the species harvested are top predators (Pimm & Hyman 1987). Striking examples are the long-term effects of perch and pike removal on the fish populations of Lake Windermere, UK (Mills &

Hurley 1990), and the effects of fishery on the pelagic fish community in Lake Tanganyika (Coulter 1991).

16.2 INTERACTIONS WITHIN FISH COMMUNITIES

Complex biotic interactions within and between size-structured populations are generally assumed to organise many natural assemblages (Larkin 1956). Competition and predation are commonly considered the most important interspecific interactions affecting community structure and dynamics (e.g. Begon *et al.* 1996), although the spectrum of possible interactions is much more varied (Abrams 1987). In the fish communities of tropical lakes, documented interspecific relationships are indeed diverse, including the commensalistic and even mutualistic associations observed in the cichlid communities of Lakes Tanganyika and Malawi (Nakai *et al.* 1994; Stauffer *et al.* 1996). Two of the sand-dwelling haplochromine cichlids of Lake Malawi forage on benthic invertebrates suspended by the feeding activity of a third, whose gill rakers are more sparse than those of the other two. Of these, one species takes small arthropods and cladocerans, whilst the other feeds on larger chironomid larvae (Stauffer *et al.* 1996).

16.2.1 Competition

In nature, competition is difficult to demonstrate convincingly. Fish are commonly thought to avoid competition via spatial, behavioural and temporal partitioning of resources (Ross 1986). Numerous examples of food and habitat segregation have been documented, but fewer exist of temporal segregation, except the seasonal succession of fish fry. The problem is that it is impossible to know whether an observed habitat or diet segregation is due to competition at an earlier time ('the ghost of competition past'), or to other reasons. High resource-use overlap is a necessary condition for present competition, but is not sufficient in itself: high overlap may also mean that there is in fact no competition; i.e. resources are so abundant that there has been no need for resource division. For example, in Võrtsjärv (Estonia), although ruffe (*Gymnocephalus cernuus*), bream (*Abramis brama*) and eels (*Anguilla anguilla*) all feed largely on larvae of *Chironomus plumosus*, these remain abundant, and there are few indications of competition between the above species (Kangur & Kangur 1996).

High overlap, combined with proven scarcity of the resource, is better evidence for competition. For example, depressed growth can be used to indicate food shortage. The almost universal occurrence of density-dependent growth in lake fish populations shows that resources are often limiting to fish and that, besides intraspecific competition, interspecific competition is also thus likely to occur, at least periodically. The clearest evidence for interspecific competition comes from circumstances in which the presence of a potential competitor causes a shift in the diet or habitat use of another species, or depresses its growth. Rigorous proofs of competition may only be derived from experiments, but suggestive evidence can also sometimes be obtained from field data.

Direct experimental evidence of interspecific food competition between yellow perch and pumpkinseed (*Lepomis gibbosus*) was obtained from manipulation of littoral enclosures (Hanson & Leggett 1985). In both species, growth was dependent on conspecific density. Interspecific competition was asymmetrical, so that yellow perch growth was depressed when pumpkinseed were present, whilst pumpkinseed growth was not affected by yellow perch. Yellow perch growth was also markedly reduced in Lake Memphremagog when pumpkinseed abundance increased.

Diet overlap between perch and roach in North European lakes is normally low (Persson 1983). The larvae and the juveniles of both species feed on zooplankton, but on reaching the length of about 10 cm, perch normally switch to zoobenthos, and then, at 15–20 cm, to piscivory. In productive environments, roach may remain planktivorous as adults, but if plankton is scarce they feed on various benthic animals, macrophytes and even on cyanobacteria (Persson 1983, 1987). In spite of low overlap, in productive lakes, the competitive effect of roach on perch may still be substantial, because roach may be ten times more abundant than perch (Persson 1983). Indeed, the resource partitioning observed seems to result from competition. Roach are more successful in competition at least partly because they are able to feed on smaller zooplanktonic prey than larger perch. In the absence of roach, juvenile perch may remain zooplanktivorous until they grow to piscivorous size (Persson 1986; Persson & Greenberg 1990).

In a whole-lake experiment, perch shifted their diet from zooplankton to benthic macroinvertebrates at a younger age when abundance of roach was high (Persson 1986). In field enclosures, increased roach density raised the proportion of macroinvertebrates eaten by young perch (Persson & Greenberg 1990). However, in littoral enclosures, no interspecific effects were detected, in spite of lower abundance of zooplankton food (Persson 1987), probably because perch was more efficient than roach at feeding on chironomid larvae abundant in the littoral.

In accordance with these experimental results, perch in a nutrient-rich lake were restricted to the littoral habitat for most of the summer, whilst roach, to a larger extent, utilised the open water habitat (Persson 1987). Further experiments by Bergman (1990), with perch, roach and ruffe, likewise suggested that perch growth is negatively correlated with roach abundance (Fig. 16.5), but no shift to benthic feeding was found, probably because of the presence of ruffe, a more specialised benthivore. Ruffe growth was less affected by the presence of roach, and roach growth was independent of that of either perch or ruffe. Increasing ruffe density reduced perch growth (Fig. 16.5), and led to an increased proportion of zooplankton in the diet of perch, at the same time as the biomass of the favoured macroinvertebrate prey decreased (Bergman & Greenberg 1994). The well-known decreasing abundance of perch along a productivity gradient (e.g. Persson *et al.* 1991) may occur because perch, which exhibit ontogenetic diet shifts, are competitively sandwiched between more specialised planktivores and benthivores which lack such shifts. In its planktivorous stage, perch needs to compete with a planktivore (roach), and in its

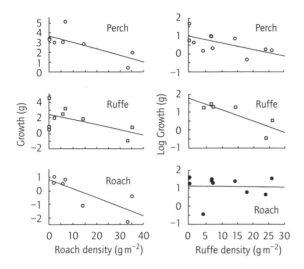

Fig. 16.5 Effects of roach (left) or ruffe (right) density on growth of perch, ruffe and roach in experimental enclosures. (Based on Bergman 1990; Bergman & Greenberg 1994.)

benthivorous stage, with a benthivore (ruffe) (Bergman & Greenberg 1994).

Roach also indirectly affect perch adversely. When abundant, roach tend to exacerbate eutrophication, increasing water turbidity (e.g. Andersson *et al.* 1978), thus impairing foraging success of the visually feeding perch. On the other hand, as indicated above, roach is a superior competitor in productive lakes, whilst perch is favoured by oligotrophic conditions, under which roach must meet competition from specialist planktivorous coregonids (especially vendace (*Coregonus albula*)) in the open water (Svärdson 1976). There thus seem to be positive feedbacks tending to accentuate roach dominance in the fish communities of lakes whose productivity is increasing.

In nature, especially in seasonally fluctuating environments, it is likely that resources become limiting only for restricted periods. Competition may then appear insignificant for most of the time (when resources are abundant relative to the needs of competitors), but yet be strong during periods of resource scarcity (or during population peaks of competitors). In a study of four tropical

floodplain fish assemblages, statistically significant guild structure was evident in all but one low-diversity assemblage (Winemiller & Pianka 1990). All four assemblages revealed significant partitioning of food resources. The most species-rich assemblage exhibited extremely high levels of resource segregation during the period of desiccation of aquatic habitat and increased fish densities.

In a similar way, during early summer, the majority of fish species in Pyhäjärvi (southwest Finland) feed on the same species of zooplankton (*Bosmina coregoni*). Yet populations of this cladoceran can still increase, indicating that there is overabundance of food and no competition between planktivores at that time. Later during the summer, the *Bosmina* population, and those of other crustacean zooplankton, collapsed, mainly owing to fish predation. In these circumstances, competition for food between fish species became keen and led to food segregation. Vendace continued to feed upon zooplankton, whilst whitefish (*Coregonus lavaretus*) and medium-sized perch switched to benthic food, and smelt (*Osmerus eperlanus*) and large perch to benthos and fish fry (Sarvala *et al.*, unpublished). These results are consistent with a hypothesis of niche compression in response to diffuse competition during temporal reduction in availability of preferred food resources (Winemiller & Pianka 1990).

A survey of the fish fauna and zooplankton in northern Swedish lakes showed the importance of food competition in moulding the fish community. The results of species introductions were especially illuminative (Nilsson & Pejler 1973; Svärdson 1976; Nilsson 1978). Competition between brown trout and Arctic charr led to interactive segregation in respect of both food and habitat (Nilsson 1978). Introduced whitefish have often eliminated native Arctic charr (Nilsson 1978; Svärdson 1976). Nilsson & Pejler (1973) classified these lakes according to their fish fauna as (i) brown trout (*Salmo trutta*), (ii) Arctic charr (*Salvelinus alpinus*), (iii) charr–trout, (iv) charr–trout–whitefish and (v) whitefish lakes. This series of fish assemblages represents increasing efficiency of planktivory, which is clearly reflected in the zooplankton species composition and size distributions in these lakes. The sequence also roughly corresponds to an altitudinal zonation, to which lowland lakes, with more diverse fauna, should be added.

Whitefish lakes normally contain many other species, whilst typical species in lowland lakes are pike, perch and roach, and often also burbot (*Lota lota*), smelt and pikeperch (*Stizostedion lucioperca*). When present by themselves (i.e. with no other fish species), brown trout can be planktivorous even after the juvenile phase, but when sympatric with Arctic charr their feeding niche is compressed, and their food consists of benthic items and small fish (e.g. charr: Damsgård & Langeland 1994). Charr also prefer benthic food when by themselves. Whitefish are dominant over charr as planktivores, eventually leading to decline or exclusion of charr, whilst brown trout are little affected.

Specialisation to pelagic life seems to be linked to competitive dominance (Svärdson 1976). Capability of feeding on smaller and, perhaps, also on sparser food items than competitors is probably the basic element of the competitive ability. In Claytor Lake (Virginia, USA), planktivory by a population of dense alewife (*Alosa pseudoharengus*) shifted zooplankton composition towards smaller forms, with clear adverse effects on the growth of young-of-the-year walleyes and white bass (Kohler & Ney 1981). In an oligotrophic Norwegian lake, introduced roach were more efficient zooplankton feeders than whitefish, and suppressed the native whitefish population (Langeland & Nøst 1994). Vendace, which is perhaps the most specialised planktivore in northern lakes, is considered competitively superior to whitefish in Swedish lakes (Svärdson 1976), and a similar conclusion seems to emerge from recent Finnish experience, although circumstances are not as simple as expected.

In several lakes, whitefish growth improved in years with low vendace stocks (Huhmarniemi *et al.* 1985; Heikinheimo-Schmid 1992; Raitaniemi *et al.* 1999; Sarvala *et al.*, unpublished). However, in Pyhäjärvi (southwest Finland), although diet overlap between vendace and whitefish was largest in the 0+ age group, whitefish growth improved particularly in the older age groups, which despite strong vendace stock possessed little food overlap with vendace (Sarvala *et al.*, unpublished). Diet and habitat shifts may be involved, as exemplified by some northern Norwegian lakes in the Paatsjoki River, which vendace has recently invaded from Lake Inari (Finland). In these lakes, plankton-feeding whitefish originally lived in the whole pelagic area as the only planktivorous fish, but after vendace invasion whitefish were restricted to the littoral zone and switched to predominantly benthic feeding (Bøhn *et al.* 1996). Similar differences in whitefish habitat use between pure whitefish and whitefish-vendace lakes from Sweden were mentioned by Svärdson (1976).

Several other cases of resource partitioning, presumably caused by competition, have been described in lake fish communities. In many European lakes, where rudd (*Scardinius erythrophthalmus*) and roach coexist, distribution of rudd is more littoral, whilst roach are more abundant in open water. Enclosure experiments by Johansson (1987) suggested that this habitat segregation may be interactive. Growth rates of both species were lowered when both were present, indicating interspecific competition. Both species were feeding mainly on zooplankton, but the impact of roach on zooplankton abundance and size structure was larger, suggesting that it is the superior competitor (Johansson 1987).

Partial habitat segregation between roach and ide (*Leuciscus idus*) in a Norwegian lake was accompanied by diet differences (Brabrand 1985). Compensatory changes in diet and habitat segregation were observed for roach and bleak in two Norwegian lakes (Fig. 16.6; Vøllestad 1985). During spring, habitat overlap was high but there was no diet overlap; later in summer, diet overlap increased but habitat overlap simultaneously declined.

A combination of diet and spatial segregation likewise enabled the coexistence of two planktivorous fish species in Lago di Maggiore, Italy (Berg & Grimaldi 1966). During spring and autumn, the native agone (*Alosa fallax lacustris*) fed more on copepods, whilst introduced whitefish 'bondella'

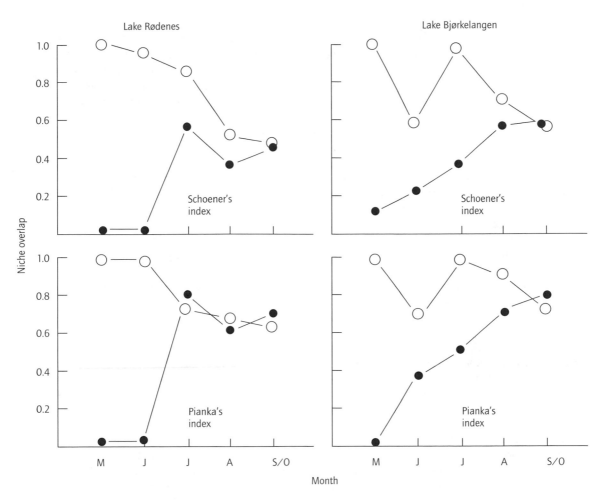

Fig. 16.6   Diet (●) and habitat (○) overlap of roach and bleak over the summer season in two Norwegian lakes. (Redrawn from Vøllestad 1985; reproduced with permission of *Holarctic Ecology*.)

(*C. lavaretus*) depended on cladocerans, especially *Daphnia*. During the zooplankton maximum (July–August) *Daphnia* was the main food for both species. The two species were also spatially segregated. During winter, when both species fed mainly on copepods, *Alosa* lived offshore and whitefish in inshore areas; during summer *Alosa* remained in the warm epilimnion, whilst the cold-stenothermic coregonids stayed deeper.

Such spatial segregation does not necessarily result from competitive relationships: differences in the vertical distributions of cisco (*C. artedii*) and yellow perch in Wisconsin (USA) lakes were predictable from their preferred temperatures derived from single-species laboratory experiments, accounting for light avoidance (Rudstam & Magnuson 1985). However, in Lake Michigan, adult alewife shifted their thermal distribution by 7–8°C after a notable increase of small bloaters with similar zooplankton diet (Crowder & Magnuson 1982). This habitat shift was probably due to competitive interactions.

16.2.2 Predation

Predation (including fishing by humans) is a very influential factor in moulding fish communities. Piscivorous fish biomass may vary within wide limits, from zero to more than 50% of total biomass; piscivores are particularly abundant in many tropical communities. Owing to morphological, physiological and behavioural constraints on predator–prey relationships, species interactions are size-dependent, leading to hierarchical organisation (Evans *et al.* 1987). Predators always take prey smaller than themselves. Capture success therefore varies according to the sizes of the predator and the prey. In piscivorous fish, optimal prey length seems to be about 10% of predator length (Popova 1978; Paradis *et al.* 1996).

In laboratory and field experiments, pike preferred the soft-rayed rudd over the spiny-rayed perch in open water, but in dense vegetation, perch became more available, and were selected over rudd. Habitat distribution of perch and rudd in a lake was influenced by predation risk. Accordingly, rudd were mainly found in the inner part of the littoral zone, whilst perch were encountered in the outer part and in the pelagic zone (Eklöv & Hamrin 1989). The antipredatory spines located on the midbody and the gillcover of perch reduced predation pressure by pike, so that perch could stay in open water. In southern Finland, the planktivorous fish community in the pelagic zone of large, oligotrophic lakes usually consists of the coregonids vendace and whitefish, and the smelt. Some perch, roach and bleak (*Alburnus alburnus*) may also move into the pelagic zone, especially in more productive lakes. Native predators in such systems are large smelt, perch, pike, burbot and brown trout but, owing to fishing and other anthropogenic effects, large piscivores have become scarce. Stocking of brown trout increased during the 1980s, elevating trout abundances above natural levels. Later, pikeperch stocking also became common, especially in more productive lakes. As a result of overstocking (relative to food resources), trout growth has deteriorated in many lakes.

Increased piscivore abundances are also suspected as one reason for an exceptionally long decline phase of vendace populations from the mid-1980s to the mid-1990s. In Finland, trout are mostly stocked in lakes as 2-year old, 25-cm long fish, which normally attain a length of >50 cm after two lake summers. Diet studies show that stocked brown trout always prefer small vendace (less than about 12 cm) if available (Helminen *et al.* 1997a; Niva & Julkunen 1998). Lacking suitably sized vendace, trout feed on smelt, and eventually on perch and roach, or even insects (mainly from the surface). Trout growth is most rapid on vendace food, and on invertebrate food especially trout grow slowly. Bioenergetic calculations suggest that at stocking densities exceeding 0.5 fish ha^{-1} brown trout are able to consume a major part of the vendace population. Accordingly, vendace stocks recovered in lakes where trout stocking rates declined below this limit (Helminen *et al.* 1997a).

Fisheries management relying on high piscivore stocking rates has also run into problems in the lake Inarijärvi in northernmost Finland. In Inarijärvi, salmonid stocking, aimed at compensating fishery losses due to water level regulation for hydropower production, exceeded prey fish production potential. Initial delays in the stocking programme were compensated by overstocking during the 1980s. The main piscivorous species stocked were brown trout, land-locked salmon (*Salmo salar*, f. *sebago*) and lake trout (*Salvelinus namaycush*) (Mutenia & Salonen 1994). Also, a new prey fish, the vendace, invaded the lake from introductions in small lakes draining into Inarijärvi.

As stated above, vendace is the preferred prey for the piscivorous brown trout in southern Finnish lakes (Helminen *et al.* 1997a). In Inarijärvi, the vendace year class 1986 was exceptionally strong and the year class 1989 was good, and as a consequence of the abundance of prey fish the brown trout stocking episodes of the late 1980s were very successful. However, by the early 1990s, the strong vendace year class was reduced by the fishery and by piscivores, and since then the stocked trout have exhibited poor growth and survival. A multispecies fish stock model, taking into account environmental variation, and the intra- and interspecific competitive and predatory relationships of vendace, benthic whitefish and brown trout, showed that in order to improve survival, growth and quality of trout the piscivore stocking rates should be decreased (Marttunen & Kylmälä 1997). High predation pressure increased the risk of small year classes of vendace. To improve the efficiency of whitefish stocking, stocking rates should be lowered, or fishing effort on whitefish increased.

The effects of each piscivore may be specific, depending on prey and habitat preferences. In Oulujärvi (northern Finland), in which both trout and pikeperch are being stocked in great numbers, diet studies show that trout mostly feed on vendace and pikeperch on smelt (Vehanen *et al.* 1998). Irrespective of their own size, both piscivores favour small prey fish, 4–10 cm in length. Small pikeperch feed more on vendace than do large ones. Of the predators capable of taking larger prey, pike tend to be littoral and burbot swim close to the bottom. Predation by perch, and especially by smelt, focuses on fish larvae and on fingerlings; perch themselves are preyed upon by the larger piscivores, especially pikeperch.

Experimental studies have shown that the presence of a predator can evoke dramatic changes in abundance and species composition in the fish community (Kitchell *et al.* 1994). In this process, the behavioural responses of the prey to predators can be as effective as, and more rapid than, the actual predation (Kitchell *et al.* 1994). Predator avoidance through habitat selection often produces size segregation of prey. For example, introduction of piscivorous pikeperch in the eutrophic Gjersjøen (Norway) led to an almost complete disappearance of juvenile roach from the pelagic zone (fish densities declined from 12,000–15,000 to 250 fish ha^{-1}), whilst total fish density remained unchanged in the littoral zone (Brabrand & Faafeng 1993). Prior to pikeperch introduction, pelagic roach (age 1+ and 2+ juveniles) underwent diel migration, forming dense littoral shoals by day, and dispersing quite evenly throughout the lake by night. After pikeperch introduction, juvenile roach displayed behavioural habitat shift, from the pelagic to the littoral zone, whilst larger roach were now found in the pelagic. In the littoral, perch

became the most abundant species. In neighbouring lakes without pikeperch, no such changes were observed.

Very similar relationships between pikeperch and different size-classes of roach were observed in The Netherlands by Lammens *et al.* (1992). Such strong interactions are made possible by the behaviour of roach and pikeperch, which lead to large temporal and spatial overlaps between the predator and the prey. In deep stratified lakes with smelt, these fish take refuge from pikeperch below the thermocline, and thus predation is limited to nighttime when smelt migrate into the epilimnion (Northcote & Rundberg 1970). For roach, the refuge is probably the littoral (Brabrand & Faafeng 1993). Similar size-dependent responses to piscivory have been reported in other species; the pelagic behaviour of prey species is largely dependent on predator presence (Arctic charr versus piscivores: L'Abée-Lund *et al.* 1993; stickleback versus rainbow trout (*Onchorhynchus mykiss*): Jakobsen *et al.* 1988; smelt versus potential piscivores: Gliwicz & Jachner 1992; minnows versus largemouth bass: Kitchell *et al.* 1994). In Michigan (USA) lakes, observed habitat shifts of bluegills (*Lepomis macrochirus*) may be predicted from a dynamic optimisation model of habitat choice, which took into account the trade-off between growth rate (food availability) and risk of predation by largemouth bass (Werner & Hall 1988). Small bluegill were confined within the vegetation, and the size at which fish moved into the pelagic zone was directly correlated with the density of largemouth bass (Werner & Hall 1988).

16.2.3 Interactions of competition and predation

Predator avoidance by habitat shifts often leads to increased intraspecific competition (Persson 1993). In Finnish lakes, small crucian carp were shown to find partial refuge from perch predation in littoral vegetation, whilst larger individuals attained a size refuge (Holopainen *et al.* 1997). Crowding in the littoral refuge led to slower growth, poorer condition and higher overwintering mortality.

More complex interactions arise in size-structured populations of predators and prey because of ontogenetic diet shifts. At some stage in their lifecycles, predators and prey may be competitors. Most fish larvae initially feed on zooplankton, switching successively to larger prey whilst growing. Pikeperch is a predator on small perch, but also a food competitor of larger piscivorous perch. Pikeperch and perch abundances are often inversely correlated, as are growth rates of walleye and yellow perch (Rudstam *et al.* 1996). Ruffe (*Gymnocephalus cernuus*) are food competitors to young burbot, whilst adult burbot prey upon ruffe (Svärdson 1976).

A useful example of the complexity of fish species interactions is the relationship between European perch and roach (Sumari 1971; Persson 1986, 1997). As described above, juvenile perch and roach are competitors, whilst larger perch may be predators on roach, and also cannibalistic on conspecifics. Competitive interactions between perch and roach are highly asymmetric, roach strongly affecting perch, with no evidence of inverse influence (Persson 1987, 1997; Persson & Greenberg 1990). In accordance with experimental results, a survey of fish populations in small Finnish lakes showed that the main factors affecting perch populations are interspecific competition with roach, and predation by piscivorous pike, burbot and perch (cannibalism), and possibly also by roach (on larval perch) (Sumari 1971). Biomass of roach is also significantly related to abiotic factors (conductivity, pH and total hardness; Persson 1997), reflecting known sensitivity of roach to acidity (e.g. Rask *et al.* 1995) as well as the observation that more eutrophic conditions favour roach (e.g. Tonn *et al.* 1990; Persson *et al.* 1991). Persson (1997) suggested that abiotic factors affected perch biomass mainly indirectly through the performance of roach. Piscivore biomass was positively related to roach biomass, and negatively related to perch biomass. The relationship between perch and roach thus may involve so-called **apparent competition**, arising when competitors share one or more predators.

On the other hand, the switch by perch to a piscivorous diet is partly regulated by food competition with roach at the juvenile stage (Persson & Greenberg 1990). Thus, in spite of food competition between the piscivorous pikeperch and perch, predation by pikeperch on roach may partially release perch from roach competition, and lead to increased numbers of piscivorous perch. The latter are then better able to control numbers of juvenile roach in the littoral zone, thus enhancing even more the possibilities for successful recruitment of perch to piscivorous size. Such **intraguild predation** (Polis & Holt 1992) may lead to a positive feedback loop, although cannibalism by piscivorous perch tends to cancel it out.

In perch–roach interaction, the prey species is competitively superior over juveniles of the predator. In contrast, central mudminnows in Wisconsin (USA) lakes, which compete with yearling yellow perch, and are simultaneously controlled by piscivorous yellow perch, are competitively inferior to yearling yellow perch (Tonn *et al.* 1986).

Relationships between vendace and smelt also involve both competition and predation. Vendace fry are more efficient planktivores than those of smelt (Karjalainen *et al.* 1997) and, owing to their fast growth, they are available as prey even for large smelt only for a short period during early summer. However, if a vendace population has declined, e.g. for climatic reasons (Helminen *et al.* 1997b), smelt may increase and, whilst abundant, large smelt exert considerable predation pressure upon late larvae and juveniles of vendace (Sterligova 1979), and may even prevent vendace stock from recovering.

16.2.4 Interactions in juvenile fish communities

Because cohort success in many fish species is determined during the early stages of life (e.g. Miller *et al.* 1988; Garvey *et al.* 1998), which therefore also affect community composition, interactions in larval and early juvenile assemblages deserve special attention. During summer, huge numbers of larval fish appear in the pelagic and littoral zones of temperate and subarctic lakes. Location of spawning sites and differences in spawning behaviour affect the distribution of larval fish between the pelagic and the littoral zones, but altogether millions of young fish of several species form the summer community of larval and juvenile fish (0+ fish), which utilise zooplankton as the main food resource (Keast 1980, 1985; Karjalainen *et al.* 1998). Therefore, intense interactions between fish species are likely to occur during the early stages of life (Colby *et al.* 1987).

In large temperate lakes, owing to high species diversity, larval fish can be sampled almost throughout the three summer months (Fig. 16.7; Karjalainen *et al.* 1998). In Puruvesi (Finland), off-spring of winter-spawning (e.g. burbot) and of autumn-spawning species (e.g. coregonids, *Coregonus* sp.) are hatched as early as April or May, whilst those of spring-spawning species (e.g. pike, perch, roach) begin to appear in larval fish samples at the beginning of June. Thus, composition of the 0+ fish community varies seasonally, exhibiting a succession of various species. Timing of spawning in different species partially explains the changing structure of this community (Keast 1980), whilst habitat shifts during development regulate species composition in a restricted area. As fish larvae develop, and their swimming capabilities improve, they are able actively to begin to select their micro-

habitat, although avoidance of unfavourable conditions (e.g. currents) may remain the major factor controlling habitat use and selection (Garner 1996). Water temperature largely regulates development rates of fish (e.g. Luczynski 1991), and temporal variation of the 0+ fish community. In tropical lakes, in which temperatures do not limit fish spawning, some aspect of seasonality is still imposed by environmental factors affecting food production, or by biotic pressures, such as competition for spawning grounds, or living space (Lowe-McConnell 1987). Many cichlid species seem to reproduce continuously throughout the year, but staggering of peak spawning times in sympatric cichlids has been observed, and the pelagic clupeids and cyprinids tend to breed seasonally (Lowe-McConnell 1987).

As well as temporally, composition of larval and juvenile fish communities varies also spatially. A simple division is into the littoral and pelagic communities. Spatial variation is also affected by specific spawning behaviour, and development of early life-stages. The yolk-sac larvae of several species are positively phototactic, and they swim or float near the surface after hatching, as though a part of the plankton community. The

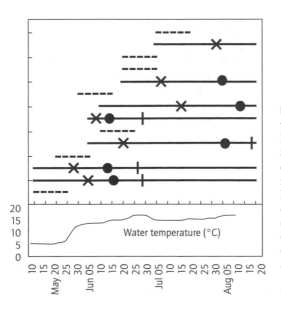

Fig. 16.7 Seasonal succession of young-of-the-year fish in Lake Puruvesi, eastern Finland in 1992. Solid lines indicate the occurrence of common species and dashed lines represent the occurrence of newly hatched larvae of other species. For common fish species, crosses, filled circles and vertical lines show the dates when on average a total length of 15 mm, 30 mm or 45 mm was attained. (From Karjalainen *et al.* 1998; reproduced with permission of the University of Joensuu.)

larvae of some species (e.g. smelt, vendace and perch; Jachner 1989; Mooij 1996; Karjalainen *et al.* 1997) may disperse throughout the open-water area of lakes, both in the pelagic and littoral zones, whilst larvae of some other species (e.g. roach, bream (*Abramis brama*), whitefish; Jachner 1989; Viljanen & Karjalainen 1992; Garner 1996; Mooij 1996; Urho 1996) are restricted to areas surrounding the spawning sites, near the shoreline or river banks. Dispersion of larval fish may positively affect the survival of fish larvae, e.g. by decreasing predation, by providing transport to areas of favourable feeding, or by increasing the areas of species distribution (Economou 1991; Fuiman & Magurran 1994). Another **endpoint strategy** of reproduction (Winemiller & Rose 1992) is based on retention of larvae in a narrow, favourable habitat (e.g. among vegetation), thus avoiding unsuitable conditions.

In many lakes, the greatest densities of newly hatched larvae are generally observed near the water surface (Viljanen *et al.* 1995) and in the shallow littoral zone (Viljanen & Karjalainen 1992; Mooij 1996). Owing to differences in vertical distribution, pelagic 0+ fish schools are often well segregated from schools of adult individuals (Hamrin 1986; Hamrin & Persson 1986; Gliwicz & Jachner 1992; Sydänoja *et al.* 1995). Young fish may stay in the warmer surface layers, whilst adult fish often live in the deeper waters. Interspecific differences of vertical distribution have also been observed in the juvenile stages of fish. Towards the end of summer, the 0+ fish begin to migrate vertically and their depth of occurrence now varies: for instance roach seem to occur in the upper layers, and perch and smelt in deeper waters (Jachner 1991; Appenzeller & Leggett 1995; Karjalainen *et al.* 1997).

During larval and juvenile development, the pelagic and littoral communities are in close interaction. Many larval fish species undergo considerable ontogenetic habitat shifts (Keast 1985; Post & McQueen 1988; Copp & Kováč 1996). Some fish species migrate from the littoral to the pelagic zone, whilst others move in the opposite direction. Migration to open water may occur soon after hatching (e.g. in smelt *O. mordax* and *O. eperlanus*;

Rupp 1965; Jachner 1989), the larval period (e.g. in vendace; Viljanen & Karjalainen 1992) or in the juvenile stage (e.g. *Labidesthes sicculus*, *Pimephales notatus*, *Fundulus diaphanus*; Keast 1985).

In contrast, in Lake Opinicon (Ontario, Canada), species diversity and abundance of juvenile fish in summer weedbed communities increased towards the end of the summer, with increasing invertebrate abundance. Most fish species entering the community had spent their larval period elsewhere (Keast 1985). Active selection of habitats is based on maximising growth rate, by increasing food intake and metabolic efficiency, and by minimising death rate by reducing predation pressure (Garner 1996), although in individual cases it may be difficult to separate the effective factors (Post & McQueen 1988; Urho 1996).

Most larval fish, and many juveniles, utilise zooplankton as their main food resource. Small larvae are gape-limited, and the size of the mouth opening affects the upper limit of ingestible prey size during larval development (e.g. Keast 1980; Hartmann 1986; Ponton & Müller 1990; Schael *et al.* 1991; Bremigan & Stein 1994; Mayer & Wahl 1997). Because all larval fish prey mainly on zooplankton, gape-limited feeding of small fish basically leads to similar diet if they spend their early life in the same habitat. Size of prey items and availability of food seem to be the most important factors regulating the diet of larval fish (Mayer & Wahl 1997; Karjalainen *et al.* 1998). Thus, simultaneous production of young by several lake fish species during summer may lead to high interspecific food competition, especially in the most favourable, sheltered areas such as amongst vegetation (Keast 1985).

On the other hand, when fish grow, variations in the size range and composition of food will increase (Keast 1980, 1985; Hammer 1985; Mayer & Wahl 1997, Karjalainen *et al.* 1998) and diets become different. Diet shift of young fish, from small to large prey items, and from zooplankton ('baby food') to benthic invertebrates, insects and fish larvae ('adult type food') reflects specific changes in swimming ability, feeding behaviour and habitat selection. In all, temporal and spatial variation, as well as ontogenetic diet shifts of various fish species in the 0+ fish community, make it possible to harvest the same planktonic food resources, and to produce high growth of young fish. This is a key factor regulating species diversity and abundance in the whole fish community.

Besides sufficiency of food, predation is another main factor which has been identified as determining abundance and survival of young fish, and to further recruitment of a year class (e.g. Miller *et al.* 1988; Luecke *et al.* 1990; Pepin & Myers 1991; Helminen *et al.* 1997b). Larval fish are vulnerable to predation by various piscivorous fish (e.g. perch), and even by species not generally piscivores (e.g. whitefish and minnow; Huusko & Sutela 1997). Owing to late spawning and hatching, the larvae of some cyprinid species are efficiently preyed upon by young-of-the-year predators such as pike or pikeperch (Mehner *et al.* 1996; Mooij 1996). When larval fish grow, their risk of falling prey to invertebrate predators (Hartig & Jude 1988) or small-sized fish decreases (Miller *et al.* 1988), but at the same time they appear in the diet of larger piscivores. Predator–prey interactions in fish communities are not only affected by size of prey, and predator, but also by their behaviour and by spatial heterogeneity in the environment. Prey fish may exhibit antipredatory behaviours; reduced activity and formation of schools are two general behavioural patterns of juvenile fish against predators (Fuiman & Magurran 1994; Eklöv & Persson 1995).

Young fish of several species are dependent on refuges, which provide shelter from predators (Eklöv & Persson 1995; Garner 1996). Habitat heterogeneity and different forms of structural complexity produce higher species diversity by maintaining specific refuges and food resources for 0+ fish (Copp 1992; Eklöv 1997). Specific selection of microhabitats is influenced by type of substrate, morphology of lake basin, currents, vegetation, temperature, water quality and nutrient concentrations. Water quality may also directly regulate abundance of some species. In eutrophied lakes, abundance of autumn- and winter-spawning species tends to decline, mainly because their eggs are unable to survive prolonged incubation in deteriorating oxygen conditions caused by excessive

influx of organic matter to the sediments (e.g. Müller 1992).

Comprehensive analysis of larval fish communities and their microhabitats may serve as a versatile bioindicator, which describes biotic processes and abiotic environment in various fresh waters (Copp *et al.* 1991). Knowledge of fish microhabitat use is also essential in the design and implementation of conservation or restoration of aquatic ecosystems (Copp 1992).

16.2.5 A case study: species interactions in North American Great Lakes

The history of fish community change in the North American Great Lakes, recently summarised by Kitchell *et al.* (1994), provides illuminating examples of the importance of species interactions. Overfishing, in combination with invasion by the sea lamprey (*Petromyzon marinus*), caused collapse of the native piscivore populations. In Lakes Michigan and Ontario, this allowed the increase of an exotic planktivore (the alewife), eventually up to more than 90% of total fish biomass. Ensuing food shortage led to massive die-offs of alewife. In the cold waters of Lake Superior, the alewife was unable to survive, but exotic rainbow smelt (*Osmerus mordax*) became abundant instead. Many native species declined (six of the seven coregonids endemic to the Great Lakes became extinct) as a result of competition–predation interaction with alewife or smelt (Crowder 1980; Eck & Wells 1987).

During the 1960s and early 1970s a chemical control programme was established in order to suppress lamprey recruitment. This was followed by intensive stocking with Pacific salmonids (*Oncorhynchus* spp.) and native lake trout. Lake Superior now contains an almost self-sustaining population of lake trout. Smelt populations have declined to 10% of former abundance, and populations of native coregonids (e.g. lake herring, *C. artedii*, and a deepwater cisco, the bloater, *C. hoyi*) have recovered locally. Bloaters now possess significantly fewer and shorter gillrakers, and shift from pelagic zooplanktivory to benthic foraging at least 2 years earlier in their life-history than before the increase of alewife (Crowder 1986). Alewife are more efficient at feeding on small zooplankton, but bloaters seem to be more efficient benthic foragers. These observations strongly support the idea of the importance of competition in the structure of the fish community.

In Lake Michigan, salmonid stocking rates increased predation rates to values which exceeded alewife replacement potential and, as a consequence, alewife declined during the early 1980s, to 10–20% of their former abundance. Several colder than normal years may also have contributed to alewife decline (Eck & Wells 1987). Palaeolimnological evidence indicates a greater intensity of piscivory in Lake Michigan than ever before (Kitchell *et al.* 1994). As anticipated on the basis of bioenergetic analyses (Stewart *et al.* 1981), native species (e.g., deepwater cisco, yellow perch and several sculpin species, *Cottus* spp.) recovered, and now comprise 80% or more of total fish biomass. Compensatory responses by native zooplanktivorous fish have now exceeded the historical evidence of steady-state behaviour. In Lake Ontario, the fish community has followed a similar development, but with a 10-year lag. During peak abundance of alewife and smelt, several native species became extinct, which diminished the recovery potential of the fish community. Decline of the alewife populations in Lakes Michigan and Ontario has caused managers to reduce predator stocking rates (Jones *et al.* 1993).

Although some time-series analyses of commercial catches in Lakes Huron and Superior provide little evidence for interspecific interactions (Henderson & Fry 1987; Stone & Cohen 1990), bioenergetic estimates of food consumption based on observed growth and mortality rates have convincingly shown the significance of introduced piscivores in these lakes (Kitchell *et al.* 1994). Prognoses based on the bioenergetic calculations have largely been borne out (Kitchell & Crowder 1986). Predator–prey interactions, and the carrying capacity for stocked salmonids, are thus the main concerns of the present-day fisheries in these lakes (Kitchell *et al.* 1994). These systems possess strong predatory inertia (Stewart *et al.* 1981); stocking decisions are made 3 to 5 years before a cohort of

predators exerts its maximum effect. The key prey species (alewife and smelt) exhibit highly variable recruitment success, which includes the density-independent effects of variable weather. Unexpectedly, the predation rate of piscivores exhibited little response to lake-wide averages of prey fish abundance, although these varied over two orders of magnitude (Kitchell *et al.* 1994). In schooling species, effective prey density encountered by the predator may well be very different from lake-wide average abundance. Then, predation and growth rates will remain similar until prey densities reach some critical lower limit. They will then fall sharply, unless the predator switches to alternative prey.

16.2.6　Another case study: effects of predation in Lake Victoria

Development of the fish community in Lake Victoria during recent decades provides a candid example of the potential of piscivory in modifying a fish community. The original fish community was diverse (>550 species), including perhaps 500 species (>95% endemic) within the haplochromines of the family Cichlidae (Seehausen *et al.* 1997). An indigenous fishery existed in the littoral zone for many of these species. Fishing intensified during the 1950s, with the advent of outboard motors and nylon nets; large members of the piscivorous guilds and potamodromous species were heavily exploited. Many haplochromine species remained abundant, but supported fisheries of only modest economic value. In order to increase the commercial value of the fishery, the piscivorous Nile perch (*Lates niloticus*) and the omnivorous, rapidly growing Nile tilapia (*Oreochromis niloticus*) were introduced from their native habitats in Lake Albert and Lake Turkana (Ogutu-Ohwayo 1990a). Populations of the introduced species increased slowly at first, but then expanded exponentially during the 1970s and throughout the 1980s. Major fisheries developed for utilising Nile perch and total yield increased fourfold compared with previous values.

However, the Nile perch is a voracious predator, which grows to 1 kg during its first year of life, and which can reach 100 kg, although the most commonly harvested sizes are 2–4 kg at ages 2+ or 3+. Smaller juveniles are restricted to shallow and/or nearshore environments, but larger juvenile and adult Nile perch occupy all habitats of the lake where oxygen concentrations are sufficient. Recently, eutrophication has led to seasonal anoxia, which restricts the use of deeper waters for all fishes (Hecky *et al.* 1994).

Initially, haplochromine species comprised 80–90% of the Nile perch diet (Ogutu-Ohwayo 1990b). Many haplochromines are trophic specialists; most mature at small size and possess low reproductive potential, owing to small clutch sizes and extended parental care. Such a life-history makes them particularly vulnerable to predation. Accordingly, most haplochromine stocks declined to low numbers as the Nile perch population increased. Although lake-wide extinction is difficult to demonstrate, local decline has been revealing. In the Mwanza Gulf (Tanzania), whilst Nile perch were increasing, experimental trawl catches of haplochromines declined from 70–75 to +1 kg ha in 1 year. At the same time, the diet of Nile perch gradually shifted from almost exclusively haplochromine prey to other fish, as haplochromines declined.

Juvenile Nile perch feed largely on invertebrates, mainly on freshwater shrimps, which have considerably increased; large Nile perch feed on cyprinids, small *Lates*, tilapia and haplochromines. By 1987, Nile perch in Tanzanian waters had reached a biomass of 150 kg ha^{-1} (Kitchell *et al.* 1997). Similar changes were recorded in Ugandan waters, throughout the lake and around its many islands. It has been estimated that 150–200 of the endemic fish species have already been lost (Kaufman 1992; Witte *et al.* 1992). Both littoral trophic specialists and pelagic zooplanktivores and insectivores declined. However, these population collapses can only be partly explained by Nile perch predation. For example, stenotopic rock-dwelling cichlids, of which, in Lake Victoria, there are more than 200 species, are rarely eaten by Nile perch. Yet many such species have disappeared during the past 10 years. During the 1990s, Nile perch catches began to decline, whilst fishing

effort is still increasing. Fishermen are switching to smaller mesh sizes, and juvenile Nile perch are increasingly caught by beach seines. As a result, in areas of highest fishing intensity, some haplochromine populations are recovering (Kitchell *et al.* 1997; Witte *et al.* 2000).

The main components of the fish community are now Nile perch, Nile tilapia and a small zooplanktivorous cyprinid, *Rastrineobola argentea*. Nile tilapia was sympatric with Nile perch, and appears resistant to predation owing to its rapid growth rate and large adult size (>30 cm total length). Gill net fisheries based on this species are now very productive and profitable (Kitchell *et al.* 1997). In contrast, *Rastrineobola* matures at a small size (4.5–5.0 cm). Its evasive and schooling behaviour allow it to occupy the surface waters, where it has increased four to fivefold after the appearance of Nile perch. Fisheries based on this species are expanding.

From a purely socioeconomic perspective, the Lake Victoria fisheries seem to be in an acceptable state (Kitchell *et al.* 1997). However, from an ecological or biodiversity point of view, the situation is far from satisfactory and is probably unstable. The original, very diverse trophic system has become extremely simplified. Recent studies also show that the consequences of the increase in turbidity of the lake due to eutrophication (see above) are dramatic for, and detrimental to, the evolution and persistence of the cichlid species. Moreover, fish community changes are due not only to introduction of an alien predator, but also of exotic tilapiine competitors, some of which may hybridise with the two native tilapiines (Craig 1992). Previous experience of Nile perch introduction into Lake Kyoga (Uganda) also shows a rapid expansion of the fishery, overexploitation and gradual recovery of populations of many prey species (Okemwa & Ogari 1994).

Kitchell *et al.* (1997) analysed the community effects of Nile perch in Lake Victoria by combining simple simulation models and a bioenergetic model. Analyses showed that development of a large-mesh gill net fishery reduced total predation by Nile perch to 40% of the levels experienced during the late 1970s, when the densities of this species were greatest. Expansion of beach seine and small-mesh gill net fisheries could reduce the total predation to 25%, and a combination of all fishing methods would lead to a total predation impact of only 10% of the peak level. Because of extensive cannibalism by Nile perch, the effects of moderate fishery on the haplochromines are not simple, whilst more intensive exploitation of the predator always improves the recovery potential of the prey. Intensified fishery of large predatory species in other African lakes has always resulted in decline of the target species, and shifted community composition towards smaller taxa (Coulter 1991; Craig 1992).

16.3   FISH COMMUNITIES AND ENVIRONMENTAL CHANGE

16.3.1   Eutrophication

Eutrophication of lake ecosystems is a consequence of the disturbed nutrient balance of such water bodies. It is caused by increased external nutrient load, in most cases due to human activities,

which then often leads to increased biological production. Along with increasing productivity of the entire ecosystem, fish production and biomass tend also to increase. Fish accelerate nutrient dynamics in eutrophic lakes, and thus increase the effects of internal nutrient load, which further accentuates the eutrophication process.

When lakes change from oligotrophic to mesotrophic, and then to eutrophic, major changes occur in the structure of the fish community. A general trend in the large, temperate lakes of Europe is the change from coregonids to percids, and, along with increasing nutrient load, to cyprinids (Fig. 16.8; Leach *et al.* 1977; Persson *et al.* 1991). A similar pattern prevails in North America, except that centrarchids, which are naturally lacking from Europe, increase with nutrient load in a similar way to cyprinids (Colby *et al.* 1972).

In a recent survey based on stratified random sampling, information on fish status of 710 Finnish lakes was obtained by questionnaires, with special emphasis on changes caused by eutrophication (Tammi *et al.* 1999). The most striking of the reported changes were increased abundance of roach and bream, and decrease of

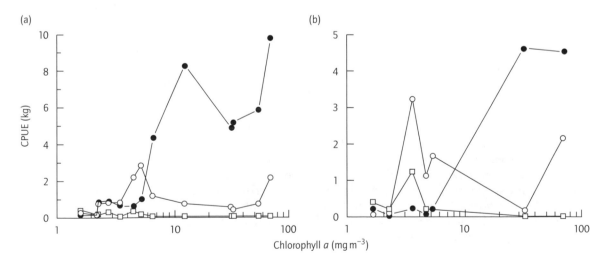

Fig. 16.8 Biomass (catch per unit effort) of Salmoniformes (mainly coregonids, □), percids (○) and cyprinids (●) along the productivity gradient (measured as phytoplankton chlorophyll) in the benthic habitat (left) and the pelagic habitat (right): CPUE, catch-per-unit-effort. (Redrawn from Persson *et al.* 1991; reproduced with permission from Elsevier Science.)

burbot and vendace. These observations were highly consistent with long-term records from eutrophied lakes (Svärdson & Molin 1981; Keto & Sammalkorpi 1988).

During eutrophication, various changes in the interactions within the fish communities of lakes take place. The increase of roach and bream biomass to hundreds of kilograms per hectare (Keto & Sammalkorpi 1988; Berg *et al.* 1997; Sarvala *et al.* 1998b) leads to high intra- and interspecific competition, which is indicated by the slow growth of these fish (Horppila 1994, Sarvala *et al.*, unpublished). Decline of perch (Svärdson & Molin 1981) may be due to competition with roach (Persson 1983) and with ruffe, which may be favoured by the low transparency of the waters (Bergman 1988). For burbot, and the coregonids, vendace and whitefish, reductions in hypolimnetic oxygen concentrations may limit access to cool water. Increased sediment accumulation smothers developing eggs of the latter (Sterligova *et al.* 1988; Müller 1992).

Hypolimnetic oxygen deficits have probably contributed to the disappearance of four bottom-living fish species from several eutrophicated lowland German lakes (Eckmann 1995). Reduced interspecific competition allows several cyprinid species to occupy the pelagic zone in deep and meso/eutrophic lakes, and these are able to invade the pelagic zone, even in oligotrophic lakes, if warm-water planktivorous competitors are absent (Svärdson 1976). Among predatory fish species, pikeperch is favoured by the turbid conditions of eutrophic lakes (Svärdson & Molin 1981). The possibilities of mitigating harmful effects of eutrophication by stocking with predatory fish, mostly pikeperch or pike, have been examined recently in several studies (Salonen *et al.* 1996; Berg *et al.* 1997).

16.3.2 *Acidification*

Acidification owing to air pollution is an environmental problem at the opposite end of the trophic gradient of lakes compared with eutrophication. It has been a considerable problem of oligotrophic surface waters in catchments with low acid neutralising capacity in Norway, Sweden, Finland, the USA and Canada. On the basis of historical catch statistics, interviews and other documents it has been suggested that fish damage by acidification began in Norway during the late 1800s and early 1900s (Hesthagen & Hansen 1991), in Sweden during the early 1900s (Hultberg 1985) and in Finland after World War II at the latest (Rask & Tuunainen 1990). In Norway, it has been assessed (on the basis of several questionnaires) that the number of fish populations recently lost and affected in lakes is, respectively, 9600 and 5400, most of them (8200 and 3900) being brown trout populations (Hesthagen 1997). In Finland, a corresponding assessment based on water chemistry, and test fishing data from lakes of the National Acidification Research Programme, led to an estimate of 1000–2100 fish populations lost and 1200–2400 affected, most of them (800–1700 lost, 500–1000 affected) being roach (Rask *et al.* 1995).

Because most acidified lakes are relatively small and oligotrophic, their fish communities were, in most cases, simple, consisting of only a few species. Therefore, interspecific interactions in these fish communities were simple as well. This is especially the case in Norway, where in most acid sensitive lakes brown trout is the only fish species. In Finland, the five most common fish species in small lakes (Tonn *et al.* 1990)—perch, roach, pike, ruffe and burbot—can be commonly found in acid sensitive lakes. In the course of acidification, roach and burbot are affected, and rapidly lost, because they are the most acid sensitive, with pH 5.5 being critical for reproduction (Almer *et al.* 1974; Magnuson *et al.* 1984). For ruffe the critical pH is 5.0; pike and perch are most tolerant with a critical pH of around 4.5 (Almer *et al.* 1974). In North America, where diversity of the fish communities tends to be higher than in Europe, corresponding gradual extinctions of fish species in acidifying lakes have also been reported (Jackson & Harvey 1995). Below pH 6.5, numbers of fish species in Ontario lakes declined linearly with decreasing pH (Matuszek & Beggs 1988).

Because the two species are food competitors (Svärdson 1976; Persson 1983), the decline of roach as a result of acidification is beneficial to perch (Appelberg *et al.* 1993). Similarly, disappearance of ruffe at a later stage of acidification is beneficial for perch. Still, it must be kept in mind that conditions other than acidity, such as oligotrophy and clear water, favour perch in coexistence with roach (Svärdson 1976) and ruffe (Bergman 1988), whereas eutrophication and increasing turbidity favour roach and ruffe against perch (Svärdson & Molin 1981).

For predatory fish, problems of prey availability occur with acidification. This is due to decrease and disappearance of prey populations, especially roach. Furthermore, owing to repeated failures of reproduction (Almer *et al.* 1974; Rosseland *et al.* 1980) those fish populations most greatly affected often consist entirely of old and large individuals, so that young pike cannot find suitable prey. Pike and perch commonly coexist as the only fish species in acidifying lakes, but there are a few documented cases where perch have disappeared before pike (Hultberg & Stenson 1970; Raitaniemi 1995). What remains is a fish community composed solely of slowly growing pike, feeding mostly on invertebrates.

In acid sensitive headwater lakes, perch is often the only fish species present with high population density, but slow growth (Raitaniemi *et al.* 1988). During the course of acidification, perch densities have declined in many lakes owing to reproduction failures. In some of these lakes, growth rates of the remaining perch have clearly accelerated, owing to decreased intraspecific food competition (Hultberg 1985; Raitaniemi *et al.* 1988). This also shows that, despite possible physiological stress, perch can grow well feeding entirely on invertebrate food. In lakes recovering from acidification, reproduction of perch has begun again after intervals even of 10 years. Population density has increased sharply, and consequently growth rates have declined drastically (Nyberg *et al.* 1995).

16.3.3 *Climatic change*

Temperature is an important regulator of fish biology and distribution. Thus, significant effects of climate change on fish communities are to be expected (Shuter & Post 1990). Predicted warming of the climate will probably expand the distribution of warm-water fish species in lakes, whereas for cold-water species, especially salmonids, an opposite trend has been suggested. Probably, responses will not be so simple, because during their early life most fish species prefer warm water, and because the changes in competitive and predator–prey relations are difficult to forecast. Furthermore, in temperate regions, amounts of precipitation should increase, which, together with mild winters, is presently causing changes in hydrology and increasing external loads of substances to lakes. This may enhance eutrophication processes in lakes and, consequently, accelerate eutrophication-related changes in fish communities and interactions within them.

In northern Europe, warming of climate probably may not cause any remarkable changes in the fish species composition of lakes. However, this statement may not apply to all regions. For example, according to Mandrak (1989), 27 new species may migrate into the area of North American Great Lakes, whilst at same time some cold-water species will disappear. In Finland, many species, such as perch and especially pikeperch, live close to their northern extremes of distribution. During warm summers this leads to the appearance of strong year classes (Lappalainen *et al.* 1996; Sarvala & Helminen 1996). According to model predictions (Lappalainen & Lehtonen 1997), warm-water species such as roach, bream, perch and pikeperch will all benefit from warming, whereas in southern Finland the habitats of cold-water taxa such as brown trout and whitefish will shrink. Similar predictions for future trends in distribution of salmonids have also been presented for North America (Meisner 1990; Schindler *et al.* 1990). However, effects of changing species interactions may add considerable uncertainty to such predictions (Davis *et al.* 1998).

16.4   FISH PRODUCTION AND FISHERIES YIELD

Amongst the products of aquatic ecosystems, the fisheries harvest has long been of most interest to humanity. Fish production is ultimately dependent on aquatic primary production, driven by solar radiation and nutrient inputs from the land, although complex interactions between biological components of the system can considerably affect transfer efficiencies. Fish harvesting may exert a very strong influence upon the whole ecosystem: according to a recent estimate, about 24% of fresh-water primary production is required in order to maintain present global levels of lacustrine and riverine fish catches (Pauly & Christensen 1995).

Several attempts have been made to develop simple methods to predict fish yield from water bodies (Leach *et al.* 1987). Regression and correlation techniques have been used in order to relate potential fish harvest to a variety of physical and biological variables. One of the earliest empirical models was based on the finding that the total fish yield from a lake is closely correlated with lake area (Rounsefell 1946). This apparently quite trivial relationship is still one of the best predictors of fish yield in lakes (Youngs & Heimbuch 1982; Ranta & Lindström 1989). Usually, catch per unit area decreases with increasing lake area (Schneider & Haedrich 1989).

Rawson (1952) noted that long-term average catches in large lakes that are intensively fished are negatively correlated with mean depth. More generally, Rawson claimed that the factors affecting lake productivity may be divided into three groups, namely morphometric, edaphic and climatic. Mean depth can be used as an indicator of morphometry within a lake basin. Edaphic factors affect lake chemistry, which regulates general productivity, and total dissolved solids or conductivity can be used to indicate the edaphic component. For a simplified comparison with fish yields, indices of lake morphometry and chemistry were combined into the so-called morphoedaphic index (Ryder 1965, 1982; Ryder *et al.* 1974; Leach *et al.* 1987), obtained by dividing total dissolved solids or conductivity by mean depth (or sometimes maximum depth). Fish yields generally increased with increasing values of morphoedaphic index (Ryder 1982).

The basic morphoedaphic index was thought to be most useful at a regional level. Leach *et al.* (1987) found that amongst existing methods for

prediction of fish yield the morphoedaphic index gave values most closely in accordance with historical yields in the North American Great Lakes and Lake Winnipeg (Canada). However, on a global scale, differences in climate exert far greater effect on fish yields than water chemistry and mean depth (e.g. Kitaev 1994). Thus, in any global model of fish harvest, climatic factors, light and temperature must be taken into account. Indeed, in two world-wide data sets of 43 and of 80 lakes, mean annual air temperature accounted for 74% and 62% of the variability of maximum sustained fish yields, whilst the morphoedaphic index accounted only for an additional 7–9% of yield variability (Schlesinger & Regier 1982).

In principle, because of the overwhelming importance of phosphorus in regulating freshwater productivity, fish yield should be better predictable from total phosphorus concentration or load (Hrbáček 1969; Hanson & Leggett 1982), or some other nutrient variable, or, even better, directly from phytoplankton, zooplankton or zoobenthos abundance or production (Hrbáček 1969; Oglesby 1977; Matuszek 1978; Liang *et al.* 1981; Hanson & Leggett 1982; Biró & Vörös 1982). If the range of the independent variable is wide enough, maximum yields may be found at intermediate values (e.g. Biró & Vörös 1982).

Applications of morphoedaphic indices have been strongly criticised (Youngs & Heimbuch 1982; Ranta & Lindström 1989, 1993; Schneider & Haedrich 1989; Jackson *et al.* 1990). The original correlation between fish yield and morphoedaphic index was partly spurious, because the term for lake area occurred on both sides of the equation (the dependent variable was catch per unit area, and the independent variable contained mean depth, which is derived from lake volume and area; Jackson *et al.* 1990). This led to inflated percentages of explained variation. In a more correct form (relating total catch to surface area and total dissolved solids), confidence limits were extremely wide. Theoretical scaling of the fish harvest on lake volume, which is consistent with the known tendency of small lakes to yield more fish per unit area, reduced the upper limits to 28 times the lower limit (Schneider & Haedrich 1989). In general, after accounting for the effect of lake area or volume, the increase in explained catch variance due to water quality variables was negligible.

Besides these simple, empirically derived estimators, methods used for predicting fish production in lakes also encompass more theoretically based models, and even complex ecosystem simulation models. The biomass-size spectrum models of Borgmann (1982, 1987) or Thiebaux & Dickie (1992) provide useful information for moderate expenditures of time and effort. Their underlying assumption is that, in aquatic systems, phytoplankton, zooplankton and fish will reach about equal biomass, allowing prediction of fish biomass from phytoplankton or zooplankton data. Size-specific rate constants can then be used in order to derive potential fish production. However, these predictions may not approximate actual yields very well. In a comparison of different methods applied to the North American Great Lakes, the potential production estimated from invertebrate production with the Borgmann method was considerably greater than historical yields, or yield estimates from the empirical models (Leach et al. 1987). In the same vein, when applied to worldwide lake and reservoir data, predictions of fish biomass from Borgmann's models were only weakly correlated to measured fish biomass (Cyr & Peters 1996).

A new approach combines life-history information of small and large fish with (i) Borgmann's production model, (ii) empirical information on population production to biomass ratios, and (iii) classic population dynamics theory, in order to estimate potential production and optimum sustained yield for each group. Historical sustained yield, as a percentage of optimum sustained yield, varies from a low of 6% for small fish to 100% for large species (Leach et al. 1987). Another recent method of estimating fish production begins from estimated biomass for a given region, and then utilises a general relationship between the production:biomass ratio and body size to predict production (Boudreau & Dickie 1989).

These simple methods, based on morphoedaphic indices, total nutrient concentrations or primary production, are attractive to fisheries managers because they are easy and relatively inexpensive to apply and interpret. Application of complex models has been limited because their data inputs are often prohibitively difficult, and expensive to obtain. However, approaches utilising information on the life-history features and ecological requirements of the fish species exploited (including interactions with the food) are likely to provide more realistic predictions under changing fishing pressures and environmental conditions.

All analyses of fish yield versus water quality are confounded by the strong influence of fishing intensity on the level of fish yield, and also on fish production. Although the productivity of fish communities is determined by energy inputs, nutrients, edaphic factors and habitat variables, the distribution of production by species is strongly influenced by interactions between them (Evans et al. 1987), and total catch levels depend on the intensity of fishing. Annual fish catches are often only around 20–25% of total fish production (Borgmann et al. 1984; Houde & Rutherford 1993). Several published data sets exhibit a tight positive relationship between fish yield and fishing effort, sometimes with a postulated asymptote, or decreasing catch, at very high fishing effort (e.g. Bayley 1988), sometimes with apparently continuously increasing catches with increasing effort (e.g. Schlesinger & McCombie 1983; Ranta & Lindström 1989, 1993). In the latter case, recorded fish yields probably did not even approach production potential, explaining the poor correlations with water quality. It should also be noted that fishing effort probably varies according to fish community composition: abundance of commercially desirable species is likely to elicit higher fishing effort. Other factors affecting yield include fish community structure (Carlander 1955: total biomass is maximal when species richness is high, but maximum biomass of individual species is greatest when few species are present) and predator–prey interactions.

Increasing fishing effort raises fishing mortality, but not in a linear fashion. At low fishing effort, fish yield is solely a function of effort (e.g. Ranta & Lindström 1989). At higher effort, fish biomass begins to decrease, but yield remains constant, because increasing production compensates for lowered biomass. Increasing fishing effort further leads to overfishing and rapid collapse of yield. Most reported lake fisheries seem to be located on the left ascending part of the fish-yield–fishing-effort curve, and thus do not tell us anything about relations between sustained fish yield and general productivity of the system. In fact, as noted by, for example, Ryder *et al.* (1974), even the original application of the morphoedaphic principle required that lakes be fished intensively enough to reveal their fish production potential.

Although the principle behind the morphoedaphic index was sound, the route from the physical and chemical basis of productivity to fish yields seems to be too long to retain much predictive power. Fish production should be more predictable than yield, and models describing fish production as a function of phytoplankton production or biomass are inherently more accurate, and subject to fewer exceptions, than are those related to morphoedaphic factors (Oglesby 1977). Indeed, fish production in lakes exhibits close correlation with primary production (Morgan 1980; Downing *et al.* 1990), but not with the morphoedaphic index (Downing *et al.* 1990). Fisheries yield is also predictable from primary production, but in relation to that factor, yield from lakes is almost an order of magnitude lower than that from marine systems (Fig. 16.9; Nixon 1988).

This difference may simply be due to the more efficient fishing methods utilised in coastal seas. This interpretation is supported by examination of the few lake fisheries which are really efficient. In Pyhäjärvi in southwest Finland, sustainable fish catches have been 0.7–0.8% of primary production, i.e. quite comparable to, or greater than, those in coastal seas (Sarvala *et al.* 1998a). Moreover, most recent estimates of fish yield in large European lakes also imply efficiencies close to those reported from marine localities (range 0.02–0.46%, averages 0.09–0.29% of primary production; Lavrentyeva & Lavrentyev 1996). In addition, a large part of traditional fish catches in lakes consist of piscivorous species, whilst the most productive sea fisheries are often based on planktivorous or demersal species. The lower the trophic

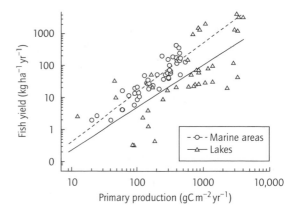

Fig. 16.9 Fish yield versus phytoplankton primary production in lakes (data from Oglesby 1977; Sarvala *et al.* 1998a, 1999) and estuarine and marine systems (data from Nixon 1988).

position of the exploited species in the food web, the higher may be the efficiency of the fishery relative to primary production. The productive fishery of Pyhäjärvi is mainly based on zooplanktivorous species (Sarvala *et al.* 1994) and, likewise, the high catch rates from Lake Tanganyika derive two-thirds from zooplanktivorous clupeids (Sarvala *et al.* 1999).

16.5 REFERENCES

Abrams, P.A. (1987) On classifying interactions between populations. *Oecologia*, **73**, 272–81.

Almer, B., Dickson, W., Ekström, C., Hörnström, E. & Miller, U. (1974) Effects of acidification on Swedish lakes. *Ambio*, **3**, 30–6.

Andersson, G., Berggren, H., Cronberg, G. & Gelin, C. (1978) Effects of planktivorous and benthivorous fish on organisms and water chemistry. *Hydrobiologia*, **59**, 9–15.

Angermeier, P.L. & Winston, M.R. (1998) Local vs regional influences on local diversity in stream fish communities of Virginia. *Ecology*, **79**, 911–27.

Appelberg, M., Henrikson, B.-I., Henrikson, L. & Svedäng, M. (1993) Biotic interactions within the littoral community of Swedish forest lakes during acidification. *Ambio*, **22**, 290–7.

Appenzeller, A.R. & Leggett, W.C. (1995) An evaluation of light-mediated vertical migration of fish based on hydroacoustic analysis of the diel vertical movements of rainbow smelt (*Osmerus mordax*). *Canadian Journal of Fisheries and Aquatic Sciences*, **52**, 504–11.

Austen, D.J., Bayley, P.B. & Menzel, B.W. (1994) Importance of the guild concept to fisheries research and management. *Fisheries*, **19**, 12–20.

Barbour, C.D. & Brown, J.H. (1974) Fish species diversity in lakes. *American Naturalist*, **108**, 473–89.

Bayley, P.B. (1988) Accounting for effort when comparing tropical fisheries in lakes, river-floodplains, and lagoons. *Limnology and Oceanography*, **33**, 963–72.

Begon, M., Harper, J.L. & Townsend, C.R. (1996) *Ecology* (3rd edn). Blackwell, Oxford, 1068 pp.

Berg, A. & Grimaldi, E. (1966) Ecological relationships between planktophagic fish species in the Lago Maggiore. *Verhandlungen Internationale Vereinigung Limnologie*, **16**, 1065–73.

Berg, S., Jeppesen, E. & Sondergaard, M. (1997) Pike (*Esox lucius* L.) stocking as a biomanipulation tool. 1. Effects on the fish population in Lake Lyng, Denmark. *Hydrobiologia*, **342/343**, 311–8.

Bergman, E. (1988) Foraging abilities and niche breadths of two percids, *Perca fluviatilis* and *Gymnocephalus cernua*, under different environmental conditions. *Journal of Animal Ecology*, **57**, 443–53.

Bergman, E. (1990) Effects of roach *Rutilus rutilus* on two percids, *Perca fluviatilis* and *Gymnocephalus cernua*: importance of species interactions for diet shifts. *Oikos*, **57**, 241–9.

Bergman, E. & Greenberg, L. (1994) Competition between a planktivore, a benthivore, and a species with ontogenetic diet shifts. *Ecology*, **75**, 1233–45.

Biró, P. & Vörös, L. (1982) Relationships between phytoplankton and fish yields in Lake Balaton. *Hydrobiologia*, **97**, 3–7.

Bøhn, T., Amundsen, P.-A. & Staldvik, F. (1996) *Invasjon av lagesild i Pasvikvassdraget—status og konsekvenser pr. 1995*. Mimeo, Norges Fiskerihøgskole, Universitetet i Tromsø, Norway, 42 pp.

Borgmann, U. (1982) Particle-size conversion efficiency and total animal production in pelagic ecosystems. *Canadian Journal of Fisheries and Aquatic Sciences*, **39**, 668–74.

Borgmann, U. (1987) Models on the slope of, and biomass flow up, the biomass–size spectrum. *Canadian Journal of Fisheries and Aquatic Sciences*, **44** (Supplement 2), 136–40.

Borgmann, U., Shear, H. & Moore, J. (1984) Zooplankton and potential fish production in Lake Ontario. *Canadian Journal of Fisheries and Aquatic Sciences*, **41**, 1303–9.

Boudreau, P.R. & Dickie, L.M. (1989) Biological model of fisheries production based on physiological and ecological scalings of body size. *Canadian Journal of Fisheries and Aquatic Sciences*, **46**, 614–23.

Brabrand, Å. (1985) Food of roach (*Rutilus rutilus*) and ide (*Leuciscus idus*): significance of diet shift for interspecific competition in omnivorous fishes. *Oecologia*, **66**, 461–7.

Brabrand, Å. & Faafeng, B. (1993) Habitat shift in roach (*Rutilus rutilus*) induced by pikeperch (*Stizostedion lucioperca*) introduction: predation risk versus pelagic behaviour. *Oecologia*, **95**, 38–46.

Bremigan, M.T. & Stein, R.A. (1994) Gape-dependent larval foraging and zooplankton size: implications for fish recruitment across systems. *Canadian Journal of Fisheries and Aquatic Sciences*, **51**, 913–22.

Browne, R.A. (1981) Lakes as islands: biogeographic distribution, turnover rates, and species composition in the lakes of central New York. *Journal of Biogeography*, **8**, 75–83.

Caley, M.J. & Schluter, D. (1997) The relationship between local and regional diversity. *Ecology*, **78**, 70–80.

Carlander, K.D. (1955) The standing crop of fish in lakes. *Journal of the Fisheries Research Board of Canada*, **12**, 543–70.

Colby, P.J., Ryan, P.A., Schupp, D.H. & Serns, S.L. (1987) Interactions in north-temperate lake fish communities. *Canadian Journal of Fisheries and Aquatic Sciences*, **44** (Supplement 2), 104–28.

Colby, P.J., Spagler, G.R., Hurley, D.A. & McCombie, A.M. (1972) Effects of eutrophication on salmonid communities in oligotrophic lakes. *Journal of the Fisheries Research Board of Canada*, **29**, 1602–12.

Copp, G.H. (1992) An empirical model for predicting microhabitat of 0+ juvenile fishes in a lowland river catchment. *Oecologia*, **91**, 338–45.

Copp, G.H. & Kováč, V. (1996) When do fish with indirect development become juveniles? *Canadian Journal of Fisheries and Aquatic Sciences*, **53**, 746–52.

Copp, G.H., Oliver, J.M., Penáz, M. & Roux, A.L. (1991) Juvenile fishes as functional describers of fluvial ecosystem dynamics: applications on the River Rhône, France. *Regulated Rivers: Research & Management*, **6**, 135–45.

Cornell, H.V. & Lawton, J.H. (1992) Species interactions, local and regional processes, and limits to the richness of ecological communities: a theoretical perspective. *Journal of Animal Ecology*, **61**, 1–12.

Coulter, G.W. (ed.) (1991) *Lake Tanganyika and its Life*. British Museum (Natural History), Oxford University Press, Oxford, 354 pp.

Coulter, G.W. (1994) Lake Tanganyika. *Archiv für Hydrobiologie Beiheft Ergebnisse Limnologie*, **44**, 13–18.

Craig, J.F. (1992) Human-induced changes in the composition of fish communities in the African Great Lakes. *Reviews in Fish Biology and Fisheries*, **2**, 93–124.

Crowder, L.B. (1980) Alewife, rainbow smelt and native fishes in Lake Michigan: competition or predation? *Environmental Biology of Fishes*, **5**, 225–33.

Crowder, L.B. (1986) Ecological and morphological shifts in Lake Michigan fishes: glimpses of the ghost of competition past. *Environmental Biology of Fishes*, **16**, 147–57.

Crowder, L.B. & Magnuson, J.J. (1982) Thermal habitat shifts by fishes at the thermocline in Lake Michigan. *Canadian Journal of Fisheries and Aquatic Sciences*, **39**, 1046–50.

Cyr, H. & Peters, R.H. (1996) Biomass–size spectra and the prediction of fish biomass in lakes. *Canadian Journal of Fisheries and Aquatic Sciences*, **53**, 994–1006.

Damsgård, B. & Langeland, A. (1994) Effects of stocking of piscivorous brown trout, *Salmo trutta* L, on stunted Arctic charr, *Salvelinus alpinus* (L.). *Ecology of Freshwater Fishes*, **3**, 59–66.

Davis, A.J., Lawton, J.H., Shorrocks, B. & Jenkinson, L.S. (1998) Individualistic species responses invalidate simple physiological models of community dynamics under global environmental change. *Journal of Animal Ecology*, **67**, 600–12.

Dejoux, C. (1994) Lake Titicaca. *Archiv für Hydrobiologie Beiheft Ergebnisse Limnologie*, **44**, 35–42.

Downing, J.A., Plante, C. & Lalonde, S. (1990) Fish production correlated with primary productivity, not the morphoedaphic index. *Canadian Journal of Fisheries and Aquatic Sciences*, **47**, 1929–36.

Eadie, J.McA. & Keast, A. (1984) Resource heterogeneity and fish species diversity in lakes. *Canadian Journal of Zoology*, **62**, 1689–95.

Eadie, J.McA., Hurly, T.A., Montgomerie, R.D. & Teather, K.L. (1986) Lakes and rivers as islands: species-area relationships in the fish faunas of Ontario. *Environmental Biology of Fishes*, **15**, 81–9.

Eck, G.W. & Wells, L. (1987) Recent changes in Lake Michigan's fish community and their probable causes, with emphasis on the role of the alewife (*Alosa pseudoharengus*). *Canadian Journal of Fisheries and Aquatic Sciences*, **44** (Supplement 2), 53–60.

Eckmann, R. (1995) Fish species richness in lakes of the northeastern lowlands of Germany. *Ecology of Freshwater Fishes*, **4**, 62–9.

Economou, N. (1991) Is dispersal of fish eggs, embryos and larvae an insurance against density dependence? *Environmental Biology of Fishes*, **31**, 313–21.

Eklöv, P. (1997) Effects of habitat complexity and prey abundance on the spatial and temporal distributions of perch (*Perca fluviatilis*) and pike (*Esox lucius*). *Canadian Journal of Fisheries and Aquatic Sciences*, **54**, 1520–31.

Eklöv, P. & Hamrin, S.F. (1989) Predatory efficiency and prey selection: interactions between pike *Esox lucius*, perch *Perca fluviatilis* and rudd *Scardinus erythrophthalmus*. *Oikos*, **56**, 149–56.

Eklöv, P. & Persson, L. (1995) Species-specific antipredator capacities and prey refuges: interactions between piscivorous perch (*Perca fluviatilis*) and juvenile perch and roach (*Rutilus rutilus*). *Behavioural Ecology and Sociobiology*, **37**, 169–78.

Eronen, M., Glückert, G., Hatakka, L., van de Plassche, O., van der Plicht, J. & Rantala, P. (2001) Rates of Holocene isostatic uplift and relative sea-level lowering of the Baltic in SW Finland based on studies of isolation contacts. *Boreas*, **30**, 17–30.

Evans, D.O., Henderson, B.A., Bax, N.J., Marshall, T.R., Oglesby, R.T. & Christie, W.J. (1987) Concepts and methods of community ecology applied to freshwater fisheries management. *Canadian Journal of Fisheries and Aquatic Sciences*, **44** (Supplement 2), 448–70.

Fernando, C.H. (1994) Zooplankton, fish and fisheries in tropical freshwaters. *Hydrobiologia*, **272**, 105–23.

Fernando, C.H. & Holcik, J. (1989) Origin, composition, and yield of fish in reservoirs. *Archiv für Hydrobiologie Beiheft Ergebnisse Limnologie*, **33**, 637–41.

Fuiman, L.A. & Magurran, A.E. (1994) Development of predator defences in fishes. *Reviews in Fish Biology and Fisheries*, **4**, 145–83.

Galis, F. & Metz, J.A.J. (1998) Why are there so many cichlid species? *Trends in Ecology and Evolution*, **13**, 1–2.

Garner, P. (1996) Microhabitat use and diet of 0+ cyprinid fishes in a lentic, regulated reach of the River Great Ouse, England. *Journal of Fish Biology*, **48**, 367–82.

Garvey, J.E., Dingledine, N.A., Donovan, N.S. & Stein, R.A. (1998) Exploring spatial and temporal variation within reservoir food webs: predictions for fish assemblages. *Ecological Applications*, **8**, 104–20.

Gashagaza, M.M. (1991) Diversity of breeding habits in lamprologine cichlids in Lake Tanganyika. *Physiological Ecology Japan*, **28**, 29–65.

Gliwicz, Z.M. & Jachner, A. (1992) Diel migrations of juvenile fish: a ghost of predation past or present? *Archiv für Hydrobiologie*, **124**, 385–410.

Goldschmidt, T. & Witte, F. (1992) Explosive speciation and adaptive radiation of haplochromine cichlids from Lake Victoria: an illustration of the scientific value of a lost species flock. *Mitteilungen Internationale Vereinigung Limnologie*, **23**, 101–7.

Greenwood, P.H. (1994). Lake Victoria. *Archiv für Hydrobiologie Beiheft Ergebnisse Limnologie*, **44**, 19–26.

Griffiths, D. (1997). Local and regional species richness in North American lacustrine fish. *Journal of Animal Ecology*, **66**, 49–56.

Guégan, J.-F., Lek, S. & Oberdorff, T. (1998) Energy availability and habitat heterogeneity predict global riverine fish diversity. *Nature*, **391**, 382–4.

Hammer, C. (1985) Feeding behaviour of roach (*Rutilus rutilus*) larvae and the fry of perch (*Perca fluviatilis*) in Lake Lankau. *Archiv für Hydrobiologie*, **103**, 61–74.

Hamrin, S.F. (1986) Vertical distribution and habitat partitioning between different size classes of vendace, *Coregonus albula*, in thermally stratified lakes. *Canadian Journal of Fisheries and Aquatic Sciences*, **43**, 1617–25.

Hamrin, S.F. & Persson, L. (1986) Asymmetrical competition between age classes as a factor causing population oscillations in an obligate planktivorous fish species. *Oikos*, **47**, 223–32.

Hanson, J.M. & Leggett, W.C. (1982) Empirical prediction of fish biomass and yield. *Canadian Journal of Fisheries and Aquatic Sciences*, **39**, 257–63.

Hanson, J.M. & Leggett, W.C. (1985) Experimental and field evidence for inter- and intraspecific competition in two freshwater fishes. *Canadian Journal of Fisheries and Aquatic Sciences*, **42**, 280–6.

Hartig, J.H. & Jude, D.J. (1988) Ecological and evolutionary significance of cyclopoid predation on fish larvae. *Journal of Plankton Research*, **10**, 573–7.

Hartmann, J. (1986) Interspecific predictors of selected prey of young fishes. *Archiv für Hydrobiologie Beiheft Ergebnisse Limnologie*, **22**, 373–86.

Hecky, R.E., Bugenyi, F.W.B., Ochumba, P., *et al.* (1994) Deoxygenation of the deep water of Lake Victoria, East Africa. *Limnology and Oceanography*, **39**, 1476–81.

Heikinheimo-Schmid, O. (1992) Management of European whitefish (*Coregonus lavaretus* L. s.l.) stocks in Lake Paasivesi, eastern Finland. *Polskie Archiwum Hydrobiologii*, **39**, 827–35.

Helminen, H., Marjomäki, T.J., Koivurinta, M. & Valkeajärvi, P. (1997a) Taimenistutusten väheneminen elvytti osaltaan muikkukantoja. *Suomen Kalastuslehti* **104**, 38–43. (In Finnish.)

Helminen, H., Sarvala, J. & Karjalainen, J. (1997b) Patterns in vendace recruitment in Lake Pyhäjärvi, southwest Finland. *Journal of Fish Biology*, **51** (Supplement A), 303–16.

Helminen, H., Karjalainen, J., Kurkilahti, M., Rask, M. & Sarvala, J. (2000) Eutrophication and fish biodiversity in Finnish lakes. *Verhandlungen Internationale Vereinigung Limnologie*, **27**, 194–9.

Henderson, B.A. & Fry, F.E.J. (1987) Interspecific relations among fish species in South Bay, Lake Huron, 1949–84. *Canadian Journal of Fisheries and Aquatic Sciences*, **44** (Supplement 2), 10–14.

Hert, E. (1990) Factors in habitat partitioning in *Pseudotropheus aurora* (Pisces: Cichlidae), an introduced species to a species-rich community of Lake Malawi. *Journal of Fish Biology*, **36**, 853–65.

Hesthagen, T. (1997) *Population responses of arctic charr (Salvelinus alpinus (L.)) and brown trout (Salmo trutta L.) to acidification in Norwegian inland waters.* PhD thesis, Norwegian University of Science and Technology, Department of Zoology, Trondheim, Norway, 56 pp.

Hesthagen, T. & Hansen, L.P. (1991) Estimates of annual loss of Atlantic salmon, *Salmo salar* L., in Norway due to acidification. *Aquaculture and Fisheries Management*, **22**, 85–91.

Hinch, S.G., Collins, N.C. & Harvey, H.H. (1991) Relative abundance of littoral zone fishes: biotic interactions, abiotic factors, and postglacial colonization. *Ecology*, **72**, 1314–24.

Holopainen, I.J., Tonn, W.M. & Paszkowski, C.A. (1997) Tales of two fish: the dichotomous biology of crucian carp (*Carassius carassius* (L.)) in northern Europe. *Annales Zoologici Fennici*, **34**, 1–22.

Hori, M., Gashagaza, M.M., Nshombo, M. & Kawanabe, H. (1993) Littoral fish communities in Lake Tanganyika: irreplaceable diversity supported by intricate interactions among species. *Conservation Biology*, **7**, 657–66.

Horppila, J. (1994) The diet and growth of roach in Lake Vesijärvi and possible changes in the course of biomanipulation. *Hydrobiologia*, **294**, 35–41.

Houde, E.D. & Rutherford, E.S. (1993) Recent trends in estuarine fisheries: predictions of fish production and yield. *Estuaries*, **16**, 161–76.

Hrbáček, J. (1969) Relations between some environmental parameters and the fish yield as a basis for a predictive model. *Verhandlungen internationale Vereinigung für theoretische und angewandte Limnologie*, **17**, 1069–81.

Hugueny, B. & Paugy, D. (1995) Unsaturated fish communities in African rivers. *American Naturalist*, **146**, 162–9.

Huhmarniemi, A., Niemi, A. & Palomäki, R. (1985) Whitefish and vendace stocks in the regulated Lake Pyhäjärvi, central Finland. In: Alabaster, J.S. (ed.), *Habitat Modification and Freshwater Fisheries*. Food and Agriculture Organisation and Butterworths, London, 165–72.

Hultberg, H. (1985) Changes in fish population and water chemistry in Lakes Gårdsjön and neighbouring lakes during the last century. *Ecological Bulletin*, **37**, 64–72.

Hultberg, H. & Stenson, J. (1970) Försurningens effekter på fiskfaunan i två bohuslänska småsjöar. *Fauna och flor*, **65**, 11–20.

Huusko, A. & Sutela, T. (1997) Minnow predation on vendace larvae: intersection of alternative prey phenologies and size-based vulnerability. *Journal of Fish Biology*, **50**, 965–77.

Inskip, P.D. & Magnuson, J.J. (1983) Changes in fish populations over an 80-year period: Big Pine Lake, Wisconsin. *Transactions of the American Fisheries Society*, **112**, 378–89.

Jachner, A. (1989) Growth of fry of three fish species from pelagial of mesotrophic lake. *Polskie Archiwum Hydrobiologii*, **36**, 359–71.

Jachner, A. (1991) Food and habitat partitioning among juveniles of three fish species in the pelagial of a mesotrophic lake. *Hydrobiologia*, **226**, 81–9.

Jackson, D.A. & Harvey, H.H. (1993) Fish and benthic invertebrates: community concordance and community-environment relationships. *Canadian Journal of Fisheries and Aquatic Sciences*, **50**, 2641–51.

Jackson, D.A. & Harvey, H.H. (1995) Gradual reduction and extinction of fish populations in acid lakes. *Water Air and Soil Pollution*, **85**, 389–94.

Jackson, D.A., Harvey, H.H. & Somers, K.M. (1990) Ratios in aquatic sciences: statistical shortcomings with mean depth and the morphoedaphic index. *Canadian Journal of Fisheries and Aquatic Sciences*, **47**, 1788–95.

Jakobsen, P.J., Johnsen, G.H. & Larsson, P. (1988) Effect of predation risk and parasitism on the feeding ecology, habitat use and abundance of lacustrine three spine stickleback (*Gasterosteus aculeatus*). *Canadian Journal of Fisheries and Aquatic Sciences*, **45**, 426–431.

Johansson, L. (1987) Experimental evidence for interactive habitat segregation between roach (*Rutilus rutilus*) and rudd (*Scardinius erythrophthalmus*) in a shallow eutrophic lake. *Oecologia*, **73**, 21–27.

Johnson, T.B. & Kitchell, J.F. (1996) Long-term changes in zooplanktivorous fish community composition: implications for food webs. *Canadian Journal of Fisheries and Aquatic Sciences*, **53**, 2792–803.

Johnson, T.C., Scholz, C.A., Talbot, M.R., *et al.* (1996) Late Pleistocene desiccation of Lake Victoria and rapid evolution of cichlid fishes. *Science*, **273**, 1091–3.

Jones, M., Koonce, J. & O'Gorman, R. (1993) Sustainability of hatchery-dependent salmonine fisheries in Lake Ontario: the conflict between predator demand and prey supply. *Transactions of the American Fisheries Society*, **122**, 1002–18.

Kangur, K. & Kangur, A. (1996) Feeding of ruffe (*Gymnocephalus cernuus*) in relation to the abundance of benthic organisms in Lake Võrtsjärv (Estonia). *Annales Zoologici Fennici* **33**, 473–80.

Karjalainen, J., Turunen, T., Helminen, H., Sarvala, J. & Huuskonen, H. (1997) Food selection and consumption of 0+ smelt (*Osmerus eperlanus* (L.)) and vendace (*Coregonus albula* (L.)) in the pelagial zone of Finnish lakes. *Archiv für Hydrobiologie Special Issues Advances in Limnology*, **49**, 37–49.

Karjalainen, J., Ollikainen, S., Staff, S., Viljanen, M. & Väisänen, P. (1998) Larval fish communities in Lake Puruvesi: species composition and diet. *Publications of Karelian Institute, University of Joensuu*, **122**, 52–5. (In Finnish.)

Kaufman, L. (1992) Catastrophic change in species-rich freshwater ecosystems. The lessons of Lake Victoria. *BioScience*, **42**, 846–58.

Keast, A. (1980) Food and feeding relationships of young fish in the first weeks after the beginning of exogenous feeding in Lake Opinicon, Ontario. *Environmental Biology of Fishes*, **5**, 305–14.

Keast, A. (1985) Development of dietary specializations in a summer community of juvenile fishes. *Environmental Biology of Fishes*, **13**, 211–24.

Keto, J. & Sammalkorpi, I. (1988) A fading recovery: a conceptual model for Lake Vesijärvi management and research. *Aqua Fennica*, **18**, 193–204.

Kitaev, S.P. (1994) *Ikhtiomassa i ryboprodukciya malykh i srednikh ozer i sposoby ikh opredeleniya*. Nauka, Sankt-Peterburg, 177 pp. (In Russian.)

Kitchell, J.F. & Crowder, L.B. (1986) Predator–prey interactions in Lake Michigan: model predictions and recent dynamics. *Environmental Biology of Fishes*, **16**, 205–11.

Kitchell, J.F., Eby, L.A., He, X., Schindler, D.E. & Wright, R.A. (1994) Predator–prey dynamics in an ecosystem context. *Journal of Fish Biology*, **45** (Supplement A), 209–26.

Kitchell, J.F., Schindler, D.E., Ogutu-Ohwayo, R. & Reinthal, P.N. (1997) The Nile perch in Lake Victoria: interactions between predation and fisheries. *Ecological Applications*, **7**, 653–64.

Kohda, M., Yanagisawa, Y., Sato, T., *et al.* (1996) Geographical colour variation in cichlid fishes at the southern end of Lake Tanganyika. *Environmental Biology of Fishes*, **45**, 237–48.

Kohler, C.C. & Ney, J.J. (1981) Consequences of an alewife die-off to fish and zooplankton in a reservoir. *Transactions of the American Fisheries Society*, **110**, 360–9.

Kornfield, I. & Carpenter, K.E. (1984) Cyprinids of Lake Lanao, Philippines: taxonomic validity, evolutionary rates and speciation scenarios. In: Echelle, A.A. & Kornfield, I. (eds), *Evolution of Fish Species Flocks.* University of Maine Press, Orono, ME, 69–84.

L'Abée-Lund, J.H., Langeland, A., Jonsson, B. & Ugedal, O. (1993) Spatial segregation by age and size in Arctic charr: a trade-off between feeding possibility and risk of predation. *Journal of Animal Ecology*, **62**, 160–8.

Lammens, E.H.H.R., Landman, A.F., McGillavry, P.J. & Vlink, B. (1992) The role of predation and competition in determining the distribution of common bream, roach and white bream in Dutch eutrophic lakes. *Environmental Biology of Fishes*, **33**, 195–205.

Langeland, A. & Nøst, T. (1994) Introduction of roach (*Rutilus rutilus*) in an oligohumic lake. 1. Competition impacts on whitefish (*Coregonus lavaretus*). *Verhandlungen Internationale Vereinigung Limnologie*, **25**, 2113–7.

Lappalainen, J. & Lehtonen, H. (1997) Temperature habitats for freshwater fishes in a warming climate. *Boreal Environment Research*, **2**, 69–84.

Lappalainen, J., Lehtonen, H., Böhling, P. & Erm, V. (1996) Covariation in year-class strength of perch, *Perca fluviatilis* L. and pikeperch, *Stizostedion lucioperca* (L.). *Annales Zoologici Fennici*, **33**, 421–6.

Larkin, P.A. (1956) Interspecific competition and population control in freshwater fish. *Journal of the Fisheries Research Board of Canada*, **13**, 327–42.

Lavrentyeva, G.M. & Lavrentyev, P.J. (1996) The relationship between fish yield and primary production in large European freshwater lakes. *Hydrobiologia*, **322**, 261–6.

Leach, J.H., Johnson, M.G., Kelso, J.R.M., Hartmann, J., Nümann, W. & Entz, B. (1977) Responses of percids and their habitats to eutrophication. *Journal of the Fisheries Research Board of Canada*, **34**, 1964–71.

Leach, J.H., Dickie, L.M., Shuter, B., Borgmann, U., Hyman, J. & Lysack, W. (1987) A review of methods for prediction of potential fish production with application to the Great Lakes and Lake Winnipeg. *Canadian Journal of Fisheries and Aquatic Sciences*, **44** (Supplement 2), 471–85.

Liang, Y., Melack, J.M. & Wang, J. (1981) Primary production and fish yields in Chinese ponds and lakes. *Transactions of the American Fisheries Society*, **110**, 346–50.

Liem, K.F. (1973) Evolutionary strategies and morphological innovations: cichlid pharyngeal jaws. *Systematic Zoology*, **22**, 425–41.

Lowe-McConnell, R. (1987) *Ecological Studies in Tropical Fish Communities.* Cambridge University Press, Cambridge, 382 pp.

Lowe-McConnell, R. (1996) Fish communities in the African Great Lakes. *Environmental Biology of Fishes*, **45**, 219–35.

Luczynski, M. (1991) Temperature requirement for growth and survival of larval vendace, *Coregonus albula* (L.). *Journal of Fish Biology*, **38**, 29–35.

Luecke, C., Rice, J., Crowder, L.B., Yeo, S.E. & Binkowski, F.P. (1990) Recruitment mechanisms of bloater in Lake Michigan: an analysis of the predatory gauntlet. *Canadian Journal of Fisheries and Aquatic Sciences*, **47**, 524–32.

MacArthur, R.H. & Wilson, E.O. (1967) *The Theory of Island Biogeography.* Princeton University Press, Princeton, NJ, 203 pp.

Magnuson, J.J. & Lathrop, R.C. (1992) Historical changes in the fish community. In: Kitchell, J.F. (ed.), *Food Web Management: a Case Study of Lake Mendota.* Springer-Verlag, New York, 193–231.

Magnuson, J.J., Baker, J.P. & Rahel, F.J. (1984) A critical assessment of effects of acidification on fisheries in North America. *Philosophical Transactions of the Royal Society of London, Series B*, **305**, 501–16.

Magnuson, J.J., Tonn, W.M., Banerjee, A., Toivonen, J., Sanchez, O. & Rask, M. (1998) Isolation vs. extinction in the assembly of fishes in small northern lakes. *Ecology*, **79**, 2941–56.

Mahon, R. (1984) Divergent structure of fish taxocenes of north temperate streams. *Canadian Journal of Fisheries and Aquatic Sciences*, **41**, 330–50.

Mandrak, N.E. (1989) Potential invasion of the Great Lakes by fish species associated with climatic warming. *Journal of Great Lakes Research*, **15**, 306–16.

Marttunen, M. & Kylmälä, P. (1997) Kalakantojen hoitomalli Inarijärven kalaistutusten vaikutusten arvioinnissa (Evaluation of the effects of fish stockings by fish stock model in Lake Inarijärvi). *Suomen Ympäristö*, **117**, 1–83. (In Finnish.)

Matuszek, J.E. (1978) Empirical predictions of fish yields of large North American lakes. *Transactions of the American Fisheries Society*, **107**, 385–94.

Matuszek, J.E. & Beggs, G.L. (1988) Fish species richness in relation to lake area, pH, and other abiotic factors in Ontario lakes. *Canadian Journal of Fisheries and Aquatic Sciences*, **45**, 1931–41.

Mayer, C.M. & Wahl, D.H. (1997) The relationship between prey selectivity and growth and survival in a larval fish. *Canadian Journal of Fisheries and Aquatic Sciences*, **54**, 1504–12.

McDowall, R.M. (1987a) The native fish. In: Viner, A.B. (ed.), *Inland Waters of New Zealand*. DSIR Bulletin 241. DSIR Science Information Publishing Centre, Wellington, New Zealand, 291–306.

McDowall, R.M. (1987b) Impacts of exotic fishes on the native fauna. In: Viner, A.B. (ed.), *Inland waters of New Zealand*. DSIR Bulletin 241. DSIR Science Information Publishing Centre, Wellington, New Zealand, 333–47.

Mehner, T., Schultz, H., Bauer, D., Herbst, R., Voigt, H. & Benndorf, J. (1996) Intraguild predation and cannibalism in age-0 perch (*Perca fluviatilis*) and age-0 zander (*Stizostedion lucioperca*): interactions with zooplankton succession, prey fish availability and temperature. *Annales Zoologici Fennici*, **33**, 353–61.

Meisner, J.D. (1990) Effect of climatic warming on the southern margins of the native range of brook trout, *Salvelinus fontinalis*. *Canadian Journal of Fisheries and Aquatic Sciences*, **47**, 1065–70.

Meyer, A. (1993) Phylogenetic relationships and evolutionary processes in East African cichlid fishes. *Trends in Ecology and Evolution*, **8**, 279–84.

Meyer, A., Kocher, T.D., Basasibwaki, P. & Wilson, A.C. (1990) Monophyletic origin of Lake Victoria cichlid fishes suggested by mitochondrial DNA sequences. *Nature*, **347**, 550–3.

Meyer, A., Montero, C. & Spreinat, A. (1994) Evolutionary history of the cichlid fish species flocks of the East African great lakes inferred from molecular phylogenetic data. *Archiv für Hydrobiologie Beiheft Ergebnisse Limnologie*, **44**, 407–23.

Milbrink, G. & Holmgren, S. (1981) Fish species interactions in a fertilised reservoir. *Report Institute of Freshwater Research Drottningholm*, **59**, 121–7.

Miller, T.J., Crowder, L.B., Rice, J.A. & Marschall, E.A. (1988) Larval size and recruitment mechanism in fishes: toward a conceptual framework. *Canadian Journal of Fisheries and Aquatic Sciences*, **45**, 1657–70.

Mills, C.A. & Hurley, M.A. (1990) Long-term studies on the Windermere populations of perch (*Perca fluviatilis*), pike (*Esox lucius*) and Arctic charr (*Salvelinus alpinus*). *Freshwater Biology*, **23**, 119–36.

Minns, C.K. (1989) Factors affecting fish species richness in Ontario lakes. *Transactions of the American Fisheries Society*, **118**, 533–45.

Mooij, W.M. (1996) Variation in abundance and survival of fish larvae in shallow eutrophic lake Tjeukemeer. *Environmental Biology of Fishes*, **46**, 265–79.

Morgan, N. (1980) Secondary production. In: Le Cren, E.D. & Lowe-McConnell, R.H. (eds), *The Functioning of Freshwater Ecosystems*. Cambridge University Press, Cambridge, 247–340.

Mori, S. & Miura, T. (1980) List of plant and animal species living in Lake Biwa. *Memoirs of the Faculty of Sciences Kyoto University Series Biology*, **8**, 1–33.

Müller, R. (1992) Trophic state and its implications for natural reproduction of salmonid fish. *Hydrobiologia*, **243/244**, 261–8.

Mutenia, A. & Salonen, E. (1994) Rehabilitation of the fisheries of Lake Inari, northern Finland. In: Cowx, I.G. (ed.), *Rehabilitation of Freshwater Fisheries*. Fishing News Books, Oxford, 280–8.

Nakai, K., Kawanabe, H. & Gashagaza, M.M. (1994) Ecological studies on the littoral cichlid communities of Lake Tanganyika: the coexistence of many endemic species. *Archiv für Hydrobiologie Beiheft Ergebnisse Limnologie*, **44**, 373–89.

Nakashima, B.S., Gascon, D. & Leggett, W.C. (1977) Species diversity of littoral zone fishes along a phosphorus-production gradient in Lake Memphremagog, Quebec–Vermont. *Journal of the Fisheries Research Board of Canada*, **34**, 167–70.

Nilsson, N.-A. (1978) The role of size-biased predation in competition and interactive segregation in fish. In: Gerking, S.D. (ed.), *Ecology of Freshwater Fish Production*. Blackwell Scientific Publishers, Oxford, 303–25.

Nilsson, N.-A. & Pejler, B. (1973) On the relation between fish fauna and zooplankton composition in North Swedish lakes. *Report Institute of Freshwater Research Drottningholm*, **53**, 51–77.

Nishida, M. (1991) Lake Tanganyika as an evolutionary reservoir of old lineages of East African cichlids: inferences from allozyme data. *Experientia*, **47**, 974–9.

Niva, T. & Julkunen, M. (1998) Effect of population fluctuation of vendace (*Coregonus albula*) on the diet and growth of stocked brown trout (*Salmo trutta*). *Archiv für Hydrobiologie Special Issues Advances in Limnology*, **50**, 295–303.

Nixon, S.W. (1988) Physical energy inputs and the comparative ecology of lake and marine ecosystems. *Limnology and Oceanography*, **33**, 1005–25.

Northcote, T.G. & Rundberg, H. (1970) Spatial distribution of pelagic fishes in Lambarfjärden (Mälaren, Sweden) with particular reference to interaction between Coregonus albula and Osmerus eperlanus. *Report Institute of Freshwater Research Drottningholm*, **50**, 133–67.

Nyberg, K., Raitaniemi, J., Rask, M., Mannio, J. & Vuorenmaa, J. (1995) What can perch population data tell us about the acidification history of a lake? *Water, Air and Soil Pollution*, **85**, 395–400.

Oberdorff, T., Guégan, J.-F. & Hugueny, B. (1995) Global scale patterns of fish species richness in rivers. *Ecography*, **18**, 345–52.

Oberdorff, T., Hugueny, B., Compin, A. & Belkessam, D. (1998) Non-interactive fish communities in the coastal streams of North-western France. *Journal of Animal Ecology*, **67**, 472–84.

Oglesby, R.T. (1977) Relationships of fish yield to lake phytoplankton standing crop, production, and morphoedaphic factors. *Journal of the Fisheries Research Board of Canada*, **34**, 2271–9.

Ogutu-Ohwayo, R. (1990a) The decline of the native fishes of lakes Victoria and Kyoga (East Africa) and the impact of introduced species, especially the Nile perch, *Lates niloticus*, and the Nile tilapia, *Oreochromis niloticus*. *Environmental Biology of Fishes*, **27**, 81–96.

Ogutu-Ohwayo, R. (1990b) Changes in the prey ingested and the variations in the Nile perch and other fish stocks in Lake Kyoga and the northern waters of Lake Victoria (Uganda). *Journal of Fish Biology*, **37**, 55–63.

Okemwa, E. & Ogari, J. (1994) Introductions and extinction of fish in Lake Victoria. In: Cowx, I.G. (ed.), *Rehabilitation of Freshwater Fisheries*. Fishing News Books, Oxford, 326–37.

Owen, R.B., Crossley, R., Johnson, T.C., *et al.* (1990) Major low levels of Lake Malawi and their implications for speciation rates in cichlid fishes. *Proceedings of the Royal Society of London, Series B*, **240**, 519–53.

Paradis, A.R., Pepin, P. Brown, J.A. (1996) Vulnerability of fish eggs and larvae to predation: review of the influence of the relative size of prey and predator. *Canadian Journal of Fisheries and Aquatic Sciences*, **53**, 1226–35.

Pauly, D. & Christensen, V. (1995) Primary production required to sustain global fisheries. *Nature*, **374**, 255–7.

Pepin, P. & Myers, R.A. (1991) Significance of egg and larval size to recruitment variability of temperate marine fish. *Canadian Journal of Fisheries and Aquatic Sciences*, **48**, 1820–8.

Persson, L. (1983) Effects of intra- and interspecific competition on dynamics and size structure of a perch *Perca fluviatilis* and a roach *Rutilus rutilus* population. *Oikos*, **41**, 126–32.

Persson, L. (1986) Effects of reduced interspecific competition on resource utilisation in perch (*Perca fluviatilis*). *Ecology*, **67**, 355–64.

Persson, L. (1987) Effects of habitat and season on competitive interactions between roach (*Rutilus rutilus*) and perch (*Perca fluviatilis*). *Oecologia*, **73**, 170–7.

Persson, L. (1993) Predator-mediated competition in prey refuges: the importance of habitat dependent prey resources. *Oikos*, **68**, 12–22.

Persson, L. (1994) Natural shifts in the structure of fish communities: mechanisms and constraints on perturbation sustenance. In: Cowx, I.G. (ed.), *Rehabilitation of Freshwater Fisheries*. Fishing News Books, Oxford, 421–34.

Persson, L. (1997) Competition, predation and environmental factors as structuring forces in freshwater fish communities: Sumari (1971) revisited. *Canadian Journal of Fisheries and Aquatic Sciences*, **54**, 85–8.

Persson, L. & Greenberg, L.A. (1990) Competitive juvenile bottlenecks: the perch (*Perca fluviatilis*)—roach (*Rutilus rutilus*) interaction. *Ecology*, **71**, 44–56.

Persson, L., Diehl, S., Johansson, L., Andersson, G. & Hamrin, S.F. (1991) Shifts in fish communities along the productivity gradient of temperate lakes—patterns and the importance of size-structured interactions. *Journal of Fish Biology*, **38**, 281–93.

Peterson, R.H. & Martin-Robichaud, D.J. (1988) Community analysis of fish populations in headwater lakes of New Brunswick and Nova Scotia. *Proceedings of the Nova Scotia Institute of Science*, **38**, 55–72.

Pimm, S.L. & Hyman, J.B. (1987) Ecological stability in the context of multispecies fisheries. *Canadian Journal of Fisheries and Aquatic Sciences*, **44** (Supplement 2), 84–94.

Polis, G.A. & Holt, R.D. (1992) Intraguild predation: the dynamics of complex trophic interactions. *Trends in Ecology and Evolution*, **7**, 151–4.

Ponton, D. & Müller, R. (1990) Size of prey ingested by whitefish, *Coregonus* sp. larvae. Are *Coregonus* larvae gape-limited predators? *Journal of Fish Biology*, **36**, 67–72.

Popova, O.A. (1978) The role of predaceous fish in ecosystems. In: Gerking, S.D. (ed.), *Ecology of Freshwater Fish Production*. Blackwell Scientific Publications, Oxford, 215–49.

Post, J.R. & McQueen, D.J. (1988) Ontogenetic changes in the distribution of larval and juvenile yellow perch (*Perca flavescens*): a response to prey or predators? *Canadian Journal of Fisheries and Aquatic Sciences*, **45**, 1820–6.

Qizhe, L. & Qiuling, Y. (1994) Present status and development of lake fisheries in Jiangsu Province, China. In: Cowx, I.G. (ed.), *Rehabilitation of Freshwater Fisheries*. Fishing News Books, Oxford, 48–56.

Radomski, P.J. & Goeman, T.J. (1995) The homogenising of Minnesota lake fish assemblages. *Fisheries*, **20**, 20–23.

Rahel, F.J. (1984) Factors structuring fish assemblages along a bog lake successional gradient. *Ecology*, **65**, 1276–89.

Rahel, F.J. (1986) Biogeographic influences on fish species composition of northern Wisconsin lakes with applications for lake acidification studies. *Canadian Journal of Fisheries and Aquatic Sciences*, **43**, 124–34.

Raitaniemi, J. (1995) The growth of young pike in small Finnish lakes with different acidity-related water properties and fish species composition. *Journal of Fish Biology*, **47**, 115–25.

Raitaniemi, J., Rask, M. & Vuorinen, P.J. (1988) The growth of perch, *Perca fluviatilis* L, in small Finnish lakes at different stages of acidification. *Annales Zoologici Fennici*, **25**, 209–19.

Raitaniemi, J., Malinen, T., Nyberg, K. & Rask, M. (1999) The growth of whitefish in relation to water quality and fish species composition. *Journal of Fish Biology*, **54**, 741–56.

Ranta, E. & Lindström, K. (1989) Prediction of lake-specific fish yield. *Fisheries Research*, **8**, 113–28.

Ranta, E. & Lindström, K. (1993) Theory on fish yield versus water quality in lakes. *Annales Zoologici Fennici*, **30**, 71–5.

Rask, M. & Tuunainen, P. (1990) Acid-induced changes in fish populations of small Finnish lakes. In: Kauppi, P., Kenttämies, K. & Anttila, P. (eds), *Acidification in Finland*. Springer-Verlag, Berlin, 911–27.

Rask, M., Mannio, J., Forsius, M., Posch, M. & Vuorinen, P.J. (1995) How many fish populations in Finland are affected by acid precipitation? *Environmental Biology of Fishes*, **42**, 51–63.

Rawson, D.S. (1952) Mean depth and the fish production of large lakes. *Ecology*, **33**, 513–21.

Ribbink, A.J. (1994a) Lake Malawi. *Archiv für Hydrobiologie Beiheft Ergebnisse Limnologie*, **44**, 27–33.

Ribbink, A.J. (1994b) Biodiversity and speciation in freshwater fishes with particular reference to African cichlids. In: Giller, P.S., Hildrew, A.G. & Raffaelli, D.G.

(eds), *Aquatic Ecology. Scale, Pattern and Process*. Symposia of the British Ecological Society 34, Blackwell Science, Oxford, 261–88.

Ricklefs, R.E. (1987) Community diversity: relative roles of local and regional processes. *Science*, **235**, 167–71.

Robinson, C.L.K. & Tonn, W.M. (1989) Influence of environmental factors and piscivory in structuring fish assemblages of small Alberta lakes. *Canadian Journal of Fisheries and Aquatic Sciences*, **46**, 81–9.

Rodríguez, M.A. & Lewis, W.M., Jr. (1994) Regulation and stability in fish assemblages of neotropical floodplain lakes. *Oecologia*, **99**, 166–80.

Rohde, K. (1998) Latitudinal gradients in species diversity. Area matters, but how much? *Oikos*, **82**, 184–90.

Rosenzweig, M.L. & Sandlin, E.A. (1997) Species diversity and latitudes: listening to area's signal. *Oikos*, **80**, 172–6.

Ross, S.T. (1986) Resource partitioning in fish assemblages: a review of field studies. *Copeia*, **1986**, 352–88.

Rosseland, B.O., Sevaldrud, I., Svalastog, D. & Muniz, I.P. (1980) Studies on freshwater fish populations — effects of acidification on reproduction, population structure, growth and food selection. In: Drabløs, D. & Tollan, A. (eds.) *Ecological Impact of Acid Precipitation*. Proceedings of an International Conference, Sandefjord, Norway, 11–14 March, 336–7.

Rounsefell, G.A. (1946) Fish production in lakes as a guide for estimating production in proposed reservoirs. *Copeia*, **1**, 29–40.

Rudstam, L.G. & Magnuson, J.J. (1985) Predicting the vertical distribution of fish populations: analysis of cisco, *Coregonus artedii*, and yellow perch, *Perca flavescens*. *Canadian Journal of Fisheries and Aquatic Sciences*, **42**, 1178–88.

Rudstam, L.G., Green, D.M., Forney, J.L., Stang, D.L. & Evans, J.T. (1996) Evidence of interactions between walleye and yellow perch in New York State lakes. *Annales Zoologici Fennici*, **33**, 443–9.

Rupp, R.S. (1965) Shore-spawning and survival of eggs of the American smelt. *Transactions of the American Fisheries Society*, **94**, 160–8.

Ryder, R.A. (1965) A method for estimating the potential fish production of north-temperate lakes. *Transactions of the American Fisheries Society*, **94**, 214–8.

Ryder, R.A. (1982) The morphoedaphic index — use, abuse, and fundamental concepts. *Transactions of the American Fisheries Society*, **111**, 154–64.

Ryder, R.A. & Kerr, S.R. (1990) Harmonic communities in aquatic ecosystems: a management perspective. In: van Densen, W.L.T., Steinmetz, B. & Hughes, R.H. (eds), *Management of Freshwater Fisheries*. Proceedings of a Symposium Organized by the European Inland Fisheries Advisory Commission, Göteborg, Sweden, 31 May–3 June 1988, PUDOC (Publishing House of the Wageningen Agricultural University), Wageningen, 594–23.

Ryder, R.A., Kerr, S.R., Loftus, K.H. & Regier, H.A. (1974) The morphoedaphic index, a fish yield estimator—review and evaluation. *Journal of the Fisheries Research Board of Canada*, **31**, 663–88.

Salonen, S., Helminen, H. & Sarvala, J. (1996) Feasibility of controlling coarse fish populations through pikeperch (*Stizostedion lucioperca*) stocking in Lake Köyliönjärvi, SW Finland. *Annales Zoologici Fennici*, **33**, 451–7.

Sandlund, O.T., Gunnarsson, K., Jónasson, P.M., *et al.* (1992) The arctic charr *Salvelinus alpinus* in Thingvallavatn. *Oikos*, **64**, 305–51.

Sarvala, J. (1993) Utilisation of eutrophication for fish production. *Memorie dell'Istituto Italiano di Idrobiologia*, **52**, 171–90.

Sarvala, J. & Helminen, H. (1996) Year-class fluctuations of perch (*Perca fluviatilis*) in Lake Pyhäjärvi, Southwest Finland. *Annales Zoologici Fennici*, **33**, 389–96.

Sarvala, J., Helminen, H. & Hirvonen, A. (1994) The effect of intensive fishing on fish populations in Lake Pyhäjärvi, south-west Finland. In: Cowx, I.G. (ed.), *Rehabilitation of Freshwater Fisheries*. Fishing News Books, Oxford, 77–89.

Sarvala, J., Helminen, H. & Auvinen, H. (1998a) Portrait of a flourishing freshwater fishery: Pyhäjärvi, a lake in SW-Finland. *Boreal Environment Research*, **3**, 329–45.

Sarvala, J., Helminen, H., Saarikari, V., Salonen, S. & Vuorio, K. (1998b) Relations between planktivorous fish abundance, zooplankton and phytoplankton in three lakes of differing productivity. *Hydrobiologia*, **363**, 81–95.

Sarvala, J., Salonen, K., Järvinen, M., *et al.* (1999) Trophic structure of Lake Tanganyika: carbon flows in the pelagic food web. *Hydrobiologia*, **407**, 149–73.

Schael, D.M., Rudstam, L.G. & Post, J.R. (1991) Gape limitation and prey selection in larval yellow perch (*Perca flavescens*), freshwater drum (*Aplodinotus grunniens*), and black grappie (*Pomoxis nigromaculatus*). *Canadian Journal of Fisheries and Aquatic Sciences*, **48**, 1919–25.

Schindler, D.W., Beaty, K.G., Fee, E.J., *et al.* (1990) Effects of climatic warming on lakes of the central boreal forest. *Science*, **250**, 967–70.

Schlesinger, D.A. & McCombie, A.M. (1983) An evaluation of climatic, morphoedaphic, and effort data as pre-dictors of yields from Ontario sport fisheries. *Ontario Fisheries Technical Report Series*, **10**, 1–14.

Schlesinger, D.A. & Regier, H.A. (1982) Climatic and morphoedaphic indices of fish yields from natural lakes. *Transactions of the American Fisheries Society*, **111**, 141–50.

Schliewen, U.K., Tautz, D. & Pääbo, S. (1994) Sympatric speciation suggested by monophyly of crater lake cichlids. *Nature*, **368**, 629–32.

Schneider, D.C. & Haedrich, R.L. (1989) Prediction limits of allometric equations: a reanalysis of Ryder's morphoedaphic index. *Canadian Journal of Fisheries and Aquatic Sciences*, **46**, 503–8.

Seehausen, O. (1996) *Lake Victoria Rock Cichlids. Taxonomy, Ecology and Distribution*. Verduijn Cichlids, Zevenhuizen, 304 pp.

Seehausen, O., van Alphen, J.J.M. & Witte, F. (1997) Cichlid fish diversity threatened by eutrophication that curbs sexual selection. *Science*, **277**, 1808–11.

Shuter, B.J. & Post, J.R. (1990) Climate, population viability, and the zoogeography of temperate fishes. *Transactions of the American Fisheries Society*, **119**, 314–36.

Snoeks, J. (1994) The haplochromine fishes (Teleostei: Cichlidae) of Lake Kivu, East Africa: a taxonomic revision with notes on their ecology. *Annales de la Musée royale Afrique centrale Tervuren, sciences zoologique*, **270**, 1–221.

Snoeks, J., Rüber, L. & Verheyen, E. (1994) The Tanganyika problem: comments on the taxonomy and distribution patterns of its cichlid fauna. *Archiv für Hydrobiologie Beiheft Ergebnisse Limnologie*, **44**, 355–72.

Snoeks, J., De Vos, L. & Thys van den Audenaerde, D. (1997) The ichthyography of Lake Kivu. *South African Journal of Science*, **93**, 579–84.

Stauffer, J.R.J., Jr., LoVullo, T.J. & Han, H.Y. (1996) Commensalistic feeding relationships of three Lake Malawi fish species. *Transactions of the American Fisheries Society*, **125**, 224–9.

Sterligova, O.P. (1979) Koryushka *Osmerus eperlanus* (L.) i ee rol v ikhtiofaune Syamozera. *Voprosy Ikhtiologii*, **19**, 793–800. (In Russian.)

Sterligova, O.P., Pavlovskij, S.A. & Komulainen, S.F. (1988) Reproduction of coregonids in the eutrophicated Lake Sjamozero, Karelian ASSR. *Finnish Fisheries Research*, **9**, 485–8.

Stewart, D.J., Kitchell, J.F. & Crowder, L.B. (1981) Forage fishes and their salmonid predators in Lake Michigan. *Transactions of the American Fisheries Society*, **110**, 751–63.

Stone, J.N. & Cohen, Y. (1990) Changes in species inter-actions of the Lake Superior fisheries system after the control of sea lamprey as indicated by time series models. *Canadian Journal of Fisheries and Aquatic Sciences*, **47**, 251–61.

Sturmbauer, C. & Meyer, A. (1992) Genetic divergence, speciation and morphological stasis in a lineage of African cichlid fishes. *Nature*, **358**, 578–81.

Sumari, O. (1971) Structure of the perch populations of some ponds in Finland. *Annales Zoologici Fennici*, **8**, 406–21.

Svärdson, G. (1976) Interspecific population dominance in fish communities of Scandinavian lakes. *Report Institute of Freshwater Research Drottningholm*, **55**, 144–71.

Svärdson, G. & Molin, G. (1981) The impact of eutrophication and climate on a warmwater fish community. *Report Institute of Freshwater Research Drottningholm*, **59**, 142–51.

Sydänoja, A., Helminen, H. & Sarvala, J. (1995) Vertical migration of vendace (*Coregonus albula*) in a thermally unstratified lake (Pyhäjärvi, SW Finland). *Archiv für Hydrobiologie Special Issues Advances in Limnology*, **46**, 77–286.

Tammi, J., Lappalainen, A., Mannio, J., Rask, M. & Vuorenmaa, J. (1999) Effects of eutrophication on fish and fisheries in Finnish lakes: a survey based on random sampling. *Fisheries Management and Ecology*, **6**, 173–86.

Thiebaux, M.L. & Dickie, L.M. (1992) Models of aquatic biomass size spectra and the common structure of their solutions. *Journal of Theoretical Biology*, **159**, 147–61.

Tonn, W.M. & Magnuson, J.J. (1982) Patterns in the species composition and richness of fish assemblages in northern Wisconsin lakes. *Ecology*, **63**, 1149–66.

Tonn, W.M., Magnuson, J.J. & Forbes, A.M. (1983) Community analysis in fishery management: An application with northern Wisconsin lakes. *Transactions of the American Fisheries Society*, **112**, 368–77.

Tonn, W.M., Paszkowski, C.A. & Moermond, T.C. (1986) Competition in *Umbra–Perca* fish assemblages: experimental and field evidence. *Oecologia*, **69**, 126–33.

Tonn, W.M., Paszkowski, C.A. & Holopainen, I.J. (1989) Responses of crucian carp populations to differential predation pressure in a manipulated pond. *Canadian Journal of Zoology*, **67**, 2841–9.

Tonn, W.M., Magnuson, J.J., Rask, M. & Toivonen, J. (1990) Intercontinental comparison of small-lake fish assemblages: the balance between local and regional processes. *American Naturalist*, **136**, 345–75.

Trendall, J. (1988) The distribution and dispersal of introduced fish at Thumbi West Island in Lake Malawi. *Journal of Fish Biology*, **33**, 357–69.

Turner, G.F. (1994) Speciation mechanisms in Malawi cichlids: a critical review. *Archiv für Hydrobiologie Beiheft Ergebnisse Limnologie*, **44**, 139–60.

Turner, G.F. (1999) Explosive speciation of African cichlid fishes. In: Magurran, A.E. & May, R.M. (eds), *Evolution of Biological Diversity*. Oxford University Press, Oxford, 113–29.

Turner, G.F., Seehausen, O., Knight, M.E., Allender, C.J. & Robinson, R.L. (2001) How many species of cichlid fishes are there in African lakes? *Molecular Ecology*, **10**, 793–806.

Urho, L. (1996) Habitat shifts of perch larvae as survival strategy. *Annales Zoologici Fennici*, **33**, 329–40.

Van Oppen, M.J.H., Turner, G.F., Rico, C., *et al.* (1997) Unusually fine-scale genetic structuring found in rapidly speciating Malawi cichlid fishes. *Proceedings of the Royal Society of London, Series B*, **264**, 1803–12.

Vehanen, T., Hyvärinen, P. & Huusko, A. (1998) Food consumption and prey orientation of piscivorous brown trout (*Salmo trutta*) and pikeperch (*Stizostedion lucioperca*) in a large regulated lake. *Journal of Applied Ichtyology*, **14**, 15–22.

Viljanen, M. & Karjalainen, J. (1992) Comparison of sampling techniques for vendace (*Coregonus albula*) and European whitefish (*Coregonus lavaretus*) larvae in large Finnish lakes. *Polskie Archiwum Hydrobiologii*, **39**, 361–9.

Viljanen, M., Karjalainen, J., Helminen, H., Sarvala, J. & Sydänoja, A. (1995) Night-day catch ratios of coregonid larvae in three large lakes in Finland. *Archiv für Hydrobiologie Special Issues Advances in Limnology*, **46**, 195–201.

Vøllestad, L.A. (1985) Resource partitioning of roach *Rutilus rutilus* and bleak *Alburnus alburnus* in two eutrophic lakes in SE Norway. *Holarctic Ecology*, **8**, 88–92.

Welcomme, R.L. (1985) *River Fisheries*. FAO Fisheries Technical Paper 262. Food and Agriculture Organisation, Rome, 330 pp.

Werner, E.E. & Hall, D.J. (1988) Ontogenetic habitat shifts in bluegill: the foraging rate–predation risk trade-off. *Ecology*, **69**, 1352–66.

Winemiller, K.O. & Pianka, E.R. (1990) Organization of natural assemblages of desert lizards and tropical fishes. *Ecological Monographs*, **60**, 27–55.

Winemiller, K. & Rose, K. (1992) Patterns of life

history diversification in North American fishes: implications for population regulation. *Canadian Journal of Fisheries and Aquatic Sciences*, **49**, 2196–218.

Witte, F. & van Oijen, M.J.P. (1990) Taxonomy, ecology and fishery of haplochromine trophic groups. *Zoologische Verhandlungen Leiden*, **262**, 1–47.

Witte, F., Goldschmidt, T., Wanink, J., van Oijen, K., Witte-Maas, E. & Bouton, N. (1992) The destruction of an endemic species flock: quantitative data on the decline of the haplochromine cichlids of Lake Victoria. *Environmental Biology of Fishes*, **34**, 1–28.

Witte, F., Msuku, B.S., Wanink, J.H., *et al.* (2000) Recovery of cichlid species in Lake Victoria: an examination of factors leading to differential extinction. *Reviews in Fish Biology and Fisheries*, **10**, 233–41.

Worthington, E.B. & Lowe-McConnell, R.H. (1994) African lakes reviewed: creation and destruction of biodiversity. *Environmental Conservation*, **21**, 199–213.

Wright, D.H. (1983) Species-energy theory: an extension of species-area theory. *Oikos*, **41**, 496–506.

Youngs, W.D. & Heimbuch, D.G. (1982) Another consideration of the morphoedaphic index. *Transactions of the American Fisheries Society*, **111**, 151–3.

17 Self-regulation of Limnetic Ecosystems

CLAUDIA PAHL-WOSTL

17.1 INTRODUCTION

Lakes played an important role in development of the ecosystem concept. Owing to their relatively discrete existence, they were perceived as microcosms (Forbes 1887; Thienemann 1925) and thus, in the sense that the influence of internal processes is greater than that of external factors, as autonomous systems. However, this statement applies predominantly to flows of nutrients and energy. At the same time, lakes are also highly influenced by environmental fluctuations. Owing to the relative weakness of differences within their spatial structure, patterns of organisation in space are usually also weak. Therefore, organisation of the biological community in lakes has mainly been perceived as determined by the vagaries of the physical environment. In order to understand the importance of self-regulation in lacustrine ecosystems, one must consider dynamic, rather than static, patterns of organisation. This chapter therefore introduces a conceptual framework designed to account for self-regulation and structural complexity in dynamic ecological networks, and discusses its implications for our understanding of ecosystem function in lakes and elsewhere.

17.2 CONCEPTS OF REGULATION

Self-regulation may be defined as control of a system's properties by its own internal feedback mechanisms. Figure 17.1 summarises the essential elements required in order to account for self-regulation. Regulation can only be explained within a system perspective, and in relation to descriptors of a system's state—e.g., total biomass, productivity and stability. The state of a system may be changed both by endogenous and exogenous sources of influence. The former are derived from processes such as nutrient recycling, or species succession, whereas exogenous influence stems from processes that originate outside the system, such as the seasonal dynamics of environmental variables, or nutrient inputs. Self-regulation is only meaningful in ecosystems that possess at least a minimal degree of autonomy. Which concepts of autonomy are useful in this context, and what is required for a lake to be autonomous?

The concepts of autonomy (and thus, of self-regulation) are dependent on scales in time and space. For example, when the water body is sufficiently large for interactions within the habitat (material exchanges) to be more powerful than those that operate across habitat boundaries, especially with respect to the littoral zone and the sediments, pelagic ecosystems may be thought of as units of self-regulation. Considerations of scale are of vital importance in determining the characteristics of lacustrine habitats and their relationships (e.g. volume/surface area, water retention time and epilimnion depth/total depth determine the influence of sediments and nutrient gradients). Thus, the difference between small ponds and large systems such as Lake Baikal may be greater than that between Lake Baikal and some marine systems. The smaller the lake, the more the benthic and the pelagic food webs need to be considered as a single unit of organisation and regulation (see Chapter 9). However, this statement does not imply that links between the pelagic and the benthos, and/or the profundal zones, are entirely absent, or are of no importance in large lakes (Lodge *et al.* 1988). The benthos is, for example, the major spawning ground for many pelagic fish. Similarly, resting stages, which may remain in the sediments, are of vital importance for long-term survival of many planktonic organisms.

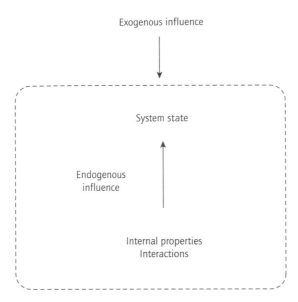

Fig. 17.1 Essential requirements to account for self-regulation.

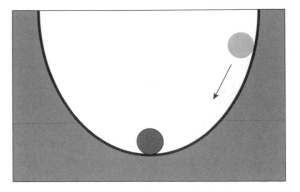

Fig. 17.2 Mechanical analogue for system stability.

This chapter therefore focuses on the understanding of general patterns of pelagic community organisation and their importance in self-regulation. Before discussing community structure and organisation, it is useful to consider a plausible reference state of an ecosystem, maintained by regulatory processes.

One of the major founding principles of ecology is the belief in 'the balance of nature', with its strong emphasis on equilibrium and stability in terms of complete absence of variation. The principle is based on the assumption that the properties of systems are conserved, and temporal variability is suppressed. The hypothesis underlying many concepts of stability may be summarised as: ecosystems are organised in such a way that any deviations from the equilibrium state are counteracted by negative feedback forces which prevent the system from deviating from a preferred state.

This hypothesis is illustrated in Fig. 17.2 by a mechanical analogue. Like a ball in a hollow, the state of an ecosystem may be described by the degree to which it is embedded in a well representing the combined influence of opposing interactive forces. If the system is in its time-invariant, equilibrium state—e.g. if grazing equals primary production—the opposed forces are balanced. If the system deviates from this equilibrium state, negative feedback effects are assumed to pull it back to equilibrium, just as the force of gravitation will pull the ball back into the well, and finally to rest.

Traditional concepts of ecosystem structure and regulation emanate from such thinking. The entire field of food web theory is, to a large extent, motivated by the hypothesis that patterns that are observed in nature ought to correspond to those that are shown to yield stable equilibrium points in model investigations. The finding of an inverse relationship between number of species and food-web connectance has been taken as support for the theoretically derived trade-off between food web complexity and stability. Deterministic models, mainly of the Lotka-Volterra type, have been used to derive the further structural properties to be expected in systems attaining stability, as in the absence of positive omnivory or the absence of positive feedback. The inability to find these properties in datasets from natural food webs is taken as further evidence that natural systems exhibit configurations assuring stability (Lawton & Warren 1988; Cohen et al. 1990; Pimm et al. 1991).

However, food web theory may be seriously compromised: (i) by ignoring largely spatial and temporal scales; (ii) by weak data exhibiting, for example, a highly heterogeneous level of aggregation (taxonomic and trophic species, functional groups); and (iii) by the absence of criteria to record a link or not (e.g. Hastings 1988; Paine 1988). The stability–diversity (complexity) debate has always suffered from a lack of coherent definitions for these two properties, and the resulting heterogeneity and incompleteness both of theoretical arguments and of field studies. It is therefore useful first to examine the concepts of stability more closely.

Pimm (1984) and Pimm *et al.* (1991) identified five concepts which have been referred to by theoretical and empirical ecologists as stability. In these statements, the term equilibrium refers to an equilibrium point, where all variables are time invariant. The statements are:

1 In the mathematical sense, a system is considered to be **stable** if and only if its variables return to equilibrium conditions after displacement from them.

2 **Resilience** corresponds to the rate at which variables return to their equilibrium after having been displaced from it. The greater the resilience, the faster the recovery following perturbation. Systems with high resilience may therefore be said to be 'stable'.

3 **Persistence** is a measure of the length of time a variable lasts before it is changed to a new value. Systems with high persistence may also be called stable.

4 **Resistance** measures the degree to which a variable is changed, following a perturbation. Systems with high resistance may be described as stable.

5 **Variability** is the degree to which a variable varies over time. Systems exhibiting little variability may be described as stable.

There does not yet seem to be general agreement on the consistent use of these various terms. In addition, depending on one's research interests, individual species abundances, species composition or trophic level abundance may be thought of as variables of interest. Hence, when discussing stability and equilibrium of ecosystems, one should always be explicit regarding the variables, the spatiotemporal scales and the concept one refers to.

It is doubtful whether, in a variable environment, the reference state of time-invariant ecosystems is appropriate. Such ecosystems would be static, rigid entities without any potential for adaptation and change. Traditional concepts of stability have been developed for ecosystems in a homogeneous environment and for single levels of ecological organisation. If these assumptions are relaxed, one may come to quite different conclusions. It has, for example, been assumed that species fluctuations lead to extinction. A plausible conclusion is therefore to claim that species able to maintain stable biomass levels are superior in evolutionary terms to those whose numbers fluctuate, and whose populations therefore come close to the threshold of extinction. However, such an argument is only valid in a perfectly homogeneous equilibrium world. Simulations with metapopulation models show that low densities also lead to more frequent extinction at the local level, whilst chaotic oscillations reduce the degree of synchronicity among populations, and thus also the risk of simultaneous extinction (Allen *et al.* 1993). A desynchronised, spatially distributed pattern of species may therefore, in an evolutionary sense, be superior to a single, stable, spatially homogeneous population.

Discussions about stability have also often been limited to an isolated view of one level of ecological organisation. However, regulation in ecological systems is decentralised, and needs to invoke several levels of organisation. Maintenance of ecosystem function depends on the diversity and flexibility of the component populations. The hypothesis can be stated that the trade-off between global and local variability is the stabilising principle of ecosystem organisation (Fig. 17.3). This is supported by empirical evidence in work reported by Tilman and co-workers (Tilman & Downing 1994; Simon Moffat 1996; Tilman 1996) which indicates that, at the system level, species-rich grassland systems possess greater resistance to disturbance with respect to functional properties, whereas biomass of individual species exhibited higher fluctuations than in species-poor systems. Reynolds (1994) argued that, during the course of ecosystem succession, similar patterns can be observed in pelagic systems. Yet it seems quite astonishing that there is not more empirical

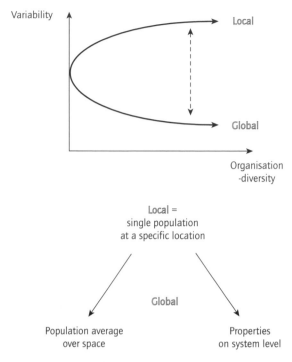

Local =
single population
at a specific location

Global

Population average   Properties
over space   on system level

Fig. 17.3 Trade-off between local and global system properties as an essential element of ecosystem dynamics.

evidence on this issue from aquatic ecosystems, where the shorter generation times of the component organisms would favour such experimental approaches.

In conclusion, one may state that stabilisation does not imply time invariance. On the contrary, for ecosystems to maintain a balance between flexibility and function, a delicate balance is required between internal degrees of freedom and constraints, a combination of positive and negative feedback effects. Positive feedback leads to a constant tendency of a system to explore the limits of its boundary conditions, and to use opportunities. Negative feedbacks keep the system within delimited bounds. Positive feedback is thus an essential ingredient for adaptive ecological and evolutionary change.

In order further to improve our understanding of ecosystem organisation along these lines of rea-

soning, investigation of the relationship between levels of ecosystem organisation and their functional and structural properties is required. Figure 17.4 depicts ecological systems as multilevel dynamic networks of energy and matter flows. At the macrolevel, an ecosystem can be described by extensive variables, such as biomass and productivity, or by intensive variables such as adaptability and structure. At the microlevel, organisms differ in phylogeny and function. Theoretical concepts of relationships among levels of organisation require a consistent definition of functional diversity. For a start, it is useful to distinguish between horizontal and vertical dimensions of organisation along the trophic gradient (Fig. 17.5). The horizontal axis refers to ensembles of species at the same trophic level—e.g., various algae. Individual species are linked by competitive interactions. The vertical axis refers to different trophic positions—the simplest case would separate primary producers, herbivores and carnivores, for example. I will first discuss self-regulation in such simple food chains, before proceeding to more complex food webs.

17.3   SELF-REGULATION IN SIMPLE FOOD CHAINS

On the one hand, because of the prevalence of stochastic environmental perturbations, biological regulation in limnetic systems may be considered inherently weak. On the other, there exists a major field of limnetic research based upon the biotic regulation of trophic structure, and on deriving applications for management purposes (see Volume 2, Chapter 18). However, interpretation of empirical data in this respect is quite ambiguous. This can only partly be explained by environmental perturbations which render every experiment quite unique. Another important reason is that every concept of biotic regulation is based on simplistic assumptions regarding community structure.

The trophic-level concept constitutes an important abstraction in the clarification and organisation of our understanding of energy transfer in ecosystems. It is often assumed that the structure

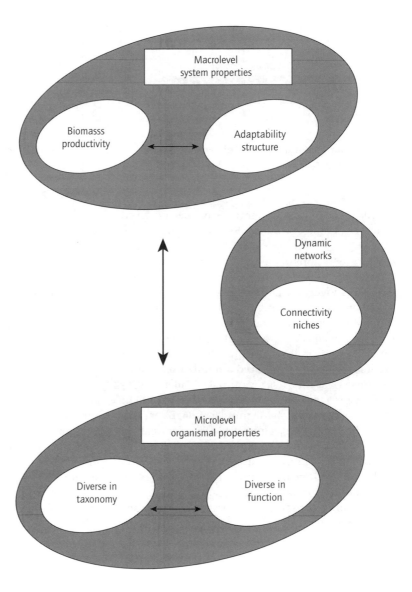

Fig. 17.4 Levels of ecosystem organisation and their interdependence. Functional properties at the level of the ecosystem as a whole are stabilised and maintained in a variable environment by changes in the ecological network comprising the component species and functional groups. A high diversity in functional pathways thus endows an ecosystem with a high degree of flexibility and adaptability. At the same time, such organisation renders ecosystem development difficult to predict and control.

of complex food webs can be reduced to a sequence of discrete trophic levels which include trophically homogeneous groups. Concepts of trophic structure have emphasised analyses of time-invariant equilibrium states in linear food-chain models. Hairston, Slobodkin and Smith viewed whole trophic levels as dynamically equivalent to single species, and assumed there to be no direct effects of population density or a given trophic level on per capita growth rate of that level (Hairston *et al.* 1960; Slobodkin *et al.* 1967). In this way, increased nutrient input raises the density both of the top level and of those that lie an even number of levels below it. The abundance of other levels remains unchanged. Figure 17.6 shows that this leads to a change in the pattern of the relative distribution of total biomass among different trophic levels. Subsequently a number of related concepts

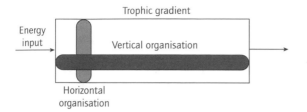

Fig. 17.5 Horizontal and vertical dimensions of ecosystem organisation.

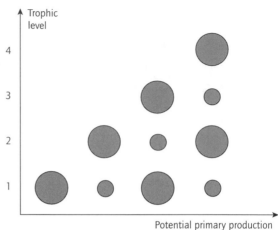

Fig. 17.6 Change in biomass distributions with increasing nutrient input and thus increasing potential primary production. The size of respective spheres indicates the relative contribution of a certain trophic level to total biomass. Such patterns are central to concepts of biomanipulation.

based on food chain dynamics have been derived (Fretwell 1987; Oksanen 1991; Carpenter *et al.* 1993).

Considerations of food chain dynamics have enhanced our understanding of the multiple causal pathways governing community dynamics. Traditional emphasis on food- or habitat-limited populations has gradually been replaced by a more balanced perspective, accounting for the importance of varying levels of predation. Theoretical predictions have stimulated a number of empirical investigations (e.g. Vanni & Findlay 1990; McQueen *et al.* 1992; Persson *et al.* 1992; Diehl 1993; Brett *et al.* 1994; Mazmuder 1994; Elser *et al.* 1995; Mateev 1995). However, the empirical results are by no means unequivocal. Since indirect evidence must often be used to support the hypotheses developed, results may be explained by different interpretations, depending on the perspective of the observer. The presence or absence of planktivorous or piscivorous fish in a lake is, for example, used to classify lakes as being two-, three- or four-link systems, respectively (e.g. Hansson 1992; Persson *et al.* 1992). Owing to lack of data, classification cannot be based on quantitative analyses of trophic structure itself.

Recent debate on the validity of biomanipulation as a management tool exemplifies some of the current controversies which hinge upon trophic structure (Carpenter & Kitchell 1992; DeMelo *et al.* 1992). The concept of cascading trophic interactions is basic to deriving those predictable patterns for regulatory processes which have been used in order to devise schemes of biomanipulation. By specific removal and/or addition of trophic

levels, one sets out to control algal biomass in eutrophic lakes. For example, reducing predation pressure of herbivores by removal of planktivorous fish is assumed to increase grazing on algae and thus to reduce algal biomass. However, a recent evaluation of a set of whole-lake experiments has revealed that food web complexity may lead to quite unexpected outcomes (Carpenter & Kitchell 1993).

Owing to the sensitivity of predictions to underlying assumptions, the practical problems of testing hypotheses about trophic level dynamics are unsurprising. Below, I briefly summarise important concerns about this approach. A comprehensive and balanced overview of the current discussion concerning structure and dynamics of food webs can be found in Polis & Winemiller (1996).

Analyses based on equilibrium assumptions have been challenged by investigations of two- and three-level food chain models with complex dynamics. These may provide results different from, or even contradicting, those obtained in models with a time-invariant steady state (Hastings & Powell 1991; Abrams 1994a,b; Hastings & Higgins 1994; Hastings 1996). Enrichment, for example, leads in general to increase in the bottom level of a food chain, with unstable dynamics. However, it may even lead to extinction of the top predator, owing to excessive fluctuations.

An important argument against the use of integer trophic levels is the ubiquitous presence of omnivory. In many practical situations, it is not possible to assign organisms with different food sources consistently to a specific trophic level. Cousins (1985, 1987) even suggested replacing the trophic level concept by a trophic continuum. Abrams (1993) showed that heterogeneity within trophic levels influences the relationship between productivity and the biomass of higher trophic levels in models of the Lotka-Volterra type. Polis & Holt (1992) reviewed the concept of **intraguild predation** (IGP). They concluded that IGP systems cannot be collapsed to shorter webs characterised by cascading trophic interactions. Recent empirical results (e.g. Carpenter & Kitchell 1993; Brett *et al.* 1994; Elser *et al.* 1995) support the importance of food web complexity.

Consumers may affect primary producers by means other than direct consumption. The importance of consumer-mediated nutrient recycling and transport has been emphasised in empirical and theoretical investigations (DeAngelis *et al.* 1989; Sterner 1990; Vanni & Findlay 1990; DeAngelis 1992; Vanni 1996; Vanni & Layne 1997; Vanni *et al.* 1997). The effects of accounting for nutrient recycling are summarised in Fig. 17.7. Whereas predation imposes a direct negative feedback on prey populations, the effect of nutrient re-

cycling is indirectly positive. In plant–herbivore systems, grazing has been observed first to raise net primary production for moderate levels of grazing, and then to lead to a decrease for a further increase of the grazing pressure (review in DeAngelis 1992).

In addition to the effects of nutrient recycling, it is now well established that the microbial loop (Stone & Weisburd 1992; Stone & Berman 1993) constitutes an important positive feedback pathway (see also Chapter 13). The importance of channelling along the traditional food chain versus the microbial loop, and thus the regulatory structure of the food web, varies over the seasonal cycle (Gaedke *et al.* 1996).

Abrams (1996) emphasises that theory dealing with response of trophic levels to nutrient input is largely based on models which assume trophic levels to be homogeneous, non-adaptive entities. He states that it is not a simple matter to predict the direction of change in a particular trophic level in response to nutrient enrichment, even when trophic levels are homogeneous. It will require more study of the types of changes which occur in the composition and functional response of the trophic levels before anything beyond speculation is possible.

We may deduce that interpretation of experimental results and empirical support of hypotheses on regulation based on the paradigm of a simple food chain are difficult, in that a range of processes is of importance. These processes are:
• resource limitation leading to bottom-up effects on consuming populations;
• competition within groups of functionally similar populations;

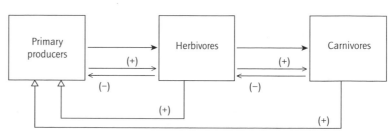

Fig. 17.7 Types of feedback effects in a food chain with nutrient recycling. The signs indicate whether an interaction is negative or positive.

• predation leading to top-down effects on prey populations;
• positive feedback via nutrient recycling and mutualistic interactions.
Even when these processes all act in concert, their relative importance varies over the seasonal cycle, which further complicates unequivocal interpretation.

In addition to the varying importance of the processes, we must consider that the simple model of a food chain may be appropriate only in exceptional cases. Real food webs are more complex in structure. The next section will discuss a systematic approach to explaining food web complexity and its consequences.

17.4   SELF-REGULATION IN COMPLEX FOOD WEBS

One of the major topics in current ecological research is investigation of the relationship between ecosystem function and diversity (Schulze & Mooney 1993). Pelagic ecosystems are ideal candidates for developing conceptual models where these relationships are addressed in a systematic fashion (Pahl-Wostl 1995).

The structural complexity of ecological networks can be approached systematically by making use of the observation that body weight to a large extent determines an organism's physiological and ecological characteristics (Peters 1983; Pahl-Wostl 1995). In most food webs, owing to the decrease in metabolic rates and the increase in generation time and radius of activity which are associated with increasing body weight (see also Chapter 9), energy flow is thus directed along a gradient of increasing scales in time and space. Pelagic systems constitute a prime example for organisation along a continuum in both trophic function and spatio-temporal scales. Figure 17.8 sketches in a stepwise fashion the transition from food chain to food web. The diagrams on the left of each figure illustrate the distribution of biomass in a two-dimensional framework of trophic level and body weight which correspond to the model of trophic transfers depicted on the right. A circle always de-

notes a weight class along a logarithmic body weight axis.

The simplest, and most often used, model structure corresponds to that of the food chain, as depicted in Fig. 17.8a. This model is based on the assumption that a trophic level is a dynamic entity, which can be adequately represented by an averaged body weight and thus by a state variable with well-defined dynamic properties. The regulatory concepts which can be derived using such a model structure have been discussed in the previous section. Food web complexity can be increased in two ways—by increasing diversity (i) along the body weight axis (and thus in time; **horizontal organisation**) and (ii) along the axis of trophic function (**vertical organisation**).

The food chain depicted in Fig. 17.8a generates a sharply peaked biomass distribution along the body weight axis. Numerous investigations have provided empirical evidence for an even rather than a peaked distribution (e.g. Sheldon *et al.* 1972, 1977; Sprules & Munawar 1986; Witek & Krajewaska-Soltys 1989; Sprules *et al.* 1991; Gaedke 1992a). Such findings may be accounted for by the trophic structure depicted in Fig. 17.8b: each trophic level is still well defined but now comprises a wide range in body weight. Such a pattern could be explained by an ensemble of food chains that are shifted along the body weight axis generating a continuum. The figure also indicates that, by taking into account the observation that a consumer's prey window extends over more than one body weight class, independent food chains may become linked. However, in the case presented, one can still discern well-defined trophic levels, which are distinguished by different shadings.

Figure 17.8c explains schematically a gradual transition to a continuum along the axis of trophic function. The food web fragment given on the right illustrates the gradual breakdown of defined trophic levels. The rectangular box denotes the window of a consumer in a higher weight class selecting organisms according to their size irrespective of their being autotrophs or heterotrophs. Such feeding strategies are common for planktonic predators

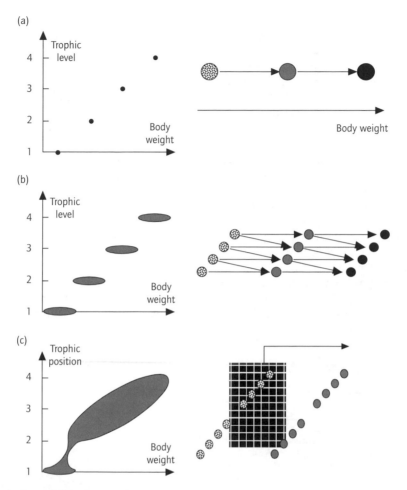

Fig. 17.8 (Left) Hypothetical distribution of biomass in a two-dimensional framework of trophic level and body weight for different models describing the energy transfer along a gradient of increasing body weight. (Right) Illustration of the transition from a food chain with discrete trophic levels to a trophic continuum. Dotted circles always refer to autotrophs, light shaded circles to herbivores, dark shaded circles to carnivores or omnivores. (a) Distribution is discrete in body weight and trophic function — food chain with discrete trophic levels. (b) Distribution with a continuum in body weight but still discrete in trophic function — ensemble of food chains which differ in their location along the body weight axis. Chains may be linked owing to overlapping prey windows. However, discrete trophic levels are still preserved. (c) Distribution with a continuum both in body weight and trophic function. The notional trophic level which can only take discrete integer values is replaced by a trophic position which may take any value — omnivorous consumer whose prey window comprises both primary producers and herbivores. The consumer's trophic position may thus range between 2 and 3. (Pahl-Wostl 1997. Reprinted from *Ecological Modelling*, **100**, 100–23. Copyright (1997). Reproduced with permission of Elsevier Science.)

(e.g. Brett *et al.* 1994; Straile 1994; Gaedke *et al.* 1996). The trophic position of such a consumer ranges somewhere between two and three. The diagram on the left depicts the continuum of trophic positions which may be generated by shifts in the prey window, and variations in the relative contribution of different kinds of prey to a consumer's overall diet. The bottleneck between primary producers and herbivores (levels 1 and 2) indicates the lack of knowledge regarding the quantitative importance of mixotrophy. Even when the existence of switching between autotrophic and heterotrophic modes of production is well documented (e.g. Porter *et al.* 1988; Brett *et al.* 1994), it would not be empirically well grounded yet to assume a continuum.

Models reflecting as a first approximation the configuration depicted in Fig. 17.8b have been investigated previously (Pahl-Wostl 1993, 1995). Figure 17.9 shows the structure of such a model comprising an ensemble of predator (P) and prey (B) pairs. Diversity in timescales increases with the number of (P)–(B) pairs differing in body weight. Figure 17.9a shows the network of nutrient flows. The size of the arrows representing internal exchanges indicates that most nutrients are recycled within the system. A (P)–(B) pair comprises two pathways of recycling: a fast, short route, deriving

from direct losses from the prey species to the nutrient pool and a slower, longer path, deriving from recycling via predators.

As represented in Fig. 17.9b, predator–prey pairs are distributed along the body weight axis which is partitioned into weight classes equally spaced on a logarithmic scale. The average body weight in the kth weight class, w_k, is expressed in fraction of w_0, the weight in class 0: $w_k = 2kw_0$. The weight ratio between neighbouring classes, w_{k+1}/w_k, equals two. The choice of a logarithmic instead of a linear scale becomes more intelligible when we consider that the weight of a predator's prey is determined by the predator–prey weight ratio, rather than by the predator–prey weight difference. A ratio is constant on a logarithmic scale. Similar considerations may be given with respect to the timescales associated with body weight. One may conceive of the body-weight axis representing that of a one-dimensional niche space,

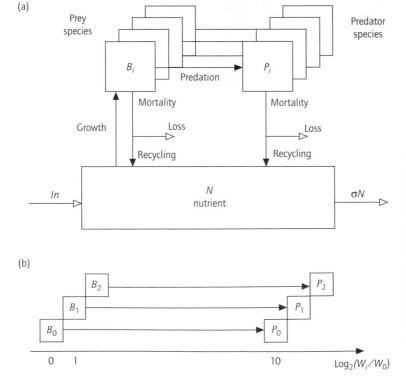

Fig. 17.9 (a) Network representation of the basic model components. The internal exchanges are shown by bold arrows in order to emphasise the quantitative preponderance of recycling over external exchanges. Exchanges with the environment are denoted by open arrow heads. (b) Arrangement of predator–prey pairs along the body weight axis. Weight is expressed as a fraction of w_0, the average weight in class 0. The weight class difference between a predator and its prey was chosen to be equal to 10, corresponding to a weight ratio of 1000.

along which species occupy niches according to their dynamic characteristics. Temporal organisation implies that species organise themselves along this axis. Model equations and parameter values are listed in Box 17.1.

A rise in the number of pairs leads to an increase in temporal organisation. The principle underlying temporal organisation for two predator–prey pairs sharing a common nutrient pool is shown in Fig. 17.10. At first glance, pairs compete for the

---

Box 17.1   Model equations and parameter values for the predator–prey pair model

Mathematical equations

$$\frac{\mathrm{d}B_i}{\mathrm{d}t} = \Psi_i\{f(N)B_i - \rho B_i - \Omega h(B_i)P_i\} \tag{17.1}$$

$$\frac{\mathrm{d}P_i}{\mathrm{d}t} = \Psi_{i+q}\{h(B_i)P_i - (\rho + \lambda P_i)P_i\} \tag{17.2}$$

$$\frac{\mathrm{d}N}{\mathrm{d}t} = In - \sigma N - \sum_{t=0}^{n} \Psi_i f(N)B_i + rec\sum_{i=0}^{n} \Psi_{i\rho}(B_i + \Omega P_i) \tag{17.3}$$

where

$$f(N) = \frac{N}{N+1}, \, h(B_i) = \frac{B_i}{B_i + K}, \quad i = 0, 1, \ldots, n; \, n \leq q - 1.$$

The quadratic loss terms of the predators simulate density-dependent limitations of growth.

Symbols and parameters

n total number of predator–prey pairs in a simulation run
i index of a predator–prey pair, with B_i being in weight class i and P_i being in weight class $i + q$

Model parameters and numerical values used in the model simulations presented here:

ρ = 0.4 rate of respiration
σ = 0.10 rate of loss from the nutrient pool
rec = 0.90 degree of recycling
K = 2.55 half-saturation constant for predation
In = 1.00 external input of nutrient
λ = 0.05 quadratic loss term of predator
ε = 0.25 allometric exponent
$\Psi_k = 2^{-\varepsilon k}$ allometric factor for the kth weight class
q = 10 predator–prey weight class difference
Ω = $2^{-\varepsilon q}$ allometric factor for a predator relative to its prey

Allometric factors are denoted with capital Greek letters; rates are referred to with small Greek letters.

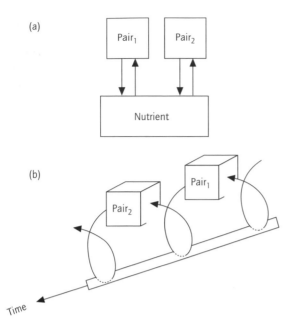

Fig. 17.10 (a) Pair$_1$ and Pair$_2$ share a common nutrient pool. (b) The activity of each pair is confined to a different time interval. The rectangle along the time axis denotes the nutrient pool. What appears as two separate feedback cycles in the time-averaged representation is revealed to be a feedback spiral linking the pairs across time.

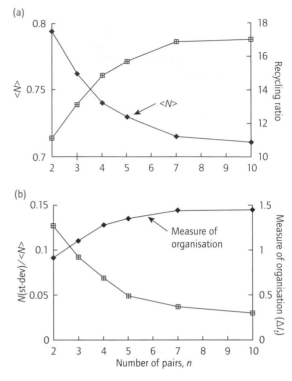

Fig. 17.11 Global system parameters as a function of the number of pairs. (a) <N>, time average of nutrient concentration in the pool and the recycling ratio, expressed as the ratio of nutrient recycled within the system to the external nutrient input. (b) Ratio of the standard deviation to <N> and the temporal organisation quantified by the measure ΔI_t (Pahl-Wostl 1995).

limiting nutrient, and are therefore linked by a negative feedback interaction. However, the nutrient is again recycled, which gives rise to positive feedback. Figure 17.10b shows that interpretation of the type of interaction may vary when the pattern of temporal activity is accounted for. Pairs engage in temporal resource partitioning, which leads to a temporal shift between nutrient utilisation and recycling. As a consequence, interaction between the two pairs may be mainly controlled by the effects of positive rather than negative feedback.

Model simulations show that an increase in the number of different food chains, and thus in dynamic diversity, produces major effects on system dynamics, with a shift from a time-invariant stable state to periodic and, finally, chaotic oscillations. However, increased variability at the level of indi-

vidual species leads to decreased variability of functional properties at the level of the system as a whole. As an example, Fig. 17.11 shows results obtained in model simulations with ensembles of predator–prey pairs. Despite chaotic fluctuations at the species level, such behaviour is an expression of temporal organisation of the ecological network. These simulations support the hypothesis of trade-off between global and local variability (cf. Fig. 17.3).

Let us now focus on the effects of increasing diversity along both the horizontal and the vertical dimensions, by comparing a simple food chain

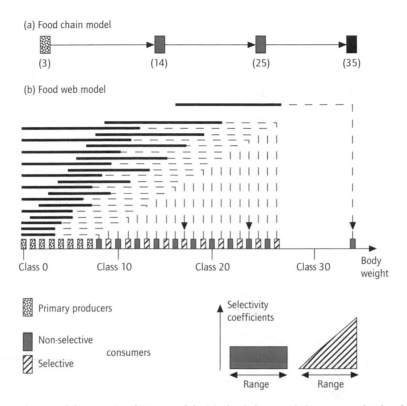

Fig. 17.12 Trophic pathways of the two simulation models: (a) a food chain with discrete trophic levels, corresponding to the model depicted in Fig. 17.8a; (b) a food web with a high diversity in dynamic and trophic function, corresponding to the model depicted in Fig. 17.8c. The body weight axis is divided into weight classes which are equally spaced on a logarithmic scale: $w_{i+1} = 2w_i$ where w_i denotes the average weight in class i. Shadings denote different functional groups. In (b) each consumer weight class is linked to a box denoting the range of the consumer's prey window. The width of a predator's prey window increases with increasing body weight. The rectangular and triangular shape of the window denote differences in feeding modes. Non-selective filter feeders prey indiscriminately on all classes in their prey window; selective raptorial feeders prey preferably on the higher classes in their prey window. The top predator in weight class 35 is optional.

with a complex food web model (cf. Pahl-Wostl 1997). Figure 17.12 depicts the pathways of trophic transfers in the food chain model with discrete trophic levels (Fig. 17.12a), and the food web model characterised by a complex trophic network (Fig. 17.12b). In the following, these two models will be referred to respectively as the **chain** and the **web** model. The models differ only in the complexity of the trophic transfers depicted. Otherwise, model parameters and the chosen type of functional response are equal. Both models account for the dy-

namics of nutrients which are recycled within the system. Both include density dependence of the predators' mortality rates. The 'species' (and thus the dynamic state variables) correspond to weight classes. Model equations and parameter values are listed in Box 17.2.

In both the chain model and the web model, the top predator in class 35 is optional. Comparisons of simulations with and without this additional top predator serve to investigate the influence of structural changes on the dynamics of the network as a

Box 17.2   Equations and parameter values for the food chain and the food web models
(cf. Pahl-Wostl 1997)

Growth (γ_i) and loss (λ_i) rates for a species in weight class i are derived from the rates of the species in weight class 0 by allometric relationships:

$$\gamma_i = k\gamma_0 2^{-\varepsilon i} \quad \text{and} \quad \lambda_i = k\lambda_0 2^{-\varepsilon i} \tag{17.4}$$

where $k = 1$ for autotrophs and $k = het$ for heterotrophs (Moloney & Field 1989).

The models are made dimensionless by expressing all rates in multiples of γ_0, the growth rate of class 0, and by expressing biomass in multiples of K_7, the half-saturation constant of the largest phytoplankton weight class. The maximum net growth rates range thus from 0.8 (class 0) to 0.0019 (class 35). In both models, the state variables comprise the nutrient in the pool (N) and the biomasses (B_i) in the weight classes.

The web model

In the web model the dynamics of the nutrient pool are described by

$$\frac{dN}{dt} = N_{in} + rec \sum_i \lambda_i B_i - \sum_{i=0}^{7} \gamma_i \frac{N}{N + K_i} B_i - \lambda_N N \tag{17.5}$$

The basic web model comprises eight classes of primary producers (classes 0–7) and 20 classes of heterotrophic consumers (8–27) which feed on the prey classes within the range of their prey windows. The growth rates of primary producers depend on nutrient availability according to a Michaelis–Menten relationship. The half-saturation constant, K_i, is assumed to decrease linearly with increasing weight class. If the constant of class 7 is normalised to 1 the constant in class 0 equals 2.75. Larger phytoplankton species are thus more efficient in utilising nutrients. This pattern accounts for the combined effects of uptake dynamics and storage capacities (Wirtz & Eckhardt 1996; Gaedke, pers. comm.).

The dynamics of phytoplankton species $i(0 \le i \le 7)$ follow

$$\frac{dB_i}{dt} = \left(\gamma_i \frac{N}{N + K_i} - \lambda_i - \gamma_i \sum_{j=8}^{26} pr_{ji} B_j \right) B_i \tag{17.6}$$

where

$$K_i = 1.0 + (7 - i)\Delta K \tag{17.7}$$

and pr_{ji}, the relative contribution of prey i to the total predation of predator j, is calculated as:

$$pr_{ji} = \frac{sel_{ji}}{\sum_k sel_{jk} B_k + K_p} \tag{17.8}$$

The dynamics of heterotrophic species i $(8 \le i \le 27)$ follows:

$$\frac{dB_i}{dt} = \left(\gamma_i \sum_{j<i} pr_{ij}B_j - \lambda_i - q_iB_i - \gamma_j \sum_{j>i} pr_{ji}B_j \right)B_i \qquad (17.9)$$

where

$$\gamma_i = het2^{-\varepsilon i}, \quad \lambda_i = het\lambda_0 2^{-\varepsilon i}, \quad q_i = hetq_0 2^{-\varepsilon i}$$

The predators in the web model are characterised by weight class and by range, shape and location of their prey window along the body weight axis. A predator i feeds on all classes j within the range of its prey window, which is defined by: $low_i = j = high_i$. Range and selectivity coefficients are a function of class and feeding mode:

$$\text{filter feeders at classes } i = \{8, 10, \ldots, 26\}$$
$$low_i = 0, \qquad high_i = 3 + (i - 8)/2 \qquad (17.10a)$$

with all $sel_{ij} = 0.45$ for all j, and

$$\text{raptorial feeders at classes } i = \{9, 11, \ldots, 27; 35\}$$
$$low_i = (i - 9)/2, \qquad high_i = 3 + (i - 9) \qquad (17.10b)$$

with $sel_{ilow_i} = 0.1$, $sel_{ihigh_i} = 1.0$ and linear interpolation in between the limits.
    An additional selective top predator $(i = 35)$ may be introduced:

$$\frac{dB_i}{dt} = \left(\gamma_i \sum_j pr_{ij}B_j - \lambda_i - q_iB_i \right)B_i \qquad (17.11)$$

The chain model

Nutrient dynamics:

$$\frac{dN}{dt} = N_{in} + rec\sum_i \lambda_iB_i - \gamma_3 \frac{N}{N + K_n}B_3 - \lambda_N N \qquad (17.12)$$

Dynamics of the phytoplankton species in class 3:

$$\frac{dB_3}{dt} = \left(\gamma_3 \frac{N}{N + K_n} - \lambda_3 - \gamma_{14} \frac{B_{14}}{B_3 + K_p} \right)B_3 \qquad (17.13)$$

Dynamics of the herbivorous consumer in class 14:

$$\frac{dB_{14}}{dt} = \left(\gamma_{14} \frac{B_3}{B_3 + K_p} - \lambda_{14} - 0.1q_{14}B_{14} - \gamma_{25} \frac{B_{25}}{B_{14} + K_p} \right)B_{14} \qquad (17.14)$$

Dynamics of the carnivorous consumer in class 25:

$$\frac{dB_{25}}{dt} = \left(\gamma_{25} \frac{B_{14}}{B_{14} + K_p} - \lambda_{25} - 0.1q_{25}B_{25} - \gamma_{35} \frac{B_{35}}{B_{25} + K_p} \right) B_{25} \qquad (17.15)$$

A secondary carnivorous consumer in class 35 may be introduced as a fourth trophic level:

$$\frac{dB_{35}}{dt} = \left(\gamma_{35} \frac{B_{25}}{B_{25} + K_p} - \lambda_{35} - 0.1q_{35}B_{35} \right) B_{35} \qquad (17.16)$$

To be able to make quantitative comparisons, the sum of the carrying capacities of the consumers in the second and third trophic levels of the chain model was set equal to the sum of the carrying capacities of the consumers in classes 8–27 of the web model. The aggregated carrying capacities of the total heterotrophic biomass are thus equal in the two models.

List of parameters and default numerical values

Symbol	Numerical value	Meaning
ε	0.25	allometric exponent
λ_0	0.3	loss rate in class 0
het	1.5	ratio of heterotrophic/autotrophic rates
K_i	f(class)	half-saturation nutrient concentration (web)
ΔK	0.25	incremental change in K_i for $\Delta i = 1$
K_p	2.0	half-saturation prey concentration
K_n	1.75	half-saturation nutrient concentration (chain)
q_0	0.05	quadratic loss term of heterotrophs
s_{ji}	$0 \le s_{ji} \le 1$	selectivity coefficient of predator j for prey i
rec	0.8	fraction of nutrient losses which is recycled
N_{in}	0.25	external nutrient input
λ_N	0.01	rate of losses from the pool

whole. In the chain model, adding this predator corresponds to a shift from a food chain with three trophic levels to one with four. In the web model, a species does not occupy a defined trophic level, but is characterised by a trophic position determined as the weighted average over all feeding pathways. Therefore, if the distribution of food intake over the various feeding pathways is not constant, a species' trophic position may vary over time. The trophic position of the additional top predator with raptorial feeding mode may range somewhere between three and four. Model versions where the predator in class 35 is absent are labelled I, whereas those where it is present are labelled II.

Figure 17.13 shows the results of simulations for the web and the chain models, with and without the additional top predator. In all simulations, nutrient input was the same and did not vary over time. Comparison of the results obtained for the chain (Fig. 17.13a) and the web (Fig. 17.13b) models shows that the former exhibits the expected effect of cascading trophic interactions. However, in the web model, the additional top predator leads to hardly any change in the trophic structure. It is further evident that total biomass and the ratio of herbivorous to autotrophic biomass (Fig. 17.13c) are greater in the chain model than in the web model. Energy transfer along the trophic gradient seems far more efficient in a model with a web type structure than in a simple food chain.

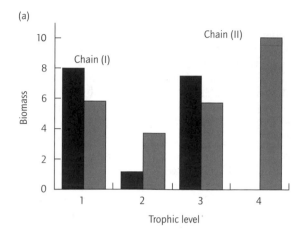

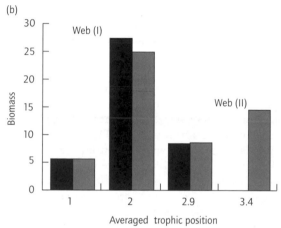

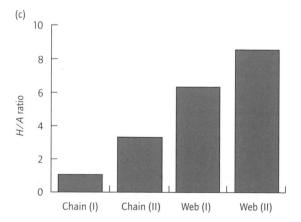

Another major difference between the chain and the web models is the effect of fluctuations in the environment, which is represented only by nutrient exchanges. The influence of a variable environment can be investigated by performing model simulations with a pulsed nutrient regime, reflecting annual seasonality in a temperate climate. Nutrient input was delivered in sine pulses over a period of 40 days, at intervals of 400 days. This may be regarded as a coarse approximation for onset of the growth period in a large, deep oligotrophic lake where winter overturn leads to high nutrient concentration in spring, and where onset of seasonal stratification isolates the photic zone of the epilimnion from the nutrient reservoir of the hypolimnion. The results obtained are shown in Fig. 17.14.

In order to facilitate comparisons, biomasses were aggregated over all primary producers, and over all heterotrophic consumers, for both the chain and the web models. The ratio of herbivorous (H) to autotrophic (A) biomass is given at the top of each figure. Figure 17.14a shows that environmental variations fluctuating on a frequency such as the annual cycle, interacting with a food chain oscillating on one or a few major frequencies, generates major effects. Pulsing of nutrient inputs leads nearly to extinction of the predator on level two, and to a very low efficiency of transfer along the trophic gradient. The result is different for the web model, with a range of internal frequencies and redundant pathways (Fig. 17.14b).

Fig. 17.13 Distribution of biomass as a function of trophic level obtained (a) with the food chain model and (b) as a function of trophic position obtained for the food web model. For the web model, the biomasses of the planktonic species classes were aggregated into three groups: all phytoplankton species with trophic position = 1, all consumers with 2 ≥ trophic position <2.5 (predominantly filter feeders), all consumers with trophic positions ≥2.5 (predominantly raptorial feeders). (c) The ratio of heterotrophic to autotrophic biomass obtained in the different simulations. Chain (I) refers to the three-level chain model, chain (II) refers to the four-level chain model, web (I) refers to the web model without top predator and web (II) to the web model with top predator.

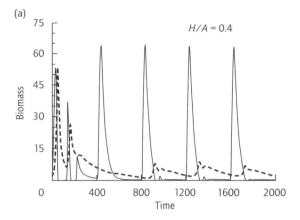

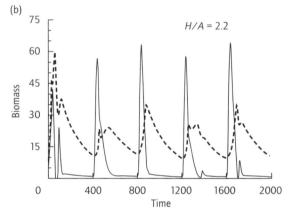

Fig. 17.14 Temporal variations of the aggregated biomasses of all primary producers (continuous lines) and of all heterotrophic consumers (dashed lines), obtained for a pulsed nutrient input (sine pulses of a duration of 40 days in intervals of 400 days). (a) Three-level chain model, (b) web model without top predator. The H/A (herbivorous/autotrophic) ratio was determined as average over the period 1200–2000.

In conclusion, one can state that, because of its dynamic and flexible network, the web model buffers the effects of structural changes such as adding a top predator, and also of changes in the environment, such as changing nutrient input. This behaviour is in sharp contrast to the sensitivity of the chain model. A more detailed comparison can be found in Pahl-Wostl (1997).

Despite the extreme simplification of the seasonal environment, the web model reproduces essential patterns of a seasonal succession which seem to be determined by the internal dynamics and the structure of the plankton web. Figure 17.15 shows the relative heterotrophic biomass as a function of weight class obtained for different model versions and for empirical data from Lake Constance (the Bodensee). Alternation between raptorial and filter feeders with each weight class (cf. Fig. 17.12) is an artefact of the model. In reality, one may expect a more continuous distribution for both functional groups. In order to eliminate the effects of this special model feature in a comparison with real data, biomasses of adjacent weight classes were aggregated. Introduction of a top predator, which can be interpreted as introduction of fish predation, leads to decline of large species and an increase in small taxa. This effect, which is much more pronounced for pulsed than for constant nutrient input, agrees with results reported in empirical studies (e.g., Vanni 1987; Kerfoot & DeAngelis 1989). Note also that the bimodal pattern obtained for version II (Fig. 17.15b) corresponds to the pattern obtained from the empirical data shown in Fig. 17.15d. The observation that similar distributions were obtained for empirical data in four successive years indicates that this pattern is a regular feature of seasonal succession. Similar patterns have been reported in other lakes (Vanni *et al.* 1997; Gaedke, pers. comm.).

In order to obtain such a bimodal pattern in model simulations, the presence of the top predator is required (cf. Fig. 17.15a & b). Otherwise, predation pressure imposed by large species is too great for smaller species to build up significant biomass. However, the presence of fish predation alone is not sufficient. Figure 17.15c shows the results obtained for a constant nutrient input, in the absence (I) and in the presence (II) of fish predation. One notes that, for a constant nutrient input, introducing fish predation does not lead to a bimodal pattern. Presence of a seasonal cycle as introduced by the pulsed nutrient input is another essential requirement. In addition, further model results have shown that both filter feeders and raptors must be present for a bimodal pattern to be obtained. This

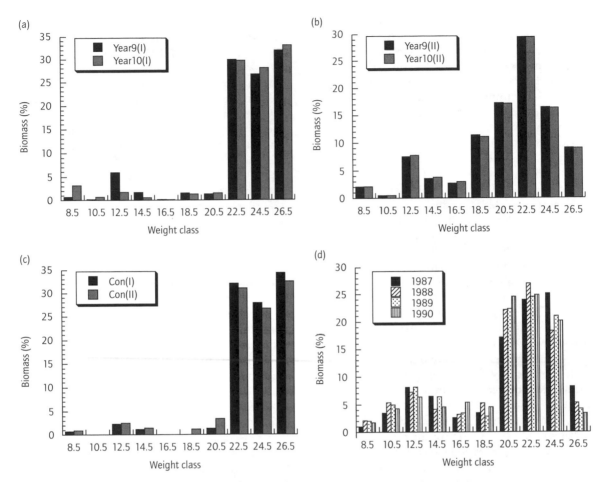

Fig. 17.15 Relative heterotrophic biomass averaged over one seasonal cycle as a function of weight class. Raptorial (Rap) and filter (Fil) feeders in adjacent classes are aggregated and their combined biomass is assigned to the class between, e.g. the united biomass of Fil(8) and Rap(9) is assigned to class 8.5. (a, b) Pulsed web model for the 9th and 10th year: (a) version I without fish and (b) version II with fish. (c) Web model for a constant nutrient input, versions I and II. (d) Biomass of heterotrophs obtained for Lake Constance (the Bodensee) in different years. The classes were aggregated accordingly to facilitate the comparison with model results. Body mass was expressed in units of carbon [pg C] and absolute biomass was measured in units of carbon per volume [pg C mL^{-1}] (Gaedke 1992b).

finding confirms observations by Brett *et al.* (1994), who emphasised the importance of the functional heterogeneity among zooplankton species for food web structure and regulation.

The framework outlined above allows us to investigate the structural complexity of pelagic networks in a systematic fashion. In extension to the issues already addressed, one may for example account for differences in life-history strategies. A comparison between daphnids and cyclopoid copepods illustrates how this can be accomplished. Daphnids are not selective and feed on a wide range of weight classes. In aggregate, the cyclopoid copepods possess a similarly large prey window. However, its width derives from the large ontogenetic shift in body weight which takes place

during maturation. The range of individuals of a
given size is much smaller in these raptorial feeders than it is for filter-feeding daphnids feeding on a
large size range of food. For daphnids, the ontogenetic shift in size (and hence the range of the prey
window) is of minor importance. Cyclopoid copepods effectively shift one trophic level owing to a
more than 100-fold increase in body size during
maturation from nauplii to adult stage.

The effects of these two types of behaviour on
the temporal organisation of an ecological network are very different (Pahl-Wostl 1995). Daphnids couple scales, thereby reducing variability
and dynamic diversity. This leads to stronger coupling between trophic levels. Hence, it is not surprising that the success of biomanipulation is
observed to be highly correlated with the influence
of large daphnids as major predators (Hosper &
Meijer 1993; Reynolds 1994). Simulation results
support statements by Strong (1992), that true
trophic cascades are restricted to sites with fairly
low-diversity, where great influence can issue
from a single functional component.

In contrast, cyclopoid copepods may increase
rather than decrease dynamic richness. Such an
increase in variability is even more pronounced for
fish, which, during ontogenetic development,
cover an even wider range of the trophic gradient,
and where the generation times exceed the length
of the annual cycle. Carpenter & Kitchell (1993)
discuss several examples of pronounced interannual effects caused by waxing and waning of fish
cohorts. Such effects add considerably to the un-
predictability of ecosystem dynamics, especially
when we take into account the extreme variability
in reproduction. Reproductive success depends on
the vagaries of the physical environment (Cushing
1990; George *et al.* 1990; Bollens *et al.* 1992) and a
favourable timing with respect to overall community dynamics—whether the temporal niche is
favourable for juveniles to thrive. Better knowledge, and models of this temporal organisation,
will allow us to derive the effects to be expected
from climate change, which may largely alter seasonal dynamics in lakes.

17.5 SELF-REGULATION AND THE RESISTANCE TO ENVIRONMENTAL AND ANTHROPOGENIC STRESS

The multilevel structure of ecosystems must also
be taken into account when we go on to consider
their response to environmental and anthropogenic stress. Structural properties of ecosystems, such as species composition, seem to be
more sensitive to stress than to functional properties. Whole-lake experiments demonstrate that
functional properties such as primary production
or respiration are rather too insensitive to monitor
the effects of a continued exposure to stress induced by acidification (Schindler 1987, 1988,
1990). Early signs of warning may be detected at
the level of species composition and morphologies, whereas functional properties exhibit a
lagged but abrupt response. These observations
suggest that the effects of environmental stress are
first buffered by structural rearrangements, but
lead finally to an abrupt decline in ecosystem function. Such an effect is illustrated by simulation results from modified versions of the chain and web
models discussed in the previous section.

In order to investigate the influence of stress,
I simply introduced a variable which represents
its effects, caused, for example, by a change in
climate, or in another environmental variable.
Stress is assumed to reduce the growth rate of algal
species. In the web model, algal species differ in
their sensitivity to stress. There is a trade-off between species growth rates in the absence of stress
and their resistance to stress. The stress sensitivity
of the single algal species in the chain model corresponds to the average of the algal ensemble in the
web model. Simulations of the effects of a gradually increasing stress on total biomass show that, in
the web model, biomass is maintained constant
over a wide range of stress, owing to replacement of
sensitive species by less sensitive taxa (Fig. 17.16).
This leads to an abrupt change in total biomass, despite a gradually increasing stress level. Structural
complexity may thus complicate monitoring, interpreting and predicting the effects of stress on
ecosystems, in particular if attention is directed
only at functional properties.

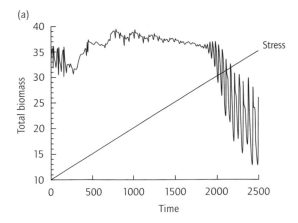

(a)

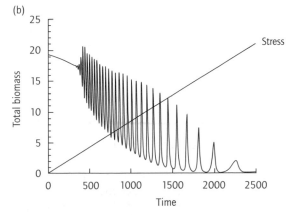

(b)

Fig. 17.16 Results from model simulations of modified versions of (a) the web model and (b) the chain model, for their response to increasing levels of stress.

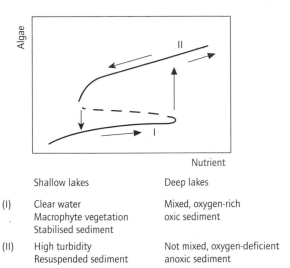

	Shallow lakes	Deep lakes
(I)	Clear water	Mixed, oxygen-rich
	Macrophyte vegetation	oxic sediment
	Stabilised sediment	
(II)	High turbidity	Not mixed, oxygen-deficient
	Resuspended sediment	anoxic sediment

Fig. 17.17 Alternative states in lacustrine ecosystems as a function of the external nutrient load. The arrows indicate that recovery for a decreasing nutrient load does not follow the same path obtained for an increasing nutrient load.

Such an abrupt decline may generate several interpretations. Extreme stress, such as severe acidification, may finally lead to an impoverished ecosystem, where only basic function is maintained. However, change in one environmental variable may cause transition to another state, and this may be abrupt rather than smooth. Even though data on the relationship between structure and function are still scarce (and even more so on their response to environmental change), it is well known that smaller lakes possess alternative states, and that the transition between these is abrupt (see also Chapter 9). A well investigated ex-

ample, of interest to ecological research and management alike, is the behaviour of lakes undergoing eutrophication.

Figure 17.17 depicts the existence of alternative states in lacustrine ecosystems as a function of the external nutrient load for two different types of lake. With increasing nutrient load, a discontinuity occurs, where a system change from one state to another occurs. Once a critical threshold is reached, a cascade of interrelated effects is triggered, amplifying the rate of change. Rapid transitions from oxic to anoxic states were monitored in cores of lake sediments (e.g. Niessen & Sturm 1987). Transition may be abrupt—from one year to the next—despite continuous variation in other lake parameters.

At intermediate nutrient levels, two alternative stable states exist. Recovery for a decreasing nutrient load does not follow the same path as that for an increase. This effect is especially pronounced in shallow lakes (Hosper & Meijer 1993). Algae render the water turbid and cause the disappearance of submerged plants. The turbid-water state is further stabilised by planktivorous fish, which suppress the larger (and more efficient) filter feeders. The clear-water state is stabilised by macrophytes which compete with algae for nutrients, and which offer spawning grounds for predatory fish such as pike. The effects are thus transmitted throughout the entire ecosystem.

Which measurements give reliable, empirical information on structural complexity? Taxonomic diversity is very difficult to determine empirically and may not always yield the appropriate information. In the models discussed here, a functional concept for diversity based on body size was chosen. Non-taxonomic, size based approaches may be of general interest in this respect. They are also more easily applied. A useful analytical tool is provided by biomass size spectra and their interpretation.

Because energy flow is directed along a gradient of increasing body weight (see section 17.4), biomass-weight spectra are a convenient tool for characterising pelagic communities. In order to obtain such spectra, organisms are aggregated into logarithmic weight classes, and the total biomass is determined for each weight class. Rendering biomass spectra more amenable to quantitative analysis, and suitable for cross-system comparisons, requires some type of standardisation. Figure 17.18 shows an example for a normalised biomass spectrum derived for the plankton community of Lake Constance (the Bodensee; Gaedke 1992b). Such normalised spectra are obtained by dividing B_i (the biomass in the ith weight class) by Δw_i (the weight range of the ith weight class). The resulting biomass density corresponds approximately to N_i (the numerical abundance of individuals in a weight class). Two attributes of these spectra are of major interest: the slope and the evenness of the distribution. Both may be quantitatively assessed by a linear regression, where a decrease in the coefficient of determination, r^2, indicates an increase in the deviation of the data points from a straight line ($r^2 = 1$ is obtained for an ideal even distribution). A slope of −1 is obtained when the same biomass is found in the different weight classes. ($N \propto W^{-1} - N_i \times W_i = B_i = \mathrm{con}$-

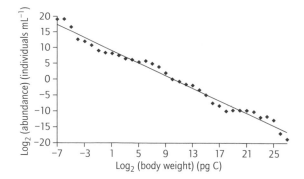

Fig. 17.18 Normalised spectrum obtained for the annual average of the plankton community's biomass in Lake Constance. The approximate abundance is obtained by dividing the biomass in a weight class by the weight range of this class. The table shows the parameters for the linear regression. A detailed discussion is given by Gaedke (1992b).

stant.) We note that the biomass distribution in Lake Constance conforms closely to the requirements of uniformity.

Sheldon *et al.* (1972) were the first to analyse particle size distributions in the open oceans. They observed a uniform distribution of particles over logarithmic size classes. Since then, further investigations confirmed the initial observations (e.g. Rodríguez & Mullin 1986; Ahrens & Peters 1991; Gaedke 1992b). It seems that large lakes and marine systems are characterised by even distributions, with an absence of major gaps in the spectrum. In these systems, the pelagic habitat is large enough to be looked upon as an autonomous ecosystem (see also Chapter 9).

Continuous biomass distributions in pelagic systems seem to reflect the homogeneous nature of the physical environment. In contrast, biomass size distributions of marine benthic environments exhibit pronounced modes corresponding to the sizes of benthic bacteria, meiofauna and macrofauna. Schwinghamer (1981) explained such patterns by arguing that organisms perceive the sedimentary environment on three scales: they may either (i) colonise particle surfaces, (ii) inhabit interstices between particles, or (iii) respond to the sediment as if it were effectively non-particulate. The transition from one scale to another is discontinuous rather than gradual. Warwick (1984) and Warwick & Joint (1987) found that species size distribution followed a similar modal pattern to the biomass spectrum. They invoked evolutionary optimisation of size-related life-history and spatial resource partitioning to explain the pattern.

Interestingly, Rodríguez *et al.* (1990) detected major gaps in the spectrum of small mountain lakes. In these systems, where the pelagic environment is relatively small, we need to take coupling between pelagic and benthos zones into account. The presence of such coupling is further supported by the observation that pelagic larvae of the macrobenthos just fit in size in the trough in the biomass spectrum (Warwick 1984; Warwick & Joint 1987).

Biomass size spectra may prove to be useful operational tools for investigating regularities and responses to environmental stress in such distributions, and whether they are related to the nature of the physical environment, in particular to the power spectrum of its variance in space and time. Systematic investigation of body size spectra may be of interest, especially for habitat characteristics and disturbance regimes. The slope of biomass size spectra (which is a measure for the efficiency of energy transfer) may be a sensitive indicator for stress. Acidification in mountain streams results in a shift of biomass distribution towards small opportunistic species (Seifried 1993), a finding which has also been reported for lakes (Ahrens & Peters 1991). Uniform distributions of biomass across the body weight gradient seem to indicate an undisturbed community which organises itself on a continuum of dynamic properties and function.

17.6 CONCLUDING REMARKS

Ecosystems are not static entities. Therefore one should not seek to preserve their present state but the goals for protection must be to maintain an ecosystem's potential for self-regulation and adaptation. Understanding the relationship between the diversity of ecological networks and ecosystem properties is essential for bridging different levels of ecological organisation. Concepts for describing the complexity of ecological networks are thus a prerequisite to defining an 'ecosystem's state of health', a reference state against which to contrast the effect of anthropogenic influence. Systematic approaches for investigating the relationship between structural and functional properties, such as that outlined in this chapter, are required. Pelagic ecosystems may prove to be an ideal system for the investigation of general ecological patterns which are of relevance far beyond the realm of lacustrine systems.

17.7 REFERENCES

Abrams, P.A. (1993) Effect of increased productivity on the abundances of trophic levels. *American Naturalist*, **141**, 351–71.

Abrams, P.A. (1996) Dynamics and interactions in food webs with adaptive foragers. In: Polis, G. & Winemiller, K. (eds), *Food Webs: Integration of Pattern and Dynamics.* Chapman & Hall, New York, 109–12.

Abrams, P. & Roth, J. (1994a) The effects of enrichment of three-species food chains with nonlinear functional response. *Ecology*, **75**, 1118–30.

Abrams, P. & Roth, J. (1994b) The response of unstable food chains to enrichment. *Evolutionary Ecology*, **8**, 150–71.

Ahrens, M. & Peters, R.H. (1991) Patterns and limitations in limnoplankton size spectra. *Canadian Journal of Fisheries and Aquatic Sciences*, **48**, 1967–78.

Allen, J.C., Schaffer, W.M. & Rosko, D. (1993) Chaos reduces species extinction by amplifying local population noise. *Nature*, **364**, 229–32.

Bollens, S., Frost, B., Schwaninger, H., Davis, C., Way, K. & Landsteiner, M. (1992) Seasonal plankton cycles in a temperate fjord and comments on the match-mismatch hypothesis. *Journal Plankton Research*, **14**, 1279–305.

Brett, M., Wiackowski, K., Lubnow, F., Mueller-Solger, A., Elser, J. & Goldman, C. (1994) Species-dependent effects of zooplankton on planktonic ecosystem processes in Castle Lake, California. *Ecology*, **75**, 2243–54.

Carpenter, S. & Kitchell, J. (1992) Trophic cascade and biomanipulation: Interface of research and management—a reply to the comment of DeMelo *et al. Limnology and Oceanography*, **37**, 208–13.

Carpenter, S.R. & Kitchell, J.F. (eds) (1993) *The Trophic Cascade in Lakes.* Cambridge University Press, Cambridge, 385 pp.

Carpenter, S.R., Frost, T., Kitchell, J.F. & Kratz, T. (1993) Species dynamics and global environmental change: a perspective from ecosystem experiments. In: Kareiva, P., Kingsolver, J. & Huey, R. (eds), *Biotic Interactions and Global Change.* Sinauer Associates, Sunderland, 267–79.

Cohen, J., Briand, F. & Newman, C. (1990) *Community Food Webs: Data and Theory.* Springer-Verlag, New York, 308 pp.

Cousins, S. (1985) The trophic continuum in marine ecosystems: structure and equations for a predictive model. *Canadian Bulletin on Fisheries and Aquatic Science,* **213,** 76–93.

Cousins, S. (1987) The decline of the trophic level concept. *Trends in Ecology and Evolution,* **2,** 312–6.

Cushing, D. (1990) Plankton production and year-class strength in fish populations: an update of the match/mismatch hypothesis. *Advances in Marine Biology,* **26,** 250–90.

DeAngelis, D.L. (1992) *Dynamics of Nutrient Cycling and Food Webs.* Chapman & Hall, London, 270 pp.

DeAngelis, D.L., Bartell, S.M. & Brenkert, A.L. (1989) Effects of nutrient recycling and food-chain length on resilience. *American Naturalist,* **134,** 778–805.

DeMelo, R., France, R. & McQueen, D. (1992) Biomanipulation: hit or myth? *Limnology and Oceanography,* **37,** 192–207.

Diehl, S. (1993) Relative consumer sizes and the strengths of direct and indirect interactions in omnivorous feeding relationships. *Oikos,* **68,** 151–7.

Elser, J., Luecke, C., Brett, M. & Goldman, C. (1995) Effects of food web compensation after manipulation of rainbow trout in an oligotrophic lake. *Ecology,* **76,** 52–69.

Forbes, S.A. (1887) The lake as a microcosm. *Illinois Natural History Survey Bulletin,* **15,** 537–50.

Fretwell, S. (1987) Food chain dynamics: the central theory of ecology? *Oikos,* **50,** 291–301.

Gaedke, U. (1992a) Identifying ecosystem properties: a case study using plankton biomass size distributions. *Ecological Modelling,* **63,** 277–98.

Gaedke, U. (1992b) The size distribution of plankton biomass in a large lake and its seasonal variability. *Limnology and Oceanography,* **37,** 1202–20.

Gaedke, U., Straile, D. & Pahl-Wostl, C. (1996) Trophic Structure and Carbon Flow Dynamics in the Pelagic Community of a Large Lake. In: Polis, G. & Winemiller, K. (eds), *Food Webs: Integration of Pattern and Dynamics.* Chapman & Hall, New York, 109–12.

George, D.G., Hewitt, D.P., Lund, J.W.G. & Smyly, W.J. (1990) The relative effects of enrichment and climate change on the long-term dynamics of *Daphnia* in Esthwaite Water, Cumbria. *Freshwater Biology,* **23,** 55–70.

Hairston, N.F., Smith, F. & Slobodkin, L. (1960) Community structure, population control and competition. *American Naturalist,* **94,** 421–5.

Hansson, L.-A. (1992) The role of food chain composition and nutrient availability in shaping algal biomass development. *Ecology,* **73,** 241–7.

Hastings, A. (1988) Food web theory and stability. *Ecology,* **69,** 1665–8.

Hastings, A. & Higgins, K. (1994) Persistence of transients in spatially structured ecological models. *Science,* **263,** 1133–6.

Hastings, A. & Powell, T. (1991) Chaos in a three-species food chain. *Ecology,* **72,** 896–903.

Hastings, P.A. (1996) What equilibrium behaviour of Lotka-Volterra models does not tell us about food webs. In: Polis, G. & Winemiller, K. (eds), *Food Webs: Integration of Pattern and Dynamics.* Chapman & Hall, New York, 211–7.

Hosper, H. & Meijer, M.-L. (1993) Biomanipulation, will it work for your lake? A simple test for the assessment of chances for clear water, following drastic fish-stock reduction in shallow, eutrophic lakes. *Ecological Engineering,* **2,** 63–72.

Kerfoot, W.C. & DeAngelis, D.L. (1989) Scale-dependent dynamics: zooplankton and the stability of freshwater food webs. *Trends in Ecology and Evolution,* **4,** 167–71.

Lawton, J. & Warren, P. (1988) Static and dynamic explanations for patterns in food webs. *Trends in Ecology and Evolution,* **3,** 242–5.

Lodge, D., Barko, J., Strayer, D., *et al.* (1988) Spatial heterogeneity and habitat interactions in lake communities. In: Carpenter, S.R. (ed.), *Complex Interactions in Lake Communities.* Springer-Verlag, New York, 229–60.

Mateev, V. (1995) The dynamics and relative strengths of bottom-up vs topdown impacts in a community of subtropical lake plankton. *Oikos,* **73,** 104–8.

Mazmuder, A. (1994) Patterns of algal biomass in dominant odd- vs even-link lake ecosystems. *Ecology,* **75,** 1141–9.

McQueen, D., Mills, E., Forney, J., Johannes, M. & Post, J. (1992) Trophic level relationships in pelagic food webs: comparisons from long-term data sets for Oneida Lake, New York (USA), and Lake St. George, Ontario (Canada). *Canadian Journal of Fisheries and Aquatic Sciences,* **49,** 1588–96.

Moloney, C. & Field, J. (1989) General allometric equations for rates of nutrient uptake, ingestion, and respiration in plankton organisms. *Limnology and Oceanography*, **34**, 1290–9.

Niessen, F. & Sturm, M. (1987) Die Sedimente des Baldeggersees (Schweiz)—Ablagerungsraum und Eutrophierungsentwicklung während der letzten 100 Jahre. *Archiv für Hydrobiologie*, **108**, 365–83.

Oksanen, L. (1991) Trophic levels and trophic dynamics: a consensus emerging. *Trends in Ecology and Evolution*, **6**, 58–60.

Pahl-Wostl, C. (1993) Food webs and ecological networks across spatial and temporal scales. *Oikos*, **66**, 415–32.

Pahl-Wostl, C. (1995) *The Dynamic Nature of Ecosystems: Chaos and Order Entwined*. Wiley, Chichester, 267 pp.

Pahl-Wostl, C. (1997) Dynamic structure of a food-web model: comparison with a food-chain model. *Ecological Modelling*, **100**, 103–23.

Paine, R.T. (1988) Food webs: road maps of interaction or grist for theoretical development? *Ecology*, **69**, 1648–54.

Persson, L., Bengtsson, J., Menge, B. & Power, M. (1996) Productivity and consumer regulation—concepts, patterns, and mechanisms. In: Polis, G. & Winemiller, K. (eds), *Food Webs: Integration of Pattern and Dynamics*. Chapman & Hall, New York, 109–12.

Persson, L., Diehl, S., Johansson, L., Andersson, G. & Hamrin, S. (1992) Trophic interactions in temperate lake ecosystems: a test of food chain theory. *American Naturalist*, **140**, 59–84.

Peters, R. (1983) *The Implications of Body Size*. Cambridge University Press, Cambridge, 329 pp.

Pimm, S.L. (1984) The complexity and stability of ecosystems. *Nature*, **307**, 321–6.

Pimm, S.L., Lawton, J. & Cohen, J. (1991) Food web patterns and their consequences. *Nature*, **350**, 669–74.

Polis, G. & Holt, R. (1992) Intraguild predation: the dynamics of complex trophic interactions. *Trends in Ecology and Evolution*, **7**, 151–4.

Polis, G. & Winemiller, K. (eds) (1996) *Food Webs: Integration of Pattern and Dynamics*. Chapman & Hall, New York, 472 pp.

Porter, K.G., Paerl, H., Hodson, R., *et al.* (1988) Microbial interactions in lake communities. In: Carpenter, S.R. (ed.), *Complex Interactions in Lake Communities*. Springer-Verlag, New York, 109–227.

Reynolds, C.S. (1994) The ecological base for the successful biomanipulation of aquatic communities. *Archiv für Hydrobiologie*, **130**, 1–33.

Reynolds, C.S. (1997) *Vegetation Processes in the Pelagic: a Model for Ecosystem Theory*. Ecology Institute, Oldendorf, 371 pp.

Rodríguez, J.F. & Mullin, M.M. (1986) Relation between biomass and body weight of plankton in a steady state oceanic ecosystem. *Limnology and Oceanography*, **31**, 361–70.

Rodríguez, J.F., Echevarria, F. & Jímenez-Gomez, F. (1990) Physiological and ecological scalings of body size in an oligotrophic, high mountain lake. *Journal of Plankton Research*, **12**, 593–9.

Schindler, D.W. (1987) Detecting ecosystem responses to anthropogenic stress. *Canadian Journal of Fisheries and Aquatic Sciences*, **44** (Supplement), 6–25.

Schindler, D.W. (1988) Effects of acid rain on freshwater ecosystems. *Science*, **239**, 149–57.

Schindler, D.W. (1990) Experimental perturbations of whole lakes as tests of hypotheses concerning ecosystem structure and function. *Oikos*, **55**, 25–41.

Schulze, E.-D. & Mooney, H.A. (eds) (1993) *Biodiversity and Ecosystem Function*. Springer-Verlag, Berlin, 558 pp.

Schwinghamer, P. (1981) Characteristic size distributions of integral benthic communities. *Canadian Journal of Fisheries and Aquatic Sciences*, **38**, 1255–63.

Seifried, A. (1993) *Die bentische Besiedlung der obersten Gauchach unter besonderer Berücksichtigung des Versauerungsgradienten*. Diploma thesis, University of Konstanz, Konstanz.

Sheldon, R., Prakash, A. & Sutcliffe, W. (1972) The size distribution of particles in the ocean. *Limnology and Oceanography*, **17**, 327–40.

Sheldon, R., Sutcliffe, W. & Paranjape, M. (1977) Structure of the pelagic food chain and relationship between plankton and fish production. *Journal of the Fisheries Research Board of Canada*, **34**, 2344–53.

Simon Moffat, A. (1996) Biodiversity is a boon to ecosystems, not species. *Science*, **271**, 1497.

Slobodkin, L., Smith, F. & Hairston, N.F. (1967) Regulation in terrestrial ecosystems, and the implied balance of nature. *American Naturalist*, **101**, 109–24.

Sprules, W. & Munawar, M. (1986) Plankton size spectra in relation to ecosystem productivity, size, and perturbation. *Canadian Journal of Fisheries and Aquatic Sciences*, **43**, 1789–94.

Sprules, W.G., Brandt, S.B., Stewart, D.J., Munawar, M., Jin, E.H. & Love, J. (1991) Biomass size spectrum of the Lake Michigan pelagic food web. *Canadian Journal of Fisheries and Aquatic Sciences*, **48**, 105–15.

Sterner, R.W. (1990) The ratio of nitrogen to phosphorus resupplied by herbivores: zooplankton and the algal competitive arena. *American Naturalist*, **136**, 209–29.

Stone, L. & Berman, T. (1993) Positive feedback in aquatic ecosystems: the case of the microbial loop. *Bulletin of Mathematical Biology*, **55**, 919–36.

Stone, L. & Weisburd, R. (1992) Positive feedback in

aquatic ecosystems. *Trends in Ecology and Evolution*, 7, 263–7.

Straile, D. (1994) *Die saisonale Entwicklung des Kohlenstoffkreislaufes im pelagischen Nahrungsnetz des Bodensees*. PhD thesis, University of Konstanz.

Strong, D. (1992) Are trophic cascades all wet? Differentiation and donor-control in speciose ecosystems. *Ecology*, 73, 747–54.

Thienemann, A. (1925) *Die Binnengewässer Mitteleuropas*. Schweizerbart'sche Verlagsbuchhandlung, Stuttgart, 255 pp.

Tilman, D. (1996) Biodiversity: population versus ecosystem stability. *Ecology*, 77, 350–63.

Tilman, S. & Downing, J. (1994) Biodiversity and stability in grasslands. *Nature*, 367, 363–5.

Vanni, M. (1987) Effects of nutrients and zooplankton size on the structure of a phytoplankton community. *Ecology*, 68, 624–35.

Vanni, M. (1996) Nutrient transport and recycling by consumers in lake food webs: implications for algal communities. In: Polis, G. & Winemiller, K. *Food Webs: Integration of Pattern and Dynamics.* Chapman & Hall, New York, 109–12.

Vanni, M. & Findlay, D. (1990) Trophic cascades and phytoplankton community structure. *Ecology*, 71, 921–37.

Vanni, M. & Layne, C.D. (1997) Nutrient recycling and herbivory as mechanisms in the top-down effect of fish on algae in lakes. *Ecology*, 78(1), 21–40.

Vanni, M., Layne, C.D. & Arnott, S.E. (1997) Top-down trophic interactions in lakes: effects of fish on nutrient dynamics. *Ecology*, 78, 1–20.

Warwick, R.M. (1984) Species size distributions in marine benthic communities. *Oecologia*, 61, 32–40.

Warwick, R.M. & Joint, R.I. (1987) The size distribution of organisms in the Celtic Sea: from bacteria to metazoa. *Oecologia*, 73, 185–91.

Wirtz, K.-W. & Eckhardt, B. (1996) Effective variables in ecosystem models with an application to phytoplankton succession. *Ecological Modelling*, 92, 33–54.

Witek, Z. & Krajewaska-Soltys, A. (1989) Some examples of the epipelagic plankton size structure in high latitude oceans. *Journal of Plankton Research*, 11, 1143–55.

18 Palaeolimnology

PATRICK O'SULLIVAN

18.1 INTRODUCTION

Palaeolimnology is the study of the ontogeny of lake and watershed ecosystems, via analysis of lake sediments. The term ontogeny (development through time) is used here in order to signify that the palaeolimnological approach encompasses **all** of the processes which affect lakes and their drainage basins throughout their lifetime as systems (e.g. succession, infilling, paludification, siltation, acidification, eutrophication), even though these are promoted by a number of different agencies (climatic, geological, successional, anthropogenic), without any implication of direction or goal.

The sediments of lakes are composed of material which originates both within lake basins themselves, and in their watershed, **and** in the atmosphere above the lake and that watershed (the 'airshed': Likens 1979). The information which lake sediments contain therefore records not only the ontogeny of lakes themselves, but also of the terrestrial areas which drain into them, and of atmospheric deposition over the lake and its watershed. Palaeolimnology may thus be used to reconstruct not only the ontogeny of lake ecosystems themselves, but also that of their catchments, and of the fundamentally important interactions between lakes and their drainage basins, especially those promoted by human action (O'Sullivan 1979).

Lake sediments therefore provide an important time perspective on a vast range of processes and 'environmental problems', enabling us to pinpoint their onset, their physical, chemical and biological manifestations, and, ultimately, their cause(s). They also allow us at least partly to reconstruct the original condition of a lake and its catchment, information which is invaluable when restoring lakes from pollution and other damage (Moss *et al.* 1996). The solutions to such problems are generally political or economic, however, and have not yet been identified using the palaeolimnological approach.

Lake sediments represent an archive of environmental change and variability unrivalled in the rest of nature. The timescales over which they accumulate lend themselves almost perfectly to the study of ecological change and of ecosystem processes, as well as of human impact on nature, including even that of modern, industrial society. Long sequences of lacustrine sediments are the only continental records which rival polar and tropical ice-caps as archives of long-term climatic variability, in terms of continuity, precision and detail (ELDP 1995).

Some pioneers of limnology (e.g. Nipkow 1920; Eggleton 1931) commented on the importance of lake sediments as a store of information. Palaeolimnology *sensu stricto* probably began at Yale during the late 1930s amongst Hutchinson and his students, notably Deevey (1942), who initially termed the subject **biostratonomy**, a word with a (characteristically) more elegant etymology, but one not widely adopted by the scientific community. It probably also does not reflect the scope of modern palaeolimnology, which is a truly interdisciplinary (i.e. **not** multidisciplinary; Emmelin 1975) subject, involving strong and explicit interactions between lake physics and chemistry (organic and inorganic), and also lake biology and ecology, as well as sediment chronology. It also calls for a working knowledge of prehistory, and of the political and economic history of human societies, all of which have modified lake-watershed ecosystems for their own purposes, a process which continues at an ever-increasing pace.

Increasingly, the emphasis of palaeolimnology

has shifted from studying processes operating in lake-watershed ecosystems themselves (succession, eutrophication) to using lake sediments as a source of proxy signals of phenomena influencing lakes and their catchments from 'outside', particularly surface water acidification and, more recently, climate change. Modern palaeolimnological research is increasingly gravitating towards teamwork, and a 'multi-core, multi-site, multi-proxy approach' (ELDP 1995), rather than restricting itself to a single microfossil taxon from one core from a single lake.

Until recently, palaeolimnology was approached via a wide range of scientific journals, only one of which (the *Journal of Paleolimnology*) specialised in the subject. A substantial number of review articles and collections of essays has appeared (e.g. Oldfield 1977, 1981; Binford *et al.* 1983; Haworth & Lund 1984; Berglund 1986; Smol 1989a,b; Davis 1990; Battarbee 1991; Charles *et al.* 1994; O'Sullivan 1995a; Talbot & Allen 1996) but none of these aspired to be comprehensive. This circumstance has now changed, with the publication of a five volume survey of palaeolimnological methods and their use in reconstructing environmental change (Last & Smol 2001a,b; Smol *et al.* 2001a,b; Birks *et al.*, in press). Collectively, these provide a comprehensive and detailed introduction to many of the techniques used in modern palaeolimnology (and some whose time is yet to come), and to the field and laboratory methods and instrumentation which they use. There are also two recent dedicated texts (Cohen 2002; Smol 2002), neither of which was I able to consult whilst preparing this chapter.

This article is not intended to compete with any of the above volumes, or in any way to be a comprehensive review of palaeolimnology. For reasons of space, it deliberately avoids practical issues, and concentrates instead on the theoretical and conceptual framework in which palaeolimnologists operate. Neither, for the same reason, is there room for specific case studies. As such, it is not meant to be an introduction to palaeolimnology for palaeolimnologists, but for those new to the subject who would like to know a little more. Sincere apologies are offered to any of my (erstwhile?)

colleagues whom I may have misrepresented or misquoted.

18.2 THE APPLICATIONS OF PALAEOLIMNOLOGY

Broadly, palaeolimnology is used to reconstruct:
1 lake ontogeny, which includes aquatic succession (maximisation of lake productivity; Odum 1969), hydroseral succession (infilling with autochthonous sediment; Walker 1970), paludification (infilling with allochthonous organic sediment; Deevey 1984), siltation and eutrophication (increase in nutrient loading; Edmondson 1991);
2 long-term and recent human impact on lakes and their catchments, including the effects of agriculture, mining, industrialisation and urbanisation on catchment soils and vegetation, erosion rates, water quality in inflowing streams, lake nutrient concentrations and other water chemistry, and lake biota—such processes normally lead to eutrophication, heavy metal and organic pollution, or lake acidification;
3 long- and short-term climatic and other changes (e.g. surface water acidification), the record of whose impact on lakes and their sediments is then used as a 'proxy' of wider regional, hemispheric or global changes.

Palaeolimnology began with the study of lake succession. In particular, Deevey's (1942) study of Linsley Pond appeared to vindicate the view that succession in lakes involves a change from oligotrophic (nutrient poor) to eutrophic (nutrient rich) conditions, as the lake fills with sediment, and the sediment–water interface (SWI) becomes deoxygenated (Deevey 1955). Thus began the confusion between eutrophication and hydroseral succession, and between nutrient concentration and productivity, which, despite Deevey's later reservations (1984), has permeated the North American literature especially, ever since (Edmondson 1991; O'Sullivan 1995b).

During the 1970s, as 'environmental problems' became apparent, palaeolimnologists joined in the study of processes such as eutrophication and, during the 1980s, surface water acidification by 'acid rain' (Battarbee *et al.* 1999). This phase led to the development of large, inter-institutional, international programmes such as SWAP (Battarbee *et al.* 1990) and PIRLA (Charles & Whitehead 1986). During the 1990s, as funding agencies at last began to believe in global warming, emphasis shifted to climatic reconstruction, although studies of long-term lake ontogeny (Likens 1985; Ralska-Jasiewiczowa *et al.* 1988; Johnson & Odada 1996) and of human impact on lakes (e.g. Verschuren *et al.* 2002) continue to appear. Over the past 20 years, palaeolimnology has changed from a minority scientific interest to a highly technical, multidisciplinary subject, generating quantitative reconstructions of numerous environmental factors (Last & Smol 2001c).

18.3   THE SOURCES AND ORIGINS OF LAKE SEDIMENTS

The material which contributes to the formation of lake sediments (Tables 18.1 and 18.2), and the processes by which it reaches the lake bottom, are also very diverse. Even now, after more than 75 years of scientific attention, they still contain much more information than can be presently extracted from them, even using the most modern analytical techniques (Moss 1980).

Allochthonous matter (Table 18.1) is that which originates outside the lake basin, in the catchment of the lake, and in the atmosphere above. It consists both of organic and of inorganic material, which enters the lake in dissolved, colloidal or suspended form, either by direct precipi-

Table 18.1   The main types of materials contributed to lake sediments by catchments. (After O'Sullivan 1995a.)

	Inorganic	Organic
Dissolved	Gases – O_2, CO_2, N_2 Major ions – Na^+, K^+, Ca^{2+}, Mg^{2+}, Cl^-, CO_3^-, SO_4^{2-}, $[HCO_3]^-$ Minor ionic constituents (NO_3^-, NH_4^+, PO_4^-) from natural vegetation, farmland, sewage treatment works, industry, etc. Other metal ions, e.g. Fe_3^+, heavy metals	Organic products of the terrestrial environment (e.g. sugars, alcohols) Other water soluble lipids such as those found in soluble animal and human wastes, e.g. urea, uric acid, etc.
Colloidal	Clay minerals from eroding soils Industrial wastes of clay mineral size (e.g. china clay waste)	Humic and fulvic acids Organic wastes from farms, sewage and other treatment works Petrochemicals
Suspended	Sand, silt and clay from stream banks and bed Eroded soil particles Solid waste, e.g. from mines, spoil heaps, quarries Atmospheric fallout	Particulate organic wastes from eroding soils, farm drains, sewage outfalls, etc. Detrital plant matter (e.g. leaf litter)
Bedload	Pebbles, cobbles, boulders Discarded consumer durables	Branches, logs, tree trunks, animal carcasses

Table 18.2   The main types of materials contributed to lake sediments by lake ecosystems. (After O'Sullivan 1995a.)

Biogenic skeletal matter	Biogenic silica (from diatoms, chrysophytes, sponges, other organisms) Biogenic carbonates (calcite, aragonite from molluscs, ostracods) Polysaccharides (e.g. chitin)
Plant and animal waste products	Organic copropel (faeces of lake animal biota) Vegetable matter – dead and decaying algal cells and colonies, macrophyte tissues
Chemically precipitated and other 'scavenged' material	Flocculated organic (i.e. humic and fulvic) and inorganic matter (e.g. metal ions and phosphorus adsorbed onto sinking particles of humic matter and/or clay)

tation onto the lake surface or via inflowing streams. Substances of biotic origin which contribute to this particulate allochthonous fraction include leaves, stems, twigs, and fruits and seeds of terrestrial plants, and their pollen and spores.

These are joined by suspended sand, silt, clay and humic particles eroded from catchment soils, and from the channels of inflowing rivers. Animal matter does not normally contribute significantly to allochthonous input to lakes. Dissolved in the stream will be important inorganic chemical species such as major ions (Na^-, K^+, Ca^{2-}, Mg^{2-}, Cl^-, CO_3^-, $[HCO_3]^-$ and SO_4^{2-}), and ecologically significant minor constituents (nitrogen (mainly NO_3^-) and phosphorus (mostly PO_4^-)). Major biomolecules generated by terrestrial ecosystems — sugars, alcohols, monocarboxylic acids, amino acids — are also present in dissolved organic form.

Diagenesis of particulate organic matter begins before detachment from the parent plant (de Leeuw & Largeau 1993). Substances may begin transport to the lake as particulate (bound) terrestrial organic matter, but arrive there as dissolved organic or (having been mineralised) inorganic material, or (after mineralisation and subsequent uptake by the river ecosystem), as particulate aquatic organic matter. Other organic material may be humified, or further humified.

Inorganic substances are also mineralised and converted from particulate to dissolved form, and may also be recycled in inflowing streams. Atmospheric precipitation also brings major ions, sometimes enhanced by 'acid rain', as well as dust and other particles from wind-blown soils and from industrial processes, especially power generation, in the form of 'fly ash'.

Autochthonous matter (Table 18.2) originates in the lake basin itself, and consists partly of plant matter (the dead and decaying cells of aquatic algae, fragments of leaves and other tissues from the periphyton). Other biogenic matter includes the siliceous **frustules** of diatoms, and the scales and cysts of chrysophytes, the calcareous shells of molluscs and some crustaceans, the chitinous skeletal components (**exuviae**) shed by other crustaceans and insect larvae, and the faecal pellets of copepods and other planktonic invertebrates.

Animal matter therefore does make significant contribution to the autochthonous component.

Precipitated chemically are substances derived from solutes with which the water column periodically becomes saturated (Chapter 4). These include calcite ($CaCO_3$) and aragonite ($CaCO_3$) flocculated during periods of high rates of photosynthesis in warm surface layers, ferric oxyhydroxide ($FeOOH$) and hydroxide [$Fe(OH)_3$], which often give lake sediments their characteristically brown colour, and black iron sulphides (mackinawite (FeS), pyrite (FeS_2), greigite (Fe_3S_4)) deposited in deoxygenated bottom waters. Clay–humic colloids are either brought in by inflowing rivers and precipitated by changing lakewater chemistry, or formed within the lake itself.

Precipitation of flocculated organic and inorganic substances is an important process for lake biogeochemistry (and hence ecology), in that as they sink through the water column they scavenge metal ions (major ions, manganese (as Mn^{4-}), iron (as Fe^{3-})) and phosphorus. Thus important nutrients are lost to the lake ecosystem. These may only be recovered if deoxygenation of bottom waters, and hence the SWI, later develops (Deevey 1984).

18.4   SEDIMENTATION PROCESSES IN LAKES

The character of any given fraction of lake sediment is as closely related to its depositional history as to its original composition and place of origin within the lake watershed system. Interpretation of palaeolimnological results is as much concerned with the **pathways** by which particular substances reach the sediment as with their original character and provenance.

Material entering the lake from its catchment, or generated in the water column, still passes through many processes before it reaches the sediments (Fig. 18.1). Some allochthonous matter (e.g., quartz (SiO_2)) takes no significant part in biological or chemical processes, but proceeds by 'route one', directly to the minerogenic matrix of the sediments. Dissolved allochthonous matter is usually

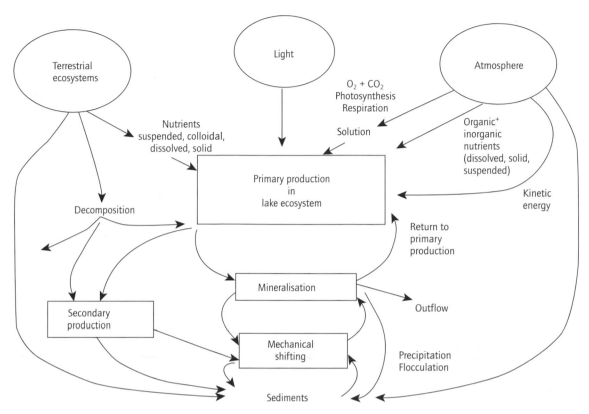

Fig. 18.1 Processes and routes by which material reaches lake sediments. (From O'Sullivan 1995a; reproduced with permission of Elsevier Science. Based on an original unpublished diagram by Dr Maureen Longmore.)

taken up by primary production, passed along the food chains, and ultimately returned to the water as dead algal and other plant matter, or as excretion products.

It then falls as detritus to the SWI, which is that part of the lake where most decomposition and mineralisation of organic matter takes place. Elements bound in dead organic matter are recovered by microbial and chemical breakdown, and reused by the ecosystem. In this sense, the SWI fulfils the same role in lakes as the soil surface in the terrestrial environment. Detritus is remineralised and reclaimed by lake biota. Alternatively, it may be resuspended by lake circulation, shifted physically ('reworked') elsewhere within the lake, further mineralised, reabsorbed, passed round the

ecosystem, and returned to the SWI. Some material follows this sequence many times before finally reaching the sediments. Other allochthonous material is not taken up immediately by primary producers, but is utilised by decomposers and/or detritus feeders. It therefore becomes incorporated in the lake ecosystem via secondary production. Alternatively, it may be decomposed, or otherwise transformed diagenetically (physically or chemically), and then taken up by the lake ecosystem. However, it may also be rendered chemically inert, and so pass directly to the SWI.

In many lakes, fine material is reworked from shallower waters into the profundal zone, a process known as **sediment focusing** (Davis *et al.* 1984; Hilton 1985). This takes place constantly in shallow, turbulent, isothermal lakes, but reaches a maximum in deeper lakes during 'overturn', or circulation after periods of stratification. Changes in lake level give rise to, and are often identified by, layers of reworked, older material embedded in younger sediments. Sediment accumulation rates are seldom even across the bottom of lakes (Dearing 1983), except in small basins which are steep-sided and deep for their surface area, in which sediment settles evenly across the lake bottom. Such lakes are therefore favoured for palaeolimnological investigations (Bennett & Willis 2001; Birks 2001), and are also more likely to contain annually laminated ('varved') sediments (O'Sullivan 1983; see 18.6.5.8).

18.5   THE OPERATIONAL FRACTIONS OF LAKE SEDIMENTS

Lake sediments therefore represent a mixture of a very wide range of substances indeed, most of which experience a complex history of formation, mobilisation, transport and deposition. Although a basic subdivision into allochthonous and autochthonous matter is generally sufficient for discussion of sediment origins, for analytical purposes further fractions are defined.

The model of lake sediment origins used by palaeolimnologists was initially developed by Mackereth (1965, 1966), in the English Lake District. He regarded sediments as a series of soils washed into lakes from the catchment surface, composed of chemically stable, highly oxidised matter. He confined his analyses to inorganic components and to total element concentrations ($mg\,g^{-1}$ dry matter (DM), or $mg\,g^{-1}$ mineral matter). Mackereth's model was refined by Engstrom & Wright (1984), since when it has become customary for palaeolimnologists to distinguish three major fractions of sediment, namely the **authigenic**, **biogenic** and **allogenic** components (Fig. 18.2).

Authigenic matter is produced by chemical precipitation and flocculation in the lakewater column (see above), and consists of the most mobile (often the ionic) forms of each element present. It is frequently adsorbed onto colloidal particles (clay, humus, other substances) in the form of ligands, and is the fraction most easily leached from the sediment matrix by weak acids or other reagents. Much authigenic material first enters the lake in solution via precipitation, or stream water, so that it is in fact ultimately allochthonous in origin. It consists, however, of material **fixed as sediment**, by chemical and physical processes within the lake-water column itself.

Biogenic material, in contrast, has passed through food chains, either in the lake ecosystem itself or the catchment. It has therefore been fixed by organisms, of whose partly decayed tissues it is largely composed, and records the history of the biota of the lake and its watershed. It consists mainly of alkali soluble material, and in many lakes biogenic silica, fixed largely by diatoms, forms a significant part of this component (Conley & Schelske 2001). **Allogenic** material is mainly minerogenic, largely crystalline and refractory, and thus much less easily extracted chemically from the sediment. Its presence constitutes a record of events taking place in the catchment (and, in certain lakes, the atmosphere).

In one sense, the above divisions are somewhat arbitrary. Movement of material between the biogenic and authigenic fractions is frequent (Fig. 18.2) and, in the case of elements such as phosphorus, rapid. In many cases, diagenesis continues to operate after burial, creating geochemically important changes in structure and composition of organic molecules and those inorganic compounds sensitive to changes in redox. It may therefore be necessary to define a further fraction (here termed the **endogenic** component), to allow for the production of new molecules within the sediments.

A note of caution: the terms authigenic and endogenic are used by mineralogists and sedimentary petrologists in a rather different way to that used here and by Engstrom & Wright (1984). Last (2001b, pp. 143–4) states that, in mineralogy, **allogenic** refers to a detrital component originating outside the lake, whereas **authigenic** signifies formation **within the sediment**, a process which

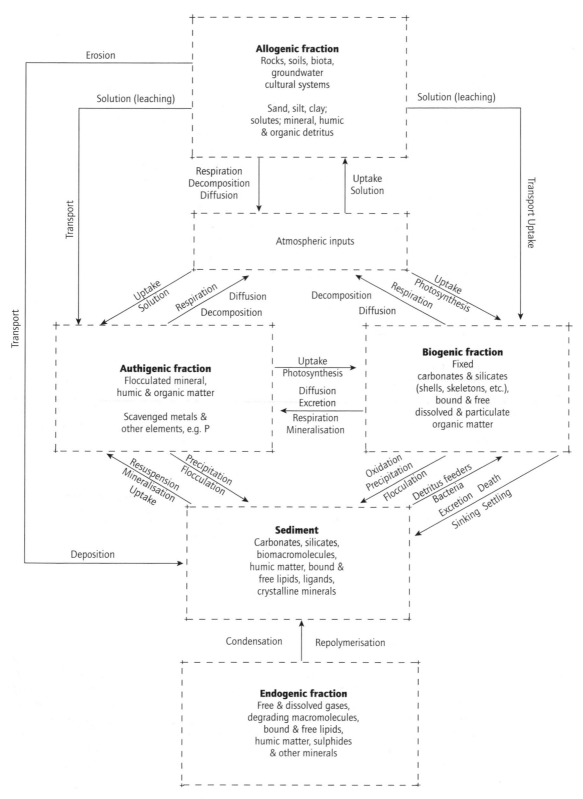

Fig. 18.2 Relationships between the various operational fractions of lake sediments, as defined by Engstrom & Wright (1984). (From O'Sullivan 1995a; reproduced with permission of Elsevier Science.)

continues long after deposition. In contrast, minerals formed in the water column of the lake are deemed **endogenic**, a term used here to signify the production of chemical species by post-depositional diagenesis. Thus, whereas allogenic is used in basically the same way in either context, authigenic and endogenic are given more or less opposite meanings.

Boyle (2001) states that, in fact, lake sediments contain only two fractions of significantly different origin, an **extractable** (or labile) component and the **residual**. The second consists mainly of inert, insoluble silicates originating in the soils, subsoils and bedrock of the catchment, and therefore corresponds exactly to the allogenic fraction of Engstrom & Wright (1984). In contrast, the extractable component is made up of porewater ions, sorbed ions, decaying lacustrine plant and animal matter, biogenic silica, carbonates, sulphides, humic compounds, iron and manganese oxyhydroxides, and trace elements, and is therefore authigenic, biogenic **and** allogenic in origin.

Engstrom & Wright's model still represents a useful framework, in that it allows more sophisticated identification of sediment composition than total element chemistry, and is therefore sensitive to changing sediment **source**. It enables investigators to distinguish sedimentary material originating exclusively in the catchment from that fixed (in its present form) almost entirely within the lake. The model therefore allows us to reconstruct not only lake and catchment ontogeny but also the interaction between processes on the catchment surface and their effects 'downstream' within the lake. Along with techniques for dating near-surface sediments (section 18.6.5), it enables results to be expressed as fractional sediment **influx** ($mg DM cm^{-2} yr^{-1}$), which, unlike concentration (see above), is not subject to perturbation by changes in sediment accumulation rate. Other processes (e.g. reworking, dilution by influx of detrital, catchment material) can also contribute to such change, however. In deep lakes with large volumes, dilution factors are very large, rendering the sedimentary record insensitive to some short-term changes (Boyle 2001).

18.6 METHODS OF ANALYSING LAKE SEDIMENTS

Detailed discussion and explanation of palaeolimnological methods developed over the past 75 years occupy more than 2000 pages of the recently published state-of-the-art summary (Last & Smol 2001a,b; Smol et al. 2001a,b), so they are clearly far too numerous to describe in detail here. Instead, Table 18.3 sets out **in outline** the main types of analysis used by palaeolimnologists. These consist mainly of techniques which:
• prepare the way for coring;
• address the physical, chemical or biological properties of lake sediments;
• determine the rates at which they accumulate;
• provide for statistical and numerical analysis of the data produced.
Many of these, especially the physical and chemical methods (Boyle 2001; Leavitt & Hodgson 2001; Zolitschka et al. 2001), have recently become automated and often require expensive equipment (Sandgren & Snowball 2001). In the absence of funds, simpler, more traditional methods may be used. Statistical and numerical methods of analysing the data are the subject of a forthcoming volume (Birks et al., in press).

18.6.1 Pre-coring exercises

Site information gathered before coring helps to identify appropriate coring locations and allows more precise interpretation of the sedimentary record. Bathymetric surveys give first approximations as to likely places to core, and those to avoid, although factors such as sediment focusing still need to be considered. Lake sediments are more variable in space and time than those found in the oceans, so that, particularly in larger lakes, especially those of tectonic origin, seismic surveys are essential (Scholz 2001). For many lakes, a fund of detailed limnological observations (water residence time, nutrient concentrations, phytoplankton succession) may exist spanning several decades (Maberly et al. 1994), and can be used to characterise their seasonal hydrological, chemical and biological regime. Catchment land-use data

Table 18.3 Outline of main methods of lake sediment analysis used by palaeolimnologists. (Based on O'Sullivan 1995a; Last & Smol 2001a,b; Smol et al. 2001a,b.)

Purpose	Proxy/determinand/method	Information obtained	Reference(s)
Pre-coring exercises:			
bathymetric survey	Lake depth, slope, morphometry, etc.	3D image of lake basin. Identification of coring locations	
seismic survey	Sediment thickness, distribution, and 'macrostructure' – location of bedding, unconformities, etc.	3D image of sediment thickness and distribution. Indications of past lake level changes	Scholz 2001
limnological data collection	Data on modern limnology of site, and on catchment land use	Identification and calibration of recent sedimentary events. Identification of nutrient and other sources. Calibration of microfossil transfer functions	O'Sullivan et al. 2001; Battarbee et al. 2001
collection of historical records of lake and drainage basin	Data on written history of lake and its drainage basin (preferably quantitative)	Changes in lake level, water residence time. Incidence of floods and other hazards. Catchment hydrology, land use and other economic history. Nutrient loadings. Pre-settlement, 'baseline' conditions	O'Sullivan 1992, 1994b, 1998
Physical analyses:			
core logging	Visual inspection and recording, photography. Digital colour imaging, (automated) whole core logging (water content, grain size, χ (below), XRF)	Log of core properties and characteristics. Automated, digital, 'objective' record of whole core properties	Nowaczyk 2001; Troels-Smith 1955; Zolitschka et al. 2001
analysis of sediment bedding and (microscopic) sediment fabric features	Digital imaging, X-radiography, CT scans, analysis of thin-sections, SEM, BSEI (BSEM), SEI	Recognition of bedding features, internal sediment structure and composition. Characterisation of laminations and sedimentary markers	Merkt & Muller 1999; Kemp et al. 2001
	Image analysis	Recording and quantification of sediment structure. Thickness and numbers of laminae, etc.	Cooper 1997; Saarinen & Peterson 2001
	Geochemical analysis	Identification and provenance of 'invisible' microtephras. Tephrochronology (see below)	Turney & Lowe 2001; van den Bogaard & Schminke 2002
textural analysis	Grain size, particle size, particle shape, sediment form and fabric	Sediment source and transport mechanisms. Depositional environment. Lake and catchment palaeohydrology	Last 2001a; Talbot & Allen 1996
mineralogy (mainly XRD, XRF plus SEM, optical methods)	Mineral composition of inorganic fraction, clay mineral content	Catchment soil profile development. Catchment erosion. Water column chemistry. Sedimentary diagenesis	Last 2001b

Table 18.3 *Continued*

Purpose	Proxy/determinand/method	Information obtained	Reference(s)
mineral magnetism	Whole core, volume susceptibility (κ)	Core logging, inter-core correlation, identification of key horizons	Nowaczyk 2001
	Low and high frequency single sample susceptibility (χ_{LF}/χ_{HF})	Magnetic mineral concentration, form and grain size. Sediment source	Dearing 1999b; Sandgren & Snowball 2001
	Frequency-dependent susceptibility $[(\chi_{LF} - \chi_{HF})/\chi_{LF}]\%$	Magnetic grain size in very fine (topsoil) range	Dearing 1999a
	Anhysteretic remanent magnetisation (ARM)	Magnetic grain size	Snowball 1999; Walden 1999a
	Isothermal remanent magnetisation (IRM, SIRM, forward and reversed IRMs)	Magnetic mineral concentration, form and grain size. Sediment source	Sandgren & Snowball 2001; Walden 1999b
	Ratios (e.g. χ_{fd}, ARM/χ, SIRM/χ. IRM/SIRM, 'S')	Composition of magnetic mineral assemblage. Component source	Dearing 1999a; Maher et al. 1999; Oldfield 1991; 1999a,b
wet/dry mass ratio (110°C, 24 h)	Water/dry matter content	Compaction/fluidity. Identification of key horizons	
ignition (550°C, 4 h)	Ash content/ignition loss	Mineral matter/ organic matter content. Identification of key horizons	Heiri et al. 2000
Chemical analyses: sequential inorganic chemistry	Allogenic Na, K, Mg (also Al, Si, Fe, Mn)	Erosion indicators	Bengtsson & Enell 1986; Engstrom & Wright 1984; Mackereth 1965, 1966
	Authigenic Mg, Ca	Palaeosalinity	Dean 1999
	Authigenic Fe, Mn oxygenated SWI periodically deoxygenated SWI	Palaeoredox catchment soils lake palaeoredox	Boyle 2001; Engstrom & Wright 1984; Mackereth 1965, 1966
	Authigenic C, N, P, biogenic N, P, biogenic silica	Eutrophication, lake silica depletion	Boyle 2001; Conley & Schelske 2001
	Allogenic N, P	Inwash of inorganic nutrients	Holloway et al. 1998; O'Sullivan 1994a
	Heavy metals	Industrialisation	Boyle 2001; Renberg 1986
	Authigenic Al, Mn	Acidification	Boyle 2001
	Sr/Mg, Sr/Ca	Palaeosalinity, palaeohydrology	Holmes 2001; Ito 2001
X-ray fluorescence (XRF)	Bulk elemental composition of solid sediment	As above, but totals only. Useful for some elements not easily dissolved (e.g. Sn)	Boyle 2001; Cousen & O'Sullivan 1984
organic geochemistry	Total organic carbon (TOC).	Lake productivity	Meyers & Teranes 2001;
	Bulk C/N ratio	Aquatic/terrestrial inputs	Meyers & Teranes 2001;
	Lipid 'biomarkers' (e.g. *n*-alkanes, fatty acids, sterols/stanols, hopanoids)	Aquatic/terrestrial inputs, hypolimnetic oxygen concentration, bacterial input	Meyers 1997; Meyers & Ishiwatari 1993; Meyers & Teranes 2001;
	Persistent organic pollutants – POPs (chlorinated pesticides, PCBs, PAHs)	Industrial pollution	Blais & Muir 2001

	Petroleum hydrocarbons	Industrialisation	Meyers & Teranes 2001
	Lignin oxidation products (C/V, S/V ratios)	Source of terrestrial plant inputs	Meyers & Teranes 2001
	Macromolecules (HI, OI)	Source of components of high molecular weight (HMW) fraction (UCM)	De Leeuw & Largeau 1993; Meyers & Teranes 2001; Tegelaar et al. 1989
	Sedimentary pigments	Algal and bacterial biomarkers, food-web interactions, UV anthropogenic impact	Leavitt 1993; Leavitt & Hodgson 2001; Sanger 1988; Züllig 1990
stable isotope chemistry	$\delta^{13}C$, freshwater carbonates	Lake productivity, atmospheric CO_2 concentration	Ito 2001; Siegenthaler & Eicher 1986
	$\delta^{13}C$, bulk organic matter	Lake productivity, changes in nutrient availability	Meyers & Teranes 2001
	$\delta^{15}N$, bulk organic matter	Lake nutrient status, nutrient cycling, nitrogen fixation, (cyano)bacterial input to sediment	Meyers & Teranes 2001; Talbot 2001
	$\delta^{18}O$, freshwater carbonates, biogenic silica, sedimentary cellulose	Source of lake waters, precipitation, lake palaeohydrology	Ito 2001; Barker et al. 2001; Conley & Schelske 2001; Wolfe et al. 2001
	Compound specific $\delta^{15}N$, $\delta^{18}O$, δD, $\delta^{14}C$	Source of lipid biomarkers (aquatic/terrestrial, C_3/C_4 plants)	Meyers & Teranes 2001
Biological methods: plant microfossils	Pollen and spores	Catchment, lake and/or regional vegetation history	Bennett & Willis 2001; Berglund & Ralska-Jasiewiczowa 1986; Prentice 1988
	Other palynomorphs	Catchment, lake, and/or regional vegetation history	Cronberg 1986; van Geel 2001
	Plant phytoliths	Catchment, lake, and/or regional vegetation history	Piperno 2001
	Microscopic charcoal	Regional fire history	Tolonen 1986a; Whitlock & Larsen 2001
	Conifer stomata	Catchment vegetation history	MacDonald 2001
plant macrofossils	Plant macrofossils	Lake basin vegetation history	Birks 2001; Collinson 1988; Wasylikowa 1986
autochthonous microfossils	Macroscopic charcoal	Catchment fire history	Tolonen 1986a; Whitlock & Larsen 2001
	Diatoms	Lake pH, salinity, nutrient status, DOC	Battarbee et al. 2001; Bradbury 1988; Stoermer & Smol 1999
	Chrysophyte scales, cysts	pH, eutrophication, salinity, climate	Cronberg 1986; Smol 1988; Zeeb & Smol 2001
	Sponge spikules	Salinity, silica availability	Frost 2001
	Protozoans	pH, trophic status	Tolonen 1986b; Douglas & Smol 2001
	Ostracod(e)s	Lake trophic history, palaeosalinity, Climate change	Carbonel et al. 1988; De Deckker 1988; Holmes 2001; Loeffler 1986
	Cladoceran remains	Lake trophic status, pH	Frey 1986, 1990; Korhola & Rautio 2001; Whiteside & Swindoll 1988

continued on p. 620

Table 18.3 *Continued*

Purpose	Proxy/determinand/method	Information obtained	Reference(s)
	Chironomid remains	Lake productivity, lake hypolimnetic oxygen status	Frey 1990; Hofmann 1986 1988; Walker 2001
Chronological methods:			
artificial fallout radionuclides	^{137}Cs, ^{241}Am	Last 50 years	Appleby 2001; Appleby et al. 1979, 1986, 1991; Oldfield & Appleby 1984; Pennington et al. 1973, 1976
naturally occurring radioactive isotopes	^{210}Pb	Last 150 years	Appleby 2001; Oldfield & Appleby 1984 Nijampurkar et al. 1998
	^{32}Si	ca 1000 years	
	^{14}C	500–50,000 years BP	Björck & Wohlfarth 2001; Hedges 1991 King & Peck 2001; Thompson 1973, 1991; Thompson & Oldfield 1986
palaeomagnetism (NRM)	Declination/inclination/Intensity	0–10,000 BP	
varves	Ferrogenic	Period of formation of deposit ($\langle 10^1\rangle$)	Anderson & Dean 1988; Dean et al. 1999; Lamoureux 2001; O'Sullivan 1983; Saarnisto 1986; Simola 1992; Sturm 1979
	Biogenic	10^5 years)	
	Calcareous		
	Clastic		
tephras	Geochemical 'fingerprinting'	Instantaneous marker horizons	Turney & Lowe 2001;
fly ash	Spheroidal carbonaceous particles (SCPs), inorganic ash spheres (IASs), 'soot'	Increase, expansion and decline based on local/national/regional historical emission record	Renberg & Wik 1984; Wik & Renberg 1996; Rose 2001
historical records	Economic and social records, especially quantitative land use data, mining and logging records, tax returns (of course!)	Duration of record	Bradbury & Waddington 1973; Davis et al. 1973; Dearing et al. 1987, 1990; Krug 1993; O'Sullivan 1992, 1994b, 1998
Data analysis:			
'unmixing'	Cluster analysis, principal components analysis (PCA), (Q-mode) factor analysis	Identification of sediment source and composition	Lees 1999; Boyle 2001
zonation of microfossil diagrams, definition of subgroups, comparison of sequences	CONISS	'Objective' description of sequences, identification of major components of assemblages, comparison between sites	Bennett & Willis 2001; Birks & Gordon 1985; Birks 1986; Grimm 1993; Prentice 1988
	PCA, rarefaction analysis, sequence slotting		
transfer functions	Canonical correspondence analysis (CCA)	Identification of species-environment relationships	ter Braak 1986, 1987
	Weighted averaging (WA), WA-(partial) least squares (WA-PLS, WA-LS)	Calculation of species optimal ranges, 'downweighting', identification of key environmental variables	Birks et al. 1990; Birks 1995; ter Braak & Juggins 1993
	Root mean squared error of prediction (RMSEP), 'jackknifing', 'bootstrapping'	Predictive ability of transfer functions	ter Braak & Juggins 1993

can be used to model past and contemporary nutrient loads (O'Sullivan 1992, 1998). Such information is not available everywhere but, where it exists, clearly enhances the potential of a particular site for palaeolimnological study.

Information on past events which have influenced the lake and its catchment (changes in hydrological regime and catchment land use, often leading to changes in nutrient loading and sediment inputs) are immensely helpful in interpreting the palaeolimnological record, especially if extended series of quantitative data are available (e.g. Dearing *et al.* 1987, 1990; Krug 1993; O'Sullivan 1998). Historical information also facilitates identification of pre-settlement or 'baseline' conditions, and the state to which lakes **could** be restored if so desired (O'Sullivan 1992, 1994a).

18.6.2 *Physical analyses of lake sediment*

Techniques for core logging, analysis of sediment bedding and fabrics, and analysis of resultant data are amongst those which have recently been automated and digitised; they are also greatly enhanced in terms of speed, power and resolution. Their objective is to identify 'the sequence and recurrence of depositional processes and events which have produced the sediment record', and to 'identify and interpret the bedding features within the core' (Kemp *et al.* 2001, p. 8), with a view to 'developing an analytical strategy for the investigation' (Kemp *et al.* 2001, p. 7). Many were developed for analysis of annually laminated sediments (Cooper 1997; Dean *et al.* 1999) but can identify fine structure even in apparently homogeneous sediment (Kemp *et al.* 2001). Determination of **whole core volume susceptibility** (κ) has now become routine in many laboratories (Nowaczyk 2001; Sandgren & Snowball 2001), and can be especially useful for initial identification of changes in gross sediment stratigraphy and for intercore correlation based on peaks produced by horizons of greater magnetisability. This is generally followed by core logging, documentation and imaging (see above).

Studies of sediment texture and mineralogy are used to investigate sediment source and transport mechanisms, past site conditions and climate, and

lake palaeohydrology (Last 2001a,b), especially in saline bodies, where they produce powerful records of lake ontogeny. Mineralogy of inorganic, detrital, allochthonous matter is used to study sediment origin, transport and deposition, intensity of catchment weathering and mass movement, catchment hydrology, floods, shoreline erosion and aeolian processes, past changes in catchment and lake basin size and morphometry, and tectonic and climatic regime.

The sediments of marl lakes, and brackish, saline and hypersaline bodies, contain very complex endogenic mineral assemblages. Combined with stable isotope studies, either of bulk sediments, or inorganic carbonates, or carbonate skeletons of certain microfossils (section 18.6.4.10), these are used to reconstruct lake palaeoalkalinity and palaeosalinity (Ito 2001). Studies of sedimentary carbonates are used to reconstruct past lake productivity, and also atmospheric CO_2 concentration. In saline lakes (because of extreme conditions for biota), these techniques offer the main basis for palaeoenvironmental reconstruction, recording changes in lakewater chemistry, salinity, lake level change, and hydrological and climatic regime (Last 2001b).

More rapid physical analyses of lake sediments, which are also applicable to many other environmental materials, are based on the study of their mineral magnetic properties (Thompson & Oldfield 1986; Dearing 1991a,b, 1999a,b; Oldfield 1991, 1999a,b; Dekkers 1997; Maher *et al.* 1999; Walden *et al.* 1999; Sandgren & Snowball 2001), the more commonly used of which (to date) are set out in Table 18.3. The theory behind environmental magnetism, the terms used to describe various types of magnetic behaviour and the use of hysteresis measurements in order to define them are discussed by Thompson & Oldfield (1986), Oldfield (1991) and Walden *et al.* (1999). **Mineral magnetism** is allied to, but distinct from, **palaeomagnetism** (the study of natural remanent magnetisation (NRM), the signal of the Earth's magnetic field; section 18.6.5.4).

Lake sediments contain a complex mixture of primary and secondary magnetic minerals, with such diverse origins as volcanic activity, natural

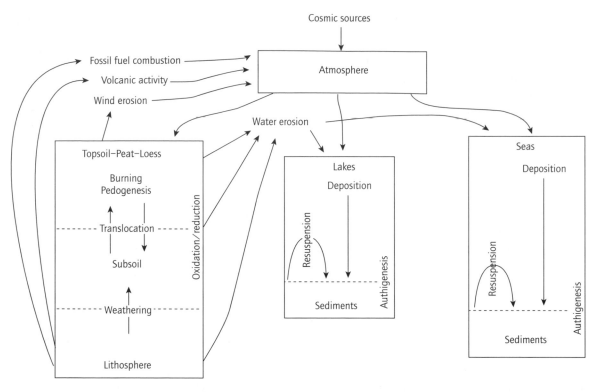

Fig. 18.3 Sources and pathways of transportation of magnetic minerals in nature. (From Thompson & Oldfield 1986; reproduced with permission of Harper Collins.)

fires, anthropogenic combustion, atmospheric pollution, catchment lithology, catchment weathering and land-use regime, lake bottom water and lake sediment oxygen availability, and lacustrine bacterial activity (Fig. 18.3; Dearing 1999a; Oldfield 1999a). Unravelling the origins and composition of this mixture is the basis of the technique. Much early work (Thompson & Oldfield 1986) assumed that the mineral magnetic signal of lake sediments is carried by detrital allochthonous material. Authigenic contributions are now being given greater emphasis, especially in cores from the deep parts of productive lakes, from lakes whose water column is incompletely mixed, or from sites wherever accumulation rates are sufficiently rapid to bury labile material before it has been acted on by oxidation processes at the SWI (Fig. 18.4; Hilton 1987; Oldfield 1999a).

The strategy of the mineral magnetic approach (Fig. 18.5; Oldfield 1991; Dearing 1999a) is
• to characterise the forms of magnetic minerals present;
• to identify their sources and origins;
• to interpret this information in terms of lake and catchment processes and their history.
The majority of minerals are not strongly magnetic, however, but **diamagnetic** (silicates and carbonates) or **paramagnetic** (iron-silicates), and only contribute to bulk magnetic signals if ferrimagnetic mineral concentrations are low (Fig. 18.6). Water and organic matter are **diamagnetic**, and dilute bulk mineral magnetic signals in fluid material close to the SWI, or highly organic sediments. Strong magnetic behaviour is exhibited by **ferrimagnetic** (magnetite (Fe_3O_4), titanomagnetite $(Fe_3O_4–Fe_2TiO_4)$, maghaemite $(\gamma\text{-}Fe_2O_3)$ and greig-

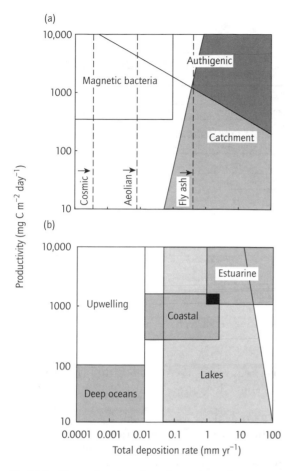

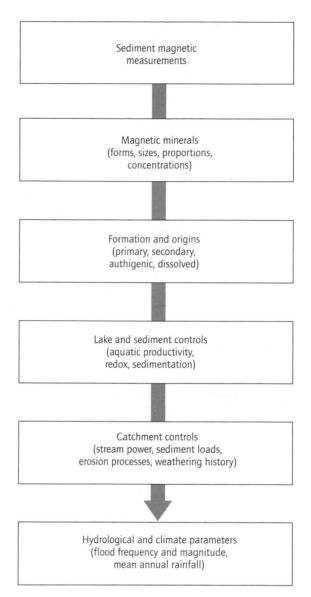

Fig. 18.4 Magnetic minerals and aquatic environments. (a) Aquatic environments and magnetic mineral source. (b) Productivity and deposition rates for various aquatic environments. Lacustrine mineral magnetic assemblages (b, right-hand side) are therefore mainly composed of material from catchment, authigenic and bacterial sources, along with a small atmospheric component, mainly 'fly ash'. (From Walden *et al.* 1999; reproduced with permission of the Quaternary Research Association.)

Fig. 18.5 Methodological strategy for mineral magnetic analysis of lake sediments. (From Dearing 1999a; reproduced with permission of Cambridge University Press.)

ite (Fe_3S_4)) and **canted antiferromagnetic** minerals (haematite (αFe_2O_3) and goethite $(\alpha FeOOH)$).

Forms of magnetic minerals are therefore identified via their response to magnetic measurements (Table 18.3). Response also depends on magnetic mineral grain size, sediment particle size, magnetic mineral concentrations and the proportions of different minerals present (see below). Ferrimagnetic minerals possess high χ (single-sample susceptibility) and low SIRM (saturated isothermal remanent magnetisation; IRM_{1T}), and are easily magnetised and saturated. In reversed fields, they lose remanence before

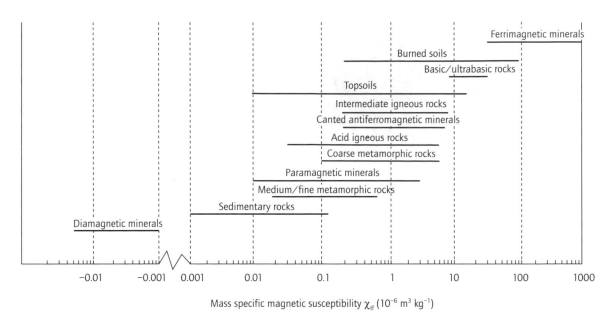

Fig. 18.6 Ranges of χ for selected environmental materials. (From Dearing 1999b; reproduced with permission of the Quaternary Research Association.)

Table 18.4 Size ranges of mineral magnetic grains of different behaviour. (From Smith 1999.)

Grain type	Symbol	Size range (μm)
Superparamagnetic	SP	<0.05
Viscous single domain	VSD	0.05–0.07
Stable single domain	SSD	0.07–0.7
Pseudo-single domain	PSD	0.7–10
Multidomain	MD	>10*

*>1, Maher et al. (1999).

IRM$_{-200mT}$ and are therefore termed magnetically 'soft'. Canted antiferromagnetic minerals possess lower χ, but higher SIRM, are less easily magnetised, or saturated, and therefore termed magnetically 'hard'. Paramagnetic and diamagnetic minerals contribute to χ, but not to IRMs.

Grain size is studied using χ (Smith 1999; Dearing 1999b, Table 18.4), although exact response depends on mineral type and grain shape. Dearing (1999b) states that χ relates first to concentration (χ_{LF} varies by 2–300; where LF refers to

low frequency), then composition (χ_{LF} varies by c.3–4) and then crystal size and shape (χ_{LF} varies by <2).

Particle size varies as much according to sorting during transport and deposition, as to the properties of magnetic minerals themselves (Dearing 1999b). Aeolian and fluvial sediments are winnowed in favour of smaller sizes, but preferentially transported in larger sizes during extreme events such as storms and floods. Offshore cores contain finer particles than those from marginal locations.

Concentration: small concentrations of ferrimagnetic minerals (e.g. stable single domain (SSD) magnetite) completely override paramagnetic components, and SIRM is also influenced by the presence of small amounts of ferrimagnetic minerals.

Proportions are studied using ratios and 'backfields' (Table 18.3). Ultrafine superparamagnetic (SP) grains (<0.03 μm), contributed mainly by topsoil, are detected by frequency dependent susceptibility (χ_{fd}, Table 18.3). Multidomain (MD) and SSD grains possess high χ_{LF}, but no χ_{fd}. Discrimination between SP grains (<0.05 μm) and bacterial

Table 18.5 Applications of lake sediment based studies of environmental magnetism. (After Dearing 1999a; Maher *et al.* 1999; Oldfield 1999b.)

Long-term (glacial–interglacial) and short-term (last 2000 yr) climate change

Identification of tephras, history of volcanic eruptions

Atmospheric pollution, surface water acidification

Detrital input to lakes from catchments – palaeohydrology, erosion, sediment source, climatic and land-use change

History of flood events

Fire histories

Lake-level changes, basin size

Whole-lake sediment budgets, patterns of sedimentation in small lake basins

Eutrophication by eroded soil inputs (edaphic, edaphic/cultural)

Changes in lake productivity and oxygen regime resulting from eutrophication, climate change

magnetosomes (0.02–0.4 µm) is possible using ARM/χ_{LF}. 'Soft' and 'hard' components are separated using $SIRM/\chi$; the higher the ratio, the 'harder' the mineral assemblage. $SIRM/\chi$ is also influenced by grain size, with SSD magnetite possessing high SIRM. The proportions of 'hard' and 'soft' components can also be studied using '*S*' ($IRM_{-100mT}/SIRM$). Any remanence due to the presence of 'magnetite' is removed before IRM_{-100mT}, whereas that resulting from haematite/goethite persists well beyond that value.

Table 18.5 lists just some of the many applications of lake-sediment-based studies of environmental magnetism. Many of these provide strong contributions to multiproxy palaeoenvironmental studies which are complementary to the results of chemical and biological analyses of sediments considered below.

More traditional means of identifying inwashed mineral or other detrital matter involve the determination of water or dry matter content (Table 18.3). Samples for mineral magnetic measurements may be used, provided that the drying temperature does not exceed 50°C. Ignition in air of dried sediment removes combustible, mainly organic matter (**loss on ignition**, LOI; Heiri *et al.* 2000) and leaves **ash** or **mineral content** of the sediment, also related to influx of mineral matter, and

trends in sedimentary organic/mineral matter content. Re-ignition for two further hours at 950°C (1000°C, Boyle 2001) allows the determination of carbonates. Where organic matter content is <10% DM, this method underestimates the true carbon content (Mackereth 1966; Boyle 2001). In many lacustrine sediments, LOI = c.2 × TOC (total organic carbon; Meyers & Teranes 2001).

18.6.3 *Chemical analyses*

18.6.3.1 Inorganic chemistry

Mackereth (1965, 1966) noted that the Late Pleistocene and early Holocene sediments of the English Lake District contained high concentrations [$mg\,g^{-1}$ DM] of the highly mobile metals sodium and potassium. These then declined during the mid-Holocene, as forests developed and soils matured, and increased again when the human cultures of that area began practising agriculture. He concluded that lake sediments consist mainly of stable, oxidised matter, and essentially represent a series of soils washed into lakes, thus providing a record of erosion of detrital matter from catchment to lake. He also proposed that iron and manganese, whose mobility responds to changing redox, could be used to study the Holocene ontogeny of catchment soils and the onset of podsolisation, provided that concentrations are determined on cores from lakes whose SWI remains permanently oxygenated. Once the SWI becomes deoxygenated (an important stage in lake ontogeny; Deevey 1955), sedimentary iron and manganese concentrations refer to the oxygen status of lake bottom waters, rather than that of catchment soils. A key diagnostic feature is departure of Fe/Mn ratios from the lithospheric values we would expect if removal from catchment to lake was entirely by erosion (Fig. 18.7; Boyle 2001).

Engstrom & Wright's (1984) model improves our ability to interpret the changes in lake sediment profiles studied by Mackereth. During periods of erosion, **influx** of allogenic sodium and potassium increase, but **concentrations** do not vary from lithospheric values. Influx of authigenic sodium and potassium may not vary at all. In-

Case 1. Biologically impoverished lake and catchment   Case 2. Productive soil; lake impoverished and permanently aerobic

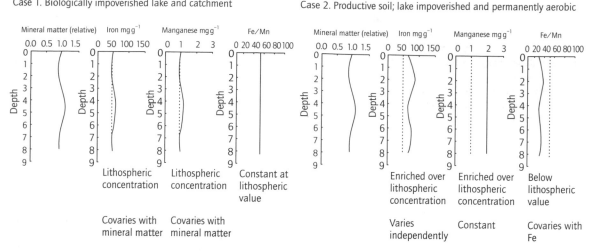

Case 3. Biologically rich soil and lake.   Case 4. Lake highly enriched, and permanently anaerobic
Sediment sensitive to hypolimnetic O_2

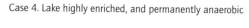

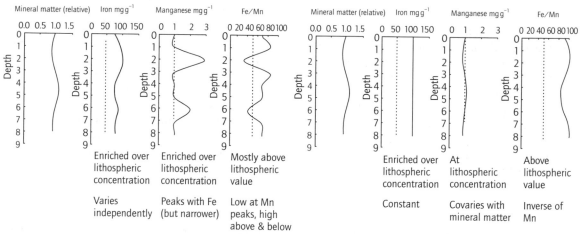

Fig. 18.7 Cartoon of Mackereth's (1966) hypothesis for interpretation of iron and manganese profiles from lake sediments. Vertical dashed lines indicate lithospheric concentrations. (From Boyle 2001; reproduced with kind permission of Kluwer Academic Publishers.)

creased mobilisation (in solution) of soil manganese, and then iron, affects the Fe/Mn ratio in the authigenic fraction, but not in the allogenic component. Phases of soil erosion are denoted by increases in influx of both allogenic iron and manganese, not in lithospheric ratios.

Boyle (2001) questions this approach on the grounds that operational distinction between the allogenic component and the authigenic and biogenic fractions is not fully justifiable. Scavenging of metal ions (and phosphorus) by humic and/or clay particles of catchment origin sinking through the water column (section 18.3) produces material which may be allogenic, authigenic and biogenic in origin. Instead, he suggests that only two chemical components of lake sediments should be distinguished, the extractable or labile fraction and the residual. Other schemes (e.g. compositional (organic, sorbed, lattice) and operational (acid extractable, reducible, oxidisable, residual)) could also be considered, although these have their own limitations. Relating chemical composition to sediment origin, or operational fractions to sediment source, is not easy.

In practice, Engstrom & Wright's (1984) method separates the sediment into three fractions, effectively (i) the acid soluble ('authigenic') component, (ii) biogenic silica (that part of the extractable fraction soluble in dilute alkali), and (iii) the minerogenic, crystalline, residual, 'allogenic' component. Certain other species (e.g. nutrients, heavy metals) sometimes also appear in fraction (ii). Such data are best used as supporting information for biological and other proxies, rather than by themselves (Boyle 2001). Total sediment chemistry may also be studied in the solid state (e.g. using X-ray fluorescence (XRF) spectroscopy). This is useful for elements such as tin (Sn) which do not easily dissolve in common reagents (e.g. Cousen & O'Sullivan 1984), or as a complement to mineral magnetic studies (Smith 1985).

Mackereth also used calcium and magnesium profiles as erosion indicators, but sedimentary calcium concentrations are nowadays related to lake productivity rather than catchment history (Dean 1999), especially in calcareous systems. Determination of the authigenic Ca/Mg ratio is increasingly used to reconstruct palaeosalinity, an important variable in the study of climatic change (section 18.6.3.3), although Boyle (2001) states that porewater diffusion means that this is not valid.

Elements used to reconstruct lake trophic status and productivity are carbon, nitrogen and phosphorus. Fractionation is used to assess the proportion of each element entering the lake in solution, and that present in the detrital, allogenic component. This distinction is used to discriminate between phosphorus (and nitrogen) supplied to the lake in solution (e.g. by sewage treatment works) and that eroded, transported and deposited in solid form (e.g. from agricultural land). In this way, it is possible to identify the **source** of nutrients contributing to different phases of eutrophica-

tion (O'Sullivan 1994a). Holloway *et al.* (1998) recently reported substantial quantities of bedrock nitrogen in some environments.

Allogenic aluminium and silicon are proxies of erosion. Authigenic aluminium may be a proxy for mobilisation of this element in catchment soils, under the influence of acidification. Biogenic silica is a proxy for lake productivity, and is also used to identify lake silica depletion and the switch from diatom to other algal production (e.g. cyanobacteria) which often takes place during advanced eutrophication (Conley & Schelske 2001). Biogenic silica as measured by these authors is not the same as determined by Engstrom & Wright's method.

Heavy metal profiles are used to study the effects of mining on lake sediments (Henon *et al.* 1999; O'Sullivan 1999), and atmospheric pollution, over the last 100–200 years. They are also used in the study of lake acidification, as indicators both of atmospheric pollution and of increased mobilisation of heavy metals, as a result of acidification. Acidification, diffusion, and surface enrichment all produce distortions of the sedimentary heavy metal record (Boyle 2001).

18.6.3.2 Organic chemistry

Organic matter in lake sediments is a complex mixture of carbohydrates, proteins, humic compounds, lipids and other biochemicals (Robinson *et al.* 1984; Cranwell *et al.* 1987; Meyers & Ishiwatari 1993; Meyers 1997; Meyers & Lallier-Vergès 1999; Meyers & Teranes 2001), derived mainly from three sources: terrestrial plants, aquatic macrophytes and algae, and aquatic bacteria. Animal material normally contributes <10% sedimentary organic matter (SOM), except perhaps in lakes with substantial bird populations. Diagenesis begins on the parent organism, and continues during transport, so that >80% of aquatic material may be remineralised before leaving the epilimnion. More resistant compounds are gradually selected, at the expense of labile matter. New compounds (degradation products) are formed by breakdown and/or alteration of the original substances, a process which continues after burial. Interpretation of sedimentary profiles is as much a question of reconstructing the diagenetic history of compounds, as of tracing their origin.

The majority of compounds present (carbohydrates, proteins, polypeptides) are easily degraded by bacterial action and chemical oxidation. Other high molecular weight (HMW) compounds (e.g. lignin (which may contribute 33% of SOM) and humic substances) are more resistant. Humic substances constitute c.60–70% of geologically 'young' lake sediment, or 90% of more 'mature' material. HMW compounds make up the major part of 'protokerogen' (Tegelaar *et al.* 1989; de Leeuw & Largeau 1993).

Specific origins of HMW compounds are more difficult to trace than those of **lipids**, which, although they make up only c.1–5% of SOM, have been studied much more extensively. The sediments of deeper, unproductive, well-oxygenated lakes are rich in terrestrial compounds which survive best in oxidising conditions, especially 'long-chain' lipids (see below). In shallow, productive lakes, a greater proportion of more mobile types of aquatic organic material is sedimented, even though bacterial activity is greater. Many 'short-chain' terrestrial compounds are mostly destroyed before ever reaching the lake.

Total organic carbon (TOC) in lake sediments is broadly related to lake productivity. Concentrations vary with distance from the shore, and are associated with particle size, clastic input and carbonate concentration. Finer sediments offshore are generally richer in SOM. The source of SOM can be broadly identified using bulk sedimentary carbon/nitrogen (C/N) ratios (Table 18.3; Fig. 18.8). Values for fresh algal material usually lie between 4 and 10, whereas those for C_3 terrestrial plant matter generally exceed 17, and for C_4 plant matter, 35. C/N ratios in lake sediments rise as inputs of terrestrial plant matter increase, and therefore can be used to identify the relative contribution of inwashed terrestrial material. It should not be forgotten that the nitrogen in this ratio is normally both organic and inorganic (Meyers & Teranes 2001).

'Biomarkers' are organic compounds with distinctive biotic sources (Meyers & Teranes 2001), which retain their molecular identity after sedimentation, or change diagenetically in ways

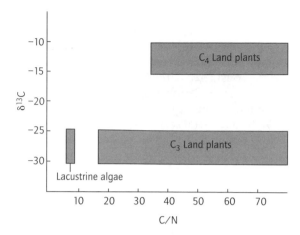

Fig. 18.8 Carbon/nitogen (C/N) ratios, and values of $\delta^{13}C$, for sedimentary organic matter (SOM) from various sources. (From Meyers & Teranes 2001; reproduced with kind permission of Kluwer Academic Publishers.)

which can be clearly detected. Many organic biomarker compounds are lipids—relatively small, simple, stable organic molecules, most of which are soluble in common organic solvents (ethanol, dichloromethane (DCM), acetone, etc.)—but these represent only a small fraction of total SOM. Diagenetic destruction of more labile compounds tends to emphasise their apparent importance.

The main sources of lipids in lake sediments are terrestrial plants, aquatic algae and macrophytes, and aquatic bacteria. Animal and human matter (sterols, stanols) may also be significant. Normal (*n*)-alkanes synthesised by aquatic algae and photosynthetic bacteria are rich in 'short chain' C_{17} molecules, whereas those produced by terrestrial vascular plants contain abundant 'long chain' C_{27}, C_{29} and C_{31} compounds, especially those which originate as epicuticular plant waxes. Chain length of fatty acids is also related to source, longer chains (24:0–28:0) being terrestrial in origin, shorter chains (15:0–18:0) being produced endogenetically (within sediments). Other means of differentiating aquatic SOM from terrestrial include determination of $\delta^{13}C$ (section 18.6.3.3) and carbon preference index (CPI), with terrestrial plants exhibiting a strong preference for odd numbered carbon chains, aquatic and other bacteria little or none. Long-chain C_{27} sterols are produced by aquatic organisms, C_{29} by terrestrial biota.

Other lipids are formed diagenetically (Meyers & Ishiwatari 1993). In deeper sediments, sterols decline relative to total SOM, as they are transformed into stanols and, ultimately, to steranes. The 5α stanols are characteristic of oxidising environments, 5β stanols of reducing conditions. Some specific compounds, e.g. cholesterol and coprostanol, also signify (respectively) inputs to lakes from higher animal and human sources. Hopanoids (tricyclic and pentacyclic triterpenoids) are formed within sediments, mainly by bacterial action upon other compounds, and are useful as bacterial markers.

Also formed naturally in sediments by diagenesis, but virtually absent from the rest of the biosphere until very recent times, are polycyclic aromatic hydrocarbons (PAHs). These are combustion products, especially of fossil fuels, both in power stations and internal combustion engines, and increase in sediments deposited since the growth of industrial society. Other organic indicators of industrial pollution are petroleum hydrocarbons and POPs (persistent organic products; Blais & Muir 2001), which include many chlorinated pesticides (aldrin, dieldrin, DDT—dichlorodiphenyltrichloroethane), polychlorinated biphenyls (PCBs) and PAHs.

Lignins, being confined almost completely to land plants, are used as tracers of inputs of terrestrial plant material. Gymnosperm lignin is differentiated from that of angiosperms by its C/V (cinnamyl/vanillyl) and S/V (syringyl/vanillyl) ratios (Meyers & Teranes 2001; Fig. 18.9). These properties are used in the reconstruction of catchment vegetation changes.

Most SOM, however, is made up of 'complex mixtures' of HMW compounds less easily isolated than lipids or lignin phenols. Rock-Eval pyrolysis of HMW matter (progressive heating at 200 to 600°C) generates hydrocarbon-like substances, which are then more easily characterised. Compounds are identified via their hydrogen (HI) and oxygen (OI) indices (Fig. 18.10). Type I OM is rich in microbial matter and higher plant waxes, whereas Type II is algal in origin, and made up

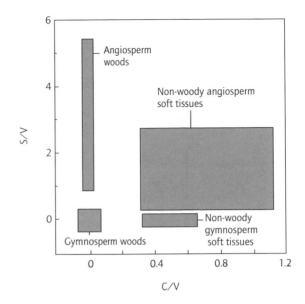

Fig. 18.9 C/V (cinnamyl/vanillyl) versus S/V (syringyl/vanillyl) ratios for lignins from various plant sources. (From Meyers & Teranes 2001; reproduced with kind permission of Kluwer Academic Publishers.)

mainly of hydrocarbons. Type III OM is poor in hydrocarbons, but rich in carbohydrates, and represents woody plant matter. Talbot & Livingstone (1989) give typical HI values of major types of SOM in lake–watershed systems. Elevated TOC concentrations and HI values are related to greater algal productivity, including diatoms, and especially cyanobacteria (Ariztegui *et al.* 1996).

18.6.3.3 Stable isotopes

Stable isotope ratios may be determined on bulk sediments, fractions (carbonates, biogenic silica, OM, cellulose, lipid biomarkers), or specific microfossils (section 18.6.4). The main ratios used are $^{13}C/^{12}C$ ($\delta^{13}C$), $^{15}N/^{14}N$ ($\delta^{15}N$), $^{18}O/^{16}O$ ($\delta^{18}O$) and D/H (δ^2H, or δD).

Sedimentary carbonates. The forms of carbonates present in lake sediments, and their potential sources (Dean 1999; Last 2001b), are very diverse, and the chemistry of the relationships between

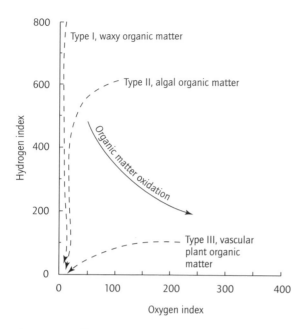

Fig. 18.10 Hydrogen index and oxygen index values for various types of high molecular weight organic matter, following Rock-Eval pyrolysis. (From Meyers & Teranes 2001; reproduced with kind permission of Kluwer Academic Publishers.)

them are highly complex (Fig. 18.11). Broadly, in lake sediments, they may be detrital (supplied by the catchment), endogenic (formed in the lake-water column), authigenic (produced *in situ* by sedimentary pore waters), or diagenetic (formed by *in situ* diagenesis of other carbonates). Biogenic and bio-induced carbonates (Ito 2001) are synthesised by biota (ostracods, molluscs), or produced by biotic influence on water chemistry (e.g. by charophytes). These are discussed in section 18.6.4.10.

Like many other components of lake sediments, therefore, sedimentary carbonates are a mixture, each component possessing a different source and $\delta^{13}C$ (Fig. 18.10). Exchange between different fractions in the lakewater column also involves isotopic fractionation, denoted in the figure by ε. Endogenic carbonate is formed in isotopic equilibrium with lake waters, and, in theory, is a proxy for lake palaeohydrology. Unfortunately, the bulk non-detrital stable isotope signal is perturbed by authigenic and diagenetic carbonates, so that identifying sources of sedimentary carbonates is a key issue in palaeolimnology. Lakes are more diverse and more complex environments than

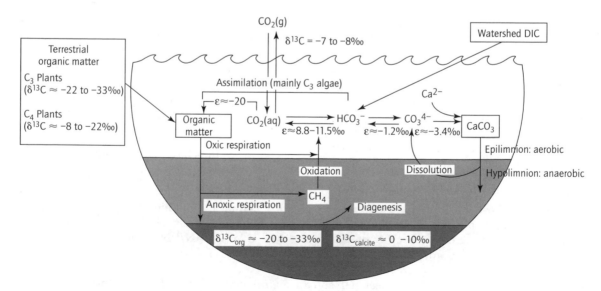

Fig. 18.11 Idealised carbon isotope cycle in a small, stratifying freshwater lake: ε, isotopic enrichment factor; DIC, dissolved inorganic carbon. (From Ito 2001; reproduced with kind permission of Kluwer Academic Publishers.)

oceans (see Chapters 2 and 3), and sedimentary carbonate records are more difficult to interpret (Ito 2001). Palaeolimnological analyses based on sedimentary carbonates require supporting information from other proxies (modern lake hydrology and chemistry) to compare with the carbonate record, and the critical identification of carbonate by scanning electron microscopy (SEM) and X-ray diffraction (XRD) (Last 2001b).

The $\delta^{18}O$ and δD contents of sedimentary carbonates are used to study lake palaeohydrology and climate change, mainly in low and mid-latitude lakes, or lakes of continental interiors. The $\delta^{18}O$ content of meteoric waters is a function of moisture source, airmass trajectory and distance from the sea (Ito 2001). The $\delta^{18}O$ and δD contents in precipitation decline owing to decreasing temperature at the site of condensation and to increasing latitude, altitude and distance from the sea (Wolfe *et al.* 2001). Evaporation from lake waters preferentially removes ^{16}O and ^1H, enriching the remaining body with ^{18}O and D. The $\delta^{18}O$ content of aragonite is greater than that of calcite, so that changes in mineralogy generate false signals (Ito 2001). Owing to evaporation, and the influence of runoff and groundwater, lake waters are rarely in isotopic equilibrium with precipitation.

The $\delta^{13}C$ content of sedimentary carbonate is a proxy of lake productivity and of atmospheric CO_2. Lacustrine carbonates (endogenic, biogenic) are synthesised from the epilimnetic dissolved inorganic carbon (DIC) pool (Chapter 4), whose ultimate source is atmospheric CO_2. Phytoplankton and macrophytes use $^{12}CO_2$ as first preference (Chapters 9, 10 and 11). Thus, $\delta^{13}C$ of endogenic carbonates reflects ^{12}C depletion of epilimnetic DIC, or 'drawdown' of atmospheric CO_2, by algal photosynthesis. Seasonal mixing and DIC inputs from rivers and/or groundwater may distort the signal (Wolfe *et al.* 2001).

Seasonal cycles of $\delta^{18}O$ and $\delta^{13}C$ in lake waters are not entirely separate. Depletion of epilimnetic $DI^{12}C$ during peak seasonal growth may be accompanied by maximum $\delta^{18}O$ (and δD) of lake waters, owing to increased evaporation and inputs of seasonally $\delta^{18}O$ enriched precipitation. This may be reflected in $\delta^{13}C$ and $\delta^{18}O$ in sedimentary carbonates.

Biogenic silica. Oxygen isotopes in biogenic silica are being developed as proxies of lake palaeohydrology and climate change. Microfossils available for this purpose include diatoms (Conley & Schelske 2001) and plant phytoliths (Piperno 2001). Studies of diatom $\delta^{18}O$, χ, diatom frequencies, pollen and green algae were used by Barker *et al.* (2001) to reconstruct lake moisture balance for the last 14,000 years for two tarns on Mt Kenya (Chapter 2). Periods of stable, enriched $\delta^{18}O$ (+27 to +37‰) coincide with phases of drier conditions, and weakening of the African and Asian monsoons. More variable, depleted $\delta^{18}O$ (<20‰ in an enclosed lake, <26‰ in an open basin) indicates increased precipitation and strengthening of the monsoon.

Sedimentary organic matter. The $\delta^{18}O$ and δD contents of bulk SOM have not been much used in palaeolimnology but they offer considerable potential for the study of past climates (Meyers & Teranes 2001). The $\delta^{13}C$ content in bulk SOM is used to reconstruct lake productivity and changes in water-column nutrient availability. As phytoplankton absorb DIC from lake waters (Chapter 4), they preferentially utilise ^{12}C, producing OM whose $\delta^{12}C$ is c.20‰ lighter than its inorganic source.

Sedimentation of dead phytoplankton depletes ^{12}C of surface waters, leading to net change in water column $\delta^{13}C$. Changes in $\delta^{13}C$ in SOM relate to past variations in phytoplankton production (e.g. due to eutrophication), although other factors (pH, temperature, nutrient limitation, growth rate) are also involved (Chapter 10). Use of bicarbonate (from lake waters or catchment runoff), or diagenetic processes, may distort the ratio (Meyers & Teranes 2001).

Nitrogen isotope ratios in SOM also contribute to the identification of source and the reconstruction of past nutrient availability and lake productivity, especially in organic-rich sediments. Interpretation is less easy than with $\delta^{13}C$ (Meyers & Teranes 2001; Talbot 2001). The $\delta^{15}N$ content of aquatic algal and macrophytic material is generally greater than that of terrestrial plant matter but this may be obscured by isotopic fractionation of DIN by phytoplankton (and heterotrophs) and by in-lake denitrification.

Most nitrogen in lacustrine sediments is located within autochthonous algal material (Talbot 2001), and reflects the $\delta^{15}N$ of the DIN from which it is synthesised. Cyanobacteria, which may operate as nitrogen fixers and utilise atmospheric nitrogen, produce a different $\delta^{15}N$ signal, as does inwashed soil material, which contains organic nitrogen fixed by bacteria. Problems of interpretation occur where diagenesis takes place and may be detected where nitrogen concentration falls and the C/N ratio rises, as more labile, nitrogen-rich, algal compounds are destroyed. Where exchangeable and bound sedimentary nitrogen occur together (e.g. after soil inwash), bulk determinations may not separate them. Effects of nitrogen inputs to sediments from diatoms have yet to be completely assessed (Talbot 2001).

Changes in sedimentary $\delta^{15}N$ over the last 100–200 years are associated with anthropogenic perturbation and acceleration of the nitrogen cycle. In catchments where modernisation of agriculture, industrialisation and urbanisation have occurred, eutrophication has led to draw down of DIN into sediments and to an increase in sedimentary $\delta^{15}N$ (Talbot 2001). In phosphorus-deficient lakes, $\delta^{15}N$ signals may be unchanged (Meyers & Teranes 2001). Shifts to low sedimentary $\delta^{15}N$ indicate an increase in nitrogen-fixing cyanobacteria. In some tropical catchments (e.g. Lake Victoria), increase of agricultural intensity using swidden ('slash-burn') has led to rising sedimentary $\delta^{15}N$.

Finney et al. (2002) used a combination of sedimentary $\delta^{15}N$ concentration and diatom indicator species (section 18.6.4.4) to study climate-driven population changes of sockeye salmon (Oncorhyncus nerka (Walb.), Gill & Jordan) in Alaskan lakes. Increases in $\delta^{15}N$ concentration and in the abundance of diatoms characteristic of more eutrophic waters denote expansion in numbers of spawning salmon bringing more nutrients into these lakes. Proxies were calibrated against observational data for recent decades.

Cellulose. The $\delta^{18}O$ content of sedimentary cellulose (SC) present in dead algal cells, zooplankton faecal pellets or amorphous plant matter is a useful proxy for lake palaeohydrology and past climates in carbonate-free systems (Wolfe et al. 2001). Although the compound is ubiquitous in the plant kingdom, most cellulose in offshore lacustrine sediments is of algal origin. Influx of allochthonous organic matter and its diagenesis may influence results. Cyanobacteria do not contain cellulose.

The $\delta^{13}C$ and δ^8O contents are determined but not δD, as some H atoms present are exchangeable. The $\delta^{18}O$ contents in SC and in tropical and temperate lake waters are practically identical, suggesting no post-depositional fractionation. Where $\delta^{13}C$ of bulk SOM and SC vary together, most of the latter is aquatic in origin. Where they vary against each other, inputs of terrestrial cellulose may have occurred. Studies of HI and OI (section 18.6.3.2) may be used further to investigate origin (Wolfe et al. 2001).

Lipids. Compound-specific (capillary) gas chromatography–isotope ratio mass spectrometry (GC–IRMS) enables isolation of individual compounds from SOM. The determination of their individual isotopic ratios gives a more precise means of recognising the origin of specific sedimentary organic compounds, their diagenetic transformation and the depositional environment than do bulk SOM or conventional biomarker analyses (Meyers & Teranes 2001). Separate organisms may synthesise the same lipid biomarker but with quite different $\delta^{13}C$. n-Alkanes from bulk organic matter, C_3, C_4 and CAM plants all possess quite different $\delta^{13}C$. The ratios $\delta^{15}N$, δ^8O, δD and $\delta^{14}C$ are also determined, although less commonly so far. The last plays a key role in compound specific accelerator mass spectrometer (AMS) radiocarbon dating ('molecular dating'; section 18.6.5).

18.6.3.4 Pigments

Plant pigments and their degradation products are non-lipid organic compounds, extractable from

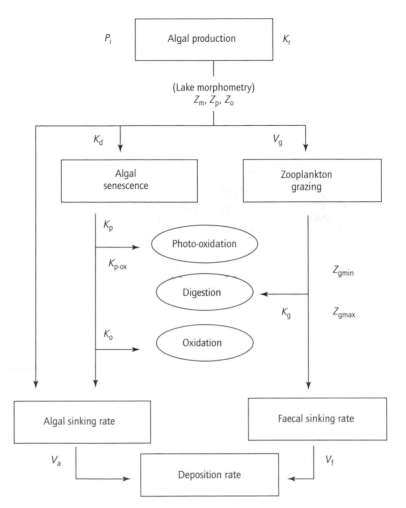

Fig. 18.12 Pathways of pigment production, diagenesis and degradation within freshwater lakes: K_d, digestion by herbivores; K_g, grazing; K_o, oxidation; K_p, depth of light penetration; K_{p-ox}, rates of photooxidation; K_r, algal cell recruitment rate; P_i, population size; V_a, algal sinking rate; V_f, faecal sinking rate; V_g, rate of ingestion; Z_{gmax}, maximum depth of grazers; Z_{gmin}, minimum depth of grazers; Z_m, maximum depth of lake; Z_o, depth of oxygen penetration; Z_p, light extinction coefficient. (Modified from Cuddington & Leavitt 1999; reproduced with permission of the *Canadian Journal of Fisheries and Aquatic Sciences*.)

lake sediments often long after all traces of their parent organisms have disappeared (Sanger 1988; Züllig, 1990; Leavitt & Hodgson 2001). Their main sources are planktonic and benthic algae, phototrophic bacteria, macrophytes, and terrestrial or other undegraded detritus. Compounds such as β-carotene, chlorophyll (Chl) *a* and phaeophytin *a* are indicators of general algal abundance, whereas several carotenoids are specific to particular groups of organisms.

Derivatives of chlorophylls and carotenoids are indicators of diagenetic pigment transformation (by grazing, anoxia, light, etc.). Degradation occurs mainly in the water column. Once buried in sediments, pigments are degraded more slowly, or they persist in stable, derivative form. Generally, as with lipids (section 18.6.3.2), simpler compounds degrade much less rapidly than complex molecules. Comparison between sites is more difficult than with other proxies, in that conditions of pigment formation (lake depth, transparency, temperature and community structure; Fig. 18.12) differ markedly from lake to lake (Leavitt & Hodgson 2001).

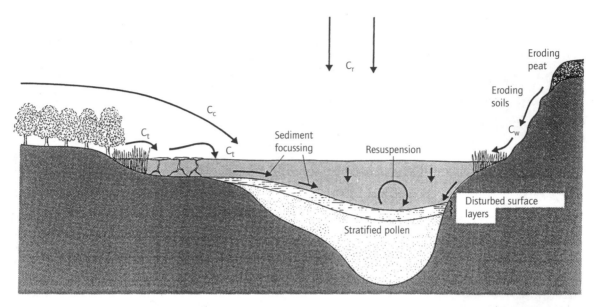

Fig. 18.13 Sources and pathways of pollen at a small lake. See text for explanation of C_r, C_c, C_t, C_l and C_w. (Modified after Moore *et al.* 1991; reproduced with permission of Blackwell Publishing Ltd.)

Fossil pigments are used to reconstruct algal and bacterial community composition, food web interactions, changes in UV radiation and anthropogenic impact (eutrophication, acidification, fisheries management, land-use practise). Certain carotenoids (echinone, zeaxathin, oscillaxanthin, myxoxanthopyll, apahanizophyll, alloxanthin, lutein) originate from specific groups of bacteria and algae (respectively, Cyanobacteria, Cryptophyta and Euglenophyta) which generally lack other fossil proxies. They are also used, as well as biogenic silica (section 18.6.3.1), to identify the switch from diatoms to cyanobacteria which often occurs during advanced eutrophication (Züllig 1982), and to reconstruct changes in food webs, especially the development of 'fish–cyanobacteria' associations characteristic of highly eutrophicated lakes (Ryding & Rast 1989). They are therefore applied to the reconstruction of events higher up the food chain than most proxies, including lake community structure and food web interactions (Leavitt *et al.* 1994).

18.6.4 *Biological analyses*

18.6.4.1 Pollen and spores

The main plant microfossils used in palaeoenvironmental reconstruction are pollen and spores of higher plants (angiosperms, gymnosperms, pteridophytes). These are washed into lakes by surface waters, blown in by the wind, or contributed locally by aquatic taxa (Fig. 18.13). Owing to their great resistance to chemical and bacterial attack (except oxidation), pollen and spores are readily sedimented and preserved in most waterlogged sediments.

Pollen analysis (or **palynology**) is the main palaeoenvironmental technique for the reconstruction of catchment vegetation history and other related change (climate, human impact, succession; Berglund & Ralska-Jasiewiczowa 1986; Prentice 1988; Faegri & Iversen 1989; Moore *et al.* 1991; Bennett & Willis 2001), and is used by palaeolimnologists to investigate terrestrial events and changes which influence lake ecosystems. Results may be expressed as pollen **percent-**

ages (of a pollen **sum**), **concentration** (grains cm^{-3} or grains mL^{-1}), or **influx** (grains cm^{-2} yr^{-1}). Most palynologists use percentage diagrams as routine, but concentration and influx give insights into vegetation changes to which these are insensitive.

Bennett & Willis (2001) recommend that sites most suitable for producing pollen diagrams which record local vegetation changes, and sedimentation of pollen in an undisturbed environment, with minimal mixing and reworking, are small lakes (<100 m diameter), deep for their surface area, with little or no stream input. Detailed studies of **aerial** dispersal of pollen in temperate forest environments (Tauber 1960; Jacobson & Bradshaw 1981) indicate that 80% of pollen collected by such sites comes from a source area <c.300 m in diameter (C_t in Fig. 18.13), and therefore records **local** vegetation changes. In basins >100 m across, 50% of pollen (C_c in Fig. 18.13) originates in an area several kilometres in diameter, and beyond 300 m, 90%. A small percentage (C_r in Fig. 18.13; c.10%?) is very far-travelled pollen (up to hundreds of kilometres), termed 'long distance transport'. Pollen diagrams from larger sites therefore record **regional** vegetation changes.

These principles apply, however, only to aerial transport of pollen in temperate and boreal forests, where most plant taxa are anemophilous ('wind pollinated'), whereas in some environments (e.g. tropical moist forests) the taxa are mainly entomophilous ('insect pollinated') and their pollen is much less easily dispersed. Such environments are palynological 'blind spots'. Within given environments, 'a single pollen sample is a representation of . . . abundance of plants with poorly dispersed pollen close to the site, and [those] with well-dispersed pollen [at] much greater distances' (Bennett & Willis 2001, p. 24). Pollen source area increases in open vegetation, so that small lakes in grassland, steppe, desert and tundra collect pollen from much wider areas. Interpretation of pollen diagrams from patchy vegetation (Sugita *et al.* 1999) is a highly complex subject.

Where lakes receive waters from inflowing streams, riverine inputs (C_w in Fig. 18.13) may reach 85–97% of total pollen, and completely 'swamp' aerial and other sources (Peck 1973; Bonny 1978). Floods bring in pollen and spores from catchment soils, with emphasis on entomophilous taxa, whose pollen falls directly on to their surface, and highly resistant pteridophyte spores. Inputs to 'enclosed' (seepage) lakes (drainage basin/lake area (D, *sensu* Hutchinson 1957) <10 : 1) are therefore more representative of aerial transport of pollen and spores, whilst 'unenclosed' drainage lakes (D > 40 : 1) are heavily influenced by streams (Bonny 1976). In temperate forested environments, c.50% of pollen input to drainage lakes is contributed by streams, but after forest clearance this rises to >80% (Pennington 1979).

Resuspension and 'reworking' of pollen (Fig. 18.13), from the littoral to the profundal zone, takes place continually in exposed or shallow sites, but operates mainly during overturn in deeper lakes (Davis 1973; Peck 1973; Bonny & Allen 1984), which leads to sediment 'focusing' (Davis & Ford 1982). Traditionally, cores for palynological investigation are collected from the deepest part of the lake, which is still 'probably the best strategy' (Davis *et al.* 1984; p. 288). However, deposition of sediment, and hence of pollen, across basins, is not even, either in time or space, and what is now the deepest part of the lake may not always have been in the past (Dearing 1983; Davis *et al.* 1984).

Differential preservation, or **deterioration** of pollen (Havinga 1984; Cushing 1964), through microbiological, chemical and physical means, occurs in catchment soils and during transport by rivers. Sometimes, and at some sites, it leads to the presence of large amounts of unclassifiable, 'deteriorated' pollen. **Redeposition** of pollen takes place from older deposits upstream. The presence of reworked pollen from the lake margin in cores from deeper parts of the lake is often a sign of changing lake levels (Digerfeldt 1986).

Small, deep lakes with no stream input therefore minimise the number of factors palynologists need to evaluate when interpreting lacustrine pollen diagrams, and produce results which record the vegetation history of specific, local areas. Large lakes with substantial drainage basins generate results of regional significance, or beyond, but are inherently complex, and the data they produce are proportionally more difficult to interpret. In multidisciplinary investigations, choice of site is not always dictated by requirements of single techniques, however, and it is often necessary to produce pollen diagrams from lakes with significant riverine inputs, where the pollen source area potentially constitutes the entire catchment, however large that may be.

18.6.4.2   Other 'palynomorphs'

Other microfossils ('palynomorphs') occur on pollen slides, including the remains of bacteria, algae (*Scenedesmus, Pediastrum, Botryococcus*, akinetes of *Anabaena* and *Aphanizomenon*), algal spores (Chrysophyta (section 18.6.4.5), Chlorophyta), invertebrate body parts (Cladocera, section 18.6.4.7) and eggs (Cladocera, Rotifera). More details of these, and the factors controlling their distribution and abundance, are discussed by Cronberg (1986), van Geel (1986, 2001) and Zeeb & Smol (2001).

The lignified stomata of conifers (and other higher plants) occur as palynomorphs (MacDonald 2001) and also as macrofossils (see next section). These are also used to reconstruct vegetation change, especially the arrival of particular species at treelines, where percentage pollen diagrams may not detect subtle changes. Also present are microscopic charcoal particles (<100 μm), used by many investigators to reconstruct **regional** fire histories (see below). Siliceous phytoliths of certain terrestrial plant taxa (grasses, sedges, gymnosperms, pteridophytes, Fagaceae, Magnoliaceae, Cucurbitaceae, Dipterocarpaceae, Bambusodeae (bamboos), *Oryza sativa* (rice), *Zea mays* (maize)), are also used to reconstruct vegetation change, especially in palynological 'blind spots' (section 18.6.4.1) such as grasslands or tropical forests (Piperno 2001). Plant phytoliths normally contain sufficient residual organic matter to be AMS radiocarbon dated (section 18.6.5). This is an important property, as the organic matter in question is synthesised from atmospheric carbon dioxide, so avoiding the pernicious 'hard water' reservoir effect (section 18.6.5).

18.6.4.3   Plant and other macrofossils

Macrofossils are mainly plant remnants (seeds, fruits and spores of higher plants, leaves, stomata, bulbscales, flowers, bulbils, rhizomes, roots, twigs, bark, remains of mosses, lichens and liverworts, oospores of Charophyta, remains of Cyanobacteria and algae) large enough to be seen by the naked human eye (Wasylikowa 1986; Collinson 1988; Birks 2001). Animal remains (Cladocera, insects (Trichoptera, Coleoptera), molluscs, acarid mites, ostracods, Foramininfera, fish scales, otoliths, bones (section 18.6.4.10)) also occur (Birks 2001). Results are presented as concentration (numbers cm^{-3}) or influx (numbers $cm^{-2}\,yr^{-1}$), but not percentages, as there is no 'macrofossil rain', and no 'macrofossil sum'.

Macrofossils are more precisely identifiable than pollen and spores (usually to species) and include taxa (e.g. *Juncus*) not usually represented in pollen diagrams. They are also much less readily dispersed than pollen and, so, reflect the specific existence of plants and other organisms in and around the lake itself, and in its catchment, rather than in a wider source areas (Fig. 18.14). With some exceptions (e.g. stomata (MacDonald 2001), *Betula* remains used in AMS radiocarbon dating (Björck & Wohlfarth 2001), section 18.6.5), macrofossils are not transported to the deep parts of lakes. In forested areas, lakes <500 m in diameter, up to 10 m deep and with no inflowing streams are favourable sites (compare pollen, section 18.6.4.1) but, in open environments, lakes with inflows are preferred (Birks 2001).

Plant macrofossils in lake sediments are mostly autochthonous and their main use is the reconstruction of lake ontogeny, rather than watershed vegetation (Collinson 1988). Macrofossils are indicators of presence where palynology is too insensitive to record vegetation change (section 18.6.4.1) or in treeless environments, where local pollen production is low, and long distance transport most pronounced. Leaf fragments containing conifer stomata, or stomata themselves (MacDonald 2001), are more accurate indicators of species presence at boreal and alpine tree lines than pollen. Macrofossils are also used, with paly-

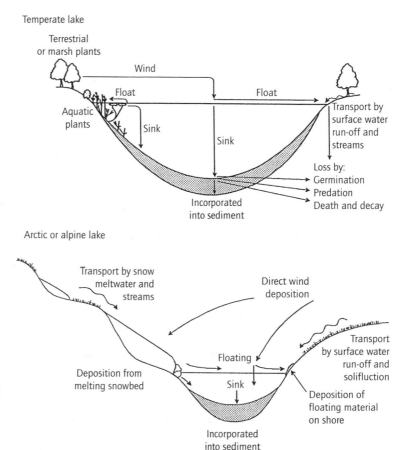

Fig. 18.14 Processes leading to preservation of macrofossil assemblages in (top) a small temperate lake and (bottom) an arctic or alpine lake. (Modified after Birks 2001; reproduced with kind permission of Kluwer Academic Publishers.)

nology, to reconstruct Holocene and other lake-level changes (Digerfeldt 1986).

Macrofossils are increasingly used for AMS radiocarbon dating (Björck & Wohlfarth 2001, section 18.6.5). Terrestrial plant macrofossils are preferred because they contain organic carbon synthesised from atmospheric CO_2, and therefore are not subject to the 'hard-water' reservoir effect which distorts dates from aquatic plant matter. Birks (2001) cautions against use of other kinds of macrofossils for this purpose, even aquatic mosses, although aquatic macrofossils from soft water and/or acid lakes, where atmospheric CO_2 is a major source of carbon, may be an exception. Stomatal density on leaves of conifers and other species to reconstruct past concentrations of at-

mospheric carbon dioxide offers a neat co-proxy for AMS radiocarbon dating of plant macrofossils.

Charcoal macrofossils (>200 μm) in lake sediments are used to reconstruct **local** (as opposed to regional, see section 18.6.4.2) fire histories (Tolonen 1986a; Whitlock & Larsen 2001). Sources may be regional (distant fires), extralocal (outside the watershed) and local (within). Most macroscopic charcoal is not transported far, so that its presence in lake sediments usually indicates the incidence of local or, perhaps, extralocal fires (within <6 km). Site selection is crucial. Large watersheds magnify allochthonous inputs and increase the potential contribution of secondary charcoal. Steep terrain also produces a strong secondary component. Small lakes, with no inflowing streams (section 18.6.4), offer the best prospect of detecting fires of local, specific origin. Estimates of fire frequency are only possible if contiguous samples are used and high-precision chronology is available. Varved sediments (section 18.6.5) are often preferred. Recent sedimentary charcoal peaks are calibrated against known fires, although dendrochronological reconstructions may include low-intensity, local events, and documentary records are often imprecise and non-quantitative.

18.6.4.4 Diatoms

Amongst the most important lacustrine microfossils used in 'palaeoenvironmental reconstruction' (Battarbee 1986, 1991; Bradbury 1988; Dixit *et al.* 1992; Fritz *et al.* 1999; Hall & Smol 1999; Stoermer & Smol 1999; Battarbee *et al.* 2001) are the siliceous **frustules** (cell walls) of diatoms (Algae: Bacillariophyta). These are initially classified according to habitat (into planktonic or benthic forms), although some taxa are common to both.

Benthic diatoms inhabit the littoral zone, where light penetrates to the lake bottom, and are classified as being **epilithic** (living attached to stones), **epiphytic** (attached to plants), **epipsammic** (attached to sand grains) and **epipelic** (living in organic lake mud). Many taxa are both epilithic and epiphytic, whereas the epipsammic and epipelic habitats are more exacting, especially with respect to light and oxygen. Benthic diatoms are often also found suspended in the water column, particularly in small, shallow, turbulent lakes.

The factors controlling the distribution of diatoms are very numerous. Battarbee *et al.* (2001) list those of significance in palaeolimnology as:
1 physical factors—temperature, light, mixing, turbulence, ice cover;
2 chemical factors—pH, dissolved organic carbon (DOC), salinity and conductivity, nutrient concentration.
Biological factors (grazing, mutualism, parasitism) are also important, but more difficult to evaluate. Physical factors, especially temperature and mixing, are of importance to climate reconstruction but underlie so many others (lake circulation, pH,

photosynthesis, productivity, water chemistry) that they are difficult to reconstruct, **of themselves**, although ice-cover may be an exception (Battarbee *et al.* 2001). Attention has therefore focused on chemical factors, of which pH is the most important (Battarbee *et al.* 2001), although Battarbee (1986) earlier emphasised salinity.

pH. Diatoms were initially classified (Hustedt 1937–1939) according to modern distributions and pH of surrounding waters, using the terms **alkalibiontic** (living only at pH > 7), **alkaliphilous** (more widely distributed at pH > 7), **indifferent** (circumneutral), **acidophilous** (more widely distributed at pH < 7) and **acidobiontic** (living at pH 5.5 and below). The Hustedt system was used in early pH reconstructions of surface waters, especially acidified lakes (Battarbee *et al.* 1986), even though allocation of some taxa to particular pH categories was always ambiguous. It has now been replaced by pH reconstruction using transfer functions based on estimates of the optimal ranges of individual diatom taxa along the gradient of natural pH (Battarbee *et al.* 2001). This approach is also based on observations of modern diatom distribution ('training sets') but emphasises continuity of species turnover along a gradient, rather than individual pH categories (section 18.6.6). pH training sets are used to reconstruct lake ontogeny, surface water acidification (Charles & Whitehead 1986; Battarbee *et al.* 1990) and climate change.

DOC. Studies of acidic, brown-water ('dystrophic') lakes show that DOC rather than pH itself explains most of the variation between samples, although results are not always repeatable (Battarbee *et al.* 2001). This offers the potential of reconstructing not only climate change, especially in boreal regions, but also optical properties of the water column.

Salinity. Diatoms were also formerly classified according to salinity (Hustedt 1957), using terms such as **polyhalobous** (widely distributed), **mesohalobous** (brackish) and **oligohalobous** (fresh), although within each category species with widely varying tolerance limits (**euryhaline** taxa) occurred. This system has also now been replaced by quantitative salinity reconstruction using transfer functions (Battarbee *et al.* 2001) based on training sets of modern lakes used to define salinity and other optima (e.g. brine type) for particular taxa (Fritz *et al.* 1999; section 18.6.6).

Two types of saline system are distinguished, with characteristic ionic concentrations. **Thalassic** systems (estuaries, bays, lagoons) are coastal, with salinity therefore based on Na^+ and Cl^-. **Athalassic** systems are inland salt-lake environments, where the range of salinity fluctuations and the ions on which salinity/conductivity is based (e.g. Na^+, Ca^{2+}, Mg^{2+}; Cl^-, CO_3^-, $[HCO_3]^-$, SO_4^{2-}) are much greater (Fritz *et al.* 1999; Last 2001b). Transfer-function-based studies of athalassic systems are increasingly used to reconstruct lake hydrology and salinity, and hence climate change, in continental interiors (the North American Great Plains, Africa) and other regions where inland saline lakes occur (Fritz *et al.* 1999). Studies of thalassic systems are traditionally used to reconstruct shoreline displacement, sea-level change (eustatic and isostatic) and isolation of estuaries or lagoons from the sea (Vos & de Wolff 1993), although problems are encountered in separating allochthonous from autochthonous diatom input (Battarbee *et al.* 2001).

Nutrients. The main nutrients limiting diatom growth in surface waters are silicon, nitrogen and phosphorus. Of these, phosphorus appears to be the key element, as it produces the greatest impact on diatom communities. An increase in the ratio of planktonic to benthic taxa is a clear sign, especially in shallow lakes, of phosphorus-led eutrophication, and is a result of decline in freshwater macrophytes (a major habitat for epiphytic diatoms) and the effects of shading of benthos by increased phytoplankton standing crop (see also Chapter 9). Silicon limits the growth of diatoms in temperate lakes in late spring, when they are replaced by other algae (Moss 1980), and also influences diatom species composition in tropical lakes (Kilham *et al.* 1986; Kilham & Kilham 1990).

Diatom-based reconstruction of changing nutrient loadings on standing waters has been used very widely indeed to investigate lake eutrophication and there are by now literally thousands of these studies, including many 'classic' examples. The use of transfer functions to reconstruct lake nutrient loadings, mainly total phosphorus (Hall & Smol 1992; Bennion *et al.* 1996), is further reviewed by Hall & Smol (1999).

Results are expressed as **percentages** (relative counts, 300–600 valves), **concentrations** (numbers of frustules unit vol.$^{-1}$), or **biovolume accumulation rates** (concentration divided by the sediment accumulation rate, and adjusted for variation in frustule size between taxa). Potential sources of error, which are responsible for differences between living diatom communities and fossil assemblages, include:

1 spatial and temporal homogenisation by bioturbation, or 'reworking' from marginal or other sources within the lake (compare pollen, section 18.6.4.1);

2 inwash of allochthonous diatoms from upstream, grazing by zooplankton, removal of lacustrine taxa via the outflow;

3 dissolution or partial dissolution of diatoms.

Fossil assemblages thus contain diatoms supplied by the entire range of habitats present in the lake, integrating information over a number of spatial and temporal scales (seasonal, annual, multiannual). However, they also represent diatom mixtures from different biotopes and successional communities, whose representation in sediments depends on relative productivities. Such problems are addressed via sediment-trap studies, and comparing living and fossil assemblages. Considerable in-basin variations in **diatom accumulation rates** between cores are observed, therefore normally precluding reconstruction of diatom **productivity**. Between-core variations in the **composition** of subfossil diatom assemblages (percentages) within single lake basins are relatively small, however.

Diatom dissolution occurs in the lakewater column and in the sediments, but mainly at the SWI (Battarbee *et al.* 2001). In warm alkaline or saline lakes, especially at low latitude, diatoms may be absent from the sediment record. Preservation is best in the cold, soft-water lakes of the boreal zone. Dissolution is usually partial, affecting individual taxa, leading to distortions in composition of fossil assemblages, in favour of more robust, heavily silicified forms. Where opal silica becomes limiting (e.g. at high nutrient concentrations), diatoms are replaced by other taxa, especially cyanobacteria (sections 18.6.3.1, 18.6.3.4), and in acid lakes by Chrysophyta (see next section).

18.6.4.6   Other siliceous microfossils

Other siliceous lacustrine microfossils occur on slides prepared for diatom analysis, including chrysophyte scales and cysts (Cronberg 1986; Smol 1988; Zeeb & Smol 2001), ebridian endoskeletons (Korhola & Smol 2001), freshwater sponge spicules (Frost 2001) and the plates and scales of protozoans ('rhizopods', Tolonen 1986b; Douglas & Smol 2001). Tests of testate amoebae are isolated separately (Beyens & Meisterfeld 2001).

Chrysophytes ('golden brown algae') occur mainly in soft-water lakes, where they sometimes replace diatoms as the main indicator group. A few taxa (e.g. *Mallomonas*, *Synura*) produce scales, bristles and spines, but all make cysts (resting stages). Knowledge of taxonomy, provenance and identification of these is still being developed. Surface sample training sets exist for pH (scales and cysts), eutrophication (scales and cysts decline as nutrient loadings rise), salinity (cysts), climate and other factors. The ratio of cysts to total diatoms is a rough measure of the relative importance of the two groups in any particular sequence.

Ebridians (Korhola & Smol 2001) are marine planktonic organisms not yet much used in palaeolimnology, but which are thought to possess potential for studies of marine influx, isolation and sea-level change. Width of skeletal spikules produced by freshwater sponges (Porifera) is related to salinity and silica availability (Frost 2001). Plates, scales and tests of protozoans, especially testate amoebae, are often only identified to genus or above, but are more resistant to dissolution than other siliceous microfossils (Douglas & Smol 2001). Shells (or **tests**) of testate amoebae are identifiable to species (Beyens & Meisterfeld 2001) and, in lake sediments, are indicators for pH and/or

trophic status, and also for catchment inputs, as some prominent taxa are terrestrial. They are therefore also used to reconstruct isolation from the sea of formerly marine bodies, and the hydroseral transition from lake to bog.

18.6.4.7   Cladocera

Amongst the most important microscopic animal remains recoverable from lake sediments are those of the Cladocera (Frey 1986, 1988, 1990; Hofmann 1987, 1998; Whiteside & Swindoll 1988; Korhola & Rautio 2001). These are crustacean brachiopods (water fleas, shrimps), which possess exoskeletons of chitin. On death or moulting, these become disarticulated into **exuviae**, several thousands per millimetre of which may become embedded in sediments. The main parts fossilised are headshields, shells (carapaces), abdomens, postabdomens, postabdominal claws, antennae, mandibles, trunks and copulating hooks. Also present are ephippia ('egg cases'), which are transportable (by birds and insects) over large distances.

Cladocera may be littoral, planktonic, meiobenthic or even neustonic (Fig. 18.15). The majority of taxa live either amongst the littoral vegetation or in the open water. Only the Bosminidae (which are planktonic filter feeders, grazing the phytoplankton) and the Chydoridae (chydorids; littoral taxa, living at the SWI, or associated with macrophytes) are well preserved. Despite often being the most important cladocerans in lakes, the Daphnidae (daphnids), apart from their postabdominal claws, mandibles and ephippia, are mostly not. Copepods, which can often be more numerous than Cladocera, are also not generally preserved in lake sediments. Korhola & Rautio (2001) also discuss various other crustaceans.

The factors which control cladoceran distribution include substrate and water depth (littoral taxa), temperature and species interactions (food availability, predators). Most Cladocera prefer oligotrophic, dilute waters (Korhola & Rautio 2001; but see Chapter 14), and species diversity declines as conditions vary outside optimal ranges, either towards lower pH (<5), alkalinity and con-

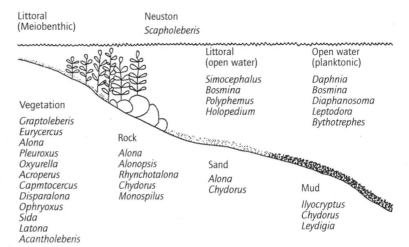

Fig. 18.15 Major ecological niches of main genera of Cladocera. (From Korhola & Rautio 2001; reproduced with kind permission of Kluwer Academic Publishers.)

ductivity, or towards increased nutrient concentrations. Changes in food availability (from other types of plankton to cyanobacteria), or in predators (from invertebrates to fish), lead (respectively) to a switch in the plankton (from Bosminidae to daphnids), or an increase in smaller Cladocera versus larger taxa. Decline of habitat (macrophytes and other substrate) is also significant (Whiteside & Swindoll 1988).

Although chydorids are mostly littoral, profundal cores provide an accurate assessment of composition of cladoceran communities in any given lake (Frey 1988; Korhola & Rautio 2001). Relative abundance of Cladocera is a useful descriptor of lake condition, from which modern surface-sample training sets are developed. Concentration of chydorids decreases markedly offshore, whilst abundance of planktonic taxa reaches a maximum in deeper waters (Frey 1986; Hofmann 1987, 1998). Variations in concentration are therefore difficult to relate to changes in lake productivity. Lake ontogeny can be successfully reconstructed using single, profundal cores, but changes in water level must be studied using multiple cores, including at least some from the littoral (Korhola & Rautio 2001).

Studies of Cladocera in sediment cores are traditionally used to reconstruct lake depth and water level, trophic status and productivity (Korhola &

Rautio 2001). Early approaches involved the use of indicator species but these have now also been replaced by transfer functions, with training sets for temperature, water level, nutrients, pH and perhaps water colour (Huttunen *et al.* 1988; Lotter *et al.* 1997; Korhola 1999). Water-level changes were formerly studied using planktonic/littoral (P/L) ratios (Frey 1986) but increased nutrient concentrations also favour planktonic against littoral taxa, whereas shifts in the plankton (from bosminids to daphnids; see above) produce false signals of chydorid 'increase' (Hofmann 1998). Lake acidification may lead to decline in fish, allowing an increase in Cladocera, a response which may also be mistaken for increased productivity.

Littoral Cladocera respond to increased nutrient loading only if their macrophyte substrate is reduced (section 18.6.4.1). Changes in planktonic species, including the classic shift from *Bosmina longirostris* to *Bosmina longispina* observed by Deevey (1942), are used as indicators of changing lake trophic status and productivity, but such changes may also be due to species interactions. This approach has therefore also now been replaced by the use of nutrient transfer functions (Lotter *et al.* 1997).

Acidification of lakes can also lead to changes in cladoceran community interactions, with loss of acid-sensitive taxa, decline in species richness and changes in total cladoceran biomass and abundance (Korhola & Rautio 2001). Shifts in cladoceran assemblages in acid lakes may also be due to changes in predation, aquatic vegetation and heavy metal concentrations. As fish generally prefer larger species and individuals as food (see Chapter 15), cladoceran size may be a factor used to reconstruct fish presence or absence in or from acidified lakes.

18.6.4.8   Chironomids

Chironomids (Chironomidae) are Diptera (true flies) whose chitinous head capsules also occur widely in lake sediments (Hofmann 1986, 1988, 1998; Walker 1987, 2001; Crisman 1988; Frey 1990), along with larval mandibles of Chaoboridae (phantom midges). Chironomid larvae experience four instars, the first planktonic, the second, third and fourth, benthic. Later instars are more robust, and more easily preserved. Larvae are mostly detritivores or filter feeders, although some are grazers or omnivores. Most are littoral benthos, but some profundal. A few (e.g. *Chironomus plumosus* L., *Chironomus anthracinus* Zetterstedt) contain sufficient haemoglobin to colour them bright red, an adaptation to life in the deoxygenated profundal of productive lakes (see Chapter 12).

The potential of chironomids for palaeoenvironmental reconstruction is considerable, especially in respect of climate (water temperature, air temperature). Walker (2001) lists more than 30 chironomid temperature training sets developed since 1990, mostly for Canada, Sweden, Finland and Switzerland, at the boreal or alpine treeline. Other factors include water depth, salinity, productivity, hypolimnetic oxygen concentration and, in the case of the Chaoboridae, fish presence or absence in acid lakes. Water depth and lake level are best reconstructed for closed basins (e.g. in East Africa or British Columbia), where changes are large enough to be registered by training sets. Productivity and hypolimnetic oxygen are studied by comparing chironomid training sets with Chl *a*, oxygen or total phosphorus concentrations.

18.6.4.9   Ostracods

Ostracods (sometimes ostracodes), or Ostracoda, are small crustacean bivalves, producing carapaces composed of low Mg calcite, 0.5–3 mm in length in the adult (Loeffler 1986; Carbonel *et al.* 1988; De Deckker 1988; De Deckker & Forester 1988; Holmes & Horne 1999; Griffiths & Holmes 2000; Holmes 2001). They occur ubiquitously in marine environments, in non-marine waters with pH neutral to alkaline, and also some acidic waters, where, however, shells are not usually preserved, owing to dissolution. Environmental factors which influence ostracod abundance include habitat type (lake size, energy level, permanence of the water body, depth, macrophyte substrate, food supply, predation), and nutrient status, salinity, temperature, dissolved oxygen and chemistry of waters. Those most easily reconstructed include habitat, temperature, salinity and ionic composition of the water (Holmes 2001).

Ostracods are classified according to habitat-preference groups (springs, streams, ponds, lakes, groundwater, ephemeral bodies, etc.) used for palaeoenvironmental reconstruction. They withstand desiccation by using resistant eggs, whose presence in the fossil record is therefore a signal of seasonal dryness. Most ostracods are benthic or epiphytic, with no truly planktonic species, but some are nektonic. Factors such as water depth and availability of macrophytes may be linked.

Some ostracods are euryhaline, others sensitive to narrow salinity ranges. In non-marine waters, diversity and abundance increase with salinity, up to the 'calcite branchpoint' (Fig. 18.16), at which $CaCO_3$ is precipitated, broadly marking the junction between fresh and saline conditions. Beyond this value, ostracod abundance increases, but diversity declines. Rather than salinity, ionic composition of waters, particularly alkalinity, Cl^- and SO_4^{2-}, may be important in controlling ostracod distribution.

Many processes contribute to differences between living ostracod communities and subfossil 'death' assemblages (Holmes 2001). Those containing both adults and the full range of instars (normally eight) have probably been deposited *in*

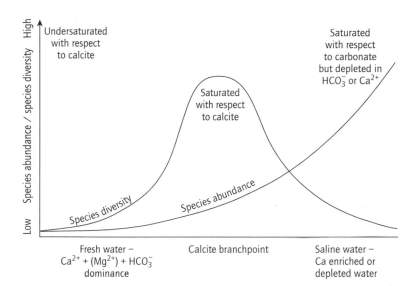

Fig. 18.16 Relationship between ostracod species diversity and abundance, and salinity, in non-marine waters. (From Holmes 2001; reproduced with kind permission of Kluwer Academic Publishers.)

situ, whereas absence either of adults, or of early instars, suggests winnowing of ostracod particle-size during transportation and redeposition. 'Valve/carapace' ratios are used as estimates of disarticulation of carapaces (into their two constituent shells) and thus energy environment. Large numbers of juvenile carapaces denote high and rapid mortality, in a suddenly unfavourable environment. Breakage, dissolution or 'overgrowth' (of new calcite) indicate reworking and/or diagenesis.

Subfossil ostracods have been used to reconstruct salinity (and, hence, effective precipitation) in closed basins in semi-arid/sub-humid regions of North America, Southern Europe, Africa, Australia and China. So far, training sets are available for North America only (Holmes 2001). Marine ostracod faunas are distinctive, so that reconstruction of palaeosalinity and marine intrusion in coastal environments represents a powerful use of this technique (Holmes 2001; Bruce *et al.*, in press). Other factors reconstructed using ostracods include temperature, habitat and lake level, especially that of Lake Titicaca, using a training set from lakes on the Bolivian Altiplano (Mourguiart & Carbonel 1994).

18.6.4.10 Stable isotope ratios and trace element chemistry of ostracod shells

Reconstructions can also be based on geochemical and stable isotope ratios of ostracod shells (Holmes 1996). Holmes (2001) lists the advantages of this approach over analysis of bulk or authigenic carbonates (sections 18.6.2 and 18.6.3), as:

- allowing precise identification of mineral phases (e.g. using XRF, SEM);
- avoiding the use of detrital carbonates not in equilibrium with lake waters;
- generating time-specific data, even to the actual season of growth;
- identifying (possibly) the location and depth of carbonate formation.

The $\delta^{18}O$ in ostracod shells is used to reconstruct $\delta^{18}O$ of lake waters at the time of carbonate formation, and hence water temperature. Ostracod carbonates are not in equilibrium with lake waters, offsets of 0.75 to 2.2‰ being common. The $\delta^{18}O$ of lake waters reflects $\delta^{18}O$ of precipitation (which is, in turn, related to distance from source; Ito 2001), basin hydrology and lake residence time. In closed basins in semi-arid and sub-humid regions, P/E ratio is the controlling factor, but, in deep lakes with isothermal hypolimnia, $\delta^{18}O$ of precipitation and, hence, its temperature, is more crucial (Holmes 2001).

The $\delta^{13}C$ of freshwater carbonates, including those present in ostracod shells, is related to $\delta^{13}DIC$ of the water column, and to the uptake of atmospheric carbon dioxide during photosynthesis (see section 18.6.3.3). The $\delta^{13}C$ of carbonates in ostracods, and in other lacustrine calcareous fossils and microfossils, is therefore a proxy of lake productivity. Trace-element (TE) studies, often of individual ostracod shells, are focused on Sr/Ca and Mg/Ca ratios as indicators of palaeosalinity (Holmes 1996). The Sr/Ca ratios in ostracod shells are positively related to those of lake waters at the time of shell formation, although Sr is strongly discriminated against during uptake. The Mg/Ca ratio of shells is also positively related to Mg/Ca of surrounding waters, although values vary between genera, and between lakes (Holmes 2001; Wansard et al. 1998). The $^{87}Sr/^{86}Sr$ ratios may be used to identify the source of carbonates present in waters in which the shells are produced.

18.6.4.11   Fish

Subfossil and fossil fish are used to reconstruct past biogeographical distributions and evolutionary relationships (see Chapter 16), as well as environmental factors. Attention focuses mainly on taphonomic studies of whole body fossilisation and, in more recent sediments, on **otoliths** (small particles of aragonite ($CaCO_3$) deposited in the ears of teleosts, which are identifiable to species), and bones and other skeletal parts (which contain apatite [$Ca_2(PO_4)_3(OH,F)$]). Both otoliths and bones possess annual or even subannual growth increments, from which determination of trace element ratios (Sr/Ca) and $\delta^{18}O$ and $\delta^{13}C$ are used to reconstruct palaeosalinity, P/E ratios, water temperature, oxygen concentration and temperature (Patterson & Smith 2001).

Studies of past fish populations are also used to reconstruct (indirectly) habitat factors such as energy environment, site elevation, lake temperature, oxygen concentration, salinity and alkalinity, and routes of fish migration (compare Chap-

ter 16). These are based on studies of modern species ranges, the study of fish **ecomorphology** (the relationship between body shape and habitat) and again on $^{87}Sr/^{86}Sr$ ratios, which indicate species' original habitat. Studies of sedimentary ^{15}N and diatom analysis were recently used to identify climatically driven changes in populations of sockeye salmon (*Oncorhyncus nerka*) in salmon nursery lakes in Alaska (Finney et al. 2002; section 18.6.3.3).

18.6.5   Chronology

Information on the rate at which sediments accumulate in lakes is fundamental to palaeolimnology. Without it, neither the **timing** of particular palaeolimnological events, nor the **rates** at which processes have operated during the recent or more distant past, nor the **duration** of particular states or episodes, can be evaluated. Processes which operate on timescales measured in millennia are studied using the standard dating techniques of Pleistocene geology (Williams et al. 1998). Dating more recent phenomena, especially the impact of industrial society on lakes and their watersheds, has required the development of techniques peculiar to palaeolimnology, which have then spread to other disciplines.

18.6.5.1   Artificial, 'fallout' radionuclides

A major technique for dating sediment profiles deposited over the past half century involves identifying the record of deposition of ^{137}Cs (half-life $(T/2)^*TH = 33$ yr) associated with testing of nuclear weapons in the atmosphere, particularly during the period AD 1954–1963 (Pennington et al. 1973). Dating is based on the known history of addition of the isotope to the environment (Fig. 18.17), rather than its decay rate (see below). In the Northern Hemisphere, ^{137}Cs began to increase in 1954, reached a minor maximum in 1959, and a major peak in 1963. Deposition then declined, owing to implementation of the Test Ban Treaty. Limits of detection were reached by 1983. Patterns in the Southern Hemisphere are similar, although less sharply defined.

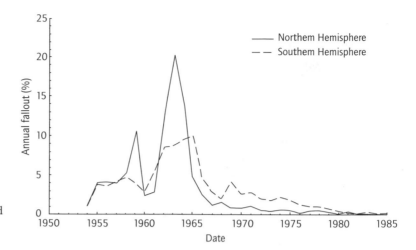

Fig. 18.17 Fallout of ^{137}Cs in the Northern (solid line) and Southern Hemispheres (dashed line) for the period AD 1954–1985. (From Appleby 2001; reproduced with kind permission of Kluwer Academic Publishers.)

Amounts deposited vary considerably from region to region, depending on latitude and annual precipitation (Appleby 2001). In most recent profiles, 1963, the peak year of ^{137}Cs deposition, can be readily identified and, in some, 1954, and also 1959. In many lakes, detail is lacking, owing to mixing at the SWI, vertical diffusion of the isotope, or (in rapidly accumulating sediments) dilution by other material. In profiles accumulating very slowly, records are often 'telescoped' and therefore also lacking in detail.

In parts of central, northern and northwest Europe, a further peak, produced by fallout from the Chernobyl accident of 1986 (originally distinguishable from 'weapons' ^{137}Cs by its ^{134}Cs /^{137}Cs ratio) is observed (Appleby *et al.* 1991). This will form a second ^{137}Cs marker in future decades, although only within the region contaminated by fallout from the accident. In some locations it has already 'swamped' the earlier (1963) peak by downward diffusion (Appleby 2001). This process is especially pronounced in lakes whose waters are poorly buffered, where ^{137}Cs should therefore probably not be used as a dating tool. Other isotopes also added to nature during the mid-twentieth century phase of atmospheric weapons testing include ^{90}Sr ($T/2*$TH=28 yr), ^{241}Am (432 yr), and $^{239, 240}$Pu and ^{241}Pu (Appleby 2001). Americium-241 is now used alongside, or instead

of, ^{137}Cs, especially in soft-water lakes, or where Chernobyl effects are pronounced (Appleby *et al.* 1991; Fig. 18.18).

18.6.5.2 Lead-210

A naturally occurring isotope formed near the end of the uranium/lead series is ^{210}Pb ($T/2*$TH=22.3 yr; Appleby 2001). This is formed (Fig. 18.18) via a series of short-lived progeny, by disintegration of atmospheric ^{222}Rn ($T/2*$TH=4 days), which is in turn the daughter of ^{226}Ra ($T/2*$TH=1620 yr). Lead-210, a solid, is deposited all over Earth's surface, from where it is washed into lakes, joining that directly deposited onto lake surfaces. Time–depth relationships are based on the decay rate of the isotope, and on measurement of activity/concentration with depth. *In situ* decay of ^{226}Ra also occurs within sediments, however, so that this component (the **supported** ^{210}Pb) must be subtracted from total ^{210}Pb, in order to leave the **unsupported** ^{210}Pb (Fig. 18.19). Dates are then calculated from changes in concentration of this component with depth.

Accurate dates generally are obtained for the last 150 years (Appleby 2001). Lead-210 is therefore another isotope suitable for dating the environmental impact of modern, industrial society. However, as the '^{210}Pb dating horizon' continues

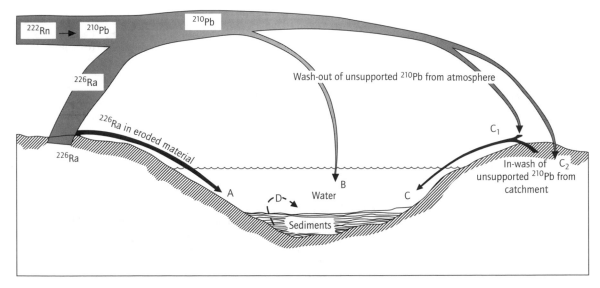

Fig. 18.18 Environmental sources and pathways of ^{210}Pb. (From Oldfield & Appleby 1984; reproduced with permission of Continuum International Publishing.)

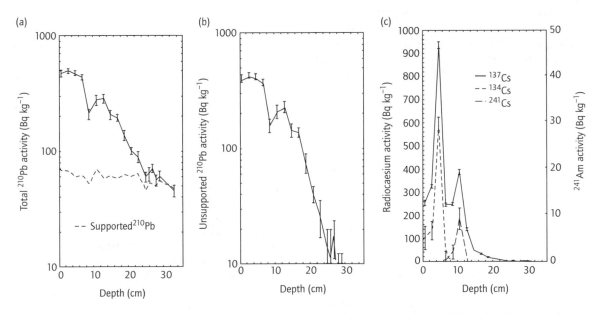

Fig. 18.19 Fallout radionuclides in a recent sediment core from Windermere (UK) showing (a) total and supported ^{210}Pb, (b) unsupported ^{210}Pb and (c) ^{137}Cs, ^{134}Cs and ^{241}Am concentration versus depth. (From Appleby 2001; reproduced with kind permission of Kluwer Academic Publishers.)

to 'rise' through the sediment profile, other techniques will need to be found with which to date the 19th century (Rose 2001). There is also a problematical 'gap' between ^{210}Pb and other isotopes, especially [14]C (see next section), which is not usually used to date material younger than about AD 1500. Other techniques, with which to bridge this gap, are therefore being sought (Nijampurkar *et al.* 1998).

Lead-210 profiles may be subject to error, mainly from mixing at the SWI, sediment focusing, inwash of radiometrically inert or [226]Ra-rich detrital material, or sediment hiatuses. Several models of accumulation of ^{210}Pb in lake sediments are used, involving assumptions either of **constant initial concentration** (Pennington *et al.* 1976) or **constant rate of supply** (Oldfield & Appleby 1984). For discussion of the applicability of these models to specific ^{210}Pb profiles, see Appleby (2001).

18.6.5.3 Radiocarbon dating of lake sediments

Longer term processes, or earlier events, are dated using the naturally occurring isotope [14]C (internationally agreed $T/2 = 5568 \pm 30$ yr; Olsson 1986, 1991; Björck & Wohlfarth 2001), which is formed in Earth's upper atmosphere by action of free neutrons on ^{14}N, forming ^{14}C, plus a displaced proton. Carbon-14 is then rapidly oxidised, mixed with the rest of atmospheric carbon dioxide, and eventually incorporated into the biosphere, or into oceanic and other carbonates. Living organisms remain in isotopic equilibrium via exchange of carbon with the rest of nature, but after death their ^{14}C content declines at a rate controlled by the half-life of the isotope.

Lake sediments, which are normally rich in organic matter (2–20% DM), have been very widely dated by this method. However, a number of key problems are encountered when dating lake sediment profiles, some of which ensue from using ^{14}C to date that particular material, and others which are due to shortcomings of the method in general (O'Sullivan 1994b; Björck & Wohlfarth 2001). The former stem from the fact that lake sediments are a mixture, and their radiocarbon content the product of their history. The carbon they contain is supplied by a considerable range of sources (autochthonous organic matter, lacustrine carbonates, detrital terrestrial and reworked organic and inorganic matter), all of which are potentially of widely differing age, each of which therefore potentially carries very different ^{14}C signals.

Sources of error which apply particularly to ^{14}C dating of **lacustrine** sediments (Fig. 18.20) include:
1 contamination by penetration of younger roots, downward percolation of humic acids, or bioturbation, all producing dates which are 'too young';
2 inclusion, even in moderately alkaline waters, of 'old', dissolved, radiometrically dead carbon, which is then incorporated in sedimentary carbonates, and taken up by aquatic biota, and sedimented as autochthonous lacustrine organic matter, both leading to the pernicious 'hard water' reservoir effect, producing ^{14}C dates which are 'too old';
3 inwash of radiometrically dead, detrital allochthonous carbon, from soils, peats, graphite, limestones and other deposits in the catchment, a factor especially likely to influence ^{14}C dates from lake sediments once human cultures practising plough agriculture inhabit a region, again producing dates which are 'too old', often identifiable, however, via reversal of time–depth relationships (O'Sullivan *et al.* 1973);
4 in coastal lakes, influence of the marine reservoir effect (Björck & Wohlfarth 2001), which also produces dates which are 'too old'.
'Bulk' radiocarbon dates from lacustrine sediment are therefore nowadays less often used than heretofore. Determinations are mainly carried out, after carbonate removal, on the alkali-soluble fraction of the sediment. This at least avoids contamination by minerogenic material, but not the 'hard water' reservoir effect, as autochthonous organic matter synthesised from water column DIC is still included in the date. Results therefore often relate to the type of pretreatment used, and to the proportions of various kinds of organic matter in the sample.

Nowadays, many ^{14}C dates are instead determined on terrestrial macrofossils (e.g. fruits,

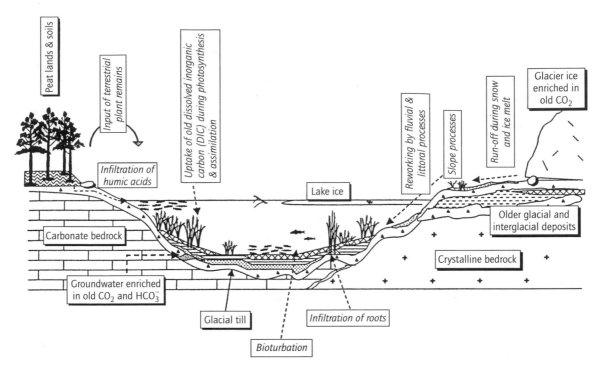

Fig. 18.20 Potential sources of error (italic font) affecting bulk radiocarbon dates from hard water (left of sketch) and soft water (right) lakes. (From Björck & Wohlfarth 2001; reproduced with kind permission of Kluwer Academic Publishers.)

twigs, scales, leaves of deciduous trees, conifer needles (but **not** wood), insect remains; section 18.6.4.3). This procedure has been greatly facilitated by the development of AMS radiocarbon dating (Hedges 1991), which requires much smaller amounts of carbon (often ≪ 1 mg) than conventional radiometric methods (usually > 0.5–1 g). Terrestrial macrofossils also contain only organic matter synthesised from atmospheric carbon dioxide, thus possessing the dual advantage of avoiding the 'hard water' reservoir effect, and giving estimates of atmospheric radiocarbon activity at the time of formation. Birks (2001) warns against the use of other kinds of macrofossil for dating, especially aquatic taxa, even from apparently carbonate-free lakes.

Macrofossils do not occur everywhere, however, so that in their absence dates must be determined on fractions which still potentially include autochthonous lacustrine carbon, or on some

other terrestrial material. One approach is to date pollen (Brown *et al.* 1989), although in any sediment sample this is itself a mixture of terrestrial and aquatic taxa, potentially of different ages (O'Sullivan 1994b). A further possibility is to date organic compounds specifically of terrestrial origin ('molecular dating'), which again have been synthesised using only atmospheric carbon dioxide. These are extracted using preparative capillary gas chromatography (PCGC), an approach initially developed for marine sediments (Eglinton *et al.* 1997).

Radiocarbon dating may be safely used to date material up to c.45,000 years in age (Björck & Wohlfarth 2001), although uncertainties increase beyond 20,000 years before present (BP). Results given in radiocarbon years BP are converted to calendar years using the (mainly tree ring) calibration curve of atmospheric radiocarbon (Stuiver *et al.* 1998). Emphasis has shifted over the past 20 years towards high-resolution sequences with multiple dates (Björck & Wohlfarth 2001). Owing to the introduction to the atmosphere, since about AD 1750, of large amounts of radiometrically inert carbon by combustion of fossil fuels (the 'Suess effect'), ^{14}C dating is not much used for the period after AD 1700. Highly radioactive 'bomb' radiocarbon, produced during the same period as ^{137}Cs and other artificial radionuclide emissions from atmospheric weapons testing (section 18.6.5.1), also affects dates on modern materials. A 'plateau' on the calibration curve for the later Middle Ages means that, from c.500 years BP, radiocarbon dates are not easily converted unambiguously into calendar years.

18.6.5.4 Palaeomagnetism/natural remanent magnetisation (NRM)

Records of Earth's magnetic field (natural remanent magnetisation—NRM) are also preserved in lake sediments (King & Peck 2001). Several types of variation occur, including **reversals**, **transitions** (during which polarity of the global field reverses, and intensity declines), **excursions** and **events**. Such features are used mainly to date events from the Pleistocene, or earlier.

Between transitions, the field is subject to geomagnetic secular variation (SV), which is measured using **declination** (D, its horizontal component), **inclination** (I, the vertical component) and **intensity** (F, its magnitude). Determination of SV has been developed as a dating technique for Holocene and Late Pleistocene lake and other sediments by Thompson (1973, 1991) and others (Thompson & Oldfield 1986), although it was originally investigated by Mackereth (1971). Patterns of SV are defined for regions of Earth's surface c.3000–5000 km in diameter (**geomagnetic provinces**), and master curves developed for each. This is not an 'absolute' dating technique, however, as master curves must be calibrated against other chronologies (e.g. radiocarbon dates, varves) and, for recent centuries, direct observations of SV. Declination reached a major westerly maximum in AD 1820, but during the early Middle Ages lay much further east than at present.

18.6.5.5 Tephras

Isochronous marker horizons are formed in lake sediments by **tephras**—material ejected during volcanic eruptions and transported by wind and washed out of the atmosphere into lakes and their catchments by rainfall. If these can be dated, and their source identified, they form very precise, practically instantaneous markers which can then be used as regional dating horizons. Regional tephrochronologies (e.g. van den Bogaard & Schminke 2002) exist for parts of all continents except Antarctica (Turney & Lowe 2001).

Tephras may be visible to the naked eye, or in thin-sections (Merkt & Muller 1999). Detection is difficult in minerogenic sediments. Also present are 'invisible tephras', or **cryptotephras**, whose identification and use as dating markers is on the increase. Both are identified as to source ('fingerprinted') by various geochemical and other methods (the respective merits of which are discussed by Turney & Lowe (2001)), and dated either directly, using fission-track dating, thermoluminescence, $^{40}K/^{40}Ar$, or $^{39}A/^{40}A$, or indirectly, with ^{14}C, artificial fallout radionuclides, ^{210}Pb or varve counts. Tephras are assigned to various eruptions known from documentary records for recent centuries if these can be matched geochemically. They are especially useful chronological tools for those parts of the palaeoenvironmental timescale where radiocarbon dating is not available (e.g. recent centuries), or where it is unable to date the record in much detail (e.g. Late Pleistocene calibration 'plateaux').

18.6.5.6 Fly ash chronologies

Chronologies for the last 150 years have also been developed using 'fly ash' particles produced by fossil fuel combustion. These are readily transported by the atmosphere, and eventually deposited in lake sediments (Wik & Renberg 1996; Rose 2001). Early sources (e.g. Renberg & Wik 1984) refer to 'soot' chronologies, but Rose (2001) states that this term is ambiguous, as fly ash is actually composed of:

- spheroidal carbonaceous particles (SCPs), which

are **never** spherical, and which are composed of un-burned combustible material;

• inorganic ash spheres (IASs), made up of non-combustible material, and **always** spherical;

• soot (particles of combusted fuel), which bears no resemblance to SCPs.

SCPs and IASs are generally > 1 μm in diameter, but soot particles are <1 μm. Up to 25% of coal is non-combustible, but the corresponding proportion for fuel-oil is only c.0.1%. The IASs therefore make up 95–98% of fly ash from coal combustion, but <20% from oil. Source is therefore investigated via relative composition. The IASs also possess natural sources (e.g. volcanoes). Fly ash is not produced by burning wood, so chronologies represent only the period since the industrialisation of power generation. Most palaeoenvironmental work focuses on SCPs.

Sedimentary records of SCP deposition (Fig. 18.21) relate to human activity, specifically to the economic history of particular industrialised countries and regions. In the UK, they begin circa AD 1850, and for most other nations, the period AD 1850–1900. IAS records possess no 'beginning', as they also come from natural sources. A rapid increase in SCPs is recorded at sites in Europe after 1945, as post-war reconstruction, increasing demand for electricity, and a switch from coal to oil all took effect. This increase occurred during the 1920s in the region of the North American Great Lakes, reflecting the earlier modernisation of industries in that area.

Deposition of SCPs in North America and most of northern and western Europe then reached a peak during the late 1960s and early 1970s. Thereafter, emission controls, taller stacks and greater efficiency led to a reduction of output, although this feature is delayed by a decade in Ireland, northern Britain and Slovenia. Since then, there has been widespread reduction in SCP sedimentation, caused by further increases in efficiency, and the so-called 'decline' of heavy industry in North America and Europe (which is, in fact, its globalisation and export to countries with lower wages and environmental standards). Recent records are complicated by use of multifuels, which means that sources may be less easily identified in future.

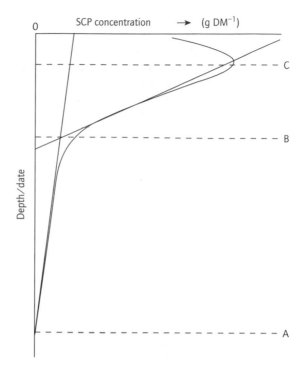

Fig. 18.21 Schematic profile of spheroidal carbonaceous particle (SCP) concentration from a lake sediment, showing (A) beginning of record, (B) rapid increase and (C) concentration peak. (From Rose *et al.* 1995; reproduced with permission of Arnold Publishers.)

In some countries (e.g. China, India), industrialised combustion of coal is still expanding.

Fly ash chronologies are used to date records of acid deposition from power generation since the mid-19th century (particularly its impact on pH of surface waters during the mid- to late 20th century), and to identify the sources of those emissions. They provide chronologies for reconstruction of lake acidification, especially where ^{137}Cs cannot be employed (section 18.6.5.1), and are also used, along with heavy metals and sulphur concentrations, in order to identify atmospheric transport and deposition of pollutants across regions and within lakes, and their sources. Chronologies must be calibrated against independent dating techniques, e.g. ^{210}Pb. As pointed out by Rose

(2001), they will eventually replace that technique for the 19th century, as the '^{210}Pb dating horizon' continues to 'rise' through the sediment profile.

18.6.5.7 Historical records

In many sediments, horizons are identified which are related to specific episodes in the history of lakes and their watersheds. Several examples have already been given in this chapter of techniques partly calibrated against documentary records. Reference has also been made (section 18.6.1) to the importance of assembling as much quantitative information as possible on the past and present condition of lake sites chosen for study. Dating is achieved by matching historical records with specific sedimentary horizons, preferably using cross-calibration with other dating techniques (e.g. ^{137}Cs, ^{210}Pb).

In some countries, detailed, quantitative agricultural records exist spanning several centuries, which can thus be used for detailed reconstruction of changing land use, sediment and/or nutrient loads on lakes. Each society develops its own administrative procedures, but this approach has been used in Sweden (Dearing *et al.* 1987, 1990; Krug 1993) and in the UK (O'Sullivan 1992, 1999 (along with catchment models; Jørgensen (1974 and Volume 2, Chapter 16)) in order to calculate changing nutrient loadings since AD 1866 on lakes in southwest England. In some parts of the USA, expansion of ragweed (*Ambrosia*) pollen in the uppermost sedimentary record provides a marker horizon for the arrival of European agriculture (Davis 1976). This feature can be dated by reference to local land registration records. Bradbury & Waddington (1973) and Engstrom *et al.* (1985) used (respectively) documentary records of mining and logging operations in order to calibrate dates from recent sections of sediment cores from lakes in Wisconsin and Vermont (both USA).

18.6.5.8 Annually laminated (varved) sediments

Early limnologists (e.g. Eggleton 1931; Nipkow 1920) noted the presence, in some lake sediments, of regular laminations. Even then, these were thought to be annual, and to be examples of (non-glacial) **varves**—sedimentary structures originally associated with near-glacial environments (de Geer 1912). During the 1970s, palaeolimnologists conducted a wide search for these deposits, as their full scientific potential came to be more fully realised. Many hundreds are now known, with more being discovered annually (Saarnisto 1986; Anderson & Dean 1988; Simola 1992), although some palaeolimnologists still regard them as rare. At one time (O'Sullivan 1983), they appeared to be confined to the North American Great Lakes region, Fennoscandia, the European Alps and some of the African Great Lakes, but this non-random distribution was merely a function of where palaeolimnologists had chosen to look for laminated sediments. Since then, many other varved sequences have been found, in North America, on the North European plain, in the *maar* lakes of western and southern Europe, and in semi-arid regions. The power of modern methods of recording and analysing varved sediments, especially those transferred from palaeoceanography (Dean *et al.* 1999; Lotter & Lemcke 1999; Kemp *et al.* 2001; Lamoureux 2001; Saarinen & Petterson 2001), has also greatly increased.

The value of annually laminated sediments partly lies in their great chronological precision, which allows specific years in which particular events are recorded in the sediment profile to be identified, even over several millennia. On geological timescales, varved sediments offer one of the few opportunities to study continental records of Pleistocene and other long-term environmental change deposited over the same time periods as those spanned by ocean sediments, but with similar or greater resolution than ice-cores (ELDP 1995). They are not just chronological tools, however, as they also enable layers of sediment deposited during specific seasons of the year to be identified. Much stronger links than normal, between the sedimentary record and observational climate or limnological data, are therefore possible.

In lakes with 'normal', unlaminated sediments (Fig. 18.22), circulation and bioturbation

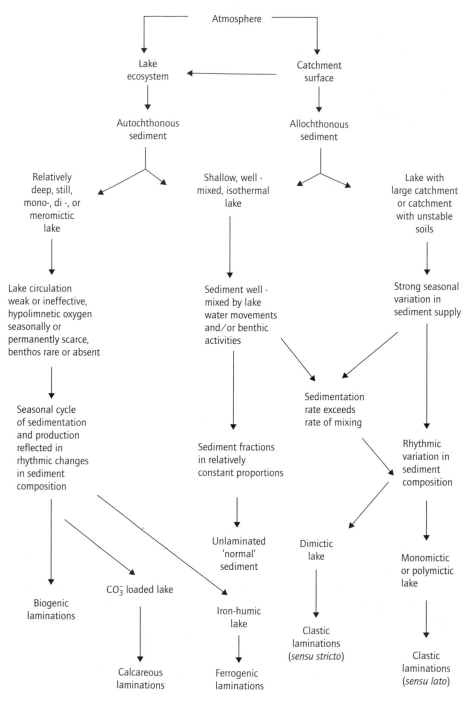

Fig. 18.22   Processes leading to varve formation in freshwater lakes. (From O'Sullivan 1983; reproduced with permission of Elsevier Science.)

generalise the sedimentary record. Layers representing individual events, or seasons, are rare. In lakes which are deep for their surface area, however, or in the deep basins of large lakes, circulation processes are weak or ineffective, and bioturbation is slight or absent. Seasonal cycles of particulate matter composition (**seston**) in the water column of such lakes are therefore reflected in rhythmic changes in sediment composition, which produce a sequence of laminae, each of which represents a season (O'Sullivan 1983; Simola 1992). If the processes generating this sequence can be identified, and the number of laminae representing an annual cycle (or **varve**) determined, laminations can be counted, and used to develop very precise chronologies of sedimentation. Although these may extend over many millennia, their potential resolution remains annual or even subannual.

In some temperate lakes, varves contain spring layers rich in diatoms, and summer laminae composed largely of other algal material, and are termed **biogenic laminations** (Fig. 18.22; O'Sullivan 1983). Layers of other materials make up the rest of the year. In boreal regions, varves are often composed of alternating black and brown layers, rich respectively in ferrous sulphide and ferric hydroxide (**ferrogenic laminations**), formed under seasonally variable lake redox. In calcareous lakes, springs are often marked by layers of sedimentary calcite, produced during periods of high rates of photosynthesis. Here, varves are termed **calcareous laminations**. Several of these (e.g. Elk Lake, Minnesota; Bradbury & Dean 1993; Dean 2002; Dean *et al.* 2002; *www.ngdc.noaa.gov/paleo/paleolim_elklake.ht ml*) have produced some of the most powerful palaeoenvironmental reconstructions. Stable isotope and geochemical ratios (section 18.6.3) are determined on calcareous laminae, whilst layers of mainly biogenic material making up the rest of the year are analysed for other proxies.

In other lakes, **clastic** varves, composed mainly of detrital, minerogenic, allochthonous material, are formed (Sturm 1979). Laminations are produced by seasonal variations in sediment composition and supply from the catchment. They are characteristic of near-glacial (e.g. alpine) environ-ments, where seasonal discharge of streams varies strongly according to freezing and melting of surface water, and also semi-arid regions, where discharge also varies seasonally, and loose soil and other material are available for transport as stream sediment. They also appear to form both in large and small lakes in humid regions, where seasonally variable sediment supply is from a relatively large drainage basin (Cooper & O'Sullivan 1998; O'Sullivan 1999).

Studies of varved sediments possess a wide range of palaeoenvironmental applications, including:

• reconstruction of rates of species immigration and duration of periods of forest stability during Pleistocene interglacials;
• estimation of rates of shoreline displacement;
• evaluating the effects of fire frequency on forest composition in boreal environments.

They also provide chronologies for studies of changing frequency of **swidden** cultivation in boreal forests (and its relation to population pressure), and eutrophication of lakes by human impact (O'Sullivan 1983). One of their main uses is calibrating other dating techniques, including ^{210}Pb (Appleby 2001), ^{14}C (Björck & Wohlfarth 2001), palaeomagnetism (King & Peck 2001) and tephrochronology (Turney & Lowe 2001). Paradoxically, varve chronologies themselves are also not 'absolute', in that the hypothesis that they contain signals representing seasonality must always be tested by some 'independent' dating method (^{210}Pb, ^{14}C, documentary records). Studies of varve **thickness** are increasingly being used to reconstruct climate, using this signal as a proxy record of lake temperature, palaeohydrology, catchment discharge or some other factor (Dean *et al.* 2002; O'Sullivan *et al.* 2002). One of the earliest studies of non-glacial varves (Lake Saki, Crimea; Shostakovich 1934; Lamb 1977) was also used for this purpose.

18.6.6 *Data analysis*

Methods of analysing palaeolimnological data, statistically or numerically, in order to improve the power of interpretations are only briefly discussed here. Such methods are described at length by Birks (1986, 1995, 1998), Birks & Gordon (1985), Maddy & Brew (1995) and by Birks *et al.* (in press).

18.6.6.1 Unmixing

Lees (1999) discusses the use of this approach in order to identify sediment source, to discriminate between sedimentary components, and to model movement of magnetic materials through the environment. Standard statistical tests (coefficient of variation (CV), Spearman's rank correlation) are used to identify the optimum number of samples needed in order to identify sediment source and to discriminate between potential source materials. Multivariate methods (cluster analysis (CA), principal components analysis (PCA), factor analysis) are used in order to group samples according to properties, to reduce the number of variables needed to characterise the dataset, and to explain (and quantify, using eigenvalues) the most important factors controlling its properties. They also identify the best interparametric ratios needed in order to discriminate further between sediment sources (section 18.6.2). Simultaneous equations are used to quantify sediment delivery from tributaries into larger rivers, and linear modelling to identify the proportions of various magnetic minerals in sediment samples. Lees (1999) concludes that, using mineral magnetic measurements, four sediment sources can be unmixed, at most.

Boyle (2001) discusses the use of element ratios for unmixing geochemical data, and identifying 'end members' and trace element contribution. Principal components analysis can be used to unmix geochemical data and to measure dissimilarity between samples using Euclidean distance. As samples are generally a mixture of n independent, multi-element components, sediment composition is generally represented by $(n-1)$ axes, usually 3–4. Sediment source is studied using Q-mode factor analysis, a 'varimax' rotation method like that applied to PCA by Lees (1999). End members are also identified using normative calculations, whereby the theoretical mineralogical composition of a sample is estimated from elemental proportions. Last (2001b), however, maintains that this method is difficult to apply to 'real

world' data, as, in nature, many minerals exist in non-equilibrium states.

18.6.6.2 Microfossil analyses

For purposes of stratigraphic description, pollen and other microfossil diagrams are divided into (microfossil) assemblage zones (Bennett & Willis 2001). These **may** be defined by eye (!), or by numerical methods (Birks & Gordon 1985). Programs for drawing microfossil diagrams (e.g. TILIAGRAPH; Grimm 1993) usually contain a facility for zoning based on a choice of several of these methods (e.g. CONISS—constrained incremental sum of squares). Trends and subgroups within the data are identified using ordination (e.g. PCA; see above) and diversity (or 'richness'), both within and between samples, by **rarefaction analysis**, which requires counts of standard size, however (Bennett & Willis 2001). Sequences may be compared with each other using 'sequence slotting'. Rates of change may be studied using measures of dissimilarity between samples (Prentice 1988), and periodicities within the data identified using various methods of time-series analysis (O'Sullivan *et al.* 2002). Results are drawn up using programs such as TILIA, psimpoll or POLPAL (Bennett & Willis 2001; *http://www.kv.geo.uu.se/datah*). Databases with which to compare one's own results are located at *http://www.imep-cnrs.com/pages/EPD.htm* and at *http://www.ngdc.noaa.gov/paleo/pollen.html*.

18.6.6.3 Transfer functions

An empirical approach to quantitative reconstruction of changes in lake and other communities is based on the development of **ecological response-** or **transfer functions** (Battarbee *et al.* 2001). Transfer functions derived from diatom training sets have been applied to the reconstruction of lake palaeosalinity, hydrology and climate (Fritz *et al.* 1999), surface water acidification (Battarbee *et al.* 1999), nutrients and eutrophication (Hall & Smol 1999), and other variables (Stoermer & Smol 1999). 'Training sets' for other microfossils (e.g. chrysophytes, cladocera, chironomids, ostracods) have also been compiled (see sections 18.6.4.5–18.6.4.8). Multifactor and multiproxy analyses, separating climate from other variables (e.g. anthropogenic influence), have also been developed (Lotter *et al.* 1997, 1998).

The training set of modern lakes, arranged along a particular environmental gradient (pH, salinity), is used to identify, and to quantify, the optimal range of particular (diatom) species, and then to reconstruct past conditions based on quantitative comparison of modern and fossil microfossil assemblages. Cores are collected from a range of lakes and the composition of the microfossil assemblage present in the uppermost 0.5–1 cm of sediment is quantified. This assemblage is deemed to represent an integrated sample of all habitats within the lake, and of seasonal variations in physical, chemical and biological factors operating within it, and its catchment, over the past few years, although the actual number of years is not usually known. Environmental variables for each lake (pH, salinity, conductivity, nutrient concentration) are then determined seasonally (monthly for nutrients), in order to derive mean or median values. If only one visit per year is possible, this should take place during circulation, rather than stratification. Representativity and quality of samples are key factors in the development of training sets.

Training sets are analysed using many different methods (Birks 1995, 1998), including canonical correspondence analysis (CCA; ter Braak 1986, 1987), in which scores for samples and taxa are arranged along ordination axes for environmental variables, the length of which indicates their contribution to explaining the variance presence in the dataset. These diagrams are used to summarise species–environment relationships, and to identify taxa characteristic of particular environmental conditions (Battarbee *et al.* 2001). The contribution of each variable to total variance is quantified using partial CCA, and a Monte Carlo permutation test which measures the independence, and the relative strength, of each variable in the dataset.

Transfer functions are developed using weighted averaging (WA), which assumes that, at particular values of environmental variables, taxa whose optimal ranges are closest to those values are most abundant. The **weighted averaging optimum** (u_k), or 'centroid', for each taxon is then cal-

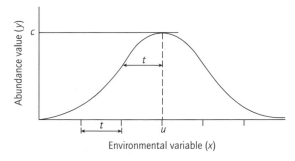

Fig. 18.23 Response curve for changing species abundance along an environmental gradient. Key: u, optimum; t, tolerance (one standard deviation); c, maximum. (From Battarbee *et al.* 2001; reproduced with kind permission of Kluwer Academic Publishers.)

culated. Weighted averaging is used to generate tolerance values for particular taxa (Fig. 18.23), and to 'down-weight' those with wide tolerance (Birks *et al.* 1990). Calculated optima are then used to quantify the value of the environmental variable at each site, by taking the average of species optima in the sample, weighted by their abundance. This approach may neglect correlations which are due to the influence of environmental variables other than those being studied, a problem overcome by using weighted averaging–partial least squares (WA–PLS; ter Braak & Juggins 1993; Birks 1995).

The ability of transfer functions to predict reconstructed environmental variables is then assessed by calculating the root mean squared error of prediction (RMSEP), which examines the correlation between the observed and the predicted values using techniques such as jack-knifing and bootstrapping (ter Braak & Juggins 1993). In the former, also called 'leave-one-out substitution' (Fritz *et al.* 1999), new training sets are formed, in each of which one sample is omitted from the original set, and its ability to measure species response and to predict lake conditions is re-evaluated. In bootstrapping, new training sets, the same size as the original set, are randomly selected from the data in the set, and the samples not selected form an independent 'test' set. This process is repeated many times, so that many combinations of training and test sets are evaluated. Software for developing transfer functions, or plotting stratigraphic data, is available for download at: http://

www.campus.ncl.ac.uk/staff/Stephen.Juggins/
software/c2home.htm

An alternative to the numerical approach is to
use an independent calibration set of the same size,
and arranged along a similar gradient (Fritz *et al.*
1999). Such sets are rarely available, however,
although the European Diatom Database (EDDI;
Battarbee *et al.* 2000) and the Diatom Paleolimnology Data (DPDC) website (http://diatom.
acnatsci.org/dpdc) are intended as web-based
sources, from which sets may be compiled. Where
fossil assemblages contain taxa not well represented in training sets, reconstructions are less
valid. Similarity between fossil assemblages and
training sets is also evaluated via modern analogue
techniques (MAT; Birks 1995), using chi-squared
distance as a measure of similarity/dissimilarity,
but this has yet to be widely applied in palaeolimnology (Battarbee *et al.* 2001).

18.7   PROSPECTS

Over the past 70 years, palaeolimnology has indeed developed from a minority scientific interest
to the multidisciplinary, multiproxy, highly automated, highly technical discipline described in
their Introduction by Last & Smol (2001a). Palaeolimnology clearly identifies the origin, manifestation and ecological consequences of many
environmental problems, and can also contribute
significantly to the evaluation of various goals
for lake and watershed rehabilitation (Battarbee
1997), but I am not aware that it has yet been used
to discuss the relative merits of proposed solutions
to such problems, which possess, of course, important economic, political and social dimensions.
Although scientists trained in the reductionist
paradigm of the mid-20th century will no doubt
feel uneasy about the use of scientific knowledge
for such purposes, palaeolimnology is uniquely positioned to develop the study of the relative impact
of different socio-economic systems on the rest
of nature (Messerli *et al.* 2000), so that this may
possibly be a direction in which the subject could
conceivably develop. These and other similar
themes will be taken up in Volume 2 of this book.

18.8   ACKNOWLEDGEMENTS

I would like to thank Frank Oldfield, Steve Row-
land and Helen Bennion for help with certain sections of this chapter, and Kimmo Tolonen and
Heikki Simola for introducing me to laminated
sediments. *Kiitos hyvää, ystavanit!*

18.9   REFERENCES

Anderson, R.Y. & Dean, W.A. (1988) Lacustrine varve
formation through time. In: Gray, J. (ed.), *Palaeolimnology: Aspects of Palaeoecology and Biogeography.*
Elsevier, Amsterdam, 215–35.

Appleby, P.G. (2001) Chronostratigraphic techniques
in recent sediments. In: Last, W.M. & Smol, J.P. (eds),
*Tracking Environmental Change using Lake Sedi-
ments*, Vol. 1, *Basin Analysis, Coring and Chronological Techniques*. Kluwer, Dordrecht, 171–203.

Appleby, P.G., Oldfield, F., Thompson, R., Huttunen, P. &
Tolonen, K. (1979) ^{210}Pb dating of annually laminated
lake sediments from Finland. *Nature*, **280**, 53–5.

Appleby, P.G., Nolan, P.J., Gifford, D.W., et al. (1986)
[210]Pb dating by low background gamma counting.
Hydrobiologia, **143**, 21–7.

Appleby, P.G., Richardson, N. & Nolan, P.J. (1991) [241]Am
dating of lake sediments. *Hydrobiologia*, **214**, 35–42.

Ariztegui, D., Hollander, D.J. & McKenzie, J.A. (1996)
Algal dominated lacustrine organic matter can be
either Type I or Type II: evidence for biological, chemi-
cal and physical controls on organic matter quality. In:
Mello, M.R.L., Trinidade, L.A.F. & Hessel, M.H.R.
(eds), *ALAGO Special Publication: Selected Papers
from the 4th Latin American Congress on Organic
Geochemistry, Bucaramanga, Colombia*. Associación
Latino-Americana de Geoquimica Orgânica (ALAGO),
Bogotá, Colombia, 12–16.

Barker, P.A., Street-Perrott, F.A., Leng, M.J., et al. (2001)
A 14,000-year oxygen isotope record from diatom
silica in two alpine lakes on Mt. Kenya. *Science*, **292**,
2307–10.

Battarbee, R.W. (1986) Diatom analysis. In: Berglund, B.E.
(ed.), *Handbook of Holocene Palaeoecology and
Palaeohydrology*. Wiley, Chichester, 527–70.

Battarbee, R.W. (1991) Recent palaeolimnology and
diatom-based environmental reconstruction. In:
Shane, L.C.K. & Cushing, E.J. (eds), *Quaternary Landscapes*. Belhaven Press, New York, 129–74.

Battarbee, R.W. (1997) Freshwater quality, naturalness and palaeolimnology. In: Boon, P.J. & Howell, D.L. (eds), *Freshwater Quality: Defining the Indefinable.* Scottish Natural Heritage, The Stationery Office, Edinburgh, 155–71.

Battarbee, R.W., Smol, J.P. & Meriläinen, J. (1986) Diatoms as indicators of pH: a historical review. In: Smol, J.P., Battarbee, R.W., Davis, R.B. & Meriläinen, J. (eds), *Diatoms and Lake Acidity: the Use of Siliceous Microfossils in the Reconstruction of pH.* Dr W. Junk, The Hague, 5–14.

Battarbee, R.W., Mason, J., Renberg, I., *et al.* (eds) (1990) Palaeolimnology and lake acidification. *Philosophical Transactions of the Royal Society of London, Series B,* **327**, 219 pp.

Battarbee, R.W., Charles, D.F., Dixit, S.S., *et al.* (1999) Diatoms as indicators of surface water acidity. In: Stoermer, E.F. & Smol, J.P. (eds), *The Diatoms: Applications for the Environmental and Earth Sciences.* Cambridge University Press, Cambridge, 85–127.

Battarbee, R.W., Juggins, S., Gasse, F., *et al.* (2000) European Diatom Database (EDDI): an information system for palaeoenvironmental reconstruction. *European Climate Science Conference, Vienna, October 1998,* 1–10.

Battarbee, R.W., Jones, V.J., Flower, R.J., *et al.* (2001) Diatoms. In: Smol, J.P., Birks, H.J.B. & Last, W.M. (eds), *Tracking Environmental Change using Lake Sediments,* Vol. 3, *Terrestrial, Algal and Siliceous Indicators.* Kluwer, Dordrecht, 155–202.

Bengtsson, L. & Enell, M. (1986) Chemical analysis. In: Berglund, B.E. (ed.), *Handbook of Holocene Palaeoecology and Palaeohydrology.* Wiley, Chichester, 423–51.

Bennett, K.D. & Willis, K.J. (2001) Pollen. In: Smol, J.P., Birks, H.J.B. & Last, W.M. (eds), *Tracking Environmental Change using Lake Sediments,* Vol. 3, *Terrestrial, Algal and Siliceous Indicators.* Kluwer, Dordrecht, 5–32.

Bennion, H., Juggins, S. & Anderson, N.J. (1996) Predicting epilimnetic phosphorus concentrations using an improved diatom-based transfer function, and its application to lake eutrophication management. *Environmental Science and Technology,* **30**, 2004–27.

Berglund, B.E. (ed.) (1986) *Handbook of Holocene Palaeoecology and Palaeohydrology.* Wiley, Chichester, 869 pp.

Berglund, B.E. & Ralska-Jasiewiczowa, M. (1986) Pollen analysis and pollen diagrams In: Berglund, B.E. (ed.), *Handbook of Holocene Palaeoecology and Palaeohydrology.* Wiley, Chichester, 455–84.

Beyens, L. & Meisterfeld, R. (2001) Protozoa: testate amoebae. In: Smol, J.P., Birks, H.J.B. & Last, W.M. (eds), *Tracking Environmental Change using Lake Sediments,* Vol. 3, *Terrestrial, Algal and Siliceous Indicators.* Kluwer, Dordrecht, 121–53,

Binford, M.W., Deevey, E.S., Jr. & Crisman, T.L. (1983) Palaeolimnology: a historical perspective on lacustrine ecosystems. *Annual Review of Ecology and Systematics,* **14**, 255–86.

Birks, H.H. (2001) Plant macrofossils. In: Smol, J.P., Birks, H.J.B. & Last, W.M. (eds), *Tracking Environmental Change using Lake Sediments,* Vol. 3, *Terrestrial, Algal and Siliceous Indicators.* Kluwer, Dordrecht, 49–74.

Birks, H.J.B. (1986) Numerical zonation, comparison and correlation of Quaternary pollen-stratigraphical data. In: Berglund, B.E. (ed.), *Handbook of Holocene Palaeoecology and Palaeohydrology.* Wiley, Chichester, 743–74.

Birks, H.J.B. (1995) Quantitative palaeoenvironmental reconstructions. In: Maddy, D.S. & Brew, J.S. (eds), *Statistical Modelling of Quaternary Data.* Technical Guide No. 5, Quaternary Research Association, Cambridge, 161–254.

Birks, H.J.B. (1998) Numerical tools in paleolimnology — progress, potentialities, and problems. *Journal of Paleolimnology,* **20**, 307–32.

Birks, H.J.B. & Gordon, A.D. (1985) *Numerical Methods in Quaternary Pollen Analysis.* Academic Press, London, 317 pp.

Birks, H.J.B., Juggins, S., Lotter, A.F., *et al.* (eds) (in press). *Tracking Environmental Change using Lake Sediments,* Vol. 5, *Data Handling and Statistical Techniques.* Kluwer, Dordtrecht.

Birks, H.J.B., Line, J.M., Juggins, S.M., *et al.* (1990) Diatoms and pH reconstruction. *Philosophical Transactions of the Royal Society of London, Series B,* **327**, 263–78.

Björck, S. & Wohlfarth, B. (2001) ^{14}C chronostratigraphic techniques in paleolimnology. In: Last, W.M. & Smol, J.P. (eds), *Tracking Environmental Change using Lake Sediments,* Vol. 1, *Basin Analysis, Coring and Chronological Techniques.* Kluwer, Dordrecht, 205–45.

Blais, J.M. & Muir, D.C.G. (2001) Paleolimnological methods and applications for persistent organic pollutants. In: Last, W.M. & Smol, J.P. (eds), *Tracking Environmental Change using Lake Sediments,* Vol. 2, *Physical and Geochemical Methods.* Kluwer, Dordrecht, 271–98.

Bonny, A.P. (1976) Recruitment of pollen to the seston

and sediment of some Lake District lakes. *Journal of Ecology*, **64**, 859–87.

Bonny, A.P. (1978) The effect of pollen recruitment processes on pollen distribution over the sediment surface of a small lake. *Journal of Ecology*, **66**, 385–416.

Bonny, A.P. & Allen, P.V. (1984) Pollen recruitment to the sediments of an enclosed lake in Shropshire, England. In: Haworth, E.Y. & Lund, J.W.G. (eds), *Lake Sediments and Environmental History*. Leicester University Press, Leicester, 231–59.

Boyle, J.F. (2001) Inorganic geochemical methods in paleolimnology. In: Last, W.M. & Smol, J.P. (eds), *Tracking Environmental Change using Lake Sediments*, Vol. 2, *Physical and Geochemical Methods*. Kluwer, Dordrecht, 83–141.

Bradbury, J.P. (1988) Fossil diatoms and Neogene palaeolimnology. In: Gray, J. (ed.), *Palaeolimnology: Aspects of Palaeoecology and Biogeography*. Elsevier, Amsterdam, 299–316.

Bradbury, J.P. & Dean, W.E. (eds) (1993) Elk Lake Minnesota: evidence for rapid climate change in the North-Central United States. *Geological Society of America Special Paper*, **276**, 336 pp.

Bradbury, J.P. & Waddington, J.C.B. (1973) The impact of European settlement on Shagawa Lake, Northeastern Minnesota, USA In: Birks, H.J.B. & West, R.G. (eds), *Quaternary Plant Ecology*. Blackwell, Oxford, 289–307.

Brown, T.A., Nelson, D.E., Mathewes, R.W., *et al.* (1989) Radiocarbon dating of pollen by accelerator mass spectrometry. *Quaternary Research*, **32**, 205–12.

Bruce, A., Horne, D.J. & Whittaker, J.E. (in press) Ostracoda as a tool for the assessment and monitoring of the aquatic environment of the Fleet Lagoon, Dorset, UK. *Aquatic Conservation*.

Carbonel, P., Colin, J-P., Danielpol, D.L., *et al.* (1988) Palaeoecology of limnic Ostracodes: a review of some major topics. In: Gray, J. (ed.), *Palaeolimnology: Aspects of Palaeoecology and Biogeography*. Elsevier, Amsterdam, 413–62.

Charles, D.F. & Whitehead, D.R. (1986) The PIRLA project: paleoecological investigation of recent acidification. *Hydrobiologia*, **143**, 3–20.

Charles, D.F., Smol, J.P. & Engstrom, D.R (1994) Paleolimnological approaches to biomonitoring. In: Loeb, S. & Spacie, A. (eds), *Biological Monitoring of Aquatic Systems*. Lewis Press, Ann Arbor, MI, 233–93.

Cohen, A.S. (2002) *Paleolimnology. History and Evolution of Lake Ecosystems*. Oxford University Press, Oxford, 350 pp.

Collinson, M.E. (1988) Freshwater macrophytes in palaeolimnology In: Gray, J. (ed.), *Palaeolimnology: Aspects of Palaeoecology and Biogeography*. Elsevier, Amsterdam, 317–42.

Conley, D.J. & Schelske, C.L. (2001) Biogenic silica. In: Smol, J.P., Birks, H.J.B. & Last, W.M. (eds), *Tracking Environmental Change using Lake Sediments*, Vol. 3, *Terrestrial, Algal and Siliceous Indicators*. Kluwer, Dordrecht, 281–93.

Cooper, M.C. (1997) The use of digital image analysis in the study of laminated sediments. *Journal of Paleolimnology*, **19**, 33–40.

Cooper, M.C. & O'Sullivan, P.E. (1998) The laminated sediments of Loch Ness: preliminary construction of a chronology of sedimentation, and its use in assessing Holocene climatic variability. *Palaeogeography, Palaeoclimatology, Palaeoecology*, **140**, 23–31.

Cousen, S.M. & O'Sullivan, P.E. (1984) The use of X-ray fluorescence spectrometry (XRFS) in palaeolimnology. *Verhandlungen Vereinigung Internationale Limnologie*, **23**, 1388–90.

Cranwell, P.A., Eglinton, G. & Robinson, N. (1987) Lipids of aquatic organisms as potential contributors to lacustrine sediments, II. *Organic Geochemistry*, **11**, 513–27.

Crisman, T.C. (1988) The use of subfossil benthic invertebrates in aquatic resource management. In: Adams, W.J., Chapman, G.A. & Landis, W.G. (eds), *Aquatic Toxicology and Hazard Assessment*, Vol. 10. American Society for Testing Materials, Philadelphia, 71–88.

Cronberg, G. (1986) Blue-green algae, green algae and Chrysophyceae in sediments In: Berglund, B.E. (ed.), *Handbook of Holocene Palaeoecology and Palaeohydrology*. Wiley, Chichester, 507–26.

Cuddington, K. & Leavitt, P.R. (1999) An individual-based model of pigment flux in lakes: implications for organic biogeochemistry and paleoecology. *Canadian Journal of Fisheries and Aquatic Sciences*, **56**, 1964–77.

Cushing, E.J. (1964) Re-deposited pollen spectra in Late-Wisconsin pollen spectra from East-Central Minnesota. *American Journal of Science*, **262**, 1075–88.

Davis, M.B. (1973) Redeposition of pollen grains in lake sediment. *Limnology and Oceanography*, **18**, 44–52.

Davis, M.B. (1976) Erosion rates and land use history in southern Michigan. *Environmental Conservation*, **3**, 139–48.

Davis, M.B. & Ford, M.S., Jr. (1982) Sediment focusing in Mirror Lake, New Hampshire. *Limnology and Oceanography*, **27**, 147–50.

Davis, M.B., Brubaker, L.B. & Webb, T. III (1973) Calibration of absolute pollen influx In: Birks, H.J.B. & West,

R.G. (eds), *Quaternary Plant Ecology.* Blackwell, Oxford, 9–25.

Davis, M.B., Moeller, R.E. & Ford, J. (1984) Sediment focusing and pollen influx In: Haworth, E.Y. & Lund, J.W.G. (eds), *Lake Sediments and Environmental History.* Leicester University Press, Leicester, 261–93.

Davis, R.B. (ed.) (1990) *Palaeolimnology and the Reconstruction of Ancient Environments.* Kluwer, Dordrecht, 254 pp.

Dean, J.M., Kemp, A.E.S., Bull, D., *et al.* (1999) Taking varves to bits: Scanning electron microscopy in the study of laminated sediments and varves. *Journal of Paleolimnology,* **22**, 121–36.

Dean, W.E. (1999) The carbon cycle and biogeochemical dynamics in lake sediments. *Journal of Paleolimnology,* **21**, 375–93.

Dean, W.E. (2002) A 1500-year record of climatic and environmental change in Elk Lake, Clearwater county, Minnesota II: Geochemistry, mineralogy, and stable isotopes. *Journal of Paleolimnology,* **27**, 301–19

Dean, W.E., Anderson, R.Y., Bradbury, J.P., *et al.* (2002) A 1500-year record of climatic and environmental change in Elk Lake, Minnesota I: varve thickness and gray-scale density. *Journal of Paleolimnology,* **27**, 287–99

Dearing, J.A. (1983) Changing patterns of sediment accumulation in a small lake in Scania, southern Sweden. *Hydrobiologia,* **103**, 59–64.

Dearing, J.A. (1991a) Lake sediment records of erosional processes. *Hydrobiologia,* **214**, 99–106.

Dearing, J.A. (1991b) Erosion and land use. In: Berglund, B.E. (ed.), *The Cultural Landscape in Sweden during the past 6000 Years. Ecological Bulletin,* **41**, 283–92.

Dearing, J.A. (1999a) Holocene environmental change from magnetic proxies in lake sediments. In: Thompson, R. & Maher, B. (eds), *Quaternary Climates, Environments and Magnetism.* Cambridge University Press, Cambridge, 231–78.

Dearing, J.A. (1999b) Magnetic susceptibility. In: Walden, J., Oldfield, F. & Smith, J.P. (eds), *Environmental Magnetism: a Practical Guide.* Technical Guide No. 6, Quaternary Research Association, Cambridge, 35–62.

Dearing, J.A., Alström, K., Bergman, A., *et al.* (1990) Recent and long term records of soil erosion from southern Sweden In: Boardman, J., Foster, I.D.L. & Dearing, J.A. (eds), *Soil Erosion on Agricultural Land.* Wiley, Chichester, 173–91.

Dearing, J.A., Håkansson, H-L., Liedberg-Jönsson, *et al.* (1987) Lake sediments used to quantify the erosional

response to land use change in southern Sweden. *Oikos,* **50**, 60–78.

De Deckker, P. (1988) An account of techniques using Ostracodes in palaeolimnology in Australia. In: Gray, J. (ed.), *Palaeolimnology: Aspects of Palaeoecology and Biogeography.* Elsevier, Amsterdam, 463–76.

De Deckker, P. & Forester, R.M. (1988) The use of ostracods to reconstruct palaeoenvironmental records. In: De Deckker, P., Colin, J-P. & Peypouquet, J-P. (eds), *Ostracoda in the Earth Sciences.* Elsevier, Amsterdam, 175–99.

Deevey, E.S., Jr. (1942) Studies on Connecticut lake sediments. III—the biostratonomy of Linsley Pond. *American Journal of Science,* **240**, 233–64.

Deevey, E.S., Jr. (1955) The obliteration of the hypolimnion. *Memorie dell'Istituto Italiano di Idrobiologie, Supplement,* **8**, 9–38.

Deevey, E.S., Jr. (1984) Stress, strain, and stability of lacustrine ecosystems. In: Haworth, E.Y. & Lund, J.W.G. (eds), *Lake Sediments and Environmental History.* Leicester University Press, Leicester, 203–29.

De Geer, G. (1912) A geochronology of the last 12,000 years. *Proceedings of the International Geological Congress, Stockholm,* 241–53.

Dekkers, M.J. (1997) Environmental magnetism: an introduction. *Geologie en Mijnbouw,* **76**, 163–82.

De Leeuw, J.W. & Largeau, C. (1993) A review of macromolecular organic compounds that comprise living organisms, and their role in kerogen, coal and petroleum formation. In: Engel, M.H. & Macko, S.A. *Organic Geochemistry: Principles and Applications.* Plenum, New York, 23–72.

Digerfeldt, G. (1986) Studies on lake level changes. In: Berglund, B.E. (ed.), *Handbook of Holocene Palaeoecology and Palaeohydrology.* Wiley, Chichester, 127–43.

Dixit, S.S., Smol, J.P., Kingston, J.C., *et al.* (1992) Diatoms: powerful indicators of environmental change. *Environmental Science and Technology,* **26**, 23–33.

Douglas, M.S.V. & Smol, J.P. (2001) Siliceous protozoan plates and scales. In: Smol, J.P., Birks, H.J.B. & Last, W.M. (eds), *Tracking Environmental Change using Lake Sediments,* Vol. 3, *Terrestrial, Algal and Siliceous Indicators.* Kluwer, Dordrecht, 265–79.

Edmondson, W.T. (1991) *The Uses of Ecology: Lake Washington and Beyond.* University of Washington Press, Seattle, 329 pp.

Eggleton, F.E. (1931) A limnological study of the profundal bottom fauna of certain freshwater lakes. *Ecological Monographs,* **1**, 231–301.

Eglinton, T.I., Benitez Nelson, B.C., Pearson, A., *et al.* (1997) Variability in radiocarbon ages of individual organic compounds from marine sediments. *Science*, **277**, 796–9.

ELDP (1995) *Climate Prediction Based on the Study of Climate Dynamics on an Annual Basis during the Last Glacial/Interglacial Cycle.* European Lakes Drilling Program, Preliminary Project Outline, May 1994, GeoForschungZentrum, Potsdam, 35 pp. (www.gfz-potsdam.de)

Emmelin, L. (1975) *Environmental Education at University Level.* Council of Europe, Strasbourg, 150 pp.

Engstrom, D.R. & Wright, H.E., Jr. (1984) Chemical stratigraphy of lake sediments as a record of environmental change. In: Haworth, E.Y. & Lund, J.W.G. (eds), *Lake Sediments and Environmental History.* Leicester University Press, Leicester, 11–69.

Engstrom, D.R., Swain. E.B. & Kingston, J.C. (1985) A paleolimnological record of human disturbance from Harvey Lake, Vermont—geochemistry, pigments and diatoms. *Freshwater Biology*, **15**, 261–88.

Faegri, K. & Iversen, J. (1989) *Textbook of Pollen Analysis* (4th edition). Wiley, Chichester, 328 pp.

Finney, B.P., Gregory-Eaves, I., Douglas, M.S.V., *et al.* (2002) Fisheries productivity in the northeastern Pacific Ocean over the past 2,200 years. *Nature*, **416**, 729–33

Frey, D.G. (1986) Cladocera analysis In: Berglund, B.E. (ed.), *Handbook of Holocene Palaeoecology and Palaeohydrology.* Wiley, Chichester, 667–92.

Frey, D.G. (1988) Littoral and offshore communities of diatoms, cladocerans and dipterous larvae, and their interpretation in paleolimnology. *Journal of Paleolimnology*, **1**, 179–91.

Frey, D.G. (1990) Littoral and offshore communities of diatoms, cladocerans, and dipterous larvae, and their interpretation in palaeolimnology. In: Davis, R.B. (ed.) *Palaeolimnology and the Reconstruction of Ancient Environments.* Kluwer, Dordrecht, 133–45.

Fritz, S.C., Cumming, B.C., Gasse, F., *et al.* (1999) Diatoms as indicators of hydrologic and climatic change in saline lakes. In: Stoermer, E.F. & Smol, J.P. (eds), *The Diatoms: Applications for the Environmental and Earth sciences.* Cambridge University Press, Cambridge, 41–72.

Frost, T.M. (2001) Freshwater sponges. In: Smol, J.P., Birks, H.J.B. & Last, W.M. (eds), *Tracking Environmental Change using Lake Sediments*, Vol. 3, *Terrestrial, Algal and Siliceous Indicators.* Kluwer, Dordrecht, 253–63.

Griffiths, H.I. & Holmes, J.A. (2000) *Non-marine Ostracods and Quaternary Palaeoenvironments.* Quaternary Research Association, London, 188 pp.

Grimm, E. (1993) *TILIA 2.0* (Software). Illinois State Museum, Springfield, IL.

Hall, R.I & Smol, J.P. (1992) A weighted averaging regression and calibration model for inferring total phosphorus concentration from diatoms in British Columbia (Canada) lakes. *Freshwater Biology*, **27**, 417–34.

Hall, R.I. & Smol, J.P. (1999) Diatoms as indicators of lake eutrophication. In: Stoermer, E.F. & Smol, J.P. (eds), *The Diatoms: Applications for the Environmental and Earth Sciences.* Cambridge University Press, Cambridge, 128–68.

Havinga, A.J. (1984) A twenty year experimental investigation into the differential corrosion susceptibility of pollen and spores in various soil types. *Pollen et Spores*, **26**, 541–58.

Haworth, E.Y. & Lund, J.W.G. (eds) (1984) *Lake sediments and Environmental History.* Leicester University Press, Leicester, 411 pp.

Hedges, R.E.M. (1991) AMS dating: present status and potential applications In: Lowe, J.J. (ed.), *Radiocarbon Dating: Recent Applications and Future Potential.* Quaternary Proceedings No. 1, Quaternary Research Association, Cambridge, 5–10.

Heiri, O., Lotter, A.F. & Lemcke, G. (2000) Loss on ignition as method for estimating organic and carbonate contents in sediments: reproducibility and comparability of results. *Journal of Paleolimnology*, **25**, 101–10.

Henon, D.N., O'Sullivan, P.E., Matthews, N.M., *et al.* (1999) Swanpool, Falmouth. In: Scourse, J.D. & Furze, M.F.A. (eds), *The Quaternary of West Cornwall.* Quaternary Research Association, London, 122–35.

Hilton, J. (1985) A conceptual framework for predicting the occurrence of sediment focusing and sediment redistribution in small lakes. *Limnology and Oceanography*, **30**, 1131–43.

Hilton, J. (1987) A simple model for the interpretation of magnetic records in lacustrine and ocean sediments. *Quaternary Research*, **27**, 160–6.

Hofmann, W. (1986) Chironomid analysis. In: Berglund, B.E. (ed.), *Handbook of Holocene Palaeoecology and Palaeohydrology.* Wiley, Chichester, 715–27.

Hofmann, W. (1987) Cladocera in space and time: analysis of lake sediments. *Hydrobiologia*, **145**, 315–21.

Hofmann, W. (1988) The significance of chironomid analysis (Insecta: Diptera) for palaeolimnological research. In: Gray, J. (ed.), *Palaeolimnology: Aspects of Palaeoecology and Biogeography.* Elsevier, Amsterdam, 501–9.

Hofmann, W. (1998) Cladocerans and chironomids as indicators of lake level changes in north temperate lakes. *Journal of Paleolimnology*, **19**, 55–62.

Holloway, J.M., Dahlgren, R.A., Hansen, B., *et al.* (1998) Contribution of bedrock nitrogen to high nitrate concentrations in stream water. *Nature*, **395**, 785–8.

Holmes, J.A. (1996) Trace element and stable isotope geochemistry of non-marine ostracod shells in Quaternary palaeoenvironmental reconstruction. *Journal of Paleolimnology*, **15**, 223–35.

Holmes, J.A. (2001) Ostracoda. In: Smol, J.P., Birks, H.J.B. & Last, W.M. (eds), *Tracking Environmental Change using Lake Sediments*, Vol. 4, *Zoological Indicators.* Kluwer, Dordrecht, 125–51.

Holmes, J.A. & Horne, D.J. (eds) (1999) Non-marine ostracods: evolution and environment. *Palaeogeography, Palaeoclimatology, Palaeoecology*, **148**, 185 pp.

Hustedt, F. (1937–1939) Systematische und ökologische Untersuchungen über den Diatomeen-Flora von Java, Bali, und Sumatra. *Archiv für Hydrobiologie Supplement Band* **15 & 16**.

Hustedt, F. (1957) Die Diatomeenflora des Fluss-systems der Weser im Gebiet der Hansestadt Bremen. *Abhlangerungen Naturwissenschaften Vereinigung Bremen*, **34**, 181–440.

Hutchinson, G.E. (1957) *A Treatise on Limnology*, Vol. 1, *Geography, Physics, Chemistry.* Wiley, New York, 1016 pp.

Hutchinson, G.E. (1973) Eutrophication. *American Scientist*, **61**, 269–79.

Huttunen, P., Meriläinen, J., Cotton, C., *et al.* (1988) Attempts to reconstruct lake water pH and colour from sedimentary diatoms and Cladocera. *Verhandlungen Internationale Vereinigung Limnologie*, **23**, 870–3.

Ito, E. (2001) Application of stable isotope techniques to inorganic and biogenic carbonates. In: Last, W.M. & Smol, J.P. (eds), *Tracking Environmental Change using Lake Sediments*, Vol. 2, *Physical and Geochemical Methods.* Kluwer, Dordrecht, 351–71.

Jacobson, G.L. & Bradshaw, R.H.W. (1981) The selection of sites for palaeovegetational studies. *Quaternary Research*, **16**, 80–96.

Johnson, T.C. & Odada, E.O. (eds) (1996) *The Limnology, Climatology and Paleoclimatology of the East African Lakes.* Gordon & Breach, Amsterdam, 456 pp.

Jørgensen, S-E. (1974) *Lake Management.* Pergamon, Oxford, 167 pp.

Kemp, A.E.S., Dean, J., Pearce, R.B., *et al.* (2001) Recognition and analysis of bedding and sediment fabric features. In: Last, W.M. & Smol, J.P. (eds), *Tracking Environmental Change using Lake Sediments*, Vol. 2, *Physical and Geochemical Methods.* Kluwer, Dordrecht, 7–22.

Kilham, P. & Kilham, S.S. (1990) Endless summer—internal loading processes dominate nutrient cycling in tropical lakes. *Freshwater Biology*, **23**, 379–89.

Kilham, P., Kilham, S.S. & Hecky, R.E. (1986) Hypothesised resource relationships among African planktonic diatoms. *Limnology and Oceanography*, **41**, 1052–62.

King, J. & Peck, J. (2001) Use of paleomagnetism in studies of lake sediments. In: Last, W.M. & Smol, J.P. (eds), *Tracking Environmental Change using Lake Sediments*, Vol. 1, *Basin Analysis, Coring and Chronological Techniques.* Kluwer, Dordrecht, 371–89.

Korhola, A. (1999) Distribution patterns of Cladocera in subarctic Fennoscandian lakes, and their potential in environmental reconstruction. *Ecography*, **22**, 357–73.

Korhola, A. & Rautio, M. (2001) Cladocera and other branchiopod crustaceans. In: Smol, J.P., Birks, H.J.B. & Last, W.M. (eds), *Tracking Environmental Change using Lake Sediments*, Vol. 4, *Zoological Indicators.* Kluwer, Dordrecht, 5–41.

Korhola, A. & Smol, J.P. (2001) Ebridians. In: Smol, J.P., Birks, H.J.B. & Last, W.M. (eds), *Tracking Environmental Change using Lake Sediments*, Vol. 3, *Terrestrial, Algal and Siliceous Indicators.* Kluwer, Dordrecht, 225–34.

Krug, A. (1993) Drainage history and land use pattern of a Swedish river system—their importance for understanding nitrogen and phosphorus load. *Hydrobiologia*, **251**, 285–96.

Lamb, H.H. (1977) *Climate: Past, Present and Future*, Vol. 2, *Climate History and the Future.* Appendix v.29. Methuen, London, 613–17.

Lamoureux, S. (2001) Varve chronology techniques. In: Last, W.M. & Smol, J.P. (eds), *Tracking Environmental Change using Lake Sediments*, Vol. 1, *Basin Analysis, Coring and Chronological Techniques.* Kluwer, Dordrecht, 247–60.

Last, W.M. (2001a) Textural analysis of lake sediments. In: Last, W.M. & Smol, J.P. (eds), *Tracking Environmental Change using Lake Sediments*, Vol. 2, *Physical and Geochemical Methods.* Kluwer, Dordrecht, 41–81.

Last, W.M. (2001b) Mineralogical analysis of lake sediments. In: Last, W.M. & Smol, J.P. (eds), *Tracking Environmental Change using Lake Sediments*, Vol. 2, *Physical and Geochemical Methods.* Kluwer, Dordrecht, 143–87.

Last, W.M. & Smol, J.P. (eds) (2001a) *Tracking Environmental Change using Lake Sediments*, Vol. 1, *Basin Analysis, Coring and Chronological Techniques.* Kluwer, Dordrecht, 548 pp.

Last, W.M. & Smol, J.P. (eds) (2001b) *Tracking Environmental Change using Lake Sediments*, Vol. 2, *Physical and Geochemical Methods.* Kluwer, Dordrecht, 504 pp.

Last, W.M. & Smol, J.P. (2001c) Introduction. In: Last, W.M. & Smol, J.P. (eds), *Tracking Environmental Change using Lake Sediments*, Vol. 1, *Basin Analysis, Coring and Chronological Techniques*. Kluwer, Dordrecht, 1–5.

Leavitt, P.R. (1993) A review of factors that regulate carotenoid and chlorophyll deposition and fossil pigment abundance. *Journal of Paleolimnology*, **9**, 109–27.

Leavitt, P.R. & Hodgson, D.A. (2001) Sedimentary pigments. In: Smol, J.P., Birks, H.J.B. & Last, W.M. (eds), *Tracking Environmental Change using Lake Sediments*, Vol. 3, *Terrestrial, Algal and Siliceous Indicators*. Kluwer, Dordrecht, 295–325.

Leavitt, P.R., Sandford, P.R., Carpenter, S.R., *et al.* (1994) An annual fossil record of production, planktivory and piscivory during whole-lake experiments. *Journal of Paleolimnology*, **11**, 133–49.

Lees, J. (1999) Evaluating magnetic parameters for use in source identification, classification and modelling of natural and environmental materials. In: Walden, J., Oldfield, F. & Smith, J.P. (eds), *Environmental Magnetism: a Practical Guide*. Technical Guide No. 6, Quaternary Research Association, Cambridge, 113–38.

Likens, G.E. (1979) The role of watershed and airshed in lake metabolism. *Archiv für Hydrobiologie, Beihefte Ergebnisse der Limnologie*, **13**, 195–211.

Likens, G.E. (ed.) (1985) *An ecosystem approach to aquatic ecology: Mirror Lake and its Environment*. Springer-Verlag, New York, 526 pp.

Loeffler, H. (1986) Ostracod analysis. In: Berglund, B.E. (ed.), *Handbook of Holocene Palaeoecology and Palaeohydrology*. Wiley, Chichester, 693–702.

Lotter, A.F. & Lemcke, G. (1999) Methods for preparing and counting biochemical varves. *Boreas*, **28**, 243–52.

Lotter, A.F., Birks, H.J.B., Hofmann, W., *et al.* (1997) Modern diatom, cladocera, chironomid and chrysophyte cyst assemblages as quantitative indicators for the reconstruction of past environmental conditions in the Alps. I. Climate. *Journal of Paleolimnology*, **18**, 395–420.

Lotter, A.F., Birks, H.J.B., Hofmann, W., *et al.* (1998) Modern diatom, cladocera, chironomid and chrysophyte cyst assemblages as quantitative indicators for the reconstruction of past environmental conditions in the Alps. II. Nutrients. *Journal of Paleolimnology*, **19**, 443–63.

Maberly, S.C., Hurley, M.A., Butterwick, C., *et al.* (1994) The rise and fall of *Asterionella formosa* in the south basin of Windermere: analysis of a 45-year series of data. *Freshwater Biology*, **31**, 19–34.

MacDonald, G.M. (2001) Conifer stomata. In: Smol, J.P., Birks, H.J.B. & Last, W.M. (eds), *Tracking Environmental Change using Lake Sediments*, Vol. 3, *Terrestrial, Algal and Siliceous Indicators*. Kluwer, Dordrecht, 33–47.

Mackereth, F.J.H. (1965) Chemical investigation of lake sediments and their interpretation. *Proceedings of the Royal Society of London*, **161**, 295–309.

Mackereth, F.J.H. (1966) Some chemical observations on Post-Glacial lake sediments. *Philosophical Transactions of the Royal Society of London, Series B*, **250**, 165–213.

Mackereth, F.J.H. (1971) On the variation in direction of the horizontal component of remanent magnetisation in lake sediments. *Earth and Planetary Science Letters*, **12**, 332–8.

Maddy, D. & Brew, J.S. (1995) *Statistical Modelling of Quaternary Science Data*. Technical Guide No. 5, Quaternary Research Association, London, 271 pp.

Maher, B.A., Thompson, R. & Hounslow, M.W. (1999) Introduction. In: Thompson, R. & Maher, B. (eds), *Quaternary Climates, Environments and Magnetism*. Cambridge University Press, Cambridge, 1–48.

Merkt, J. & Muller, H. (1999) Varve chronology and palynology of the Lateglacial in Northwest Germany from lacustrine sediments of Hamelsee in Lower Saxony. *Quaternary International*, **61**, 41–59.

Messerli, B., Grosjean, M., Hofer, T., *et al.* (2000) From nature-dominated to human-dominated environmental changes. *Quaternary Science Reviews*, **19**, 459–79.

Meyers, P.A. (1997) Organic geochemical proxies of paleoceanographic, paleolimnologic, and paleoclimatic processes. *Organic Geochemistry*, **27**, 213–50.

Meyers, P.A. & Ishiwatari, R. (1993) The early diagenesis of organic matter in lacustrine sediments. In: Engel, M.H. & Macko, S.A. *Organic Geochemistry: Principles and Applications*. Plenum, New York, 185–296.

Meyers, P.A. & Lallier-Vergès, E. (1999) Lacustrine sedimentary organic matter records of Late Quaternary paleoclimates. *Journal of Paleolimnology*, **21**, 345–72.

Meyers, P.A. & Teranes, J.L. (2001) Sediment organic matter. In: Last, W.M. & Smol, J.P. (eds), *Tracking Environmental Change using Lake Sediments*, Vol. 2, *Physical and Geochemical Methods*. Kluwer, Dordrecht, 239–69.

Moore, P.D., Webb, J.A. & Collinson, M.E. (1991) *Pollen Analysis* (2nd edition). Blackwell, Oxford, 216 pp.

Moss, B. (1980) *The Ecology of Fresh Waters*. Blackwell, Oxford, 332 pp.

Moss, B., Madgwick, J. & Phillips, G.W. (1996) *A Guide to the Restoration of Nutrient-enriched Shallow Lakes.* Broads Authority, Norwich, 179 pp.

Mourguiart, P. & Carbonel, P. (1994) A quantitative method of paleo-lake level reconstruction using ostracod assemblages: an example from the Bolivian Altiplano. *Hydrobiologia,* **288**, 183–93.

Nijampurkar, V.N., Rao, D.K., Oldfield, F. & Renberg, I. (1998) The half-life of Si-32: a new estimate based on varved sediments. *Earth and Planetary Science Letters,* **163**, 191–6.

Nipkow, F. (1920) Vorfläuge Mitteilungen über untersuchungen des Schlammbatzes in Zürichsee. *Zeitschrift für Hydrologie,* **1**, 100–22.

Nowaczyk, N.R. (2001) Logging of magnetic susceptibility. In: Last, W.M. & Smol, J.P. (eds), *Tracking Environmental Change using Lake Sediments,* Vol. 1, *Basin Analysis, Coring and Chronological Techniques.* Kluwer, Dordrecht, 155–70.

Odum, E.P. (1969) The strategy of ecosystem development. *Science,* **164**, 262–70.

Oldfield, F. (1977) Lakes and their drainage basins as units of sediment-based ecological study. *Progress in Physical Geography,* **3**, 460–504.

Oldfield, F. (1981) Peats and lake sediments. In: Goudie, A.S. (ed.), *Geomorphological Techniques.* Allen & Unwin, London, 306–26.

Oldfield, F. (1991) Environmental magnetism—a personal perspective. *Quaternary Science Reviews,* **10**, 73–85.

Oldfield, F. (1999a) The rock magnetic identification of magnetic mineral and grain size assemblages. In: Walden, J., Oldfield, F. & Smith, J.P. (eds), *Environmental Magnetism: a Practical Guide.* Technical Guide No. 6, Quaternary Research Association, Cambridge, 98–112.

Oldfield, F. (1999b) Environmental magnetism: the range of applications. In: Walden, J., Oldfield, F. & Smith, J.P. (eds), *Environmental Magnetism: a Practical Guide.* Technical Guide No. 6, Quaternary Research Association, Cambridge, 212–22.

Oldfield, F. & Appleby, P.G. (1984) Empirical testing of ^{210}Pb dating models for lake sediments In: Haworth, E.Y. & Lund, J.W.G. (eds), *Lake Sediments and Environmental History.* Leicester University Press, Leicester, 93–124.

Olsson, I. (1986) Radiometric dating. In: Berglund, B.E. (ed.), *Handbook of Holocene Palaeoecology and Palaeohydrology.* Wiley, Chichester, 273–312.

Olsson, I. (1991) Accuracy and precision in sediment chronology. *Hydrobiologia,* **214**, 25–34.

O'Sullivan, P.E. (1979) The ecosystem–watershed concept in the environmental sciences—a review. *International Journal of Environmental Studies,* **13**, 273–81.

O'Sullivan, P.E. (1983) Annually laminated sediments and the study of Quaternary environmental changes—a review. *Quaternary Science Reviews,* **1**, 245–313.

O'Sullivan, P.E. (1992) The eutrophication of shallow coastal lakes in Southwest England—understanding, and recommendations for restoration, based on palaeolimnology, historical records, and the modelling of changing phosphorus loads. *Hydrobiologia,* **243/244**, 421–34.

O'Sullivan, P.E. (1994a) The Natural History of Slapton Ley Nature Reserve. XXI: the palaeolimnology of the uppermost sediments of the Lower Ley, with interpretations based on ^{210}Pb dating and the historical record. *Field Studies,* **8**, 405–49.

O'Sullivan, P.E. (1994b) Improving the accuracy of radiocarbon dates using annually laminated sediments In: Hicks, S.P., Miller, U. & Saarnisto, M. (eds), *Laminated Sediments.* Council of Europe, Rixensart, Belgium, 63–88.

O'Sullivan, P.E. (1995a) Palaeolimnology. *Encyclopaedia of Environmental Biology,* Vol. 3. Academic Press, San Diego, 37–60.

O'Sullivan, P.E. (1995b) Eutrophication—a review. *International Journal of Environmental Studies,* **47**, 173–95.

O'Sullivan, P.E. (1998) Cows, pigs, war, but (so far as is known) no witches—the historical ecology of the Lower Ley, Slapton, Devon, and its catchment, 1840 to the present. In: Blacksell, M., Matthews, J.M. & Sims, P.C. (eds), *Environmental Management and Change in Plymouth and the South West. Essays in Honour of Dr John C. Goodridge.* University of Plymouth, Plymouth, 53–72.

O'Sullivan, P.E. (1999) Loe Pool and Loe Bar. In: Scourse, J.D. & Furze, M.F.A. (eds), *The Quaternary of West Cornwall.* Quaternary Research Association, London, 141–55.

O'Sullivan, P.E., Oldfield, F. & Battarbee, R.W. (1973) Preliminary studies of Lough Neagh sediments. I, stratigraphy, chronology and pollen analysis. In: Birks, H.J.B. & West, R.G. (eds), *Quaternary Plant Ecology.* Blackwell, Oxford, 267–78.

O'Sullivan, P.E., Heathwaite, A.L., Appleby, P.G., *et al.* (1991) Palaeolimnology of Slapton Ley, Devon, UK. *Hydrobiologia,* **214**, 115–24.

O'Sullivan, P.E., Glaser, R., Jacobeit, J., *et al.* (2001) Natural variability of climate on the Atlantic margin of Europe over the past two millennia—a proposal for further study. *Terra Nostra*, **2001/3**, 33–42.

O'Sullivan, P.E., Moyeed, R., Nicholson. M.J., *et al.* (2002) Comparison between instrumental, observational and high resolution proxy sedimentary records of Late Holocene climatic change on the Atlantic seaboard of Europe—a discussion of possibilities. *Quaternary International*, **88**, 27–44.

Patterson, W.P. & Smith, G.R. (2001) Fish. In: Smol, J.P., Birks, H.J.B. & Last, W.M. (eds), *Tracking Environmental Change using Lake Sediments*, Vol. 4, *Zoological Indicators*. Kluwer, Dordrecht, 173–87.

Peck, R.M. (1973) Pollen budget studies in a small Yorkshire catchment. In: Birks, H.J.B. & West, R.G. (eds), *Quaternary Plant Ecology*. Blackwell, Oxford, 43–60.

Pennington, W. (1979) The origin of pollen in lake sediments: an enclosed lake compared with one receiving inflowing streams. *New Phytologist*, **83**, 189–213.

Pennington, W., Cambray, R.S. & Fisher, E.M. (1973) Observations on lake sediments using fall-out ^{137}Cs as a tracer. *Nature*, **242**, 324–6.

Pennington, W., Cambray, R.S., Eakins, J.D., *et al.* (1976) Radionuclide dating of the recent sediments of Blelham Tarn. *Freshwater Biology*, **6**, 317–31.

Piperno, D.R. (2001) Phytoliths. In: Smol, J.P., Birks, H.J.B. & Last, W.M. (eds), *Tracking Environmental Change using Lake Sediments*, Vol. 3, *Terrestrial, Algal and Siliceous Indicators*. Kluwer, Dordrecht, 235–51.

Prentice, I.C. (1988) Records of vegetation in time and space: the principles of pollen analysis. In: Huntley, B. & Webb, T., III (eds), *Vegetation History*: Handbook of Vegetation Science 7, Kluwer, Dordrecht, 17–42.

Ralska-Jasiewiczowa, M., Goslar, T., Madeyska, T., *et al.* (eds) (1988) *Lake Gosciaz, central Poland: a Monographic Study*, Part 1. W. Szafer Institute of Botany, Polish Academy of Sciences, Krakow, 339 pp.

Renberg, I. (1986) Concentration and annual accumulation values of heavy metals in lake sediments: their significance in studies of the history of heavy metal pollution. *Hydrobiologia*, **143**, 379–85.

Renberg, I. & Wik, M. (1984) Dating of recent lake sediments by soot particle counting. *Verhandlungen Vereinigung Internationale Limnologie*, **22**, 712–18.

Robinson, N., Cranwell, P.A., Finlay, B.J., *et al.* (1984) Lipids of aquatic organisms as potential contributors to lacustrine sediments. *Organic Geochemistry*, **6**, 143–52.

Rose, N.L. (2001) Fly ash particles. In: Last, W.M. & Smol, J.P. (eds), *Tracking Environmental Change using Lake Sediments*, Vol. 2, *Physical and Geochemical Methods*. Kluwer, Dordrecht, 319–49.

Rose, N.L., Harlosck, S., Appleby, P.G. & Battarbee, R.W. (1995) The dating of recent lake sediments in the United Kingdom and Ireland using spheroidal carbonaceous particle concentration profiles. *The Holocene*, **5**, 328–35.

Ryding, S-O. & Rast, W. (eds) (1989) *The Control of Eutrophication of Lakes and Reservoirs*. Parthenon, Carnforth, 314 pp.

Saarinen, T. & Petterson, G. (2001) Image analysis techniques. In: Last, W.M. & Smol, J.P. (eds), *Tracking Environmental Change using Lake Sediments*, Vol. 2, *Physical and Geochemical Methods*. Kluwer, Dordrecht, 23–39.

Saarnisto, M. (1986) Annually laminated lake sediments. In: Berglund, B.E. (ed.), *Handbook of Holocene Palaeoecology and Palaeohydrology*. Wiley, Chichester, 693–702.

Sandgren, P. & Snowball, I. (2001) Application of mineral magnetic techniques to paleolimnology. In: Last, W.M. & Smol, J.P. (eds), *Tracking Environmental Change using Lake Sediments*, Vol. 2, *Physical and Geochemical Methods*. Kluwer, Dordrecht, 217–37.

Sanger, J.E. (1988) Fossil pigments in palaeoecology and palaeolimnology. In: Gray, J. (ed.), *Palaeolimnology: Aspects of Palaeoecology and Biogeography*. Elsevier, Amsterdam, 343–59.

Scholz, C.A. (2001) Applications of seismic stratigraphy in lake basins. In: Last, W.M. & Smol, J.P. (eds), *Tracking Environmental Change using Lake Sediments*, Vol. 1, *Basin Analysis, Coring and Chronological Techniques*. Kluwer, Dordrecht, 7–22.

Shostakovich, V.B. (1934) Ilovye otlozhenia ozer i periodicheskie kolebania v yavleniyakh prirody (Clayey lake deposits and natural periodical variations). *Zapiski Gosudarstvennogo Gidrologicheskogo Instituta*, **13**, 95–140.

Siegenthaler, U. & Eicher, U. (1986) Stable oxygen and carbon isotope analyses. In: Berglund, B.E. (ed.), *Handbook of Holocene Palaeoecology and Palaeohydrology*. Wiley, Chichester, 407–22.

Simola, H. (1992) Structural elements in varved lake sediments. *Geological Survey of Finland, Special Paper* **14**, 5–9.

Smith, J.P. (1985) *Mineral magnetic studies on two Shropshire–Cheshire meres*. Unpublished PhD thesis, University of Liverpool.

Smith, J.P. (1999) An introduction to the magnetic properties of natural materials. In: Walden, J., Oldfield, F. & Smith, J.P. (eds), *Environmental Magnetism: a*

Practical Guide. Technical Guide No. 6, Quaternary Research Association, Cambridge, 5–26.

Smol, J.P. (1988) Chrysophycaean microfossils in palaeolimnological studies. In: Gray, J. (ed.), *Palaeolimnology: Aspects of Palaeoecology and Biogeography.* Elsevier, Amsterdam, 287–98.

Smol, J.P. (1989a) Paleolimnology—recent advances and future challenges. In: De Bernardi, R. (ed.), *Scientific Perspectives in Theoretical and Applied Limnology. Memorie dell'Istituto Italiano Idrobiologia,* **47**, 253–76.

Smol. J.P. (1989b) Paleolimnology: an important tool for effective ecosystem management. *Journal of Aquatic Ecosystem Health,* **1**, 49–58.

Smol, J.P. (2002) *Pollution of Lakes and Rivers: a Palaeoenvironmental Perspective.* Arnold, London, 280 pp.

Smol, J.P., Birks, H.J.B. & Last, W.M. (eds) (2001a) *Tracking Environmental Change using Lake Sediments,* Vol. 3, *Terrestrial, Algal and Siliceous Indicators.* Kluwer, Dordrecht, 371 pp.

Smol, J.P., Birks, H.J.B. & Last, W.M. (eds) (2001b) *Tracking Environmental Change using Lake Sediments,* Vol. 4, *Zoological Indicators.* Kluwer, Dordrecht, 217 pp.

Snowball, I. (1999) Electromagnetic units and their use in environmental magnetic studies. In: Walden, J., Oldfield, F. & Smith, J.P. (eds), *Environmental Magnetism: a Practical Guide.* Technical Guide 6, Quaternary Research Association, London, 89–97.

Stoermer, E.F. & Smol, J.P. (eds) (1999) *The Diatoms: Applications for the Environmental and Earth Sciences.* Cambridge University Press, Cambridge, 469 pp.

Stuiver, M., Reimer, P.J., Bard, E., *et al.* (1998) INTCAL98 radiocarbon age calibration, 24,000–0 cal BP. *Radiocarbon,* **40**, 1041–83.

Sturm, M. (1979) Origin of clastic varves. In: Schlüchter, C. (ed.), *Moraines and Varves.* Balkema, Rotterdam, 281–5.

Sugita, S., Gaillard, M-J. & Bröström, A. (1999) Landscape openness and pollen records: a simulation approach. *The Holocene,* **9**, 409–21.

Talbot, M.R. (2001) Nitrogen isotopes in paleolimnology. In: Last, W.M. & Smol, J.P. (eds), *Tracking Environmental Change using Lake Sediments,* Vol. 2, *Physical and Geochemical Methods.* Kluwer, Dordrecht, 401–39.

Talbot, M.R. & Allen, P.A. (1996) Lakes. In: Reading, H.G. (ed.), *Sedimentary Environments: Processes, Facies and Stratigraphy* (3rd edition). Blackwell, Oxford, 83–124.

Talbot, M.R. & Livingstone, D.A. (1989) Hydrogen index and carbon isotopes of lacustrine organic matter as lake level indicators. *Palaeogeography, Palaeoclimatology, Palaeoecology,* **70**, 121–37.

Tegelaar, E.W., de Leeuw, J.W., Derenne, S. & Largeau, C. (1989) A reappraisal of kerogen formation. *Geochimica Cosmochimica Acta,* **53**, 3103–6.

Ter Braak, C.J.F. (1986) Canonical correspondence analysis; a new eigenvector method for multivariate direct gradient analysis. *Ecology,* **67**, 1167–79.

Ter Braak, C.J.F. (1987) *CANOCO—a FORTRAN Program for Canonical Community Ordination by [Partial] [Detrended] [Canonical] Correspondence Analysis, Principal Components Analysis, and Redundancy Analysis (Version 2.1).* ITI-TNO, Wageningen, Netherlands, 95 pp.

Ter Braak, C.J.F. & Juggins, S. (1993) Weighted averaging partial least squares regression (WA-PLS): an improved method for reconstructing environmental variables from species assemblages. *Hydrobiologia,* **269/270**, 485–502.

Thompson, R. (1973) Palaeolimnology and palaeomagnetism. *Nature,* **242**, 182–4.

Thompson, R. (1991) Palaeomagnetic dating. In: Smart, P. & Frances, P.D. (eds), *Quaternary Dating Methods— a User's Guide.* Quaternary Research Association, Cambridge, 177–98.

Thompson, R. & Oldfield, F. (1986) *Environmental Magnetism.* George Allen & Unwin, London, 227 pp.

Tolonen, K. (1986a) Charred particle analysis. In: Berglund, B.E. (ed.), *Handbook of Holocene Palaeoecology and Palaeohydrology.* Wiley, Chichester, 485–96.

Tolonen, K. (1986b) Rhizopod analysis. In: Berglund, B.E. (ed.), *Handbook of Holocene Palaeoecology and Palaeohydrology.* Wiley, Chichester, 645–66.

Troels-Smith, J. (1955) Characterisation of unconsolidated sediments. *Danmarks Geologiske Undersøgelse* IV, **3**, 73 pp.

Turney, C.S.M. & Lowe, J.J. (2001) Tephrochronology. In: Last, W.M. & Smol, J.P. (eds), *Tracking Environmental Change using Lake Sediments,* Vol. 1, *Basin Analysis, Coring and Chronological Techniques.* Kluwer, Dordrecht, 451–69.

Van den Bogaard C. & Schmincke H.U. (2002) Linking the North Atlantic to central Europe: a high-resolution Holocene tephrochronological record from northern Germany. *Journal of Quaternary Science,* **17**, 3–20.

Van Geel, B. (1986) Application of fungal and algal remains and other microfossils in palynological analyses. In: Berglund, B.E. (ed.), *Handbook of Holocene*

Palaeoecology and Palaeohydrology. Wiley, Chichester, 497–505.

Van Geel, B. (2001) Non-pollen palynomorphs. In: Smol, J.P., Birks, H.J.B. & Last, W.M. (eds), *Tracking Environmental Change using Lake Sediments*, Vol. 3, *Terrestrial, Algal and Siliceous Indicators.* Kluwer, Dordrecht, 99–119.

Verschuren, D., Johnson, T.C., Kling, H.J., *et al.* (2002) History and timing of human impact on Lake Victoria. *Proceedings of the Royal Society of London*, **B269**, 289–94.

Vos, P. & De Wolf, H. (1993) Diatoms as a tool for reconstructing sediment environments in coastal wetlands *Hydrobiologia*, **269/270**, 285–6.

Walden, J. (1999a) Remanence measurements. In: Walden, J., Oldfield, F. & Smith, J.P. (eds), *Environmental Magnetism: a Practical Guide.* Technical Guide No. 6, Quaternary Research Association, Cambridge, 63–88.

Walden, J. (1999b) Sample collection and preparation. In: Walden, J., Oldfield, F. & Smith, J.P. (eds), *Environmental Magnetism: a Practical Guide.* Technical Guide No. 6, Quaternary Research Association, Cambridge, 25–34.

Walden, J., Oldfield, F. & Smith, J.P. (eds) (1999) *Environmental Magnetism: a Practical Guide.* Technical Guide No. 6, Quaternary Research Association, Cambridge, 243 pp.

Walker, D. (1970) Direction and rate in some British postglacial hydroseres. In: Walker, D. & West, R.G. (eds), *Studies in the Vegetational History of the British Isles.* Cambridge University Press, Cambridge, 117–39.

Walker, I. (1987) Chironomidae (Diptera) in palaeoecology. *Quaternary Science Reviews*, **6**, 29–40.

Walker, I. (2001) Midges: Chironomidae and related taxa. In: Smol, J.P., Birks, H.J.B. & Last, W.M. (eds), *Tracking Environmental Change using Lake Sediments*, Vol. 4, *Zoological Indicators.* Kluwer, Dordrecht, 43–65.

Wansard, G., De Deckker, P. & Julià, R. (1998) Variability in ostracod partition coefficients $D(Sr)$ and $D(Mg)$. Implications for lacustrine palaeoenvironmental reconstructions. *Chemical Geology*, **146**, 39–54.

Wasylikowa, K. (1986) Analysis of fossil fruits and seeds. In: Berglund, B.E. (ed.), *Handbook of Holocene Palaeoecology and Palaeohydrology.* Wiley, Chichester, 571–90.

Whiteside, M.C. & Swindoll, M.R. (1988) Guidelines and limitations to cladoceran palaeoecological interpretations. In: Gray, J. (ed.), *Palaeolimnology: Aspects of Palaeoecology and Biogeography.* Elsevier, Amsterdam, 405–12.

Whitlock, C. & Larsen, C. (2001) Charcoal as a fire proxy. In: Smol, J.P., Birks, H.J.B. & Last, W.M. (eds), *Tracking Environmental Change using Lake Sediments*, Vol. 3, *Terrestrial, Algal and Siliceous Indicators.* Kluwer, Dordrecht, 75–97.

Wik, M. & Renberg, I. (1996) Environmental records of carbonaceous fly-ash particles from fossil fuel combustion. A summary. *Journal of Paleolimnology*, **15**, 193–206.

Williams, M.A.J., Dunkerley, D., De Deckker, P., *et al.* (1998) *Quaternary Environments* (2nd edition). Edward Arnold, London, 329 pp.

Wolfe, B.B., Edwards, T.W.D., Elgood, R.J., *et al.* (2001) Carbon and oxygen isotope analysis of lake sediment cellulose: methods and applications. In: Last, W.M. & Smol, J.P. (eds), *Tracking Environmental Change using Lake Sediments*, Vol. 2, *Physical and Geochemical Methods.* Kluwer, Dordrecht, 373–400.

Zeeb, B.A. & Smol, J.P. (2001) Chrysophyte scales and cysts. In: Smol, J.P., Birks, H.J.B. & Last, W.M. (eds), *Tracking Environmental Change using Lake Sediments*, Vol. 3, *Terrestrial, Algal and Siliceous Indicators.* Kluwer, Dordrecht, 203–23.

Zolitschka, B., Mingram, J., van der Gaast, S., *et al.* (2001) Sediment logging techniques. In: Last, W.M. & Smol, J.P. (eds), *Tracking Environmental Change using Lake Sediments*, Vol. 1, *Basin Analysis, Coring and Chronological Techniques.* Kluwer, Dordrecht, 137–53.

Züllig, H. (1982) Untersuchungen über die Stratigraphie von Carotinoiden im geschichteten Sediment von tsein Schweizer Seen zur Erkundung früherer Phytoplankton-Entfaltungen. *Schweizerisch Zeitschrift für Hydrologie*, **44**, 1–98.

Züllig, H. (1990) Role of carotenoids in lake sediments for reconstructing trophic history during the late Quaternary In: Davis, R.B. (ed.), *Palaeolimnology and the Reconstruction of Ancient Environments.* Kluwer, Dordrecht, 191–209.

Index